Formulas/Equations

Distance Formula

If $P_1 = (x_1, y_1)$ and $P_2 = (x_2, y_2)$, the distance from P_1 to P_2 is

$$d(P_1, P_2) = \sqrt{(x_2 - x_1)^2 + (y_2 - y_1)^2}$$

Standard Equation of a Circle

The standard equation of a circle of radius r with center at (h, k) is

$$(x - h)^2 + (y - k)^2 = r^2$$

Slope Formula

The slope m of the line containing the points $P_1 = (x_1, y_1)$ and $P_2 = (x_2, y_2)$ is

$$m = \frac{y_2 - y_1}{x_2 - x_1} \qquad \text{if } x_1 \neq x_2$$

$$m \text{ is undefined} \qquad \text{if } x_1 = x_2$$

Point–Slope Equation of a Line

The equation of a line with slope m containing the point (x_1, y_1) is

$$y - y_1 = m(x - x_1)$$

Slope–Intercept Equation of a Line

The equation of a line with slope m and y-intercept b is

$$y = mx + b$$

Quadratic Formula

The solutions of the equation $ax^2 + bx + c = 0, a \neq 0$, are

$$x = \frac{-b \pm \sqrt{b^2 - 4ac}}{2a}$$

If $b^2 - 4ac > 0$, there are two real unequal solutions.

If $b^2 - 4ac = 0$, there is a repeated real solution.

If $b^2 - 4ac < 0$, there are two complex solutions that are not real.

Geometry Formulas

Circle

r = Radius, A = Area, C = Circumference

$A = \pi r^2 \qquad C = 2\pi r$

Triangle

b = Base, h = Altitude (Height), A = area

$A = \frac{1}{2}bh$

Rectangle

l = Length, w = Width, A = area, P = perimeter

$A = lw \qquad P = 2l + 2w$

Rectangular Box

l = Length, w = Width, h = Height, V = Volume

$V = lwh$

Sphere

r = Radius, V = Volume, S = Surface area

$V = \frac{4}{3}\pi r^3 \qquad S = 4\pi r^2$

INSTRUCTOR'S EDITION
College Algebra

Fifth Edition

MICHAEL SULLIVAN
Chicago State University

Prentice Hall, Upper Saddle River, NJ 07458

Acquisitions Editor: *Sally Simpson*
Editor-in-Chief: *Jerome Grant*
Editorial Director: *Tim Bozik*
Associate Editor-in-Chief, Development: *Carol Trueheart*
Production Editor: *Robert C. Walters*
Senior Managing Editor: *Linda Mihatov Behrens*
Executive Managing Editor: *Kathleen Schiaparelli*
Assistant Vice President of Production and Manufacturing: *David W. Riccardi*
Marketing Manager: *Patrice L. Jones*
Manufacturing Buyer: *Alan Fischer*
Manufacturing Manager: *Trudy Pisciotti*
Supplements Editor/Editorial Assistant: *Joanne Wendelken*
Art Director: *Maureen Eide*
Associate Creative Director: *Amy Rosen*
Director of Creative Services: *Paula Maylahn*
Art Manager: *Gus Vibal*
Interior Designer: *Elm Street Publishing*
Cover Designer: *Maureen Eide*
Cover Photo: *Kids Hiking Together, Red Butte, Aspen, Colorado; Team Russell/Adventure Photo and Film*
Photo Researcher: *Beth Boyd*
Photo Research Administrator: *Melinda Reo*
Art Studio: *Academy Artworks*

 © 1999, 1996, 1993, 1990, 1987 by Prentice-Hall, Inc.
Simon & Schuster/A Viacom Company
Upper Saddle River, NJ 07458

Printed in the United States of America

10 9 8 7 6 5 4 3 2 1

ISBN 0-13-081009-6

PRENTICE-HALL INTERNATIONAL (UK) LIMITED, LONDON
PRENTICE-HALL OF AUSTRALIA PTY. LIMITED, SYDNEY
PRENTICE-HALL CANADA INC. TORONTO
PRENTICE-HALL HISPANOAMERICANA, S.A., MEXICO
PRENTICE-HALL OF INDIA PRIVATE LIMITED, NEW DELHI
PRENTICE-HALL OF JAPAN, INC., TOKYO
SIMON & SCHUSTER ASIA PTE, LTD., SINGAPORE
EDITORA PRENTICE-HALL DO BRASIL, LTDA., RIO DE JANEIRO

In Memory of Mary

CONTENTS

CHAPTER 8   SYSTEMS OF EQUATIONS AND INEQUALITIES   533

CHAPTER 9   SEQUENCES; INDUCTION; COUNTING; PROBABILITY   633

APPENDIX   GRAPHING UTILITIES   709

As a professor at an urban public university for over 30 years, I am aware of the varied needs of college algebra students who range from having little mathematical background and a fear of mathematics courses to those who have had a strong education and are extremely motivated. For some of your students, this will be their last course in mathematics, while others may decide to further their mathematical education. I have written this text for both groups. As the author of precalculus, engineering calculus, finite math and business calculus texts, and, as a teacher, I understand what students must know if they are to be focused and successful in upper level mathematics courses. However, as a father of four, I also understand the realities of college life. I have taken great pains to insure that the text contains solid, student-friendly examples and problems, as well as a clear writing style.

In the Fifth Edition

The Fifth Edition builds upon a solid foundation by integrating new features and techniques that further enhance student interest and involvement. The elements of previous editions that have proved successful remain, while many changes, some obvious, others subtle, have been made. A huge benefit of authoring a successful series is the broad-based feedback upon which improvements and additions are ultimately based. Virtually every change to this edition is the result of thoughtful comments and suggestions made from colleagues and students who have used previous editions. I am sincerely grateful for this feedback and have tried to make changes that improve the flow and usability of the text. For example, some topics have been moved to better reflect the way teachers approach the course. In other places, problems have been added where more practice was needed. The testing package has been significantly upgraded with the addition of Testpro 3 online testing. The addition of Mathpro Explorer tutorial software and the Sullivan web site, www.prenhall.com/sullivan, represent a truly useful integration of learning technology.

Changes to the Fifth Edition

Each chapter now begins with an Internet Excursion involving a real world problem that can be analyzed utilizing the material found in the chapter and information found on the Web.

Many new exercises, problems, and examples utilize real data in table form.

Specific Organizational Changes
- Chapter 1: The section on Complex Numbers (1.9) has been removed and now appears as part of Section 5.6.
- Chapter 2: The section on Quadratic Equations with a Negative Discriminant (2.4) has been removed. The two sections (1.9) and (2.4) are combined and appear in Chapter 5 (5.6). [Section 5.6 can be covered at anytime after 2.3.]

- Chapter 4: Section 4.5, One-to-One Functions; Inverse Functions, has been removed and now appears as the first section of Chapter 6, Exponential and Logarithmic Functions.

  The above changes are predicated on the belief that it is better to place material as close as possible to where it is actually needed than it is to cover it, set it aside, and then expect the student to remember it when it is needed.

- Chapter 5: Sections 5.4–5.7 have been rewritten in a more concise way.
- Chapter 10: Section 10.1, Matrix Algebra, and Section 10.2, Partial Fraction Decomposition, now appear in Chapter 8, Systems of Equations and Inequalities. Section 10.3, Vectors, has been removed entirely.

Specific Content Changes

- Chapter 2: Problems have been added to the Chapter Review to more accurately represent the material of the chapter.
- Chapter 3: Scatter diagrams are used in Section 3.1 to further motivate the idea of rectangular coordinates. A new Section, 3.5 Linear Curve Fitting, explores the idea of linearly related data and fitting lines to scatter diagrams. This section is optional and does not require graphing calculators.
- Chapter 4: In Section 4.2, more emphasis is placed on the average rate of change of a function.
- Chapter 5: Section 5.1 contains a brief discussion of data that fits a quadratic function.
- Chapter 6: Section 6.4, Properties of Logarithms, contains a brief discussion of data that fits exponential or logarithmic functions. Section 6.7, Growth and Decay, now contains an example of Logistic Growth.

As a result of these changes, this edition will be an improved teaching device for professors and a better learning tool for students.

Acknowledgments

Textbooks are written by authors, but evolve from an idea into final form through the efforts of many people. Special thanks to Don Dellen, who first suggested this book and the other books in this series. Don's extensive contributions to publishing and mathematics are well known; we all miss him dearly.

There are many people we would like to thank for their input, encouragement, patience, and support. They have our deepest thanks and appreciation. We apologize for any omissions . . .

James Africh, *College of DuPage*
Steve Agronsky, *Cal Poly State University*
Dave Anderson, *South Suburban College*
Joby Milo Anthony, *University of Central Florida*
James E. Arnold, *University of Wisconsin-Milwaukee*
Agnes Azzolino, *Middlesex County College*
Wilson P. Banks, *Illinois State University*
Dale R. Bedgood, *East Texas State University*
Beth Beno, *South Suburban College*

Carolyn Bernath, *Tallahassee Community College*
William H. Beyer, *University of Akron*
Richelle Blair, *Lakeland Community College*
Trudy Bratten, *Grossmont College*
William J. Cable, *University of Wisconsin-Stevens Point*
Lois Calamia, *Brookdale Community College*
Roger Carlsen, *Moraine Valley Community College*
John Collado, *South Suburban College*

Denise Corbett, *East Carolina University*

Theodore C. Coskey, *South Seattle Community College*

John Davenport, *East Texas State University*

Duane E. Deal, *Ball State University*

Vivian Dennis, *Eastfield College*

Guesna Dohrman, *Tallahassee Community College*

Karen R. Dougan, *University of Florida*

Louise Dyson, *Clark College*

Paul D. East, *Lexington Community College*

Don Edmondson, *University of Texas-Austin*

Christopher Ennis, *University of Minnesota*

Ralph Esparza, Jr., *Richland College*

Garret J. Etgen, *University of Houston*

W. A. Ferguson, *University of Illinois-Urbana/Champaign*

Iris B. Fetts, *Clemson University*

Mason Flake, *student at Edison Community College*

Merle Friel, *Humboldt State University*

Richard A. Fritz, *Moraine Valley Community College*

Carolyn Funk, *South Suburban College*

Dewey Furness, *Ricke College*

Dawit Getachew, *Chicago State University*

Wayne Gibson, *Rancho Santiago College*

Sudhir Kumar Goel, *Valdosta State University*

Joan Goliday, *Sante Fe Community College*

Frederic Gooding, *Goucher College*

Ken Gurganus, *University of North Carolina*

James E. Hall, *University of Wisconsin-Madison*

Judy Hall, *West Virginia University*

Edward R. Hancock, *DeVry Institute of Technology*

Julia Hassett, *DeVry Institute-Dupage*

Brother Herron, *Brother Rice High School*

Kim Hughes, *California State College-San Bernardino*

Ron Jamison, *Brigham Young University*

Richard A. Jensen, *Manatee Community College*

Sandra G. Johnson, *St. Cloud State University*

Moana H. Karsteter, *Tallahassee Community College*

Arthur Kaufman, *College of Staten Island*

Thomas Kearns, *North Kentucky University*

Keith Kuchar, *Manatee Community College*

Tor Kwembe, *Chicago State University*

Linda J. Kyle, *Tarrant Country Jr. College*

H. E. Lacey, *Texas A & M University*

Christopher Lattin, *Oakton Community College*

Adele LeGere, *Oakton Community College*

Stanley Lukawecki, *Clemson University*

Janice C. Lyon, *Tallahassee Community College*

Virginia McCarthy, *Iowa State University*

James McCollow, *DeVry Institute of Technology*

Laurence Maher, *North Texas State University*

Jay A. Malmstrom, *Oklahoma City Community College*

James Maxwell, *Oklahoma State University-Stillwater*

Carolyn Meitler, *Concordia University*

Eldon Miller, *University of Mississippi*

James Miller, *West Virginia University*

Michael Miller, *Iowa State University*

Kathleen Miranda, *SUNY at Old Westbury*

A. Muhundan, *Manatee Community College*

Jane Murphy, *Middlesex Community College*

Bill Naegele, *South Suburban College*

James Nymann, *University of Texas-El Paso*

Sharon O'Donnell, *Chicago State University*

Seth F. Oppenheimer, *Mississippi State University*

E. James Peake, *Iowa State University*

Thomas Radin, *San Joaquin Delta College*

Ken A. Rager, *Metropolitan State College*

Elsi Reinhardt, *Truckee Meadows Community College*

Jane Ringwald, *Iowa State University*

Stephen Rodi, *Austin Community College*

Howard L. Rolf, *Baylor University*

Edward Rozema, *University of Tennessee at Chattanooga*

Dennis C. Runde, *Manatee Community College*

John Sanders, *Chicago State University*

Susan Sandmeyer, *Jamestown Community College*

A. K. Shamma, *University of West Florida*

Martin Sherry, *Lower Columbia College*

Anita Sikes, *Delgado Community College*

Timothy Sipka, *Alma College*

Lori Smellegar, *Manatee Community College*

John Spellman, *Southwest Texas State University*

Becky Stamper, *Western Kentucky University*

Judy Staver, *Florida Community College-South*

Neil Stephens, *Hinsdale South High School*

Diane Tesar, *South Suburban College*

Tommy Thompson, *Brookhaven College*

Richard J. Tondra, *Iowa State University*

Marvel Townsend, *University of Florida*

Jim Trudnowski, *Carroll College*

Richard G. Vinson, *University of Southern Alabama*

Mary Voxman, *University of Idaho*

Darlene Whitkenack, *Northern Illinois University*

Christine Wilson, *West Virginia University*

Carlton Woods, *Auburn University*

George Zazi, *Chicago State University*

Recognition and thanks are due particularly to the following individuals for their valuable assistance in the preparation of this edition: Jerome Grant for his support and commitment; Sally Simpson, for her genuine interest and insightful direction; Bob Walters for his organizational skill as production supervisor; Patrice Lumumba Jones for his innovative marketing efforts; Tony Palermino for his specific editorial comments; the entire Prentice Hall sales staff for their confidence; and to Katy Murphy and Michael Sullivan, III for checking the answers to all the exercises.

Michael Sullivan

As you begin your study of College Algebra, you may feel overwhelmed by the number of theorems, definitions, procedures, and equations that confront you. You may even wonder whether or not you can learn all of this material in a single course. These concerns are normal. Keep in mind that the concepts of College Algebra are all around us as we go through our daily routines. Many of the concepts you will learn to express mathematically, you already know intuitively. For many of you, this may be your last math course, while for others, just the first in a series of many. Either way, this text was written with you in mind. I have taught college algebra courses for over thirty years. I am also the father of four college students who called home, from time to time, frustrated and with questions. I know what you're going through. So I have written a text that doesn't overwhelm, or unnecessarily complicate the concepts of College Algebra, but at the same time gives you the skills and practice you need to be successful.

This text is designed to help you, the student, master the terminology and basic concepts of College Algebra. These aims have helped to shape every aspect of the book. Many learning aids are built into the format of the text to make your study of the material easier and more rewarding. This book is meant to be a "machine for learning," one that can help you focus your efforts and get the most from the time and energy you invest.

Please do not hesitate to contact me, through Prentice Hall, with any suggestions or comments that would improve this text.

Best Wishes!

Michael Sullivan

CHAPTER 4

Functions and Their Graphs

I magine yourself as an expert for the EPA and sitting in congressional policy meetings having to explain the importance of balancing energy resources with environmental protection. On the following page is the Internet Excursion placing you in exactly that position and asking the tough questions a member of Congress might ask. Use the Sullivan website at:

www.prenhall.com/sullivan

to link to the Internet resources needed to answer the questions asked.

PREPARING FOR THIS CHAPTER

Before getting started on this chapter, review the following concepts:

Domain of a Variable *(p. 17)*

Graphs of Certain Equations *(Example 2, p. 175; Example 3, p. 176; Example 4, p. 176; Example 11, p. 182)*

Tests for Symmetry of an Equation *(p. 180)*

Procedure for Finding Intercepts of an Equation *(p. 178)*

Steps for Setting up Applied Problems *(p. 99)*

OUTLINE

4.1 Functions
4.2 More about Functions
4.3 Graphing Techniques: Transformations
4.4 Operations on Functions; Composite Functions
4.5 Mathematical Models: Constructing Functions
Chapter Review

PAGE 227

CHAPTER OPENERS

Each chapter begins with Preparing for this Chapter, which serves as a "just-in-time" review, while Chapter Outlines provide students with a basic road map of the chapter.

Students will already have an understanding of which concepts are most important before beginning the chapter.

INTERNET PROJECTS

Exploration in each chapter starts right from the beginning with optional Internet Projects inviting students to use the Sullivan Web site to gather information and real data to solve mathematical problems.

Students can see what a useful and accessible tool the Internet is. Likewise, they will also be able to see algebra as a valuable tool for understanding the world outside the classroom.

Internet Projects also help foster active learning and participation in lecture.

Netsite: http://www.prenhall.com/sullivan/

HOW LONG WILL THE OIL LAST?

The global supply of oil and other sources of energy is more than adequate to meet present needs, but most of this supply is outside the United States. Currently, oil supplies about 40 percent of the world's energy, with the United States being the biggest consumer. Since oil is a nonrenewable energy resource, plans must be made for the eventuality of diminishing supply.

Suppose that you were a consultant for the EPA and members of Congress asked you to sit on the Energy Policy Steering Committee as an expert analyst. You must convince the committee of the importance of environmental concerns in planning the global energy system. You are very aware of the fact that present modes of energy use and production threaten serious environmental deterioration. Your plan is to create a set of "what if" functions that will allow your committee to model many possible scenarios for their policy guidelines.

1. $E(t)$, the first function that you create, will allow you to model United States oil consumption over a period of time. Make a scatter diagram using the EPA Data on United States oil consumption per capita from 1950–1990. Since oil provides 40 percent of the energy consumed, adjust your figures to get oil consumption per capita. Use the LINear REGression tool on your graphing utility to find the linear function $E(t)$ of best fit for modeling the data. Does this seem like a good model?

2. Your second function, $P(t)$, is a population modeling tool. From the United States Census Data, make a scatter diagram, and then use a QUADratic REGression to model the data. Next, let $O(t) = E(t) \cdot P(t)$. What does $O(t)$ model? Graph $O(t)$. Compare to the actual figures from BP Petroleum.

PAGE 228

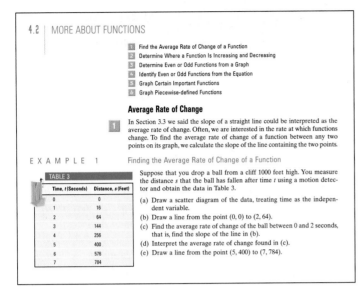

4.2 | MORE ABOUT FUNCTIONS

1. Find the Average Rate of Change of a Function
2. Determine Where a Function Is Increasing and Decreasing
3. Determine Even or Odd Functions from a Graph
4. Identify Even or Odd Functions from the Equation
5. Graph Certain Important Functions
6. Graph Piecewise-defined Functions

Average Rate of Change

1. In Section 3.3 we said the slope of a straight line could be interpreted as the average rate of change. Often, we are interested in the rate at which functions change. To find the average rate of change of a function between any two points on its graph, we calculate the slope of the line containing the two points.

EXAMPLE 1   Finding the Average Rate of Change of a Function

TABLE 3

Time, t (Seconds)	Distance, s (Feet)
0	0
1	16
2	64
3	144
4	256
5	400
6	576
7	784

Suppose that you drop a ball from a cliff 1000 feet high. You measure the distance s that the ball has fallen after time t using a motion detector and obtain the data in Table 3.

(a) Draw a scatter diagram of the data, treating time as the independent variable.
(b) Draw a line from the point $(0, 0)$ to $(2, 64)$.
(c) Find the average rate of change of the ball between 0 and 2 seconds, that is, find the slope of the line in (b).
(d) Interpret the average rate of change found in (c).
(e) Draw a line from the point $(5, 400)$ to $(7, 784)$.

PAGE 248

SECTION OBJECTIVES

Each section begins with a bulleted synopsis of the topics it contains.

Students have a clear idea of which concepts they will cover, enhancing their comfort level as new topics are introduced.

OPTIONAL GRAPHING UTILITIES AND TECHNIQUES

Graphing utilities are optional in this text and their use is clearly identified by the use of a graphing utility icon ▦ . At appropriate places, graphing utilities are used for Exploration, Seeing the Concept, and Checks, often using TI-83 screen shots. Optional graphing utility exercises and examples also appear.

Graphing utilities allow students to explore algebraic concepts visually, promoting more thorough understanding of key objectives.

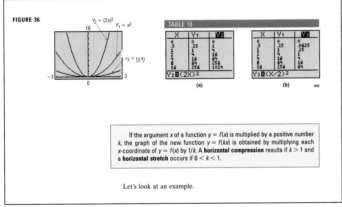

FIGURE 36

If the argument x of a function $y = f(x)$ is multiplied by a positive number k, the graph of the new function $y = f(kx)$ is obtained by multiplying each x-coordinate of $y = f(x)$ by $1/k$. A **horizontal compression** results if $k > 1$ and a **horizontal stretch** occurs if $0 < k < 1$.

Let's look at an example.

PAGE 270

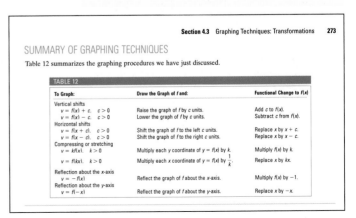

Section 4.3   Graphing Techniques: Transformations   273

SUMMARY OF GRAPHING TECHNIQUES

Table 12 summarizes the graphing procedures we have just discussed.

TABLE 12

To Graph:	Draw the Graph of f and:	Functional Change to $f(x)$
Vertical shifts		
$y = f(x) + c,\quad c > 0$	Raise the graph of f by c units.	Add c to $f(x)$.
$y = f(x) - c,\quad c > 0$	Lower the graph of f by c units.	Subtract c from $f(x)$.
Horizontal shifts		
$y = f(x + c),\quad c > 0$	Shift the graph of f to the left c units.	Replace x by $x + c$.
$y = f(x - c),\quad c > 0$	Shift the graph of f to the right c units.	Replace x by $x - c$.
Compressing or stretching		
$y = kf(x),\quad k > 0$	Multiply each y coordinate of $y = f(x)$ by k.	Multiply $f(x)$ by k.
$y = f(kx),\quad k > 0$	Multiply each x coordinate of $y = f(x)$ by $\frac{1}{k}$.	Replace x by kx.
Reflection about the x-axis		
$y = -f(x)$	Reflect the graph of f about the x-axis.	Multiply $f(x)$ by -1.
Reflection about the y-axis		
$y = f(-x)$	Reflect the graph of f about the y-axis.	Replace x by $-x$.

PAGE 273

270   Chapter 4   Functions and Their Graphs

 Now work Problem 33.

What happens if the argument x of a function $y = f(x)$ is multiplied by a positive number k, creating a new function $y = f(kx)$. To find the answer, we look at the following Exploration.

 Exploration   On the same screen, graph each of the following functions:

$$Y_1 = f(x) = x^2$$
$$Y_2 = f(2x) = (2x)^2$$
$$Y_3 = f\left(\frac{1}{2}x\right) = \left(\frac{1}{2}x\right)^2$$

You should have obtained the graphs shown in Figure 36. The graph of $Y_2 = (2x)^2$ is the graph of $Y_1 = x^2$ compressed horizontally. Look at Table 10(a). Notice that $(1, 1)$, $(2, 4)$, $(4, 16)$, and $(16, 256)$ are points on the graph of $Y_1 = x^2$. Also, $(0.5, 1)$, $(1, 4)$, $(2, 16)$, and $(8, 256)$ are points on the graph of $Y_2 = (2x)^2$. Thus, the graph of $Y_2 = (2x)^2$ is obtained by multiplying the x-coordinate of each point on the graph of $Y_1 = x^2$ by $\frac{1}{2}$. The graph of $Y_3 = \left(\frac{1}{2}x\right)^2$ is the graph of $Y_1 = x^2$ stretched horizontally. Look at Table 10(b). Notice that $(0.5, 0.25)$, $(1, 1)$, $(2, 4)$, $(4, 16)$ are points on the graph of $Y_1 = x^2$. Also, $(1, 0.25)$, $(2, 1)$, $(4, 4)$, $(8, 16)$ are points on the graph of $Y_3 = \left(\frac{1}{2}x\right)^2$. Thus, the graph of $Y_3 = \left(\frac{1}{2}x\right)^2$ is obtained by multiplying the x-coordinate of each point on the graph of $Y_1 = x^2$ by a factor of 2.

FIGURE 36

TABLE 10

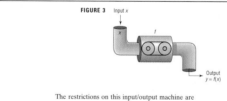

(a)   (b)

If the argument x of a function $y = f(x)$ is multiplied by a positive number k, the graph of the new function $y = f(kx)$ is obtained by multiplying each x-coordinate of $y = f(x)$ by $1/k$. A **horizontal compression** results if $k > 1$ and a **horizontal stretch** occurs if $0 < k < 1$.

Let's look at an example.

PAGE 270

"NOW WORK" PROBLEMS

Blue "Now Work" icons appear at appropriate places within each section, sending the student to work specific exercises as topics are introduced.

Students are able to test their understanding as they study, rather than waiting until the end of the section and trying to determine where they may have missed a concept.

A hidden benefit of the "Now Work" problems is that they enable students to ask more specific questions in class, because they know exactly in which areas they need help.

A CLEAR WRITING STYLE

Sullivan's accessible writing style is apparent throughout.

An author who writes clearly makes potentially difficult concepts intuitive. Class time is infinitely more productive when students have the support of a well-written text.

FIGURE 3   Input x

The restrictions on this input/output machine are

1. It only accepts numbers from the domain of the function.
2. For each input, there is exactly one output (which may be repeated for different inputs).

For a function $y = f(x)$, the variable x is called the **independent variable**, because it can be assigned any of the permissible numbers from the domain. The variable y is called the **dependent variable**, because its value depends on x.

EXAMPLE 4   Finding Values of a Function

For the function G defined by $G(x) = 2x^2 - 3x$, evaluate:

(a) $G(3)$ (b) $G(x) + G(3)$ (c) $G(-x)$
(d) $-G(x)$ (e) $G(x + 3)$

PAGE 232

"Saving the Economic Future of Krispy Krunchy Candy Co."

The Krispy Krunchy Candy Company is facing a financial crisis because the cost of shipping cocoa beans from Ghana has increased. The CEO has decided that the way to stay afloat would be to reduce the size of their GIANT KRISPY KRUNCHY BAR by 10% but keep the price the same. He doesn't want to lose any customers, however. Therefore, he wants the change in size to be as unobtrusive as possible. The present dimensions of the GIANT KRISPY KRUNCHY BAR are 12 centimeters (cm) in length, 7 cm in width, and 3 cm in thickness. The CEO has asked your consulting firm to come up with the best way to shrink the candy bar. Because millions of dollars are riding on this decision, you will need to find all answers in centimeters to 3 decimal places and all percents to 5 significant digits.

1. Make a sketch of the candy bar, roughly to scale, and label it.
2. What is the present volume of the candy bar?
3. What would be the new volume after a 10% reduction?
4. What would be the new volume if each dimension were reduced by 10%. Is this the same as your answer to question 3? (It shouldn't be.) Explain the difference.
5. Consider reducing two of the dimensions by the same number of centimeters, while holding the third dimension fixed. (There are three possibilities; in each case, the new volume is to be 90% of the original volume.)
6. Consider reducing two (but not three) of the dimensions by the same percent. What would the new dimensions be in each case? (There are three possibilities; in each case you want the new volume to be 90% of the original volume.)
7. Consider reducing all three of the dimensions by the same percent. What would the new dimensions be?
8. Within your group, make a decision about which of the seven possibilities that you found would be the best one to recommend to the CEO of Krispy Krunchy. Write out two or three sentences to justify your choice.
9. Would you mind if you discovered that your favorite candy bar had been reduced in size while the price stayed the same? Do you think a candy company might actually do this to improve their financial standing? What is your protection as a consumer from being fooled?

In every chapter, there is a full-page collaborative project, a Mission Possible. These multi-tasked projects, written by Hester Lewellen, one of the co-authors of the University of Chicago High School Mathematics Project, will help your students learn through and with each other.

The collaborative format has been shown to increase student motivation and interest.

---

36. **Federal Income Tax**   Two 1997 Tax Rate Schedules are given in the accompanying table. If x equals the amount on Form 1040, line 37, and y equals the tax due, construct a function f for each schedule.

1997 TAX RATE SCHEDULES

SCHEDULE X—IF YOUR FILING STATUS IS SINGLE				SCHEDULE Y-1—USE IF YOUR FILING STATUS IS MARRIED FILING JOINTLY OR QUALIFYING WIDOW(ER)			
If the amount on Form 1040, line 37, is: Over—	But not over—	Enter on Form 1040, line 38	of the amount over—	If the amount on Form 1040, line 37, is: Over—	But not over—	Enter on Form 1040, line 38	of the amount over—
$ 0	$ 24,650	$ 0 + 15%	$ 0	$ 0	$ 41,200	$ 0 + 15%	$ 0
24,650	59,750	3,698 + 28	24,650	41,200	99,600	6,180 + 28	41,200
59,750	124,650	13,526 + 31	59,750	99,600	151,750	22,532 + 31	99,600
124,650	271,050	33,645 + 36	124,650	151,750	271,050	38,699 + 36	151,750
271,050		86,349 + 39.6	271,050	271,050		81,647 + 39.6	271,050

37. **Inscribing a Cylinder in a Sphere**   Inscribe a right circular cylinder of height h and radius r in a sphere of fixed radius R. See the illustra-

38. **Inscribing a Cylinder in a Cone**   Inscribe a right circular cylinder of height h and radius r in a cone of fixed radius R and fixed height H.

Sullivan provides students with applications that employ sourced real-world data.

Students find it difficult to get into hypothetical "widget" problems. Real data allows students to see the relevancy of algebra outside the classroom.

(b) Draw a line from the point (0, 24,000) to (25, 27,750) on the scatter diagram found in (a).
(c) Find the average rate of change of the cost between 0 and 25 bicycles.
(d) Interpret the average rate of change found in (c). In economics, this is called **marginal cost.**
(e) Draw a line from the point (190, 42,750) to (223, 46,500) on the scatter diagram found in (a).
(f) Find the average rate of change of the cost between 190 and 223 bicycles.
(g) Interpret the average rate of change found in (f).

79. **Revenue from Selling Bikes**   The following data represent the total revenue that would be received from selling x bicycles at Tunney's Bi-cycle Shop.

output does Tunney's Bicycle Shop's profits stop increasing?

80. **Cost of Tuition**   The following data represent the average cost of tuition and required fees (in dollars) at public four-year colleges.

Year	Average Cost of Tuition
1990	2,035
1991	2,159
1992	2,410
1993	2,604
1994	2,822

Source: U.S. National Center for Education Statistics.

OVERVIEW

4.5 | MATHEMATICAL MODELS: CONSTRUCTING FUNCTIONS

> 1 | Construct and Analyze Functions

1 Real-world problems often result in mathematical models that involve functions. These functions need to be constructed or built based on the information given. In constructing functions, we must be able to translate the verbal description into the language of mathematics. We do this by assigning symbols to represent the independent and dependent variables and then finding the function or rule that relates these variables.

EXAMPLE 1   Area of a Rectangle with Fixed Perimeter

The perimeter of a rectangle is 50 feet. Express its area A as a function of the length x of a side.

FIGURE 46

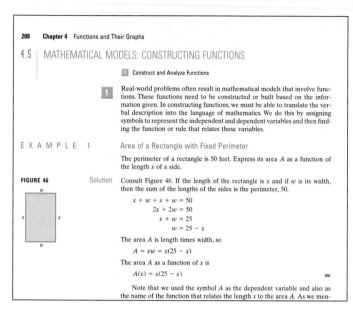

Solution   Consult Figure 46. If the length of the rectangle is x and if w is its width, then the sum of the lengths of the sides is the perimeter, 50.

$$x + w + x + w = 50$$
$$2x + 2w = 50$$
$$x + w = 25$$
$$w = 25 - x$$

The area A is length times width, so

$$A = xw = x(25 - x)$$

The area A as a function of x is

$$A(x) = x(25 - x)$$

Note that we used the symbol A as the dependent variable and also as the name of the function that relates the length x to the area A. As we men-

MODELING

Students are asked to create and manipulate mathematical models.

The construction and interpretation of mathematical models is essential to better understanding of not only precalculus, but virtually any mathematics course that your students are likely to take during their college careers.

END-OF-SECTION EXERCISES

Sullivan's exercises are unparalleled in terms of thorough coverage and accuracy.

Well-executed problem sets better prepare students for exams.

Each end-of-section exercise set begins with visual and concept-based problems, starting students out with the basics of the section.

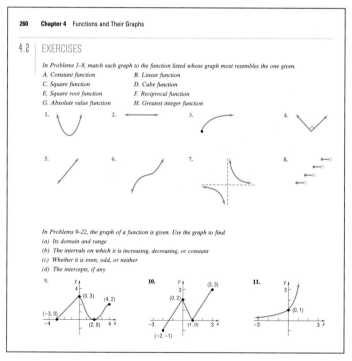

CHAPTER REVIEW

THINGS TO KNOW

Function

A relation between two sets of real numbers so that each number x in the first set, the domain, has corresponding to it exactly one number y in the second set. The range is the set of y values of the function for the x values in the domain.

x is the independent variable; y is the dependent variable.

A function f may be defined implicitly by an equation involving x and y or explicitly by writing $y = f(x)$.

A function can also be characterized as a set of ordered pairs (x, y) or $(x, f(x))$ in which no two distinct pairs have the same first element.

Function notation

$y = f(x)$

f is a symbol for the function.

x is the argument, or independent variable.

y is the dependent variable.

$f(x)$ is the value of the function at x, or the image of x.

HOW TO

Determine whether a relation represents a function

Find the domain and range of a function from its graph

Find the domain of a function given its equation

Determine algebraically whether a function is even or odd

Graph certain functions by shifting, compressing, stretching, and/or reflecting (see Table 12)

Given the graph of a function, determine where it is increasing or decreasing

Find the composite of two functions and its domain

Construct functions in applications, including piecewise-defined functions

FILL-IN-THE-BLANK ITEMS

1. If f is a function defined by the equation $y = f(x)$, then x is called the _____ variable and y is the _____ variable.

2. A set of points in the xy-plane is the graph of a function if and only if no _____ line contains more than one point of the set.

3. A(n) _____ function f is one for which $f(-x) = f(x)$ for every x in the domain of f; a(n) _____ function f is one for which $f(-x) = -f(x)$ for every x in the domain of f.

4. Suppose that the graph of a function f is known. Then the graph of $y = f(x - 2)$ may be obtained by a(n) _____ shift of the graph of f to the _____ a distance of 2 units.

5. If $f(x) = x + 1$ and $g(x) = x^3$, then _____ $= (x + 1)^3$.

6. For two functions f and g, $(g \circ f)(x) = $ _____.

REVIEW EXERCISES

Blue problem numbers indicate the author's suggestions for use in a Practice Test.

1. Given that f is a linear function, $f(4) = -5$, and $f(0) = 3$, write the equation that defines f.

2. Given that g is a linear function with slope $= -4$ and $g(-2) = 2$, write the equation that defines g.

3. A function f is defined by

$$f(x) = \frac{Ax + 5}{6x - 2}$$

If $f(1) = 4$, find A.

4. A function g is defined by

$$g(x) = \frac{A}{x} + \frac{8}{x^2}$$

If $g(-1) = 0$, find A.

5. Tell which of the following graphs are graphs of functions.

(a)

(b)

(c)

(d)

The Chapter Review checks students' understanding of the chapter material in several ways. "Things to Know" gives a general overview of review topics. The "How To" section gives the student a concept-by-concept listing of the operations they are expected to perform. The "Review Exercises" then serve as a chance to practice the concepts presented within the chapter.

Students master and feel confident in knowing the chapter concepts before moving on to tackle more difficult mathematics. The Chapter Review section is also another excellent review for students when preparing for an exam or final.

GIVE YOUR STUDENTS THE POWER TO EXPLORE AND DISCOVER MATHEMATICAL CONCEPTS.

MATHPRO 3.0

- An interactive and networkable text-specific CD-ROM tutorial.
- Algorithmically generates exercises to help students gain confidence and successfully progress through their course.
- Features problem-specific step-by-step tutorial help.

For Windows and Power Macintosh.

MATHPRO EXPLORER CD

MathPro Explorer offers a unique dimension to mathematics that no other tutorial software can match: Exploratory Math Labs. Now, your students can interact with and see dynamic fundamental operations such as coordinating graphs, modeling data, testing equation solutions, and experimenting with equations and inequalities.

Other MathPro Explorer features:

- Includes all the capabilities of MathPro 3.0.
- Twenty pre-formatted explorations include Toolkit tools that students may link to share and explore data and concepts.
- Available to students as an upgrade via the bookstore.

For Windows and Power Macintosh.

TESTPRO 3.0

- Algorithmically generates multiple forms of different but equivalent tests by chapter, section, or objective.
- Includes multiple-choice and free-response options.
- Features editing and formatting capabilities.

For Windows and Macintosh.

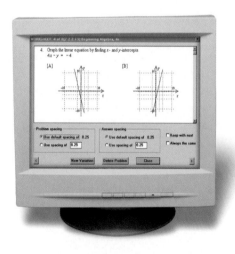

THE SULLIVAN WEB SITE

For students to succeed in mathematics, they need practice in problem solving and the opportunity to explore. Prentice Hall's Companion Website for Sullivan provides both—and much more:

- Each text-specific site features easy-to-use study and exploration aids.
- Organized by chapter
- **Destinations.** Links to interesting and relevant mathematics sites.
- **Communication Center.** Links to online chat board and messaging functions.
- **Problem Solving Center.** Offers both Practice Exercises and Quizzes in multiple choice forms. Hints and page references are included. Quizzes and exercises offer students immediate feedback and the opportunity to send completed quizzes to their instructors.
- **Chapter Projects.** Linked to the Internet Projects at the beginning of each chapter.

Check out the site today!

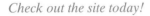

SUPPLEMENTS

STUDENT SUPPLEMENTS

- **Student Solutions Manual**
  Worked solutions to all odd-numbered exercises from the text and complete solutions for chapter review problems and chapter tests.
  Student Solutions Manual for *College Algebra*
  ISBN: 0-13-081012-6
- **Lecture Videos**
  The instructional tapes, in a lecture format, feature worked-out examples and exercises, taken from each section of the text.
  Lecture Videos for *College Algebra*
  ISBN: 0-13-081017-7
- *New York Times* **Themes of the Times**
  A free newspaper from Prentice Hall and *The New York Times*. Interesting and current articles on mathematics which invite discussion and writing about mathematics.
- **Mathematics on the Internet**
  Free guide providing a brief history of the Internet, discussing the use of the World Wide Web, and describing how to find your way within the Internet and how to find others on it.

INSTRUCTOR SUPPLEMENTS

- **Instructor's Edition**
  Includes answers to all exercises at the end of the book. Instructor's Edition for *College Algebra*
  ISBN: 0-13-081009-6
- **Instructor's Solutions Manual**
  Contains complete step-by-step worked-out solutions to all even-numbered exercises in the textbook. Also included are strategies for using the Collaborative Learning projects found in each chapter.
  Instructor's Solutions Manual for *College Algebra*
  ISBN: 0-13-081011-8
- **Test Item File**
  Hard copy of the algorithmic computerized testing materials.
  Test Item File for *College Algebra*
  ISBN: 0-13-081014-2

LIST OF APPLICATIONS

Acoustics

amplifying sound, 490–91
distance measured with sound, 127
intensity of sound, 484–85, 487
loudness, 485, 487, 491

Agriculture

crop selection, 631
farm management, 625, 626
fencing of field, 610
watering field, 120, 226

Archaeology

dating
of ancient tools, 478
of fossil, 482
time of death of prehistoric man, 492

Architecture

parabolic arch, 322
windows
design, 85
dimensions of, 120, 631
Norman, 80, 322
special, 322, 409

Art

framing a painting, 160
page design, 247

Astronomy

Kepler's Third Law of Planetary
Motion, 224
light speed, 33
light-year, 33
planetary orbits, 513, 516
reflecting telescopes, 506
satellites, 220, 434, 445

Automotive. *See* Motor Vehicles

Banking. *See also* Finance

loans, 106

Biology. *See also* Medicine

bacterial growth, 263, 436, 481, 482,
483, 492
cell growth, 476–77

fruit fly population, 480–81
insect population growth, 481
mosquito colony growth, 481–82
tree age, 482

Business. *See also* Economics; Finance

advertising, 323
automobile production, 286
blending coffee, 106, 160, 618, 631
candy bar size reduction, 121
car rentals, 323
constant rate jobs, 104, 632
cookie orders, 631
copying machine utilization, 160
cost
average, 365
of can, 365
of car rental, 297
of charter, 160
of commodity, 286
of manufacturing, 20, 85, 105,
263–64, 342
minimizing, 409, 627, 632
of natural gas, 264–65
of printing, 342
of production, 286, 593
of transporting goods, 297
demand for corn, 423
depreciation, 660
discounts, 106, 121
employment and labor force, 214
expenses, 107
farm management, 625, 626
gross salary, 246
hourly wages, 100, 105
productivity and, 173, 215
manufacturing trucks, 618
market penetration of Intel's coprocessor, 483
meat market management, 626
mixing candy, 106
mixing nuts, 106, 546, 619, 631
price and quantity demanded, 458–59
price markups, 106
product demand, 491
product design, 626
production scheduling, 625
productivity vs. earnings, 302–3
profit
annual, 435
of cigarette company, 278

computation of, 196, 593
maximization of, 623–24, 626, 632
monthly, 151
reduction from sale, 108
quarterly corporate earnings, 85
restaurant management, 546
revenue
airline, 626–27
from bicycle sales, 264
maximization of, 320
theater, 547
total, 105
salary increases, 660, 707
sales commission, 143
sales data vs. income, 214
salvage value, 491
theater attendance, 106
transporting goods, 619
working together, 106, 121, 159

Chemistry

carbon monoxide concentration, 225
drug concentration, 365
gas laws, 220
laboratory work stations, 547–48
mixture problems, 64, 102–3
acids, 106, 159, 559–60, 562, 631
cement, 108
salt solutions, 107, 160
water and antifreeze, 107
Newton's Law of Cooling, 479–80, 482
pH, 444
purity of gold, 108
radioactivity, 457–58, 477
in Chernobyl, 482
decay, 481, 482
salt decomposition in water, 482
self-catalytic reactions, 325
sugar molecules, 108
volume of gas, 143

Combinatorics

book arrangement, 689, 705
codes, 682, 688
coin toss, 697, 700
combination locks, 689
committee formation, 686, 689, 690,
706
flag arrangements, 688, 706
home choices, 705
letters formation, 688
license plate possibilities, 705

Geometry

area
- of circle, 105, 238, 297
- of equilateral triangle, 224, 297
- of isosceles right triangle, 246
- of isosceles triangle, 290–91
- of rectangle, 20, 246, 296, 297
- of square, 80, 105, 117–18, 296, 297
- of triangle, 20, 80, 219
- of window, 85

circumference of circle, 20, 105
collinear points, 575
cylinder inscribed in cone, 298
cylinder inscribed in sphere, 298
dimensions
- of isosceles triangle, 610
- of lot, 631
- of rectangle, 120, 122, 409, 610
- of right triangle, 631

equation of line, 575
interior angles of convex polygon, 666
legs of triangle, 122
perimeter
- of equilateral triangle, 20
- of rectangle, 20, 105, 219, 321, 546
- of square, 105

polygon sides, 122
Pythagorean theorem, 219
slope, 206–7
surface area
- of balloon, 286
- of cube, 20
- of sphere, 20

tangent line, 207
volume
- of balloon, 286
- of cone, 294
- of cube, 20
- of cylinder, 294
- of right circular cone, 219
- of right circular cylinder, 219
- of sphere, 20, 219, 296

Insurance

life expectancy, 137

Landscaping

fencing, 546
pond enclosure, 409

Media

response to TV advertising, 434

Medicine

AIDS cases in U.S., 409–10
blood type classification, 680
drug concentration in bloodstream, 434, 445
healing of wounds, 434, 445
pharmacy prescription, 547, 563
red cell count, 24

Miscellaneous

arithmetic mean, 136–37
bending a wire, 610
Challenger disaster, 435–36
charter club memberships, 323
disseminating information, 491
filling swimming pool, 289–90
geometric mean, 136–37
harmonic mean, 136–37
height vs. weight, 214
IQ tests, 143
life expectancy, 137
magnitude of telescope, 491
painting house, 562
postage stamps, 436
rumor spreading, 445
spreading rumors, 434
suspension bridge cables, 317–18, 322

Motion. *See also* Physics

circular, 80
rolling circle, 21
uniform, 103–4, 107, 108, 118–19, 122, 159, 160

Motor vehicles. *See also* Energy

alcohol and driving, 445
car rentals, 323
cost of car, 106
miles per gallon, 324
production of, 286
thefts of, 341
tune-up and brake job probability, 707

Navigation

air traffic control, 322
closeness of two ships, 316–17, 321, 322
from island to town, 240–41, 247
two ships approaching, 409

Nutrition

animal, 626
dietary requirements, 625

Optics, 433

light through glass, 444
magnitude of telescope, 491
mirrors, 531
reflecting telescopes, 506

Physics. *See also* Motion

beam strength, 302
bouncing balls, 660, 706–7
falling objects, 219
force, 105, 221
force of wind on window, 219
gravity
- acceleration due to, 364
- on Earth, 246
- on Jupiter, 246
- on Moon, 410

height of ball, 325
horsepower, 219
kinetic energy, 105, 220
light intensity, 159
Newton's Law, 219
pendulum motion, 219
- arc length of, 657, 660
- period, 247, 277, 423
- total swings before stopping, 657–58

pressure, 105
projectile motion
- analysis of, 315–16, 321, 322
- vertically-thrown object, 121, 122, 151

safe load for beam, 220
sound to measure distance, 127
stress of materials, 220
stretching spring, 219
vibrating strings, 219, 224
weight of body, 219, 220
- elevation and, 247–48

work, 105

Population. *See also* Biology

of endangered species, 483
of fruit fly, 480–81
growth of, 482
rabbit colony growth, 643
of rare insect, 364
of VCR owners, 483

Rate and time, 302

current of stream, 547
distance and, 245–46
of emptying
- oil tankers, 108
- pool, 108
- tank, 121, 159
- tub, 108

PHOTO CREDITS

Preliminaries

On the page following each chapter opener is an Internet Excursion. These Excursions will place you in intriguing situations where information from various Internet resources are needed in order to answer thought-provoking questions. All of the links and even more information on these Excursions can be found on your Sullivan website.

The URL address is:

www.prenhall.com/sullivan

Once at the website, find the appropriate text and explore the student's pages of the site. You will find an interactive version of the Excursion as well as many other useful and helpful materials.

Turn the page and learn all about . . . .

OUTLINE

FIBONACCI NUMBERS AND THE GOLDEN RATIO

As a biologist, part of your job is to observe the patterns that occur in nature. You are currently studying a colony of bees. Each male worker bee has just one parent (a queen bee), while each female bee has two parents. As a result, the male worker bee's family tree has the following pattern:

A Male Bee's Family Tree	
The bee itself	1
Number of parents	1
Number of grandparents	2
Number of great-grandparents	3
Number of great-great-grandparents	5
Number of great-great-great-grandparents	8
Number of great-great-great-great-grandparents	13

Suddenly, you recall a college class in which you studied the *Fibonacci sequence*, a famous sequence of numbers discovered by an Italian mathematician named *Leonardo of Pisa*, also known as Fibonacci. You realize that you have just found yet another occurence of the Fibonacci numbers in nature.

1. *The Rabbit's Story* is an entertaining account of how Fibonacci might have discovered his sequence of numbers. Can you see the *pattern* in these numbers? Explain the pattern in your own words. Look also at *The First 100 Fibonacci Numbers*.
2. Give some other examples of how the Fibonacci numbers occur in *nature*.
3. Part of your success in the study of mathematics depends on your ability to recognize and use patterns. But you must be careful when doing so! For example, try finding the value of $(x^3 - 6x^2 + 17x - 6)/6$ for $x = 1, 2, 3, \ldots$. Is this a formula for the Fibonacci numbers? Read about *"Fibonacci forgeries"* to find out.
4. What happens when you divide a Fibonacci number by the previous Fibonacci number? For example, look at the following ratios:

    $$1/1 = 1 \quad 2/1 = 2 \quad 3/2 = 1.5 \quad 5/3 \approx 1.667 \quad 8/5 = 1.6 \quad 13/8 = 1.625$$

    As you use larger and larger Fibonacci numbers, the ratio will get closer and closer to a very special number known as the *Golden Ratio* (also called the Golden Mean or the Golden Section). Describe how this number occurs in nature and how it can be used in *art, architecture, geometric designs, music,* and even *floral arrangements*. Also look at *The Golden Ratio—Calculator Investigation* and *How to Draw a Golden Rectangle*.

The word *algebra* is derived from the Arabic word *al-jabr*. This word is a part of the title of a ninth century work, "Hisâb al-jabr w'al-muqâbalah," written by Mohammed ibn Mûsâ al-Khowârizmî. The word *al-jabr* means "a restoration," a reference to the fact that, if a number is added to one side of an equation, then it must also be added to the other side in order to "restore" the equality. The title of the work, freely translated, means "The science of reduction and cancellation." Of course, today, algebra has come to mean a great deal more.

1.1 | REAL NUMBERS

1 Classify Numbers
2 Perform Arithmetic Operations
3 Know Properties of Real Numbers
4 Determine the Domain of a Variable

Algebra can be thought of as a generalization of arithmetic in which letters may be used to represent numbers.

Operations

We will use letters such as *x, y, a, b,* and *c* to represent numbers. The symbols used in algebra for the operations of addition, subtraction, multiplication, and division are $+$, $-$, $\cdot$, and $/$. The words used to describe the results of these operations are **sum, difference, product,** and **quotient.** Table 1 summarizes these ideas.

TABLE 1		
Operation	**Symbol**	**Words**
Addition	$a + b$	Sum: *a* plus *b*
Subtraction	$a - b$	Difference: *a* less *b*
Multiplication	$a \cdot b$, $(a) \cdot b$, $a \cdot (b)$, $(a) \cdot (b)$, ab, $(a)b$, $a(b)$, $(a)(b)$	Product: *a* times *b*
Division	a/b or $\dfrac{a}{b}$	Quotient: *a* divided by *b*

In algebra, we generally avoid using the multiplication sign $\times$ and the division sign $\div$ so familiar in arithmetic. Notice also that when two expressions are placed next to each other without an operation symbol, as in *ab*, or in parentheses, as in $(a)(b)$, it is understood that the expressions, called **factors,** are to be multiplied.

The four operations of addition, subtraction, multiplication, and division are called **binary operations,** since each is performed on a pair of numbers.

The symbol $=$, called an **equal sign** and read as "equals" or "is," is used to express the idea that the number or expression on the left of the equal sign is equivalent to the number or expression on the right.

E X A M P L E   1   Writing Statements Using Symbols

(a) The sum of 2 and 7 equals 9. In symbols, this statement is written as $2 + 7 = 9$.

(b) The product of 3 and 5 is 15. In symbols, this statement is written as $3 \cdot 5 = 15$. ∎

Equality

We have used the equal sign to convey the idea that one expression is equivalent to another. Three important properties of equality are listed next. In this list, *a* and *b* represent numbers.

1. The **reflexive property** states that a number always equals itself; that is, $a = a$.

Although this result seems obvious, it forms the basis for much of what we do in algebra.

2. The **symmetric property** states that, if $a = b$, then $b = a$.

3. The **principle of substitution** states that, if $a = b$, then we may substitute b for a in any expression containing a.

Sets

When we want to treat a collection of similar but distinct objects as a whole, we use the idea of a **set.** For example, the set of *digits* consists of the collection of numbers 0, 1, 2, 3, 4, 5, 6, 7, 8, and 9. If we use the symbol D to denote the set of digits, then we can write

$$D = \{0, 1, 2, 3, 4, 5, 6, 7, 8, 9\}$$

In this notation, the braces { } are used to enclose the objects, or **elements,** in the set. This method of denoting a set is called the **roster method.** A second way to denote a set is to use **set-builder notation,** where the set D of digits is written as

$$D = \{\quad x \quad | \quad x \text{ is a digit}\}$$

Read as "D is the set of all x such that x is a digit."

EXAMPLE 2   Using Set-builder Notation and the Roster Method

(a) $E = \{x | x \text{ is an even digit}\} = \{0, 2, 4, 6, 8\}$
(b) $O = \{x | x \text{ is an odd digit}\} = \{1, 3, 5, 7, 9\}$ ▬

In listing the elements of a set, we do not list an element more than once because the elements of a set are distinct. Also, the order in which the elements are listed is not relevant. Thus, for example, $\{2, 3\}$ and $\{3, 2\}$ both represent the same set.

If every element of a set A is also an element of a set B, then we say that A is a **subset** of B. If two sets A and B have the same elements, then we say that A **equals** B. For example, $\{1, 2, 3\}$ is a subset of $\{1, 2, 3, 4, 5\}$, and $\{1, 2, 3\}$ equals $\{2, 3, 1\}$.

Classification of Numbers

It is helpful to classify the various kinds of numbers that we deal with in algebra. The **counting numbers,** or **natural numbers,** are the numbers 1, 2, 3, 4, ... (The three dots, called an **ellipsis,** indicate that the pattern continues indefinitely.) As their name implies, these numbers are often used to count things. For example, there are 26 letters in our alphabet; there are 100 cents in a dollar. The **whole numbers** are the numbers 0, 1, 2, 3, . . . , that is, the counting numbers together with 0.

The **integers** are the numbers $\ldots, -3, -2, -1, 0, 1, 2, 3, \ldots$.

These numbers prove useful in many situations. For example, if your checking account has $10 in it and you write a check for $15, you can represent the current balance as $-$5.

Notice that the counting numbers are included among the whole numbers. Each time we expand a number system, such as from the whole numbers to the integers, we do so in order to be able to handle new, and usually more complicated, problems. Thus, the integers allow us to solve problems requiring both positive and negative counting numbers, such as profit/loss, height above/below sea level, temperature above/below 0°F, and so on.

But integers alone are not sufficient for *all* problems. For example, they do not answer the questions "What part of a dollar is 38 cents?" or "What part of a pound is 5 ounces?" To answer such questions, we enlarge our number system to include *rational numbers*. For example, $\frac{38}{100}$ answers the question "What part of a dollar is 38 cents?" and $\frac{5}{16}$ answers the question "What part of a pound is 5 ounces?"

A **rational number** is a number that can be expressed as a quotient a/b of two integers. The integer a is called the **numerator,** and the integer b, which cannot be 0, is called the **denominator.**

Examples of rational numbers are $\frac{3}{4}, \frac{5}{2}, \frac{0}{4}, -\frac{2}{3}$, and $\frac{100}{3}$. Since $a/1 = a$ for any integer a, it follows that the rational numbers contain the integers as a special case.

Real Numbers

Rational numbers may be represented as **decimals.** For example, the rational numbers $\frac{3}{4}, \frac{5}{2}, -\frac{2}{3}$, and $\frac{7}{66}$ may be represented as decimals by merely carrying out the indicated division:

$$\frac{3}{4} = 0.75 \qquad \frac{5}{2} = 2.5 \qquad -\frac{2}{3} = -0.666\ldots \qquad \frac{7}{66} = 0.1060606\ldots$$

Notice that the decimal representations of $\frac{3}{4}$ and $\frac{5}{2}$ terminate, or end. The decimal representations of $-\frac{2}{3}$ and $\frac{7}{66}$ do not terminate, but they do exhibit a pattern of repetition. For $-\frac{2}{3}$, the 6 repeats indefinitely; for $\frac{7}{66}$, the block 06 repeats indefinitely. It can be shown that every rational number may be represented by a decimal that either terminates or is nonterminating with a repeating block of digits, and vice versa.

On the other hand, there are decimals that do not fit into either of these categories. Such decimals represent **irrational numbers.** For example, the decimal $0.1234567891011121314\ldots$, in which we write down the positive integers successively one after the other, will neither repeat (think about it) nor terminate and so represents an irrational number. Thus, every irrational number may be represented by a decimal that neither repeats nor terminates.

Irrational numbers occur naturally. For example, consider the isosceles right triangle whose legs are each of length 1. See Figure 1. The number that equals the length of the hypotenuse is the positive number whose square

FIGURE 1

is 2. It can be shown that this number, symbolized by $\sqrt{2}$, is an irrational number.

Also, the number that equals the ratio of the circumference C to the diameter d of any circle, denoted by the symbol π (the Greek letter pi), is an irrational number. See Figure 2.

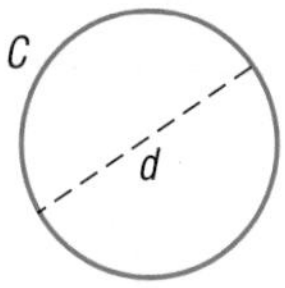

The irrational numbers $\sqrt{2}$ and π have decimal representations that begin as follows:

$$\sqrt{2} = 1.414213\ldots \qquad \pi = 3.14159\ldots$$

In practice, irrational numbers are generally represented by approximations. For example, using the symbol $\approx$ (read as "approximately equal to"), we can write

$$\sqrt{2} \approx 1.4142 \qquad \pi \approx 3.1416$$

We shall discuss the idea of approximation in more detail in the next section.

Other examples of irrational numbers are $\sqrt{3}, \sqrt{5}, \sqrt{7}, \sqrt[3]{2}, \sqrt[3]{3}$ and so on.

FIGURE 2

$\pi = \dfrac{C}{d}$

Together, the rational numbers and irrational numbers form the **real numbers.**

Thus, every decimal may be represented by a real number (either rational or irrational) and, conversely, every real number may be represented by a decimal. It is this feature of real numbers that gives them their practicality. In the physical world, many changing quantities, such as the length of a heated rod, the velocity of a falling object, and so on, are assumed to pass through every possible magnitude from the initial one to the final one as they change. Real numbers in the form of decimals provide a convenient way to measure such quantities as they change. Figure 3 shows the relationship of various types of numbers.

Some Conventions

In algebra, we agree in advance to the following three conventions:

1. Given $a \cdot b + c$ or $c + a \cdot b$, we agree to multiply a by b first and then add the result to c. Thus, we have $2 \cdot 4 + 5 = 8 + 5 = 13$ and $3 + 4 \cdot 5 = 3 + 20 = 23$.
2. In arithmetic, when mixed numbers are used, addition is understood; for example, $2\frac{3}{4}$ means $2 + \frac{3}{4}$. In algebra, use of a mixed number may be confusing because the absence of an operation symbol between two terms is generally taken to mean multiplication. Thus, we prefer not to use mixed numbers in algebra. The expression $2\frac{3}{4}$ is written instead as 2.75 or as $\frac{11}{4}$.
3. Parentheses are used in algebra to group terms and indicate precedence of operations. Thus, $(2 + 3) \cdot 4$ is an instruction to add 2 and 3 and then multiply by 4, so $(2 + 3) \cdot 4 = 5 \cdot 4 = 20$. When parentheses are nested,

FIGURE 3

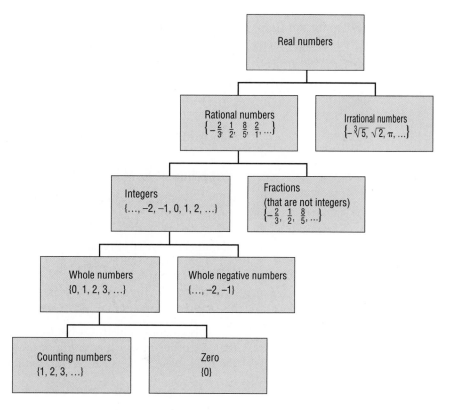

the inside parentheses are evaluated first, followed in order by the outer ones. For example,

$$3 - \{2 - [4 + 5 \cdot (8 - 2)]\} = 3 - \{2 - [4 + 5 \cdot 6]\}$$
$$= 3 - \{2 - [4 + 30]\}$$
$$= 3 - \{2 - 34\}$$
$$= 3 - \{-32\}$$
$$= 3 + 32 = 35$$

Inside first
Second
Third

Notice in the preceding expression that we used braces { } and brackets [ ] to assist us in seeing the relative order of operations. Calculators and computers, needing no such help, will merely nest parentheses. Thus, a computer "sees" the expression

$$3 - \{2 - [4 + 5 \cdot (8 - 2)]\}$$

as

$$3 - (2 - (4 + 5 \cdot (8 - 2)))$$

and will perform the operations from inside to outside.

When we divide two expressions, as in

$$\frac{2 + 3}{4 + 8}$$

it is understood that the division bar acts like parentheses; that is,

$$\frac{2 + 3}{4 + 8} = \frac{(2 + 3)}{(4 + 8)}$$

The following list gives the rules for the order of operations.

Rules for the Order
of Operations

1. Begin with the innermost parentheses and work outward. Remember that in dividing two expressions the numerator and denominator are treated as if they were enclosed in parentheses.
2. Perform multiplications and divisions, working from left to right.
3. Perform additions and subtractions, working from left to right.

Now work Problem 21 in the exercise set at the end of this section.

Properties of Real Numbers

We begin with an example.

EXAMPLE   3

Commutative Properties

(a) $3 + 5 = 8$ (b) $2 \cdot 3 = 6$
    $5 + 3 = 8$ $3 \cdot 2 = 6$
    $3 + 5 = 5 + 3$ $2 \cdot 3 = 3 \cdot 2$ ▬

This example illustrates the first property of real numbers, which states that the order in which addition or multiplication takes place will not affect the final result. This is called the **commutative property.**

Commutative Properties

$$a + b = b + a \qquad a \cdot b = b \cdot a \qquad (1)$$

Here, and in the properties listed below and on pages 9–12, a, b, and c represent real numbers.

EXAMPLE   4

Associative Properties

(a) $2 + (3 + 4) = 2 + 7 = 9$ (b) $2 \cdot (3 \cdot 4) = 2 \cdot 12 = 24$
    $(2 + 3) + 4 = 5 + 4 = 9$ $(2 \cdot 3) \cdot 4 = 6 \cdot 4 = 24$
    $2 + (3 + 4) = (2 + 3) + 4$ $2 \cdot (3 \cdot 4) = (2 \cdot 3) \cdot 4$ ▬

The way we add or multiply three real numbers will not affect the final result. Thus, expressions such as $2 + 3 + 4$ and $3 \cdot 4 \cdot 5$ present no ambiguity, even though addition and multiplication are performed on one pair of numbers at a time. This property is called the **associative property.**

Associative Properties

$$a + (b + c) = (a + b) + c = a + b + c \qquad (2a)$$
$$a \cdot (b \cdot c) = (a \cdot b) \cdot c = a \cdot b \cdot c \qquad (2b)$$

The next property is perhaps the most important.

Distributive Property

$$a \cdot (b + c) = a \cdot b + a \cdot c \qquad (3a)$$
$$(a + b) \cdot c = a \cdot c + b \cdot c \qquad (3b)$$

The **distributive property** may be used in two different ways.

E X A M P L E 5

Distributive Property

(a) $2 \cdot (x + 3) = 2 \cdot x + 2 \cdot 3 = 2x + 6$ Use to remove parentheses.
(b) $3x + 5x = (3 + 5)x = 8x$ Use to combine two expressions.   ▬

Now work Problem 35.

The real numbers 0 and 1 have unique properties.

E X A M P L E 6

Identity Properties

(a) $4 + 0 = 0 + 4 = 4$ (b) $3 \cdot 1 = 1 \cdot 3 = 3$ ▬

The properties of 0 and 1 illustrated in Example 6 are called the **identity properties.**

Identity Properties

$$0 + a = a + 0 = a \qquad a \cdot 1 = 1 \cdot a = a \qquad (4)$$

We call 0 the **additive identity** and 1 the **multiplicative identity.**
For each real number a, there is a real number $-a$, called the **additive inverse** of a, having the following property:

Additive Inverse Property

$$a + (-a) = -a + a = 0 \qquad (5a)$$

E X A M P L E 7

Finding an Additive Inverse

(a) The additive inverse of 6 is -6, because $6 + (-6) = 0$.
(b) The additive inverse of -8 is $-(-8) = 8$, because $-8 + 8 = 0$.   ▬

The additive inverse of a, that is, $-a$, is often called the *negative* of a or the *opposite* of a. The use of such terms can be dangerous because they suggest that the additive inverse is a negative number, which it may not be. For example, the additive inverse of -3, or $-(-3)$, equals 3, a positive number.

For each *nonzero* real number a, there is a real number $1/a$, called the **multiplicative inverse** of a, having the following property:

Multiplicative Inverse Property

$$a \cdot \frac{1}{a} = \frac{1}{a} \cdot a = 1 \quad \text{if } a \neq 0 \qquad (5b)$$

The multiplicative inverse $1/a$ of a nonzero real number a is also referred to as the **reciprocal** of a.

E X A M P L E   8   Finding a Reciprocal

(a) The reciprocal of 6 is $\frac{1}{6}$, because $6 \cdot \frac{1}{6} = 1$.

(b) The reciprocal of -3 is $\frac{1}{-3}$, because $-3 \cdot \frac{1}{-3} = 1$.

(c) The reciprocal of $\frac{2}{3}$ is $\frac{3}{2}$, because $\frac{2}{3} \cdot \frac{3}{2} = 1$.   ∎

With these properties for adding and multiplying real numbers, we can now define the operations of subtraction and division as follows:

The **difference** $a - b$, also read "a less b" or "a minus b," is defined as

$$a - b = a + (-b) \qquad (6)$$

Thus, to subtract b from a, add the opposite of b to a.

If b is a nonzero real number, the **quotient** a/b, also read as "a divided by b" or "the ratio of a to b," is defined as

$$\frac{a}{b} = a \cdot \frac{1}{b} \quad \text{if } b \neq 0 \qquad (7)$$

E X A M P L E   9   Working with Differences and Quotients

(a) $8 - 5 = 8 + (-5) = 3$ (b) $4 - 9 = 4 + (-9) = -5$

(c) $\frac{5}{8} = (5)\left(\frac{1}{8}\right)$ (d) $\frac{20}{-5} = (20)\left(\frac{1}{-5}\right) = -4$ ∎

The additional properties of real numbers given next will also be useful. Although these can be derived or proved using equations (1)–(5), we shall not attempt to prove them here. As before, a, b, c, and d denote real numbers in the properties listed.

For any number a, the product of a times 0 is always 0; that is,

Multiplication by Zero

$$a \cdot 0 = 0 \qquad\qquad (8)$$

For a nonzero number a,

Division Properties

$$\frac{0}{a} = 0 \qquad \frac{a}{a} = 1 \qquad \text{if } a \neq 0 \qquad\qquad (9)$$

Note: Division by 0 is *not allowed*. One reason is to avoid the following difficulty: $\frac{2}{0} = x$ means to find x such that $0 \cdot x = 2$. But $0 \cdot x$ equals 0 for all x, so there is *no* number x such that $\frac{2}{0} = x$.

Rules of Signs

$$a(-b) = -(ab) \qquad (-a)b = -(ab) \qquad (-a)(-b) = ab$$

$$-(-a) = a \qquad\qquad \frac{a}{-b} = \frac{-a}{b} = -\frac{a}{b} \qquad \frac{-a}{-b} = \frac{a}{b} \qquad (10)$$

E X A M P L E   10

Applying the Rules of Signs

(a) $2(-3) = -(2 \cdot 3) = -6$ (b) $(-3)(-5) = 3 \cdot 5 = 15$

(c) $\dfrac{3}{-2} = \dfrac{-3}{2} = -\dfrac{3}{2}$ (d) $\dfrac{-4}{-9} = \dfrac{4}{9}$ (e) $\dfrac{x}{2} = \dfrac{1}{2} \cdot x$ ▬

If c is a nonzero number, then

Cancellation Properties

$$ac = bc \quad \text{implies} \quad a = b \qquad \text{if } c \neq 0$$

$$\frac{ac}{bc} = \frac{a}{b} \qquad\qquad\qquad\qquad \text{if } b \neq 0, c \neq 0 \qquad (11)$$

Zero-Product Property

$$\text{If } ab = 0, \text{ then } a = 0 \text{ or } b = 0, \text{ or both.} \qquad (12)$$

The Zero-Product Property is used often to draw conclusions. For example, if twice a number x is 0 (that is, if $2x = 0$), then the number x must be 0.

Arithmetic of Quotients

$$\frac{a}{b} + \frac{c}{d} = \frac{ad + bc}{bd} \qquad\qquad \text{if } b \neq 0, d \neq 0 \qquad\qquad (13)$$

$$\frac{a}{b} \cdot \frac{c}{d} = \frac{ac}{bd} \qquad\qquad \text{if } b \neq 0, d \neq 0 \qquad\qquad (14)$$

$$\frac{\dfrac{a}{b}}{\dfrac{c}{d}} = \frac{a}{b} \cdot \frac{d}{c} = \frac{ad}{bc} = \qquad \text{if } b \neq 0, c \neq 0, d \neq 0 \qquad (15)$$

E X A M P L E   11   Adding Quotients

$$\frac{2}{3} + \frac{5}{2} \underset{\substack{\uparrow \\ \text{By equation (13)}}}{=} \frac{2 \cdot 2 + 3 \cdot 5}{3 \cdot 2} = \frac{4 + 15}{6} = \frac{19}{6}$$

 Now work Problem 29.

E X A M P L E   12   Subtracting Quotients

$$\frac{3}{5} - \frac{2}{3} \underset{\substack{\uparrow \\ \text{By equation (6)}}}{=} \frac{3}{5} + \left(-\frac{2}{3}\right) \underset{\substack{\uparrow \\ \text{By equation (10)}}}{=} \frac{3}{5} + \frac{-2}{3}$$

$$\underset{\substack{\uparrow \\ \text{By equation (13)}}}{=} \frac{3 \cdot 3 + 5 \cdot (-2)}{5 \cdot 3} = \frac{9 + (-10)}{15} = \frac{-1}{15}$$

E X A M P L E   13   Multiplying Quotients

$$\frac{8}{3} \cdot \frac{15}{4} \underset{\substack{\uparrow \\ \text{By equation (14)}}}{=} \frac{8 \cdot 15}{3 \cdot 4} = \frac{2 \cdot 4 \cdot 3 \cdot 5}{3 \cdot 4 \cdot 1} \underset{\substack{\uparrow \\ \text{By equation (11)}}}{=} \frac{2 \cdot 5}{1} = 10$$

Note: We follow the common practice of using slash marks to indicate cancellation. Slanting the cancellation marks in different directions for different factors, as in Example 13, is a good practice to follow, since it will help in checking for errors.

EXAMPLE 14

Dividing Quotients

$$\frac{\dfrac{3}{5}}{\dfrac{7}{9}} = \frac{3}{5} \cdot \frac{9}{7} = \frac{3 \cdot 9}{5 \cdot 7} = \frac{27}{35}$$

By equation (14)

By equation (15)

Note: In writing quotients, we shall follow the usual convention and write the quotient in **lowest terms;** that is, we write it so that any common divisors of the numerator and the denominator have been removed using the cancellation properties, equation (11). Thus,

$$\frac{90}{24} = \frac{15 \cdot 6}{4 \cdot 6} = \frac{15}{4}$$

$$\frac{24x^2}{18x} = \frac{4 \cdot 6 \cdot x \cdot x}{3 \cdot 6 \cdot x} = \frac{4x}{3} \qquad x \neq 0$$

Now work Problem 31.

The Real Number Line

It can be shown that there is a one-to-one correspondence between real numbers and points on a line. That is, every real number corresponds to a point on the line and, conversely, each point on the line has a unique real number associated with it. We establish this correspondence of real numbers with points on a line in the following manner.

We start with a line that is, for convenience, drawn horizontally. Pick a point on the line and label it *O*, for **origin.** Then pick another point some fixed distance to the right of *O* and label it *U*, for **unit,** as shown in Figure 4.

The fixed distance, which may be 1 inch, 1 centimeter, 1 light-year, or any unit distance, determines the **scale.** We associate the real number 0 with the origin *O* and the number 1 with the point *U*. Refer now to Figure 5. The point to the right of *U* that is twice as far from *O* as *U* is associated with the number 2. The point to the right of *U* that is three times as far from *O* as *U* is associated with the number 3. The point midway between *O* and *U* is assigned the number 0.5, or $\frac{1}{2}$. Corresponding points to the left of the origin *O* are assigned the numbers $-\frac{1}{2}, -1, -2, -3$, and so on, depending on how far they are from *O*. Notice in Figure 5 that we placed an arrowhead on the right end of the line to indicate the direction in which the assigned numbers increase. Figure 5 also shows the points associated with the irrational numbers $\sqrt{2}$ and π.

FIGURE 4
Horizontal line

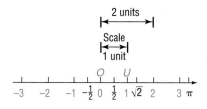

FIGURE 5
Real number line

The real number associated with a point *P* is called the **coordinate** of *P*, and the line whose points have been assigned coordinates is called the **real number line.**

The real number line divides the real numbers into three classes, as shown in Figure 6:

FIGURE 6

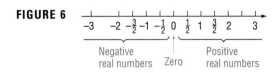

1. The **negative real numbers** are the coordinates of points to the left of the origin O.
2. The real number **zero** is the coordinate of the origin O.
3. The **positive real numbers** are the coordinates of points to the right of the origin O.

Negative and positive numbers have the following multiplication properties:

Multiplication Properties of Positive and Negative Numbers

1. The product of two positive numbers is a positive number.
2. The product of two negative numbers is a positive number.
3. The product of a positive number and a negative number is a negative number.

Inequalities

FIGURE 7

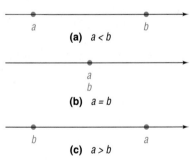

(a) $a < b$

(b) $a = b$

(c) $a > b$

An important property of the real number line follows from the fact that, given two numbers (points) a and b, either a is to the left of b, a equals b, or a is to the right of b. See Figure 7.

If a is to the left of b, we say "a is less than b" and write $a < b$. If a is to the right of b, we say "a is greater than b" and write $a > b$. Of course, if a equals b, we write $a = b$. If a is either less than or equal to b, we write $a \leq b$. Similarly, $a \geq b$ means that a is either greater than or equal to b. Collectively, the symbols $<$, $>$, $\leq$, and $\geq$ are called **inequality symbols.** A formal definition of *less than* and *greater than* follows.

Let a and b be two real numbers. We say that a **is less than** b, written $a < b$, or, equivalently, that b **is greater than** a, written $b > a$, if the difference $b - a$ is positive.

Note that $a < b$ and $b > a$ means the same thing. Thus, it does not matter whether we write $2 < 3$ or $3 > 2$.

E X A M P L E   15

Using Inequality Symbols

(a) $3 < 7$ (b) $-8 > -16$ (c) $-6 < 0$

(d) $-8 < -4$ (e) $4 > -1$ (f) $8 > 0$ ▬

In Example 15(a), we conclude that $3 < 7$ either because 3 is to the left of 7 on the real number line or because the difference $7 - 3 = 4$, a positive real number.

Similarly, we conclude in Example 15(b) that $-8 > -16$ either because -8 lies to the right of -16 on the real number line or because the difference $-8 - (-16) = -8 + 16 = 8$, a positive real number.

Look again at Example 15. You may find it useful to observe that the inequality symbol always points in the direction of the smaller number.

Statements of the form $a < b$ or $b > a$ are called **strict inequalities,** while statements of the form $a \leq b$ or $b \geq a$ are called **nonstrict inequalities.** An **inequality** is a statement in which two expressions are related by an inequality symbol. The expressions are referred to as the **sides** of the inequality.

Based on the discussion thus far, we conclude that

> $a > 0$ is equivalent to a is positive
>
> $a < 0$ is equivalent to a is negative

Thus, we sometimes read $a > 0$ by saying that "a is positive." If $a \geq 0$, then either $a > 0$ or $a = 0$, and we may read this as "a is nonnegative."

 Now work Problem 59.

Absolute Value

The *absolute value* of a number a is the distance to the origin from the point whose coordinate is a. For example, the point whose coordinate is -4 is 4 units from the origin. The point whose coordinate is 3 is 3 units from the origin. See Figure 8. Thus, the absolute value of -4 is 4, and the absolute value of 3 is 3.

A more formal definition of absolute value is given next.

FIGURE 8

4 units   3 units

-5 -4 -3 -2 -1 0 1 2 3 4

> The **absolute value** of a real number *a,* denoted by the symbol $|a|$, is defined by the rules
>
> $$|a| = a \quad \text{if } a \geq 0 \qquad \text{and} \qquad |a| = -a \quad \text{if } a < 0$$

For example, since $-4 < 0$, the second rule must be used to get $|-4| = -(-4) = 4$.

E X A M P L E   16

Computing Absolute Value

(a) $|8| = 8$ (b) $|0| = 0$ (c) $|-15| = 15$ ■

Now work Problem 75.

Look again at Figure 8. The distance from the point whose coordinate is -4 to the point whose coordinate is 3 is 7 units. This distance is the difference $3 - (-4)$, obtained by subtracting the smaller coordinate from the larger. However, since $|3 - (-4)| = |7| = 7$ and $|-4 - 3| = |-7| = 7$, we can use absolute value to calculate the distance between two points without being concerned about which coordinate is smaller.

If P and Q are two points on a real number line with coordinates a and b, respectively, the **distance between P and Q**, denoted by $d(P, Q)$, is

$$d(P, Q) = |b - a|$$

Since $|b - a| = |a - b|$, it follows that $d(P, Q) = d(Q, P)$.

E X A M P L E   17

Finding Distance on a Number Line

Let P, Q, and R be points on a real number line with coordinates -5, 7, and -3, respectively. Find the distance

(a) Between P and Q (b) Between Q and R

Solution   (a) $d(P, Q) = |7 - (-5)| = |12| = 12$ (see Figure 9)
(b) $d(Q, R) = |-3 - 7| = |-10| = 10$

FIGURE 9

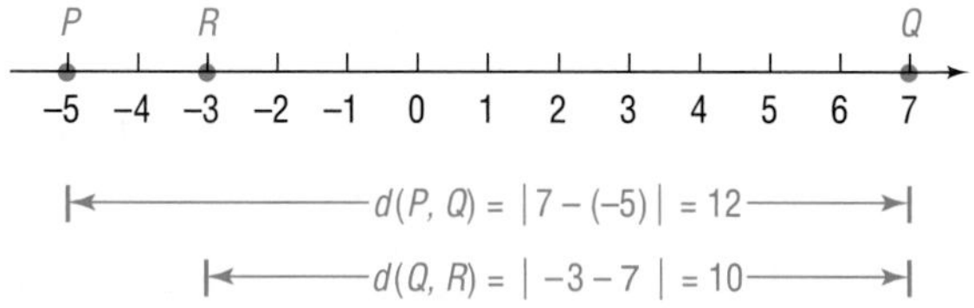

$$d(P, Q) = |7 - (-5)| = 12$$
$$d(Q, R) = |-3 - 7| = 10$$

■

Constants and Variables

We said earlier that in algebra we use letters such as x, y, a, b, and c to represent numbers. If the letter used is to represent *any* number from a given set of numbers, it is called a **variable.** A **constant** is either a fixed number, such as 5 or $\sqrt{3}$, or a letter that represents a fixed (possibly unspecified) number.

E X A M P L E   18

Area of a Circle

The formula for the area of a circle is

$$A = \pi r^2$$

FIGURE 10
$A = \pi r^2$

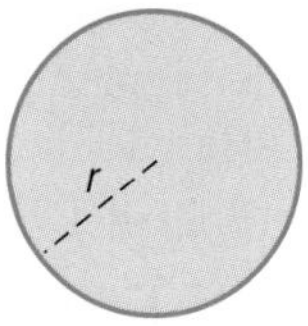

In this formula, r is a variable representing the radius of the circle, A is a variable representing the area of the circle, and π is a constant (≈ 3.14). See Figure 10. ∎

4

In working with expressions or formulas involving variables, the variables may be allowed to take on values from only a certain set of numbers. For example, in the formula for the area A of a circle of radius r, $A = \pi r^2$, the variable r is necessarily restricted to the positive real numbers. In the expression $1/x$, the variable x cannot take on the value 0, since division by 0 is not allowed.

> The set of values that a variable may assume is called the **domain of the variable.**

E X A M P L E   19   Finding the Domain of a Variable

The domain of the variable x in the expression

$$\frac{5}{x - 2}$$

is $\{x | x \neq 2\}$, since, if $x = 2$, the denominator becomes 0, which is not allowed. ∎

E X A M P L E   20   Circumference of a Circle

In the formula for the circumference C of a circle of radius r,

$$C = 2\pi r$$

the domain of the variable r, representing the radius of the circle, is the set of positive real numbers. The domain of the variable C, representing the circumference of the circle, is also the set of positive real numbers. ∎

In describing the domain of a variable, we may use either set notation or words, whichever is more convenient.

Now work Problems 85 and 95.

HISTORICAL FEATURE   The real number system has a history that stretches back at least to the ancient Babylonians (1800 BC). It is remarkable how much the ancient Babylonian attitudes resemble our own. As we stated in the text, the fundamental difficulty with irrational numbers is that they cannot be written as quotients of integers or, equivalently, as repeating or terminating decimals. The Babylonians wrote their numbers in a system based on 60 in the same way that we write ours based on 10. They would carry as many places for π as the accuracy of the problem demanded, just as we now use

$$\pi \approx 3\frac{1}{7} \quad \text{or} \quad \pi \approx 3.1416 \quad \text{or} \quad \pi \approx 3.14159 \quad \text{or} \quad \pi \approx 3.141592653589$$

depending on how accurate we need to be.

Things were very different for the Greeks, whose number system allowed only rational numbers. When it was discovered that $\sqrt{2}$ was not a rational number, this was regarded as a fundamental flaw in the number concept. So serious was the matter that the Pythagorean Brotherhood (an early mathematical society) is said to have drowned one of its members for revealing this terrible secret. Greek mathematicians then turned away from the number concept, expressing facts about whole numbers in terms of line segments.

In astronomy, however, Babylonian methods, including the Babylonian number system, continued to be used. Simon Stevin (1548–1620), probably using the Babylonian system as a model, invented the decimal system, complete with rules of calculation, in 1585. [Others, for example, al-Kashi of Samarkand (d. 1424), had made some progress in the same direction.] The decimal system so effectively conceals the difficulties that the need for more logical precision began to be felt only in the early 1800s. Around 1880, Georg Cantor (1845–1918) and Richard Dedekind (1831–1916) gave precise definitions of real numbers. Cantor's definition, although more abstract and precise, has its roots in the decimal (and hence Babylonian) numerical system.

Sets and set theory were a spin-off of the research that went into clarifying the foundations of the real number system. Set theory has developed into a large discipline of its own, and many mathematicians regard it as the foundation upon which modern mathematics is built. Cantor's discoveries that infinite sets can also be counted and that there are different sizes of infinite sets are among the most astounding results of modern mathematics.

HISTORICAL PROBLEMS *The Babylonian number system was based on* 60. *Thus* 2,30 *means* $2 + \frac{30}{60} = 2.5$, *and* 4,25,14 *means*

$$4 + \frac{25}{60} + \frac{14}{60^2} = 4 + \frac{1514}{3600} = 4.42055555\ldots$$

1. What are the following numbers in Babylonian notation?
    (a) $1\frac{1}{3}$ (b) $2\frac{5}{6}$
2. What are the following Babylonian numbers when written as fractions and as decimals?
    (a) 2,20   (b) 4,52,30   (c) 3,8,29,44

1.1 | EXERCISES

In Problems 1–12, express each rational number as a decimal, either repeating or terminating.

1. $\dfrac{1}{3}$ **2.** $\dfrac{3}{5}$ **3.** $\dfrac{1}{8}$ **4.** $\dfrac{10}{3}$ **5.** $-\dfrac{8}{5}$ **6.** $-\dfrac{16}{3}$

7. $\dfrac{1}{9}$ **8.** $\dfrac{1}{10}$ **9.** $\dfrac{4}{25}$ **10.** $-\dfrac{5}{6}$ **11.** $-\dfrac{3}{7}$ **12.** $\dfrac{8}{11}$

In Problems 13–32, evaluate each expression.

13. $9 - 4 + 2$ **14.** $6 - 4 + 3$ **15.** $-6 + 4 \cdot 3$ **16.** $8 - 4 \cdot 2$

17. $4 + 5 - 8$ **18.** $8 - 3 - 4$ **19.** $4 + \dfrac{1}{3}$ **20.** $2 - \dfrac{1}{2}$

21. $6 - [3 \cdot 5 + 2 \cdot (3 - 2)]$ **22.** $2 \cdot [8 - 3(4 + 2)] - 3$ **23.** $2 \cdot (3 - 5) + 8 \cdot 2 - 1$

24. $1 - (4 \cdot 3 - 2 + 2)$ **25.** $10 - [6 - 2 \cdot 2 + (8 - 3)] \cdot 2$ **26.** $2 - 5 \cdot 4 - [6 \cdot (3 - 4)]$

27. $5 - 3 \cdot \dfrac{1}{2}$

28. $5 + 4 \cdot \dfrac{1}{3}$

29. $\dfrac{4}{5} + \dfrac{8}{3}$

30. $\dfrac{2}{5} - \dfrac{4}{3}$

31. $\dfrac{\frac{16}{3} - \frac{1}{2}}{\frac{1}{3}}$

32. $\dfrac{\frac{14}{3} + \frac{1}{2}}{\frac{5}{4}}$

In Problems 33–44, use the distributive property to remove the parentheses.

33. $6(x + 4)$

34. $4(2x - 1)$

35. $x(x - 4)$

36. $4x(x + 3)$

37. $(x + 2)(x + 4)$

38. $(x + 5)(x + 1)$

39. $(x - 2)(x + 1)$

40. $(x - 4)(x + 1)$

41. $(x - 8)(x - 2)$

42. $(x - 4)(x - 2)$

43. $(x + 2)(x - 2)$

44. $(x - 3)(x + 3)$

45. Explain to a friend how the distributive property is used to justify the fact that $2x + 3x = 5x$.

46. Explain to a friend why $2 + 3 \cdot 4 = 14$, while $(2 + 3) \cdot 4 = 20$.

47. Explain why $2(3 \cdot 4)$ is not equal to $(2 \cdot 3) \cdot (2 \cdot 4)$.

48. Explain why $\dfrac{4 + 3}{2 + 5}$ is not equal to $\dfrac{4}{2} + \dfrac{3}{5}$.

49. Does $\frac{1}{3}$ equal 0.333? If not, which is larger? By how much?

50. Does $\frac{2}{3}$ equal 0.666? If not, which is larger? By how much?

51. Is subtraction commutative? Support your conclusion with an example.

52. Is subtraction associative? Support your conclusion with an example.

53. Is division commutative? Support your conclusion with an example.

54. Is division associative? Support your conclusion with an example.

55. If $2 = x$, why does $x = 2$?

56. If $x = 5$, why does $x^2 + x = 30$?

57. On a real number line, label the points with coordinates of $0, 1, -1, \frac{5}{2}, -2.5, \frac{3}{4}$, and 0.25.

58. Repeat Problem 57 for the coordinates $0, -2, 2, -1.5, \frac{3}{2}, \frac{1}{3}$, and $\frac{2}{3}$.

In Problems 59–64, write each statement as an inequality.

59. x is positive

60. z is negative

61. x is less than 3

62. y is greater than -4

63. The absolute value of x is less than or equal to 1.

64. The reciprocal of x is greater than or equal to 2.

In Problems 65–74, $a > 0$ and $b < 0$. Tell whether the given expression is positive, negative, or 0.

65. $|b|$

66. $a \cdot b$

67. b^2

68. a/b

69. $a - b$

70. $b - a$

71. $|a| + |b|$

72. $|ab|$

73. $b + |b|$

74. $a + |a|$

In Problems 75–84, find the value of each expression if $x = 3$ and $y = -2$.

75. $|x + y|$

76. $|x - y|$

77. $|x| + |y|$

78. $|x| - |y|$

79. $\dfrac{|x|}{x}$

80. $\dfrac{|y|}{y}$

81. $|4x - 5y|$

82. $|3x + 2y|$

83. $||4x| - |5y||$

84. $3|x| + 2|y|$

In Problems 85–94, determine the domain of the variable x in each expression.

85. $\dfrac{4}{x - 5}$

86. $\dfrac{-6}{x + 4}$

87. $\dfrac{x}{x + 4}$

88. $\dfrac{x - 2}{x - 6}$

89. $\dfrac{1}{x} + \dfrac{1}{x - 1}$

90. $\dfrac{3}{x(x - 4)}$

91. $\dfrac{x + 1}{x(x + 2)}$

92. $\dfrac{2}{x + 1} + \dfrac{3}{x + 2}$

93. $\dfrac{x + 1}{x^2 + 4}$

94. $\dfrac{x - 3}{x^2 + 6}$

In Problems 95–104, express each statement as an equation involving the indicated variables and constant. What is the domain of each variable?

95. Area of a Rectangle The area A of a rectangle is the product of its length l times its width w.

96. Perimeter of a Rectangle The perimeter P of a rectangle is twice the sum of its length l and its width w.

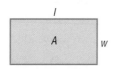

97. Circumference of a Circle The circumference C of a circle is the product of π times its diameter d.

98. Area of a Triangle The area A of a triangle is one-half the product of its base b times its height h.

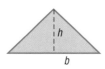

99. Area of an Equilateral Triangle The area A of an equilateral triangle is $\sqrt{3}/4$ times the square of the length x of one side.

100. Perimeter of an Equilateral Triangle The perimeter P of an equilateral triangle is 3 times the length x of one side.

101. Volume of a Sphere The volume V of a sphere is the product of $\frac{4}{3}$ times π times the cube of the radius r.

102. Surface Area of a Sphere The surface area S of a sphere is the product of 4 times π times the square of the radius r.

103. Volume of a Cube The volume V of a cube is the cube of the length x of a side.

104. Surface Area of a Cube The surface area S of a cube is 6 times the square of the length x of a side.

In Problems 105–108, use the formula $C = \frac{5}{9}(F - 32)$ for converting degrees Fahrenheit into degrees Celsius to find the Celsius measure of each Fahrenheit temperature.

105. $F = 32°$ **106.** $F = 212°$ **107.** $F = 77°$ **108.** $F = -4°$

109. Manufacturing Cost The weekly production cost C of manufacturing x watches is given by the formula $C = 4000 + 2x$, where the variable C is in dollars.
  (a) What is the cost of producing 1000 watches?
  (b) What is the cost of producing 2000 watches?

110. Balancing a Checkbook At the beginning of the month, Mike had a balance of $210 in his checking account. During the next month, he deposited $80, wrote a check for $120, made an-

other deposit of $25, wrote two checks for $60 and $32, and was assessed a monthly service charge of $5. What was his balance at the end of the month?

111. Find the value of $1 + 2 + 3 + 4 + \cdots + 99$. [**Hint:** Regroup the terms as $(1 + 99) + (2 + 98) + \cdots$ and proceed from there.]

112. Find the value of $1 + 3 + 5 + 7 + \cdots + 99$.

113. Find the value of $-2 - 4 - 6 - 8 - \cdots - 98$.

114. Find the value of $1 - 2 + 3 - 4 + 5 - 6 + \cdots + 97 - 98 + 99$.

115. Consumer Survey A recent Gallup poll on "body image" found that two-thirds of college-educated women want a more muscular, "hard-bodied" look. If 300 college-educated women were involved in this poll, how many would have answered "yes" to the question "Do you want a more muscular, hard-bodied look?"

116. On the real number line in the figure, we have placed a circle of radius 1 so that the circle touches the line at 0. This point on the circle is labeled *P*. If the circle now rolls without slippage in the positive direction along the real number line, at what coordinates will the point *P* again come into contact with the line?

[**Hint:** $C = 2\pi r$.]

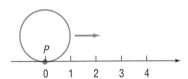

117. Are there any real numbers that are both rational and irrational? Are there any real numbers that are neither? Explain your reasoning.

118. Explain why the sum of a rational number and an irrational number must be irrational.

119. What rational number does the repeating decimal 0.9999 . . . equal?

120. Is there a positive real number "closest" to 0?

121. I'm thinking of a number! It lies between 1 and 10; its square is rational and lies between 1 and 10. The number is larger than π. Correct to two decimal places, name the number. Now think of your own number, describe it, and challenge a fellow student to name it.

122. Write a brief paragraph that illustrates the similarities and differences between "less than" ($<$) and "less than or equal" ($\leq$).

1.2 APPROXIMATIONS; CALCULATORS

1 Distinguish between Rounding and Truncating a Decimal
2 Perform Arithmetic Operations on a Calculator

1 In approximating decimals, we either *round off* or *truncate* to a given number of decimal places. The number of places establishes the location of the *final digit* in the decimal approximation.

> **Truncation:** Drop all the digits that follow the specified final digit in the decimal.
> **Rounding:** Identify the specified final digit in the decimal. If the next digit is 5 or more, add 1 to the final digit; if the next digit is 4 or less, leave the final digit as it is. Now truncate following the final digit.

E X A M P L E 1

Approximating a Decimal to Two Places

Approximate 20.98752 to two decimal places by

(a) Truncating   (b) Rounding

Solution   For 20.98752, the final digit is 8, since it is two decimal places from the decimal point.

(a) To truncate, we remove all digits following the final digit 8. Thus, the truncation of 20.98752 to two decimal places is 20.98.

(b) The digit following the final digit 8 is the digit 7. Since 7 is 5 or more, we add 1 to the final digit 8 and truncate. The rounded form of 20.98752 to two decimal places is 20.99.

E X A M P L E   2   Approximating a Decimal to Two and Four Places

Number	Rounded to Two Decimal Places	Rounded to Four Decimal Places	Truncated to Two Decimal Places	Truncated to Four Decimal Places
(a) 3.14159	3.14	3.1416	3.14	3.1415
(b) 0.056128	0.06	0.0561	0.05	0.0561
(c) 893.46125	893.46	893.4613	893.46	893.4612	■

 Now work Problem 1.

Calculators

Calculators are finite machines. As a result, they are incapable of displaying decimals that contain a large number of digits. For example, some calculators are capable of displaying only eight digits. When a number requires more than eight digits, the calculator either truncates or rounds. To see how your calculator handles decimals, divide 2 by 3. How many digits do you see? Is the last digit a 6 or a 7? If it is a 6, your calculator truncates; if it is a 7, your calculator rounds.

There are different kinds of calculators. An **arithmetic** calculator can only add, subtract, multiply, and divide numbers; therefore, this type is not adequate for this course. **Scientific** calculators have all the capabilities of arithmetic calculators and also contain **function keys** labeled ln, log, sin, cos, tan, x^y, inv, and so on. As you proceed through this text, you will discover how to use many of the function keys. **Graphing** calculators have all the capabilities of scientific calculators and contain a screen on which graphs can be displayed.

For those who have access to a graphing calculator, we have included comments, examples, and exercises marked with a 🖩 , indicating that a graphing calculator is required. We have also included an appendix that explains some of the capabilities of a graphing calculator. The 🖩 comments, examples, and exercises may be omitted without loss of continuity, if so desired.

Calculator Operations

2   Arithmetic operations on your calculator are generally shown as keys labeled as follows:

$\boxed{+}$ Addition $\boxed{-}$ Subtraction $\boxed{\times}$ Multiplication $\boxed{\div}$ Division

Your calculator also has the following key:

$\boxed{AC}$ or $\boxed{C}$ To clear the memory and the display window of previous entries.

Now, let's look at a few examples.

E X A M P L E 3   Using a Calculator

Evaluate: $8 + 9 \cdot 6$

Solution   Keystrokes: 8 + 9 × 6 =

Display: 8   9   6   62

E X A M P L E 4   Using a Calculator

Evaluate: $\dfrac{12}{6} + 5$

Solution   Keystrokes: 12 ÷ 6 + 5 =

Display: 12   6   2   5   7

E X A M P L E 5   Using a Calculator

Evaluate: $3 + 2 \cdot \dfrac{9}{3} - 6$

Solution   Keystrokes: 3 + 2 × 9 ÷

Display: 3   2   9   18

Keystrokes: 3 − 6 =

Display: 3   9   6   3

These examples should convince you that a calculator that uses the algebraic system does arithmetic according to the agreed-upon conventions regarding order of operations. However, there are times when you need to be careful.

The next example illustrates a use of the keys for parentheses,

(and   )

E X A M P L E 6   Using a Calculator

Evaluate: $\dfrac{22 + 8}{8 + 2}$

Solution   Remember, we treat this expression as if parentheses enclose the numerator and the denominator.

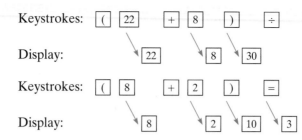

Be careful! If you work the problem in Example 6 without using parentheses, you obtain 22 + 8/8 + 2 = 22 + 1 + 2 = 25.

 Now work Problem 25.

1.2 | EXERCISES

In Problems 1–12, approximate each number (a) rounded and (b) truncated to three decimal places.

1. 18.9526   2. 25.86134   3. 28.65319   4. 99.05249   5. 0.06291   6. 0.05388

7. 9.9985   8. 1.0006   9. $\dfrac{3}{7}$ 10. $\dfrac{5}{9}$ 11. $\dfrac{521}{15}$ 12. $\dfrac{81}{5}$

In Problems 13–36, use a calculator to approximate each expression. Round off your answer to two decimal places.

13. $(8.51)^2$ 14. $(9.62)^2$ 15. $4.1 + (3.2)(8.3)$

16. $(8.1)(4.2) + 6.1$ 17. $(9.6)^2 + (6.4)^2$ 18. $(4.1)^2 + (9.8)^2$

19. $8.6 + \dfrac{10.2}{4.2}$ 20. $9.1 - \dfrac{8.2}{10.2}$ 21. $\dfrac{2.3 - 9.25}{8.91 + 5.4}$

22. $\dfrac{4.73 - 2.4}{81 + 6.39}$ 23. $\dfrac{\pi + 8}{10.2 + 8.6}$ 24. $\dfrac{21.3 - \pi}{6.1 + 8.8}$

25. $(22.6 + 8.5)/81.3 + 21.2$ 26. $(16.5 - 7.2) \cdot 51.2 - 18.6$ 27. $22.6 + 8.5/81.3 + 21.2$

28. $16.5 - 7.2/51.2 - 18.6$ 29. $(22.6 + 8.5)/(81.3 + 21.2)$ 30. $(16.5 - 7.2)/(51.2 - 18.6)$

31. $22.6 + 8.5/(81.3 + 21.2)$ 32. $16.5 - 7.2/(51.2 - 18.6)$ 33. $[42.1(4.11 + 8.6) - 25.3]/7.6$

34. $8.2/[19.3 - 2.2(8.3 + 11.6)]$ 35. $[(4.2 - 8.7)^2 + (5.1 - 2.77)^2]^2$ 36. $[(4.1 - 18.22)/(9.3 - 45.12)]^2$

In Problems 37–40, place the correct symbol, <, =, or >, in each box.

37. $\dfrac{5}{9}\ \square\ 0.5555$ 38. $\dfrac{5}{8}\ \square\ 0.6255$ 39. $\sqrt{3}\ \square\ 1.732$ 40. $\pi\ \square\ \dfrac{22}{7}$

41. **Ecology** In 1900, the level of carbon dioxide in the atmosphere was 290 parts per million parts of atmosphere.* What fraction of the atmosphere was carbon dioxide in 1900? Express this fraction as a decimal.

42. **Ecology** In 1988, the level of carbon dioxide in the atmosphere was 350 parts per million parts of atmosphere.* What fraction of the atmosphere was carbon dioxide in 1988? Express this fraction as a decimal.

43. **Medicine** Women have only four-fifths as many red cells in each drop of their blood as men do. If a typical man has 1000 red cells in a drop of blood, how many red cells does a typical woman have in a drop of blood?

44. On your calculator, divide 8 by 0. What happens?

 45. Compare your calculator to a calculator of a different kind. What similarities do you see? What differences?

46. When do truncation and rounding yield the same result? Explain.

*Source: Testimony of Syukuro Manabe, an atmospheric scientist at Princeton's Geophysical Fluid Dynamics Laboratory, before the U.S. Senate Energy and Natural Resources Committee, June 1988.

1.3 | INTEGER EXPONENTS

1 Evaluate Expressions Containing Exponents
2 Work with the Laws of Exponents
3 Use a Calculator to Evaluate Exponents
4 Use Scientific Notation

Integer exponents provide a shorthand device for representing repeated multiplications of a real number.

If a is a real number and n is a positive integer, then the **symbol a^n** represents the product of n factors of a. That is,

$$a^n = \underbrace{a \cdot a \cdot \ldots \cdot a}_{n \text{ factors}}$$

(1)

Here it is understood that $a^1 = a$.

In particular, we have

$a^1 = a$
$a^2 = a \cdot a$
$a^3 = a \cdot a \cdot a$

and so on.

1

In the expression a^n, a is called the **base** and n is called the **exponent,** or **power.** We read a^n as "a raised to the power n" or as "a to the nth power." We usually read a^2 as "a squared" and a^3 as "a cubed."

E X A M P L E 1   Evaluating Expressions Containing Exponents

(a) $2^3 = 2 \cdot 2 \cdot 2 = 8$
(b) $5^2 = 5 \cdot 5 = 25$
(c) $10^1 = 10$
(d) $(-2)^4 = (-2)(-2)(-2)(-2) = 16$
(e) $-2^4 = -(2 \cdot 2 \cdot 2 \cdot 2) = -16$

In words, 2 cubed equals 8; 5 squared equals 25; 10 to the first power equals 10; and so on. ∎

Notice the difference between Examples 1(d) and 1(e): The exponent applies only to the symbol or parenthetical expression immediately preceding it.

We define a raised to a negative power as follows:

If n is a positive integer and if a is a nonzero real number, then we define

$$a^{-n} = \frac{1}{a^n} \qquad \text{if } a \neq 0 \qquad\qquad (2)$$

Thus, whenever you encounter a negative exponent, think "reciprocal."

EXAMPLE 2   Evaluating Expressions Containing Negative Exponents

(a) $2^{-3} = \dfrac{1}{2^3} = \dfrac{1}{8}$

(b) $x^{-4} = \dfrac{1}{x^4}$

(c) $\left(\dfrac{1}{5}\right)^{-2} = \dfrac{1}{\left(\dfrac{1}{5}\right)^2} = \dfrac{1}{\dfrac{1}{25}} = 25$

  Now work Problem 3.

If a is a nonzero number, we define

$$a^0 = 1 \qquad \text{if } a \neq 0 \qquad\qquad (3)$$

Notice that we do not allow the base a to be 0 in a^{-n} or in a^0.

Laws of Exponents

Several general rules can be used when dealing with exponents. The first rule we consider is used when multiplying two expressions that have the same base:

$$\underbrace{3^2}_{\substack{\uparrow \\ \text{Same base}}} \cdot \underbrace{3^4}_{} = \underbrace{(3 \cdot 3)}_{2 \text{ factors}} \underbrace{(3 \cdot 3 \cdot 3 \cdot 3)}_{4 \text{ factors}} = \underbrace{3 \cdot 3 \cdot 3 \cdot 3 \cdot 3 \cdot 3}_{6 \text{ factors}} = 3^6$$

Multiplying ↓ ; Sum of powers 2 and 4; Same base

In general, if a is any real number and if m and n are positive integers, we have

$$a^m \cdot a^n = \underbrace{(a \cdot a \cdot \ldots \cdot a)}_{m \text{ factors}}\underbrace{(a \cdot a \cdot \ldots \cdot a)}_{n \text{ factors}} = \underbrace{a \cdot a \cdot \ldots \cdot a}_{m + n \text{ factors}} = a^{m+n}$$

Thus, to multiply two expressions having the same base, retain the base and add the exponents. That is, if a is a real number and m, n are positive integers, then

$$a^m a^n = a^{m+n} \qquad (4)$$

It can be shown that equation (4) is true whether the integers m and n are positive, negative, or 0. Of course, if m, n, or $m + n$ is 0 or negative, then a cannot be 0, as stated earlier.

E X A M P L E   3

Using Equation (4)

(a) $2 \cdot 2^4 = 2^5 = 32$ (b) $(-3)^2(-3)^{-3} = (-3)^{-1} = \dfrac{1}{-3}$

(c) $x^{-3} \cdot x^5 = x^2$ (d) $(-y)^{-1}(-y)^{-3} = (-y)^{-4} = \dfrac{1}{(-y)^4} = \dfrac{1}{y^4}$ ▬

 Now work Problem 11.

Another law of exponents applies when an expression containing a power is itself raised to a power:

$$(2^3)^4 = \underbrace{2^3 \cdot 2^3 \cdot 2^3 \cdot 2^3}_{\text{4 factors}} = \underbrace{(\underbrace{2 \cdot 2 \cdot 2}_{\text{3 factors}})(\underbrace{2 \cdot 2 \cdot 2}_{\text{3 factors}})(\underbrace{2 \cdot 2 \cdot 2}_{\text{3 factors}})(\underbrace{2 \cdot 2 \cdot 2}_{\text{3 factors}})}_{3 \cdot 4 = 12 \text{ factors}} = 2^{12}$$

In general, if a is any real number and if m and n are positive integers, we have

$$(a^m)^n = \underbrace{a^m \cdot a^m \cdot \ldots \cdot a^m}_{n \text{ factors}} = \underbrace{(\underbrace{a \cdot a \cdot \ldots \cdot a}_{m \text{ factors}})(\underbrace{a \cdot a \cdot \ldots \cdot a}_{m \text{ factors}}) \ldots (\underbrace{a \cdot a \cdot \ldots \cdot a}_{m \text{ factors}})}_{m \cdot n \text{ factors}} = a^{mn}$$

Thus, if an expression containing a power is raised to a power, retain the base and multiply the powers. That is, if a is a real number and m, n are positive integers, then

$$(a^m)^n = a^{mn} \qquad (5)$$

It can be shown that equation (5) is true whether the integers m and n are positive, negative, or 0. Again, if m or n is 0 or negative, then a must not be 0.

E X A M P L E   4

Using Equation (5)

(a) $[(-2)^3]^2 = (-2)^6 = 64$ (b) $(3^{-4})^0 = 3^{(-4)(0)} = 3^0 = 1$

(c) $(x^{-3})^2 = x^{-6} = \dfrac{1}{x^6}$ (d) $(y^{-1})^{-2} = y^{(-1)(-2)} = y^2$ ▬

The next law of exponents involves raising a product to a power:

$$(2 \cdot 5)^3 = (2 \cdot 5)(2 \cdot 5)(2 \cdot 5) = (2 \cdot 2 \cdot 2)(5 \cdot 5 \cdot 5) = 2^3 \cdot 5^3$$

In general, if a and b are real numbers and if n is a positive integer, we have

$$(a \cdot b)^n = \underbrace{(ab)(ab) \cdot \ldots \cdot (ab)}_{n \text{ factors}} = \underbrace{(a \cdot a \cdot \ldots \cdot a)}_{n \text{ factors}}\underbrace{(b \cdot b \cdot \ldots \cdot b)}_{n \text{ factors}} = a^n \cdot b^n$$

Thus, if a product is raised to a power, the result equals the product of each factor raised to that power. That is, if a, b are real numbers and n is a positive integer, then

$$(a \cdot b)^n = a^n \cdot b^n \qquad (6)$$

It can be shown that equation (6) is true whether the integer n is positive, negative, or 0. If n is 0 or negative, neither a nor b can be 0.

E X A M P L E   5

Using Equation (6)

(a) $(2x)^3 = 2^3 \cdot x^3 = 8x^3$

(b) $(-2x)^0 = (-2)^0 \cdot x^0 = 1 \cdot 1 = 1 \qquad x \neq 0$

(c) $(ax)^{-2} = a^{-2}x^{-2} = \dfrac{1}{a^2} \cdot \dfrac{1}{x^2} = \dfrac{1}{a^2x^2} \qquad x \neq 0, a \neq 0$

(d) $(ax)^{-2} = \dfrac{1}{(ax)^2} = \dfrac{1}{a^2x^2} \qquad x \neq 0, a \neq 0$

Now work Problem 29.

Equations (4) and (6) both involve products. Two analogous laws involve quotients.

If a and b are real numbers and if m and n are integers, then

$$\frac{a^m}{a^n} = a^{m-n} = \frac{1}{a^{n-m}} \qquad \text{if } a \neq 0$$

$$\left(\frac{a}{b}\right)^n = \frac{a^n}{b^n} \qquad \text{if } b \neq 0 \qquad\qquad (7)$$

E X A M P L E   6

Using Equation (7)

(a) $\dfrac{2^6}{2^4} = 2^{6-4} = 2^2 = 4$

(b) $\left(\dfrac{2}{3}\right)^4 = \dfrac{2^4}{3^4} = \dfrac{16}{81}$

(c) $\dfrac{x^{-2}}{x^{-5}} = x^{-2-(-5)} = x^3$

(d) $\left(\dfrac{5}{2}\right)^{-2} = \dfrac{1}{\left(\dfrac{5}{2}\right)^2} = \dfrac{1}{\dfrac{5^2}{2^2}} = \dfrac{1}{\dfrac{25}{4}} = \dfrac{4}{25}$

You may have observed the following shortcut for problems like Example 6(d):

$$\left(\frac{a}{b}\right)^{-n} = \left(\frac{b}{a}\right)^{n} \qquad \text{if } a \neq 0, b \neq 0 \qquad (8)$$

E X A M P L E   7

Using Equation (8)

$$\left(\frac{2}{3}\right)^{-3} = \left(\frac{3}{2}\right)^{3} = \frac{27}{8} \qquad \left(\frac{3}{2}\right)^{-3} = \left(\frac{2}{3}\right)^{3} = \frac{8}{27}$$

■

  Now work Problem 19.

E X A M P L E   8

Using the Laws of Exponents

Write each expression so that all exponents are positive.

(a) $\dfrac{x^5 y^{-2}}{(x^3 y)^2}, \quad x \neq 0, y \neq 0$ (b) $\left(\dfrac{x^{-3}}{3y^{-1}}\right)^{-2}, \quad x \neq 0, y \neq 0$

Solution (a) $\dfrac{x^5 y^{-2}}{(x^3 y)^2} = \dfrac{x^5 y^{-2}}{(x^3)^2 y^2} = \dfrac{x^5 y^{-2}}{x^6 y^2} = \dfrac{x^5}{x^6} \cdot \dfrac{y^{-2}}{y^2}$

$$= x^{5-6} y^{-2-2} = x^{-1} y^{-4} = \frac{1}{x} \cdot \frac{1}{y^4} = \frac{1}{xy^4}$$

(b) $\left(\dfrac{x^{-3}}{3y^{-1}}\right)^{-2} = \dfrac{(x^{-3})^{-2}}{(3y^{-1})^{-2}} = \dfrac{x^6}{3^{-2}(y^{-1})^{-2}} = \dfrac{x^6}{\frac{1}{9}y^2} = \dfrac{9x^6}{y^2}$

■

  Now work Problem 47.

Calculator Use

3 Your calculator should have the key $\boxed{x^y}$ or $\boxed{y^x}$ or $\boxed{\wedge}$, which is used for computations involving exponents. The next example shows how this key is used.

E X A M P L E   9

Exponents on a Calculator

Evaluate: $(2.3)^5$

Solution Keystrokes: $\boxed{2.3}$ $\boxed{x^y}$ $\boxed{5}$ $\boxed{=}$

Display: $\boxed{2.3}$ $\boxed{5}$ $\boxed{64.36343}$

■

Negative exponents may be handled in a variety of ways on your calculator.

E X A M P L E   10

Negative Exponents on a Calculator

Evaluate: $(8.1)^{-3}$

Solution A   For this solution, we use the $\boxed{\pm}$ key, which is used to change the sign of a number.

Keystrokes: $\boxed{8.1}$ $\boxed{x^y}$ $\boxed{3}$ $\boxed{\pm}$ $\boxed{=}$

Display: $\boxed{8.1}$ $\boxed{3}$ $\boxed{-3}$ $\boxed{0.00188}$ ▬

Solution B   For this solution, we use the reciprocal key $\boxed{1/x}$.

Keystrokes: $\boxed{8.1}$ $\boxed{x^y}$ $\boxed{3}$ $\boxed{=}$ $\boxed{1/x}$

Display: $\boxed{8.1}$ $\boxed{3}$ $\boxed{531.441}$ $\boxed{0.00188}$ ▬

 Solution C   Keystrokes: $\boxed{8.1}$ $\boxed{\wedge}$ $\boxed{(-)}$ $\boxed{3}$ $\boxed{EXE}$

Display: $\boxed{8.1}$ $\boxed{x^y}$ $\boxed{-}$ $\boxed{3}$ $\boxed{0.00188}$ ▬

Notice that Solution B uses the fact that $a^{-n} = 1/a^n$.

 Now work Problem 63.

Scientific Notation

Measurements of physical quantities can range from very small to very large. For example, the mass of a proton is approximately 0.00000000000000000000000000167 kilogram and the mass of Earth is about 5,980,000,000,000,000,000,000,000 kilograms. These numbers obviously are tedious to write down and difficult to read, so we use exponents to rewrite each. When a number has been written as the product of a number x, where $1 \le x < 10$, times a power of 10, it is said to be written in **scientific notation.** In scientific notation,

Mass of a proton $= 1.67 \times 10^{-27}$ kilogram

Mass of Earth $= 5.98 \times 10^{24}$ kilograms

Converting a Decimal to Scientific Notation

> To change a positive number into scientific notation, do the following:
>
> 1. Count the number N of places that the decimal point must be moved in order to arrive at a number x, where $1 \le x < 10$.
> 2. If the original number is greater than or equal to 1, the scientific notation is $x \times 10^N$. If the original number is between 0 and 1, the scientific notation is $x \times 10^{-N}$.

E X A M P L E   11   Using Scientific Notation

Write each number in scientific notation.

(a) 9582   (b) 1.245   (c) 0.285   (d) 0.000561

Solution   (a) The decimal point in 9582 follows the 2. Thus, we count

$$9 \; 5 \; 8 \; 2 \; .$$
$$\begin{array}{cccc} \uparrow & \uparrow & \uparrow & \\ 3 & 2 & 1 & \end{array}$$

stopping after three moves because 9.582 is a number between 1 and 10. Since 9582 is greater than 1, we write

$$9582 = 9.582 \times 10^3$$

(b) The decimal point in 1.245 is between the 1 and 2. Since the number is already between 1 and 10, the scientific notation for it is $1.245 \times 10^0 = 1.245$.

(c) The decimal point in 0.285 is between the 0 and the 2. Thus, we count

$$0 \quad . \quad 2 \quad 8 \quad 5$$
$$1$$

stopping after one move because 2.85 is a number between 1 and 10. Since 0.285 is between 0 and 1, we write

$$0.285 = 2.85 \times 10^{-1}$$

(d) The decimal point in 0.000561 is moved as follows:

$$0 \quad . \quad 0 \quad 0 \quad 0 \quad 5 \quad 6 \quad 1$$
$$1 \quad 2 \quad 3 \quad 4$$

Thus,

$$0.000561 = 5.61 \times 10^{-4}$$

 Now work Problem 69.

E X A M P L E   12   Changing from Scientific Notation to Decimals

Write each number as a decimal.

(a) 2.1×10^4 (b) 3.26×10^{-5} (c) 1×10^{-2}

Solution   (a) $2.1 \times 10^4 = 2 \quad . \quad 1 \quad 0 \quad 0 \quad 0 \quad \times 10^4 = 21,000$
$$1 \quad 2 \quad 3 \quad 4$$

(b) $3.26 \times 10^{-5} = 0 \quad 0 \quad 0 \quad 0 \quad 0 \quad 3 \quad . \quad 2 \quad 6 \times 10^{-5} = 0.0000326$
$$5 \quad 4 \quad 3 \quad 2 \quad 1$$

(c) $1 \times 10^{-2} = 0 \quad 0 \quad 1 \quad . \quad \times 10^{-2} = 0.01$
$$2 \quad 1$$

On a calculator, a number such as 3.615×10^{12} is displayed either as $\boxed{3.615 \quad 12}$ or as $\boxed{3.615 \quad \text{E}12}$, depending on the calculator.

 Now work Problem 77.

SUMMARY

We close this section by summarizing the Laws of Exponents. In the list that follows, a and b are real numbers and m and n are integers. Also, we assume that no denominator is 0 and that all expressions are defined.

Laws of Exponents

$$a^{-n} = \frac{1}{a^n} \qquad a^0 = 1$$

$$a^m a^n = a^{m+n} \qquad (a^m)^n = a^{mn} \qquad (ab)^n = a^n b^n$$

$$\frac{a^m}{a^n} = a^{m-n} = \frac{1}{a^{n-m}} \qquad \left(\frac{a}{b}\right)^n = \frac{a^n}{b^n}$$

HISTORICAL FEATURE   Our method of writing exponents originated with René Descartes (1596–1650), although the concept goes back in various forms to the ancient Babylonians. Even after its introduction by Descartes in 1637, the method took a remarkable amount of time to become completely standardized, and expressions like $aaaaa + 3aaaa + 2aaa - 4aa + 2a + 1$ remained common until 1750. The concept of a rational exponent (see Section 1.8) was known by 1400, although inconvenient notation prevented the development of any extensive theory. John Wallis (1616–1703), in 1655, was the first to give a fairly complete explanation of negative and rational exponents, and Sir Isaac Newton's (1642–1727) use of them made exponents standard in their current form.

1.3 | EXERCISES

In Problems 1–24, simplify each expression.

1. 4^2

2. -4^2

3. 4^{-2}

4. $(-4)^2$

5. -4^{-2}

6. $(-4)^{-2}$

7. $4^0 \cdot 2^{-3}$

8. $(-2)^{-3} \cdot 3^0$

9. $2^{-3} + \left(\frac{1}{2}\right)^3$

10. $3^{-2} + \left(\frac{1}{3}\right)^2$

11. $3^{-6} \cdot 3^4$

12. $4^{-2} \cdot 4^3$

13. $\frac{(3^2)^2}{(2^3)^2}$

14. $\frac{(2^3)^3}{(2^2)^3}$

15. $\left(\frac{2}{3}\right)^{-3}$

16. $\left(\frac{3}{2}\right)^{-2}$

17. $\frac{2^3 \cdot 3^2}{2^4 \cdot 3^{-2}}$

18. $\frac{3^{-2} \cdot 5^3}{3^2 \cdot 5}$

19. $\left(\frac{9}{2}\right)^{-2}$

20. $\left(\frac{6}{5}\right)^{-3}$

21. $\frac{2^{-2}}{3}$

22. $\frac{3^{-2}}{2}$

23. $\frac{-3^{-1}}{2^{-1}}$

24. $\frac{-2^{-3}}{-1}$

In Problems 25–56, simplify each expression so that all exponents are positive. Whenever an exponent is negative or 0, we assume that the base does not equal 0.

25. $x^0 y^2$

26. $x^{-1} y$

27. xy^{-2}

28. $x^0 y^4$

29. $(8x^3)^{-2}$

30. $(-8x^3)^{-2}$

31. $-4x^{-1}$

32. $(-4x)^{-1}$

33. $3x^0$

34. $(3x)^0$

35. $\frac{x^{-2} y^3}{xy^4}$

36. $\frac{x^{-2} y}{xy^2}$

37. $x^{-1} y^{-1}$

38. $\frac{x^{-2} y^{-3}}{x}$

39. $\frac{x^{-1}}{y^{-1}}$

40. $\left(\dfrac{2x}{3}\right)^{-1}$ **41.** $\left(\dfrac{4y}{5x}\right)^{-2}$ **42.** $(x^2y)^{-2}$ **43.** $x^{-2}y^{-2}$ **44.** $x^{-1}y^{-1}$

45. $\dfrac{x^{-1}y^{-2}z^3}{x^2yz^3}$ **46.** $\dfrac{3x^{-2}yz^2}{x^4y^{-3}z^2}$ **47.** $\dfrac{(-2)^3x^4(yz)^2}{3^2xy^3z^4}$ **48.** $\dfrac{4x^{-2}(yz)^{-1}}{(-5)^2x^4y^2z^{-2}}$ **49.** $\dfrac{\left(\dfrac{x}{y}\right)^{-2}\cdot\left(\dfrac{y}{x}\right)^4}{x^2y^3}$

50. $\dfrac{\left(\dfrac{y}{x}\right)^2}{x^{-2}y}$ **51.** $\left(\dfrac{3x^{-1}}{4y^{-1}}\right)^{-2}$ **52.** $\left(\dfrac{5x^{-2}}{6y^{-2}}\right)^{-3}$ **53.** $\dfrac{(xy^{-1})^{-2}}{xy^3}$ **54.** $\dfrac{(3xy^{-1})^2}{(2x^{-1}y)^3}$

55. $\left(\dfrac{x}{y^2}\right)^{-2}\cdot(y^2)^{-1}$ **56.** $\dfrac{(x^2)^{-3}y^3}{(x^3y)^{-2}}$

57. Find the value of the expression $2x^3 - 3x^2 + 5x - 4$ if $x = 2$. If $x = 1$.

58. Find the value of the expression $4x^3 + 3x^2 - x + 2$ if $x = 1$. If $x = 2$.

59. What is the value of $\dfrac{(666)^4}{(222)^4}$? **60.** What is the value of $(0.1)^3(20)^3$?

In Problems 61–68, evaluate each expression. Round your answers to three decimal places.

61. $(8.2)^6$ **62.** $(3.7)^5$ **63.** $(6.1)^{-3}$ **64.** $(2.2)^{-5}$
65. $(-2.8)^6$ **66.** $-(2.8)^6$ **67.** $(-8.11)^{-4}$ **68.** $-(8.11)^{-4}$

In Problems 69–76, write each number in scientific notation.

69. 454.2 **70.** 32.14 **71.** 0.013 **72.** 0.00421
73. 32,155 **74.** 21,210 **75.** 0.000423 **76.** 0.0514

In Problems 77–84, write each number as a decimal.

77. 6.15×10^4 **78.** 9.7×10^3 **79.** 1.214×10^{-3} **80.** 9.88×10^{-4}
81. 1.1×10^8 **82.** 4.112×10^2 **83.** 8.1×10^{-2} **84.** 6.453×10^{-1}

85. Astronomy One light-year is defined by astronomers to be the distance a beam of light will travel in 1 year (365 days). If the speed of light is 186,000 miles per second, how many miles are in a light-year? Express your answer in scientific notation.

86. Astronomy How long does it take a beam of light to reach Earth from the Sun, when the Sun is 93,000,000 miles from Earth? Express your answer in seconds, using scientific notation.

 87. Look at the summary box where the Laws of Exponents are given. List them in the order of most importance to you. Write a brief position paper defending your ordering.

88. Write a paragraph to justify the definition given in the text that $a^0 = 1$, $a \neq 0$.

1.4 POLYNOMIALS

1 Add and Subtract Polynomials
2 Multiply Polynomials
3 Know Formulas for Special Products
4 Divide Polynomials

We have described algebra as a generalization of arithmetic in which letters are used to represent real numbers. From now on, we shall use the letters at the end of the alphabet, such as x, y, and z, to represent variables and the letters at the beginning of the alphabet, such as a, b, and c, to represent constants. Thus, in the expressions $3x + 5$ and $ax + b$, it is understood that x is a variable and that a and b are constants, even though the constants a and b are unspecified. As you will find out, the context usually makes the intended meaning clear.

Now we introduce some basic vocabulary.

A **monomial** in one variable is the product of a constant times a variable raised to a nonnegative integer power. Thus, a monomial is of the form

$$ax^k$$

where a is a constant, x is a variable, and $k \geq 0$ is an integer. The constant a is called the **coefficient** of the monomial. If $a \neq 0$, then k is called the **degree** of the monomial.

Examples of monomials are:

Monomial	Coefficient	Degree	
$6x^2$	6	2	
$-\sqrt{2}x^3$	$-\sqrt{2}$	3	
3	3	0	Since $3 = 3 \cdot 1 = 3x^0$
$-5x$	-5	1	Since $-5x = -5x^1$
x^4	1	4	Since $x^4 = 1 \cdot x^4$

Two monomials ax^k and bx^k with the same degree and the same variable are called **like terms.** Such monomials, when added or subtracted, can be combined into a single monomial by using the distributive property. For example,

$$2x^2 + 5x^2 = (2 + 5)x^2 = 7x^2 \quad \text{and} \quad 8x^3 - 5x^3 = (8 - 5)x^3 = 3x^3$$

The sum or difference of two monomials having different degrees is called a **binomial.** The sum or difference of three monomials with three different degrees is called a **trinomial.** For example,

$x^2 - 2$ is a binomial.
$x^3 - 3x + 5$ is a trinomial.
$2x^2 + 5x^2 + 2 = 7x^2 + 2$ is a binomial.

A **polynomial** in one variable is an algebraic expression of the form

$$a_n x^n + a_{n-1} x^{n-1} + \cdots + a_1 x + a_0 \tag{1}$$

where $a_n, a_{n-1}, \ldots, a_1, a_0$ are constants,* called the **coefficients** of the polynomial, $n \geq 0$ is an integer, and x is a variable. If $a_n \neq 0$, it is called the **leading coefficient**, and n is called the **degree** of the polynomial.

The monomials that make up a polynomial are called its **terms.** If all the coefficients are 0, the polynomial is called the **zero polynomial,** which has no degree.

Polynomials are usually written in **standard form,** beginning with the nonzero term of highest degree and continuing with terms in descending order according to degree. Examples of polynomials are:

---

*The notation a_n is read as "a sub n." The number n is called a **subscript** and should not be confused with an exponent. We use subscripts in order to distinguish one constant from another when a large or undetermined number of constants is required.

Polynomial	Coefficients	Degree
$3x^2 - 5 = 3x^2 + 0 \cdot x + (-5)$	$3, 0, -5$	2
$8 - 2x + x^2 = 1 \cdot x^2 - 2x + 8$	$1, -2, 8$	2
$5x + \sqrt{2} = 5x^1 + \sqrt{2}$	$5, \sqrt{2}$	1
$3 = 3 \cdot 1 = 3 \cdot x^0$	3	0
0	0	No degree

Although we have been using x to represent the variable, letters such as y or z are also commonly used. Thus,

$3x^4 - x^2 + 2$ is a polynomial (in x) of degree 4.
$9y^3 - 2y^2 + y - 3$ is a polynomial (in y) of degree 3.
$z^5 + \pi$ is a polynomial (in z) of degree 5.

Algebraic expressions such as

$$\frac{1}{x} \quad \text{and} \quad \frac{x^2 + 1}{x + 5}$$

are not polynomials. The first is not a polynomial because $1/x = x^{-1}$ has an exponent that is not a nonnegative integer. Although the second expression is the quotient of two polynomials, the polynomial in the denominator has degree greater than 0, so the expression cannot be a polynomial.

Adding and Subtracting Polynomials

1 Polynomials are added and subtracted by combining like terms.

E X A M P L E   1   Adding Polynomials

Find the sum of the polynomials

$$8x^3 - 2x^2 + 6x - 2 \quad \text{and} \quad 3x^4 - 2x^3 + x^2 + x$$

Solution   We shall find the sum in two ways.

Horizontal Addition: The idea here is to group the like terms and then combine them.

$$(8x^3 - 2x^2 + 6x - 2) + (3x^4 - 2x^3 + x^2 + x)$$
$$= 3x^4 + (8x^3 - 2x^3) + (-2x^2 + x^2) + (6x + x) - 2$$
$$= 3x^4 + 6x^3 - x^2 + 7x - 2$$

Vertical Addition: The idea here is to vertically line up the like terms in each polynomial and then add the coefficients.

$$
\begin{array}{ccccc}
x^4 & x^3 & x^2 & x^1 & x^0 \\
& 8x^3 - & 2x^2 + & 6x - & 2 \\
(+) \quad 3x^4 - & 2x^3 + & x^2 + & x & \\
\hline
3x^4 + & 6x^3 - & x^2 + & 7x - & 2
\end{array}
$$

The difference of two polynomials also can be found in either of the ways just outlined.

E X A M P L E   2

Subtracting Polynomials

Find the difference: $(3x^4 - 4x^3 + 6x^2 - 1) - (2x^4 - 8x^2 - 6x + 5)$

Solution   *Horizontal Subtraction:*

$$(3x^4 - 4x^3 + 6x^2 - 1) - (2x^4 - 8x^2 - 6x + 5)$$
$$= 3x^4 - 4x^3 + 6x^2 - 1 + \underbrace{(-2x^4 + 8x^2 + 6x - 5)}$$

Be sure to change the sign of each term in the second polynomial.

$$= \underset{\uparrow}{(3x^4 - 2x^4)} + (-4x^3) + (6x^2 + 8x^2) + 6x + (-1 - 5)$$

Group like terms.

$$= x^4 - 4x^3 + 14x^2 + 6x - 6$$

Vertical Subtraction: We line up like terms, change the sign of each coefficient of the second polynomial, and add.

x^4	x^3	x^2	x^1	x^0		x^4	x^3	x^2	x^1	x^0

$$3x^4 - 4x^3 + 6x^2 \qquad - 1 = \qquad 3x^4 - 4x^3 + \; 6x^2 \qquad - 1$$
$$(-)[2x^4 \qquad - 8x^2 - 6x + 5] = (+)\underline{-2x^4 \qquad + \; 8x^2 + 6x - 5}$$
$$\qquad\qquad\qquad\qquad\qquad x^4 - 4x^3 + 14x^2 + 6x - 6 \quad \blacksquare$$

The choice of which of these methods to use for adding and subtracting polynomials is left to you. To save space, we shall most often use the horizontal format.

Now work Problem 13.

Multiplying Polynomials

The product of two monomials ax^n and bx^m is obtained using the laws of exponents and commutative and associative properties. Thus,

$$ax^n \cdot bx^m = abx^{n+m}$$

Products of polynomials are found by repeated use of the distributive property and the laws of exponents. Again, there is a choice of horizontal or vertical format.

E X A M P L E   3

Multiplying Polynomials

Find the product: $(2x + 5)(x^2 - x + 2)$

Solution   *Horizontal Multiplication:*

$$(2x + 5)(x^2 - x + 2) \underset{\uparrow}{=} 2x(x^2 - x + 2) + 5(x^2 - x + 2)$$

Distributive property

$$\underset{\uparrow}{=} 2x \cdot x^2 - 2x \cdot x + 2x \cdot 2 + 5 \cdot x^2 - 5 \cdot x + 5 \cdot 2$$

Distributive property

$$= 2x^3 - 2x^2 + 4x + 5x^2 - 5x + 10$$

↑
Law of exponents

$$= 2x^3 + 3x^2 - x + 10$$

↑
Combine like terms

Vertical Multiplication: The idea here is very much like multiplying a two-digit number by a three-digit number.

$$
\begin{array}{r}
x^2 - x + 2 \\
2x + 5 \\
\hline
2x^3 - 2x^2 + 4x \\
5x^2 - 5x + 10 \\
\hline
2x^3 + 3x^2 - x + 10
\end{array}
$$

(+)

This line is $2x(x^2 - x + 2)$.
This line is $5(x^2 - x + 2)$.
Sum of the above two lines.   ■

Now work Problem 45.

Special Products

3

Certain products, which we call **special products,** occur frequently in algebra. We can calculate them easily using the **FOIL** (First, Outer, Inner, Last) method of multiplying two binomials:

```
          ┌──── Outer ────┐
          │ ┌─ First ─┐   │
```
$$(ax + b)(cx + d) = ax(cx + d) + b(cx + d)$$
```
          │ │         │
          │ └─ Inner  │
          └─── Last ──┘
```

$$
\begin{aligned}
&= \overset{\text{First}}{ax \cdot cx} + \overset{\text{Outer}}{ax \cdot d} + \overset{\text{Inner}}{b \cdot cx} + \overset{\text{Last}}{b \cdot d} \\
&= acx^2 + adx + bcx + bd \\
&= acx^2 + (ad + bc)x + bd
\end{aligned}
$$

E X A M P L E   4

Using FOIL to Find Products of the Form $(x - a)(x + a)$

$$(x - 3)(x + 3) = x^2 + 3x - 3x - 9 = x^2 - 9$$ ■

E X A M P L E   5

Using FOIL to Find the Square of a Binomial

(a) $(x + 2)^2 = (x + 2)(x + 2) = x^2 + 2x + 2x + 4 = x^2 + 4x + 4$
(b) $(x - 3)^2 = (x - 3)(x - 3) = x^2 - 3x - 3x + 9 = x^2 - 6x + 9$ ■

E X A M P L E   6

Using FOIL to Find the Product of Two Binomials

(a) $(x + 3)(x + 1) = x^2 + x + 3x + 3 = x^2 + 4x + 3$
(b) $(2x + 1)(3x + 4) = 6x^2 + 8x + 3x + 4 = 6x^2 + 11x + 4$ ■

Now work Problems 29 and 31.

Each of the preceding examples can be calculated using a general formula. In the equations that follow, x, a, b, c, and d are real numbers:

Difference of Two Squares

$$(x - a)(x + a) = x^2 - a^2 \tag{2}$$

Squares of Binomials, or Perfect Squares

$$(x + a)^2 = x^2 + 2ax + a^2 \tag{3a}$$
$$(x - a)^2 = x^2 - 2ax + a^2 \tag{3b}$$

Miscellaneous Trinomials

$$(x + a)(x + b) = x^2 + (a + b)x + ab \tag{4a}$$
$$(ax + b)(cx + d) = acx^2 + (ad + bc)x + bd \tag{4b}$$

Let's look at some more examples that lead to general formulas.

E X A M P L E 7   Cubing a Binomial

$$(a) \ (x + 2)^3 = (x + 2)(x + 2)^2 = (x + 2)(x^2 + 4x + 4)$$
$$= (x^3 + 4x^2 + 4x) + (2x^2 + 8x + 8)$$
$$= x^3 + 6x^2 + 12x + 8$$
$$(b) \ (x - 1)^3 = (x - 1)(x - 1)^2 = (x - 1)(x^2 - 2x + 1)$$
$$= (x^3 - 2x^2 + x) - (x^2 - 2x + 1)$$
$$= x^3 - 3x^2 + 3x - 1 \quad \blacksquare$$

E X A M P L E 8   Forming the Difference of Two Cubes

$$(x - 1)(x^2 + x + 1) = x(x^2 + x + 1) - 1(x^2 + x + 1)$$
$$= x^3 + x^2 + x - x^2 - x - 1$$
$$= x^3 - 1 \quad \blacksquare$$

E X A M P L E 9   Forming the Sum of Two Cubes

$$(x + 2)(x^2 - 2x + 4) = x(x^2 - 2x + 4) + 2(x^2 - 2x + 4)$$
$$= (x^3 - 2x^2 + 4x) + (2x^2 - 4x + 8)$$
$$= x^3 + 8 \quad \blacksquare$$

Cubes of Binomials, or Perfect Cubes

$$(x + a)^3 = x^3 + 3ax^2 + 3a^2x + a^3 \qquad (5a)$$
$$(x - a)^3 = x^3 - 3ax^2 + 3a^2x - a^3 \qquad (5b)$$

Difference of Two Cubes

$$(x - a)(x^2 + ax + a^2) = x^3 - a^3 \qquad (6)$$

Sum of Two Cubes

$$(x + a)(x^2 - ax + a^2) = x^3 + a^3 \qquad (7)$$

The formulas in equations (2)–(7) are used often, and their patterns should be committed to memory. But if you forget one or are unsure of its form, you should be able to derive it as needed.

We derived equation (4b) earlier. Now we will verify equation (5b); derivations of the remaining rules are left as exercises.

Equation (5b)
$$\begin{aligned}
(x - a)^3 &= (x - a)(x - a)^2 = (x - a)(x^2 - 2ax + a^2) \\
&= x(x^2 - 2ax + a^2) - a(x^2 - 2ax + a^2) \\
&= x^3 - 2ax^2 + a^2x - ax^2 + 2a^2x - a^3 \\
&= x^3 - 3ax^2 + 3a^2x - a^3
\end{aligned}$$

 Now work Problems 41 and 71.

Dividing Polynomials

 The procedure for dividing two polynomials is similar to the procedure for dividing two integers. This process should be familiar to you, but we review it briefly next.

E X A M P L E   10   Dividing Two Integers

Divide 842 by 15.

Solution

$$
\begin{array}{r}
56 \\
15\overline{)842} \\
\underline{75} \\
92 \\
\underline{90} \\
2
\end{array}
$$

← Quotient
← Dividend
← 5 · 15 (Subtract)

← 6 · 15 (Subtract)
← Remainder

Thus, $\dfrac{842}{15} = 56 + \dfrac{2}{15}$.

In the long division process detailed in Example 10, the number 15 is called the **divisor,** the number 842 is called the **dividend,** the number 56 is called the **quotient,** and the number 2 is called the **remainder.**

To check the answer obtained in a division problem, multiply the quotient by the divisor and add the remainder. The answer should be the dividend.

$$(\text{Quotient})(\text{Divisor}) + \text{Remainder} = \text{Dividend}$$

For example, we can check the results obtained in Example 10 as follows:

$$(56)(15) + 2 = 840 + 2 = 842$$

To divide two polynomials, we first must write each polynomial in standard form. The process then follows a pattern similar to that of Example 10. The next example illustrates the procedure.

E X A M P L E   11   Dividing Two Polynomials

Find the quotient and the remainder when

$$3x^3 + 4x^2 + x + 7 \quad \text{is divided by} \quad x^2 + 1$$

Solution   Each polynomial is in standard form. The dividend is $3x^3 + 4x^2 + x + 7$, and the divisor is $x^2 + 1$.

STEP 1: Divide the leading term of the dividend, $3x^3$, by the leading term of the divisor, x^2. Enter the result, $3x$, over the term $3x^3$, as follows:

$$\begin{array}{r} 3x \\ x^2 + 1 \overline{)3x^3 + 4x^2 + x + 7} \end{array}$$

STEP 2: Multiply $3x$ by $x^2 + 1$ and enter the result below the dividend.

$$\begin{array}{r} 3x \\ x^2 + 1 \overline{)3x^3 + 4x^2 + x + 7} \\ \underline{3x^3 + 3x} \end{array}$$ ← $3x \cdot (x^2 + 1) = 3x^3 + 3x$

↑

Notice that we align the $3x$ term under the x to make the next step easier.

STEP 3: Subtract and bring down the remaining terms.

$$\begin{array}{r} 3x \\ x^2 + 1 \overline{)3x^3 + 4x^2 + x + 7} \\ \underline{3x^3 + 3x} \\ 4x^2 - 2x + 7 \end{array}$$ ← Subtract.
← Bring down the $4x^2$ and the 7.

STEP 4: Repeat Steps 1–3 using $4x^2 - 2x + 7$ as the dividend.

$$
\begin{array}{r}
3x + 4 \\
x^2 + 1\overline{)3x^3 + 4x^2 + x + 7} \\
\underline{3x^3 + 3x} \\
4x^2 - 2x + 7 \\
\underline{4x^2 + 4} \\
- 2x + 3
\end{array}
$$

← Divide $4x^2$ by x^2 to get 4.
← Multiply $x^2 + 1$ by 4; subtract.

Since x^2 does not divide $-2x$ evenly (that is, the result is not a monomial), the process ends. The quotient is $3x + 4$, and the remainder is $-2x + 3$.

Check: (Quotient)(Divisor) + Remainder

$$
\begin{aligned}
&= (3x + 4)(x^2 + 1) + (-2x + 3) \\
&= 3x^3 + 4x^2 + 3x + 4 + (-2x + 3) \\
&= 3x^3 + 4x^2 + x + 7 = \text{Dividend}
\end{aligned}
$$

Thus,

$$
\frac{3x^3 + 4x^2 + x + 7}{x^2 + 1} = 3x + 4 + \frac{-2x + 3}{x^2 + 1}
$$

The next example combines the steps involved in long division.

E X A M P L E   12   Dividing Two Polynomials

Find the quotient and the remainder when

$$x^4 - 3x^3 + 2x - 5 \quad \text{is divided by} \quad x^2 - x + 1$$

Solution   In setting up this division problem, it is necessary to leave a space for the missing x^2 term in the dividend.

$$
\begin{array}{r}
x^2 - 2x - 3 \quad \text{← Quotient}\\
x^2 - x + 1\overline{)x^4 - 3x^3 + 2x - 5} \quad \text{← Dividend}\\
\underline{x^4 - x^3 + x^2} \\
-2x^3 - x^2 + 2x - 5 \\
\underline{-2x^3 + 2x^2 - 2x} \\
-3x^2 + 4x - 5 \\
\underline{-3x^2 + 3x - 3} \\
x - 2 \quad \text{← Remainder}
\end{array}
$$

Divisor →
Subtract →
Subtract →
Subtract →

Check: (Quotient)(Divisor) + Remainder

$$
\begin{aligned}
&= (x^2 - 2x - 3)(x^2 - x + 1) + x - 2 \\
&= x^4 - x^3 + x^2 - 2x^3 + 2x^2 - 2x - 3x^2 + 3x - 3 + x - 2 \\
&= x^4 - 3x^3 + 2x - 5 = \text{Dividend}
\end{aligned}
$$

Thus,

$$\frac{x^4 - 3x^3 + 2x - 5}{x^2 - x + 1} = x^2 - 2x - 3 + \frac{x - 2}{x^2 - x + 1}$$

The process of dividing two polynomials leads to the following result:

Theorem

The remainder after dividing two polynomials is either the zero polynomial or a polynomial of degree less than the degree of the divisor.

 Now work Problem 85.

Polynomials in Two or More Variables

Our discussion thus far has involved only polynomials in a single variable. A **polynomial in two variables** x and y is the sum of one or more monomials of the form $ax^n y^m$, where a is a constant called the **coefficient,** x and y are variables, and n and m are nonnegative integers. The **degree of the monomial** $ax^n y^m$ is $n + m$. The **degree of a polynomial** in two variables x and y is the highest degree of all the monomials with nonzero coefficients that appear.

Polynomials in three variables x, y, and z and polynomials in more than three variables are defined in a similar way. Here are some examples:

$$3x^2 + 2x^3y + 5 \qquad \pi x^3 - y^2 \qquad x^4 + 4x^3y - xy^3 + y^4$$

Two variables,   Two variables,   Two variables,
degree is 4   degree is 3   degree is 4

$$x^2 + y^2 - z^2 + 4 \qquad x^3 y^2 z \qquad 5x^2 - 4y^2 + z^3 y + 2w^2 x$$

Three variables,   Three variables,   Four variables,
degree is 2   degree is 6   degree is 4

Adding, subtracting, and multiplying polynomials in two or more variables are handled in the same way as for polynomials in one variable.

E X A M P L E   13   Subtracting Polynomials in Two Variables

$$(x^3 - 3x^2y + 3xy^2 - y^3) - (x^3 - x^2y + y^3)$$
$$= (x^3 - x^3) + (-3x^2y + x^2y) + 3xy^2 + (-y^3 - y^3)$$
$$= -2x^2y + 3xy^2 - 2y^3$$

E X A M P L E   14   Multiplying Polynomials in Two Variables

$$(xy - 2)(2x^2 - xy + y^2) = xy(2x^2 - xy + y^2) - 2(2x^2 - xy + y^2)$$
$$= 2x^3y - x^2y^2 + xy^3 - 4x^2 + 2xy - 2y^2$$

1.4 | EXERCISES

In Problems 1–10, tell whether the expression is a polynomial. If it is, give its degree.

1. $3x^2 - 5$ **2.** $1 - 4x$ **3.** 5 **4.** $-\pi$ **5.** $3x^2 - \dfrac{5}{x}$

6. $\dfrac{3}{x} + 2$ **7.** $2y^3 - \sqrt{2}$ **8.** $10z^2 + z$ **9.** $\dfrac{x^2 + 5}{x^3 - 1}$ **10.** $\dfrac{3x^3 + 2x - 1}{x^2 + x + 1}$

In Problems 11–76, perform the indicated operations. Express each answer as a polynomial.

11. $(x^2 + 4x + 5) + (3x - 3)$
12. $(x^3 + 3x^2 + 2) + (x^2 - 4x + 4)$
13. $(x^3 - 2x^2 + 5x + 10) - (2x^2 - 4x + 3)$
14. $(x^2 - 3x - 4) - (x^3 - 3x^2 + x + 5)$
15. $(6x^5 + x^3 + x) + (5x^4 - x^3 + 3x^2)$
16. $(10x^5 - 8x^2) + (3x^3 - 2x^2 + 6)$
17. $3(x^2 - 3x + 1) + 2(3x^2 + x - 4)$
18. $-2(x^2 + x + 1) + 6(-5x^2 - x + 2)$
19. $6(x^3 + x^2 - 3) - 4(2x^3 - 3x^2)$
20. $8(4x^3 - 3x^2 - 1) - 6(4x^3 + 8x - 2)$
21. $(x^2 - x + 2) + (2x^2 - 3x + 5) - (x^2 + 1)$
22. $(x^2 + 1) - (4x^2 + 5) + (x^2 + x - 2)$
23. $9(y^2 - 3y + 4) - 6(1 - y^2)$
24. $8(1 - y^3) + 4(1 + y + y^2 + y^3)$
25. $(x + a)^2 - x^2$
26. $(x - a)^2 - x^2$
27. $(x + a)^3 - x^3$
28. $(x - a)^3 - x^3$
29. $(x + 5)(x - 3)$
30. $(x + 4)(x - 2)$
31. $(x + 8)(2x + 1)$
32. $(2x - 1)(x + 2)$
33. $(-3x + 1)(x + 4)$
34. $(-2x - 1)(x + 1)$
35. $(1 - 4x)(2 - 3x)$
36. $(2 - x)(2 - 3x)$
37. $(2x - 5)(2x + 5)$
38. $(3x + 1)(3x - 1)$
39. $(2x - 5)^2$
40. $(3x + 1)^2$
41. $(x - 1)(x^2 + x + 1)$
42. $(x + 1)(x^2 - x + 1)$
43. $(2x + 5)(x^2 - 2)$
44. $(3x - 5)(x^2 + 1)$
45. $(2x - 3)(3x^2 - 2x + 1)$
46. $(2x + 3)(3x^2 - 2x + 4)$
47. $(3x + 4y)(3x - 4y)$
48. $(2x + 3y)^2$
49. $(x + y)(x - 2y)$
50. $(x - 2y)(x - y)$
51. $(x^2 + 2x + y^2) + (x + y - 3y^2)$
52. $(x^2 - 2xy + y^2) - (x^2 - xy)$
53. $(x - y)^2 - (x + y)^2$
54. $(x - y)^2 - (x^2 + y^2)$
55. $(x + 2y)^2 + (x - 3y)^2$
56. $(2x - y)^2 - (x - y)^2$
57. $[(x + y)^2 + z^2] + [x^2 + (y + z)^2]$
58. $[(x - y)^2 + z^2] + [x^2 + (y - z)^2]$
59. $(x - y)(x^2 + xy + y^2)$
60. $(x + y)(x^2 - xy + y^2)$
61. $(x + 2)^2(x - 2)$
62. $(x - 3)^2(x + 3)$
63. $(x - 1)^2(2x + 3)$
64. $(x + 5)^2(3x - 1)$
65. $(x - 1)(x - 2)(x - 3)$
66. $(x + 1)(x + 2)(x + 3)$
67. $(x + 1)^3 - (x - 1)^3$
68. $(x - 2)^3 - (x + 2)^3$
69. $(x - 1)^2(x + 1)^2$
70. $(x - 1)^3(x + 1)$
71. $(2x + 3)^3$
72. $(3x + 2)^3$
73. $(2x - 3a)^3$
74. $(3x - 2a)^3$
75. $(x + y + z)(x - y - z)$
76. $(x + y - z)(x - y + z)$

In Problems 77–106, find the quotient and the remainder. Check your work by verifying that

  (Quotient)(Divisor) + Remainder = Dividend

77. $4x^3 - 3x^2 + x + 1$ divided by x
78. $3x^3 - x^2 + x - 2$ divided by x
79. $4x^3 - 3x^2 + x + 1$ divided by $x + 2$
80. $3x^3 - x^2 + x - 2$ divided by $x + 2$
81. $4x^3 - 3x^2 + x + 1$ divided by $x - 4$
82. $3x^3 - x^2 + x - 2$ divided by $x - 4$
83. $4x^3 - 3x^2 + x + 1$ divided by x^2
84. $3x^3 - x^2 + x - 2$ divided by x^2
85. $4x^3 - 3x^2 + x + 1$ divided by $x^2 + 2$
86. $3x^3 - x^2 + x - 2$ divided by $x^2 + 2$
87. $4x^3 - 3x^2 + x + 1$ divided by $2x^3 - 1$
88. $3x^3 - x^2 + x - 2$ divided by $3x^3 - 1$
89. $4x^3 - 3x^2 + x + 1$ divided by $2x^2 + x + 1$
90. $3x^3 - x^2 + x - 2$ divided by $3x^2 + x + 1$

91. $4x^3 - 3x^2 + x + 1$ divided by $4x^2 + 1$

92. $3x^3 - x^2 + x - 2$ divided by $3x - 1$

93. $x^4 - 1$ divided by $x - 1$

94. $x^4 - 1$ divided by $x + 1$

95. $x^4 - 1$ divided by $x^2 - 1$

96. $x^4 - 1$ divided by $x^2 + 1$

97. $-4x^3 + x^2 - 4$ divided by $x - 1$

98. $-3x^4 - 2x - 1$ divided by $x - 1$

99. $1 - x^2 + x^4$ divided by $x^2 + x + 1$

100. $1 - x^2 + x^4$ divided by $x^2 - x + 1$

101. $1 - x^2 + x^4$ divided by $1 - x^2$

102. $1 - x^2 + x^4$ divided by $1 + x^2$

103. $x^3 - a^3$ divided by $x - a$

104. $x^3 + a^3$ divided by $x + a$

105. $x^4 - a^4$ divided by $x - a$

106. $x^5 - a^5$ divided by $x - a$

107. Explain why the degree of the product of two polynomials equals the sum of their degrees.

108. Explain why the degree of the sum of two polynomials equals the larger of their degrees.

109. Derive equations (2), (3a), (3b), and (6) (pp. 38 and 39).

110. Derive equations (4a), (5a), and (7) (pp. 38 and 39).

111. Develop a formula for $(x + a)^4$.

112. Develop a formula for $(x + a)^5$. Now compare $(x + a)^2$, $(x + a)^3$, $(x + a)^4$, and $(x + a)^5$. What do you notice?

113. Do you prefer adding two polynomials using the horizontal method or the vertical method? Write a brief position paper defending your choice.

114. Do you prefer to memorize the rule for the square of a binomial $(x + a)^2$ or to use FOIL to obtain the product? Write a brief position paper defending your choice.

115. Find the sum of a, b, c, and d if

$$\frac{x^3 - 2x^2 + 3x + 5}{x + 2} = ax^2 + bx + c + \frac{d}{x + 2}$$

1.5   FACTORING POLYNOMIALS

> **1** Factor the Difference of Two Squares and the Sum and the Difference of Two Cubes
> **2** Factor Perfect Squares
> **3** Factor Second-degree Polynomials (Trinomials)
> **4** Factor by Grouping

Consider the following product:

$$(2x + 3)(x - 4) = 2x^2 - 5x - 12$$

The two polynomials on the left side are called **factors** of the polynomial on the right side. Expressing a given polynomial as a product of other polynomials, that is, finding the factors of a polynomial, is called **factoring.**

We shall restrict our discussion here to factoring polynomials in one variable into products of polynomials in one variable, where all coefficients are integers. We call this **factoring over the integers.** There will be times, though, when we will want to **factor over the rational numbers** and even **factor over the real numbers.** Factoring over the rational numbers means to write a given polynomial whose coefficients are rational numbers as a product of polynomials whose coefficients are also rational numbers. Factoring over the real numbers means to write a given polynomial whose coefficients are real numbers as a product of polynomials whose coefficients are also real numbers. Unless specified otherwise, we will be factoring over the integers.

Any polynomial can be written as the product of 1 times itself or as -1 times its additive inverse. If a polynomial cannot be written as the product of two other polynomials (excluding 1 and -1), then the polynomial is said to be **prime.** When a polynomial has been written as a product consisting only of prime factors, it is said to be **factored completely.** Examples of prime polynomials are

$$2, \quad 3, \quad 5, \quad x, \quad x + 1, \quad x - 1, \quad 3x + 4$$

The first factor to look for in a factoring problem is a common monomial factor present in each term of the polynomial. If one is present, use the distributive property to factor it out. For example:

Polynomial	Common Monomial Factor	Remaining Factor	Factored Form
$2x + 4$	2	$x + 2$	$2x + 4 = 2(x + 2)$
$3x - 6$	3	$x - 2$	$3x - 6 = 3(x - 2)$
$2x^2 - 4x + 8$	2	$x^2 - 2x + 4$	$2x^2 - 4x + 8 = 2(x^2 - 2x + 4)$
$8x - 12$	4	$2x - 3$	$8x - 12 = 4(2x - 3)$
$x^2 + x$	x	$x + 1$	$x^2 + x = x(x + 1)$
$x^3 - 3x^2$	x^2	$x - 3$	$x^3 - 3x^2 = x^2(x - 3)$
$6x^2 + 9x$	$3x$	$2x + 3$	$6x^2 + 9x = 3x(2x + 3)$

Notice that, once all common monomial factors have been removed from a polynomial, the remaining factor is either a prime polynomial of degree 1 or a polynomial of degree 2 or higher. (Do you see why?) Thus, we concentrate on techniques for factoring polynomials of degree 2 or higher that contain no monomial factors.

Special Factors

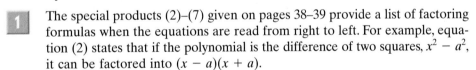

The special products (2)–(7) given on pages 38–39 provide a list of factoring formulas when the equations are read from right to left. For example, equation (2) states that if the polynomial is the difference of two squares, $x^2 - a^2$, it can be factored into $(x - a)(x + a)$.

E X A M P L E   1   Factoring the Difference of Two Squares

Factor completely: $x^2 - 4$

Solution   We notice that $x^2 - 4$ is the difference of two squares, x^2 and 2^2. Thus, using equation (2), we find that

$$x^2 - 4 = (x - 2)(x + 2)$$

E X A M P L E   2   Factoring the Difference of Two Cubes

Factor completely: $x^3 - 1$

Solution   Equation (6) states that the difference of two cubes, $x^3 - a^3$, can be factored as $(x - a)(x^2 + ax + a^2)$. Because $x^3 - 1$ is the difference of two cubes, x^3 and 1^3, we find that

$$x^3 - 1 = (x - 1)(x^2 + x + 1)$$

EXAMPLE 3   Factoring the Sum of Two Cubes

Factor completely: $x^3 + 8$

Solution   Equation (7) states that the sum of two cubes, $x^3 + a^3$, can be factored as $(x + a)(x^2 - ax + a^2)$. Because $x^3 + 8$ is the sum of two cubes, x^3 and 2^3, we have

$$x^3 + 8 = (x + 2)(x^2 - 2x + 4)$$

EXAMPLE 4   Factoring the Difference of Two Squares

Factor completely: $x^4 - 16$

Solution   Using equation (2) for the difference of two squares, $x^4 = (x^2)^2$ and $16 = 4^2$, we have

$$x^4 - 16 = (x^2 - 4)(x^2 + 4)$$

But $x^2 - 4$ is also the difference of two squares. Thus,

$$x^4 - 16 = (x^2 - 4)(x^2 + 4) = (x - 2)(x + 2)(x^2 + 4)$$

 Now work Problems 11 and 37.

2   Whenever the first term and third term of a trinomial are both positive and are perfect squares, such as $x^2, 9x^2, 1$, and 4, check to see whether either of the special products (3a) or (3b) applies.

EXAMPLE 5   Factoring Perfect Squares

Factor completely: $x^2 + 6x + 9$

Solution   The first term, x^2, and the third term, $9 = 3^2$, are perfect squares. Because the middle term is twice the product of x and 3, we use equation (3a) to obtain

$$x^2 + 6x + 9 = (x + 3)^2$$

EXAMPLE 6   Factoring Perfect Squares

Factor completely: $9x^2 - 6x + 1$

Solution   The first term, $9x^2 = (3x)^2$, and the third term, $1 = 1^2$, are perfect squares. Because the middle term is twice the product of $3x$ and 1, we use equation (3b) to obtain

$$9x^2 - 6x + 1 = (3x - 1)^2$$

EXAMPLE 7   Factoring Perfect Squares

Factor completely: $25x^2 + 30x + 9$

Solution   The first term, $25x^2 = (5x)^2$, and the third term, $9 = 3^2$, are perfect squares. Because the middle term is twice the product of $5x$ and 3, we use a form of equation (3a) to obtain

$$25x^2 + 30x + 9 = (5x + 3)^2$$

Now work Problems 21 and 33.

If a trinomial is not a perfect square, it may be possible to factor it using the technique discussed next.

Factoring Second-degree Polynomials

Factoring a second-degree polynomial, $Ax^2 + Bx + C$, where A, B, and C are integers, is a matter of skill, experience, and often some trial and error. The idea behind factoring $Ax^2 + Bx + C$ is to see whether it can be made equal to the product of two, possibly equal, first-degree polynomials. Thus, we want to see whether there are integers a, b, c, and d so that

$$Ax^2 + Bx + C = (ax + b)(cx + d)$$

We start with second-degree polynomials that have a leading coefficient of 1. Such a polynomial, if it can be factored, must follow the form of the special product (4a):

$$x^2 + Bx + C = (x + a)(x + b) = x^2 + (a + b)x + ab \qquad B = a + b, C = ab$$

Note the pattern in this formula: The correct choice of the factors a and b of the constant term C must add up to the coefficient of the middle term B.

E X A M P L E   8   Factoring Trinomials

Factor completely: $x^2 + 7x + 12$

Solution   First, determine all possible integral factors of the constant term 12 and then compute their sums:

Factors of 12	1, 12	−1, −12	2, 6	−2, −6	3, 4	−3, −4
Sum	13	−13	8	−8	7	−7

The factors of 12 that add up to 7, the coefficient of the middle term, are 3 and 4. Thus,

$$x^2 + 7x + 12 = (x + 3)(x + 4)$$

E X A M P L E   9   Factoring Trinomials

Factor completely: $x^2 - 6x + 8$

Solution   First, determine all possible integral factors of the constant term 8 and then compute each sum:

Factors of 8	1, 8	−1, −8	2, 4	−2, −4
Sum	9	−9	6	−6

Since −6 is the coefficient of the middle term, we have

$$x^2 - 6x + 8 = (x - 2)(x - 4)$$

E X A M P L E   10   Factoring Trinomials

Factor completely: $x^2 - x - 12$

Solution   First, determine all possible integral factors of -12 and then compute each sum:

Factors of -12	1, -12	-1, 12	2, -6	-2, 6	3, -4	-3, 4
Sum	-11	11	-4	4	-1	1

Since -1 is the coefficient of the middle term, we have

$$x^2 - x - 12 = (x + 3)(x - 4)$$

E X A M P L E   11   Factoring Trinomials

Factor completely: $x^2 + 4x - 12$

Solution   The factors -2 and 6 of -12 have the sum 4. Thus,

$$x^2 + 4x - 12 = (x - 2)(x + 6)$$

To avoid errors in factoring, always check your answer by multiplying it out to see if the result equals the original expression.
When none of the possibilities works, the polynomial is prime.

E X A M P L E   12   Identifying Prime Polynomials

Show that $x^2 + 9$ is prime.

Solution   First, list the integral factors of 9 and then compute their sums:

Factors of 9	1, 9	-1, -9	3, 3	-3, -3
Sum	10	-10	6	-6

Since the coefficient of the middle term in $x^2 + 9$ is 0 and none of the preceding sums equals 0, we conclude that $x^2 + 9$ is prime.

Example 12 demonstrates a more general result:

Theorem

Any polynomial of the form $x^2 + a^2$, a real, is prime.

 Now work Problems 15 and 25.

When the leading coefficient is not 1, a somewhat longer list of possibilities may be required. Observe the pattern in the following formula:

$$Ax^2 + Bx + C = (ax + b)(cx + d) = acx^2 + (ad + bc)x + bd$$

$$A = ac, B = ad + bc, C = bd$$

Our task is to find factors a and c of the leading coefficient A and factors b and d of the constant term C so that the expression $ad + bc$ equals the coefficient B of the middle term.

E X A M P L E   13

Factoring Trinomials

Factor completely: $2x^2 + 5x + 3$

Solution   The positive integral factors of the leading coefficient $ac = 2$ are $a = 2$, $c = 1$. We begin the factorization by writing

$$2x^2 + 5x + 3 = (2x \quad)(x \quad)$$

The positive integral factors of the constant term $bd = 3$ are $b = 1, d = 3$, or $b = 3, d = 1$. This suggests the following possibilities:

$$2x^2 + 5x + 3 \begin{cases} (2x \quad 1)(x \quad 3) \\ (2x \quad 3)(x \quad 1) \end{cases}$$

Next, we select the signs to be placed inside the factors. Since the constant term is positive and the coefficient of the middle term is positive, the only signs that can possibly work are + signs. (Do you see why?) Thus,

$$2x^2 + 5x + 3 \begin{cases} (2x + 1)(x + 3) = 2x^2 + 7x + 3 \\ (2x + 3)(x + 1) = 2x^2 + 5x + 3 \end{cases}$$

We conclude that

$$2x^2 + 5x + 3 = (2x + 3)(x + 1)$$

E X A M P L E   14

Factoring Trinomials

Factor completely: $2x^2 - x - 6$

Solution   The positive integral factors of 2 are 2, 1 and the positive integral factors of 6 are 1, 6, and 2, 3. This suggests the following possibilities:

$$2x^2 - x - 6 \begin{cases} (2x \quad 1)(x \quad 6) \\ (2x \quad 6)(x \quad 1) \\ (2x \quad 2)(x \quad 3) \\ (2x \quad 3)(x \quad 2) \end{cases}$$

Since the constant term is negative, the signs chosen for each possible factor must be opposite. This leads to the possibilities

$$(2x - 1)(x + 6) = 2x^2 + 11x - 6 \qquad (2x - 2)(x + 3) = 2x^2 + 4x - 6$$
$$(2x + 1)(x - 6) = 2x^2 - 11x - 6 \qquad (2x + 2)(x - 3) = 2x^2 - 4x - 6$$
$$(2x - 6)(x + 1) = 2x^2 - 4x - 6 \qquad (2x - 3)(x + 2) = 2x^2 + x - 6$$
$$(2x + 6)(x - 1) = 2x^2 + 4x - 6 \qquad (2x + 3)(x - 2) = 2x^2 - x - 6$$

Thus, $2x^2 - x - 6 = (2x + 3)(x - 2)$.

Note: $(2x \pm 3)(x \mp 2)$ is a notation that can be used to save space. It actually represents two products: $(2x + 3)(x - 2)$ and $(2x - 3)(x + 2)$.

Study the patterns illustrated in these examples carefully. Practice will give you the experience needed to use this factoring technique skillfully and efficiently.*

Now work Problem 51.

Other Factoring Techniques

 Sometimes a common factor does not occur in every term of the polynomial, but in each of several groups of terms that together make up the polynomial. When this happens, the common factor can be factored out of each group by means of the distributive property. This technique is called **factoring by grouping.**

E X A M P L E   15   Factoring by Grouping

Factor completely by grouping: $(x^2 + 2)x + (x^2 + 2) \cdot 3$

Solution   Notice the common factor $x^2 + 2$. By applying the distributive property, we have

$$(x^2 + 2)x + (x^2 + 2) \cdot 3 = (x^2 + 2)(x + 3)$$

Since $x^2 + 2$ and $x + 3$ are prime, the factorization is complete. ■

E X A M P L E   16   Factoring by Grouping

Factor completely by grouping: $x^3 - 4x^2 + 2x - 8$

Solution   To see if factoring by grouping will work, group the first two terms and the last two terms. Then look for a common factor in each group. In this example, we can factor x^2 from $x^3 - 4x^2$ and 2 from $2x - 8$. The remaining factor in each case is the same, $x - 4$. This means factoring by grouping will work, as follows:

$$\begin{aligned}
x^3 - 4x^2 + 2x - 8 &= (x^3 - 4x^2) + (2x - 8) \\
&= x^2(x - 4) + 2(x - 4) \\
&= (x^2 + 2)(x - 4)
\end{aligned}$$

Since $x^2 + 2$ and $x - 4$ are prime, the factorization is complete. ■

E X A M P L E   17   Factoring by Grouping

Factor completely by grouping: $3x^3 + 4x^2 - 6x - 8$

---

*There is another method for factoring trinomials, whose leading coefficient is not 1, that does not require trial and error. It is especially efficient when trial and error lead to many possibilities. See Problem 93 in Exercises 1.5 for the procedure.

Solution   Here, $3x + 4$ is a common factor of $3x^3 + 4x^2$ and of $-6x - 8$. Hence,

$$3x^3 + 4x^2 - 6x - 8 = (3x^3 + 4x^2) - (6x + 8)$$
$$= x^2(3x + 4) - 2(3x + 4)$$
$$= (x^2 - 2)(3x + 4)$$

Since $x^2 - 2$ and $3x + 4$ are prime (over the integers), the factorization is complete.   ▬

 Now work Problem 83.

SUMMARY

We close this section with a capsule summary.

Type of Polynomial	Method	Example
Any polynomial	Look for common monomial factors. (Always do this first!)	$6x^2 + 9x = 3x(2x + 3)$
Binomials of degree 2 or higher	Check for a special product: Difference of two squares, $x^2 - a^2$ Difference of two cubes, $x^3 - a^3$ Sum of two cubes, $x^3 + a^3$	$x^2 - 16 = (x - 4)(x + 4)$ $x^3 - 64 = (x - 4)(x^2 + 4x + 16)$ $x^3 + 27 = (x + 3)(x^2 - 3x + 9)$
Trinomials of degree 2	Check for a perfect square, $(x \pm a)^2$. List possibilities.	$x^2 + 8x + 16 = (x + 4)^2$ $6x^2 + x - 1 = (2x + 1)(3x - 1)$
Three or more terms	Grouping	$2x^3 - 3x^2 + 4x - 6 = (2x - 3)(x^2 + 2)$

1.5   EXERCISES

In Problems 1–10, factor each polynomial by removing the common monomial factor.

1. $3x + 6$ **2.** $7x - 14$ **3.** $ax^2 + a$ **4.** $ax - a$

5. $x^3 + x^2 + x$ **6.** $x^3 - x^2 + x$ **7.** $2x^2 - 2x$ **8.** $3x^2 - 3x$

9. $3x^2y - 6xy^2 + 12xy$ **10.** $60x^2y - 48xy^2 + 72x^3y$

In Problems 11–88, factor completely each polynomial. If the polynomial cannot be factored, say it is prime.

11. $x^2 - 1$ **12.** $x^2 - 9$ **13.** $1 - 4x^2$ **14.** $1 - 9x^2$

15. $x^2 + 7x + 10$ **16.** $x^2 + 5x + 4$ **17.** $x^2 - 10x + 21$ **18.** $x^2 - 6x + 8$

19. $x^2 - 2x - 8$ **20.** $x^2 - 4x - 5$ **21.** $x^2 + 2x + 1$ **22.** $x^2 + 4x + 4$

23. $x^2 - 4x + 4$ **24.** $x^2 - 2x + 1$ **25.** $15 + 2x - x^2$ **26.** $14 + 6x - x^2$

27. $3x^2 - 12x - 36$ **28.** $x^3 + 8x^2 - 20x$ **29.** $y^4 + 11y^3 + 30y^2$

30. $3y^3 - 18y^2 - 48y$ **31.** $16x^2 + 8x + 1$ **32.** $25x^2 + 10x + 1$

33. $4x^2 + 12x + 9$ **34.** $9x^2 - 12x + 4$ **35.** $ax^2 - 4a^2x - 45a^3$

36. $bx^2 + 14b^2x + 45b^3$ **37.** $x^3 - 125$ **38.** $x^3 + 125$

39. $27 - 8x^3$ **40.** $64 - 8x^3$ **41.** $2x^4 + 16x$

42. $3x^5 - 3x^2$ **43.** $3x^2 + 4x + 1$ **44.** $4x^2 + 3x - 1$

45. $x^4 - 81$

46. $x^4 - 1$

47. $x^6 - 2x^3 + 1$

48. $x^6 + 2x^3 + 1$

49. $x^7 - x^5$

50. $x^8 - x^5$

51. $2z^2 + 5z + 3$

52. $6z^2 - z - 1$

53. $16x^2 + 24x + 9$

54. $9x^2 - 24x + 16$

55. $5 + 16x - 16x^2$

56. $5 + 11x - 16x^2$

57. $4y^2 - 16y + 15$

58. $9y^2 + 9y - 4$

59. $18x^2 - 9x - 27$

60. $8x^2 - 6x - 2$

61. $8x^2 + 2x + 6$

62. $9x^2 - 3x + 3$

63. $x^2 - x + 4$

64. $x^2 + 6x + 9$

65. $4x^3 - 10x^2 - 6x$

66. $27x^3 - 9x^2 - 6x$

67. $x^4 + 2x^2 + 1$

68. $x^4 - 6x^2 + 9$

69. $1 - 8x^2 - 9x^4$

70. $4 - 14x^2 - 8x^4$

71. $8 - 64x^3$

72. $8 + 64x^3$

73. $x(x + 3) - 6(x + 3)$

74. $5(3x - 7) + x(3x - 7)$

75. $(x + 2)^2 - 5(x + 2)$

76. $(x - 1)^2 - 2(x - 1)$

77. $(2x - 1)^2 - 25$

78. $(5x - 3)^2 - 9$

79. $(3x - 2)^3 - 27$

80. $(5x + 1)^3 - 1$

81. $3(x^2 + 10x + 25) - 4(x + 5)$

82. $7(x^2 - 6x + 9) + 5(x - 3)$

83. $x^3 + 2x^2 - x - 2$

84. $x^3 - 3x^2 - x + 3$

85. $x^4 - x^3 + x - 1$

86. $x^4 + x^3 + x + 1$

87. $x^5 + x^3 + 8x^2 + 8$

88. $x^5 - x^3 + 8x^2 - 8$

89. Show that $x^2 + 4$ is prime.

90. Show that $x^2 + x + 1$ is prime.

91. Make up a polynomial that factors into a perfect square.

92. Explain to a fellow student what you look for first when presented with a factoring problem. What do you do next?

93. Another method* for factoring a trinomial $ax^2 + bx + c$ is given next:

STEP 1: Multiply the last coefficient, c, by the first coefficient, a. Replace the coefficient c by ac and a by 1.

$6x^2 - x - 12$ $\quad a = 6, c = -12$

$x^2 - x - 72$ $\quad ac = -72$

STEP 2: Factor this polynomial

$(x - 9)(x + 8)$

STEP 3: Using the original coefficient a, replace x by ax in each factor. Thus, we replace x by $6x$.

$(6x - 9)(6x + 8)$

STEP 4: Factor out all common factors and discard each number.

$3(2x - 3) \cdot 2(3x + 4)$

$(2x - 3)(3x + 4) = 6x^2 - x - 12$

Use this method to solve Examples 13 and 14. Which method do you prefer, the one in the text or the one given here? Which method would you use to teach factoring to someone for the first time? How does this method detect primes?

1.6 RATIONAL EXPRESSIONS

- **1** Reduce a Rational Expression to Lowest Terms
- **2** Multiply and Divide Rational Expressions
- **3** Add and Subtract Rational Expressions
- **4** Use the Least Common Multiple Method
- **5** Simplify Mixed Quotients

If we form the quotient of two polynomials, the result is called a **rational expression.** Some examples of rational expressions are

(a) $\dfrac{x^3 + 1}{x}$ (b) $\dfrac{3x^2 + x - 2}{x^2 + 5}$ (c) $\dfrac{x}{x^2 - 1}$ (d) $\dfrac{xy^2}{(x - y)^2}$

Expressions (a), (b), and (c) are rational expressions in one variable, x, whereas (d) is a rational expression in two variables, x and y.

*Robert Atkin, *Factoring Trinomials,* Association of Mathematics Teachers of New York, Summer 1991.

Rational expressions are described in the same manner as rational numbers. Thus, in expression (a), the polynomial $x^3 + 1$ is called the **numerator,** and x is called the **denominator.** When the numerator and denominator of a rational expression contain no common factors (except 1 and -1), we say the rational expression is **reduced to lowest terms,** or **simplified.** Remember that the denominator of a rational expression can never be 0. Thus, the domain of the variable x in expression (a) is $x \neq 0$. For expression (b), x can be any real number since $x^2 + 5$ is never 0. In (c), $x \neq -1, x \neq 1$; and in (d), $x \neq y$.

A rational expression is reduced to lowest terms by factoring completely the numerator and the denominator and canceling any common factors by using the cancellation property,

$$\frac{a\cancel{c}}{b\cancel{c}} = \frac{a}{b} \qquad \text{if } b \neq 0, c \neq 0 \tag{1}$$

E X A M P L E   1   **Reducing to Lowest Terms**

Reduce each rational expression to lowest terms.

(a) $\dfrac{x^2 + 4x + 4}{x^2 + 3x + 2}$ (b) $\dfrac{x^3 - 8}{x^3 - 2x^2}$ (c) $\dfrac{8 - 2x}{x^2 - x - 12}$

Solution (a) $\dfrac{x^2 + 4x + 4}{x^2 + 3x + 2} = \dfrac{\cancel{(x+2)}(x+2)}{\cancel{(x+2)}(x+1)} = \dfrac{x+2}{x+1}, \qquad x \neq -2, -1$

(b) $\dfrac{x^3 - 8}{x^3 - 2x^2} = \dfrac{\cancel{(x-2)}(x^2 + 2x + 4)}{x^2\cancel{(x-2)}} = \dfrac{x^2 + 2x + 4}{x^2}, \qquad x \neq 0, 2$

(c) $\dfrac{8 - 2x}{x^2 - x - 12} = \dfrac{2(4 - x)}{(x - 4)(x + 3)} = \dfrac{2(-1)\cancel{(x-4)}}{\cancel{(x-4)}(x + 3)} = \dfrac{-2}{x + 3}, x \neq -3, 4$ ∎

Now work Problem 3.

Multiplying and Dividing Rational Expressions

The rules for multiplying and dividing rational expressions are the same as the rules for multiplying and dividing rational numbers. If $\dfrac{a}{b}$ and $\dfrac{c}{d}, b \neq 0,$ $d \neq 0$, are two rational expressions, then

$$\frac{a}{b} \cdot \frac{c}{d} = \frac{ac}{bd} \qquad \text{if } b \neq 0, d \neq 0 \tag{2}$$

$$\frac{\dfrac{a}{b}}{\dfrac{c}{d}} = \frac{a}{b} \cdot \frac{d}{c} = \frac{ad}{bc} \qquad \text{if } b \neq 0, c \neq 0, d \neq 0 \tag{3}$$

In using equations (2) and (3) with rational expressions, be sure first to factor each polynomial completely so that common factors can be canceled. We shall follow the practice of leaving our answers in factored form.

E X A M P L E   2   Multiplying and Dividing Rational Expressions

Perform the indicated operation and simplify the result. Leave your answer in factored form.

(a) $\dfrac{x^2 - 2x + 1}{x^3 + x} \cdot \dfrac{4x^2 + 4}{x^2 + x - 2}$ (b) $\dfrac{\dfrac{x + 3}{x^2 - 4}}{\dfrac{x^2 - x - 12}{x^3 - 8}}$

Solution   (a) $\dfrac{x^2 - 2x + 1}{x^3 + x} \cdot \dfrac{4x^2 + 4}{x^2 + x - 2} = \dfrac{(x - 1)^2}{x(x^2 + 1)} \cdot \dfrac{4(x^2 + 1)}{(x + 2)(x - 1)}$

$$= \dfrac{(x - 1)^2(4)(x^2 + 1)}{x(x^2 + 1)(x + 2)(x - 1)} = \dfrac{4(x - 1)}{x(x + 2)},$$

$$x \neq -2, 0, 1$$

(b) $\dfrac{\dfrac{x + 3}{x^2 - 4}}{\dfrac{x^2 - x - 12}{x^3 - 8}} = \dfrac{x + 3}{x^2 - 4} \cdot \dfrac{x^3 - 8}{x^2 - x - 12}$

$$= \dfrac{x + 3}{(x - 2)(x + 2)} \cdot \dfrac{(x - 2)(x^2 + 2x + 4)}{(x - 4)(x + 3)}$$

$$= \dfrac{(x + 3)(x - 2)(x^2 + 2x + 4)}{(x - 2)(x + 2)(x - 4)(x + 3)} = \dfrac{x^2 + 2x + 4}{(x + 2)(x - 4)},$$

$$x \neq -3, -2, 2, 4 \quad \blacksquare$$

 Now work Problems 23 and 29.

Adding and Subtracting Rational Expressions

3   The rules for adding and subtracting rational expressions are the same as the rules for adding and subtracting rational numbers. Thus, if the denominators of two rational expressions to be added (or subtracted) are equal, we add (or subtract) the numerators and keep the common denominator.

If $\dfrac{a}{b}$ and $\dfrac{c}{b}$ are two rational expressions, then

$$\dfrac{a}{b} + \dfrac{c}{b} = \dfrac{a + c}{b} \qquad \dfrac{a}{b} - \dfrac{c}{b} = \dfrac{a - c}{b} \qquad \text{if } b \neq 0 \qquad (4)$$

E X A M P L E   3   Adding and Subtracting Rational Expressions
with Equal Denominators

Perform the indicated operation and simplify the result. Leave your answer in factored form.

(a) $\dfrac{2x^2 - 4}{2x + 5} + \dfrac{x + 3}{2x + 5}$, $x \neq -\dfrac{5}{2}$ (b) $\dfrac{x}{x - 3} - \dfrac{3x + 2}{x - 3}$, $x \neq 3$

Solution   (a) $\dfrac{2x^2 - 4}{2x + 5} + \dfrac{x + 3}{2x + 5} = \dfrac{(2x^2 - 4) + (x + 3)}{2x + 5}$

$$= \dfrac{2x^2 + x - 1}{2x + 5} = \dfrac{(2x - 1)(x + 1)}{2x + 5}$$

(b) $\dfrac{x}{x - 3} - \dfrac{3x + 2}{x - 3} = \dfrac{x - (3x + 2)}{x - 3} = \dfrac{x - 3x - 2}{x - 3}$

$$= \dfrac{-2x - 2}{x - 3} = \dfrac{-2(x + 1)}{x - 3}$$ ■

E X A M P L E   4   Adding Rational Expressions Whose Denominators Are Additive
Inverses of Each Other

Perform the indicated operation and simplify the result. Leave your answer in factored form.

$$\dfrac{2x}{x - 3} + \dfrac{5}{3 - x},   x \neq 3$$

Solution   Notice that the denominators of the two rational expressions are different. However, the denominator of the second expression is just the additive inverse of the denominator of the first. That is,

$$3 - x = -x + 3 = -1 \cdot (x - 3) = -(x - 3)$$

Thus,

$$\dfrac{2x}{x - 3} + \dfrac{5}{3 - x} = \dfrac{2x}{x - 3} + \dfrac{5}{-(x - 3)} = \dfrac{2x}{x - 3} + \dfrac{-5}{x - 3}$$

$$\underset{3 - x = -(x - 3)}{\uparrow} \qquad \underset{\frac{a}{-b} = \frac{-a}{b}}{\uparrow}$$

$$= \dfrac{2x + (-5)}{x - 3} = \dfrac{2x - 5}{x - 3}$$ ■

If the denominators of two rational expressions to be added or subtracted are not equal, we can use the general formulas for adding and subtracting quotients:

$$\dfrac{a}{b} + \dfrac{c}{d} = \dfrac{a}{b} \cdot \dfrac{d}{d} + \dfrac{b}{b} \cdot \dfrac{c}{d} = \dfrac{ad + bc}{bd} \qquad \text{if } b \neq 0, d \neq 0$$

$$\dfrac{a}{b} - \dfrac{c}{d} = \dfrac{a \cdot d}{b \cdot d} - \dfrac{b \cdot c}{b \cdot d} = \dfrac{ad - bc}{bd} \qquad \text{if } b \neq 0, d \neq 0$$ (5)

E X A M P L E   5   Adding and Subtracting Rational Expressions
                         with Unequal Denominators

Perform the indicated operation and simplify the result. Leave your answer
in factored form.

(a) $\dfrac{x-3}{x+4} + \dfrac{x}{x-2}$, $x \neq -4, 2$ (b) $\dfrac{x^2}{x^2-4} - \dfrac{1}{x}$, $x \neq -2, 0, 2$

Solution   (a) $\dfrac{x-3}{x+4} + \dfrac{x}{x-2} \underset{\underset{(5)}{\uparrow}}{=} \dfrac{(x-3)(x-2) + (x+4)(x)}{(x+4)(x-2)}$

$$= \dfrac{x^2 - 5x + 6 + x^2 + 4x}{(x+4)(x-2)} = \dfrac{2x^2 - x + 6}{(x+4)(x-2)}$$

(b) $\dfrac{x^2}{x^2-4} - \dfrac{1}{x} = \dfrac{x^2(x) - (x^2-4)(1)}{(x^2-4)(x)} = \dfrac{x^3 - x^2 + 4}{(x-2)(x+2)(x)}$ ∎

Now work Problem 49.

Least Common Multiple (LCM)

 If the denominators of two rational expressions to be added (or subtracted)
have common factors, we usually do not use the general rules given by equation (5), since, in doing so, we make the problem more complicated than it
needs to be. Instead, just as with fractions, we apply the **least common multiple (LCM) method** by using the polynomial of least degree that contains
each denominator polynomial as a factor. Then we rewrite each rational expression using the LCM as the common denominator and use equation (4)
to do the addition (or subtraction).

To find the least common multiple of two or more polynomials, first factor completely each polynomial. The LCM is the product of the different
prime factors of each polynomial, each factor appearing the greatest number of times it occurs in each polynomial. The next two examples will give
you the idea.

E X A M P L E   6   Finding the Least Common Multiple

Find the least common multiple of the following pair of polynomials:

$x(x-1)^2(x+1)$ and $4(x-1)(x+1)^3$

Solution   The polynomials are already factored completely as

$x(x-1)^2(x+1)$ and $4(x-1)(x+1)^3$

Start by writing the factors of the left-hand polynomial. (Or you could start
with the one on the right.)

$x(x-1)^2(x+1)$

Now look at the right-hand polynomial. Its first factor, 4, does not appear in our list, so we insert it:

$4x(x-1)^2(x+1)$

The next factor, $x - 1$, is already in our list, so no change is necessary. The final factor is $(x + 1)^3$. Since our list has $x + 1$ to the first power only, we replace $x + 1$ in the list by $(x + 1)^3$. The LCM is

$$4x(x - 1)^2(x + 1)^3$$

Notice that the LCM is, in fact, the polynomial of least degree that contains $x(x - 1)^2(x + 1)$ and $4(x - 1)(x + 1)^3$ as factors. ▬

E X A M P L E   7   Finding the Least Common Multiple

Find the least common multiple of the following pair of polynomials:

$$2x^2 - 2x - 12 \quad \text{and} \quad x^3 - 3x^2$$

Solution   First, we factor completely each polynomial:

$$2x^2 - 2x - 12 = 2(x^2 - x - 6) = 2(x - 3)(x + 2)$$
$$x^3 - 3x^2 = x^2(x - 3)$$

Now we write the factors of the first polynomial:

$$2(x - 3)(x + 2)$$

Looking at the factors that appear in the second polynomial, we see that the first factor, x^2, does not appear in our list, so we insert it:

$$2x^2(x - 3)(x + 2)$$

The remaining factor, $x - 3$, is already in our list. Thus, the LCM of $2x^2 - 2x - 12$ and $x^3 - 3x^2$ is $2x^2(x - 3)(x + 2)$. ▬

 Now work Problem 61.

The next example illustrates how the LCM is used for adding and subtracting rational expressions.

E X A M P L E   8   Using the Least Common Multiple to Add Rational Expressions

Perform the indicated operation and simplify the result. Leave your answer in factored form.

$$\frac{x}{x^2 + 3x + 2} + \frac{2x - 3}{x^2 - 1} \quad x \neq -2, -1, 1$$

Solution   First, we find the LCM of the denominators:

$$x^2 + 3x + 2 = (x + 2)(x + 1)$$
$$x^2 - 1 = (x - 1)(x + 1)$$

The LCM is $(x + 2)(x + 1)(x - 1)$. Next, we rewrite each rational expression using the LCM as the denominator:

$$\frac{x}{x^2 + 3x + 2} = \frac{x}{(x + 2)(x + 1)} = \frac{x(x - 1)}{(x + 2)(x + 1)(x - 1)}$$

↑

Multiply numerator and denominator by $x - 1$ to get the LCM in the denominator.

$$\frac{2x - 3}{x^2 - 1} = \frac{2x - 3}{(x - 1)(x + 1)} = \frac{(2x - 3)(x + 2)}{(x - 1)(x + 1)(x + 2)}$$

↑

Multiply numerator and denominator by $x + 2$ to get the LCM in the denominator.

Now we can add by using equation (4).

$$\frac{x}{x^2 + 3x + 2} + \frac{2x - 3}{x^2 - 1} = \frac{x(x - 1)}{(x + 2)(x + 1)(x - 1)} + \frac{(2x - 3)(x + 2)}{(x + 2)(x + 1)(x - 1)}$$

$$= \frac{(x^2 - x) + (2x^2 + x - 6)}{(x + 2)(x + 1)(x - 1)}$$

$$= \frac{3x^2 - 6}{(x + 2)(x + 1)(x - 1)} = \frac{3(x^2 - 2)}{(x + 2)(x + 1)(x - 1)}$$

∎

If we had not used the LCM technique to add the quotients in Example 8, but decided instead to use the general rule of equation (5), we would have obtained a more complicated expression, as follows:

$$\frac{x}{x^2 + 3x + 2} + \frac{2x - 3}{x^2 - 1} = \frac{x(x^2 - 1) + (x^2 + 3x + 2)(2x - 3)}{(x^2 + 3x + 2)(x^2 - 1)}$$

$$= \frac{3x^3 + 3x^2 - 6x - 6}{(x^2 + 3x + 2)(x^2 - 1)} = \frac{3(x^3 + x^2 - 2x - 2)}{(x^2 + 3x + 2)(x^2 - 1)}$$

Now we are faced with a more complicated problem of expressing this quotient in lowest terms. It is always best to first look for common factors in the denominators of expressions to be added or subtracted and use the LCM if any common factors are found.

E X A M P L E   9

Using the Least Common Multiple to Combine Three Rational Expressions

Perform the indicated operations and simplify the result. Leave your answer in factored form.

$$\frac{1}{x^2 + x} - \frac{x + 4}{x^3 + 1} + \frac{3}{x^2} \qquad x \neq -1, 0$$

Solution   Again, we start by finding the LCM of the denominators:

$$x^2 + x = x(x + 1)$$
$$x^3 + 1 = (x + 1)(x^2 - x + 1)$$
$$x^2 = x^2$$

The LCM is $x^2(x + 1)(x^2 - x + 1)$. Thus,

$$\frac{1}{x^2 + x} - \frac{x + 4}{x^3 + 1} + \frac{3}{x^2}$$

$$= \frac{1}{x(x + 1)} - \frac{x + 4}{(x + 1)(x^2 - x + 1)} + \frac{3}{x^2}$$

$$= \frac{x(x^2 - x + 1)}{x^2(x + 1)(x^2 - x + 1)} - \frac{x^2(x + 4)}{x^2(x + 1)(x^2 - x + 1)} + \frac{3(x + 1)(x^2 - x + 1)}{x^2(x + 1)(x^2 - x + 1)}$$

$$= \frac{(x^3 - x^2 + x) - (x^3 + 4x^2) + 3(x^3 + 1)}{x^2(x + 1)(x^2 - x + 1)} = \frac{3x^3 - 5x^2 + x + 3}{x^2(x + 1)(x^2 - x + 1)} \quad \blacksquare$$

 Now work Problem 69.

Mixed Quotients

5 When sums and/or differences of rational expressions appear as the numerator and/or denominator of a quotient, the quotient is called a **mixed quotient.*** For example,

$$\frac{1 + \dfrac{1}{x}}{1 - \dfrac{1}{x}} \quad \text{and} \quad \frac{\dfrac{x^2}{x^2 - 4} - 3}{\dfrac{x - 3}{x + 2} - 1}$$

are mixed quotients. To **simplify** a mixed quotient means to write it as a rational expression reduced to lowest terms. This can be accomplished in either of two ways:

Simplifying a Mixed Quotient

METHOD 1:	Treat the numerator and denominator of the mixed quotient separately, performing whatever operations are indicated and simplifying the results. Follow this by simplifying the resulting rational expression.
METHOD 2:	Find the LCM of the denominators of all rational expressions that appear in the mixed quotient. Multiply the numerator and denominator of the mixed quotient by the LCM and simplify the result.

We will use both methods in the next example. By carefully studying each method, you can discover in which situations one method may be easier to use than the other.

E X A M P L E 10

Simplifying a Mixed Quotient

Simplify: $\dfrac{\dfrac{1}{2} + \dfrac{3}{x}}{\dfrac{x + 3}{4}}$ $x \neq -3, 0$

Solution METHOD 1: First, we perform the indicated operation in the numerator, and then we divide:

*Some texts use the term **complex fraction.**

$$\frac{\dfrac{1}{2} + \dfrac{3}{x}}{\dfrac{x+3}{4}} = \frac{\dfrac{1 \cdot x + 2 \cdot 3}{2 \cdot x}}{\dfrac{x+3}{4}} = \frac{\dfrac{x+6}{2x}}{\dfrac{x+3}{4}} = \frac{x+6}{2x} \cdot \frac{4}{x+3}$$

$$\uparrow \qquad\qquad\qquad\qquad\qquad\qquad\qquad \uparrow$$

Rule for adding quotients　　　　　Rule for dividing quotients

$$= \frac{(x+6) \cdot 4}{2 \cdot x \cdot (x+3)} = \frac{2 \cdot 2 \cdot (x+6)}{2 \cdot x \cdot (x+3)} = \frac{2(x+6)}{x(x+3)}$$

$$\uparrow$$

Rule for multiplying quotients

METHOD 2: The rational expressions that appear in the mixed quotient are

$$\frac{1}{2}, \quad \frac{3}{x}, \quad \frac{x+3}{4}$$

The LCM of their denominators is $4x$. Thus, we multiply the numerator and denominator of the mixed quotient by $4x$ and then simplify:

$$\frac{\dfrac{1}{2} + \dfrac{3}{x}}{\dfrac{x+3}{4}} = \frac{4x \cdot \left(\dfrac{1}{2} + \dfrac{3}{x}\right)}{4x \cdot \left(\dfrac{x+3}{4}\right)} = \frac{4x \cdot \dfrac{1}{2} + 4x \cdot \dfrac{3}{x}}{\dfrac{4x \cdot (x+3)}{4}}$$

$$\uparrow \qquad\qquad\qquad\qquad \uparrow$$

Multiply by $4x$　　　　Distributive property in numerator

$$= \frac{2 \cdot 2x \cdot \dfrac{1}{2} + 4x \cdot \dfrac{3}{x}}{\dfrac{4x \cdot (x+3)}{4}} = \frac{2x + 12}{x(x+3)} = \frac{2(x+6)}{x(x+3)}$$

$$\qquad\qquad\qquad\qquad\qquad \uparrow \qquad\qquad \uparrow$$

Simplify　　　Factor

E X A M P L E　11 Simplifying a Mixed Quotient

Simplify: $\dfrac{\dfrac{x^2}{x^2-4} - 3}{\dfrac{x-3}{x+2} - 1}$ $\qquad x \neq -2, 2$

Solution We shall use Method 1:

$$\frac{\dfrac{x^2}{x^2-4} - 3}{\dfrac{x-3}{x+2} - 1} = \frac{\dfrac{x^2}{x^2-4} - \dfrac{3(x^2-4)}{x^2-4}}{\dfrac{x-3}{x+2} - \dfrac{x+2}{x+2}} = \frac{\dfrac{x^2 - 3(x^2-4)}{x^2-4}}{\dfrac{(x-3) - (x+2)}{x+2}} = \frac{\dfrac{-2x^2 + 12}{x^2-4}}{\dfrac{-5}{x+2}}$$

$$= \frac{\dfrac{-2(x^2-6)}{(x-2)(x+2)}}{\dfrac{-5}{x+2}} = \frac{-2(x^2-6)}{(x-2)(x+2)} \cdot \frac{x+2}{-5}$$

$$= \frac{-2(x^2-6)(x+2)}{(x-2)(x+2)(-5)} = \frac{2(x^2-6)}{5(x-2)}$$

1.6 | EXERCISES

In Problems 1–20, reduce each rational expression to lowest terms.

1. $\dfrac{3x + 9}{x^2 - 9}$

2. $\dfrac{4x^2 + 8x}{12x + 24}$

3. $\dfrac{x^2 - 2x}{3x - 6}$

4. $\dfrac{15x^2 + 24x}{3x^2}$

5. $\dfrac{24x^2}{12x^2 - 6x}$

6. $\dfrac{x^2 + 4x + 4}{x^2 - 16}$

7. $\dfrac{y^2 - 25}{2y^2 - 8y - 10}$

8. $\dfrac{3y^2 - y - 2}{3y^2 + 5y + 2}$

9. $\dfrac{x^2 + 4x - 5}{x^2 - 2x + 1}$

10. $\dfrac{x - x^2}{x^2 + x - 2}$

11. $\dfrac{x^2 - 4}{x^2 + 5x + 6}$

12. $\dfrac{x^2 + x - 6}{9 - x^2}$

13. $\dfrac{x^2 + 5x - 14}{2 - x}$

14. $\dfrac{2x^2 + 5x - 3}{1 - 2x}$

15. $\dfrac{2x^3 - x^2 - 10x}{x^3 - 2x^2 - 8x}$

16. $\dfrac{4x^4 + 2x^3 - 6x^2}{4x^4 + 26x^3 + 30x^2}$

17. $\dfrac{(x - 4)^2 - 9}{(x + 3)^2 - 16}$

18. $\dfrac{(x + 2)^2 - 8x}{(x - 2)^2}$

19. $\dfrac{6x(x - 1) - 12}{x^3 - 8}$

20. $\dfrac{3(x - 2)^2 + 17(x - 2)}{2(x - 2)^2 + 7(x - 2)}$

In Problems 21–34, perform the indicated operation and simplify the result. Leave your answer in factored form.

21. $\dfrac{3x + 6}{5x^2} \cdot \dfrac{x^2 - x - 6}{x^2 - 4}$

22. $\dfrac{9x^2 - 25}{2x - 2} \cdot \dfrac{1 - x^2}{6x + 10}$

23. $\dfrac{4x^2 - 1}{x^2 - 16} \cdot \dfrac{x^2 - 4x}{2x + 1}$

24. $\dfrac{12}{x^2 - x} \cdot \dfrac{x^2 - 1}{4x - 2}$

25. $\dfrac{4x - 8}{-3x} \cdot \dfrac{12}{12 - 6x}$

26. $\dfrac{6x - 27}{5x} \cdot \dfrac{2}{4x - 18}$

27. $\dfrac{x^2 - 3x - 10}{x^2 + 2x - 35} \cdot \dfrac{x^2 + 4x - 21}{x^2 + 9x + 14}$

28. $\dfrac{x^2 + x - 6}{x^2 + 4x - 5} \cdot \dfrac{x^2 - 25}{x^2 + 2x - 15}$

29. $\dfrac{\dfrac{2x^2 - x - 28}{3x^2 - x - 2}}{\dfrac{4x^2 + 16x + 7}{3x^2 + 11x + 6}}$

30. $\dfrac{\dfrac{9x^2 - 3x - 2}{12x^2 + 5x - 2}}{\dfrac{9x^2 - 6x + 1}{8x^2 + 10x - 3}}$

31. $\dfrac{\dfrac{8x^2 - 6x + 1}{4x^2 - 1}}{\dfrac{12x^2 - 5x - 2}{6x^2 - x - 2}}$

32. $\dfrac{\dfrac{3x^2 + 2x - 1}{5x^2 - 9x - 2}}{\dfrac{2x^2 - x - 3}{10x^2 - 13x - 3}}$

33. $\dfrac{5x^2 - x}{3x + 2} \cdot \dfrac{2x^2 - x}{2x^2 - x - 1} \cdot \dfrac{10x^2 + 3x - 1}{6x^2 + x - 2}$

34. $\dfrac{x^2 - 7x - 8}{x^2 + 2x - 15} \cdot \dfrac{x^2 + 8x + 12}{x^2 + 6x - 7} \cdot \dfrac{x^2 + 4x - 5}{x^2 + 11x + 18}$

In Problems 35–58, perform the indicated operations and simplify the result. Leave your answer in factored form.

35. $\dfrac{x}{2} + \dfrac{5}{2}$

36. $\dfrac{3}{x} - \dfrac{6}{x}$

37. $\dfrac{x^2}{2x - 3} - \dfrac{4}{2x - 3}$

38. $\dfrac{3x^2}{2x - 1} - \dfrac{9}{2x - 1}$

39. $\dfrac{x + 1}{x - 3} + \dfrac{2x - 3}{x - 3}$

40. $\dfrac{2x - 5}{3x + 2} + \dfrac{x + 4}{3x + 2}$

41. $\dfrac{3x + 5}{2x - 1} - \dfrac{2x - 4}{2x - 1}$

42. $\dfrac{5x - 4}{3x + 4} - \dfrac{x + 1}{3x + 4}$

43. $\dfrac{4}{x - 2} + \dfrac{x}{2 - x}$

44. $\dfrac{6}{x - 1} - \dfrac{x}{1 - x}$

45. $\dfrac{4}{x - 1} - \dfrac{2}{x + 2}$

46. $\dfrac{2}{x + 5} - \dfrac{5}{x - 5}$

47. $\dfrac{x}{x + 1} + \dfrac{2x - 3}{x - 1}$

48. $\dfrac{3x}{x - 4} + \dfrac{2x}{x + 3}$

49. $\dfrac{x - 3}{x + 2} - \dfrac{x + 4}{x - 2}$

50. $\dfrac{2x - 3}{x - 1} - \dfrac{2x + 1}{x + 1}$

51. $\dfrac{x}{x^2 - 4} + \dfrac{1}{x}$

52. $\dfrac{x - 1}{x^3} + \dfrac{x}{x^2 + 1}$

53. $\dfrac{x^3}{(x-1)^2} - \dfrac{x^2+1}{x}$

54. $\dfrac{3x^2}{4} - \dfrac{x^3}{x^2-1}$

55. $\dfrac{x}{x+1} + \dfrac{x-2}{x-1} - \dfrac{x+1}{x-2}$

56. $\dfrac{3x+1}{x} + \dfrac{x}{x-1} - \dfrac{2x}{x+1}$

57. $\dfrac{1}{x} + \dfrac{1}{x+1} - \dfrac{1}{x-1}$

58. $\dfrac{1}{x} - \dfrac{1}{x-1} - \dfrac{1}{x-2}$

In Problems 59–66, find the LCM of the given polynomials.

59. $x^2 - 4, \quad x^2 - x - 2$

60. $x^2 - x - 12, \quad x^2 - 8x + 16$

61. $x^3 - x, \quad x^2 - x$

62. $3x^2 - 27, \quad 2x^2 - x - 15$

63. $4x^3 - 4x^2 + x, \quad 2x^3 - x^2, \quad x^3$

64. $x - 3, \quad x^2 + 3x, \quad x^3 - 9x$

65. $x^3 - x, \quad x^3 - 2x^2 + x, \quad x^3 - 1$

66. $x^2 + 4x + 4, \quad x^3 + 2x^2, \quad (x+2)^3$

In Problems 67–78, perform the indicated operations and simplify the result. Leave your answer in factored form.

67. $\dfrac{x}{x^2 - 7x + 6} - \dfrac{x}{x^2 - 2x - 24}$

68. $\dfrac{x}{x-3} - \dfrac{x+1}{x^2 + 5x - 24}$

69. $\dfrac{4x}{x^2 - 4} - \dfrac{2}{x^2 + x - 6}$

70. $\dfrac{3x}{x-1} - \dfrac{x-4}{x^2 - 2x + 1}$

71. $\dfrac{3}{(x-1)^2(x+1)} + \dfrac{2}{(x-1)(x+1)^2}$

72. $\dfrac{2}{(x+2)^2(x-1)} - \dfrac{6}{(x+2)(x-1)^2}$

73. $\dfrac{x+4}{x^2 - x - 2} - \dfrac{2x+3}{x^2 + 2x - 8}$

74. $\dfrac{2x-3}{x^2 + 8x + 7} - \dfrac{x-2}{(x+1)^2}$

75. $\dfrac{1}{x} - \dfrac{2}{x^2 + x} + \dfrac{3}{x^3 - x^2}$

76. $\dfrac{x}{(x-1)^2} + \dfrac{2}{x} - \dfrac{x+1}{x^3 - x^2}$

77. $\dfrac{1}{h}\left(\dfrac{1}{x+h} - \dfrac{1}{x}\right)$

78. $\dfrac{1}{h}\left[\dfrac{1}{(x+h)^2} - \dfrac{1}{x^2}\right]$

In Problems 79–90, perform the indicated operations and simplify the result. Leave your answer in factored form.

79. $\dfrac{1 + \dfrac{1}{x}}{1 - \dfrac{1}{x}}$

80. $\dfrac{4 + \dfrac{1}{x^2}}{3 - \dfrac{1}{x^2}}$

81. $\dfrac{x - \dfrac{1}{x}}{x + \dfrac{1}{x}}$

82. $\dfrac{1 - \dfrac{x}{x+1}}{2 - \dfrac{x-1}{x}}$

83. $\dfrac{\dfrac{x+4}{x-2} - \dfrac{x-3}{x+1}}{x+1}$

84. $\dfrac{\dfrac{x-2}{x+1} - \dfrac{x}{x-2}}{x+3}$

85. $\dfrac{\dfrac{x-2}{x+2} + \dfrac{x-1}{x+1}}{\dfrac{x}{x+1} - \dfrac{2x-3}{x}}$

86. $\dfrac{\dfrac{2x+5}{x} - \dfrac{x}{x-3}}{\dfrac{x^2}{x-3} - \dfrac{(x+1)^2}{x+3}}$

87. $1 - \dfrac{1}{1 - \dfrac{1}{x}}$

88. $1 - \dfrac{1}{1 - \dfrac{1}{1-x}}$

89. $\dfrac{\dfrac{x+h-2}{x+h+2} - \dfrac{x-2}{x+2}}{h}$

90. $\dfrac{\dfrac{x+h+1}{x+h-1} - \dfrac{x+1}{x-1}}{h}$

91. The following expressions are called **continued fractions:**

$$1 + \frac{1}{x}, \quad 1 + \frac{1}{1 + \frac{1}{x}}, \quad 1 + \frac{1}{1 + \frac{1}{1 + \frac{1}{x}}}, \quad 1 + \frac{1}{1 + \frac{1}{1 + \frac{1}{1 + \frac{1}{x}}}}, \dots$$

Each simplifies to an expression of the form

$$\frac{ax + b}{bx + c}$$

Trace the successive values of a, b, and c as you "continue" the fraction. Can you discover the patterns that these values follow? Go to the library and research Fibonacci numbers. Write a report on your findings.

92. Explain to a fellow student when you would use the LCM method to add two rational expressions. Give two examples of adding two rational expressions, one in which you use the LCM and the other in which you do not.

93. Which of the two methods given in the text for simplifying mixed quotients do you prefer? Write a brief paragraph stating the reasons for your choice.

1.7 | SQUARE ROOTS; RADICALS

> **1** Work with Properties of Square Roots
> **2** Rationalize the Denominator
> **3** Simplify nth Roots
> **4** Simplify Radicals

Square Roots

A real number is squared when it is raised to the power 2. The inverse of squaring is finding a **square root.** For example, since $6^2 = 36$ and $(-6)^2 = 36$, the numbers 6 and -6 are square roots of 36.

The symbol $\sqrt{}$, called a **radical sign,** is used to denote the **principal,** or nonnegative, square root. Thus, $\sqrt{36} = 6$.

> In general, if a is a nonnegative real number, the nonnegative number b such that $b^2 = a$ is the **principal square root** of a and is denoted by $b = \sqrt{a}$.

The following comments are noteworthy:

1. Negative numbers do not have square roots (in the real number system), because the square of any real number is *nonnegative*. For example, $\sqrt{-4}$ is not a real number, because there is no real number whose square is -4.
2. The principal square root of 0 is 0, since $0^2 = 0$. That is, $\sqrt{0} = 0$.
3. The principal square root of a positive number is positive.
4. If $c \geq 0$, then $(\sqrt{c})^2 = c$. For example, $(\sqrt{2})^2 = 2$ and $(\sqrt{3})^2 = 3$.

E X A M P L E 1

Evaluating Square Roots

(a) $\sqrt{64} = 8$ (b) $\sqrt{\dfrac{1}{16}} = \dfrac{1}{4}$ (c) $(\sqrt{1.4})^2 = 1.4$ ▬

Examples 1(a) and (b) are examples of **perfect square roots.** Thus, 64 is a **perfect square,** since $64 = 8^2$; and $\dfrac{1}{16}$ is a perfect square, since $\dfrac{1}{16} = \left(\dfrac{1}{4}\right)^2$.

In general, we have the rule

$$\sqrt{a^2} = |a| \tag{1}$$

Notice the need for the absolute value in equation (1). Since $a^2 \geq 0$, the principal square root of a^2 is defined whether $a > 0$ or $a < 0$. However, since the principal square root is nonnegative, we need the absolute value to ensure the nonnegative result.

E X A M P L E 2

Square Roots of Perfect Squares

(a) $\sqrt{(2.3)^2} = |2.3| = 2.3$ (b) $\sqrt{(-2.3)^2} = |-2.3| = 2.3$

(c) $\sqrt{x^2} = |x|$ ▬

Properties of Square Roots

We begin with the following observation:

$$\sqrt{4 \cdot 25} = \sqrt{100} = 10 \quad \text{and} \quad \sqrt{4}\sqrt{25} = 2 \cdot 5 = 10$$

This suggests the following property of square roots:
If a and b are each nonnegative real numbers, then

Product Property of Square Roots

$$\sqrt{ab} = \sqrt{a}\sqrt{b} \tag{2}$$

When used in connection with square roots, the direction "simplify" means to remove from the square root any perfect squares that occur as factors. We can use equation (2) to simplify a square root that contains a perfect square as a factor, as illustrated by the following examples.

E X A M P L E 3

Simplifying Square Roots

(a) $\sqrt{32} = \sqrt{16 \cdot 2} = \sqrt{16}\sqrt{2} = 4\sqrt{2}$

     ↑   ↑

   16 is a   (2)

  perfect square.

(b) $\sqrt{135} = \sqrt{9 \cdot 15} = \sqrt{9}\sqrt{15} = 3\sqrt{15}$

$\qquad\uparrow\qquad\qquad\qquad\uparrow$

$\quad$ Factor out the $\qquad$ (2)
$\quad$ perfect square.

(c) $\sqrt{5}\sqrt{10} = \sqrt{5 \cdot 10} = \sqrt{50} = \sqrt{25 \cdot 2} = \sqrt{25}\sqrt{2} = 5\sqrt{2}$ ∎

E X A M P L E   4

Simplifying Square Roots

Simplify:

(a) $-3\sqrt{72}$ $\qquad$ (b) $\sqrt{75x^2}$ $\qquad$ (c) $\sqrt{2x}\sqrt{6x}$

Solution $\quad$ (a) $-3\sqrt{72} = -3\sqrt{36 \cdot 2} = -3\sqrt{36}\sqrt{2} = -3 \cdot 6\sqrt{2} = -18\sqrt{2}$

$\qquad$ (b) $\sqrt{75x^2} = \sqrt{(25x^2)(3)} = \sqrt{25x^2}\sqrt{3} = \sqrt{25}\sqrt{x^2}\sqrt{3} = 5|x|\sqrt{3}$

$\qquad\qquad\qquad\qquad\qquad\qquad\qquad\qquad\qquad\qquad\qquad\uparrow$
$\qquad\qquad\qquad\qquad\qquad\qquad\qquad\qquad\qquad\qquad\quad$ (1)

$\qquad$ (c) $\sqrt{2x}\sqrt{6x} = \sqrt{2x \cdot 6x} = \sqrt{12x^2} = \sqrt{3 \cdot 4x^2} = \sqrt{3}\sqrt{4x^2} = \sqrt{3} \cdot 2x$ ∎

 Now work Problem 9.

Another property of square roots is suggested by the following examples:

$$\sqrt{\frac{36}{9}} = \sqrt{4} = 2 \quad \text{and} \quad \frac{\sqrt{36}}{\sqrt{9}} = \frac{6}{3} = 2$$

If a is a nonnegative real number and b is a positive real number, then

Quotient Property of Square Roots

$$\sqrt{\frac{a}{b}} = \frac{\sqrt{a}}{\sqrt{b}} \qquad\qquad (3)$$

E X A M P L E   5

Simplifying Square Roots

(a) $\sqrt{\frac{81}{25}} = \frac{\sqrt{81}}{\sqrt{25}} = \frac{9}{5}$

(b) $\frac{\sqrt{24}}{\sqrt{3}} = \sqrt{\frac{24}{3}} = \sqrt{8} = \sqrt{4 \cdot 2} = \sqrt{4}\sqrt{2} = 2\sqrt{2}$

(c) $\frac{\sqrt{x^5y}}{\sqrt{x^3y^3}} = \sqrt{\frac{x^5y}{x^3y^3}} = \sqrt{\frac{x^2}{y^2}} = \sqrt{\left(\frac{x}{y}\right)^2} = \left|\frac{x}{y}\right|$ ∎

 Now work Problem 31.

Rationalizing

When square roots occur in quotients, it has become common practice to rewrite the quotient so that the denominator contains no square roots. This process is referred to as **rationalizing the denominator.**

$\qquad$ The idea is to find an appropriate expression so that, when it is multiplied by the square root in the denominator, the new denominator that results contains no square roots. For example:

If Denominator Contains the Factor	Multiply By	To Obtain Denominator Free of Radicals
$\sqrt{3}$	$\sqrt{3}$	$(\sqrt{3})^2 = 3$
$\sqrt{3} + 1$	$\sqrt{3} - 1$	$(\sqrt{3})^2 - 1^2 = 3 - 1 = 2$
$\sqrt{2} - 3$	$\sqrt{2} + 3$	$(\sqrt{2})^2 - 3^2 = 2 - 9 = -7$
$\sqrt{5} - \sqrt{3}$	$\sqrt{5} + \sqrt{3}$	$(\sqrt{5})^2 - (\sqrt{3})^2 = 5 - 3 = 2$

In rationalizing the denominator of a quotient, be sure to multiply both the numerator and the denominator by the appropriate expression so that an equivalent quotient is obtained.

E X A M P L E   6

Rationalizing Denominators

Rationalize the denominator: $\dfrac{1}{\sqrt{3}}$

Solution   The denominator contains the factor $\sqrt{3}$, so we multiply the numerator and denominator by $\sqrt{3}$ to obtain

$$\frac{1}{\sqrt{3}} = \frac{1}{\sqrt{3}} \cdot \frac{\sqrt{3}}{\sqrt{3}} = \frac{\sqrt{3}}{(\sqrt{3})^2} = \frac{\sqrt{3}}{3}$$

The new denominator, 3, contains no square roots, so the original quotient has been rationalized.   ∎

E X A M P L E   7

Rationalizing Denominators

Rationalize the denominator: $\dfrac{5}{4\sqrt{2}}$

Solution   The denominator contains the factor $\sqrt{2}$, so we multiply the numerator and denominator by $\sqrt{2}$ to obtain

$$\frac{5}{4\sqrt{2}} = \frac{5}{4\sqrt{2}} \cdot \frac{\sqrt{2}}{\sqrt{2}} = \frac{5\sqrt{2}}{4(\sqrt{2})^2} = \frac{5\sqrt{2}}{4 \cdot 2} = \frac{5\sqrt{2}}{8}$$ ∎

E X A M P L E   8

Rationalizing Denominators

Rationalize the denominator: $\dfrac{\sqrt{2}}{\sqrt{3} - \sqrt{2}}$

Solution   The denominator contains the factor $\sqrt{3} - \sqrt{2}$, so we multiply the numerator and denominator by $\sqrt{3} + \sqrt{2}$ to obtain

$$\frac{\sqrt{2}}{\sqrt{3} - \sqrt{2}} = \frac{\sqrt{2}}{\sqrt{3} - \sqrt{2}} \cdot \frac{\sqrt{3} + \sqrt{2}}{\sqrt{3} + \sqrt{2}} = \frac{\sqrt{2}(\sqrt{3} + \sqrt{2})}{(\sqrt{3})^2 - (\sqrt{2})^2}$$

$$= \frac{\sqrt{2}\sqrt{3} + (\sqrt{2})^2}{3 - 2} = \sqrt{6} + 2$$ ∎

Now work Problem 71.

In calculus, sometimes the numerator must be rationalized.

E X A M P L E 9

Rationalizing Numerators

Rationalize the numerator: $\dfrac{\sqrt{x} - 2}{\sqrt{x} + 1}$, $x \geq 0$

Solution We multiply by $\dfrac{\sqrt{x} + 2}{\sqrt{x} + 2}$:

$$\frac{\sqrt{x} - 2}{\sqrt{x} + 1} = \frac{\sqrt{x} - 2}{\sqrt{x} + 1} \cdot \frac{\sqrt{x} + 2}{\sqrt{x} + 2} = \frac{(\sqrt{x})^2 - 2^2}{(\sqrt{x} + 1)(\sqrt{x} + 2)} = \frac{x - 4}{x + 3\sqrt{x} + 2}$$

▬

*n*th Roots

The **principal *n*th root of a real number *a*,** symbolized by $\sqrt[n]{a}$, is defined as follows:

$\sqrt[n]{a} = b$ means $a = b^n$ where $a \geq 0$ and $b \geq 0$ if *n* is even and *a, b* are any real numbers if *n* is odd

Notice that if *a* is negative and *n* is even then $\sqrt[n]{a}$ is not defined. When it is defined, the principal *n*th root of a number is unique.

The symbol $\sqrt[n]{a}$ for the principal *n*th root of *a* is sometimes called a **radical;** the integer *n* is called the **index,** and *a* is called the **radicand.** If the index of a radical is 2, we call $\sqrt[2]{a}$ the **square root** of *a* and omit the index 2 by simply writing $\sqrt{a}$. If the index is 3, we call $\sqrt[3]{a}$ the **cube root** of *a*.

E X A M P L E 10

Simplifying Principal *n*th Roots

(a) $\sqrt[3]{8} = \sqrt[3]{2^3} = 2$ (b) $\sqrt[3]{-64} = \sqrt[3]{(-4)^3} = -4$

(c) $\sqrt[4]{1/16} = \sqrt[4]{(1/2)^4} = 1/2$ (d) $\sqrt[6]{(-2)^6} = |-2| = 2$

▬

(a), (b), and (c) are examples of **perfect roots.** Thus, 8 and -64 are perfect cube roots, since $8 = 2^3$ and $-64 = (-4)^3$; $\frac{1}{2}$ is a perfect fourth root of $\frac{1}{16}$, since $\frac{1}{16} = (\frac{1}{2})^4$.

Notice the need for the absolute value in Example 10(d). If *n* is even, then a^n is positive whether $a > 0$ or $a < 0$. But if *n* is even, the principal *n*th root must be nonnegative. Hence, the reason for using the absolute value: it gives a nonnegative result.

In general, if $n \geq 2$ is a positive integer and *a* is a real number, we have

$$\sqrt[n]{a^n} = a \quad \text{if } n \text{ is odd} \tag{4a}$$
$$\sqrt[n]{a^n} = |a| \quad \text{if } n \text{ is even} \tag{4b}$$

Radicals provide a way of representing many irrational real numbers. For example, there is no rational number whose square is 2. Thus, using decimals, we can only approximate the positive number whose square is 2. Using radicals, we can say that $\sqrt{2}$ *is* the positive number whose square is 2. Of course, to obtain a decimal approximation for a radical, we can use a calculator.

E X A M P L E   11

Using a Calculator to Approximate Roots

Use a calculator to approximate $\sqrt[3]{16}$.

Keystrokes: $\boxed{16}$ $\boxed{\text{SHIFT}}$ $\boxed{x^y}$ $\boxed{3}$ $\boxed{=}$

Display: $\boxed{16}$ $\boxed{3}$ $\boxed{2.5198421}$ ▬

Note: On some calculators the inverse key $\boxed{\text{INV}}$ is used with the $\boxed{x^y}$ key. In either case, y equals the index of the radical.

Now work Problem 87.

Properties of Radicals

Let $n \geq 2$ and $m \geq 2$ denote positive integers, and let a and b represent real numbers. Assuming all radicals are defined, we have the following properties:

Properties of Radicals

$$\sqrt[n]{ab} = \sqrt[n]{a}\sqrt[n]{b} \tag{5a}$$

$$\sqrt[n]{\frac{a}{b}} = \frac{\sqrt[n]{a}}{\sqrt[n]{b}} \tag{5b}$$

$$\sqrt[n]{a^m} = (\sqrt[n]{a})^m \tag{5c}$$

$$\sqrt[m]{\sqrt[n]{a}} = \sqrt[mn]{a} \tag{5d}$$

---

4   When used in reference to radicals, the direction to "simplify" will mean to remove from the radicals any perfect roots that occur as factors. Let's look at some examples of how the preceding rules are applied to simplify radicals.

E X A M P L E   12

Simplifying Radicals

$$\sqrt[3]{16} = \sqrt[3]{8 \cdot 2} = \sqrt[3]{8} \cdot \sqrt[3]{2} = 2\sqrt[3]{2}$$

↑   ↑

Factor out (5a)
perfect cube.   ▬

E X A M P L E   13

Simplifying Radicals

$$\sqrt[3]{-16x^4} = \sqrt[3]{-8 \cdot 2 \cdot x^3 \cdot x} = \sqrt[3]{(-8x^3)(2x)}$$

↑   ↑

Factor perfect   Combine perfect
cubes inside radical.   cubes.

$$= \sqrt[3]{(-2x)^3 \cdot 2x} = \sqrt[3]{(-2x)^3} \cdot \sqrt[3]{2x}$$

$$= -2x\sqrt[3]{2x}$$

↑
(5a)   ▬

E X A M P L E   14

Simplifying Radicals

$$\sqrt{\sqrt[3]{x^7}} = \sqrt[6]{x^7} = \sqrt[6]{x^6 \cdot x} = \sqrt[6]{x^6} \cdot \sqrt[6]{x} = |x|\sqrt[6]{x}$$
$\quad\uparrow$ $\qquad\qquad\qquad\qquad\qquad\qquad\uparrow$
$\quad$(5d) $\qquad\qquad\qquad\qquad\qquad\qquad$(4b)

E X A M P L E   15

Simplifying Radicals

$$\sqrt[3]{\frac{8x^5}{27}} = \sqrt[3]{\frac{2^3 x^3 x^2}{3^3}} = \sqrt[3]{\left(\frac{2x}{3}\right)^3 \cdot x^2} = \sqrt[3]{\left(\frac{2x}{3}\right)^3} \cdot \sqrt[3]{x^2} = \frac{2x}{3}\sqrt[3]{x^2}$$

Now work Problem 37.

Two or more radicals can be combined, provided they have the same index and the same radicand. Such radicals are called **like radicals.**

E X A M P L E   16

Combining Like Radicals

$$\begin{aligned}
4\sqrt{27} - 8\sqrt{12} + \sqrt{3} &= 4\sqrt{9 \cdot 3} - 8\sqrt{4 \cdot 3} + \sqrt{3} \\
&= 4 \cdot \sqrt{9}\sqrt{3} - 8 \cdot \sqrt{4}\sqrt{3} + \sqrt{3} \\
&= 12\sqrt{3} - 16\sqrt{3} + \sqrt{3} \qquad \text{Factor out } \sqrt{3}. \\
&= (12 - 16 + 1)\sqrt{3} = -3\sqrt{3}
\end{aligned}$$

E X A M P L E   17

Combining Like Radicals

$$\begin{aligned}
\sqrt[3]{8x^4} + \sqrt[3]{-x} + 4\sqrt[3]{27x} &= \sqrt[3]{2^3 x^3 x} + \sqrt[3]{-1 \cdot x} + 4\sqrt[3]{3^3 x} \\
&= \sqrt[3]{(2x)^3} \cdot \sqrt[3]{x} + \sqrt[3]{-1} \cdot \sqrt[3]{x} + 4\sqrt[3]{3^3} \cdot \sqrt[3]{x} \\
&= 2x\sqrt[3]{x} - 1 \cdot \sqrt[3]{x} + 12\sqrt[3]{x} \\
&= (2x + 11)\sqrt[3]{x}
\end{aligned}$$

Now work Problem 45.

HISTORICAL FEATURE   The radical sign, $\sqrt{}$, was first used in print by Coss in 1525. It is thought to be the manuscript form of the letter r (for the Latin word *radix = root*), although this is not quite conclusively proved. It took a long time for $\sqrt{}$ to become the standard symbol for a square root and much longer to standardize $\sqrt[3]{}, \sqrt[4]{}, \sqrt[5]{}$, and so on. The indices of the root were placed in every conceivable position, with

$$\sqrt{}\,{\overset{3}{}}8, \quad \sqrt{}\,\text{③}8, \quad \text{and} \quad \underset{3}{\sqrt{}}8$$

all being variants for $\sqrt[3]{8}$. The notation $\sqrt{}\sqrt{}16$ was popular for $\sqrt[4]{16}$. By the 1700s, the index had settled where we now put it.

The bar on top of the present radical symbol, as follows,

$$\sqrt{a^2 + 2ab + b^2}$$

is the last survivor of the **vinculum,** a bar placed atop an expression to indicate what we would now indicate with parentheses. For example,

$$a\overline{b + c} = a(b + c)$$

HISTORICAL PROBLEMS

1. Christian Dibuadius, a Dane, in 1605 used the early symbols $\sqrt{}$ for square root and $\sqrt{C}$ for cube root. Change the following into modern notation:

(a) $\sqrt{C8}$ (b) $\sqrt{C}\sqrt{C}512$ (c) $\sqrt{}\sqrt{C64}$ (d) $\sqrt{C}\sqrt{64}$

(e) $\sqrt{}\sqrt{81}$ (f) $\sqrt{}\sqrt{256x^8}$ (g) $\sqrt{C-4}+\sqrt{8}$

2. Oresme (1360) used

$$\frac{1}{2}2^p \text{ for } 2^{1/2} \quad \text{and} \quad \boxed{1\frac{1}{2}}2^p \text{ for } 2^{1+1/2} = 2^{3/2}.$$

What are the following in modern notation?

(a) $\frac{1}{3}8^p$ (b) $\frac{1}{4}256^p$ (c) $\boxed{2\frac{1}{3}}8^p$ (d) $\boxed{-1\frac{1}{4}}64^p$

1.7 | EXERCISES

In Problems 1–12, evaluate each perfect root.

1. $\sqrt{25}$ 2. $\sqrt{81}$ 3. $\sqrt[3]{27}$ 4. $\sqrt[3]{125}$ 5. $\sqrt[3]{-64}$

6. $\sqrt[3]{-8}$ 7. $\sqrt{\dfrac{1}{9}}$ 8. $\sqrt[3]{\dfrac{27}{8}}$ 9. $\sqrt{25x^4}$ 10. $\sqrt[3]{64x^6}$

11. $\sqrt[3]{8(1+x)^3}$ 12. $\sqrt{4(x+4)^2}$

In Problems 13–40, simplify each expression. Assume that all radicals containing variables are defined.

13. $\sqrt{8}$ 14. $\sqrt[4]{32}$ 15. $\sqrt[3]{16x^4}$ 16. $\sqrt{27x^3}$ 17. $\sqrt{\sqrt{x^6}}$

18. $\sqrt{\sqrt{x^6}}$ 19. $\sqrt{\dfrac{25x^3}{9x}}$ 20. $\sqrt[3]{\dfrac{x}{8x^4}}$ 21. $\sqrt[4]{x^{12}y^8}$ 22. $\sqrt[5]{x^{10}y^5}$

23. $\sqrt[4]{\dfrac{x^9y^8}{xy^{12}}}$ 24. $\sqrt[3]{\dfrac{3xy^5}{81x^4y^2}}$ 25. $\sqrt{36x}$ 26. $\sqrt{9x^5}$ 27. $\sqrt{3x^2}\sqrt{12x}$

28. $\sqrt{5x}\sqrt{20x^3}$ 29. $\dfrac{\sqrt{3xy^3}\sqrt{2x^2y}}{\sqrt{6x^3y^4}}$ 30. $\dfrac{\sqrt[3]{x^2y}\sqrt[3]{125x^3}}{\sqrt[3]{8x^3y^4}}$

31. $\sqrt{\dfrac{16y^4}{9x^2}}$ 32. $\sqrt{\dfrac{9x^4}{16y^6}}$ 33. $(\sqrt{5}\sqrt[3]{9})^2$

34. $(\sqrt[3]{3}\sqrt{10})^4$ 35. $\sqrt{\dfrac{2x-3}{2x^4+3x^3}}\sqrt{\dfrac{x}{4x^2-9}}$ 36. $\sqrt[3]{\dfrac{x-1}{x^2+2x+1}}\sqrt[3]{\dfrac{(x-1)^2}{x+1}}$

37. $\sqrt{\sqrt[3]{\sqrt[4]{x}}}$ 38. $\sqrt[4]{\sqrt[3]{x}}$ 39. $\sqrt{\dfrac{x-1}{x+1}}\sqrt{\dfrac{x^2+2x+1}{x^2-1}}$ 40. $\sqrt{\dfrac{x^2+4}{x(x^2-4)}}\sqrt{\dfrac{4x^2}{x^4-16}}$

In Problems 41–50, simplify each expression.

41. $3\sqrt{2}+4\sqrt{2}-\sqrt{2}$ 42. $6\sqrt{5}+\sqrt{5}-4\sqrt{5}$ 43. $3\sqrt{2}-\sqrt{18}+2\sqrt{8}$

44. $5\sqrt{3}+2\sqrt{12}-3\sqrt{27}$ 45. $\sqrt[3]{16}+5\sqrt[3]{2}-2\sqrt[3]{54}$ 46. $9\sqrt[3]{24}-\sqrt[3]{81}$

47. $\sqrt{8x^3}-3\sqrt{50x}+\sqrt{2x^5}$, $\;x\geq 0$ 48. $\sqrt{x^2y}-3x\sqrt{9y}+4\sqrt{25y}$, $\;x\geq 0, y\geq 0$

49. $\sqrt[3]{16x^4y}-3x\sqrt[3]{2xy}+5\sqrt[3]{-2xy^4}$ 50. $8xy-\sqrt{25x^2y^2}+\sqrt[3]{8x^3y^3}$, $\;x\geq 0, y\geq 0$

In Problems 51–64, perform the indicated operation and simplify the results.

51. $(3\sqrt{6})(4\sqrt{3})$

52. $(5\sqrt{8})(-3\sqrt{6})$

53. $(\sqrt{3}+3)(\sqrt{3}-4)$

54. $(\sqrt{5}-2)(\sqrt{5}+6)$

55. $(3\sqrt{7}+4)(2\sqrt{7}+3)$

56. $(2\sqrt{6}+3)^2$

57. $(\sqrt{x}-1)^2, \quad x \geq 0$

58. $(\sqrt{x}+\sqrt{5})^2$

59. $(\sqrt[3]{x}-1)^3$

60. $(\sqrt[3]{x}+2)^3$

61. $(2\sqrt{x}-3)(2\sqrt{x}+5)$

62. $(4\sqrt{x}-3)(\sqrt{x}+3)$

63. $\sqrt{1-x^2} - \dfrac{1}{\sqrt{1-x^2}}, \quad -1 < x < 1$

64. $\sqrt{1-x^2} + \dfrac{x^2}{\sqrt{1-x^2}}, \quad -1 < x < 1$

In Problems 65–82, rationalize the denominator of each expression.

65. $\dfrac{2}{\sqrt{5}}$

66. $\dfrac{\sqrt{3}}{\sqrt{5}}$

67. $\dfrac{8}{\sqrt{6}}$

68. $\dfrac{5}{\sqrt{10}}$

69. $\dfrac{1}{\sqrt{x}}, \quad x > 0$

70. $\dfrac{x}{\sqrt{x^2+4}}$

71. $\dfrac{3}{5+\sqrt{2}}$

72. $\dfrac{2}{\sqrt{7}-2}$

73. $\dfrac{3}{4+\sqrt{7}}$

74. $\dfrac{10}{4-\sqrt{2}}$

75. $\dfrac{\sqrt{5}}{2+3\sqrt{5}}$

76. $\dfrac{\sqrt{3}}{2\sqrt{3}+3}$

77. $\dfrac{\sqrt{3}-\sqrt{2}}{\sqrt{3}+\sqrt{2}}$

78. $\dfrac{\sqrt{5}+\sqrt{3}}{\sqrt{5}-\sqrt{3}}$

79. $\dfrac{1}{\sqrt{x}+2}, \quad x \geq 0$

80. $\dfrac{1}{\sqrt{x}-3}, \quad x \geq 0, x \neq 9$

81. $\dfrac{\sqrt{x+h}-\sqrt{x}}{\sqrt{x+h}+\sqrt{x}}$

82. $\dfrac{\sqrt{x+h}+\sqrt{x-h}}{\sqrt{x+h}-\sqrt{x-h}}$

In Problems 83–86, rationalize the numerator of each expression.

83. $\dfrac{2-\sqrt{5}}{3+2\sqrt{5}}$

84. $\dfrac{4+\sqrt{3}}{2+3\sqrt{3}}$

85. $\dfrac{\sqrt{x+h}-\sqrt{x}}{h}$

86. $\dfrac{\dfrac{1}{\sqrt{x+h}}-\dfrac{1}{\sqrt{x}}}{h}$

In Problems 87–94, use a calculator to approximate each radical. Round your answer to two decimal places.

87. $\sqrt{2}$

88. $\sqrt{7}$

89. $\sqrt[3]{4}$

90. $\sqrt[3]{-5}$

91. $\dfrac{2+\sqrt{3}}{3-\sqrt{5}}$

92. $\dfrac{\sqrt{5}-2}{\sqrt{2}+4}$

93. $\dfrac{3\sqrt[3]{5}-\sqrt{2}}{\sqrt{3}}$

94. $\dfrac{2\sqrt{3}-\sqrt[3]{4}}{\sqrt{2}}$

95. Give an example to show that $\sqrt{a^2}$ is not equal to a. Use it to explain why $\sqrt{a^2} = |a|$.

1.8 | RATIONAL EXPONENTS

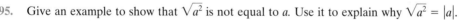 Evaluate Expressions with Fractional Exponents

[2] Simplify Radicals Using Rational Exponents

Our purpose in this section is to give a definition for "*a* raised to the power *m/n*," where *a* is a real number and *m/n* is a rational number. However, we want the definition we give to ensure that the laws of exponents stated for integer exponents in Section 1.3 remain true for rational exponents. For example, if the law of exponents $(a^r)^s = a^{rs}$ is to hold, then it must be true that

$$(a^{1/n})^n = a^{(1/n)n} = a^1 = a$$

That is, $a^{1/n}$ is a number that, when raised to the power *n*, is *a*. But this was the definition we gave of the principal *n*th root of *a* in Section 1.7. Thus,

$$a^{1/2} = \sqrt{a} \qquad a^{1/3} = \sqrt[3]{a} \qquad a^{1/4} = \sqrt[4]{a}$$

and we can state the following definition:

If a is a real number and $n \geq 2$ is an integer, then

$$a^{1/n} = \sqrt[n]{a} \qquad \qquad (1)$$

provided $\sqrt[n]{a}$ exists.

E X A M P L E   1   Evaluating Expressions Containing Fractional Exponents

(a) $4^{1/2} = \sqrt{4} = 2$ (b) $(-27)^{1/3} = \sqrt[3]{-27} = -3$

(c) $8^{1/2} = \sqrt{8} = 2\sqrt{2}$ (d) $16^{1/3} = \sqrt[3]{16} = 2\sqrt[3]{2}$ ■

Note that if n is even and $a < 0$ then $\sqrt[n]{a}$ and $a^{1/n}$ do not exist.

We now seek a definition for $a^{m/n}$, where m and n are integers containing no common factors (except 1 and -1) and $n \geq 2$. Again, we want the definition to obey the laws of exponents stated earlier. For example,

$$a^{m/n} = a^{m(1/n)} = (a^m)^{1/n} \quad \text{and} \quad a^{m/n} = a^{(1/n)m} = (a^{1/n})^m$$

If a is a real number and m and n are integers containing no common factors with $n \geq 2$, then

$$a^{m/n} = \sqrt[n]{a^m} = (\sqrt[n]{a})^m \qquad \qquad (2)$$

provided $\sqrt[n]{a}$ exists.

We have two comments about equation (2):

1. The exponent m/n must be in lowest terms and n must be positive.
2. In simplifying the rational expression $a^{m/n}$, either $\sqrt[n]{a^m}$ or $(\sqrt[n]{a})^m$ may be used, the choice depending on which is easier to simplify. Generally, taking the root first, as in $(\sqrt[n]{a})^m$, is preferred.

E X A M P L E   2   Using Equation (2)

(a) $4^{3/2} = (\sqrt{4})^3 = 2^3 = 8$

(b) $(-8)^{4/3} = (\sqrt[3]{-8})^4 = (-2)^4 = 16$

(c) $(32)^{-2/5} = (\sqrt[5]{32})^{-2} = 2^{-2} = \dfrac{1}{4}$

(d) $4^{6/4} = 4^{3/2} = (\sqrt{4})^3 = 2^3 = 8$ ■

Now work Problem 7.

Based on the definition of $a^{m/n}$, no meaning is given to $a^{m/n}$ if a is a negative real number and n is an even integer.

The definitions in equations (1) and (2) were stated so that the laws of exponents would remain true for rational exponents. For convenience, we list again the Laws of Exponents.

M I S S I O N   P O S S I B L E

Buying a Used Car

Your team has decided to begin their association by purchasing a good used car that can take them to the various locations they'll need to visit during this course. They have found four possibilities at four different dealerships. The cars seem to be more or less equivalent in price. However, the dealerships are offering different loan packages, and your team needs to sit down and compare the financial advantages and disadvantages of each.

The formula that you need to work out the details follows. For the time being, you do not need to know how and why the formula works; you only need to be able to work with the numbers. In the formula, the A stands for the amount that you would have to borrow from the dealer. (If the car costs $9000, and you can only put down $500, then you would need to borrow $8500, which would eventually be paid back to the dealer plus interest.) The R represents the amount you would have to pay back per month. The i represents the annual percentage rate (APR) that you would pay in interest. The n represents the number of times per year that the interest would be compounded. The t represents the number of years that you would be making payments on the car.

$$A = R \left[\frac{1 - \left(1 + \frac{i}{n}\right)^{-nt}}{\frac{i}{n}} \right]$$

Here are the deals that you were offered:

	Amt. Borrowed	APR	# Times Compounded Per Year
Dealer 1	$10,000	12.3%	12
Dealer 2	$9500	12.6%	4
Dealer 3	$9000	13%	6
Dealer 4	$10,500	12.2%	12

1. Suppose that you want to have the loan paid off in 3 years. How much would your car payment be in each case? Assume that the number of car payments that you make per year equals the number of interest periods per year. (Remember that, when substituting a percent, it must be changed to decimal form; for example, 5.3% would become 0.053 in the formula.)

2. In each case, how much money will you pay back altogether? How much of the amount that you paid the dealer was interest?

3. If it is true that the four cars are more or less interchangeable in quality, drivability, and the like, which car would your team purchase? What factor would you consider most important?

If a and b are real numbers and r and s are rational numbers, then

Laws of Exponents

$$a^r a^s = a^{r+s} \qquad (a^r)^s = a^{rs} \qquad (ab)^r = a^r \cdot b^r$$

$$a^{-r} = \frac{1}{a^r} \qquad \left(\frac{a}{b}\right)^r = \frac{a^r}{b^r} \qquad \frac{a^r}{a^s} = a^{r-s} = \frac{1}{a^{s-r}}$$

where it is assumed that all expressions used are defined.
Rational exponents can sometimes be used to simplify radicals.

2

EXAMPLE 3 Simplifying Radicals Using Rational Exponents

Simplify each expression.

(a) $\sqrt[9]{x^3}$ (b) $\sqrt[3]{4}\sqrt{2}$ (c) $\sqrt{a}\sqrt[3]{a}$

Solution (a) $\sqrt[9]{x^3} = x^{3/9} = x^{1/3} = \sqrt[3]{x}$

(b) $\sqrt[3]{4}\sqrt{2} = 4^{1/3} \cdot 2^{1/2} = (2^2)^{1/3} \cdot 2^{1/2} = 2^{2/3} \cdot 2^{1/2} = 2^{7/6} = 2 \cdot 2^{1/6} = 2\sqrt[6]{2}$

(c) $\sqrt{a}\sqrt[3]{a} = (a \cdot a^{1/3})^{1/2} = (a^{4/3})^{1/2} = a^{2/3} = \sqrt[3]{a^2}$ ∎

EXAMPLE 4 Simplifying Expressions Containing Rational Exponents

Simplify each expression. Express your answer so that only positive exponents occur. Assume that the variables are positive.

(a) $\left(\dfrac{2x^{1/3}}{y^{2/3}}\right)^{-3}$ (b) $\left(\dfrac{9x^2 y^{1/3}}{x^{1/3}y}\right)^{1/2}$ (c) $\dfrac{(2x+5)^{1/3}(2x+5)^{-1/2}}{(2x+5)^{-3/4}}$

Solution (a) $\left(\dfrac{2x^{1/3}}{y^{2/3}}\right)^{-3} \underset{\substack{\uparrow \\ \text{Equation (8), page 29}}}{=} \left(\dfrac{y^{2/3}}{2x^{1/3}}\right)^3 = \dfrac{(y^{2/3})^3}{(2x^{1/3})^3} = \dfrac{y^2}{2^3(x^{1/3})^3} = \dfrac{y^2}{8x}$

(b) $\left(\dfrac{9x^2 y^{1/3}}{x^{1/3}y}\right)^{1/2} = \left(\dfrac{9x^{2-(1/3)}}{y^{1-(1/3)}}\right)^{1/2} = \left(\dfrac{9x^{5/3}}{y^{2/3}}\right)^{1/2} = \dfrac{9^{1/2}(x^{5/3})^{1/2}}{(y^{2/3})^{1/2}} = \dfrac{3x^{5/6}}{y^{1/3}}$

(c) $\dfrac{(2x+5)^{1/3}(2x+5)^{-1/2}}{(2x+5)^{-3/4}} = (2x+5)^{(1/3)-(1/2)-(-3/4)}$

$= (2x+5)^{(4-6+9)/12} = (2x+5)^{7/12}$ ∎

Now work Problem 25.

The next three examples illustrate some algebra that you will need to know for certain calculus problems.

EXAMPLE 5 Writing an Expression as a Single Quotient

Write the following expression as a single quotient in which only positive exponents appear.

$$(x^2 + 1)^{1/2} + x \cdot \frac{1}{2}(x^2 + 1)^{-1/2} \cdot 2x$$

Solution

$$(x^2 + 1)^{1/2} + x \cdot \frac{1}{2}(x^2 + 1)^{-1/2} \cdot 2x = (x^2 + 1)^{1/2} + \frac{x^2}{(x^2 + 1)^{1/2}}$$

$$= \frac{(x^2 + 1)^{1/2}(x^2 + 1)^{1/2} + x^2}{(x^2 + 1)^{1/2}} = \frac{(x^2 + 1) + x^2}{(x^2 + 1)^{1/2}} = \frac{2x^2 + 1}{(x^2 + 1)^{1/2}}$$

We can also solve this problem by factoring out the common term with the smallest negative power, that is, $(x^2 + 1)^{-1/2}$:

$$(x^2 + 1)^{1/2} + x \cdot \frac{1}{2}(x^2 + 1)^{-1/2} \cdot 2x = (x^2 + 1)^{-1/2}[(x^2 + 1) + x^2]$$

$$= (x^2 + 1)^{-1/2}[2x^2 + 1] = \frac{2x^2 + 1}{(x^2 + 1)^{1/2}}$$

E X A M P L E   6

Writing an Expression as a Single Quotient

Write the following expression as a single quotient in which only positive exponents appear.

$$\frac{(x^2 + 4)^{1/2} \cdot 2x - x^2 \cdot \frac{1}{2}(x^2 + 4)^{-1/2} \cdot 2x}{x^2 + 4}$$

Solution

$$\frac{(x^2 + 4)^{1/2} \cdot 2 - x^2 \cdot \frac{1}{2}(x^2 + 4)^{-1/2} \cdot 2x}{x^2 + 4} = \frac{2x(x^2 + 4)^{1/2} - x^3(x^2 + 4)^{-1/2}}{x^2 + 4}$$

$$= \frac{2x(x^2 + 4)^{1/2} - \frac{x^3}{(x^2 + 4)^{1/2}}}{x^2 + 4}$$

$$= \frac{\frac{2x(x^2 + 4)^{1/2}(x^2 + 4)^{1/2} - x^3}{(x^2 + 4)^{1/2}}}{x^2 + 4} = \frac{2x(x^2 + 4) - x^3}{(x^2 + 4)^{1/2}} \cdot \frac{1}{x^2 + 4}$$

$$= \frac{2x^3 + 8x - x^3}{(x^2 + 4)^{3/2}} = \frac{x^3 + 8x}{(x^2 + 4)^{3/2}} = \frac{x(x^2 + 8)}{(x^2 + 4)^{3/2}}$$

Now work Problem 33.

E X A M P L E   7

Factoring an Expression Containing Rational Exponents

Factor: $4x^{1/3}(2x + 1) + 2x^{4/3}$

Solution

We begin by looking for factors that are common to the two terms. Notice that 2 and $x^{1/3}$ are common factors. Thus,

$$4x^{1/3}(2x + 1) + 2x^{4/3} = 2x^{1/3}[2(2x + 1) + x] = 2x^{1/3}(5x + 2)$$

1.8 | EXERCISES

In Problems 1–20, simplify each expression.

1. $8^{2/3}$

2. $4^{3/2}$

3. $(-27)^{2/3}$

4. $(-64)^{2/3}$

5. $4^{-3/2}$

6. $(-8)^{-5/3}$

7. $9^{-3/2}$

8. $25^{-5/2}$

9. $\left(\dfrac{9}{4}\right)^{3/2}$

10. $\left(\dfrac{27}{8}\right)^{2/3}$

11. $\left(\dfrac{4}{9}\right)^{-3/2}$

12. $\left(\dfrac{8}{27}\right)^{-2/3}$

13. $4^{1.5}$

14. $16^{-1.5}$

15. $\left(\dfrac{1}{4}\right)^{-1.5}$

16. $\left(\dfrac{1}{9}\right)^{1.5}$

17. $(\sqrt{3})^6$

18. $(\sqrt[3]{4})^6$

19. $(\sqrt{5})^{-2}$

20. $(\sqrt[4]{3})^{-8}$

In Problems 21–30, simplify each expression. Express your answer so that only positive exponents occur. Assume that the variables are positive.

21. $x^{2/3}x^{-1/2}$

22. $x^{4/3}x^{-1/4}$

23. $(x^3y^6)^{2/3}$

24. $(x^4y^8)^{5/4}$

25. $(x^2y)^{1/3}(xy^2)^{2/3}$

26. $(xy)^{1/4}(x^2y^2)^{1/2}$

27. $(16x^2y^{-1/3})^{3/4}$

28. $(4x^{-1}y^{1/3})^{3/2}$

29. $\left(\dfrac{x^{2/5}y^{-1/5}}{x^{-1/3}}\right)^{15}$

30. $\left(\dfrac{x^{1/2}}{y^2}\right)^4\left(\dfrac{y^{1/3}}{x^{-2/3}}\right)^3$

In Problems 31–44, write each expression as a single quotient in which only positive exponents and/or radicals appear.

31. $\dfrac{x}{(1+x)^{1/2}} + 2(1+x)^{1/2}$

32. $\dfrac{1+x}{2x^{1/2}} + x^{1/2}$

33. $2x(x^2+1)^{1/2} + x^2 \cdot \dfrac{1}{2}(x^2+1)^{-1/2} \cdot 2x$

34. $(x+1)^{1/3} + x \cdot \dfrac{1}{3}(x+1)^{-2/3}$

35. $\sqrt{4x+3} \cdot \dfrac{1}{2\sqrt{x-5}} + \sqrt{x-5} \cdot \dfrac{1}{5\sqrt{4x+3}}$

36. $\dfrac{\sqrt[3]{8x+1}}{3\sqrt[3]{(x-2)^2}} + \dfrac{\sqrt[3]{x-2}}{24\sqrt[3]{(8x+1)^2}}$

37. $\dfrac{\sqrt{1+x} - x \cdot \dfrac{1}{2\sqrt{1+x}}}{1+x}$

38. $\dfrac{\sqrt{x^2+1} - x \cdot \dfrac{2x}{2\sqrt{x^2+1}}}{x^2+1}$

39. $\dfrac{(x+4)^{1/2} - 2x(x+4)^{-1/2}}{x+4}$

40. $\dfrac{(9-x^2)^{1/2} + x^2(9-x^2)^{-1/2}}{9-x^2}$

41. $\dfrac{\dfrac{x^2}{(x^2-1)^{1/2}} - (x^2-1)^{1/2}}{x^2}$

42. $\dfrac{(x^2+4)^{1/2} - x^2(x^2+4)^{-1/2}}{x^2+4}$

43. $\dfrac{\dfrac{1+x^2}{2\sqrt{x}} - 2x\sqrt{x}}{(1+x^2)^2}$

44. $\dfrac{2x(1-x^2)^{1/3} + \dfrac{2}{3}x^3(1-x^2)^{-2/3}}{(1-x^2)^{2/3}}$

In Problems 45–52, factor each expression.

45. $(x+1)^{3/2} + x \cdot \dfrac{3}{2}(x+1)^{1/2}$

46. $(x^2+4)^{4/3} + x \cdot \dfrac{4}{3}(x^2+4)^{1/3} \cdot 2x$

47. $6x^{1/2}(x^2+x) - 8x^{3/2} - 8x^{1/2}$

48. $6x^{1/2}(2x+3) + x^{3/2} \cdot 8$

49. $3(x^2+4)^{4/3} + x \cdot 4(x^2+4)^{1/3} \cdot 2x$

50. $2x(3x+4)^{4/3} + x^2 \cdot 4(3x+4)^{1/3}$

51. $4(3x+5)^{1/3}(2x+3)^{3/2} + 3(3x+5)^{4/3}(2x+3)^{1/2}$

52. $6(6x+1)^{1/3}(4x-3)^{3/2} + 6(6x+1)^{4/3}(4x-3)^{1/2}$

53. If $u = \dfrac{1}{2}\left(x^3 - \dfrac{1}{x^3}\right)$, show that $\sqrt{1+u^2} = \dfrac{1}{2}\left(x^3 + \dfrac{1}{x^3}\right)$.

54. If $u = 2\sqrt{3}x\sqrt{3x^2+1}$, show that $1 + u^2 = (6x^2+1)^2$.

55. If $u = \dfrac{1}{2}\left(x^2 - \dfrac{1}{x^2}\right)$, show that $1 + u^2 = \dfrac{1}{4}\left(x^2 + \dfrac{1}{x^2}\right)^2$.

1.9 | GEOMETRY TOPICS

1 Use the Pythagorean Theorem and Its Converse

2 Know Geometry Formulas

In this section we review some topics studied in geometry that we shall need for our study of algebra.

Pythagorean Theorem

FIGURE 11

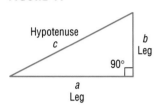

1 The *Pythagorean Theorem* is a statement about *right triangles.* A **right triangle** is one that contains a **right angle,** that is, an angle of 90°. The side of the triangle opposite the 90° angle is called the **hypotenuse;** the remaining two sides are called **legs.** In Figure 11 we have used c to represent the length of the hypotenuse and a and b to represent the lengths of the legs. Notice the use of the symbol $\ulcorner$ to show the 90° angle. We now state the Pythagorean Theorem.

> Pythagorean Theorem
>
> In a right triangle, the square of the length of the hypotenuse is equal to the sum of the squares of the lengths of the legs. That is, in the right triangle shown in Figure 11,
>
> $$c^2 = a^2 + b^2 \tag{1}$$

We shall prove this result at the end of this section. ▬

E X A M P L E 1   Finding the Hypotenuse of a Right Triangle

In a right triangle, one leg is of length 4 and the other is of length 3. What is the length of the hypotenuse?

Solution   Since the triangle is a right triangle, we use the Pythagorean Theorem with $a = 4$ and $b = 3$ to find the length c of the hypotenuse. Thus, from equation (1), we have

$$c^2 = a^2 + b^2$$
$$c^2 = 4^2 + 3^2 = 16 + 9 = 25$$
$$c = 5$$

▬

Now work Problem 3.

The converse of the Pythagorean Theorem is also true.

> Converse of the Pythagorean Theorem
>
> In a triangle, if the square of the length of one side equals the sum of the squares of the lengths of the other two sides, then the triangle is a right triangle. The 90° angle is opposite the longest side.

We shall prove this result at the end of this section. ▬

E X A M P L E   2   Verifying That a Triangle Is a Right Triangle

Show that a triangle whose sides are of lengths 5, 12, and 13 is a right triangle. Identify the hypotenuse.

Solution   We square the lengths of the sides:

FIGURE 12

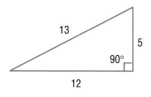

25,   144,   169

Notice that the sum of the first two squares (25 and 144) equals the third square (169). Hence, the triangle is a right triangle. The longest side, 13, is the hypotenuse. See Figure 12.   ∎

   Now work Problem 11.

Geometry Formulas

Certain formulas from geometry are useful in solving algebra problems. We list some of these formulas next.

For a rectangle of length l and width w,

$$\text{Area} = lw \qquad \text{Perimeter} = 2l + 2w$$

For a triangle with base b and altitude h,

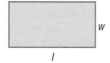

$$\text{Area} = \frac{1}{2}bh$$

For a circle of radius r (diameter $d = 2r$),

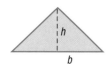

$$\text{Area} = \pi r^2 \qquad \text{Circumference} = 2\pi r = \pi d$$

For a rectangular box of length l, width w, and height h,

$$\text{Volume} = lwh$$

Proof of the   We begin with a square, each side of length $a + b$. In this square, we can form
Pythagorean Theorem   four right triangles, each having legs equal in length to a and b. See Figure 13.

FIGURE 13

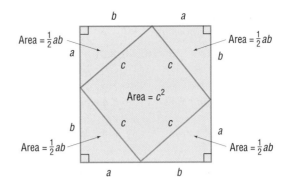

All of these triangles are **congruent** (two sides and their included angle are equal). As a result, the hypotenuse of each is the same, say c, and the color shading in Figure 13 indicates a square with an area equal to c^2. The area of the original square with side $a + b$ equals the sum of the areas of the four triangles (each of area $\frac{1}{2}ab$) plus the area of the square with side c. Thus,

$$(a + b)^2 = \tfrac{1}{2}ab + \tfrac{1}{2}ab + \tfrac{1}{2}ab + \tfrac{1}{2}ab + c^2$$
$$a^2 + 2ab + b^2 = 2ab + c^2$$
$$a^2 + b^2 = c^2$$

The proof is complete. ▄

Proof of the Converse of the Pythagorean Theorem

FIGURE 14

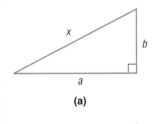

(a)

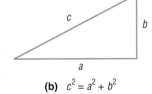

(b) $c^2 = a^2 + b^2$

We begin with two triangles: one a right triangle with legs a and b and the other a triangle with sides a, b, and c for which $c^2 = a^2 + b^2$ (see Figure 14). By the Pythagorean Theorem, the length x of the third side of the first triangle is

$$x^2 = a^2 + b^2$$

But $c^2 = a^2 + b^2$. Hence,

$$x^2 = c^2$$
$$x = c$$

The two triangles have the same sides and are therefore congruent; hence, corresponding angles are equal. Thus, the angle opposite side c of the second triangle equals 90°.

The proof is complete. ▄

1.9 | EXERCISES

In Problems 1–6, the lengths of the legs of a right triangle are given. Find the hypotenuse.

1. $a = 5$, $b = 12$ **2.** $a = 6$, $b = 8$ **3.** $a = 10$, $b = 24$ **4.** $a = 4$, $b = 3$

5. $a = 7$, $b = 24$ **6.** $a = 14$, $b = 48$

In Problems 7–14, the lengths of the sides of a triangle are given. Determine which are right triangles. For those that are, identify the hypotenuse.

7. 3, 4, 5   **8.** 6, 8, 10   **9.** 4, 5, 6   **10.** 2, 2, 3

11. 7, 24, 25   **12.** 10, 24, 26   **13.** 6, 4, 3   **14.** 5, 4, 7

In Problems 15–18, find the area A and the perimeter P of a rectangle with length l and width w.

15. $l = 5$, $w = 3$ **16.** $l = 4$, $w = 2$ **17.** $l = \dfrac{1}{2}$, $w = \dfrac{1}{3}$ **18.** $l = \dfrac{3}{4}$, $w = \dfrac{4}{3}$

In Problems 19–22, find the area A of a triangle with base b and altitude h.

19. $b = 4$, $h = 3$ **20.** $b = 3$, $h = 6$ **21.** $b = \dfrac{1}{2}$, $h = \dfrac{3}{4}$ **22.** $b = \dfrac{3}{4}$, $h = \dfrac{4}{3}$

In Problems 23–26, find the area A and circumference C of a circle with radius r (use $\pi \approx 3.14$).

23. $r = 1$ **24.** $r = 2$ **25.** $r = \dfrac{3}{2}$ **26.** $r = \dfrac{5}{4}$

In Problems 27–30, find the volume V of a rectangular box with length l, width w, and height h.

27. $l = 4$, $w = 2$, $h = 6$ **28.** $l = 5$, $w = 2$, $h = 4$

29. $l = \dfrac{1}{3}$, $w = \dfrac{5}{2}$, $h = 4$ **30.** $l = \dfrac{1}{4}$, $w = \dfrac{1}{2}$, $h = \dfrac{4}{3}$

31. How many feet does a wheel with a diameter of 16 inches travel after four revolutions?

32. How many revolutions will a circular disk with a diameter of 4 feet have completed after it has rolled 20 feet?

33. In the figure below, *ABCD* is a square, with each side of length 6 feet. The width of the border (shaded portion) between the outer square *EFGH* and *ABCD* is 2 feet. Find the area of the border.

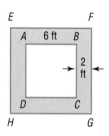

34. Refer to the figure below. Square *ABCD* has an area of 100 square feet; square *BEFG* has an area of 16 square feet. What is the area of the triangle *CGF*?

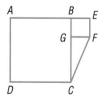

35. Architecture A Norman window consists of a rectangle surmounted by a semicircle. Find the area of the Norman window shown in the illustration. How much wood frame is needed to enclose the window?

36. Construction A circular swimming pool, 20 feet in diameter, is enclosed by a wooden deck that is 3 feet wide. What is the area of the deck? How much fence is required to enclose the deck?

37. How far can you see? The tallest inhabited building in the world is the Sears Tower in Chicago.* If the observation tower is 1450 feet above ground level, use the figure to determine how far a person standing in the observation tower can see (with the aid of a telescope). Use 3960 miles for the radius of Earth.
[**Note:** 1 mile = 5280 feet]

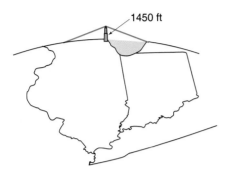

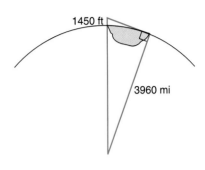

In Problems 38–40, use the fact that the radius of Earth is 3960 miles.

38. How far can you see? The conning tower of the USS *Silversides,* a World War II submarine now permanently stationed in Muskegon, Michigan, is approximately 20 feet above sea level. How far can one see from the conning tower?

39. How far can you see? A person who is 6 feet tall is standing on the beach in Fort Lauderdale, Florida, and looks out onto the Atlantic Ocean. Suddenly, a ship appears on the horizon. How far is the ship from shore?

40. How far can you see? The deck of a destroyer is 100 feet above sea level. How far can a person see from the deck? How far can a person see from the bridge, which is 150 feet above sea level?

41. Suppose that m and n are positive integers with $m > n$. If $a = m^2 - n^2$, $b = 2mn$, and $c = m^2 + n^2$, show that a, b, and c are the lengths of the sides of a right triangle. (This formula can be used to find the sides of a right triangle that are integers, such as 3, 4, 5; 5, 12, 13; and so on. Such triplets of integers are called *Pythagorean triples.*)

 42. You have 1000 feet of flexible pool siding and wish to construct a swimming pool. Experiment with rectangular-shaped pools with perimeters of 1000 feet. How do their areas vary? What is the shape of the rectangle with the largest area? Now compute the area enclosed by a circular pool with a perimeter (circumference) of 1000 feet. What would be your choice of shape for the pool? If rectangular, what is your preference for dimensions? Justify your choice. If your only consideration is to have a pool that encloses the most area, what shape should you use?

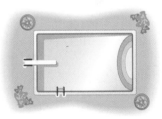

Source: Council on Tall Buildings and Urban Habitat (1997): Sears Tower No. 1 for tallest roof (1450 ft) and tallest occupied floor (1431 ft).

43. **The Gibb's Hill lighthouse, Southampton, Bermuda,** in operation since 1846, stands 117 feet high on a hill 245 feet high, so its beam of light is 362 feet above sea level. A brochure states that the light itself can be seen on the horizon about 26 miles distant. Verify the correctness of this information. The brochure further states that ships 40 miles away can see the light and planes flying at 10,000 feet can see it 120 miles away. Verify the accuracy of these statements. What assumption did the brochure make about the height of the ship?

CHAPTER REVIEW

THINGS TO KNOW

Classification of numbers

Counting numbers	$1, 2, 3 \ldots$
Whole numbers	$0, 1, 2, 3 \ldots$
Integers	$\ldots, -2, -1, 0, 1, 2, \ldots$
Rational numbers	Quotients of two integers (denominators not equal to 0); terminating or infinite, repeating decimals
Irrational numbers	Infinite, nonrepeating decimals
Real numbers	Rational or irrational numbers

Properties of real numbers

Commutative properties	$a + b = b + a, \quad a \cdot b = b \cdot a$
Associative properties	$a + (b + c) = (a + b) + c,$ $a \cdot (b \cdot c) = (a \cdot b) \cdot c$
Distributive property	$a \cdot (b + c) = a \cdot b + a \cdot c$
Identity properties	$a + 0 = a, \quad a \cdot 1 = a$
Inverse properties	$a + (-a) = 0, \quad a \cdot \dfrac{1}{a} = 1,$ where $a \neq 0$
Zero-product property	If $ab = 0$, then $a = 0$ or $b = 0$ or both.

Absolute value

$|a| = a$ if $a \geq 0, \quad |a| = -a$ if $a < 0$

Exponents

$a^n = a \cdot a \cdot \ldots \cdot a$, n a positive integer $\qquad a^0 = 1, \ a \neq 0 \qquad a^{-n} = \dfrac{1}{a^n}, \ a \neq 0, n$ a positive integer

$\underbrace{}_{n \text{ factors}}$

Laws of exponents

$a^m \cdot a^n = a^{m+n}, \qquad (a^m)^n = a^{mn}, \qquad (a \cdot b)^n = a^n \cdot b^n \qquad \dfrac{a^m}{a^n} = a^{m-n} = \dfrac{1}{a^{n-m}}, \qquad \left(\dfrac{a}{b}\right)^n = \dfrac{a^n}{b^n}$

Polynomial

Algebraic expression of the form $a_n x^n + a_{n-1}x^{n-1} + \cdots + a_1 x + a_0$, n a nonnegative integer

Special products/factoring formulas

$(x - a)(x + a) = x^2 - a^2$

$(x + a)^2 = x^2 + 2ax + a^2$, $(x - a)^2 = x^2 - 2ax + a^2$

$(x + a)(x + b) = x^2 + (a + b)x + ab$

$(ax + b)(cx + d) = acx^2 + (ad + bc)x + bd$

$(x + a)^3 = x^3 + 3ax^2 + 3a^2x + a^3$, $(x - a)^3 = x^3 - 3ax^2 + 3a^2x - a^3$

$(x - a)(x^2 + ax + a^2) = x^3 - a^3$, $(x + a)(x^2 - ax + a^2) = x^3 + a^3$

Radicals; Fractional exponents

$\sqrt[n]{a} = b$ means $a = b^n$, where $a \geq 0$ and $b \geq 0$ if n is even and a, b are any real numbers if n is odd

$a^{m/n} = \sqrt[n]{a^m} = (\sqrt[n]{a})^m$

Pythagorean Theorem

In a right triangle, the square of the length of the hypotenuse is equal to the sum of the squares of the lengths of the legs.

Converse of a Pythagorean Theorem

In a triangle, if the square of the length of one side equals the sum of the squares of the lengths of the other two sides, then the triangle is a right triangle.

HOW TO

Add, subtract, multiply, divide, and factor polynomials

Add, subtract, multiply, divide, and simplify rational expressions

Rationalize denominators of quotients containing radicals

FILL-IN-THE-BLANK ITEMS

1. The _____ property of equality states that if $a = b$ then $b = a$.

2. The _____ properties state that $a + b = b + a$ and $ab = ba$.

3. The _____ property states that $a \cdot (b + c) = a \cdot b + a \cdot c$.

4. In the expression 2^4, the number 2 is called the _____, and 4 is called the _____.

5. The polynomial $5x^3 - 3x^2 + 2x - 4$ is of degree _____; the _____ coefficient is 5.

6. In the expression $\sqrt[3]{25}$, the number 3 is called the _____; the number 25 is called the _____.

7. In a right triangle the side that is longest is called the _____.

TRUE/FALSE ITEMS

T F   **1.** Rational numbers are also real numbers.

T F   **2.** Irrational numbers have decimals that neither repeat nor terminate.

T F   **3.** $(a + b)^2 = a^2 + b^2$

T F   **4.** The polynomial $x^2 + 9$ is prime.

T F   **5.** To rationalize $\dfrac{3}{2 - \sqrt{3}}$, you would multiply the numerator and the denominator by $-2 + \sqrt{3}$.

T F   **6.** The circumference of a circle of diameter d is πd.

REVIEW EXERCISES

Blue problem numbers indicate the author's suggestions for use in a Practice Test.

In Problems 1–20, perform the indicated operations.

1. $3 - 4 \cdot 5 + 6$ **2.** $8 + 4 \cdot 2 - 6$ **3.** $\dfrac{3}{4} - \dfrac{7}{12}$ **4.** $\dfrac{9}{8} - \dfrac{3}{4}$

5. $\dfrac{\frac{15}{2} + \frac{1}{4}}{\frac{2}{3}}$ **6.** $\dfrac{\frac{9}{2} + \frac{3}{4}}{\frac{4}{3}}$ **7.** $5^2 - 3^3 \cdot 2$ **8.** $2^3 + 3^2 \cdot 4$

9. $\dfrac{2^{-3} \cdot 5^0}{4^2}$ **10.** $\dfrac{3^{-2} - 4}{2^{-2}}$ **11.** $(2\sqrt{5} - 2)(2\sqrt{5} + 2)$ **12.** $(2\sqrt{3} + \sqrt{2})(2\sqrt{3} - \sqrt{2})$

13. $\left(\dfrac{8}{27}\right)^{-2/3}$ **14.** $\left(\dfrac{4}{9}\right)^{-3/2}$ **15.** $(\sqrt[3]{2})^{-3}$ **16.** $(4\sqrt{32})^{-1/2}$

17. $|6 - 8^{1/3}|$ **18.** $|25^{1/2} - 27^{2/3}|$ **19.** $\sqrt{|3^2 - 5^2|}$ **20.** $\sqrt[3]{-|4^2 - 2^3|}$

In Problems 21–32, write each expression so that all exponents are positive. Assume that $x > 0$ and $y > 0$.

21. $\dfrac{x^{-2}}{y^{-2}}$ **22.** $\left(\dfrac{x^{-1}}{y^{-3}}\right)^2$ **23.** $\dfrac{(x^2 y)^{-4}}{(xy)^{-3}}$ **24.** $\dfrac{\left(\dfrac{x}{y}\right)^2}{\left(\dfrac{x}{y}\right)^{-1}}$

25. $\dfrac{\left(\dfrac{x^2}{y}\right)^2}{\left(\dfrac{x}{y^2}\right)^3}$ **26.** $\dfrac{\left(\dfrac{2x}{3y^2}\right)^{-1}}{\dfrac{4}{y^3}}$ **27.** $\dfrac{x^{-2}}{x^{-2} + y^{-2}}$ **28.** $\dfrac{x^{-1} + y^{-1}}{x^{-1} - y^{-1}}$

29. $(25x^{-4/3}y^{-2/3})^{3/2}$ **30.** $(16x^{-2/3}y^{4/3})^{-3/2}$ **31.** $\left(\dfrac{2x^{-1/2}}{y^{-3/4}}\right)^{-4}$ **32.** $\left(\dfrac{8x^{-3/2}}{y^{-3}}\right)^{-2/3}$

In Problems 33–40, perform the indicated operations. Express your answer as a polynomial.

33. $(2x - 3)(-4x + 2)$ **34.** $(3x + 4)(-8x - 2)$

35. $4(3x^3 - 2x^2 + 1) - 3(x^3 + 4x^2 - 2x - 3)$ **36.** $8(1 - x^2 + x^3) - 4(1 + 2x^2 - 4x^4)$

37. $(2x - 5)(3x^2 + 2)$ **38.** $(1 - 2x^3)(1 - 4x)$

39. $(x + 1)(x + 2)(x - 3)$ **40.** $(x + 1)(x + 3)(x - 5)$

In Problems 41–50, find the quotient and remainder. Check your work by verifying that

(Quotient)(Divisor) + Remainder = Dividend

41. $3x^3 - x^2 + x + 4$ divided by $x - 3$ **42.** $2x^3 - 3x^2 + x + 1$ divided by $x - 2$

43. $-3x^4 + x^2 + 2$ divided by $x^2 + 1$ **44.** $-4x^3 + x^2 - 2$ divided by $x^2 - 1$

45. $8x^4 - 2x^2 + 5x + 1$ divided by $x^2 - 3x + 1$ **46.** $3x^4 - x^3 - 8x + 4$ divided by $x^2 + 3x - 2$

47. $x^5 + 1$ divided by $x + 1$ **48.** $x^5 - 1$ divided by $x - 1$

49. $6x^5 + 3x^4 - 4x^3 - 2x^2 + 2x + 1$ divided by $2x + 1$

50. $6x^5 - 3x^4 - 4x^3 + 2x^2 + 2x - 1$ divided by $2x - 1$

In Problems 51–64, factor each polynomial completely (over the integers). If the polynomial cannot be factored, say it is prime.

51. $x^2 + 5x - 14$ **52.** $x^2 - 9x + 14$ **53.** $6x^2 - 5x - 6$ **54.** $6x^2 + x - 2$

55. $3x^2 - 15x - 42$ **56.** $2x^3 + 18x^2 + 28x$ **57.** $8x^3 + 1$ **58.** $27x^3 - 8$

59. $2x^3 + 3x^2 - 2x - 3$ **60.** $2x^3 + 3x^2 + 2x + 3$ **61.** $25x^2 - 4$ **62.** $16x^2 - 1$

63. $9x^2 + 1$ **64.** $x^2 - x + 1$

In Problems 65–74, perform the indicated operation and simplify.

65. $\dfrac{2x^2 + 11x + 14}{x^2 - 4}$

66. $\dfrac{x^2 - 5x - 14}{4 - x^2}$

67. $\dfrac{9x^2 - 1}{x^2 - 9} \cdot \dfrac{3x - 9}{9x^2 + 6x + 1}$

68. $\dfrac{x^2 - 25}{x^3 - 4x^2 - 5x} \cdot \dfrac{x^2 + x}{1 - x^2}$

69. $\dfrac{x + 1}{x - 1} - \dfrac{x - 1}{x + 1}$

70. $\dfrac{x}{x + 1} - \dfrac{2x}{x + 2}$

71. $\dfrac{3x + 4}{x^2 - 4} - \dfrac{2x - 3}{x^2 + 4x + 4}$

72. $\dfrac{x^2}{2x^2 + 5x - 3} + \dfrac{x^2}{2x^2 - 5x + 2}$

73. $1 + \dfrac{1}{1 + \dfrac{1}{x}}$

74. $1 - \dfrac{1}{1 + \dfrac{1}{x}}$

In Problems 75–80, rationalize the denominator of each expression.

75. $\dfrac{4}{\sqrt{5}}$

76. $\dfrac{-2}{\sqrt{3}}$

77. $\dfrac{2}{1 - \sqrt{2}}$

78. $\dfrac{-4}{1 + \sqrt{3}}$

79. $\dfrac{1 + \sqrt{5}}{1 - \sqrt{5}}$

80. $\dfrac{4\sqrt{3} + 2}{2\sqrt{3} + 1}$

In Problems 81–84, write each expression as a single quotient in which only positive exponents and/or radicals appear.

81. $(2 + x^2)^{1/2} + x \cdot \frac{1}{2}(2 + x^2)^{-1/2} \cdot 2x$

82. $(x^2 + 4)^{2/3} + x \cdot \frac{2}{3}(x^2 + 4)^{-1/3} \cdot 2x$

83. $\dfrac{(x + 4)^{1/2} \cdot 2x - x^2 \cdot \frac{1}{2}(x + 4)^{-1/2}}{x + 4}$

84. $\dfrac{(x^2 + 4)^{1/2} \cdot 2x - x^2 \cdot \frac{1}{2}(x^2 + 4)^{-1/2} \cdot 2x}{x^2 + 4}$

85. Manufacturing Cost The weekly production cost C of manufacturing x hand calculators is given by the formula

$$C = 3000 + 6x - \dfrac{x^2}{1000}$$

    (a) What is the cost of producing 1000 hand calculators?

    (b) What is the cost of producing 3000 hand calculators?

86. Quarterly Corporate Earnings In the first quarter of its fiscal year, a company posted earnings of $1.20 per share. During the second and third quarters, it posted losses of $0.75 per share and $0.30 per share, respectively. In the fourth quarter, it earned a modest $0.20 per share. What were the annual earnings per share of this company?

87. Design A window consists of a rectangle surmounted by a triangle. Find the area of the window shown in the illustration. How much wood frame is needed to enclose the window?

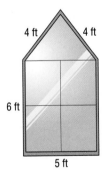

88. Construction A rectangular swimming pool, 20 feet long and 10 feet wide, is enclosed by a wooden deck that is 3 feet wide. What is the area of the deck? How much fence is required to enclose the deck?

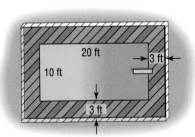

89. **Construction** A statue with a circular base of radius 3 feet is enclosed by a circular pond as shown in the illustration. What is the area of the pond? How much fence is required to enclose the pond?

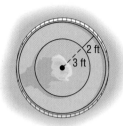

90. **How far can a pilot see?** On a recent flight to San Francisco, the pilot announced we were 139 miles from the city, flying at an altitude of 35,000 feet. The pilot claimed he could see the Golden Gate bridge and beyond. Was he telling the truth? How far could he see?

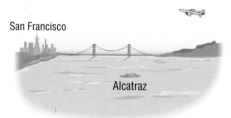

91. Use the material in this chapter to create a problem that uses each of the following words:
(a) Simplify (b) Factor (c) Reduce

92. A rational number is defined as the quotient of two integers. When written as a decimal, the decimal will either repeat or terminate. By looking at the denominator of the rational number, there is a way to tell in advance whether its decimal representation will repeat or terminate. Make a list of rational numbers and their decimals. See if you can discover the pattern. Confirm your conclusion by consulting books on number theory at the library. Write a brief essay on your findings.

93. The current time is 12 noon CST. What time (CST) will it be 12,997 hours from now?

94. Both $a/0$ ($a \neq 0$) and $0/0$ are undefined, but for different reasons. Write a paragraph or two explaining the different reasons.

Equations and Inequalities

Economists are fascinated with the consumption habits of individuals. Theories abound regarding how people decide to spend their money. On the following page, you will be asked to verify the hypothesis that income rises with age, reaches a peak, and then declines. Use the Sullivan website at:

www.prenhall.com/sullivan

to help link you to the Internet resources you will need to complete your assignment.

Netsite: http://www.prenhall.com/sullivan/

LIFE CYCLE HYPOTHESIS

When John Maynard Keynes wrote his *General Theory* in 1936, he stated there was a direct relation between the current consumption of an individual and his/her current income. Since this revolutionary thought first appeared, there have been many new theories regarding the consumption of individuals. One such theory submitted by Franco Modigliani and Richard Brumberg in 1955 is the life-cycle theory of consumption. These individuals felt that consumption was constant and, therefore, not a function of current income, but rather of expected lifetime income. Incomes start at low levels during the early earning years, increase until the earner reaches his/her peak earning potential, and then begin to decline as an individual approaches retirement age.

1. Find the quadratic equation of best fit for men's income, using the *median income data* for the most recent year available, in constant (this year's) dollars. Treat age as x and income as y. Use the average age of each age group. For example, for the age group 15–24, let $x = (15 + 24)/2 = 19.5$.
2. Find the quadratic equation of best fit for women's income, using the *median income data* for the most recent year available, in constant dollars. Treat age as x and income as y.
3. At what age can a man expect to earn $25,000? Can a woman ever expect to earn $25,000?
4. Determine the median income of a man 45 years old. Determine the median income of a woman 45 years old.
5. Find the quadratic equation of best fit for men's income, using the *median income data* for 1980, in constant dollars.
6. Find the quadratic equation of best fit for women's income, using the *median income data* for 1980, in constant dollars.
7. At what age did a man earn $25,000 in this years dollars? Could a woman ever expect to earn $25,000?
8. Determine the median income of a man 45 year's old in 1980, using constant dollars. Determine the median income of a woman 45 years old in 1980, using constant dollars.
9. Compare the median income of a man 45 years old in 1980 with the median income of a 45 year old man today. Compare the median income of a woman 45 years old in 1980 with the median income of a 45 year old woman today.

The investigation of equations and their solutions has played a central role in algebra for many hundreds of years. In fact as early as 2000 BC, the Babylonians had a well-developed algebra that included a solution for quadratic equations (see the Historical Feature in Section 2.3). Of late, the study of inequalities (Sections 2.6–2.8) has become equally important. For example, the problem of optimizing airline scheduling leads to linear inequalities. As you go through this chapter, you will see evidence of the importance of the topics presented here for solving problems that occur in business, engineering, and the social and natural sciences.

We shall limit our discussion in this chapter to equations and inequalities containing a single variable. Later, we will study equations and inequalities containing more than one variable.

2.1 | EQUATIONS

1. Solve an Equation in One Variable
2. Solve a Linear Equation
3. Solve Equations That Lead to Linear Equations

An **equation in one variable** is a statement in which two expressions, at least one containing the variable, are equal. The expressions are called the **sides** of the equation. Since an equation is a statement, it may not be true, depending on the value of the variable. Unless otherwise restricted, the admissible values of the variable are those in the domain of the variable. Those admissible values of the variable, if any, that result in a true statement are called **solutions,** or **roots,** of the equation. To **solve an equation** means to find all the solutions of the equation.

For example, the following are all equations in one variable, x:

$$x + 5 = 9 \qquad x^2 + 5x = 2x - 2 \qquad \frac{x^2 - 4}{x + 1} = 0 \qquad x^2 + 9 = 5$$

The first of these statements, $x + 5 = 9$, is true when $x = 4$ and false for any other choice of x. Thus, 4 is a solution of the equation $x + 5 = 9$. We also say that 4 **satisfies** the equation $x + 5 = 9$, because, when we substitute 4 for x, a true statement results.

Sometimes an equation will have more than one solution. For example, the equation

$$\frac{x^2 - 4}{x + 1} = 0$$

has either $x = -2$ or $x = 2$ as a solution.

Sometimes we will write the solution of an equation in set notation. This set is called the **solution set** of the equation. For example, the solution set of the equation $x^2 - 9 = 0$ is $\{-3, 3\}$.

Unless indicated otherwise, we will limit ourselves to real solutions. Some equations have no real solution. For example, $x^2 + 9 = 5$ has no real solution, because there is no real number whose square when added to 9 equals 5.

An equation that is satisfied for every choice of the variable for which both sides are defined is called an **identity.** For example, the equation

$$3x + 5 = x + 3 + 2x + 2$$

is an identity, because this statement is true for any real number x.

1. Two or more equations that have precisely the same solutions are called **equivalent equations.** For example, all the following equations are equivalent, because each has only the solution $x = 5$:

$$2x + 3 = 13$$
$$2x = 10$$
$$x = 5$$

These three equations illustrate one method for solving many types of equations: Replace the original equation by an equivalent equation, and continue until an equation with an obvious solution, such as $x = 5$, is reached. The question though, is "How do I obtain an equivalent equation?" In general, there are five ways to do so.

Procedures That Result in Equivalent Equations

1. Interchange the two sides of the equation:

    Replace $\quad 3 = x \quad$ by $\quad x = 3$

2. Simplify the sides of the equation by combining like terms, eliminating parentheses, and so on:

    Replace $\quad (x + 2) + 6 = 2x + (x + 1)$

    by $\qquad\quad x + 8 = 3x + 1$

3. Add or subtract the same expression on both sides of the equation:

    Replace $\qquad\quad 3x - 5 = 4$

    by $\qquad (3x - 5) + 5 = 4 + 5$

4. Multiply or divide both sides of the equation by the same nonzero expression:

    Replace $\qquad \dfrac{3x}{x - 1} = \dfrac{6}{x - 1} \qquad x \neq 1$

    by $\qquad \dfrac{3x}{x - 1} \cdot (x - 1) = \dfrac{6}{x - 1} \cdot (x - 1)$

5. If one side of the equation is 0 and the other side can be factored, then we may use the Zero-Product Property* and set each factor equal to 0:

    Replace $\qquad x(x - 3) = 0$

    by $\qquad x = 0 \quad$ or $\quad x - 3 = 0$

Whenever it is possible to solve an equation in your head, do so. For example:

The solution of $2x = 8$ is $x = 4$.

The solution of $3x - 15 = 0$ is $x = 5$.

Often, though, some rearrangement is necessary.

E X A M P L E 1

Solving an Equation

Solve the equation: $\quad 3x - 5 = 4$

Solution

We replace the original equation by a succession of equivalent equations:

$$3x - 5 = 4$$
$$(3x - 5) + 5 = 4 + 5 \qquad \text{Add 5 to both sides.}$$
$$3x = 9 \qquad \text{Simplify.}$$
$$\frac{3x}{3} = \frac{9}{3} \qquad \text{Divide both sides by 3.}$$
$$x = 3 \qquad \text{Simplify.}$$

The last equation, $x = 3$, has the single solution 3. All these equations are equivalent, so 3 is the only solution of the original equation $3x - 5 = 4$.

*The Zero-Product Property says that if $ab = 0$ then $a = 0$ or $b = 0$ or both equal 0.

Check: It is a good practice to check the solution by substituting 3 for x in the original equation:

$$3x - 5 = 4$$
$$3(3) - 5 \stackrel{?}{=} 4$$
$$9 - 5 \stackrel{?}{=} 4$$
$$4 = 4$$

The solution checks.

 Now work Problem 15.

E X A M P L E   2   Solving an Equation

Solve the equation: $\dfrac{3x}{x - 1} + 2 = \dfrac{3}{x - 1}$

Solution   First, we note that the domain of the variable is $\{x \mid x \neq 1\}$. Since the two quotients in the equation have the same denominator, $x - 1$, we can simplify by multiplying both sides by $x - 1$. The resulting equation is equivalent to the original equation, since we are multiplying by $x - 1$, which is not 0 (remember, $x \neq 1$).

$$\frac{3x}{x - 1} + 2 = \frac{3}{x - 1}$$

$$\left(\frac{3x}{x - 1} + 2\right) \cdot (x - 1) = \frac{3}{x - 1} \cdot (x - 1) \qquad \text{Multiply both sides by } x - 1; \text{ cancel on the right.}$$

$$\frac{3x}{x - 1} \cdot (x - 1) + 2 \cdot (x - 1) = 3 \qquad \text{Use the Distributive Property on the left side; cancel on the left.}$$

$$3x + (2x - 2) = 3 \qquad \text{Simplify.}$$
$$5x - 2 = 3$$
$$5x = 5 \qquad \text{Add 2 to each side.}$$
$$x = 1 \qquad \text{Divide both sides by 5.}$$

The solution appears to be 1. But recall that $x = 1$ is not in the domain of the variable. Thus, the equation has no solution.

 Now work Problem 41.

In the next examples, we use the Zero-Product Property (procedure 5, listed in the box on p. 90).

E X A M P L E   3   Solving Equations by Factoring

Solve the equations: (a) $x^2 = 4x$ (b) $x^3 - x^2 - 4x + 4 = 0$

Solution   (a) We begin by collecting all terms on one side. This results in 0 on one side and an expression to be factored on the other.

$$x^2 = 4x$$
$$x^2 - 4x = 0$$
$$x(x - 4) = 0 \qquad \text{Factor.}$$
$$x = 0 \quad \text{or} \quad x - 4 = 0 \qquad \text{Apply the Zero-Product Property.}$$
$$x = 4$$

The solution set is $\{0, 4\}$.

Check: $x = 0$: $0^2 = 4 \cdot 0$ So 0 is a solution.

$x = 4$: $4^2 = 4 \cdot 4$ So 4 is a solution.

(b) Do you recall the method of factoring by grouping from Chapter 1? We group the terms of $x^3 - x^2 - 4x + 4 = 0$ as follows:

$$(x^3 - x^2) - (4x - 4) = 0$$

Factor out x^2 from the first grouping and 4 from the second:

$$x^2(x - 1) - 4(x - 1) = 0$$

This reveals the common factor $(x - 1)$, so we have

$$(x^2 - 4)(x - 1) = 0$$

$$(x - 2)(x + 2)(x - 1) = 0 \qquad \text{Factor again.}$$

$(x - 2) = 0$ or $x + 2 = 0$ or $x - 1 = 0$ Set each factor equal to 0.

$\qquad x = 2 \qquad\qquad x = -2 \qquad\qquad x = 1$ Solve.

The solution set is $\{-2, 1, 2\}$.

Check: $x = 2$: $2^3 - 2^2 - 4(2) + 4 = 8 - 4 - 8 + 4 = 0$ 2 is a solution.

$x = -2$: $(-2)^3 - (-2)^2 - 4(-2) + 4 = -8 - (4) + 8 + 4 = 0$ −2 is a solution.

$x = 1$: $1^3 - 1^2 - 4(1) + 4 = 1 - 1 - 4 + 4 = 0$ 1 is a solution. ∎

Steps for Solving Equations

> STEP 1: List any restrictions on the domain of the variable.
> STEP 2: Simplify the equation by replacing the original equation by a succession of equivalent equations following the procedures listed earlier.
> STEP 3: If the result of Step 2 is a product of factors equal to 0, use the Zero-Product Property and set each factor equal to 0 (procedure 5).
> STEP 4: Check your solution(s).

Now work Problems 43 and 73.

Linear Equations

Linear equations are equations such as

$$3x + 12 = 0 \qquad -2x + 5 = 0 \qquad 4x - 3 = 0$$

A general definition is given next.

> A **linear equation in one variable** is equivalent to an equation of the form
>
> $$ax + b = 0$$
>
> where a and b are real numbers and $a \neq 0$.

Sometimes, a linear equation is called a **first-degree equation,** because the left side is a polynomial in x of degree 1.

It is relatively easy to solve a linear equation:

$$ax + b = 0$$

$$ax = -b \qquad \text{Subtract } b \text{ from both sides.}$$

$$x = \frac{-b}{a} \qquad \text{Divide both sides by } a, a \neq 0.$$

The linear equation $ax + b = 0$ has the single solution given by the formula $x = -b/a$.

E X A M P L E   4   Solving a Linear Equation

Solve the equation: $\frac{1}{2}(p + 5) - 4 = \frac{1}{3}(2p - 1)$

Solution   To clear the equation of fractions, we multiply both sides by 6, the least common multiple of the denominators of the fractions $\frac{1}{2}$ and $\frac{1}{3}$.

$$\frac{1}{2}(p + 5) - 4 = \frac{1}{3}(2p - 1)$$

$$6\left[\frac{1}{2}(p + 5) - 4\right] = 6\left[\frac{1}{3}(2p - 1)\right] \qquad \text{Multiply both sides by 6, the LCM of 2 and 3.}$$

$$3(p + 5) - 24 = 2(2p - 1) \qquad \text{Use the Distributive Property on the left and the Associative Property on the right.}$$

$$3p + 15 - 24 = 4p - 2 \qquad \text{Use the Distributive Property.}$$

$$3p - 9 = 4p - 2 \qquad \text{Combine like terms.}$$

$$3p = 4p + 7 \qquad \text{Add 9 to each side.}$$

$$-p = 7 \qquad \text{Subtract } 4p \text{ from each side.}$$

$$p = -7 \qquad \text{Multiply both sides by } -1.$$

Check: $\frac{1}{2}(p + 5) - 4 = \frac{1}{2}(-7 + 5) - 4 = \frac{1}{2}(-2) - 4 = -1 - 4 = -5$

$\frac{1}{3}(2p - 1) = \frac{1}{3}(-14 - 1) = \frac{1}{3}(-15) = -5$

Since the two expressions are equal, the solution $p = -7$ checks.   ▬

E X A M P L E   5   Solving a Linear Equation Using a Calculator

Solve the equation: $2.78x + \dfrac{2}{17.931} = 54.06$

Round the answer to two decimal places.

Solution   To avoid rounding errors, we solve for x before using the calculator. We proceed as in Example 4.

$$2.78x + \frac{2}{17.931} = 54.06$$

$$2.78x = 54.06 - \frac{2}{17.931} \qquad \text{Subtract 2/17.931 from each side.}$$

$$x = \frac{54.06 - (2/17.931)}{2.78} \qquad \text{Divide each side by 2.78.}$$

Now use your calculator.

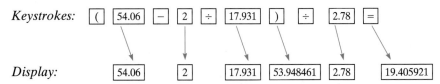

Keystrokes: $\boxed{(}$ $\boxed{54.06}$ $\boxed{-}$ $\boxed{2}$ $\boxed{\div}$ $\boxed{17.931}$ $\boxed{)}$ $\boxed{\div}$ $\boxed{2.78}$ $\boxed{=}$

Display: $\boxed{54.06}$ $\boxed{2}$ $\boxed{17.931}$ $\boxed{53.948461}$ $\boxed{2.78}$ $\boxed{19.405921}$

The solution, rounded to two decimal places, is 19.41.

Check: We store the unrounded solution in memory and proceed to check.

$$(2.78)(19.405921) + (2/17.931) = 54.06$$ ▬

Now work Problems 25 and 61.

Equations That Lead to Linear Equations

The next two examples illustrate the solution of equations that appear to be not linear, but lead to linear equations upon simplification.

E X A M P L E   6
Solving Equations

Solve the equation: $(2y + 1)(y - 1) = (y + 5)(2y - 5)$

Solution

$$
\begin{aligned}
(2y + 1)(y - 1) &= (y + 5)(2y - 5) \\
2y^2 - y - 1 &= 2y^2 + 5y - 25 &&\text{Multiply and combine like terms.}\\
-y - 1 &= 5y - 25 &&\text{Subtract } 2y^2 \text{ from each side.}\\
-y - 1 &= 5y - 24 &&\text{Add 1 to each side.}\\
-6y &= -24 &&\text{Subtract } 5y \text{ from each side.}\\
y &= 4 &&\text{Divide both sides by } -6.
\end{aligned}
$$

Check: $(2y + 1)(y - 1) = (8 + 1)(3) = (9)(3) = 27$
$(y + 5)(2y - 5) = (4 + 5)(8 - 5) = (9)(3) = 27$

Since the two expressions are equal, the solution $y = 4$ checks.   ▬

E X A M P L E   7
Solving Equations

Solve the equation: $\dfrac{3}{x - 2} = \dfrac{1}{x - 1} + \dfrac{7}{(x - 1)(x - 2)}$

Solution   First, we note that the domain of the variable is $\{x \mid x \neq 1, x \neq 2\}$. As we did in Example 4, we clear the equation of fractions by multiplying both sides by the least common multiple of the denominators of the three fractions, $(x - 1)(x - 2)$:

$$\frac{3}{x - 2} = \frac{1}{x - 1} + \frac{7}{(x - 1)(x - 2)}$$

$$(x - 1)(x - 2)\frac{3}{x - 2} = (x - 1)(x - 2)\left[\frac{1}{x - 1} + \frac{7}{(x - 1)(x - 2)}\right] \quad \begin{array}{l}\text{Multiply both sides by } (x - 1)(x - 2).\\ \text{Cancel on the left.}\end{array}$$

$$3x - 3 = (x - 1)(x - 2)\frac{1}{x - 1} + (x - 1)(x - 2)\frac{7}{(x - 1)(x - 2)} \quad \begin{array}{l}\text{Use the Distributive Property on}\\ \text{each side; cancel on the right.}\end{array}$$

$$3x - 3 = (x - 2) + 7$$
$$3x - 3 = x + 5 \qquad\qquad\qquad\qquad\qquad \text{Combine like terms.}$$
$$2x = 8 \qquad\qquad\qquad\qquad\qquad\quad \begin{array}{l}\text{Add 3 to each side. Subtract } x \text{ from}\\ \text{each side.}\end{array}$$
$$x = 4 \qquad\qquad\qquad\qquad\qquad\qquad \text{Divide by 2.}$$

Check: $\dfrac{3}{x - 2} = \dfrac{3}{4 - 2} = \dfrac{3}{2}$

$\dfrac{1}{x - 1} + \dfrac{7}{(x - 1)(x - 2)} = \dfrac{1}{3} + \dfrac{7}{(3)(2)} = \dfrac{1}{3} + \dfrac{7}{6} = \dfrac{9}{6} = \dfrac{3}{2}$

Since the two expressions are equal, the solution $x = 4$ checks.   ▬

Now work Problem 55.

E X A M P L E 8   Converting to Fahrenheit from Celsius

In the United States we measure temperature both in degrees Fahrenheit (°F) and in degrees Celsius (°C), which are related by the formula $C = \frac{5}{9}(F - 32)$. What are the Fahrenheit temperatures corresponding to Celsius temperatures of 0°, 10°, 20°, and 30°C?

Solution   We could solve four equations for F by replacing C each time by 0, 10, 20, and 30. Instead, it is much easier and faster first to solve the equation $C = \frac{5}{9}(F - 32)$ for F and then substitute in the values of C:

$$C = \frac{5}{9}(F - 32)$$

$$9C = 5(F - 32) \qquad \text{Multiply by 9.}$$

$$9C = 5F - 160 \qquad \text{Use the Distributive Property.}$$

$$5F - 160 = 9C \qquad \text{Interchange sides.}$$

$$5F = 9C + 160 \qquad \text{Add 160 to each side.}$$

$$F = \frac{9}{5}C + 32 \qquad \text{Divide both sides by 5.}$$

We can now do the required arithmetic:

$$0°C: \quad F = \frac{9}{5}(0) + 32 = 32°F$$

$$10°C: \quad F = \frac{9}{5}(10) + 32 = 50°F$$

$$20°C: \quad F = \frac{9}{5}(20) + 32 = 68°F$$

$$30°C: \quad F = \frac{9}{5}(30) + 32 = 86°F$$

SUMMARY

To solve a linear equation, follow these steps:

Steps for Solving a Linear Equation

STEP 1: If necessary, clear the equation of fractions by multiplying both sides by the least common multiple (LCM) of the denominators of all the fractions.

STEP 2: Remove all parentheses and simplify.

STEP 3: Collect all terms containing the variable on one side and all remaining terms on the other side.

STEP 4: Check your solution(s).

HISTORICAL FEATURE   The solution of equations is among the oldest of mathematical activities, and efforts to systematize this activity determined much of the shape of modern mathematics.

Consider the following problem, and its solution, using only words: Solve the problem of how many apples Jim has, given that

"Bob's five apples and Jim's apples together make twelve apples" by thinking,

"Jim's apples are all twelve apples less Bob's five apples" and then concluding,

"Jim has seven apples."

The mental steps translated into algebra are

$$5 + x = 12$$
$$x = 12 - 5$$
$$x = 7$$

The solution of this problem using only words is the earliest form of algebra. Such problems were solved exactly this way in Babylonia in 1800 BC. We know almost nothing of mathematical work before this date, although most authorities believe the sophistication of the earliest known texts indicates that a long period of previous development must have occurred. The method of writing out equations in words persisted for thousands of years, and although it now seems extremely cumbersome, it was used very effectively by many generations of mathematicians. The Arabs developed a good deal of the theory of cubic equations while writing out all the equations in words. About AD 1500, the tendency to abbreviate words in the written equations began to lead in the direction of modern notation; for example, the Latin word *et* (meaning *and*) developed into the plus sign, +. Although the occasional use of letters to represent variables dates back to AD 1200, the practice did not become common until about AD 1600. Development thereafter was rapid, and by 1635 algebraic notation did not differ essentially from what we use now.

2.1 | EXERCISES

In Problems 1–8, mentally solve each equation.

1. $7x = 21$ **2.** $6x = -24$ **3.** $5x + 15 = 0$ **4.** $3x + 18 = 0$

5. $2x - 3 = 0$ **6.** $3x + 4 = 0$ **7.** $\frac{1}{3}x = \frac{5}{12}$ **8.** $\frac{2}{3}x = \frac{9}{2}$

In Problems 9–60, solve each equation.

9. $3x + 2 = x$ **10.** $2x + 7 = 3x$ **11.** $2t - 6 = 3 - t$

12. $5y + 6 = -18 - y$ **13.** $6 - x = 2x + 9$ **14.** $3 - 2x = 2 - x$

15. $3 + 2n = 5n + 7$ **16.** $3 - 2m = 3m + 1$ **17.** $2(3 + 2x) = 3(x - 4)$

18. $3(2 - x) = 2x - 1$ **19.** $8x - (3x + 2) = 3x - 10$ **20.** $7 - (2x - 1) = 10$

21. $\frac{3}{2}x + 2 = \frac{1}{2} - \frac{1}{2}x$ **22.** $\frac{1}{3}x = 2 - \frac{2}{3}x$ **23.** $\frac{1}{2}x - 5 = \frac{3}{4}x$

24. $1 - \frac{1}{2}x = 6$ **25.** $\frac{2}{3}p = \frac{1}{2}p + \frac{1}{3}$ **26.** $\frac{1}{2} - \frac{1}{3}p = \frac{4}{3}$

27. $0.9t = 0.4 + 0.1t$ **28.** $0.9t = 1 + t$ **29.** $\dfrac{x + 1}{3} + \dfrac{x + 2}{7} = 5$

30. $\dfrac{2x + 1}{3} + 16 = 3x$ **31.** $\dfrac{2}{y} + \dfrac{4}{y} = 3$ **32.** $\dfrac{4}{y} - 5 = \dfrac{5}{2y}$

33. $\dfrac{1}{2} + \dfrac{2}{x} = \dfrac{3}{4}$ **34.** $\dfrac{3}{x} - \dfrac{1}{3} = \dfrac{1}{6}$ **35.** $(x + 7)(x - 1) = (x + 1)^2$

36. $(x + 2)(x - 3) = (x + 3)^2$ **37.** $x(2x - 3) = (2x + 1)(x - 4)$ **38.** $x(1 + 2x) = (2x - 1)(x - 2)$

39. $z(z^2 + 1) = 3 + z^3$ **40.** $w(4 - w^2) = 8 - w^3$ **41.** $\dfrac{x}{x - 2} + 3 = \dfrac{2}{x - 2}$

42. $\dfrac{2x}{x + 3} = \dfrac{-6}{x + 3} - 2$ **43.** $x^2 = 9x$ **44.** $x^3 = x^2$

45. $t^3 - 9t^2 = 0$ **46.** $4z^3 - 8z^2 = 0$ **47.** $\dfrac{2x}{x^2 - 4} = \dfrac{4}{x^2 - 4} - \dfrac{3}{x + 2}$

48. $\dfrac{x}{x^2 - 9} + \dfrac{4}{x + 3} = \dfrac{3}{x^2 - 9}$

49. $\dfrac{x}{x + 2} = \dfrac{1}{2}$

50. $\dfrac{3x}{x - 1} = 2$

51. $\dfrac{5}{2x - 3} = \dfrac{2}{x + 5}$

52. $\dfrac{-2}{x + 4} = \dfrac{-3}{x + 6}$

53. $\dfrac{6t + 7}{4t - 1} = \dfrac{3t + 8}{2t - 4}$

54. $\dfrac{8w + 5}{10w - 7} = \dfrac{4w - 3}{5w + 7}$

55. $\dfrac{4}{x - 2} = \dfrac{-3}{x + 5} + \dfrac{7}{(x + 5)(x - 2)}$

56. $\dfrac{-4}{2x + 3} + \dfrac{1}{x - 1} = \dfrac{1}{(2x + 3)(x - 1)}$

57. $\dfrac{2}{y + 3} + \dfrac{3}{y - 4} = \dfrac{5}{y + 6}$

58. $\dfrac{5}{5z - 11} + \dfrac{4}{2z - 3} = \dfrac{-3}{5 - z}$

59. $\dfrac{x}{x^2 - 1} - \dfrac{x + 3}{x^2 - x} = \dfrac{-3}{x^2 + x}$

60. $\dfrac{x + 1}{x^2 + 2x} - \dfrac{x + 4}{x^2 + x} = \dfrac{-3}{x^2 + 3x + 2}$

In Problems 61–64, use a calculator to solve each equation. Round the solution to two decimal places.

61. $3.2x + \dfrac{21.3}{65.871} = 19.23$

62. $6.2x - \dfrac{19.1}{83.72} = 0.195$

63. $14.72 - 21.58x = \dfrac{18}{2.11}x + 2.4$

64. $18.63x - \dfrac{21.2}{2.6} = \dfrac{14x}{2.32} - 20$

In Problems 65–76, find the real solutions of each equation by factoring.

65. $x^2 - 7x + 12 = 0$

66. $x^2 - x - 6 = 0$

67. $2x^2 + 5x - 3 = 0$

68. $3x^2 + 5x - 2 = 0$

69. $x^3 = 9x$

70. $x^4 = x^2$

71. $x^3 + x^2 - 20x = 0$

72. $x^3 + 6x^2 - 7x = 0$

73. $x^3 + x^2 - x - 1 = 0$

74. $x^3 + 4x^2 - x - 4 = 0$

75. $x^3 - 3x^2 - 4x + 12 = 0$

76. $x^3 - 3x^2 - x + 3 = 0$

In Problems 77–82, solve each equation. The letters a, b, and c are constants.

77. $ax - b = c, \quad a \neq 0$

78. $1 - ax = b, \quad a \neq 0$

79. $\dfrac{x}{a} + \dfrac{x}{b} = c, \quad a \neq 0, b \neq 0, a \neq -b$

80. $\dfrac{a}{x} + \dfrac{b}{x} = c, \quad c \neq 0$

81. $\dfrac{1}{x - a} + \dfrac{1}{x + a} = \dfrac{2}{x - 1}$

82. $\dfrac{b + c}{x + a} = \dfrac{b - c}{x - a}, \quad c \neq 0, a \neq 0$

83. Find the number a for which $x = 4$ is a solution of the equation

$$x + 2a = 16 + ax - 6a$$

84. Find the number b for which $x = 2$ is a solution of the equation

$$x + 2b = x - 4 + 2bx$$

Problems 85–90, list some formulas that occur in applications. Solve each formula for the indicated variable.

85. **Electricity** $\dfrac{1}{R} = \dfrac{1}{R_1} + \dfrac{1}{R_2}$ for R

86. **Finance** $A = P(1 + rt)$ for r

87. **Mechanics** $F = \dfrac{mv^2}{R}$ for R

88. **Chemistry** $PV = nRT$ for T

89. **Mathematics** $S = \dfrac{a}{1 - r}$ for r

90. **Mechanics** $v = -gt + v_0$ for t

 91. One of the steps in the following list contains an error. Identify it and explain what is wrong.

$$x = 2 \tag{1}$$
$$3x - 2x = 2 \tag{2}$$
$$3x = 2x + 2 \tag{3}$$
$$x^2 + 3x = x^2 + 2x + 2 \tag{4}$$
$$x^2 + 3x - 10 = x^2 + 2x - 8 \tag{5}$$
$$(x - 2)(x + 5) = (x - 2)(x + 4) \tag{6}$$
$$x + 5 = x + 4 \tag{7}$$
$$1 = 0 \tag{8}$$

92. Which of the following pairs of equations are equivalent? Explain.
    (a) $x^2 = 9; x = 3$
    (b) $x = \sqrt{9}; x = 3$
    (c) $(x - 1)(x - 2) = (x - 1)^2; x - 2 = x - 1$

93. The equation
$$\frac{5}{x + 3} + 3 = \frac{8 + x}{x + 3}$$

has no solution, yet when we go through the process of solving it we obtain $x = -3$. Write a brief paragraph to explain what causes this to happen.

94. Make up an equation that has no solution and give it to a fellow student to solve. Ask the fellow student to write a critique of your equation.

2.2 | SETTING UP EQUATIONS: APPLICATIONS

1 Translate Verbal Descriptions into Mathematical Expressions
2 Set Up Applied Problems
3 Solve Interest Problems
4 Solve Mixture Problems
5 Solve Uniform Motion Problems
6 Solve Constant Rate Jobs Problems

The previous section provides the tools for solving equations. But, unfortunately, applied problems do not come in the form "Solve the equation . . . ." Instead, they are narratives that supply information—hopefully, enough to answer the question that inevitably arises. Thus, to solve applied problems we must be able to translate the verbal description into the language of mathematics. We do this by using symbols (usually letters of the alphabet) to represent unknown quantities and then finding relationships (such as equations) that involve these symbols. The process of doing this is called **mathematical modeling.**

Any solution to the mathematical problem must be checked against the mathematical problem, the verbal description, and the real problem. See Figure 1 for an illustration of the modeling process.

FIGURE 1

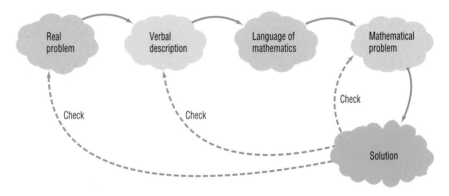

1 Let's look at a few examples that will help you to translate certain words into mathematical symbols.

E X A M P L E   1   Translating Verbal Descriptions into Mathematical Expressions

(a) The area of a rectangle is the product of its length times its width.
    Translation: If A is used to represent the area, l the length, and w the width, then $A = lw$.

(b) For uniform motion, the velocity of an object equals the distance traveled divided by the time required.

Translation: If v is the velocity, s the distance, and t the time, then $v = s/t$.

(c) A total of \$5000 is invested, some in stocks and some in bonds. If the amount invested in stocks is x, express the amount invested in bonds in terms of x.

Translation: If x is the amount invested in stocks, then the amount invested in bonds is $5000 - x$, since their sum is $x + (5000 - x) = 5000$.

(d) Let x denote a number.

The number 5 times as large as x is $5x$.

The number 3 less than x is $x - 3$.

The number that exceeds x by 4 is $x + 4$.

The number that, when added to x, gives 5 is $5 - x$. ∎

 Now work Problem 1.

 Mathematical equations that represent real situations should be consistent in terms of the units used. In Example 1(a), if l is measured in feet, then w also must be expressed in feet, and A will be expressed in square feet. In Example 1(b), if v is measured in miles per hour, then the distance s must be expressed in miles and the time t must be expressed in hours. It is a good practice to check units to be sure that they are consistent and make sense.

Although each situation has its own unique features, we can provide an outline of the steps to follow in setting up applied problems.

Steps for Setting Up Applied Problems

> STEP 1: Read the problem carefully, perhaps two or three times. Pay particular attention to the question being asked in order to identify what you are looking for. If you can, determine realistic possibilities for the answer.
>
> STEP 2: Assign a letter (variable) to represent what you are looking for, and, if necessary, express any remaining unknown quantities in terms of this variable.
>
> STEP 3: Make a list of all the known facts, and write down any relationships among them, especially any that involve the variable. These may take the form of an equation (or, later, an inequality) involving the variable. If possible, draw an appropriately labeled diagram to assist you. Sometimes, a table or chart helps.
>
> STEP 4: Solve the equation for the variable, and then answer the question asked in the problem.
>
> STEP 5: Check the answer with the facts in the problem. If it agrees, congratulations! If it does not agree, try again.

Let's look at an example.

E X A M P L E 2   Investments

A total of \$18,000 is invested, some in stocks and some in bonds. If the amount invested in bonds is half that invested in stocks, how much is invested in each category?

Solution   STEP 1: We are being asked to find the amount of two investments. These amounts must total \$18,000. (Do you see why?)

STEP 2: If we let x equal the amount invested in stocks, then $18,000 - x$ is the amount invested in bonds. [Look back at Example 1(c) to see why.]

STEP 3: We set up a table:

Amount in Stocks	Amount in Bonds	Reason
x	$18,000 - x$	Total invested is $18,000

From the table and the given fact that the total amount invested in bonds $(18,000 - x)$ is half that in stocks (x), we obtain the equation $18,000 - x = \frac{1}{2}x$.

STEP 4:
$$18,000 - x = \tfrac{1}{2}x$$
$$18,000 = x + \tfrac{1}{2}x$$
$$18,000 = \tfrac{3}{2}x$$
$$\left(\tfrac{2}{3}\right)18,000 = \left(\tfrac{2}{3}\right)\left(\tfrac{3}{2}x\right)$$
$$12,000 = x$$

Thus, $12,000 is invested in stocks and $18,000 - $12,000 = $6000 is invested in bonds.

STEP 5: The total invested is $12,000 + $6000 = $18,000, and the amount in bonds ($6000) is half that in stocks ($12,000). ∎

 Now work Problem 11.

E X A M P L E   3   Determining an Hourly Wage

Shannon grossed $435 one week by working 52 hours. Her employer pays time-and-a-half for all hours worked in excess of 40 hours. With this information, can you determine Shannon's regular hourly wage?

Solution   STEP 1: We are looking for an hourly wage. Our answer will be in dollars per hour.

STEP 2: Let x represent the regular hourly wage; x is measured in dollars per hour.

STEP 3: We set up a table:

	Hours Worked	Hourly Wage	Salary
Regular	40	x	$40x$
Overtime	12	$1.5x$	$12(1.5x) = 18x$

The sum of regular salary plus overtime salary will equal $435. Thus, from the table, $40x + 18x = 435$.

STEP 4:
$$40x + 18x = 435$$
$$58x = 435$$
$$x = 7.50$$

Thus, Shannon's regular hourly wage is $7.50 per hour.

STEP 5: Forty hours yields a salary of $40(7.50) = $300, and 12 hours of overtime yields a salary of $12(1.5)(7.50) = $135, for a total of $435. ∎

 Now work Problem 15.

Interest

3

The next example involves **interest.** Interest is money paid for the use of money. The total amount borrowed (whether by an individual from a bank in the form of a loan or by a bank from an individual in the form of a savings account) is called the **principal.** The **rate of interest,** expressed as a percent, is the amount charged for the use of the principal for a given period of time, usually on a yearly (that is, per annum) basis.

Simple Interest Formula

    If a principal of P dollars is borrowed for a period of t years at a per annum interest rate r, expressed as a decimal, the interest I charged is

$$I = Prt \qquad\qquad (1)$$

Interest charged according to formula (1) is called **simple interest.**

E X A M P L E   4   Finance: Computing Interest on a Loan

Suppose that Juanita borrows $500 for 6 months at the simple interest rate of 9% per annum. What is the interest Juanita will be charged on this loan? How much does Juanita owe after 6 months?

Solution   The rate of interest is given per annum, so the actual time the money is borrowed must be expressed in years. Thus, the interest charged would be the principal ($500) times the rate of interest (9% = 0.09) times the time in years $\left(\frac{1}{2}\right)$:

$$\text{Interest charged} = I = Prt = (500)(0.09)\left(\tfrac{1}{2}\right) = \$22.50$$

Juanita will owe what she borrowed, plus interest; that is, $500 + $22.50 = $522.50.

—

E X A M P L E   5   Financial Planning

Candy has $70,000 to invest and requires an overall rate of return of 9%. She can invest in a safe, government-insured Certificate of Deposit, but it only pays 8%. To obtain 9%, she agrees to invest some of her money in non-insured corporate bonds paying 12%. How much should be placed in each investment to achieve her goals?

Solution   STEP 1: The question is asking for two dollar amounts: the principal to invest in the corporate bonds and the principal to invest in the certificate of deposit.

STEP 2: We let x represent the amount (in dollars) to be invested in the bonds. Then $70{,}000 - x$ is the amount that will be invested in the certificate. (Do you see why?)

STEP 3: We set up a table:

	Principal $	Rate	Time yr	Interest $
Bonds	x	12% = 0.12	1	$0.12x$
Certificate	$70{,}000 - x$	8% = 0.08	1	$0.08(70{,}000 - x)$
Total	$70{,}000$	9% = 0.09	1	$0.09(70{,}000) = 6300$

Since the total interest from the investments is equal to $0.09(70,000) = 6300$, we must have the equation

$$0.12x + 0.08(70,000 - x) = 6300$$

(Note that the units are consistent: the unit is dollars on each side.)

STEP 4: $0.12x + 5600 - 0.08x = 6300$
$$0.04x = 700$$
$$x = 17,500$$

Thus, Candy should place $17,500 in the bonds and $70,000 − $17,500 = $52,500 in the certificate.

STEP 5: The interest on the bonds after 1 year is $0.12(\$17,500) = \$2100$; the interest on the certificate after 1 year is $0.08(\$52,500) = \$4200$. The total annual interest is $6300, the required amount. ■

 Now work Problem 21.

Mixture Problems

 The next example is a type usually referred to as a **mixture problem.**

EXAMPLE 6   Chemistry: Mixing Acids

In a chemistry laboratory the concentration of one solution is 10% hydrochloric acid (HCl) and that of a second solution is 60% HCl. How many milliliters (mL) of each should be mixed to obtain 50 mL of a 30% HCl solution?

Solution   Let x represent the number of milliliters of the 10% solution. Then $50 - x$ equals the number of milliliters of the 60% solution. See Figure 2.

FIGURE 2

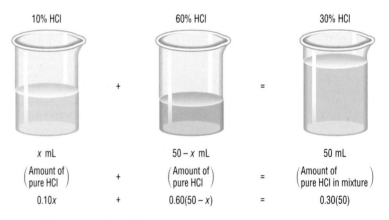

Based on the figure, we form a table:

	Amount mL	Concentration of HCl	Amount of Pure Acid mL
10% HCl	x	10% = 0.10	0.10x
60% HCl	$50 - x$	60% = 0.60	0.60(50 − x)
30% HCl	50	30% = 0.30	0.30(50) = 15

The amount of HCl in the 30% solution (15 milliliters) must equal the sum of the amounts of HCl found in the 10% solution and the 60% solution. Thus, we have the equation

$$0.10x + 0.60(50 - x) = 15$$
$$0.10x + 30 - 0.60x = 15$$
$$-0.50x = -15$$
$$x = 30 \text{ mL}$$

Thus, 30 mL of the 10% acid solution, when mixed with 20 mL of the 60% acid solution, yields 50 mL of a 30% acid solution.

Check: To check this answer, we note that there are $0.10(30) = 3$ mL of acid in the 10% solution and $0.60(20) = 12$ mL of acid in the 60% solution. The 50 mL mixture therefore contains 15 mL of acid, for an acid concentration of $\frac{15}{50} = 0.30 = 30\%$ acid solution. ■

Now work Problem 25.

Uniform Motion

5 The next example deals with moving objects.

> **Uniform Motion Formula**
> If an object moves at an average velocity *v*, the distance *s* covered in time *t* is given by the formula
>
> $$s = vt \qquad\qquad (2)$$

That is, Distance = Velocity · Time. Objects that are moving in accordance with formula (2) are said to be in **uniform motion.**

E X A M P L E   7   Physics: Uniform Motion

Tanya, who is a long-distance runner, runs at an average velocity of 8 miles per hour (mph). Two hours after Tanya leaves your house, you leave in your Honda and follow the same route. If your average velocity is 40 mph, how long will it be before you catch up to Tanya? How far will each of you be from your house?

Solution   Refer to Figure 3. We use *t* to represent the time (in hours) that it takes the Honda to catch up with Tanya. When this occurs, the total time elapsed for Tanya is $t + 2$ hours.

FIGURE 3

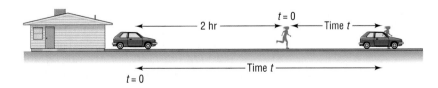

Set up the following table:

	Velocity mph	Time hr	Distance mi
Tanya	8	$t + 2$	$8(t + 2)$
Honda	40	t	$40t$

Since the distance traveled is the same, we are led to the following equation:

$$8(t + 2) = 40t$$
$$8t + 16 = 40t$$
$$32t = 16$$
$$t = \frac{1}{2} \text{ hour}$$

It will take the Honda $\frac{1}{2}$ hour to catch up to Tanya. Each of you will have gone 20 miles.

Check: In 2.5 hours, Tanya travels a distance of $(2.5)(8) = 20$ miles. In $\frac{1}{2}$ hour, the Honda travels a distance of $(\frac{1}{2})(40) = 20$ miles. ▬

 Now work Problem 41.

Constant Rate Jobs

 This section involves jobs that are performed at a **constant rate.** Our assumption is that, if a job can be done in t units of time, $1/t$ of the job is done in 1 unit of time. Let's look at an example.

E X A M P L E   8   Working Together to Do a Job

At 10 AM Danny is asked by his father to weed the garden. From past experience Danny knows this will take him 4 hours, working alone. His older brother, Mike, when it is his turn to do this job, requires 6 hours. Since Mike wants to go golfing with Danny and has a reservation for 1 PM, he agrees to help Danny. Assuming no gain or loss of efficiency, when will they finish if they work together? Can they make the golf date?

Solution   In 1 hour, Danny does $\frac{1}{4}$ of the job, and in 1 hour, Mike does $\frac{1}{6}$ of the job. Let t be the time (in hours) it takes them to do the job together. In 1 hour, then $1/t$ of the job is completed. We reason as follows:

$$\left(\begin{array}{c}\text{Part done by Danny}\\\text{in 1 hour}\end{array}\right) + \left(\begin{array}{c}\text{Part done by Mike}\\\text{in 1 hour}\end{array}\right) = \left(\begin{array}{c}\text{Part done together}\\\text{in 1 hour}\end{array}\right)$$

Now we set up the table in the margin. From the table,

	Hours to Do Job	Part of Job Done in 1 Hour
Danny	4	$\frac{1}{4}$
Mike	6	$\frac{1}{6}$
Together	t	$\frac{1}{t}$

$$\frac{1}{4} + \frac{1}{6} = \frac{1}{t}$$
$$\frac{3 + 2}{12} = \frac{1}{t}$$
$$t = \frac{12}{5}$$

Working together, the job can be done in $\frac{12}{5}$ hours, or 2 hours, 24 minutes. They should make the golf date, since they will finish at 12:24 PM. ▬

 Now work Problem 37.

2.2 EXERCISES

In Problems 1–10, translate each sentence into a mathematical equation. Be sure to identify the meaning of all symbols.

1. **Geometry**   The area of a circle is the product of the number π times the square of the radius.

2. **Geometry**   The circumference of a circle is the product of the number π times twice the radius.

3. **Geometry**   The area of a square is the square of the length of a side.

4. **Geometry**   The perimeter of a square is four times the length of a side.

5. **Physics**   Force equals the product of mass times acceleration.

6. **Physics**   Pressure is force per unit area.

7. **Physics**   Work equals force times distance.

8. **Physics**   Kinetic energy is one-half the product of the mass times the square of the velocity.

9. **Business**   The total variable cost of manufacturing x dishwashers is $150 per dishwasher times the number of dishwashers manufactured.

10. **Business**   The total revenue derived from selling x dishwashers is $250 per dishwasher times the number of dishwashers sold.

11. **Finance**   A total of $20,000 is to be invested, some in bonds and some in Certificates of Deposit (CDs). If the amount invested in bonds is to exceed that in CDs by $2000, how much will be invested in each type of investment?

12. **Finance**   A total of $10,000 is to be divided between Sean and George, with George to receive $2000 less than Sean. How much will each receive?

13. **Finance**   An inheritance of $900,000 is to be divided among Scott, Alice, and Tricia in the following manner: Alice is to receive $\frac{3}{4}$ of what Scott gets, while Tricia gets $\frac{1}{2}$ of what Scott gets. How much does each receive?

14. **Sharing the Cost of a Pizza**   Canter and Carole agree to share the cost of an $18 pizza based on how much each ate. If Carole ate $\frac{2}{3}$ the amount Canter ate, how much should each pay? [**Hint:** Some pizza may be left.]

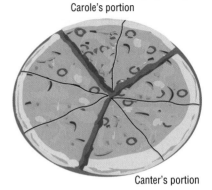

Carole's portion

Canter's portion

15. **Computing Hourly Wages**   Sandra, who is paid time-and-a-half for hours worked in excess of 40 hours, had gross weekly wages of $442 for 48 hours worked. What is the regular hourly rate?

16. **Computing Hourly Wages**   Leigh is paid time-and-a-half for hours worked in excess of 40 hours and double-time for hours worked on Sunday. If Leigh had gross weekly wages of $342 for working 50 hours, 4 of which were on Sunday, what is her regular hourly rate?

17. **Football**   In an NFL football game, the Bears scored a total of 41 points, including one safety (2 points) and two field goals (3 points each). After scoring a touchdown (6 points), a team is given the chance to score 1 or 2 extra points. The Bears, in trying to score 1 extra point after each touchdown, missed 2 extra points. How many touchdowns did they get?

18. **Basketball**   In a basketball game, the Bulls scored a total of 103 points and made three times as many field goals (2 points each) as free throws (1 point each). They also made eleven 3 point baskets. How many field goals did they have?

19. **Geometry**   The perimeter of a rectangle is 60 feet. Find its length and width if the length is 8 feet longer than the width.

20. **Geometry**   The perimeter of a rectangle is 42 meters. Find its length and width if the length is twice the width.

21. **Financial Planning**   A recent retiree, Kate requires $6000 per year in extra income. She has $50,000 to invest and can invest in noninsured B-rated bonds paying 15% per year or in a government-insured Certificate of Deposit (CD) paying 7% per year. How much money should be invested in each to realize exactly $6000 in interest per year?

22. **Financial Planning** After 2 years, Kate (see Problem 21) finds that she now will require $7000 per year. Assuming that the remaining information is the same, how should the money be reinvested?

23. **Banking** A bank loaned out $12,000, part of it at the rate of 8% per year and the rest at the rate of 18% per year. If the interest received totaled $1000, how much was loaned at 8%?

24. **Banking** Scott, a loan officer at a bank, has $1,000,000 to lend and is required to obtain an average return of 18% per year. If he can lend at the rate of 19% or the rate of 16%, how much can he lend at the 16% rate and still meet his requirement?

25. **Chemistry: Mixing Acids** For a certain experiment, a student requires 100 cubic centimeters (cc) of a solution that is 8% HCl. The storeroom has only solutions that are 15% HCl and 5% HCl. How many cubic centimeters of each available solution should be mixed to get 100 cc of 8% HCl?

26. **Business: Blending Coffee** A coffee manufacturer wants to market a new blend of coffee that will cost $3.90 per pound by mixing two coffees that sell for $2.75 and $5 per pound, respectively. What amounts of each coffee should be blended to obtain the desired mixture?

    [**Hint:** Assume that the total weight of the desired blend is 100 pounds.]

27. **Business: Mixing Nuts** A nut store normally sells cashews for $4.00 per pound and peanuts for $1.50 per pound. But at the end of the month, the peanuts had not sold well, so in order to sell 60 pounds of peanuts, the manager decided to mix the 60 pounds of peanuts with some cashews and sell the mixture for $2.50 per pound. How many pounds of cashews should be mixed with the peanuts to ensure no change in the profit?

28. **Business: Mixing Candy** A candy store sells boxes of candy containing caramels and cremes. Each box sells for $12.50 and holds 30 pieces of candy (all pieces are the same size). If the caramels cost $0.25 to produce and the cremes cost $0.45 to produce, how many of each should be in a box to make a profit of $3?

29. **Chemistry: Mixing Acids** How many ounces of pure water should be added to 20 ounces of a 40% solution of muriatic acid to obtain a 30% solution of muriatic acid?

30. **Chemistry: Mixing Acids** How many cubic centimeters of pure hydrochloric acid should be added to 20 cc of a 30% solution of hydrochloric acid to obtain a 50% solution?

31. **Business: Theater Attendance** The manager of the Coral Theater wants to know whether the majority of its patrons are adults or children. During a week in July, 5200 tickets were sold and the receipts totaled $20,335. The adult admission is $4.75, and the children's admission is $2.50. How many adult patrons were there?

32. **Business: Discount Pricing** A wool suit, discounted by 30% for a clearance sale, has a price tag of $399. What was the suit's original price?

33. **Business: Discount Pricing** A builder of tract homes reduced the price of a model by 15%. If the new price is $125,000, what was its original price? How much can be saved by purchasing the model?

34. **Business: Discount Pricing** A car dealer, at a year-end clearance, reduces the list price of last year's models by 15%. If a certain four-door model has a discounted price of $8000, what was its list price? How much can be saved by purchasing last year's model?

35. **Business: Marking Up the Price of Books** A college book store marks up the price it pays the publisher for a book by 25%. If the selling price of a book is $56.00, how much did the book store pay for the book?

36. **Personal Finance: Cost of a Car** The suggested list price of a new car is $12,000. The dealer's cost is 85% of list. How much will you pay if the dealer is willing to accept $100 over cost for the car?

37. **Working Together on a Job** Trent can deliver his newspapers in 30 minutes. It takes Lois 20 minutes to do the same route. How long would it take them to deliver the newspapers if they work together?

38. **Working Together on a Job** Patrice, by himself, can paint four rooms in 10 hours. If he hires April to help, they can do the same job together in 6 hours. If he lets April work alone, how long will it take for her to paint four rooms?

39. **Computing Grades** Going into the final exam, which will count as two tests, Sarah has test scores of 80, 83, 71, 61, and 95. What score does Sarah need on the final in order to have an average score of 80?

40. Computing Grades Going into the final exam, which will count as two-thirds of the final grade, Mark has test scores of 86, 80, 84, and 90. What score does Mark need on the final in order to earn a B, which requires an average score of 80? What does he need to earn an A, which requires an average of 90?

41. Physics: Uniform Motion A motorboat can maintain a constant speed of 16 miles per hour relative to the water. The boat makes a trip upstream to a certain point in 20 minutes; the return trip takes 15 minutes. What is the speed of the current? (See the figure.)

42. Physics: Uniform Motion A motorboat heads upstream on a river that has a current of 3 miles per hour. The trip upstream takes 5 hours, while the return trip takes 2.5 hours. What is the speed of the motorboat? (Assume that the motorboat maintains a constant speed relative to the water.)

43. Enclosing a Garden A gardener has 46 feet of fencing to be used to enclose a rectangular garden that has a border 2 feet wide surrounding it (see the figure).
(a) If the length of the garden is to be twice its width, what will be the dimensions of the garden?
(b) What is the area of the garden?
(c) If the length and width of the garden were to be the same, what would be the dimensions of the garden?
(d) What would be the area of the square garden?

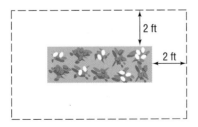

44. Construction A pond is enclosed by a wooden deck that is 3 feet wide. The fence surrounding the deck is 100 feet long.

(a) If the pond is square, what are its dimensions?
(b) If the pond is rectangular and the length of the pond is to be three times its width, what are the dimensions of the pond?
(c) If the pond is circular, what is the diameter of the pond?
(d) Which pond has the most area?

45. Football A tight end can run the 100 yard dash in 12 seconds. A defensive back can do it in 10 seconds. The tight end catches a pass at his own 20 yard line with the defensive back at the 15 yard line. (See the figure.) If no other players are nearby, at what yard line will the defensive back catch up to the tight end?
[**Hint:** At time $t = 0$, the defensive back is 5 yards behind the tight end.]

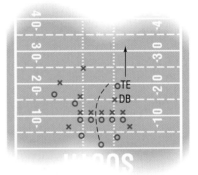

46. Computing Business Expenses Therese, an outside saleswoman, uses her car for both business and pleasure. Last year, she traveled 30,000 miles, using 900 gallons of gasoline. Her car gets 40 miles per gallon on the highway and 25 in the city. She can deduct all highway travel, but no city travel, on her taxes. How many miles should Therese be allowed as a business expense?

47. Mixing Water and Antifreeze How much water should be added to 1 gallon of pure antifreeze to obtain a solution that is 60% antifreeze?

48. Mixing Water and Antifreeze The cooling system of a certain foreign-made car has a capacity of 15 liters. If the system is filled with a mixture that is 40% antifreeze, how much of this mixture should be drained and replaced by pure antifreeze so that the system is filled with a solution that is 60% antifreeze?

49. Chemistry: Salt Solutions How much water must be evaporated from 32 ounces of a 4% salt solution to make a 6% salt solution?

50. Chemistry: Salt Solutions How much water must be evaporated from 240 gallons of a 3% salt solution to produce a 5% salt solution?

51. Physics: Uniform Motion A Metra commuter train leaves Union Station in Chicago at 12 noon. Two hours later, an Amtrak train leaves on the same track, traveling at an average speed that is 50 miles per hour faster than the Metra train. At 3 PM, the Amtrak train is 10 miles behind the commuter train. How fast is each going?

52. Physics: Uniform Motion Two cars enter the Florida Turnpike at Commercial Boulevard at 8:00 AM, each heading for Wildwood. One car's average speed is 10 miles per hour more than the other's. The faster car arrives at Wildwood at 11:00 AM, $\frac{1}{2}$ hour before the other car. What is the average speed of each car? How far did each travel?

53. Purity of Gold The purity of gold is measured in karats, with pure gold being 24 karats. Other purities of gold are expressed as proportional parts of pure gold. Thus, 18 karat gold is $\frac{18}{24}$, or 75% pure gold; 12 karat gold is $\frac{12}{24}$, or 50% pure gold; and so on. How much 12 karat gold should be mixed with pure gold to obtain 60 grams of 16 karat gold?

54. Chemistry: Sugar Molecules A sugar molecule has twice as many atoms of hydrogen as it does oxygen and one more atom of carbon than oxygen. If a sugar molecule has a total of 45 atoms, how many are oxygen? How many are hydrogen?

55. Running a Race Mike can run the mile in 6 minutes, and Dan can run the mile in 9 minutes. If Mike gives Dan a head start of 1 minute, how far from the start will Mike pass Dan? (See the figure.) How long does it take?

56. Range of an Airplane An air rescue plane averages 300 miles per hour in still air. It carries enough fuel for 5 hours of flying time. If, upon takeoff, it encounters a wind of 30 mph, how far can it fly and return safely? (Assume that the wind remains constant.)

57. Emptying Oil Tankers An oil tanker can be emptied by the main pump in 4 hours. An auxiliary pump can empty the tanker in 9 hours. If the main pump is started at 9 AM, when should the auxiliary pump be started so that the tanker is emptied by noon?

58. Cement Mix A 20 pound bag of Economy brand cement mix contains 25% cement and 75% sand. How much pure cement must be added to produce a cement mix that is 40% cement?

59. Emptying a Tub A bathroom tub will fill in 15 minutes with both faucets open and the stopper in place. With both faucets closed and the stopper removed, the tub will empty in 20 minutes. How long will it take for the tub to fill if both faucets are open and the stopper is removed?

60. Using Two Pumps A 5 horsepower (hp) pump can empty a pool in 5 hours. A smaller, 2 hp pump empties the same pool in 8 hours. The pumps are used together to begin emptying this pool. After two hours, the 2 hp pump breaks down. How long will it take the larger pump to empty the pool?

61. Comparing Olympic Heroes In the 1984 Olympics, Carl Lewis of the United States won the gold medal in the 100 meter race with a time of 9.99 seconds. In the 1896 Olympics, Thomas Burke, also of the United States, won the gold medal in the 100 meter race in 12.0 seconds. If they ran in the same race repeating their respective times, by how many meters would Lewis beat Burke?

 62. Critical Thinking You are the manager of a clothing store and have just purchased 100 dress shirts for $20.00 each. After 1 month of selling the shirts at the regular price, you plan to have a sale giving 40% off the original selling price. However, you still want to make a profit of $4 on each shirt at the sale price. What should you price the shirts at initially to ensure this? If, instead of 40% off at the sale, you give 50% off, by how much is your profit reduced?

63. Critical Thinking Make up a word problem that requires solving a linear equation as part of its solution. Exchange problems with a friend. Write a critique of your friend's problem.

64. Critical Thinking Without solving, explain what is wrong with the following mixture problem: How many liters of 25% ethanol should be added to 20 liters of 48% ethanol to obtain a solution of 58% ethanol? Now go through an algebraic solution. What happens?

2.3 | QUADRATIC EQUATIONS

1. Solve a Quadratic Equation by Factoring
2. Know How to Complete the Square
3. Solve a Quadratic Equation by Completing the Square
4. Solve a Quadratic Equation Using the Quadratic Formula
5. Solve Applied Problems

Quadratic equations are equations such as

$$2x^2 + x + 8 = 0 \qquad 3x^2 - 5x + 6 = 0 \qquad x^2 - 9 = 0$$

A general definition is given next.

A **quadratic equation** is an equation equivalent to one of the form

$$ax^2 + bx + c = 0 \tag{1}$$

where *a*, *b*, and *c* are real numbers and $a \neq 0$.

A quadratic equation written in the form $ax^2 + bx + c = 0$ is said to be in **standard form.**

Sometimes, a quadratic equation is called a **second-degree equation,** because the left side is a polynomial of degree 2. We shall discuss three ways of solving quadratic equations: by factoring, by completing the square, and by using the quadratic formula.

Factoring

1. We have already used factoring to solve equations. In particular, when a quadratic equation is written in standard form, $ax^2 + bx + c = 0$, it may be possible to factor the expression on the left side as the product of two first-degree polynomials. Then, by using the Zero-Product Property and setting each factor equal to 0, we can solve the resulting linear equations and obtain the solutions of the quadratic equation.

Let's look at an example.

EXAMPLE 1 Solving a Quadratic Equation by Factoring

Solve the equation: $x^2 - 5x + 6 = 0$

Solution The equation is in the standard form specified in equation (1). The left side may be factored as

$$x^2 - 5x + 6 = 0$$
$$(x - 2)(x - 3) = 0$$

We set each factor equal to 0 and solve the resulting first-degree equations.

$$x - 2 = 0 \quad \text{or} \quad x - 3 = 0$$
$$x = 2 \qquad\qquad x = 3$$

The solution set is $\{2, 3\}$.

E X A M P L E   2   Solving a Quadratic Equation by Factoring

Solve the equation: $x^2 = 12 - x$

Solution   We put the equation in standard form by adding $x - 12$ to each side:

$$x^2 = 12 - x$$
$$x^2 + x - 12 = 0$$

The left side may now be factored as

$$(x + 4)(x - 3) = 0$$

so that

$$x + 4 = 0 \quad \text{or} \quad x - 3 = 0$$
$$x = -4 \qquad\qquad x = 3$$

The solution set is $\{-4, 3\}$.   ▬

When the left side factors into two linear equations with the same solution, the quadratic equation is said to have a **repeated solution.** We also call this solution a **root of multiplicity 2,** or a **double root.**

E X A M P L E   3   Solving a Quadratic Equation by Factoring

Solve the equation: $x^2 - 6x + 9 = 0$

Solution   This equation is already in standard form, and the left side can be factored:

$$x^2 - 6x + 9 = 0$$
$$(x - 3)(x - 3) = 0$$

so

$$x = 3 \quad \text{or} \quad x = 3$$

This equation has only the repeated solution 3.   ▬

Now work Problems 3 and 13.

Completing the Square

We begin with a preliminary result. Suppose that we wish to solve the quadratic equation

$$x^2 = p \tag{2}$$

where $p \geq 0$ is a nonnegative number. We proceed as in the earlier examples:

$$x^2 - p = 0 \qquad \text{Put in standard form.}$$
$$(x - \sqrt{p})(x + \sqrt{p}) = 0 \qquad \text{Factor (over the real numbers).}$$
$$x = \sqrt{p} \quad \text{or} \quad x = -\sqrt{p} \qquad \text{Solve.}$$

Thus, we have the following result:

$$\text{If } x^2 = p \text{ and } p \geq 0, \text{ then } x = \sqrt{p} \text{ or } x = -\sqrt{p}. \tag{3}$$

Note that if $p > 0$ the equation $x^2 = p$ has two solutions, $x = \sqrt{p}$ and $x = -\sqrt{p}$. We usually abbreviate these solutions as $x = \pm\sqrt{p}$, read as "x equals plus or minus the square root of p." For example, the two solutions of the equation

$$x^2 = 4$$

are

$$x = \pm\sqrt{4}$$

and since $\sqrt{4} = 2$, we have

$$x = \pm 2$$

The solution set is $\{-2, 2\}$.

E X A M P L E   4   Solving Quadratic Equations Using Equation (3)

Solve each equation.

(a) $x^2 = 5$ (b) $(x - 2)^2 = 16$

Solution   (a) We use the result in equation (3) to get

$$x^2 = 5$$
$$x = \pm\sqrt{5}$$
$$x = \sqrt{5} \quad \text{or} \quad x = -\sqrt{5}$$

The solution set is $\{-\sqrt{5}, \sqrt{5}\}$.

(b) We use the result in equation (3) to get

$$(x - 2)^2 = 16$$
$$x - 2 = \pm\sqrt{16}$$
$$x - 2 = \sqrt{16} \quad \text{or} \quad x - 2 = -\sqrt{16}$$
$$x - 2 = 4 \qquad\qquad x - 2 = -4$$
$$x = 6 \qquad\qquad\quad x = -2$$

The solution set is $\{-2, 6\}$.   ▬

We now introduce the method of **completing the square.** The idea behind this method is to "adjust" the left side of a quadratic equation, $ax^2 + bx + c = 0$, so that it becomes a perfect square, that is, the square of a first-degree polynomial. For example, $x^2 + 6x + 9$ and $x^2 - 4x + 4$ are perfect squares because

$$x^2 + 6x + 9 = (x + 3)^2 \quad \text{and} \quad x^2 - 4x + 4 = (x - 2)^2$$

How do we "adjust" the left side? We do it by adding the appropriate number to the left side to create a perfect square. For example, to make $x^2 + 6x$ a perfect square, we add 9.

Let's look at several examples of completing the square when the coefficient of x^2 is 1:

Start	**Add**	**Result**
$x^2 + 4x$	4	$x^2 + 4x + 4 = (x + 2)^2$
$x^2 + 12x$	36	$x^2 + 12x + 36 = (x + 6)^2$
$x^2 - 6x$	9	$x^2 - 6x + 9 = (x - 3)^2$
$x^2 + x$	$\frac{1}{4}$	$x^2 + x + \frac{1}{4} = (x + \frac{1}{2})^2$

Do you see the pattern? Provided the coefficient of x^2 is 1, we complete the square by adding the square of $\frac{1}{2}$ the coefficient of x:

Start	**Add**	**Result**
$x^2 + mx$	$\left(\dfrac{m}{2}\right)^2$	$x^2 + mx + \left(\dfrac{m}{2}\right)^2 = \left(x + \dfrac{m}{2}\right)^2$

Now work Problem 67.

The next example illustrates how the procedure of completing the square can be used to solve a quadratic equation.

E X A M P L E 5   Solving Quadratic Equations by Completing the Square

Solve by completing the square: $x^2 + 5x + 4 = 0$

Solution   We always begin this procedure by rearranging the equation so that the constant is on the right side:

$$x^2 + 5x + 4 = 0$$
$$x^2 + 5x = -4$$

Since the coefficient of x^2 is 1, we can complete the square on the left side by adding $\left(\frac{1}{2} \cdot 5\right)^2 = \frac{25}{4}$. Of course, in an equation, whatever we add to the left side also must be added to the right side. Thus, we add $\frac{25}{4}$ to *both* sides:

$$x^2 + 5x + \frac{25}{4} = -4 + \frac{25}{4} \qquad \text{Add } \tfrac{25}{4} \text{ to each side}$$

$$\left(x + \frac{5}{2}\right)^2 = \frac{9}{4} \qquad \text{Complete the square}$$

$$x + \frac{5}{2} = \pm\sqrt{\frac{9}{4}} \qquad \text{Apply formula (3)}$$

$$x + \frac{5}{2} = \pm\frac{3}{2}$$

$$x = -\frac{5}{2} \pm \frac{3}{2}$$

$$x = -\frac{5}{2} + \frac{3}{2} = -1 \quad \text{or} \quad x = -\frac{5}{2} - \frac{3}{2} = -4$$

The solution set is $\{-4, -1\}$. ▬

The solution of the equation in Example 5 also can be obtained by factoring. Rework Example 5 using this technique.

The next example illustrates an equation that cannot be solved by factoring.

E X A M P L E 6   Solving Quadratic Equations by Completing the Square

Solve by completing the square: $2x^2 - 8x - 5 = 0$

Solution   First, we rewrite the equation:

$$2x^2 - 8x - 5 = 0$$
$$2x^2 - 8x = 5$$

Next, we divide by 2 so that the coefficient of x^2 is 1. (This enables us to complete the square at the next step.)

$$x^2 - 4x = \frac{5}{2}$$

Finally, we complete the square by adding 4 to each side:

$$x^2 - 4x + 4 = \frac{5}{2} + 4$$

$$(x - 2)^2 = \frac{13}{2}$$

$$x - 2 = \pm\sqrt{\frac{13}{2}}$$

$$x = 2 \pm \sqrt{\frac{13}{2}}$$

We choose to leave our answer in this compact form. Thus, the solution set is $\{2 - \sqrt{\frac{13}{2}}, 2 + \sqrt{\frac{13}{2}}\}$. ∎

Note: If we wanted an approximation, say to two decimal places, of these solutions, we would use a calculator to get $\{-0.55, 4.55\}$.

 Now work Problem 73.

The Quadratic Formula

 We can use the method of completing the square to obtain a general formula for solving the quadratic equation:

$$ax^2 + bx + c = 0 \qquad a \neq 0$$

As in Examples 5 and 6, we rearrange the terms as

$$ax^2 + bx = -c$$

Since $a \neq 0$, we can divide both sides by a to get

$$x^2 + \frac{b}{a}x = -\frac{c}{a}$$

Now the coefficient of x^2 is 1. To complete the square on the left side, add the square of $\frac{1}{2}$ the coefficient of x; that is, add

$$\left(\frac{1}{2}\cdot\frac{b}{a}\right)^2 = \frac{b^2}{4a^2}$$

to each side. Then

$$x^2 + \frac{b}{a}x + \frac{b^2}{4a^2} = \frac{b^2}{4a^2} - \frac{c}{a}$$

$$\left(x + \frac{b}{2a}\right)^2 = \frac{b^2 - 4ac}{4a^2} \tag{4}$$

Provided $b^2 - 4ac \geq 0$, we now can apply the result in equation (3) to get

$$x + \frac{b}{2a} = \pm\sqrt{\frac{b^2 - 4ac}{4a^2}}$$

$$x = -\frac{b}{2a} \pm \frac{\sqrt{b^2 - 4ac}}{2a} = \frac{-b \pm \sqrt{b^2 - 4ac}}{2a}$$

What if $b^2 - 4ac$ is negative? Then equation (4) states that the left expression (a real number squared) equals the right expression (a negative number). Since this occurrence is impossible for real numbers, we conclude that if $b^2 - 4ac < 0$ the quadratic equation has no *real* solution. (We discuss quadratic equations for which the quantity $b^2 - 4ac < 0$ in detail in Section 5.6.)*

We now state the *quadratic formula*.

Theorem

Consider the quadratic equation

$$ax^2 + bx + c = 0 \qquad a \neq 0$$

If $b^2 - 4ac < 0$, this equation has no real solution.
If $b^2 - 4ac \geq 0$, the real solution(s) of this equation is (are) given by the **quadratic formula:**

Quadratic Formula

$$x = \frac{-b \pm \sqrt{b^2 - 4ac}}{2a} \tag{5}$$

The quantity **$b^2 - 4ac$** is called the **discriminant** of the quadratic equation, because its value tells us whether the equation has real solutions. In fact, it also tells us how many solutions to expect.

Discriminant of a Quadratic Equation

For a quadratic equation $ax^2 + bx + c = 0$:

1. If $b^2 - 4ac > 0$, there are two unequal real solutions.
2. If $b^2 - 4ac = 0$, there is a repeated solution, a root of multiplicity 2.
3. If $b^2 - 4ac < 0$, there is no real solution.

Thus, when asked to find the real solutions, if any, of a quadratic equation, always evaluate the discriminant first to see how many real solutions there are.

EXAMPLE 7

Solving a Quadratic Equation Using the Quadratic Formula

Use the quadratic formula to find the real solutions, if any, of the equation:

$$3x^2 - 5x + 1 = 0$$

Solution   The equation is in standard form, so we compare it to $ax^2 + bx + c = 0$ to find a, b, and c:

$$3x^2 - 5x + 1 = 0$$
$$ax^2 + bx + c = 0$$

*Section 5.6 may be covered anytime after completing this section without any loss of continuity.

With $a = 3$, $b = -5$, and $c = 1$, we evaluate the discriminant $b^2 - 4ac$:

$$b^2 - 4ac = (-5)^2 - 4(3)(1) = 25 - 12 = 13$$

Since $b^2 - 4ac > 0$, there are two real solutions, which can be found using the quadratic formula:

$$x = \frac{-b \pm \sqrt{b^2 - 4ac}}{2a} = \frac{5 \pm \sqrt{13}}{6}$$

The solution set is $\left\{ \dfrac{5 - \sqrt{13}}{6}, \dfrac{5 + \sqrt{13}}{6} \right\}$.

E X A M P L E 8

Solving Quadratic Equations Using the Quadratic Formula

Use the quadratic formula to find the real solutions, if any, of the equation:

$$\tfrac{25}{2}x^2 - 30x + 18 = 0$$

Solution
The equation is given in standard form. However, to simplify the arithmetic, we clear the fractions:

$$\tfrac{25}{2}x^2 - 30x + 18 = 0$$
$$25x^2 - 60x + 36 = 0 \qquad \text{Clear fractions; multiply by 2.}$$
$$ax^2 + bx + c = 0 \qquad \text{Compare to standard form.}$$

With $a = 25$, $b = -60$, and $c = 36$, we evaluate the discriminant:

$$b^2 - 4ac = (-60)^2 - 4(25)(36) = 3600 - 3600 = 0$$

The equation has a repeated solution, which we find by using the quadratic formula:

$$x = \frac{-b \pm \sqrt{b^2 - 4ac}}{2a} = \frac{60 \pm \sqrt{0}}{50} = \frac{60}{50} = \frac{6}{5}$$

The repeated solution is $\frac{6}{5}$.

E X A M P L E 9

Solving Quadratic Equations Using the Quadratic Formula

Use the quadratic formula to find the real solutions, if any, of the equation:

$$3x^2 + 2 = 4x$$

Solution
The equation, as given, is not in standard form.

$$3x^2 + 2 = 4x$$
$$3x^2 - 4x + 2 = 0 \qquad \text{Put in standard form.}$$
$$ax^2 + bx + c = 0 \qquad \text{Compare to standard form.}$$

With $a = 3$, $b = -4$, and $c = 2$, we find

$$b^2 - 4ac = 16 - 24 = -8$$

Since $b^2 - 4ac < 0$, the equation has no real solution.

 Now work Problems 25 and 35.

Sometimes a given equation can be transformed into a quadratic equation so that it can be solved using the quadratic formula.

E X A M P L E   10   Solving Quadratic Equations Using the Quadratic Formula

Find the real solutions, if any, of the equation: $9 + \dfrac{3}{x} - \dfrac{2}{x^2} = 0, x \neq 0$

Solution   In its present form, the equation

$$9 + \frac{3}{x} - \frac{2}{x^2} = 0$$

is not a quadratic equation. However, it can be transformed into one by multiplying each side by x^2. The result is

$$9x^2 + 3x - 2 = 0$$

Although we multiplied each side by x^2, we know that $x^2 \neq 0$ (do you see why?), so this quadratic equation is equivalent to the original equation.

Using $a = 9$, $b = 3$, and $c = -2$, the discriminant is

$$b^2 - 4ac = 9 + 72 = 81$$

Since $b^2 - 4ac > 0$, the new equation has two real solutions:

$$x = \frac{-b \pm \sqrt{b^2 - 4ac}}{2a} = \frac{-3 \pm \sqrt{81}}{18} = \frac{-3 \pm 9}{18}$$

$$x = \frac{-3 + 9}{18} = \frac{6}{18} = \frac{1}{3} \quad \text{or} \quad x = \frac{-3 - 9}{18} = \frac{-12}{18} = \frac{-2}{3}$$

The solution set is $\{-\frac{2}{3}, \frac{1}{3}\}$.

SUMMARY

Procedure for Solving a Quadratic Equation
To solve a quadratic equation, first put it in standard form:

$$ax^2 + bx + c = 0$$

Then:

STEP 1: Identify a, b, and c.
STEP 2: Evaluate the discriminant, $b^2 - 4ac$.
STEP 3: (a) If the discriminant is negative, the equation has no real solution.
       (b) If the discriminant is nonnegative, determine whether the left side can be factored. If you can easily spot factors, use the factoring method to solve the equation. Otherwise, use the quadratic formula or the method of completing the square.

Applications

5 Many applied problems require the solution of a quadratic equation. Let's look at one that you will probably see again in a slightly different form if you study calculus.

E X A M P L E   11

Preview of a Calculus Problem

From each corner of a square piece of sheet metal, remove a square of side 9 centimeters. Turn up the edges to form an open box. If the box is to hold 144 cubic centimeters (cc), what should be the dimensions of the piece of sheet metal?

Solution   We use Figure 4 as a guide. We have labeled by x the length of a side of the square piece of sheet metal. The box will be of height 9 centimeters and its square base will have $x - 18$ as the length of a side. The volume (Length × Width × Height) of the box is therefore

$$9(x - 18)(x - 18) = 9(x - 18)^2$$

FIGURE 4

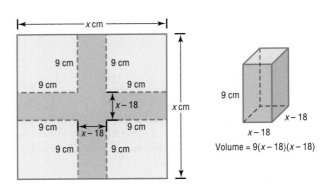

Since the volume of the box is to be 144 cc, we have

$$9(x - 18)^2 = 144$$
$$(x - 18)^2 = 16 \quad \text{Divide each side by 9}$$
$$x - 18 = \pm 4 \quad \text{Apply formula (3)}$$
$$x = 18 \pm 4$$
$$x = 22 \quad \text{or} \quad x = 14$$

We discard the solution $x = 14$ (do you see why?) and conclude that the sheet metal should be 22 centimeters by 22 centimeters.

Check: If we begin with a piece of sheet metal 22 centimeters by 22 centimeters, cut out a 9 centimeter square from each corner, and fold up the edges, we get a box whose dimensions are 9 by 4 by 4, with volume $9 \times 4 \times 4 = 144$ cc, as required.   ▬

E X A M P L E   12

Preview of a Calculus Problem

A piece of wire 8 feet in length is to be cut into two pieces. Each piece will then be bent into a square. Where should the cut in the wire be made if the sum of the areas of these squares is to be 2 square feet?

FIGURE 5

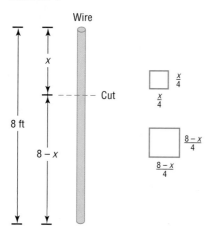

Solution   We use Figure 5 as a guide. We have labeled by x the length of one of the pieces of wire after it has been cut. The remaining piece will be of length $8 - x$. If each length is bent into a square, then one of the squares has a side of length $x/4$ and the other a side of length $(8 - x)/4$. Since the sum of the areas of these two squares is 2, we have the equation

$$\left(\frac{x}{4}\right)^2 + \left(\frac{8 - x}{4}\right)^2 = 2$$

$$\frac{x^2}{16} + \frac{64 - 16x + x^2}{16} = 2 \qquad \text{Remove parentheses}$$

$$2x^2 - 16x + 64 = 32 \qquad \text{Clear fractions and simplify}$$

$$2x^2 - 16x + 32 = 0 \qquad \text{Put in standard form.}$$

$$x^2 - 8x + 16 = 0 \qquad \text{Divide by 2.}$$

$$(x - 4)^2 = 0 \qquad \text{Factor.}$$

$$x = 4$$

Since $x = 4, 8 - x = 4$, and the original piece of wire should be cut into two pieces, each of length 4 feet.

Check: If the length of each piece of wire is 4 feet, then each piece can be formed into a square whose side is 1 foot. The area of each square is then 1 square foot, so the sum of the areas is 2 square feet, as required.   ▄

E X A M P L E   13   Physics: Uniform Motion

A motorboat heads upstream a distance of 24 miles on a river whose current is running at 3 miles per hour (mph). The trip up and back takes 6 hours. Assuming that the motorboat maintained a constant speed relative to the water, what was its speed?

FIGURE 6

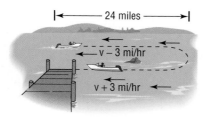

Solution   See Figure 6. We use v to represent the constant speed of the motorboat relative to the water. Then the true speed going upstream is $v - 3$ mph, and the true speed going downstream is $v + 3$ mph. Since Distance = Velocity × Time, then Time = Distance/Velocity. We set up a table.

	Velocity mi/hr	Distance mi	Time = Distance/Velocity hr
Upstream	$v - 3$	24	$\dfrac{24}{v - 3}$
Downstream	$v + 3$	24	$\dfrac{24}{v + 3}$

Since the total time up and back is 6 hours, we have

$$\frac{24}{v - 3} + \frac{24}{v + 3} = 6$$

$$\frac{24(v + 3) + 24(v - 3)}{(v - 3)(v + 3)} = 6 \qquad \text{Add the quotients on the left}$$

$$\frac{48v}{v^2 - 9} = 6 \qquad \text{Simplify}$$

$$48v = 6(v^2 - 9)$$
$$6v^2 - 48v - 54 = 0$$
$$v^2 - 8v - 9 = 0$$
$$(v - 9)(v + 1) = 0$$
$$v = 9 \quad \text{or} \quad v = -1$$

We discard the solution $v = -1$ mph, so the speed of the motorboat relative to the water is 9 mph. ∎

HISTORICAL FEATURE   Problems using quadratic equations are found in the oldest known mathematical literature. Babylonians and Egyptians were solving such problems before 1800 BC. Euclid solved quadratic equations geometrically in his *Data* (300 BC), and the Hindus and Arabs gave rules for solving any quadratic equation with real roots. Because negative numbers were not freely used before AD 1500, there were several different types of quadratic equations, each with its own rule. Thomas Harriot (1560–1621) introduced the method of factoring to obtain solutions, and François Viète (1540–1603) introduced a method that is essentially completing the square.

Until modern times it was usual to neglect the negative roots (if there were any), and equations involving square roots of negative quantities were regarded as unsolvable until the 1500s.

HISTORICAL PROBLEMS   1. *One of al-Khowârizmî's solutions*   We solve $x^2 + 12x = 85$ by drawing the square shown. The area of the four white rectangles and the yellow square is $x^2 + 12x$. We then set this expression equal to 85 to get the equation $x^2 + 12x = 85$. If we add the four blue squares, we will have a larger square of known area. Complete the solution.

2. *Viète's method*   We solve $x^2 + 12x - 85 = 0$ by letting $x = u + z$. Then

$$(u + z)^2 + 12(u + z) - 85 = 0$$
$$u^2 + (2z + 12)u + (z^2 + 12z - 85) = 0$$

Now select z so that $2z + 12 = 0$ and finish the solution.

3. *Another method to get the quadratic formula*   Look at equation (4) on page 113. Rewrite the right side as $(\sqrt{b^2 - 4ac}/2a)^2$ and then subtract it from each side. The right side is now 0 and the left side is a difference of two squares. If you factor this difference of two squares, you will easily be able to get the quadratic formula, and, moreover, the quadratic expression is factored, which is sometimes useful.

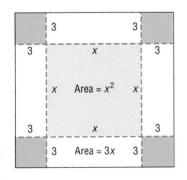

2.3 | EXERCISES

In Problems 1–20, solve each equation by factoring.

1. $x^2 = 9x$

2. $x^2 = -4x$

3. $x^2 - 25 = 0$

4. $x^2 - 9 = 0$

5. $z^2 + z - 6 = 0$

6. $v^2 + 7v + 6 = 0$

7. $2x^2 - 5x - 3 = 0$

8. $3x^2 + 5x + 2 = 0$

9. $3t^2 - 48 = 0$

10. $2y^2 - 50 = 0$

11. $x(x - 8) + 12 = 0$

12. $x(x + 4) = 12$

13. $4x^2 + 9 = 12x$

14. $25x^2 + 16 = 40x$

15. $6(p^2 - 1) = 5p$

16. $2(2u^2 - 4u) + 3 = 0$

17. $6x - 5 = \dfrac{6}{x}$

18. $x + \dfrac{12}{x} = 7$

19. $\dfrac{4(x - 2)}{x - 3} + \dfrac{3}{x} = \dfrac{-3}{x(x - 3)}$

20. $\dfrac{5}{x + 4} = 4 + \dfrac{3}{x - 2}$

In Problems 21–40, find the real solutions, if any, of each equation. Use the quadratic formula.

21. $x^2 - 4x + 2 = 0$

22. $x^2 + 4x + 2 = 0$

23. $x^2 - 4x - 1 = 0$

24. $x^2 + 6x + 1 = 0$

25. $2x^2 - 5x + 3 = 0$

26. $2x^2 + 5x + 3 = 0$

27. $4y^2 - y + 2 = 0$

28. $4t^2 + t + 1 = 0$

29. $4x^2 = 1 - 2x$

30. $2x^2 = 1 - 2x$

31. $4x^2 = 9x$

32. $5x = 4x^2$

33. $9t^2 - 6t + 1 = 0$

34. $4u^2 - 6u + 9 = 0$

35. $\dfrac{3}{4}x^2 - \dfrac{1}{4}x - \dfrac{1}{2} = 0$

36. $\dfrac{2}{3}x^2 - x - 3 = 0$

37. $4 - \dfrac{1}{x} - \dfrac{2}{x^2} = 0$

38. $4 + \dfrac{1}{x} - \dfrac{1}{x^2} = 0$

39. $3x = 1 - \dfrac{1}{x}$

40. $x = 1 - \dfrac{4}{x}$

In Problems 41–48, find the real solutions, if any, of each equation. Use the quadratic formula and a calculator. Express any solutions rounded to two decimal places.

41. $x^2 - 4.1x + 2.2 = 0$

42. $x^2 + 3.9x + 1.8 = 0$

43. $x^2 + \sqrt{3}x - 3 = 0$

44. $x^2 + \sqrt{2}x - 2 = 0$

45. $\pi x^2 - x - \pi = 0$

46. $\pi x^2 + \pi x - 2 = 0$

47. $3x^2 + 8\pi x + \sqrt{29} = 0$

48. $\pi x^2 - 15\sqrt{2}x + 20 = 0$

In Problems 49–60, find the real solutions, if any, of each quadratic equation. Use any method.

49. $x^2 - 5 = 0$

50. $x^2 - 6 = 0$

51. $16x^2 - 8x + 1 = 0$

52. $9x^2 - 6x + 1 = 0$

53. $10x^2 - 19x - 15 = 0$

54. $6x^2 + 7x - 20 = 0$

55. $2 + z = 6z^2$

56. $2 = y + 6y^2$

57. $x^2 + \sqrt{2}x = \dfrac{1}{2}$

58. $\dfrac{1}{2}x^2 = \sqrt{2}x + 1$

59. $x^2 + x = 4$

60. $x^2 + x = 1$

In Problems 61–66, use the discriminant to determine whether each quadratic equation has two unequal real solutions, a repeated real solution, or no real solution, without solving the equation.

61. $2x^2 - 6x + 7 = 0$

62. $x^2 + 4x + 7 = 0$

63. $9x^2 - 30x + 25 = 0$

64. $25x^2 - 20x + 4 = 0$

65. $3x^2 + 5x - 8 = 0$

66. $2x^2 - 3x - 7 = 0$

In Problems 67–72, what number should be added to complete the square of each expression?

67. $x^2 + 8x$

68. $x^2 - 4x$

69. $x^2 + \dfrac{1}{2}x$

70. $x^2 - \dfrac{1}{3}x$

71. $x^2 - \dfrac{2}{3}x$

72. $x^2 - \dfrac{2}{5}x$

In Problems 73–78, solve each equation by completing the square.

73. $x^2 + 4x - 21 = 0$

74. $x^2 - 6x = 13$

75. $x^2 - \dfrac{1}{2}x = \dfrac{3}{16}$

76. $x^2 + \dfrac{2}{3}x = \dfrac{1}{3}$

77. $3x^2 + x - \dfrac{1}{2} = 0$

78. $2x^2 - 3x = 1$

79. Dimensions of a Window The area of the opening of a rectangular window is to be 143 square feet. If the length is to be 2 feet more than the width, what are the dimensions?

80. Dimensions of a Window The area of a rectangular window is to be 306 square centimeters. If the length exceeds the width by 1 centimeter, what are the dimensions?

81. Geometry Find the dimensions of a rectangle whose perimeter is 26 meters and whose area is 40 square meters.

82. Watering a Field An adjustable water sprinkler that sprays water in a circular pattern is placed at the center of a square field whose area is 1250 square feet (see the figure). What is the shortest radius setting that can be used if the field is to be completely enclosed within the circle?

83. **Constructing a Box** An open box is to be constructed from a square piece of sheet metal by removing a square of side 1 foot from each corner and turning up the edges. If the box is to hold 4 cubic feet, what should be the dimensions of the sheet metal?

84. **Constructing a Box** Rework Problem 83 if the piece of sheet metal is a rectangle whose length is twice its width.

85. **Physics** A ball is thrown vertically upward from the top of a building 96 feet tall with an initial velocity of 80 feet per second. The distance s (in feet) of the ball from the ground after t seconds is $s = 96 + 80t - 16t^2$.
    (a) After how many seconds does the ball strike the ground?
    (b) After how many seconds will the ball pass the top of the building on its way down?

86. **Constructing a Coffee Can** A 39 ounce can of Hills Bros.® coffee requires 188.5 square inches of aluminum. If its height is 7 inches, what is its radius? (The surface area A of a right cylinder is $A = 2\pi r^2 + 2\pi rh$, where r is the radius and h is the height.)

87. **Working Together to Get a Job Done** Mike and Dan, working together, can paint the exterior of a house in 6 days. Mike, by himself, can complete this job in 5 days less than Dan. How long will it take Mike to complete the job by himself?

88. **Emptying a Tank** Two pumps of different sizes, working together, can empty a fuel tank in 5 hours. The larger pump can empty this tank in 4 hours less than the smaller one. If the larger one is out of order, how long will it take the smaller one to do the job alone?

89. **Business: Large Order Discounts** A company charges $200 for each box of tools on orders of 150 or fewer boxes. If a customer orders x boxes in excess of 150, the cost for each box ordered is reduced by x dollars. If a customer's bill came to $30,625, how many boxes were ordered?

90. **Dimensions of a Patio** A contractor orders 8 cubic yards of premixed cement, all of which is to be used to pour a patio that will be 4 inches thick. If the length of the patio is specified to be twice the width, what will be the patio dimensions? (1 cubic yard = 27 cubic feet)

91. **Constructing a Border around a Garden** A landscaper, who just completed a rectangular flower garden measuring 6 feet by 10 feet, orders 1 cubic yard of premixed cement, all of which is to be used to create a border of uniform width around the garden. If the border is to have a depth of 3 inches, how wide will the border be?

92. **Physics** An object is propelled vertically upward with an initial velocity of 20 meters per second. The distance s (in meters) of the object from the ground after t seconds is $s = -4.9t^2 + 20t$.
    (a) When will the object be 15 meters above the ground?
    (b) When will it strike the ground?
    (c) Will the object reach a height of 100 meters?
    (d) What is the maximum height?

93. **Reducing the Size of a Candy Bar** A jumbo chocolate bar with a rectangular shape measures 12 centimeters in length, 7 centimeters in width, and 3 centimeters in thickness. Due to escalating costs of cocoa, management decides to reduce the volume of the bar by 10%. To accomplish this reduction, management decides the new bar should have the same 3 centimeter thickness, but the length and width each should be reduced an equal number of centimeters. What should be the dimensions of the new candy bar?

94. **Reducing the Size of a Candy Bar** Rework Problem 93 if the reduction is to be 20%.

95. **Constructing a Border around a Pool** A pool in the shape of a circle measures 10 feet across. One cubic yard of concrete is to be used to create a circular border of uniform width around the pool. If the border is to have a depth of 3 inches, how wide will the border be? (1 cubic yard = 27 cubic feet). See the illustration on p. 122.

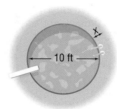

← 10 ft →

96. Constructing a Border around a Pool Rework Problem 95 if the depth of the border is 4 inches.

97. Physics: Uniform Motion A motorboat maintained a constant speed of 15 miles per hour relative to the water in going 10 miles upstream and then returning. The total time for the trip was 1.5 hours. Use this information to find the speed of the current.

98. Geometry The hypotenuse of a right triangle measures 13 centimeters. Find the lengths of the legs if their sum is 17 centimeters.

99. Geometry The diagonal of a rectangle measures 10 inches. If the length is 2 inches more than the width, find the dimensions of the rectangle.

100. Physics An object is thrown down from the top of a building 1280 feet tall with an initial velocity of 32 feet per second. The distance s (in feet) of the object from the ground after t seconds is $s = 1280 - 32t - 16t^2$.
(a) When will the object strike ground?
(b) What is the height of the object after 4 seconds?

101. Show that the sum of the roots of a quadratic equation is $-b/a$.

102. Show that the product of the roots of a quadratic equation is c/a.

103. Find k such that the equation $kx^2 + x + k = 0$ has a repeated real solution.

104. Find k such that the equation $x^2 - kx + 4 = 0$ has a repeated real solution.

105. Show that the real solutions of the equation $ax^2 + bx + c = 0$ are the negatives of the real solutions of the equation $ax^2 - bx + c = 0$. Assume that $b^2 - 4ac \geq 0$.

106. Show that the real solutions of the equation $ax^2 + bx + c = 0$ are the reciprocals of the real solutions of the equation $cx^2 + bx + a = 0$. Assume that $b^2 - 4ac \geq 0$.

107. The sum of the consecutive integers $1, 2, 3, \ldots, n$ is given by the formula $\frac{1}{2}n(n + 1)$. How many consecutive integers, starting with 1, must be added to get a sum of 666?

108. Geometry If a polygon of n sides has $\frac{1}{2}n(n - 3)$ diagonals, how many sides will a polygon with 65 diagonals have? Is there a polygon with 80 diagonals?

 109. Computing Average Speed In going from Chicago to Atlanta, a car averages 45 miles per hour, and in going from Atlanta to Miami, it averages 55 miles per hour. If Atlanta is halfway between Chicago and Miami, what is the average speed from Chicago to Miami? Discuss an intuitive solution. Write a paragraph defending your intuitive solution. Then solve the problem algebraically. Is your intuitive solution the same as the algebraic one? If not, find the flaw.

110. Speed of a Plane On a recent flight from Phoenix to Kansas City, a distance of 919 nautical miles, the plane arrived 20 minutes early. On leaving the aircraft, I asked the captain, "What was our tail wind?" He replied "I don't know, but our ground speed was 550 knots." How can you determine if enough information is provided to find the tail wind? If possible, find the tail wind. (1 knot = 1 nautical mile per hour)

111. Describe three ways you might solve a quadratic equation. State your preferred method; explain why you chose it.

112. Explain the benefits of evaluating the discriminant of a quadratic equation before attempting to solve it.

113. Make up three quadratic equations: one having two distinct solutions, one having no real solution, and one having exactly one real solution.

114. The word *quadratic* seems to imply four (*quad*), yet a quadratic equation is an equation that involves a polynomial of degree 2. Investigate the origin of the term *quadratic* as it is used in the expression *quadratic equation*. Write a brief essay on your findings.

115. Write a program that will solve a quadratic equation:
{Enter the coefficient of x squared}   READ (a);
{Enter the coefficient of x}   READ (b);
{Enter the constant term}   READ (c);
IF $b^2 - 4ac < 0$
THEN {write no real solution}
ELSE IF $b^2 - 4ac = 0$
    THEN {write $-b/2a$ is a double root}
    ELSE {write $(-b + \text{SQRT}(b^2 - 4ac)/2a$
        or $(-b - \text{SQRT}(b^2 - 4ac))/2a$
        is a solution};

2.4 | OTHER TYPES OF EQUATIONS

> **1** Solve an Equation Containing Radicals
> **2** Solve Equations Quadratic in Form

In this section we look at other types of equations, most of which can be solved using variations of the techniques already discussed.

Equations Containing Radicals

1 When the variable in an equation occurs in a square root, cube root, and so on, that is, when it occurs in a radical, the equation is called a **radical equation.** Sometimes, a suitable operation will change a radical equation to one that is linear or quadratic. A commonly used procedure is to isolate the most complicated radical on one side of the equation and then eliminate it by raising each side to a power equal to the index of the radical. Care must be taken, however, because apparent solutions that are not, in fact, solutions of the original equation may result. These are called **extraneous solutions.** Thus, we need to check all answers when working with radical equations.

E X A M P L E 1

Solving a Radical Equation

Find the real solutions of the equation: $\sqrt[3]{2x - 4} - 2 = 0$

Solution

The equation contains a radical whose index is 3. We isolate it on the left side:

$$\sqrt[3]{2x - 4} - 2 = 0$$
$$\sqrt[3]{2x - 4} = 2$$

Now raise each side to the third power (the index of the radical is 3) and solve:

$$(\sqrt[3]{2x - 4})^3 = 2^3$$
$$2x - 4 = 8$$
$$2x = 12$$
$$x = 6$$

Check: $\sqrt[3]{2(6) - 4} - 2 = \sqrt[3]{12 - 4} - 2 = \sqrt[3]{8} - 2 = 2 - 2 = 0$

The solution is $x = 6$.

Sometimes, we need to raise each side to a power more than once in order to solve a radical equation.

E X A M P L E 2

Solving a Radical Equation

Find the real solutions of the equation: $\sqrt{2x + 3} - \sqrt{x + 2} = 2$

Solution

First, we choose to isolate the more complicated radical expression (in this case, $\sqrt{2x + 3}$) on the left side:

$$\sqrt{2x + 3} = \sqrt{x + 2} + 2$$

Now square both sides (the index of the radical is 2):

$$(\sqrt{2x + 3})^2 = (\sqrt{x + 2} + 2)^2$$
$$2x + 3 = (\sqrt{x + 2})^2 + 4\sqrt{x + 2} + 4$$
$$2x + 3 = x + 2 + 4\sqrt{x + 2} + 4$$

Because the equation still contains a radical, we combine like terms, isolate the remaining radical on the right side, and again square both sides:

$$x - 3 = 4\sqrt{x + 2}$$
$$(x - 3)^2 = 16(x + 2) \qquad \text{Square each side}$$
$$x^2 - 6x + 9 = 16x + 32 \qquad \text{Remove parentheses}$$
$$x^2 - 22x - 23 = 0 \qquad \text{Put in standard form}$$
$$(x - 23)(x + 1) = 0 \qquad \text{Factor}$$
$$x = 23 \quad \text{or} \quad x = -1$$

The original equation appears to have the solution set $\{-1, 23\}$. However, we have not yet checked.

Check: $\sqrt{2(23) + 3} - \sqrt{23 + 2} = \sqrt{49} - \sqrt{25} = 7 - 5 = 2$
$$\sqrt{2(-1) + 3} - \sqrt{-1 + 2} = \sqrt{1} - \sqrt{1} = 1 - 1 = 0$$

Thus, the equation has only one real solution, 23; the solution -1 is extraneous.   ▬

Now work Problem 9.

Equations Quadratic in Form

2

The equation $x^4 + x^2 - 12 = 0$ is not quadratic in x, but it is quadratic in x^2. That is, if we let $u = x^2$, we get $u^2 + u - 12 = 0$, a quadratic equation. This equation can be solved for u and, in turn, by using $u = x^2$, we can find the solutions x of the original equation.

In general, if an appropriate substitution u transforms an equation into one of the form

$$au^2 + bu + c = 0 \qquad a \neq 0$$

then the original equation is called an **equation of the quadratic type** or an **equation quadratic in form.**

The difficulty of solving such an equation lies in the determination that the equation is, in fact, quadratic in form. After you are told an equation is quadratic in form, it is easy enough to see it, but some practice is needed to enable you to recognize them on your own.

E X A M P L E   3

Solving Equations That Are Quadratic in Form

Find the real solutions of the equation: $(x + 2)^2 + 11(x + 2) - 12 = 0$

Solution   For this equation, let $u = x + 2$. Then $u^2 = (x + 2)^2$, and the original equation,

$$(x + 2)^2 + 11(x + 2) - 12 = 0$$

becomes

$$u^2 + 11u - 12 = 0 \quad \text{\small } u = x + 2$$
$$(u + 12)(u - 1) = 0 \quad \text{\small Factor.}$$
$$u = -12 \quad \text{or} \quad u = 1 \quad \text{\small Solve.}$$

But we want to solve for x. Because $u = x + 2$, we have

$$x + 2 = -12 \quad \text{or} \quad x + 2 = 1$$
$$x = -14 \qquad\qquad x = -1$$

Check: $x = -14$: $(-14 + 2)^2 + 11(-14 + 2) - 12 = (-12)^2 + 11(-12) - 12 = 144 - 132 - 12 = 0$

$x = -1$: $(-1 + 2)^2 + 11(-1 + 2) - 12 = 1 + 11 - 12 = 0$

The original equation has the solution set $\{-14, -1\}$. ▬

E X A M P L E 4

Solving Equations That Are Quadratic in Form

Find the real solutions of the equation: $(x^2 - 1)^2 + (x^2 - 1) - 12 = 0$

Solution For the equation $(x^2 - 1)^2 + (x^2 - 1) - 12 = 0$, we let $u = x^2 - 1$ so that $u^2 = (x^2 - 1)^2$. Then the original equation

$$(x^2 - 1)^2 + (x^2 - 1) - 12 = 0$$

becomes

$$u^2 + u - 12 = 0 \quad \text{\small } u = x^2 - 1$$
$$(u + 4)(u - 3) = 0 \quad \text{\small Factor.}$$
$$u = -4 \quad \text{or} \quad u = 3 \quad \text{\small Solve.}$$

But remember that we want to solve for x. Because $u = x^2 - 1$, we have

$$x^2 - 1 = -4 \quad \text{or} \quad x^2 - 1 = 3$$
$$x^2 = -3 \qquad\qquad x^2 = 4$$

The first of these has no real solution; the second has the solution set $\{-2, 2\}$.

Check: $x = -2$: $(4 - 1)^2 + (4 - 1) - 12 = 9 + 3 - 12 = 0$

$x = 2$: $(4 - 1)^2 + (4 - 1) - 12 = 9 + 3 - 12 = 0$

Thus, $\{-2, 2\}$ is the solution set of the original equation. ▬

E X A M P L E 5

Solving Equations That Are Quadratic in Form

Find the real solutions of the equation: $x + 2\sqrt{x} - 3 = 0$

Solution For the equation $x + 2\sqrt{x} - 3 = 0$, let $u = \sqrt{x}$. Then $u^2 = x$, and the original equation,

$$x + 2\sqrt{x} - 3 = 0$$

becomes

$$u^2 + 2u - 3 = 0 \quad \text{\small } u = \sqrt{x}$$
$$(u + 3)(u - 1) = 0 \quad \text{\small Factor.}$$
$$u = -3 \quad \text{or} \quad u = 1 \quad \text{\small Solve.}$$

Since $u = \sqrt{x}$, we have $\sqrt{x} = -3$ or $\sqrt{x} = 1$. The first of these, $\sqrt{x} = -3$, has no real solution, since the square root of a real number is never negative. The second one, $\sqrt{x} = 1$, has the solution $x = 1$.

Check: $1 + 2\sqrt{1} - 3 = 1 + 2 - 3 = 0$

Thus, $x = 1$ is the only solution of the original equation. ∎

 Another method for solving Example 5 would be to treat it as a radical equation. Solve it this way for practice.

The idea should now be clear. If an equation contains an expression and that same expression squared, make a substitution for the expression. You may get a quadratic equation.

 Now work Problem 41.

2.4 | **EXERCISES**

In Problems 1–30, find the real solutions of each equation.

1. $\sqrt{2t - 1} = 1$

2. $\sqrt{3t + 4} = 2$

3. $\sqrt{3t + 1} = -6$

4. $\sqrt{5t + 4} = -2$

5. $\sqrt[3]{1 - 2x} - 3 = 0$

6. $\sqrt[3]{1 - 2x} - 1 = 0$

7. $x = 6\sqrt{x}$

8. $x = 4\sqrt{x}$

9. $\sqrt{15 - 2x} = x$

10. $\sqrt{12 - x} = x$

11. $x = 2\sqrt{x - 1}$

12. $x = 2\sqrt{-x - 1}$

13. $\sqrt{x^2 - x - 4} = x + 2$

14. $\sqrt{3 - x + x^2} = x - 2$

15. $3 + \sqrt{3x + 1} = x$

16. $2 + \sqrt{12 - 2x} = x$

17. $\sqrt{2x + 3} - \sqrt{x + 1} = 1$

18. $\sqrt{3x + 7} + \sqrt{x + 2} = 1$

19. $\sqrt{3x + 1} - \sqrt{x - 1} = 2$

20. $\sqrt{3x - 5} - \sqrt{x + 7} = 2$

21. $\sqrt{3 - 2\sqrt{x}} = \sqrt{x}$

22. $\sqrt{10 + 3\sqrt{x}} = \sqrt{x}$

23. $(3x + 1)^{1/2} = 4$

24. $(3x - 5)^{1/2} = 2$

25. $(5x - 2)^{1/3} = 2$

26. $(2x + 1)^{1/3} = -1$

27. $(x^2 + 9)^{1/2} = 5$

28. $(x^2 - 16)^{1/2} = 9$

29. $x^{3/2} - 2x^{1/2} = 0$

30. $x^{3/4} - 4x^{1/4} = 0$

In Problems 31–62, find the real solutions of each equation.

31. $x^4 - 5x^2 + 4 = 0$

32. $x^4 - 10x^2 + 25 = 0$

33. $3x^4 - 2x^2 - 1 = 0$

34. $2x^4 - 5x^2 - 12 = 0$

35. $x^6 + 7x^3 - 8 = 0$

36. $x^6 - 7x^3 - 8 = 0$

37. $(x + 2)^2 + 7(x + 2) + 12 = 0$

38. $(2x + 5)^2 - (2x + 5) - 6 = 0$

39. $(3x + 4)^2 - 6(3x + 4) + 9 = 0$

40. $(2 - x)^2 + (2 - x) - 20 = 0$

41. $2(s + 1)^2 - 5(s + 1) = 3$

42. $3(1 - y)^2 + 5(1 - y) + 2 = 0$

43. $x - 4\sqrt{x} = 0$

44. $x + 8\sqrt{x} = 0$

45. $x + \sqrt{x} = 20$

46. $x + \sqrt{x} = 6$

47. $t^{1/2} - 2t^{1/4} + 1 = 0$

48. $z^{1/2} - 4z^{1/4} + 4 = 0$

49. $4x^{1/2} - 9x^{1/4} + 4 = 0$

50. $x^{1/2} - 3x^{1/4} + 2 = 0$

51. $\sqrt[4]{5x^2 - 6} = x$

52. $\sqrt[4]{4 - 5x^2} = x$

53. $x^2 + 3x + \sqrt{x^2 + 3x} = 6$

54. $x^2 - 3x - \sqrt{x^2 - 3x} = 2$

55. $\dfrac{1}{(x + 1)^2} = \dfrac{1}{x + 1} + 2$

56. $\dfrac{1}{(x - 1)^2} + \dfrac{1}{x - 1} = 12$

57. $3x^{-2} - 7x^{-1} - 6 = 0$

58. $2x^{-2} - 3x^{-1} - 4 = 0$

59. $2x^{2/3} - 5x^{1/3} - 3 = 0$

60. $3x^{4/3} + 5x^{2/3} - 2 = 0$

61. $\left(\dfrac{v}{v + 1}\right)^2 + \dfrac{2v}{v + 1} = 8$

62. $\left(\dfrac{y}{y - 1}\right)^2 = 6\left(\dfrac{y}{y - 1}\right) + 7$

In Problems 63–68, find the real solutions of each equation. Use a calculator to express solutions correct to two decimal places.

63. $x - 4x^{1/2} + 2 = 0$

64. $x^{2/3} + 4x^{1/3} + 2 = 0$

65. $x^4 + \sqrt{3}x^2 - 3 = 0$

66. $x^4 + \sqrt{2}x^2 - 2 = 0$

67. $\pi(1 + t)^2 = \pi + 1 + t$

68. $\pi(1 + r)^2 = 2 + \pi(1 + r)$

69. If $k = \dfrac{x + 3}{x - 3}$ and $k^2 - k = 12$, find x.

70. If $k = \dfrac{x + 3}{x - 4}$ and $k^2 - 3k = 28$, find x.

71. **Physics: Using Sound to Measure Distance** The depth of a well can sometimes be found by dropping an object into the well and measuring the time elapsed until a sound is heard. If t_1 is the time (measured in seconds) it takes for the object to strike the water in the well, then t_1 will obey the equation $s = 16t_1^2$, where s is the distance (measured in feet). It follows that $t_1 = \sqrt{s}/4$. Suppose that t_2 is the time it takes for the sound of the impact to reach your ears. Because sound waves are known to travel at a speed of approximately 1100 feet per second, the time t_2 to travel the distance s will be $t_2 = s/1100$. Now $t_1 + t_2$ is the total time that elapses from the moment the object is dropped to the moment a sound is heard. Thus, we have the equation

$$\text{Total time elapsed} = \frac{\sqrt{s}}{4} + \frac{s}{1100}$$

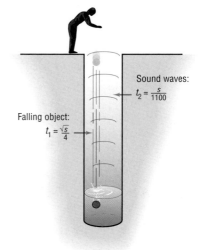

Sound waves:
$t_2 = \frac{s}{1100}$

Falling object:
$t_1 = \frac{\sqrt{s}}{4}$

Find the depth of a well if the total time elapsed from dropping a rock to hearing it hit the water is 4 seconds. Express the answer correct to the nearest integer.

72. Make up a radical equation that has no solution.

73. Make up a radical equation that has an extraneous solution.

74. Discuss what there is in the solving process for radical equations that leads to the possibility of extraneous solutions. Why is there no such possibility for linear and quadratic equations?

2.5 INEQUALITIES

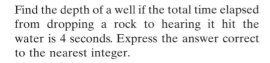

1 Use Properties of Inequalities
2 Graph Inequalities
3 Use Interval Notation

Properties of Inequalities

1 In working with inequalities, we will need to know certain properties that they obey.

We begin with the **trichotomy property,** which states that either two numbers are equal or one of them is less than the other.

For any pair of numbers a and b,

Trichotomy Property

$$a < b \quad \text{or} \quad a = b \quad \text{or} \quad b < a$$

If $b = 0$, the trichotomy property states that, for any real number a,

$$a < 0 \quad \text{or} \quad a = 0 \quad \text{or} \quad a > 0$$

That is, any real number is negative or 0 or positive, a fact we have already noted.

The product of two positive real numbers is positive, the product of two negative real numbers is positive, and the product of 0 and 0 is 0. Thus, for any real number a, the value of a^2 is 0 or positive; that is, a^2 is nonnegative. This is called the **nonnegative property.**

For any real number a, we have

Nonnegative Property

$$a^2 \geq 0$$

FIGURE 7

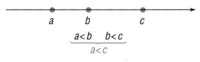

In Figure 7, we can see that, if a lies to the left of b and b lies to the left of c, it follows that a must also lie to the left of c. This is called the **transitive property** for the inequality $<$. There is also a corresponding property for the inequality $>$.

Transitive Property of Inequalities

If $a < b$ and $b < c$, then $a < c$.	(1a)
If $a > b$ and $b > c$, then $a > c$.	(1b)

 Draw an illustration similar to Figure 7 that depicts the transitive property (1b) for the inequality $>$.

If we add the same number to both sides of an inequality, we obtain an equivalent inequality. For example, since $3 < 5$, then $3 + 4 < 5 + 4$ or $7 < 9$. This is called the **addition property** of inequalities.

MISSION POSSIBLE

"Saving the Economic Future of Krispy Krunchy Candy Co."

The Krispy Krunchy Candy Company is facing a financial crisis because the cost of shipping cocoa beans from Ghana has increased. The CEO has decided that the way to stay afloat would be to reduce the size of their GIANT KRISPY KRUNCHY BAR by 10% but keep the price the same. He doesn't want to lose any customers, however. Therefore, he wants the change in size to be as unobtrusive as possible. The present dimensions of the GIANT KRISPY KRUNCHY BAR are 12 centimeters (cm) in length, 7 cm in width, and 3 cm in thickness. The CEO has asked your consulting firm to come up with the best way to shrink the candy bar. Because millions of dollars are riding on this decision, you will need to find all answers in centimeters to 3 decimal places and all percents to 5 significant digits.

1.   Make a sketch of the candy bar, roughly to scale, and label it.
2.   What is the present volume of the candy bar?
3.   What would be the new volume after a 10% reduction?
4.   What would be the new volume if each dimension were reduced by 10%. Is this the same as your answer to question 3? (It shouldn't be.) Explain the difference.
5.   Consider reducing two of the dimensions by the same number of centimeters, while holding the third dimension fixed. (There are three possibilities; in each case, the new volume is to be 90% of the original volume.)
6.   Consider reducing two (but not three) of the dimensions by the same percent. What would the new dimensions be in each case? (There are three possibilities; in each case you want the new volume to be 90% of the original volume.)
7.   Consider reducing all three of the dimensions by the same percent. What would the new dimensions be?
8.   Within your group, make a decision about which of the seven possibilities that you found would be the best one to recommend to the CEO of Krispy Krunchy. Write out two or three sentences to justify your choice.
9.   Would you mind if you discovered that your favorite candy bar had been reduced in size while the price stayed the same? Do you think a candy company might actually do this to improve their financial standing? What is your protection as a consumer from being fooled?

Addition Property of Inequalities

If $a < b$, then $a + c < b + c$.   (2a)

If $a > b$, then $a + c > b + c$.   (2b)

The addition property states that the sense, or direction, of an inequality remains unchanged if the same number is added to each side. Figure 8 illustrates the addition property (2a). In Figure 8(a), we see that a lies to the left of b. If c is positive, then $a + c$ and $b + c$ each lie c units to the right of a and b, respectively. Consequently, $a + c$ must lie to the left of $b + c$; that is, $a + c < b + c$. Figure 8(b) illustrates the situation if c is negative.

FIGURE 8

c units

c units

a b $a+c$ $b+c$

(a) If $a < b$ and $c > 0$,
then $a + c < b + c$.

$-c$ units

$-c$ units

$a+c$ $b+c$ a b

(b) If $a < b$ and $c < 0$,
then $a + c < b + c$.

 Draw an illustration similar to Figure 8 that illustrates the addition property (2b).

E X A M P L E   1

Addition Property of Inequalities

(a) If $x < -5$, then $x + 5 < -5 + 5$ or $x + 5 < 0$.
(b) If $x > 2$, then $x + (-2) > 2 + (-2)$ or $x - 2 > 0$.

 Now work Problem 9.

We will use two examples to arrive at our next property.

E X A M P L E   2

Multiplying an Inequality by a Positive Number

Express as an inequality the result of multiplying each side of the inequality $3 < 7$ by 2.

Solution   We begin with

$$3 < 7$$

Multiplying each side by 2 yields the numbers 6 and 14, so we have

$$6 < 14$$

E X A M P L E   3

Multiplying an Inequality by Negative Number

Express as an inequality the result of multiplying each side of the inequality $9 > 2$ by -4.

Solution   We begin with

$$9 > 2$$

Multiplying each side by -4 yields the numbers -36 and -8, so we have

$$-36 < -8$$

    Note that the effect of multiplying both sides of $9 > 2$ by the negative number -4 is that the direction of the inequality symbol is reversed.
    Examples 2 and 3 illustrate the following general **multiplication properties** for inequalities:

Multiplication Properties for Inequalities

> If $a < b$ and if $c > 0$, then $ac < bc$.
> If $a < b$ and if $c < 0$, then $ac > bc$.   (3a)
>
> If $a > b$ and if $c > 0$, then $ac > bc$.
> If $a > b$ and if $c < 0$, then $ac < bc$.   (3b)

    The multiplication properties state that the sense, or direction, of an inequality *remains the same* if each side is multiplied by a *positive* real number, whereas the direction is *reversed* if each side is multiplied by a *negative* real number.

E X A M P L E   4   Multiplication Property of Inequalities

(a) If $2x < 6$, then $\frac{1}{2}(2x) < \frac{1}{2}(6)$ or $x < 3$.

(b) If $\dfrac{x}{-3} > 12$, then $-3\left(\dfrac{x}{-3}\right) < -3(12)$ or $x < -36$.

(c) If $-4x > -8$, then $\dfrac{-4x}{-4} < \dfrac{-8}{-4}$ or $x < 2$.

(d) If $-x < 8$, then $(-1)(-x) > (-1)(8)$ or $x > -8$.

Now work Problem 15.

    The **reciprocal property** states that the reciprocal of a positive real number is positive and that the reciprocal of a negative real number is negative.

Reciprocal Property for Inequalities

> If $a > 0$, then $\dfrac{1}{a} > 0$.   (4a)
>
> If $a < 0$, then $\dfrac{1}{a} < 0$.   (4b)

Graphing Inequalities

2 We shall find it useful in later work to graph inequalities on the real number line.

EXAMPLE 5 Graphing Inequalities

(a) On the real number line, graph all numbers x for which $x > 4$.
(b) On the real number line, graph all numbers x for which $x \leq 5$.

Solution (a) See Figure 9. Notice that we use a left parenthesis to indicate that the number 4 is *not* part of the graph.

FIGURE 9

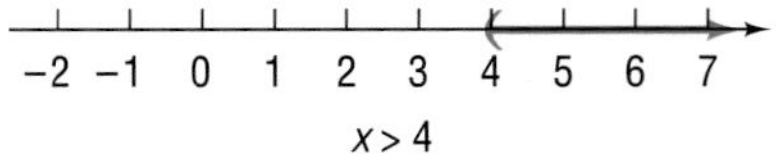

−2 −1 0 1 2 3 4 5 6 7
x > 4

FIGURE 10

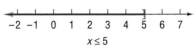

−2 −1 0 1 2 3 4 5 6 7
x ≤ 5

(b) See Figure 10. Notice that we use a right bracket to indicate that the number 5 *is* part of the graph.

Inequalities are often combined.

EXAMPLE 6 Graphing Combined Inequalities

On the real number line, graph all numbers x for which $x > 4$ and $x < 6$.

Solution We first graph each inequality separately, as illustrated in Figure 11(a). Then it is easy to see that the numbers that belong to *both* the graph of $x > 4$ *and* the graph of $x < 6$ are as shown in Figure 11(b). For example, 5 is part of the graph because $5 > 4$ and $5 < 6$; 7 is not on the graph because 7 is not less than 6.

FIGURE 11

−2 −1 0 1 2 3 4 5 6 7
x > 4
−2 −1 0 1 2 3 4 5 6 7
x < 6
(a)
−2 −1 0 1 2 3 4 5 6 7
x > 4 and x < 6
(b)

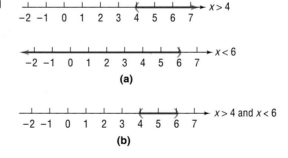

EXAMPLE 7 Graphing Combined Inequalities

On the real number line, graph all numbers x for which $x > 4$ or $x \leq -1$.

Solution See Figure 12(a), where each inequality is graphed separately. The numbers that belong to *either* the graph of $x > 4$ *or* the graph of $x \leq -1$ are graphed in Figure 12(b). For example, 5 is part of the graph because $5 > 4$; 2 is not on the graph because 2 is not greater than 4 nor is 2 less than or equal to −1.

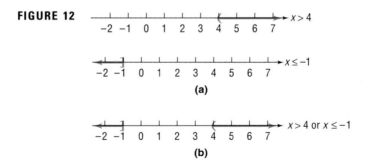

FIGURE 12

(a)

(b)

EXAMPLE 8

Graphing Combined Inequalities

On the real number line, graph all numbers x for which $x > 4$ and $x > 6$.

Solution   Figure 13(a) shows the graphs of the separate inequalities, $x > 4$, $x > 6$. Figure 13(b) shows the numbers for which *both* $x > 4$ *and* $x > 6$.

FIGURE 13

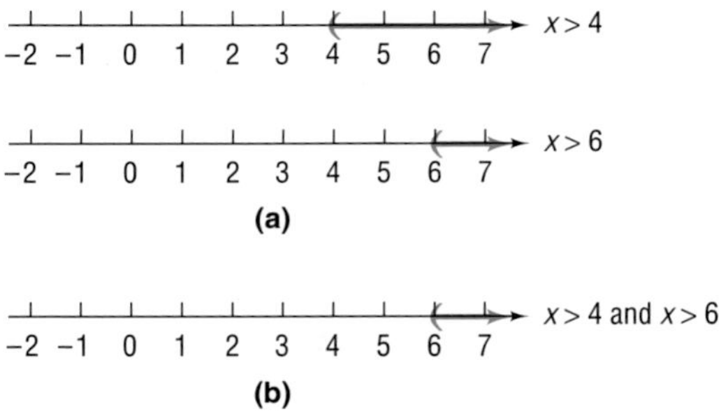

(a)

(b)

EXAMPLE 9

Graphing Combined Inequalities

On the real number line, graph all numbers x for which $x \geq 4$ or $x \geq 1$.

Solution   Figure 14(a) shows the graph of the separate inequalities, $x \geq 4$, $x \geq 1$. Figure 14(b) shows the numbers for which *either* $x \geq 4$ *or* $x \geq 1$.

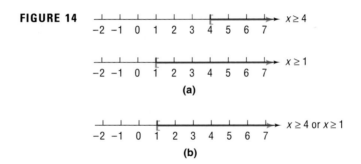

FIGURE 14

(a)

(b)

E X A M P L E   10   Graphing Combined Inequalities

On the real number line, graph all numbers x for which $x \le 4$ and $x \ge 6$.

Solution   Figure 15(a) shows the graphs of the separate inequalities, $x \le 4$, $x \ge 6$. Figure 15(b) illustrates that there are no numbers for which *both* $x \le 4$ *and* $x \ge 6$.

FIGURE 15

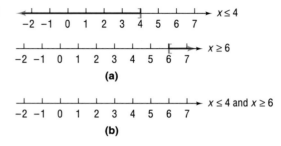

(a)

(b)

 Now work Problem 19.

Suppose that a and b are two real numbers and $a < b$. We shall use the notation $a < x < b$ to mean that x is a number *between* a and b. Thus, the expression $a < x < b$ is equivalent to the two inequalities $a < x$ and $x < b$. Similarly, the expression $a \le x \le b$ is equivalent to the two inequalities $a \le x$ and $x \le b$. The remaining two possibilities, $a \le x < b$ and $a < x \le b$, are defined analogously.

Although it is acceptable to write $3 \ge x \ge 2$, it is preferable to reverse the inequality symbols and write instead $2 \le x \le 3$ so that, as you read from left to right, the values go from smaller to larger.

A statement such as $2 \le x \le 1$ is false because there is no number x for which $2 \le x$ and $x \le 1$. Finally, we never mix inequality symbols, as in $2 \le x \ge 3$.

Intervals

 Let a and b represent two real numbers with $a < b$:

A **closed interval**, denoted by **[a, b]**, consists of all real numbers x for which $a \le x \le b$.

An **open interval**, denoted by **(a, b)**, consists of all real numbers x for which $a < x < b$.

The **half-open**, or **half-closed**, **intervals** are **(a, b]**, consisting of all real numbers x for which $a < x \le b$, and **[a, b)**, consisting of all real numbers x for which $a \le x < b$.

In each of these definitions, a is called the **left endpoint** and b the **right endpoint** of the interval. Figure 16 illustrates each type of interval.

The symbol ∞ (read as "infinity") is not a real number but a notational device used to indicate unboundedness in the positive direction. The symbol $-\infty$ (read as "minus infinity") also is not a real number, but a notational device used to indicate unboundedness in the negative

FIGURE 16

(a) Closed interval (b) Open interval (c) Half-open (half-closed) intervals

direction. Using the symbols ∞ and $-\infty$, we can define five other kinds of intervals:

$[a, \infty)$ consists of all real numbers x for which $x \geq a$ ($a \leq x < \infty$)
(a, ∞) consists of all real numbers x for which $x > a$ ($a < x < \infty$)
$(-\infty, a]$ consists of all real numbers x for which $x \leq a$ ($-\infty < x \leq a$)
$(-\infty, a)$ consists of all real numbers x for which $x < a$ ($-\infty < x < a$)
$(-\infty, \infty)$ consists of all real numbers x ($-\infty < x < \infty$)

Figure 17 illustrates these types of intervals.

FIGURE 17

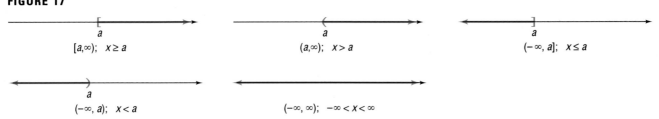

Note that ∞ (and $-\infty$) is never included as an endpoint, since it is not a real number.

E X A M P L E   11

Writing Inequalities Using Interval Notation

Write each inequality using interval notation.

(a) $1 \leq x \leq 3$ (b) $-4 < x < 0$ (c) $x > 5$ (d) $x \leq 1$

Solution

(a) $1 \leq x \leq 3$ describes all numbers x between 1 and 3, inclusive. In interval notation, we write $[1, 3]$.
(b) In interval notation, $-4 < x < 0$ is written $(-4, 0)$.
(c) $x > 5$ consists of all numbers x greater than 5. In interval notation, we write $(5, \infty)$.
(d) In interval notation, $x \leq 1$ is written $(-\infty, 1]$. ▬

E X A M P L E   12

Writing Intervals Using Inequality Notation

Write each interval as an inequality involving x.

(a) $[1, 4)$ (b) $(2, \infty)$ (c) $[2, 3]$ (d) $(-\infty, -3]$

Solution

(a) $[1, 4)$ consists of all numbers x for which $1 \leq x < 4$.
(b) $(2, \infty)$ consists of all numbers x for which $x > 2$ ($2 < x < \infty$).
(c) $[2, 3]$ consists of all numbers x for which $2 \leq x \leq 3$.
(d) $(-\infty, -3]$ consists of all numbers x for which $x \leq -3$ ($-\infty < x \leq -3$). ▬

2.5 | EXERCISES

In Problems 1–8, express the graph shown in color using interval notation. Also express each as an inequality involving x.

1.

2.

3.

4.

5.

6.

7.

8.

In Problems 9–16, fill in the blank with the correct inequality symbol.

9. If $x < 5$, then $x - 5$ _____ 0.

10. If $x < -4$, then $x + 4$ _____ 0.

11. If $x > -4$, then $x + 4$ _____ 0.

12. If $x > 6$, then $x - 6$ _____ 0.

13. If $x > -4$, then $3x$ _____ -12.

14. If $x < 3$, then $2x$ _____ 6.

15. If $x < 6$, then $-2x$ _____ -12.

16. If $x > -2$, then $-4x$ _____ 8.

In Problems 17–26, graph the numbers x, if any, on the real number line.

17. $x \geq -2$

18. $x < 4$

19. $x \geq 4$ and $x < 6$

20. $x > 3$ and $x \leq 7$

21. $x \leq 0$ or $x < 6$

22. $x > 0$ or $x \geq 5$

23. $x \leq -2$ and $x > 1$

24. $x \geq 4$ and $x < -2$

25. $x \leq -2$ or $x > 1$

26. $x \geq 4$ or $x < -2$

In Problems 27–34, write each inequality using interval notation, and illustrate each inequality using the real number line.

27. $0 \leq x \leq 4$

28. $-1 < x < 5$

29. $4 \leq x < 6$

30. $-2 < x < 0$

31. $x \geq 4$

32. $x \leq 5$

33. $x < -4$

34. $x > 1$

In Problems 35–42, write each interval as an inequality involving x, and illustrate each inequality using the real number line.

35. $[2, 5]$

36. $(1, 2)$

37. $(-3, -2)$

38. $[0, 1)$

39. $[4, \infty)$

40. $(-\infty, 2]$

41. $(-\infty, -3)$

42. $(-8, \infty)$

43. If $a \leq b$ and $c > 0$, show that $ac \leq bc$.
[**Hint:** Since $a \leq b$, it follows that $a - b \leq 0$. Now multiply each side by c.]

44. If $a \leq b$ and $c < 0$, show that $ac \geq bc$.

45. **Arithmetic Mean** If $a < b$, show that $a < (a + b)/2 < b$. The number $(a + b)/2$ is called the **arithmetic mean** of a and b.

46. Refer to Problem 45. Show that the arithmetic mean of a and b is equidistant from a and b.

47. **Geometric Mean** If $0 < a < b$, show that $a < \sqrt{ab} < b$. The number $\sqrt{ab}$ is called the **geometric mean** of a and b.

48. Refer to Problems 45 and 47. Show that the geometric mean of a and b is less than the arithmetic mean of a and b.

49. **Harmonic Mean** For $0 < a < b$, let h be defined by

$$\frac{1}{h} = \frac{1}{2}\left(\frac{1}{a} + \frac{1}{b}\right)$$

Show that $a < h < b$. The number h is called the **harmonic mean** of a and b.

50. Refer to Problems 45, 47, and 49. Show that the harmonic mean of a and b equals the geometric mean squared divided by the arithmetic mean.

51. A young adult may be defined as someone older than 21, but less than 30 years of age. Express this statement using inequalities.

52. Middle-aged may be defined as being 40 or more and less than 60. Express this statement using inequalities.

53. Life Expectancy Metropolitan Life Insurance Co. reported that an average 25-year-old male in 1996 could expect to live at least 48.4 more years, and an average 25-year-old female in 1996 could expect to live at least 54.7 more years.
(a) To what age can an average 25-year-old male expect to live? Express your answer as an inequality.

(b) To what age can an average 25-year-old female expect to live? Express your answer as an inequality.
(c) Who can expect to live longer, a male or a female? By how many years?

 54. How would you explain to a fellow student the underlying reason for the multiplication property for inequalities (page 131); that is, the sense or direction of an inequality remains the same if each side is multiplied by a positive real number, whereas the direction is reversed if each side is multiplied by a negative real number?

2.6 | LINEAR INEQUALITIES

> **1** Solve Linear Inequalities
> **2** Solve Combined Inequalities

An **inequality in one variable** is a statement involving two expressions, at least one containing the variable, separated by one of the inequality symbols, $<$, $\leq$, $>$, or $\geq$. To **solve an inequality** means to find all values of the variable for which the statement is true. These values are called **solutions** of the inequality.

For example, the following are all inequalities involving one variable, x:

$$x + 5 < 8 \qquad 2x - 3 \geq 4 \qquad x^2 - 1 \leq 3 \qquad \frac{x + 1}{x - 2} > 0$$

Two inequalities having exactly the same solution set are called **equivalent inequalities.**

As with equations, one method for solving an inequality is to replace it by a series of equivalent inequalities, until an inequality with an obvious solution, such as $x < 3$, is obtained. We obtain equivalent inequalities by applying some of the same operations as those used to find equivalent equations. The addition property and the multiplication properties form the basis for the following procedures.

Procedures That Leave the Inequality Symbol Unchanged	1. Simplify both sides of the inequality by combining like terms and eliminating parentheses: $\quad$ Replace $\qquad (x + 2) + 6 > 2x + (x + 1)$ $\quad$ by $\qquad\qquad\qquad x + 8 > 3x + 1$

2. Add or subtract the same expression on both sides of the inequality:

$$\text{Replace} \qquad 3x - 5 < 4$$
$$\text{by} \quad (3x - 5) + 5 < 4 + 5$$

3. Multiply or divide both sides of the inequality by the same *positive* expression:

$$\text{Replace} \quad 4x > 16 \quad \text{by} \quad \frac{4x}{4} > \frac{16}{4}$$

Procedures That Reverse the Sense or Direction of the Inequality Symbol

1. Interchange the two sides of the inequality

$$\text{Replace} \qquad 3 < x \quad \text{by} \quad x > 3$$

2. Multiply or divide both sides of the inequality by the same *negative* expression:

$$\text{Replace} \qquad -2x > 6 \quad \text{by} \quad \frac{-2x}{-2} < \frac{6}{-2}$$

1 A **linear inequality in one variable** is an inequality equivalent to one of the forms

$$ax + b < 0 \qquad ax + b > 0$$
$$ax + b \leq 0 \qquad ax + b \geq 0$$

where a and b are real numbers and $a \neq 0$.

The remainder of this section deals with solving linear inequalities. In the next section, we discuss the solution of other types of inequalities. As the examples that follow illustrate, we solve linear inequalities using many of the same steps that we would use to solve a linear equation. In writing the solution of an inequality, we may use either set notation or interval notation, whichever is more convenient.

E X A M P L E   1   Solving Linear Inequalities

Solve the inequality $3 - 2x < 5$, and draw a graph to illustrate the solution.

Solution

$$3 - 2x < 5$$
$$3 - 2x - 3 < 5 - 3 \qquad \text{Subtract 3 from both sides.}$$
$$-2x < 2 \qquad \text{Simplify.}$$
$$\frac{-2x}{-2} > \frac{2}{-2} \qquad \begin{array}{l}\text{Divide both sides by } -2. \text{ (The sense of} \\ \text{the inequality symbol is reversed.)}\end{array}$$
$$x > -1 \qquad \text{Simplify.}$$

FIGURE 18
$x > -1$ or $(-1, \infty)$

The solution set is $\{x | x > -1\}$ or, using interval notation, all numbers in the interval $(-1, \infty)$. See Figure 18 for the graph. ∎

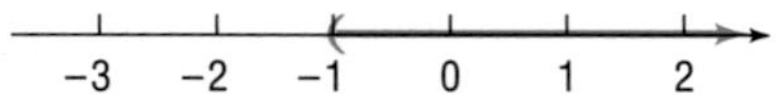

E X A M P L E   2   Solving Linear Inequalities

Solve the inequality $4x + 7 \geq 2x - 3$, and draw a graph to illustrate the solution.

Solution

$$4x + 7 \geq 2x - 3$$

$$4x + 7 - 7 \geq 2x - 3 - 7 \qquad \text{Subtract 7 from both sides.}$$

$$4x \geq 2x - 10 \qquad \text{Simplify.}$$

$$4x - 2x \geq 2x - 10 - 2x \qquad \text{Subtract } 2x \text{ from both sides.}$$

$$2x \geq -10 \qquad \text{Simplify.}$$

$$\frac{2x}{2} \geq \frac{-10}{2} \qquad \text{Divide both sides by 2. (The sense of the inequality symbol is unchanged.)}$$

$$x \geq -5 \qquad \text{Simplify.}$$

FIGURE 19
$x \geq -5$ or $[-5, \infty)$

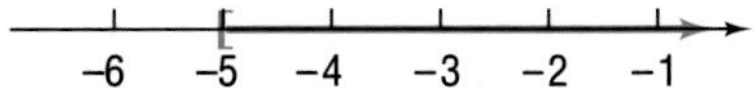

The solution set is $\{x \mid x \geq -5\}$ or, using interval notation, all numbers in the interval $[-5, \infty)$.

See Figure 19 for the graph. ▬

Now work Problem 11.

E X A M P L E   3   Solving Combined Inequalities

Solve the inequality $-5 < 3x - 2 < 1$, and draw a graph to illustrate the solution.

Solution   Recall that the inequality

$$-5 < 3x - 2 < 1$$

is equivalent to the two inequalities

$$-5 < 3x - 2 \quad \text{and} \quad 3x - 2 < 1$$

We will solve each of these inequalities separately. For the first inequality,

$$-5 < 3x - 2$$

$$-5 + 2 < 3x - 2 + 2 \qquad \text{Add 2 to both sides.}$$

$$-3 < 3x \qquad \text{Simplify.}$$

$$\frac{-3}{3} < \frac{3x}{3} \qquad \text{Divide both sides by 3.}$$

$$-1 < x \qquad \text{Simplify.}$$

The second inequality is solved as follows:

$$3x - 2 < 1$$

$$3x - 2 + 2 < 1 + 2 \qquad \text{Add 2 to both sides.}$$

$$3x < 3 \qquad \text{Simplify.}$$

$$\frac{3x}{3} < \frac{3}{3} \qquad \text{Divide both sides by 3.}$$

$$x < 1 \qquad \text{Simplify.}$$

The solution set of the original pair of inequalities consists of all x for which

$$-1 < x \quad \text{and} \quad x < 1$$

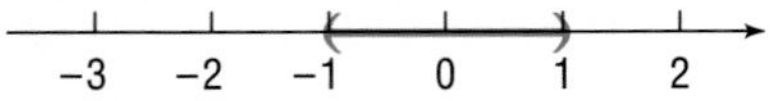

This may be written more compactly as $\{x \mid -1 < x < 1\}$. In interval notation, the solution is $(-1, 1)$. See Figure 20 for the graph. ▬

FIGURE 20
$-1 < x < 1$ or $(-1, 1)$

We observe in the preceding process that the two inequalities we solved required exactly the same steps. A shortcut to solving the original inequality is to deal with the two inequalities at the same time, as follows:

$$
\begin{array}{lll}
-5 < 3x - 2 & < 1 & \\
-5 + 2 < 3x - 2 + 2 < 1 + 2 & & \text{Add 2 to each part.} \\
-3 < \quad 3x \quad < 3 & & \text{Simplify.} \\
\dfrac{-3}{3} < \quad \dfrac{3x}{3} \quad < \dfrac{3}{3} & & \text{Divide each part by 3.} \\
-1 < \quad x \quad < 1 & & \text{Simplify.}
\end{array}
$$

We use this shortcut in the next example.

EXAMPLE 4

Solving Combined Inequalities

Solve the inequality:

$$-1 \le \frac{3 - 5x}{2} \le 9$$

Graph the solution set.

Solution

$$
\begin{array}{lll}
-1 \le \dfrac{3 - 5x}{2} \le 9 & & \\
2(-1) \le 2\left(\dfrac{3 - 5x}{2}\right) \le 2(9) & & \text{Multiply each part by 2 to remove the denominator.} \\
-2 \le \quad 3 - 5x \quad \le 18 & & \text{Simplify.} \\
-2 - 3 \le 3 - 5x - 3 \le 18 - 3 & & \text{Subtract 3 from each part to isolate the term containing } x. \\
-5 \le \quad -5x \quad \le 15 & & \text{Simplify.} \\
\dfrac{-5}{-5} \ge \quad \dfrac{-5x}{-5} \quad \ge \dfrac{15}{-5} & & \text{Divide each part by } -5 \text{ (change the sense of each inequality symbol).} \\
1 \ge \quad x \quad \ge -3 & & \text{Simplify.} \\
-3 \le \quad x \quad \le 1 & & \text{Reverse the order so that the numbers get larger as you read from left to right.}
\end{array}
$$

FIGURE 21
$-3 \le x \le 1$ or $[-3, 1]$

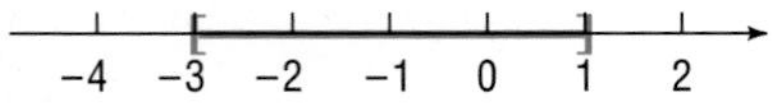

The solution set is $\{x \mid -3 \le x \le 1\}$, that is, all x in $[-3, 1]$. Figure 21 illustrates the graph. ▬

Now work Problem 23.

EXAMPLE 5

Using the Reciprocal Property to Solve an Inequality

Solve the inequality: $(4x - 1)^{-1} > 0$

FIGURE 22
$x > \frac{1}{4}$ or $(\frac{1}{4}, \infty)$

Solution   Since $(4x - 1)^{-1} = 1/(4x - 1)$ and since the Reciprocal Property states that when $1/a > 0$ then $a > 0$, we have

$$(4x - 1)^{-1} > 0$$

$$\frac{1}{4x - 1} > 0$$

$$4x - 1 > 0 \quad \text{Reciprocal Property}$$

$$4x > 1$$

$$x > 1/4$$

The solution set is $\{x | x > 1/4\}$; that is, all x in $(1/4, \infty)$. Figure 22 illustrates the graph. ∎

Now work Problem 35.

E X A M P L E   6   Creating Equivalent Inequalities

If $-1 < x < 4$, find a and b so that $a < 2x + 1 < b$.

Solution   The idea here is to change the middle part of the combined inequality from x to $2x + 1$, using properties of inequalities.

$$
\begin{aligned}
-1 < \quad & x \quad < 4 \\
-2 < \quad & 2x \quad < 8 \qquad \text{Multiply each part by 2.} \\
-1 < \quad & 2x + 1 < 9 \qquad \text{Add 1 to each part.}
\end{aligned}
$$

Thus, $a = -1$ and $b = 9$. ∎

Now work Problem 43.

Let's look at an applied problem involving linear inequalities.

E X A M P L E   7   Physics: Ohm's Law

In electricity, Ohm's law states that $E = IR$, where E is the voltage (in volts), I is the current (in amperes), and R is the resistance (in ohms). An air-conditioning unit is rated at a resistance of 10 ohms. If the voltage varies from 110 to 120 volts, inclusive, what corresponding range of current will the air conditioner draw?

Solution   The voltage lies between 110 and 120, inclusive, so

$$
\begin{aligned}
110 \le \quad & E \quad \le 120 \\
110 \le \quad & IR \quad \le 120 \qquad \text{Ohm's law, } E = IR \\
110 \le \quad & I(10) \le 120 \qquad R = 10 \\
\frac{110}{10} \le \quad & \frac{I(10)}{10} \le \frac{120}{10} \qquad \text{Divide each part by 10.} \\
11 \le \quad & I \quad \le 12 \qquad \text{Simplify.}
\end{aligned}
$$

The air conditioner will draw between 11 and 12 amperes of current, inclusive.

∎

HISTORICAL FEATURE Inequalities are a relatively new component of the algebra curriculum. They have been introduced in the last 25 years for two important reasons.

First, if approximations are made in a problem, inequalities allow calculation of how serious the error is likely to be. This use is important for practical applications of mathematics and is also critical for the understanding of calculus (in its modern version, due principally to Augustin Louis Cauchy, 1789–1857, and Karl Weierstrass, 1815–1897).

Second, linear inequalities are the basis for solving a kind of problem that involves finding a maximum or minimum of a quantity, depending on variables that are subject to certain constraints. For example, we might wish to ship several products by truck from several different factories, each factory having a limited supply of each product, and to get enough of the products to a central point within three days using the minimum amount of gas possible. Such problems, called *linear programming problems,* are discussed later in this text.

2.6 | EXERCISES

In Problems 1–4, an inequality is given. Write the equivalent inequality obtained by:
(a) Adding −3 to each side of the given inequality.
(b) Subtracting 5 from each side of the given inequality.
(c) Multiplying each side of the given inequality by 3.
(d) Multiplying each side of the given inequality by −2.

1. $3 < 5$ **2.** $2 > 1$ **3.** $2x + 1 < 2$ **4.** $1 - 2x > 5$

In Problems 5–40, solve each inequality. Graph the solution set.

5. $x + 1 < 5$

6. $x - 6 < 1$

7. $1 - 2x \leq 3$

8. $2 - 3x \leq 5$

9. $3x - 7 > 2$

10. $2x + 5 > 1$

11. $3x - 1 \geq 3 + 5x$

12. $2x - 2 \geq 3 + 3x$

13. $-2(x + 3) < 8$

14. $-3(1 - x) < 12$

15. $4 - 3(1 - x) \leq 3$

16. $8 - 4(2 - x) \leq -2x$

17. $\frac{1}{2}(x - 4) > x + 8$

18. $3x + 4 > \frac{1}{3}(x - 2)$

19. $\frac{x}{2} \geq 1 - \frac{x}{4}$

20. $\frac{x}{3} \geq 2 + \frac{x}{6}$

21. $0 \leq 2x - 6 \leq 4$

22. $4 \leq 2x + 2 \leq 10$

23. $-5 \leq 4 - 3x \leq 2$

24. $-3 \leq 3 - 2x \leq 9$

25. $-3 < \frac{2x - 1}{3} < 0$

26. $0 < \frac{3x + 2}{2} < 7$

27. $1 < 1 - \frac{1}{2}x < 4$

28. $0 < 1 - \frac{1}{3}x < 1$

29. $(x + 2)(x - 3) > (x - 1)(x + 1)$

30. $(x - 1)(x + 1) > (x - 3)(x + 4)$

31. $x(4x + 3) \leq (2x + 1)^2$

32. $x(9x - 5) \leq (3x - 1)^2$

33. $\frac{1}{2} \leq \frac{x + 1}{3} < \frac{3}{4}$

34. $\frac{1}{3} < \frac{x + 1}{2} \leq \frac{2}{3}$

35. $(4x + 2)^{-1} < 0$

36. $(2x - 1)^{-1} > 0$

37. $0 < \frac{2}{x} < \frac{3}{5}$

38. $0 < \frac{4}{x} < \frac{2}{3}$

39. $0 < (2x - 4)^{-1} < \frac{1}{2}$

40. $0 < (3x + 6)^{-1} < \frac{1}{3}$

In Problems 41–50, find a and b.

41. If $-1 < x < 1$, then $a < x + 4 < b$.

42. If $-3 < x < 2$, then $a < x - 6 < b$.

43. If $2 < x < 3$, then $a < -4x < b$.

44. If $-4 < x < 0$, then $a < \frac{1}{2}x < b$.

45. If $0 < x < 4$, then $a < 2x + 3 < b$.

46. If $-3 < x < 3$, then $a < 1 - 2x < b$.

47. If $-3 < x < 0$, then $a < \dfrac{1}{x + 4} < b$.

48. If $2 < x < 4$, then $a < \dfrac{1}{x - 6} < b$.

49. If $6 < 3x < 12$, then $a < x^2 < b$.

50. If $0 < 2x < 6$, then $a < x^2 < b$.

51. If $-10 < x < 5$, solve: $\dfrac{2}{x + 10} > \dfrac{-1}{x - 5}$.

52. If $x > 3$, solve: $\dfrac{x}{x - 3} > 2$.

53. What is the domain of the variable in the expression $\sqrt{3x + 6}$?

54. What is the domain of the variable in the expression $\sqrt{8 + 2x}$?

55. **General Chemistry** For a certain ideal gas, the volume V (in cubic centimeters) equals 20 times the temperature T (in degrees Celsius). If the temperature varies from 80° to 120°C, inclusive, what is the corresponding range of the volume of the gas?

56. An investor has $1000 to invest for a period of 1 year. What range of per annum simple interest rates are needed to obtain interest that varies from $90 to $110, inclusive?

57. **Real Estate** A real estate agent agrees to sell a large apartment complex according to the following commission schedule: $45,000 plus 25% of the selling price in excess of $900,000. Assuming that the complex will sell at some price between $900,000 and $1,100,000, inclusive, over what range does the agent's commission vary? How does the commission vary as a percent of selling price?

58. **Sales Commission** A used car salesperson is paid a commission of $25 plus 40% of the selling price in excess of owner's cost. The owner claims that used cars typically sell for at least owner's cost plus $70 and at most owner's cost plus $300. For each sale made, over what range can the salesperson expect the commission to vary?

59. **Federal Tax Withholding** The percentage method of withholding for federal income tax (1998)* states that a single person whose weekly wages, after subtracting withholding allowances, are over $517, but not over $1105, shall have $69.90 plus 28% of the excess over $517 withheld. Over what range does the amount withheld vary if the weekly wages vary from $525 to $600, inclusive?

60. **Federal Tax Withholding** Rework Problem 59 if the weekly wages vary from $600 to $700, inclusive.

61. **Electricity Rates** Commonwealth Edison Company's energy charge for electricity is 10.494¢ per kilowatt-hour.† In addition, each monthly bill contains a customer charge of $9.36. If your bill ranged from a low of $80.24 to a high of $271.80, over what range did usage vary (in kilowatt-hours)?

62. **Water Bills** The Village of Oak Lawn charges homeowners $21.60 per quarter-year plus $1.70 per 1000 gallons for water usage in excess of 12,000 gallons.‡ In 1998, one homeowner's quarterly bill ranged from a high of $65.80 to a low of $28.40. Over what range did water usage vary?

63. **Markup of a New Car** The markup over dealer's cost of a new car ranges from 12% to 18%. If the sticker price is $8800, over what range will the dealer's cost vary?

64. **IQ Tests** A standard intelligence test has an average score of 100. According to statistical theory, of the people who take the test, the 2.5% with the highest scores will have scores of more than 1.95σ above the average, where σ (sigma, a number called the *standard deviation*) depends on the nature of the test. If $\sigma = 12$ for this test and there is (in principle) no upper limit to the score possible on the test, write the interval of possible test scores of the people in the top 2.5%.

Source: Employer's Tax Guide, Department of the Treasury, Internal Revenue Service, 1998.

†*Source:* Commonwealth Edison Co., Chicago, Illinois, 1998.

‡*Source:* Village of Oak Lawn, Illinois, 1998.

65. In your Economics 101 class, you have scores of 68, 82, 87, and 89 on the first four of five tests. To get a grade of B, the average of the five test scores must be greater than or equal to 80 and less than 90. Solve an inequality to find the range of the score that you need on the last test to get a B.

What do I need to get a B?

66. Repeat Problem 65 if the fifth test counts double.

67. A car that averages 25 miles per gallon has a tank that holds 20 gallons of gasoline. After a trip that covered at least 300 miles, the car ran out of gasoline. What is the range of the amount of gasoline (in gallons) that was in the tank at the start of the trip?

68. Repeat Problem 67 if the same car runs out of gasoline after a trip of no more than 250 miles.

 69. Make up a linear inequality that has $-3 < x < 4$ as a solution.

70. Make up a linear inequality that has $x < -3$ as a solution.

71. Do you prefer to use inequality notation or interval notation to express the solution to an inequality? Give your reasons. Are there particular circumstances when you prefer one to the other? Cite examples.

2.7 | POLYNOMIAL AND RATIONAL INEQUALITIES

> **1** Solve Polynomial Inequalities
> **2** Solve Rational Inequalities

1 In this section, we solve inequalities that contain polynomials of degree 2 and higher, as well as some that contain rational expressions. Before we can solve such inequalities, we rearrange them so that the polynomial or rational expression is on the left side and 0 is on the right side. Then we factor the left side. An example will show you why.

EXAMPLE 1   Solving Quadratic Inequalities

Solve the inequality $x^2 + x - 12 > 0$, and graph the solution set.

Solution   We factor the left side, obtaining

$$x^2 + x - 12 > 0$$
$$(x + 4)(x - 3) > 0$$

The product of two real numbers is positive either when both factors are positive or when both factors are negative.

Both Negative			or	**Both Positive**		
$x + 4 < 0$	and	$x - 3 < 0$		$x + 4 > 0$	and	$x - 3 > 0$
$x < -4$	and	$x < 3$		$x > -4$	and	$x > 3$

The numbers x that are less than -4 and at the same time less than 3 are simply

$$x < -4$$

The numbers x that are greater than -4 and at the same time greater than 3 are simply

$$x > 3$$

or

The solution set is $\{x | x < -4 \text{ or } x > 3\}$. In interval notation, we write the solution as $(-\infty, -4)$ or $(3, \infty)$. See Figure 23. ■

FIGURE 23
$x < -4$ or $x > 3$; $(-\infty, -4)$ or $(3, \infty)$

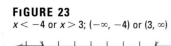

We can also obtain the solution to the inequality of Example 1 by another method. The left-hand side of the inequality is factored so that it becomes $(x + 4)(x - 3) > 0$, as before. We then use the real number line to construct a graph that uses the solutions to the equation

$$x^2 + x - 12 = (x + 4)(x - 3) = 0$$

which are $x = -4$ and $x = 3$. These numbers separate the real number line into three intervals: $-\infty < x < -4, -4 < x < 3,$ and $3 < x < \infty$. See Figure 24(a).

Now, if $x < -4$, then $x + 4 < 0$. We indicate this fact about the expression $x + 4$ by placing minus signs $(- - -)$ to the left of -4. See Figure 24(b). If $x > -4$, then $x + 4 > 0$. We indicate this fact about $x + 4$ by placing plus signs $(+ + +)$ to the right of -4.

Similarly, if $x < 3$, then $x - 3 < 0$. We indicate this fact about $x - 3$ by placing minus signs to the left of 3. If $x > 3$, then $x - 3 > 0$. We indicate this fact about $x - 3$ by placing plus signs to the right of 3. See Figure 24(b).

Next we prepare Figure 24(c) as follows: Since we know that the expressions $x + 4$ and $x - 3$ are both negative for $x < -4$, it follows that their product will be positive for $x < -4$. Since we know that $x - 3$ is negative and $x + 4$ is positive for $-4 < x < 3$, it follows that their product is negative for $-4 < x < 3$. Finally, since both expressions are positive for $x > 3$, their product is positive for $x > 3$. We place plus and minus signs as shown in Figure 24(c) to indicate these facts.

Finally, in Figure 24(d) we show on the number line where the expression $(x + 4)(x - 3)$ is positive. The solution to the original inequality $(x + 4)(x - 3) = x^2 + x - 12 > 0$ is $\{x | x < -4 \text{ or } x > 3\}$, as before.

FIGURE 24

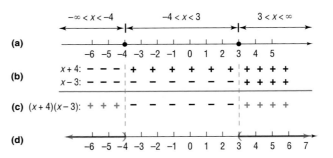

The preceding discussion demonstrates that the sign of each factor of an expression is the same on each interval that the real number line was divided into. Consequently, an alternative, and simpler, approach to obtaining Figure 24(c) would be to select a **test number** in each interval and use it to evaluate each factor to see if it is positive or negative. You may choose any number in the interval as a test number. In Figure 25(a), the test numbers that we selected, $-5, 1,$ and 4 are in red.

FIGURE 25

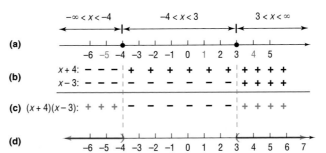

For $x + 4$, we insert $- - -$ under $-\infty < x < -4$ because, for the test number -5, the value of $x + 4$ is $-5 + 4 = -1$, a negative number. Continuing across, we insert $+ + +$ under $-4 < x < 3$ because, for the test number 1, the value of $x + 4$ is $1 + 4 = 5$, a positive number. This process is continued for each factor and for each interval to obtain Figure 25(b). The rest of Figure 25 is obtained as before.

We shall employ the method of using a test number to solve inequalities. Here is another example showing all the details.

E X A M P L E 2

Solving Quadratic Inequalities

Solve the inequality $x^2 \leq 4x + 12$, and graph the solution set.

Solution

First, we rearrange the inequality so that 0 is on the right side:

$$x^2 \leq 4x + 12$$
$$x^2 - 4x - 12 \leq 0$$
$$(x + 2)(x - 6) \leq 0 \qquad \text{Factor.}$$

Next, we set the left side equal to 0 and solve the resulting equation:

$$(x + 2)(x - 6) = 0$$

The solutions of the equation are -2 and 6, and they separate the real number line into three intervals:

$$-\infty < x < -2 \qquad -2 < x < 6 \qquad 6 < x < \infty$$

We choose -3, 1, and 7 as test numbers. See Figure 26(a).

Now, for the test number -3, we find that $x + 2 = -3 + 2 = -1$, a negative number, so we place minus signs under the interval $-\infty < x < -2$. For the test number 1, we find $x + 2 = 1 + 2 = 3$, a positive number, so we place plus signs under the interval $-2 < x < 6$. Continuing in this fashion, we obtain Figure 26(b).

Next we enter the signs of the product $(x + 2)(x - 6)$ in Figure 26(c). Since we want to know where the product $(x + 2)(x - 6)$ is negative, we conclude that the solutions are numbers x for which $-2 < x < 6$. However, because the original inequality is nonstrict, numbers x that satisfy the equation $x^2 = 4x + 12$ are also solutions of the inequality $x^2 \leq 4x + 12$. Thus, we include -2 and 6, and the solution set of the given inequality is $\{x | -2 \leq x \leq 6\}$, that is, all x in $[-2, 6]$. See Figure 26(d).

FIGURE 26

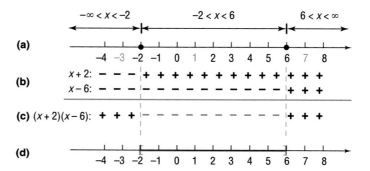

Now work Problems 3 and 7.

We have been solving inequalities by rearranging the inequality so that 0 is on the right side, setting the left side equal to 0, and solving the resulting equation. The solutions are then used to separate the real number line into intervals. But what if the resulting equation has no real solution? In this case, we rely on the following result.

> **Theorem**
>
> If a polynomial equation has no real solutions, the polynomial is either always positive or always negative.

For example, the equation

$$x^2 + 5x + 8 = 0$$

has no real solutions. (Do you see why? Its discriminant, $b^2 - 4ac = 25 - 32 = -7$, is negative.) The value of $x^2 + 5x + 8$ is therefore always positive or always negative. To see which is true, we test its value at some number (0 is the easiest). Because $0^2 + 5(0) + 8 = 8$ is positive, we conclude that $x^2 + 5x + 8 > 0$ for all x.

E X A M P L E   3

Solving a Polynomial Inequality

Solve the inequality $x^4 \leq x$, and graph the solution set.

Solution   We rewrite the inequality so that 0 is on the right side:

$$x^4 \leq x$$
$$x^4 - x \leq 0$$

Then we proceed to factor the left side:

$$x^4 - x \leq 0$$
$$x(x^3 - 1) \leq 0$$
$$x(x - 1)(x^2 + x + 1) \leq 0$$

The solutions of the equation

$$x(x - 1)(x^2 + x + 1) = 0$$

are just 0 and 1, since the equation $x^2 + x + 1 = 0$ has no real solutions. We note that the expression $x^2 + x + 1$ is always positive. (Do you see why?) Next, we use 0 and 1 to separate the real number line into three intervals:

$$-\infty < x < 0 \qquad 0 < x < 1 \qquad 1 < x < \infty$$

Then we construct Figure 27, showing the signs of x, $x - 1$, and $x^2 + x + 1$. Note that, since $x^2 + x + 1 > 0$ for all x, we enter plus signs next to it.

FIGURE 27

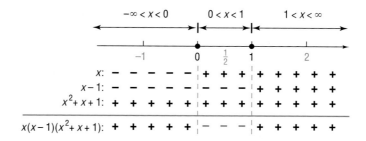

FIGURE 28
$0 \leq x \leq 1$ or $[0, 1]$

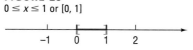

Since we want to know where $x^4 \leq x$ or, equivalently, where $x(x - 1)(x^2 + x + 1) \leq 0$, we conclude from Figure 27 that the solution set is $\{x | 0 \leq x \leq 1\}$, that is, all x in $[0, 1]$. See Figure 28. ■

Now work Problem 17.

Let's solve a rational inequality.

E X A M P L E 4   Solving a Rational Inequality

Solve the inequality $\dfrac{(x + 3)(2 - x)}{(x - 1)^2} > 0$ and graph the solution set.

Solution   The domain of the variable x is $\{x | x \neq 1\}$. The rational expression is factored with zero on the right-hand side. Since the sign of a rational expression depends on the sign of its numerator and the sign of its denominator, we separate the real number line into intervals using the numbers obtained by setting the numerator and the denominator equal to zero. For this example, these numbers are: -3, 1, and 2. See Figure 29. Notice that since $(x - 1)^2 > 0$ for all $x \neq 1$, we can enter plus signs next to it and avoid using test numbers.

FIGURE 29

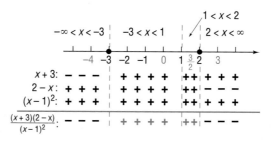

FIGURE 30
$-3 < x < 2$ or $(-3, 2)$, $x \neq 1$

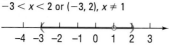

The bottom row in Figure 29 reveals the numbers x for which $\dfrac{(x + 3)(2 - x)}{(x - 1)^2} > 0$. We conclude that the solution set is $\{x | -3 < x < 2, x \neq 1\}$; that is, all x in $(-3, 2)$, except $x = 1$. See Figure 30 for the graph of the solution set. Notice the hole at $x = 1$ to indicate 1 is to be excluded. ■

E X A M P L E 5   Solving a Rational Inequality

Solve the inequality $\dfrac{4x + 5}{x + 2} \geq 3$, and graph the solution set.

Solution We first note that the domain of the variable consists of all real numbers except -2. We rearrange terms so that 0 is on the right side:

$$\frac{4x+5}{x+2} \geq 3$$

$$\frac{4x+5}{x+2} - 3 \geq 0$$

$$\frac{4x+5 - 3(x+2)}{x+2} \geq 0 \qquad \text{Rewrite using } x+2 \text{ as the denominator.}$$

$$\frac{x-1}{x+2} \geq 0 \qquad \text{Simplify.}$$

The sign of a rational expression depends on the sign of its numerator and the sign of its denominator. Thus, for a rational expression, we separate the real number line into intervals using the numbers obtained by setting the numerator and the denominator equal to 0. For this example, they are -2 and 1. See Figure 31.

FIGURE 31

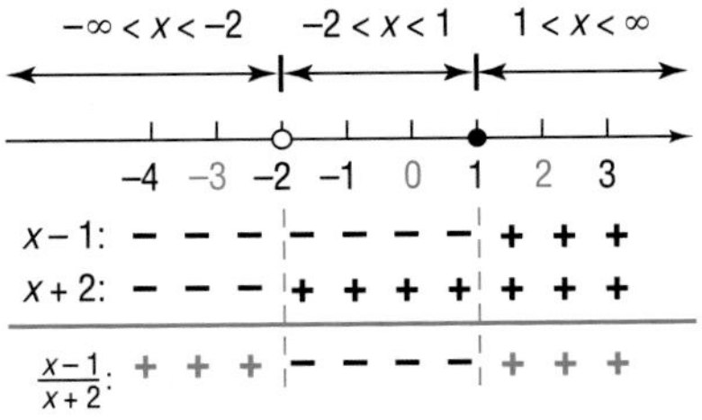

$$-\infty < x < -2 \quad -2 < x < 1 \quad 1 < x < \infty$$

$x-1$:	$-\ -\ -$	$-\ -\ -\ -$	$+\ +\ +$
$x+2$:	$-\ -\ -$	$+\ +\ +\ +$	$+\ +\ +$
$\frac{x-1}{x+2}$:	$+\ +\ +$	$-\ -\ -\ -$	$+\ +\ +$

-4 -3 -2 -1   0   1   2   3

FIGURE 32
$x < -2$ or $x \geq 1$ $(-\infty, -2)$ or $[1, \infty)$

-5 -4 -3 -2 -1   0   1   2   3   4

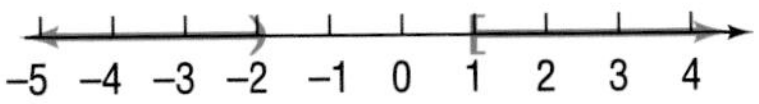

The bottom line in Figure 31 reveals the number x for which $(x-1)/(x+2)$ is positive. However, we want to know where the expression $(x-1)/(x+2)$ is positive or 0. Since $(x-1)/(x+2) = 0$ only if $x = 1$, we conclude that the solution set is $\{x \mid x < -2 \text{ or } x \geq 1\}$, that is, all x in $(-\infty, -2)$ or $[1, \infty)$. See Figure 32. ∎

In Example 5, you may wonder why we did not first multiply both sides of the inequality by $x + 2$ to clear the denominator. The reason is that we do not know whether $x + 2$ is positive or negative and, as a result, we do not know whether to reverse the sense of the inequality symbol after multiplying $x + 2$. However, there is nothing to prevent us from multiplying both sides by $(x + 2)^2$ which is always positive, since $x \neq -2$. (Do you see why?) Then

$$\frac{4x+5}{x+2} \geq 3 \qquad x \neq -2$$

$$\frac{4x+5}{x+2}(x+2)^2 \geq 3(x+2)^2 \qquad \text{Multiply by } (x+2)^2 > 0$$

$$(4x+5)(x+2) \geq 3(x^2 + 4x + 4) \qquad \text{Simplify}$$

$$4x^2 + 13x + 10 \geq 3x^2 + 12x + 12$$

$$x^2 + x - 2 \geq 0 \qquad \text{Obtain zero on the right}$$

$$(x+2)(x-1) \geq 0 \qquad x \neq -2 \qquad \text{Factor}$$

This last expression leads to the same solution set obtained in Example 5.

 Now work Problem 29.

For our next example, we need the following equivalence property of inequalities.

If a and b are real numbers and if $a \geq 0$ and $b \geq 0$, then

$$a \leq b \quad \text{is equivalent to} \quad \sqrt{a} \leq \sqrt{b} \qquad (1)$$

The proof of this property is left as an exercise.

EXAMPLE 6 **Physics**

An object is dropped from the roof of a building 150 feet tall. After t seconds, the object will be $150 - 16t^2$ feet above the ground. During what interval of time will the object be between 54 and 118 feet above the ground?

Solution The inequality to be solved has the form

$$
\begin{array}{rcll}
54 \leq 150 - 16t^2 &\leq& 118 & \\
-96 \leq \quad -16t^2 &\leq& -32 & \text{Subtract 150 from each part.} \\
6 \geq \qquad t^2 &\geq& 2 & \text{Divide each part by } -16. \\
2 \leq \qquad t^2 &\leq& 6 & \text{Reverse inequalities}
\end{array}
$$

For the situation described, the time t is assumed to be a nonnegative real number. With this restriction, by equation (1), the inequality

$$2 \leq t^2 \leq 6 \qquad t \geq 0$$

is equivalent to

$$\sqrt{2} \leq t \leq \sqrt{6}$$

which may be approximated by

$$1.41 \leq t \leq 2.45$$

For times ranging from approximately 1.41 to 2.45 seconds, the object will be between 54 and 118 feet above the ground. ▬

2.7 | EXERCISES

In Problems 1–46, solve each inequality.

1. $(x - 5)(x + 2) < 0$
2. $(x - 5)(x + 2) > 0$
3. $x^2 - 4x > 0$
4. $x^2 + 8x > 0$
5. $x^2 - 9 < 0$
6. $x^2 - 1 < 0$
7. $x^2 + x > 2$
8. $x^2 + 7x < -12$
9. $2x^2 \leq 5x + 3$
10. $6x^2 \leq 6 + 5x$
11. $x(x - 7) > 8$
12. $x(x + 1) > 20$
13. $4x^2 + 9 < 6x$
14. $25x^2 + 16 < 40x$
15. $6(x^2 - 1) > 5x$
16. $2(2x^2 - 3x) > -9$
17. $(x - 1)(x^2 + x + 4) > 0$
18. $(x + 2)(x^2 - x + 1) > 0$
19. $(x - 1)(x - 2)(x - 3) \leq 0$
20. $(x + 1)(x + 2)(x + 3) \leq 0$
21. $x^3 - 2x^2 - 3x > 0$

22. $x^3 + 2x^2 - 3x > 0$

23. $x^4 > x^2$

24. $x^4 < 4x^2$

25. $x^3 > 4x^2$

26. $x^3 < 9x^2$

27. $x^4 > 1$

28. $x^3 > 1$

29. $\dfrac{x+1}{x-1} > 0$

30. $\dfrac{x-3}{x+1} > 0$

31. $\dfrac{(x-1)(x+1)}{x} < 0$

32. $\dfrac{(x-3)(x+2)}{x-1} < 0$

33. $\dfrac{(x-2)^2}{x^2-1} \geq 0$

34. $\dfrac{(x+5)^2}{x^2-4} \geq 0$

35. $6x - 5 < \dfrac{6}{x}$

36. $x + \dfrac{12}{x} < 7$

37. $\dfrac{x+4}{x-2} \leq 1$

38. $\dfrac{x+2}{x-4} \geq 1$

39. $\dfrac{3x-5}{x+2} \leq 2$

40. $\dfrac{x-4}{2x+4} \geq 1$

41. $\dfrac{1}{x-2} < \dfrac{2}{3x-9}$

42. $\dfrac{5}{x-3} > \dfrac{3}{x+1}$

43. $\dfrac{2x+5}{x+1} > \dfrac{x+1}{x-1}$

44. $\dfrac{1}{x+2} > \dfrac{3}{x+1}$

45. $\dfrac{x^2(3+x)(x+4)}{(x+5)(x-1)} \geq 0$

46. $\dfrac{x(x^2+1)(x-2)}{(x-1)(x+1)} \geq 0$

47. For what positive numbers will the cube of a number exceed 4 times its square?

48. For what positive numbers will the square of a number exceed twice the number?

49. What is the domain of the variable in the expression $\sqrt{x^2 - 16}$?

50. What is the domain of the variable in the expression $\sqrt{x^3 - 3x^2}$?

51. What is the domain of the variable in the expression $\sqrt{\dfrac{x-2}{x+4}}$?

52. What is the domain of the variable in the expression $\sqrt{\dfrac{x-1}{x+4}}$?

53. **Physics** A ball is thrown vertically upward with an initial velocity of 80 feet per second. The distance s (in feet) of the ball from the ground after t seconds is $s = 80t - 16t^2$. For what time interval is the ball more than 96 feet above the ground? (See the figure.)

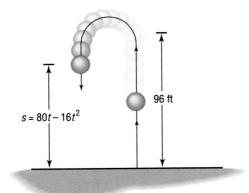

$s = 80t - 16t^2$

96 ft

54. **Physics** Rework Problem 53 to find when the ball is less than 64 feet above the ground.

55. **Business** The monthly revenue achieved by selling x wristwatches is figured to be $x(40 - 0.2x)$ dollars. The wholesale cost of each watch is

$28. How many watches must be sold each month to achieve a profit (revenue − cost) of at least $100?

56. **Business** The monthly revenue achieved by selling x boxes of candy is figured to be $x(5 - 0.05x)$ dollars. The wholesale cost of each box of candy is $1.50. How many boxes must be sold each month to achieve a profit of at least $60?

57. Prove that if a, b are real numbers and $a \geq 0$, $b \geq 0$ then

$a \leq b$ is equivalent to $\sqrt{a} \leq \sqrt{b}$

[**Hint:** $b - a = (\sqrt{b} - \sqrt{a})(\sqrt{b} + \sqrt{a})$]

58. Find k such that the equation $x^2 + kx + 1 = 0$ has no real solution.

59. Find k such that the equation $kx^2 + 2x + 1 = 0$ has two distinct real solutions.

60. Make up an inequality that has no solution. Make up one that has exactly one solution.

61. The inequality $x^2 + 1 < -5$ has no solution. Explain why.

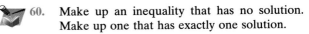

2.8 │ EQUATIONS AND INEQUALITIES INVOLVING ABSOLUTE VALUE

    1 Solve Equations Involving Absolute Value
    2 Solve Inequalities Involving Absolute Value

The absolute value of a real number a has been defined as

$$|a| = \begin{cases} a & \text{if } a \geq 0 \\ -a & \text{if } a < 0 \end{cases}$$

Also, recall that, geometrically, the absolute value of a equals the distance from the origin to the point whose coordinate is a.

In this section, we shall discuss equations and inequalities involving absolute value. In solving such equations and inequalities, the following properties will prove useful.

Theorem   Properties of Absolute Value

1. The absolute value of any real number a is always nonnegative; that is,

$$|a| \geq 0 \tag{1}$$

2. The absolute value of any real number a equals the principal square root of the number squared; that is,

$$|a| = \sqrt{a^2} \tag{2}$$

3. The absolute value of the product of two real numbers a and b equals the product of their absolute values; that is,

$$|ab| = |a| \cdot |b| \tag{3}$$

Proof Property (1) follows directly from the definition of absolute value. Property (2) was given earlier in Section 1.7. Property (3) is proved by using property (2):

$$|ab| = \sqrt{(ab)^2} = \sqrt{a^2 b^2} = \sqrt{a^2} \cdot \sqrt{b^2} = |a| \cdot |b|$$

    1 We shall now take up the problem of solving equations and inequalities that contain absolute values.

Because there are two points whose distance from the origin is 5 units, -5 and 5, the equation $|x| = 5$ will have the solution set $\{-5, 5\}$. We are thus led to the following result:

> **Theorem** Equations Involving Absolute Value
>
> If the absolute value of an expression equals some positive number a, then the expression itself equals either a or $-a$. Thus,
>
> $$|u| = a \quad \text{is equivalent to} \quad u = a \quad \text{or} \quad u = -a \qquad (4)$$

E X A M P L E   1

Solving an Equation Involving Absolute Value

Solve the equation: $|x + 4| = 13$.

Solution

This follows the form of equation (4), where $u = x + 4$. Thus, there are two possibilities:

$$x + 4 = 13 \quad \text{or} \quad x + 4 = -13$$
$$x = 9 \qquad\qquad x = -17$$

Thus, the solution set is $\{-17, 9\}$.

Now work Problem 3.

Let's look at an inequality involving absolute value.

E X A M P L E   2

Solving an Inequality Involving Absolute Value

Solve the inequality: $|x| < 4$.

FIGURE 33
$-4 < x < 4$ or $(-4, 4)$

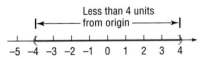

Solution

We are looking for all points whose coordinate x is a distance less than 4 units from the origin. See Figure 33 for an illustration. Because any x between -4 and 4 satisfies the condition $|x| < 4$, the solution set consists of all numbers x for which $-4 < x < 4$, that is, all x in $(-4, 4)$.

We are led to the following results:

> **Theorem**
>
> If a is any positive number, then
>
> $$|u| < a \quad \text{is equivalent to} \quad -a < u < a \qquad (5)$$
> $$|u| \leq a \quad \text{is equivalent to} \quad -a \leq u \leq a \qquad (6)$$
>
> In other words, $|u| < a$ is equivalent to $-a < u$ and $u < a$.

E X A M P L E   3

Solving an Inequality Involving Absolute Value

Solve the inequality $|2x + 4| \leq 3$, and graph the solution set.

Solution

$$\begin{array}{rcll} |2x+4| & \leq 3 & & \text{This follows the form of statement (6); the expression } u = 2x+4 \text{ is inside the absolute value bars.} \\ -3 \leq & 2x+4 & \leq 3 & \text{Apply statement (6).} \\ -3-4 \leq & 2x+4-4 & \leq 3-4 & \text{Subtract 4 from each part.} \\ -7 \leq & 2x & \leq -1 & \text{Simplify.} \\ \frac{-7}{2} \leq & \frac{2x}{2} & \leq \frac{-1}{2} & \text{Divide each part by 2.} \\ -\frac{7}{2} \leq & x & \leq -\frac{1}{2} & \text{Simplify.} \end{array}$$

FIGURE 34
$-\frac{7}{2} \leq x \leq \frac{1}{2}$ or $[-\frac{7}{2}, -\frac{1}{2}]$

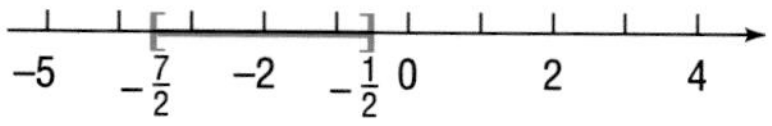

The solution set is $\{x \mid -\frac{7}{2} \leq x \leq -\frac{1}{2}\}$, that is, all x in $[-\frac{7}{2}, -\frac{1}{2}]$. See Figure 34.

EXAMPLE 4

Solving an Inequality Involving Absolute Value

Solve the inequality $|1 - 4x| < 5$, and graph the solution set.

Solution

$$\begin{array}{rcll} |1-4x| & < 5 & & \text{This expression follows the form of statement (5); the expression } u = 1-4x \text{ is inside the absolute value bars.} \\ -5 < & 1-4x & < 5 & \text{Apply statement (5).} \\ -5-1 < & 1-4x-1 & < 5-1 & \text{Subtract 1 from each part.} \\ -6 < & -4x & < 4 & \text{Simplify.} \\ \frac{-6}{-4} > & \frac{-4x}{-4} & > \frac{4}{-4} & \text{Divide each part by } -4 \text{, which reverses the sense of the inequality symbols.} \\ \frac{3}{2} > & x & > -1 & \text{Simplify.} \\ -1 < & x & < \frac{3}{2} & \text{Rearrange the ordering.} \end{array}$$

FIGURE 35
$-1 < x < \frac{3}{2}$ or $(-1, \frac{3}{2})$

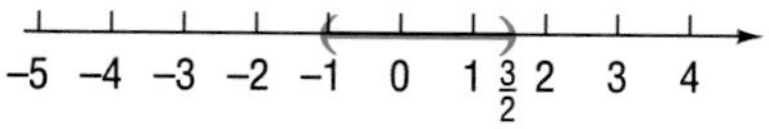

The solution set is $\{x \mid -1 < x < \frac{3}{2}\}$, that is, all x in $(-1, \frac{3}{2})$. See Figure 35.

Now work Problem 27.

EXAMPLE 5

Solving an Inequality Involving Absolute Value

Solve the inequality $|x| > 3$, and graph the solution set.

Solution

FIGURE 36
$x < -3$ or $x > 3$; $(-\infty, -3)$ or $(3, \infty)$

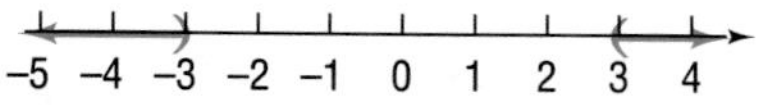

We are looking for all points whose coordinate x is a distance greater than 3 units from the origin. Figure 36 illustrates the situation. We conclude that any x less than -3 or greater than 3 satisfies the condition $|x| > 3$. Consequently, the solution set consists of all numbers x for which $x < -3$ or $x > 3$, that is, all x in $(-\infty, -3)$ or $(3, \infty)$.

This leads to the following results:

Theorem

If a is any positive number, then

$$|u| > a \quad \text{is equivalent to} \quad u < -a \quad \text{or} \quad u > a \qquad (7)$$
$$|u| \geq a \quad \text{is equivalent to} \quad u \leq -a \quad \text{or} \quad u \geq a \qquad (8)$$

E X A M P L E   6   Solving an Inequality Involving Absolute Value

Solve the inequality $|2x - 5| > 3$, and graph the solution set.

Solution

$|2x - 5| > 3$ This follows the form of statement (7); the expression
$u = 2x - 5$ is inside the absolute value bars.

$$2x - 5 < -3 \qquad \text{or} \qquad 2x - 5 > 3 \qquad \text{Apply statement (7).}$$

$$2x - 5 + 5 < -3 + 5 \quad \text{or} \quad 2x - 5 + 5 > 3 + 5 \qquad \text{Add 5 to each part.}$$

$$2x < 2 \qquad \text{or} \qquad 2x > 8 \qquad \text{Simplify.}$$

$$\frac{2x}{2} < \frac{2}{2} \qquad \text{or} \qquad \frac{2x}{2} > \frac{8}{2} \qquad \text{Divide each part by 2.}$$

$$x < 1 \qquad \text{or} \qquad x > 4 \qquad \text{Simplify.}$$

FIGURE 37
$x < 1$ or $x > 4$; $(-\infty, 1)$ or $(4, \infty)$

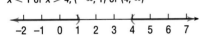

The solution set is $\{x | x < 1 \text{ or } x > 4\}$, that is, all x in $(-\infty, 1)$ or $(4, \infty)$. See Figure 37.

Warning: A common error to be avoided is to attempt to write the solution $x < 1$ or $x > 4$ as $1 > x > 4$, which is incorrect, since there are no numbers x for which $1 > x$ *and* $x > 4$. Another common error is to "mix" the symbols and write $1 < x > 4$, which, of course, makes no sense.

2.8 | EXERCISES

In Problems 1–22, solve each equation.

1. $|2x| = 6$

2. $|3x| = 12$

3. $|2x + 3| = 5$

4. $|3x - 1| = 2$

5. $|1 - 4t| = 5$

6. $|1 - 2z| = 3$

7. $|-2x| = 8$

8. $|-x| = 1$

9. $|-2|x = 4$

10. $|3|x = 9$

11. $\frac{2}{3}|x| = 8$

12. $\frac{3}{4}|x| = 9$

13. $\left|\frac{x}{3} + \frac{2}{5}\right| = 2$

14. $\left|\frac{x}{2} - \frac{1}{3}\right| = 1$

15. $|u - 2| = -\frac{1}{2}$

16. $|2 - v| = -1$

17. $|x^2 - 9| = 0$

18. $|x^2 - 16| = 0$

19. $|x^2 - 2x| = 3$

20. $|x^2 + x| = 12$

21. $|x^2 + x - 1| = 1$

22. $|x^2 + 3x - 2| = 2$

In Problems 23–42, solve each inequality.

23. $|2x| < 8$

24. $|3x| < 15$

25. $|3x| > 12$

26. $|2x| > 6$

27. $|x - 2| < 1$

28. $|x + 4| < 2$

29. $|3t - 2| \leq 4$

30. $|2u + 5| \leq 7$

31. $|x - 3| \geq 2$ **32.** $|x + 4| \geq 2$ **33.** $|1 - 4x| < 5$ **34.** $|1 - 2x| < 3$

35. $|1 - 2x| > 3$ **36.** $|2 - 3x| > 1$ **37.** $|x + 3| > 0$ **38.** $|3 - x| > 0$

39. $|x + 2| > -3$ **40.** $|2 - x| > -2$ **41.** $|2x - 1| < 0.02$ **42.** $|3x - 2| < 0.02$

In Problems 43–48, find a and b.

43. If $|x - 1| < 3$, then $a < x + 4 < b$.

44. If $|x + 2| < 5$, then $a < x - 2 < b$.

45. If $|x + 4| \leq 2$, then $a \leq 2x - 3 \leq b$.

46. If $|x - 3| \leq 1$, then $a \leq 3x + 1 \leq b$.

47. If $|x - 2| \leq 7$, then $a \leq \dfrac{1}{x - 10} \leq b$.

48. If $|x + 1| \leq 3$, then $a \leq \dfrac{1}{x + 5} \leq b$.

49. If $b \neq 0$, prove that $\left| \dfrac{a}{b} \right| = \dfrac{|a|}{|b|}$.

50. Show that $a \leq |a|$.

51. Prove the *triangle inequality* $|a + b| \leq |a| + |b|$.

[**Hint:** Expand $|a + b|^2 = (a + b)^2$, and use the result of Problem 50.]

52. Prove that $|a - b| \geq |a| - |b|$.

[**Hint:** Apply the triangle inequality from Problem 51 to $|a| = |(a - b) + b|$.]

53. Express the fact that x differs from 3 by less than $\frac{1}{2}$ as an inequality involving an absolute value. Solve for x.

54. Express the fact that x differs from -4 by less than 1 as an inequality involving an absolute value. Solve for x.

55. Express the fact that x differs from -3 by more than 2 as an inequality involving an absolute value. Solve for x.

56. Express the fact that x differs from 2 by more than 3 as an inequality involving an absolute value. Solve for x.

57. **Body Temperature** "Normal" human body temperature is 98.6°F. If a temperature x that differs from normal by at least 1.5° is considered unhealthy, write the condition for an unhealthy temperature x as an inequality involving an absolute value, and solve for x.

58. **Household Voltage** In the United States, normal household voltage is 115 volts. However, it is not uncommon for actual voltage to differ from normal voltage by at most 5 volts. Express this situation as an inequality involving an absolute value. Use x as the actual voltage and solve for x.

59. If $a > 0$, show that the solution set of the inequality

$$x^2 < a$$

consists of all numbers x for which

$$-\sqrt{a} < x < \sqrt{a}$$

60. If $a > 0$, show that the solution set of the inequality

$$x^2 > a$$

consists of all numbers x for which

$$x < -\sqrt{a} \quad \text{or} \quad x > \sqrt{a}$$

In Problems 61–68, use the results found in Problems 59 and 60 to solve each inequality.

61. $x^2 < 1$ **62.** $x^2 < 4$ **63.** $x^2 \geq 9$ **64.** $x^2 \geq 1$

65. $x^2 \leq 16$ **66.** $x^2 \leq 9$ **67.** $x^2 > 4$ **68.** $x^2 > 16$

69. Solve $|3x - |2x + 1|| = 4$.

70. Solve $|x + |3x - 2|| = 2$.

71. The equation $|x| = -2$ has no solution. Explain why.

72. The inequality $|x| > -0.5$ has all real numbers as a solution. Explain why.

CHAPTER REVIEW

THINGS TO KNOW

Quadratic equation and quadratic formula

If $ax^2 + bx + c = 0$, $a \neq 0$, and if $b^2 - 4ac \geq 0$, then $x = \dfrac{-b \pm \sqrt{b^2 - 4ac}}{2a}$.

Discriminant

If $b^2 - 4ac > 0$, there are two distinct real solutions.
If $b^2 - 4ac = 0$, there is one repeated real solution.
If $b^2 - 4ac < 0$, there are no real solutions.

Inequality properties

Trichotomy property	$a < b$ or $a = b$ or $b < a$
Transitive property	If $a < b$ and $b < c$, then $a < c$.
	If $a > b$ and $b > c$, then $a > c$.
Addition property	If $a < b$, then $a + c < b + c$.
	If $a > b$, then $a + c > b + c$.
Multiplication properties	(a) If $a < b$ and if $c > 0$, then $ac < bc$.
	If $a < b$ and if $c < 0$, then $ac > bc$.
	(b) If $a > b$ and if $c > 0$, then $ac > bc$.
	If $a > b$ and if $c < 0$, then $ac < bc$.
Reciprocal property	If $a > 0$, then $\dfrac{1}{a} > 0$.
	If $a < 0$, then $\dfrac{1}{a} < 0$.

Interval notation

$[a, b]$	$\{x \mid a \leq x \leq b\}$	$(-\infty, a]$	$\{x \mid x \leq a\}$
$[a, b)$	$\{x \mid a \leq x < b\}$	$(-\infty, a)$	$\{x \mid x < a\}$
$(a, b]$	$\{x \mid a < x \leq b\}$	$(-\infty, \infty)$	All real numbers.
(a, b)	$\{x \mid a < x < b\}$		
$[a, \infty)$	$\{x \mid x \geq a\}$		
(a, ∞)	$\{x \mid x > a\}$		

Absolute value

If $|u| = a$, $a > 0$, then $u = -a$ or $u = a$.
If $|u| \leq a$, $a > 0$, then $-a \leq u \leq a$.
If $|u| \geq a$, $a > 0$, then $u \leq -a$ or $u \geq a$.

HOW TO

Solve linear equations in one variable
Solve quadratic equations
Solve radical equations
Solve linear inequalities in one variable

Solve polynomial and rational inequalities
Solve equations and inequalities involving absolute value
Solve applied problems

FILL-IN-THE-BLANK ITEMS

1. Two equations (or inequalities) that have precisely the same solution set are called _____.
2. An equation that is satisfied for every choice of the variable for which both sides are meaningful is called a(n) _____.
3. To complete the square of the expression $x^2 + 5x$, you would _____ the number _____.
4. The quantity $b^2 - 4ac$ is called the _____ of a quadratic equation. If it is _____, the equation has no real solution.

5. If $a < 0$, then $|a| = $ _____.

6. When a quadratic equation has a repeated solution, it is called a(n) _____ root or a root of _____ _____.

7. When an apparent solution does not satisfy the original equation, it is called a(n) _____ solution.

8. If each side of an inequality is multiplied by a(n) _____ number, then the sense of the inequality symbol is reversed.

9. The equation $|x^2| = 4$ has two real solutions, _____ and _____.

TRUE/FALSE ITEMS

T F **1.** Equations can have no solutions, one solution, or more than one solution.

T F **2.** Quadratic equations always have two real solutions.

T F **3.** If the discriminant of a quadratic equation is positive, then the equation has two solutions that are negatives of each other.

T F **4.** The square of any real number is always nonnegative.

T F **5.** The expression $x^2 + x + 1$ is positive for any real number x.

6. If $a < b$ and $c < 0$, which of the following statements are true?

T F   (a) $a \pm c < b \pm c$

T F   (b) $a \cdot c < b \cdot c$

T F   (c) $a/c > b/c$

7. If $a^2 < b^2$, $b \neq 0$, which of the following statements are true?

T F   (a) $a < b$

T F   (b) $a^2/b^2 < 1$

T F   (c) $b^2 - a^2 > 0$

REVIEW EXERCISES

Blue problem numbers indicate the author's suggestions for use in a Practice Test.

In Problems 1–42, find all real solutions, if any, of each equation. (Where they appear, a, b, m, and n are positive constants.)

1. $2 - \dfrac{x}{3} = 8$

2. $\dfrac{x}{4} - 2 = 4$

3. $-2(5 - 3x) + 8 = 4 + 5x$

4. $(6 - 3x) - 2(1 + x) = 6x$

5. $\dfrac{3x}{4} - \dfrac{x}{3} = \dfrac{1}{12}$

6. $\dfrac{4 - 2x}{3} + \dfrac{1}{6} = 2x$

7. $\dfrac{x}{x - 1} = \dfrac{6}{5}, \quad x \neq 1$

8. $\dfrac{4x - 5}{3 - 7x} = 2, \quad x \neq \frac{3}{7}$

9. $x(1 - x) = 6$

10. $x(1 + x) = 6$

11. $\dfrac{1}{2}\left(x - \dfrac{1}{3}\right) = \dfrac{3}{4} - \dfrac{x}{6}$

12. $\dfrac{1 - 3x}{4} = \dfrac{x + 6}{3} + \dfrac{1}{2}$

13. $(x - 1)(2x + 3) = 3$

14. $x(2 - x) = 3(x - 4)$

15. $2x + 3 = 4x^2$

16. $1 + 6x = 4x^2$

17. $\sqrt[3]{x^2 - 1} = 2$

18. $\sqrt{1 + x^3} = 3$

19. $x(x + 1) + 2 = 0$

20. $3x^2 - x + 1 = 0$

21. $x^4 - 5x^2 + 4 = 0$

22. $3x^4 + 4x^2 + 1 = 0$

23. $\sqrt{2x - 3} + x = 3$

24. $\sqrt{2x - 1} = x - 2$

25. $x^{3/2} + 5x^{1/2} = 0$

26. $x^{2/3} + x = 0$

27. $\sqrt{x + 1} + \sqrt{x - 1} = \sqrt{2x + 1}$

28. $\sqrt{2x - 1} - \sqrt{x - 5} = 3$

29. $2\sqrt[3]{x^2} - \sqrt[3]{x} = 1$

30. $4\sqrt[3]{x^2} = 1$

31. $x^{-6} - 7x^{-3} - 8 = 0$

32. $6x^{-1} - 5x^{-1/2} + 1 = 0$

33. $x^2 + m^2 = 2mx + (nx)^2$

34. $b^2x^2 + 2ax = x^2 + a^2$

35. $10a^2x^2 - 2abx - 36b^2 = 0$

36. $\dfrac{1}{x - m} + \dfrac{1}{x - n} = \dfrac{2}{x}, \quad x \neq 0, m, n$

37. $\sqrt{x^2 + 3x + 7} - \sqrt{x^2 - 3x + 9} + 2 = 0$

38. $\sqrt{x^2 + 3x + 7} - \sqrt{x^2 + 3x + 9} = 2$

39. $|2x + 3| = 7$

40. $|3x - 1| = 5$

41. $|2 - 3x| = 7$

42. $|1 - 2x| = 3$

In Problems 43–62, solve each inequality.

43. $\dfrac{2x-3}{5} + 2 \le \dfrac{x}{2}$

44. $\dfrac{5-x}{3} \le 6x - 4$

45. $-9 \le \dfrac{2x+3}{-4} \le 7$

46. $-4 < \dfrac{2x-2}{3} < 6$

47. $6 > \dfrac{3-3x}{12} > 2$

48. $6 > \dfrac{5-3x}{2} \ge -3$

49. $2x^2 + 5x - 12 < 0$

50. $3x^2 - 2x - 1 \ge 0$

51. $\dfrac{6}{x+3} \ge 1$

52. $\dfrac{-2}{1-3x} < 1$

53. $\dfrac{2x-6}{1-x} < 2$

54. $\dfrac{3-2x}{2x+5} \ge 2$

55. $\dfrac{(x-2)(x-1)}{x-3} > 0$

56. $\dfrac{x+1}{x(x-5)} \le 0$

57. $\dfrac{x^2 - 8x + 12}{x^2 - 16} > 0$

58. $\dfrac{x(x^2 + x - 2)}{x^2 + 9x + 20} \le 0$

59. $|3x + 4| < \frac{1}{2}$

60. $|1 - 2x| < \frac{1}{3}$

61. $|2x - 5| \ge 9$

62. $|3x + 1| \ge 10$

In Problems 63–66, what number should be added to complete the square of each expression.

63. $x^2 + 6x$

64. $x^2 - 10x$

65. $x^2 - \dfrac{4}{3}x$

66. $x^2 + \dfrac{4}{5}x$

67. Lightning and Thunder A flash of lightning is seen, and the resulting thunderclap is heard 3 seconds later. If the speed of sound averages 1100 feet per second, how far away is the storm?

68. Physics: Intensity of Light The intensity I (in candlepower) of a certain light source obeys the equation $I = 900/x^2$, where x is the distance (in meters) from the light. Over what range of distances can an object be placed from this light source so that the range of intensity of light is from 1600 to 3600 candlepower, inclusive?

69. Extent of Search and Rescue A search plane has a cruising speed of 250 miles per hour and carries enough fuel for at most 5 hours of flying. If there is a wind that averages 30 miles per hour and the direction of search is with the wind one way and against it the other, how far can the search plane travel?

70. Extent of Search and Rescue If the search plane described in Problem 69 is able to add a supplementary fuel tank that allows for an additional 2 hours of flying, how much farther can the plane extend its search?

71. Rescue at Sea A life raft, set adrift from a sinking ship 150 miles offshore, travels directly toward a Coast Guard station at the rate of 5 miles per hour. At the time the raft is set adrift, a rescue helicopter is dispatched from the Coast Guard station. If the helicopter's average speed is 90 miles per hour, how long will it take the helicopter to reach the life raft?

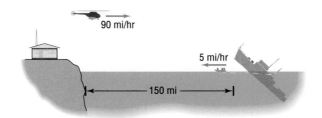

72. Physics: Uniform Motion Two bees leave two locations 150 meters apart and fly, without stopping, back and forth between these two locations at average speeds of 3 meters per second and 5 meters per second, respectively. How long is it until the bees meet for the first time? How long is it until they meet for the second time?

73. Working Together to Get a Job Done Clarissa and Shawna, working together, can paint the exterior of a house in 6 days. Clarissa by herself can complete this job in 5 days less than Shawna. How long will it take Clarissa to complete the job by herself?

74. Emptying a Tank Two pumps of different sizes, working together, can empty a fuel tank in 5 hours. The larger pump can empty this tank in 4 hours less than the smaller one. If the larger one is out of order, how long will it take the smaller one to do the job alone?

75. Chemistry: Mixing Acids A laboratory has 60 cubic centimeters (cc) of a solution that is 40% HCl acid. How many cc of a 15% solution of HCl acid should be mixed with the 60 cc of 40% acid to obtain a solution of 25% HCl? How much of the 25% solution is there?

76. Business: Blending Coffee A coffee house has 20 pounds of a coffee that sells for $4 per pound. How many pounds of a coffee that sells for $8/lb should be mixed with the 20 pounds of $4 coffee to obtain a blend that will sell for $5 per pound? How much of the $5/lb coffee is there to sell?

77. Chemistry: Salt Solutions How much water should be added to 64 ounces of a 10% salt solution to make a 2% salt solution?

78. Chemistry: Salt Solutions How much water must be evaporated from 64 ounces of a 2% salt solution to make a 10% salt solution?

79. Physics: Uniform Motion A man is walking at an average speed of 4 miles per hour alongside a railroad track. A freight train, going in the same direction at an average speed of 30 miles per hour, requires 5 seconds to pass the man. How long is the freight train? Give your answer in feet.

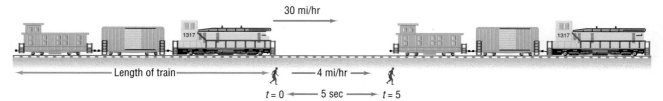

80. Framing a Painting An artist has 50 inches of oak trim to frame a painting. The frame is to have a border 3 inches wide surrounding the painting.
(a) If the painting is square, what are its dimensions? What are the dimensions of the frame?
(b) If the painting is rectangular with a length twice its width, what are the dimensions of the painting? What are the dimensions of the frame?

81. Using Two Pumps An 8 horsepower (hp) pump can fill a tank in 8 hours. A smaller, 3 hp pump fills the same tank in 12 hours. The pumps are used together to begin filling this tank. After four hours, the 8 hp pump breaks down. How long will it take the smaller pump to fill the tank?

82. Pleasing Proportions One formula stating the relationship between the length l and width w of a rectangle of "pleasing proportion" is $l^2 = w(l + w)$. How should a 4 foot by 8 foot sheet of plasterboard be cut so that the result is a rectangle of "pleasing proportion" with a width of 4 feet?

83. Business: Determining the Cost of a Charter A group of 20 senior citizens can charter a bus for a one-day excursion trip for $15 per person. The charter company agrees to reduce the price of each ticket by 10¢ for each additional passenger in excess of 20 who goes on the trip, up to a maximum of 44 passengers (the capacity of the bus). If the final bill from the charter company was $482.40, how many seniors went on the trip, and how much did each pay?

84. Utilizing Copying Machines A new copying machine can do a certain job in 1 hour less than an older copier. Together they can do this job in 72 minutes. How long would it take the older copier by itself to do the job?

85. In a 100-meter race, Todd crosses the finish line 5 meters ahead of Scott. To even things up, Todd suggests to Scott that they race again, this time with Todd lining up 5 meters behind the start.
(a) Assuming that Todd and Scott run the same pace as before, does the second race end in a tie?
(b) If not, who wins?
(c) By how many meters does he win?
(d) How far back should Todd start so that the race ends in a tie?

After running the race a second time, Scott, to even things up, suggests to Todd that he (Scott) line up 5 meters in front of the start.
(e) Assuming again that they run at the same pace as in the first race, does the third race result in a tie?
(f) If not, who wins?
(g) By how many meters?
(h) How far up should Scott start so that the race ends in a tie?

86. Explain the difference between the following three problems. Are there any similarities in their solution?
(a) Write the expression as a single quotient:
$$\frac{x}{x - 2} + \frac{x}{x^2 - 4}$$
(b) Solve: $\dfrac{x}{x - 2} + \dfrac{x}{x^2 - 4} = 0$
(c) Solve: $\dfrac{x}{x - 2} + \dfrac{x}{x^2 - 4} < 0$

Graphs

Sir Isaac Newton (1642–1727), an Englishman, and Gottfried Wilhelm von Leibniz, a German, share the honor of founding calculus. Newton, also famous for his contributions to physics, taught at Trinity College. What would it have been like to be one of his students? Use the URL address

www.prenhall.com/sullivan

and experience first-hand one of Newton's projects.

PREPARING FOR THIS CHAPTER

Before getting started on this chapter, review the following concepts:

Real Number Line *(p. 13)*
Distance on the Number Line *(p. 16)*
Pythagorean Theorem *(p. 77)*
Completing the Square *(p. 111)*

OUTLINE

THE CATALOG OF EQUATIONS

Trinity College, 1667. You have just begun your college work and one of your teachers is a strange fellow named *Isaac Newton*. His ideas are difficult to understand and very radical. Yet, when you listen to him speak, his passion makes a profound impression on you. His curious experiments and clever devices attract you to him.

After class one day, he asks you to help him with his Catalog of Equations. He quickly explains about the coordinate geometry of Descartes and how algebra and geometry are now joined together. He assigns to you the equation

$$(-2x + x^2 + y^2)^2 = x^2 + y^2$$

He gives you a manuscript for *The Witch,* and asks you to do a similar work for your assigned curve. As you walk away, you feel a great sense of honor in being selected to do research for Isaac Newton.

1. Have you ever thought about such a Catalog of Equations? Take a quick tour of the famous curves at the *MacTutor Website.* Notice the Cartesian formulas (x and y) for the curves. How would you organize a catalog for the classification of algebraic curves?
2. Today a student does not hear much about a Catalog of Equations. Why do you suppose that is?
3. From looking at your equation, can you use any of the symmetry principles to get some idea of its graph? Can you graph this equation on your calculator? Are you able to graph it by making a chart of x and y values?
4. Visit The Visual Dictionary of Plane Curves Website and read about catacaustic curves. Can you determine the catacaustic of your curve with a light source on the edge? How could you graph it?
5. Tie together your explorations into a "manuscript."

The idea of using a system of rectangular coordinates dates back to ancient times, when such a system was used for surveying and city planning. Apollonius of Perga, in 200 BC, used a form of rectangular coordinates in his work on conics, although this use does not stand out as clearly as it does in modern treatments. Sporadic use of rectangular coordinates continued until the 1600's. By that time, algebra had developed sufficiently so that René Descartes (1596–1650) and Pierre de Fermat (1601–1665) could take the crucial step, which was the use of rectangular coordinates to translate geometry problems into algebra problems, and vice versa. This step was supremely important for two reasons. First, it allowed both geometers and algebraists to gain critical new insights into their subjects, which previously had been regarded as separate but now were seen to be connected in many important ways. Second, the insights gained made possible the development of calculus, which greatly enlarged the number of areas in which mathematics could be applied and made possible a much deeper understanding of these areas.

Information, often in the form of data, is the basis for making informed decisions. In this chapter we discuss data that can be represented using rectangular coordinates.

3.1 | RECTANGULAR COORDINATES; SCATTER DIAGRAMS

1 Distance Formula
2 Midpoint Formula
3 Draw and Interpret Scatter Diagrams

We locate a point on the real number line by assigning it a single real number, called the *coordinate of the point.* For work in a two-dimensional plane, we locate points by using two numbers.

FIGURE 1

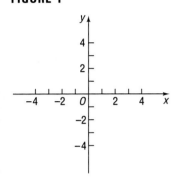

We begin with two real number lines located in the same plane: one horizontal and the other vertical. We call the horizontal line the **x-axis,** the vertical line the **y-axis,** and the point of intersection the **origin O.** We assign coordinates to every point on these number lines as shown in Figure 1, using a convenient scale. In mathematics, we usually use the same scale on each axis; in applications, a different scale is often used on each axis.

The origin O has a value of 0 on both the x-axis and the y-axis. We follow the usual convention that points on the x-axis to the right of O are associated with positive real numbers, and those to the left of O are associated with negative real numbers. Those on the y-axis above O are associated with positive real numbers, and those below O are associated with negative real numbers. In Figure 1, the x-axis and y-axis are labeled as x and y, respectively, and we have used an arrow at the end of each axis to denote the positive direction.

The coordinate system described here is called a **rectangular,** or **Cartesian*** **coordinate system.** The plane formed by the x-axis and y-axis is sometimes called the **xy-plane,** and the x-axis and y-axis are referred to as the **coordinate axes.**

FIGURE 2

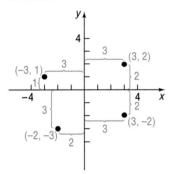

Any point P in the xy-plane can then be located by using an **ordered pair** (x, y) of real numbers. Let x denote the signed distance of P from the y-axis (*signed* in the sense that, if P is to the right of the y-axis, then $x > 0$, and if P is to the left of the y-axis, then $x < 0$); and let y denote the signed distance of P from the x-axis. The ordered pair (x, y), also called the **coordinates** of P, then gives us enough information to locate the point P in the plane.

For example, to locate the point whose coordinates are $(-3, 1)$, go 3 units along the x-axis to the left of O and then go straight up 1 unit. We **plot** this point by placing a dot at this location. See Figure 2 in which the points with coordinates $(-3, 1)$, $(-2, -3)$, $(3, -2)$, and $(3, 2)$ are plotted.

The origin has coordinates $(0, 0)$. Any point on the x-axis has coordinates of the form $(x, 0)$, and any point on the y-axis has coordinates of the form $(0, y)$.

If (x, y) are the coordinates of a point P, then x is called the **x-coordinate,** or **abscissa,** of P, and y is called the **y-coordinate,** or **ordinate,** of P. We identify the point P by its coordinates (x, y) by writing $P = (x, y)$. Usually, we will simply say "the point (x, y)" rather than "the point whose coordinates are (x, y)."

FIGURE 3

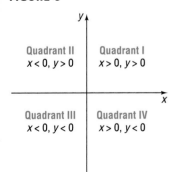

The coordinate axes divide the xy-plane into four sections, called **quadrants,** as shown in Figure 3. In quadrant I, both the x-coordinate and the y-coordinate of all points are positive; in quadrant II, x is negative and y is positive; in quadrant III, both x and y are negative; and in quadrant IV, x is positive and y is negative. Points on the coordinate axes belong to no quadrant.

Now work Problem 1.

---

*Named after René Descartes (1596–1650), a French mathematician, philosopher, and theologian.

FIGURE 4

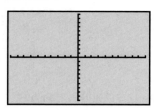

Comment   On a graphing calculator, you can set the scale on each axis. Once this has been done, you obtain the viewing rectangle. See Figure 4 for a typical **viewing rectangle.** You should now read Section 1, The Viewing Rectangle, in the Appendix, page 709.   ▬

Distance between Points

If the same units of measurement, such as inches, centimeters, and so on, are used for both the x-axis and the y-axis, then all distances in the xy-plane can be measured using this unit of measurement.

E X A M P L E   1

Finding the Distance between Two Points

Find the distance d between the points $(1, 3)$ and $(5, 6)$.

Solution   First we plot the points $(1, 3)$ and $(5, 6)$ as shown in Figure 5(a). Then we draw a horizontal line from $(1, 3)$ to $(5, 3)$ and a vertical line from $(5, 3)$ to $(5, 6)$, forming a right triangle, as in Figure 5(b). One leg of the triangle is of length 4 and the other is of length 3. By the Pythagorean Theorem (see Section 1.9), the square of the distance d that we seek is

$$d^2 = 4^2 + 3^2 = 16 + 9 = 25$$
$$d = 5$$

FIGURE 5

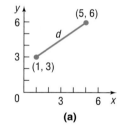

(a)

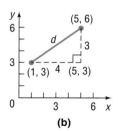

(b)   ▬

The **distance formula** provides a straightforward method for computing the distance between two points.

Theorem   Distance Formula

The distance between two points $P_1 = (x_1, y_1)$ and $P_2 = (x_2, y_2)$, denoted by $d(P_1, P_2)$, is

$$d(P_1, P_2) = \sqrt{(x_2 - x_1)^2 + (y_2 - y_1)^2} \qquad (1)$$

▬

That is, to compute the distance between two points, find the difference of the x-coordinates, square it, and add this to the square of the difference of the y-coordinates. The square root of this sum is the distance. See Figure 6.

FIGURE 6

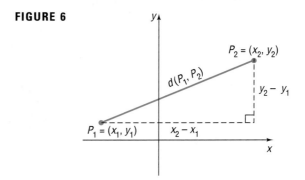

Proof of the Distance Formula   Let (x_1, y_1) denote the coordinates of point P_1, and let (x_2, y_2) denote the coordinates of point P_2. Assume that the line joining P_1 and P_2 is neither horizontal nor vertical. Refer to Figure 7(a). The coordinates of P_3 are (x_2, y_1). The horizontal distance from P_1 to P_3 is the absolute value of the difference of the x-coordinates, $|x_2 - x_1|$. The vertical distance from P_3 to P_2 is the absolute value of the difference of the y-coordinates, $|y_2 - y_1|$. See Figure 7(b). The distance $d(P_1, P_2)$ that we seek is the length of the hypotenuse of the right triangle, so, by the Pythagorean Theorem, it follows that

$$[d(P_1, P_2)]^2 = |x_2 - x_1|^2 + |y_2 - y_1|^2$$
$$= (x_2 - x_1)^2 + (y_2 - y_1)^2$$
$$d(P_1, P_2) = \sqrt{(x_2 - x_1)^2 + (y_2 - y_1)^2}$$

FIGURE 7

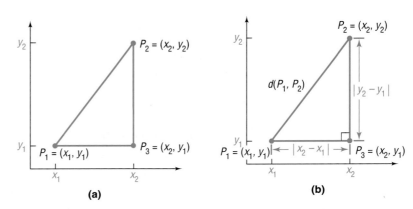

(a)

(b)

Now, if the line joining P_1 and P_2 is horizontal, then the y-coordinate of P_1 equals the y-coordinate of P_2; that is, $y_1 = y_2$. Refer to Figure 8(a). In this case, the distance formula (1) still works, because, for $y_1 = y_2$, it reduces to

$$d(P_1, P_2) = \sqrt{(x_2 - x_1)^2 + 0^2} = \sqrt{(x_2 - x_1)^2} = |x_2 - x_1|$$

A similar argument holds if the line joining P_1 and P_2 is vertical. See Figure 8(b). Thus, the distance formula is valid in all cases.

FIGURE 8

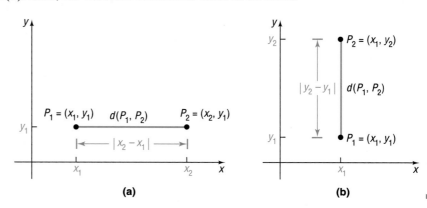

(a)   (b)

EXAMPLE 2

Finding the Distance between Two Points

Find the distance d between the points $(-4, 5)$ and $(3, 2)$.

Solution   Using the distance formula (1), the solution is obtained as follows:

$$d = \sqrt{[3 - (-4)]^2 + (2 - 5)^2} = \sqrt{7^2 + (-3)^2}$$
$$= \sqrt{49 + 9} = \sqrt{58} \approx 7.62$$

Now work Problem 5.

The distance between two points $P_1 = (x_1, y_1)$ and $P_2 = (x_2, y_2)$ is never a negative number. Furthermore, the distance between two points is 0 only when the points are identical, that is, when $x_1 = x_2$ and $y_1 = y_2$. Also, because $(x_2 - x_1)^2 = (x_1 - x_2)^2$ and $(y_2 - y_1)^2 = (y_1 - y_2)^2$, it makes no difference whether the distance is computed from P_1 to P_2 or from P_2 to P_1; that is, $d(P_1, P_2) = d(P_2, P_1)$.

The introduction to this chapter mentioned that rectangular coordinates enable us to translate geometry problems into algebra problems, and vice versa. The next example shows how algebra (the distance formula) can be used to solve geometry problems.

EXAMPLE 3

Using Algebra to Solve Geometry Problems

Consider the three points $A = (-2, 1)$, $B = (2, 3)$, and $C = (3, 1)$.

(a) Plot each point and form the triangle ABC.
(b) Find the length of each side of the triangle.
(c) Verify that the triangle is a right triangle.
(d) Find the area of the triangle.

Solution   (a) Points A, B, and C and triangle ABC are plotted in Figure 9.

(b) $d(A, B) = \sqrt{[2 - (-2)]^2 + (3 - 1)^2} = \sqrt{16 + 4} = \sqrt{20} = 2\sqrt{5}$
$d(B, C) = \sqrt{(3 - 2)^2 + (1 - 3)^2} = \sqrt{1 + 4} = \sqrt{5}$
$d(A, C) = \sqrt{[3 - (-2)]^2 + (1 - 1)^2} = \sqrt{25 + 0} = 5$

FIGURE 9

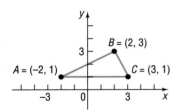

(c) To show that the triangle is a right triangle, we need to show that the sum of the squares of the lengths of two of the sides equals the square of the length of the third side. (Why is this sufficient?) Looking at Figure 9, it seems reasonable to conjecture that the right angle is at vertex B. Thus, we shall check to see whether

$$[d(A, B)]^2 + [d(B, C)]^2 = [d(A, C)]^2$$

We find that

$$[d(A, B)]^2 + [d(B, C)]^2 = (2\sqrt{5})^2 + (\sqrt{5})^2$$
$$= 20 + 5 = 25 = [d(A, C)]^2$$

so it follows from the converse of the Pythagorean Theorem that triangle ABC is a right triangle.

(d) Because the right angle is at B, the sides AB and BC form the base and altitude of the triangle. Its area is therefore

$$\text{Area} = \frac{1}{2}(\text{Base})(\text{Altitude}) = \frac{1}{2}(2\sqrt{5})(\sqrt{5}) = 5 \text{ square units}$$

 Now work Problem 19.

Midpoint Formula

FIGURE 10

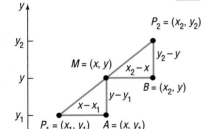

We now derive a formula for the coordinates of the **midpoint of a line segment**. Let $P_1 = (x_1, y_1)$ and $P_2 = (x_2, y_2)$ be the endpoints of a line segment, and let $M = (x, y)$ be the point on the line segment that is the same distance from P_1 as it is from P_2. See Figure 10. The triangles P_1AM and MBP_2 are congruent.* [Do you see why? Angle $AP_1M =$ Angle BMP_2.† Angle $P_1MA =$ Angle MP_2B, and $d(P_1, M) = d(M, P_2)$ is given. Thus, we have Angle–Side–Angle.] Hence, corresponding sides are equal in length. That is,

$$x - x_1 = x_2 - x \quad \text{and} \quad y - y_1 = y_2 - y$$
$$2x = x_1 + x_2 \qquad\qquad 2y = y_1 + y_2$$
$$x = \frac{x_1 + x_2}{2} \qquad\qquad y = \frac{y_1 + y_2}{2}$$

> **Theorem** Midpoint Formula
>
> The midpoint (x, y) of the line segment from $P_1 = (x_1, y_1)$ to $P_2 = (x_2, y_2)$ is
>
> $$(x, y) = \left(\frac{x_1 + x_2}{2}, \frac{y_1 + y_1}{2} \right) \qquad (2)$$

*The following statement is a postulate from geometry. Two triangles are congruent if their sides are the same length (SSS), or if two sides and the included angle are the same (SAS), or if two angles and the included side are the same (ASA).

†Another postulate from geometry states that the transversal $\overline{P_1P_2}$, forms equal corresponding angles with the parallel lines $\overline{P_1A}$ and $\overline{MB}$.

Thus, to find the midpoint of a line segment, we average the *x*-coordinates and the *y*-coordinates of the endpoints.

EXAMPLE 4

Finding the Midpoint of a Line Segment

Find the midpoint of a line segment from $P_1 = (-5, 3)$ to $P_2 = (3, 1)$. Plot the points P_1 and P_2 and their midpoint. Check your answer.

Solution

We apply the midpoint formula (2) using $x_1 = -5, x_2 = 3, y_1 = 3,$ and $y_2 = 1$. Then the coordinates (x, y) of the midpoint M are

$$x = \frac{x_1 + x_2}{2} = \frac{-5 + 3}{2} = -1 \quad \text{and} \quad y = \frac{y_1 + y_2}{2} = \frac{3 + 1}{2} = 2$$

That is, $M = (-1, 2)$. See Figure 11.

FIGURE 11

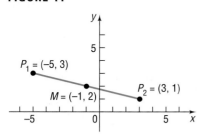

Check: Because M is the midpoint, we check the answer by verifying that $d(P_1, M) = d(M, P_2)$:

$$d(P_1, M) = \sqrt{[-1 - (-5)]^2 + (2 - 3)^2} = \sqrt{16 + 1} = \sqrt{17}$$
$$d(M, P_2) = \sqrt{[3 - (-1)]^2 + (1 - 2)^2} = \sqrt{16 + 1} = \sqrt{17}$$

 Now work Problem 29.

Scatter Diagrams

 A **relation** is a correspondence between two variables, say, *x* and *y*. When relations are written as ordered pairs (x, y), we say that *x* is related to *y*. Often, we are interested in specifying the type of relation (such as an equation) that might exist between two variables. The first step in finding this relation is to plot the ordered pairs using rectangular coordinates. The resulting graph is called a **scatter diagram.**

EXAMPLE 5

Drawing a Scatter Diagram

The data in Table 1 represent the total number *x* of accidents for all U.S. General Aviation Flying and the corresponding total number *y* of fatalities as a result of the accidents for the years 1984–1993.

(a) Draw a scatter diagram.

(b) Use a graphing utility to draw a scatter diagram.

(c) Describe what happens to the total number of fatalities as the number of accidents increases.

TABLE 1		
Total Number of Accidents, x	**Total Fatalities, y**	**(x, y)**
3016	1042	(3016, 1042)
2738	955	(2738, 955)
2582	967	(2582, 967)
2494	838	(2494, 838)
2386	800	(2386, 800)
2230	768	(2230, 768)
2214	766	(2214, 766)
2170	781	(2170, 781)
2074	862	(2074, 862)
2022	715	(2022, 715)

Source: National Transportation Safety Board.

Solution (a) To draw a scatter diagram, we plot the ordered pairs listed in Table 1, using total number of accidents as the *x*-coordinate and total number of fatalities as the *y*-coordinate. See Figure 12(a). Notice that the points in a scatter diagram are not connected.

(b) Figure 12(b) shows the scatter diagram of the data using a graphing utility.

FIGURE 12

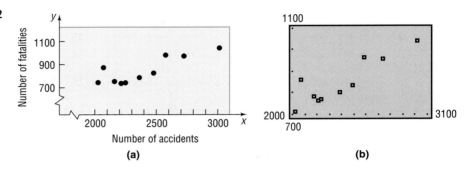

(a)

(b)

(c) From the scatter diagram, as the number of accidents increases, we see that the number of fatalities increases. ▬

Now work Problem 49.

3.1 | EXERCISES

In Problems 1 and 2, plot each point in the xy-plane. Tell in which quadrant or on what coordinate axis each point lies.

1. (a) $A = (-3, 2)$ (b) $B = (6, 0)$ (c) $C = (-2, -2)$
    (d) $D = (6, 5)$ (e) $E = (0, -3)$ (f) $F = (6, -3)$

2. (a) $A = (1, 4)$ (b) $B = (-3, -4)$ (c) $C = (-3, 4)$
    (d) $D = (4, 1)$ (e) $E = (0, 1)$ (f) $F = (-3, 0)$

3. Plot the points $(2, 0)$, $(2, -3)$, $(2, 4)$, $(2, 1)$, and $(2, -1)$. Describe the set of all points of the form $(2, y)$, where y is a real number.

4. Plot the points $(0, 3)$, $(1, 3)$, $(-2, 3)$, $(5, 3)$, and $(-4, 3)$. Describe the set of all points of the form $(x, 3)$, where x is a real number.

In Problems 5–18, find the distance $d(P_1, P_2)$ between the points P_1 and P_2.

5.
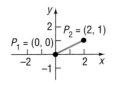

6.
7.
8.
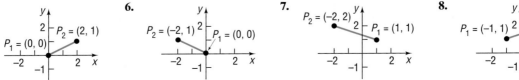
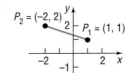
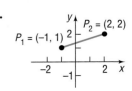

9. $P_1 = (3, -4); \quad P_2 = (5, 4)$

10. $P_1 = (-1, 0); \quad P_2 = (2, 4)$

11. $P_1 = (-3, 2); \quad P_2 = (6, 0)$

12. $P_1 = (2, -3); \quad P_2 = (4, 2)$

13. $P_1 = (4, -3); \quad P_2 = (6, 4)$

14. $P_1 = (-4, -3); \quad P_2 = (6, 2)$

15. $P_1 = (-0.2, 0.3); \quad P_2 = (2.3, 1.1)$

16. $P_1 = (1.2, 2.3); \quad P_2 = (-0.3, 1.1)$

17. $P_1 = (a, b); \quad P_2 = (0, 0)$

18. $P_1 = (a, a); \quad P_2 = (0, 0)$

In Problems 19–28, plot each point and form the triangle ABC. Verify that the triangle is a right triangle. Find its area.

19. $A = (-2, 5); \quad B = (1, 3); \quad C = (-1, 0)$

20. $A = (-2, 5); \quad B = (12, 3); \quad C = (10, -11)$

21. $A = (-5, 3); \quad B = (6, 0); \quad C = (5, 5)$

22. $A = (-6, 3); \quad B = (3, -5); \quad C = (-1, 5)$

23. $A = (4, -3); \quad B = (0, -3); \quad C = (4, 2)$

24. $A = (4, -3); \quad B = (4, 1); \quad C = (2, 1)$

25. Find all points having an x-coordinate of 2 whose distance from the point $(-2, -1)$ is 5.

26. Find all points having a y-coordinate of -3 whose distance from the point $(1, 2)$ is 13.

27. Find all points on the x-axis that are 5 units from the point $(4, -3)$.

28. Find all points on the y-axis that are 5 units from the point $(4, 4)$.

In Problems 29–38, find the midpoint of the line segment joining the points P_1 and P_2.

29. $P_1 = (5, -4); \quad P_2 = (3, 2)$

30. $P_1 = (-1, 0); \quad P_2 = (2, 4)$

31. $P_1 = (-3, 2); \quad P_2 = (6, 0)$

32. $P_1 = (2, -3); \quad P_2 = (4, 2)$

33. $P_1 = (4, -3); \quad P_2 = (6, 1)$

34. $P_1 = (-4, -3); \quad P_2 = (2, 2)$

35. $P_1 = (-0.2, 0.3); \quad P_2 = (2.3, 1.1)$

36. $P_1 = (1.2, 2.3); \quad P_2 = (-0.3, 1.1)$

37. $P_1 = (a, b); \quad P_2 = (0, 0)$

38. $P_1 = (a, a); \quad P_2 = (0, 0)$

39. The **medians** of a triangle are the line segments from each vertex to the midpoint of the opposite side (see the figure). Find the lengths of the medians of the triangle with vertices at $A = (0, 0)$, $B = (0, 6)$, and $C = (4, 4)$.

40. An **equilateral triangle** is one in which all three sides are of equal length. If two vertices of an equilateral triangle are $(0, 4)$ and $(0, 0)$, find the third vertex. How many of these triangles are possible?

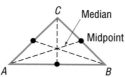

*In Problems 41–44, find the length of each side of the triangle determined by the three points P_1, P_2, and P_3. State whether the triangle is an isosceles triangle, a right triangle, neither of these, or both. (An **isosceles triangle** is one in which at least two of the sides are of equal length.)*

41. $P_1 = (2, 1); \quad P_2 = (-4, 1); \quad P_3 = (-4, -3)$

42. $P_1 = (-1, 4); \quad P_2 = (6, 2); \quad P_3 = (4, -5)$

43. $P_1 = (-2, -1); \quad P_2 = (0, 7); \quad P_3 = (3, 2)$

44. $P_1 = (7, 2); \quad P_2 = (-4, 0); \quad P_3 = (4, 6)$

In Problems 45–48, find the length of the line segment. Assume that the endpoints of each line segment have integer coordinates.

45.

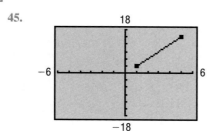

46.

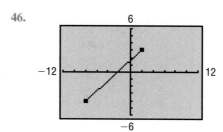

47.

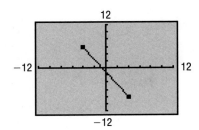

48.

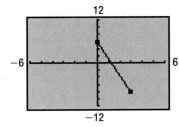

49. Prices of Motorola Stock The following data represent the closing price of Motorola, Inc., stock at the end of each month in 1997.
(a) Draw a scatter diagram.
(b) Using a graphing utility, draw a scatter diagram.
(c) How would you describe the trend in this data over time?

Month	Closing Price
January, 1997	$68\frac{1}{4}$
February, 1997	56
March, 1997	$60\frac{1}{2}$
April, 1997	$57\frac{1}{8}$
May, 1997	$66\frac{3}{8}$
June, 1997	$76\frac{1}{8}$
July, 1997	$80\frac{3}{8}$
August, 1997	$73\frac{3}{8}$
September, 1997	$71\frac{7}{8}$
October, 1997	62
November, 1997	$62\frac{7}{8}$
December, 1997	57

Courtesy of A. G. Edwards & Sons, Inc.

50. Price of Microsoft Corp. Stock The following data represent the closing price of Microsoft Corporation stock at the end of each month in 1997.
(a) Draw a scatter diagram.
(b) Using a graphing utility, draw a scatter diagram.
(c) How would you describe the trend in this data over time?

Month	Closing Price
January, 1997	102
February, 1997	$97\frac{1}{2}$
March, 1997	$91\frac{1}{2}$
April, 1997	$121\frac{1}{2}$
May, 1997	124
June, 1997	$128\frac{3}{8}$
July, 1997	$141\frac{3}{8}$
August, 1997	$132\frac{1}{8}$
September, 1997	$132\frac{1}{4}$
October, 1997	138
November, 1997	$141\frac{1}{2}$
December, 1997	$129\frac{3}{4}$

Courtesy of A. G. Edwards & Sons, Inc.

51. Fuel Consumption The following data represent average fuel consumption per automobile for the years 1984–1993.
(a) Draw a scatter diagram.
(b) Using a graphing utility, draw a scatter diagram.
(c) How would you describe the trend in this data over time?

Year	Average Fuel Consumption
1984	536
1985	525
1986	526
1987	514
1988	509
1989	509
1990	502
1991	496
1992	512
1993	513

Source: U.S. Federal Highway Administration.

52. Gross Domestic Product The following data represent per capita Gross Domestic Product (GDP) of the United States for the years 1985–1994 in constant 1987 dollars (dollars adjusted for inflation). GDP is the total value of all goods and services produced in the United States.

Year	Per Capita Gross Domestic Product
1985	17,944
1986	18,299
1987	18,694
1988	19,252
1989	19,556
1990	19,593
1991	19,263
1992	19,490
1993	19,879
1994	20,476

Source: U.S. Bureau of Economic Analysis.

(a) Draw a scatter diagram.
(b) Using a graphing utility, draw a scatter diagram.
(c) How would you describe the trend in this data over time?

53. Disposable Personal Income The following data represent per capita disposable personal income x and per capita personal consumption expenditures y for the years 1985–1994 in current dollars (dollars adjusted for inflation).
(a) Draw a scatter diagram.
(b) Using a graphing utility, draw a scatter diagram.
(c) What happens to per capita personal consumption as per capita disposable income increases?

Per Capita Disposable Income, x	Per Capita Personal Consumption, y
13,258	12,013
13,552	12,336
13,545	12,568
13,890	12,903
14,005	13,025
14,101	13,063
14,003	12,860
14,279	13,110
14,341	13,361
14,696	13,711

Source: U.S. Bureau of Economic Analysis.

54. Natural Gas versus Coal The following data represent the amount of energy provided by natural gas and coal measured in trillions of Btu's (British Thermal Units) for the New England states. Let the amount of energy provided by natural gas be x and the amount of energy provided by coal be y.

State	Natural Gas, x	Coal, y
Maine	5	22
New Hampshire	17	35
Vermont	8	1
Massachusetts	306	111
Rhode Island	79	0
Connecticut	114	22

Source: U.S. Energy Information Administration.

(a) Draw a scatter diagram.
(b) Using a graphing utility, draw a scatter diagram.
(c) What happens to the amount of energy provided by coal as the amount of energy provided by natural gas increases?

55. Fuel Consumption The data on p. 173 represent the average fuel consumption per car (gallons) and the pollutant emissions of carbon monoxide (thousands of tons) for 1983–1993. Let x represent the average fuel consumption per car and let y represent the pollutant emissions of carbon monoxide.

Average Fuel Consumption Per Car, x	Pollutant Emissions of Carbon Monoxide, y
553	115,334
536	114,262
525	112,072
526	108,070
514	105,117
509	106,100
509	100,806
502	103,753
496	99,898
512	96,368
513	97,208

Source: Statistical Abstract of the United States, 1995.

(a) Draw a scatter diagram.

(b) Using a graphing utility, draw a scatter diagram.

(c) What happens to the level of carbon monoxide as the average fuel consumption per car increases?

56. Hourly Pay and Productivity   The following data represent the average hourly earnings of production workers and productivity (output per hour) for the years 1986–1995. Let x represent productivity and let y represent average hourly earnings.

Productivity, x	Average Hourly Earnings, y
94.2	8.76
94.1	8.98
94.6	9.28
95.3	9.66
96.1	10.01
96.7	10.32
100	10.57
100.2	10.83
100.7	11.12
100.8	11.44

Source: Bureau of Labor Statistics.

(a) Draw a scatter diagram.

(b) Using a graphing utility, draw a scatter diagram.

(c) What happens to the average hourly earnings as productivity increases?

57. Baseball   A major league baseball "diamond" is actually a square, 90 feet on a side (see the figure). What is the distance directly from home plate to second base (the diagonal of the square)?

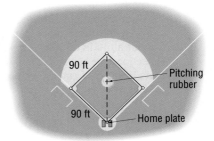

58. Little League Baseball   The layout of a Little League playing field is a square, 60 feet on a side.* How far is it directly from home plate to second base (the diagonal of the square)?

59. Baseball   Refer to Problem 57. Overlay a rectangular coordinate system on a major league baseball diamond so that the origin is at home plate, the positive x-axis lies in the direction from home plate to first base, and the positive y-axis lies in the direction from home plate to third base.

(a) What are the coordinates of first base, second base, and third base? Use feet as the unit of measurement.

(b) If the right fielder is located at (310, 15), how far is it from there to second base?

(c) If the center fielder is located at (300, 300), how far is it from there to third base?

60. Little League Baseball   Refer to Problem 58. Overlay a rectangular coordinate system on a Little League baseball diamond so that the origin is at home plate, the positive x-axis lies in the direction from home plate to first base, and the positive y-axis lies in the direction from home plate to third base.

(a) What are the coordinates of first base, second base, and third base? Use feet as the unit of measurement.

(b) If the right fielder is located at (180, 20), how far is it from there to second base?

(c) If the center fielder is located at (220, 220), how far is it from there to third base?

Source: Little League Baseball, Official Regulations and Playing Rules, 1997.

61. A Dodge Intrepid and a Mack truck leave an intersection at the same time. The Intrepid heads east at an average speed of 30 miles per hour, while the truck heads south at an average speed of 40 miles per hour. Find an expression for their distance apart d (in miles) at the end of t hours.

62. A hot-air balloon, headed due east at an average speed of 15 miles per hour and at a constant altitude of 100 feet, passes over an intersection (see the figure). Find an expression for its distance d (measured in feet) from the intersection t seconds later.

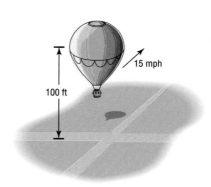

3.2 | GRAPHS OF EQUATIONS

1. Graph Equations
2. Find the Intercepts on a Graph
3. Find the Intercepts from an Equation
4. Test an Equation for Symmetry with Respect to the (a) x-axis, (b) y-axis, and (c) origin

1 Illustrations play an important role in helping us to visualize the relationships that exist between two variable quantities. Figure 13 shows the weekly chart of interest rates for 30-year Treasury bonds from November 29, 1996, through December 31, 1997. For example, the interest rate for January 17, 1997, was about 6.825% (682.5 basis points). Such illustrations are usually referred to as *graphs*. The **graph of an equation** in two variables x and y consists of the set of points in the xy-plane whose coordinates (x, y) satisfy the equation. Let's look at some examples.

FIGURE 13
Thirty-year Treasury rates

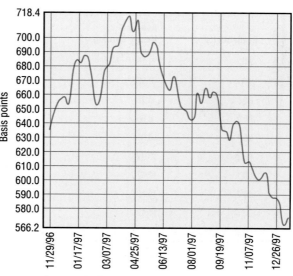

Courtesy of A. G. Edwards & Sons, Inc.

E X A M P L E   1   Graphing an Equation

Graph the equation: $y = 2x + 5$.

Solution   We want to find all points (x, y) that satisfy the equation. To locate some of these points (and thus get an idea of the pattern of the graph), we assign some numbers to x and find corresponding values for y:

If	Then	Point on Graph
$x = 0$	$y = 2(0) + 5 = 5$	$(0, 5)$
$x = 1$	$y = 2(1) + 5 = 7$	$(1, 7)$
$x = -5$	$y = 2(-5) + 5 = -5$	$(-5, -5)$
$x = 10$	$y = 2(10) + 5 = 25$	$(10, 25)$

By plotting these points and then connecting them, we obtain the graph of the equation (a line), as shown in Figure 14.

FIGURE 14
$y = 2x + 5$

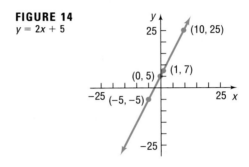

EXAMPLE 2

Graphing an Equation

Graph the equation: $y = x^2$

Solution   Table 2 provides several points on the graph. In Figure 15 we plot these points and connect them with a smooth curve to obtain the graph (a *parabola*).

TABLE 2		
x	$y = x^2$	(x, y)
-4	16	$(-4, 16)$
-3	9	$(-3, 9)$
-2	4	$(-2, 4)$
-1	1	$(-1, 1)$
0	0	$(0, 0)$
1	1	$(1, 1)$
2	4	$(2, 4)$
3	9	$(3, 9)$
4	16	$(4, 16)$

FIGURE 15
$y = x^2$

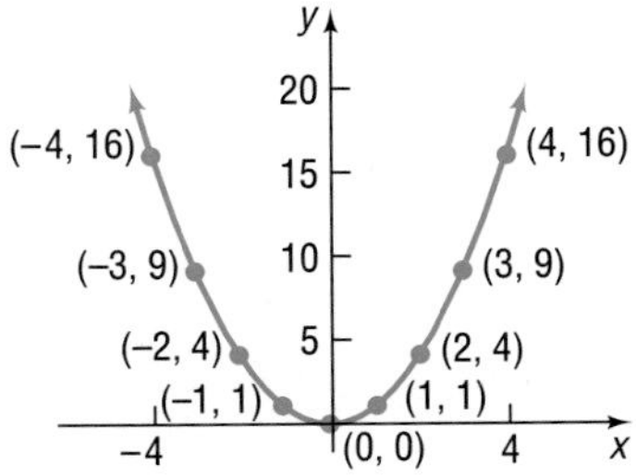

The graphs of the equations shown in Figures 14 and 15 do not show all points. For example, in Figure 14, the point $(20, 45)$ is a part of the graph of $y = 2x + 5$, but it is not shown. Since the graph of $y = 2x + 5$ could be extended out as far as we please, we use arrows to indicate that the pattern shown continues. Thus, it is important when illustrating a graph to present enough of the graph so that any viewer of the illustration will "see" the rest of it as an obvious continuation of what is actually there. This is referred to as a **complete graph.**

So, one way to obtain a complete graph of an equation is to plot a sufficient number of points on the graph until a pattern becomes evident. Then these points are connected with a smooth curve following the suggested pattern. But how many points are sufficient? Sometimes knowledge about the equation tells us. For example, we will learn in the next section that, if an equation is of the form $y = mx + b$, then its graph is a line. In this case, two points would suffice to obtain the graph.

One of the purposes of this book is to investigate the properties of equations in order to decide whether a graph is complete. Sometimes we shall graph equations by plotting a sufficient number of points on the graph until a pattern becomes evident; then we connect these points with a smooth curve following the suggested pattern. (Shortly, we shall investigate various techniques that will enable us to graph an equation without plotting so many points.) Other times we shall graph equations based on properties of the equation.

 Comment   Another way to obtain the graph of an equation is to use a graphing utility. Read Section 2, Using a Graphing Utility to Graph Equations, in the Appendix, page 712.   ▄

E X A M P L E   3   Graphing an Equation

Graph the equation: $y = x^3$

Solution   We set up Table 3, listing several points on the graph. Figure 16 illustrates some of these points and the graph of $y = x^3$.

TABLE 3		
x	**y = x³**	**(x, y)**
−3	−27	(−3, −27)
−2	−8	(−2, −8)
−1	−1	(−1, −1)
0	0	(0, 0)
1	1	(1, 1)
2	8	(2, 8)
3	27	(3, 27)

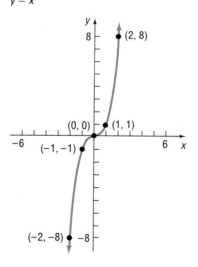

FIGURE 16
$y = x^3$

E X A M P L E   4   Graphing an Equation

Graph the equation: $x = y^2$

Solution   We set up Table 4, listing several points on the graph. In this case, because of the form of the equation, we assign some numbers to y and find corresponding values of x. Figure 17 illustrates some of these points and the graph of $x = y^2$.

y	$x = y^2$	(x, y)
-3	9	$(9, -3)$
-2	4	$(4, -2)$
-1	1	$(1, -1)$
0	0	$(0, 0)$
1	1	$(1, 1)$
2	4	$(4, 2)$
3	9	$(9, 3)$
4	16	$(16, 4)$

TABLE 4

FIGURE 17
$x = y^2$

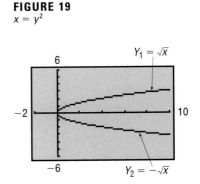

If we restrict y so that $y \geq 0$, the equation $x = y^2$, $y \geq 0$, may be written equivalently as $y = \sqrt{x}$. The portion of the graph of $x = y^2$ in quadrant I is therefore the graph of $y = \sqrt{x}$. See Figure 18.

Comment: To see the graph of the equation $x = y^2$ on a graphing calculator, you will need to graph two equations $Y_1 = \sqrt{x}$ and $Y_2 = -\sqrt{x}$. We discuss why in the next chapter. See Figure 19.

FIGURE 18
$y = \sqrt{x}$

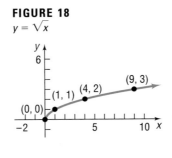

FIGURE 19
$x = y^2$

Now work Problem 27.

We said earlier that we would discuss techniques that reduce the number of points required to graph an equation. Two such techniques involve finding *intercepts* and checking for *symmetry*.

FIGURE 20

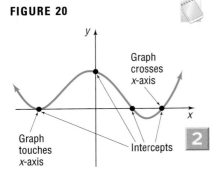

Intercepts

The points, if any, at which a graph crosses or touches the coordinate axes are called the **intercepts**. See Figure 20. The x-coordinate of a point at which the graph crosses or touches the x-axis is an **x-intercept**, and the y-coordinate of a point at which the graph crosses or touches the y-axis is a **y-intercept**.

E X A M P L E   5

Finding Intercepts on a Graph

Find the intercepts of the graph in Figure 21. What are its x-intercepts? What are its y-intercepts?

FIGURE 21

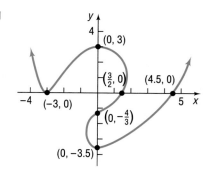

Solution   The intercepts of the graph are the points

$$(-3, 0),\quad (0, 3),\quad \left(\frac{3}{2}, 0\right),\quad \left(0, -\frac{4}{3}\right),\quad (0, -3.5),\quad (4.5, 0)$$

The x-intercepts are $-3, \frac{3}{2}, 4.5$; the y-intercepts are $-3.5, -\frac{4}{3}, 3$. ▬

3   The intercepts of the graph of an equation can be found by using the fact that points on the x-axis have y-coordinates equal to 0, and points on the y-axis have x-coordinates equal to 0.

Procedure
for Finding Intercepts

1. To find the x-intercept(s), if any, of the graph of an equation, let $y = 0$ in the equation and solve for x.
2. To find the y-intercept(s), if any, of the graph of an equation, let $x = 0$ in the equation and solve for y.

E X A M P L E   6

Finding Intercepts from an Equation

Find the x-intercept(s) and the y-intercept(s) of the graph of $y = x^2 - 4$. Graph $y = x^2 - 4$.

Solution   To find the x-intercept(s), we let $y = 0$ and obtain the equation

$$x^2 - 4 = 0$$

The equation has two solutions, -2 and 2. Thus, the x-intercepts are -2 and 2. To find the y-intercept(s), we let $x = 0$ and obtain the equation

$$y = -4$$

Thus, the y-intercept is -4.

Since $x^2 \geq 0$ for all x, we deduce from the equation $y = x^2 - 4$ that $y \geq -4$ for all x. This information, plus the points from Table 5, provide enough information to graph $y = x^2 - 4$. See Figure 22.

TABLE 5		
x	$y = x^2 - 4$	(x, y)
-3	5	$(-3, 5)$
-1	-3	$(-1, -3)$
1	-3	$(1, -3)$
3	5	$(3, 5)$

FIGURE 22
$y = x^2 - 4$

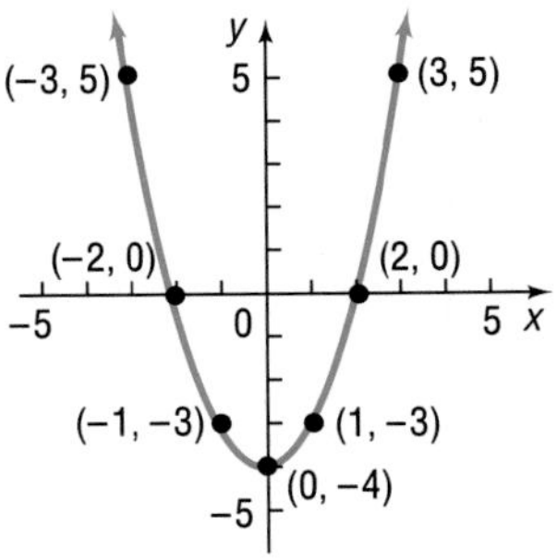

Now work Problem 13(a).

Comment For many equations, finding intercepts may not be so easy. In such cases, a graphing utility can be used. Read Section 3, Using a Graphing Utility to Locate Intercepts, in the Appendix, page 716, to find out how a graphing utility locates intercepts.

Symmetry

We have just seen the role intercepts play in obtaining key points on the graph of an equation. Another helpful tool for graphing equations involves *symmetry*, particularly symmetry with respect to the *x*-axis, the *y*-axis, and the origin.

FIGURE 23
Symmetric with respect to the *x*-axis

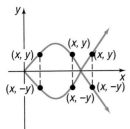

> A graph is said to be **symmetric with respect to the *x*-axis** if, for every point (x, y) on the graph, the point $(x, -y)$ is also on the graph.

Figure 23 illustrates the definition. Notice that, when a graph is symmetric with respect to the *x*-axis, the part of the graph above the *x*-axis is a reflection or mirror image of the part below it, and vice versa.

E X A M P L E 7

Points Symmetric with Respect to the *x*-Axis

If a graph is symmetric with respect to the *x*-axis and the point $(3, 2)$ is on the graph, then the point $(3, -2)$ is also on the graph.

> A graph is said to be **symmetric with respect to the *y*-axis** if, for every point (x, y) on the graph, the point $(-x, y)$ is also on the graph.

Figure 24 illustrates the definition. Notice that, when a graph is symmetric with respect to the y-axis, the part of the graph to the right of the y-axis is a reflection of the part to the left of it, and vice versa.

FIGURE 24
Symmetric with respect to the y-axis

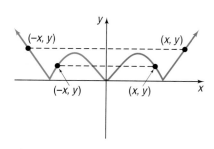

E X A M P L E   8

Points Symmetric with Respect to the y-Axis

If a graph is symmetric with respect to the y-axis and the point (5, 8) is on the graph, then the point (−5, 8) is also on the graph. ▬

FIGURE 25
Symmetric with respect to the origin

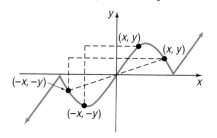

> A graph is said to be **symmetric with respect to the origin** if, for every point (x, y) on the graph, the point (−x, −y) is also on the graph.

Figure 25 illustrates the definition. Notice that symmetry with respect to the origin may be viewed in two ways:

1.   As a reflection about the y-axis, followed by a reflection about the x-axis.
2.   As a projection along a line through the origin so that the distances from the origin are equal.

E X A M P L E   9

Points Symmetric with Respect to the Origin

If a graph is symmetric with respect to the origin and the point (4, 2) is on the graph, then the point (−4, −2) is also on the graph. ▬

Now work Problems 1 and 13(b).

When the graph of an equation is symmetric with respect to a coordinate axis or the origin, the number of points that you need to plot in order to see the pattern is reduced. For example, if the graph of an equation is symmetric with respect to the y-axis, then, once points to the right of the y-axis are plotted, an equal number of points on the graph can be obtained by reflecting them about the y-axis. Thus, before we graph an equation, we first want to determine whether it has any symmetry. The following tests are used for this purpose.

Tests for Symmetry

> To test the graph of an equation for symmetry with respect to the
>
> **x-Axis**   Replace y by −y in the equation. If an equivalent equation results, the graph of the equation is symmetric with respect to the x-axis.
>
> **y-Axis**   Replace x by −x in the equation. If an equivalent equation results, the graph of the equation is symmetric with respect to the y-axis.
>
> **Origin**   Replace x by −x and y by −y in the equation. If an equivalent equation results, the graph of the equation is symmetric with respect to the origin.

Let's look at an equation that we have already graphed to see how these tests are used.

EXAMPLE   10   Testing An Equation for Symmetry $(x = y^2)$

(a) To test the graph of the equation $x = y^2$ for symmetry with respect to the x-axis, we replace y by $-y$ in the equation, as follows:

$$x = y^2 \qquad \text{Original equation.}$$
$$x = (-y)^2 \qquad \text{Replace } y \text{ by } -y.$$
$$x = y^2 \qquad \text{Simplify.}$$

When we replace y by $-y$, the result is the same equation. Thus, the graph is symmetric with respect to the x-axis.

(b) To test the graph of the equation $x = y^2$ for symmetry with respect to the y-axis, we replace x by $-x$ in the equation:

$$x = y^2 \qquad \text{Original equation.}$$
$$-x = y^2 \qquad \text{Replace } x \text{ by } -x.$$

Because we arrive at the equation $-x = y^2$, which is not equivalent to the original equation, we conclude that the graph is not symmetric with respect to the y-axis.

(c) To test for symmetry with respect to the origin, we replace x by $-x$ and y by $-y$:

$$x = y^2 \qquad \text{Original equation.}$$
$$-x = (-y)^2 \qquad \text{Replace } x \text{ by } -x \text{ and } y \text{ by } -y.$$
$$-x = y^2 \qquad \text{Simplify.}$$

The resulting equation, $-x = y^2$, is not equivalent to the original equation. We conclude that the graph is not symmetric with respect to the origin. ∎

Figure 26(a) illustrates the graph of $x = y^2$. In forming a table of points on the graph of $x = y^2$, we can restrict ourselves to points whose y-coordinates are positive. Once these are plotted and connected, a reflection about the x-axis (because of the symmetry) provides the rest of the graph.

Figures 26(b) and (c) illustrate two other equations that we graphed earlier. Notice how the existence of symmetry reduces the number of points that we need to plot.

FIGURE 26

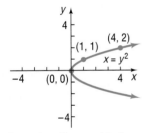

(a) Symmetry with respect to the x-axis

(b) Symmetry with respect to the y-axis

(c) Symmetry with respect to the origin

    Now work Problem 49.

E X A M P L E   11   Graphing the Equation $y = 1/x$

Graph the equation: $y = \dfrac{1}{x}$

Find any intercepts and check for symmetry first.

Solution   We check for intercepts first. If we let $x = 0$, we obtain a 0 denominator, which is not allowed. Hence, there is no y-intercept. If we let $y = 0$, we get the equation $1/x = 0$, which has no solution. Hence, there is no x-intercept. Thus, the graph of $y = 1/x$ does not cross or touch the coordinate axes.

Next we check for symmetry:

x-Axis: Replacing y by $-y$ yields $-y = 1/x$, which is not equivalent to $y = 1/x$.

y-Axis: Replacing x by $-x$ yields $y = -1/x$, which is not equivalent to $y = 1/x$.

Origin: Replacing x by $-x$ and y by $-y$ yields $-y = -1/x$, which is equivalent to $y = 1/x$.

The graph is symmetric with respect to the origin.

Finally, we set up Table 6, listing several points on the graph. Because of the symmetry with respect to the origin, we use only positive values of x. From Table 6, we infer that if x is a large and positive number then $y = 1/x$ is a positive number close to 0. We also infer that if x is a positive number close to 0 then $y = 1/x$ is a large and positive number. Armed with this information, we can graph the equation. Figure 27 illustrates some of these points and the graph of $y = 1/x$. Observe how the absence of intercepts and the existence of symmetry with respect to the origin were utilized.

TABLE 6		
x	**y = 1/x**	**(x, y)**
$\frac{1}{10}$	10	$(\frac{1}{10}, 10)$
$\frac{1}{3}$	3	$(\frac{1}{3}, 3)$
$\frac{1}{2}$	2	$(\frac{1}{2}, 2)$
1	1	$(1, 1)$
2	$\frac{1}{2}$	$(2, \frac{1}{2})$
3	$\frac{1}{3}$	$(3, \frac{1}{3})$
10	$\frac{1}{10}$	$(10, \frac{1}{10})$

FIGURE 27

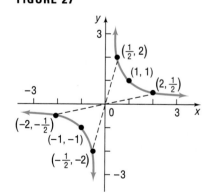

FIGURE 28

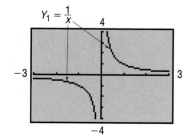

Comment   Figure 28 shows the graph of $y = 1/x$ using a graphing utility. We infer from the graph that there are no intercepts. We may also infer from the graph that there is possible symmetry with respect to the origin. The TRACE feature provides further evidence of this symmetry. As the graph is TRACEd, the points $(-x, -y)$ and (x, y) are obtained. For example, the pair of points $(-0.95238, -1.05)$ and $(0.95238, 1.05)$ both lie on the graph. ▪

3.2 | EXERCISES

In Problems 1–10, plot each point. Then plot the point that is symmetric to it with respect to (a) the x-axis; (b) the y-axis; (c) the origin.

1. $(3, 4)$ 2. $(5, 3)$ 3. $(-2, 1)$ 4. $(4, -2)$ 5. $(1, 1)$

6. $(-1, -1)$ 7. $(-3, -4)$ 8. $(4, 0)$ 9. $(0, -3)$ 10. $(-3, 0)$

In Problems 11–26, the graph of an equation is given.

(a) List the intercepts of the graph.

(b) Based on the graph, tell whether the graph is symmetric with respect to the x-axis, y-axis, and/or origin.

11.
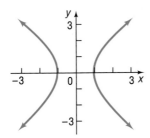

12.
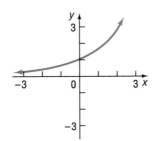

13.
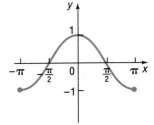

14.
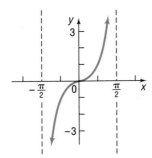

15.
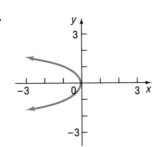

16.
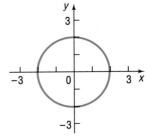

17.
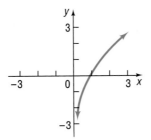

18.
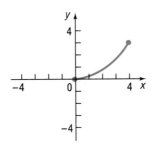

19.
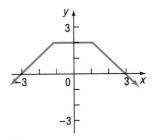

20.
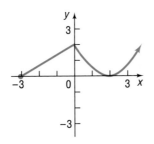

21.
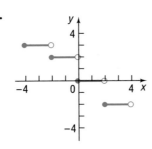

22.
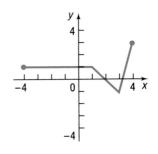

23.

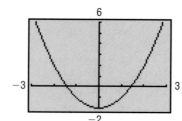

24.

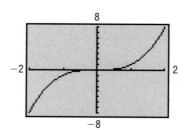

25.

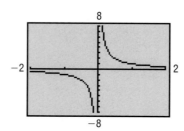

26.

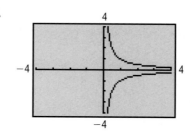

In Problems 27–32, tell whether the given points are on the graph of the equation.

27. Equation: $y = x^4 - \sqrt{x}$
Points: $(0, 0)$; $(1, 1)$; $(-1, 0)$

28. Equation: $y = x^3 - 2\sqrt{x}$
Points: $(0, 0)$; $(1, 1)$; $(1, -1)$

29. Equation: $y^2 = x^2 + 9$
Points: $(0, 3)$; $(3, 0)$; $(-3, 0)$

30. Equation: $y^3 = x + 1$
Points: $(1, 2)$; $(0, 1)$; $(-1, 0)$

31. Equation: $x^2 + y^2 = 4$
Points: $(0, 2)$; $(-2, 2)$; $(\sqrt{2}, \sqrt{2})$

32. Equation: $x^2 + 4y^2 = 4$
Points: $(0, 1)$; $(2, 0)$; $(2, \frac{1}{2})$

33. If $(a, 2)$ is a point on the graph of $y = 3x + 5$, what is a?

34. If $(2, b)$ is a point on the graph of $y = x^2 + 4x$, what is b?

35. If (a, b) is a point on the graph of $2x + 3y = 6$, write an equation that relates a to b.

36. If $(2, 0)$ and $(0, 5)$ are points on the graph of $y = mx + b$, what are m and b?

In Problems 37–40, use the graph on the right.

37. Draw the graph to make it symmetric with respect to the *x*-axis.

38. Draw the graph to make it symmetric with respect to the *y*-axis.

39. Draw the graph to make it symmetric with respect to the origin.

40. Draw the graph to make it symmetric with respect to the *x*-axis, *y*-axis, and origin.

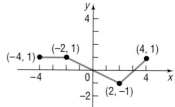

In Problems 41–44, use the graph on the right.

41. Draw the graph to make it symmetric with respect to the *x*-axis.

42. Draw the graph to make it symmetric with respect to the *y*-axis.

43. Draw the graph to make it symmetric with respect to the origin.

44. Draw the graph to make it symmetric with respect to the *x*-axis, *y*-axis, and origin.

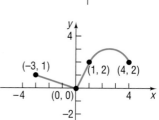

In Problems 45–58, list the intercepts and test for symmetry.

45. $x^2 = y$

46. $y^2 = x$

47. $y = 3x$

48. $y = -5x$

49. $x^2 + y - 9 = 0$

50. $y^2 - x - 4 = 0$

51. $9x^2 + 4y^2 = 36$

52. $4x^2 + y^2 = 4$

53. $y = x^3 - 27$

54. $y = x^4 - 1$

55. $y = x^2 - 3x - 4$

56. $y = x^2 + 4$

57. $y = \dfrac{x}{x^2 + 9}$

58. $y = \dfrac{x^2 - 4}{x}$

In Problem 59, you may use a graphing utility, but it is not required.

59. (a) Graph $y = \sqrt{x^2}, y = x, y = |x|,$ and $y = (\sqrt{x})^2,$ noting which graphs are the same.

   (b) Explain why the graphs of $y = \sqrt{x^2}$ and $y = |x|$ are the same.

   (c) Explain why the graphs of $y = x$ and $y = (\sqrt{x})^2$ are not the same.

   (d) Explain why the graphs of $y = \sqrt{x^2}$ and $y = x$ are not the same.

60. Make up an equation with the intercepts $(2, 0)$, $(4, 0)$, and $(0, 1)$. Compare your equation with a friend's equation. Comment on any similarities.

61. An equation is being tested for symmetry with respect to the x-axis, the y-axis, and the origin. Explain why, if two of these symmetries are present, the remaining one must also be present.

62. Draw a graph that contains the points $(-2, -1)$, $(0, 1)$, $(1, 3)$, and $(3, 5)$. Compare your graph with those of other students. Are most of the graphs almost straight lines? How many are "curved"? Discuss the various ways that these points might be connected.

3.3 | LINES

1. Calculate and Interpret the Slope of a Line
2. Graph Lines
3. Find the Equation of Vertical Lines
4. Use the Point–Slope Form of a Line; Identify Horizontal Lines
5. Find the Equation of a Line from Two Points
6. Write the Equation of a Line in Slope–Intercept Form
7. Identify the Slope and y-Intercept of a Line from its Equation
8. Write the Equation of a Line in General Form

In this section we study a certain type of equation that contains two variables, called a *linear equation,* and its graph, a *line.*

Slope of a Line

FIGURE 29

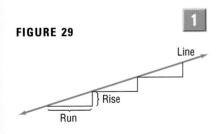
Line
} Rise
Run

1 Consider the staircase illustrated in Figure 29. Each step contains exactly the same horizontal **run** and the same vertical **rise.** The ratio of the rise to the run, called the *slope,* is a numerical measure of the steepness of the staircase. For example, if the run is increased and the rise remains the same, the staircase becomes less steep. If the run is kept the same, but the rise is increased, the staircase becomes more steep. This important characteristic of a line is best defined using rectangular coordinates.

Let $P = (x_1, y_1)$ and $Q = (x_2, y_2)$ be two distinct points with $x_1 \neq x_2$. The **slope** m of the nonvertical line L containing P and Q is defined by the formula

$$m = \frac{y_2 - y_1}{x_2 - x_1} \qquad x_1 \neq x_2 \qquad \text{(1)}$$

If $x_1 = x_2,$ L is a **vertical line** and the slope m of L is **undefined** (since this results in division by 0).

Figure 30(a) provides an illustration of the slope of a nonvertical line; Figure 30(b) illustrates a vertical line.

FIGURE 30

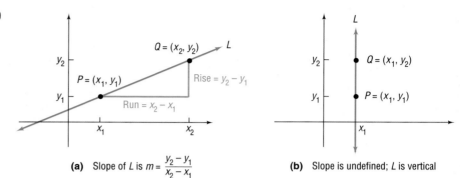

(a) Slope of L is $m = \dfrac{y_2 - y_1}{x_2 - x_1}$

(b) Slope is undefined; L is vertical

As Figure 30(a) illustrates, the slope m of a nonvertical line may be viewed as

$$m = \frac{y_2 - y_1}{x_2 - x_1} = \frac{\text{Rise}}{\text{Run}}$$

We can also express the slope m of a nonvertical line as

$$m = \frac{y_2 - y_1}{x_2 - x_1} = \frac{\text{Change in } y}{\text{Change in } x} = \frac{\Delta y}{\Delta x}$$

That is, the slope m of a nonvertical line L measures the amount y changes as x changes from x_1 to x_2. This is called the **average rate of change** of y with respect to x.

Two comments about computing the slope of a nonvertical line may prove helpful:

1. Any two distinct points on the line can be used to compute the slope of the line. (See Figure 31 for justification.)

FIGURE 31
Triangles ABC and PQR are similar (equal angles). Hence, ratios of corresponding sides are proportional. Thus:

Slope using P and $Q = \dfrac{y_2 - y_1}{x_2 - x_1}$

$\qquad = $ Slope using A and $B = \dfrac{d(B, C)}{d(A, C)}$

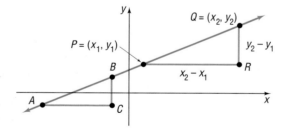

2. The slope of a line may be computed from $P = (x_1, y_1)$ to $Q = (x_2, y_2)$ or from Q to P, because

$$\frac{y_2 - y_1}{x_2 - x_1} = \frac{y_1 - y_2}{x_1 - x_2}$$

E X A M P L E   1   Finding and Interpreting the Slope of a Line Joining Two Points

The slope m of the line joining the points $(1, 2)$ and $(5, -3)$ may be computed as

$$m = \frac{-3 - 2}{5 - 1} = \frac{-5}{4} \quad \text{or as} \quad m = \frac{2 - (-3)}{1 - 5} = \frac{5}{-4} = \frac{-5}{4}$$

For every 4-unit change in x, y will change by -5 units. That is, if x increases by 4 units, then y decreases by 5 units. The average rate of change of y with respect to x is $\frac{-5}{4}$.

Now work Problem 7.

To get a better idea of the meaning of the slope m of a line L, consider the following example.

E X A M P L E   2

Finding the Slopes of Various Lines Containing the Same Point (2, 3)

Compute the slopes of the lines L_1, L_2, L_3, and L_4 containing the following pairs of points. Graph all four lines on the same set of coordinate axes.

$$
\begin{array}{lll}
L_1: & P = (2, 3) & Q_1 = (-1, -2) \\
L_2: & P = (2, 3) & Q_2 = (3, -1) \\
L_3: & P = (2, 3) & Q_3 = (5, 3) \\
L_4: & P = (2, 3) & Q_4 = (2, 5)
\end{array}
$$

Solution

FIGURE 32

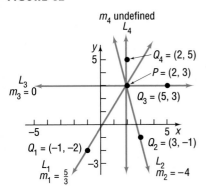

Let m_1, m_2, m_3, and m_4 denote the slopes of the lines L_1, L_2, L_3, and L_4, respectively. Then

$$m_1 = \frac{-2 - 3}{-1 - 2} = \frac{-5}{-3} = \frac{5}{3} \quad \text{A rise of 5 divided by a run of 3.}$$

$$m_2 = \frac{-1 - 3}{3 - 2} = \frac{-4}{1} = -4$$

$$m_3 = \frac{3 - 3}{5 - 2} = \frac{0}{3} = 0$$

m_4 is undefined

The graphs of these lines are given in Figure 32.

Figure 32 illustrates the following facts:

FIGURE 33

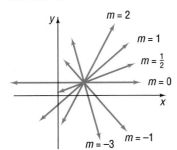

1. When the slope of a line is positive, the line slants upward from left to right (L_1).
2. When the slope of a line is negative, the line slants downward from left to right (L_2).
3. When the slope is 0, the line is horizontal (L_3).
4. When the slope is undefined, the line is vertical (L_4).

Figure 33 illustrates some additional facts about the slope of a line. Note that the closer the line is to the vertical position, the greater the magnitude of the slope.

Comment Now read Section 4, Square Screens, in the Appendix, page 720.

FIGURE 34

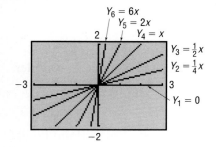

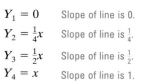 **Seeing the Concept** On the same square screen, graph the following equations:

$Y_1 = 0$ Slope of line is 0.

$Y_2 = \frac{1}{4}x$ Slope of line is $\frac{1}{4}$.

$Y_3 = \frac{1}{2}x$ Slope of line is $\frac{1}{2}$.

$Y_4 = x$ Slope of line is 1.

$Y_5 = 2x$ Slope of line is 2.

$Y_6 = 6x$ Slope of line is 6.

See Figure 34. ∎

FIGURE 35

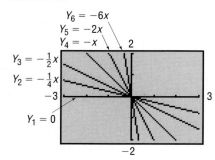

Seeing the Concept On the same square screen, graph the following equations:

$Y_1 = 0$ Slope of line is 0.

$Y_2 = -\frac{1}{4}x$ Slope of line is $-\frac{1}{4}$.

$Y_3 = -\frac{1}{2}x$ Slope of line is $-\frac{1}{2}$.

$Y_4 = -x$ Slope of line is -1.

$Y_5 = -2x$ Slope of line is -2.

$Y_6 = -6x$ Slope of line is -6.

See Figure 35. ∎

2   The next example illustrates how the slope of a line can be used to graph the line.

E X A M P L E   3   Graphing a Line Given a Point and a Slope

Draw a graph of the line that passes through the point $(3, 2)$ and has a slope of:

(a) $\frac{3}{4}$ (b) $-\frac{4}{5}$

Solution   (a) Slope = Rise/Run. The fact that the slope is $\frac{3}{4}$ means that for every horizontal movement (run) of 4 units to the right there will be a vertical movement (rise) of 3 units. If we start at the given point $(3, 2)$ and move 4 units to the right and 3 units up, we reach the point $(7, 5)$. By drawing the line through this point and the point $(3, 2)$, we have the graph. See Figure 36.

(b) The fact that the slope is

FIGURE 36

Slope = $\frac{3}{4}$

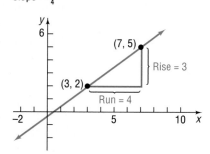

$$-\frac{4}{5} = \frac{-4}{5} = \frac{\text{Rise}}{\text{Run}}$$

means that for every horizontal movement of 5 units to the right there will be a corresponding vertical movement of -4 units (a downward movement). If we start at the given point $(3, 2)$ and move 5 units to the right and then 4 units down, we arrive at the point $(8, -2)$. By drawing the line through these points, we have the graph. See Figure 37.

FIGURE 37
Slope $= -\dfrac{4}{5}$

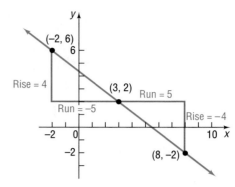

Alternatively, we can set

$$-\frac{4}{5} = \frac{4}{-5} = \frac{\text{Rise}}{\text{Run}}$$

so that for every horizontal movement of -5 units (a movement to the left) there will be a corresponding vertical movement of 4 units (upward). This approach brings us to the point $(-2, 6)$, which is also on the graph shown in Figure 37. ▬

 Now work Problem 17.

Equations of Lines

 Now that we have discussed the slope of a line, we are ready to derive equations of lines. As we shall see, there are several forms of the equation of a line. Let's start with an example.

E X A M P L E   4

Graphing a Line

Graph the equation: $x = 3$.

Solution

FIGURE 38

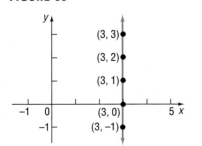

We are looking for all points (x, y) in the plane for which $x = 3$. Thus, no matter what y-coordinate is used, the corresponding x-coordinate always equals 3. Consequently, the graph of the equation $x = 3$ is a vertical line with x-intercept 3 and undefined slope. See Figure 38. ▬

As suggested by Example 4, we have the following result:

Theorem   Equation of a Vertical Line

A vertical line is given by the equation of the form

$$x = a$$

where a is the x-intercept.

▬

Comment   To graph an equation using a graphing utility, we need to express the equation in the form $y =$ expression in x. But $x = 3$ cannot be put in this form. To overcome this, most graphing utilities have special ways for drawing vertical lines. LINE, PLOT, and VERT are among the more common ones. Consult your manual to determine the correct methodology for your graphing utility.

FIGURE 39

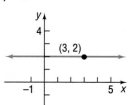

4

Now let L be a nonvertical line with slope m containing the point (x_1, y_1). See Figure 39. For any other point (x, y) on L, we have

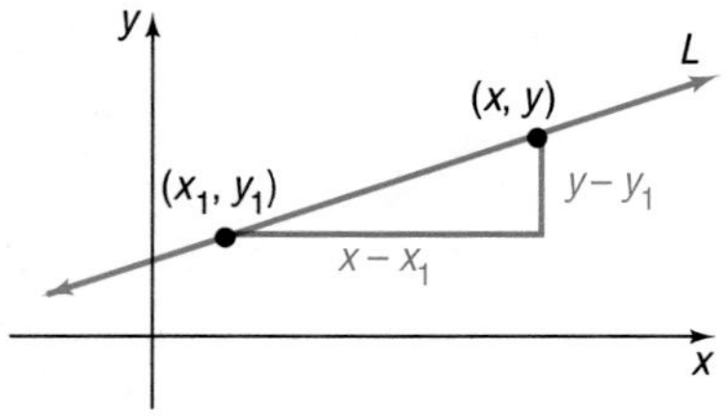

$$m = \frac{y - y_1}{x - x_1} \quad \text{or} \quad y - y_1 = m(x - x_1)$$

Theorem   Point–Slope Form of an Equation of a Line

An equation of a nonvertical line of slope m that passes through the point (x_1, y_1) is

$$y - y_1 = m(x - x_1) \tag{2}$$

E X A M P L E   5

Using the Point–Slope Form of a Line

An equation of the line with slope 4 and passing through the point $(1, 2)$ can be found by using the point–slope form with $m = 4$, $x_1 = 1$, and $y_1 = 2$:

$$y - y_1 = m(x - x_1)$$
$$y - 2 = 4(x - 1)$$
$$y = 4x - 2$$

See Figure 40.

FIGURE 40
$y = 4x - 2$

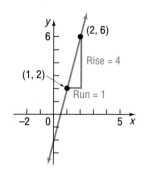

E X A M P L E   6

Finding the Equation of a Horizontal Line

Find an equation of the horizontal line passing through the point $(3, 2)$.

FIGURE 41
$y = 2$

Solution   The slope of a horizontal line is 0. To get an equation, we use the point–slope form with $m = 0$, $x_1 = 3$, and $y_1 = 2$:

$$y - y_1 = m(x - x_1)$$
$$y - 2 = 0 \cdot (x - 3)$$
$$y - 2 = 0$$
$$y = 2$$

See Figure 41 for the graph.

As suggested by Example 6, we have the following result:

> **Theorem** Equation of a Horizontal Line
>
> A horizontal line is given by an equation of the form
>
> $$y = b$$
>
> where b is the y-intercept.

EXAMPLE 7

Finding an Equation of a Line Given Two Points

Find an equation of the line L passing through the points $(2, 3)$ and $(-4, 5)$. Graph the line L.

Solution Since two points are given, we first compute the slope of the line:

$$m = \frac{5 - 3}{-4 - 2} = \frac{2}{-6} = -\frac{1}{3}$$

FIGURE 42
$y - 3 = -\dfrac{1}{3}(x - 2)$

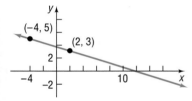

We use the point $(2, 3)$ and the fact that the slope $m = -\dfrac{1}{3}$ to get the point–slope form of the equation of the line:

$$y - 3 = -\frac{1}{3}(x - 2)$$

See Figure 42 for the graph.

In the solution to Example 7, we could have used the other point, $(-4, 5)$, instead of the point $(2, 3)$. The equation that results, although it looks different, is equivalent to the equation we obtained in the example. (Try it for yourself.)

Another useful equation of a line is obtained when the slope m and y-intercept b are known. In this event, we know both the slope m of the line and a point $(0, b)$ on the line; thus, we may use the point–slope form, equation (2), to obtain the following equation:

$$y - b = m(x - 0) \quad \text{or} \quad y = mx + b$$

> **Theorem** Slope–Intercept Form of an Equation of a Line
>
> An equation of a line L with slope m and y-intercept b is
>
> $$y = mx + b \tag{3}$$

FIGURE 43
$y = mx + 2$

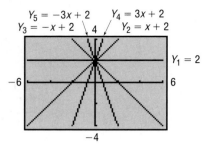

Seeing the Concept To see the role that the slope m plays, graph the following lines on the same square screen.

$$Y_1 = 2$$
$$Y_2 = x + 2$$
$$Y_3 = -x + 2$$
$$Y_4 = 3x + 2$$
$$Y_5 = -3x + 2$$

See Figure 43. What do you conclude about the lines $y = mx + 2$?   ▄

FIGURE 44
$y = 2x + b$

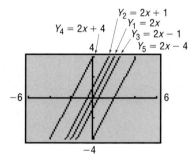

Seeing the Concept To see the role of the y-intercept b, graph the following lines on the same square screen.

$$Y_1 = 2x$$
$$Y_2 = 2x + 1$$
$$Y_3 = 2x - 1$$
$$Y_4 = 2x + 4$$
$$Y_5 = 2x - 4$$

See Figure 44. What do you conclude about the lines $y = 2x + b$?   ▄

7   When the equation of a line is written in slope–intercept form, it is easy to find the slope m and y-intercept b of the line. For example, suppose that the equation of a line is

$$y = -2x + 3$$

Compare it to $y = mx + b$:

$$y = -2x + 3$$
$$\uparrow \qquad \uparrow$$
$$y = \quad mx + b$$

The slope of this line is -2 and its y-intercept is 3.

E X A M P L E   8   Finding the Slope and y-Intercept

Find the slope m and y-intercept b of the equation $2x + 4y - 8 = 0$. Graph the equation.

Solution   To obtain the slope and y-intercept, we transform the equation into its slope–intercept form. Thus, we need to solve for y:

$$2x + 4y - 8 = 0$$
$$4y = -2x + 8$$
$$y = -\tfrac{1}{2}x + 2 \qquad y = mx + b$$

The coefficient of x, $-\tfrac{1}{2}$, is the slope, and the y-intercept is 2. We can graph the line in two ways:

1.   Use the fact that the y-intercept is 2 and the slope is $-\tfrac{1}{2}$. Then, starting at the point $(0, 2)$, go to the right 2 units and then down 1 unit to the point $(2, 1)$. See Figure 45.

Or:

FIGURE 45
$2x + 4y - 8 = 0$

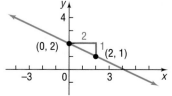

FIGURE 46
$2x + 4y - 8 = 0$

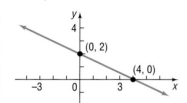

2. Locate the intercepts. Because the y-intercept is 2, we know one intercept is $(0, 2)$. To obtain the x-intercept, let $y = 0$ and solve for x. When $y = 0$, we have

$$2x + 4 \cdot 0 - 8 = 0$$
$$2x - 8 = 0$$
$$x = 4$$

Thus, the intercepts are $(4, 0)$ and $(0, 2)$. See Figure 46. ∎

 Now work Problem 45.

 The form of the equation of the line in Example 8, $2x + 4y - 8 = 0$, is called the *general form*.

> The equation of a line L is in **general form** when it is written as
>
> $$Ax + By + C = 0 \qquad (4)$$
>
> where A, B, and C are three real numbers and A and B are not both 0.

Every line has an equation that is equivalent to an equation written in general form. For example, a vertical line whose equation is

$$x = a$$

can be written in the general form

$$1 \cdot x + 0 \cdot y - a = 0 \quad A = 1, B = 0, C = -a.$$

A horizontal line whose equation is

$$y = b$$

can be written in the general form

$$0 \cdot x + 1 \cdot y - b = 0 \quad A = 0, B = 1, C = -b.$$

Lines that are neither vertical nor horizontal have general equations of the form

$$Ax + By + C = 0 \quad A \neq 0 \text{ and } B \neq 0.$$

Because the equation of every line can be written in general form, any equation equivalent to (4) is called a **linear equation.**

The next example illustrates one way of graphing a linear equation.

E X A M P L E   9   Finding the Intercepts of a Line

Find the intercepts of the line $2x + 3y - 6 = 0$. Graph this line.

Solution   To find the point at which the graph crosses the x-axis, that is, to find the x-intercept, we need to find the number x for which $y = 0$. Thus, we let $y = 0$ to get

$$2x + 3(0) - 6 = 0$$
$$2x - 6 = 0$$
$$x = 3$$

FIGURE 47
$2x + 3y - 6 = 0$

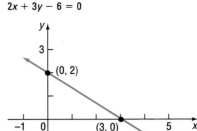

The x-intercept is 3. To find the y-intercept, we let $x = 0$ and solve for y:

$$2(0) + 3y - 6 = 0$$
$$3y - 6 = 0$$
$$y = 2$$

The y-intercept is 2.

To graph the line, we use the intercepts. Since the x-intercept is 3 and the y-intercept is 2, we know two points on the line: $(3, 0)$ and $(0, 2)$. Because two points determine a unique line, we do not need further information to graph the line. See Figure 47. ∎

The next example illustrates a typical situation that requires the use of linear equations.

E X A M P L E   10

Computing the Cost of Operating a Car

The National Car Rental Company has determined that the cumulative cost of operating a vehicle is \$0.41 per mile.

(a) Write an equation that relates the cumulative cost C, in dollars, of operating a car and the number x of miles it has been driven.
(b) What is the cost of operating a car with 1000 miles on it?
(c) What is the cost of operating a car with 2000 miles on it?

Solution (a) If x is the number of miles that the car has been driven, then the cumulative cost C, in dollars, is $0.41x$. Thus, an equation relating C and x is

$$C = 0.41x \qquad x \geq 0$$

The average cost per mile, \$0.41, is the slope of the line $C = 0.41x$. In other words, the cost increases by \$0.41 for each additional mile driven. See Figure 48.

(b) The cost of operating a car with 1000 miles on it is

$$C = 0.41x = 0.41(1000) = \$410$$

(c) The cost of operating a car with 2000 miles on it is

$$C = 0.41x = 0.41(2000) = \$820$$

FIGURE 48
$C = 0.41x$

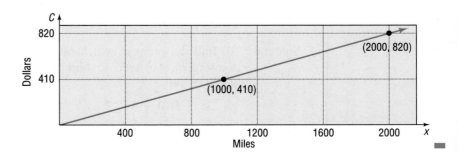

∎

3.3 | EXERCISES

In Problems 1–4; (a) find the slope of the line; (b) interpret the slope.

1.
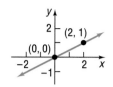

2.
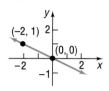

3.
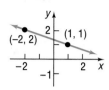

4.
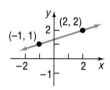

In Problems 5–14, plot each pair of points and determine the slope of the line containing them. Graph the line.

5. $(2, 3); (4, 0)$ **6.** $(4, 2); (3, 4)$ **7.** $(-2, 3); (2, 1)$ **8.** $(-1, 1); (2, 3)$

9. $(-3, -1); (2, -1)$ **10.** $(4, 2); (-5, 2)$ **11.** $(-1, 2); (-1, -2)$ **12.** $(2, 0); (2, 2)$

13. $(\sqrt{2}, 3); (1, \sqrt{3})$ **14.** $(-2\sqrt{2}, 0); (4, \sqrt{5})$

In Problems 15–22, graph the line passing through the point P and having slope m.

15. $P = (1, 2);\quad m = 3$ **16.** $P = (2, 1);\quad m = 4$ **17.** $P = (2, 4);\quad m = \frac{-3}{4}$

18. $P = (1, 3);\quad m = \frac{-2}{5}$ **19.** $P = (-1, 3);\quad m = 0$ **20.** $P = (2, -4);\quad m = 0$

21. $P = (0, 3);\quad$ slope undefined **22.** $P = (-2, 0);\quad$ slope undefined

In Problems 23–26, find an equation of each line. Express your answer using either the general form or the slope–intercept form of the equation of a line, whichever you prefer.

23.
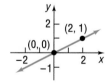

24.
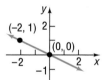

25.
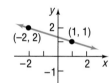

26.
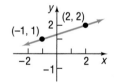

In Problems 27–38, find an equation for the line with the given properties. Express your answer using either the general form or the slope–intercept form of the equation of a line, whichever you prefer.

27. Slope $= 4$; passing through $(-2, 5)$ **28.** Slope $= 3$; passing through $(2, -3)$

29. Slope $= -\frac{2}{3}$; passing through $(1, -1)$ **30.** Slope $= \frac{1}{2}$; passing through $(3, 1)$

31. Passing through $(1, 3)$ and $(-1, 2)$ **32.** Passing through $(-3, 4)$ and $(2, 5)$

33. Slope $= -4$; y-intercept $= 3$ **34.** Slope $= -2$; y-intercept $= -3$

35. x-intercept $= 2$; y-intercept $= -1$ **36.** x-intercept $= -4$; y-intercept $= 4$

37. Slope undefined; passing through $(2, -4)$ **38.** Slope undefined; passing through $(3, -8)$

In Problems 39–58, find the slope and y-intercept of each line. Graph the line.

39. $y = 2x + 3$ **40.** $y = -3x + 4$ **41.** $\frac{1}{2}y = x - 1$ **42.** $\frac{1}{3}x + y = 2$

43. $y = \frac{1}{2}x + 2$ **44.** $y = 2x + \frac{1}{2}$ **45.** $x + 2y = 4$ **46.** $-x + 3y = 6$

47. $2x - 3y = 6$ **48.** $3x + 2y = 6$ **49.** $x + y = 1$ **50.** $x - y = 2$

51. $x = -4$ **52.** $y = -1$ **53.** $y = 5$ **54.** $x = 2$

55. $y - x = 0$ **56.** $x + y = 0$ **57.** $2y - 3x = 0$ **58.** $3x + 2y = 0$

59. Find an equation of the *x*-axis.

60. Find an equation of the *y*-axis.

61. **Measuring Temperature** The relationship between Celsius (°C) and Fahrenheit (°F) degrees for measuring temperature is linear. Find an equation relating °C and °F if 0°C corresponds to 32°F and 100°C corresponds to 212°F. Use the equation to find the Celsius measure of 70°F.

62. **Measuring Temperature** The Kelvin (K) scale for measuring temperature is obtained by adding 273 to the Celsius temperature.
(a) Write an equation relating K and °C.
(b) Write an equation relating K and °F (see Problem 61).

63. **Business: Computing Profit** Each Sunday a newspaper agency sells *x* copies of a newspaper for $1.00 per copy. The cost to the agency of each newspaper is $0.50. The agency pays a fixed cost for storage, delivery, and so on, of $100 per Sunday.
(a) Write an equation that relates the profit *P*, in dollars, to the number *x* of copies sold. Graph this equation.
(b) What is the profit to the agency if 1,000 copies are sold?
(c) What is the profit to the agency if 5,000 copies are sold?

64. **Business: Computing Profit** Repeat Problem 63 if the cost to the agency is $0.45 per copy and the fixed cost is $125 per Sunday.

65. **Cost of Electricity** In 1997, Florida Power and Light Company supplied electricity to residential customers for a monthly customer charge of $5.65 plus 6.543¢ per kilowatt-hour supplied in the month for the first 750 kilowatt-hours used.* Write an equation that relates the monthly charge C, in dollars, to the number *x* of kilowatt-hours used in the month. Graph this equation. What is the monthly charge for using 300 kilowatt-hours? For using 750 kilowatt-hours?

66. Show that an equation for a line with nonzero *x*- and *y*-intercepts can be written as

$$\frac{x}{a} + \frac{y}{b} = 1$$

where *a* is the *x*-intercept and *b* is the *y*-intercept. This is called the **intercept form** of the equation of a line.

* *Source:* Florida Power and Light Co., Miami, Florida, 1997.

In Problems 67–70, match each graph with the correct equation:
(a) y = x;   (b) y = 2x;   (c) y = x/2;   (d) y = 4x.

67.

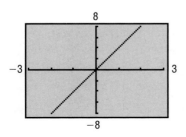

68.

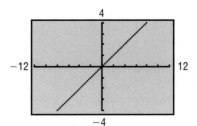

69.

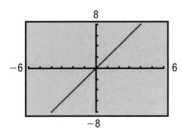

70.

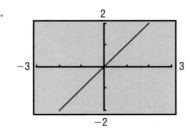

In Problems 71–76, write an equation of each line. Express your answer using either the general form or the slope–intercept form of the equation of a line, whichever you prefer.

71.

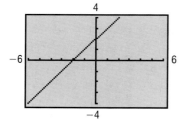

72.

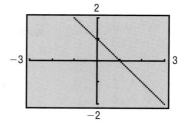

73.

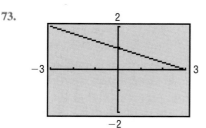

74.

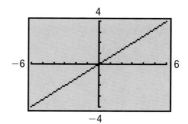

75.

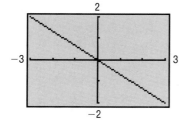

76.

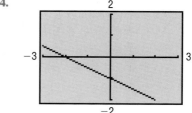

77. Which form of the equation of a line do you prefer to use? Justify your position with an example that shows that your choice is better than another. Have reasons.

78. Can every line be written in slope–intercept form? Explain.

79. Does every line have two distinct intercepts? Explain. Are there lines that have no intercepts? Explain.

80. What can you say about two lines that have equal slopes and equal y-intercepts?

81. What can you say about two lines with the same x-intercept and the same y-intercept? Assume that the x-intercept is not 0.

82. If two lines have the same slope, but different x-intercepts, can they have the same y-intercept?

83. If two lines have the same y-intercept, but different slopes, can they have the same x-intercept? What is the only way that this can happen?

84. The accepted symbol used to denote the slope of a line is the letter m. Investigate the origin of this symbolism. Begin by consulting a French dictionary and looking up the French word *monter*. Write a brief essay on your findings.

85. **Grade of a Road** The term *grade* is used to describe the inclination of a road. How does this term relate to the notion of slope of a line? Is a 4% grade very steep? Investigate the grades of some mountainous roads and determine their slopes. Write a brief essay on your findings.

86. **Carpentry** Carpenters use the term pitch to describe the steepness of staircases and roofs. How does pitch relate to slope? Investigate typical pitches used for stairs and for roofs. Write a brief essay on your findings.

3.4 | PARALLEL AND PERPENDICULAR LINES; CIRCLES

> **1** Define Parallel Lines
> **2** Find Equations of Parallel Lines
> **3** Define Perpendicular Lines
> **4** Find Equations of Perpendicular Lines
> **5** Graph a Circle
> **6** Write the Standard Form of the Equation of a Circle
> **7** Identify the Center and Radius of a Circle in General Form and Graph It
> **8** Write the General Form of the Equation of a Circle

Parallel and Perpendicular Lines

FIGURE 49
The lines are parallel if and only if their slopes are equal.

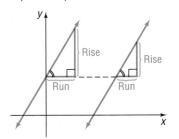

1 When two lines (in the plane) have no points in common, they are said to be **parallel.** Look at Figure 49. There we have drawn two lines and have constructed two right triangles by drawing sides parallel to the coordinate axes. These lines are parallel if and only if the right triangles are similar. (Do you see why? Two angles are equal.) But the triangles are similar if and only if the ratios of corresponding sides are equal.

This suggests the following result:

> **Theorem**   Criterion for Parallel Lines
>
> Two distinct nonvertical lines are parallel if and only if their slopes are equal.

The use of the words "if and only if" in the preceding theorem means that actually two statements are being made, one the converse of the other.

If two distinct nonvertical lines are parallel, then their slopes are equal.
If two distinct nonvertical lines have equal slopes, then they are parallel.

 Now work Problem 1(a).

E X A M · P L E   1

FIGURE 50
Parallel lines

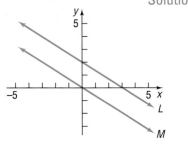

Showing That Two Lines Are Parallel

Show that the lines given by the following equations are parallel:

$$L: \quad 2x + 3y - 6 = 0 \qquad M: \quad 4x + 6y = 0$$

Solution   To determine whether these lines have equal slopes, we write each equation in slope–intercept form:

$$L: \quad 2x + 3y - 6 = 0 \qquad\qquad M: \quad 4x + 6y = 0$$
$$3y = -2x + 6 \qquad\qquad\qquad 6y = -4x$$
$$y = -\tfrac{2}{3}x + 2 \qquad\qquad\qquad y = -\tfrac{2}{3}x$$
$$\text{Slope} = -\tfrac{2}{3} \qquad\qquad\qquad \text{Slope} = -\tfrac{2}{3}$$

Because these lines have the same slope, $-\tfrac{2}{3}$, but different y-intercepts, the lines are parallel. See Figure 50.

E X A M P L E   2   Finding a Line That Is Parallel to a Given Line

Find an equation for the line that contains the point $(2, -3)$ and is parallel to the line $2x + y - 6 = 0$.

FIGURE 51

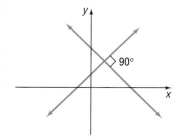

Solution   The slope of the line that we seek equals the slope of the line $2x + y - 6 = 0$, since the two lines are to be parallel. Thus, we begin by writing the equation of the line $2x + y - 6 = 0$ in slope–intercept form:

$$2x + y - 6 = 0$$
$$y = -2x + 6$$

The slope is -2. Since the line we seek contains the point $(2, -3)$, we use the point–slope form to obtain

$$y + 3 = -2(x - 2) \qquad \text{Point-slope form.}$$
$$2x + y - 1 = 0 \qquad \text{General form.}$$
$$y = -2x + 1 \qquad \text{Slope–intercept form.}$$

This line is parallel to the line $2x + y - 6 = 0$ and contains the point $(2, -3)$. See Figure 51. ∎

Now work Problem 15.

3   When two lines intersect at a right angle (90°), they are said to be **perpendicular.** See Figure 52.
The following result gives a condition, in terms of their slopes, for two lines to be perpendicular:

FIGURE 52
Perpendicular lines

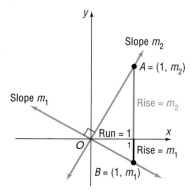

> **Theorem   Criterion for Perpendicular Lines**
>
> Two nonvertical lines are perpendicular if and only if the product of their slopes is -1. ∎

Here, we shall prove the "only if" part of the statement:

> If two nonvertical lines are perpendicular, then the product of their slopes is -1.

In Problem 69, you are asked to prove the "if" part of the theorem; that is:

> If two nonvertical lines have slopes whose product is -1, then the lines are perpendicular.

FIGURE 53

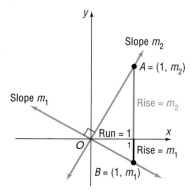

Proof   Let m_1 and m_2 denote the slopes of the two lines. There is no loss in generality (that is, neither the angle nor the slopes are affected) if we situate the lines so that they meet at the origin. See Figure 53. The point $A = (1, m_2)$ is on the line having slope m_2, and the point $B = (1, m_1)$ is on the line having slope m_1. (Do you see why this must be true?)
Suppose that the lines are perpendicular. Then triangle OAB is a right triangle. As a result of the Pythagorean Theorem, it follows that

$$[d(O, A)]^2 + [d(O, B)]^2 = [d(A, B)]^2 \qquad (1)$$

By the distance formula, we can write each of these distances as

$$[d(O, A)]^2 = (1 - 0)^2 + (m_2 - 0)^2 = 1 + m_2^2$$
$$[d(O, B)]^2 = (1 - 0)^2 + (m_1 - 0)^2 = 1 + m_1^2$$
$$[d(A, B)]^2 = (1 - 1)^2 + (m_2 - m_1)^2 = m_2^2 - 2m_1m_2 + m_1^2$$

Using these facts in equation (1), we get

$$(1 + m_2^2) + (1 + m_1^2) = m_2^2 - 2m_1m_2 + m_1^2$$

which, upon simplification, can be written as

$$m_1m_2 = -1$$

Thus, if the lines are perpendicular, the product of their slopes is -1.   ▄

You may find it easier to remember the condition for two nonvertical lines to be perpendicular by observing that the equality $m_1m_2 = -1$ means that m_1 and m_2 are negative reciprocals of each other; that is, either $m_1 = -1/m_2$ or $m_2 = -1/m_1$.

E X A M P L E   3   Finding the Slope of a Line Perpendicular to a Given Line

If a line has slope $\frac{3}{2}$, any line having slope $-\frac{2}{3}$ is perpendicular to it.   ▄

 Now work Problem 1(b).

E X A M P L E   4   Finding the Equation of a Line Perpendicular to a Given Line

 Find an equation of the line passing through the point $(1, -2)$ and perpendicular to the line $x + 3y - 6 = 0$. Graph the two lines.

Solution   We first write the equation of the given line in slope–intercept form to find its slope:

$$x + 3y - 6 = 0$$
$$3y = -x + 6$$
$$y = -\tfrac{1}{3}x + 2$$

FIGURE 54

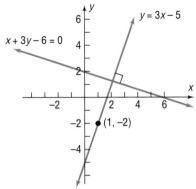

The given line has slope $-\frac{1}{3}$. Any line perpendicular to this line will have slope 3. Because we require the point $(1, -2)$ to be on this line with slope 3, we use the point–slope form of the equation of a line:

$$y - (-2) = 3(x - 1) \quad \text{Point-slope form.}$$
$$y + 2 = 3(x - 1)$$

This equation is equivalent to the forms

$$3x - y - 5 = 0 \qquad \text{General form.}$$
$$y = 3x - 5 \quad \text{Slope–intercept form.}$$

Figure 54 shows the graphs.   ▄

Now work Problem 23.

⌨ **Warning** Be sure to use a square screen when you graph perpendicular lines. Otherwise, the angle between the two lines will appear distorted. ∎

Circles

One advantage of a coordinate system is that it enables us to translate a geometric statement into an algebraic statement, and vice versa. Consider, for example, the following geometric statement that defines a circle.

> A **circle** is a set of points in the *xy*-plane that are a fixed distance *r* from a fixed point (*h, k*). The fixed distance *r* is called the **radius,** and the fixed point (*h, k*) is called the **center** of the circle.

FIGURE 55

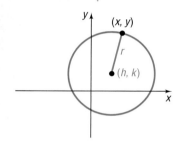

Figure 55 shows the graph of a circle. Is there an equation having this graph? If so, what is the equation? To find the equation, we let (*x, y*) represent the coordinates of any point on a circle with radius *r* and center (*h, k*). Then the distance between the points (*x, y*) and (*h, k*) must always equal *r*. That is, by the distance formula,

$$\sqrt{(x - h)^2 + (y - k)^2} = r$$

or, equivalently,

$$(x - h)^2 + (y - k)^2 = r^2$$

The **standard form of an equation of a circle** with radius *r* and center (*h, k*) is

$$(x - h)^2 + (y - k)^2 = r^2 \qquad (2)$$

Conversely, by reversing the steps, we conclude: The graph of any equation of the form of equation (2) is that of a circle with radius *r* and center (*h, k*).

E X A M P L E   5

Graphing a Circle

Graph the equation: $(x + 3)^2 + (y - 2)^2 = 16$

Solution By comparing the given equation to the standard form of the equation of a circle, equation (2), we conclude that the graph of the given equation is a circle. Moreover, the comparison yields information about the circle:

$$(x + 3)^2 + (y - 2)^2 = 16$$
$$(x - (-3))^2 + (y - 2)^2 = 4^2$$
$$\uparrow \qquad\qquad \uparrow \qquad\quad \uparrow$$
$$(x - h)^2 \quad + \quad (y - k)^2 \ = \ r^2$$

We see that $h = -3$, $k = 2$, and $r = 4$. Hence, the circle has center $(-3, 2)$ and a radius of 4 units. To graph this circle, we first plot the center $(-3, 2)$. Since the radius is 4, we can locate four points on the circle by going out

4 units to the left, to the right, up, and down from the center. These four points can then be used as guides to obtain the graph. See Figure 56.

FIGURE 56

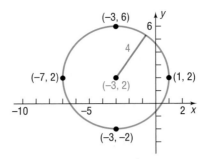

(−3, 6)
(−7, 2)
(−3, 2)
(1, 2)
(−3, −2)

Now work Problem 43.

E X A M P L E   6

Using a Graphing Utility to Graph a Circle

Graph the equation: $x^2 + y^2 = 4$

FIGURE 57

Solution

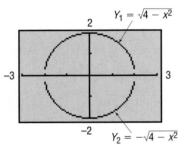

$Y_1 = \sqrt{4 - x^2}$

$Y_2 = -\sqrt{4 - x^2}$

This is the equation of a circle with center at the origin and radius 2. In order to graph this equation, we must first solve for y.*

$$x^2 + y^2 = 4$$
$$y^2 = 4 - x^2$$
$$y = \pm\sqrt{4 - x^2}$$

There are two equations to graph: first, we graph $Y_1 = \sqrt{4 - x^2}$ and then $Y_2 = -\sqrt{4 - x^2}$ on the same square screen. (Your circle will appear oval if you do not use a square screen.) See Figure 57.

E X A M P L E   7

Writing the Standard Form of the Equation of a Circle

Write the standard form of the equation of the circle with radius 3 and center (1, −2).

Solution

Using the form of equation (2) and substituting the values $r = 3$, $h = 1$, and $k = -2$, we have

$$(x - h)^2 + (y - k)^2 = r^2$$
$$(x - 1)^2 + (y + 2)^2 = 9$$

Now work Problem 27.

The standard form of an equation of a circle of radius r with center at the origin (0, 0) is

$$x^2 + y^2 = r^2$$

*Some graphing utilities (e.g., TI-82, TI-83, TI-85, TI-86) have a CIRCLE feature that allows the user to enter only the coordinates of the center of the circle and its radius to graph the circle.

If the radius $r = 1$, the circle whose center is at the origin is called the **unit circle** and has the equation

$$x^2 + y^2 = 1$$

FIGURE 58
Unit circle $x^2 + y^2 = 1$

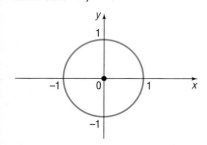

See Figure 58.

If we eliminate the parentheses from the standard form of the equation of the circle obtained in Example 7, we get

$$(x - 1)^2 + (y + 2)^2 = 9$$
$$x^2 - 2x + 1 + y^2 + 4y + 4 = 9$$

which we find, upon simplifying, is equivalent to

$$x^2 + y^2 - 2x + 4y - 4 = 0$$

By completing the squares on both the x- and y-terms, it can be shown that any equation of the form

$$x^2 + y^2 + ax + by + c = 0$$

has a graph that is a circle, or a point, or has no graph at all. For example, the graph of the equation $x^2 + y^2 = 0$ is the single point $(0, 0)$. The equation $x^2 + y^2 + 5 = 0$, or $x^2 + y^2 = -5$, has no graph, because sums of squares of real numbers are never negative. When its graph is a circle, the equation

$$x^2 + y^2 + ax + by + c = 0$$

is referred to as the **general form of the equation of a circle.**

7   The next example shows how to transform an equation in the general form to an equivalent equation in standard form. As we said earlier, the idea is to use the method of completing the square on both the x- and y-terms.

E X A M P L E   8   Graphing a Circle Whose Equation Is in General Form

Graph the equation: $x^2 + y^2 + 4x - 6y + 12 = 0$

Solution   We rearrange the equation as follows:

$$(x^2 + 4x) + (y^2 - 6y) = -12$$

Next, we complete the square of each expression in parentheses. Remember that any number added on the left also must be added on the right:

$$(x^2 + 4x + 4) + (y^2 - 6y + 9) = -12 + 4 + 9$$
$$(x + 2)^2 + (y - 3)^2 = 1$$

We recognize this equation as the standard form of the equation of a circle with radius 1 and center $(-2, 3)$.

To graph the equation, we use the center $(-2, 3)$ and the radius 1. Figure 59 illustrates the graph.

FIGURE 59 $x^2 + y^2 + 4x - 6y + 12 = 0$

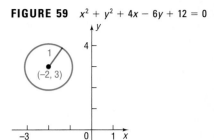

Now work Problem 45.

E X A M P L E   9

Finding the General Equation of a Circle

Find the general equation of the circle whose center is $(1, -2)$ and whose graph contains the point $(4, -2)$.

Solution

To find the equation of a circle, we need to know its center and its radius. Here, we know that the center is $(1, -2)$. Since the point $(4, -2)$ is on the graph, the radius r will equal the distance from $(4, -2)$ to the center $(1, -2)$. See Figure 60. Thus,

$$r = \sqrt{(4 - 1)^2 + [-2 - (-2)]^2}$$
$$= \sqrt{9} = 3$$

FIGURE 60
$x^2 + y^2 - 2x + 4y - 4 = 0$

The standard form of the equation of the circle is

$$(x - 1)^2 + (y + 2)^2 = 9$$

Eliminating the parentheses and rearranging terms, we get the general form of the equation

$$x^2 + y^2 - 2x + 4y - 4 = 0$$

Overview

The preceding discussion about lines and circles dealt with two main types of problems that can be generalized as follows:

1.   Given an equation, classify it and graph it.
2.   Given a graph, or information about a graph, find its equation.

This text deals with both types of problems. We shall study various equations, classify them, and graph them. Although the second type of problem is usually more difficult to solve than the first, in many instances a graphing utility can be used to solve such problems.

3.4 | EXERCISES

In Problems 1–10, the equation of a line L is given. Find the slope of a line that is (a) parallel to L; (b) perpendicular to L.

1. $y = 6x$

2. $y = -3x$

3. $y = -\frac{1}{2}x + 2$

4. $y = \frac{2}{3}x - 1$

5. $2x - 4y + 5 = 0$

6. $3x + y = 4$

7. $3x + 5y - 10 = 0$

8. $4x - 3y + 7 = 0$

9. $x = 7$

10. $y = 8$

In Problems 11–14, find an equation for the line L. Express your answer using either the general form or the slope–intercept form of the equation of a line, whichever you prefer.

11.
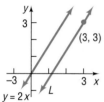
L is parallel to y = 2x

12.
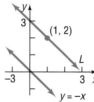
L is parallel to y = −x

13.
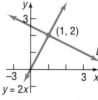
L is perpendicular to y = 2x

14.
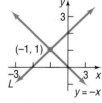
L is perpendicular to y = −x

In Problems 15–26, find an equation for the line with the given properties. Express your answer using either the general form or the slope–intercept form of the equation of a line, whichever you prefer.

15. Parallel to the line $y = 4x$; passing through $(-1, 2)$

16. Parallel to the line $y = -5x$; passing through $(-1, 2)$

17. Parallel to the line $2x - y + 2 = 0$; passing through $(0, 0)$

18. Parallel to the line $x - 2y + 5 = 0$; passing through $(0, 0)$

19. Parallel to the line $x = 5$; passing through $(4, 2)$

20. Parallel to the line $y = 5$; passing through $(4, 2)$

21. Perpendicular to the line $y = \frac{1}{2}x + 4$; passing through $(1, -2)$

22. Perpendicular to the line $y = 2x - 3$; passing through $(1, -2)$

23. Perpendicular to the line $2x + y - 2 = 0$; passing through $(-3, 0)$

24. Perpendicular to the line $x - 2y + 5 = 0$; passing through $(0, 4)$

25. Perpendicular to the line $x = -3$; passing through $(3, 4)$

26. Perpendicular to the line $y = -3$; passing through $(3, 4)$

In Problems 27–30, find the center and radius of each circle. Write the standard form of the equation.

27.
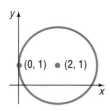

28.
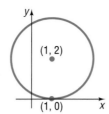

29.
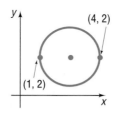

30.
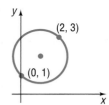

In Problems 31–40, write the standard form of the equation and the general form of the equation of each circle of radius r and center (h, k). Graph each circle.

31. $r = 1$; $(h, k) = (1, -1)$

32. $r = 2$; $(h, k) = (-2, 1)$

33. $r = 2$; $(h, k) = (0, 2)$

34. $r = 3$; $(h, k) = (1, 0)$

35. $r = 5$; $(h, k) = (4, -3)$

36. $r = 4$; $(h, k) = (2, -3)$

37. $r = 2$; $(h, k) = (0, 0)$

38. $r = 3$; $(h, k) = (0, 0)$

39. $r = \frac{1}{2}$; $(h, k) = (\frac{1}{2}, 0)$

40. $r = \frac{1}{2}$; $(h, k) = (0, -\frac{1}{2})$

In Problems 41–50, find the center (h, k) and radius r of each circle. Graph each circle.

41. $x^2 + y^2 = 4$

42. $x^2 + (y - 1)^2 = 1$

43. $2(x - 3)^2 + 2y^2 = 8$

44. $3(x + 1)^2 + 3(y - 1)^2 = 6$

45. $x^2 + y^2 + 4x - 4y - 1 = 0$

46. $x^2 + y^2 - 6x + 2y + 9 = 0$

47. $x^2 + y^2 - x + 2y + 1 = 0$

48. $x^2 + y^2 + x + y - \frac{1}{2} = 0$

49. $2x^2 + 2y^2 - 12x + 8y - 24 = 0$

50. $2x^2 + 2y^2 + 8x + 7 = 0$

In Problems 51–56, find the general form of the equation of each circle.

51. Center at the origin and passing through the point $(-3, 2)$

52. Center $(1, 0)$ and passing through the point $(-2, 3)$

53. Center $(2, 3)$ and tangent to the x-axis

54. Center $(-3, 1)$ and tangent to the y-axis

55. With endpoints of a diameter at $(1, 4)$ and $(-3, 2)$

56. With endpoints of a diameter at $(4, 3)$ and $(0, 1)$

57. **Geometry** Use slopes to show that the triangle whose vertices are $(-2, 5), (1, 3)$, and $(-1, 0)$ is a right triangle.

58. **Geometry** Use slopes to show that the quadrilateral whose vertices are $(1, -1), (4, 1), (2, 2)$, and $(5, 4)$ is a parallelogram.

59. **Geometry** Use slopes to show that the quadrilateral whose vertices are $(-1, 0), (2, 3), (1, -2)$, and $(4, 1)$ is a rectangle.

60. **Geometry** Use slopes and the distance formula to show that the quadrilateral whose vertices are $(0, 0), (1, 3), (4, 2)$, and $(3, -1)$ is a square.

 In Problems 61–64, match each graph with the correct equation.

(a) $(x - 3)^2 + (y + 3)^2 = 9$

(c) $(x - 1)^2 + (y + 2)^2 = 4$

(b) $(x + 1)^2 + (y - 2)^2 = 4$

(d) $(x + 3)^2 + (y - 3)^2 = 9$

61.

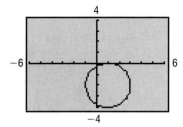

62.

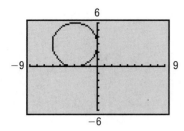

63.

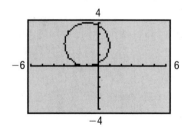

64.

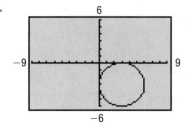

In Problems 65–68, find the standard form of the equation of each circle.

65.

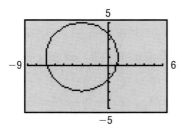

66.

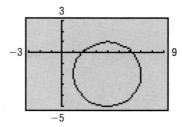

67.

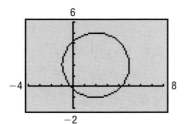

68.

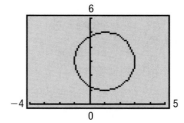

69. Prove that if two nonvertical lines have slopes whose product is -1 then the lines are perpendicular.

[**Hint:** Refer to Figure 53, and use the converse of the Pythagorean Theorem.]

70. **Weather Satellites**   Earth is represented on a map of a portion of the solar system so that its surface is the circle with equation $x^2 + y^2 + 2x + 4y - 4091 = 0$. A weather satellite circles 0.6 units above Earth with the center of its circular orbit at the center of Earth. Find the equation for the orbit of the satellite on this map.

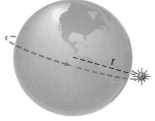

71. The **tangent line** to a circle may be defined as the line that intersects the circle in a single point, called the **point of tangency** (see the figure). If the equation of the circle is $x^2 + y^2 = r^2$ and the equation of the tangent line is $y = mx + b$, show that:
(a) $r^2(1 + m^2) = b^2$

[**Hint:** The quadratic equation $x^2 + (mx + b)^2 = r^2$ has exactly one solution.]

(b) The point of tangency is $(-r^2 m/b, r^2/b)$.
(c) The tangent line is perpendicular to the line containing the center of the circle and the point of tangency.

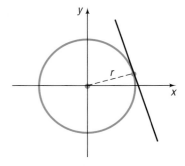

72. The **Greek method** for finding the equation of the tangent line to a circle used the fact that at any point on a circle the line containing the radius and the tangent line are perpendicular (see Problem 71). Use this method to find an equation of the tangent line to the circle $x^2 + y^2 = 9$ at the point $(1, 2\sqrt{2})$.

73. Use the Greek method described in Problem 72 to find an equation of the tangent line to the circle $x^2 + y^2 - 4x + 6y + 4 = 0$ at the point $(3, 2\sqrt{2} - 3)$.

74. Refer to Problem 71. The line $x - 2y + 4 = 0$ is tangent to a circle at $(0, 2)$. The line $y = 2x - 7$ is tangent to the same circle at $(3, -1)$. Find the center of the circle.

75. Find an equation of the line containing the centers of the two circles

$$x^2 + y^2 - 4x + 6y + 4 = 0$$
and $x^2 + y^2 + 6x + 4y + 9 = 0$

76. Show that the line containing the points (a, b) and (b, a) is perpendicular to the line $y = x$. Also show that the midpoint of (a, b) and (b, a) lies on the line $y = x$.

77. The equation $2x - y + C = 0$ defines a **family of lines,** one line for each value of C. On one set of coordinate axes, graph the members of the family when $C = -4, C = 0,$ and $C = 2$. Can you draw a conclusion from the graph about each member of the family?

78. Rework Problem 77 for the family of lines $Cx + y + 4 = 0$.

79. If a circle of radius 2 is made to roll along the x-axis, what is an equation for the path of the center of the circle?

3.5 | LINEAR CURVE FITTING

1 Distinguish between Linear and Nonlinear Relations
2 Use a Graphing Utility to Find the Line of Best Fit

Curve fitting is an area of statistics in which a relation between two or more variables is explained through an equation. For example, the equation $S = \$100,000 + 12A$, where S represents sales (in dollars) and A is the advertising expenditure (in dollars), implies that if advertising expenditures were \$0, sales would be $\$100,000 + 12(0) = \$100,000$ and if advertising expenditures were \$10,000, sales would be $\$100,000 + 12(\$10,000) = \$220,000$. In this model, the variable A is called the predictor (independent) variable and S is called the response (dependent) variable because, if the level of advertising is known, it can be used to predict sales. Curve fitting is used to find an equation that relates two or more variables, using observed or experimental data.

Steps for Curve Fitting

> STEP 1: Obtain data that relate two variables. Then plot ordered pairs of the variables as points to obtain a **scatter diagram.**
> STEP 2: Find an equation that describes this relation.

1 Scatter diagrams are used to help us to see the type of relation that exists between two variables. In this text, we will discuss a variety of different relations that may exist between two variables. For now, we concentrate on distinguishing between linear and nonlinear relations. See Figure 61.

FIGURE 61
Linear Relations

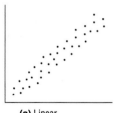

(a) Linear
$y = mx + b, m > 0$

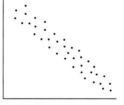

(b) Linear
$y = mx + b, m < 0$

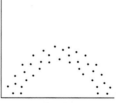

(c) Nonlinear

(d) Nonlinear

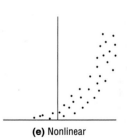

(e) Nonlinear

Predicting the Future of Olympic Track Events

Some people think that women athletes are beginning to catch up to men in the Olympic track events. In fact, researchers at UCLA published data in 1992 that seemed to indicate that within 65 years the top men and women runners would be able to compete on an equal basis. Other researchers disagree. Here are some data on the 200-meter run with winning times for the Olympics in the year given.

	Men's Time in Seconds	Women's Time in Seconds
1948	21.1	24.4
1952	20.7	23.7
1956	20.6	23.4
1960	20.5	24.0
1964	20.3	23.0
1968	19.83	22.5
1972	20.00	22.40
1976	20.23	22.37
1980	20.19	22.03
1984	19.80	21.81
1988	19.75	21.34
1992	19.73	21.72
1996	19.32	22.12

(You may notice that the introduction of better timing devices meant more accurate measures starting in 1968 for the men and 1972 for the women.)

1. To begin your investigation of the UCLA conjecture, make a graph of these data. To get the kind of accuracy that you need, you should use graph paper. You will need to use only the first quadrant. On your x-axis, place the Olympic years from 1948 to 2048, counting by 4's. On your y-axis place the numbers from 15 to 25, which represent the seconds; if you are using graph paper, allow about four squares per one second. Use X's to represent the men's times on the graph and O's to represent the women's times. These should form what we call a scatter plot, not a neat line or curve.

2. Using a ruler or straightedge, draw a line that represents roughly the slope and direction indicated by the men's scores. Do the same for the women's scores. Write a sentence or two explaining why you think your line is a good representation of the scatter plot.

3. Next find the equation for each line, using your graph to estimate the slope and y-intercept for each. Try to be as accurate as possible, remembering your units and using the points where the line crosses intersections of the grid. The slope will be in seconds per year.

4. Do your two lines appear to cross? In what year do they cross? If you solve the two equations algebraically, do you get the same answer?

5. Some graphing calculators enable you do this problem in a statistics mode. They will plot the points that you type in and find a line that represents the data. Use your graphing calculator to find out whether its equations match yours.

6. Make a group decision about whether you think that women's times will catch up to men's times in the future. Write two to three sentences explaining why you believe they will or will not.

EXAMPLE 1   Distinguishing between Linear and Nonlinear Relations

Determine whether the relation between the two variables shown in Figure 62 is linear or nonlinear.

FIGURE 62

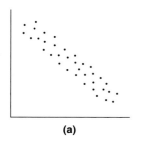
(a)

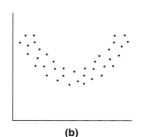
(b)

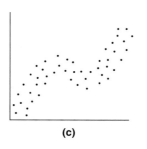
(c)

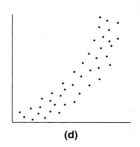
(d)

Solution   (a) Linear   (b) Nonlinear   (c) Nonlinear   (d) Nonlinear

We will study linearly related data in this section. Fitting an equation to nonlinear data will be discussed in later chapters.

Suppose that the scatter diagram of a set of data appears to be linearly related. One way to obtain an equation for such data is to draw a line through two points on the scatter diagram and estimate the equation of the line.

EXAMPLE 2   Finding an Equation for Linearly Related Data

A farmer collected the following data, which show crop yields for various amounts of fertilizer used.

Plot	1	2	3	4	5	6	7	8	9	10	11	12
Fertilizer, X (pounds/100 ft^2)	0	0	5	5	10	10	15	15	20	20	25	25
Yield, Y (bushels)	4	6	10	7	12	10	15	17	18	21	23	22

(a) Draw a scatter diagram of the data.
(b) Select two points from the data and find an equation of the line containing the points.
(c) Graph the line on the scatter diagram.

Solution   (a) The data collected indicate that a relation exists between the amount of fertilizer used and crop yield. To draw a scatter diagram, we plot points, using fertilizer as the x-coordinate and yield as the y-coordinate. See Figure 63. From the scatter diagram, it appears that a linear relation exists between the amount of fertilizer used and yield.

FIGURE 63

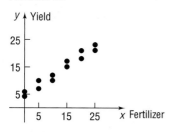

(b) Select two points, say $(0, 4)$ and $(25, 23)$. (You should select your own two points and complete the solution.) The slope of the line joining the points $(0, 4)$ and $(25, 23)$ is

$$m = \frac{23 - 4}{25 - 0} = \frac{19}{25} = 0.76$$

FIGURE 64

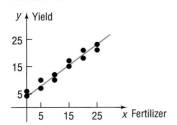

The equation of the line with slope 0.76 and passing through (0, 4) is found using the point–slope form with $m = 0.76$, $x_1 = 0$, and $y_1 = 4$:

$$y - y_1 = m(x - x_1)$$
$$y - 4 = 0.76(x - 0)$$
$$y = 0.76x + 4$$

(c) Figure 64 shows the scatter diagram with the graph of the line found in (b).

 Now work Problem 7(a), (b), and (c).

The line obtained in Example 2 depends on the selection of points, which will vary from person to person. So the line we found might be different from the line you found. Although the line we found in Example 2 appears to "fit" the data well, there may be a line that "fits it better." Do you think your line "fits" the data better? Is there a line of "best fit"? As it turns out, there is a method for finding the line that best fits linearly related data.* This line is called the **line of best fit.**

 ## Lines of Best Fit

E X A M P L E   3   Finding the Line of Best Fit

 With the data from Example 2, find the line of best fit using a graphing utility.

Solution   Graphing utilities contain built-in programs that find the linear equation of "best fit" for a collection of points in a scatter diagram. (Look in your owner's manual under Linear Regression or Line of Best Fit for details on how to execute the program.) Upon executing the LINear REGression program, we obtain the results shown in Figure 65. The output that the utility provides shows us the equation, $y = ax + b$, where a is the slope of the line and b is the y-intercept. The line of best fit that relates fertilizer and yield is

Yield = 0.717(fertilizer) + 4.786

Figure 66 shows the graph of the line of best fit, along with the scatter diagram.

FIGURE 65   **FIGURE 66**

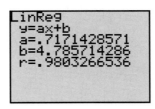   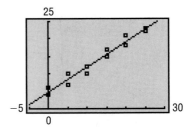

 Now work Problem 7(d) and (e).

*We shall not discuss in this book the underlying mathematics of lines of best fit. Most books in statistics and many in linear algebra discuss this topic.

Does the line of best fit appear to be a good "fit"? In other words, does the line appear to accurately describe the relation between yield and fertilizer?

And just how "good" is this line of "best fit"? The answers are given by what is called the *correlation coefficient*. Look again at Figure 65. The last line of output is $r = 0.98$. This number, called the **correlation coefficient, r,** $0 \leq |r| \leq 1$, is a measure of the strength of the *linear relation* that exists between two variables. The closer $|r|$ is to 1, the more perfect the linear relationship is. If r is close to 0, there is little or no *linear* relationship between the variables. A negative value of r, $r < 0$, indicates that as x increases y decreases; a positive value of r, $r > 0$, indicates that as x increases y does also. Thus, the data given in Example 2, having a correlation coefficient of 0.98, are strongly indicative of a linear relationship.

3.5 | EXERCISES

In Problems 1–6, examine the scatter diagram and determine whether the type of relation, if any, that may exist is linear or nonlinear.

1.
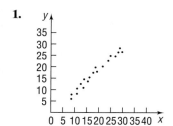

2.
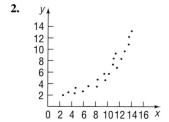

3.
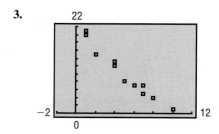

4.
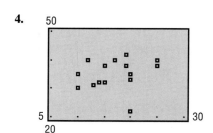

5.
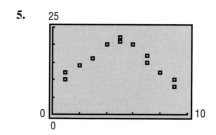

6.
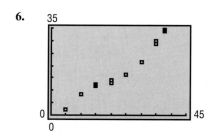

In Problems 7–14:

(a) Draw a scatter diagram.

(b) Select two points from the scatter diagram and find the equation of the line through the points selected.*

(c) Graph the line found in (b) on the scatter diagram.

(d) Use a graphing utility to find the line of best fit.

(e) Use a graphing utility to graph the line of best fit on the scatter diagram.

7.
x	3	4	5	6	7	8	9
y	4	6	7	10	12	14	16

8.
x	3	5	7	9	11	13
y	0	2	3	6	9	11

9.
x	-2	-1	0	1	2
y	-4	0	1	4	5

10.
x	-2	-1	0	1	2
y	7	6	3	2	0

11.
x	20	30	40	50	60
y	100	95	91	83	70

12.
x	5	10	15	20	25
y	2	4	7	11	18

13.
x	-20	-17	-15	-14	-10
y	100	120	118	130	140

14.
x	-30	-27	-25	-20	-14
y	10	12	13	13	18

15. **Consumption and Disposable Income** An economist wishes to estimate a line that relates personal consumption expenditures (C) and disposable income (I). Both C and I are in thousands of dollars. She interviews 8 heads of households for families of size 4 and obtains the following data:

C (000)	I (000)
16	20
18	20
13	18
21	27
27	36
26	37
36	45
39	50

Let I represent the independent variable and C the dependent variable.
(a) Draw a scatter diagram.
(b) Find a line that fits the data.*
(c) Interpret the slope. The slope of this line is called the **marginal propensity to consume.**
(d) Predict the consumption of a family whose disposable income is $42,000.
(e) Use a graphing utility to find the line of best fit to the data.

16. **Marginal Propensity to Save** The same economist as the one in Problem 15 wants to estimate a line that relates savings (S) and disposable income (I). Let $S = I - C$ be the dependent variable and I the independent variable. The slope of this line is called the **marginal propensity to save.**
(a) Draw a scatter diagram.
(b) Find a line that fits the data.
(c) Interpret the slope.
(d) Predict the savings of a family whose income is $42,000.
(e) Use a graphing utility to find the line of best fit to the data.

17. **Average Speed of a Car** An individual wanted to determine the relation that might exist between speed and miles per gallon of an automobile. Let X be the average speed of a car on the highway measured in miles per hour, and let Y represent the miles per gallon of the automobile. The following data are collected:

X	50	55	55	60	60	62	65	65
Y	28	26	25	22	20	20	17	15

(a) Draw a scatter diagram.
(b) Find a line that fits the data.*
(c) Interpret the slope.
(d) Predict the miles per gallon of a car traveling 61 miles per hour.
(e) Use a graphing utility to find the line of best fit to the data.

*Answers will vary. We'll use the first and last data points in the answer section.

18. **Height versus Weight**   A doctor wished to determine whether a relation exists between the height of a female and her weight. She obtained the heights and weights of 10 females aged 18 to 24. Let height be the independent variable, X, measured in inches, and weight be the dependent variable, Y, measured in pounds.

X	60	61	62	62	64	65	65	67	68	68
Y	105	110	115	120	120	125	130	135	135	145

(a) Draw a scatter diagram.
(b) Find a line that fits the data.
(c) Interpret the slope.
(d) Predict the weight of a female aged 18 to 24 whose height is 66 inches.
(e) Use a graphing utility to find the line of best fit to the data.

19. **Sales Data versus Income**   The following data represent sales and net income before taxes (both are in billions of dollars) for all manufacturing firms within the United States for 1980–1989. Treat sales as the independent variable and net income before taxes as the dependent variable.

Year	Sales	Net Income Before Taxes
1980	1912.8	92.6
1981	2144.7	101.3
1982	2039.4	70.9
1983	2114.3	85.8
1984	2335.0	107.6
1985	2331.4	87.6
1986	2220.9	83.1
1987	2378.2	115.6
1988	2596.2	154.6
1989	2745.1	136.3

Source: Economic Report of the President, February 1995.

(a) Draw a scatter diagram.
(b) Find a line that fits the data.
(c) Interpret the slope.
(d) Predict the net income before taxes of manufacturing firms in 1990 if sales are $2456.4 billion.
(e) Use a graphing utility to find the line of best fit to the data.

20. **Employment and the Labor Force**   The following data represent the civilian labor force (people aged 16 years and older, excluding those serving in the military) and the number of employed people in the United States for the years 1981–1991. Treat the size of the labor force as the independent variable and the number employed as the dependent variable. Both the size of the labor force and the number employed are measured in thousands of people.

Year	Civilian Labor Force	Number Employed
1981	108,670	100,397
1982	110,204	99,526
1983	111,550	100,834
1984	113,544	105,005
1985	115,461	107,150
1986	117,834	109,597
1987	119,865	112,440
1988	121,669	114,968
1989	123,869	117,342
1990	124,787	117,914
1991	125,303	116,877

Source: Business Statistics, 1963–1991, U.S. Department of Commerce, Economics and Statistics Administration, Bureau of Economic Analysis, June 1992.

(a) Draw a scatter diagram.
(b) Find a line that fits the data.
(c) Interpret the slope.
(d) Predict the number of employed people if the civilian labor force is 122,340,000 people.
(e) Use a graphing utility to find the line of best fit to the data.

21. **Disposable Personal Income**   Use the data from Problem 53, Section 3.1.
(a) With a graphing utility, find the line of best fit, letting per capita disposable income represent the independent variable.
(b) Interpret the slope.
(c) Predict per capita personal consumption if per capita disposable income is $14,989.

22. **Natural Gas versus Coal**   Use the data from Problem 54, Section 3.1.
(a) With a graphing utility, find the line of best fit, letting the amount of natural gas represent the independent variable.
(b) Interpret the slope.
(c) Predict the amount of energy provided by coal if the amount of energy provided by natural gas is 67 billion BTUs.

23. **Fuel Consumption** Use the data from Problem 55, Section 3.1.
    (a) With a graphing utility, find the line of best fit, letting the average fuel consumption per car represent the independent variable.
    (b) Interpret the slope.
    (c) Predict the pollutant emissions of carbon monoxide if the average fuel consumption per car is 505 gallons.

24. **Hourly Pay and Productivity** Use the data from Problem 56, Section 3.1.
    (a) With a graphing utility, find the line of best fit, letting productivity represent the independent variable.
    (b) Interpret the slope.
    (c) Predict average hourly earnings if productivity is 101.3.

3.6 | VARIATION

 Construct a Model Using Direct Variation
2 Construct a Model Using Inverse Variation
3 Construct a Model Using Joint or Combined Variation

When a mathematical model is developed for a real-world problem, it often involves relationships between quantities that are expressed in terms of proportionality:

Force is proportional to acceleration.

For an ideal gas held at a constant temperature, pressure and volume are inversely proportional.

The force of attraction between two heavenly bodies is inversely proportional to the square of the distance between them.

Revenue is directly proportional to sales.

Each of these statements illustrates the idea of **variation,** or how one quantity varies in relation to another quantity. Quantities may vary *directly, indirectly,* or *jointly.*

Direct Variation

1 Let x and y denote two quantities. Then y **varies directly** with x, or y is **directly proportional to** x, if there is a nonzero number k such that

$$y = kx$$

FIGURE 67
$y = kx, k > 0, x \geq 0$

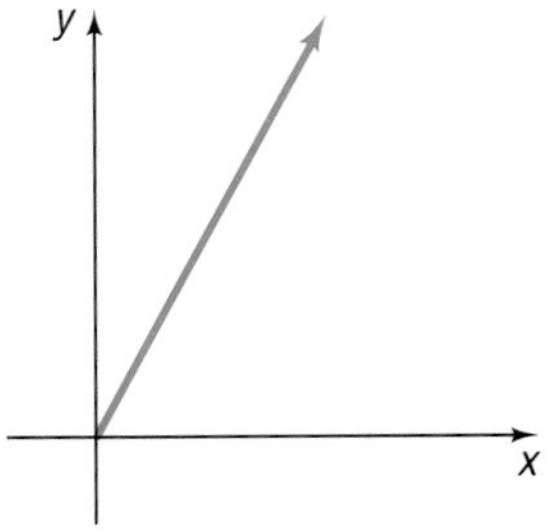

The number k is called the **constant of proportionality.**

The graph in Figure 67 illustrates the relationship between y and x if y varies directly with x and $k > 0$, $x \geq 0$. Note that the constant of proportionality is, in fact, the slope of the line.

If we know that two quantities vary directly, then knowing the value of each quantity in one instance enables us to write a formula that is true in all cases.

E X A M P L E   1   Chemistry: Gas Law

For a certain gas enclosed in a container of fixed volume, the pressure P (in newtons per square meter) varies directly with temperature T (in kelvins). If the pressure is found to be 20 newtons per square meter at a temperature of 60 K, find a formula that relates pressure P to temperature T. Then find the pressure P when $T = 120$ K.

Solution   Because P varies directly with T, we know that

$$P = kT$$

for some constant k. Because $P = 20$ when $T = 60$,

$$20 = k(60)$$
$$k = \tfrac{1}{3}$$

Thus, in all cases,

$$P = \tfrac{1}{3}T$$

In particular, when $T = 120$ K, we find

$$P = \tfrac{1}{3}(120) = 40 \text{ newtons per square meter}$$

Figure 68 illustrates the relationship between the pressure P and the temperature T.   ▬

FIGURE 68

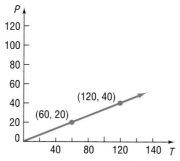

Now work Problem 1.

FIGURE 69

$y = \dfrac{k}{x}; k > 0, x > 0$

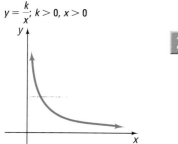

Inverse Variation

2 Let x and y denote two quantities. Then y **varies inversely** with x, or y is **inversely proportional to** x, if there is a nonzero constant k such that

$$y = \frac{k}{x}$$

The graph in Figure 69 illustrates the relationship between y and x if y varies inversely with x and $k > 0, x > 0$.

E X A M P L E   2   Safe Weight That Can Be Supported by a Piece of Pine

The weight W that can be safely sup-   **FIGURE 70**
ported by a 2-inch by 4-inch piece of
lumber varies inversely with its length l.
See Figure 70. Experiments indicate that
the maximum weight a 10-foot long pine
2-by-4 can support is 500 pounds. Write
a general formula relating the safe weight
W (in pounds) to length l (in feet). Find
the maximum weight W that can be
safely supported by a length of 25 feet.

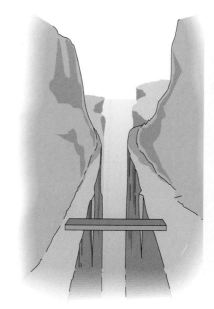

Solution   Because W varies inversely with l, we
know that

$$W = \frac{k}{l}$$

for some constant k. Because $W = 500$
when $l = 10$, we have

$$500 = \frac{k}{10}$$
$$k = 5000$$

FIGURE 71

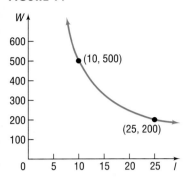

Thus, in all cases,

$$W = \frac{5000}{l}$$

In particular, the maximum weight W that can be safely supported by a piece of pine 25 feet in length is

$$W = \frac{5000}{25} = 200 \text{ pounds}$$

Figure 71 illustrates the relationship between the weight W and the length l.

In direct or inverse variation, the quantities that vary may be raised to powers. For example, in the early seventeenth century, Johannes Kepler (1571–1630) discovered that the square of the period T of a planet varies directly with the cube of its mean distance a from the Sun. That is, $T^2 = ka^3$, where k is the constant of proportionality.

Joint Variation and Combined Variation

When a variable quantity Q is proportional to the product of two or more other variables, we say that Q **varies jointly** with these quantities. Finally, combinations of direct and/or inverse variation may occur. This is usually referred to as **combined variation.**

Let's look at an example.

E X A M P L E 3

Loss of Heat through a Wall

The loss of heat through a wall varies jointly with the area of the wall and the difference between the inside and outside temperatures and varies inversely with the thickness of the wall. Write an equation that relates these quantities.

Solution

We begin by assigning symbols to represent the quantities:

L = Heat loss T = Temperature difference
A = Area of wall d = Thickness of wall

Then

$$L = k\frac{AT}{d}$$

where k is the constant of proportionality.

E X A M P L E 4

Force of the Wind on a Window

The force F of the wind on a flat surface positioned at a right angle to the direction of the wind varies jointly with the area A of the surface and the square of the speed v of the wind. A wind of 30 miles per hour blowing on a window measuring 4 feet by 5 feet has a force of 150 pounds. (See Figure 72.) What is the force on a window measuring 3 feet by 4 feet caused by a wind of 50 miles per hour?

FIGURE 72

Solution   Since F varies jointly with A and v^2, we have

$$F = kAv^2$$

where k is the constant of proportionality. We are told that $F = 150$ when $v = 30$ and $A = 4 \cdot 5 = 20$. Thus,

$$150 = k(20)(900)$$

$$k = \frac{1}{120}$$

The general formula is therefore

$$F = \frac{1}{120}Av^2$$

For a wind of 50 miles per hour blowing on a window whose area is $A = 3 \cdot 4 = 12$ square feet, the force F is

$$F = \frac{1}{120}(12)(2500) = 250 \text{ pounds}$$

3.6 | EXERCISES

In Problems 1–12, write a general formula to describe each variation.

1. y varies directly with x; $y = 2$ when $x = 10$
2. v varies directly with t; $v = 16$ when $t = 2$
3. A varies directly with x^2; $A = 4\pi$ when $x = 2$
4. V varies directly with x^3; $V = 36\pi$ when $x = 3$
5. F varies inversely with d^2; $F = 10$ when $d = 5$
6. y varies inversely with $\sqrt{x}$; $y = 4$ when $x = 9$
7. z varies directly with the sum of the squares of x and y; $z = 5$ when $x = 3$ and $y = 4$
8. T varies jointly with the cube root of x and the square of d; $T = 18$ when $x = 8$ and $d = 3$
9. M varies directly with the square of d and inversely with the square root of x; $M = 24$ when $x = 9$ and $d = 4$
10. z varies directly with the sum of the cube of x and the square of y; $z = 1$ when $x = 2$ and $y = 3$
11. The square of T varies directly with the cube of a and inversely with the square of d; $T = 2$ when $a = 2$ and $d = 4$.
12. The cube of z varies directly with the sum of the squares of x and y; $z = 2$ when $x = 9$ and $y = 4$.

In Problems 13–20, write an equation that relates the quantities.

13. **Geometry** The volume V of a sphere varies directly with the cube of its radius r. The constant of proportionality is $4\pi/3$.

14. **Geometry** The square of the hypotenuse c of a right triangle varies directly with the sum of the squares of the legs a and b. The constant of proportionality is 1.

15. **Geometry** The area A of a triangle varies jointly with the lengths of the base b and the height h. The constant of proportionality is $\frac{1}{2}$.

16. **Geometry** The perimeter p of a rectangle varies directly with the sum of the lengths of its sides l and w. The constant of proportionality is 2.

17. **Geometry** The volume V of a right circular cylinder varies jointly with the square of its radius r and its height h. The constant of proportionality is π. (See the figure.)

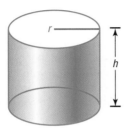

18. **Geometry** The volume V of a right circular cone varies jointly with the square of its radius r and its height h. The constant of proportionality is $\pi/3$. (See the figure.)

19. **Physics: Newton's Law** The force F (in newtons) of attraction between two bodies varies jointly with their masses m and M (in kilograms) and inversely with the square of the distance d (in meters) between them. The constant of proportionality is $G = 6.67 \times 10^{-11}$.

20. **Physics: Simple Pendulum** The *period* of a pendulum is the time required for one oscillation; the pendulum is usually referred to as *simple* when the angle made to the vertical is less than 5°. The period T of a simple pendulum (in seconds) varies directly with the square root of its length l (in feet). The constant of proportionality is $2\pi/\sqrt{32}$.

21. **Physics: Falling Objects** The distance s that an object falls is directly proportional to the square of the time t of the fall. If an object falls 16 feet in 1 second, how far will it fall in 3 seconds? How long will it take an object to fall 64 feet?

22. **Physics: Falling Objects** The velocity v of a falling object is directly proportional to the time t of the fall. If, after 2 seconds, the velocity of the object is 64 feet per second, what will its velocity be after 3 seconds?

23. **Physics: Stretching a Spring** The elongation E of a spring balance varies directly with the applied weight W (see the figure). If $E = 3$ when $W = 20$, find E when $W = 15$.

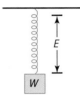

24. **Physics: Vibrating String** The rate of vibration of a string under constant tension varies inversely with the length of the string. If a string is 48 inches long and vibrates 256 times per second, what is the length of a string that vibrates 576 times per second?

25. **Weight of a Body** The weight of a body above the surface of Earth varies inversely with the square of the distance from the center of Earth. If a certain body weighs 55 pounds when it is 3960 miles from the center of Earth, how much will it weigh when it is 3965 miles from the center?

26. **Force of the Wind on a Window** The force exerted by the wind on a plane surface varies jointly with the area of the surface and the square of the velocity of the wind. If the force on an area of 20 square feet is 11 pounds when the wind velocity is 22 miles per hour, find the force on a surface area of 47.125 square feet when the wind velocity is 36.5 miles per hour.

27. **Horsepower** The horsepower (hp) that a shaft can safely transmit varies jointly with its speed (in revolutions per minute, rpm) and the cube of its diameter. If a shaft of a certain material 2 inches in diameter can transmit 36 hp at 75 rpm, what diameter must the shaft have in order to transmit 45 hp at 125 rpm?

28. **Weight of a Body**   The weight of a body varies inversely with the square of its distance from the center of Earth. Assuming that the radius of Earth is 3960 miles, how much would a man weigh at an altitude of 1 mile above Earth's surface if he weighs 200 pounds on Earth's surface?

29. **Physics: Kinetic Energy**   The kinetic energy K of a moving object varies jointly with its mass m and the square of its velocity v. If an object weighing 25 pounds and moving with a velocity of 100 feet per second has a kinetic energy of 400 foot-pounds, find its kinetic energy when the velocity is 150 feet per second.

30. **Electrical Resistance of a Wire**   The electrical resistance of a wire varies directly with the length of the wire and inversely with the square of the diameter of the wire. If a wire 432 feet long and 4 millimeters in diameter has a resistance of 1.24 ohms, find the length of a wire of the same material whose resistance is 1.44 ohms and whose diameter is 3 millimeters.

31. **Measuring the Stress of Materials**   The stress in the material of a pipe subject to internal pressure varies jointly with the internal pressure and the internal diameter of the pipe and inversely with the thickness of the pipe. The stress is 100 pounds per square inch when the diameter is 5 inches, the thickness is 0.75 inch, and the internal pressure is 25 pounds per square inch. Find the stress when the internal pressure is 40 pounds per square inch if the diameter is 8 inches and the thickness is 0.50 inch.

32. **Safe Load for a Beam**   The maximum safe load for a horizontal rectangular beam varies jointly with the width of the beam and the square of the thickness of the beam and inversely with its length. If an 8-foot beam will support up to 750 pounds when the beam is 4 inches wide and 2 inches thick, what is the maximum safe load in a similar beam 10 feet long, 6 inches wide, and 2 inches thick?

33. **Resistance due to a Conductor**   The resistance (in ohms) of a circular conductor varies directly with the length of the conductor and inversely with the square of the radius of the conductor. If 50 feet of wire with a radius of 6×10^{-3} inch has a resistance of 10 ohms, what would be the resistance of 100 feet of the same wire if the radius is increased to 7×10^{-3} inch?

34. **Chemistry: Gas Laws**   The volume V of an ideal gas varies directly with the temperature T and inversely with the pressure P. Write an equation relating V, T, and P using k as the constant of proportionality. If a cylinder contains oxygen at a temperature of 300 K and a pressure of 15 atmospheres in a volume of 100 liters, what is the constant of proportionality k? If a piston is lowered into the cylinder, decreasing the volume occupied by the gas to 80 liters and raising the temperature to 310 K, what is the gas pressure?

In Problems 36–40, use the result obtained in Problem 35.

35. **Satellites in Orbit**   The speed v required of a satellite to maintain a near-Earth circular orbit is directly proportional to the square root of the distance r of the satellite from the center of Earth.* The constant of proportionality is $\sqrt{g}$, where g is the acceleration of gravity for Earth. Write an equation that shows the relationship between v and r. (The radius of Earth is approximately 3960 miles.)

*Near-Earth orbits are at least 100 miles above Earth's surface (out of Earth's atmosphere) and up to an altitude of approximately 15,000 miles. The effect of the gravitational attraction of other bodies is ignored. Although the acceleration of gravity at such altitudes is somewhat less than $g \approx 32$ feet per second per second $\approx 79,036$ miles per hour per hour, we shall ignore this discrepancy in our calculations.

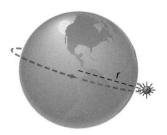

36. **Speed Required to Stay in Orbit**   What speed is required to maintain a communications satellite in a circular orbit 500 miles above Earth's surface?

37. **Speed Required to Stay in Orbit**   Find the speed of a satellite that moves in a circular orbit 140 miles above Earth's surface.

38. Distance of a Satellite from Earth Find the distance of a satellite from the surface of Earth as it moves around Earth in a circular orbit at a constant speed of 18,630 miles per hour.

39. Distance of a Satellite from Earth A weather satellite orbits Earth in a circle every 1.5 hours. How high is it above Earth?

40. Distance and Speed of a Military Satellite A military satellite orbits Earth every 2 hours. How high is this satellite and what is its speed?

In Problems 42–46, use the result obtained in Problem 41.

41. Physical Force The force *F* (in newtons) required to maintain an object in a circular path varies jointly with the mass *m* (in kilograms) of the object and the square of its speed *v* (in meters per second) and inversely with the radius *r* (in meters) of the circular path. The constant of proportionality is 1. Write an equation relating *F, m, v,* and *r.*

42. Physical Forces A motorcycle with mass 150 kilograms is driven at a constant speed of 120 kilometers per hour on a circular track with a radius of 100 meters. To keep the motorcycle from skidding, what frictional force must be exerted by the tires on the track?

43. Physical Forces If the speed of the motorcycle described in Problem 42 is increased by 10%, by how much is the frictional force of the tires increased?

44. Physical Forces If the radius of the track described in Problem 42 is cut in half, how much slower should the motorcycle be driven to maintain the same frictional force?

45. Physical Forces Judy is spinning a bucket of water in a horizontal plane at the end of a rope

of length *L* (see the figure). If she triples the speed of the bucket, how many times as hard must she pull on the rope?

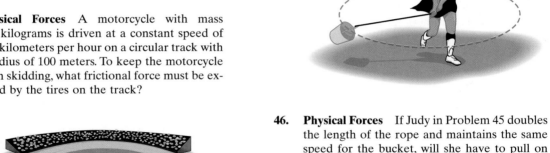

46. Physical Forces If Judy in Problem 45 doubles the length of the rope and maintains the same speed for the bucket, will she have to pull on the rope more or less? How much?

47. The formula on page 217 attributed to Johannes Kepler is one of the famous three Keplerian Laws of Planetary Motion. Go to the library and research these laws. Write a brief paper about these laws and Kepler's place in history.

48. Make up a real-world problem different from any in the text that you think involves two variables that vary directly. Exchange your problem with another student's to solve and critique.

49. Make up a real-world problem different from any in the text that you think involves two variables that vary inversely. Exchange your problem with another student's to solve and critique.

50. Make up a real-world problem different from any in the text that you think involves three variables that vary jointly. Exchange your problem with another student's to solve and critique.

CHAPTER REVIEW

THINGS TO KNOW

Formulas

Distance formula	$d = \sqrt{(x_2 - x_1)^2 + (y_2 - y_1)^2}$
Midpoint formula	$(x, y) = \left(\dfrac{x_1 + x_2}{2}, \dfrac{y_1 + y_2}{2} \right)$
Slope	$m = \dfrac{y_2 - y_1}{x_2 - x_1}$, if $x_1 \neq x_2$; undefined if $x_1 = x_2$
Parallel lines	Equal slopes $(m_1 = m_2)$
Perpendicular lines	Product of slopes is -1 $(m_1 \cdot m_2 = -1)$
Direct variation	$y = kx$
Inverse variation	$y = \dfrac{k}{x}$

Equations

Vertical line	$x = a$
Horizontal line	$y = b$
Point–slope form of the equation of a line	$y - y_1 = m(x - x_1)$; m is the slope of the line, (x_1, y_1) is a point on the line
Slope–intercept form of the equation of a line	$y = mx + b$; m is the slope of the line, b is the y-intercept
General form of the equation of a line	$Ax + By + C = 0$; A, B, not both 0
Standard form of the equation of a circle	$(x - h)^2 + (y - k)^2 = r^2$; r is the radius of the circle, (h, k) is the center of the circle
General form of the equation of a circle	$x^2 + y^2 + ax + by + c = 0$
Equation of the unit circle	$x^2 + y^2 = 1$

HOW TO:

Use the distance formula	Draw scatter diagrams
Graph equations by plotting points	Obtain the equation of a line
Find the intercepts of a graph	Obtain the equation of a circle
Find the intercepts of an equation	Find the center and radius of a circle, given the equation
Test an equation for symmetry	Graph circles
Find the slope and intercepts of a line, given the equation	Solve variation problems
Graph lines	Find the equation of the line of best fit

FILL-IN-THE-BLANK ITEMS

1. If (x, y) are the coordinates of a point P in the xy-plane, then x is called the _____ of P and y is the _____ of P.

2. If three distinct points P, Q, and R all lie on a line and if $d(P, Q) = d(Q, R)$, then Q is called the _____ of the line segment from P to R.

3. If for every point (x, y) on a graph the point $(-x, y)$ is also on the graph, then the graph is symmetric with respect to the _____.

4. The set of points in the xy-plane that are a fixed distance from a fixed point is called a(n) _____. The fixed distance is called the _____; the fixed point is called the _____.

5. The slope of a vertical line is _____; the slope of a horizontal line is _____.

6. Two nonvertical lines have slopes m_1 and m_2, respectively. The lines are parallel if _____; the lines are perpendicular if _____.

7. If z varies jointly as x^2 and y^3 and inversely as $\sqrt{t}$, then $z =$ _____, where k is the constant of proportionality.

TRUE/FALSE ITEMS

T F **1.** The distance between two points is sometimes a negative number.

T F **2.** The graph of the equation $y = x^4 + x^2 + 1$ is symmetric with respect to the y-axis.

T F **3.** Vertical lines have undefined slope.

T F **4.** The slope of the line $2y = 3x + 5$ is 3.

T F **5.** Perpendicular lines have slopes that are reciprocals of one another.

T F **6.** The radius of the circle $x^2 + y^2 = 9$ is 3.

T F **7.** If y varies inversely with x, then as x increases in value, y will decrease in value.

REVIEW EXERCISES

Blue problem numbers indicate the author's suggestions for use in a Practice Test.

In Problems 1–10, find an equation of the line having the given characteristics. Express your answer using either the general form or the slope–intercept form of the equation of a line, whichever you prefer.

1. Slope $= -2$;   passing through $(3, -1)$
2. Slope $= 0$;   passing through $(-5, 4)$
3. Slope undefined;   passing through $(-3, 4)$
4. x-intercept $= 2$;   passing through $(4, -5)$
5. y-intercept $= -2$;   passing through $(5, -3)$
6. Passing through $(3, -4)$ and $(2, 1)$
7. Parallel to the line $2x - 3y + 4 = 0$;   passing through $(-5, 3)$
8. Parallel to the line $x + y - 2 = 0$;   passing through $(1, -3)$
9. Perpendicular to the line $x + y - 2 = 0$;   passing through $(4, -3)$
10. Perpendicular to the line $3x - y + 4 = 0$;   passing through $(-2, 4)$

In Problems 11–16, graph each line, and label the x-intercept and y-intercept.

11. $4x - 5y + 20 = 0$

12. $3x + 4y - 12 = 0$

13. $\dfrac{1}{2}x - \dfrac{1}{3}y + \dfrac{1}{6} = 0$

14. $-\dfrac{3}{4}x + \dfrac{1}{2}y = 0$

15. $\sqrt{2}x + \sqrt{3}y = \sqrt{6}$

16. $\dfrac{x}{3} + \dfrac{y}{4} = 1$

In Problems 17–22, find the center and radius of each circle. Graph each circle.

17. $x^2 + (y - 1)^2 = 4$

18. $(x + 2)^2 + y^2 = 9$

19. $x^2 + y^2 - 2x + 4y - 4 = 0$

20. $x^2 + y^2 + 4x - 4y - 1 = 0$

21. $3x^2 + 3y^2 - 6x + 12y = 0$

22. $2x^2 + 2y^2 - 4x = 0$

23. Find the slope of the line containing the points $(7, 4)$ and $(-3, 2)$. What is the distance between these points? What is their midpoint?

24. Find the slope of the line containing the points $(2, 5)$ and $(6, -3)$. What is the distance between these points? What is their midpoint?

25. The figure on the right shows the graph of two parallel lines. Which of the following pairs of equations might have such a graph?

(a) $x - 2y = 3$
$x + 2y = 7$
(b) $x + y = 2$
$x + y = -1$
(c) $x - y = -2$
$x - y = 1$
(d) $x - y = -2$
$2x - 2y = -4$
(e) $x + 2y = 2$
$x + 2y = -1$

26. The figure on the right shows the graph of two perpendicular lines. Which of the following pairs of equations might have such a graph?

(a) $y - 2x = 2$
$y + 2x = -1$
(b) $y - 2x = 0$
$2y + x = 0$
(c) $2y - x = 2$
$2y + x = -2$
(d) $y - 2x = 2$
$x + 2y = -1$
(e) $2x + y = -2$
$2y + x = -2$

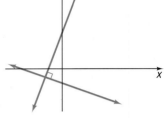

In Problems 27–34, list the intercepts and test for symmetry.

27. $2x = 3y^2$ **28.** $y = 5x$ **29.** $x^2 + 4y^2 = 16$ **30.** $9x^2 - y^2 = 9$

31. $y = x^4 + 2x^2 + 1$ **32.** $y = x^3 - x$ **33.** $x^2 + x + y^2 + 2y = 0$ **34.** $x^2 + 4x + y^2 - 2y = 0$

35. **Geometry** The area of an equilateral triangle varies directly with the square of the length of a side. If the area of the equilateral triangle whose sides are of length 1 centimeter is $\sqrt{3}/4$, find the length s of each side of an equilateral triangle whose area A is 16 square centimeters.

36. **Vibrating Strings** In a vibrating string, the pitch (measured in vibrations per second) varies directly with the square root of the tension of the string (measured in pounds). If a certain string vibrates 300 times per second under a tension of 9 pounds, find the tension required to cause the string to vibrate 400 times per second.

37. **Kepler's Third Law of Planetary Motion** Kepler's Third Law of Planetary Motion states that the square of the period T of revolution of a planet is proportional to the cube of its mean distance from the Sun. If the mean distance of Earth from the Sun is 93 million miles, what is the mean distance a of the planet Mercury from the Sun, given that Mercury has a "year" of 88 days?

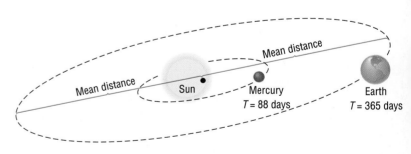

38. Use Problem 37 to find the mean distance of the planet Jupiter from the Sun, given that Jupiter circles the Sun every $5\sqrt{5}$ years.

39. Show that the points $A = (3, 4)$, $B = (1, 1)$, and $C = (-2, 3)$ are the vertices of an isosceles triangle.

40. Show that the points $A = (-2, 0)$, $B = (-4, 4)$, and $C = (8, 5)$ are the vertices of a right triangle in two ways:

(a) By using the converse of the Pythagorean Theorem
(b) By using the slopes of the lines joining the vertices

41. Show that the points $A = (2, 5)$, $B = (6, 1)$, and $C = (8, -1)$ lie on a straight line by using slopes.

42. Show that the points $A = (1, 5)$, $B (2, 4)$, and $C = (-3, 5)$ lie on a circle with center $(-1, 2)$. What is the radius of this circle?

43. The endpoints of the diameter of a circle are $(-3, 2)$ and $(5, -6)$. Find the center and radius of the circle. Write the general equation of this circle.

44. Find two numbers y such that the distance from $(-3, 2)$ to $(5, y)$ is 10.

45. Concentration of Carbon Monoxide in the Air The following data represent the average concentration of carbon monoxide in parts per million (ppm) in the air for 1987–1993.

Year	Concentration of Carbon Monoxide (ppm)
1987	6.69
1988	6.38
1989	6.34
1990	5.87
1991	5.55
1992	5.18
1993	4.88

Source: U.S. Environmental Protection Agency.

(a) Treating the year as the x-coordinate and the average level of carbon monoxide as the y-coordinate, draw a scatter diagram of the data.

(b) What is the slope of the line joining the points (1987, 6.69) and (1990, 5.87)?

(c) Interpret this slope.

(d) What is the slope of the line joining the points (1990, 5.87) and (1993, 4.88)?

(e) Interpret this slope.

 (f) Use a graphing utility to find the slope of the line of best fit for these data.

(g) Interpret this slope.

 (h) How do you explain the differences among the three slopes obtained?

(i) What is the trend in the data? In other words, as time passes, what is happening to the average level of carbon monoxide in the air? Why do you think this is happening?

46. Value of a Portfolio The following data represent the value of the Vanguard Index Trust-500 Portfolio for 1987–1995.

Year	Value (Dollars)
1987	54.26
1988	63.06
1989	82.84
1990	80.08
1991	104.28
1992	112.03
1993	123.11
1994	124.56
1995	171.20

Source: Vanguard Index Trust, Annual Report 1995.

(a) Treating the year as the x-coordinate and the value of the Vanguard Index Trust-500 Portfolio as the y-coordinate, draw a scatter diagram of the data.

(b) What is the slope of the line joining the points (1987, 54.26) and (1991, 104.28)?

(c) Interpret this slope.

(d) What is the slope of the line joining the points (1991, 104.28) and (1995, 171.20)?

(e) Interpret this slope.

(f) Use a graphing utility to find the slope of the line of best fit for these data.

(g) Interpret this slope.

(h) If you were managing this trust, which of the three slopes would you use to convince someone to invest? Why?

(i) What is the trend in the data? In other words, as time passes, what is happening to the value of the Vanguard Index Trust-500 Portfolio?

47. Make up four problems that you might be asked to do given the two points $(-3, 4)$ and $(6, 1)$. Each problem should involve a different concept. Be sure your directions are clearly stated.

48. Describe each of the following graphs in the xy-plane. Give justification.
(a) $x = 0$
(b) $y = 0$
(c) $x + y = 0$
(d) $xy = 0$
(e) $x^2 + y^2 = 0$

49. Suppose that you have a rectangular field that requires watering. Your watering system consists of an arm of variable length that rotates so that the watering pattern is a circle. Decide where to position the arm and what length it should be so that the entire field is watered most efficiently. When does it become desirable to use more than one arm?

[**Hint:** Use a rectangular coordinate system positioned so that the axes bisect the rectangle (see the figure). Write equations for the circle(s) swept out by the watering arm(s).]

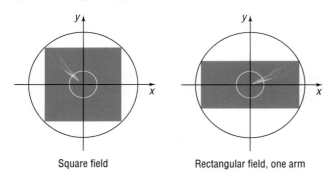

Square field   Rectangular field, one arm

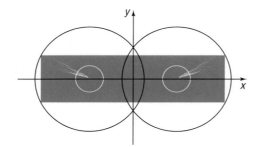

Rectangular field, two arms

 50. Graph $Y_1 = x$ and $Y_2 = 0.99x + 0.01$ on the same screen. It appears the graphs are identical. Explain why this happens. Provide a way to correct this situation.

51. Why does the graph of $Y_1 = x/6$, a straight line, appear to consist of a collection of tiny horizontal line segments when drawn using a graphing utility?

Functions and Their Graphs

I magine yourself as an expert for the EPA and sitting in congressional policy meetings having to explain the importance of balancing energy resources with environmental protection. On the following page is the Internet Excursion placing you in exactly that position and asking the tough questions a member of Congress might ask. Use the Sullivan website at:

www.prenhall.com/sullivan

to link to the Internet resources needed to answer the questions asked.

PREPARING FOR THIS CHAPTER

Before getting started on this chapter, review the following concepts:

Domain of a Variable *(p. 17)*

Graphs of Certain Equations *(Example 2, p. 175; Example 3, p. 176; Example 4, p. 176; Example 11, p. 182)*

Tests for Symmetry of an Equation *(p. 180)*

Procedure for Finding Intercepts of an Equation *(p. 178)*

Steps for Setting up Applied Problems *(p. 99)*

OUTLINE

HOW LONG WILL THE OIL LAST?

The global supply of oil and other sources of energy is more than adequate to meet present needs, but most of this supply is outside the United States. Currently, oil supplies about 40 percent of the world's energy, with the United States being the biggest consumer. Since oil is a nonrenewable energy resource, plans must be made for the eventuality of diminishing supply.

Suppose that you were a consultant for the EPA and members of Congress asked you to sit on the Energy Policy Steering Committee as an expert analyst. You must convince the committee of the importance of environmental concerns in planning the global energy system. You are very aware of the fact that present modes of energy use and production threaten serious environmental deterioration. Your plan is to create a set of "what if" functions that will allow your committee to model many possible scenarios for their policy guidelines.

1. $E(t)$, the first function that you create, will allow you to model United States oil consumption over a period of time. Make a scatter diagram using the EPA Data on United States energy consumption per capita from 1950–1990. Since oil provides 40 percent of the energy consumed, adjust your figures to get oil consumption per capita. Use the LINear REGression tool on your graphing utility to find the linear function $E(t)$ of best fit for modeling the data. Does this seem like a good model?

2. Your second function, $P(t)$, is a population modeling tool. From the United States Census Data, make a scatter diagram, and then use a QUADratic REGression to model the data. Next, let $O(t) = E(t) \cdot P(t)$. What does $O(t)$ model? Graph $O(t)$. Compare to the actual figures from BP Petroleum.

3. Using the function $O(t)$, estimate the area under the graph for the years 1997 through 2000. To do this make a trapezoid and find its area. Now estimate the areas under $O(t)$ using 10-year intervals. For what year does the total of all the trapezoid areas equal the remaining United States oil reserve?

4. Could this type of analysis be used to predict when the world's supply of oil will run out? Can your committee think of any methods that might improve your predictions?

Perhaps the most central idea in mathematics is the notion of a *function*. This important chapter deals with what a function is, how to graph functions, how to perform operations on functions, and how functions are used in applications.

The word *function* apparently was introduced by René Descartes in 1637. For him, a function simply meant any positive integral power of a variable *x*. Gottfried Wilhelm Leibniz (1646–1716), who always emphasized the geometric side of mathematics, used the word function to denote any quantity associated with a curve, such as the coordinates of a point on the curve. Leonhard Euler (1707–1783) employed the word to mean any equation or formula involving variables and constants. His idea of a function is similar to the one most often seen in courses that precede calculus. Later, the use of functions in investigating heat flow equations led to a very broad definition, due to Lejeune Dirichlet (1805–1859), which describes a function as a rule or correspondence between two sets. It is his definition that we use here.

4.1 | FUNCTIONS

- 1 Determine Whether a Relation Represents a Function
- 2 Find the Value of a Function
- 3 Find the Domain of a Function
- 4 Identify the Graph of a Function
- 5 Obtain Information from or about the Graph of a Function

In Section 3.1, we said that a **relation** is a correspondence between two variables, say x and y. When relations are written as ordered pairs (x, y), we say that x is related to y. Often, we are interested in specifying the type of relation (such as an equation) that might exist between the two variables.

For example, the relation between the revenue R resulting from the sale of x items selling for \$10 each may be expressed by the equation $R = 10x$. If we know how many items have been sold, then we can calculate the revenue by using the equation $R = 10x$. This equation is an example of a *function*.

As another example, suppose that an icicle falls off a building from a height of 64 feet above the ground. According to a law of physics, the distance s (in feet) of the icicle from the ground after t seconds is given (approximately) by the formula $s = 64 - 16t^2$. When $t = 0$ seconds, the icicle is $s = 64$ feet above the ground. After 1 second, the icicle is $s = 64 - 16(1)^2 = 48$ feet above the ground. After 2 seconds, the icicle strikes the ground. The formula $s = 64 - 16t^2$ provides a way of finding the distance s when the time t $(0 \le t \le 2)$ is prescribed. There is a correspondence between each time t in the interval $0 \le t \le 2$ and the distance s. We say that the distance s is a *function* of the time t because

1. There is a correspondence between the set of times and the set of distances.
2. There is exactly one distance s obtained for a prescribed time t in the interval $0 \le t \le 2$.

Let's now look at the definition of a function.

Definition of Function

FIGURE 1

Let X and Y be two nonempty sets of real numbers.* A **function** from X into Y is a relation that associates with each element of X a unique element of Y.

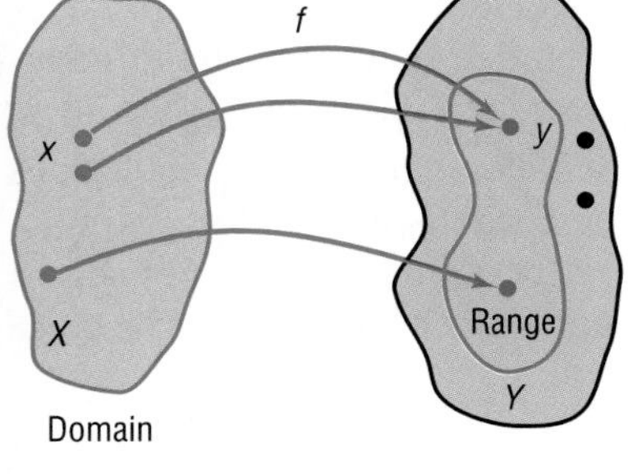

The set X is called the **domain** of the function. For each element x in X, the corresponding element y in Y is called the **value** of the function at x, or the **image** of x. The set of all images of the elements of the domain is called the **range** of the function. See Figure 1.

Since there may be some elements in Y that are not the image of some x in X, it follows that the range of a function may be a subset of Y, as shown in Figure 1.

---

*The two sets X and Y can also be sets of complex numbers (discussed in Section 5.6), and then we have defined a complex function. In the broad definition (due to Lejeune Dirichlet), X and Y can be any two sets.

Warning Do not confuse the two meanings given for the word range. When used in connection with a function, it means the set of all images of the elements of the domain of the function. When the word RANGE is used in connection with a graphing utility, it means the settings used for the viewing rectangle. ■

Not all relations between two sets are functions.

E X A M P L E   1

Determining Whether a Relation Represents a Function

Determine whether the following relations represent functions.

(a) For this relation, the domain represents the employees of Sara's Pre-Owned Car Mart and the range represents their base salary.

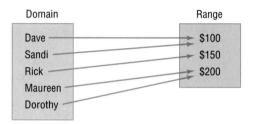

(b) For this relation, the domain represents the employees of Sara's Pre-Owned Car Mart and the range represents their phone number(s).

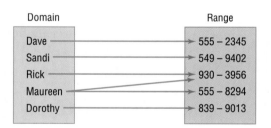

Solution (a) The relation is a function because each element in the domain corresponds to a unique element in the range. Notice that more than one element in the domain can correspond to the same element in the range.

(b) The relation is not a function because each element in the domain does not correspond to a unique element in the range. Maureen has two telephone numbers; therefore, if Maureen is chosen from the domain, a unique telephone number cannot be assigned to her. ■

We may think of a function as a set of ordered pairs (x, y) in which no two distinct pairs have the same first element. The set of all first elements x is the domain of the function, and the set of all second elements y is its range. Thus, there is associated with each element x in the domain a unique element y in the range.

E X A M P L E   2   Determining Whether a Relation Represents a Function

Determine whether each relation represents a function.

(a) $\{(1, 4), (2, 5), (3, 6), (4, 7)\}$
(b) $\{(1, 4), (2, 4), (3, 5), (6, 10)\}$
(c) $\{(-3, 9), (-2, 4), (0, 0), (1, 1), (-3, 8)\}$

Solution   (a) This relation is a function because there are no distinct ordered pairs with the same first element.
(b) This relation is a function because there are no distinct ordered pairs with the same first element.
(c) This relation is not a function because there is a first element, -3, that corresponds to two different second elements, 9 and 8.   ∎

In Example 2(b), notice that 1 and 2 in the domain each have the same image in the range. This does not violate the definition of a function; two different first elements can have the same second element. A violation of the definition occurs when two ordered pairs have the same first element and different second elements, as in Example 2(c).

The relation referred to in the definition of a function is most often given as an equation in two variables, usually denoted x and y.

E X A M P L E   3   Example of a Function

Consider the function defined by the equation

$$y = 2x - 5 \qquad 1 \le x \le 6$$

The domain $1 \le x \le 6$ specifies that the number x is restricted to the real numbers from 1 to 6, inclusive. The equation $y = 2x - 5$ specifies that the number x is to be multiplied by 2 and then 5 is to be subtracted from the result to get y. For example, if $x = \frac{3}{2}$, then $y = 2 \cdot \frac{3}{2} - 5 = -2$.   ∎

 Now work Problems 1 and 5.

Function Notation

Functions are often denoted by letters such as f, F, g, G and so on. If f is a function, then for each number x in its domain the corresponding image in the range is designated by the symbol $f(x)$, read as "f of x" or as "f at x." We refer to $f(x)$ as the **value of f at the number x.** Thus, $f(x)$ is the number that results when x is given and the function f is applied; $f(x)$ does *not* mean "f times x." For example, the function given in Example 3 may be written as $y = f(x) = 2x - 5, 1 \le x \le 6$. Then $f(\frac{3}{2}) = -2$.

Figure 2 illustrates some other functions. Note that for each function illustrated, to each x in the domain, there is one value in the range.

FIGURE 2

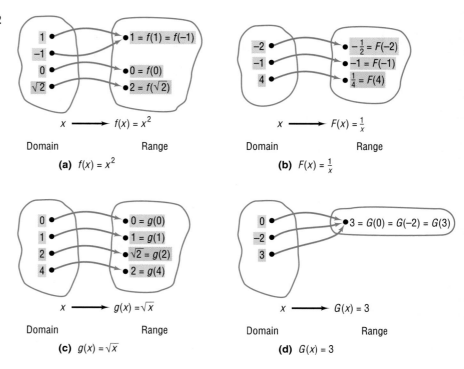

(a) $f(x) = x^2$

(b) $F(x) = \frac{1}{x}$

(c) $g(x) = \sqrt{x}$

(d) $G(x) = 3$

Sometimes it is helpful to think of a function f as a machine that receives as input a number from the domain, manipulates it, and outputs the value. See Figure 3.

FIGURE 3 Input x

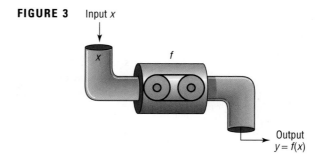

The restrictions on this input/output machine are

1. It only accepts numbers from the domain of the function.
2. For each input, there is exactly one output (which may be repeated for different inputs).

For a function $y = f(x)$, the variable x is called the **independent variable,** because it can be assigned any of the permissible numbers from the domain. The variable y is called the **dependent variable,** because its value depends on x.

Any symbol can be used to represent the independent and dependent variables. For example, if f is the *cube function,* then f can be defined by $f(x) = x^3$ or $f(t) = t^3$ or $f(z) = z^3$. All three functions are the same: Each tells us to cube the independent variable. In practice, the symbols used for the independent and dependent variables are based on common usage.

The variable x is also called the **argument** of the function. Thinking of the independent variable as an argument sometimes can make it easier to find the value of a function. For example, if f is the function defined by $f(x) = x^3$, then f tells us to cube the argument. Thus, $f(2)$ means to cube 2, $f(a)$ means to cube the number a, and $f(x + h)$ means to cube the quantity $x + h$.

E X A M P L E   4   Finding Values of a Function

For the function G defined by $G(x) = 2x^2 - 3x$, evaluate:

(a) $G(3)$ (b) $G(x) + G(3)$ (c) $G(-x)$

(d) $-G(x)$ (e) $G(x + 3)$

Solution   (a) We substitute 3 for x in the equation for G to get

$$G(3) = 2(3)^2 - 3(3) = 18 - 9 = 9$$

(b) $G(x) + G(3) = (2x^2 - 3x) + (9) = 2x^2 - 3x + 9$

(c) We substitute $-x$ for x in the equation for G:

$$G(-x) = 2(-x)^2 - 3(-x) = 2x^2 + 3x$$

(d) $-G(x) = -(2x^2 - 3x) = -2x^2 + 3x$

(e) $G(x + 3) = 2(x + 3)^2 - 3(x + 3)$ Notice the use of parentheses here.

$= 2(x^2 + 6x + 9) - 3x - 9$

$= 2x^2 + 12x + 18 - 3x - 9$

$= 2x^2 + 9x + 9$ ∎

Notice in this example that $G(x + 3) \neq G(x) + G(3)$ and $G(-x) \neq -G(x)$.

 Now work Problem 13.

Most calculators have special keys that enable you to find the value of certain commonly used functions. For example, you should be able to find the square function, $f(x) = x^2$; the square root function, $f(x) = \sqrt{x}$; the reciprocal function, $f(x) = 1/x = x^{-1}$; and many others that will be discussed later in this book (such as $\ln x$, $\log x$, and so on). Verify the results of Example 5 on your calculator.

E X A M P L E   5   Finding Values of a Function on a Calculator

(a) $f(x) = x^2$; $f(1.234) = 1.522756$

(b) $F(x) = 1/x$; $F(1.234) = 0.8103727715$

(c) $g(x) = \sqrt{x}$; $g(1.234) = 1.110855526$ ∎

In general, when a function f is defined by an equation in x and y, we say that the function f is given **implicitly.** If it is possible to solve the equation for y in terms of x, then we write $y = f(x)$ and say that the function is given **explicitly.** In fact, we usually write "the function $y = f(x)$" when we mean "the function f defined by the equation $y = f(x)$." Although this usage is not entirely correct, it is rather common and should not cause any confusion. For example,

Implicit Form	**Explicit Form**
$3x + y = 5$	$y = f(x) = -3x + 5$
$x^2 - y = 6$	$y = f(x) = x^2 - 6$
$xy = 4$	$y = f(x) = 4/x$

Not all equations in x and y define a function $y = f(x)$. If an equation is solved for y and two or more values of y can be obtained for a given x, then the equation does not define a function. For example, consider the equation $x^2 + y^2 = 1$, which defines a circle. If we solve for y, we obtain $y = \pm\sqrt{1 - x^2}$, so two values of y result for numbers x between -1 and 1. Thus, $x^2 + y^2 = 1$ does not define a function.

 Comment The explicit form of a function is the form required by a graphing calculator. Now do you see why it is necessary to graph a circle in two "pieces"? ▬

We list below a summary of some important facts to remember about a function f.

SUMMARY OF IMPORTANT FACTS ABOUT FUNCTIONS

1.   To each x in the domain of f, there is one and only one image $f(x)$ in the range.
2. f is the symbol we use to denote the function. It is symbolic of the domain and the equation that we use to get from an x in the domain to $f(x)$ in the range.
3.   If $y = f(x)$, then x is called the independent variable or argument of f and y is called the dependent variable or the value of f at x.

Domain of a Function

 Often, the domain of a function f is not specified; instead, only the equation defining the function is given. In such cases, we agree that the domain of f is the largest set of real numbers for which the value $f(x)$ is a real number. Thus, the domain of f is the same as the domain of the variable x in the expression $f(x)$.

E X A M P L E   6

Finding the Domain of a Function

Find the domain of each of the following functions:

(a) $f(x) = x^2 + 5x$ (b) $g(x) = \dfrac{3x}{x^2 - 4}$ (c) $h(x) = \sqrt{4 - 3x}$

Solution

(a) The function tells us to square a number and then add five times the number. Since these operations can be performed on any real number, we conclude that the domain of f is all real numbers.
(b) The function g tells us to divide $3x$ by $x^2 - 4$. Since division by 0 is not allowed, the denominator $x^2 - 4$ can never be 0. Thus, x can never equal 2 or -2. The domain of the function g is $\{x \mid x \neq -2, x \neq 2\}$.

(c) The function h tells us to take the square root of $4 - 3x$. But only non-negative numbers have real square roots. Hence, we require that

$$4 - 3x \geq 0$$
$$-3x \geq -4$$
$$x \leq \tfrac{4}{3}$$

The domain of h is $\{x \mid x \leq \tfrac{4}{3}\}$ or the interval $(-\infty, \tfrac{4}{3}]$. ▬

Now work Problem 53.

If x is in the domain of a function f, we shall say that **f is defined at x,** or **$f(x)$ exists.** If x is not in the domain of f, we say that **f is not defined at x,** or **$f(x)$ does not exist.** For example, if $f(x) = x/(x^2 - 1)$, then $f(0)$ exists, but $f(1)$ and $f(-1)$ do not exist. (Do you see why?)

We have not said much about finding the range of a function. The reason is that when a function is defined by an equation it is often difficult to find the range. Therefore, we shall usually be content to find just the domain of a function when only the rule for the function is given. We shall express the domain of a function using inequalities, interval notation, set notation, or words, whichever is most convenient.

The Graph of a Function

In applications, a graph often demonstrates more clearly the relationship between two variables than, say, an equation or table would. For example, Table 1 shows the price per share of Disney stock at the end of each month during 1997.

If we plot the data in Table 1, using the date as the x-coordinate and the price as the y-coordinate, and then connect the points, we obtain the graph in Figure 4.

TABLE 1			
Date	**Closing Price**	**Date**	**Closing Price**
1/31/97	$72\tfrac{7}{8}$	7/31/97	$80\tfrac{3}{4}$
2/28/97	$74\tfrac{1}{4}$	8/31/97	$76\tfrac{3}{4}$
3/31/97	$72\tfrac{7}{8}$	9/30/97	$80\tfrac{5}{8}$
4/30/97	$81\tfrac{3}{4}$	10/31/97	$82\tfrac{3}{8}$
5/31/97	$81\tfrac{7}{8}$	11/30/97	95
6/30/97	$80\tfrac{1}{4}$	12/31/97	99

Courtesy of A.G. Edwards & Sons, Inc.

FIGURE 4
Monthly closing prices of Disney stock in 1997

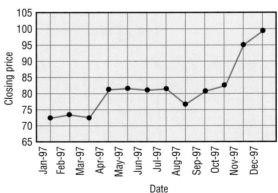

We can see from the graph that the price of the stock was fairly constant during May, June, and July and was rising during November and December. The graph also shows that the lowest price during this period occurred at the end of January and March, while the highest occurred at the end of December. Equations and tables, on the other hand, usually require some calculations and interpretation before this kind of information can be "seen."

Look again at Figure 4. The graph shows that for each time on the horizontal axis there is only one price on the vertical axis. Thus, the graph represents a function, although the exact rule for getting from time to price is not given.

When a function is defined by an equation in x and y, the **graph** of the function is the graph of the equation, that is, the set of points (x, y) in the xy-plane that satisfies the equation.

 Comment   When we select a viewing rectangle to graph a function, the values of Xmin, Xmax give the domain that we wish to view, while Ymin, Ymax give the range that we wish to view. These settings usually do not represent the actual domain and range of the function.   ▬

 Not every collection of points in the xy-plane represents the graph of a function. Remember, for a function, each number x in the domain has one and only one image y. Thus, the graph of a function cannot contain two points with the same x-coordinate and different y-coordinates. Therefore, the graph of a function must satisfy the following **vertical-line test.**

> **Theorem**   Vertical-line Test
>
> A set of points in the xy-plane is the graph of a function if and only if every vertical line intersects the graph in at most one point.

▬

It follows that, if any vertical line intersects a graph at more than one point, the graph is not the graph of a function.

E X A M P L E   7   Identifying the Graph of a Function

Which of the graphs in Figure 5 are graphs of functions?

FIGURE 5

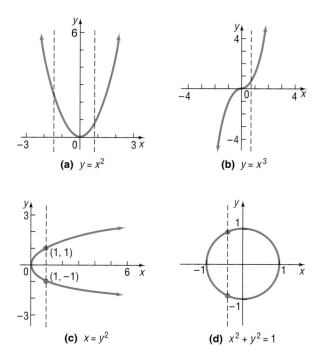

(a) $y = x^2$

(b) $y = x^3$

(c) $x = y^2$

(d) $x^2 + y^2 = 1$

Solution   The graphs in Figures 5(a) and 5(b) are graphs of functions, because a vertical line intersects each graph in at most one point. The graphs in Figures 5(c) and 5(d) are not graphs of functions, because some vertical line intersects each graph in more than one point. ▬

Now work Problem 41.

If (x, y) is a point on the graph of a function f, then y is the value of f at x, that is, $y = f(x)$. The next example illustrates how to obtain information about a function if its graph is given.

E X A M P L E   8   Obtaining Information from the Graph of a Function

FIGURE 6

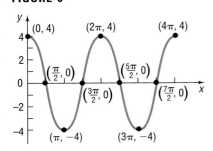

Let f be the function whose graph is given in Figure 6. The graph of f represents the distance that the bob of a pendulum is from its "at rest" position. Negative values of y mean that the pendulum is to the left of the "at rest" position and positive values of y mean that the pendulum is to the right of the "at rest" position.

(a) What is the value of the function when $x = 0$, $x = 3\pi/2$, and $x = 3\pi$?
(b) What is the domain of f?
(c) What is the range of f?
(d) List the intercepts. (Recall that these are the points, if any, where the graph crosses or touches the coordinate axes.)

Solution   (a) Since $(0, 4)$ is on the graph of f, the y-coordinate 4 is the value of f at the x-coordinate 0; that is, $f(0) = 4$. In a similar way, we find that when $x = 3\pi/2$ then $y = 0$; or $f(3\pi/2) = 0$. When $x = 3\pi$, then $y = -4$; or $f(3\pi) = -4$.
(b) To determine the domain of f, we notice that the points on the graph of f will have x-coordinates between 0 and 4π, inclusive; and, for each number x between 0 and 4π, there is a point $(x, f(x))$ on the graph. Thus, the domain of f is $\{x | 0 \leq x \leq 4\pi\}$ or the interval $[0, 4\pi]$.
(c) The points on the graph all have y-coordinates between -4 and 4, inclusive; and, for each such number y, there is at least one number x in the domain. Hence, the range of f is $\{y | -4 \leq y \leq 4\}$ or the interval $[-4, 4]$.
(d) The intercepts are $(0, 4)$, $(\pi/2, 0)$, $(3\pi/2, 0)$, $(5\pi/2, 0)$, and $(7\pi/2, 0)$. ▬

When the graph of a function is given, its domain may be viewed as the shadow created by the graph on the x-axis by vertical beams of light. Its range can be viewed as the shadow created by the graph on the y-axis by horizontal beams of light. Try this technique with the graph given in Figure 6.

Now work Problems 37 and 39.

E X A M P L E   9   Obtaining Information about the Graph of a Function

Consider the function: $f(x) = \dfrac{x}{x + 2}$

(a) Is the point $(1, 1/2)$ on the graph of f?
(b) If $x = -1$, what is $f(x)$? What point is on the graph of f?
(c) If $f(x) = 2$, what is x? What point is on the graph of f?

Solution   (a) When $x = 1$, then $f(x) = f(1) = 1/(1 + 2) = 1/3$. The point $(1, 1/2)$ is therefore not on the graph of f.

(b) If $x = -1$, then $f(x) = f(-1) = -1/(-1 + 2) = -1$, so the point $(-1, -1)$ is on the graph of f.

(c) If $f(x) = 2$, then

$$\frac{x}{x + 2} = 2$$
$$x = 2(x + 2)$$
$$x = 2x + 4$$
$$x = -4$$

The point $(-4, 2)$ is on the graph of f. ▬

Now work Problem 33.

Applications

When we use functions in applications, the domain may be restricted by physical or geometric considerations. For example, the domain of the function f defined by $f(x) = x^2$ is the set of all real numbers. However, if f is used to obtain the area of a square when the length x of a side is known, then we must restrict the domain of f to the positive real numbers, since the length of a side can never be 0 or negative.

E X A M P L E   10   Area of a Circle

Express the area of a circle as a function of its radius.

Solution   We know that the formula for the area A of a circle of radius r is $A = \pi r^2$. If we use r to represent the independent variable and A to represent the dependent variable, the function expressing this relationship is

$$A(r) = \pi r^2$$

In this setting, the domain is $\{r | r > 0\}$. (Do you see why?) ▬

E X A M P L E   11   Construction Cost

Sally, a builder of homes, is interested in finding a function that relates the cost C of building a house and x, the number of square feet of the house, so that she can easily provide estimates to clients for various sized homes. The data she obtained from constructing homes last year are listed in Table 2.

TABLE 2		
	Square Feet, x	Cost, C
	1500	165,000
	1600	176,300
	1700	183,000
	1700	189,500
	1830	201,700
	1970	217,000
	2050	220,000
	2100	237,400

(a) Does the relation defined by the set of ordered pairs (x, C) represent a function?

(b) Draw a scatter diagram of the data.

(c) Select two points from the data and find an equation of the line containing the points.

(d) Interpret the slope.

(e) Express the relationship found in (c) using function notation.

(f) What is the domain of the function?

(g) Use the function to find the cost to build a 2000 square foot house.

(h) Using a graphing utility, find the line of best fit relating square feet and cost. Interpret the slope.

(i) Give some reasons to explain why the cost of building a 1700 square foot house might cost $183,000 in one case and $189,500 in another.

Solution (a) No, because the first element 1700 has, corresponding to it, two second elements, 183,000 and 189,500.

(b) See Figure 7.

FIGURE 7

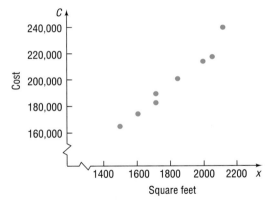

(c) We will choose the points (1500, 165,000) and (2050, 220,000). (You should select your own two points and complete the solution.) The slope of the line containing these points is

$$m = \frac{220,000 - 165,000}{2050 - 1500} = \frac{55,000}{550} = 100$$

The equation of the line containing these points is

$$y - 165,000 = 100(x - 1500)$$
$$y = 100x + 15,000$$

(d) The cost per square foot to build the house is $100. For each additional square foot of house, the cost increases by $100.

(e) Using function notation, $C(x) = 100x + 15,000$.

(f) The domain is $\{x \mid x > 0\}$ since a house cannot have 0 or negative square feet.

(g) The cost to build a 2000 square foot house is

$$C(2000) = 100(2000) + 15,000 = \$215,000$$

(h) The line of best fit is $y = 112.09x - 3734.02$. The cost per square foot to build a house is approximately $112.09. Thus, for each additional square foot of house being built, the cost increases by $112.09.

(i) One explanation is that the fixtures used in one house are more expensive than those used in the other house. ▬

Observe in the solution to Example 11 that we used the symbol C in two ways: It is used to name the function, and it is used to symbolize the dependent variable. This double use is common in applications and should not cause any difficulty.

Refer back to Example 11. Notice that the data in Table 2 does not represent a function since the same first element is paired with two different second elements ((1700, 183,000), (1700, 189,500)). However, when we find a line that fits these data, we find a function that relates the data. Thus, **curve fitting** is a process whereby we find a functional relationship between two or more variables even though the data, themselves, may not represent a function.

Now work Problems 73 and 81.

Examples 10 and 11 demonstrate that some functions are determined from data, while others are based on geometric, physical, or other relations.

EXAMPLE 12

Getting from an Island to Town

An island is 2 miles from the nearest point P on a straight shoreline. A town is 12 miles down the shore from P.

(a) If a person can row a boat at an average speed of 3 miles per hour and the same person can walk 5 miles per hour, express the time T that it takes to go from the island to town as a function of the distance x from P to where the person lands the boat. See Figure 8.

(b) What is the domain of T?

(c) How long will it take to travel from the island to town if the person lands the boat 4 miles from P?

(d) How long will it take if the person lands the boat 8 miles from P?

Solution

(a) Figure 8 illustrates the situation. The distance d_1 from the island to the landing point is the hypotenuse of a right triangle. Using the Pythagorean Theorem, d_1 satisfies the equation

$$d_1^2 = 4 + x^2 \quad \text{or} \quad d_1 = \sqrt{4 + x^2}$$

FIGURE 8

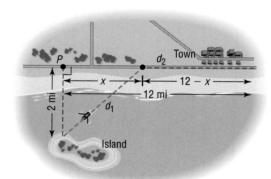

21. Find $f(0)$ and $f(-6)$.

22. Find $f(6)$ and $f(11)$.

23. Is $f(2)$ positive or negative?

24. Is $f(8)$ positive or negative?

25. For what numbers x is $f(x) = 0$?

26. For what numbers x is $f(x) > 0$?

27. What is the domain of f?

28. What is the range of f?

29. What are the x-intercepts?

30. What are the y-intercepts?

31. How often does the line $y = \frac{1}{2}$ intersect the the graph?

32. How often does the line $y = 3$ intersect the graph?

In Problems 33–36, answer the questions about the given function.

33. $f(x) = \dfrac{x + 2}{x - 6}$

    (a) Is the point $(3, 14)$ on the graph of f?

    (b) If $x = 4$, what is $f(x)$? What point is on the graph of f?

    (c) If $f(x) = 2$, what is x? What point is on the graph of f?

    (d) What is the domain of f?

34. $f(x) = \dfrac{x^2 + 2}{x + 4}$

    (a) Is the point $(1, \frac{3}{5})$ on the graph of f?

    (b) If $x = 0$, what is $f(x)$? What point is on the graph of f?

    (c) If $f(x) = \frac{1}{2}$, what is x? What point is on the graph of f?

    (d) What is the domain of f?

35. $f(x) = \dfrac{2x^2}{x^4 + 1}$

    (a) Is the point $(-1, 1)$ on the graph of f?

    (b) If $x = 2$, what is $f(x)$? What point is on the graph of f?

    (c) If $f(x) = 1$, what is x? What point is on the graph of f?

    (d) What is the domain of f?

36. $f(x) = \dfrac{2x}{x - 2}$

    (a) Is the point $(\frac{1}{2}, -\frac{2}{3})$ on the graph of f?

    (b) If $x = 4$, what is $f(x)$? What point is on the graph of f?

    (c) If $f(x) = 1$, what is x? What point is on the graph of f?

    (d) What is the domain of f?

In Problems 37–48, determine whether the graph is that of a function by using the vertical-line test. If it is, use the graph to find:

(a) Its domain and range

(b) The intercepts, if any

(c) Any symmetry with respect to the x-axis, y-axis, or origin

37.

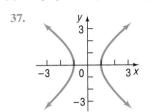

38.

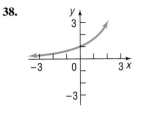

39.

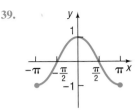

40.

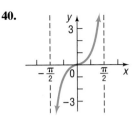

41.

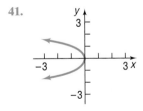

42.

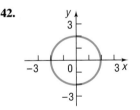

43.

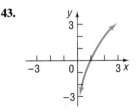

44.

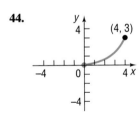

45.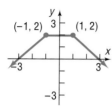

46.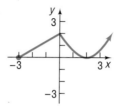

47.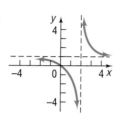

48.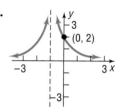

In Problems 49–62, find the domain of each function.

49. $f(x) = 3x + 4$

50. $f(x) = 5x^2 + 2$

51. $f(x) = \dfrac{x}{x^2 + 1}$

52. $f(x) = \dfrac{x^2}{x^2 + 1}$

53. $g(x) = \dfrac{x}{x^2 - 1}$

54. $h(x) = \dfrac{x}{x - 1}$

55. $F(x) = \dfrac{x - 2}{x^3 + x}$

56. $G(x) = \dfrac{x + 4}{x^3 - 4x}$

57. $h(x) = \sqrt{3x - 12}$

58. $G(x) = \sqrt{1 - x}$

59. $f(x) = \dfrac{4}{\sqrt{x - 9}}$

60. $f(x) = \dfrac{x}{\sqrt{x - 4}}$

61. $p(x) = \sqrt{\dfrac{x - 2}{x - 1}}$

62. $q(x) = \sqrt{x^2 - x - 2}$

 63. Match each of the following functions with the graphs that best describe the situation.
  (a) The cost of building a house as a function of its square footage.
  (b) The height of an egg dropped from a 300-foot building as a function of time.
  (c) The height of a human as a function of time.
  (d) The demand for Big Macs as a function of price.
  (e) The height of a child on a swing as a function of time.

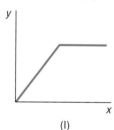

(I)

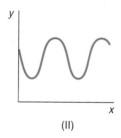

(II)

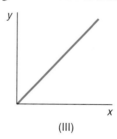

(III)

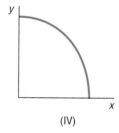

(IV)

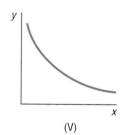

(V)

64. Match each of the following functions with the graph that best describes the situation.
  (a) The temperature of a bowl of soup as a function of time.
  (b) The number of hours of daylight during two years.
  (c) The population of Florida as a function of time.
  (d) The distance of a car traveling at a constant velocity as a function of time.
  (e) The height of a golf ball hit with a 7-iron as a function of time.

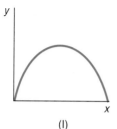

(I)

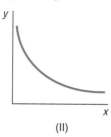

(II)

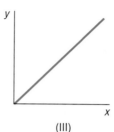

(III)

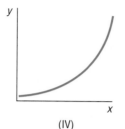

(IV)

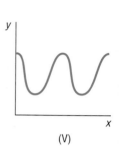

(V)

65. Consider the following scenario: Barbara decides to take a walk. She leaves home, walks 2 blocks in 5 minutes at a constant speed, and realizes that she forgot to lock the door. So Barbara runs home in 1 minute. While at her doorstep, it takes her 1 minute to find her keys and lock the door. Barbara walks 5 blocks in 15 minutes and then decides to jog home. It takes her 7 minutes to get home. Draw a graph of Barbara's distance from home (in blocks) as a function of time.

66. Consider the following scenario: Jayne enjoys riding her bicycle through the woods. At the forest preserve, she gets on her bicycle and rides up a 2,000-foot hill in 10 minutes. She then travels down the hill in 3 minutes. The next 5,000 feet is level terrain and she covers the distance in 20 minutes. She rests for 15 minutes. Jayne then travels 10,000 feet in 30 minutes. Draw a graph of Jayne's distance traveled (in feet) as a function of time.

67. If $f(x) = 2x^3 + Ax^2 + 4x - 5$ and $f(2) = 5$, what is the value of A?

68. If $f(x) = 3x^2 - Bx + 4$ and $f(-1) = 12$, what is the value of B?

69. If $f(x) = (3x + 8)/(2x - A)$ and $f(0) = 2$, what is the value of A?

70. If $f(x) = (2x - B)/(3x + 4)$ and $f(2) = \frac{1}{2}$, what is the value of B?

71. If $f(x) = (2x - A)/(x - 3)$ and $f(4) = 0$, what is the value of A? Where is f not defined?

72. If $f(x) = (x - B)/(x - A)$, $f(2) = 0$, and $f(1)$ is undefined, what are the values of A and B?

73. Demand for Jeans The marketing manager at Levi–Strauss wishes to find a function that relates the demand D of men's jeans and p, the price of the jeans. The following data were obtained based on a price history of the jeans.

Price ($/Pair), p	Demand (Pairs of Jeans Sold per Day), D
20	60
22	57
23	56
23	53
27	52
29	49
30	44

(a) Does the relation defined by the set of ordered pairs (p, D) represent a function?
(b) Draw a scatter diagram of the data.
(c) Select two points from the data and find an equation of the line containing the points. (The answer in the back of the book uses the first and last data points.)
(d) Interpret the slope.
(e) Express the relationship found in (c) using function notation.
(f) What is the domain of the function?
(g) How many jeans will be demanded if the price is $28 a pair?
(h) Using a graphing utility, find the line of best fit relating price and quantity demanded.

74. Advertising and Sales Revenue A marketing firm wishes to find a function that relates the sales S of a product and A, the amount spent on advertising the product. The data are obtained from past experience. Advertising and sales are measured in thousands of dollars.

Advertising Expenditures, A	Sales, S
20	335
22	339
22.5	338
24	343
24	341
27	350
28.3	351

(a) Does the relation defined by the set of ordered pairs (A, S) represent a function?
(b) Draw a scatter diagram of the data.
(c) Select two points from the data and find an equation of the line containing the points.
(d) Interpret the slope.
(e) Express the relationship found in (c) using function notation.
(f) What is the domain of the function?
(g) Predict sales if advertising expenditures are $25,000.
(h) Using a graphing utility, find the line of best fit relating advertising expenditures and sales.

75. Distance and Time Using the following data, find a function that relates the distance, s, in miles, driven by a Ford Taurus and t, the time the Taurus has been driven.

Time (Hours), t	Distance (Miles), s
0	0
1	30
2	55
3	83
4	100
5	150
6	210
7	260
8	300

(a) Does the relation defined by the set of ordered pairs (t, s) represent a function?

(b) Draw a scatter diagram of the data.

(c) Select two points from the data and find an equation of the line containing the points. (The answer in the back of the book uses the first and last data points.)

(d) Interpret the slope.

(e) Express the relationship found in (c) using function notation.

(f) What is the domain of the function?

(g) Predict the distance the car is driven after 11 hours.

 (h) Using a graphing utility, find the line of best fit relating time and distance.

76. **High School versus College GPA** An administrator at Southern Illinois University wants to find a function that relates a student's college grade point average G to the high school grade point average, x. She randomly selects 8 students and obtains the following data:

High School GPA, x	College GPA, G
2.73	2.43
2.92	2.97
3.45	3.63
3.78	3.81
2.56	2.83
2.98	2.81
3.67	3.45
3.10	2.93

(a) Does the relation defined by the set of ordered pairs (x, G) represent a function?

(b) Draw a scatter diagram of the data.

(c) Select two points from the data and find an equation of the line containing the points.

(d) Interpret the slope.

(e) Express the relationship found in (c) using function notation.

(f) What is the domain of the function?

(g) Predict a student's college GPA if her high school GPA is 3.23.

(h) Using a graphing utility, find the line of best fit relating high school GPA and college GPA.

77. **Effect of Gravity on Earth** If a rock falls from a height of 20 meters on Earth, the height H (in meters) after x seconds is approximately

$$H(x) = 20 - 4.9x^2$$

(a) What is the height of the rock when $x = 1$ second? $x = 1.1$ seconds? $x = 1.2$ seconds? $x = 1.3$ seconds?

(b) When is the height of the rock 15 meters? When is it 10 meters? When is it 5 meters?

(c) When does the rock strike the ground?

78. **Effect of Gravity on Jupiter** If a rock falls from a height of 20 meters on the planet Jupiter, its height H (in meters) after x seconds is approximately

$$H(x) = 20 - 13x^2$$

(a) What is the height of the rock when $x = 1$ second? $x = 1.1$ seconds? $x = 1.2$ seconds?

(b) When is the height of the rock 15 meters? When is it 10 meters? When is it 5 meters?

(c) When does the rock strike the ground?

79. **Geometry** Express the area A of a rectangle as a function of the length x if the length is twice the width of the rectangle.

80. **Geometry** Express the area A of an isosceles right triangle as a function of the length x of one of the two equal sides.

81. Express the gross salary G of a person who earns $10 per hour as a function of the number x of hours worked.

82. Tiffany, a commissioned salesperson, earns $100 base pay plus $10 per item sold. Express her gross salary G as a function of the number x of items sold.

83. Refer to Example 12. Is there a place to land the boat so that the travel time is least? Do you think this place is closer to town or closer to P? Discuss the possibilities. Give reasons.

84. Refer to Example 12. Graph the function $T = T(x)$. Use TRACE to see how the time T varies as x changes from 0 to 12. What value of x results in the least time?

85. **Installing Cable TV** MetroMedia Cable is asked to provide service to a customer whose house is located 2 miles from the road along which the cable is buried. The nearest connection box for the cable is located 5 miles down the road (see the figure).

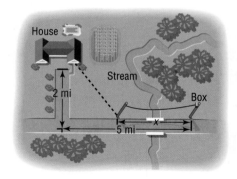

(a) If the installation cost is $10 per mile along the road and $14 per mile off the road, express the total cost C of installation as a function of the distance x (in miles) from the connection box to the point where the cable installation turns off the road. Give the domain.

(b) Compute the cost if $x = 1$ mile.

(c) Compute the cost if $x = 3$ miles.

 (d) Graph the function $C = C(x)$. Use TRACE to see how the cost C varies as x changes from 0 to 5.

(e) What value of x results in the least cost?

86. **Time Required to Go from an Island to a Town** An island is 3 miles from the nearest point P on a straight shoreline. A town is located 20 miles down the shore from P. (Refer to Figure 8 for a similar situation.)

(a) If a person has a boat that averages 12 miles per hour and the same person can run 5 miles per hour, express the time T that it takes to go from the island to town as a function of x, where x is the distance from P to where the person lands the boat. Give the domain.

(b) How long will it take to travel from the island to town if you land the boat 8 miles from P?

(c) How long will it take if you land the boat 12 miles from P?

(d) Graph the function $T = T(x)$. Use TRACE to see how the time T varies as x changes from 0 to 20.

(e) What value of x results in the least time?

 (f) The least time occurs by heading directly to town from the island. Explain why this solution makes sense.

87. **Page Design** A page with dimensions of $8\frac{1}{2}$ inches by 11 inches has a border of uniform width x surrounding the printed matter of the page, as shown in the figure.

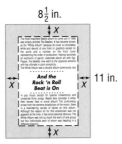

(a) Write a formula for the area A of the printed part of the page as a function of the width x of the border.

(b) Give the domain and range of A.

(c) Find the area of the printed page for borders of widths 1 inch, 1.2 inches, and 1.5 inches.

(d) Graph the function $A = A(x)$.

(e) Use TRACE to determine what margin should be used to obtain an area of 70 square inches and of 50 square inches.

88. **Cost of Trans-Atlantic Travel** A Boeing 747 crosses the Atlantic Ocean (3000 miles) with an airspeed of 500 miles per hour. The cost C (in dollars) per passenger is given by

$$C(x) = 100 + \frac{x}{10} + \frac{36,000}{x}$$

where x is the ground speed (airspeed $\pm$ wind).

(a) What is the cost per passenger for quiescent (no wind) conditions?

(b) What is the cost per passenger with a head wind of 50 miles per hour?

(c) What is the cost per passenger with a tail wind of 100 miles per hour?

(d) What is the cost per passenger with a head wind of 100 miles per hour?

(e) Graph the function $C = C(x)$.

(f) As x varies from 400 to 600 miles per hour, how does the cost vary?

89. **Period of a Pendulum** The period T (in seconds) of a simple pendulum is a function of its length l (in feet) defined by the equation

$$T(l) = 2\pi\sqrt{\frac{l}{g}}$$

where $g \approx 32.2$ feet per second is the acceleration due to gravity.

(a) Use a graphing utility to graph the function $T = T(l)$.

(b) Use the TRACE function to see how the period T varies as l changes from 1 to 10.

(c) What length should be used if a period of 10 seconds is required?

90. **Effect of Elevation on Weight** If an object weighs m pounds at sea level, then its weight W (in pounds) at a height of h miles above sea level is given approximately by

$$W(h) = m\left(\frac{4000}{4000 + h}\right)^2$$

(a) If Amy weighs 120 pounds at sea level, how much will she weigh on Pike's Peak, which is 14,110 feet above sea level?

(b) Use a graphing utility to graph the function $W = W(h)$. Use $m = 120$ pounds.

(c) Use the TRACE function to see how weight W varies as h changes from 0 to 5 miles.

(d) At what height will Amy weigh 121 pounds?

(e) Does your answer to part (d) seem reasonable?

In Problems 91–98, tell whether the set of ordered pairs (x, y) defined by each equation is a function.

91. $y = x^2 + 2x$

92. $y = x^3 - 3x$

93. $y = \dfrac{2}{x}$

94. $y = \dfrac{3}{x} - 3$

95. $y^2 = 1 - x^2$

96. $y = \pm\sqrt{1 - 2x}$

97. $x^2 + y = 1$

98. $x + 2y^2 = 1$

99. Some functions f have the property that $f(a + b) = f(a) + f(b)$ for all real numbers a and b. Which of the following functions have this property?
(a) $h(x) = 2x$ (b) $g(x) = x^2$
(c) $F(x) = 5x - 2$ (d) $G(x) = 1/x$

100. Draw the graph of a function whose domain is $\{x \mid -3 \le x \le 8,\ x \ne 5\}$ and whose range is $\{y \mid -1 \le y \le 2,\ y \ne 0\}$. What point(s) in the rectangle $-3 \le x \le 8$, $-1 \le y \le 2$ cannot be on the graph? Compare your graph with those of other students. What differences do you see?

101. Are the functions $f(x) = x - 1$ and $g(x) = (x^2 - 1)/(x + 1)$ the same? Explain.

102. Describe how you would proceed to find the domain and range of a function if you were

given its graph. How would your strategy change if, instead, you were given the equation defining the function?

103. How many x-intercepts can the graph of a function have? How many y-intercepts can it have?

104. Is a graph that consists of a single point the graph of a function? Can you write the equation of such a function?

105. Is there a function whose graph is symmetric with respect to the x-axis?

106. Investigate when, historically, the use of function notation $y = f(x)$ first appeared.

4.2   MORE ABOUT FUNCTIONS

1	Find the Average Rate of Change of a Function
2	Determine Where a Function Is Increasing and Decreasing
3	Determine Even or Odd Functions from a Graph
4	Identify Even or Odd Functions from the Equation
5	Graph Certain Important Functions
6	Graph Piecewise-defined Functions

Average Rate of Change

1   In Section 3.3 we said the slope of a straight line could be interpreted as the average rate of change. Often, we are interested in the rate at which functions change. To find the average rate of change of a function between any two points on its graph, we calculate the slope of the line containing the two points.

E X A M P L E   1   Finding the Average Rate of Change of a Function

Suppose that you drop a ball from a cliff 1000 feet high. You measure the distance s that the ball has fallen after time t using a motion detector and obtain the data in Table 3.

(a) Draw a scatter diagram of the data, treating time as the independent variable.

(b) Draw a line from the point $(0, 0)$ to $(2, 64)$.

(c) Find the average rate of change of the ball between 0 and 2 seconds, that is, find the slope of the line in (b).

(d) Interpret the average rate of change found in (c).

(e) Draw a line from the point $(5, 400)$ to $(7, 784)$.

TABLE 3

Time, t (Seconds)	Distance, s (Feet)
0	0
1	16
2	64
3	144
4	256
5	400
6	576
7	784

(f) Find the average rate of change of the ball between 5 and 7 seconds.

(g) Interpret the average rate of change found in (f).

(h) What is happening to the average rate of change as time passes?

Solution

(a) We plot the ordered pairs $(0, 0)$, $(1, 16)$, and so on, using rectangular co-ordinates. See Figure 9.

(b) Draw the line from $(0, 0)$ to $(2, 64)$. See Figure 10.

(c) The average rate of change is found by computing the slope of the line joining the points $(0, 0)$ and $(2, 64)$.

$$\text{Average Rate of Change} = \frac{\Delta s}{\Delta t} = \frac{64 - 0}{2 - 0} = \frac{64}{2} = 32 \text{ ft/sec}$$

(d) Since the change in distance divided by the change in time represents a speed, we would interpret the slope of the line as an average speed. The average speed of the ball between 0 and 2 seconds is 32 feet per second.

(e) Draw the line from $(5, 400)$ to $(7, 784)$. See Figure 11.

FIGURE 9

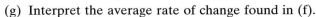

FIGURE 10

FIGURE 11

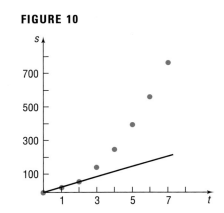

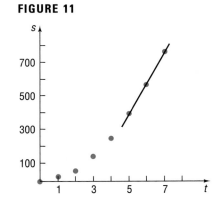

(f) The average rate of change of the ball between 5 and 7 seconds is

$$\text{Average Rate of Change} = \frac{\Delta s}{\Delta t} = \frac{784 - 400}{7 - 5} = \frac{384}{2} = 192 \text{ ft/sec}$$

(g) The average speed of the ball between 5 and 7 seconds is 192 feet per second.

(h) The average speed of the ball is increasing as time passes since the ball is accelerating due to the effect of gravity.   ▬

If we know the function f that relates the time t to the distance s that the ball has fallen, $s = f(t)$, then the average speed of the ball between 0 and 2 seconds is

$$\frac{f(2) - f(0)}{2 - 0} = \frac{64 - 0}{2} = 32 \text{ ft/sec}$$

The average speed of the ball between 5 and 7 seconds is

$$\frac{f(7) - f(5)}{7 - 5} = \frac{784 - 400}{2} = 192 \text{ ft/sec}$$

Expressions like these occur frequently in calculus.

If c is in the domain of a function $y = f(x)$, the **average rate of change of f** between c and x is defined as

$$\text{Average rate of change} = \frac{\Delta y}{\Delta x} = \frac{f(x) - f(c)}{x - c} \qquad x \neq c \qquad (1)$$

This expression is also called the **difference quotient** of f at c.

The average rate of change of a function has an important geometric interpretation. Look at the graph of $y = f(x)$ in Figure 12. We have labeled two points on the graph: $(c, f(c))$ and $(x, f(x))$. The slope of the line containing these two points is

$$\frac{f(x) - f(c)}{x - c}$$

This line is called a **secant line.**

FIGURE 12

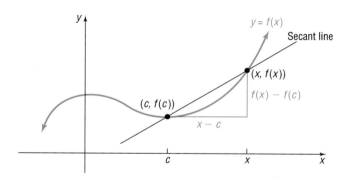

Theorem   Slope of a Secant Line

The average rate of change of a function equals the slope of a secant line containing two points on its graph.

E X A M P L E   2   Finding the Average Rate of Change

The function $s(t) = -4.9t^2 + 100t + 5$ gives the height s (in meters) of a bullet fired straight up as a function of time t (in seconds).

(a) Find the average rate of change of the height of the bullet between 2 and t seconds.

(b) Using the result found in (a), find the average rate of change of the height of the bullet between 2 and 3 seconds.

Solution   (a) From expression (1), we seek

$$\frac{\Delta s}{\Delta t} = \frac{s(t) - s(2)}{t - 2} \qquad t \neq 2$$

We begin by finding $s(2)$:

$$s(2) = -4.9(2)^2 + 100(2) + 5 = -19.6 + 200 + 5 = 185.4$$

Then the average rate of change of s between 2 and t seconds is

$$\frac{s(t) - s(2)}{t - 2} = \frac{-4.9t^2 + 100t + 5 - 185.4}{t - 2} = \frac{-4.9t^2 + 100t - 180.4}{t - 2}$$

$$= \frac{(-4.9t + 90.2)(t - 2)}{t - 2} = -4.9t + 90.2$$

(b) The average rate of change between 2 and 3 seconds is found by letting $t = 3$ in the expression found in (a). Thus, the average rate of change of the bullet between 2 and 3 seconds is $-4.9(3) + 90.2 = 75.5$ meters per second.

  Now work Problem 29.

Increasing and Decreasing Functions

2 Consider the graph given in Figure 13. If you look from left to right along the graph of the function, you will notice that parts of the graph are rising, parts are falling, and parts are horizontal. In such cases, the function is described as *increasing, decreasing,* and *constant,* respectively.

FIGURE 13

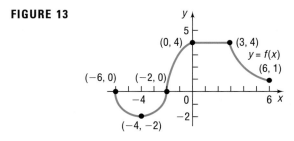

E X A M P L E   3   Determining Where a Function Is Increasing, Decreasing, or Constant

Where is the function in Figure 13 increasing? Where is it decreasing? Where is it constant?

Solution   To answer the question of where a function is increasing, where it is decreasing, and where it is constant, we use inequalities involving the independent variable x or we use open intervals* of x-coordinates. The graph in Figure 13 is rising (increasing) from the point $(-4, -2)$ to the point $(0, 4)$, so we conclude that it is increasing on the open interval $(-4, 0)$ (or for $-4 < x < 0$). The graph is falling (decreasing) from the point $(-6, 0)$ to the point $(-4, -2)$ and from the point $(3, 4)$ to the point $(6, 1)$. We conclude that the graph is decreasing on the open intervals $(-6, -4)$ and $(3, 6)$ (or for $-6 < x < -4$ and $3 < x < 6$). The graph is constant on the open interval $(0, 3)$ (or for $0 < x < 3$).

More precise definitions follow.

*The open interval (a, b) consists of all real numbers x for which $a < x < b$. Refer to Section 2.5, if necessary.

A function f is **increasing** on an open interval I if, for any choice of x_1 and x_2 in I, with $x_1 < x_2$, we have $f(x_1) < f(x_2)$.

A function f is **decreasing** on an open interval I if, for any choice of x_1 and x_2 in I, with $x_1 < x_2$, we have $f(x_1) > f(x_2)$.

A function f is **constant** on an open interval I if, for all choices of x in I, the values $f(x)$ are equal.

Thus, the graph of an increasing function goes up from left to right, the graph of a decreasing function goes down from left to right, and the graph of a constant function remains at a fixed height. Figure 14 illustrates the definitions.

FIGURE 14

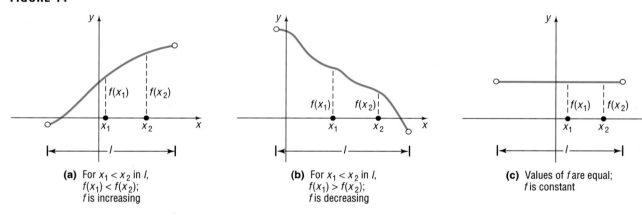

(a) For $x_1 < x_2$ in I, $f(x_1) < f(x_2)$; f is increasing

(b) For $x_1 < x_2$ in I, $f(x_1) > f(x_2)$; f is decreasing

(c) Values of f are equal; f is constant

Even and Odd Functions

A function f is **even** if for every number x in its domain the number $-x$ is also in the domain and

$$f(-x) = f(x)$$

In other words, if $y = f(x)$, then f is even if and only if, whenever the point (x, y) is on the graph of f, the point $(-x, y)$ is also on the graph.

> A function f is **odd** if for every number x in its domain the number $-x$ is also in the domain and
>
> $$f(-x) = -f(x)$$

In other words, if $y = f(x)$, then f is odd if and only if, whenever the point (x, y) is on the graph of f, the point $(-x, -y)$ is also on the graph.

Refer to Section 3.2, where the tests for symmetry are listed. The following results are then evident.

Theorem

A function is even if and only if its graph is symmetric with respect to the y-axis. A function is odd if and only if its graph is symmetric with respect to the origin.

E X A M P L E   4

Determining Even and Odd Functions from the Graph

Determine whether each graph in Figure 15 is the graph of an even function, an odd function, or a function that is neither even nor odd.

FIGURE 15

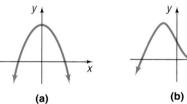

(a) (b)

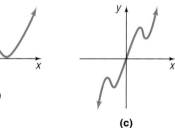

(c)

Solution   The graph in Figure 15(a) is that of an even function, because the graph is symmetric with respect to the y-axis. The function whose graph is given in Figure 15(b) is neither even nor odd, because the graph is neither symmetric with respect to the y-axis nor symmetric with respect to the origin. The function whose graph is given in Figure 15(c) is odd, because its graph is symmetric with respect to the origin.

Now work Problem 9.

In the next example, we show how to verify algebraically whether a function is even, odd, or neither.

E X A M P L E   5

Identifying Even and Odd Functions Algebraically

Determine whether each of the following functions is even, odd, or neither. Then determine whether the graph is symmetric with respect to the y-axis or with respect to the origin.

(a) $f(x) = x^2 - 5$ (b) $g(x) = x^3 - 1$
(c) $h(x) = 5x^3 - x$ (d) $F(x) = |x|$

Solution   (a) We replace x by $-x$ in $f(x) = x^2 - 5$. Then

$$f(-x) = (-x)^2 - 5 = x^2 - 5 = f(x)$$

Since $f(-x) = f(x)$, we conclude that f is an even function, and the graph is symmetric with respect to the y-axis.

(b) We replace x by $-x$. Then

$$g(-x) = (-x)^3 - 1 = -x^3 - 1$$

Since $g(-x) \neq g(x)$ and $g(-x) \neq -g(x)$, we conclude that g is neither even nor odd. The graph is not symmetric with respect to the y-axis nor with respect to the origin.

(c) We replace x by $-x$ in $h(x) = 5x^3 - x$. Then

$$h(-x) = 5(-x)^3 - (-x) = -5x^3 + x = -(5x^3 - x) = -h(x)$$

Since $h(-x) = -h(x)$, h is an odd function, and the graph of h is symmetric with respect to the origin.

(d) We replace x by $-x$ in $F(x) = |x|$. Then

$$F(-x) = |-x| = |x| = F(x)$$

Since $F(-x) = F(x)$, F is an even function, and the graph of F is symmetric with respect to the y-axis.   ▬

Now work Problem 39.

Library of Functions

⑤   We now give names to some of the functions that we have encountered. In going through this list, pay special attention to the characteristics of each function, particularly to the shape of each graph. Knowing these graphs will lay the foundation for later graphing techniques.

Linear Functions

$$f(x) = mx + b \qquad m \text{ and } b \text{ are real numbers.}$$

The domain of a **linear function** f consists of all real numbers. The graph of this function is a nonvertical line with slope m and y-intercept b. A linear function is increasing if $m > 0$, decreasing if $m < 0$, and constant if $m = 0$.

FIGURE 16
Constant Function

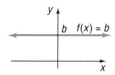

Constant Function

$$f(x) = b \qquad b \text{ is a real number}$$

See Figure 16.

A **constant function** is a special linear function ($m = 0$). Its domain is the set of all real numbers; its range is the set consisting of a single number b. Its graph is a horizontal line whose y-intercept is b. The constant function is an even function whose graph is constant over its domain.

FIGURE 17
Identity Function

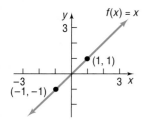

Identity Function

$$f(x) = x$$

See Figure 17.

The **identity function** is also a special linear function. Its domain and its range are the set of all real numbers. Its graph is a line whose slope is $m = 1$ and whose y-intercept is 0. The line consists of all points for which the x-coordinate equals the y-coordinate. The identity function is an odd function that is increasing over its domain. Note that the graph bisects quadrants I and III.

FIGURE 18
Square Function

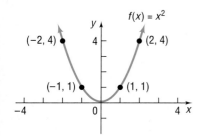

Square Function

$$f(x) = x^2$$

See Figure 18.

The domain of the **square function** f is the set of all real numbers; its range is the set of nonnegative real numbers. The graph of this function is a parabola, whose intercept is at $(0, 0)$. The square function is an even function that is decreasing on the interval $(-\infty, 0)$ and increasing on the interval $(0, \infty)$.

FIGURE 19
Cube Function

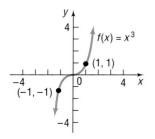

Cube Function

$$f(x) = x^3$$

See Figure 19.

The domain and range of the **cube function** are the set of all real numbers. The intercept of the graph is at $(0, 0)$. The cube function is odd and is increasing on the interval $(-\infty, \infty)$.

FIGURE 20
Square Root Function

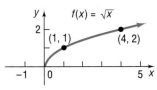

Square Root Function

$$f(x) = \sqrt{x}$$

See Figure 20.

The domain and range of the **square root function** are the set of non-negative real numbers. The intercept of the graph is at $(0, 0)$. The square root function is neither even nor odd and is increasing on the interval $(0, \infty)$.

FIGURE 21
Reciprocal Function

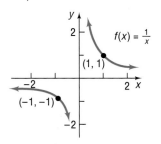

Reciprocal Function

$$f(x) = \frac{1}{x}$$

Refer to Example 11, p. 182, for a discussion of the equation $y = 1/x$. See Figure 21.

The domain and range of the **reciprocal function** are the set of all nonzero real numbers. The graph has no intercepts. The reciprocal function is decreasing on the intervals $(-\infty, 0)$ and $(0, \infty)$ and is an odd function.

FIGURE 22
Absolute Value Function

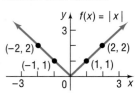

Absolute Value Function

$$f(x) = |x|$$

See Figure 22.

The domain of the **absolute value function** is the set of all real numbers; its range is the set of nonnegative real numbers. The intercept of the graph is at $(0, 0)$. If $x \geq 0$, then $f(x) = x$ and the graph of f is part of the line $y = x$; if $x < 0$, then $f(x) = -x$ and the graph of f is part of the line $y = -x$. The absolute value function is an even function; it is decreasing on the interval $(-\infty, 0)$ and increasing on the interval $(0, \infty)$.

 Check: Graph $y = |x|$ on a square screen and compare what you see with Figure 22. Note that some graphing calculators use the symbols abs (x) for absolute value. If your utility has no built-in absolute value function, you can still graph $y = |x|$ by using the fact that $|x| = \sqrt{x^2}$.   ▄

The notation $\text{int}(x)$ stands for the largest integer less than or equal to x. For example,

$$\text{int}(1) = 1 \quad \text{int}(2.5) = 2 \quad \text{int}(\tfrac{1}{2}) = 0 \quad \text{int}(\tfrac{-3}{4}) = -1 \quad \text{int}(\pi) = 3$$

This type of correspondence occurs frequently enough in mathematics that we give it a name.

Greatest Integer Function

$$f(x) = \text{int}(x) = \text{Greatest integer less than or equal to } x$$

We obtain the graph of $f(x) = \text{int}(x)$ by plotting several points. See Table 4. For values of x, $-1 \le x < 0$, the value of $f(x) = \text{int}(x)$ is -1; for values of x, $0 \le x < 1$, the value of f is 0. See Figure 23 for the graph.

	TABLE 4	
x	**y = f(x) = int(x)**	**(x, y)**
-1	-1	$(-1, -1)$
$-\frac{1}{2}$	-1	$(-\frac{1}{2}, -1)$
$-\frac{1}{4}$	-1	$(-\frac{1}{4}, -1)$
0	0	$(0, 0)$
$\frac{1}{4}$	0	$(\frac{1}{4}, 0)$
$\frac{1}{2}$	0	$(\frac{1}{2}, 0)$
$\frac{3}{4}$	0	$(\frac{3}{4}, 0)$

FIGURE 23
Greatest Integer Function

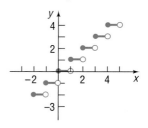

The domain of the **greatest integer function** is the set of all real numbers; its range is the set of integers. The y-intercept of the graph is at 0. The x-intercepts lie in the interval $[0, 1)$. The greatest integer function is neither even nor odd. It is constant on every interval of the form $[k, k + 1)$, for k an integer. In Figure 23, we use a solid dot to indicate, for example, that at $x = 1$ the value of f is $f(1) = 1$; we use an open circle to illustrate that the function does not assume the value of 0 at $x = 1$.

From the graph of the greatest integer function, we can see why it is also called a **step function.** At $x = 0, x = \pm1, x = \pm2$, and so on, this function exhibits what is called a *discontinuity;* that is, at integer values, the graph suddenly "steps" from one value to another without taking on any of the intermediate values. For example, to the immediate left of $x = 3$, the y-coordinates are 2, and to the immediate right of $x = 3$, the y-coordinates are 3.

Comment   When graphing a function, you can choose either the **connected mode,** in which points plotted on the screen are connected, making the graph appear without any breaks, or the **dot mode,** in which only the points plotted appear. When graphing the greatest integer function with a graphing utility, it is necessary to be in the **dot mode.** This is to prevent the utility from "connecting the dots" when $f(x)$ changes from one integer value to the next. See Figure 24.

FIGURE 24
$f(x) = \text{int}(x)$

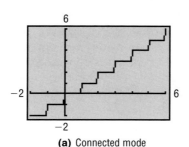

(a) Connected mode

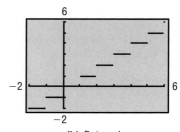

(b) Dot mode

The functions that we have discussed so far are basic. Whenever you encounter one of them, you should see a mental picture of its graph. For example, if you encounter the function $f(x) = x^2$, you should see in your mind's eye a picture like Figure 18.

Now work Problems 1–8.

Piecewise-defined Functions

Sometimes, a function is defined differently on different parts of its domain. For example, the absolute value function $f(x) = |x|$ is actually defined by two equations: $f(x) = x$ if $x \geq 0$ and $f(x) = -x$ if $x < 0$. For convenience, we generally combine these equations into one expression as

$$f(x) = |x| = \begin{cases} x & \text{if } x \geq 0 \\ -x & \text{if } x < 0 \end{cases}$$

When functions are defined by more than one equation, they are called **piecewise-defined** functions.

Let's look at another example of a piecewise-defined function.

E X A M P L E   6   Analyzing a Piecewise-defined Function

The function f is defined as:

$$f(x) = \begin{cases} -x + 1 & \text{if } -1 \leq x < 1 \\ 2 & \text{if } x = 1 \\ x^2 & \text{if } x > 1 \end{cases}$$

(a) Find $f(0)$, $f(1)$, and $f(2)$.   (b) Determine the domain of f.
(c) Graph f.   (d) Use the graph to find the range of f.

Solution   (a) To find $f(0)$, we observe that when $x = 0$ the equation for f is given by $f(x) = -x + 1$. So we have

$$f(0) = -0 + 1 = 1$$

When $x = 1$, the equation for f is $f(x) = 2$. Thus,

$$f(1) = 2$$

When $x = 2$, the equation for f is $f(x) = x^2$. So

$$f(2) = 2^2 = 4$$

(b) To find the domain of f, we look at its definition. We conclude that the domain of f is $\{x \mid x \geq -1\}$, or $[-1, \infty)$.

(c) To graph f, we graph "each piece." Thus, we first graph the line $y = -x + 1$ and keep only the part for which $-1 \leq x < 1$. Then we plot the point $(1, 2)$ because, when $x = 1$, $f(x) = 2$. Finally, we graph the parabola $y = x^2$ and keep only the part for which $x > 1$. See Figure 25.

(d) From the graph, we conclude that the range of f is $\{y \mid y > 0\}$, or $(0, \infty)$.

FIGURE 25

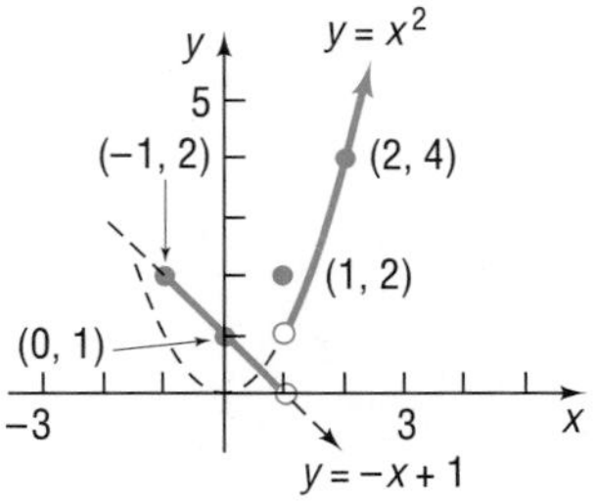

(Graph showing y = x², with points (−1, 2), (2, 4), (1, 2), (0, 1), axes labeled x and y, marks at 5, −3, 3, and line y = −x + 1)

Now work Problem 53.

E X A M P L E   7   Cost of Electricity

In the winter, Commonwealth Edison Company supplies electricity to residences for a monthly customer charge of $8.91 plus 10.494¢ per kilowatt-
hour (kWhr) for the first 400 kWhr supplied in the month and 7.91¢ per
kWhr for all usage over 400 kWhr in the month.*

(a) What is the charge for using 300 kWhr in a month?
(b) What is the charge for using 700 kWhr in a month?
(c) If C is the monthly charge for x kWhr, express C as a function of x.

Solution   (a) For 300 kWhr, the charge is $8.91 plus 10.494¢ = $0.10494 per kWhr.
Thus,

$$\text{Charge} = \$8.91 + \$0.10494(300) = \$40.39$$

(b) For 700 kWhr, the charge is $8.91 plus 10.494¢ for the first 400 kWhr
plus 7.91¢ for the 300 kWhr in excess of 400. Thus,

$$\text{Charge} = \$8.91 + \$0.10494(400) + \$0.0791(300) = \$74.62$$

(c) If $0 \le x \le 400$, the monthly charge C (in dollars) can be found by multiplying x times $0.10494 and adding the monthly customer charge of
$8.91. Thus, if $0 \le x \le 400$, then $C(x) = 0.10494x + 8.91$. For $x > 400$, the
charge is $0.10494(400) + 8.91 + 0.0791(x - 400)$, since $x - 400$ equals
the usage in excess of 400 kWhr, which costs $0.0791 per kWhr. Thus, if
$x > 400$, then

$$\begin{aligned}
C(x) &= 0.10494(400) + 8.91 + 0.0791(x - 400) \\
&= 50.89 + 0.0791(x - 400) \\
&= 0.0791x + 19.25
\end{aligned}$$

The rule for computing C follows two equations:

$$C(x) = \begin{cases} 0.10494x + 8.91 & \text{if } 0 \le x \le 400 \\ 0.0791x + 19.25 & \text{if } x > 400 \end{cases}$$

See Figure 26 for the graph.

FIGURE 26

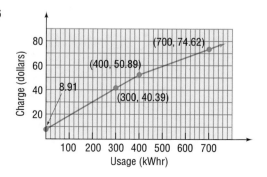

*Source: Commonwealth Edison Co., Chicago, Illinois, 1997.

4.2 │ EXERCISES

In Problems 1–8, match each graph to the function listed whose graph most resembles the one given.

A. *Constant function*   B. *Linear function*

C. *Square function*   D. *Cube function*

E. *Square root function*   F. *Reciprocal function*

G. *Absolute value function* H. *Greatest integer function*

1.   **2.**   **3.**   **4.**

5.   **6.** 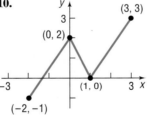   **7.** 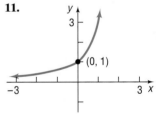   **8.**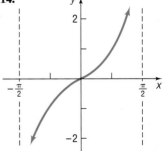

In Problems 9–22, the graph of a function is given. Use the graph to find

(a) Its domain and range

(b) The intervals on which it is increasing, decreasing, or constant

(c) Whether it is even, odd, or neither

(d) The intercepts, if any

9.

10.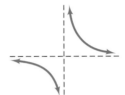

11.

12.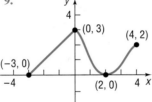

13.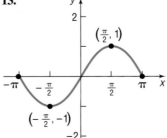

14.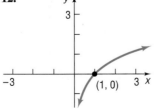

15.

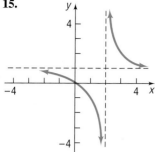

16.

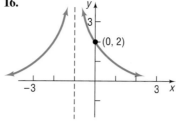

17.

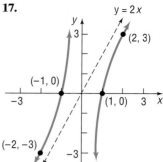

18.

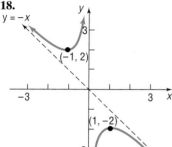

19.

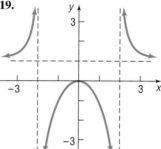

20.

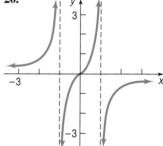

In Problems 21–22, assume that the entire graph is shown.

21.

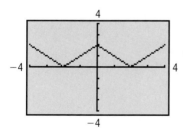

22.

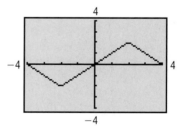

23. If

$$f(x) = \begin{cases} x^2 & \text{if } x < 0 \\ 2 & \text{if } x = 0 \\ 2x + 1 & \text{if } x > 0 \end{cases}$$

find: (a) $f(-2)$ (b) $f(0)$ (c) $f(2)$

24. If

$$f(x) = \begin{cases} x^3 & \text{if } x < 0 \\ 3x + 2 & \text{if } x \geq 0 \end{cases}$$

find: (a) $f(-1)$ (b) $f(0)$ (c) $f(1)$

25. If $f(x) = \text{int}(2x)$, find: (a) $f(1.2)$ (b) $f(1.6)$ (c) $f(-1.8)$

26. If $f(x) = \text{int}(x/2)$, find: (a) $f(1.2)$ (b) $f(1.6)$ (c) $f(-1.8)$

In Problems 27–38, find the average rate of change of f between 1 and x,

$$\frac{f(x) - f(1)}{x - 1} \qquad x \neq 1$$

for each function f. Be sure to simplify.

27. $f(x) = 3x$ **28.** $f(x) = -2x$ **29.** $f(x) = 1 - 3x$ **30.** $f(x) = x^2 + 1$

31. $f(x) = 3x^2 - 2x$ **32.** $f(x) = 4x - 2x^2$ **33.** $f(x) = x^3 - x$ **34.** $f(x) = x^3 + x$

35. $f(x) = \dfrac{2}{x + 1}$ **36.** $f(x) = \dfrac{1}{x^2}$ **37.** $f(x) = \sqrt{x}$ **38.** $f(x) = \sqrt{x + 3}$

In Problems 39–50, determine (algebraically) whether each function is even, odd, or neither.

39. $f(x) = 4x^3$ **40.** $f(x) = 2x^4 - x^2$ **41.** $g(x) = 2x^2 - 5$ **42.** $h(x) = 3x^3 + 2$

43. $F(x) = \sqrt[3]{x}$ **44.** $G(x) = \sqrt{x}$ **45.** $f(x) = x + |x|$ **46.** $f(x) = \sqrt[3]{2x^2 + 1}$

47. $g(x) = \dfrac{1}{x^2}$ **48.** $h(x) = \dfrac{x}{x^2 - 1}$ **49.** $h(x) = \dfrac{x^3}{3x^2 - 9}$ **50.** $F(x) = \dfrac{x}{|x|}$

 51. How many *x*-intercepts can a function defined on an interval have if it is increasing on that interval? Explain.

52. How many *y*-intercepts can a function have? Explain.

In Problems 53–58:

(a) Find the domain of each function.   *(b) Locate any intercepts.*

(c) Graph each function.   *(d) Based on the graph, find the range.*

53. $f(x) = \begin{cases} 2x & \text{if } x \neq 0 \\ 0 & \text{if } x = 0 \end{cases}$ **54.** $f(x) = \begin{cases} 3x & \text{if } x \neq 0 \\ 4 & \text{if } x = 0 \end{cases}$

55. $f(x) = \begin{cases} 1 + x & \text{if } x < 0 \\ x^2 & \text{if } x \geq 0 \end{cases}$ **56.** $f(x) = \begin{cases} 1/x & \text{if } x < 0 \\ \sqrt{x} & \text{if } x \geq 0 \end{cases}$

57. $f(x) = \begin{cases} |x| & \text{if } -2 \leq x < 0 \\ 1 & \text{if } x = 0 \\ x^3 & \text{if } x > 0 \end{cases}$ **58.** $f(x) = \begin{cases} 3 + x & \text{if } -3 \leq x < 0 \\ 3 & \text{if } x = 0 \\ \sqrt{x} & \text{if } x > 0 \end{cases}$

In Problems 59–66, find the following for each function:

(a) f(-x)   *(b) -f(x)*   *(c) f(2x)*   *(d) f(x - 3)*   *(e) f(1/x)*   *(f) 1/f(x)*

59. $f(x) = 2x + 5$ **60.** $f(x) = 3 - x$ **61.** $f(x) = 2x^2 - 4$ **62.** $f(x) = x^3 + 1$

63. $f(x) = x^3 - 3x$ **64.** $f(x) = x^2 + x$ **65.** $f(x) = |x|$ **66.** $f(x) = \dfrac{1}{x}$

Problems 67–70 require the following definition. **Secant Line:** *The slope of the secant line containing the two points (x, f(x)) and (x + h, f(x + h)) on the graph of a function y = f(x) may be given as*

$$\frac{f(x + h) - f(x)}{h}$$

In Problems 67–70, express the slope of the secant line of each function in terms of x and h. Be sure to simplify your answer.

67. $f(x) = 2x + 5$ **68.** $f(x) = -3x + 2$ **69.** $f(x) = x^2 + 2x$ **70.** $f(x) = 1/x$

In Problems 71–74, the graph of a piecewise-defined function is given. Write a definition for each function.

71.

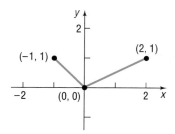

72.

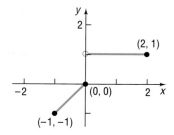

73.

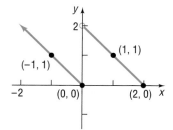

74.

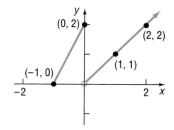

In Problems 75 and 76, decide whether each function is even. Give a reason.

75. $f(x) = \begin{cases} x^2 + 4 & \text{if } x \neq 2 \\ 6 & \text{if } x = 2 \end{cases}$

76. $f(x) = \begin{cases} x^2 + 4 & \text{if } x \neq 2 \\ 5 & \text{if } x = 2 \end{cases}$

77. Growth of Bacteria The following data represents the population of an unknown bacteria.

Time (Days)	Population
0	50
1	153
2	234
3	357
4	547
5	839
6	1280

(a) Draw a scatter diagram of the data, treating the time as the independent variable.
(b) Draw a line from the point $(0, 50)$ to $(1, 153)$ on the scatter diagram found in (a).
(c) Find the average rate of change of the population between 0 and 1 days.
(d) Interpret the average rate of change found in (c).
(e) Draw a line from the point $(5, 839)$ to $(6, 1280)$ on the scatter diagram found in (a).
(f) Find the average rate of change of the population between 5 and 6 days.

(g) Interpret the average rate of change found in (f).
(h) What is happening to the average rate of change of the population as time passes?

78. Cost of Manufacturing Bikes The following data represents the monthly cost of producing bicycles at Tunney's Bicycle Shop.

Number of Bicycles, x	Total Cost of Production, C (Dollars)
0	24,000
25	27,750
60	31,500
102	35,250
150	39,000
190	42,750
223	46,500
249	50,250

(a) Draw a scatter diagram of the data, treating the number of bicycles produced as the independent variable.

(b) Draw a line from the point (0, 24,000) to (25,27,750) on the scatter diagram found in (a).

(c) Find the average rate of change of the cost between 0 and 25 bicycles.

(d) Interpret the average rate of change found in (c). In economics, this is called **marginal cost.**

(e) Draw a line from the point (190, 42,750) to (223, 46,500) on the scatter diagram found in (a).

(f) Find the average rate of change of the cost between 190 and 223 bicycles.

(g) Interpret the average rate of change found in (f).

79. **Revenue from Selling Bikes** The following data represent the total revenue that would be received from selling x bicycles at Tunney's Bicycle Shop.

Number of Bicycles, x	Total Revenue, R (Dollars)
0	0
25	28,000
60	45,000
102	53,400
150	59,160
190	62,360
223	64,835
249	66,525

(a) Draw a scatter diagram of the data, treating the number of bicycles produced as the independent variable.

(b) Draw a line from the point (0, 0) to (25, 28,000) on the scatter diagram found in (a).

(c) Find the average rate of change of revenue between 0 and 25 bicycles.

(d) Interpret the average rate of change found in (c). In economics, this is called **marginal revenue.**

(e) Draw a line from the point (190, 62,360) to (223, 64,835) on the scatter diagram found in (a).

(f) Find the average rate of change of revenue between 190 and 223 bicycles.

(g) Interpret the average rate of change found in (f).

(h) What is happening to marginal revenue as the number of bicycles sold increases?

(i) Refer to Problem 78(d). A firm's profit will increase as long as marginal revenue is greater than marginal cost. At what level of output does Tunney's Bicycle Shop's profits stop increasing?

80. **Cost of Tuition** The following data represent the average cost of tuition and required fees (in dollars) at public four-year colleges.

Year	Average Cost of Tuition
1990	2,035
1991	2,159
1992	2,410
1993	2,604
1994	2,822

Source: U.S. National Center for Education Statistics.

(a) Draw a scatter diagram of the data, treating the year as the independent variable.

(b) Draw a line from the point (1990, 2035) to (1992, 2410) on the scatter diagram found in (a).

(c) Find the average rate of change of the cost of tuition and required fees between 1990 and 1992.

(d) Interpret the average rate of change found in (c).

(e) Draw a line from the point (1992, 2410) to (1994, 2822) on the scatter diagram found in (a).

(f) Find the average rate of change of the cost of tuition and required fees between 1992 and 1994.

(g) Interpret the average rate of change found in (f).

(h) What is happening to the average rate of change for the cost of tuition and required fees as time passes?

81. **Cost of Natural Gas** In January 1997, the Peoples Gas Company had the following rate schedule* for natural gas usage in single-family residences:

Monthly service charge	$9.00
Per therm service charge,	
1st 50 therms	$0.36375/therm
Over 50 therms	$0.11445/therm
Gas charge	$0.3256/therm

(a) What is the charge for using 50 therms in a month?

(b) What is the charge for using 500 therms in a month?

Source: The Peoples Gas Company, Chicago, Illinois.

(c) Construct a function that relates the monthly charge C for x therms of gas.

(d) Graph this function.

82. **Cost of Natural Gas** In January 1997, Northern Illinois Gas Company had the following rate schedule* for natural gas usage in single-family residences:

Monthly customer charge	$6.00
Distribution charge,	
1st 20 therms	$0.2012/therm
Next 30 therms	$0.1117/therm
Over 50 therms	$0.0374/therm
Gas supply charge	$0.3113/therm

(a) What is the charge for using 40 therms in a month?

(b) What is the charge for using 202 therms in a month?

(c) Construct a function that gives the monthly charge C for x therms of gas.

(d) Graph this function.

83. Graph $y = x^2$. Then on the same screen graph $y = x^2 + 2$, followed by $y = x^2 + 4$, followed by $y = x^2 - 2$. What pattern do you observe? Can you predict the graph of $y = x^2 - 4$? Of $y = x^2 + 5$?

84. Graph $y = x^2$. Then on the same screen graph $y = (x - 2)^2$, followed by $y = (x - 4)^2$, followed by $y = (x + 2)^2$. What pattern do you observe? Can you predict the graph of $y = (x + 4)^2$? Of $y = (x - 5)^2$?

85. Graph $y = |x|$. Then on the same screen graph $y = 2|x|$, followed by $y = 4|x|$, followed by $y = \frac{1}{2}|x|$. What pattern do you observe? Can you predict the graph of $y = \frac{1}{4}|x|$? Of $y = 5|x|$?

86. Graph $y = x^2$. Then on the same screen graph $y = -x^2$. What pattern do you observe? Now try $y = |x|$ and $y = -|x|$. What do you conclude?

87. Graph $y = \sqrt{x}$. Then on the same screen graph $y = \sqrt{-x}$. What pattern do you observe? Now try $y = 2x + 1$ and $y = 2(-x) + 1$. What do you conclude?

88. Graph $y = x^3$. Then on the same screen graph $y = (x - 1)^3 + 2$. Could you have predicted the result?

89. Graph $y = x^2$, $y = x^4$, and $y = x^6$ on the same screen. What do you notice that is the same about each graph? What do you notice that is different?

90. Graph $y = x^3$, $y = x^5$, and $y = x^7$ on the same screen. What do you notice that is the same about each graph? What do you notice that is different?

91. Consider the equation

$$y = \begin{cases} 1 & \text{if } x \text{ is rational} \\ 0 & \text{if } x \text{ is irrational} \end{cases}$$
Is this a function? What is its domain? What is its range? What is its y-intercept, if any? What are its x-intercepts, if any? Is it even, odd, or neither? How would you describe its graph?

92. Define some functions that pass through $(0, 0)$ and $(1, 1)$ and are increasing for $x \geq 0$. Begin your list with $y = \sqrt{x}, y = x$, and $y = x^2$. Can you propose a general result about such functions?

93. Can you think of a function that is both even and odd?

Source: Northern Illinois Gas Company, Naperville, Illinois.

4.3 | GRAPHING TECHNIQUES: TRANSFORMATIONS

1   Graph Functions Using Horizontal and Vertical Shifts
2   Graph Functions Using Compressions and Stretches
3   Graph Functions Using Reflections about the x-Axis or y-Axis

At this stage, if you were asked to graph any of the functions defined by $y = x$, $y = x^2$, $y = x^3$, $y = \sqrt{x}$, $y = |x|$, or $y = 1/x$, your response should be, "Yes, I recognize these functions and know the general shapes of their graphs." (If this is not your answer, review the previous section and Figures 17 through 22.)

Sometimes, we are asked to graph a function that is "almost" like one that we already know how to graph. In this section, we look at some of these functions and develop techniques for graphing them. Collectively, these techniques are referred to as **transformations.**

| 1 | **Vertical Shifts** |

E X A M P L E   1

Vertical Shift Up

Use the graph of $f(x) = x^2$ to obtain the graph of $g(x) = x^2 + 3$.

Solution   We begin by obtaining some points on the graphs of f and g. For example, when $x = 0$, then $y = f(0) = 0$ and $y = g(0) = 3$. When $x = 1$, then $y = f(1) = 1$ and $y = g(1) = 4$. Table 5 lists these and a few other points on each graph. We conclude that the graph of g is identical to that of f, except that it is shifted vertically up 3 units. See Figure 27.

TABLE 5		
x	**$y = f(x)$** $= x^2$	**$y = g(x)$** $= x^2 + 3$
-2	4	7
-1	1	4
0	0	3
1	1	4
2	4	7

FIGURE 27

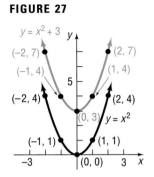

FIGURE 28

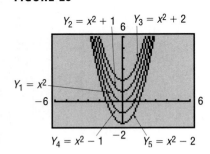

Seeing the Concept   On the same screen, graph each of the following functions:

$$Y_1 = x^2$$
$$Y_2 = x^2 + 1$$
$$Y_3 = x^2 + 2$$
$$Y_4 = x^2 - 1$$
$$Y_5 = x^2 - 2$$

Figure 28 illustrates the graphs. You should have observed a general pattern. With $Y_1 = x^2$ on the screen, the graph of $Y_2 = x^2 + 1$ is identical to that of $Y_1 = x^2$, except that it is shifted vertically up 1 unit. Similarly, $Y_3 = x^2 + 2$ is identical to that of $Y_1 = x^2$, except that it is shifted vertically up 2 units. The graph of $Y_4 = x^2 - 1$ is identical to that of $Y_1 = x^2$, except that it is shifted vertically down 1 unit.

We are led to the following conclusion:

> If a real number c is added to the right side of a function $y = f(x)$, the graph of the new function $y = f(x) + c$ is the graph of f **shifted vertically** up (if $c > 0$) or down (if $c < 0$).

Let's look at another example.

EXAMPLE 2

Vertical Shift Down

Use the graph of $f(x) = x^2$ to obtain the graph of $h(x) = x^2 - 4$.

Solution   Table 6 lists some points on the graphs of f and h. The graph of h is identical to that of f, except that is shifted down 4 units. See Figure 29.

TABLE 6		
x	$y = f(x)$ $= x^2$	$y = h(x)$ $= x^2 - 4$
-2	4	0
-1	1	-3
0	0	-4
1	1	-3
2	4	0

FIGURE 29

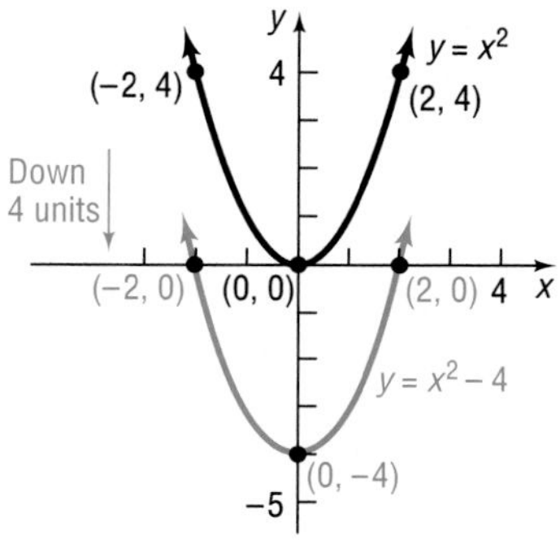

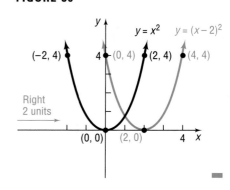

 Now work Problem 25.

Horizontal Shifts

EXAMPLE 3

Horizontal Shift to the Right

Use the graph of $f(x) = x^2$ to obtain the graph of $g(x) = (x - 2)^2$.

Solution   The function $g(x) = (x - 2)^2$ is basically a square function. Table 7 lists some points on the graphs of f and g. Note that when $f(x) = 0$ then $x = 0$, and when $g(x) = 0$, then $x = 2$. Also, when $f(x) = 4$, then $x = -2$ or 2, and when $g(x) = 4$, then $x = 0$ or 4. We conclude that the graph of g is identical to that of f, except that it is shifted 2 units to the right. See Figure 30.

TABLE 7		
x	$y = f(x)$ $= x^2$	$y = g(x)$ $= (x - 2)^2$
-2	4	16
0	0	4
2	4	0
4	16	4

FIGURE 30

FIGURE 31

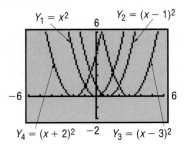

Seeing the Concept   On the same screen, graph each of the following functions:

$$Y_1 = x^2$$
$$Y_2 = (x - 1)^2$$
$$Y_3 = (x - 3)^2$$
$$Y_4 = (x + 2)^2$$

Figure 31 illustrates the graphs.

You should have observed the following pattern. With the graph of $Y_1 = x^2$ on the screen, the graph of $Y_2 = (x - 1)^2$ is identical to that of $Y = x^2$, except that it is shifted horizontally to the right 1 unit. Similarly, the graph of $Y_3 = (x - 3)^2$ is identical to that of $Y_1 = x^2$, except that it is shifted horizontally to the right 3 units. Finally, the graph of $Y_4 = (x + 2)^2$ is identical to that of $Y_1 = x^2$, except that it is shifted horizontally to the left 2 units. ▬

We are led to the following conclusion.

If the argument x of a function f is replaced by $x - c$, c a real number, the graph of the new function $g(x) = f(x - c)$ is the graph of f **shifted horizontally** left (if $c < 0$) or right (if $c > 0$).

E X A M P L E   4   Horizontal Shift to the Left

Use the graph of $f(x) = x^2$ to obtain the graph of $h(x) = (x + 4)^2$.

Solution   Again, the function $h(x) = (x + 4)^2$ is basically a square function. Thus, its graph is the same as that of f, except that it is shifted 4 units to the left. (Do you see why? $(x + 4)^2 = [x - (-4)]^2$) See Figure 32.

FIGURE 32

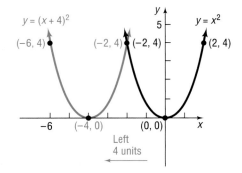

Now work Problem 31.

Vertical and horizontal shifts are sometimes combined.

E X A M P L E   5   Combining Vertical and Horizontal Shifts

FIGURE 33

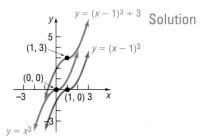

Graph the function $f(x) = (x - 1)^3 + 3$.

Solution   We graph f in steps. First, we note that the rule for f is basically a cube function. Thus, we begin with the graph of $y = x^3$, shown in blue in Figure 33. Next, to get the graph of $y = (x - 1)^3$, we shift the graph of $y = x^3$ horizontally 1 unit to the right. See the graph shown in red in Figure 33. Finally, to get the graph of $y = (x - 1)^3 + 3$, we shift the graph of $y = (x - 1)^3$ vertically up 3 units. See the graph shown in green in Figure 33. Note the points that have been plotted on each graph. Using key points such as these can be helpful in keeping track of just what is taking place. ▬

In Example 5, if the vertical shift had been done first, followed by the horizontal shift, the final graph would have been the same. (Try it for yourself.)

Now work Problem 43.

 Compressions and Stretches

E X A M P L E   6

Vertical Stretch

Use the graph of $f(x) = |x|$ to obtain the graph of $g(x) = 2|x|$.

Solution   To see the relationship between the graphs of f and g, we form Table 8, listing points on each graph. For each x, the y-coordinate of a point on the graph of g is 2 times as large as the corresponding y-coordinate on the graph of f. The graph of $f(x) = |x|$ is vertically stretched by a factor of 2 [for example, from $(1, 1)$ to $(1, 2)$] to obtain the graph of $g(x) = 2|x|$. See Figure 34.

TABLE 8

x	$y = f(x)$ $= |x|$	$y = g(x)$ $= 2|x|$
-2	2	4
-1	1	2
0	0	0
1	1	2
2	2	4

FIGURE 34

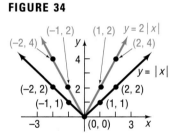

E X A M P L E   7

Vertical Compression

Use the graph of $f(x) = |x|$ to obtain the graph of $h(x) = \frac{1}{2}|x|$.

Solution   For each x, the y-coordinate of a point on the graph of h is $\frac{1}{2}$ as large as the corresponding y-coordinate on the graph of f. The graph of $f(x) = |x|$ is vertically compressed by a factor of $\frac{1}{2}$ [for example, from $(2, 2)$ to $(2, 1)$] to obtain the graph of $h(x) = \frac{1}{2}|x|$. See Table 9 and Figure 35.

TABLE 9

x	$y = f(x)$ $= |x|$	$y = h(x)$ $= \frac{1}{2}|x|$
-2	2	1
-1	1	$\frac{1}{2}$
0	0	0
1	1	$\frac{1}{2}$
2	2	1

FIGURE 35

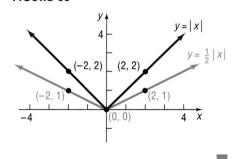

When the right side of a function $y = f(x)$ is multiplied by a positive number k, the graph of the new function $y = kf(x)$ is a **vertically compressed** (if $0 < k < 1$) or **stretched** (if $k > 1$) version of the graph of $y = f(x)$.

Now work Problem 33.

What happens if the argument x of a function $y = f(x)$ is multiplied by a positive number k, creating a new function $y = f(kx)$. To find the answer, we look at the following Exploration.

Exploration   On the same screen, graph each of the following functions:

$$Y_1 = f(x) = x^2$$
$$Y_2 = f(2x) = (2x)^2$$
$$Y_3 = f\left(\frac{1}{2}x\right) = \left(\frac{1}{2}x\right)^2$$

You should have obtained the graphs shown in Figure 36. The graph of $Y_2 = (2x)^2$ is the graph of $Y_1 = x^2$ compressed horizontally. Look at Table 10(a). Notice that $(1, 1)$, $(2, 4)$, $(4, 16)$, and $(16, 256)$ are points on the graph of $Y_1 = x^2$. Also, $(0.5, 1)$, $(1, 4)$, $(2, 16)$, and $(8, 256)$ are points on the graph of $Y_2 = (2x)^2$. Thus, the graph of $Y_2 = (2x)^2$ is obtained by multiplying the x-coordinate of each point on the graph of $Y_1 = x^2$ by $\frac{1}{2}$. The graph of $Y_3 = \left(\frac{1}{2}x\right)^2$ is the graph of $Y_1 = x^2$ stretched horizontally. Look at Table 10(b). Notice that $(0.5, 0.25)$, $(1, 1)$, $(2, 4)$, $(4, 16)$ are points on the graph of $Y_1 = x^2$. Also, $(1, 0.25)$, $(2, 1)$, $(4, 4)$, $(8, 16)$ are points on the graph of $Y_3 = \left(\frac{1}{2}x\right)^2$. Thus, the graph of $Y_3 = \left(\frac{1}{2}x\right)^2$ is obtained by multiplying the x-coordinate of each point on the graph of $Y_1 = x^2$ by a factor of 2.

FIGURE 36

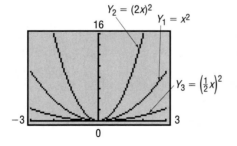

TABLE 10

X	Y_1	Y_2
0	0	0
.5	.25	1
1	1	4
2	4	16
4	16	64
8	64	256
16	256	1024

$Y_2 \boxminus (2X)^2$

(a)

X	Y_1	Y_3
0	0	0
.5	.25	.0625
1	1	.25
2	4	1
4	16	4
8	64	16
16	256	64

$Y_3 \boxminus (X/2)^2$

(b)

> If the argument x of a function $y = f(x)$ is multiplied by a positive number k, the graph of the new function $y = f(kx)$ is obtained by multiplying each x-coordinate of $y = f(x)$ by $1/k$. A **horizontal compression** results if $k > 1$ and a **horizontal stretch** occurs if $0 < k < 1$.

Let's look at an example.

E X A M P L E   8

Graphing Using Stretches and Compressions

The graph of $y = f(x)$ is given in Figure 37. Use this graph to find the graphs of

(a) $y = 3f(x)$ (b) $y = f(3x)$

FIGURE 37

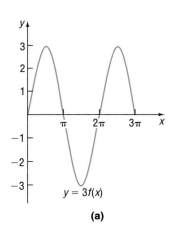

Solution   (a) The graph of $y = 3f(x)$ is obtained by multiplying each y-coordinate of $y = f(x)$ by a factor of 3. See Figure 38(a).

(b) The graph of $y = f(3x)$ is obtained from the graph of $y = f(x)$ by multiplying each x-coordinate of $y = f(x)$ by a factor of $\frac{1}{3}$. See Figure 38(b).

FIGURE 38

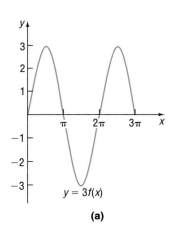

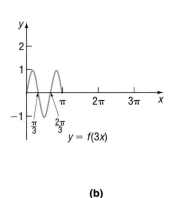

(a)

(b)

 Now work Problems 55(d) and 55(f).

3 **Reflections about the x-Axis and the y-Axis**

E X A M P L E   9

Reflection about the x-Axis

Graph the function: $f(x) = -x^2$

Solution   We begin with the graph of $y = x^2$, as shown in Figure 39. For each point (x, y) on the graph of $y = x^2$, the point $(x, -y)$ is on the graph of $y = -x^2$, as indicated in Table 11. Thus, we can draw the graph of $y = -x^2$ by reflecting the graph of $y = x^2$ about the x-axis. See Figure 39.

	TABLE 11		
x	$y = x^2$	$y = -x^2$
-2	4	-4
-1	1	-1
0	0	0
1	1	-1
2	4	-4

FIGURE 39

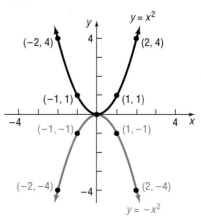

When the right side of the equation $y = f(x)$ is multiplied by -1, the graph of the new function $y = -f(x)$ is the **reflection about the x-axis** of the graph of the function $y = f(x)$.

 Now work Problem 37.

E X A M P L E   10

Reflection about the y-Axis

Graph the function: $f(x) = \sqrt{-x}$

Solution   First, notice that the domain of f consists of all real numbers x for which $-x \geq 0$ or equivalently, $x \leq 0$. To get the graph of $f(x) = \sqrt{-x}$, we begin with the graph of $y = \sqrt{x}$, as shown in Figure 40. For each point (x, y) on the graph of $y = \sqrt{x}$, the point $(-x, y)$ is on the graph of $y = \sqrt{-x}$. Thus, we get the graph of $y = \sqrt{-x}$ by reflecting the graph of $y = \sqrt{x}$ about the y-axis. See Figure 40.

FIGURE 40

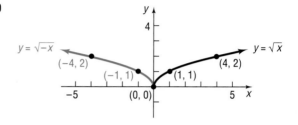

When the graph of the function $y = f(x)$ is known, the graph of the new function $y = f(-x)$ is the **reflection about the y-axis** of the graph of the function $y = f(x)$.

SUMMARY OF GRAPHING TECHNIQUES

Table 12 summarizes the graphing procedures we have just discussed.

TABLE 12		
To Graph:	**Draw the Graph of f and:**	**Functional Change to $f(x)$**
Vertical shifts		
$\quad y = f(x) + c,\quad c > 0$	Raise the graph of f by c units.	Add c to $f(x)$.
$\quad y = f(x) - c,\quad c > 0$	Lower the graph of f by c units.	Subtract c from $f(x)$.
Horizontal shifts		
$\quad y = f(x + c),\quad c > 0$	Shift the graph of f to the left c units.	Replace x by $x + c$.
$\quad y = f(x - c),\quad c > 0$	Shift the graph of f to the right c units.	Replace x by $x - c$.
Compressing or stretching		
$\quad y = kf(x),\quad k > 0$	Multiply each y coordinate of $y = f(x)$ by k.	Multiply $f(x)$ by k.
$\quad y = f(kx),\quad k > 0$	Multiply each x coordinate of $y = f(x)$ by $\dfrac{1}{k}$.	Replace x by kx.
Reflection about the x-axis		
$\quad y = -f(x)$	Reflect the graph of f about the x-axis.	Multiply $f(x)$ by -1.
Reflection about the y-axis		
$\quad y = f(-x)$	Reflect the graph of f about the y-axis.	Replace x by $-x$.

The examples that follow combine some of the procedures outlined in this section to get the required graph.

EXAMPLE 11

Determining the Function Obtained from a Series of Transformations

Find the function that is finally graphed after the following three transformations are applied to the graph of $y = |x|$.

1. Shift left 2 units
2. Shift up 3 units
3. Reflect about the y-axis

Solution
1. Shift left 2 units: Replace x by $x + 2$ $y = |x + 2|$
2. Shift up 3 units: Add 3 $y = |x + 2| + 3$
3. Reflect about the y-axis: Replace x by $-x$ $y = |-x + 2| + 3$ ■

 Now work Problem 21.

EXAMPLE 12

Combining Graphing Procedures

Graph the function: $f(x) = \dfrac{3}{x - 2} + 1$.

Solution   We use the following steps to obtain the graph of f:

STEP 1: $y = \dfrac{1}{x}$ Reciprocal function.

STEP 2: $y = \dfrac{3}{x}$ Multiply by 3; vertical stretch of the graph of $y = \dfrac{1}{x}$ by a factor of 3.

STEP 3: $y = \dfrac{3}{x - 2}$ Replace x by $x - 2$; horizontal shift to the right 2 units.

STEP 4: $y = \dfrac{3}{x - 2} + 1$ Add 1; vertical shift up 1 unit.

See Figure 41.

FIGURE 41

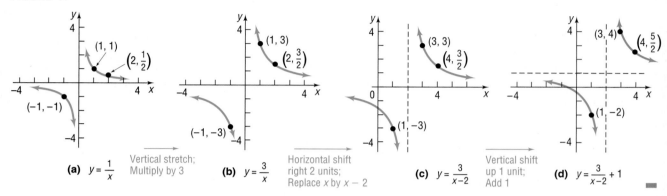

There are other orderings of the steps shown in Example 12 that would also result in the graph of f. For example, try this one:

STEP 1: $y = \dfrac{1}{x}$ Reciprocal function.

STEP 2: $y = \dfrac{1}{x - 2}$ Replace x by $x - 2$; horizontal shift to the right 2 units.

STEP 3: $y = \dfrac{3}{x - 2}$ Multiply by 3; vertical stretch of the graph of $y = \dfrac{1}{x - 2}$ by a factor of 3.

STEP 4: $y = \dfrac{3}{x - 2} + 1$ Add 1; vertical shift up 1 unit.

E X A M P L E   13   **Combining Graphing Procedures**

Graph the function: $f(x) = \sqrt{1 - x} + 2$.

Solution   We use the following steps to get the graph of $y = \sqrt{1 - x} + 2$:

STEP 1: $y = \sqrt{x}$ Square root function.

STEP 2: $y = \sqrt{x + 1}$ Replace x by $x + 1$; horizontal shift left 1 unit.

STEP 3: $y = \sqrt{-x + 1} = \sqrt{1 - x}$ Replace x by $-x$; reflect about y-axis.

STEP 4: $y = \sqrt{1 - x} + 2$ Add 2; vertical shift up 2 units.

See Figure 42.

FIGURE 42

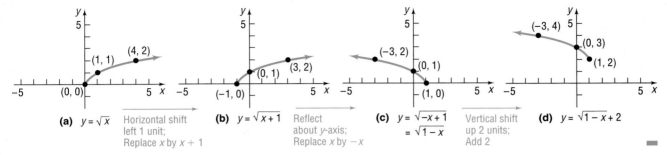

(a) $y = \sqrt{x}$ Horizontal shift
left 1 unit;
Replace x by $x + 1$

(b) $y = \sqrt{x+1}$ Reflect
about y-axis;
Replace x by $-x$

(c) $y = \sqrt{-x+1}$ Vertical shift
$= \sqrt{1-x}$ up 2 units;
Add 2

(d) $y = \sqrt{1-x} + 2$

4.3 EXERCISES

In Problems 1–12, match each graph to one of the following functions:

A. $y = x^2 + 2$ B. $y = -x^2 + 2$ C. $y = |x| + 2$ D. $y = -|x| + 2$
E. $y = (x - 2)^2$ F. $y = -(x + 2)^2$ G. $y = |x - 2|$ H. $y = -|x + 2|$
I. $y = 2x^2$ J. $y = -2x^2$ K. $y = 2|x|$ L. $y = -2|x|$

1.
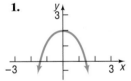

2.
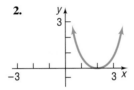

3.
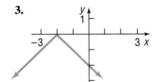

4.
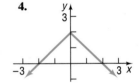

5.
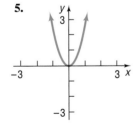

6.
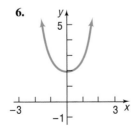

7.
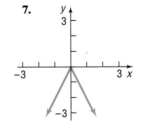

8.
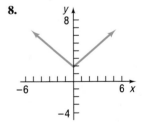

9.
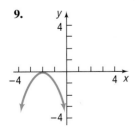

10.
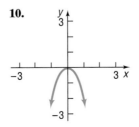

11.
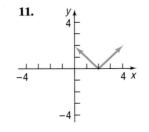

12.
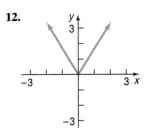

In Problems 13–20, write the function whose graph is the graph of $y = x^3$*, but is*

13. Shifted to the right 4 units

14. Shifted to the left 4 units

15. Shifted up 4 units

16. Shifted down 4 units

17. Reflected about the *y*-axis

18. Reflected about the *x*-axis

19. Vertically stretched by a factor of 4

20. Horizontally stretched by a factor of 4

In Problems 21–24, find the function that is finally graphed after the following transformations are applied to the graph of $y = \sqrt{x}$.

21. (1) Shift up 2 units
(2) Reflect about the *x*-axis
(3) Reflect about the *y*-axis

22. (1) Reflect about the *x*-axis
(2) Shift right 3 units
(3) Shift down 2 units

23. (1) Reflect about the *x*-axis
(2) Shift up 2 units
(3) Shift left 3 units

24. (1) Shift up 2 units
(2) Reflect about the *y*-axis
(3) Shift left 3 units

In Problems 25–54, graph each function using the techniques of shifting, compressing, stretching, and/or reflecting. Start with the graph of the basic function (for example, $y = x^2$*) and show all stages.*

25. $f(x) = x^2 - 1$ **26.** $f(x) = x^2 + 4$ **27.** $g(x) = x^3 + 1$ **28.** $g(x) = x^3 - 1$

29. $h(x) = \sqrt{x - 2}$ **30.** $h(x) = \sqrt{x + 1}$ **31.** $f(x) = (x - 1)^3$ **32.** $f(x) = (x + 2)^3$

33. $g(x) = 4\sqrt{x}$ **34.** $g(x) = \frac{1}{2}\sqrt{x}$ **35.** $h(x) = \dfrac{1}{2x}$ **36.** $h(x) = \dfrac{4}{x}$

37. $f(x) = -|x|$ **38.** $f(x) = -\sqrt{x}$ **39.** $g(x) = -\dfrac{1}{x}$ **40.** $g(x) = -x^3$

41. $h(x) = \text{int}(-x)$ **42.** $h(x) = \dfrac{1}{-x}$ **43.** $f(x) = (x + 1)^2 - 3$ **44.** $f(x) = (x - 2)^2 + 1$

45. $g(x) = \sqrt{x - 2} + 1$ **46.** $g(x) = |x + 1| - 3$ **47.** $h(x) = \sqrt{-x} - 2$ **48.** $h(x) = \dfrac{4}{x} + 2$

49. $f(x) = (x + 1)^3 - 1$ **50.** $f(x) = 4\sqrt{x - 1}$ **51.** $g(x) = 2|1 - x|$ **52.** $g(x) = 4\sqrt{2 - x}$

53. $h(x) = 2\,\text{int}(x - 1)$ **54.** $h(x) = -x^3 + 2$

In Problems 55–60, the graph of a function f is illustrated. Use the graph of f as the first step toward graphing each of the following functions:

(a) $F(x) = f(x) + 3$ (b) $G(x) = f(x + 2)$ (c) $P(x) = -f(x)$

(d) $Q(x) = \frac{1}{2}f(x)$ (e) $g(x) = f(-x)$ (f) $h(x) = f(2x)$

55.
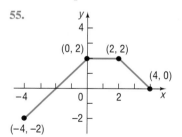

56.
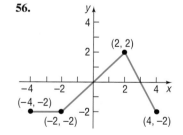

57.
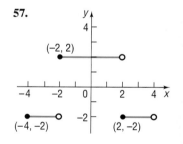

58.
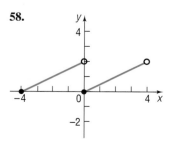

59.
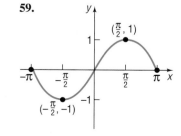

60.
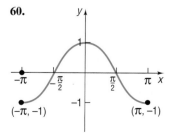

61. Exploration
(a) Use a graphing utility to graph $y = x + 1$ and $y = |x + 1|$.
(b) Graph $y = 4 - x^2$ and $y = |4 - x^2|$.
(c) Graph $y = x^3 + x$ and $y = |x^3 + x|$.
(d) What do you conclude about the relationship between the graphs of $y = f(x)$ and $y = |f(x)|$?

62. Exploration
(a) Use a graphing utility to graph $y = x + 1$ and $y = |x| + 1$.
(b) Graph $y = 4 - x^2$ and $y = 4 - |x|^2$.
(c) Graph $y = x^3 + x$ and $y = |x|^3 + |x|$.
(d) What do you conclude about the relationship between the graphs of $y = f(x)$ and $y = f(|x|)$?

63. The graph of a function f is illustrated in the figure.
(a) Draw the graph of $y = |f(x)|$.
(b) Draw the graph of $y = f(|x|)$.

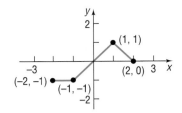

64. The graph of a function f is illustrated in the figure.
(a) Draw the graph of $y = |f(x)|$.
(b) Draw the graph of $y = f(|x|)$.

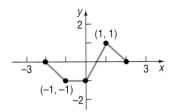

In Problems 65–70, complete the square of each quadratic expression. Then graph each function using the technique of shifting.
Hint: If $f(x) = x^2 - 2x$, then $f(x) = (x^2 - 2x + 1) - 1 = (x - 1)^2 + 1$.

65. $f(x) = x^2 + 2x$

66. $f(x) = x^2 - 6x$

67. $f(x) = x^2 - 8x + 1$

68. $f(x) = x^2 + 4x + 2$

69. $f(x) = x^2 + x + 1$

70. $f(x) = x^2 - x + 1$

71. The equation $y = (x - c)^2$ defines a *family of parabolas*, one parabola for each value of c. On one set of coordinate axes, graph the members of the family for $c = 0, c = 3, c = -2$.

72. Repeat Problem 71 for the family of parabolas $y = x^2 + c$.

73. Temperature Measurements The relationship between the Celsius (°C) and Fahrenheit (°F) scales for measuring temperature is given by the equation

$$F = \frac{9}{5}C + 32$$

The relationship between the Celsius (°C) and Kelvin (K) scales is $K = C + 273$. Graph the equation $F = \frac{9}{5}C + 32$ using degrees Fahrenheit on the y-axis and degrees Celsius on the x-axis. Use the techniques introduced in this section to obtain the graph showing the relationship between Kelvin and Fahrenheit temperatures.

74. Period of a Pendulum The period T (in seconds) of a simple pendulum is a function of its length l (in feet) defined by the equation

$$T = 2\pi\sqrt{\frac{l}{g}}$$

where $g \approx 32.2$ feet per second per second is the acceleration of gravity.
(a) Use a graphing utility to graph the function $T = T(l)$.
(b) Now graph the functions $T = T(l + 1)$, $T = T(l + 2)$, and $T = T(l + 3)$.
(c) Discuss how adding to the length l changes the period T.
(d) Now graph the functions $T = T(2l)$, $T = T(3l)$, and $T = T(4l)$.
(e) Discuss how multiplying the length l by factors of 2, 3, and 4 changes the period T.

75. **Cigarette Company Profits**   The daily profits of a cigarette company from selling x packs of cigarettes are given by

$$p(x) = -0.05x^2 + 100x - 2000$$

The government wishes to impose a tax on cigarettes (sometimes called a sin tax) that gives the company the option of either paying a flat tax of \$10,000 per day or a tax of 10% on profits. As Chief Financial Officer (CFO) of the company, you need to decide which tax is the better option for the company.

(a) On the same viewing rectangle graph $Y_1 = p(x) - 10,000$ and $Y_2 = (1 - 0.10)p(x)$.

(b) Based on the graph, which option would you select? Why?

(c) Using the terminology learned in this section, describe each graph in terms of the graph of $p(x)$.

(d) Suppose that the government offered the option of a flat tax of \$4800 or a tax of 10% on profits. Which option would you select? Why?

4.4 | OPERATIONS ON FUNCTIONS; COMPOSITE FUNCTIONS

> **1** Form the Sum, Difference, Product, and Quotient of Two Functions
>
> **2** Form the Composite Function and Find its Domain

In this section, we introduce some operations on functions. We shall see that functions, like numbers, can be added, subtracted, multiplied, and divided. For example, if $f(x) = x^2 + 9$ and $g(x) = 3x + 5$, then

$$f(x) + g(x) = (x^2 + 9) + (3x + 5) = x^2 + 3x + 14$$

The new function $y = x^2 + 3x + 14$ is called the *sum function* $f + g$. Similarly,

$$f(x) \cdot g(x) = (x^2 + 9)(3x + 5) = 3x^3 + 5x^2 + 27x + 45$$

The new function $y = 3x^3 + 5x^2 + 27x + 45$ is called the *product function* $f \cdot g$.

The general definitions are given next.

If f and g are functions:
Their **sum $f + g$** is the function defined by

$$(f + g)(x) = f(x) + g(x)$$

The domain of $f + g$ consists of the numbers x that are in the domain of f and in the domain of g.

Their **difference $f - g$** is the function defined by

$$(f - g)(x) = f(x) - g(x)$$

The domain of $f - g$ consists of the numbers x that are in the domain of f and in the domain of g.

Their **product f · g** is the function defined by

$$(f \cdot g)(x) = f(x) \cdot g(x)$$

The domain of $f \cdot g$ consists of the numbers x that are in the domain of f and in the domain of g.

Their **quotient f/g** is the function defined by

$$\frac{f}{g}(x) = \frac{f(x)}{g(x)} \qquad g(x) \neq 0$$

The domain of f/g consists of the numbers x for which $g(x) \neq 0$ that are in the domain of f and in the domain of g.

E X A M P L E   1

Operations on Functions

Let f and g be two functions defined as

$$f(x) = \sqrt{x + 2} \quad \text{and} \quad g(x) = \sqrt{3 - x}$$

Find the following, and determine the domain in each case:

(a) $(f + g)(x)$ (b) $(f - g)(x)$ (c) $(f \cdot g)(x)$ (d) $(f/g)(x)$

Solution

(a) $(f + g)(x) = f(x) + g(x) = \sqrt{x + 2} + \sqrt{3 - x}$

(b) $(f - g)(x) = f(x) - g(x) = \sqrt{x + 2} - \sqrt{3 - x}$

(c) $(f \cdot g)(x) = f(x) \cdot g(x) = (\sqrt{x + 2})(\sqrt{3 - x}) = \sqrt{(x + 2)(3 - x)}$

(d) $\left(\dfrac{f}{g}\right)(x) = \dfrac{f(x)}{g(x)} = \dfrac{\sqrt{x + 2}}{\sqrt{3 - x}} = \dfrac{\sqrt{(x + 2)(3 - x)}}{3 - x}$

The domain of f consists of all numbers x for which $x \geq -2$; the domain of g consists of all numbers x for which $x \leq 3$. The numbers x common to both these domains are those for which $-2 \leq x \leq 3$. As a result, the numbers x for which $-2 \leq x \leq 3$ comprise the domain of the sum function $f + g$, the difference function $f - g$, and the product function $f \cdot g$. For the quotient function f/g, we must exclude from this set the number 3, because the denominator, g, has the value 0 when $x = 3$. Thus, the domain of f/g consists of all x for which $-2 \leq x < 3$. ■

Now work Problem 1.

In calculus, it is sometimes helpful to view a complicated function as the sum, difference, product, or quotient of simpler functions. For example,

$$F(x) = x^2 + \sqrt{x} \text{ is the sum of } f(x) = x^2 \text{ and } g(x) = \sqrt{x}.$$

$$H(x) = \frac{(x^2 - 1)}{(x^2 + 1)} \text{ is the quotient of } f(x) = x^2 - 1 \text{ and } g(x) = x^2 + 1.$$

Composite Functions

2 Consider the function $y = (2x + 3)^2$. If we write $y = f(u) = u^2$ and $u = g(x) = 2x + 3$, then, by a substitution process, we can obtain the original function: $y = f(u) = f(g(x)) = (2x + 3)^2$. This process is called **composition.**

In general, suppose that f and g are two functions, and suppose that x is a number in the domain of g. By evaluating g at x, we get $g(x)$. If $g(x)$ is in the domain of f, then we may evaluate f at $g(x)$ and thereby obtain the expression $f(g(x))$. The correspondence from x to $f(g(x))$ is called the *composite function $f \circ g$.*

Look carefully at Figure 43. Only those x's in the domain of g for which $g(x)$ is in the domain of f can be in the domain of $f \circ g$. The reason is that if $g(x)$ is not in the domain of f then $f(g(x))$ is not defined. Thus, the domain of $f \circ g$ is a subset of the domain of g; the range of $f \circ g$ is a subset of the range of f.

FIGURE 43

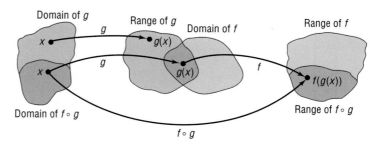

Given two functions f and g, the **composite function,** denoted by $f \circ g$ (read as "f composed with g"), is defined by

$$(f \circ g)(x) = f(g(x))$$

The domain of $f \circ g$ is the set of all numbers x in the domain of g such that $g(x)$ is in the domain of f.

Figure 44 provides a second illustration of the definition. Notice that the "inside" function g in $f(g(x))$ is done first.

FIGURE 44

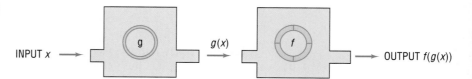

Let's look at some examples.

EXAMPLE 2   Evaluating a Composite Function

Suppose that $f(x) = 2x^2 - 3$ and $g(x) = 4x$. Find

(a) $(f \circ g)(1)$ (b) $(g \circ f)(1)$ (c) $(f \circ f)(-2)$ (d) $(g \circ g)(-1)$

Solution (a) $(f \circ g)(1) = f(g(1)) = f(4) = 2 \cdot 16 - 3 = 29$

$$\uparrow \qquad \uparrow$$
$$g(x) = 4x \quad f(x) = 2x^2 - 3$$
$$g(1) = 4$$

(b) $(g \circ f)(1) = g(f(1)) = g(-1) = 4 \cdot (-1) = -4$

$$\uparrow \qquad \uparrow$$
$$f(x) = 2x^2 - 3 \quad g(x) = 4x$$
$$f(1) = -1$$

(c) $(f \circ f)(-2) = f(f(-2)) = f(5) = 2 \cdot 25 - 3 = 47$

$$\uparrow$$
$$f(-2) = 5$$

(d) $(g \circ g)(-1) = g(g(-1)) = g(-4) = 4 \cdot (-4) = -16$

$$\uparrow$$
$$g(-1) = -4$$

■

Now work Problem 13.

Look back at Figure 43. In determining the domain of the composite function $(f \circ g)(x) = f(g(x))$, keep the following two thoughts in mind about the input x:

1. $g(x)$ must be defined so any x not in the domain of g must be excluded.
2. $f(g(x))$ must be defined so any x for which $g(x)$ is not in the domain of f must be excluded.

E X A M P L E   3

Finding the Domain of $f \circ g$

Find the domain of $(f \circ g)(x)$ if $f(x) = \dfrac{1}{x + 2}$ and $g(x) = \dfrac{4}{x - 1}$.

Solution For $(f \circ g)(x) = f(g(x))$, we first note that the domain of g is $\{x | x \neq 1\}$, so we exclude 1 from the domain of $f \circ g$. Next, we note that the domain of f is $\{x | x \neq -2\}$. Thus, $g(x)$ cannot equal -2, so we solve the equation

$$\frac{4}{x - 1} = -2$$
$$4 = -2(x - 1)$$
$$4 = -2x + 2$$
$$2x = -2$$
$$x = -1$$

We also exclude -1 from the domain of $f \circ g$. The domain of $f \circ g$ is $\{x | x \neq -1, x \neq 1\}$.

Check: For $x = 1$, $g(x) = \dfrac{4}{x - 1}$ is not defined, so $(f \circ g)(x) = f(g(x))$ is not defined.

For $x = -1$, $g(-1) = 4/-2 = -2$ and $(f \circ g)(-1) = f(g(-1)) = f(-2)$ is not defined.

■

Now work Problem 23.

E X A M P L E   4   Finding a Composite Function

Suppose that $f(x) = \dfrac{1}{x + 2}$ and $g(x) = \dfrac{4}{x - 1}$. Find the following composite functions, and then find the domain of each composite function.

(a) $f \circ g$ (b) $g \circ f$ (c) $f \circ f$ (d) $g \circ g$

Solution   The domain of f is $\{x | x \neq -2\}$ and the domain of g is $\{x | x \neq 1\}$.

(a) $(f \circ g)(x) = f(g(x)) = f\left(\dfrac{4}{x - 1}\right) = \dfrac{1}{\dfrac{4}{x - 1} + 2} = \dfrac{x - 1}{4 + 2(x - 1)} = \dfrac{x - 1}{2x + 2}$

$\uparrow$
Multiply by $\frac{x-1}{x-1}$

The domain of $f \circ g$ consists of those x in the domain of g ($x \neq 1$) for which $g(x) = 4/(x - 1) \neq -2$ or, equivalently, $x \neq -1$. Thus, the domain of $f \circ g$ is $\{x | x \neq -1, x \neq 1\}$. (See Example 3.)

(b) $(g \circ f)(x) = g(f(x)) = g\left(\dfrac{1}{x + 2}\right) = \dfrac{4}{\dfrac{1}{x + 2} - 1} = \dfrac{4(x + 2)}{1 - (x + 2)} = \dfrac{4(x + 2)}{-x - 1}$

The domain of $g \circ f$ consists of those x in the domain of f ($x \neq -2$) for which

$$f(x) = \dfrac{1}{x + 2} \neq 1 \qquad \begin{array}{l} \frac{1}{x+2} = 1 \\[4pt] 1 = x + 2 \\[2pt] x = -1 \end{array}$$

or equivalently, $x \neq -1$.

The domain of $g \circ f$ is $\{x | x \neq -1, x \neq -2\}$.

(c) $(f \circ f)(x) = f(f(x)) = f\left(\dfrac{1}{x + 2}\right) = \dfrac{1}{\dfrac{1}{x + 2} + 2} = \dfrac{x + 2}{1 + 2(x + 2)} = \dfrac{x + 2}{2x + 5}$

The domain of $f \circ f$ consists of those x in the domain of f ($x \neq -2$) for which

$$f(x) = \dfrac{1}{x + 2} \neq -2 \qquad \begin{array}{l} \frac{1}{x+2} = -2 \\[4pt] 1 = -2(x + 2) \\[2pt] 1 = -2x - 4 \\[2pt] 2x = -5 \\[2pt] x = -\frac{5}{2} \end{array}$$

or, equivalently,

$$x \neq -\dfrac{5}{2}$$

The domain of $f \circ f$ is $\{x | x \neq -\frac{5}{2}, x \neq -2\}$.

(d) $(g \circ g)(x) = g(g(x)) = g\left(\dfrac{4}{x - 1}\right) = \dfrac{4}{\dfrac{4}{x - 1} - 1} = \dfrac{4(x - 1)}{4 - (x - 1)} = \dfrac{4(x - 1)}{-x + 5}$

The domain of $g \circ g$ consists of those x in the domain of $g(x \neq 1)$ for which

$$g(x) = \frac{4}{x - 1} \neq 1 \qquad \frac{4}{x-1} = 1$$

$$4 = x - 1$$

or, equivalently, $x \neq 5$. $\qquad x = 5$

Thus, the domain of $g \circ g$ is $\{x | x \neq 1, x \neq 5\}$.

 Now work Problems 35 and 37.

Examples 4(a) and 4(b) illustrate that, in general, $f \circ g \neq g \circ f$. However, sometimes $f \circ g$ does equal $g \circ f$, as shown in the next example.

E X A M P L E   5   Showing Two Composite Functions Equal

If $f(x) = 3x - 4$ and $g(x) = \frac{1}{3}(x + 4)$, show that $(f \circ g)(x) = (g \circ f)(x) = x$ for every x.

Solution   The domain of f and g is all real numbers, so the domain of $f \circ g$ and $g \circ f$ is also all real numbers.

$$(f \circ g)(x) = f(g(x))$$

$$= f\left(\frac{x + 4}{3}\right) \qquad g(x) = \frac{1}{3}(x + 4) = \frac{x+4}{3}.$$

$$= 3\left(\frac{x + 4}{3}\right) - 4 \qquad \begin{array}{l} \text{Substitute } g(x) \text{ into the equation for} \\ f, f(x) = 3x - 4. \end{array}$$

$$= x + 4 - 4 = x$$

$$(g \circ f)(x) = g(f(x))$$

$$= g(3x - 4) \qquad f(x) = 3x - 4.$$

$$= \frac{1}{3}[(3x - 4) + 4] \qquad \text{Substitute } f(x) \text{ into the equation for } g,$$

$$g(x) = \frac{1}{3}(x + 4).$$

$$= \frac{1}{3}(3x) = x$$

Thus, $(f \circ g)(x) = (g \circ f)(x) = x$.

In Section 6.1, we shall see that there is an important relationship between functions f and g for which $(f \circ g)(x) = (g \circ f)(x) = x$.

 Now work Problem 47.

Calculus Application

Some techniques in calculus require that we be able to determine the components of a composite function. For example, the function $H(x) = \sqrt{x + 1}$ is the composition of the functions f and g, where $f(x) = \sqrt{x}$ and $g(x) = x + 1$, because $H(x) = (f \circ g)(x) = f(g(x)) = f(x + 1) = \sqrt{x + 1}$.

EXAMPLE 6   Finding the Components of a Composite Function

Find functions f and g such that $f \circ g = H$ if $H(x) = (x^2 + 1)^{50}$.

Solution   The function H takes $x^2 + 1$ and raises it to the power 50. A natural way to decompose H is to raise the function $g(x) = x^2 + 1$ to the power 50. Thus, if we let $f(x) = x^{50}$ and $g(x) = x^2 + 1$, then

$$(f \circ g)(x) = f(g(x))$$
$$= f(x^2 + 1)$$
$$= (x^2 + 1)^{50} = H(x)$$

FIGURE 45

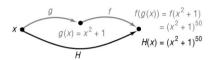

See Figure 45.

Other functions f and g may be found for which $f \circ g = H$ in Example 6. For example, if $f(x) = x^2$ and $g(x) = (x^2 + 1)^{25}$, then

$$(f \circ g)(x) = f(g(x)) = f((x^2 + 1)^{25}) = [(x^2 + 1)^{25}]^2 = (x^2 + 1)^{50}$$

Thus, although the functions f and g found as a solution to Example 6 are not unique, there is usually a "natural" selection for f and g that comes to mind first. This natural selection usually will enable you to use your calculator most efficiently. Let's look at Example 6. To calculate the value of H at, say, 2, we proceed as follows:

Keystrokes: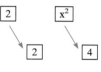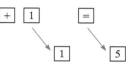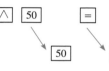

Display:

EXAMPLE 7   Finding the Components of a Composite Function

Find functions f and g such that $f \circ g = H$ if $H(x) = 1/(x + 1)$.

Solution   Here, H is the reciprocal of $g(x) = x + 1$. Thus, if we let $f(x) = 1/x$ and $g(x) = x + 1$, we find that

$$(f \circ g)(x) = f(g(x)) = f(x + 1) = \frac{1}{x + 1} = H(x)$$

4.4 | EXERCISES

In Problems 1–10, for the given functions f and g, find the following functions and state the domain of each:

(a) $f + g$ (b) $f - g$ (c) $f \cdot g$ (d) f/g

1. $f(x) = 3x + 4$; $g(x) = 2x - 3$
2. $f(x) = 2x + 1$; $g(x) = 3x - 2$
3. $f(x) = x - 1$; $g(x) = 2x^2$
4. $f(x) = 2x^2 + 3$; $g(x) = 4x^3 + 1$
5. $f(x) = \sqrt{x}$; $g(x) = 3x - 5$
6. $f(x) = |x|$; $g(x) = x$
7. $f(x) = 1 + \dfrac{1}{x}$; $g(x) = \dfrac{1}{x}$
8. $f(x) = 2x^2 - x$; $g(x) = 2x^2 + x$
9. $f(x) = \dfrac{2x + 3}{3x - 2}$; $g(x) = \dfrac{4x}{3x - 2}$
10. $f(x) = \sqrt{x + 1}$; $g(x) = \dfrac{2}{x}$
11. Given $f(x) = 3x + 1$ and $(f + g)(x) = 6 - \frac{1}{2}x$, find the function g.
12. Given $f(x) = 1/x$ and $(f/g)(x) = (x + 1)/(x^2 - x)$, find the function g.

In Problems 13–22, for the given functions f and g, find

(a) $(f \circ g)(4)$ (b) $(g \circ f)(2)$ (c) $(f \circ f)(1)$ (d) $(g \circ g)(0)$

13. $f(x) = 2x;\ \ g(x) = 3x^2 + 1$

14. $f(x) = 3x + 2;\ \ g(x) = 2x^2 - 1$

15. $f(x) = 4x^2 - 3;\ \ g(x) = 3 - \frac{1}{2}x^2$

16. $f(x) = 2x^2;\ \ g(x) = 1 - 3x^2$

17. $f(x) = \sqrt{x};\ \ g(x) = 2x$

18. $f(x) = \sqrt{x + 1};\ \ g(x) = 3x$

19. $f(x) = |x|;\ \ g(x) = \dfrac{1}{x^2 + 1}$

20. $f(x) = |x - 2|;\ \ g(x) = \dfrac{3}{x^2 + 2}$

21. $f(x) = \dfrac{3}{x^2 + 1};\ \ g(x) = \sqrt{x}$

22. $f(x) = x^3;\ \ g(x) = \dfrac{2}{x^2 + 1}$

In Problems 23–30, find the domain of the composite function f ∘ g.

23. $f(x) = \dfrac{3}{x - 1};\ \ g(x) = \dfrac{2}{x}$

24. $f(x) = \dfrac{1}{x + 3};\ \ g(x) = \dfrac{-2}{x}$

25. $f(x) = \dfrac{x}{x - 1};\ \ g(x) = \dfrac{-4}{x}$

26. $f(x) = \dfrac{x}{x + 3};\ \ g(x) = \dfrac{2}{x}$

27. $f(x) = \sqrt{x};\ \ g(x) = 2x + 3$

28. $f(x) = \sqrt{x - 2};\ \ g(x) = 1 - 2x$

29. $f(x) = \sqrt{x + 1};\ \ g(x) = \dfrac{2}{x - 1}$

30. $f(x) = \sqrt{3 - x};\ \ g(x) = \dfrac{2}{x - 2}$

In Problems 31–46, for the given functions f and g, find

(a) $f \circ g$ (b) $g \circ f$ (c) $f \circ f$ (d) $g \circ g$

State the domain of each composite function.

31. $f(x) = 2x + 3;\ \ g(x) = 3x$

32. $f(x) = -x;\ \ g(x) = 2x - 4$

33. $f(x) = 3x + 1;\ \ g(x) = x^2$

34. $f(x) = x + 1;\ \ g(x) = x^2 + 4$

35. $f(x) = x^2;\ \ g(x) = x^2 + 4$

36. $f(x) = x^2 + 1;\ \ g(x) = 2x^2 + 3$

37. $f(x) = \dfrac{3}{x - 1};\ \ g(x) = \dfrac{2}{x}$

38. $f(x) = \dfrac{1}{x + 3};\ \ g(x) = \dfrac{-2}{x}$

39. $f(x) = \dfrac{x}{x - 1};\ \ g(x) = \dfrac{-4}{x}$

40. $f(x) = \dfrac{x}{x + 3};\ \ g(x) = \dfrac{2}{x}$

41. $f(x) = \sqrt{x};\ \ g(x) = 2x + 3$

42. $f(x) = \sqrt{x - 2};\ \ g(x) = 1 - 2x$

43. $f(x) = \sqrt{x + 1};\ \ g(x) = \dfrac{2}{x - 1}$

44. $f(x) = \sqrt{3 - x};\ \ g(x) = \dfrac{2}{x - 2}$

45. $f(x) = ax + b;\ \ g(x) = cx + d$

46. $f(x) = \dfrac{ax + b}{cx + d};\ \ g(x) = mx$

In Problems 47–54, show that $(f \circ g)(x) = (g \circ f)(x) = x$.

47. $f(x) = 2x;\ \ g(x) = \frac{1}{2}x$

48. $f(x) = 4x;\ \ g(x) = \frac{1}{4}x$

49. $f(x) = x^3;\ \ g(x) = \sqrt[3]{x}$

50. $f(x) = x + 5;\ \ g(x) = x - 5$

51. $f(x) = 2x - 6;\ \ g(x) = \frac{1}{2}(x + 6)$

52. $f(x) = 4 - 3x;\ \ g(x) = \frac{1}{3}(4 - x)$

53. $f(x) = ax + b;\ \ g(x) = \dfrac{1}{a}(x - b), a \neq 0$

54. $f(x) = \dfrac{1}{x};\ \ g(x) = \dfrac{1}{x}$

In Problems 55–62, let $f(x) = x^2$, $g(x) = 3x$, and $h(x) = \sqrt{x} + 1$. Express each function as a composite of f, g, and/or h.

55. $F(x) = 9x^2$

56. $G(x) = 3x^2$

57. $H(x) = |x| + 1$

58. $p(x) = 3\sqrt{x} + 3$

59. $q(x) = x + 2\sqrt{x} + 1$

60. $R(x) = 9x$

61. $P(x) = x^4$

62. $Q(x) = \sqrt{\sqrt{x} + 1} + 1$

In Problems 63–70, find functions f and g so that $f \circ g = H$.

63. $H(x) = (2x + 3)^4$

64. $H(x) = (1 + x^2)^{3/2}$

65. $H(x) = \sqrt{x^2 + x + 1}$

66. $H(x) = \dfrac{1}{1 + x^2}$

67. $H(x) = \left(1 - \dfrac{1}{x^2}\right)^2$

68. $H(x) = |2x^2 + 3|$

69. $H(x) = \text{int}(x^2 + 1)$

70. $H(x) = (4 - x^2)^{-4}$

71. If $f(x) = 2x^3 - 3x^2 + 4x - 1$ and $g(x) = 2$, find $(f \circ g)(x)$ and $(g \circ f)(x)$.

72. If $f(x) = x/(x - 1)$, find $(f \circ f)(x)$.

73. If $f(x) = 2x^2 + 5$ and $g(x) = 3x + a$ find a so that the graph of $f \circ g$ crosses the y-axis at 23.

74. If $f(x) = 3x^2 - 7$ and $g(x) = 2x + a$, find a so that the graph of $f \circ g$ crosses the y-axis at 68.

75. **Surface Area of a Balloon** The surface area S (in square meters) of a hot-air balloon is given by

$$S(r) = 4\pi r^2$$

where r is the radius of the balloon (in meters). If the radius r is increasing with time t (in seconds) according to the formula $r(t) = \frac{2}{3}t^3$, $t \geq 0$, find the surface area S of the balloon as a function of the time t.

76. **Volume of a Balloon** The volume V (in cubic meters) of the hot-air balloon described in Problem 75 is given by $V(r) = \frac{4}{3}\pi r^3$. If the radius r is the same function of t as in Problem 75, find the volume V as a function of the time t.

77. **Automobile Production** The number N of cars produced at a certain factory in 1 day after t hours of operation is given by $N(t) = 100t - 5t^2$, $0 \leq t \leq 10$. If the cost C (in dollars) of producing N cars is $C(N) = 15,000 + 8000N$, find the cost C as a function of the time t of operation of the factory.

78. **Environmental Concerns** The spread of oil leaking from a tanker is in the shape of a circle. If the radius r (in feet) of the spread after t hours is $r(t) = 200\sqrt{t}$, find the area A of the oil slick as a function of the time t.

79. **Production Cost** The price p of a certain product and the quantity x sold obey the demand equation

$$p = -\frac{1}{4}x + 100 \qquad 0 \leq x \leq 400$$

Suppose that the cost C of producing x units is

$$C = \frac{\sqrt{x}}{25} + 600$$

Assuming that all items produced are sold, find the cost C as a function of the price p.
[**Hint:** Solve for x in the demand equation, and then form the composite.]

80. **Cost of a Commodity** The price p of a certain commodity and the quantity x sold obey the demand equation

$$p = -\frac{1}{5}x + 200 \qquad 0 \leq x \leq 1000$$

Suppose that the cost C of producing x units is

$$C = \frac{\sqrt{x}}{10} + 400$$

Assuming that all items produced are sold, find the cost C as a function of the price p.

81. If f and g are odd functions, show that the composite function $f \circ g$ is also odd.

82. If f is an odd function and g is an even function, show that the composite functions $f \circ g$ and $g \circ f$ are also even.

MISSION POSSIBLE

Optimal Hiring Practices at Platinum Chip, Incorporated

Your team works in the marketing department of Platinum Chip, Incorporated. You've been asked to perform some research concerning the firm's labor options. Upon consulting with the company's engineers, it has been determined that the number Q of units produced is given as a function of labor, L, in hours, by $Q = 8L^{\frac{1}{2}}$. For example, if an employee works 9 hours, then there will be $Q = 8(9)^{\frac{1}{2}} = 24$ units produced. In addition, the firm's engineers provide you with the following cost function that relates the cost per month, C, to Q, the units produced by the firm per month:

$$C(Q) = 0.002Q^3 + 0.5Q^2 - Q + 1000.$$

1. Management wishes to know cost as a function of labor.
2. A microeconomist hired by Platinum found that demand is given by $Q(P) = 5000 - 20P$, where P is the price per unit. Find revenue R as a function of units produced, Q. (Hint: Revenue equals the number of units produced times the price per unit.)
3. The Chief Financial Officer (CFO) has asked your team to find revenue as a function of labor.
4. The CFO has also asked you to determine profit as a function of labor.
5. Your team decides to provide a graphical solution to management. Graph profit as a function of labor. What is the implied domain of this function?
6. Human resources wishes to know the optimal level of labor for the company.
7. Investor relations is about to hold a conference call with institutional investors and needs to know the monthly maximum profit of the firm.
8. The production department needs to know the optimal level of units produced each month.
9. Determine the optimal price to charge for each part.
10. The production function $Q = 8L^{\frac{1}{2}}$ is a variation of the Cobb-Douglas production function, $Q = K^\alpha L^\beta$ where K is capital (machinery, factory, etc.) and L is labor. In the short-run, capital is assumed constant. In the long-run, all inputs are variable. If $\alpha + \beta = 1$, the firm experiences constant returns to scale which means if all inputs increase $x\%$, then output will also increase $x\%$. What must α be in order for Platinum Chip, Inc. to experience constant returns to scale?
11. Graph the production function $Q = 8L^{\frac{1}{2}}$. Explain why the shape of the curve reasonably portrays what actually occurs in the production process as labor increases while all other inputs are fixed.

4.5 | MATHEMATICAL MODELS: CONSTRUCTING FUNCTIONS

1 Construct and Analyze Functions

1 Real-world problems often result in mathematical models that involve functions. These functions need to be constructed or built based on the information given. In constructing functions, we must be able to translate the verbal description into the language of mathematics. We do this by assigning symbols to represent the independent and dependent variables and then finding the function or rule that relates these variables.

E X A M P L E   1   Area of a Rectangle with Fixed Perimeter

The perimeter of a rectangle is 50 feet. Express its area A as a function of the length x of a side.

FIGURE 46

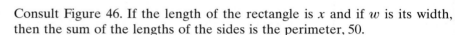

Solution   Consult Figure 46. If the length of the rectangle is x and if w is its width, then the sum of the lengths of the sides is the perimeter, 50.

$$x + w + x + w = 50$$
$$2x + 2w = 50$$
$$x + w = 25$$
$$w = 25 - x$$

The area A is length times width, so

$$A = xw = x(25 - x)$$

The area A as a function of x is

$$A(x) = x(25 - x)$$

Note that we used the symbol A as the dependent variable and also as the name of the function that relates the length x to the area A. As we mentioned earlier, this double usage is common in applications and should cause no difficulties.

E X A M P L E   2   Economics: Demand Equations

In economics, revenue R is defined as the amount of money derived from the sale of a product and is equal to the unit selling price p of the product times the number x of units actually sold. That is,

$$R = xp$$

In economics, the Law of Demand states that p and x are related: As one increases, the other decreases. Suppose that p and x are related by the following **demand equation:**

$$p = -\tfrac{1}{10}x + 20 \qquad 0 \le x \le 200$$

Express the revenue R as a function of the number x of units sold.

Solution   Since $R = xp$ and $p = -\tfrac{1}{10}x + 20$, it follows that

$$R(x) = xp = x(-\tfrac{1}{10}x + 20) = -\tfrac{1}{10}x^2 + 20x$$

 Now work Problem 3.

E X A M P L E   3

Finding the Distance from the Origin to a Point on a Graph

Let $P = (x, y)$ be a point on the graph of $y = x^2 - 1$.

(a) Express the distance d from P to the origin O as a function of x.
(b) What is d if $x = 0$?
(c) What is d if $x = 1$?
(d) What is d if $x = \sqrt{2}/2$?

(e) Use a graphing utility to graph the function $d = d(x)$, $x \geq 0$. Correct to two decimal places, find the value of x at which d has a local minimum. [This gives the point(s) on the graph of $y = x^2 - 1$ closest to the origin.]

FIGURE 47

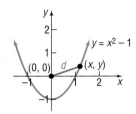

Solution (a) Figure 47 illustrates the graph. The distance d from P to O is

$$d = \sqrt{x^2 + y^2}$$

Since P is a point on the graph of $y = x^2 - 1$, we have

$$d(x) = \sqrt{x^2 + (x^2 - 1)^2} = \sqrt{x^4 - x^2 + 1}$$

Thus, we have expressed the distance d as a function of x.

(b) If $x = 0$, the distance d is

$$d(0) = \sqrt{1} = 1$$

(c) If $x = 1$, the distance d is

FIGURE 48

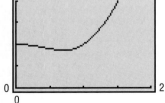

$$d(1) = \sqrt{1 - 1 + 1} = 1$$

(d) If $x = \sqrt{2}/2$, the distance d is

$$d\left(\frac{\sqrt{2}}{2}\right) = \sqrt{\left(\frac{\sqrt{2}}{2}\right)^4 - \left(\frac{\sqrt{2}}{2}\right)^2 + 1} = \sqrt{\frac{1}{4} - \frac{1}{2} + 1} = \frac{\sqrt{3}}{2}$$

(e) Figure 48 shows the graph of $Y_1 = \sqrt{x^4 - x^2 + 1}$. Using the MINIMUM feature on a graphing utility, we find that, when $x \approx 0.70$ the value of d is smallest ($d \approx 0.86$) correct to two decimal places. ▬

Now work Problem 13.

E X A M P L E   4

Filling a Swimming Pool

A rectangular swimming pool 20 meters long and 10 meters wide is 4 meters deep at one end and 1 meter deep at the other. Figure 49 illustrates a cross-sectional view of the pool. Water is being pumped into the pool at the deep end.

FIGURE 49

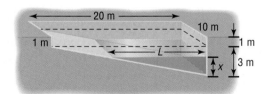

(a) Find a function that expresses the volume V of water in the pool as a function of the height x of the water at the deep end.

(b) Find the volume when the height is 1 meter.

(c) Find the volume when the height is 2 meters.

 (d) Use a graphing utility to graph the function $V = V(x)$. At what height is the volume 20 cubic meters? 100 cubic meters?

Solution (a) Let L denote the distance (in meters) measured at water level from the deep end to the short end. Notice that L and x form the sides of a triangle that is similar to the triangle whose sides are 20 meters by 3 meters. Thus, L and x are related by the equation

$$\frac{L}{x} = \frac{20}{3} \quad \text{or} \quad L = \frac{20x}{3} \quad 0 \le x \le 3$$

The volume V of water in the pool at any time is

$$V = \binom{\text{cross-sectional}}{\text{triangular area}}(\text{width}) = (\tfrac{1}{2}Lx)(10) \quad \text{cubic meters}$$

Since $L = 20x/3$, we have

$$V(x) = \left(\frac{1}{2} \cdot \frac{20x}{3} \cdot x\right)(10) = \frac{100}{3}x^2 \quad \text{cubic meters}$$

(b) When the height x of the water is 1 meter, the volume $V = V(x)$ is

$$V(1) = \frac{100}{3} \cdot 1^2 = 33.3 \text{ cubic meters}$$

FIGURE 50

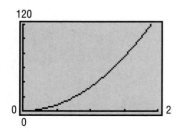

(c) When the height x of the water is 2 meters, the volume $V = V(x)$ is

$$V(2) = \frac{100}{3} \cdot 2^2 = \frac{400}{3} = 133.3 \text{ cubic meters}$$

(d) See Figure 50.

When $x \approx 0.77$ meter, the volume is 20 cubic meters. When $x \approx 1.73$ meters, the volume is 100 cubic meters. ∎

E X A M P L E 5 Area of an Isosceles Triangle

Consider an isosceles triangle of fixed perimeter p.

(a) If x equals the length of one of the two equal sides, express the area A as a function of x.

(b) What is the domain of A?

(c) Use a graphing utility to graph $A = A(x)$ for $p = 4$.

(d) For what value of x is the area largest?

FIGURE 51

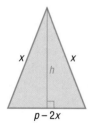

Solution (a) Look at Figure 51. Since the equal sides are of length x, the third side must be of length $p - 2x$. (Do you see why?) We know that the area A is

$$A = \tfrac{1}{2}(\text{base})(\text{height})$$

To find the height h, we drop the perpendicular to the base of length $p - 2x$ and use the fact that the perpendicular bisects the base. Then, by the Pythagorean Theorem, we have

$$h^2 = x^2 - \left(\frac{p - 2x}{2}\right)^2 = x^2 - \frac{1}{4}(p^2 - 4px + 4x^2)$$

$$= px - \frac{1}{4}p^2 = \frac{4px - p^2}{4}$$

$$h = \sqrt{\frac{4px - p^2}{4}} = \frac{\sqrt{p}}{2}\sqrt{4x - p}$$

The area A is given by

$$A = \frac{1}{2} \cdot (p - 2x)\frac{\sqrt{p}}{2}\sqrt{4x - p} = \frac{\sqrt{p}}{4}(p - 2x)\sqrt{4x - p}$$

(b) The domain of A is found as follows. Because of the expression $\sqrt{4x - p}$, we require that

$$4x - p > 0$$

$$x > \frac{p}{4}$$

Since $p - 2x$ is a side of the triangle, we also require that

$$p - 2x > 0$$

$$-2x > -p$$

$$x < \frac{p}{2}$$

FIGURE 52

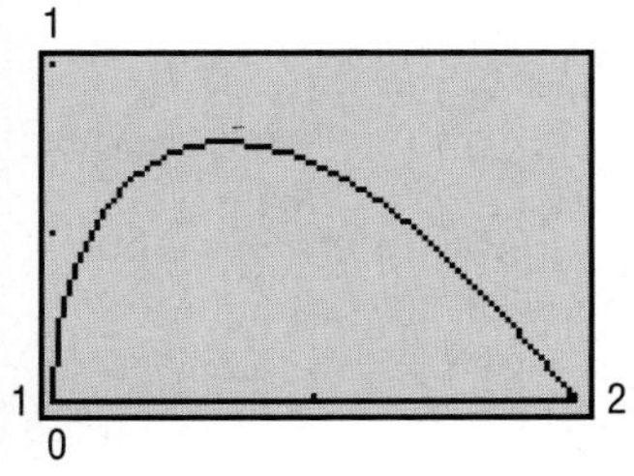

Thus, the domain of A is $p/4 < x < p/2$, or $(p/4, p/2)$, and we state the function A as

$$A(x) = \frac{\sqrt{p}}{4}(p - 2x)\sqrt{4x - p} \qquad \frac{p}{4} < x < \frac{p}{2}$$

(c) See Figure 52 for the graph of $A(x) = 2(2 - x)\sqrt{x - 1}$.

(d) The area is largest (approximately 0.77) when $x \approx 1.33$.

Now work Problem 7.

E X A M P L E   6   Minimum Payments and Interest Charged for Credit Cards

(a) Holders of credit cards issued by banks, department stores, oil companies, and so on, receive bills each month that state minimum amounts that must be paid by a certain due date. The minimum due depends on the total amount owed. One such credit card company uses the following rules: For a bill of less than $10, the entire amount is due. For a bill of at least $10 but less than $500, the minimum due is $10. There is a minimum of $30 due on a bill of at least $500 but less than $1000, a minimum of $50 due on a bill of at least $1000 but less than $1500, and a minimum of $70 due on bills of $1500 or more. Find the function f that describes the minimum payment due on a bill of x dollars. Graph f.

(b) The card holder may pay any amount between the minimum due and the total owed. The organization issuing the card charges the card holder interest of 1.5% per month for the first $1000 owed and 1% per month on any unpaid balance over $1000. Find the function g that gives the amount of interest charged per month on a balance of x dollars. Graph g.

Solution   (a) The function f that describes the minimum payment due on a bill of x dollars is

$$f(x) = \begin{cases} x & \text{if} & 0 \le x < 10 \\ 10 & \text{if} & 10 \le x < 500 \\ 30 & \text{if} & 500 \le x < 1000 \\ 50 & \text{if} & 1000 \le x < 1500 \\ 70 & \text{if} & 1500 \le x \end{cases}$$

To graph this function f, we proceed as follows. For $0 \le x < 10$, draw the graph of $y = x$; for $10 \le x < 500$, draw the graph of the constant function $y = 10$; for $500 \le x < 1000$, draw the graph of the constant function $y = 30$; and so on. The graph of f is given in Figure 53.

FIGURE 53

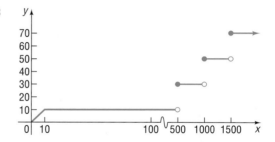

(b) If $g(x)$ is the amount of interest charged per month on a balance of x, then $g(x) = 0.015x$ for $0 \le x \le 1000$. The amount of the unpaid balance above $1000 is $x - 1000$. If the balance due is $x > 1000$, then the interest is $0.015(1000) + 0.01(x - 1000) = 15 + 0.01x - 10 = 5 + 0.01x$, so

$$g(x) = \begin{cases} 0.015x & \text{if } 0 \le x \le 1000 \\ 5 + 0.01x & \text{if } x > 1000 \end{cases}$$

See Figure 54.   ∎

FIGURE 54

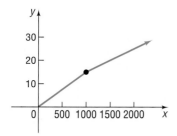

E X A M P L E   7   Making a Playpen

FIGURE 55

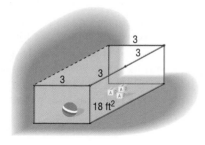

A manufacturer of children's playpens makes a square model that can be opened at one corner and attached at right angles to a wall or, perhaps, the side of a house. If each side is 3 feet in length, the open configuration doubles the available area in which the child can play from 9 square feet to 18 square feet. See Figure 55.

Now, suppose we place hinges at the outer corners to allow for a configuration like the one shown in Figure 56.

(a) Express the area A of this configuration as a function of the distance x between the two parallel sides.

(b) Find the domain of A.

(c) Find A if $x = 5$.

 (d) Graph $A = A(x)$. For what value of x is the area largest? What is the maximum area?*

FIGURE 56

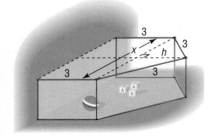

Solution   (a) Refer to Figure 56. The area A that we seek consists of the area of a rectangle (with width 3 and length x) and the area of an isosceles triangle (with base x and two equal sides of length 3). The height h of the triangle may be found using the Pythagorean Theorem.

$$h^2 = 3^2 - \left(\frac{x}{2}\right)^2 = 9 - \frac{x^2}{4} = \frac{36 - x^2}{4}$$

$$h = \tfrac{1}{2}\sqrt{36 - x^2}$$

The area A enclosed by the playpen is

$$A = \text{area of rectangle} + \text{area of triangle} = 3x + \tfrac{1}{2}x\left(\tfrac{1}{2}\sqrt{36 - x^2}\right)$$

$$A(x) = 3x + \frac{x\sqrt{36 - x^2}}{4}$$

Thus, we have expressed the area A as a function of x.

(b) To find the domain of A, we note first that $x > 0$, since x is a length. Also, the expression under the radical must be positive, so

$$36 - x^2 > 0$$
$$x^2 < 36$$
$$-6 < x < 6$$

Combining these restrictions, we find that the domain of A is $0 < x < 6$, or $(0, 6)$.

FIGURE 57

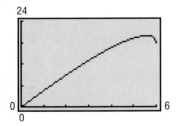

(c) If $x = 5$, the area is

$$A(5) = 3(5) + \frac{5}{4}\sqrt{36 - (5)^2} \approx 19.15 \text{ square feet}$$

Thus, if the width of the playpen is 5 feet, its area is 19.15 square feet.

(d) See Figure 57. The maximum area is about 19.81 square feet, obtained when x is about 5.58 feet.   ▬

*Adapted from *Proceedings, Summer Conference for College Teachers on Applied Mathematics* (University of Missouri, Rolla), 1971.

4.5 | EXERCISES

1. **Volume of a Cylinder**   The volume V of a right circular cylinder of height h and radius r is $V = \pi r^2 h$. If the height is twice the radius, express the volume V as a function of r.

2. **Volume of a Cone**   The volume V of a right circular cone is $V = \frac{1}{3}\pi r^2 h$. If the height is twice the radius, express the volume V as a function of r.

3. **Demand Equation**   The price p and the quantity x sold of a certain product obey the demand equation

$$p = -\tfrac{1}{6}x + 100 \qquad 0 \le x \le 600$$

   (a) Express the revenue R as a function of x. (Remember, $R = xp$.)
   (b) What is the revenue if 200 units are sold?
   (c) Graph the revenue function using a graphing utility.
   (d) What quantity x maximizes revenue? What is the maximum revenue?
   (e) What price should the company charge to maximize revenue?

4. **Demand Equation**   The price p and the quantity x sold of a certain product obey the demand equation

$$p = -\tfrac{1}{3}x + 100 \qquad 0 \le x \le 300$$

   (a) Express the revenue R as a function of x.
   (b) What is the revenue if 100 units are sold?
   (c) Graph the revenue function using a graphing utility.
   (d) What quantity x maximizes revenue? What is the maximum revenue?
   (e) What price should the company charge to maximize revenue?

5. **Demand Equation**   The price p and the quantity x sold of a certain product obey the demand equation

$$x = -5p + 100 \qquad 0 \le p \le 20$$

   (a) Express the revenue R as a function of x.
   (b) What is the revenue if 15 units are sold?
   (c) Graph the revenue function using a graphing utility.
   (d) What quantity x maximizes revenue? What is the maximum revenue?

   (e) What price should the company charge to maximize revenue?

6. **Demand Equation**   The price p and the quantity x sold of a certain product obey the demand equation

$$x = -20p + 500 \qquad 0 \le p \le 25$$

   (a) Express the revenue R as a function of x.
   (b) What is the revenue if 20 units are sold?
   (c) Graph the revenue function using a graphing utility.
   (d) What quantity x maximizes revenue? What is the maximum revenue?
   (e) What price should the company charge to maximize revenue?

7. **Enclosing a Rectangular Field**   David has available 400 yards of fencing and wishes to enclose a rectangular area.
   (a) Express the area A of the rectangle as a function of the width x of the rectangle.
   (b) What is the domain of A?
   (c) Graph $A = A(x)$ using a graphing utility. For what value of x is the area largest?

8. **Enclosing a Rectangular Field along a River**   Beth has 3000 feet of fencing available to enclose a rectangular field. One side of the field lies along a river, so only three sides require fencing.
   (a) Express the area A of the rectangle as a function of x, where x is the length of the side parallel to the river.
   (b) Graph $A = A(x)$ using a graphing utility. For what value of x is the area largest?

9. A wire of length x is bent into the shape of a circle.
   (a) Express the circumference of the circle as a function of x.
   (b) Express the area of the circle as a function of x.

10. A wire of length x is bent into the shape of a square.
   (a) Express the perimeter of the square as a function of x.
   (b) Express the area of the square as a function of x.

11. A right triangle has one vertex on the graph of $y = x^3, x > 0$, at (x, y), another at the origin, and the third on the positive y-axis at $(0, y)$, as shown

in the figure. Express the area A of the triangle as a function of x.

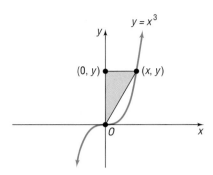

12. A right triangle has one vertex on the graph of $y = 9 - x^2$, $x > 0$, at (x, y), another at the origin, and the third on the positive x-axis at $(x, 0)$. Express the area A of the triangle as a function of x.

13. Let $P = (x, y)$ be a point on the graph of $y = x^2 - 8$.
    (a) Express the distance d from P to the origin as a function of x.
    (b) What is d if $x = 0$?
    (c) What is d if $x = 1$?
    (d) Use a graphing utility to graph $d = d(x)$.
    (e) For what values of x is d smallest?

14. Let $P = (x, y)$ be a point on the graph of $y = x^2 - 8$.
    (a) Express the distance d from P to the point $(0, -1)$ as a function of x.
    (b) What is d if $x = 0$?
    (c) What is d if $x = -1$?
    (d) Use a graphing utility to graph $d = d(x)$.
    (e) For what values of x is d smallest?

15. Let $P = (x, y)$ be a point on the graph of $y = \sqrt{x}$.
    (a) Express the distance d from P to the point $(1, 0)$ as a function of x.
    (b) Use a graphing utility to graph $d = d(x)$.
    (c) For what values of x is d smallest?

16. Let $P = (x, y)$ be a point on the graph of $y = 1/x$.
    (a) Express the distance d from P to the origin as a function of x.
    (b) Use a graphing utility to graph $d = d(x)$.
    (c) For what values of x is d smallest?

17. Two cars leave an intersection at the same time. One is headed south at a constant speed of 30 miles per hour; the other is headed west at a constant speed of 40 miles per hour (see the figure). Express the distance d between the cars as a function of the time t.

[**Hint:** At $t = 0$, the cars leave the intersection.]

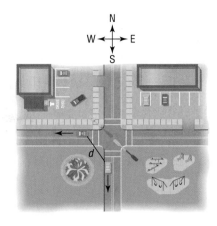

18. Two cars are approaching an intersection. One is 2 miles south of the intersection and is moving at a constant speed of 30 miles per hour. At the same time, the other car is 3 miles east of the intersection and is moving at a constant speed of 40 miles per hour.
    (a) Express the distance d between the cars as a function of time t.

[**Hint:** At $t = 0$, the cars are 2 miles south and 3 miles east of the intersection, respectively.]

    (b) Use a graphing utility to graph $d = d(t)$. For what value of t is d smallest?

19. **Constructing an Open Box** An open box with a square base is to be made from a square piece of cardboard 24 inches on a side by cutting out a square from each corner and turning up the sides (see the figure).
    (a) Express the volume V of the box as a function of the length x of the side of the square cut from each corner.
    (b) What is the volume if a 3-inch square is cut out?
    (c) What is the volume if a 10-inch square is cut out?
    (d) Graph $V = V(x)$. For what value of x is V largest?

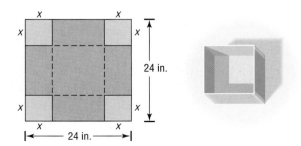

20. Constructing an Open Box An open box with a square base is required to have a volume of 10 cubic feet.
(a) Express the amount A of material used to make such a box as a function of the length x of a side of the square base.
(b) How much material is required for a base 1 foot by 1 foot?
(c) How much material is required for a base 2 feet by 2 feet?
 (d) Graph $A = A(x)$. For what value of x is A smallest?

21. Constructing a Closed Box A closed box with a square base is required to have a volume of 10 cubic feet.
(a) Express the amount A of material used to make such a box as a function of the length x of a side of the square base.
(b) How much material is required for a base 1 foot by 1 foot?
(c) How much material is required for a base 2 feet by 2 feet?
(d) Graph $A = A(x)$. For what value of x is A smallest?

22. Spheres The volume V of a sphere of radius r is $V = \frac{4}{3}\pi r^3$; the surface area S of this sphere is $S = 4\pi r^2$. Express the volume V as a function of the surface area S. If the surface area doubles, how does the volume change?

23. A rectangle has one corner on the graph of $y = 16 - x^2$, another at the origin, a third on the positive y-axis, and the fourth on the positive x-axis (see the figure).

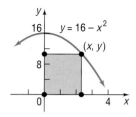

(a) Express the area A of the rectangle as a function of x.
(b) What is the domain of A?
(c) Graph $A = A(x)$. For what value of x is A largest?

24. A rectangle is inscribed in a semicircle of radius 2 (see the figure). Let $P = (x, y)$ be the

point in quadrant I that is a vertex of the rectangle and is on the circle.

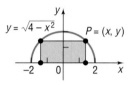

(a) Express the area A of the rectangle as a function of x.
(b) Express the perimeter p of the rectangle as a function of x.
(c) Graph $A = A(x)$. For what value of x is A largest?
(d) Graph $p = p(x)$. For what value of x is p largest?

25. A rectangle is inscribed in a circle of radius 2 (see the figure). Let $P = (x, y)$ be the point in quadrant I that is a vertex of the rectangle and is on the circle.

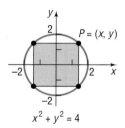

(a) Express the area A of the rectangle as a function of x.
(b) Express the perimeter p of the rectangle as a function of x.
(c) Graph $A = A(x)$. For what value of x is A largest?
(d) Graph $p = p(x)$. For what value of x is p largest?

26. A circle of radius r is inscribed in a square (see the figure).

(a) Express the area A of the square as a function of the radius r of the circle.
(b) Express the perimeter p of the square as a function of r.

27. A wire 10 meters long is to be cut into two pieces. One piece will be shaped as a square, and the other piece will be shaped as a circle (see the figure).

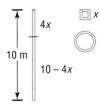

(a) Express the total area A enclosed by the pieces of wire as a function of the length x of a side of the square.
(b) What is the domain of A?
 (c) Graph $A = A(x)$. For what value of x is A smallest?

28. A wire 10 meters long is to be cut into two pieces. One piece will be shaped as an equilateral triangle, and the other piece will be shaped as a circle.
(a) Express the total area A enclosed by the pieces of wire as a function of the length x of a side of the equilateral triangle.
(b) What is the domain of A?
(c) Graph $A = A(x)$. For what value of x is A smallest?

29. A semicircle of radius r is inscribed in a rectangle so that the diameter of the semicircle is the length of the rectangle (see the figure).

(a) Express the area A of the rectangle as a function of the radius r of the semicircle.
(b) Express the perimeter p of the rectangle as a function of r.

30. An equilateral triangle is inscribed in a circle of radius r. See the figure. Express the circumference C of the circle as a function of the length x of a side of the triangle.

[**Hint:** First show that $r^2 = x^2/3$.]

31. An equilateral triangle is inscribed in a circle of radius r. See the figure. Express the area A

within the circle, but outside the triangle, as a function of the length x of a side of the triangle.

32. **Cost of Transporting Goods** A trucking company transports goods between Chicago and New York, a distance of 960 miles. The company's policy is to charge, for each pound, $0.50 per mile for the first 100 miles, $0.40 per mile for the next 300 miles, $0.25 per mile for the next 400 miles, and no charge for the remaining 160 miles.
(a) Graph the relationship between the cost of transportation in dollars and mileage over the entire 960-mile route.
(b) Find the cost as a function of mileage for hauls between 100 and 400 miles from Chicago.
(c) Find the cost as a function of mileage for hauls between 400 and 800 miles from Chicago.

33. **Car Rental Costs** An economy car rented in Florida from National Car Rental® on a weekly basis costs $95 per week.* Extra days cost $24 per day until the day rate exceeds the weekly rate, in which case the weekly rate applies. Find the cost C of renting an economy car as a piecewise-defined function of the number x of days used, where $7 \le x \le 14$. Graph this function.

[**Note:** Any part of a day counts as a full day.]

34. Rework Problem 33 for a luxury car, which costs $219 on a weekly basis with extra days at $45 per day.

35. Water is poured into a container in the shape of a right circular cone with radius 4 feet and height 16 feet (see the figure). Express the volume V of the water in the cone as a function of the height h of the water.

[**Hint:** The volume V of a cone of radius r and height h is $V = \frac{1}{3}\pi r^2 h$.]

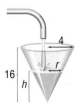

*Source: National Car Rental®, 1997.

36. Federal Income Tax Two 1997 Tax Rate Schedules are given in the accompanying table. If x equals the amount on Form 1040, line 37, and y equals the tax due, construct a function f for each schedule.

1997 TAX RATE SCHEDULES

SCHEDULE X—IF YOUR FILING STATUS IS SINGLE					SCHEDULE Y-1—USE IF YOUR FILING STATUS IS MARRIED FILING JOINTLY OR QUALIFYING WIDOW(ER)				
If the amount on Form 1040, line 37, is: Over—	But not over—	Enter on Form 1040, line 38		of the amount over—	If the amount on Form 1040, line 37, is: Over—	But not over—	Enter on Form 1040, line 38		of the amount over—
$ 0	$ 24,650	$ 0 +	15%	$ 0	$ 0	$ 41,200	$ 0 +	15%	$ 0
24,650	59,750	3,698 +	28	24,650	41,200	99,600	6,180 +	28	41,200
59,750	124,650	13,526 +	31	59,750	99,600	151,750	22,532 +	31	99,600
124,650	271,050	33,645 +	36	124,650	151,750	271,050	38,699 +	36	151,750
271,050		86,349 +	39.6	271,050	271,050		81,647 +	39.6	271,050

37. Inscribing a Cylinder in a Sphere Inscribe a right circular cylinder of height h and radius r in a sphere of fixed radius R. See the illustration. Express the volume V of the cylinder as a function of h.

[**Hint:** $V = \pi r^2 h$. Note also the right triangle.]

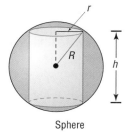

Sphere

38. Inscribing a Cylinder in a Cone Inscribe a right circular cylinder of height h and radius r in a cone of fixed radius R and fixed height H. See the illustration. Express the volume V of the cylinder as a function of r.

[**Hint:** $V = \pi r^2 h$. Note also the similar triangles.]

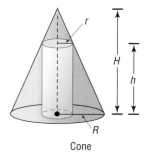

Cone

CHAPTER REVIEW

THINGS TO KNOW

Function

A relation between two sets of real numbers so that each number x in the first set, the domain, has corresponding to it exactly one number y in the second set. The range is the set of y values of the function for the x values in the domain.

x is the independent variable; y is the dependent variable.

A function f may be defined implicitly by an equation involving x and y or explicitly by writing $y = f(x)$.

A function can also be characterized as a set of ordered pairs (x, y) or $(x, f(x))$ in which no two distinct pairs have the same first element.

Function notation

$y = f(x)$

f is a symbol for the function.

x is the argument, or independent variable.

y is the dependent variable.

$f(x)$ is the value of the function at x, or the image of x.

Domain
If unspecified, the domain of a function f is the largest set of real numbers for which $f(x)$ is a real number.

Vertical line test
A set of points in the plane is the graph of a function if and only if every vertical line intersects the graph in at most one point.

Even function f
$f(-x) = f(x)$ for every x in the domain ($-x$ must also be in the domain).

Odd function f
$f(-x) = -f(x)$ for every x in the domain ($-x$ must also be in the domain).

LIBRARY OF FUNCTIONS

Linear function
$f(x) = mx + b$ Graph is a straight line with slope m and y-intercept b.

Constant function
$f(x) = b$ Graph is a horizontal line with y-intercept b (see Figure 16).

Identity function
$f(x) = x$ Graph is a straight line with slope 1 and y-intercept 0 (see Figure 17).

Square function
$f(x) = x^2$ Graph is a parabola with intercept at $(0, 0)$ (see Figure 18).

Cube function
$f(x) = x^3$ See Figure 19.

Square root function
$f(x) = \sqrt{x}$ See Figure 20.

Reciprocal function
$f(x) = 1/x$ See Figure 21.

Absolute value function
$f(x) = |x|$ See Figure 22.

HOW TO

Determine whether a relation represents a function

Find the domain and range of a function from its graph

Find the domain of a function given its equation

Determine algebraically whether a function is even or odd

Graph certain functions by shifting, compressing, stretching, and/or reflecting (see Table 12)

Given the graph of a function, determine where it is increasing or decreasing

Find the composite of two functions and its domain

Construct functions in applications, including piecewise-defined functions

FILL-IN-THE-BLANK ITEMS

1. If f is a function defined by the equation $y = f(x)$, then x is called the _____ variable and y is the _____ variable.

2. A set of points in the xy-plane is the graph of a function if and only if no _____ line contains more than one point of the set.

3. A(n) _____ function f is one for which $f(-x) = f(x)$ for every x in the domain of f; a(n) _____ function f is one for which $f(-x) = -f(x)$ for every x in the domain of f.

4. Suppose that the graph of a function f is known. Then the graph of $y = f(x - 2)$ may be obtained by a(n) _____ shift of the graph of f to the _____ a distance of 2 units.

5. If $f(x) = x + 1$ and $g(x) = x^3$, then _____ $= (x + 1)^3$.

6. For two functions f and g, $(g \circ f)(x) =$ _____.

TRUE/FALSE ITEMS

T F   **1.**   Every relation is a function.

T F   **2.**   Vertical lines intersect the graph of a function in no more than one point.

T F   **3.**   The y-intercept of the graph of the function $y = f(x)$ whose domain is all real numbers is $f(0)$.

T F   **4.**   Increasing functions always cross the y-axis.

T F   **5.**   Even functions have graphs that are symmetric with respect to the origin.

T F   **6.**   The graph of $y = f(-x)$ is the reflection about the y-axis of the graph of $y = f(x)$.

T F   **7.** $f(g(x)) = f(x) \cdot g(x)$

T F   **8.**   The domain of the composite function $(f \circ g)(x)$ is the same as that of $g(x)$.

REVIEW EXERCISES

Blue problem numbers indicate the author's suggestions for use in a Practice Test.

1.   Given that f is a linear function, $f(4) = -5$, and $f(0) = 3$, write the equation that defines f.

2.   Given that g is a linear function with slope $= -4$ and $g(-2) = 2$, write the equation that defines g.

3.   A function f is defined by

$$f(x) = \frac{Ax + 5}{6x - 2}$$

If $f(1) = 4$, find A.

4.   A function g is defined by

$$g(x) = \frac{A}{x} + \frac{8}{x^2}$$

If $g(-1) = 0$, find A.

5.   Tell which of the following graphs are graphs of functions.

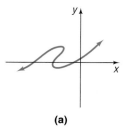

(a)

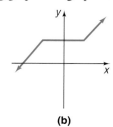

(b)

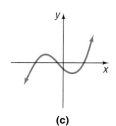

(c)

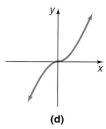

(d)

6.   Use the graph of the function f shown to find
(a) The domain and range of f
(b) The intervals on which f is increasing
(c) The intervals on which f is constant
(d) The intercepts of f

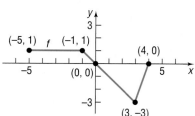

In Problems 7–12, find the following for each function:

(a) $f(-x)$ (b) $-f(x)$ (c) $f(x + 2)$ (d) $f(x - 2)$

7. $f(x) = \dfrac{3x}{x^2 - 4}$

8. $f(x) = \dfrac{x^2}{x + 2}$

9. $f(x) = \sqrt{x^2 - 4}$

10. $f(x) = |x^2 - 4|$

11. $f(x) = \dfrac{x^2 - 4}{x^2}$

12. $f(x) = \dfrac{x^3}{x^2 - 4}$

In Problems 13–18, determine (algebraically) whether the given function is even, odd, or neither.

13. $f(x) = x^3 - 4x$

14. $g(x) = \dfrac{4 + x^2}{1 + x^4}$

15. $h(x) = \dfrac{1}{x^4} + \dfrac{1}{x^2} + 1$

16. $F(x) = \sqrt{1 - x^3}$

17. $G(x) = 1 - x + x^3$ 18. $H(x) = 1 + x + x^2$

In Problems 19–30, find the domain of each function.

19. $f(x) = \dfrac{x}{x^2 - 9}$ **20.** $f(x) = \dfrac{3x^2}{x - 2}$ **21.** $f(x) = \sqrt{2 - x}$ **22.** $f(x) = \sqrt{x + 2}$

23. $h(x) = \dfrac{\sqrt{x}}{|x|}$ **24.** $g(x) = \dfrac{|x|}{x}$ **25.** $f(x) = \dfrac{x}{x^2 + 2x - 3}$ **26.** $F(x) = \dfrac{1}{x^2 - 3x - 4}$

27. $G(x) = \begin{cases} |x| & \text{if } -1 \le x \le 1 \\ 1/x & \text{if } x > 1 \end{cases}$ **28.** $H(x) = \begin{cases} 1/x & \text{if } 0 < x < 4 \\ x - 4 & \text{if } 4 \le x \le 8 \end{cases}$

29. $f(x) = \begin{cases} 1/(x - 2) & \text{if } x > 2 \\ 0 & \text{if } x = 2 \\ 3x & \text{if } 0 \le x \le 2 \end{cases}$ **30.** $g(x) = \begin{cases} |1 - x| & \text{if } x < 1 \\ 3 & \text{if } x = 1 \\ x + 1 & \text{if } 1 < x \le 3 \end{cases}$

In Problems 31–34, find the average rate of change between 0 and x for each function f. Be sure to simplify.

31. $f(x) = 2 - 5x$ **32.** $f(x) = 2x^2 + 7$ **33.** $f(x) = 3x - 4x^2$ **34.** $f(x) = x^2 - 3x + 2$

In Problems 35–54, graph each function using the techniques of shifting, compressing or stretching, and reflections. Identify any intercepts on the graph. State the domain and, based on the graph, find the range.

35. $F(x) = |x| - 4$ **36.** $f(x) = |x| + 4$ **37.** $g(x) = -|x|$ **38.** $g(x) = \frac{1}{2}|x|$

39. $h(x) = \sqrt{x - 1}$ **40.** $h(x) = \sqrt{x} - 1$ **41.** $f(x) = \sqrt{1 - x}$ **42.** $f(x) = -\sqrt{x}$

43. $F(x) = \begin{cases} x^2 + 4 & \text{if } x < 0 \\ 4 - x^2 & \text{if } x \ge 0 \end{cases}$ **44.** $H(x) = \begin{cases} |1 - x| & \text{if } 0 \le x \le 2 \\ |x - 1| & \text{if } x > 2 \end{cases}$ **45.** $h(x) = (x - 1)^2 + 2$

46. $h(x) = (x + 2)^2 - 3$ **47.** $g(x) = (x - 1)^3 + 1$ **48.** $g(x) = (x + 2)^3 - 8$

49. $f(x) = \begin{cases} 2\sqrt{x} & \text{if } x \ge 4 \\ x & \text{if } 0 < x < 4 \end{cases}$ **50.** $f(x) = \begin{cases} 3|x| & \text{if } x < 0 \\ \sqrt{1 - x} & \text{if } 0 \le x \le 1 \end{cases}$ **51.** $g(x) = \dfrac{1}{x - 1} + 1$

52. $g(x) = \dfrac{1}{x + 2} - 2$ **53.** $h(x) = \text{int}(-x)$ **54.** $h(x) = -\text{int}(x)$

In Problems 55–60, for the given functions f and g; find:
(a) $(f \circ g)(2)$ *(b)* $(g \circ f)(-2)$ *(c)* $(f \circ f)(4)$ *(d)* $(g \circ g)(-1)$

55. $f(x) = 3x - 5; \quad g(x) = 1 - 2x^2$ **56.** $f(x) = 4 - x; \quad g(x) = 1 + x^2$

57. $f(x) = \sqrt{x + 2}; \quad g(x) = 2x^2 + 1$ **58.** $f(x) = 1 - 3x^2; \quad g(x) = \sqrt{4 - x}$

59. $f(x) = \dfrac{1}{x^2 + 4}; \quad g(x) = 3x - 2$ **60.** $f(x) = \dfrac{2}{1 + 2x^2}; \quad g(x) = 3x$

In Problems 61–66, find f ∘ g, g ∘ f, f ∘ f, and g ∘ g for each pair of functions. State the domain of each.

61. $f(x) = 2 - x; \quad g(x) = 3x + 1$ **62.** $f(x) = 2x - 1; \quad g(x) = 2x + 1$

63. $f(x) = 3x^2 + x + 1; \quad g(x) = |3x|$ **64.** $f(x) = \sqrt{3x}; \quad g(x) = 1 + x + x^2$

65. $f(x) = \dfrac{x + 1}{x - 1}; \quad g(x) = \dfrac{1}{x}$ **66.** $f(x) = \sqrt{x - 3}; \quad g(x) = 3/x$

67. For the graph of the function f shown below:
    (a) Draw the graph of $y = f(-x)$.
    (b) Draw the graph of $y = -f(x)$.
    (c) Draw the graph of $y = f(x + 2)$.
    (d) Draw the graph of $y = f(x) + 2$.
    (e) Draw the graph of $y = 2f(x)$.
    (f) Draw the graph of $y = f(3x)$.

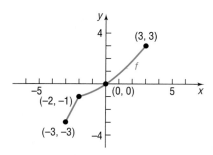

68. Repeat Problem 67 for the graph of the function g shown below.

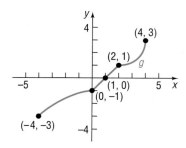

69. **Temperature Conversion** The temperature T of the air is approximately a linear function of the altitude h for altitudes within 10,000 meters of the surface of Earth. If the surface temperature is 30°C and the temperature at 10,000 meters is 5°C, find the function $T = T(h)$.

70. **Speed as a Function of Time** The speed v (in feet per second) of a car is a linear function of the time t (in seconds) for $10 \leq t \leq 30$. If after each second the speed of the car has increased by 5 feet per second and if after 20 seconds the speed is 80 feet per second, how fast is the car going after 30 seconds? Find the function $v = v(t)$.

71. **Strength of a Beam** The strength of a rectangular wooden beam is proportional to the product of the width and the cube of its depth (see the figure). If the beam is to be cut from a log in the shape of a cylinder of radius 3 feet, ex-

press the strength S of the beam as a function of the width x. What is the domain of S?

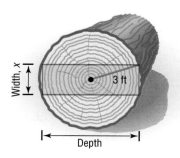

72. **Productivity versus Earnings** The following data represent the average hourly earnings and productivity (output per hour) of production workers for the years 1986–1995. Let productivity x be the independent variable and average hourly earnings y be the dependent variable.

Productivity	Average Hourly Earnings
94.2	8.76
94.1	8.98
94.6	9.28
95.3	9.66
96.1	10.01
96.7	10.32
100	10.57
100.2	10.83
100.7	11.12
100.8	11.44

Source: Bureau of Labor Statistics.

(a) Draw a scatter diagram of the data.
(b) Draw a line from the point (94.2, 8.76) to (96.7, 10.32) on the scatter diagram found in (a).
(c) Find the average rate of change of hourly earnings for productivity between 94.2 and 96.7.
(d) Interpret the average rate of change found in (c).
(e) Draw a line from the point (96.7, 10.32) to (100.8, 11.44) on the scatter diagram found in (a).
(f) Find the average rate of change of hourly earnings for productivity between 96.7 and 100.8.
(g) Interpret the average rate of change found in (f).

(h) What is happening to the average rate of change of hourly earnings as productivity increases?

73. Speed of a Parachutist The following data represent the distance that a parachutist has fallen over time.

Time (Seconds)	Distance (Meters)
0	0
5	112.5
10	490
15	1102.5
20	1960

(a) Draw a scatter diagram of the data.
(b) Draw a line from the point $(0,0)$ to $(5,112.5)$.
(c) Find the average rate of change of distance between 0 and 5 seconds.
(d) Interpret the average rate of change found in (c).
(e) Draw a line from the points $(15, 1102.5)$ to $(20, 1960)$.
(f) Find the average rate of change of distance between 15 and 20 seconds.
(g) Interpret the average rate of change found in (f).
(h) What is happening to the average rate of change of distance as time passes?

74. Material Needed to Make a Drum A steel drum in the shape of a right circular cylinder is required to have a volume of 100 cubic feet.

(a) Express the amount A of material required to make the drum as a function of the radius r to the cylinder.
(b) How much material is required if the drum is of radius 3 feet?
(c) Of radius 4 feet?
(d) Of radius 5 feet?
(e) Graph $A = A(r)$. For what value of r is A smallest?

75. Cost of a Drum A drum in the shape of a right circular cylinder is required to have a volume of 500 cubic centimeters. The top and bottom are made of material that costs 6¢ per square centimeter, while the sides are made of material that costs 4¢ per square centimeter.

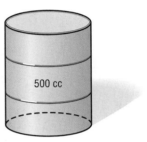

500 cc

(a) Express the total cost C of the material as a function of the radius r of the cylinder.
(b) What is the cost if the radius is 4 cm?
(c) What is the cost if the radius is 8 cm?
(d) Graph $C = C(r)$. For what value of r is the cost C least?

CHAPTER

5

Polynomial
and Rational Functions

Radio telescopes and satellite dishes have their foundations in the mathematical concepts of parabolas and the quadratic function. On the following page, you have been hired as a high school physics teacher and need to introduce your class to Galileo's *Analysis of Motion.* Use the Sullivan website at

www.prenhall.com/sullivan

to link to the Internet resources that you will need to prepare your lesson.

PREPARING FOR THIS CHAPTER

Before getting started on this chapter, review the following concepts:

Completing the Square *(p. 111)*

The Discriminant of a Quadratic Equation *(p. 114)*

Graphing Techniques: Transformations *(Section 4.3)*

Factoring Polynomials *(Section 1.5)*

Procedure for Finding Intercepts *(p. 178)*

Graphs of Certain Functions: Example 2 *(p. 175);* Example 3 *(p. 176);* Example 11 *(p. 182)*

Polynomial and Rational Inequalities *(Section 2.7)*

Dividing Polynomials *(p. 39)*

OUTLINE

THE PARABOLA

Quadratic functions, like all polynomials, are formed using only the basic rules of arithmetic: addition, subtraction, multiplication, and division. The graph of a quadratic function is called a **parabola.** The parabola has properties that make it useful for satellite dishes, car headlights, radio telescopes, and reflecting telescopes, including the *liquid mirror telescope.* Although the Greeks knew about the parabola as a conic section, it was *Galileo* who first observed that a quadratic function can be used to describe the motion of a falling body.

Suppose that you are a high school physics teacher and you want to put together an Internet lesson for your class. You would like to begin with a short lecture on Galileo's *Analysis of Motion.* His analysis was a major advance in science. The quadratic function was like his sword for fighting against the intellectual bondage of medieval times. You plan to duplicate *his experiments* with in-class, hands-on activity.

1. As you start to prepare your Internet lesson, you search the net for "parabolas." You are overwhelmed with the number of links. How might you handle this information overload?
2. How many different kinds of applications of the parabola did you find on the Web? Did you find any links that refer to suspension bridges?
3. For the math side, you want to make a list of important definitions and find some good illustrations of the essential features of parabolas. You want to try to find interactive sites. Did you find any?
4. Given the physical description of the *largest radio telescope on the planet* at Arecibo, Puerto Rico, approximate the dish with a quadratic function. Then approximate it with a circle. Graph both functions with your calculator. Can you determine the amount of difference between the two graphs? What advantages are there in making the radio telescope circular?
5. Suppose that you perform Galileo's *parabola experiment.* Model the data with a quadratic function. How well does the graph of your quadratic function fit the data? How well do your *data* fit the ideal law?

In Chapters 3 and 4, we graphed linear functions $f(x) = ax + b$, $a \neq 0$; the square function $f(x) = x^2$; and the cube function $f(x) = x^3$. Each of these functions belongs to the class of functions called *polynomial functions,* which we discuss further in this chapter. We will also discuss *rational functions,* which are ratios of polynomial functions. In this chapter, we place special emphasis on the graphs of polynomial and rational functions. This emphasis will demonstrate the importance of evaluating polynomials and solving polynomial equations (Section 5.5). Section 5.6 introduces complex numbers and the role that they play in solving quadratic equations with a negative discriminant. This paves the way for the Fundamental Theorem of Algebra (Section 5.7).

5.1 | QUADRATIC FUNCTIONS; CURVE FITTING

> **1** Graph a Quadratic Function Using Transformations
> **2** Identify the Vertex and Axis of Symmetry of a Quadratic Function
> **3** Graph a Quadratic Function Using Its Vertex, Axis, and Intercepts
> **4** Use the Maximum or Minimum Value of a Quadratic Function to Solve Applied Problems
> **5** Use a Graphing Utility to Find the Quadratic Function of Best Fit

---

A **quadratic function** is a function of the form

$$f(x) = ax^2 + bx + c \qquad (1)$$

where a, b, and c are real numbers and $a \neq 0$. The domain of a quadratic function consists of all real numbers.

Many applications require a knowledge of quadratic functions. For example, suppose that Texas Instruments collects the data shown in Table 1, which relates the number x of calculators sold at the price p per calculator. Since the price of a product determines the quantity that will be purchased, we treat price as the independent variable.

TABLE 1

Price per Calculator, p (Dollars)	Number of Calculators, x
60	11,100
65	10,115
70	9,652
75	8,731
80	8,087
85	7,205
90	6,439

A linear relationship between the number x of calculators and the price p per calculator may be given by the equation

$$x = 21,000 - 150p$$

Then the revenue R derived from selling x calculators at the price p per calculator is

$$\begin{aligned} R &= xp \\ &= (21,000 - 150p)p \\ &= -150p^2 + 21,000p \end{aligned}$$

Thus, revenue as a function of price is a quadratic function. Figure 1 illustrates the graph of this revenue function, whose domain is $0 \leq p \leq 140$, since both x and p must be nonnegative.

FIGURE 1
Graph of a revenue function:
$R = -150p^2 + 21,000p$

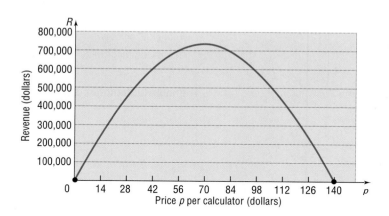

A second situation in which a quadratic function appears involves the motion of a projectile. Based on Newton's second law of motion (force equals mass times acceleration, $F = ma$), it can be shown that, ignoring air resistance, the path of a projectile propelled upward at an inclination to the horizontal is the graph of a quadratic function. See Figure 2 for an illustration.

FIGURE 2
Path of a cannonball

Graphing Quadratic Functions

We know how to graph quadratic functions of the form $f(x) = ax^2$, $a \neq 0$, based on prior discussions. Figure 3 shows the graph of three functions of the form $f(x) = ax^2$, $a > 0$, for $a = 1$, $a = \frac{1}{2}$, and $a = 3$. Notice that the larger the value of a, the "narrower" the graph is, and the smaller the value of a, the "wider" the graph is.

Figure 4 shows the graphs of $f(x) = ax^2$ for $a < 0$. Notice that these graphs are reflections about the x-axis of the graphs in Figure 3. Based on the results of these two figures, we can draw some general conclusions about the graph of $f(x) = ax^2$. First, as $|a|$ increases, the graph becomes "narrower," and as $|a|$ gets closer to zero, the graph gets "wider." Second, if a is positive, the graph opens "up," and if a is negative, the graph opens "down."

FIGURE 3

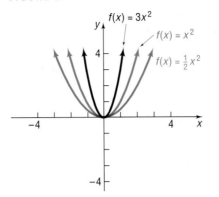

$f(x) = 3x^2$
$f(x) = x^2$
$f(x) = \frac{1}{2}x^2$

FIGURE 4

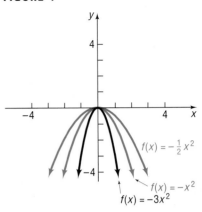

$f(x) = -\frac{1}{2}x^2$
$f(x) = -x^2$
$f(x) = -3x^2$

FIGURE 5
Graphs of a quadratic function,
$f(x) = ax^2 + bx + c, a \neq 0$

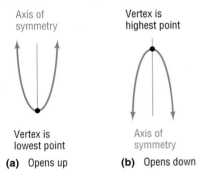

Axis of symmetry

Vertex is highest point

Vertex is lowest point

Axis of symmetry

(a) Opens up

(b) Opens down

The graphs in Figures 3 and 4 are typical of the graphs of all quadratic functions, which we call **parabolas.*** Refer to Figure 5, where two parabolas are pictured. The one on the left **opens up** and has a lowest point; the one on the right **opens down** and has a highest point. The lowest or highest point of a parabola is called the **vertex.** The vertical line passing through the vertex in each parabola in Figure 5 is called the **axis of symmetry** (sometimes abbreviated to **axis**) of the parabola. Because the parabola is symmetric about its axis, the axis of symmetry of a parabola can be used to advantage in graphing the parabola.

The parabolas shown in Figure 5 are the graphs of a quadratic function $f(x) = ax^2 + bx + c, a \neq 0$. Notice that the coordinate axes are not included in the figure. Depending on the values of a, b, and c, the axes could be placed anywhere. The important fact is that, except possibly for compression or stretching, the shape of the graph of a quadratic function will look like one of the parabolas in Figure 5.

1 In the following example, we use techniques from Section 4.3 to graph a quadratic function $f(x) = ax^2 + bx + c, a \neq 0$. In so doing, we shall complete the square and write the function f in the form $f(x) = a(x - h)^2 + k$.

E X A M P L E   1   Graphing a Quadratic Function Using Transformations

Graph the function: $f(x) = 2x^2 + 8x + 5$

Solution   We begin by completing the square on the right side:

$$f(x) = 2x^2 + 8x + 5$$
$$= 2(x^2 + 4x) + 5$$
$$= 2(x^2 + 4x + 4) + 5 - 8$$
$$= 2(x + 2)^2 - 3$$

Factor out the 2 from $2x^2 + 8x$.
Complete the square of $2(x^2 + 4x)$.
Notice that the factor of 2 requires
that 8 be added and subtracted.

(2)

*We shall study parabolas using a geometric definition later in this book.

We start with the graph of $y = 2x^2$, a stretched version of the graph of $y = x^2$. Then the graph of f can be obtained in three stages, as shown in Figure 6. Now compare this graph to the graph in Figure 5(a). The graph of $f(x) = 2x^2 + 8x + 5$ is a parabola that opens up and has its vertex (lowest point) at $(-2, -3)$. Its axis of symmetry is the line $x = -2$.

FIGURE 6

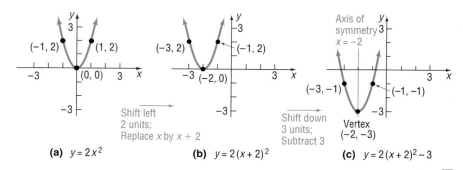

(a) $y = 2x^2$

(b) $y = 2(x + 2)^2$

(c) $y = 2(x + 2)^2 - 3$

Shift left 2 units; Replace x by $x + 2$

Shift down 3 units; Subtract 3

Check: Graph $f(x) = 2x^2 + 8x + 5$ and use the MINIMUM command to locate its vertex.

Now work Problem 17.

The method used in Example 1 can be used to graph any quadratic function $f(x) = ax^2 + bx + c, a \neq 0$, as follows:

$$f(x) = ax^2 + bx + c$$

$$= a\left(x^2 + \frac{b}{a}x\right) + c \qquad \text{Factor out } a \text{ from } ax^2 + bx.$$

$$= a\left(x^2 + \frac{b}{a}x + \frac{b^2}{4a^2}\right) + c - a\left(\frac{b^2}{4a^2}\right) \qquad \begin{array}{l}\text{Complete the square by adding and}\\ \text{subtracting } a(b^2/4a^2). \text{ Look closely at}\\ \text{this step!}\end{array}$$

$$= a\left(x + \frac{b}{2a}\right)^2 + c - \frac{b^2}{4a}$$

$$= a\left(x + \frac{b}{2a}\right)^2 + \frac{4ac - b^2}{4a}$$

Based on the above results, we conclude:

If $h = -b/2a$ and $k = (4ac - b^2)/4a$, then

$$f(x) = ax^2 + bx + c = a(x - h)^2 + k \qquad\qquad (3)$$

The graph of f is the parabola $y = ax^2$ shifted horizontally h units and vertically k units. As a result, the vertex is at (h, k), and the graph opens up if $a > 0$ and down if $a < 0$. The axis is the vertical line $x = h$.

For example, compare equation (3) with equation (2) of Example 1.

$$f(x) = 2(x + 2)^2 - 3$$
$$f(x) = a(x - h)^2 + k$$

We conclude that $a = 2$, so the graph opens up. Also, we find that $h = -2$ and $k = -3$, so its vertex is at $(-2, -3)$.

It is not required to complete the square to obtain the vertex. In almost every case, it is easier to obtain the vertex of a quadratic function f by remembering that its x-coordinate is $h = -b/2a$. The y-coordinate can then be found by evaluating f at $-b/2a$.

These results are summarized next:

Characteristics of the Graph of a Quadratic Function

$$f(x) = ax^2 + bx + c$$

$$\text{Vertex} = \left(\frac{-b}{2a}, f\left(\frac{-b}{2a}\right)\right) \qquad \text{Axis of Symmetry: the line } x = \frac{-b}{2a} \qquad (4)$$

Parabola opens up if $a > 0$. Parabola opens down if $a < 0$.

E X A M P L E 2

Locating the Vertex without Graphing

Without graphing, locate the vertex and axis of the parabola defined by $f(x) = -3x^2 + 6x + 1$. Does it open up or down?

Solution

For this quadratic function, $a = -3$, $b = 6$, and $c = 1$. The x-coordinate of the vertex is

$$h = \frac{-b}{2a} = \frac{-6}{-6} = 1$$

The y-coordinate of the vertex is therefore

$$k = f\left(\frac{-b}{2a}\right) = f(1) = -3 + 6 + 1 = 4$$

The vertex is located at the point $(1, 4)$. The axis of symmetry is the line $x = 1$. Finally, because $a = -3 < 0$, the parabola opens down.　■

The information we gathered in Example 2, together with the location of the intercepts, usually provides enough information to graph $f(x) = ax^2 + bx + c$, $a \neq 0$. The y-intercept is the value of f at $x = 0$, that is, $f(0) = c$. The x-intercepts, if there are any, are found by solving the equation

$$f(x) = ax^2 + bx + c = 0$$

This equation has two, one, or no real solutions, depending on whether the discriminant $b^2 - 4ac$ is positive, 0, or negative. Thus, it has corresponding x-intercepts, as follows:

The *x*-intercepts
of a Quadratic Function

1. If the discriminant $b^2 - 4ac > 0$, the graph of $f(x) = ax^2 + bx + c$ has two distinct *x*-intercepts and so will cross the *x*-axis in two places.
2. If the discriminant $b^2 - 4ac = 0$, the graph of $f(x) = ax^2 + bx + c$ has one *x*-intercept and touches the *x*-axis at its vertex.
3. If the discriminant $b^2 - 4ac < 0$, the graph of $f(x) = ax^2 + bx + c$ has no *x*-intercept and so will not cross or touch the *x*-axis.

Figure 7 illustrates these possibilities for parabolas that open up.

FIGURE 7
$f(x) = ax^2 + bx + c,\ a > 0$

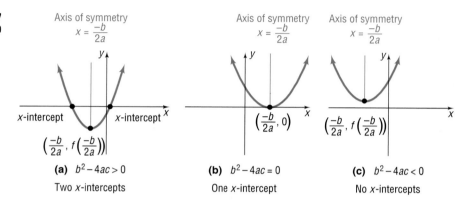

(a) $b^2 - 4ac > 0$
Two *x*-intercepts

(b) $b^2 - 4ac = 0$
One *x*-intercept

(c) $b^2 - 4ac < 0$
No *x*-intercepts

E X A M P L E   3

Graphing a Quadratic Function Using Its Vertex, Axis, and Intercepts

Use the information from Example 2 and the locations of the intercepts to graph $f(x) = -3x^2 + 6x + 1$.

Solution   In Example 2, we found the vertex to be at $(1, 4)$ and the axis of symmetry to be $x = 1$. The *y*-intercept is found by letting $x = 0$. Thus, the *y*-intercept is $f(0) = 1$. The *x*-intercepts are found by letting $f(x) = 0$. This results in the equation

$$-3x^2 + 6x + 1 = 0 \qquad a = -3, \quad b = 6, \quad c = 1$$

The discriminant $b^2 - 4ac = (6)^2 - 4(-3)(1) = 36 + 12 = 48 > 0$, so the equation has two real solutions and the graph has two *x*-intercepts. Using the quadratic formula, we find

$$x = \frac{-b + \sqrt{b^2 - 4ac}}{2a} = \frac{-6 + \sqrt{48}}{-6} = \frac{-6 + 4\sqrt{3}}{-6} \approx -0.15$$

FIGURE 8
$f(x) = -3x^2 + 6x + 1$

and

$$x = \frac{-b - \sqrt{b^2 - 4ac}}{2a} = \frac{-6 - \sqrt{48}}{-6} = \frac{-6 - 4\sqrt{3}}{-6} \approx 2.15$$

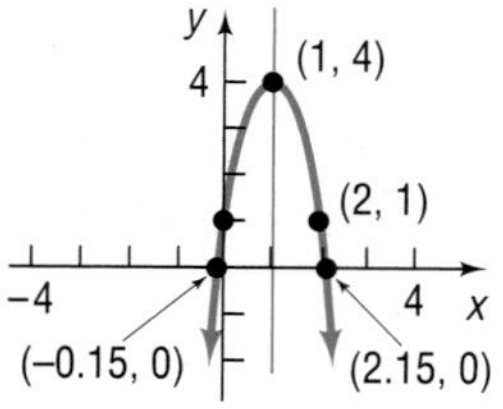

The *x*-intercepts are approximately -0.15 and 2.15.

The graph is given in Figure 8. Notice how we used the *y*-intercept and the axis of symmetry, $x = 1$, to obtain the additional point $(2, 1)$ on the graph. ∎

Check: Graph $f(x) = -3x^2 + 6x + 1$. Use ROOT or ZERO to locate the two *x*-intercepts and use MAXIMUM to locate the vertex. ∎

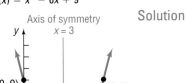

Now work Problem 25.

If the graph of a quadratic function has one x-intercept or none, it may be necessary to plot some additional points to obtain the graph.

E X A M P L E   4

Graphing a Quadratic Function Using Its Vertex, Axis, and Intercepts

Graph $f(x) = x^2 - 6x + 9$ by determining whether its graph opens up or down and by finding its vertex, axis of symmetry, y-intercept, and x-intercepts, if any.

FIGURE 9
$f(x) = x^2 - 6x + 9$

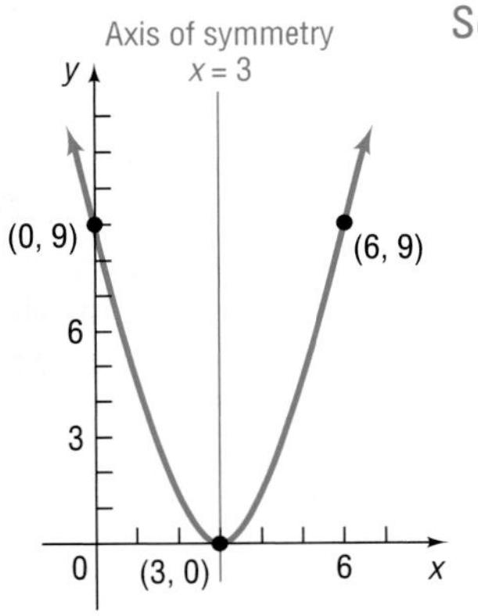

Solution   For $f(x) = x^2 - 6x + 9$, we have $a = 1$, $b = -6$, and $c = 9$. Since $a = 1 > 0$, the parabola opens up. The x-coordinate of the vertex is

$$h = \frac{-b}{2a} = \frac{-(-6)}{2 \cdot 1} = 3$$

The y-coordinate of the vertex is

$$k = f(3) = 9 - 6 \cdot 3 + 9 = 0$$

So the vertex is at $(3, 0)$. The axis of symmetry is the line $x = 3$. The y-intercept is $f(0) = 9$. Since the vertex $(3, 0)$ lies on the x-axis, the graph will touch the x-axis at the x-intercept. By using the axis of symmetry and the y-intercept at $(0, 9)$, we can locate the point $(6, 9)$ on the graph. See Figure 9.   ▬

E X A M P L E   5

Graphing a Quadratic Function Using Its Vertex, Axis, and Intercepts

Graph $f(x) = 2x^2 + x + 1$ by determining whether its graph opens up or down and by finding its vertex, axis of symmetry, y-intercept, and x-intercepts, if any.

Solution   For $f(x) = 2x^2 + x + 1$, we have $a = 2$, $b = 1$, and $c = 1$. Since $a = 2 > 0$, the parabola opens up. The x-coordinate of the vertex is

$$h = \frac{-b}{2a} = -\frac{1}{4}$$

FIGURE 10
$f(x) = 2x^2 + x + 1$

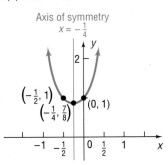

The y-coordinate of the vertex is

$$k = f\left(-\tfrac{1}{4}\right) = 2\left(\tfrac{1}{16}\right) + \left(-\tfrac{1}{4}\right) + 1 = \tfrac{7}{8}$$

So, the vertex is at $\left(-\tfrac{1}{4}, \tfrac{7}{8}\right)$. The axis of symmetry is the line $x = -\tfrac{1}{4}$. The y-intercept is $f(0) = 1$. The x-intercept(s), if any, obey the equation

$$2x^2 + x + 1 = 0$$

Since the discriminant $b^2 - 4ac = 1 - 8 = -7 < 0$, this equation has no real solution, and so the graph has no x-intercepts. We use the point $(0, 1)$ and the axis of symmetry $x = -\tfrac{1}{4}$ to locate the point $\left(-\tfrac{1}{2}, 1\right)$ on the graph. See Figure 10.   ▬

SUMMARY

We have two ways to graph a quadratic function:

1. Complete the square and apply shifting techniques (Example 1).
2. Use the results given in display (4) on page 311 to find the vertex and axis of symmetry and to determine whether the graph opens up or down. Then locate the y-intercept and the x-intercepts, if there are any (Examples 2 to 5).

 Now work Problem 31.

Applications

 We have already seen that the graph of a quadratic function $f(x) = ax^2 + bx + c$ is a parabola with vertex at $(-b/2a, f(-b/2a))$. This vertex is the highest point on the graph if $a < 0$ and the lowest point on the graph if $a > 0$. If the vertex is the highest point ($a < 0$), then $f(-b/2a)$ is the **maximum value** of f. If the vertex is the lowest point ($a > 0$), then $f(-b/2a)$ is the **minimum value** of f. These ideas give rise to many applications.

E X A M P L E   6   Maximizing Revenue

The marketing department at Texas Instruments has found that, when certain calculators are sold at a price of p dollars per unit, the revenue R (in dollars) as a function of the price p is

$$R(p) = -150p^2 + 21{,}000p$$

What unit price should be established in order to maximize revenue? If this price is charged, what is the maximum revenue?

Solution   The revenue R is

$$R(p) = -150p^2 + 21{,}000p = ap^2 + bp + c$$

The function R is a quadratic function with $a = -150$, $b = 21{,}000$, and $c = 0$. Because $a < 0$, the vertex is the highest point of the parabola. The revenue R is therefore a maximum when the price p is

$$p = \frac{-b}{2a} = \frac{-21{,}000}{2(-150)} = \frac{-21{,}000}{-300} = \$70.00$$

The maximum revenue R is

$$R(70) = -150(70)^2 + 21{,}000(70) = \$735{,}000$$

See Figure 11 for an illustration.

FIGURE 11
$R(p) = -150p^2 + 21,000p$

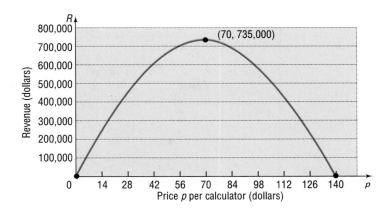

Now work **Problem 47.**

E X A M P L E   7

Analyzing the Motion of a Projectile

A projectile is fired from a cliff 500 feet above the water at an inclination of 45° to the horizontal, with a muzzle velocity of 400 feet per second. In physics, it is established that the height h of the projectile above the water is given by

$$h(x) = \frac{-32x^2}{(400)^2} + x + 500$$

where x is the horizontal distance of the projectile from the base of the cliff. See Figure 12.

FIGURE 12

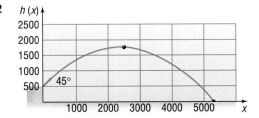

(a) Find the maximum height of the projectile.
(b) How far from the base of the cliff will the projectile strike the water?

Solution   (a) The height of the projectile is given by a quadratic function:

$$h(x) = \frac{-32x^2}{(400)^2} + x + 500 = \frac{-1}{5000}x^2 + x + 500$$

We are looking for the maximum value of h. Since the maximum value is obtained at the vertex, we compute

$$x = \frac{-b}{2a} = \frac{-1}{2(-1/5000)} = \frac{5000}{2} = 2500$$

The maximum height of the projectile is

$$h(2500) = \frac{-1}{5000}(2500)^2 + 2500 + 500 = -1250 + 2500 + 500 = 1750 \text{ ft}$$

(b) The projectile will strike the water when its height is zero. To find the distance x traveled, we need to solve the equation

$$h(x) = \frac{-1}{5000}x^2 + x + 500 = 0$$

We use the quadratic formula with

$$b^2 - 4ac = 1 - 4\left(\frac{-1}{5000}\right)(500) = 1.4$$

$$x = \frac{-1 \pm \sqrt{1.4}}{2(-1/5000)} \approx \begin{cases} -458 \\ 5458 \end{cases}$$

We discard the negative solution and find that the projectile will strike the water a distance of about 5458 feet from the base of the cliff. ▬

Exploration Graph:

$$h(x) = \frac{-1}{5000}x^2 + x + 500 \qquad 0 \le x \le 5500$$

Use MAXIMUM to find the maximum height of the projectile and use ROOT or ZERO to find the distance from the base of the cliff at which it strikes the water. Compare your results with those obtained in the text. TRACE the path of the projectile. How far from the base of the cliff is the projectile when its height is 1000 ft? 1500 ft? ▬

Now work Problem 55.

E X A M P L E 8   Navigation

A cruise ship leaves the Port of Miami heading due east at a constant speed of 5 knots (1 knot = 1 nautical mile per hour). At 5:00 PM, the cruise ship is 5 nautical miles due south of a cabin cruiser that is moving south at a constant speed of 10 knots. At what time are the two ships closest?

FIGURE 13

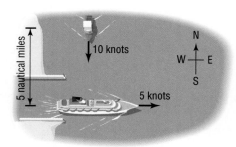
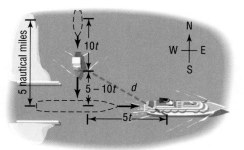

(a) Position at 5:00 PM   (b) Position at time t

FIGURE 14

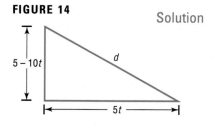

Solution   We begin with an illustration depicting the relative position of each ship at 5:00 PM. See Figure 13(a). After a time t (in hours) has passed, the cruise ship has moved east $5t$ nautical miles, and the cabin cruiser has moved south $10t$ nautical miles. Figure 13(b) illustrates the relative position of each ship after t hours. Figure 14 shows a right triangle extracted from Figure 13(b). By the Pythagorean Theorem, the square of the distance d between the ships after time t is

$$d^2 = (5 - 10t)^2 + (5t)^2$$
$$= 125t^2 - 100t + 25$$

Now, the distance d is a minimum when d^2 is a minimum. Because d^2 is a quadratic function of t, it follows that d^2, and hence d, is a minimum when

$$t = \frac{-b}{2a} = \frac{100}{2(125)} = \frac{2}{5} \text{ hour}$$

Thus, the ships are closest after $\frac{2}{5}(60) = 24$ minutes, that is, at 5:24 PM. ▬

In a suspension bridge, the main cables are of parabolic shape because, if the total weight of a bridge is uniformly distributed along its length, the only cable shape that will bear the load evenly is that of a parabola. The Golden Gate Bridge in San Francisco is an example of a suspension bridge.

E X A M P L E   9

The Golden Gate Bridge

The Golden Gate Bridge, a suspension bridge, spans the entrance to San Francisco Bay. Its 746-foot-tall towers are 4200 feet apart. The bridge is suspended from two huge cables more than 3 feet in diameter; the 90-foot-wide roadway is 220 feet above the water. The cables are parabolic in shape and touch the road surface at the center of the bridge. Find the height of the cable at a distance of 1000 feet from the center.

Solution   We begin by choosing the placement of the coordinate axes so that the x-axis coincides with the road surface and the origin coincides with the center of the bridge. As a result, the twin towers will be vertical (height $746 - 220 = 526$ feet above the road) and located 2100 feet from the center. Also, the cable, which has the shape of a parabola, will extend from the towers, open up, and have its vertex at $(0, 0)$. As illustrated in Figure 15, the choice of placement of the axes enables us to identify the equation of the parabola as $y = ax^2, a > 0$. We also can see that the points $(-2100, 526)$ and $(2100, 526)$ are on the graph.

FIGURE 15

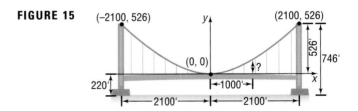

Based on these facts, we can find the value of a in $y = ax^2$:

$$y = ax^2$$
$$526 = a(2100)^2$$
$$a = \frac{526}{(2100)^2}$$

The equation of the parabola is therefore

$$y = \frac{526}{(2100)^2}x^2$$

The height of the cable when $x = 1000$ is

$$y = \frac{526}{(2100)^2}(1000)^2 \approx 119.3 \text{ feet}$$

Thus, the cable is 119.3 feet high at a distance of 1000 feet from the center of the bridge.

Now work Problem 59.

Curve Fitting

In Section 3.5, we found the line of best fit for data that appeared to be linearly related. It was noted that data may also follow a nonlinear relation. Figures 16(a) and (b) show scatter diagrams of data that follow a quadratic relation.

FIGURE 16

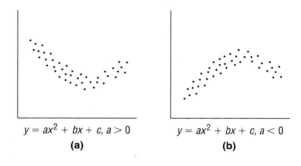

$y = ax^2 + bx + c, a > 0$ $y = ax^2 + bx + c, a < 0$

(a)   (b)

E X A M P L E   10

Plot	Fertilizer, x (Pounds/100 ft²)	Yield (Bushels)
1	0	4
2	0	6
3	5	10
4	5	7
5	10	12
6	10	10
7	15	15
8	15	17
9	20	18
10	20	21
11	25	20
12	25	21
13	30	21
14	30	22
15	35	21
16	35	20
17	40	19
18	40	19

Fitting a Quadratic Function to Data

Refer back to Example 2 in Section 3.5. Suppose that the farmer collected additional data, as shown in the table that show crop yields Y for various amounts of fertilizer used, x.

(a) Draw a scatter diagram of the data. Comment on the type of relation that may exist between the two variables.

(b) The quadratic function of best fit to these data is

$$Y(x) = -0.0171x^2 + 1.0765x + 3.8939$$

Use this function to determine the optimal amount of fertilizer to apply.

(c) Use the function to predict crop yield when the optimal amount of fertilizer is applied.

 (d) Use a graphing utility to verify that the function given in (b) is the quadratic function of best fit.

(e) With a graphing utility, draw a scatter diagram of the data and then graph the quadratic function of best fit on the scatter diagram.

Solution

(a) Figure 17 shows the scatter diagram, from which it appears that the data follow a quadratic relation, with $a < 0$.

(b) Based on the quadratic function of best fit, the optimal amount of fertilizer to apply is

FIGURE 17

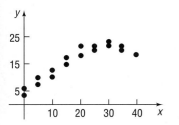

$$h = \frac{-b}{2a} = \frac{-1.0765}{2(-0.0171)} \approx 31.5 \text{ pounds of fertilizer per 100 square feet}$$

(c) We evaluate the function $Y(x)$ for $x = 31.5$:

$$Y(31.5) = -0.0171(31.5)^2 + 1.0765(31.5) + 3.8939 \approx 20.8 \text{ bushels}$$

If we apply 31.5 pounds of fertilizer per 100 square feet, the crop yield will be 20.8 bushels according to the quadratic function of best fit.

(d) Upon executing the QUADratic REGression program, we obtain the results shown in Figure 18. The output that the utility provides shows us the equation $y = ax^2 + bx + c$. The quadratic function of best fit is $Y(x) = -0.0171x^2 + 1.0765x + 3.8939$, where x represents the amount of fertilizer used and Y represents crop yield.

(e) Figure 19 shows the graph of the quadratic function found in (b) drawn on the scatter diagram.

FIGURE 18

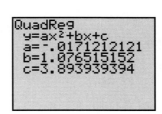

FIGURE 19

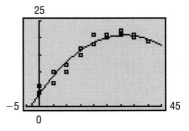

Look again at Figure 18. Notice that the output given by the graphing calculator does not include r, the correlation coefficient. Recall that the correlation coefficient is a measure of the strength of a **linear** relation that exists between two variables. The graphing calculator does not provide an indication of how well the function "fits" the data in terms of r since the function cannot be expressed as a linear function.

5.1 | EXERCISES

In Problems 1–8, match each graph to one of the following functions:

A. $y = x^2 - 1$ B. $y = -x^2 - 1$ C. $y = x^2 - 2x + 1$ D. $y = x^2 + 2x + 1$

E. $y = x^2 - 2x + 2$ F. $y = x^2 + 2x$ G. $y = x^2 - 2x$ H. $y = x^2 + 2x + 2$

1.

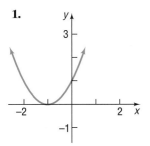

2.

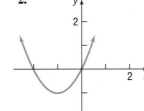

3.

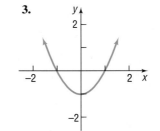

4.

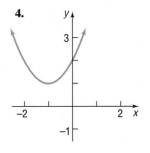

5.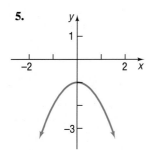

6.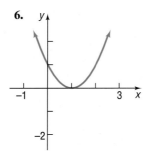

7.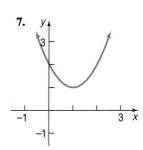

8.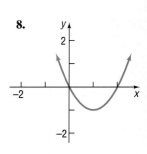

In Problems 9–24, graph the function f by starting with the graph of $y = x^2$ and using transformations (shifting, compressing, stretching, and/or reflection). [**Hint:** *If necessary, write f in the form $f(x) = a(x - h)^2 + k$.*]

9. $f(x) = \dfrac{1}{4}x^2$

10. $f(x) = 2x^2$

11. $f(x) = \dfrac{1}{4}x^2 - 2$

12. $f(x) = 2x^2 - 3$

13. $f(x) = \dfrac{1}{4}x^2 + 2$

14. $f(x) = 2x^2 + 4$

15. $f(x) = \dfrac{1}{4}x^2 + 1$

16. $f(x) = -2x^2 - 2$

17. $f(x) = x^2 + 4x + 2$

18. $f(x) = x^2 - 6x - 1$

19. $f(x) = 2x^2 - 4x + 1$

20. $f(x) = 3x^2 + 6x$

21. $f(x) = -x^2 - 2x$

22. $f(x) = -2x^2 + 6x + 2$

23. $f(x) = \dfrac{1}{2}x^2 + x - 1$

24. $f(x) = \dfrac{2}{3}x^2 + \dfrac{4}{3}x - 1$

In Problems 25–38, graph each quadratic function by determining whether its graph opens up or down and by finding its vertex, axis of symmetry, y-intercept, and x-intercepts, if any.

25. $f(x) = x^2 + 2x - 8$
26. $f(x) = x^2 - 2x - 3$
27. $f(x) = -x^2 - 3x + 4$
28. $f(x) = -x^2 + x + 2$
29. $f(x) = x^2 + 2x + 1$
30. $f(x) = -x^2 + 4x - 4$
31. $f(x) = 2x^2 - x + 2$
32. $f(x) = 4x^2 - 2x + 1$
33. $f(x) = -2x^2 + 2x - 3$
34. $f(x) = -3x^2 + 3x - 2$
35. $f(x) = 3x^2 + 6x + 2$
36. $f(x) = 2x^2 + 5x + 3$
37. $f(x) = -4x^2 - 6x + 2$
38. $f(x) = 3x^2 - 8x + 2$

In Problems 39–44, determine, without graphing, whether the given quadratic function has a maximum value or a minimum value and then find the value.

39. $f(x) = 2x^2 + 12x - 3$
40. $f(x) = 4x^2 - 8x + 3$
41. $f(x) = -x^2 + 10x - 4$
42. $f(x) = -2x^2 + 8x + 3$
43. $f(x) = -3x^2 + 12x + 1$
44. $f(x) = 4x^2 - 4x$

45. The graph of the function $f(x) = ax^2 + bx + c$ has vertex at $x = 0$ and passes through the points $(0, 2)$ and $(1, 8)$. Find a, b, and c.

46. The graph of the function $f(x) = ax^2 + bx + c$ has vertex at $x = 1$ and passes through the points $(0, 1)$ and $(-1, -8)$. Find a, b, and c.

47. **Maximizing Revenue** Suppose that the manufacturer of a gas clothes dryer has found that, when the unit price is p dollars, the revenue R (in dollars) is

$$R = -4p^2 + 4000p$$

What unit price should be established for the dryer to maximize revenue? What is the maximum revenue?

48. **Maximizing Revenue** The John Deere company has found that the revenue from sales of heavy-duty tractors is a function of the unit price p that it charges. If the revenue R is

$$R = -\dfrac{1}{2}p^2 + 1900p$$

what unit price p should be charged to maximize revenue? What is the maximum revenue?

49. Rectangles with Fixed Perimeter What is the largest rectangular area that can be enclosed with 400 feet of fencing? What are the dimensions of the rectangle?

50. Rectangles with Fixed Perimeter What are the dimensions of a rectangle of a fixed perimeter P that result in the largest area?

51. Enclosing the Most Area with a Fence A farmer with 4000 meters of fencing wants to enclose a rectangular plot that borders on a river. If the farmer does not fence the side along the river, what is the largest area that can be enclosed? (See the figure.)

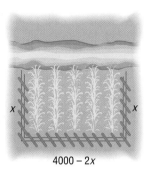

x x

$4000 - 2x$

52. Enclosing the Most Area with a Fence A farmer with 2000 meters of fencing wants to enclose a rectangular plot that borders on a straight highway. If the farmer does not fence the side along the highway, what is the largest area that can be enclosed?

53. Enclosing the Most Area with a Fence A farmer with 10,000 meters of fencing wants to enclose a rectangular field and then divide it into two plots with a fence parallel to one of the sides (see the figure). What is the largest area that can be enclosed?

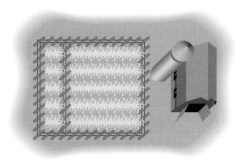

54. Enclosing the Most Area with a Fence A farmer with 10,000 meters of fencing wants to enclose a rectangular field and then divide it into three plots with two fences parallel to one of the sides. What is the largest area that can be enclosed?

55. Analyzing the Motion of a Projectile A projectile is fired from a cliff 200 feet above the water at an inclination of 45° to the horizontal, with a muzzle velocity of 50 feet per second. The height h of the projectile above the water is given by

$$h(x) = \frac{-32x^2}{(50)^2} + x + 200$$

where x is the horizontal distance of the projectile from the base of the cliff.
(a) How far from the base of the cliff is the height of the projectile a maximum?
(b) Find the maximum height of the projectile.
(c) How far from the base of the cliff will the projectile strike the water?
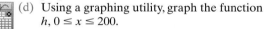 (d) Using a graphing utility, graph the function h, $0 \le x \le 200$.
(e) Use a graphing utility to verify the solutions found in (b) and (c).
(f) When the height of the projectile is 100 feet above the water, how far is it from the cliff?

56. Analyzing the Motion of a Projectile A projectile is fired at an inclination of 45° to the horizontal, with a muzzle velocity of 100 feet per second. The height h of the projectile is given by

$$h(x) = \frac{-32x^2}{(100)^2} + x$$

where x is the horizontal distance of the projectile from the firing point.
(a) How far from the firing point is the height of the projectile a maximum?
(b) Find the maximum height of the projectile.
(c) How far from the firing point will the projectile strike the ground?
(d) Using a graphing utility, graph the function h, $0 \le x \le 350$.
(e) Use a graphing utility to verify the results obtained in parts (b) and (c).
(f) When the height of the projectile is 50 feet above the ground, how far has it traveled horizontally?

57. Navigation The *U.S.S. Independence* maintains a constant speed of 10 knots heading due north. At 4:00 PM the ship's radar detects a destroyer 100 nautical miles due east of the carrier. If the destroyer is heading due west at 20 knots, when will the two ships be the closest? (1 knot = 1 nautical mile per hour)

58. Air Traffic Control An air traffic controller sees two aircraft flying at the same altitude on his screen. One, a Beechcraft Duchess, is headed due west at 150 miles per hour. The other, a Lear jet, is 15 miles due north of the Beechcraft and is headed due south at 400 miles per hour. How close will the two aircraft come to each other?

59. Suspension Bridge A suspension bridge with weight uniformly distributed along its length has twin towers that extend 75 meters above the road surface and are 400 meters apart. The cables are parabolic in shape and are suspended from the tops of the towers. The cables touch the road surface at the center of the bridge. Find the height of the cables at a point 100 meters from the center. (Assume that the road is level.)

60. Architecture A parabolic arch has a span of 120 feet and a maximum height of 25 feet. Choose suitable rectangular coordinate axes and find the equation of the parabola. Then calculate the height of the arch at points 10 feet, 20 feet, and 40 feet from the center.

61. Constructing Rain Gutters A rain gutter is to be made of aluminum sheets that are 12 inches wide by turning up the edges 90°. What depth will provide maximum cross-sectional area and hence allow the most water to flow?

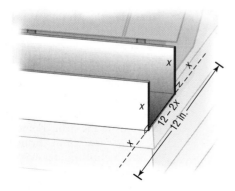

62. Navigation At 4 PM the *Ecstasy* leaves the Port of Miami heading due east at a constant speed of 15 knots. At the same time, a pleasure boat located 100 nautical miles northeast of the Port of Miami is headed due south at a constant speed of 12 knots. When are the two ships closest? How close do they get to each other? (Express your answer in nautical miles; 1 knot = 1 nautical mile per hour.)

63. Norman Windows A Norman window has the shape of a rectangle surmounted by a semicircle of diameter equal to the width of the rectangle (see the figure). If the perimeter of the window is 20 feet, what dimensions will admit the most light (maximize the area)?

[**Hint:** Circumference of circle = $2\pi r$; area of circle = πr^2, where r is the radius of the circle.]

64. Constructing a Stadium A track and field playing area is in the shape of a rectangle with semicircles at each end (see the figure). The inside perimeter of the track is to be 1500 meters. What should the dimensions of the rectangle be so that the area of the rectangle is a maximum?

65. Architecture A special window has the shape of a rectangle surmounted by an equilateral triangle (see the figure). If the perimeter of the window is 16 feet, what dimensions will admit the most light?

[**Hint:** Area of an equilateral triangle = $(\sqrt{3}/4)x^2$, where x is the length of a side of the triangle.]

66. Analyzing the Motion of a Projectile A projectile is fired at an inclination of 45° to the horizontal with an initial velocity of v_0 feet per second. If the starting point is the origin, the x-axis is horizontal, and the y-axis is vertical, then the height y (in feet) after a horizontal distance x has been traversed is approximately

$$y - \frac{-32x^2}{v_0^2} + x$$

(a) Find the maximum height in terms of the initial velocity v_0.

(b) If the initial velocity is doubled, what happens to the maximum height?

(c) Assuming that the ground is flat, how far from the starting point will the projectile land if the initial velocity is 64 feet per second?

67. **Charter Club Memberships** A charter flight club charges its members $400 per year. But, for each new member in excess of 60, the charge for every member is reduced by $5. What number of members leads to a maximum revenue?

68. **Car Rentals** A car rental agency has 24 identical cars. The owner of the agency finds that all the cars can be rented at a price of $10 per day. However, for each $2 increase in rental, one of the cars is not rented. What should be charged to maximize income?

69. **Calculus: Simpson's Rule** The figure shows the graph of $y = ax^2 + bx + c$. Suppose that the points $(-h, y_0)$, $(0, y_1)$, and (h, y_2) are on the graph. It can be shown that the area enclosed by the parabola, the x-axis, and the lines $x = -h$ and $x = h$ is

$$\text{Area} = \frac{h}{3}(2ah^2 + 6c)$$

Show that this area may also be given by

$$\text{Area} = \frac{h}{3}(y_0 + 4y_1 + y_2)$$

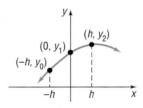

70. Let $f(x) = ax^2 + bx + c$, where a, b, and c are odd integers. If x is an integer, show that $f(x)$ must be an odd integer.

[**Hint:** x is either an even integer or an odd integer.]

71. **Life Cycle Hypothesis** An individual's income varies with his or her age. The following table shows the median income I of individuals of different age groups within the United States for 1995. For each class, let the class midpoint represent the independent variable x. For the class "65 years and older," we will assume that class midpoint is 69.5.

Age	Class Midpoint, x	Median Income, I
15–24 years	19.5	$20,979
25–34 years	29.5	$34,701
35–44 years	39.5	$43,465
45–54 years	49.5	$48,058
55–64 years	59.5	$38,077
65 years and older	69.5	$19,096

Source: U.S. Census Bureau

(a) Draw a scatter diagram of the data. Comment on the type of relation that may exist between the two variables.

(b) The quadratic function of best fit to these data is

$$I(x) = -42.6x^2 + 3805.5x - 38,526$$

Use this function to determine the age at which an individual can expect to earn the most income.

(c) Use the function to predict the peak income earned.

 (d) Use a graphing utility to verify that the function given in (b) is the quadratic function of best fit.

(e) With a graphing utility, draw a scatter diagram of the data and then graph the quadratic function of best fit on the scatter diagram.

72. **Advertising** A small manufacturing firm collected the following data on advertising expenditures A (in thousands of dollars) and total revenue R (in thousands of dollars).

Advertising	Total Revenue
20	$6101
22	$6222
25	$6350
25	$6378
27	$6453
28	$6423
29	$6360
31	$6231

(a) Draw a scatter diagram of the data. Comment on the type of relation that may exist between the two variables.

(b) The quadratic function of best fit to these data is

$$R(A) = -7.76A^2 + 411.88A + 942.72$$

Use this function to determine the optimal level of advertising for this firm.

(c) Use the function to find the revenue that the firm can expect if it uses the optimal level of advertising.

 (d) Use a graphing utility to verify that the function given in (b) is the quadratic function of best fit.

(e) With a graphing utility, draw a scatter diagram of the data and then graph the quadratic function of best fit on the scatter diagram.

73. Fuel Consumption   The following data represent the average fuel consumption C by cars (in billions of gallons) for the years 1980–1993.

Year, t	Average Fuel Consumption, C
1980	71.9
1981	71.0
1982	70.1
1983	69.9
1984	68.7
1985	69.3
1986	71.4
1987	70.6
1988	71.9
1989	72.7
1990	72.0
1991	70.7
1992	73.9
1993	75.1

Source: U.S. Federal Highway Administration

(a) Draw a scatter diagram of the data. Comment on the type of relation that may exist between the two variables.

(b) The quadratic function of best fit to these data is

$$C(t) = 0.063t^2 - 250t + 248,077$$

Use this function to determine the year in which average fuel consumption was lowest.

(c) Use the function to predict the average fuel consumption for 1994.

(d) Use a graphing utility to verify that the function given in (b) is the quadratic function of best fit.

(e) With a graphing utility, draw a scatter diagram of the data and then graph the quadratic function of best fit on the scatter diagram.

74. Miles per Gallon   An engineer collects data showing the speed s of a Ford Taurus and its average miles per gallon, M. See the table.

Speed, s	Miles per Gallon, M
30	18
35	20
40	23
40	25
45	25
50	28
55	30
60	29
65	26
65	25
70	25

(a) Draw a scatter diagram of the data. Comment on the type of relation that may exist between the two variables.

(b) The quadratic function of best fit to these data is

$$M(s) = -0.018s^2 + 1.93s - 25.34$$

Use this function to determine the speed that maximizes miles per gallon.

(c) Use the function to predict miles per gallon for a speed of 63 miles per hour.

(d) Use a graphing utility to verify that the function given in (b) is the quadratic function of best fit.

(e) With a graphing utility, draw a scatter diagram of the data and then graph the quadratic function of best fit on the scatter diagram.

75. Height of a Ball A physicist throws a ball at an inclination of 45° to the horizontal. The following data represent the height of the ball h at the instant it has traveled x feet horizontally.

Distance, x	Height, h
20	25
40	40
60	55
80	65
100	71
120	77
140	77
160	75
180	71
200	64

(a) Draw a scatter diagram of the data. Comment on the type of relation that may exist between the two variables.

(b) The quadratic function of best fit to these data is

$$h(x) = -0.0037x^2 + 1.03x + 5.7$$

Use this function to determine how far the ball will travel before it reaches its maximum height.

(c) Use the function to find the maximum height of the ball.

 (d) Use a graphing utility to verify that the function given in (b) is the quadratic function of best fit.

(e) With a graphing utility, draw a scatter diagram of the data and then graph the quadratic function of best fit on the scatter diagram.

76. Enrollment in Public Schools The following data represent the enrollment E in all public schools (both elementary and high school) for the academic years 1980–1981 to 1988–1989. Let 1 represent the academic year 1980–1981, 2 the academic year 1981–1982, and so on.

(a) Draw a scatter diagram of the data. Comment on the type of relation that may exist between the two variables.

Year, t	Enrollment, E
1	41.5
2	40.8
3	40.1
4	39.6
5	39.1
6	39.1
7	39.6
8	39.8
9	40.1

(b) The quadratic function of best fit to these data is

$$E(t) = 0.01t^2 - 1.16t + 42.62$$

Use this function to determine when enrollment was lowest.

(c) Use a graphing utility to verify that the function given in (b) is the quadratic function of best fit.

(d) With a graphing utility, draw a scatter diagram of the data and then graph the quadratic function of best fit on the scatter diagram.

77. Chemical Reactions A self-catalytic chemical reaction results in the formation of a compound that causes the formation ratio to increase. If the reaction rate V is given by

$$V(x) = kx(a - x) \qquad 0 \le x \le a$$

where k is a positive constant, a is the initial amount of the compound, and x is the variable amount of the compound, for what value of x is the reaction rate a maximum?

 78. Make up a quadratic function that opens down and has only one x-intercept. Compare yours with others in the class. What are the similarities? What are the differences?

79. On one set of coordinate axes, graph the family of parabolas $f(x) = x^2 + 2x + c$ for $c = -3$, $c = 0$, and $c = 1$. Describe the characteristics of a member of this family.

80. On one set of coordinate axes, graph the family of parabolas $f(x) = x^2 + bx + 1$ for $b = -4$, $b = 0$, and $b = 4$. Describe the general characteristics of this family.

5.2 | POLYNOMIAL FUNCTIONS

> **1** Identify Polynomials and Their Degree
> **2** Graph Polynomial Functions Using Transformations
> **3** Identify the Zeros of a Polynomial and Their Multiplicity
> **4** Analyze the Graph of a Polynomial Function

Polynomial functions are among the simplest expressions in algebra. They are easy to evaluate: only addition and repeated multiplication are required. Because of this, they are often used to approximate other, more complicated functions. In this section, we investigate characteristics of this important class of function.

A **polynomial function** is a function of the form

$$f(x) = a_n x^n + a_{n-1} x^{n-1} + \cdots + a_1 x + a_0 \qquad (1)$$

where $a_n, a_{n-1}, \ldots, a_1, a_0$ are real numbers and n is a nonnegative integer. The domain consists of all real numbers.

---

1   Thus, a polynomial function is a function whose rule is given by a polynomial in one variable (refer to Section 1.4). The **degree** of a polynomial function is the degree of the polynomial in one variable.

E X A M P L E   1   Identifying Polynomial Functions

Determine which of the following are polynomial functions. For those that are, state the degree; for those that are not, tell why not.

(a) $f(x) = 2 - 3x^4$ (b) $g(x) = \sqrt{x}$ (c) $h(x) = \dfrac{x^2 - 2}{x^3 - 1}$

(d) $F(x) = 0$ (e) $G(x) = 8$

Solution   (a) f is a polynomial function of degree 4.
       (b) g is not a polynomial function. The variable x is raised to the $\frac{1}{2}$ power, which is not a nonnegative integer.
       (c) h is not a polynomial function. It is the ratio of two polynomials, and the polynomial in the denominator is of positive degree.
       (d) F is the zero polynomial function; it is not assigned a degree.
       (e) G is a nonzero constant function, a polynomial function of degree 0. ■

 Now work Problems 1 and 5.

We have already discussed in detail polynomial functions of degrees 0, 1, and 2. See Table 2 for a summary of the characteristics of the graphs of these polynomial functions.

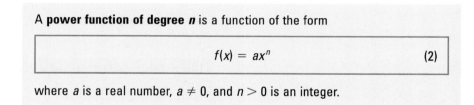

TABLE 2			
Degree	**Form**	**Name**	**Graph**
No degree	$f(x) = 0$	Zero function	The x-axis
0	$f(x) = a_0, \quad a_0 \neq 0$	Constant function	Horizontal line with y-intercept a_0
1	$f(x) = a_1 x + a_0, \quad a_1 \neq 0$	Linear function	Nonvertical, nonhorizontal line with slope a_1 and y-intercept a_0
2	$f(x) = a_2 x^2 + a_1 x + a_0, \quad a_2 \neq 0$	Quadratic function	Parabola: graph opens up if $a_2 > 0$; graph opens down if $a_2 < 0$

Power Functions

First, we consider a special kind of polynomial function called a *power function.*

> A **power function of degree *n*** is a function of the form
>
> $$f(x) = ax^n \tag{2}$$
>
> where a is a real number, $a \neq 0$, and $n > 0$ is an integer.

The graph of a power function of degree 1, $f(x) = ax$, is a straight line, with slope a, that passes through the origin. The graph of a power function of degree 2, $f(x) = ax^2$, is a parabola, with vertex at the origin, that opens up if $a > 0$ and down if $a < 0$.

If we know how to graph a power function of the form $f(x) = x^n$, then a compression or stretch and, perhaps, a reflection about the x-axis will enable us to obtain the graph of $g(x) = ax^n$. Consequently, we shall concentrate on graphing power functions of the form $f(x) = x^n$.

We begin with power functions of even degree of the form $f(x) = x^n$, $n \geq 2$ and n even. The domain of f is the set of all real numbers, and the range is the set of nonnegative real numbers. Such a power function is an even function (do you see why?), and hence its graph is symmetric with respect to the y-axis. Its graph always contains the origin and the points $(-1, 1)$ and $(1, 1)$.

If $n = 2$, the graph is the familiar parabola $y = x^2$ that opens up, with vertex at the origin. If $n \geq 4$, the graph of $f(x) = x^n$, n even, will be closer to the x-axis than the parabola $y = x^2$ if $-1 < x < 1$ and farther from the x-axis than the parabola $y = x^2$ if $x < -1$ or if $x > 1$. Figure 20(a) illustrates this conclusion. Figure 20(b) shows the graphs of $y = x^4$ and $y = x^8$ for comparison.

FIGURE 20

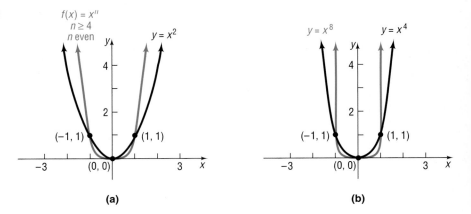

(a)

(b)

From Figure 20, we can see that as n increases, the graph of $f(x) = x^n$, $n \geq 2$ and n even, tends to flatten out near the origin and to increase very rapidly when x is far from 0. For large n, it may appear that the graph coincides with the x-axis near the origin, but it does not; the graph actually touches the x-axis only at the origin (see Table 3). Also, for large n, it may appear that for $x < -1$ or for $x > 1$ the graph is vertical, but it is not; it is only increasing very rapidly in those intervals. If the graphs were enlarged many times, these distinctions would be clear.

TABLE 3			
	$x = 0.1$	$x = 0.3$	$x = 0.5$
$f(x) = x^8$	10^{-8}	0.0000656	0.0039063
$f(x) = x^{20}$	10^{-20}	$3.487 \cdot 10^{-11}$	0.000001
$f(x) = x^{40}$	10^{-40}	$1.216 \cdot 10^{-21}$	$9.095 \cdot 10^{-13}$

Seeing the Concept Graph $Y_1 = x^4$, $Y_2 = x^8$, and $Y_3 = x^{12}$ using the viewing rectangle $-2 \leq x \leq 2$, $-4 \leq y \leq 16$. Then graph each one again using the viewing rectangle $-1 \leq x \leq 1$, $0 \leq y \leq 1$. See Figure 21.

FIGURE 21

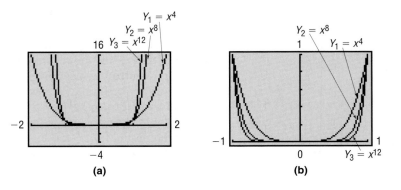

(a)

(b)

Now we consider power functions of odd degree of the form $f(x) = x^n$, $n \geq 3$ and n odd. The domain and range of f are the set of real numbers. Such a power function is an odd function (do you see why?), and hence its graph is symmetric with respect to the origin. Its graph always contains the origin and the points $(-1, -1)$ and $(1, 1)$.

The graph of $f(x) = x^n$ when $n = 3$ has been shown several times and is repeated in Figure 22. If $n \geq 5$, the graph of $f(x) = x^n$, n odd, will be closer to the x-axis than that of $y = x^3$ if $-1 < x < 1$ and farther from the x-axis than that of $y = x^3$ if $x < -1$ or if $x > 1$. Figure 22 also illustrates this conclusion.

Figure 23 shows the graph of $y = x^5$ and the graph of $y = x^9$ for further comparison.

FIGURE 22

FIGURE 23

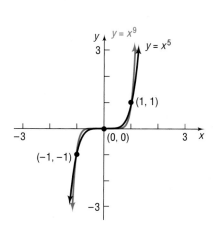

From Figures 22 and 23 we can see that as n increases, the graph of $f(x) = x^n$, $n \geq 3$ and n odd, tends to flatten out near the origin and to become nearly vertical when x is far from 0.

Seeing the Concept Graph $Y_1 = x^3$, $Y_2 = x^7$, and $Y_3 = x^{11}$ using the viewing rectangle $-2 \leq x \leq 2$, $-16 \leq y \leq 16$. Then graph each one again using the viewing rectangle $-1 \leq x \leq 1$, $-1 \leq y \leq 1$. See Figure 24.

FIGURE 24

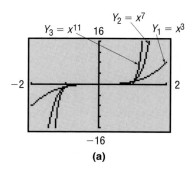 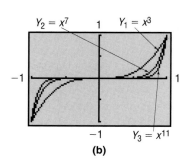

(a)   (b)

Transformations (shifting, compression, stretching, and reflection), when used in conjunction with the facts just presented, enable us to graph a variety of polynomials.

E X A M P L E 2

Graphing Polynomial Functions Using Transformations

Graph: $f(x) = 1 - x^5$

Solution Figure 25 shows the required stages.

FIGURE 25

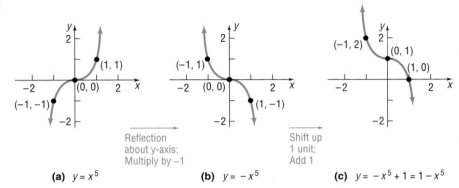

Reflection
about y-axis;
Multiply by −1

Shift up
1 unit;
Add 1

(a) $y = x^5$

(b) $y = -x^5$

(c) $y = -x^5 + 1 = 1 - x^5$

E X A M P L E 3 ·

Graphing Polynomial Functions Using Transformations

Graph: $f(x) = \frac{1}{2}(x - 1)^4$

Solution Figure 26 shows the required stages.

FIGURE 26

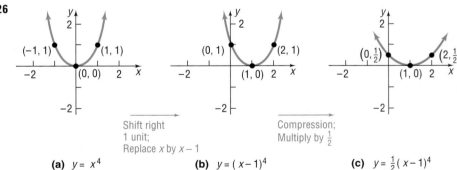

Shift right
1 unit;
Replace x by x − 1

Compression;
Multiply by $\frac{1}{2}$

(a) $y = x^4$

(b) $y = (x-1)^4$

(c) $y = \frac{1}{2}(x-1)^4$

Now work Problem 15.

Graphing Other Polynomials

To graph most polynomial functions of degree 3 or higher requires techniques beyond the scope of this text. If you take a course in calculus, you will learn that the graph of every polynomial function is both smooth and continuous. By **smooth,** we mean that the graph contains no sharp corners or cusps; by **continuous,** we mean that the graph has no gaps or holes and can be drawn without lifting pencil from paper. See Figure 27(a) and 27(b).

FIGURE 27

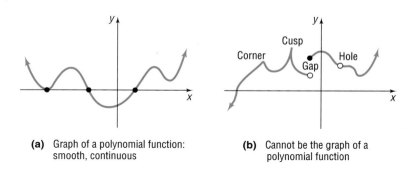

(a) Graph of a polynomial function: smooth, continuous

(b) Cannot be the graph of a polynomial function

Figure 28 shows the graph of a polynomial function with four x-intercepts. Notice that at the x-intercepts the graph must either cross the x-axis or touch the x-axis. Consequently, between consecutive x-intercepts the graph is either above the x-axis or below the x-axis. We will make use of this characteristic of the graph of a polynomial shortly.

FIGURE 28

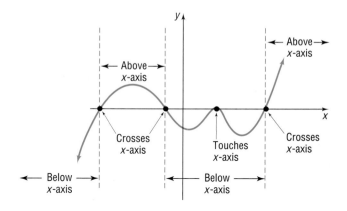

If a polynomial function f is factored completely, it is easy to solve the equation $f(x) = 0$ and locate the x-intercepts of the graph. For example, if $f(x) = (x - 1)^2(x + 3)$, then the solutions of the equation

$$f(x) = (x - 1)^2(x + 3) = 0$$

are easily identified as 1 and -3. Based on this result, we make the following observations:

If f is a polynomial function and r is a real number for which $f(r) = 0$, then r is called a (real) **zero of f**, or **root of f**. If r is a (real) zero of f, then

(a) r is an x-intercept of the graph of f.
(b) $(x - r)$ is a factor of f.

3

If the same factor $x - r$ occurs more than once, then r is called a **repeated,** or **multiple, zero of f.** More precisely, we have the following definition.

> If $(x - r)^m$ is a factor of a polynomial f and $(x - r)^{m+1}$ is not a factor of f, then r is called a **zero of multiplicity m of f.**

E X A M P L E   4   Identifying Zeros and Their Multiplicities

For the polynomial

$$f(x) = 5(x - 2)(x + 3)^2\left(x - \frac{1}{2}\right)^4$$

2 is a zero of multiplicity 1.
-3 is a zero of multiplicity 2.
$\frac{1}{2}$ is a zero of multiplicity 4.

∎

In the solution to Example 4, notice that, if you add the multiplicities $(1 + 2 + 4 = 7)$, you obtain the degree of the polynomial.

Suppose that it is possible to factor completely a polynomial function and, as a result, locate all the x-intercepts of its graph (the real zeros of the function). As mentioned earlier, these x-intercepts then divide the x-axis into open intervals and, on each such interval, the graph of the polynomial will be either above or below the x-axis. Let's look at an example.

E X A M P L E   5   Graphing a Polynomial Using Its x-Intercepts

For the polynomial: $f(x) = x^2(x - 2)$

(a) Find the x- and y-intercepts of the graph of f.
(b) Use the x-intercepts to find the intervals on which the graph of f is above the x-axis and the intervals on which the graph of f is below the x-axis.
(c) Locate other points on the graph and connect all the points plotted with a smooth curve.

Solution   (a) The y-intercept is $f(0) = 0^2(0 - 2) = 0$. The x-intercepts satisfy the equation

$$f(x) = x^2(x - 2) = 0$$

from which we find

$$x^2 = 0 \quad \text{or} \quad x - 2 = 0$$
$$x = 0 \qquad\qquad x = 2$$

The x-intercepts are 0 and 2.

(b) The two x-intercepts divide the x-axis into three intervals:

$$-\infty < x < 0 \qquad 0 < x < 2 \qquad 2 < x < \infty$$

Since the graph of f crosses (or touches) the x-axis only at $x = 0$ and $x = 2$, it follows that the graph of f is either above the x-axis $[f(x) > 0]$

or below the x-axis $[f(x) < 0]$ on each of these three intervals. Thus, we need to solve the following two inequalities:

$$f(x) = x^2(x - 2) > 0 \quad \text{and} \quad f(x) = x^2(x - 2) < 0$$

Recalling the procedure discussed in Section 2.7, we set up Figure 29(a). Thus, the graph of f is above the x-axis for $2 < x < \infty$ and below the x-axis for $-\infty < x < 0$ and $0 < x < 2$.

(c) Because we used test numbers, we know three additional points on the graph: $(-1, -3)$, $(1, -1)$, and $(3, 9)$. Figure 29(b) illustrates these points, the intercepts, and a smooth, continuous curve (the graph of f) connecting them.

FIGURE 29

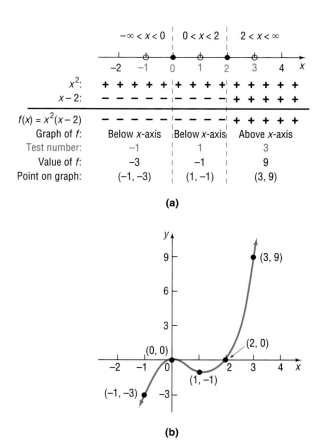

(a)

(b)

Notice that the graph of f in Figure 29(b) *crosses* the x-axis at $x = 2$, a *zero of multiplicity 1.* Notice also that the graph of f *touches* the x-axis at $x = 0$, a *zero of multiplicity 2,* since the graph is below the x-axis on both sides of $(0, 0)$.

This suggests the following result:

If r Is a Zero of Even Multiplicity	Sign of $f(x)$ does not change from one side to the other side of r.	Graph **touches** x-axis at r.
If r Is a Zero of Odd Multiplicity	Sign of $f(x)$ changes from one side to the other side of r.	Graph **crosses** x-axis at r.

Look again at Figure 29(b). We cannot be sure just how low the graph actually goes in the interval $0 < x < 2$. But we do know that somewhere in the interval $0 < x < 2$ the graph of f must change direction (from decreasing to increasing). The points at which a graph changes direction are called **turning points.** In calculus, such points are called **local maxima** or **local minima,** and techniques for locating them are given. So we shall not ask for the location of turning points in our graphs. Instead, we will use the following result from calculus, which tells us the maximum number of turning points that the graph of a polynomial function can have.

> **Theorem**
>
> If f is a polynomial function of degree n, then f has at most $n - 1$ turning points.

For example, the graph of $f(x) = x^2(x - 2)$ shown in Figure 29(b) is the graph of a polynomial of degree 3 and has $3 - 1 = 2$ turning points: one at $(0, 0)$ and the other somewhere in the interval $0 < x < 2$.

FIGURE 30

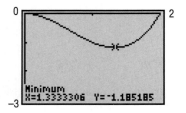

 Exploration A graphing utility can be used to locate the turning points of a graph. Graph $y = x^2(x - 2)$. Use MINIMUM to find the location of the turning point in the interval $0 < x < 2$. See Figure 30.

One last remark about Figure 29(b). Notice that the graph of $f(x) = x^2(x - 2)$ looks somewhat like the graph of $y = x^3$. In fact, for very large values of x, either positive or negative, there is little difference. To see for yourself, use your calculator to compare the values of $f(x) = x^2(x - 2)$ and $y = x^3$ for $x = -100,000$ and $x = 100,000$. The behavior of the graph of a function for large values of x, either positive or negative, is referred to as its **end behavior.**

> **Theorem** End Behavior
>
> For large values of x, either positive or negative, the graph of the polynomial
>
> $$f(x) = a_n x^n + a_{n-1} x^{n-1} + \cdots + a_1 x + a_0$$
>
> resembles the graph of the power function $y = a_n x^n$.

The following box summarizes some features of the graph of a polynomial function.

4

<table>
<tr><td rowspan="5">**Summary: Graph of a Polynomial Function**
$f(x) = a_n x^n + a_{n-1} x^{n-1} +$
$\cdots + a_1 x + a_0, \ a_n \neq 0$</td><td>Degree of the polynomial f: n
Maximum number of turning points: $n - 1$
At zero of even multiplicity: graph of f touches x-axis
At zero of odd multiplicity: graph of f crosses x-axis
Between zeros, graph of f is either above the x-axis or below it
End behavior: For large x, graph of f behaves like graph of $y = a_n x^n$</td></tr>
</table>

Now work Problem 19.

E X A M P L E   6

Analyzing the Graph of a Polynomial Function

For the polynomial: $f(x) = x^3 + x^2 - 12x$

(a) Find the x- and y-intercepts of the graph of f.

(b) Determine whether the graph crosses or touches the x-axis at each x-intercept.

(c) End behavior: find the power function that the graph of f resembles for large values of x.

(d) Determine the maximum number of turning points on the graph of f.

(e) Use the x-intercepts and test numbers to find the intervals on which the graph of f is above the x-axis and the intervals on which the graph is below the x-axis.

(f) Put all the information together, and connect the points with a smooth, continuous curve to obtain the graph of f.

Solution

(a) The y-intercept is $f(0) = 0$. To find the x-intercepts, if any, we factor f:

$$f(x) = x^3 + x^2 - 12x = x(x^2 + x - 12) = x(x + 4)(x - 3)$$

Solving the equation $f(x) = x(x + 4)(x - 3) = 0$, we find that the x-intercepts, or zeros of f, are $-4, 0$, and 3.

(b) Since each zero of f is of multiplicity 1, the graph of f will cross the x-axis at each x-intercept.

(c) End behavior: the graph of f resembles that of the power function $y = x^3$ for large values of x.

(d) The graph of f will contain at most two turning points.

(e) The three x-intercepts divide the x-axis into four intervals:

$$-\infty < x < -4 \qquad -4 < x < 0 \qquad 0 < x < 3 \qquad 3 < x < \infty$$

To determine the sign of $f(x)$ in each interval, we choose test numbers and construct Figure 31(a).

(f) The graph of f is given in Figure 31(b).

FIGURE 31
$f(x) = x^3 + x^2 - 12x$

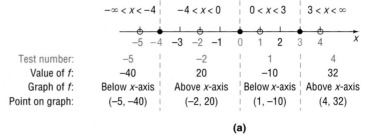

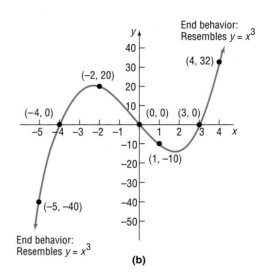

Exploration Graph $y = x^3 + x^2 - 12x$. Compare what you see with Figure 31. Use MAXIMUM/MINIMUM to locate the two turning points.

 Now work Problem 33.

E X A M P L E   7

Analyzing the Graph of a Polynomial Function

Follow the instructions for Example 6 for the following polynomial:

$$f(x) = x^2(x - 4)(x + 1)$$

Solution   (a) The y-intercept is $f(0) = 0$. The x-intercepts satisfy the equation

$$f(x) = x^2(x - 4)(x + 1) = 0$$

So

$$x^2 = 0 \quad \text{or} \quad x - 4 = 0 \quad \text{or} \quad x + 1 = 0$$
$$x = 0 \qquad\qquad x = 4 \qquad\qquad x = -1$$

The x-intercepts are $-1, 0,$ and 4.

(b) The intercept 0 is a zero of multiplicity 2, so the graph of f will touch the x-axis at 0; 4 and -1 are zeros of multiplicity 1, so the graph of f will cross the x-axis at 4 and -1.

(c) End behavior: the graph of f resembles that of the power function $y = x^4$ for large values of x.

(d) The graph of f will contain at most three turning points.

(e) The three x-intercepts divide the x-axis into four intervals:

$$-\infty < x < -1 \qquad -1 < x < 0 \qquad 0 < x < 4 \qquad 4 < x < \infty$$

To determine the sign of $f(x)$ in each interval, we choose test numbers and construct Figure 32(a).

(f) The graph of f is given in Figure 32(b).

FIGURE 32
$f(x) = x^2(x - 4)(x + 1)$

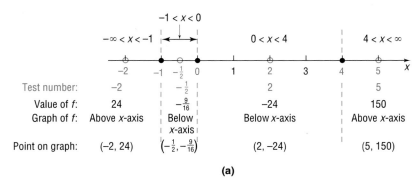

	$-\infty < x < -1$	$-1 < x < 0$	$0 < x < 4$	$4 < x < \infty$
Test number:	-2	$-\frac{1}{2}$	2	5
Value of f:	24	$-\frac{9}{16}$	-24	150
Graph of f:	Above x-axis	Below x-axis	Below x-axis	Above x-axis
Point on graph:	$(-2, 24)$	$\left(-\frac{1}{2}, -\frac{9}{16}\right)$	$(2, -24)$	$(5, 150)$

(a)

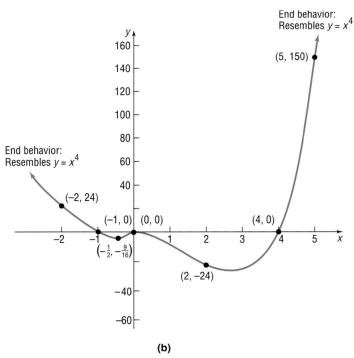

(b)

 Exploration Graph $y = x^2(x - 4)(x + 1)$. Compare what you see with Figure 32. Use MAXIMUM/MINIMUM to locate the two turning points besides $(0, 0)$.

Now work Problem 43.

E X A M P L E   8   Using a Graphing Utility to Graph a Polynomial Function

Graph: $f(x) = \pi x^4 - \sqrt{7} x^3 + 5x - 1$

Solution   We begin by making the following observations:

1.   The y-intercept is $f(0) = -1$.
2.   End behavior: the graph will behave like $y = \pi x^4$ for large values of x.
3.   The graph will have no more than four x-intercepts and no more than three turning points.

These observations help us set the viewing rectangle for our first attempt at a complete graph. See Figure 33(a). Figure 33(b) shows the complete graph.

FIGURE 33

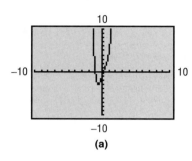

(a)

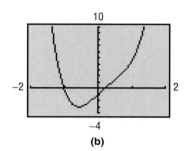

(b)

Using ROOT or ZERO, we find the x-intercepts to be -1.01 and 0.20, correct to two decimal places. Using MINIMUM, we find that the only turning point is located at $(-0.57, -3.02)$, correct to two decimal places. ■

SUMMARY

To analyze the graph of a polynomial function $y = f(x)$, follow these steps:

Steps for Analyzing
the Graph of a Polynomial

STEP 1: (a) Find the x-intercepts, if any, by solving the equation $f(x) = 0$.

(b) Find the y-intercept by letting $x = 0$ and finding the value of $f(0)$.

STEP 2: Determine whether the graph of f crosses or touches the x-axis at each x-intercept.

STEP 3: End behavior: find the power function that the graph of f resembles for large values of x.

STEP 4: Determine the maximum number of turning points on the graph of f.

STEP 5: Use the x-intercept(s) and test numbers to find the intervals on which the graph of f is above the x-axis and the intervals on which the graph is below the x-axis.

STEP 6: Plot the points obtained in Steps 1 and 5, and use the remaining information to connect them with a smooth, continuous curve.

5.2 | EXERCISES

In Problems 1–10, determine which functions are polynomial functions. For those that are, state the degree. For those that are not, tell why not.

1. $f(x) = 4x + x^3$

2. $f(x) = 5x^2 + 4x^4$

3. $g(x) = \dfrac{1 - x^2}{2}$

4. $h(x) = 3 - \dfrac{1}{2}x$

5. $f(x) = 1 - \dfrac{1}{x}$

6. $f(x) = x(x - 1)$

7. $g(x) = x^{3/2} - x^2 + 2$

8. $h(x) = \sqrt{x}(\sqrt{x} - 1)$

9. $F(x) = 5x^4 - \pi x^3 + \dfrac{1}{2}$

10. $F(x) = \dfrac{x^2 - 5}{x^3}$

In Problems 11–18, use transformations of the graph of $y = x^4$ to graph each function.

11. $f(x) = (x + 1)^4$

12. $f(x) = x^4 + 2$

13. $f(x) = \dfrac{1}{2}x^4$

14. $f(x) = -x^4$

15. $f(x) = 2(x + 1)^4 + 1$

16. $f(x) = 3 - (x + 2)^4$

17. $f(x) = -\dfrac{(x - 2)^4}{2} - 1$

18. $f(x) = 1 - \dfrac{(x + 1)^4}{3}$

In Problems 19–30, for each polynomial function, list each real zero and its multiplicity. Determine whether the graph crosses or touches the x-axis at each x-intercept. Find the power function that the graph of f resembles for large values of x.

19. $f(x) = 3(x - 7)(x + 3)^2$

20. $f(x) = 4(x + 4)(x + 3)^3$

21. $f(x) = 4(x^2 + 1)(x - 2)^3$

22. $f(x) = 2(x - 3)(x + 4)^3$

23. $f(x) = -2\left(x + \dfrac{1}{2}\right)^2(x^2 + 4)^2$

24. $f(x) = \left(x - \dfrac{1}{3}\right)^2(x - 1)^3$

25. $f(x) = (x - 5)^3(x + 4)^2$

26. $f(x) = (x + \sqrt{3})^2(x - 2)^4$

27. $f(x) = 3(x^2 + 8)(x^2 + 9)^2$

28. $f(x) = -2(x^2 + 3)^3$

29. $f(x) = -2x^2(x^2 - 2)$

30. $f(x) = 4x(x^2 - 3)$

In Problems 31–54, for each polynomial function f:
(a) Find the x- and y-intercepts of f.
(b) Determine whether the graph of f crosses or touches the x-axis at each x-intercept.
(c) End behavior: find the power function that the graph of f resembles for large values of x.
(d) Determine the maximum number of turning points on the graph of f.
(e) Use the x-intercept(s) and test numbers to find the intervals on which the graph of f is above and below the x-axis.
(f) Plot the points obtained in parts (a) and (e), and use the remaining information to connect them with a smooth, continuous curve.

31. $f(x) = (x - 1)^2$

32. $f(x) = (x - 2)^3$

33. $f(x) = x^2(x - 3)$

34. $f(x) = x(x + 2)^2$

35. $f(x) = 6x^3(x + 4)$

36. $f(x) = 5x(x - 1)^3$

37. $f(x) = -4x^2(x + 2)$

38. $f(x) = -\dfrac{1}{2}x^3(x + 4)$

39. $f(x) = x(x - 2)(x + 4)$

40. $f(x) = x(x + 4)(x - 3)$

41. $f(x) = 4x - x^3$

42. $f(x) = x - x^3$

43. $f(x) = x^2(x - 2)(x + 2)$

44. $f(x) = x^2(x - 3)(x + 4)$

45. $f(x) = x^2(x - 2)^2$

46. $f(x) = x^3(x - 3)$

47. $f(x) = x^2(x - 3)(x + 1)$

48. $f(x) = x^2(x - 3)(x - 1)$

49. $f(x) = x(x + 2)(x - 4)(x - 6)$

50. $f(x) = x(x - 2)(x + 2)(x + 4)$

51. $f(x) = x^2(x - 2)(x^2 + 3)$

52. $f(x) = x^2(x^2 + 1)(x + 4)$

53. $f(x) = -x^2(x^2 - 1)(x + 1)$

54. $f(x) = -x^2(x^2 - 4)(x - 5)$

55. Consult the illustration. Which of the following polynomial functions might have this graph? (More than one answer may be possible.)

(a) $y = -4x(x - 1)(x - 2)$
(b) $y = x^2(x - 1)^2(x - 2)$
(c) $y = 3x(x - 1)(x - 2)$
(d) $y = x(x - 1)^2(x - 2)^2$
(e) $y = x^3(x - 1)(x - 2)$
(f) $y = -x(1 - x)(x - 2)$

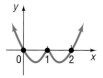

56. Consult the illustration. Which of the following polynomial functions might have this graph? (More than one answer may be possible.)

(a) $y = 2x^3(x - 1)(x - 2)^2$
(b) $y = x^2(x - 1)(x - 2)$
(c) $y = x^3(x - 1)^2(x - 2)$
(d) $y = x^2(x - 1)^2(x - 2)^2$
(e) $y = 5x(x - 1)^2(x - 2)$
(f) $y = -2x(x - 1)^2(2 - x)$

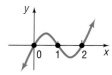

57. Consult the illustration. Which of the following polynomial functions might have this graph? (More than one answer may be possible.)

(a) $y = \frac{1}{2}(x^2 - 1)(x - 2)$

(b) $y = -\frac{1}{2}(x^2 + 1)(x - 2)$

(c) $y = (x^2 - 1)\left(1 - \frac{x}{2}\right)$

(d) $y = -\frac{1}{2}(x^2 - 1)^2(x - 2)$

(e) $y = \left(x^2 + \frac{1}{2}\right)(x^2 - 1)(2 - x)$

(f) $y = -(x - 1)(x - 2)(x + 1)$

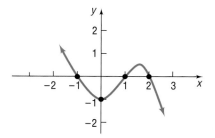

58. Consult the illustration. Which of the following polynomial functions might have this graph? (More than one answer may be possible.)

(a) $y = -\frac{1}{2}(x^2 - 1)(x - 2)(x + 1)$

(b) $y = -\frac{1}{2}(x^2 + 1)(x - 2)(x + 1)$

(c) $y = -\frac{1}{2}(x + 1)^2(x - 1)(x - 2)$

(d) $y = (x - 1)^2(x + 1)\left(1 - \frac{x}{2}\right)$

(e) $y = -(x - 1)^2(x - 2)(x + 1)$

(f) $y = -\left(x^2 + \frac{1}{2}\right)(x - 1)^2(x + 1)(x - 2)$

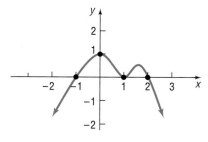

In Problems 59–72, graph each polynomial function. Approximate the x-intercepts, if any, and the turning points correct to two decimal places.

59. $f(x) = x^3 + 0.2x^2 - 1.5876x - 0.31752$

60. $f(x) = x^3 - 0.8x^2 - 4.6656x + 3.73248$

61. $f(x) = x^3 + 2.56x^2 - 3.31x + 0.89$

62. $f(x) = x^3 - 2.91x^2 - 7.668x - 3.8151$

63. $f(x) = x^4 - 2.5x^2 + 0.5625$

64. $f(x) = x^4 - 18.5x^2 + 50.2619$

65. $f(x) = x^4 + 0.65x^3 - 16.6319x^2 + 14.209335x - 3.1264785$

66. $f(x) = x^4 + 3.45x^3 - 11.6639x^2 - 5.864241x - 0.69257738$

67. $f(x) = \pi x^3 + \sqrt{2}x^2 - x - 2$

68. $f(x) = -2x^3 + \pi x^2 + \sqrt{3}x + 1$

69. $f(x) = 2x^4 - \pi x^3 + \sqrt{5}x - 4$

70. $f(x) = -1.2x^4 + 0.5x^2 - \sqrt{3}x + 2$

71. $f(x) = -2x^5 - \sqrt{2}x^2 - x - \sqrt{2}$

72. $f(x) = \pi x^5 + \pi x^4 + \sqrt{3}x + 1$

73. Motor Vehicle Thefts   The following data represent the number of motor vehicle thefts (in thousands) in the United States for the years 1983–1993, where 1 represents 1983, 2 represents 1984, and so on.

Year, x	Motor Vehicle Thefts, M
1983, 1	1008
1984, 2	1032
1985, 3	1103
1986, 4	1224
1987, 5	1289
1988, 6	1433
1989, 7	1565
1990, 8	1636
1991, 9	1662
1992, 10	1611
1993, 11	1561

Source: U.S. Federal Bureau of Investigation

(a) Draw a scatter diagram of the data. Comment on the type of relation that may exist between the two variables.

(b) The cubic function of best fit to these data is
$$M(x) = -2.4x^3 + 37.3x^2 - 70.4x + 1043.8$$
Use this function to predict the number of motor vehicle thefts in 1994.

(c) Use a graphing utility to verify that the function given in (b) is the cubic function of best fit.

(d) With a graphing utility, draw a scatter diagram of the data and then graph the cubic function of best fit on the scatter diagram.

 (e) Do you think that the function given in (b) will be useful in predicting the number of motor vehicle thefts in 1999?

74. Larceny Thefts   The following data represent the number of larceny thefts (in thousands) in the United States for the years 1983–1993, where 1 represents 1983, 2 represents 1984, and so on.

Year, x	Larceny Thefts, L
1983, 1	6713
1984, 2	6592
1985, 3	6926
1986, 4	7257
1987, 5	7500
1988, 6	7706
1989, 7	7872
1990, 8	7946
1991, 9	8142
1992, 10	7915
1993, 11	7821

Source: U.S. Federal Bureau of Investigation

(a) Draw a scatter diagram of the data. Comment on the type of relation that may exist between the two variables.

(b) The cubic function of best fit to these data is
$$L(x) = -4.37x^3 + 59.29x^2 - 14.02x + 6578$$
Use this function to predict the number of larceny thefts in 1994.

(c) Use a graphing utility to verify that the function given in (b) is the cubic function of best fit.

(d) With a graphing utility, draw a scatter diagram of the data and then graph the cubic function of best fit on the scatter diagram.

(e) Do you think the function found in (b) will be useful in predicting the number of larceny thefts in 1999?

75. Cost of Manufacturing The following data represent the cost C of manufacturing Chevy Cavaliers (in thousands of dollars) and the number x of Cavaliers produced.

Number of Cavaliers Produced, x	Cost, C
0	10
1	23
2	31
3	38
4	43
5	50
6	59
7	70
8	85
9	105
10	135

(a) Draw a scatter diagram of the data. Comment on the type of relation that may exist between the two variables.

(b) Find the average rate of change of cost between four and five Cavaliers. Economists call this the **marginal cost.**

(c) What is the marginal cost of producing between eight and nine Cavaliers?

(d) The cubic function of best fit to these data is

$$C(x) = 0.2x^3 - 2.3x^2 + 14.3x + 10.2$$

Use this function to predict the cost of manufacturing eleven Cavaliers.

 (e) Use a graphing utility to verify that the function given in (d) is the cubic function of best fit.

(f) With a graphing utility, draw a scatter diagram of the data and then graph the cubic function of best fit on the scatter diagram.

 (g) Interpret the y-intercept.

76. Cost of Printing The following data represent the weekly cost of printing textbooks C (in thousands) and the number of texts printed (in thousands), x.

(a) Draw a scatter diagram of the data. Comment on the type of relation that may exist between the two variables.

(b) Find the average rate of change of cost between 10,000 and 13,000 textbooks. Economists call this the **marginal cost.**

(c) What is the marginal cost of producing between 18,000 and 20,000 textbooks?

Number of Text Books, x	Cost, C
0	100
5	128.1
10	144
13	153.5
17	161.2
18	162.6
20	166.3
23	178.9
25	190.2
27	221.8

(d) The cubic function of best fit to these data is

$$C(x) = 0.015x^3 - 0.595x^2 + 9.15x + 98.43$$

Use this function to predict the cost of printing 22,000 texts per week.

(e) Use a graphing utility to verify that the function given in (d) is the cubic function of best fit.

(f) With a graphing utility, draw a scatter diagram of the data and then graph the cubic function of best fit on the scatter diagram.

(g) Interpret the y-intercept.

77. Can the graph of a polynomial function have no y-intercept? Can it have no x-intercepts? Explain.

78. Write a few paragraphs that provide a general strategy for graphing a polynomial function. Be sure to mention the following: degree, intercepts, and turning points.

79. Make up a polynomial that has the following characteristics: crosses the x-axis at -1 and 4, touches the x-axis at 0 and 2, and is above the x-axis between 0 and 2. Give your polynomial to a fellow classmate and ask for a written critique of your polynomial.

80. Make up two polynomials, not of the same degree, with the following characteristics: crosses the x-axis at -2, touches the x-axis at 1, and is above the x-axis between -2 and 1. Give your polynomials to a fellow classmate and ask for a written critique of your polynomials.

81. The graph of a polynomial function is always smooth and continuous. Name a function studied earlier that is smooth and not continuous. Name one that is continuous, but not smooth.

82. Which of the following statements are true regarding the graph of the cubic polynomial $f(x) = x^3 + bx^2 + cx + d$? (Give reasons for your conclusions.)

(a) It intersects the y-axis in one and only one point.

(b) It intersects the x-axis in at most three points.

(c) It intersects the x-axis at least once.

(d) For x very large, it behaves like the graph of $y = x^3$.

(e) It is symmetric with respect to the origin.

(f) It passes through the origin.

5.3 | RATIONAL FUNCTIONS

1 Find the Domain of a Rational Function

2 Determine the Vertical Asymptotes of a Rational Function

3 Determine the Horizontal or Oblique Asymptotes of a Rational Function

4 Analyze the Graph of a Rational Function

Ratios of integers are called *rational numbers.* Similarly, ratios of polynomial functions are called *rational functions.*

> A **rational function** is a function of the form
>
> $$R(x) = \frac{p(x)}{q(x)}$$
>
> where p and q are polynomial functions and q is not the zero polynomial. The domain consists of all real numbers except those for which the denominator q is 0.

E X A M P L E   1   Finding the Domain of a Rational Function

1 (a) The domain of $R(x) = \dfrac{2x^2 - 4}{x + 5}$ consists of all real numbers x except -5.

(b) The domain of $R(x) = \dfrac{1}{x^2 - 4}$ consists of all real numbers x except -2 and 2.

(c) The domain of $R(x) = \dfrac{x^3}{x^2 + 1}$ consists of all real numbers.

(d) The domain of $R(x) = \dfrac{-x^2 + 2}{3}$ consists of all real numbers.

(e) The domain of $R(x) = \dfrac{x^2 - 1}{x - 1}$ consists of all real numbers x except 1.   ∎

It is important to observe that the functions

$$R(x) = \frac{x^2 - 1}{x - 1} \quad \text{and} \quad f(x) = x + 1$$

are not equal, since the domain of R is $\{x \mid x \neq 1\}$ and the domain of f is all real numbers.

 Now work Problem 3.

If $R(x) = p(x)/q(x)$ is a rational function and if p and q have no common factors, then the rational function R is said to be in **lowest terms.** For a rational function $R(x) = p(x)/q(x)$ in lowest terms, the zeros, if any, of the numerator are the x-intercepts of the graph of R and so will play a major role in graphing R. The zeros of the denominator of R [that is, the numbers x, if any, for which $q(x) = 0$], although not in the domain of R, also play a major role in the graph of R. We will discuss this role shortly.

We have already discussed the characteristics of the rational function $f(x) = 1/x$. (Refer to Example 11, page 182.) The next rational function we take up is $H(x) = 1/x^2$.

E X A M P L E   2

Graphing $y = \dfrac{1}{x^2}$

Analyze the graph: $H(x) = \dfrac{1}{x^2}$

Solution   The domain of $H(x) = 1/x^2$ consists of all real numbers x except 0. Thus, the graph has no y-intercept, because x can never equal 0. The graph has no x-intercept, because the equation $H(x) = 0$ has no solution. Therefore, the graph of H will not cross either of the coordinate axes. Because

$$H(-x) = \frac{1}{(-x)^2} = \frac{1}{x^2} = H(x)$$

H is an even function, so its graph is symmetric with respect to the y-axis.

Table 4 shows the behavior of $H(x) = 1/x^2$ for selected positive numbers x (we will use symmetry to obtain the graph of H when $x < 0$). From Table 4, it is apparent that, as the values of x approach (get closer to) 0, the values of $H(x)$ become larger and larger positive numbers. When this happens, we say that H is **unbounded in the positive direction.** We symbolize this by writing $H \to \infty$ (read as "H **approaches infinity**"). In calculus, **limits** are used to convey these ideas. There we use the symbolism $\lim\limits_{x \to 0} H(x) = \infty$, read the limit of $H(x)$ as x approaches zero equals infinity, to mean that $H(x) \to \infty$ as $x \to 0$.

Look again at Table 4. As $x \to \infty$, the values of $H(x)$ approach 0 (the end behavior of the graph). In calculus, this is symbolized by writing $\lim\limits_{x \to \infty} H(x) = 0$. Figure 34 shows the graph.

TABLE 4

x	$H(x) = 1/x^2$
$\frac{1}{100}$	10,000
$\frac{1}{10}$	100
$\frac{1}{2}$	4
1	1
2	$\frac{1}{4}$
10	$\frac{1}{100}$
100	$\frac{1}{10,000}$

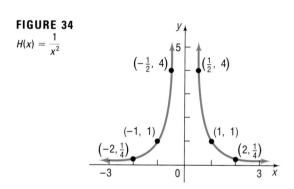

FIGURE 34
$H(x) = \dfrac{1}{x^2}$

Sometimes transformations (shifting, compressing, stretching, and reflection) can be used to graph a rational function.

E X A M P L E   3

Using Transformations to Graph a Rational Function

Graph the rational function: $R(x) = \dfrac{1}{(x-2)^2} + 1$

Solution   First, we take note of the fact that the domain of R consists of all real numbers except $x = 2$. To graph R, we start with the graph of $y = 1/x^2$. See Figure 35 for the steps.

FIGURE 35

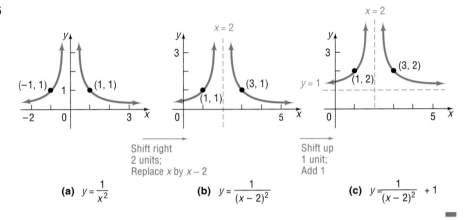

Shift right
2 units;
Replace x by $x - 2$

Shift up
1 unit;
Add 1

(a) $y = \dfrac{1}{x^2}$ (b) $y = \dfrac{1}{(x-2)^2}$ (c) $y = \dfrac{1}{(x-2)^2} + 1$

Now work Problem 21.

Asymptotes

In Figure 35(c), notice that as the values of x become more negative, that is, as x becomes **unbounded in the negative direction** ($x \to -\infty$, read as "x **approaches negative infinity**"), the values $R(x)$ approach 1. In fact, we can conclude the following from Figure 35(c):

1.   As $x \to -\infty$, the values $R(x)$ approach 1. $[\lim\limits_{x \to -\infty} R(x) = 1]$

2.   As x approaches 2, the values $R(x) \to \infty$. $[\lim\limits_{x \to 2} R(x) = \infty]$

3.   As $x \to \infty$, the values $R(x)$ approach 1. $[\lim\limits_{x \to \infty} R(x) = 1]$

This behavior of the graph is depicted by the vertical line $x = 2$ and the horizontal line $y = 1$. These lines are called *asymptotes* of the graph, which we define as follows:

Let R denote a function:

If, as $x \to -\infty$ or as $x \to \infty$, the values of $R(x)$ approach some fixed number, L, then the line $y = L$ is a **horizontal asymptote** of the graph of R.

If, as x approaches some number c, the values $|R(x)| \to \infty$, then the line $x = c$ is a **vertical asymptote** of the graph of R.

Even though asymptotes of a function are not part of the graph of the function, they provide information about the way the graph looks. Figure 36 illustrates some of the possibilities.

FIGURE 36

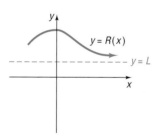

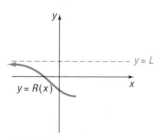

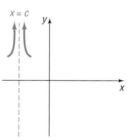

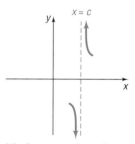

(a) End behavior: As $x \to \infty$, the values of $R(x)$ approach L. That is, the points on the graph of R are getting closer to the line $y = L$; $y = L$ is a horizontal asymptote.

(b) End behavior: As $x \to -\infty$, the values of $R(x)$ approach L. That is, the points on the graph of R are getting closer to the line $y = L$; $y = L$ is a horizontal asymptote.

(c) As x approaches c, the values of $|R(x)| \to \infty$, That is, the points on the graph of R are getting closer to the line $x = c$; $x = c$ is a vertical asymptote.

(d) As x approaches c, the values of $|R(x)| \to \infty$, That is, the points on the graph of R are getting closer to the line $x = c$; $x = c$ is a vertical asymptote.

FIGURE 37
Oblique asymptote

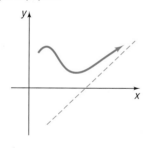

Thus, an asymptote is a line that a certain part of the graph of a function gets closer and closer to but never touches. However, other parts of the graph of the function may intersect a nonvertical asymptote. The graph of the function will never intersect a vertical asymptote. Notice that a horizontal asymptote, when it occurs, describes a certain behavior of the graph as $x \to \infty$ or as $x \to -\infty$, that is, its end behavior. A vertical asymptote, when it occurs, describes a certain behavior of the graph when x is close to some number c.

If an asymptote is neither horizontal nor vertical, it is called **oblique.** Figure 37 shows an oblique asymptote. An oblique asymptote, when it occurs, describes the end behavior of the graph.

Finding Asymptotes

The vertical asymptotes, if any, of a rational function $R(x) = p(x)/q(x)$, in lowest terms, are found by factoring the denominator of $q(x)$. Suppose that $x - r$ is a factor of the denominator. Now, as x approaches r, symbolized as $x \to r$, the values of $x - r$ approach 0, causing the ratio to become unbounded, that is, causing $|R(x)| \to \infty$. Based on the definition, we conclude that the line $x = r$ is a vertical asymptote.

> **Theorem** Locating Vertical Asymptotes
>
> A rational function $R(x) = p(x)/q(x)$, *in lowest terms,* will have a vertical asymptote $x = r$, if $x - r$ is a factor of the denominator q. That is, if r is a zero of the denominator of a rational function $R(x) = p(x)/q(x)$, in lowest terms, then R will have the vertical asymptote $x = r$.

Warning: If a rational function is not in lowest terms, an application of this theorem may result in an incorrect listing of vertical asymptotes.

E X A M P L E   4

Finding Vertical Asymptotes

Find the vertical asymptotes, if any, of the graph of each rational function.

(a) $R(x) = \dfrac{x}{x^2 - 4}$ (b) $F(x) = \dfrac{x + 3}{x - 1}$ (c) $H(x) = \dfrac{x^2}{x^2 + 1}$

(d) $G(x) = \dfrac{x^2 - 9}{x^2 + 4x - 21}$

Solution

(a) The zeros of the denominator $x^2 - 4$ are -2 and 2. Hence, the lines $x = -2$ and $x = 2$ are the vertical asymptotes of the graph of R.

(b) The only zero of the denominator is 1. Hence, the line $x = 1$ is the only vertical asymptote of the graph of F.

(c) The denominator has no zeros. Hence, the graph of H has no vertical asymptotes.

(d) Factor $G(x)$ to determine if it is in lowest terms.

$$G(x) = \frac{x^2 - 9}{x^2 + 4x - 21} = \frac{(x + 3)(x - 3)}{(x + 7)(x - 3)} = \frac{x + 3}{x + 7}$$

The only zero of the denominator of $G(x)$ in lowest terms is -7. Hence, the line $x = -7$ is the only vertical asymptote of the graph of G. ▪

As Example 4 points out, rational functions can have no vertical asymptotes, one vertical asymptote, or more than one vertical asymptote. However, the graph of a rational function will never intersect any of its vertical asymptotes. (Do you know why?)

 Exploration Graph each of the following rational functions:

$$Y_1 = \frac{1}{x - 1} \qquad Y_2 = \frac{-1}{x - 1} \qquad Y_3 = \frac{1}{(x - 1)^2} \qquad Y_4 = \frac{-1}{(x - 1)^2}$$

Each has the vertical asymptote $x = 1$. Use TRACE (or a TABLE) to see what happens as x approaches 1. Be sure to look at both sides of $x = 1$. ▪

3   The procedure for finding horizontal and oblique asymptotes is somewhat more involved. To find such asymptotes, we need to know how the values of a function behave as $x \to -\infty$ or as $x \to \infty$.

If a rational function $R(x)$ is **proper,** that is, if the degree of the numerator is less than the degree of the denominator, then, as $x \to -\infty$ or as $x \to \infty$, the values of $R(x)$ approach 0. Consequently, the line $y = 0$ (the x-axis) is a horizontal asymptote of the graph.

Theorem

If a rational function is proper, the line $y = 0$ is a horizontal asymptote of its graph.

E X A M P L E   5

Finding Horizontal Asymptotes

Find the horizontal asymptotes, if any, of the graph of

$$R(x) = \frac{x - 12}{4x^2 + x + 1}$$

Solution   The rational function R is proper, since the degree of the numerator, 1, is less than the degree of the denominator, 2. We conclude that the line $y = 0$ is a horizontal asymptote of the graph of R. ∎

To see why $y = 0$ is a horizontal asymptote of the function R in Example 5, we need to investigate the behavior of R for x unbounded. When x is unbounded, the numerator of R, which is $x - 12$, can be approximated by the power function $y = x$, while the denominator of R, which is $4x^2 + x + 1$, can be approximated by the power function $y = 4x^2$.
Thus,

$$R(x) = \underset{\substack{\uparrow \\ \text{For } x \text{ unbounded}}}{\frac{x - 12}{4x^2 + x + 1}} \approx \frac{x}{4x^2} = \underset{\substack{\uparrow \\ \text{As } x \to -\infty \text{ or } x \to \infty}}{\frac{1}{4x} \to 0}$$

This shows that the line $y = 0$ is a horizontal asymptote of the graph of R.

If a rational function $R(x)$ is **improper,** that is, if the degree of the numerator is greater than or equal to the degree of the denominator, we must use long division to write the rational function as the sum of a polynomial $f(x)$ plus a proper rational function $r(x)$. That is, we write

$$R(x) = \frac{p(x)}{q(x)} = f(x) + r(x)$$

where $f(x)$ is a polynomial and $r(x)$ is a proper rational function. Since $r(x)$ is proper, then $r(x) \to 0$ as $x \to -\infty$ or as $x \to \infty$. Thus,

$$R(x) = \frac{p(x)}{q(x)} \to f(x) \qquad \text{as} \qquad x \to -\infty \qquad \text{or as} \qquad x \to \infty$$

The possibilities are listed next:

1. If $f(x) = b$, a constant, then the line $y = b$ is a horizontal asymptote of the graph of R.
2. If $f(x) = ax + b, a \neq 0$, then the line $y = ax + b$ is an oblique asymptote of the graph of R.
3. In all other cases, the graph of R approaches the graph of f, and there are no horizontal or oblique asymptotes.

The following examples demonstrate these conclusions.

E X A M P L E   6

Finding Horizontal or Oblique Asymptotes

Find the horizontal or oblique asymptotes, if any, of the graph of

$$H(x) = \frac{3x^4 - x^2}{x^3 - x^2 + 1}$$

Solution   The rational function H is improper, since the degree of the numerator, 4, is larger than the degree of the denominator, 3. To find any horizontal or oblique asymptotes, we use long division:

$$
\begin{array}{r}
3x + 3 \\
x^3 - x^2 + 1 \overline{\smash{\big)}\, 3x^4 - x^2 } \\
\underline{3x^4 - 3x^3 + 3x} \\
3x^3 - x^2 - 3x \\
\underline{3x^3 - 3x^2 + 3} \\
2x^2 - 3x - 3
\end{array}
$$

Thus,

$$
H(x) = \frac{3x^4 - x^2}{x^3 - x^2 + 1} = 3x + 3 + \frac{2x^2 - 3x - 3}{x^3 - x^2 + 1}
$$

Then, as $x \to -\infty$ or as $x \to \infty$, the remainder will behave as follows:

$$
\frac{2x^2 - 3x - 3}{x^3 - x^2 + 1} \approx \frac{2x^2}{x^3} = \frac{2}{x} \to 0
$$

Thus, as $x \to -\infty$ or as $x \to \infty$, we have $H(x) \to 3x + 3$. We conclude that the graph of the rational function H has an oblique asymptote $y = 3x + 3$. ∎

E X A M P L E   7

Finding Horizontal or Oblique Asymptotes

Find the horizontal or oblique asymptotes, if any, of the graph of

$$
R(x) = \frac{8x^2 - x + 2}{4x^2 - 1}
$$

Solution   The rational function R is improper, since the degree of the numerator, 2, equals the degree of the denominator, 2. To find any horizontal or oblique asymptotes, we use long division:

$$
\begin{array}{r}
2 \\
4x^2 - 1 \overline{\smash{\big)}\, 8x^2 - x + 2} \\
\underline{8x^2 - 2} \\
-x + 4
\end{array}
$$

Thus,

$$
R(x) = \frac{8x^2 - x + 2}{4x^2 - 1} = 2 + \frac{-x + 4}{4x^2 - 1}
$$

Then, as $x \to -\infty$ or as $x \to \infty$, the remainder will behave as follows:

$$\frac{-x + 4}{4x^2 - 1} \approx \frac{-x}{4x^2} = \frac{-1}{4x} \to 0$$

Thus, as $x \to -\infty$ or as $x \to \infty$, we have $R(x) \to 2$. We conclude that $y = 2$ is a horizontal asymptote of the graph. ▬

In Example 7, note that the quotient 2 obtained by long division is the quotient of the leading coefficients of the numerator polynomial and the denominator polynomial $\left(\frac{8}{4}\right)$. This means we can avoid the long division process for rational functions whose numerator and denominator *are of the same degree* and conclude that the quotient of the leading coefficients will give us the horizontal asymptote.

Now work Problem 39.

E X A M P L E 8 Finding Horizontal or Oblique Asymptotes

Find the horizontal or oblique asymptotes, if any, of the graph of

$$G(x) = \frac{2x^5 - x^3 + 2}{x^3 - 1}$$

Solution The rational function G is improper, since the degree of the numerator, 5, is larger than the degree of the denominator, 3. To find any horizontal or oblique asymptotes, we use long division:

$$
\begin{array}{r}
2x^2 - 1 \\
x^3 - 1 \overline{) 2x^5 - x^3 + 2 } \\
\underline{2x^5 - 2x^2 } \\
-x^3 + 2x^2 + 2 \\
\underline{-x^3 + 1 } \\
2x^2 + 1
\end{array}
$$

Thus,

$$G(x) = \frac{2x^5 - x^3 + 2}{x^3 - 1} = 2x^2 - 1 + \frac{2x^2 + 1}{x^3 - 1}$$

Then, as $x \to -\infty$ or as $x \to \infty$, the remainder will behave as follows:

$$\frac{2x^2 + 1}{x^3 - 1} \approx \frac{2x^2}{x^3} = \frac{2}{x} \to 0$$

Thus, as $x \to -\infty$ or as $x \to \infty$, we have $G(x) \to 2x^2 - 1$. We conclude that, for large values of x, the graph of G approaches the graph of $y = 2x^2 - 1$. Since $y = 2x^2 - 1$ is not a linear function, G has no horizontal or oblique asymptotes. ▬

We summarize next the procedure for finding horizontal and oblique asymptotes.

Finding Horizontal and Oblique Asymptotes of a Rational Function *R*

Consider the rational function

$$R(x) = \frac{p(x)}{q(x)} = \frac{a_n x^n + a_{n-1}x^{n-1} + \cdots + a_1 x + a_0}{b_m x^m + b_{m-1}x^{m-1} + \cdots + b_1 x + b_0}$$

in which the degree of the numerator is n and the degree of the denominator is m.

1. If $n < m$, then R is a proper rational function, and the graph of R will have the horizontal asymptote $y = 0$ (the x-axis).
2. If $n \geq m$, then R is improper. Here long division is used.
   (a) If $n = m$, the quotient obtained will be a number $L\,(= a_n/b_m)$ and the line $y = L(= a_n/b_m)$, is a horizontal asymptote.
   (b) If $n = m + 1$, the quotient obtained is of the form $ax + b$ (a polynomial of degree 1), and the line $y = ax + b$ is an oblique asymptote.
   (c) If $n > m + 1$, the quotient obtained is a polynomial of degree 2 or higher, and R has neither a horizontal nor an oblique asymptote. In this case, for x unbounded, the graph of R will behave like the graph of the quotient.

Note: The graph of a rational function either has one horizontal or one oblique asymptote or else has no horizontal or no oblique asymptote.

Now we are ready to graph rational functions.

Graphing Rational Functions

We commented earlier that calculus provides the tools required to graph a polynomial function accurately. The same holds true for rational functions. However, we can gather together quite a bit of information about their graphs to get an idea of the general shape and position of the graph.

In the examples that follow, we will analyze the graph of a rational function $R(x) = p(x)/q(x)$ by applying the following steps:

Analyzing the Graph of a Rational Function

STEP 1: Find the domain of the rational function.
STEP 2: Locate the intercepts, if any, of the graph. The x-intercepts, if any, of $R(x) = p(x)/q(x)$ satisfy the equation $p(x) = 0$. The y-intercept, if there is one, is $R(0)$.
STEP 3: Test for symmetry. Replace x by $-x$ in $R(x)$. If $R(-x) = R(x)$, there is symmetry with respect to the y-axis; if $R(-x) = -R(x)$, there is symmetry with respect to the origin.
STEP 4: Write R in lowest terms and find the real zeros of the denominator. With R in lowest terms, each zero will give rise to a vertical asymptote.
STEP 5: Locate the horizontal or oblique asymptotes, if any, using the procedure given earlier. Determine points, if any, at which the graph of R intersects these asymptotes.
STEP 6: Determine where the graph is above the x-axis and where the graph is below the x-axis.
STEP 7: Graph the asymptotes, if any, found in Steps 4 and 5. Plot the points found in Steps 2, 5, and 6. Use all the information to connect the points with a smooth curve.

E X A M P L E   9   Analyzing the Graph of a Rational Function

Analyze the graph of the rational function: $R(x) = \dfrac{x - 1}{x^2 - 4}$.

Solution   First, we factor both the numerator and the denominator of R:

$$R(x) = \frac{x - 1}{(x + 2)(x - 2)}$$

R is in lowest terms.

STEP 1: The domain of R is $\{x \mid x \neq -2, x \neq 2\}$.

STEP 2: We locate the x-intercepts by finding the zeros of the numerator. By inspection, 1 is the only x-intercept. The y-intercept is $R(0) = \frac{1}{4}$.

STEP 3: Because

$$R(-x) = \frac{-x - 1}{x^2 - 4}$$

we conclude that R is neither even nor odd. Thus, there is no symmetry with respect to the y-axis or the origin.

STEP 4: We locate the vertical asymptotes by factoring the denominator: $x^2 - 4 = (x + 2)(x - 2)$. Since R is in lowest terms, the graph of R has two vertical asymptotes: the lines $x = -2$ and $x = 2$.

STEP 5: The degree of the numerator is less than the degree of the denominator, so R is proper and the line $y = 0$ (the x-axis) is a horizontal asymptote of the graph. To determine if the graph of R intersects the horizontal asymptote, we solve the equation $R(x) = 0$. The only solution is $x = 1$, so the graph of R intersects the horizontal asymptote at $(1, 0)$.

STEP 6: The zero of the numerator, 1, and the zeros of the denominator, -2 and 2, divide the x-axis into four intervals:

$$-\infty < x < -2 \qquad -2 < x < 1 \qquad 1 < x < 2 \qquad 2 < x < \infty$$

Now we construct Figure 38.

FIGURE 38

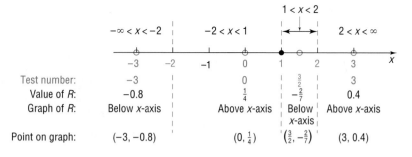

	$-\infty < x < -2$	$-2 < x < 1$	$1 < x < 2$	$2 < x < \infty$
Test number:	-3	0	$\frac{3}{2}$	3
Value of R:	-0.8	$\frac{1}{4}$	$-\frac{2}{7}$	0.4
Graph of R:	Below x-axis	Above x-axis	Below x-axis	Above x-axis
Point on graph:	$(-3, -0.8)$	$(0, \frac{1}{4})$	$(\frac{3}{2}, -\frac{2}{7})$	$(3, 0.4)$

STEP 7: We begin by graphing the asymptotes and plotting the points found in Steps 2, 5, and 6. See Figure 39(a). Next, we determine the behavior of the graph near the asymptotes. Since the x-axis is a horizontal asymptote and the graph lies below the x-axis for $-\infty < x < -2$, we can sketch a portion of the graph by placing a small arrow to the far left and under the x-axis. Since the line $x = -2$ is a vertical asymptote and the graph lies below the x-axis for $-\infty < x < -2$,

we continue the sketch with an arrow placed well below the x-axis and approaching the line $x = -2$ on the left. Similar explanations account for the positions of the other portions of the graph. In particular, note how we use the facts that the graph lies above the x-axis for $-2 < x < 1$ and below the x-axis for $1 < x < 2$ and $(1, 0)$ is an x-intercept to draw the conclusion that the graph crosses the x-axis at $(1, 0)$. Figure 39(b) shows the complete sketch.

FIGURE 39

$R(x) = \dfrac{x - 1}{x^2 - 4}$

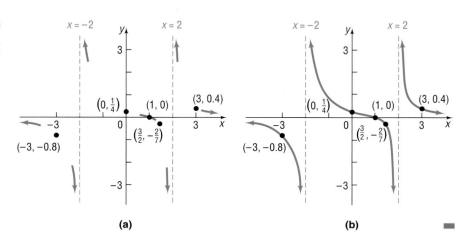

(a)

(b)

Exploration Graph $R(x) = (x - 1)/(x^2 - 4)$.

Solution The analysis just completed in Example 9 helps us to set the viewing rectangle to obtain a complete graph. Figure 40(a) shows the graph of $R(x) = \dfrac{x - 1}{x^2 - 4}$ in connected mode, and Figure 40(b) shows it in dot mode. Notice in Figure 40(a) that the graph has vertical lines at $x = -2$ and $x = 2$. This is due to the fact that, when the graphing utility is in connected mode, it will "connect the dots" between consecutive pixels. We know that the graph of R does not cross the lines $x = -2$ and $x = 2$, since R is not defined at $x = -2$ or $x = 2$. Thus, when graphing rational functions, dot mode should be used to avoid extraneous vertical lines that are not part of the graph.

FIGURE 40

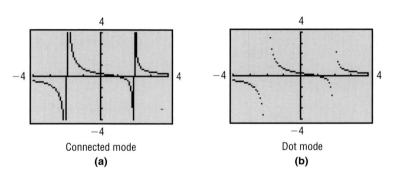

Connected mode

(a)

Dot mode

(b)

Now work Problem 47.

E X A M P L E   10   Analyzing the Graph of a Rational Function

Analyze the graph of the rational function: $R(x) = \dfrac{x^2 - 1}{x}$.

Solution   STEP 1: The domain of R is $\{x \mid x \neq 0\}$.
STEP 2: The graph has two x-intercepts: -1 and 1. There is no y-intercept.
STEP 3: Since $R(-x) = -R(x)$, the function is odd and the graph is symmetric with respect to the origin.
STEP 4: Since R is in lowest terms, the graph of $R(x)$ has the line $x = 0$ (the y-axis) as a vertical asymptote.
STEP 5: The rational function R is improper, since the degree of the numerator, 2, is larger than the degree of the denominator, 1. To find any horizontal or oblique asymptotes, we use long division:

$$
\begin{array}{r}
x \\
x{\overline{\smash{\big)}\,x^2 - 1}} \\
\underline{x^2 } \\
-1
\end{array}
$$

The quotient is x, so the line $y = x$ is an oblique asymptote of the graph. To determine whether the graph of R intersects the asymptote $y = x$, we solve the equation $R(x) = x$:

$$R(x) = \frac{x^2 - 1}{x} = x$$

$$x^2 - 1 = x^2$$

$$-1 = 0 \quad \text{Impossible.}$$

We conclude that the equation $(x^2 - 1)/x = x$ has no solution, so the graph of $R(x)$ does not intersect the line $y = x$.
STEP 6: The zeros of the numerator are -1 and 1; the denominator has the zero 0. Thus, we divide the x-axis into four intervals:

$$-\infty < x < -1 \qquad -1 < x < 0 \qquad 0 < x < 1 \qquad 1 < x < \infty$$

Now we construct Figure 41.

FIGURE 41

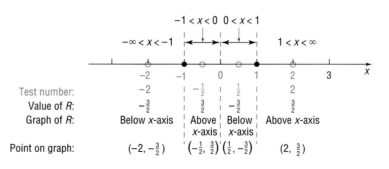

STEP 7: Figure 42(a) shows a partial graph using the facts that we have gathered. The complete graph is given in Figure 42(b).

FIGURE 42

$R(x) = \dfrac{x^2 - 1}{x}$

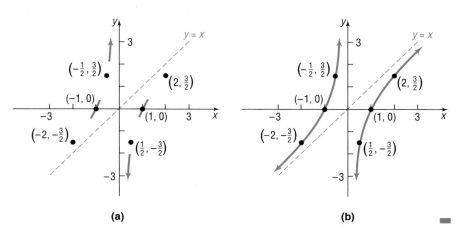

(a)   (b)

Seeing the Concept Graph $R(x) = (x^2 - 1)/x$ and compare what you see with Figure 42. Could you have predicted from the graph that $y = x$ is an oblique asymptote? Graph $y = x$ and ZOOM-OUT. What do you observe?

Now work Problem 51.

E X A M P L E   11

Analyzing the Graph of a Rational Function

Analyze the graph of the rational function: $R(x) = \dfrac{x^4 + 1}{x^2}$

Solution   STEP 1: The domain of R is $\{x | x \neq 0\}$.

STEP 2: The graph has no x-intercepts and no y-intercepts.

STEP 3: Since $R(-x) = R(x)$, the function is even and the graph is symmetric with respect to the y-axis.

STEP 4: Since R is in lowest terms, the graph of $R(x)$ has the line $x = 0$ (the y-axis) as a vertical asymptote.

STEP 5: The rational function R is improper. To find any horizontal or oblique asymptotes, we use long division:

$$
\begin{array}{r}
x^2 \\
x^2{\overline{\smash{)}\,x^4 + 1}} \\
\underline{x^4 } \\
1
\end{array}
$$

The quotient is x^2, so the graph has no horizontal or oblique asymptotes. However, the graph of R will approach the graph of $y = x^2$ as $x \to -\infty$ and as $x \to \infty$.

STEP 6: The numerator has no zeros, and the denominator has one zero at 0. Thus, we divide the x-axis into the following two intervals: $-\infty < x < 0$ and $0 < x < \infty$. Now we set up Figure 43.

FIGURE 43

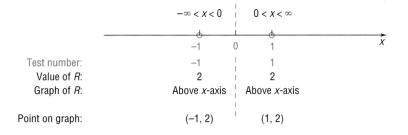

STEP 7: Figure 44 shows the graph.

FIGURE 44
$$R(x) = \frac{x^4 + 1}{x^2}$$

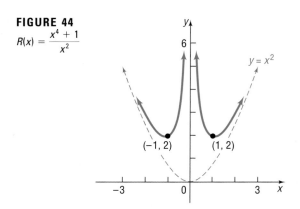

 Seeing the Concept Graph $R(x) = (x^4 + 1)/x^2$ and compare what you see with Figure 44. Use MINIMUM to find the two turning points. Enter $y = x^2$ and ZOOM-OUT. What do you see?

Now work Problem 49.

E X A M P L E   12   Analyzing the Graph of a Rational Function

Analyze the graph of the rational function: $R(x) = \dfrac{3x^2 - 3x}{x^2 + x - 12}$

Solution   We factor R to get

$$R(x) = \frac{3x(x - 1)}{(x + 4)(x - 3)}$$

STEP 1: The domain of R is $\{x | x \neq -4, x \neq 3\}$.

STEP 2: The graph has two x-intercepts: 0 and 1. The y-intercept is $R(0) = 0$.

STEP 3: There is no symmetry with respect to the y-axis or the origin.

STEP 4: Since R is in lowest terms, the graph of R has two vertical asymptotes: $x = -4$ and $x = 3$.

STEP 5: Since the degree of the numerator equals the degree of the denominator, the graph has a horizontal asymptote. To find it, we form the quotient of the leading coefficient of the numerator, 3, and the leading coefficient of the denominator, 1. Thus, the graph of R has the horizontal asymptote $y = 3$. To find out whether the graph of R intersects the asymptote, we solve the equation $R(x) = 3$.

$$R(x) = \frac{3x^2 - 3x}{x^2 + x - 12} = 3$$
$$3x^2 - 3x = 3x^2 + 3x - 36$$
$$-6x = -36$$
$$x = 6$$

Thus, the graph intersects the line $y = 3$ only at $x = 6$, and $(6, 3)$ is a point on the graph of R.

STEP 6: The zeros of the numerator, 0 and 1, and the zeros of the denominator, -4 and 3, divide the x-axis into five intervals:

$$-\infty < x < -4 \qquad -4 < x < 0 \qquad 0 < x < 1 \qquad 1 < x < 3 \qquad 3 < x < \infty$$

Now we can construct Figure 45.

FIGURE 45

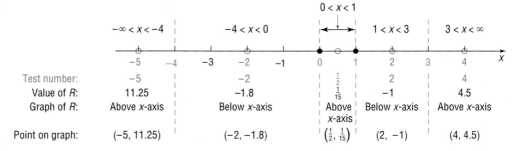

	$-\infty < x < -4$	$-4 < x < 0$	$0 < x < 1$	$1 < x < 3$	$3 < x < \infty$
Test number:	-5	-2	$\frac{1}{2}$	2	4
Value of R:	11.25	-1.8	$\frac{1}{15}$	-1	4.5
Graph of R:	Above x-axis	Below x-axis	Above x-axis	Below x-axis	Above x-axis
Point on graph:	$(-5, 11.25)$	$(-2, -1.8)$	$\left(\frac{1}{2}, \frac{1}{15}\right)$	$(2, -1)$	$(4, 4.5)$

STEP 7: Figure 46(a) shows a partial graph. Notice that we have not yet used the fact that the line $y = 3$ is a horizontal asymptote, because we do not know yet whether the graph of R crosses or touches the line $y = 3$ at $(6, 3)$. To see whether the graph, in fact, crosses or touches the line $y = 3$, we plot an additional point to the right of $(6, 3)$. We use $x = 7$ to find $R(7) = \frac{63}{22} < 3$. Thus, the graph crosses $y = 3$ at $x = 6$. Because $(6, 3)$ is the only point where the graph of R intersects the asymptote $y = 3$, the graph must approach

the line $y = 3$ from above as $x \to -\infty$ and approach the line $y = 3$ from below as $x \to \infty$. See Figure 46(b). The completed graph is shown in Figure 46(c).

FIGURE 46

$$R(x) = \frac{3x^2 - 3x}{x^2 + x - 12}$$

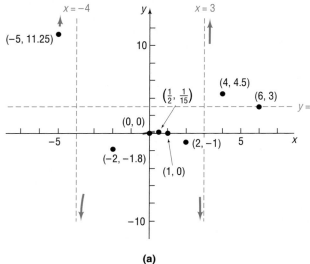

(a)

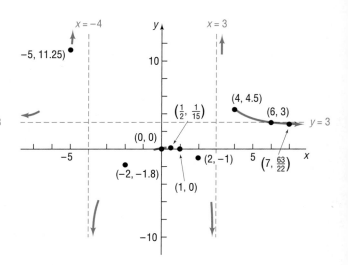

(b)

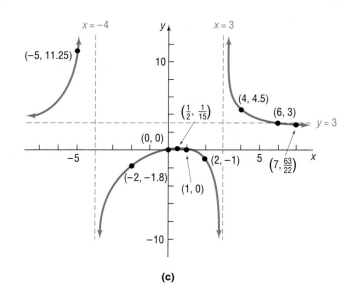

(c)

Exploration Graph $R(x) = \dfrac{3x^2 - 3x}{x^2 + x - 12}$.

Solution Figure 47 shows the graph in connected mode, and Figure 48(a) shows it in dot mode. Neither graph displays well the behavior between the two x-intercepts, 0 and 1. Nor do they display well the fact that the graph crosses the horizontal asymptote at (6, 3). To see these parts better, we use BOX and graph R for $-1 \le x \le 2$, Figure 48(b), and for $4 \le x \le 60$, Figure 49(b).

FIGURE 47

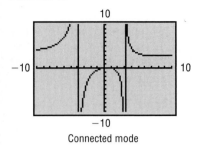

Connected mode

FIGURE 48

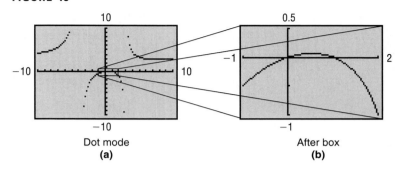

Dot mode
(a)

After box
(b)

FIGURE 49

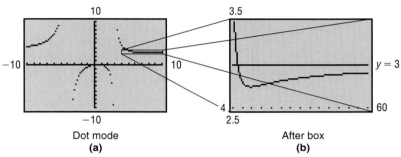

Dot mode
(a)

After box
(b)

The new graphs reflect the behavior produced by the analysis. Further, we observe two turning points, one between 0 and 1 and the other to the right of 4. Correct to two decimal places, these turning points are $(0.52, 0.06)$ and $(11.47, 2.74)$. ∎

E X A M P L E   13

Analyzing the Graph of a Rational Function with a Hole

Analyze the graph of the rational function: $R(x) = \dfrac{2x^2 - 5x + 2}{x^2 - 4}$.

Solution   We factor R and obtain

$$R(x) = \frac{(2x - 1)(x - 2)}{(x + 2)(x - 2)}$$

In lowest terms,

$$R(x) = \frac{2x - 1}{x + 2}$$

STEP 1: The domain of R is $\{x \mid x \ne -2, x \ne 2\}$.
STEP 2: The graph has one x-intercept: $1/2$. The y-intercept is $R(0) = -1/2$.
STEP 3: Because

$$R(-x) = \frac{2x^2 + 5x + 2}{x^2 - 4}$$

we conclude that R is neither even nor odd. Thus, there is no symmetry with respect to the y-axis or the origin.

STEP 4: The graph has one vertical asymptote: $x = -2$, since $x + 2$ is the only factor of the denominator of $R(x)$ *in lowest terms.* However, the rational function is undefined at both $x = 2$ and $x = -2$.

STEP 5: Since the degree of the numerator equals the degree of the denominator, the graph has a horizontal asymptote. To find it, we form the quotient of the leading coefficient of the numerator, 2, and the leading coefficient of the denominator, 1. Thus, the graph of R has the horizontal asymptote $y = 2$. To find out whether the graph of R intersects the asymptote, we solve the equation $R(x) = 2$.

$$R(x) = \frac{2x - 1}{x + 2} = 2$$
$$2x - 1 = 2(x + 2)$$
$$2x - 1 = 2x + 4$$
$$-1 = 4 \qquad \text{Impossible.}$$

Thus, the graph does not intersect the line $y = 2$.

STEP 6: The zeros of the numerator and denominator, -2, 1/2, and 2, divide the x-axis into four intervals:

$$-\infty < x < -2 \qquad -2 < x < 1/2 \qquad 1/2 < x < 2 \qquad 2 < x < \infty$$

Now we construct Figure 50.

FIGURE 50

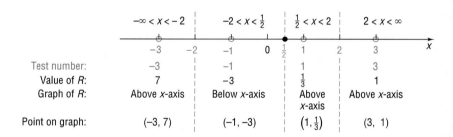

	$-\infty < x < -2$	$-2 < x < \frac{1}{2}$	$\frac{1}{2} < x < 2$	$2 < x < \infty$
Test number:	-3	-1	1	3
Value of R:	7	-3	$\frac{1}{3}$	1
Graph of R:	Above x-axis	Below x-axis	Above x-axis	Above x-axis
Point on graph:	$(-3, 7)$	$(-1, -3)$	$\left(1, \frac{1}{3}\right)$	$(3, 1)$

STEP 7: See Figure 51. Notice the vertical asymptote $x = -2$ and the hole at the point (2, 3/4). R is not defined at -2 and 2.

FIGURE 51

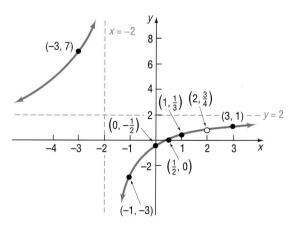

As Example 13 shows, the zeros of the denominator of a rational function give rise to either vertical asymptotes or holes on the graph.

Exploration   Graph: $R(x) = \dfrac{2x^2 - 5x + 2}{x^2 - 4}$. Do you see the hole at $(2, 3/4)$?

TRACE along the graph. Did you obtain an ERROR at $x = 2$? Are you convinced that an algebraic analysis of a rational function is required in order to accurately interpret the graph obtained with a graphing utility?

E X A M P L E   14   Finding the Least Cost of a Can

Reynolds Metal Company manufacturers aluminum cans in the shape of a cylinder with a capacity of 500 cubic centimeters ($\frac{1}{2}$ liter). The top and bottom of the can will be made of a special aluminum alloy that costs 0.05¢ per square centimeter. The sides of the can are to be made of material that costs 0.02¢ per square centimeter.

(a) Express the cost of material for the can as a function of the radius r of the can.
(b) Use a graphing utility to graph the function $C = C(r)$.
(c) What value of r will result in the least cost?
(d) What is this least cost?

FIGURE 52

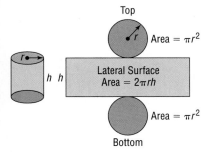

Top
r Area $= \pi r^2$

Lateral Surface
Area $= 2\pi rh$

h h

Area $= \pi r^2$

Bottom

Solution (a) Figure 52 illustrates the situation. Notice that the material required to produce a cylindrical can of height h and radius r consists of a rectangle of area $2\pi rh$ and two circles, each of area πr^2. The total cost C (in cents) of manufacturing the can is therefore

$$C = \text{cost of top and bottom} + \text{cost of side}$$
$$= \underbrace{2(\pi r^2)}_{\substack{\text{Total area} \\ \text{of top and} \\ \text{bottom}}} \underbrace{(0.05)}_{\substack{\text{Cost/unit} \\ \text{area}}} + \underbrace{(2\pi rh)}_{\substack{\text{Total} \\ \text{area of} \\ \text{side}}} \underbrace{(0.02)}_{\substack{\text{Cost/unit} \\ \text{area}}}$$
$$= 0.10\pi r^2 + 0.04\pi rh$$

But we have the additional restriction that the height h and radius r must be chosen so that the volume V of the can is 500 cubic centimeters. Since $V = \pi r^2 h$, we have

$$500 = \pi r^2 h \quad \text{or} \quad h = \frac{500}{\pi r^2}$$

FIGURE 53

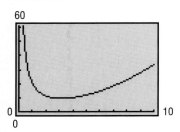

Thus, the cost C, in cents, as a function of the radius r is

$$C(r) = 0.10\pi r^2 + 0.04\pi r\left(\frac{500}{\pi r^2}\right) = 0.10\pi r^2 + \frac{20}{r} = \frac{0.10\pi r^3 + 20}{r}$$

(b) See Figure 53.
(c) Using the MINIMUM command, the cost is least for a radius of about 3.16 centimeters.
(d) The least cost is $C(3.16) = 9.46$¢.

5.3 | EXERCISES

In Problems 1–12, find the domain of each rational function.

1. $R(x) = \dfrac{4x}{x - 3}$

2. $R(x) = \dfrac{5x^2}{3 + x}$

3. $H(x) = \dfrac{-4x^2}{(x - 2)(x + 4)}$

4. $G(x) = \dfrac{6}{(x + 3)(4 - x)}$

5. $F(x) = \dfrac{3x(x - 1)}{2x^2 - 5x - 3}$

6. $Q(x) = \dfrac{-x(1 - x)}{3x^2 + 5x - 2}$

7. $R(x) = \dfrac{x}{x^3 - 8}$

8. $R(x) = \dfrac{x}{x^4 - 1}$

9. $H(x) = \dfrac{3x^2 + x}{x^2 + 4}$

10. $G(x) = \dfrac{x - 3}{x^4 + 1}$

11. $R(x) = \dfrac{3(x^2 - x - 6)}{4(x^2 - 9)}$

12. $F(x) = \dfrac{-2(x^2 - 4)}{3(x^2 + 4x + 4)}$

In Problems 13–18, use the graph shown to find:
(a) The domain and range of each function
(b) The intercepts, if any
(c) Horizontal asymptotes, if any
(d) Vertical asymptotes, if any
(e) Oblique asymptotes, if any

13.

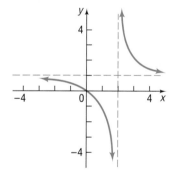

14.

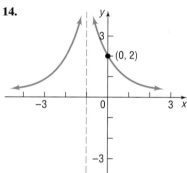

15.

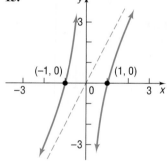

16.

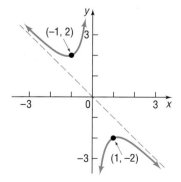

17.

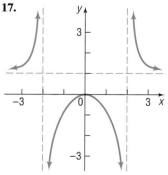

18.

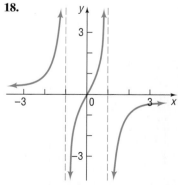

In Problems 19–30, graph each rational function using transformations (shifting, compression, stretching, and reflection).

19. $R(x) = \dfrac{1}{(x-1)^2}$

20. $R(x) = \dfrac{3}{x}$

21. $H(x) = \dfrac{-2}{x+1}$

22. $G(x) = \dfrac{2}{(x+2)^2}$

23. $R(x) = \dfrac{1}{x^2+4x+4}$

24. $R(x) = \dfrac{1}{x-1} + 1$

25. $F(x) = 1 - \dfrac{1}{x}$

26. $Q(x) = 1 + \dfrac{1}{x}$

27. $R(x) = \dfrac{x^2-4}{x^2}$

28. $R(x) = \dfrac{x-4}{x}$

29. $G(x) = 1 + \dfrac{2}{(x-3)^2}$

30. $F(x) = 2 - \dfrac{1}{(x+1)}$

In Problems 31–42, find the vertical, horizontal, and oblique asymptotes, if any, of each rational function. Do not graph.

31. $R(x) = \dfrac{3x}{x+4}$

32. $R(x) = \dfrac{3x+5}{x-6}$

33. $H(x) = \dfrac{x^4+2x^2+1}{x^2-x+1}$

34. $G(x) = \dfrac{-x^2+1}{x+5}$

35. $T(x) = \dfrac{x^2}{x^4-1}$

36. $P(x) = \dfrac{4x^5}{x^3-1}$

37. $Q(x) = \dfrac{5-x^2}{3x^4}$

38. $F(x) = \dfrac{-2x^2+1}{2x^3+4x^2}$

39. $R(x) = \dfrac{3x^4+4}{x^3+3x}$

40. $R(x) = \dfrac{6x^2+x+12}{3x^2-5x-2}$

41. $G(x) = \dfrac{x^3-1}{x-x^2}$

42. $F(x) = \dfrac{x-1}{x-x^3}$

In Problems 43–72, follow Steps 1 through 7 on page 351 to graph each rational function.

43. $R(x) = \dfrac{x+1}{x(x+4)}$

44. $R(x) = \dfrac{x}{(x-1)(x+2)}$

45. $R(x) = \dfrac{3x+3}{2x+4}$

46. $R(x) = \dfrac{2x+4}{x-1}$

47. $R(x) = \dfrac{3}{x^2-4}$

48. $R(x) = \dfrac{6}{x^2-x-6}$

49. $P(x) = \dfrac{x^4+x^2+1}{x^2-1}$

50. $Q(x) = \dfrac{x^4-1}{x^2-4}$

51. $H(x) = \dfrac{x^3-1}{x^2-9}$

52. $G(x) = \dfrac{x^3+1}{x^2+2x}$

53. $R(x) = \dfrac{x^2}{x^2+x-6}$

54. $R(x) = \dfrac{x^2+x-12}{x^2-4}$

55. $G(x) = \dfrac{x}{x^2-4}$

56. $G(x) = \dfrac{3x}{x^2-1}$

57. $R(x) = \dfrac{3}{(x-1)(x^2-4)}$

58. $R(x) = \dfrac{-4}{(x+1)(x^2-9)}$

59. $H(x) = 4\dfrac{x^2-1}{x^4-16}$

60. $H(x) = \dfrac{x^2+4}{x^4-1}$

61. $F(x) = \dfrac{x^2-3x-4}{x+2}$

62. $F(x) = \dfrac{x^2+3x+2}{x-1}$

63. $R(x) = \dfrac{x^2+x-12}{x-4}$

64. $R(x) = \dfrac{x^2-x-12}{x+5}$

65. $F(x) = \dfrac{x^2+x-12}{x+2}$

66. $G(x) = \dfrac{x^2-x-12}{x+1}$

67. $R(x) = \dfrac{x(x-1)^2}{(x+3)^3}$

68. $R(x) = \dfrac{(x-1)(x+2)(x-3)}{x(x-4)^2}$

69. $R(x) = \dfrac{x^2+x-12}{x^2-x-6}$

70. $R(x) = \dfrac{x^2+3x-10}{x^2+8x+15}$

71. $R(x) = \dfrac{6x^2-7x-3}{2x^2-7x+6}$

72. $R(x) = \dfrac{8x^2+26x+15}{2x^2-x-15}$

73. If the graph of a rational function R has the vertical asymptote $x = 4$, then the factor $x - 4$ must be present in the denominator of R. Explain why.

74. If the graph of a rational function R has the horizontal asymptote $y = 2$, then the degree of the numerator of R equals the degree of the denominator of R. Explain why.

75. Consult the illustration. Which of the following rational functions might have this graph? (More than one answer might be possible.)

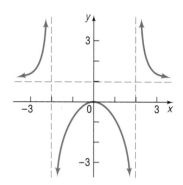

(a) $y = \dfrac{4x^2}{x^2 - 4}$

(b) $y = \dfrac{x}{x^2 - 4}$

(c) $y = \dfrac{x^2}{x^2 - 4}$

(d) $y = \dfrac{x^2(x^2 + 1)}{(x^2 + 4)(x^2 - 4)}$

(e) $y = \dfrac{x^3}{x^2 - 4}$

(f) $y = \dfrac{x^2 - 4}{x^2}$

76. Consult the illustration. Which of the following rational functions might have this graph? (More than one answer may be possible.)

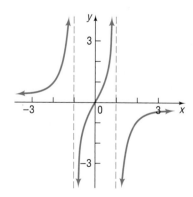

(a) $y = \dfrac{2x}{x^2 - 1}$

(b) $y = \dfrac{-3x}{x^2 - 1}$

(c) $y = \dfrac{x^3}{x^2 - 1}$

(d) $y = \dfrac{x^2 - 1}{-3x}$

(e) $y = \dfrac{-x^3}{(x^2 + 1)(x^2 - 1)}$

(f) $y = \dfrac{-x^2}{(x^2 + 1)(x^2 - 1)}$

77. **Gravity** In physics, it is established that the acceleration due to gravity, g, at a height h meters above sea level is given by

$$g(h) = \frac{3.99 \times 10^{14}}{(6.374 \times 10^6 + h)^2}$$

where 6.374×10^6 is the radius of Earth in meters.

(a) What is the acceleration due to gravity at sea level?

(b) The Sears Tower in Chicago, Illinois, is 443 meters tall. What is the acceleration due to gravity at the top of the Sears Tower?

(c) The peak of Mount Everest is 8848 meters above sea level. What is the acceleration due to gravity on the peak of Mount Everest?

(d) Find the horizontal asymptote of $g(h)$.

(e) Using your graphing utility, graph $g(h)$.

 (f) Solve $g(h) = 0$. How do you interpret your answer?

78. **Population Model** A rare species of insect was discovered in the Amazon Rain Forest. In order to protect the species, environmentalists declare the insect endangered and transplant the insects into a protected area. The population of the insect t months after being transplanted is given by P:

$$P(t) = \frac{50(1 + 0.5t)}{(2 + 0.01t)}$$

(a) How many insects were discovered? In other words, what was the population when $t = 0$?

(b) What will the population be after 5 years?

(c) Determine the horizontal asymptote of $P(t)$. What is the largest population that the protected area can sustain?

(d) Using your graphing utility, graph $P(t)$.

(e) TRACE $P(t)$ for large values of t to verify your answer to (c).

79. Drug Concentration The concentration C of a certain drug in a patient's bloodstream t hours after injection is given by

$$C(t) = \frac{t}{2t^2 + 1}$$

(a) Find the horizontal asymptote of $C(t)$. What happens to the concentration of the drug as t increases?

(b) Using your graphing utility, graph $C(t)$.

(c) Determine the time at which the concentration is highest.

80. Drug Concentration The concentration C of a certain drug in a patient's bloodstream t minutes after injection is given by

$$C(t) = \frac{50t}{(t^2 + 25)}$$

(a) Find the horizontal asymptote of $C(t)$. What happens to the concentration of the drug as t increases?

(b) Using your graphing utility, graph $C(t)$.

(c) Determine the time at which the concentration is highest.

81. Average Cost In Problem 75, Exercise 5.2, the cost function for manufacturing Chevy Cavaliers was found to be

$$C(x) = 0.2156x^3 - 2.3473x^2 + 14.3275x + 10.2238$$

Economists define the **average cost function** as

$$\overline{C}(x) = \frac{C(x)}{x}$$

(a) Find the average cost function.

(b) What is the average cost of producing six Cavaliers per hour?

(c) What is the average cost of producing nine Cavaliers per hour?

(d) Using your graphing utility, graph the average cost function.

(e) Using your graphing utility, find the number of Cavaliers that should be produced per hour to minimize average cost.

(f) What is the minimum average cost?

82. Average Cost In Problem 76, Exercise 5.2, the cost function for printing x textbooks (in thousands) was found to be

$$C(x) = 0.0155x^3 - 0.5951x^2 + 9.1502x + 98.4327$$

(a) Find the average cost function (refer to Problem 81).

(b) What is the average cost of printing 13 thousand textbooks per week?

(c) What is the average cost of printing 25 thousand textbooks per week?

(d) Using your graphing utility, graph the average cost function.

(e) Using your graphing utility, find the number of textbooks that should be printed to minimize average cost.

(f) What is the minimum average cost?

83. Minimizing Surface Area United Parcel Service has contracted you to design a closed box with a square base that has a volume of 10,000 cubic inches. See the illustration.

(a) Find a function for the surface area of the box.

(b) Using a graphing utility, graph the function found in (a).

(c) What is the minimum amount of cardboard that can be used to construct the box?

(d) What are the dimensions of the box that minimize surface area?

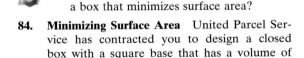
(e) Why might UPS be interested in designing a box that minimizes surface area?

84. Minimizing Surface Area United Parcel Service has contracted you to design a closed box with a square base that has a volume of 5000 cubic inches. See the illustration.

(a) Find a function for the surface area of the box.

(b) Using a graphing utility, graph the function found in (a).

(c) What is the minimum amount of cardboard that can be used to construct the box?

(d) What are the dimensions of the box that minimize surface area?

(e) Why might UPS be interested in designing a box that minimizes surface area?

85. Cost of a Can A can in the shape of a right circular cylinder is required to have a volume of 500 cubic centimeters. The top and bottom are made of material that costs 6¢ per square centimeter, while the sides are made of material that costs 4¢ per square centimeter.

(a) Express the total cost C of the material as a function of the radius r of the cylinder. (Refer to Figure 52.)

(b) Graph $C = C(r)$. For what value of r is the cost C least?

86. Material Needed to Make a Drum A steel drum in the shape of a right circular cylinder is required to have a volume of 100 cubic feet.
(a) Express the amount A of material required to make the drum as a function of the radius r of the cylinder.

(b) How much material is required if the drum is of radius 3 feet?
(c) Of radius 4 feet?
(d) Of radius 5 feet?
 (e) Graph $A = A(r)$. For what value of r is A smallest?

87. Graph each of the following functions:

$$y = \frac{x^2 - 1}{x - 1} \qquad y = \frac{x^3 - 1}{x - 1} \qquad y = \frac{x^4 - 1}{x - 1} \qquad y = \frac{x^5 - 1}{x - 1}$$

Is $x = 1$ a vertical asymptote? Why not? What is happening for $x = 1$? What do you conjecture about $y = \frac{x^n - 1}{x - 1}, n \geq 1$ an integer, for $x = 1$?

88. Graph each of the following functions:

$$y = \frac{x^2}{x - 1} \qquad y = \frac{x^4}{x - 1} \qquad y = \frac{x^6}{x - 1} \qquad y = \frac{x^8}{x - 1}$$

What similarities do you see? What differences?

In Problems 89–94, graph each function and use MINIMUM to obtain the minimum value, correct to two decimal places.

89. $f(x) = x + \dfrac{1}{x}, \quad x > 0$

90. $f(x) = 2x + \dfrac{9}{x}, \quad x > 0$

91. $f(x) = x^2 + \dfrac{1}{x}, \quad x > 0$

92. $f(x) = 2x^2 + \dfrac{9}{x}, \quad x > 0$

93. $f(x) = x + \dfrac{1}{x^3}, \quad x > 0$

94. $f(x) = 2x + \dfrac{9}{x^3}, \quad x > 0$

 95. Consult the illustration. Make up a rational function that might have this graph. Compare yours with a friend's. What similarities do you see? What differences?

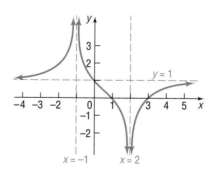

96. Can the graph of a rational function have both a horizontal and an oblique asymptote? Explain.

97. Write a few paragraphs that provide a general strategy for graphing a rational function. Be sure to mention the following: proper, improper, intercepts, and asymptotes.

98. Make up a rational function that has the following characteristics: crosses the x-axis at 3; touches the x-axis at -2; one vertical asymptote, $x = 1$; and one horizontal asymptote, $y = 2$. Give your rational function to a fellow classmate and ask for a written critique of your rational function.

99. Make up a rational function that has $y = 2x + 1$ as an oblique asymptote. Explain the methodology you used.

MISSION POSSIBLE

Responding to a Challenge by the Cardassians

While playing with virtual reality in your school's computer lab, you are suddenly faced with some very real and very nasty looking Cardassians who have burst into your school through your computer network. As usual, they are very scornful of the intelligence of Earthlings, and when you try to protest, they throw this challenge at you. "Find a rational function defined by this graph," they say, "and we promise to support your reputation for intellectual achievement in every intergalactic community this side of the Macklin Nebula. Otherwise, consider yourselves the laughing stock of the universe." And they departed back through the computer screen, laughing at what they perceived to be their extreme cleverness.

As you look at the computer screen now, all you can see is this graph:

You decide to tackle this as a team project to save the reputation of all the humans on this planet! The Cardassians mentioned "rational function," so you know that the solution will have the form

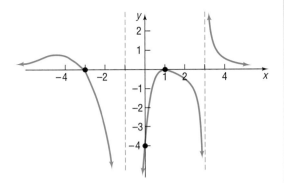

$$R(x) = \frac{a_n x^n + a_{n-1} x^{n-1} + \cdots + a_0}{b_m x^m + b_{m-1} x^{m-1} + \cdots + b_0}$$

Sometimes it is easier to think of it in factored form.

1. You notice some things right away. There are two vertical asymptotes. How do they fit into the solution? Put them in.

2. There are two x-intercepts. Where do they show up in the solution? Put them in.

3. Use a graphing utility to check what you have so far. Does it look like the original graph? Notice that 1 and −3 appear to be the only real zeros. What can you tell about the multiplicity of each? Do you need to change the multiplicity of either zero in your solution?

4. How can you account for the fact that the graph of the function goes to −∞ on both sides of one vertical asymptote and at the other vertical asymptote it goes to −∞ on one side and to ∞ on the other? Can you change the denominator in some way to account for this?

5. Is the information that you have so far consistent with the horizontal asymptote $y = 0$? If not, how should you adjust your function so that it is consistent?

6. Is the information that you have so far consistent with the y-intercept? If not, how should you adjust your function so that it is consistent?

7. Are you finished? Have you checked your solution on a graphing utility? Are you satisfied that you have it right?

8. Are there other functions whose graph might look like the one given? If so, what common characteristics do they have? Will the Cardassians be more impressed if more than one correct answer is sent?

9. Compare your graph's solution to those of other groups before typing the solution into the computer. Then send all the correct solutions off to cardass@slime.uni.

5.4 | SYNTHETIC DIVISION

1 Use Synthetic Division

To find the quotient as well as the remainder when a polynomial function f of degree 1 or higher is divided by $g(x) = x - c$, a shortened version of long division, called **synthetic division,** makes the task simpler.

To see how synthetic division works, we will use long division to divide the polynomial $f(x) = 2x^3 - x^2 + 3$ by $g(x) = x - 3$.

$$
\begin{array}{r}
2x^2 + 5x + 15 \quad \leftarrow \text{quotient} \\
x - 3 \overline{) 2x^3 - x^2 \qquad + 3} \\
\underline{2x^3 - 6x^2} \\
5x^2 \\
\underline{5x^2 - 15x} \\
15x + 3 \\
\underline{15x - 45} \\
48 \quad \leftarrow \text{remainder}
\end{array}
$$

Check: (Divisor) · (quotient) + remainder = $(x - 3)(2x^2 + 5x + 15) + 48$

$$
\begin{aligned}
&= 2x^3 + 5x^2 + 15x - 6x^2 - 15x - 45 + 48 \\
&= 2x^3 - x^2 + 3
\end{aligned}
$$

The process of synthetic division arises from rewriting the long division in a more compact form, using simpler notation. For example, in the long division above, the terms in color are not really necessary because they are identical to the terms directly above them. With these terms removed, we have

$$
\begin{array}{r}
2x^2 + 5x + 15 \\
x - 3 \overline{) 2x^3 - x^2 \qquad + 3} \\
\underline{- 6x^2} \\
5x^2 \\
\underline{- 15x} \\
15x \\
\underline{- 45} \\
48
\end{array}
$$

Most of the x's that appear in this process can also be removed, provided we are careful about positioning each coefficient. In this regard, we will need to use 0 as the coefficient of x in the dividend, because that power of x is missing. Now we have

$$
\begin{array}{r}
2x^2 + 5x + 15 \\
x - 3 \overline{) 2 \quad - 1 \qquad 0 \qquad 3} \\
\underline{- 6} \\
5 \\
\underline{- 15} \\
15 \\
\underline{- 45} \\
48
\end{array}
$$

We can make this display more compact by moving the lines up until the numbers in color align horizontally:

$$
\begin{array}{r}
2x^2 + 5x + 15 \\
x - 3\overline{\smash{\big)}2 \quad -1 \quad\;\; 0 \quad\;\; 3} \\
-6 \; -15 \; -45 \\
\hline
\bigcirc \quad\;\; 5 \quad\;\; 15 \quad\;\; 48
\end{array}
$$

Row 1
Row 2
Row 3
Row 4

Now, if we place the leading coefficient of the quotient (2) in the circled position, the first three numbers in row 4 are precisely the coefficients of the quotient, and the last number in row 4 is the remainder. Thus, row 1 is not really needed, so we can compress the process to three rows, where the bottom row contains the coefficients of both the quotient and the remainder:

$$
\begin{array}{r}
x - 3\overline{\smash{\big)}2 \quad -1 \quad\;\; 0 \quad\;\; 3} \\
-6 \; -15 \; -45 \\
\hline
2 \quad\;\; 5 \quad\;\; 15 \quad\;\; 48
\end{array}
$$

Row 1
Row 2 (subtract)
Row 3

Recall that the entries in row 3 are obtained by subtracting the entries in row 2 from those in row 1. Rather than subtracting the entries in row 2, we can change the sign of each entry and add. With this modification, our display will look like this:

$$
\begin{array}{r}
x - 3\overline{\smash{\big)}2 \quad -1 \quad\;\; 0 \quad\;\; 3} \\
6 \quad\;\; 15 \quad\;\; 45 \\
\hline
2 \quad\;\; 5 \quad\;\; 15 \quad\;\; 48
\end{array}
$$

Row 1
Row 2 (add)
Row 3

Notice that the entries in row 2 are three times the prior entries in row 3. Out last modification to the display replaces the $x - 3$ by 3. The entries in row 3 give the quotient and the remainder, as shown next.

$$
\begin{array}{r}
3\overline{\smash{\big)}2 \quad -1 \quad\;\; 0 \quad\;\; 3} \\
6 \quad\;\; 15 \quad\;\; 45 \\
\hline
2 \quad\;\; 5 \quad\;\; 15 \quad\;\; 48
\end{array}
$$

Row 1
Row 2 (add)
Row 3

Quotient   Remainder

$$2x^2 + 5x + 15 \qquad\qquad R = 48$$

Let's go through another example step by step.

EXAMPLE 1

Using Synthetic Division to Find the Quotient and Remainder

Use synthetic division to find the quotient and remainder when

$$f(x) = 3x^4 + 8x^2 - 7x + 4 \quad \text{is divided by} \quad g(x) = x - 1$$

Solution

STEP 1: Write the dividend in descending powers of x. Then copy the coefficients, remembering to insert a 0 for any missing powers of x:

$$3 \quad 0 \quad 8 \quad -7 \quad 4 \qquad \text{Row 1}$$

STEP 2: Insert the usual division symbol. Since the divisor is $x - 1$, we insert 1 to the left of the division symbol

$$1\overline{\smash{\big)}3 \quad 0 \quad 8 \quad -7 \quad 4} \qquad \text{Row 1}$$

STEP 3: Bring the 3 down two rows, and enter it in row 3:

$$
\begin{array}{r}
1)\overline{3 \quad 0 \quad 8 \quad -7 \quad 4} \\
\downarrow \qquad\qquad\qquad\qquad \\
\overline{3 \qquad\qquad\qquad\qquad}
\end{array}
\qquad
\begin{array}{l}
\text{Row 1} \\
\text{Row 2} \\
\text{Row 3}
\end{array}
$$

STEP 4: Multiply the latest entry in row 3 by 1, and place the result in row 2, but one column over to the right:

$$
\begin{array}{r}
1)\overline{3 \quad 0 \quad 8 \quad -7 \quad 4} \\
3 \qquad\qquad\qquad\qquad \\
\overline{3 \qquad\qquad\qquad\qquad}
\end{array}
\qquad
\begin{array}{l}
\text{Row 1} \\
\text{Row 2} \\
\text{Row 3}
\end{array}
$$

STEP 5: Add the entry in row 2 to the entry above it in row 1, and enter the sum to row 3:

$$
\begin{array}{r}
1)\overline{3 \quad 0 \quad 8 \quad -7 \quad 4} \\
3 \qquad\qquad\qquad\qquad \\
\overline{3 \quad 3 \qquad\qquad\qquad}
\end{array}
\qquad
\begin{array}{l}
\text{Row 1} \\
\text{Row 2} \\
\text{Row 3}
\end{array}
$$

STEP 6: Repeat steps 4 and 5 until no more entries are available in row 1:

$$
\begin{array}{r}
1)\overline{3 \quad 0 \quad 8 \quad -7 \quad 4} \\
3 \quad 3 \quad 11 \quad 4 \\
\overline{3 \quad 3 \quad 11 \quad 4 \quad 8}
\end{array}
\qquad
\begin{array}{l}
\text{Row 1} \\
\text{Row 2 (add)} \\
\text{Row 3}
\end{array}
$$

STEP 7: The final entry in row 3, an 8, is the remainder; the other entries in row 3 $(3, 3, 11,$ and $4)$ are the coefficients (in descending order) of a polynomial whose degree is 1 less than that of the dividend; this is the quotient. Thus,

$$\text{Quotient} = 3x^3 + 3x^2 + 11x + 4 \qquad \text{Remainder} = 8$$

Check: (Divisor)(quotient) + remainder

$$
\begin{aligned}
&= (x - 1)(3x^3 + 3x^2 + 11x + 4) + 8 \\
&= 3x^4 + 3x^3 + 11x^2 + 4x - 3x^3 - 3x^2 - 11x - 4 + 8 \\
&= 3x^4 + 8x^2 - 7x + 4 = \text{dividend}
\end{aligned}
$$

Let's do an example in which all seven steps are combined.

E X A M P L E   2   Using Synthetic Division to Verify a Factor

Use synthetic division to show that $g(x) = x + 3$ is a factor of

$$f(x) = 2x^5 + 5x^4 - 2x^3 + 2x^2 - 2x + 3$$

Solution   The divisor is $x + 3 = x - (-3)$, so the row 3 entries will be multiplied by -3 entered in row 2 and added to row 1:

$$
\begin{array}{r|rrrrrl}
-3) & 2 & 5 & -2 & 2 & -2 & 3 & \text{Row 1} \\
    &   & -6 & 3 & -3 & 3 & -3 & \text{Row 2} \\
\hline
    & 2 & -1 & 1 & -1 & 1 & 0 & \text{Row 3}
\end{array}
$$

Because the remainder is 0, we have

$$(\text{Divisor})(\text{quotient}) + \text{remainder} =$$
$$(x + 3)(2x^4 - x^3 + x^2 - x + 1) = 2x^5 + 5x^4 - 2x^3 + 2x^2 - 2x + 3$$

As we see, $x + 3$ is a factor of $2x^5 + 5x^4 - 2x^3 + 2x^2 - 2x + 3$.   ■

As Example 2 illustrates, the remainder after division gives information about whether the divisor is, or is not, a factor. We'll have more to say about this in the next section.

    Now work Problem 13.

5.4 | EXERCISES

In Problems 1–12, use synthetic division to find the quotient q(x) and remainder R when f(x) is divided by g(x).

1. $f(x) = x^3 - x^2 + 2x + 4$; $g(x) = x - 2$

2. $f(x) = x^3 + 2x^2 - 3x + 1$; $g(x) = x + 1$

3. $f(x) = 3x^3 + 2x^2 - x + 3$; $g(x) = x - 3$

4. $f(x) = -4x^3 + 2x^2 - x + 1$; $g(x) = x + 2$

5. $f(x) = x^5 - 4x^3 + x$; $g(x) = x + 3$

6. $f(x) = x^4 + x^2 + 2$; $g(x) = x - 2$

7. $f(x) = 4x^6 - 3x^4 + x^2 + 5$; $g(x) = x - 1$

8. $f(x) = x^5 + 5x^3 - 10$; $g(x) = x + 1$

9. $f(x) = 0.1x^3 + 0.2x$; $g(x) = x + 1.1$

10. $f(x) = 0.1x^2 - 0.2$; $g(x) = x + 2.1$

11. $f(x) = x^5 - 1$; $g(x) = x - 1$

12. $f(x) = x^5 + 1$; $g(x) = x + 1$

In Problems 13–22, use synthetic division to determine whether x − c is a factor of f(x).

13. $f(x) = 4x^3 - 3x^2 - 8x + 4$; $x - 2$

14. $f(x) = -4x^3 + 5x^2 + 8$; $x + 3$

15. $f(x) = 3x^4 - 6x^3 - 5x + 10$; $x - 2$

16. $f(x) = 4x^4 - 15x^2 - 4$; $x - 2$

17. $f(x) = 3x^6 + 82x^3 + 27$; $x + 3$

18. $f(x) = 2x^6 - 18x^4 + x^2 - 9$; $x + 3$

19. $f(x) = 4x^6 - 64x^4 + x^2 - 15$; $x + 4$

20. $f(x) = x^6 - 16x^4 + x^2 - 16$; $x + 4$

21. $f(x) = 2x^4 - x^3 + 2x - 1$; $x - 1/2$

22. $f(x) = 3x^4 + x^3 - 3x + 1$; $x + 1/3$

    23. When dividing a polynomial by $x - c$, do you prefer to use long division or synthetic division? Does the value of c make a difference to you in choosing? Give reasons.

5.5 | THE REAL ZEROS OF A POLYNOMIAL FUNCTION

> 1   Utilize the Remainder and Factor Theorems
>
> 2   Use Descartes' Rule of Signs to Determine the Number of Positive and the Number of Negative Real Zeros of a Polynomial Function
>
> 3   Use the Rational Zeros Theorem to List the Potential Rational Zeros of a Polynomial Function
>
> 4   Find the Real Zeros of a Polynomial Function
>
> 5   Solve Polynomial Equations
>
> 6   Utilize the Theorem: Bounds on Zeros
>
> 7   Use the Intermediate Value Theorem

Remainder and Factor Theorems

1   Recall that when we divide one polynomial (the dividend) by another (the divisor) we obtain a quotient polynomial and a remainder, the remainder being either the zero polynomial or a polynomial whose degree is less than the degree of the divisor. To check our work, we verify that

$$(\text{Divisor})(\text{quotient}) + \text{remainder} = \text{dividend}$$

This checking routine is the basis for a famous theorem called the **Division Algorithm* for Polynomials,** which we now state without proof.

Theorem   Division Algorithm for Polynomials

If $f(x)$ and $g(x)$ denote polynomial functions and if $g(x)$ is not the zero polynomial, then there are unique polynomial functions $q(x)$ and $r(x)$ such that

$$\frac{f(x)}{g(x)} = q(x) + \frac{r(x)}{g(x)} \quad \text{or} \quad f(x) = g(x)q(x) + r(x) \quad \quad (1)$$

$$\underset{\text{dividend}}{\uparrow} \quad \underset{\text{divisor}}{\uparrow} \ \underset{\text{quotient}}{\uparrow} \quad \underset{\text{remainder}}{\uparrow}$$

where $r(x)$ is either the zero polynomial or a polynomial of degree less than that of $g(x)$.

In equation (1), $f(x)$ is the **dividend,** $g(x)$ is the **divisor,** $q(x)$ is the **quotient,** and $r(x)$ is the **remainder.**

If the divisor $g(x)$ is a first-degree polynomial of the form

$$g(x) = x - c \quad \quad c \text{ a real number}$$

then the remainder $r(x)$ is either the zero polynomial or a polynomial of degree 0. Thus, for such divisors, the remainder is some number, say R, and we may write

---

*A systematic process in which certain steps are repeated a finite number of times is called an **algorithm.** Thus, long division is an algorithm.

$$f(x) = (x - c)q(x) + R \qquad (2)$$

This equation is an identity in x and is true for all real numbers x. In particular, it is true when $x = c$. Thus, if $x = c$, then equation (2) becomes

$$f(c) = (c - c)q(c) + R$$
$$f(c) = R$$

and equation (2) takes the form

$$f(x) = (x - c)q(x) + f(c) \qquad (3)$$

We have now proved the following result, called the **Remainder Theorem.**

> ### Remainder Theorem
>
> Let f be a polynomial function. If $f(x)$ is divided by $x - c$, then the remainder is $f(c)$.

E X A M P L E 1

Using the Remainder Theorem

Find the remainder if $f(x) = x^3 - 4x^2 + 2x - 5$ is divided by

(a) $x - 3$ (b) $x + 2$

Solution (a) The Remainder Theorem says that the remainder is $f(3)$.

$$f(3) = (3)^3 - 4(3)^2 + 2(3) - 5 = 27 - 36 + 6 - 5 = -8$$

(b) To find the remainder when $f(x)$ is divided by $x + 2 = x - (-2)$, we need to find $f(-2)$.

$$f(-2) = (-2)^3 - 4(-2)^2 + 2(-2) - 5 = -8 - 16 - 4 - 5 = -33$$

Thus, the remainder is -33. ∎

E X A M P L E 2

Using Synthetic Division to Find the Value of a Polynomial

Use synthetic division to find the value of $f(x) = -3x^4 + 2x^3 - x + 1$ at $x = -2$; that is, find $f(-2)$.

Solution The Remainder Theorem tells us that the value of a polynomial function at c equals the remainder when the polynomial is divided by $x - c$. This remainder is the final entry of the third row in the process of synthetic division. We want $f(-2)$, so we divide by $x - (-2)$:

$$
\begin{array}{r|rrrrr}
-2) & -3 & 2 & 0 & -1 & 1 \\
    &   & 6 & -16 & 32 & -62 \\
\hline
    & -3 & 8 & -16 & 31 & -61
\end{array}
$$

The quotient is $q(x) = -3x^3 + 8x^2 - 16x + 31$; the remainder is $R = -61$. Because the remainder was found to be -61, it follows from the Remainder Theorem that $f(-2) = -61$. ∎

As Example 2 illustrates, we can use the process of synthetic division to find the value of a polynomial function at a number c as an alternative to merely substituting c for x.

Comment   A graphing utility provides another way to find the value of a function, using the eVALUEate feature. Consult your manual for details. Then check the result of Example 2.   ∎

An important and useful consequence of the Remainder Theorem is the **Factor Theorem.**

> ### Factor Theorem
>
> Let f be a polynomial function. Then $x - c$ is a factor of $f(x)$ if and only if $f(c) = 0$.

∎

The Factor Theorem actually consists of two separate statements:

> 1. If $f(c) = 0$, then $x - c$ is a factor of $f(x)$.
> 2. If $x - c$ is a factor of $f(x)$, then $f(c) = 0$.

Thus, the proof requires two parts.

Proof:

1. Suppose that $f(c) = 0$. Then, by equation (3), we have

    $$f(x) = (x - c)q(x)$$

    for some polynomial $q(x)$. That is, $x - c$ is a factor of $f(x)$.
2. Suppose that $x - c$ is a factor of $f(x)$. Then there is a polynomial function q such that

    $$f(x) = (x - c)q(x)$$

    Replacing x by c, we find that

    $$f(c) = (c - c)q(c) = 0 \cdot q(c) = 0$$

This completes the proof.   ∎

One use of the Factor Theorem is to determine whether a polynomial has a particular factor.

EXAMPLE 3

Using the Factor Theorem

Use the Factor Theorem to determine whether the function $f(x) = 2x^3 - x^2 + 2x - 3$ has the factor

(a) $x - 1$ (b) $x + 3$

Solution   The Factor Theorem states that if $f(c) = 0$ then $x - c$ is a factor.

(a) Because $x - 1$ is of the form $x - c$ with $c = 1$, we find the value of $f(1)$. We choose to use substitution:

$$f(1) = 2(1)^3 - (1)^2 + 2(1) - 3 = 2 - 1 + 2 - 3 = 0$$

By the Factor Theorem, $x - 1$ is a factor of $f(x)$.

(b) To test the factor $x + 3$, we first need to write it in the form $x - c$. Since $x + 3 = x - (-3)$, we find the value of $f(-3)$; we choose to use synthetic division:

$$
\begin{array}{r|rrrr}
-3 & 2 & -1 & 2 & -3 \\
   &   & -6 & 21 & -69 \\
\hline
   & 2 & -7 & 23 & -72
\end{array}
$$

Because $f(-3) = -72 \neq 0$, we conclude from the Factor Theorem that $x - (-3) = x + 3$ is not a factor of $f(x)$.   ▬

 Now work Problem 1.

Descartes' Rule of Signs

 The real zeros of a polynomial function f are the real solutions, if any, of the equation $f(x) = 0$. They are also the x-intercepts of the graph of f. For polynomial and rational functions, we have seen the importance of the zeros for graphing. In most cases, however, the real zeros of a polynomial function are difficult to find using algebraic methods. There are no nice formulas like the quadratic formula available to help us for polynomials of degree higher than 2. Although formulas do exist for solving any third- or fourth-degree polynomial equation, they are somewhat complicated. It has been proved that no general formulas exist for polynomial equations of degree 5 or higher. In this section, we shall learn some ways of detecting information about the character of the zeros, which, in turn, may help us to find them or approximate them.

Our first theorem concerns the number of real zeros that a polynomial function may have. In counting the real zeros of a polynomial, we count each zero as many times as its multiplicity.

> Theorem   Number of Zeros
>
> A polynomial function cannot have more real zeros than its degree.
>
>   ▬

Proof   The proof is based on the Factor Theorem. If r is a real zero of a polynomial function f, then $f(r) = 0$ and, hence, $x - r$ is a factor of $f(x)$. Thus, each real zero corresponds to a factor of degree 1. Because f cannot have more first-degree factors than its degree, the result follows.   ▬

The next theorem, called **Descartes' Rule of Signs,** provides information about the number and location of the real zeros of a polynomial function, so we know where to look for zeros. Descartes' Rule of Signs assumes that the polynomial is written in descending powers of x and requires that we count the number of variations in sign of the coefficients of $f(x)$ and $f(-x)$.

For example, the following polynomial function has two variations in the signs of coefficients:

$$f(x) = -3x^7 + 4x^4 + 3x^2 - 2x - 1$$
$$= \underbrace{-3x^7 + 0x^6 + 0x^5 + 4x^4}_{- \text{ to } +} + 0x^3 + \underbrace{3x^2 - 2x}_{+ \text{ to } -} - 1$$

Notice that we ignored the zero coefficients in $0x^6$, $0x^5$, and $0x^3$ in counting the number of variations in the sign of $f(x)$. Replacing x by $-x$, we get

$$f(-x) = 3x^7 + 4x^4 + 3x^2 \underbrace{+ 2x - 1}_{+ \text{ to } -}$$

which has one variation in sign.

Theorem **Descartes' Rule of Signs**

Let f denote a polynomial function written in standard form.*

The number of positive real zeros of f either equals the number of variations in sign of the nonzero coefficients of $f(x)$ or else equals that number less an even integer.

The number of negative real zeros of f either equals the number of variations in sign of the nonzero coefficients of $f(-x)$ or else equals that number less an even integer.

We shall not prove Descartes' Rule of Signs. Let's see how it is used.

EXAMPLE 4   Using Descartes' Rule of Signs to Locate Zeros

Discuss the real zeros of $f(x) = 3x^6 - 4x^4 + 3x^3 + 2x^2 - x - 3$.

Solution   There are at most six zeros, because the polynomial is of degree 6. Since there are three variations in sign of the nonzero coefficients of $f(x)$, by Descartes' Rule of Signs we expect either three or one positive real zero(s). To continue, we look at $f(-x)$:

$$f(-x) = 3x^6 - 4x^4 - 3x^3 + 2x^2 + x - 3$$

There are three variations in sign, so we expect either three or one negative real zero(s). Equivalently, we now know that the graph of f has either three or one positive x-intercept(s) and three or one negative x-intercept(s). ▬

 Now work Problem 11.

---

*The powers of x appear in descending order.

Although we have not actually found the zeros in Example 4, we know something about the number of zeros and how many might be positive or negative. The next result, which you are asked to prove in Problem 108, is called the **Rational Zeros Theorem.** It provides information about the rational zeros of a polynomial with integer coefficients.

[3] **Rational Zeros Theorem**

> **Theorem   Rational Zeros Theorem**
>
> Let f be a polynomial function of degree 1 or higher of the form
>
> $$f(x) = a_n x^n + a_{n-1} x^{n-1} + \cdots + a_1 x + a_0 \qquad a_n \neq 0, a_0 \neq 0$$
>
> where each coefficient is an integer. If p/q, in lowest terms, is a rational zero of f, then p must be a factor of a_0 and q must be a factor of a_n.

E X A M P L E   5

Listing Potential Rational Zeros

List the potential rational zeros of

$$f(x) = 2x^3 + 11x^2 - 7x - 6$$

Solution   Because f has integer coefficients, we may use the Rational Zeros Theorem. First, we list all the integers p that are factors of $a_0 = -6$ and all the integers q that are factors of $a_3 = 2$:

p: $\pm 1, \pm 2, \pm 3, \pm 6$
q: $\pm 1, \pm 2$

Now we form all possible ratios p/q:

$$\frac{p}{q}: \quad \pm 1, \pm 2, \pm 3, \pm 6, \pm \frac{1}{2}, \pm \frac{3}{2}$$

If f has a rational zero, it will be found in this list, which contains 12 possibilities.

Now work Problem 23.

Be sure that you understand what the Rational Zeros Theorem says: For a polynomial with integer coefficients, *if* there is a rational zero, it is one of those listed. There may not be any rational zeros.

[4] Synthetic division may be used to test each potential rational zero to determine whether it is indeed a zero. To make the work easier, the integers are usually tested first. Let's continue this example.

E X A M P L E   6

Finding the Real Zeros of a Polynomial Function

Continue working with Example 5 to find the zeros of

$$f(x) = 2x^3 + 11x^2 - 7x - 6$$

Solution   We gather all the information that we can about the zeros:

STEP 1: There are at most three zeros.
STEP 2: By Descartes' Rule of Signs, there is one positive zero. Also, because

$$f(-x) = -2x^3 + 11x^2 + 7x - 6$$

there are two or no negative zeros.
STEP 3: Now we use the list of potential rational zeros obtained in Example 5: $\pm 1, \pm 2, \pm 3, \pm 6, \pm\frac{1}{2}, \pm\frac{3}{2}$. We choose to test the potential rational zero 1 using synthetic division:

$$
\begin{array}{r|rrr}
1) & 2 & 11 & -7 & -6 \\
  &   & 2 & 13 & 6 \\
\hline
  & 2 & 13 & 6 & 0
\end{array}
$$

The remainder is $f(1) = 0$. Thus, 1 is a zero and $x - 1$ is a factor of f. The entries in the bottom row of this synthetic division can be used to factor f:

$$
\begin{aligned}
f(x) &= 2x^3 + 11x^2 - 7x - 6 \\
&= (x - 1)(2x^2 + 13x + 6)
\end{aligned}
$$

Now any solution of the equation $2x^2 + 13x + 6 = 0$ will be a zero of f. Because of this, we call the equation $2x^2 + 13x + 6 = 0$ a **depressed equation** of f. Since the degree of the depressed equation of f is less than that of the original equation, we work with the depressed equation to find the zeros of f.
STEP 4: The depressed equation $2x^2 + 13x + 6 = 0$ is a quadratic equation with discriminant $b^2 - 4ac = 169 - 48 = 121 > 0$. It thus has two real solutions, which can be found by factoring:

$$2x^2 + 13x + 6 = (2x + 1)(x + 6) = 0$$
$$2x + 1 = 0 \quad \text{or} \quad x + 6 = 0$$
$$x = -\frac{1}{2} \qquad\qquad x = -6$$

The zeros of f are -6, $-\frac{1}{2}$, and 1.   ▬

Notice that the three zeros of f found in Example 6 are among those given in the list of potential rational zeros in Example 5. Also, notice that, based on the zeros, we can write f in factored form as

$$
\begin{aligned}
f(x) &= 2x^3 + 11x^2 - 7x - 6 \\
&= 2\left(x^3 + \tfrac{11}{2}x^2 - \tfrac{7}{2}x - 3\right) \\
&= 2(x + 6)\left(x + \tfrac{1}{2}\right)(x - 1)
\end{aligned}
$$

To obtain information about the real zeros of a polynomial function, follow these steps:

Steps for Finding the Real Zeros of a Polynomial Function	STEP 1: Use the degree of the polynomial to determine the maximum number of zeros.
	STEP 2: Use Descartes' Rule of Signs to determine the possible number of positive zeros and negative zeros.
	STEP 3: (a) If the polynomial has integer coefficients, use the Rational Zeros Theorem to identify those rational numbers that potentially can be zeros. (b) Use substitution or synthetic division to test each potential rational zero. (c) Each time that a zero (and thus a factor) is found, repeat Steps 2 and 3 on the depressed equation.
	STEP 4: In attempting to find the zeros, remember to use (if possible) the factoring techniques that you already know (special products, factoring by grouping, and so on).

E X A M P L E   7

Finding the Real Zeros of a Polynomial Function

Find the zeros of: $f(x) = 3x^5 - 2x^4 - 15x^3 + 10x^2 + 12x - 8$

Solution We gather all the information that we can about the zeros:

STEP 1: There are at most five zeros.

STEP 2: By Descartes' Rule of Signs, there are three or one positive zero(s). Also, because

$$f(-x) = -3x^5 - 2x^4 + 15x^3 + 10x^2 - 12x - 8$$

there are two or no negative zeros.

STEP 3: To obtain the list of potential rational zeros, we write the factors p of $a_0 = -8$ and the factors q of $a_5 = 3$:

p: $\pm 1, \pm 2, \pm 4, \pm 8$

q: $\pm 1, \pm 3$

The potential rational zeros consist of all possible quotients p/q:

$$\frac{p}{q}:\quad \pm 1, \pm 2, \pm 4, \pm 8, \pm\frac{1}{3}, \pm\frac{2}{3}, \pm\frac{4}{3}, \pm\frac{8}{3}$$

We can test the potential rational zero 1 using synthetic division:

$$
\begin{array}{r|rrrrrr}
1) & 3 & -2 & -15 & 10 & 12 & -8 \\
   &   & 3 & 1   & -14 & -4 & 8 \\
\hline
   & 3 & 1 & -14 & -4 & 8 & 0
\end{array}
$$

The remainder is $f(1) = 0$. Thus, 1 is a zero and $x - 1$ is a factor. As before, we use the entries in the bottom row of this synthetic division to factor f:

$$
\begin{aligned}
f(x) &= 3x^5 - 2x^4 - 15x^3 + 10x^2 + 12x - 8 \\
&= (x - 1)(3x^4 + x^3 - 14x^2 - 4x + 8)
\end{aligned}
$$

We now work with the first depressed equation of f:

$$3x^4 + x^3 - 14x^2 - 4x + 8 = 0$$

REPEAT STEP 2: Let $q_1(x) = 3x^4 + x^3 - 14x^2 - 4x + 8$. By Descartes' Rule of Signs, q_1 has two or no positive zeros. Also, because

$$q_1(-x) = 3x^4 - x^3 - 14x^2 + 4x + 8$$

q_1 has two or no negative zeros.

REPEAT STEP 3: The potential rational zeros of q_1 are the same as those listed earlier for f. We choose to test 1 again because it may be a repeated root:

$$
\begin{array}{r|rrrrr}
1) & 3 & 1 & -14 & -4 & 8 \\
  &   & 3 & 4 & -10 & -14 \\
\hline
  & 3 & 4 & -10 & -14 & -6
\end{array}
$$

The remainder tells us that 1 is not a zero of q_1. Now we test -1:

$$
\begin{array}{r|rrrrr}
-1) & 3 & 1 & -14 & -4 & 8 \\
   &   & -3 & 2 & 12 & -8 \\
\hline
   & 3 & -2 & -12 & 8 & 0
\end{array}
$$

We find that -1 is a zero of q_1 and therefore $x - (-1) = x + 1$ is a factor of q_1. Thus, we have

$$f(x) = (x - 1)(x + 1)(3x^3 - 2x^2 - 12x + 8)$$

REPEAT STEP 2: We work now with the second depressed equation of f:

$$3x^3 - 2x^2 - 12x + 8 = 0$$

Let $q_2(x) = 3x^3 - 2x^2 - 12x + 8$. By Descartes' Rule of Signs, q_2 has two or no positive zeros. Also, because

$$q_2(-x) = -3x^3 - 2x^2 + 12x + 8$$

q_2 has one negative zero.

REPEAT STEP 3: The list of potential rational zeros of q_2 is the same as that of f. However, because 1 was not a zero of q_1, it cannot be a zero of q_2. Also, the fact that -1 is a zero of q_1 does not mean it cannot also be a zero of q_2 (that is, it could be a repeated root of q_1). We know that there is one negative zero (which may not be rational), so we test -1 once more to determine whether it is a root of q_2:

$$
\begin{array}{r|rrrr}
-1) & 3 & -2 & -12 & 8 \\
   &   & -3 & 5 & 7 \\
\hline
   & 3 & -5 & -7 & 15
\end{array}
$$

It is not. Next, we choose to test -2:

$$
\begin{array}{r|rrrr}
-2) & 3 & -2 & -12 & 8 \\
   &   & -6 & 16 & -8 \\
\hline
   & 3 & -8 & 4 & 0
\end{array}
$$

We find that -2 is a zero, so $x - (-2) = x + 2$ is a factor. Thus, we have

$$f(x) = (x - 1)(x + 1)(x + 2)(3x^2 - 8x + 4)$$

STEP 4: The new depressed equation of f, $3x^2 - 8x + 4 = 0$, is a quadratic equation with a discriminant of $b^2 - 4ac = (-8)^2 - 4(3)(4) = 16$.

Therefore, this equation has two real solutions and, in this case, we find them by factoring:

$$3x^2 - 8x + 4 = 0$$
$$(3x - 2)(x - 2) = 0$$
$$3x - 2 = 0 \quad \text{or} \quad x - 2 = 0$$
$$x = \frac{2}{3} \qquad\qquad x = 2$$

The zeros of f are $-2, -1, \frac{2}{3}, 1$, and 2. The factored form of f is

$$f(x) = 3x^5 - 2x^4 - 15x^3 + 10x^2 + 12x - 8 = 3(x - 1)(x + 1)(x + 2)(x - \tfrac{2}{3})(x - 2)$$

Now work Problem 35.

The procedure outlined in Example 7 for finding the zeros of a polynomial can also be used to solve polynomial equations.

EXAMPLE 8   Solving a Polynomial Equation

Solve the equation: $x^5 - 5x^4 + 12x^3 - 24x^2 + 32x - 16 = 0$

Solution   The real solutions of this equation are the real zeros of the polynomial function

$$f(x) = x^5 - 5x^4 + 12x^3 - 24x^2 + 32x - 16$$

STEP 1: There are at most five real solutions.

STEP 2: By Descartes' Rule of Signs, there are five, three, or one positive solution(s). Because

$$f(-x) = -x^5 - 5x^4 - 12x^3 - 24x^2 - 32x - 16$$

there are no negative solutions.

STEP 3: Because $a_5 = 1$ and there are no negative solutions, the potential rational solutions are the positive integers $1, 2, 4, 8$, and 16. We test the potential rational solution 1 first, using synthetic division:

$$\begin{array}{r|rrrrr} 1) & 1 & -5 & 12 & -24 & 32 & -16 \\ & & 1 & -4 & 8 & -16 & 16 \\ \hline & 1 & -4 & 8 & -16 & 16 & 0 \end{array}$$

Thus, 1 is a solution and

$$x^5 - 5x^4 + 12x^3 - 24x^2 + 32x - 16 = (x - 1)(x^4 - 4x^3 + 8x^2 - 16x + 16)$$

The remaining solutions satisfy the depressed equation

$$x^4 - 4x^3 + 8x^2 - 16x + 16 = 0$$

REPEAT STEP 3: The potential rational solutions are still $1, 2, 4, 8$, and 16. We test 1 first, since it may be a repeated solution:

$$\begin{array}{r|rrrr} 1) & 1 & -4 & 8 & -16 & 16 \\ & & 1 & -3 & 5 & -11 \\ \hline & 1 & -3 & 5 & -11 & 5 \end{array}$$

Thus, 1 is not a solution of the depressed equation. We try 2 next:

$$2)\overline{1 \quad -4 \quad 8 \quad -16 \quad 16}$$
$$\underline{\quad\quad 2 \quad -4 \quad 8 \quad -16}$$
$$1 \quad -2 \quad 4 \quad -8 \quad 0$$

Thus, 2 is a solution of the depressed equation and

$$x^5 - 5x^4 + 12x^3 - 24x^2 + 32x - 16 = (x - 1)(x - 2)(x^3 - 2x^2 + 4x - 8)$$

The remaining solutions satisfy the new depressed equation

$$x^3 - 2x^2 + 4x - 8 = 0$$

REPEAT STEP 3: The potential rational solutions are now 1, 2, 4, and 8. We know 1 is not a solution (why?), so we start with 2:

$$2)\overline{1 \quad -2 \quad 4 \quad -8}$$
$$\underline{\quad\quad 2 \quad 0 \quad 8}$$
$$1 \quad 0 \quad 4 \quad 0$$

Thus, 2 is a solution of the new depressed equation and is a re-peated solution of the original equation, so

$$x^5 - 5x^4 + 12x^3 - 24x^2 + 32x - 16 = (x - 1)(x - 2)^2(x^2 + 4)$$

STEP 4: The remaining solutions satisfy the depressed equation

$$x^2 + 4 = 0$$

which has no real solutions. Thus, the real solutions are 1 and 2 (the latter being a repeated solution).   ▬

 Now work Problem 47.

The polynomial f used in Example 8 can be factored as

$$f(x) = x^5 - 5x^4 + 12x^3 - 24x^2 + 32x - 16 = (x - 1)(x - 2)^2(x^2 + 4) \quad (4)$$

The quadratic factor $x^2 + 4$ that appears in the factored form of $f(x)$ is called *irreducible,* because the polynomial $x^2 + 4$ cannot be factored over the real numbers. In general, we say that a quadratic factor $ax^2 + bx + c$ is **irreducible** if it cannot be factored over the real numbers, that is, if it is prime over the real numbers. The next result tells us what to expect when we factor a polynomial.

> ### Theorem
>
> Every polynomial function (with real coefficients) can be uniquely fac-tored into a product of linear factors and/or irreducible quadratic factors.
>
> ▬

We shall prove this result in Section 5.7, and, in fact, we shall draw sev-eral additional conclusions about the zeros of a polynomial function. One conclusion is worth noting now. If a polynomial (with real coefficients) is of odd degree, then it must contain at least one linear factor. (Do you see why?) Therefore, it will have at least one real zero.

Corollary

A polynomial function (with real coefficients) of odd degree has at least one real zero.

Bounds on Zeros

6 The search for the real zeros of a polynomial function can be reduced somewhat if *bounds* on the zeros are found. A number M is a **bound** on the zeros of a polynomial if every zero r lies between $-M$ and M, inclusive. That is, M is a bound to the zeros of a polynomial f if

$$-M \leq \text{any zero of } f \leq M$$

Theorem   Bounds on Zeros

Let f denote a polynomial function whose leading coefficient is 1:

$$f(x) = x^n + a_{n-1}x^{n-1} + \cdots + a_1 x + a_0$$

A bound M on the zeros of f is the smaller of the two numbers

$$\text{Max}\{1, |a_0| + |a_1| + \cdots + |a_{n-1}|\} \quad 1 + \text{Max}\{|a_0|, |a_1|, \ldots |a_{n-1}|\} \quad (5)$$

where Max{ } means "choose the largest entry in { }."

An example will help to make the theorem clear.

E X A M P L E   9

Using the Theorem: Bounds on Zeros

Find a bound to the zeros of each polynomial.

(a) $f(x) = x^5 + 3x^3 - 9x^2 + 5$ (b) $g(x) = 4x^5 - 2x^3 + 2x^2 + 1$

Solution   (a) The leading coefficient of f is 1.

$$f(x) = x^5 + 3x^3 - 9x^2 + 5 \quad a_4 = 0 \quad a_3 = 3 \quad a_2 = -9 \quad a_1 = 0 \quad a_0 = 5$$

We evaluate the expressions in (5).

$$\text{Max}\{1, |a_0| + |a_1| + \cdots |a_{n-1}|\} = \text{Max}\{1, |5| + |0| + |-9| + |3| + |0|\} = \text{Max}\{1, 17\} = 17$$
$$1 + \text{Max}\{|a_0|, |a_1|, \ldots, |a_{n-1}|\} = 1 + \text{Max}\{|5|, |0|, |-9|, |3|, |0|\} = 1 + 9 = 10$$

The smaller of the two numbers, 10, is the bound. Every zero of f lies between -10 and 10.

(b) First we write g so that its leading coefficient is 1.

$$g(x) = 4x^5 - 2x^3 + 2x^2 + 1 = 4(x^5 - \tfrac{1}{2}x^3 + \tfrac{1}{2}x^2 + \tfrac{1}{4})$$

Next, we evaluate the two expressions in (5) with $a_4 = 0$, $a_3 = -\tfrac{1}{2}$, $a_2 = \tfrac{1}{2}$, $a_1 = 0$, $a_0 = \tfrac{1}{4}$.

$$\text{Max}\{1, |a_0| + |a_1| + \cdots + |a_{n-1}|\} = \text{Max}\{1, |\tfrac{1}{4}| + |0| + |\tfrac{1}{2}| + |-\tfrac{1}{2}| + |0|\} = \text{Max}\{1, \tfrac{5}{4}\} = \tfrac{5}{4}$$

$$1 + \text{Max}\{|a_0|, |a_1|, |a_2|, |a_3|, |a_4|\} = 1 + \text{Max}\{|\tfrac{1}{4}|, |0|, |\tfrac{1}{2}|, |-\tfrac{1}{2}|, |0|\} = 1 + \tfrac{1}{2} = \tfrac{3}{2}$$

The smaller of the two numbers, $\tfrac{5}{4}$, is the bound. Every zero of g lies between $-\tfrac{5}{4}$ and $\tfrac{5}{4}$. ∎

Comment The bounds on the zeros of a polynomial provide good choices for setting Xmin and Xmax of the viewing rectangle. With these choices, all the x-intercepts of the graph can be seen. ∎

Intermediate Value Theorem

7 The next result, called the **Intermediate Value Theorem,** is based on the fact that the graph of a polynomial function is continuous; that is, it contains no "holes" or "gaps."

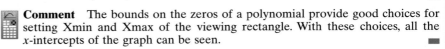

Theorem Intermediate Value Theorem

Let f denote a polynomial function. If $a < b$ and if $f(a)$ and $f(b)$ are of opposite sign, then there is at least one zero of f between a and b.

∎

Although the proof of this result requires advanced methods in calculus, it is easy to "see" why the result is true. Look at Figure 54.

FIGURE 54
If $f(a) < 0$ and $f(b) > 0$, there is a zero between a and b.

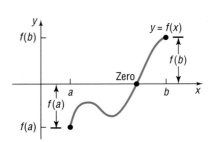

E X A M P L E 10 Using the Intermediate Value Theorem to Locate Zeros

Show that $f(x) = x^5 - x^3 - 1$ has a zero between 1 and 2.

Solution We evaluate f at 1 and at 2:

$$f(1) = -1 \quad \text{and} \quad f(2) = 23$$

Because $f(1) < 0$ and $f(2) > 0$, it follows from the Intermediate Value Theorem that f has a zero between 1 and 2. ∎

Now work Problem 79.

Let's look at the polynomial f of Example 10 more closely. Based on Descartes' Rule of Signs, f has exactly one positive real zero. Based on the Rational Zeros Theorem, 1 is the only potential positive rational zero. Since $f(1) \neq 0$, we conclude that the zero between 1 and 2 is irrational. We can use the Intermediate Value Theorem to approximate it. The steps to use follow:

Approximating the Zeros of a Polynomial Function

STEP 1: Find two consecutive integers a and $a + 1$ such that f has a zero between them.

STEP 2: Divide the interval $[a, a + 1]$ into 10 equal subintervals.

STEP 3: Evaluate f at each endpoint of the subintervals until the Intermediate Value Theorem applies; that interval then contains a zero.

STEP 4: Repeat the process starting at Step 2 until the desired accuracy is achieved.

E X A M P L E   11

Approximating the Zeros of a Polynomial Function

Find the positive zeros of $f(x) = x^5 - x^3 - 1$ correct to two decimal places.

Solution

From Example 10 we know that the positive zero is between 1 and 2. We divide the interval $[1, 2]$ into 10 equal subintervals: $[1, 1.1]$, $[1.1, 1.2]$, $[1.2, 1.3]$, $[1.3, 1.4]$, $[1.4, 1.5]$, $[1.5, 1.6]$, $[1.6, 1.7]$, $[1.7, 1.8]$, $[1.8, 1.9]$, $[1.9, 2]$. Now, we find the value of f at each endpoint until the Intermediate Value Theorem applies.

$$f(x) = x^5 - x^3 - 1$$
$$f(1.0) = -1 \qquad f(1.2) = -0.23968$$
$$f(1.1) = -0.72049 \qquad f(1.3) = 0.51593$$

We can stop here and conclude that the zero is between 1.2 and 1.3. Now we divide the interval $[1.2, 1.3]$ into 10 equal subintervals and proceed to evaluate f at each endpoint:

$$f(1.20) = -0.23968 \qquad f(1.23) = -0.0455613$$
$$f(1.21) = -0.1778185 \qquad f(1.24) = 0.025001$$
$$f(1.22) = -0.1131398$$

We conclude that the zero lies between 1.23 and 1.24, and so, correct to two decimal places, the zero is 1.23.   ∎

FIGURE 55

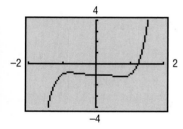

 Exploration We continue to examine the polynomial f. The Theorem Bounds on Zeros tells us that every zero is between -2 and 2. If we graph f using $-2 \leq x \leq 2$, we see that f has exactly one x-intercept. See Figure 55. Using ZERO or ROOT, we find this zero to be 1.23 correct to two decimal places.   ∎

There are many other numerical techniques for approximating the zeros of a polynomial. The one outlined in Example 11 (a variation of the *bisection method*) has the advantages that it will always work, that it can be programmed rather easily on a computer, and each time it is used another decimal place of accuracy is achieved. See Problem 113 for the bisection method, which places the zero in a succession of intervals, with each new interval being half the length of the preceding one.

HISTORICAL FEATURE Formulas for the solution of third- and fourth-degree polynomial equations exist, and, while not very practical, they do have an interesting history.

In the 1500's in Italy, mathematical contests were a popular pastime, and persons possessing methods for solving problems kept them secret. (Solutions that were published were already common knowledge.) Niccolo of Brescia (1499–1557), commonly referred to as Tartaglia ("the stammerer"), had the secret for solving cubic (third-degree) equations, which gave him a decided advantage in the contests. Girolamo Cardano (1501–1576) found out that Tartaglia had the secret, and, being interested in cubics, he requested it from Tartaglia. The reluctant Tartaglia hesitated for some time, but finally, swearing Cardano to secrecy with midnight oaths by candlelight, told him the secret. Cardano then published the solution in his book *Ars Magna* (1545), giving Tartaglia the credit but rather compromising the secrecy. Tartaglia exploded into bitter recriminations, and each wrote pamphlets that reflected on the other's mathematics, moral character, and ancestry.

The quartic (fourth-degree) equation was solved by Cardano's student Lodovico Ferrari, and this solution also was included, with credit and this time with permission, in the *Ars Magna*.

Attempts were made to solve the fifth-degree equation in similar ways, all of which failed. In the early 1800's, P. Ruffini, Niels Abel, and Evariste Galois all found ways to show that it is not possible to solve fifth-degree equations by formula, but the proofs required the introduction of new methods. Galois' methods eventually developed into a large part of modern algebra.

HISTORICAL PROBLEMS *Problems 1–8, develop the Tartaglia–Cardano solution of the cubic equation and show why it is not altogether practical.*

1. Show that the general cubic equation $y^3 + by^2 + cy + d = 0$ can be transformed into an equation of the form $x^3 + px + q = 0$ by using the substitution $y = x - b/3$.

2. In the equation $x^3 + px + q = 0$, replace x by $H + K$. Let $3HK = -p$, and show that $H^3 + K^3 = -q$.
   [**Hint:** $3H^2K + 3HK^2 = 3HKx$.]

3. Based on Problem 2, we have the two equations
   $$3HK = -p \quad \text{and} \quad H^3 + K^3 = -q$$

   Solve for K in $3HK = -p$ and substitute into $H^3 + K^3 = -q$. Then show that
   $$H = \sqrt[3]{\dfrac{-q}{2} + \sqrt{\dfrac{q^2}{4} + \dfrac{p^3}{27}}}$$

   [**Hint:** Look for an equation that is quadratic in form.]

4. Use the solution for H from Problem 3 and the equation $H^3 + K^3 = -q$ to show that
   $$K = \sqrt[3]{\dfrac{-q}{2} - \sqrt{\dfrac{q^2}{4} + \dfrac{p^3}{27}}}$$

5. Use the results from Problems 2–4 to show that the solution of $x^3 + px + q = 0$ is
   $$x = \sqrt[3]{\dfrac{-q}{2} + \sqrt{\dfrac{q^2}{4} + \dfrac{p^3}{27}}} + \sqrt[3]{\dfrac{-q}{2} - \sqrt{\dfrac{q^2}{4} + \dfrac{p^3}{27}}}$$

6. Use the result of Problem 5 to solve the equation $x^3 - 6x - 9 = 0$.

7. Use a calculator and the result of Problem 5 to solve the equation $x^3 + 3x - 14 = 0$.

8. Use the methods of this chapter to solve the equation $x^3 + 3x - 14 = 0$.

5.5 | EXERCISES

In Problems 1–10, use the Factor Theorem to determine whether $x - c$ is a factor of $f(x)$.

1. $f(x) = 4x^3 - 3x^2 - 8x + 4;\quad x - 2$

2. $f(x) = -4x^3 + 5x^2 + 8;\quad x + 3$

3. $f(x) = 3x^4 - 6x^3 - 5x + 10;\quad x - 2$

4. $f(x) = 4x^4 - 15x^2 - 4;\quad x - 2$

5. $f(x) = 3x^6 + 82x^3 + 27;\quad x + 3$

6. $f(x) = 2x^6 - 18x^4 + x^2 - 9;\quad x + 3$

7. $f(x) = 4x^6 - 64x^4 + x^2 - 15;\quad x + 4$

8. $f(x) = x^6 - 16x^4 + x^2 - 16;\quad x + 4$

9. $f(x) = 2x^4 - x^3 + 2x - 1;\quad x - \frac{1}{2}$

10. $f(x) = 3x^4 + x^3 - 3x + 1;\quad x + \frac{1}{3}$

In Problems 11–22, tell the maximum number of zeros that each polynomial function may have. Then, use Descartes' Rule of Signs to determine how many positive and how many negative zeros each polynomial function may have. Do not attempt to find the zeros.

11. $f(x) = -4x^7 + x^3 - x^2 + 2$

12. $f(x) = 5x^4 + 2x^2 - 6x - 5$

13. $f(x) = 2x^6 - 3x^2 - x + 1$

14. $f(x) = -3x^5 + 4x^4 + 2$

15. $f(x) = 3x^3 - 2x^2 + x + 2$

16. $f(x) = -x^3 - x^2 + x + 1$

17. $f(x) = -x^4 + x^2 - 1$

18. $f(x) = x^4 + 5x^3 - 2$

19. $f(x) = x^5 + x^4 + x^2 + x + 1$

20. $f(x) = x^5 - x^4 + x^3 - x^2 + x - 1$

21. $f(x) = x^6 - 1$

22. $f(x) = x^6 + 1$

In Problems 23–34, list the potential rational zeros of each polynomial function. Do not attempt to find the zeros.

23. $f(x) = 3x^4 - 3x^3 + x^2 - x + 1$

24. $f(x) = x^5 - x^4 + 2x^2 + 3$

25. $f(x) = x^5 - 6x^2 + 9x - 3$

26. $f(x) = 2x^5 - x^4 - x^2 + 1$

27. $f(x) = -4x^3 - x^2 + x + 2$

28. $f(x) = 6x^4 - x^2 + 2$

29. $f(x) = 6x^4 - x^2 + 9$

30. $f(x) = -4x^3 + x^2 + x + 6$

31. $f(x) = 2x^5 - x^3 + 2x^2 + 12$

32. $f(x) = 3x^5 - x^2 + 2x + 18$

33. $f(x) = 6x^4 + 2x^3 - x^2 + 20$

34. $f(x) = -6x^3 - x^2 + x + 10$

In Problems 35–46, use Descartes' Rule of Signs and the Rational Zeros Theorem to find all the real zeros of each polynomial function. Use the zeros to factor f over the real numbers.

35. $f(x) = x^3 + 2x^2 - 5x - 6$

36. $f(x) = x^3 + 8x^2 + 11x - 20$

37. $f(x) = 2x^3 - x^2 + 2x - 1$

38. $f(x) = 2x^3 + x^2 + 2x + 1$

39. $f(x) = x^4 + x^2 - 2$

40. $f(x) = x^4 - 3x^2 - 4$

41. $f(x) = 4x^4 + 7x^2 - 2$

42. $f(x) = 4x^4 + 15x^2 - 4$

43. $f(x) = x^4 + x^3 - 3x^2 - x + 2$

44. $f(x) = x^4 - x^3 - 6x^2 + 4x + 8$

45. $f(x) = 4x^5 - 8x^4 - x + 2$

46. $f(x) = 4x^5 + 12x^4 - x - 3$

In Problems 47–58, solve each equation in the real number system.

47. $x^4 - x^3 + 2x^2 - 4x - 8 = 0$

48. $2x^3 + 3x^2 + 2x + 3 = 0$

49. $3x^3 + 4x^2 - 7x + 2 = 0$

50. $2x^3 - 3x^2 - 3x - 5 = 0$

51. $3x^3 - x^2 - 15x + 5 = 0$

52. $2x^3 - 11x^2 + 10x + 8 = 0$

53. $x^4 + 4x^3 + 2x^2 - x + 6 = 0$

54. $x^4 - 2x^3 + 10x^2 - 18x + 9 = 0$

55. $x^3 - \frac{2}{3}x^2 + \frac{8}{3}x + 1 = 0$

56. $x^3 + \frac{3}{2}x^2 + 3x - 2 = 0$

57. $2x^4 - 19x^3 + 57x^2 - 64x + 20 = 0$

58. $2x^4 + x^3 - 24x^2 + 20x + 16 = 0$

In Problems 59–70, find the intercepts of each polynomial function f(x). Find the intervals x for which the graph of f is above the x-axis and below the x-axis. Obtain several other points on the graph, and connect them with a smooth curve. [**Hint:** Use the factored form of f (see Problems 35–46).]

59. $f(x) = x^3 + 2x^2 - 5x - 6$

60. $f(x) = x^3 + 8x^2 + 11x - 20$

61. $f(x) = 2x^3 - x^2 + 2x - 1$

62. $f(x) = 2x^3 + x^2 + 2x + 1$

63. $f(x) = x^4 + x^2 - 2$

64. $f(x) = x^4 - 3x^2 - 4$

65. $f(x) = 4x^4 + 7x^2 - 2$

66. $f(x) = 4x^4 + 15x^2 - 4$

67. $f(x) = x^4 + x^3 - 3x^2 - x + 2$

68. $f(x) = x^4 - x^3 - 6x^2 + 4x + 8$

69. $f(x) = 4x^5 - 8x^4 - x + 2$

70. $f(x) = 4x^5 + 12x^4 - x - 3$

In Problems 71–78, find bounds on the real zeros of each polynomial function.

71. $f(x) = x^4 - 3x^2 - 4$

72. $f(x) = x^4 - 5x^2 - 36$

73. $f(x) = x^4 + x^3 - x - 1$

74. $f(x) = x^4 - x^3 + x - 1$

75. $f(x) = 3x^4 + 3x^3 - x^2 - 12x - 12$

76. $f(x) = 3x^4 - 3x^3 - 5x^2 + 27x - 36$

77. $f(x) = 4x^5 - x^4 + 2x^3 - 2x^2 + x - 1$

78. $f(x) = 4x^5 + x^4 + x^3 + x^2 - 2x - 2$

In Problems 79–84, use the Intermediate Value Theorem to show that each polynomial function has a zero in the given interval.

79. $f(x) = 8x^4 - 2x^2 + 5x - 1$; $[0, 1]$

80. $f(x) = x^4 + 8x^3 - x^2 + 2$; $[-1, 0]$

81. $f(x) = 2x^3 + 6x^2 - 8x + 2$; $[-5, -4]$

82. $f(x) = 3x^3 - 10x + 9$; $[-3, -2]$

83. $f(x) = x^5 - x^4 + 7x^3 - 7x^2 - 18x + 18$; $[1.4, 1.5]$

84. $f(x) = x^5 - 3x^4 - 2x^3 + 6x^2 + x + 2$; $[1.7, 1.8]$

In Problems 85–88, each equation has a solution r in the interval indicated. Use the method of Example 11 to approximate this solution correct to two decimal places.

85. $8x^4 - 2x^2 + 5x - 1 = 0$; $0 \le r \le 1$

86. $x^4 + 8x^3 - x^2 + 2 = 0$; $-1 \le r \le 0$

87. $2x^3 + 6x^2 - 8x + 2 = 0$; $-5 \le r \le -4$

88. $3x^3 - 10x + 9 = 0$; $-3 \le r \le -2$

In Problems 89–92, each polynomial function has exactly one positive zero. Use the method of Example 11 to approximate the zero correct to two decimal places.

89. $f(x) = x^3 + x^2 + x - 4$

90. $f(x) = 2x^4 + x^2 - 1$

91. $f(x) = 2x^4 - 3x^3 - 4x^2 - 8$

92. $f(x) = 3x^3 - 2x^2 - 20$

93. Find k such that $f(x) = x^3 - kx^2 + kx + 2$ has the factor $x - 2$.

94. Find k such that $f(x) = x^4 - kx^3 + kx^2 + 1$ has the factor $x + 2$.

95. What is the remainder when $f(x) = 2x^{20} - 8x^{10} + x - 2$ is divided by $x - 1$?

96. What is the remainder when $f(x) = -3x^{17} + x^9 - x^5 + 2x$ is divided by $x + 1$?

97. Use the Factor Theorem to prove that $x - c$ is a factor of $x^n - c^n$ for any positive integer n.

98. Use the Factor Theorem to prove that $x + c$ is a factor of $x^n + c^n$ if $n \ge 1$ is an odd integer.

99. One solution of the equation $x^3 - 8x^2 + 16x - 3 = 0$ is 3. Find the sum of the remaining solutions.

100. One solution of the equation $x^3 + 5x^2 + 5x - 2 = 0$ is -2. Find the sum of the remaining solutions.

101. Is $\frac{1}{3}$ a zero of $f(x) = 2x^3 + 3x^2 - 6x + 7$? Explain.

102. Is $\frac{1}{3}$ a zero of $f(x) = 4x^3 - 5x^2 - 3x + 1$? Explain.

103. Is $\frac{3}{5}$ a zero of $f(x) = 2x^6 - 5x^4 + x^3 - x + 1$? Explain.

104. Is $\frac{2}{3}$ a zero of $f(x) = x^7 + 6x^5 - x^4 + x + 2$? Explain.

105. What is the length of the edge of a cube if, after a slice 1 inch thick is cut from one side, the volume remaining is 294 cubic inches?

106. What is the length of the edge of a cube if its volume could be doubled by an increase of 6 centimeters in one edge, an increase of 12 centimeters in a second edge, and a decrease of 4 centimeters in the third edge?

107. Let $f(x)$ be a polynomial function whose coefficients are integers. Suppose that r is a real zero of f and that the leading coefficient of f is 1. Use the Rational Zeros Theorem to show that r is either an integer or an irrational number.

108. Prove the Rational Zeros Theorem.

[**Hint:** Let p/q, where p and q have no common factors except 1 and -1, be a solution of the polynomial $f(x) = a_n x^n + a_{n-1}x^{n-1} + \cdots + a_1 x + a_0$, whose coefficients are all integers. Show that $a_n p^n + a_{n-1}p^{n-1}q + \cdots + a_1 pq^{n-1} + a_0 q^n = 0$. Now, because p is a factor of the first n terms of this equation, p must also be a factor of the term $a_0 q^n$. Since p is not a factor of q (why?), p must be a factor of a_0. Similarly, q must be a factor of a_n.]

109. Bisection Method for Approximating Zeros of a Function f We begin with two consecutive integers, a and $a + 1$, such that $f(a)$ and $f(a + 1)$ are of opposite sign. Evaluate f at the midpoint m_1 of a and $a + 1$. If $f(m_1) = 0$, then m_1 is the zero of f, and we are finished. Otherwise, $f(m_1)$ is of opposite sign to either $f(a)$ or $f(a + 1)$. Suppose that it is $f(a)$ and $f(m_1)$ that are of opposite sign. Now evaluate f at the midpoint m_2 of a and m_1. Repeat this process until the desired degree of accuracy is obtained. Note that each iteration places the zero in an interval whose length is half that of the previous interval. Use the bisection method to solve Problems 85–92.

5.6 COMPLEX NUMBERS; QUADRATIC EQUATIONS WITH A NEGATIVE DISCRIMINANT

> **1** Add, Subtract, Multiply, and Divide Complex Numbers
> **2** Solve Quadratic Equations with a Negative Discriminant

One property of a real number is that its square is nonnegative. For example, there is no real number x for which

$$x^2 = -1$$

To remedy this situation, we introduce a number called the **imaginary unit,** which we denote by i and whose square is -1. Thus,

$$i^2 = 1$$

This should not surprise you. If our universe were to consist only of integers, there would be no number x for which $2x = 1$. This unfortunate circumstance was remedied by introducing numbers such as $\frac{1}{2}$ and $\frac{2}{3}$, the *rational numbers*. If our universe were to consist only of rational numbers, there would be no number x whose square equals 2. That is, there would be no number x for which $x^2 = 2$. To remedy this, we introduced numbers such as $\sqrt{2}$ and $\sqrt[3]{5}$, the *irrational numbers*. The *real numbers,* you will recall, consist of the rational numbers and the irrational numbers. Now, if our universe were to consist only of real numbers, then there would be no number x whose square is -1. To remedy this, we introduce a number i, whose square is -1.

In the progression outlined, each time that we encountered a situation that was unsuitable, we introduced a new number system to remedy this situation. And, each new number system contained the earlier number system as a subset. The number system that results from introducing the number i is called the **complex number system.**

> **Complex numbers** are numbers of the form $a + bi$, where a and b are real numbers. The real number a is called the **real part** of the number $a + bi$; the real number b is called the **imaginary part** of $a + bi$.

For example, the complex number $-5 + 6i$ has the real part -5 and the imaginary part 6.

When a complex number is written in the form $a + bi$, where a and b are real numbers, we say it is in **standard form.** However, if the imaginary part of a complex number is negative, such as in the complex number $3 + (-2)i$, we agree to write it instead in the form $3 - 2i$.

Also, the complex number $a + 0i$ is usually written merely as a. This serves to remind us that the real numbers are a subset of the complex numbers. The complex number $0 + bi$ is usually written as bi. Sometimes the complex number bi is called a **pure imaginary number.**

1   Equality, addition, subtraction, and multiplication of complex numbers are defined so as to preserve the familiar rules of algebra for real numbers. Thus, two complex numbers are equal if and only if their real parts are equal and their imaginary parts are equal. That is,

Equality of Complex Numbers

$$a + bi = c + di \quad \text{if and only if} \quad a = c \text{ and } b = d \qquad (1)$$

Two complex numbers are added by forming the complex number whose real part is the sum of the real parts and whose imaginary part is the sum of the imaginary parts. That is,

Sum of Complex Numbers

$$(a + bi) + (c + di) = (a + c) + (b + d)i \qquad (2)$$

To subtract two complex numbers, we follow the rule

Difference of Complex Numbers

$$(a + bi) - (c + di) = (a - c) + (b - d)i \qquad (3)$$

E X A M P L E   1   Adding and Subtracting Complex Numbers

(a) $(3 + 5i) + (-2 + 3i) = [3 + (-2)] + (5 + 3)i = 1 + 8i$

(b) $(6 + 4i) - (3 + 6i) = (6 - 3) + (4 - 6)i = 3 + (-2)i = 3 - 2i$ ▬

 Now work Problem 5.

Products of complex numbers are calculated as illustrated in Example 2.

E X A M P L E   2   Multiplying Complex Numbers

$$(5 + 3i) \cdot (2 + 7i) = 5 \cdot (2 + 7i) + 3i(2 + 7i) = 10 + 35i + 6i + 21i^2$$

$\uparrow$ Distributive property $\uparrow$ Distributive property

$$= 10 + 41i + 21(-1)$$

$\uparrow$ $i^2 = -1$

$$= -11 + 41i$$ ∎

Based on the procedure of Example 2, we define the **product** of two complex numbers by the formula

Product of Complex Numbers

$$(a + bi) \cdot (c + di) = (ac - bd) + (ad + bc)i \qquad (4)$$

Do not bother to memorize formula (4). Instead, whenever it is necessary to multiply two complex numbers, follow the usual rules for multiplying two binomials, as in Example 2, remembering that $i^2 = -1$. For example,

$$(2i)(2i) = 4i^2 = -4$$
$$(2 + i)(1 - i) = 2 - 2i + i - i^2 = 3 - i$$

   Now work Problem 11.

Algebraic properties for addition and multiplication, such as the commutative, associative, and distributive properties, hold for complex numbers. Of these, the property that every nonzero complex number has a multiplicative inverse, or reciprocal, requires a closer look.

Conjugates

If $z = a + bi$ is a complex number, then its **conjugate,** denoted by $\bar{z}$ is defined as

$$\bar{z} = \overline{a + bi} = a - bi$$

For example, $\overline{2 + 3i} = 2 - 3i$ and $\overline{-6 - 2i} = -6 + 2i$.

E X A M P L E   3   Multiplying a Complex Number by Its Conjugate

Find the product of the complex number $z = 3 + 4i$ and its conjugate $\bar{z}$.

Solution   Since $\bar{z} = 3 - 4i$, we have

$$z\bar{z} = (3 + 4i)(3 - 4i) = 9 + 12i - 12i - 16i^2 = 9 + 16 = 25$$ ∎

The result obtained in Example 3 has an important generalization:

Theorem

The product of a complex number and its conjugate is a nonnegative real number. Thus, if $z = a + bi$, then

$$z\bar{z} = a^2 + b^2 \qquad (5)$$

Proof If $z = a + bi$, then

$$z\bar{z} = (a + bi)(a - bi) = a^2 - (bi)^2 = a^2 - b^2i^2 = a^2 + b^2$$

To express the reciprocal of a nonzero complex number z in standard form, multiply the numerator and denominator by its conjugate $\bar{z}$,. Thus, if $z = a + bi$ is a nonzero complex number, then

$$\frac{1}{a + bi} = \frac{1}{z} = \frac{1}{z} \cdot \frac{\bar{z}}{\bar{z}} = \frac{\bar{z}}{z\bar{z}} = \frac{a - bi}{(a + bi)(a - bi)}$$

$$= \frac{a - bi}{a^2 + b^2}$$

↑
Use (5).

$$= \frac{a}{a^2 + b^2} - \frac{b}{a^2 + b^2}i$$

E X A M P L E   4   **Writing the Reciprocal of a Complex Number in Standard Form**

Write $\dfrac{1}{3 + 4i}$ in standard form $a + bi$; that is, find the reciprocal of $3 + 4i$.

Solution   The idea is to multiply the numerator and denominator by the conjugate of $3 + 4i$, that is, the complex number $3 - 4i$. The result is

$$\frac{1}{3 + 4i} = \frac{1}{3 + 4i} \cdot \frac{3 - 4i}{3 - 4i} = \frac{3 - 4i}{9 + 16} = \frac{3}{25} - \frac{4}{25}i$$

To express the quotient of two complex numbers in standard form, we multiply the numerator and denominator of the quotient by the conjugate of the denominator.

E X A M P L E   5   **Writing the Quotient of Complex Numbers in Standard Form**

Write each of the following in standard form:

(a) $\dfrac{1 + 4i}{5 - 12i}$ (b) $\dfrac{2 - 3i}{4 - 3i}$

Solution (a) $\dfrac{1 + 4i}{5 - 12i} = \dfrac{1 + 4i}{5 - 12i} \cdot \dfrac{5 + 12i}{5 + 12i} = \dfrac{5 + 20i + 12i + 48i^2}{25 + 144}$

$= \dfrac{-43 + 32i}{169} = \dfrac{-43}{169} + \dfrac{32}{169}i$

(b) $\dfrac{2 - 3i}{4 - 3i} = \dfrac{2 - 3i}{4 - 3i} \cdot \dfrac{4 + 3i}{4 + 3i} = \dfrac{8 - 12i + 6i - 9i^2}{16 + 9} = \dfrac{17 - 6i}{25} = \dfrac{17}{25} - \dfrac{6}{25}i$

■

Now work Problem 19.

E X A M P L E 6   Writing Other Expressions in Standard Form

If $z = 2 - 3i$ and $w = 5 + 2i$, write each of the following expressions in standard form:

(a) $\dfrac{z}{w}$ (b) $\overline{z + w}$ (c) $z + \overline{z}$

Solution (a) $\dfrac{z}{w} = \dfrac{z \cdot \overline{w}}{w \cdot \overline{w}} = \dfrac{(2 - 3i)(5 - 2i)}{(5 + 2i)(5 - 2i)} = \dfrac{10 - 15i - 4i + 6i^2}{25 + 4}$

$= \dfrac{4 - 19i}{29} = \dfrac{4}{29} - \dfrac{19}{29}i$

(b) $\overline{z + w} = \overline{(2 - 3i) + (5 + 2i)} = \overline{7 - i} = 7 + i$

(c) $z + \overline{z} = (2 - 3i) + (2 + 3i) = 4$

■

The conjugate of a complex number has certain general properties that we shall find useful later.

For a real number $a = a + 0i$, the conjugate is $\overline{a} = \overline{a + 0i} = a - 0i = a$. That is,

> **Theorem**
>
> The conjugate of a real number is the real number itself.

■

Other properties of the conjugate that are direct consequences of the definition are given next. In each statement, z and w represent complex numbers.

> **Theorem**
>
> The conjugate of the conjugate of a complex number is the complex number itself:
>
> $$\overline{\overline{z}} = z \qquad\qquad (6)$$

The conjugate of the sum of two complex numbers equals the sum of their conjugates:

$$\overline{z + w} = \overline{z} + \overline{w} \tag{7}$$

The conjugate of the product of two complex numbers equals the product of their conjugates:

$$\overline{z \cdot w} = \overline{z} \cdot \overline{w} \tag{8}$$

We leave the proofs of equations (6), (7), and (8) as exercises.

Powers of i

The **powers of i** follow a pattern that is useful to know:

$$i^1 = i \qquad\qquad\qquad i^5 = i^4 \cdot i = 1 \cdot i = i$$
$$i^2 = -1 \qquad\qquad\quad i^6 = i^4 \cdot i^2 = -1$$
$$i^3 = i^2 \cdot i = -i \qquad\quad i^7 = i^4 \cdot i^3 = -i$$
$$i^4 = i^2 \cdot i^2 = (-1)(-1) = 1 \qquad i^8 = i^4 \cdot i^4 = 1$$

And so on. Thus, the powers of i repeat with every fourth power.

EXAMPLE 7

Evaluating Powers of i

(a) $i^{27} = i^{24} \cdot i^3 = (i^4)^6 \cdot i^3 = 1^6 \cdot i^3 = -i$
(b) $i^{101} = i^{100} \cdot i^1 = (i^4)^{25} \cdot i = 1^{25} \cdot i = i$

EXAMPLE 8

Writing the Power of a Complex Number in Standard Form

Write $(2 + i)^3$ in standard form.

Solution   We use the special product formula for $(x + a)^3$:

$$(x + a)^3 = x^3 + 3ax^2 + 3a^2x + a^3$$

Thus,

$$(2 + i)^3 = 2^3 + 3 \cdot i \cdot 2^2 + 3 \cdot i^2 \cdot 2 + i^3$$
$$= 8 + 12i + 6(-1) + (-i)$$
$$= 2 + 11i$$

 Now work Problem 33.

Quadratic Equations with a Negative Discriminant

2 Quadratic equations with a negative discriminant have no real number solution. However, if we extend our number system to allow complex numbers, quadratic equations will always have a solution. Since the solution to a quadratic equation involves the square root of the discriminant, we begin with a discussion of square roots of negative numbers.

If N is a positive real number, we define the **principal square root of** $-N$, denoted by $\sqrt{-N}$, as

$$\sqrt{-N} = \sqrt{N}\,i$$

where i is the imaginary unit and $i^2 = -1$.

E X A M P L E   9

Evaluating the Square Root of a Negative Number

(a) $\sqrt{-1} = \sqrt{1}\,i = i$ (b) $\sqrt{-4} = \sqrt{4}\,i = 2i$

(c) $\sqrt{-8} = \sqrt{8}\,i = 2\sqrt{2}\,i$

E X A M P L E   10

Solving Equations

Solve each equation in the complex number system.

(a) $x^2 = 4$ (b) $x^2 = -9$

Solution   (a) $x^2 = 4$

$x = \pm\sqrt{4} = \pm 2$

The equation has two solutions, -2 and 2.

(b) $x^2 = -9$

$x = \pm\sqrt{-9} = \pm\sqrt{9}\,i = \pm 3i$

The equation has two solutions, $-3i$ and $3i$.

Now work Problem 45.

Warning When working with square roots of negative numbers, do not set the square root of a product equal to the product of the square roots (which can be done with positive numbers). To see why, look at this calculation: We know that $\sqrt{100} = 10$. However, it is also true that $100 = (-25)(-4)$, so

$$10 = \sqrt{100} = \sqrt{(-25)(-4)} = \underset{\uparrow}{\sqrt{-25}}\sqrt{-4} = (\sqrt{25}i)(\sqrt{4}i) = (5i)(2i) = 10i^2 = -10$$

Here is the error.

Because we have defined the square root of a negative number, we now can restate the quadratic formula without restriction.

> **Theorem**
>
> In the complex number system, the solutions of the quadratic equation $ax^2 + bx + c = 0$, where a, b, and c are real numbers and $a \neq 0$, are given by the formula
>
> $$x = \frac{-b \pm \sqrt{b^2 - 4ac}}{2a} \qquad (9)$$

E X A M P L E   11

Solving Quadratic Equations in the Complex Number System

Solve the equation $x^2 - 4x + 8 = 0$ in the complex number system.

Solution   Here $a = 1$, $b = -4$, $c = 8$, and $b^2 - 4ac = 16 - 4(8) = -16$. Using equation (9), we find

$$x = \frac{4 \pm \sqrt{-16}}{2} = \frac{4 \pm \sqrt{16}i}{2} = \frac{4 \pm 4i}{2} = 2 \pm 2i$$

The equation has the solution set $\{2 - 2i, 2 + 2i\}$.

Check:

$$2 + 2i: \quad (2 + 2i)^2 - 4(2 + 2i) + 8 = 4 + 8i + 4i^2 - 8 - 8i + 8$$
$$= 4 - 4 = 0$$
$$2 - 2i: \quad (2 - 2i)^2 - 4(2 - 2i) + 8 = 4 - 8i + 4i^2 - 8 + 8i + 8$$
$$= 4 - 4 = 0$$

 Now work Problem 51.

The discriminant, $b^2 - 4ac$, of a quadratic equation still serves as a way to determine the character of the solutions.

Discriminant of a Quadratic Equation

> In the complex number system, consider a quadratic equation $ax^2 + bx + c = 0$ with real coefficients.
>
> 1. If $b^2 - 4ac > 0$, the equation has two unequal real solutions.
> 2. If $b^2 - 4ac = 0$, the equation has a repeated real solution, a double root.
> 3. If $b^2 - 4ac < 0$, the equation has two complex solutions that are not real. The solutions are conjugates of each other.

The third conclusion in the display is a consequence of the fact that if $b^2 - 4ac = -N < 0$ then, by the quadratic formula, the solutions are

$$x = \frac{-b + \sqrt{b^2 - 4ac}}{2a} = \frac{-b + \sqrt{-N}}{2a} = \frac{-b + \sqrt{N}i}{2a} = \frac{-b}{2a} + \frac{\sqrt{N}}{2a}i$$

and

$$x = \frac{-b - \sqrt{b^2 - 4ac}}{2a} = \frac{-b - \sqrt{-N}}{2a} = \frac{-b - \sqrt{N}i}{2a} = \frac{-b}{2a} - \frac{\sqrt{N}}{2a}i$$

which are conjugates of each other.

E X A M P L E   12   Determining the Character of the Solution of a Quadratic Equation

Without solving, determine the character of the solution of each equation in the complex number system.

(a) $3x^2 + 4x + 5 = 0$
(b) $2x^2 + 4x + 1 = 0$
(c) $9x^2 - 6x + 1 = 0$

Solution   (a) Here, $a = 3$, $b = 4$, and $c = 5$, so $b^2 - 4ac = 16 - 4(3)(5) = -44$. The solutions are complex numbers that are not real and are conjugates of each other.

(b) Here, $a = 2$, $b = 4$, and $c = 1$, so $b^2 - 4ac = 16 - 8 = 8$. The solutions are two unequal real numbers.

(c) Here, $a = 9$, $b = -6$, and $c = 1$, so $b^2 - 4ac = 36 - 4(9)(1) = 0$. The solution is a repeated real number, that is, a double root.   ∎

5.6 | EXERCISES

In Problems 1–38, write each expression in the standard form $a + bi$.

1. $(2 - 3i) + (6 + 8i)$ **2.** $(4 + 5i) + (-8 + 2i)$ **3.** $(-3 + 2i) - (4 - 4i)$ **4.** $(3 - 4i) - (-3 - 4i)$

5. $(2 - 5i) - (8 + 6i)$ **6.** $(-8 + 4i) - (2 - 2i)$ **7.** $3(2 - 6i)$ **8.** $-4(2 + 8i)$

9. $2i(2 - 3i)$ **10.** $3i(-3 + 4i)$ **11.** $(3 - 4i)(2 + i)$ **12.** $(5 + 3i)(2 - i)$

13. $(-6 + i)(-6 - i)$ **14.** $(-3 + i)(3 + i)$ **15.** $\dfrac{10}{3 - 4i}$ **16.** $\dfrac{13}{5 - 12i}$

17. $\dfrac{2 + i}{i}$ **18.** $\dfrac{2 - i}{-2i}$ **19.** $\dfrac{6 - i}{1 + i}$ **20.** $\dfrac{2 + 3i}{1 - i}$

21. $\left(\dfrac{1}{2} + \dfrac{\sqrt{3}}{2}i\right)^2$ **22.** $\left(\dfrac{\sqrt{3}}{2} - \dfrac{1}{2}i\right)^2$ **23.** $(1 + i)^2$ **24.** $(1 - i)^2$

25. i^{23} **26.** i^{14} **27.** i^{-15} **28.** i^{-23}

29. $i^6 - 5$ **30.** $4 + i^3$ **31.** $6i^3 - 4i^5$ **32.** $4i^3 - 2i^2 + 1$

33. $(1 + i)^3$ **34.** $(3i)^4 + 1$ **35.** $i^7(1 + i^2)$ **36.** $2i^4(1 + i^2)$

37. $i^6 + i^4 + i^2 + 1$ **38.** $i^7 + i^5 + i^3 + i$

In Problems 39–44, perform the indicated operations and express your answer in the form $a + bi$.

39. $\sqrt{-4}$ **40.** $\sqrt{-9}$ **41.** $\sqrt{-25}$

42. $\sqrt{-64}$ **43.** $\sqrt{(3 + 4i)(4i - 3)}$ **44.** $\sqrt{(4 + 3i)(3i - 4)}$

In Problems 45–64, solve each equation in the complex number system.

45. $x^2 + 4 = 0$ **46.** $x^2 - 4 = 0$ **47.** $x^2 - 16 = 0$ **48.** $x^2 + 25 = 0$

49. $x^2 - 6x + 13 = 0$ **50.** $x^2 + 4x + 8 = 0$ **51.** $x^2 - 6x + 10 = 0$ **52.** $x^2 - 2x + 5 = 0$

53. $8x^2 - 4x + 1 = 0$ **54.** $10x^2 + 6x + 1 = 0$ **55.** $5x^2 + 1 = 2x$ **56.** $13x^2 + 1 = 6x$

57. $x^2 + x + 1 = 0$ **58.** $x^2 - x + 1 = 0$ **59.** $x^3 - 8 = 0$ **60.** $x^3 + 27 = 0$

61. $x^4 = 16$ **62.** $x^4 = 1$ **63.** $x^4 + 13x^2 + 36 = 0$ **64.** $x^4 + 3x^2 - 4 = 0$

In Problems 65–70, without solving, determine the character of the solutions of each equation in the complex number system.

65. $3x^2 - 3x + 4 = 0$

66. $2x^2 - 4x + 1 = 0$

67. $2x^2 + 3x = 4$

68. $x^2 + 6 = 2x$

69. $9x^2 - 12x + 4 = 0$

70. $4x^2 + 12x + 9 = 0$

71. $2 + 3i$ is a solution of a quadratic equation with real coefficients. Find the other solution.

72. $4 - i$ is a solution of a quadratic equation with real coefficients. Find the other solution.

In Problems 73–76, $z = 3 - 4i$ and $w = 8 + 3i$. Write each expression in the standard form $a + bi$.

73. $z + \bar{z}$

74. $w - \bar{w}$

75. $z\bar{z}$

76. $\overline{z - w}$

77. Use $z = a + bi$ to show that $z + \bar{z} = 2a$ and that $z - \bar{z} = 2bi$.

78. Use $z = a + bi$ to show that $\bar{\bar{z}} = z$.

79. Use $z = a + bi$ and $w = c + di$ to show that $\overline{z + w} = \bar{z} + \bar{w}$.

80. Use $z = a + bi$ and $w = c + di$ to show that $\overline{z \cdot w} = \bar{z} \cdot \bar{w}$.

81. Explain to a friend how you would add two complex numbers and how you would multiply two complex numbers. Explain any differences in the two explanations.

82. Write a brief paragraph that compares the method used to rationalize the denominator of a rational expression and the method used to write a complex number in standard form.

5.7 COMPLEX ZEROS; FUNDAMENTAL THEOREM OF ALGEBRA

1   Utilize the Conjugate Pairs Theorem to Find the Complex Zeros of a Polynomial

2   Find a Polynomial Function with Specified Zeros

3   Find the Complex Zeros of a Polynomial

In Section 5.5 we found the **real** zeros of a polynomial function. In this section, we will find the **complex** zeros of a polynomial function. Since the set of real numbers is a subset of the set of complex numbers, finding the complex zeros of a function requires finding all zeros of the form $a + bi$. These zeros will be real if $b = 0$.

A variable in the complex number system is referred to as a **complex variable**. A **complex polynomial function** f of degree n is a complex function of the form

$$f(x) = a_n x^n + a_{n-1} x^{n-1} + \cdots + a_1 x + a_0 \qquad (1)$$

where $a_n, a_{n-1}, \ldots, a_1, a_0$ are complex numbers, $a_n \neq 0$, n is a nonnegative integer, and x is a complex variable. Here, a_n is called the **leading coefficient** of f. A complex number r is called a (complex) **zero** of a complex function f if $f(r) = 0$.

We have learned that some quadratic equations have no real solutions, but that in the complex number system every quadratic equation has a solution, either real or complex. The next result, proved by Karl Friedrich Gauss (1777–1855) when he was 22 years of age* gives an extension to complex polynomials. In fact, this result is so important and useful it has become known as the **Fundamental Theorem of Algebra.**

*In all, Gauss gave four different proofs of this theorem, the first one in 1799 being the subject of his doctoral dissertation.

Fundamental Theorem of Algebra

Every complex polynomial function $f(x)$ of degree $n \geq 1$ has at least one complex zero.

We shall not prove this result, as the proof is beyond the scope of this book. However, using the Fundamental Theorem of Algebra, and the Factor Theorem, we can prove the following result:

Theorem

Every complex polynomial function $f(x)$ of degree $n \geq 1$ can be factored into n linear factors (not necessarily distinct) of the form

$$f(x) = a_n(x - r_1)(x - r_2) \cdot \ldots \cdot (x - r_n) \tag{2}$$

where $a_n, r_1, r_2, \ldots, r_n$ are complex numbers.

Proof Let

$$f(x) = a_n x^n + a_{n-1}x^{n-1} + \cdots + a_1 x + a_0$$

By the Fundamental Theorem of Algebra, f has at least one zero, say, r_1. Then, by the Factor Theorem, $x - r_1$ is a factor, and

$$f(x) = (x - r_1)q_1(x)$$

where $q_1(x)$ is a complex polynomial of degree $n - 1$ whose leading coefficient is a_n. Again by the Fundamental Theorem of Algebra, the complex polynomial $q_1(x)$ has at least one zero, say, r_2. By the Factor Theorem, $q_1(x)$ has the factor $x - r_2$, so

$$q_1(x) = (x - r_2)q_2(x)$$

where $q_2(x)$ is a complex polynomial of degree $n - 2$ whose leading coefficient is a_n. Consequently,

$$f(x) = (x - r_1)(x - r_2)q_2(x)$$

Repeating this argument n times, we finally arrive at

$$f(x) = (x - r_1)(x - r_2) \cdot \ldots \cdot (x - r_n)q_n(x)$$

where $q_n(x)$ is a complex polynomial of degree $n - n = 0$ whose leading coefficient is a_n. Thus, $q_n(x) = a_n x^0 = a_n$, and so

$$f(x) = a_n(x - r_1)(x - r_2) \cdot \ldots \cdot (x - r_n)$$

Complex Polynomials with Real Coefficients

We can use the Fundamental Theorem of Algebra to obtain valuable information about the zeros of complex polynomials whose coefficients are real numbers.

Conjugate Pairs Theorem

Let $f(x)$ be a complex polynomial whose coefficients are real numbers. If $r = a + bi$ is a zero of f, then the complex conjugate $\bar{r} = a - bi$ is also a zero of f.

In other words, for complex polynomials whose coefficients are real numbers, the zeros occur in conjugate pairs.

Proof   Let

$$f(x) = a_n x^n + a_{n-1} x^{n-1} + \cdots + a_1 x + a_0$$

where $a_n, a_{n-1}, \ldots, a_1, a_0$ are real numbers and $a_n \neq 0$. If r is a zero of f, then $f(r) = 0$, so

$$a_n r^n + a_{n-1} r^{n-1} + \cdots + a_1 r + a_0 = 0$$

We take the conjugate of both sides to get

$$\overline{a_n r^n + a_{n-1} r^{n-1} + \cdots + a_1 r + a_0} = \bar{0}$$

$$\overline{a_n r^n} + \overline{a_{n-1} r^{n-1}} + \cdots + \overline{a_1 r} + \overline{a_0} = \bar{0} \quad \text{The conjugate of a sum equals the sum of the conjugates (see Section 5.6).}$$

$$\overline{a_n}(\bar{r})^n + \overline{a_{n-1}}(\bar{r})^{n-1} + \cdots + \overline{a_1}\bar{r} + \overline{a_0} = \bar{0} \quad \text{The conjugate of a product equals the product of the conjugates.}$$

$$a_n(\bar{r})^n + a_{n-1}(\bar{r})^{n-1} + \cdots + a_1\bar{r} + a_0 = 0 \quad \text{The conjugate of a real number equals the real number.}$$

This last equation states that $f(\bar{r}) = 0$; that is, $\bar{r}$, is a zero of f.

The value of this result should be clear. Once we know that, say, $3 + 4i$ is a zero of a polynomial with real coefficients then we know that $3 - 4i$ is also a zero. This result has an important corollary.

Corollary

A complex polynomial f of odd degree with real coefficients has at least one real zero.

Proof   Because complex zeros occur as conjugate pairs in a complex polynomial with real coefficients, there will always be an even number of zeros that are not real numbers. Consequently, since f is of odd degree, one of its zeros has to be a real number.

For example, the polynomial $f(x) = x^5 - 3x^4 + 4x^3 - 5$ has at least one zero that is a real number, since f is of degree 5 (odd) and has real coefficients.

E X A M P L E   1   Using the Conjugate Pairs Theorem

A polynomial f of degree 5 whose coefficients are real numbers has the zeros 1, $5i$, and $1 + i$. Find the remaining two zeros.

Solution   Since complex zeros appear as conjugate pairs, it follows that $-5i$, the conjugate of $5i$, and $1 - i$, the conjugate of $1 + i$, are the two remaining zeros. ▬

Now work Problem 1.

E X A M P L E   2   Finding a Polynomial Function Whose Zeros Are Given

Find a polynomial f of degree 4 whose coefficients are real numbers and that has the zeros 1, 1, and $-4 + i$.

Solution   Since $-4 + i$ is a zero, by the Conjugate Pairs Theorem, $-4 - i$ must also be a zero of f. Because of the Factor Theorem, if $f(c) = 0$, then $x - c$ is a factor of $f(x)$. So, we can now write f as

$$f(x) = a(x - 1)(x - 1)[x - (-4 + i)][x - (-4 - i)]$$

where a is any real number. If we let $a = 1$, we obtain

$$\begin{aligned}
f(x) &= (x - 1)(x - 1)[x - (-4 + i)][x - (-4 - i)] \\
&= (x - 1)^2[x^2 - (-4 + i)x - (-4 - i)x + (-4 + i)(-4 - i)] \\
&= (x - 1)^2(x^2 + 4x - ix + 4x + ix + 16 + 4i - 4i - i^2) \\
&= (x - 1)^2(x^2 + 8x + 17)
\end{aligned}$$

▬

 Exploration   Graph the function f found in Example 2 for $a = 1$. Does the value of a affect the zeros of f? How does the value of a affect the graph of f?

Solution   A quick analysis of the polynomial f tells us what to expect:

At most 3 turning points.
For large x, the graph will behave like $y = x^4$.
A repeated real zero at 1 so that the graph will touch the x-axis at 1.
The only x-intercept is at 1; the y-intercept is 17.

Figure 56 shows the complete graph. (Do you see why? The graph has exactly 3 turning points.) The value of a causes a stretch or compression; a reflection also occurs if $a < 0$. The zeros are not affected. ▬

FIGURE 56

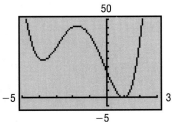

Now we can prove the theorem that we conjectured earlier in Section 5.5.

> **Theorem**
>
> Every polynomial function with real coefficients can be uniquely factored over the real numbers into a product of linear factors and/or irreducible quadratic factors.

▬

Proof   Every complex polynomial f of degree n has exactly n zeros and can be factored into a product of n linear factors. If its coefficients are real, then those zeros that are complex numbers will always occur as conjugate pairs. As a result, if $r = a + bi$ is a complex zero, then so is $\bar{r} = a - bi$. Consequently, when the linear factors $x - r$ and $x - \bar{r}$ of f are multiplied, we have

$$(x - r)(x - \bar{r}) = x^2 - (r + \bar{r})x + r\bar{r} = x^2 - 2ax + a^2 + b^2$$

This second-degree polynomial has real coefficients and is irreducible (over the real numbers). Thus, the factors of f are either linear or irreducible quadratic factors. ∎

E X A M P L E   3   Finding the Complex Zeros of a Polynomial

3   It is known that $2 + i$ is a zero of $f(x) = x^4 - 8x^3 + 64x - 105$. Find the remaining zeros.

Solution   Since $2 + i$ is zero of f and f has real coefficients, then $2 - i$ must also be a zero of f. (Complex zeros appear as conjugate pairs in a polynomial with real coefficients.) So $x - (2 + i)$ and $x - (2 - i)$ are factors of f. Therefore, their product,

$$[x - (2 + i)][x - (2 - i)] = x^2 - (2 - i)x - (2 + i)x + (2 + i)(2 - i)$$
$$= x^2 - 4x + 5$$

is also a factor of f. We use long division to obtain the other factor:

$$
\begin{array}{r}
x^2 - 4x\ -\ 21 \\
x^2 - 4x + 5 \overline{\big)\ x^4 - 8x^3 \qquad\quad + 64x - 105} \\
\underline{x^4 - 4x^3 +\ 5x^2} \\
-4x^3 -\ 5x^2 + 64x \\
\underline{-4x^3 + 16x^2 - 20x} \\
-21x^2 + 84x - 105 \\
\underline{-21x^2 + 84x - 105} \\
0
\end{array}
$$

So $f(x) = (x^2 - 4x + 5)(x^2 - 4x - 21) = (x^2 - 4x + 5)(x - 7)(x + 3)$. By the Factor Theorem, the zeros of f are $-3, 7, 2 + i$, and $2 - i$. ∎

Now work Problem 17.

The steps on page 379 Section 5.5 used to find the real zeros of a polynomial function can also be used to find the complex zeros of a polynomial.

E X A M P L E   4   Finding the Complex Zeros of a Polynomial

Find the complex zeros of the polynomial function

$$f(x) = 3x^4 + 5x^3 + 25x^2 + 45x - 18$$

STEP 1: The degree of f is 4. So f will have 4 complex zeros.

STEP 2: Descartes' Rule of Signs provides information about the real zeros. For this polynomial, there is one positive real zero. Because

$$f(-x) = 3x^4 - 5x^3 + 25x^2 - 45x - 18$$

there are three or one negative real zero(s).

STEP 3: The Rational Zeros Theorem provides information about the potential rational zeros of polynomials with integer coefficients. For this polynomial (which has integer coefficients), the potential rational zeros are

$$\pm\frac{1}{3}, \quad \pm\frac{2}{3}, \quad \pm 1, \quad \pm 2, \quad \pm 3, \quad \pm 6, \quad \pm 9, \quad \pm 18$$

We test 1 first:

$$1\overline{)\begin{array}{rrrrr} 3 & 5 & 25 & 45 & -18 \\ & 3 & 8 & 33 & 78 \\ \hline 3 & 8 & 33 & 78 & 60 \end{array}}$$

We test -1:

$$-1\overline{)\begin{array}{rrrrr} 3 & 5 & 25 & 45 & -18 \\ & -3 & -2 & -23 & -22 \\ \hline 3 & 2 & 23 & 22 & -40 \end{array}}$$

We test 2:

$$2\overline{)\begin{array}{rrrrr} 3 & 5 & 25 & 45 & -18 \\ & 6 & 22 & 94 & 278 \\ \hline 3 & 11 & 47 & 139 & 260 \end{array}}$$

We test -2:

$$-2\overline{)\begin{array}{rrrrr} 3 & 5 & 25 & 45 & -18 \\ & -6 & 2 & -54 & 18 \\ \hline 3 & -1 & 27 & -9 & 0 \end{array}}$$

Thus, $f(-2) = 0$, so -2 is a zero and $x + 2$ is a factor. The depressed equation is

$$3x^3 - x^2 + 27x - 9 = 0$$

STEP 4: We factor by grouping:

$$x^2(3x - 1) + 9(3x - 1) = 0$$
$$(x^2 + 9)(3x - 1) = 0$$

$$\begin{array}{ll} x^2 + 9 = 0 & 3x - 1 = 0 \\ x^2 = -9 & x = 1/3 \\ x = -3i \quad x = 3i & x = 1/3 \end{array}$$

The four complex zeros of f are $\{-3i, 3i, -2, 1/3\}$.

The factored form of f is

$$f(x) = 3x^4 + 5x^3 + 25x^2 + 45x - 18 = 3(x + 3i)(x - 3i)(x + 2)(x - 1/3)$$

5.7 | EXERCISES

In Problems 1–10, information is given about a complex polynomial f(x) whose coefficients are real numbers. Find the remaining zeros of f.

1. Degree 3; zeros: $3, 4 - i$
2. Degree 3; zeros: $4, 3 + i$
3. Degree 4; zeros: $i, 1 + i$
4. Degree 4; zeros: $1, 2, 2 + i$
5. Degree 5; zeros: $1, i, 2i$
6. Degree 5; zeros: $0, 1, 2, i$
7. Degree 4; zeros: $i, 2, -2$
8. Degree 4; zeros: $2 - i, -i$
9. Degree 6; zeros: $2, 2 + i, -3 - i, 0$
10. Degree 6; zeros: $i, 3 - 2i, -2 + i$

In Problems 11–16, form a polynomial f(x) with real coefficients having the given degree and zeros.

11. Degree 4; zeros: $3 + 2i$; 4, multiplicity 2
12. Degree 4; zeros: $i, 1 + 2i$
13. Degree 5; zeros: $2, -i; 1 + i$
14. Degree 6; zeros: $i, 4 - i; 2 + i$
15. Degree 4; zeros: 3, multiplicity 2; $-i$
16. Degree 5; zeros: 1, multiplicity 3; $1 + i$

In Problems 17–24, use the given zero to find the remaining zeros of each function.

17. $f(x) = x^3 - 4x^2 + 4x - 16$;   zero: $2i$
18. $g(x) = x^3 + 3x^2 + 25x + 75$;   zero: $-5i$
19. $f(x) = 2x^4 + 5x^3 + 5x^2 + 20x - 12$;   zero: $-2i$
20. $h(x) = 3x^4 + 5x^3 + 25x^2 + 45x - 18$;   zero: $3i$
21. $h(x) = x^4 - 9x^3 + 21x^2 + 21x - 130$;   zero: $3 - 2i$
22. $f(x) = x^4 - 7x^3 + 14x^2 - 38x - 60$;   zero: $1 + 3i$
23. $h(x) = 3x^5 + 2x^4 + 15x^3 + 10x^2 - 528x - 352$;   zero: $-4i$
24. $g(x) = 2x^5 - 3x^4 - 5x^3 - 15x^2 - 207x + 108$;   zero: $3i$

In Problems 25–34, find the complex zeros of each polynomial function.

25. $f(x) = x^3 - 1$
26. $f(x) = x^4 - 1$
27. $f(x) = x^3 - 8x^2 + 25x - 26$
28. $f(x) = x^3 + 13x^2 + 57x + 85$
29. $f(x) = x^4 + 5x^2 + 4$
30. $f(x) = x^4 + 13x^2 + 36$
31. $f(x) = x^4 + 2x^3 + 22x^2 + 50x - 75$
32. $f(x) = x^4 + 3x^3 - 19x^2 + 27x - 252$
33. $f(x) = 3x^4 - x^3 - 9x^2 + 159x - 52$
34. $f(x) = 2x^4 + x^3 - 35x^2 - 113x + 65$

In Problems 35 and 36, tell why the facts given are contradictory.

35. $f(x)$ is a complex polynomial of degree 3 whose coefficients are real numbers; its zeros are $4 + i, 4 - i,$ and $2 + i$.

36. $f(x)$ is a complex polynomial of degree 3 whose coefficients are real numbers; its zeros are $2, i,$ and $3 + i$.

37. $f(x)$ is a complex polynomial of degree 4 whose coefficients are real numbers; three of its zeros are $2, 1 + 2i,$ and $1 - 2i$. Explain why the remaining zero must be a real number.

38. $f(x)$ is a complex polynomial of degree 4 whose coefficients are real numbers; two of its zeros are -3 and $4 - i$. Explain why one of the remaining zeros must be a real number. Write down one of the missing zeros.

CHAPTER REVIEW

THINGS TO KNOW

Quadratic function $f(x) = ax^2 + bx + c, a \neq 0$ Vertex: $\left(\dfrac{-b}{2a}, f\left(\dfrac{-b}{2a}\right) \right)$

Axis: The line $x = \dfrac{-b}{2a}$

Parabola opens up if $a > 0$

Parabola opens down if $a < 0$

Quadratic equation and quadratic formula

If $ax^2 + bx + c = 0, a \neq 0$, then $x = \dfrac{-b \pm \sqrt{b^2 - 4ac}}{2a}$.

Discriminant

If $b^2 - 4ac > 0$, there are two distinct real solutions.

If $b^2 - 4ac = 0$, there is one repeated real solution.

If $b^2 - 4ac < 0$, there are two distinct complex solutions that are not real; the solutions are conjugates of each other.

Power function	$f(x) = x^n, n \geq 2$ even	Even function:
		Passes through $(-1, 1), (0, 0), (1, 1)$
		Opens up
	$f(x) = x^n, n \geq 3$ odd	Odd function:
		Passes through $(-1, -1), (0, 0), (1, 1)$
		Increasing
Polynomial function	$f(x) = a_n x^n + a_{n-1} x^{n-1} +$ $\cdots + a_1 x + a_0, a_n \neq 0$	At most $n - 1$ turning points; End behavior: behaves like $y = a_n x^n$ for large x.
Rational function	$R(x) = \dfrac{p(x)}{q(x)}$,	p, q are polynomial functions. See Steps 1 through 7 on page 351.

Zeros of a polynomial f   Numbers for which $f(x) = 0$; these are the x-intercepts of the graph of f.

Remainder Theorem   If a polynomial $f(x)$ is divided by $x - c$, then the remainder is $f(c)$.

Factor Theorem $x - c$ is a factor of a polynomial $f(x)$ if and only if $f(c) = 0$.

Descartes' Rule of Signs   Let f denote a polynomial function. The number of positive zeros of f either equals the number of variations in sign of the nonzero coefficients of $f(x)$ or else equals that number less some even integer. The number of negative zeros of f either equals the number of variations in sign of the nonzero coefficients of $f(-x)$ or else equals that number less some even integer.

Rational Zeros Theorem   Let f be a polynomial function of degree 1 or higher of the form

$$f(x) = a_n x^n + a_{n-1} x^{n-1} + \cdots + a_1 x + a_0, a_n \neq 0, a_0 \neq 0$$

where each coefficient is an integer. If p/q, in lowest terms, is a rational zero of f, then p must be a factor of a_0 and q must be a factor of a_n.

Fundamental Theorem of Algebra   Every complex polynomial function $f(x)$ of degree $n \geq 1$ has at least one complex zero.

Conjugate Pairs Theorem   Let $f(x)$ be a complex polynomial whose coefficients are real numbers. If $r = a + bi$ is a zero of f, then its complex conjugate $\bar{r} = a - bi$ is also a zero of f.

HOW TO

Graph quadratic functions

Graph power functions

Graph polynomial functions

Graph rational functions (see Steps 1 through 7, page 351)

Use synthetic division to divide a polynomial by $x - c$

Find the real zeros of a polynomial by using Descartes' Rule of Signs, the Rational Zeros Theorem, and depressed equations

Solve polynomial equations using Descartes' Rule of Signs, the Rational Zeros Theorem, and depressed equations

Use the theorem for Bounds on Zeros

Add, subtract, multiply, and divide complex numbers

Solve quadratic equations in the complex number system

Find the complex zeros of a polynomial

FILL-IN-THE-BLANK ITEMS

1. The graph of a quadratic function is called a(n) _____. Its lowest or highest point is called the _____.

2. In the process of long division,
(Divisor)(quotient) + _____ = _____.

3. When a polynomial function f is divided by $x - c$, the remainder is _____.

4. A polynomial function f has the factor $x - c$ if and only if _____.

5. A number r for which $f(r) = 0$ is called a(n) _____ of the function f.

6. The polynomial function $f(x) = x^5 - 2x^3 + x^2 + x - 1$ has either _____ or _____ positive real zeros; it has _____ or _____ negative real zeros.

7. The possible rational zeros of $f(x) = 2x^5 - x^3 + x^2 - x + 1$ are _____.

8. The line _____ is a horizontal asymptote of $R(x) = \dfrac{x^3 - 1}{x^3 + 1}$.

9. The line _____ is a vertical asymptote of $R(x) = \dfrac{x^3 - 1}{x^3 + 1}$.

10. In the complex number $5 + 2i$, the number 5 is called the _____ part; the number 2 is called the _____ part; the number i is called the _____ _____.

11. If $3 + 4i$ is a zero of a polynomial of degree 5 with real coefficients, then so is _____.

12. The equation $|x^2| = 4$ has four solutions: _____, _____, _____, and _____.

TRUE/FALSE ITEMS

T F **1.** Every polynomial of degree 3 with real coefficients has exactly three real zeros.

T F **2.** If $2 - 3i$ is a zero of a polynomial with real coefficients, then so is $-2 + 3i$.

T F **3.** The graph of $R(x) = \dfrac{x^2}{x - 1}$ has exactly one vertical asymptote.

T F **4.** The graph of $f(x) = x^2(x - 3)(x + 4)$ has exactly three x-intercepts.

T F **5.** If f is a polynomial function of degree 4 and if $f(2) = 5$, then

$$\frac{f(x)}{x - 2} = p(x) + \frac{5}{x - 2}$$

where $p(x)$ is a polynomial of degree 3.

T F **6.** The conjugate of $2 + \sqrt{5}i$ is $-2 - \sqrt{5}i$.

T F **7.** A polynomial of degree n with real coefficients has exactly n complex zeros. At most n of them are real numbers.

REVIEW EXERCISES

Blue problem numbers indicate the author's suggestions for use in a Practice Test.

In Problems 1–10, graph each quadratic function by determining whether its graph opens up or down and by finding its vertex, axis of symmetry, y-intercept, and x-intercepts, if any.

1. $f(x) = (x - 2)^2 + 2$

2. $f(x) = (x + 1)^2 - 4$

3. $f(x) = \frac{1}{4}x^2 - 16$

4. $f(x) = -\frac{1}{2}x^2 + 2$

5. $f(x) = -4x^2 + 4x$

6. $f(x) = 9x^2 - 6x + 3$

7. $f(x) = \frac{9}{2}x^2 + 3x + 1$

8. $f(x) = -x^2 + x + \frac{1}{2}$

9. $f(x) = 3x^2 + 4x - 1$

10. $f(x) = -2x^2 - x + 4$

In Problems 11–16, graph each function using transformations (shifting, compressing, stretching, and reflection).

11. $f(x) = (x + 2)^3$

12. $f(x) = -x^3 + 3$

13. $f(x) = -(x - 1)^4$

14. $f(x) = (x - 1)^4 - 2$

15. $f(x) = (x - 1)^4 + 2$

16. $f(x) = (1 - x)^3$

In Problems 17–22, determine whether the given quadratic function has a maximum value or a minimum value, and then find the value.

17. $f(x) = 3x^2 - 6x + 4$

18. $f(x) = 2x^2 + 8x + 5$

19. $f(x) = -x^2 + 8x - 4$

20. $f(x) = -x^2 - 10x - 3$

21. $f(x) = -3x^2 + 12x + 4$

22. $f(x) = -2x^2 + 4$

In Problems 23–30:
(a) Find the x- and y-intercepts of each polynomial function f.
(b) Determine whether the graph of f touches or crosses the x-axis at each x-intercept.
(c) End behavior: find the power function that the graph of f resembles for large values of x.
(d) Determine the maximum number of turning points of the graph of f.
(e) Use the x-intercept(s) and test numbers to find the intervals on which the graph of f is above and below the x-axis.
(f) Plot the points obtained in parts (a) and (e), and use the remaining information to connect them with a smooth curve.

23. $f(x) = x(x + 2)(x + 4)$

24. $f(x) = x(x - 2)(x - 4)$

25. $f(x) = (x - 2)^2(x + 4)$

26. $f(x) = (x - 2)(x + 4)^2$

27. $f(x) = x^3 - 4x^2$

28. $f(x) = x^3 + 4x$

29. $f(x) = (x - 1)^2(x + 3)(x + 1)$

30. $f(x) = (x - 4)(x + 2)^2(x - 2)$

In Problems 31–42, discuss each rational function following the seven steps outlined in Section 5.3.

31. $R(x) = \dfrac{2x - 6}{x}$

32. $R(x) = \dfrac{4 - x}{x}$

33. $H(x) = \dfrac{x + 2}{x(x - 2)}$

34. $H(x) = \dfrac{x}{x^2 - 1}$

35. $R(x) = \dfrac{x^2 + x - 6}{x^2 - x - 6}$

36. $R(x) = \dfrac{x^2 - 6x + 9}{x^2}$

37. $F(x) = \dfrac{x^3}{x^2 - 4}$

38. $F(x) = \dfrac{3x^3}{(x - 1)^2}$

39. $R(x) = \dfrac{2x^4}{(x - 1)^2}$

40. $R(x) = \dfrac{x^4}{x^2 - 9}$

41. $G(x) = \dfrac{x^2 - 4}{x^2 - x - 2}$

42. $F(x) = \dfrac{(x - 1)^2}{x^2 - 1}$

In Problems 43–46, use synthetic division to find the quotient q(x) and remainder R when f(x) is divided by g(x).

43. $f(x) = 8x^3 - 3x^2 + x + 4$; $g(x) = x - 1$

44. $f(x) = 2x^3 + 8x^2 - 5x + 5$; $g(x) = x - 2$

45. $f(x) = x^4 - 2x^3 + x - 1$; $g(x) = x + 2$

46. $f(x) = x^4 - x^2 + 3x$; $g(x) = x + 1$

47. Find the value of $f(x) = 12x^6 - 8x^4 + 1$ at $x = 4$.

48. Find the value of $f(x) = -16x^3 + 18x^2 - x + 2$ at $x = -2$.

In Problems 49 and 50, use Descartes' Rule of Signs to determine how many positive and negative zeros each polynomial function may have. Do not attempt to find the zeros.

49. $f(x) = 12x^8 - x^7 + 8x^4 - 2x^3 + x + 3$

50. $f(x) = -6x^5 + x^4 + 5x^3 + x + 1$

51. List all the potential rational zeros of $f(x) = 12x^8 - x^7 + 6x^4 - x^3 + x - 3$.

52. List all the potential rational zeros of $f(x) = -6x^5 + x^4 + 2x^3 - x + 1$.

In Problems 53–58, use Descartes' Rule of Signs and the Rational Zeros Theorem to find all the real zeros of each polynomial function. Use the zeros to factor f over the real numbers.

53. $f(x) = x^3 - 3x^2 - 6x + 8$

54. $f(x) = x^3 - x^2 - 10x - 8$

55. $f(x) = 4x^3 + 4x^2 - 7x + 2$

56. $f(x) = 4x^3 - 4x^2 - 7x - 2$

57. $f(x) = x^4 - 4x^3 + 9x^2 - 20x + 20$

58. $f(x) = x^4 + 6x^3 + 11x^2 + 12x + 18$

In Problems 59–62, solve each equation in the real number system.

59. $2x^4 + 2x^3 - 11x^2 + x - 6 = 0$

60. $3x^4 + 3x^3 - 17x^2 + x - 6 = 0$

61. $2x^4 + 7x^3 + x^2 - 7x - 3 = 0$

62. $2x^4 + 7x^3 - 5x^2 - 28x - 12 = 0$

In Problems 63–72, find the intercepts of each polynomial f(x). Find the intervals for which the graph of f is above and below the x-axis. Obtain several other points on the graph and connect them with a smooth curve.

63. $f(x) = x^3 - 3x^2 - 6x + 8$

64. $f(x) = x^3 - x^2 - 10x - 8$

65. $f(x) = 4x^3 + 4x^2 - 7x + 2$

66. $f(x) = 4x^3 - 4x^2 - 7x - 2$

67. $f(x) = x^4 - 4x^3 + 9x^2 - 20x + 20$

68. $f(x) = x^4 + 6x^3 + 11x^2 + 12x + 18$

69. $f(x) = 2x^4 + 2x^3 - 11x^2 + x - 6$

70. $f(x) = 3x^4 + 3x^3 - 17x^2 + x - 6$

71. $f(x) = 2x^4 + 7x^3 + x^2 - 7x - 3$

72. $f(x) = 2x^4 + 7x^3 - 5x^2 - 28x - 12$

In Problems 73–76, find bounds to the zeros of each polynomial function.

73. $f(x) = x^3 - x^2 - 4x + 2$

74. $f(x) = x^3 + x^2 - 10x - 5$

75. $f(x) = 2x^3 - 7x^2 - 10x + 35$

76. $f(x) = 3x^3 - 7x^2 - 6x + 14$

In Problems 77–80, use the Intermediate Value Theorem to show that each polynomial has a zero in the given interval.

77. $f(x) = 3x^3 - x - 1;$ $[0, 1]$

78. $f(x) = 2x^3 - x^2 - 3;$ $[1, 2]$

79. $f(x) = 8x^4 - 4x^3 - 2x - 1;$ $[0, 1]$

80. $f(x) = 3x^4 + 4x^3 - 8x - 2;$ $[1, 2]$

In Problems 81–84, each polynomial has exactly one positive zero. Approximate the zero correct to two decimal places.

81. $f(x) = x^3 - x - 2$

82. $f(x) = 2x^3 - x^2 - 3$

83. $f(x) = 8x^4 - 4x^3 - 2x - 1$

84. $f(x) = 3x^4 + 4x^3 - 8x - 2$

In Problems 85–94, use the complex number system and write each expression in the standard form a + bi.

85. $(6 + 3i) - (2 - 4i)$

86. $(8 - 3i) + (-6 + 2i)$

87. $4(3 - i) + 3(-5 + 2i)$

88. $2(1 + i) - 3(2 - 3i)$

89. $\dfrac{3}{3 + i}$

90. $\dfrac{4}{2 - i}$

91. i^{50}

92. i^{29}

93. $(2 + 3i)^3$

94. $(3 - 2i)^3$

In Problems 95–98, information is given about a complex polynomial f(x) whose coefficients are real numbers. Find the remaining zeros of f.

95. Degree 3; zeros: $4 + i, 6$

96. Degree 3; zeros: $3 + 4i, 5$

97. Degree 4; zeros: $i, 1 + i$

98. Degree 4; zeros: $1, 2, 1 + i$

In Problems 99–112, solve each equation in the complex number system.

99. $x^2 + x + 1 = 0$

100. $x^2 - x + 1 = 0$

101. $2x^2 + x - 2 = 0$

102. $3x^2 - 2x - 1 = 0$

103. $x^2 + 3 = x$

104. $2x^2 + 1 = 2x$

105. $x(1 - x) = 6$

106. $x(1 + x) = 2$

107. $x^4 + 2x^2 - 8 = 0$

108. $x^4 + 8x^2 - 9 = 0$

109. $x^3 - x^2 - 8x + 12 = 0$

110. $x^3 - 3x^2 - 4x + 12 = 0$

111. $3x^4 - 4x^3 + 4x^2 - 4x + 1 = 0$

112. $x^4 + 4x^3 + 2x^2 - 8x - 8 = 0$

113. Find the point on the line $y = x$ that is closest to the point $(3, 1)$.

[**Hint:** Find the minimum value of the function $f(x) = d^2$, where d is the distance from $(3, 1)$ to a point on the line.]

114. Find the point on the line $y = x + 1$ that is closest to the point $(4, 1)$.

115. Landscaping A landscape engineer has 200 feet of border to enclose a rectangular pond. What dimensions will result in the largest pond?

116. Geometry Find the length and width of a rectangle whose perimeter is 20 feet and whose area is 16 square feet.

117. A rectangle has one vertex on the line $y = 10 - x, x > 0$, another at the origin, one on the positive x-axis, and one on the positive y-axis. Find the largest area A that can be enclosed by the rectangle.

118. Minimizing Cost Callaway Golf Company has determined that the daily cost C of manufacturing x Big-Bertha-type gold clubs may be expressed by the quadratic function

$$C(x) = 5x^2 - 620x + 20,000$$

(a) How many clubs should be manufactured to minimize the cost?

(b) At this level of production, what is the cost per club?

(c) What are the fixed costs of production (the cost to produce 0 clubs)?

119. Minimizing Cost Scott-Jones Publishing Company has found that the cost C for paper, printing, and binding of x textbooks is given by the quadratic function

$$C(x) = 0.003x^2 - 30x + 111,800$$

(a) How many books should be manufactured for the cost to be a minimum?

(b) At this level of production, what is the cost per book?

(c) What is the cost to produce the first book $(x = 1)$?

120. Navigation At 5 PM the *Splendor of the Seas* leaves San Juan, Puerto Rico, heading due south at an average speed of 20 knots. At the same time, a Coast Guard ship located 50 nautical miles southwest of San Juan is heading due east at 25 knots. When are the two ships closest? How close do they get to each other? (Express your answer in nautical miles; 1 knot = 1 nautical mile per hour.)

121. Architecture A special window in the shape of a rectangle with semicircles at each end is to be constructed so that the outside dimensions are 100 feet in length. See the illustration. Find the dimensions of the rectangle that maximizes its area.

122. Parabolic Arch Bridges A horizontal bridge is in the shape of a parabolic arch. Given the information shown in the figure, what is the height h of the arch 2 feet from shore?

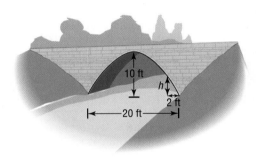

123. Aids Cases in the United States The following data represent the cumulative number of reported AIDS cases in the United States from 1983–1994.

Year, t	Number of AIDS Cases, A
1983, 1	4,589
1984, 2	10,750
1985, 3	22,399
1986, 4	41,256
1987, 5	69,592
1988, 6	104,644
1989, 7	146,574
1990, 8	193,878
1991, 9	251,638
1992, 10	326,648
1993, 11	399,613
1994, 12	457,280

Source: Center for Disease Control.

(a) Draw a scatter diagram of the data.

(b) The cubic function of best fit to these data is

$$A(t) = -68.8t^3 + 4988t^2 - 12{,}436t + 14611$$

Use this function to predict the cumulative number of AIDS cases reported in the United States in 1995.

(c) Use a graphing utility to verify that the function given in (b) is the cubic function of best fit.

(d) With a graphing utility, draw a scatter diagram of the data and then graph the cubic function of best fit on the scatter diagram.

(e) Do you think the function found in (b) will be useful in predicting the number of AIDS cases in 1999?

124. Gravity on the Moon   Neil Armstrong wishes to estimate the acceleration due to gravity on the Moon. He throws a ball (with a special chip that measures the distance to the ground in 2-second intervals) straight down into a crater that is known to be 1000 feet deep. The ball records the following data:

Time (seconds)	0	2	4	6	8	10	12	14	16
Distance (feet)	1000	969	917	844	749	633	496	337	157

(a) Draw a scatter diagram using time as the independent variable and distance as the dependent variable.

(b) The quadratic function of best fit to these data is

$$s(t) = -2.7t^2 - 10t + 1000$$

Use this function to determine the initial velocity of the ball.

(c) Use this function to predict how long it will take for the ball to strike the ground.

(d) Use a graphing utility to verify that the function given in (b) is the quadratic function of best fit.

(e) With a graphing utility, draw a scatter diagram of the data and then graph the quadratic function of best fit on the scatter diagram.

(f) According to physics theory,

$$s(t) = -\frac{1}{2}gt^2 + v_0 t + s_0$$

where $s(t)$ is the height of the object at time t, g is acceleration due to gravity, v_0 is the initial velocity of the object (this number will be negative if the object is thrown down), and s_0 is the initial height of the object. What is the acceleration due to gravity on the moon? (Compare your answer with the fact that the acceleration due to gravity on the moon is approximately -5.33 feet/sec^2.)

125. Design a polynomial function with the following characteristics: degree 6; four real zeros, one of multiplicity 3; y-intercept 3; behaves like $y = -5x^6$ for large values of x. Is this polynomial unique? Compare your polynomial with those of other students. What terms will be the same as everyone else's? Add some more characteristics, such as symmetry or naming the real zeros. How does this modify the polynomial?

126. Design a rational function with the following characteristics: three real zeros, one of multiplicity 2; y-intercept 1; vertical asymptotes $x = -2$ and $x = 3$; oblique asymptote $y = 2x + 1$. Is this rational function unique? Compare yours with those of other students. What will be the same as everyone else's? Add some more characteristics, such as symmetry or naming the real zeros. How does this modify the rational function?

127. The illustration shows the graph of a polynomial function.

(a) Is the degree of the polynomial even or odd?

(b) Is the leading coefficient positive or negative?

(c) Is the function even, odd, or neither?

(d) Why is x^2 necessarily a factor of the polynomial?

(e) What is the minimum degree of the polynomial?

(f) Formulate five different polynomials whose graphs could look like the one shown. Compare yours to those of other students. What similarities do you see? What differences?

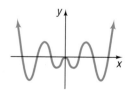

CHAPTER 6

Exponential and Logarithmic Functions

Most people view bees as pests, but farmers and scientists view bees rather differently. On the following page you will find your project as a research entomologist and will need to use the Sullivan website at:

www.prenhall.com/sullivan

to help link you to the Internet resources you will need to prepare your report.

PREPARING FOR THIS CHAPTER

Before getting started on this chapter, review the following concepts:

Exponents *(pp. 25–30 and 71–74)*

Functions *(Sections 4.1, 4.2, and 4.3)*

Simple Interest *(p. 101)*

OUTLINE

BEE PARASITIC MITE SYNDROME [BPMS]

The Bee Crisis. If you've noticed a lack of bees buzzing around your home, you're not alone. In recent years, there has been a *serious loss* of bee colonies in the United States. The cause is from two parasitic mites: Acarapis woodi and Varroa jacobsoni. Varroa first showed up in 1987. Both mites now infest bees in every state except Hawaii. There is no official wild bee census, but estimates go as high as 90 percent of the wild bee colonies have been wiped out! Scientists expect that all the bee colonies in the United States will be exposed to the Varroa mite.

Varroa mites are a serious threat to the United States honeybee industry. In 1986, more than 211,000 beekeepers produced about 200 million pounds of honey valued at $103.1 million, according to the National Honey Board. To agriculture, in general, the mite poses an even bigger threat. Crops valued at $20 billion, ranging from blueberries in Maine to almonds in California, depend on bees for pollination each year. Serious economic losses could result if Varroa mites were to devastate the United States bee population.

As a research entomologist, you want to find alternative methods of control other than the use of pesticides. Your research into Varroa resistant hybrids may offer some hope. You need to create some mathematical models that reflect possible outcomes of your research. These models will help you to determine a focus for your genetic work.

1. What are some of the *difficulties* in fighting the Varroa mite? What are likely results unless some control measures are used?
2. Pesticides are recommended to curb the Varroa epidemic. This is the main defense currently in use. How effective are the pesticides? Does the short-term gain result in any long-term costs?
3. If your research in bee hybridization allows you to breed a mite-resistant honey bee such that only one-third of your bees die from Varroa per year, would this be considered a success?
4. If you study the graphs of bee-mite interaction, you will notice some fairly well defined *logistic curves*. You search the Internet for references to logistic growth models. Using a graphing utility, determine a logistic growth function of best fit that models the situation.
5. Collect and assemble the data on *honey production* for the 1990's. Can the Varroa infestation be seen in this data?
6. Given the results of questions 4 and 5, could you justify removing the use of pesticides for the treatment of a colony?

Until now, our study of functions has concentrated primarily on polynomial and rational functions. These functions belong to the class of **algebraic functions,** that is, functions that can be expressed in terms of sums, differences, products, quotients, powers, or roots of polynomials. Functions that are not algebraic are termed **transcendental** (they transcend, or go beyond, algebraic functions).

In this chapter, we study two transcendental functions: the *exponential* and *logarithmic functions*. These functions occur frequently in a wide variety of applications, such as biology, chemistry, economics, and psychology.

The chapter begins with a discussion of inverse functions.

6.1 | ONE-TO-ONE FUNCTIONS; INVERSE FUNCTIONS

1 Determine Whether a Function Is One-to-One

2 Obtain the Graph of the Inverse Function from the Graph of the Function

3 Find an Inverse Function

In Section 4.1, we said a function f can be thought of as a machine that receives as input a number, say x, from the domain, manipulates it, and outputs the value $f(x)$. The **inverse** of f receives as input a number $f(x)$, manipulates it, and outputs the value x. If the function f is the set of ordered pairs (x, y), then the inverse of f is the set of ordered pairs (y, x).

E X A M P L E   1

Finding the Inverse of a Function

Find the inverse of the following functions.

(a) Let the domain of the function represent the employees of Yolanda's Preowned Car Mart and let the range represent their base salaries.

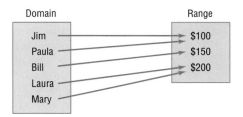

(b) Let the domain of the function represent the employees of Yolanda's Preowned Car Mart and let the range represent their spouse's names.

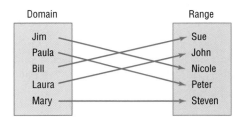

(c) $\{(-3, -27), (-2, -8), (-1, -1), (0, 0), (1, 1), (2, 8), (3, 27)\}$

(d) $\{(-3, 9), (-2, 4), (-1, 1), (0, 0), (1, 1), (2, 4), (3, 9)\}$

Solution (a) The elements in the domain represent inputs to the function, and the elements in the range represent the outputs. To find the inverse, interchange the elements in the domain with the elements in the range. For example, the function receives as input Bill and outputs $150. So the inverse receives as input $150 and outputs Bill. The inverse of the given function takes the form:

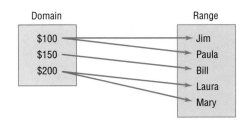

(b) The inverse of the given function is:

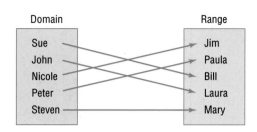

Domain   Range

(c) The inverse of the given function is found by interchanging the entries in each ordered pair and so is given by:

$$\{(-27, -3), (-8, -2), (-1, -1), (0, 0), (1, 1), (8, 2), (27, 3)\}.$$

(d) The inverse of the given function is

$$\{(9, -3), (4, -2), (1, -1), (0, 0), (1, 1), (4, 2), (9, 3)\}.$$

1

We notice that the inverses found in Examples 1(b) and (c) represent functions, since each element in the domain corresponds to a unique element in the range. The inverses found in Examples 1(a) and (d) do not represent functions, since each element in the domain does not correspond to a unique element in the range. Compare the function in Example 1(c) with the function in Example 1(d). For the function in Example 1(c), to every unique x-coordinate there corresponds a unique y-coordinate; for the function in Example 1(d), every unique x-coordinate does not correspond to a unique y-coordinate. Functions for which unique x-coordinates correspond to unique y-coordinates are called *one-to-one* functions. So, in order for the inverse of a function f to be a function itself, f must be one-to-one.

> A function f is said to be **one-to-one** if, for any choice of numbers x_1 and x_2, $x_1 \neq x_2$, in the domain of f, then $f(x_1) \neq f(x_2)$.

In other words, if f is a one-to-one function, then for each x in the domain of f there is exactly one y in the range, and no y in the range is the image of more than one x in the domain. See Figure 1.

FIGURE 1

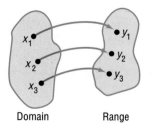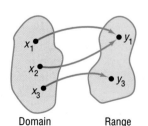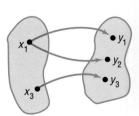

Domain   Range   Domain   Range

(a) One-to-one function: Each x in the domain has one and only one image in the range

(b) Not a one-to-one function: y_1 is the image of both x_1 and x_2

(c) Not a function: x_1 has two images, y_1 and y_2

 Now work Problem 1.

If the graph of a function f is known, there is a simple test, called the **horizontal line test,** to determine whether f is one-to-one.

FIGURE 2
$f(x_1) = f(x_2) = h$, but $x_1 \neq x_2$;
f is not a one-to-one function.

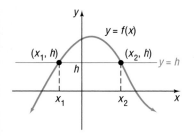

> **Theorem   Horizontal Line Test**
>
> If every horizontal line intersects the graph of a function f in at most one point, then f is one-to-one.

The reason this test works can be seen in Figure 2, where the horizontal line $y = h$ intersects the graph at two distinct points, (x_1, h) and (x_2, h), with the same second element. Thus, f is not one-to-one.

E X A M P L E   2   Using the Horizontal Line Test

For each given function, use the graph to determine whether the function is one-to-one.

(a) $f(x) = x^2$ (b) $g(x) = x^3$

Solution   (a) Figure 3(a) illustrates the horizontal line test for $f(x) = x^2$. The horizontal line $y = 1$ meets the graph of f twice, at $(1, 1)$ and at $(-1, 1)$, so f is not one-to-one.

(b) Figure 3(b) illustrates the horizontal line test for $g(x) = x^3$. Because each horizontal line will intersect the graph of g exactly once, it follows that g is one-to-one.

FIGURE 3

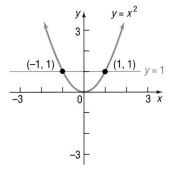

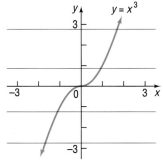

(a) A horizontal line intersects the graph twice; thus, f is not one-to-one

(b) Horizontal lines intersect the graph exactly once; thus, g is one-to-one

  Now work Problem 9.

Let's look more closely at the one-to-one function $g(x) = x^3$. This function is an increasing function. Because an increasing (or decreasing) function will always have different y values for unequal x values, it follows that a function that is increasing (or decreasing) on its domain is also a one-to-one function.

> **Theorem**
>
> An increasing (or decreasing) function is a one-to-one function.

Inverse of a Function $y = f(x)$

For a function $y = f(x)$ to have an inverse *function*, f must be one-to-one. Then for each x in its domain there is exactly one y in its range; furthermore, to each y in the range, there corresponds exactly one x in the domain. The correspondence from the range of f onto the domain of f is, therefore, also a function. It is this function that is the *inverse of f*.

We mentioned in Chapter 4 that a function $y = f(x)$ can be thought of as a rule that tells us to do something to the argument x. For example, the function $f(x) = 2x$ multiplies the argument by 2. An *inverse function* of f undoes whatever f does. Thus, the function $g(x) = \frac{1}{2}x$, which divides the argument by 2, is an inverse of $f(x) = 2x$. See Figure 4.

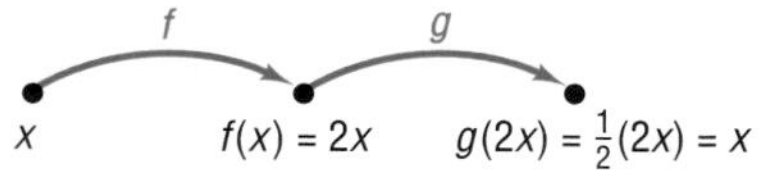

To put it another way, if we think of f as an input/output machine that processes an input x into $f(x)$, then the inverse function reverses this process, taking $f(x)$ back to x. A definition is given next.

FIGURE 4

$f(x) = 2x; \; g(x) = \dfrac{1}{2}x$

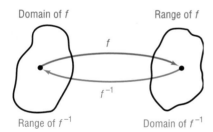

> Let f denote a one-to-one function $y = f(x)$. The **inverse of f**, denoted by f^{-1}, is a function such that $f^{-1}(f(x)) = x$ for every x in the domain of f, and $f(f^{-1}(x)) = x$ for every x in the domain of f^{-1}.

Warning: Be careful! The -1 used in f^{-1} is not an exponent. Thus, f^{-1} does *not* mean the reciprocal of f; it means the inverse of f.

FIGURE 5

Domain of f Range of f

f

f^{-1}

Range of f^{-1} Domain of f^{-1}

Figure 5 illustrates the definition.

Two facts are now apparent about a function f and its inverse f^{-1}.

> Domain of f = Range of f^{-1} Range of f = Domain of f^{-1}

Look again at Figure 5 to visualize the relationship. If we start with x, apply f, and then apply f^{-1}, we get x back again. If we start with x, apply f^{-1}, and then apply f, we get the number x back again. To put it simply, what f does, f^{-1} undoes, and vice versa:

| Input x | $\xrightarrow{\text{Apply } f.}$ | $f(x)$ | $\xrightarrow{\text{Apply } f^{-1}.}$ | $f^{-1}(f(x)) = x$ |

| Input x | $\xrightarrow{\text{Apply } f^{-1}.}$ | $f^{-1}(x)$ | $\xrightarrow{\text{Apply } f.}$ | $f(f^{-1}(x)) = x$ |

In other words,

> $$f^{-1}(f(x)) = x \quad \text{and} \quad f(f^{-1}(x)) = x$$

The preceding conditions can be used to verify that a function is, in fact, the inverse of f, as Example 3 demonstrates.

EXAMPLE 3　　　Verifying Inverse Functions

(a) We verify that the inverse of $g(x) = x^3$ is $g^{-1}(x) = \sqrt[3]{x}$ by showing that

$$g^{-1}(g(x)) = g^{-1}(x^3) = \sqrt[3]{x^3} = x$$

and

$$g(g^{-1}(x)) = g(\sqrt[3]{x}) = (\sqrt[3]{x})^3 = x$$

(b) We verify that the inverse of $h(x) = 3x$ is $h^{-1}(x) = \frac{1}{3}x$ by showing that

$$h^{-1}(h(x)) = h^{-1}(3x) = \frac{1}{3}(3x) = x$$

and

$$h(h^{-1}(x)) = h(\frac{1}{3}x) = 3(\frac{1}{3}x) = x$$

(c) We verify that the inverse of $f(x) = 2x + 3$ is $f^{-1}(x) = \frac{1}{2}(x - 3)$ by showing that

$$f^{-1}(f(x)) = f^{-1}(2x + 3) = \frac{1}{2}[(2x + 3) - 3] = \frac{1}{2}(2x) = x$$

and

$$f(f^{-1}(x)) = f(\frac{1}{2}(x - 3)) = 2[\frac{1}{2}(x - 3)] + 3 = (x - 3) + 3 = x \quad \blacksquare$$

 Now work Problem 21.

Exploration: Simultaneously graph $Y_1 = x$, $Y_2 = x^3$, and $Y_3 = \sqrt[3]{x}$ on a square screen, using the viewing rectangle $-3 \le x \le 3$, $-2 \le y \le 2$. What do you observe about the graphs of $Y_2 = x^3$, its inverse $Y_3 = \sqrt[3]{x}$, and the line $Y_1 = x$?

Repeat this experiment by simultaneously graphing $Y_1 = x$, $Y_2 = 2x + 3$, and $Y_3 = \frac{1}{2}(x - 3)$, using the viewing rectangle $-6 \le x \le 3$, $-8 \le y \le 4$. Do you see the symmetry of the graph of Y_2 and its inverse Y_3 with respect to the line $Y_1 = x$? $\quad \blacksquare$

Geometric Interpretation

FIGURE 6

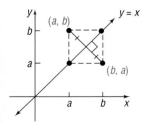

Suppose that (a, b) is a point on the graph of the one-to-one function f defined by $y = f(x)$. Then $b = f(a)$. This means that $a = f^{-1}(b)$, so (b, a) is a point on the graph of the inverse function f^{-1}. The relationship between the point (a, b) on f and the point (b, a) on f^{-1} is shown in Figure 6. The line joining (a, b) and (b, a) is perpendicular to the line $y = x$ and is bisected by the line $y = x$. (Do you see why?) It follows that the point (b, a) on f^{-1} is the reflection about the line $y = x$ of the point (a, b) on f.

Theorem

The graph of a function f and the graph of its inverse function f^{-1} are symmetric with respect to the line $y = x$.

Figure 7 illustrates this result. Notice that, once the graph of f is known, the graph of f^{-1} may be obtained by reflecting the graph of f about the line $y = x$.

FIGURE 7

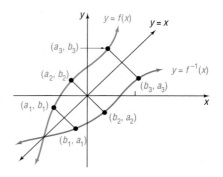

E X A M P L E   4   Graphing the Inverse Function

The graph in Figure 8(a) is that of a one-to-one function $y = f(x)$. Draw the graph of its inverse.

Solution   We begin by adding the graph of $y = x$ to Figure 8(a). Since the points $(-2, -1)$, $(-1, 0)$, and $(2, 1)$ are on the graph of f, we know that the points $(-1, -2)$, $(0, -1)$, and $(1, 2)$ must be on the graph of f^{-1}. Keeping in mind that the graph of f^{-1} is the reflection about the line $y = x$ of the graph of f, we can draw f^{-1}. See Figure 8(b).

FIGURE 8

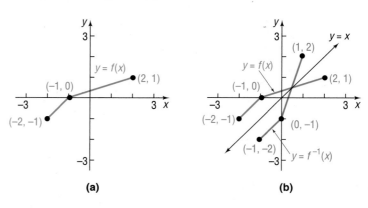

(a)   (b)

 Now work Problem 15.

Finding the Inverse Function

3 The fact that the graph of a one-to-one function f and its inverse are symmetric with respect to the line $y = x$ tells us more. It says that we can obtain f^{-1} by interchanging the roles of x and y in f. Look again at Figure 7. If f is defined by the equation

$$y = f(x)$$

then f^{-1} is defined by the equation

$$x = f(y)$$

The equation $x = f(y)$ defines f^{-1} *implicitly.* If we can solve this equation for y, we will have the *explicit* form of f^{-1}, that is,

$$y = f^{-1}(x)$$

Let's use this procedure to find the inverse of $f(x) = 2x + 3$. (Since f is a linear function and is increasing, we know that f is one-to-one.)

E X A M P L E   5

Finding the Inverse Function

Find the inverse of $f(x) = 2x + 3$. Also find the domain and range of f and f^{-1}. Graph f and f^{-1} on the same coordinate axes.

Solution   In the equation $y = 2x + 3$, interchange the variables x and y. The result,

$$x = 2y + 3$$

is an equation that defines the inverse f^{-1} implicitly. Solving for y, we obtain

$$2y + 3 = x$$
$$2y = x - 3$$
$$y = \tfrac{1}{2}(x - 3)$$

FIGURE 9

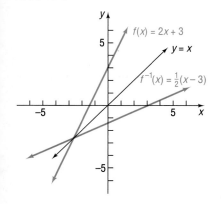

The explicit form of the inverse f^{-1} is therefore

$$f^{-1}(x) = \tfrac{1}{2}(x - 3)$$

which we verified in Example 3(c).
    Then we find

$$\text{Domain } f = \text{Range } f^{-1} = (-\infty, \infty)$$
$$\text{Range } f = \text{Domain } f^{-1} = (-\infty, \infty)$$

The graphs of $f(x) = 2x + 3$ and its inverse $f^{-1}(x) = \tfrac{1}{2}(x - 3)$ are shown in Figure 9. Note the symmetry of the graphs with respect to the line $y = x$.   ∎

We outline next the steps to follow for finding the inverse of a one-to-one function.

Procedure for Finding the Inverse of a One-to-One Function

STEP 1: In $y = f(x)$, interchange the variables x and y to obtain

$$x = f(y)$$

This equation defines the inverse function f^{-1} implicitly.

STEP 2: If possible, solve the implicit equation for y in terms of x to obtain the explicit form of f^{-1}:

$$y = f^{-1}(x)$$

STEP 3: Check the result by showing that

$$f^{-1}(f(x)) = x \quad \text{and} \quad f(f^{-1}(x)) = x$$

E X A M P L E   6

Finding the Inverse Function

The function

$$f(x) = \frac{2x + 1}{x - 1} \qquad x \neq 1$$

is one-to-one. Find its inverse and check the result.

Solution   STEP 1: Interchange the variables x and y in

$$y = \frac{2x + 1}{x - 1}$$

to obtain

$$x = \frac{2y + 1}{y - 1}$$

STEP 2: Solve for y:

$$x = \frac{2y + 1}{y - 1}$$
$$x(y - 1) = 2y + 1$$
$$xy - x = 2y + 1$$
$$xy - 2y = x + 1$$
$$(x - 2)y = x + 1$$
$$y = \frac{x + 1}{x - 2}$$

The inverse is

$$f^{-1}(x) = \frac{x + 1}{x - 2} \qquad x \neq 2$$

STEP 3: *Check:*

$$f^{-1}(f(x)) = f^{-1}\left(\frac{2x + 1}{x - 1}\right) = \frac{\dfrac{2x + 1}{x - 1} + 1}{\dfrac{2x + 1}{x - 1} - 2} = \frac{2x + 1 + x - 1}{2x + 1 - 2(x - 1)} = \frac{3x}{3} = x$$

$$f(f^{-1}(x)) = f\left(\frac{x + 1}{x - 2}\right) = \frac{2\left(\dfrac{x + 1}{x - 2}\right) + 1}{\dfrac{x + 1}{x - 2} - 1} = \frac{2(x + 1) + x - 2}{x + 1 - (x - 2)} = \frac{3x}{3} = x$$

■

Exploration   In Example 6, we found that, if $f(x) = (2x + 1)/(x - 1)$, then $f^{-1}(x) = (x + 1)/(x - 2)$. Graph $y = f(f^{-1}(x))$ on a square screen. What do you see? Are you surprised?   ■

Now work Problem 33.

If a function is not one-to-one, then it will have no inverse function. Sometimes, though, an appropriate restriction on the domain of such a function will yield a new function that is one-to-one. Let's look at an example of this common practice.

E X A M P L E   7   Finding the Inverse Function

Find the inverse of $y = f(x) = x^2$ if $x \geq 0$.

Solution   The function $f(x) = x^2$ is not one-to-one. [Refer to Example 2(a).] However, if we restrict f to only that part of its domain for which $x \geq 0$, as indicated, we have a new function that is increasing and therefore is one-to-one. As a result, the function defined by $y = x^2$, $x \geq 0$, has an inverse function, f^{-1}.

We follow the steps given previously to find f^{-1}:

STEP 1: In the equation $y = x^2$, $x \geq 0$, interchange the variables x and y. The result is

$$x = y^2 \qquad y \geq 0$$

This equation defines (implicitly) the inverse function.

STEP 2: We solve for y to get the explicit form of the inverse. Since $y \geq 0$, only one solution for y is obtained,

$$y = \sqrt{x}$$

so $f^{-1}(x) = \sqrt{x}$.

STEP 3: *Check:* $f^{-1}(f(x)) = f^{-1}(x^2) = \sqrt{x^2} = |x| = x$, since $x \geq 0$.

$$f(f^{-1}(x)) = f(\sqrt{x}) = (\sqrt{x})^2 = x$$

Figure 10 illustrates the graphs of $f(x) = x^2$, $x \geq 0$, and $f^{-1}(x) = \sqrt{x}$. ▬

FIGURE 10

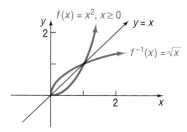

SUMMARY

1. If a function f is one-to-one, then it has an inverse function f^{-1}.
2. Domain f = Range f^{-1};   Range f = Domain f^{-1}.
3. To verify that f^{-1} is the inverse of f, show that $f^{-1}(f(x)) = x$ and $f(f^{-1}(x)) = x$.
4. The graphs of f and f^{-1} are symmetric with respect to the line $y = x$.

6.1 | EXERCISES

In Problems 1–8, (a) find the inverse and (b) determine whether the inverse represents a function.

1.

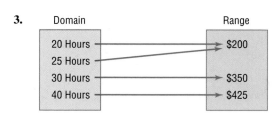

2.

3.

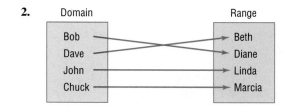

4.

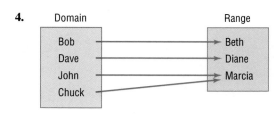

5. $\{(2, 6), (-3, 6), (4, 9), (1, 10)\}$

6. $\{(-2, 5), (-1, 3), (3, 7), (4, 12)\}$

7. $\{(0, 0), (1, 1), (2, 16), (3, 81)\}$

8. $\{(1, 2), (2, 8), (3, 18), (4, 32)\}$

In Problems 9–14, the graph of a function f is given. Use the horizontal line test to determine whether f is one-to-one.

9.

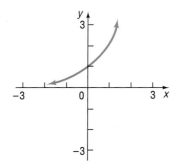

10.

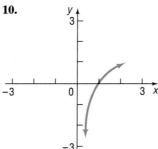

11.

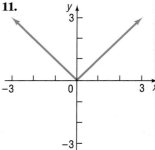

12.

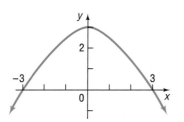

13.

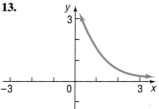

14.

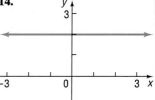

In Problems 15–20, the graph of a one-to-one function f is given. Find the graph of the inverse function f^{-1}. For convenience (and as a hint), the graph of y = x is also given.

15.

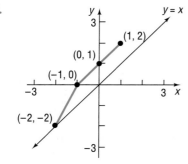

16.

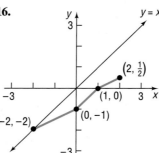

17.

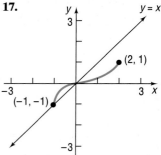

18.

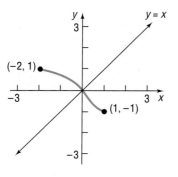

19.

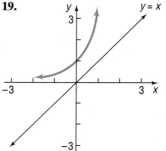

20.

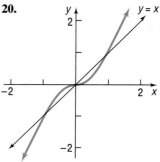

In Problems 21–30, verify that the functions f and g are inverses of each other by showing that f(g(x)) = x and g(f(x)) = x.

21. $f(x) = 3x + 4$; $g(x) = \frac{1}{3}(x - 4)$

22. $f(x) = 3 - 2x$; $g(x) = -\frac{1}{2}(x - 3)$

23. $f(x) = 4x - 8$; $g(x) = \frac{x}{4} + 2$

24. $f(x) = 2x + 6$; $g(x) = \frac{1}{2}x - 3$

25. $f(x) = x^3 - 8$; $g(x) = \sqrt[3]{x + 8}$

26. $f(x) = (x - 2)^2$, $x \ge 2$; $g(x) = \sqrt{x} + 2$, $x \ge 0$

27. $f(x) = \dfrac{1}{x}$; $g(x) = \dfrac{1}{x}$

28. $f(x) = x$; $g(x) = x$

29. $f(x) = \dfrac{2x + 3}{x + 4}$; $g(x) = \dfrac{4x - 3}{2 - x}$

30. $f(x) = \dfrac{x - 5}{2x + 3}$; $g(x) = \dfrac{3x + 5}{1 - 2x}$

In Problems 31–42, the function f is one-to-one. Find its inverse and check your answer. State the domain and range of f and f^{-1}. Graph f, f^{-1}, and y = x on the same coordinate axes.

31. $f(x) = 3x$

32. $f(x) = -4x$

33. $f(x) = 4x + 2$

34. $f(x) = 1 - 3x$

35. $f(x) = x^3 - 1$

36. $f(x) = x^3 + 1$

37. $f(x) = x^2 + 4$, $x \ge 0$

38. $f(x) = x^2 + 9$, $x \ge 0$

39. $f(x) = \dfrac{4}{x}$

40. $f(x) = -\dfrac{3}{x}$

41. $f(x) = \dfrac{1}{x - 2}$

42. $f(x) = \dfrac{4}{x + 2}$

In Problems 43–54, the function f is one-to-one. Find its inverse and check your answer. State the domain and range of f and f^{-1}.

43. $f(x) = \dfrac{2}{3 + x}$

44. $f(x) = \dfrac{4}{2 - x}$

45. $f(x) = (x + 2)^2$, $x \ge -2$

46. $f(x) = (x - 1)^2$, $x \ge 1$

47. $f(x) = \dfrac{2x}{x - 1}$

48. $f(x) = \dfrac{3x + 1}{x}$

49. $f(x) = \dfrac{3x + 4}{2x - 3}$

50. $f(x) = \dfrac{2x - 3}{x + 4}$

51. $f(x) = \dfrac{2x + 3}{x + 2}$

52. $f(x) = \dfrac{-3x - 4}{x - 2}$

53. $f(x) = 2\sqrt[3]{x}$

54. $f(x) = \dfrac{4}{\sqrt{x}}$

55. Find the inverse of the linear function:
$f(x) = mx + b$, $m \ne 0$.

56. Find the inverse of the function:
$f(x) = \sqrt{r^2 - x^2}$, $0 \le x \le r$.

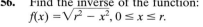 **57.** Can an even function be one-to-one? Explain.

58. Is every odd function one-to-one? Explain.

59. A function f has an inverse. If the graph of f lies in quadrant I, in which quadrant does the graph of f^{-1} lie?

60. A function f has an inverse. If the graph of f lies in quadrant II, in which quadrant does the graph of f^{-1} lie?

61. The function $f(x) = |x|$ is not one-to-one. Find a suitable restriction on the domain of f so that the new function that results is one-to-one. Then find the inverse of f.

62. The function $f(x) = x^4$ is not one-to-one. Find a suitable restriction on the domain of f so that the new function that results is one-to-one. Then find the inverse of f.

63. **Temperature Conversion** To convert from x degrees Celsius to y degrees Fahrenheit, we use the formula $y = f(x) = \frac{9}{5}x + 32$. To convert from x degrees Fahrenheit to y degrees Celsius,

we use the formula $y = g(x) = \frac{5}{9}(x - 32)$. Show that f and g are inverse functions.

64. **Demand for Corn** The demand for corn obeys the equation $p(x) = 300 - 50x$, where p is the price per bushel (in dollars) and x is the number of bushels produced, in millions. Express the production amount x as a function of the price p.

65. **Period of a Pendulum** The period T (in seconds) of a simple pendulum is a function of its length l (in feet), given by $T(l) = 2\pi\sqrt{l/g}$, where $g \approx 32.2$ feet per second per second is the acceleration of gravity. Express the length l as a function of the period T.

66. Give an example of a function whose domain is the set of real numbers and that is neither increasing nor decreasing on its domain, but is one-to-one.
[**Hint:** Use a piecewise defined function.]

67. Given
$$f(x) = \dfrac{ax + b}{cx + d}$$
find $f^{-1}(x)$. If $c \ne 0$, under what conditions on $a, b, c,$ and d is $f = f^{-1}$?

68. We said earlier that finding the range of a function f is not easy. However, if f is one-to-one, we can find its range by finding the domain of the inverse function f^{-1}. Use this technique to find the range of each of the following one-to-one functions:

(a) $f(x) = \dfrac{2x + 5}{x - 3}$

(b) $g(x) = 4 - \dfrac{2}{x}$

(c) $F(x) = \dfrac{3}{4 - x}$

69. If the graph of a function and its inverse intersect, where must this necessarily occur? Can they intersect anywhere else? Must they intersect?

70. Can a one-to-one function and its inverse be equal? What must be true about the graph of f for this to happen? Give some examples to support your conclusion.

71. Draw the graph of a one-to-one function that contains the points $(-2, -3)$, $(0, 0)$, and $(1, 5)$. Now draw the graph of its inverse. Compare your graph to those of other students. Discuss any similarities. What differences do you see?

6.2 | EXPONENTIAL FUNCTIONS

> **1** Evaluate Exponential Functions
> **2** Graph Exponential Functions
> **3** Define the Number e

1 In Chapter 1, we gave a definition for raising a real number a to a rational power. Based on that discussion, we gave meaning to expressions of the form

$$a^r$$

where the base a is a positive real number and the exponent r is a rational number.

But what is the meaning of a^x, where the base a is a positive real number and the exponent x is an irrational number? Although a rigorous definition requires methods discussed in calculus, the basis for the definition is easy to follow: Select a rational number r that is formed by truncating (removing) all but a finite number of digits from the irrational number x. Then it is reasonable to expect that

$$a^x \approx a^r$$

For example, take the irrational number $\pi = 3.14159. \ldots$. Then, an approximation to a^π is

$$a^\pi \approx a^{3.14}$$

where the digits after the hundredths position have been removed from the value for π. A better approximation would be

$$a^\pi \approx a^{3.14159}$$

where the digits after the hundred-thousandths position have been removed. Continuing this way, we can obtain approximations to a^π to any desired degree of accuracy.

Most calculators have an $\boxed{x^y}$ key or a caret key $\boxed{\wedge}$ for working with exponents. To evaluate expressions of the form a^x, enter the base a, then press the $\boxed{x^y}$ key (or the $\boxed{\wedge}$ key), enter the exponent x, and press $\boxed{=}$ (or $\boxed{\text{enter}}$).

E X A M P L E   1   Using a Calculator to Evaluate Powers of 2

Using a calculator, evaluate:

(a) $2^{1.4}$ (b) $2^{1.41}$ (c) $2^{1.414}$ (d) $2^{1.4142}$ (e) $2^{\sqrt{2}}$

Solution   (a) $2^{1.4} \approx 2.639015822$ (b) $2^{1.41} \approx 2.657371628$
(c) $2^{1.414} \approx 2.66474965$ (d) $2^{1.4142} \approx 2.665119089$
(e) $2^{\sqrt{2}} \approx 2.665144143$

   Now work Problem 1.

It can be shown that the familiar laws of exponents hold for real exponents.

Theorem   Laws of Exponents

If s, t, a, and b are real numbers with $a > 0$ and $b > 0$, then

$$a^s \cdot a^t = a^{s+t} \qquad (a^s)^t = a^{st} \qquad (ab)^s = a^s \cdot b^s$$
$$1^s = 1 \qquad a^{-s} = \frac{1}{a^s} = \left(\frac{1}{a}\right)^s \qquad a^0 = 1 \tag{1}$$

We are now ready for the following definition.

An **exponential function** is a function of the form

$$f(x) = a^x$$

where a is a positive real number ($a > 0$) and $a \neq 1$. The domain of f is the set of all real numbers.

We exclude the base $a = 1$, because this function is simply the constant function $f(x) = 1^x = 1$. We also need to exclude the bases that are negative, because, otherwise, we would have to exclude many values of x from the domain, such as $x = \frac{1}{2}$, $x = \frac{3}{4}$, and so on. [Recall that $(-2)^{1/2}$, $(-3)^{3/4}$, and so on, are not defined in the system of real numbers.]

Graphs of Exponential Functions

2 First, we graph the exponential function $f(x) = 2^x$.

E X A M P L E   2   Graphing an Exponential Function

Graph the exponential function: $f(x) = 2^x$.

Solution   The domain of $f(x) = 2^x$ consists of all real numbers. We begin by locating some points on the graph of $f(x) = 2^x$, as listed in Table 1.

Since $2^x > 0$ for all x, the range of f is $(0, \infty)$. From this, we conclude that the graph has no x-intercepts, and, in fact, the graph will lie above the x-axis. As Table 1 indicates, the y-intercept is 1. Table 1 also indicates that as $x \to -\infty$ the value of $f(x) = 2^x$ gets closer and closer to 0. Thus, the x-axis is a horizontal asymptote to the graph as $x \to -\infty$. This gives us the end behavior of the graph for x large and negative.

To determine the end behavior for x large and positive, look again at Table 1. As $x \to \infty$, $f(x) = 2^x$ grows very quickly, causing the graph of $f(x) = 2^x$ to rise very rapidly. Thus, it is apparent that f is an increasing function and hence is one-to-one.

Using all this information, we plot some of the points from Table 1 and connect them with a smooth, continuous curve, as shown in Figure 11.

TABLE 1	
x	**$f(x) = 2^x$**
-10	$2^{-10} \approx 0.00098$
-3	$2^{-3} = \frac{1}{8}$
-2	$2^{-2} = \frac{1}{4}$
-1	$2^{-1} = \frac{1}{2}$
0	$2^0 = 1$
1	$2^1 = 2$
2	$2^2 = 4$
3	$2^3 = 8$
10	$2^{10} = 1024$

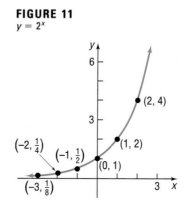

FIGURE 11
$y = 2^x$

As we shall see, graphs that look like the one in Figure 11 occur very frequently in a variety of situations. For example, look at the graph in Figure 12, which illustrates the population in Ethiopia. Researchers might conclude from this graph that the population in Ethiopia is "behaving exponentially"; that is, the graph exhibits "rapid, or exponential, growth." We shall have more to say about situations that lead to exponential growth later in this chapter. For now, we continue to seek properties of the exponential functions.

The graph of $f(x) = 2^x$ in Figure 11 is typical of all exponential functions that have a base larger than 1. Such functions are increasing functions and hence are one-to-one. Their graphs lie above the x-axis, pass through the

FIGURE 12
Ethiopia's population is growing at one of the fastest paces in the world, exacerbating continuing shortfalls in food needed to feed its millions of people. The nation, which is only about three-quarters the size of Alaska, grows with more than 2 million births each year.

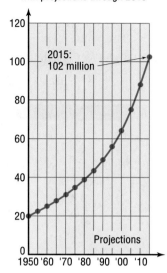

Ethiopia, in millions of people, with projections through 2015

Sources: News reports, World Bank, U.S. Agency for International Development, UN Food and Agriculture Organization, Population Reference Bureau, UNICEF, U.S. Department of Agriculture, Human Nutrition Information Service.

FIGURE 13

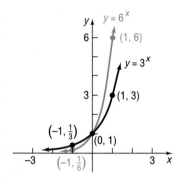

point $(0, 1)$, and thereafter rise rapidly as $x \to \infty$. As $x \to -\infty$, the x-axis is a horizontal asymptote. There are no vertical asymptotes. Finally, the graphs are smooth and continuous, with no corners or gaps.

Figure 13 illustrates the graphs of two more exponential functions whose bases are larger than 1. Notice that for the larger base the graph is steeper when $x > 0$ and is closer to the x-axis when $x < 0$.

 Seeing the Concept Graph $y = 2^x$ and compare what you see to Figure 11. Clear the screen and graph $y = 3^x$ and $y = 6^x$ and compare what you see to Figure 13. Clear the screen and graph $y = 10^x$ and $y = 100^x$. What viewing rectangle seems to work best?

The display summarizes the information we have about $f(x) = a^x, a > 1$:

$$f(x) = a^x \qquad a > 1$$
Domain: $(-\infty, \infty)$ Range: $(0, \infty)$
x-Intercepts: None y-Intercept: 1
Horizontal asymptote: x-Axis, as $x \to -\infty$
f is an increasing function
f is one-to-one
The graph of f passes through the points $(0, 1)$ and $(1, a)$

Now we consider $f(x) = a^x$ when $0 < a < 1$.

E X A M P L E   3   Graphing an Exponential Function

Graph the exponential function: $f(x) = \left(\frac{1}{2}\right)^x$.

Solution   The domain of $f(x) = \left(\frac{1}{2}\right)^x$ consists of all real numbers. As before, we locate some points on the graph, as listed in Table 2. Since $\left(\frac{1}{2}\right)^x > 0$ for all x, the range of f is $(0, \infty)$. Thus, the graph lies above the x-axis and so has no x-intercepts. The y-intercept is 1. As $x \to -\infty$, $f(x) = \left(\frac{1}{2}\right)^x$ grows very quickly. As $x \to \infty$, the values of $f(x)$ approach 0. Thus, the x-axis ($y = 0$) is a horizontal asymptote as $x \to \infty$. It is apparent that f is a decreasing function and hence is one-to-one. Figure 14 illustrates the graph.

TABLE 2	
x	$f(x) = \left(\frac{1}{2}\right)^x$
-10	$\left(\frac{1}{2}\right)^{-10} = 1024$
-3	$\left(\frac{1}{2}\right)^{-3} = 8$
-2	$\left(\frac{1}{2}\right)^{-2} = 4$
-1	$\left(\frac{1}{2}\right)^{-1} = 2$
0	$\left(\frac{1}{2}\right)^{0} = 1$
1	$\left(\frac{1}{2}\right)^{1} = \frac{1}{2}$
2	$\left(\frac{1}{2}\right)^{2} = \frac{1}{4}$
3	$\left(\frac{1}{2}\right)^{3} = \frac{1}{8}$
10	$\left(\frac{1}{2}\right)^{10} \approx 0.00098$

FIGURE 14

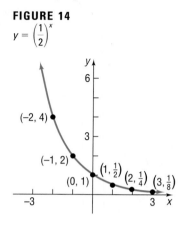

$y = \left(\frac{1}{2}\right)^x$

Note that we could have obtained the graph of $y = \left(\frac{1}{2}\right)^x$ from the graph of $y = 2^x$. If $f(x) = 2^x$, then $f(-x) = 2^{-x} = \frac{1}{2^x} = \left(\frac{1}{2}\right)^x$. Thus, the graph of $y = \left(\frac{1}{2}\right)^x = 2^{-x}$ is a reflection about the y-axis of the graph of $y = 2^x$. Compare Figures 11 and 14.

The graph of $f(x) = \left(\frac{1}{2}\right)^x$ in Figure 14 is typical of all exponential functions that have a base between 0 and 1. Such functions are decreasing, one-to-one functions. Their graphs lie above the x-axis and pass through the point $(0, 1)$. The graphs rise rapidly as $x \to -\infty$. As $x \to \infty$, the x-axis is a horizontal asymptote. There are no vertical asymptotes. Finally, the graphs are smooth and continuous, with no corners or gaps.

Figure 15 illustrates the graphs of two more exponential functions whose bases are between 0 and 1. Notice that the choice of a base closer to 0 results in a graph that is steeper when $x < 0$ and closer to the x-axis when $x > 0$.

FIGURE 15

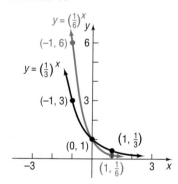

Seeing the Concept   Graph $y = \left(\frac{1}{2}\right)^x$ and compare what you see to Figure 14. Clear the screen and graph $y = \left(\frac{1}{3}\right)^x$ and $y = \left(\frac{1}{6}\right)^x$ and compare what you see to Figure 15. Clear the screen and graph $y = \left(\frac{1}{10}\right)^x$ and $y = \left(\frac{1}{100}\right)^x$. What viewing rectangle seems to work best?

The display summarizes the information we have about $f(x) = a^x$, $0 < a < 1$:

$$f(x) = a^x \quad 0 < a < 1$$
Domain: $(-\infty, \infty)$ Range: $(0, \infty)$
x-Intercepts: None y-Intercept: 1
Horizontal asymptote: x-Axis, as $x \to \infty$
f is a decreasing function
f is one-to-one
The graph of f passes through the points $(0, 1)$ and $(1, a)$

Transformations (shifting, compression, stretching, and reflection) may be used to graph many functions that are basically exponential functions.

E X A M P L E 4

Graphing Exponential Functions Using Transformations

Graph $f(x) = 2^{-x} - 3$ and determine the domain, range, and horizontal asymptote of f.

Solution   We begin with the graph of $y = 2^x$. Figure 16 shows the various steps.

FIGURE 16

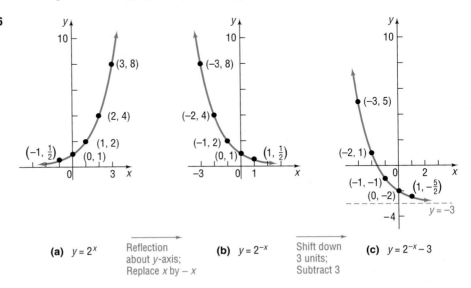

(a) $y = 2^x$ Reflection about y-axis; Replace x by $-x$ (b) $y = 2^{-x}$ Shift down 3 units; Subtract 3   (c) $y = 2^{-x} - 3$

As Figure 16(c) illustrates, the domain of $f(x) = 2^{-x} - 3$ is $(-\infty, \infty)$ and the range is $(-3, \infty)$. The horizontal asymptote of f is the line $y = -3$. ▬

E X A M P L E 5

Graphing Exponential Functions Using Transformations

Graph $f(x) = -(2^{x-3})$ and determine the domain, range, and horizontal asymptote of f.

Solution   We begin with the graph of $y = 2^x$. Figure 17 shows the various steps.

FIGURE 17

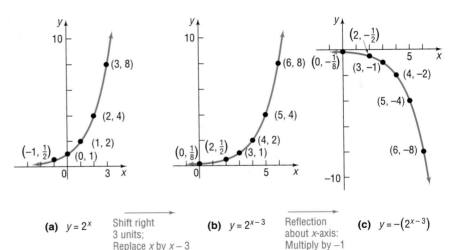

(a) $y = 2^x$ Shift right 3 units; Replace x by $x - 3$ (b) $y = 2^{x-3}$ Reflection about x-axis: Multiply by -1 (c) $y = -(2^{x-3})$

The domain of $f(x) = -(2^{x-3})$ is $(-\infty, \infty)$ and the range is $(-\infty, 0)$. The horizontal asymptote of f is $y = 0$.   ∎

Now work Problems 11 and 13.

The Base *e*

3 As we shall see shortly, many problems that occur in nature require the use of an exponential function whose base is a certain irrational number, symbolized by the letter *e*.

Let's look now at one way of arriving at this important number *e*.

> The **number *e*** is defined as the number that the expression
> $$\left(1 + \frac{1}{n}\right)^n \tag{2}$$
> approaches as $n \to \infty$. In calculus, this is expressed using limit notation as
> $$e = \lim_{n \to \infty} \left(1 + \frac{1}{n}\right)^n$$

Table 3 illustrates what happens to the defining expression (2) as *n* takes on increasingly large values. The last number in the last column in the table is correct to nine decimal places and is the same as the entry given for *e* on your calculator (if expressed correct to nine decimal places).

TABLE 3

n	$\dfrac{1}{n}$	$1 + \dfrac{1}{n}$	$\left(1 + \dfrac{1}{n}\right)^n$
1	1	2	2
2	0.5	1.5	2.25
5	0.2	1.2	2.48832
10	0.1	1.1	2.59374246
100	0.01	1.01	2.704813829
1,000	0.001	1.001	2.716923932
10,000	0.0001	1.0001	2.718145927
100,000	0.00001	1.00001	2.718268237
1,000,000	0.000001	1.000001	2.718280469
1,000,000,000	10^{-9}	$1 + 10^{-9}$	2.718281827

The exponential function $f(x) = e^x$, whose base is the number *e*, occurs with such frequency in applications that it is usually referred to as *the* exponential function. Indeed, most calculators have the key* $\boxed{e^x}$ or $\boxed{exp(x)}$, which may be used to evaluate the exponential function for a given value of *x*.

*If your calculator does not have this key but does have a $\boxed{\text{SHIFT}}$ key (or $\boxed{2^{\text{nd}}}$ key) and an $\boxed{\ln}$ key, you can display the number *e* as follows:

Keystrokes: $\boxed{1}$ $\boxed{\text{SHIFT}}$ $\boxed{\ln}$

Display: ↘ $\boxed{1}$ ↘ $\boxed{2.7182818}$

The reason this works will become clear in Section 6.3.

Now use your calculator to find e^x for $x = -2, x = -1, x = 0, x = 1$, and $x = 2$, as we have done to create Table 4. The graph of the exponential function $f(x) = e^x$ is given in Figure 18. Since $2 < e < 3$, the graph of $y = e^x$ lies between the graphs of $y = 2^x$ and $y = 3^x$. (Refer to Figures 11 and 13.)

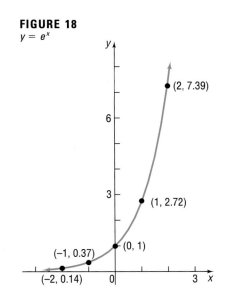

TABLE 4

x	e^x
-2	0.14
-1	0.37
0	1
1	2.72
2	7.39

FIGURE 18
$y = e^x$

 Seeing the Concept Graph $Y_1 = e^x$ and compare what you see to Figure 18. Use eVALUEate to verify the points on the graph shown in Figure 18. Now graph $Y_2 = 2^x$ and $Y_3 = 3^x$ on the same screen as $Y_1 = e^x$. Notice that the graph of $Y_1 = e^x$ lies between these two graphs. ■

 Now work Problem 19.

There are many applications involving the exponential function. Let's look at one.

E X A M P L E   6

Response to Advertising

Suppose that the percent R of people who respond to a newspaper advertisement for a new product and purchase the item advertised after t days is found using the formula

$$R = 50 - 50e^{-0.1t}$$

(a) What percent has responded and purchased after 5 days?
(b) What percent has responded and purchased after 10 days?
(c) What is the highest percent of people expected to respond and purchase?
(d) Graph $R = 50 - 50e^{-0.1t}, t > 0$. Use eVALUEate to compare the values of R for $t = 5$ and $t = 10$ to the ones obtained in parts (a) and (b). Use TRACE to determine how many days are required for R to exceed 40%?

Solution   (a) After 5 days, we have $t = 5$. The corresponding percent R of people responding and purchasing is

$$R = 50 - 50e^{(-0.1)(5)} = 50 - 50e^{-0.5}$$

We use a calculator to evaluate this expression:

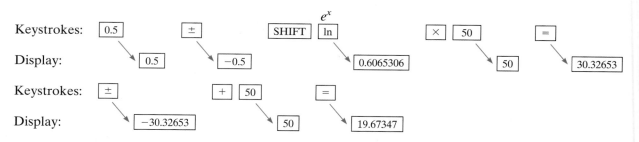

Thus, about 20% will have responded after 5 days.

(b) After 10 days, we have $t = 10$. The corresponding percent R of people responding and purchasing is

$$R = 50 - 50e^{(-0.1)(10)} = 50 - 50e^{-1} \approx 31.606$$

About 32% will have responded and purchased after 10 days.

(c) As time passes, more people are expected to respond and purchase. The highest percent expected is therefore found for the value of R as $t \to \infty$. Since $e^{-0.1t} = 1/e^{0.1t}$, it follows that $e^{-0.1t} \to 0$ as $t \to \infty$. Thus, the highest percent expected is 50%.

(d) See Figure 19. It will require just over 16 days to exceed 40%. ▬

FIGURE 19

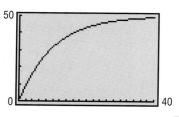

Now work Problem 43.

SUMMARY

Properties of the Exponential Function

$f(x) = a^x$, $a > 1$ Domain: $(-\infty, \infty)$; range: $(0, \infty)$; x-intercepts: none; y-intercept: 1; horizontal asymptote: x-axis as $x \to -\infty$; increasing; one-to-one
See Figure 11 for a typical graph.

$f(x) = a^x$, $0 < a < 1$ Domain: $(-\infty, \infty)$; range: $(0, \infty)$; x-intercepts: none; y-intercept: 1; horizontal asymptote: x-axis as $x \to \infty$; decreasing; one-to-one
See Figure 14 for a typical graph.

6.2 | EXERCISES

In Problems 1–10, approximate each number using a calculator. Express your answer rounded to three decimal places.

1. (a) $3^{2.2}$ (b) $3^{2.23}$ (c) $3^{2.236}$ (d) $3^{\sqrt{5}}$
2. (a) $5^{1.7}$ (b) $5^{1.73}$ (c) $5^{1.732}$ (d) $5^{\sqrt{3}}$
3. (a) $2^{3.14}$ (b) $2^{3.141}$ (c) $2^{3.1415}$ (d) 2^{π}
4. (a) $2^{2.7}$ (b) $2^{2.71}$ (c) $2^{2.718}$ (d) 2^{e}
5. (a) $3.1^{2.7}$ (b) $3.14^{2.71}$ (c) $3.141^{2.718}$ (d) π^{e}
6. (a) $2.7^{3.1}$ (b) $2.71^{3.14}$ (c) $2.718^{3.141}$ (d) e^{π}
7. $e^{1.2}$ 8. $e^{-1.3}$ 9. $e^{-0.85}$ 10. $e^{2.1}$

In Problems 11–18, the graph of an exponential function is given. Match each graph to one of the following functions:

A. $y = 3^x$
B. $y = 3^{-x}$
C. $y = -3^x$
D. $y = -3^{-x}$
E. $y = 3^x - 1$
F. $y = 3^{x-1}$
G. $y = 3^{1-x}$
H. $y = 1 - 3^x$

11.

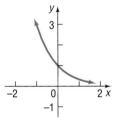

12.

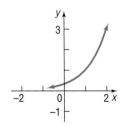

13.

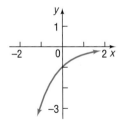

14.

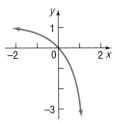

15.

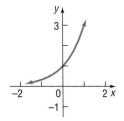

16.

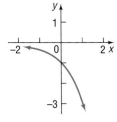

17.

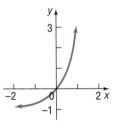

18.

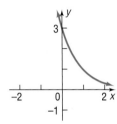

In Problems 19–26, begin with the graph of $y = e^x$ (Figure 18) and use transformations to graph each function. Determine the domain, range, and horizontal asymptote of each function.

19. $f(x) = e^{-x}$
20. $f(x) = -e^x$
21. $f(x) = e^{x+2}$
22. $f(x) = e^x - 1$
23. $f(x) = 5 - e^{-x}$
24. $f(x) = 9 - 3e^{-x}$
25. $f(x) = 2 - e^{-x/2}$
26. $f(x) = 7 - 3e^{-2x}$

In Problems 27–30, graph each function f. Based on the graph, state the domain, range, and intercepts, if any, of f.

27. $f(x) = \begin{cases} e^{-x} & \text{if } x < 0 \\ e^x & \text{if } x \geq 0 \end{cases}$

28. $f(x) = \begin{cases} e^x & \text{if } x < 0 \\ e^{-x} & \text{if } x \geq 0 \end{cases}$

29. $f(x) = \begin{cases} -e^x & \text{if } x < 0 \\ -e^{-x} & \text{if } x \geq 0 \end{cases}$

30. $f(x) = \begin{cases} -e^{-x} & \text{if } x < 0 \\ -e^x & \text{if } x \geq 0 \end{cases}$

31. If $4^x = 7$, what does 4^{-2x} equal?
32. If $2^x = 3$, what does 4^{-x} equal?
33. If $3^{-x} = 2$, what does 3^{2x} equal?
34. If $5^{-x} = 3$, what does 5^{3x} equal?

35. Optics If a single pane of glass obliterates 3% of the light passing through it, then the percent p of light that passes through n successive panes is given approximately by the equation

$$p = 100e^{-0.03n}$$

(a) What percent of light will pass through 10 panes?
(b) What percent of light will pass through 25 panes?

36. Atmospheric Pressure The atmospheric pressure p on a balloon or plane decreases with increasing height. This pressure, measured in millimeters of mercury, is related to the number of kilometers h above sea level by the formula

$$p = 760e^{-0.145h}$$

(a) Find the atmospheric pressure at a height of 2 kilometers (over a mile).

(b) What is it at a height of 10 kilometers (over 30,000 feet)?

37. Space Satellites   The number of watts w provided by a space satellite's power supply over a period of d days is given by the formula

$$w = 50e^{-0.004d}$$

(a) How much power will be available after 30 days?

(b) How much power will be available after 1 year (365 days)?

38. Healing of Wounds   The normal healing of wounds can be modeled by an exponential function. If A_0 represents the original area of the wound and if A equals the area of the wound after n days, then the formula

$$A = A_0 e^{-0.35n}$$

describes the area of a wound on the nth day following an injury when no infection is present to retard the healing. Suppose that a wound initially had an area of 100 square centimeters.

(a) If healing is taking place, how large should the area of the wound be after 3 days?

(b) How large should it be after 10 days?

39. Drug Medication   The formula

$$D = 5e^{-0.4h}$$

can be used to find the number of milligrams D of a certain drug that is in a patient's bloodstream h hours after the drug has been administered. How many milligrams will be present after 1 hour? After 6 hours?

40. Spreading of Rumors   A model for the number of people N in a college community who have heard a certain rumor is

$$N = P(1 - e^{-0.15d})$$

where P is the total population of the community and d is the number of days that have elapsed since the rumor began. In a community of 1000 students, how many students will have heard the rumor after 3 days?

41. Exponential Probability   Between 12:00 PM and 1:00 PM, cars arrive at Citibank's drive-through at the rate of 6 cars per hour (0.1 car per minute). The following formula from statistics can be used to determine the probability that a car will arrive within t minutes of 12:00 PM.

$$F(t) = 1 - e^{-0.1t}$$

(a) Determine the probability that a car will arrive within 10 minutes of 12:00 PM (that is, before 12:10 PM).

(b) Determine the probability that a car will arrive within 40 minutes of 12:00 PM (before 12:40 PM).

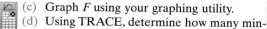
(c) Graph F using your graphing utility.

(d) Using TRACE, determine how many minutes are needed for the probability to reach 50%?

(e) What value does F approach as t becomes unbounded in the positive direction?

42. Exponential Probability   Between 5:00 PM and 6:00 PM, cars arrive at Jiffy Lube at the rate of 9 cars per hour (0.15 car per minute). The following formula from statistics can be used to determine the probability that a car will arrive within t minutes of 5:00 PM:

$$F(t) = 1 - e^{-0.15t}$$

(a) Determine the probability that a car will arrive within 15 minutes of 5:00 PM (that is, before 5:15 PM).

(b) Determine the probability that a car will arrive within 30 minutes of 5:00 PM (before 5:30 PM).

(c) Graph F using your graphing utility.

(d) Using TRACE, determine how many minutes are needed for the probability to reach 60%?

(e) What value does F approach as t becomes unbounded in the positive direction?

43. Response to TV Advertising   The percent R of viewers who respond to a television commercial for a new product after t days is found by using the formula

$$R = 70 - 70e^{-0.2t}$$

(a) What percent is expected to respond after 10 days?

(b) What percent has responded after 20 days?

(c) What is the highest percent of people expected to respond?

(d) Graph $R = 70 - 70e^{-0.2t}$, $t > 0$. Use eVALUEate to compare the values of R for $t = 10$ and $t = 20$ to the ones obtained in parts (a) and (b). Use TRACE to determine how many days are required for R to exceed 40%?

44.   Profit   The annual profit P of a company due to the sales of a particular item after it has been on the market x years is determined to be

$$P = \$100,000 - \$60,000(\tfrac{1}{2})^x$$

(a) What is the profit after 5 years?
(b) What is the profit after 10 years?
(c) What is the most profit the company can expect from this product?
 (d) Graph the profit function. Use eVALUEate to compare the values of P for $x = 5$ and $x = 10$ to the ones obtained in parts (a) and (b). Use TRACE to determine how many years it takes before a profit of \$65,000 is obtained.

45.   Alternating Current in an _RL_ Circuit   The equation governing the amount of current I (in amperes) after time t (in seconds) in a single RL circuit consisting of a resistance R (in ohms), an inductance L (in henrys), and an electromotive force E (in volts) is

$$I = \frac{E}{R}[1 - e^{-(R/L)t}]$$

(a) If $E = 120$ volts, $R = 10$ ohms, and $L = 5$ henrys, how much current I_1 is available after 0.3 second? After 0.5 second? After 1 second?
(b) What is the maximum current?
(c) Graph this function $I = I_1(t)$, measuring I along the y-axis and t along the x-axis.
(d) If $E = 120$ volts, $R = 5$ ohms, and $L = 10$ henrys, how much current I_2 is available after 0.3 second? After 0.5 second? After 1 second?
(e) What is the maximum current?
(f) Graph this function $I = I_2(t)$ on the same coordinate axes as $I_1(t)$.

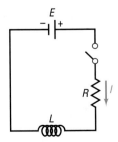

46.   Alternating Current in an _RC_ Circuit   The equation governing the amount of current I (in milliamperes) after time t (in milliseconds) in a

single RC circuit consisting of a resistance R (in ohms), a capacitance C (in microfarads), and an electromotive force E (in volts) is

$$I = \frac{E}{R}e^{-t/(RC)}$$

(a) If $E = 120$ volts, $R = 2000$ ohms, and $C = 1.0$ microfarad, how much current I_1 is available initially $(t = 0)$? After 1000 milliseconds? After 3000 milliseconds?
(b) What is the maximum current?
(c) Graph this function $I = I_1(t)$, measuring I along the y-axis and t along the x-axis.
(d) If $E = 120$ volts, $R = 1000$ ohms, and $C = 2.0$ microfarads, how much current I_2 is available initially? After 1000 milliseconds? After 3000 milliseconds?
(e) What is the maximum current?
(f) Graph this function $I = I_2(t)$ on the same coordinate axes as $I_1(t)$.

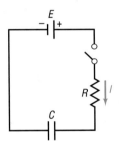

47.   The Challenger Disaster*   After the _Challenger_ disaster in 1986, a study of the 23 launches that preceded the fatal flight was made. A mathematical model was developed involving the relationship between the Fahrenheit temperature x around the O-rings and the number y of eroded or leaky primary O-rings. The model stated that

$$y = \frac{6}{1 + e^{-(5.085 - 0.1156x)}}$$

where the number 6 indicates the 6 primary O-rings on the spacecraft.

---

*Linda Tappin, "Analyzing Data Relating to the _Challenger_ Disaster," _Mathematics Teacher,_ Vol. 87, No. 6, September 1994, pp. 423–426.

(a) What is the predicted number of eroded or leaky primary O-rings at a temperature of 100°F?

(b) What is the predicted number of eroded or leaky primary O-rings at a temperature of 60°F?

(c) What is the predicted number of eroded or leaky primary O-rings at a temperature of 30°F?

 (d) Graph the equation and TRACE. At what temperature is the predicted number of eroded or leaky O-rings 1? 3? 5?

48. **Postage Stamps*** The cumulative number y of different postage stamps (regular and commemorative only) issued by the U.S. Post Office can be approximated (modeled) by the exponential function

$$y = 78e^{0.025x}$$

where x is the number of years since 1848.

(a) What is the predicted cumulative number of stamps that will have been issued by the year 1998? Check with the Postal Service and comment on the accuracy of using the function.

(b) What is the predicted cumulative number of stamps that will have been issued by the year 2000?

 (c) The cumulative number of stamps actually issued by the United States was 2 in 1848, 88 in 1868, 218 in 1888, and 341 in 1908. What conclusion can you draw about using the given function as a model over the first few decades in which stamps were issued?

*David Kullman, "Patterns of Postage-stamp Production," *Mathematics Teacher*, Vol. 85, No. 3, March 1992, pp. 188–189.

49. **Another Formula for e** Use a calculator to compute the value of

$$2 + \frac{1}{2!} + \frac{1}{3!} + \cdots + \frac{1}{n!}$$

for $n = 4$, 6, 8, and 10. Compare each result with e.

[**Hint:** $1! = 1$, $2! = 2 \cdot 1$, $3! = 3 \cdot 2 \cdot 1$, $n! = n(n-1) \cdot \ldots \cdot (3)(2)(1)$]

50. **Another Formula for e** Use a calculator to compute the various values of the expression

$$2 + \cfrac{1}{1 + \cfrac{1}{2 + \cfrac{2}{3 + \cfrac{3}{4 + 4}}}}$$

etc.

Compare the values to e.

51. If $f(x) = a^x$, show that

$$\frac{f(x+h) - f(x)}{h} = a^x \left(\frac{a^h - 1}{h} \right).$$

52. If $f(x) = a^x$, show that $f(A + B) = f(A) \cdot f(B)$.

53. If $f(x) = a^x$, show that $f(-x) = \dfrac{1}{f(x)}$.

54. If $f(x) = a^x$, show that $f(\alpha x) = [f(x)]^{\alpha}$.

55. **Historical Problem** Pierre de Fermat (1601–1665) conjectured that the function

$$f(x) = 2^{(2^x)} + 1$$

for $x = 1, 2, 3, \ldots$, would always have a value equal to a prime number. But Leonhard Euler (1707–1783) showed that this formula fails for $x = 5$. Use a calculator to determine the prime numbers produced by f for $x = 1, 2, 3, 4$. Then show that $f(5) = 641 \times 6,700,417$, which is not prime.

56. The bacteria in a 4-liter container double every minute. After 60 minutes the container is full. How long did it take to fill half the container?

57. Explain in your own words what the number e is. Provide at least two applications that require the use of this number.

58. Do you think there is a power function that increases more rapidly than an exponential function whose base is greater than 1? Explain.

6.3 | LOGARITHMIC FUNCTIONS

1 Change Exponential Expressions to Logarithmic Expressions
2 Change Logarithmic Expressions to Exponential Expressions
3 Evaluate Logarithmic Functions
4 Determine the Domain of a Logarithmic Function
5 Graph Logarithmic Functions

Recall that a one-to-one function $y = f(x)$ has an inverse that is defined (implicitly) by the equation $x = f(y)$. In particular, the exponential function $y = f(x) = a^x$, $a > 0$, $a \neq 1$, is one-to-one and hence has an inverse function that is defined implicitly by the equation

$$x = a^y \qquad a > 0, a \neq 1$$

This inverse function is so important that it is given a name, the *logarithmic function*.

> The **logarithmic function to the base a,** where $a > 0$ and $a \neq 1$, is denoted by $y = \log_a x$ (read as "y is the logarithm to the base a of x") and is defined by
>
> $$y = \log_a x \quad \text{if and only if} \quad x = a^y$$

EXAMPLE 1

Relating Logarithms to Exponents

(a) If $y = \log_3 x$, then $x = 3^y$. Thus, if $x = 9$, then $y = 2$, so $9 = 3^2$ is equivalent to $2 = \log_3 9$.

(b) If $y = \log_5 x$, then $x = 5^y$. Thus, if $x = \frac{1}{5} = 5^{-1}$, then $y = -1$, so $\frac{1}{5} = 5^{-1}$ is equivalent to $-1 = \log_5 \left(\frac{1}{5}\right)$. ∎

EXAMPLE 2

Changing Exponential Expressions to Logarithmic Expressions

1 Change each exponential expression to an equivalent expression involving a logarithm.

(a) $1.2^3 = m$ (b) $e^b = 9$ (c) $a^4 = 24$

Solution   We use the fact that $y = \log_a x$ and $x = a^y$, $a > 0$, $a \neq 1$, are equivalent.

(a) If $1.2^3 = m$, then $3 = \log_{1.2} m$.
(b) If $e^b = 9$, then $b = \log_e 9$.
(c) If $a^4 = 24$, then $4 = \log_a 24$. ∎

Now work Problem 1.

EXAMPLE 3

Changing Logarithmic Expressions to Exponential Expressions

2 Change each logarithmic expression to an equivalent expression involving an exponent.

(a) $\log_a 4 = 5$ (b) $\log_e b = -3$ (c) $\log_3 5 = c$

Solution (a) If $\log_a 4 = 5$, then $a^5 = 4$.
  (b) If $\log_e b = -3$, then $e^{-3} = b$.
  (c) If $\log_3 5 = c$, then $3^c = 5$. ▬

Now work Problem 13.

[3] To find the exact value of a logarithm, we write the logarithm in exponential notation and use the following fact:

> If $a^u = a^v$, then $u = v$. (1)

The result (1) is a consequence of the fact that exponential functions are one-to-one.

E X A M P L E 4

Finding the Exact Value of a Logarithmic Function

Find the exact value of

(a) $\log_2 16$ (b) $\log_3 \frac{1}{3}$ (c) $\log_5 25$

Solution (a) For $y = \log_2 16$, we have the equivalent exponential equation $2^y = 16 = 2^4$, so, by (1), $y = 4$. Thus, $\log_2 16 = 4$.
  (b) For $y = \log_3 \frac{1}{3}$, we have $3^y = \frac{1}{3} = 3^{-1}$, so $y = -1$. Thus, $\log_3 \frac{1}{3} = -1$.
  (c) For $y = \log_5 25$, we have $5^y = 25 = 5^2$, so $y = 2$. Thus, $\log_5 25 = 2$. ▬

Now work Problem 25.

Domain of a Logarithmic Function

[4] The logarithmic function $y = \log_a x$ has been defined as the inverse of the exponential function $y = a^x$. That is, if $f(x) = a^x$, then $f^{-1}(x) = \log_a x$. Based on the discussion given in Section 6.1 on inverse functions, we know that for a function f and its inverse f^{-1}

Domain f^{-1} = Range f and Range f^{-1} = Domain f

Consequently, it follows that

> Domain of logarithmic function = Range of exponential function = $(0, \infty)$
> Range of logarithmic function = Domain of exponential function = $(-\infty, \infty)$

In the next box, we summarize some properties of the logarithmic function:

> $y = \log_a x$ (defining equation: $x = a^y$)
> Domain: $0 < x < \infty$ Range: $-\infty < y < \infty$

Notice that the domain of a logarithmic function consists of the *positive* real numbers.

E X A M P L E   5

Finding the Domain of a Logarithmic Function

Find the domain of each logarithmic function:

(a) $F(x) = \log_2 (1 - x)$

(b) $h(x) = \log_{1/2}|x|$

Solution

(a) The domain of F consists of all x for which $(1 - x) > 0$; that is, all x for which $x < 1$, or $(-\infty, 1)$.

(b) Since $|x| > 0$ provided $x \neq 0$, the domain of h consists of all nonzero real numbers. ∎

Now work Problem 39.

Graphs of Logarithmic Functions

Since exponential functions and logarithmic functions are inverses of each other, the graph of a logarithmic function $y = \log_a x$ is the reflection about the line $y = x$ of the graph of the exponential function $y = a^x$, as shown in Figure 20.

FIGURE 20

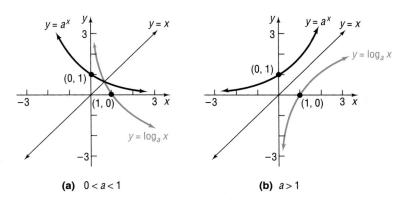

(a) $0 < a < 1$

(b) $a > 1$

Facts about the Graph of a Logarithmic Function $f(x) = \log_a x$

1. The x-intercept of the graph is 1. There is no y-intercept.
2. The y-axis is a vertical asymptote of the graph.
3. A logarithmic function is decreasing if $0 < a < 1$ and increasing if $a > 1$.
4. The graph is smooth and continuous, with no corners or gaps.

If the base of a logarithmic function is the number e, then we have the **natural logarithm function.** This function occurs so frequently in applications that it is given a special symbol, **ln** (from the Latin, *logarithmic naturalis*). Thus,

$$y = \ln x \quad \text{if and only if} \quad x = e^y$$

Since $y = \ln x$ and the exponential function $y = e^x$ are inverse functions, we can obtain the graph of $y = \ln x$ by reflecting the graph of $y = e^x$ about the line $y = x$. See Figure 21.

FIGURE 21

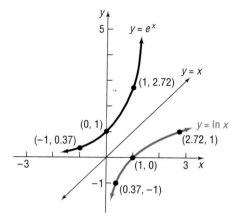

TABLE 5

x	$\ln x$
$\frac{1}{2}$	-0.69
2	0.69
3	1.10

Using a calculator with an $\boxed{\textit{ln}}$ key, we can obtain other points on the graph of $f(x) = \ln x$. See Table 5.

 Check: Graph $Y_1 = e^x$ and $Y_2 = \ln x$ on the same square screen. Use eVALUEate to verify the points on the graph given in Figure 21. Do you see the symmetry of the two graphs with respect to the line $y = x$? ▬

E X A M P L E   6

Graphing Logarithmic Functions Using Transformations

Graph $f(x) = -\ln x$ by starting with the graph of $y = \ln x$. Determine the domain, range, and vertical asymptote of f.

Solution The graph of $f(x) = -\ln x$ is obtained by a reflection about the x-axis of the graph of $y = \ln x$. See Figure 22.

FIGURE 22

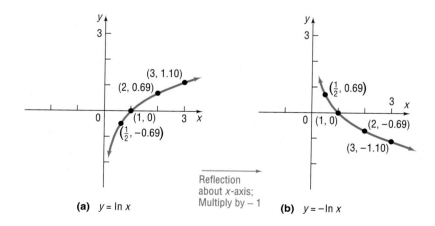

The domain of $f(x) = -\ln x$ is $(0, \infty)$, the range is $(-\infty, \infty)$, and the vertical asymptote is $x = 0$. ▬

E X A M P L E   7   Graphing Logarithmic Functions Using Transformations

Determine the domain, range, and vertical asymptote of $f(x) = \ln(x + 2)$. Graph $f(x) = \ln(x + 2)$.

Solution   The domain consists of all x for which

$$x + 2 > 0 \quad \text{or} \quad x > -2$$

The graph is obtained by applying a horizontal shift to the left 2 units, as shown in Figure 23.

FIGURE 23

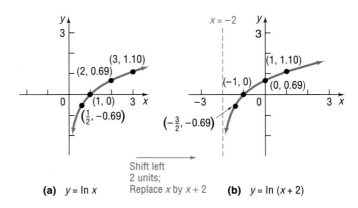

(a) $y = \ln x$ Shift left 2 units; Replace x by $x + 2$ (b) $y = \ln(x + 2)$

The range is $(-\infty, \infty)$, and $x = -2$ is the vertical asymptote.

E X A M P L E   8   Graphing Logarithmic Functions Using Transformations

Graph $f(x) = \ln(1 - x)$. Determine the domain, range, and vertical asymptote of f.

Solution   The domain consists of all x for which

$$1 - x > 0 \quad \text{or} \quad x < 1$$

To obtain the graph of $y = \ln(1 - x)$, we use the steps illustrated in Figure 24.

FIGURE 24

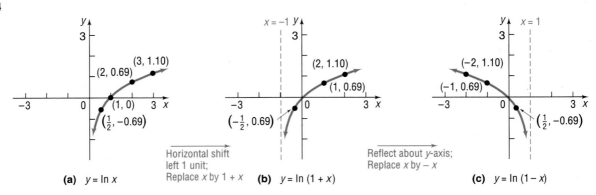

(a) $y = \ln x$ Horizontal shift left 1 unit; Replace x by $1 + x$ (b) $y = \ln(1 + x)$ Reflect about y-axis; Replace x by $-x$ (c) $y = \ln(1 - x)$

The range of $f(x) = \ln(1 - x)$ is $(-\infty, \infty)$ and the vertical asymptote is $x = 1$.

Now work Problem 67.

E X A M P L E   9   Alcohol and Driving

The concentration of alcohol in a person's blood is measurable. Recent medical research suggests that the risk R (given as a percent) of having an accident while driving a car can be modeled by the equation

$$R = 6e^{kx}$$

where x is the variable concentration of alcohol in the blood and k is a constant.

(a) Suppose that a concentration of alcohol in the blood of 0.04 results in a 10% risk ($R = 10$) of an accident. Find the constant k in the equation.

(b) Using this value of k, what is the risk if the concentration is 0.17?

(c) Using the same value of k, what concentration of alcohol corresponds to a risk of 100%?

(d) If the law asserts that anyone with a risk of having an accident of 20% or more should not have driving privileges, at what concentration of alcohol in the blood should a driver be arrested and charged with a DUI (Driving Under the Influence)?

Solution (a) For a concentration of alcohol in the blood of 0.04 and a risk of 10%, we let $x = 0.04$ and $R = 10$ in the equation and solve for k.

$$R = 6e^{kx}$$
$$10 = 6e^{k(0.04)}$$
$$\frac{10}{6} = e^{0.04k} \qquad \text{Change to a logarithmic expression.}$$

$$0.04k = \ln\frac{10}{6} = 0.5108256$$

$$k = 12.77$$

(b) Using $k = 12.77$ and $x = 0.17$ in the equation, we find the risk R to be

$$R = 6e^{kx} = 6e^{(12.77)(0.17)} = 52.6$$

For a concentration of alcohol in the blood of 0.17, the risk of an accident is about 52.6%.

(c) Using $k = 12.77$ and $R = 100$ in the equation, we find the concentration x of alcohol in the blood to be

$$R = 6e^{kx}$$
$$100 = 6e^{12.77x}$$
$$\frac{100}{6} = e^{12.77x} \qquad \text{Change to a logarithmic expression.}$$

$$12.77x = \ln\frac{100}{6} = 2.8134$$

$$x = 0.22$$

For a concentration of alcohol in the blood of 0.22, the risk of an accident is 100%.

(d) Using $k = 12.77$ and $R = 20$ in the equation, we find the concentration x of alcohol in the blood to be

$$R = 6e^{kx}$$

$$20 = 6e^{12.77x}$$

$$\frac{20}{6} = e^{12.77x}$$

$$12.77x = \ln \frac{20}{6} = 1.204$$

$$x = 0.094$$

A driver with a concentration of alcohol in the blood of 0.094 or more should be arrested and charged with DUI. ■

Note: Many states use 0.10 as the blood alcohol content at which a DUI citation is given. More and more states are lowering the blood alcohol content to 0.08.

SUMMARY

Properties of the Logarithmic Function

$f(x) = \log_a x$, $a > 1$ Domain: $(0, \infty)$;   Range: $(-\infty, \infty)$; x-intercept: 1; y-intercept: none;
($y = \log_a x$ means $x = a^y$)   vertical asymptote: y-axis;   increasing;   one-to-one
                                     See Figure 20(b) for a typical graph.

$f(x) = \log_a x$, $0 < a < 1$ Domain: $(0, \infty)$;   Range: $(-\infty, \infty)$; x-intercept: 1; y-intercept: none;
($y = \log_a x$ means $x = a^y$)   vertical asymptote: y-axis;   decreasing;   one-to-one
                                       See Figure 20(a) for a typical graph.

6.3   EXERCISES

In Problems 1–12, change each exponential expression to an equivalent expression involving a logarithm.

1. $9 = 3^2$ 2. $16 = 4^2$ 3. $a^2 = 1.6$ 4. $a^3 = 2.1$ 5. $1.1^2 = M$ 6. $2.2^3 = N$

7. $2^x = 7.2$ 8. $3^x = 4.6$ 9. $x^{\sqrt{2}} = \pi$ 10. $x^\pi = e$ 11. $e^x = 8$ 12. $e^{2.2} = M$

In Problems 13–24, change each logarithmic expression to an equivalent expression involving an exponent.

13. $\log_2 8 = 3$ 14. $\log_3 \left(\frac{1}{9}\right) = -2$ 15. $\log_a 3 = 6$ 16. $\log_b 4 = 2$

17. $\log_3 2 = x$ 18. $\log_2 6 = x$ 19. $\log_2 M = 1.3$ 20. $\log_3 N = 2.1$

21. $\log_{\sqrt{2}} \pi = x$ 22. $\log_\pi x = \frac{1}{2}$ 23. $\ln 4 = x$ 24. $\ln x = 4$

In Problems 25–36, find the exact value of each logarithm without using a calculator.

25. $\log_2 1$ 26. $\log_8 8$ 27. $\log_5 25$ 28. $\log_3 \left(\frac{1}{9}\right)$ 29. $\log_{1/2} 16$ 30. $\log_{1/3} 9$

31. $\log_{10} \sqrt{10}$ 32. $\log_5 \sqrt[3]{25}$ 33. $\log_{\sqrt{2}} 4$ 34. $\log_{\sqrt{3}} 9$ 35. $\ln \sqrt{e}$ 36. $\ln e^3$

In Problems 37–44, find the domain and the intercepts, if any, of each function.

37. $f(x) = \ln(3 - x)$ 38. $g(x) = \ln(x - 1)$ 39. $F(x) = \log_2 x^2$ 40. $H(x) = \log_5 x^3$

41. $h(x) = \log_{1/2} (x^2 - 2x + 1)$ 42. $G(x) = \log_{1/2} (x^2 - 1)$ 43. $g(x) = \log_5 \left(\frac{x + 1}{x}\right)$ 44. $h(x) = \log_3 \left(\frac{x}{x - 1}\right)$

In Problems 45–48, use a calculator to evaluate each expression. Round your answer to three decimal places.

45. $\ln \dfrac{5}{3}$

46. $\dfrac{\ln 5}{3}$

47. $\dfrac{\ln (10/3)}{0.04}$

48. $\dfrac{\ln (2/3)}{-0.1}$

49. Find a such that the graph of $f(x) = \log_a x$ contains the point $(2, 2)$.

50. Find a such that the graph of $f(x) = \log_a x$ contains the point $(\frac{1}{2}, -4)$.

In Problems 51–58, the graph of a logarithmic function is given. Match each graph to one of the following functions:

A. $y = \log_3 x$

B. $y = \log_3 (-x)$

C. $y = -\log_3 x$

D. $y = -\log_3 (-x)$

E. $y = \log_3 x - 1$

F. $y = \log_3 (x - 1)$

G. $y = \log_3 (1 - x)$

H. $y = 1 - \log_3 x$

51.
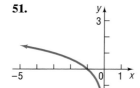

52.
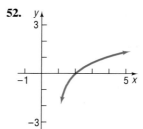

53.
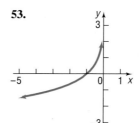

54.
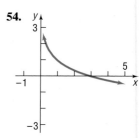

55.
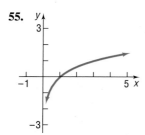

56.
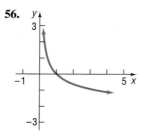

57.
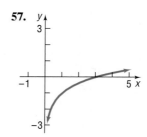

58.
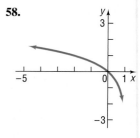

In Problems 59–70, begin with the graph of $y = \ln x$ and use transformations to graph each function. State the domain, range, and vertical asymptote of f.

59. $f(x) = \ln(x + 4)$

60. $f(x) = \ln(x - 3)$

61. $f(x) = \ln(-x)$

62. $f(x) = -\ln(-x)$

63. $g(x) = \ln 2x$

64. $h(x) = \ln \frac{1}{2}x$

65. $f(x) = 3 \ln x$

66. $f(x) = -2 \ln x$

67. $g(x) = \ln(3 - x)$

68. $h(x) = \ln(4 - x)$

69. $f(x) = -\ln(x - 1)$

70. $f(x) = 2 - \ln x$

In Problems 71–74, graph each function f. Based on the graph, state the domain, range, and intercepts, if any, of f.

71. $f(x) = \begin{cases} \ln(-x) & \text{if } x < 0 \\ \ln x & \text{if } x > 0 \end{cases}$

72. $f(x) = \begin{cases} \ln(-x) & \text{if } x \le -1 \\ -\ln(-x) & \text{if } -1 < x < 0 \end{cases}$

73. $f(x) = \begin{cases} -\ln x & \text{if } 0 < x < 1 \\ \ln x & \text{if } x \ge 1 \end{cases}$

74. $f(x) = \begin{cases} \ln x & \text{if } 0 < x < 1 \\ -\ln x & \text{if } x \ge 1 \end{cases}$

75. Optics If a single pane of glass obliterates 10% of the light passing through it, then the percent P of light that passes through n successive panes is given approximately by the equation

$$P = 100e^{-0.1n}$$

(a) How many panes are necessary to block at least 50% of the light?

(b) How many panes are necessary to block at least 75% of the light?

76. Chemistry The pH of a chemical solution is given by the formula

$$pH = -\log_{10} [H^+]$$

where $[H^+]$ is the concentration of hydrogen ions in moles per liter. Values of pH range from 0 (acidic) to 14 (alkaline).

(a) Find the pH of a 1-liter container of water with 0.0000001 mole of hydrogen ion.

(b) Find the hydrogen ion concentration of a mildly acidic solution with a pH of 4.2.

77. Space Satellites The number of watts w provided by a space satellite's power supply after d days is given by the formula

$$w = 50e^{-0.004d}$$

(a) How long will it take for the available power to drop to 30 watts?
(b) How long will it take for the available power to drop to only 5 watts?

78. Healing of Wounds The normal healing of wounds can be modeled by an exponential function. If A_0 represents the original area of the wound and if A equals the area of the wound after n days, then the formula

$$A = A_0 e^{-0.35n}$$

describes the area of a wound on the nth day following an injury when no infection is present to retard the healing. Suppose that a wound initially had an area of 100 square centimeters.

(a) If healing is taking place, how many days should pass before the wound is one-half its original size?
(b) How long before the wound is 10% of its original size?

79. Drug Medication The formula

$$D = 5e^{-0.4h}$$

can be used to find the number of milligrams D of a certain drug that is in a patient's bloodstream h hours after the drug has been administered. When the number of milligrams reaches 2, the drug is to be administered again. What is the time between injections?

80. Spreading of Rumors A model for the number of people N in a college community who have heard a certain rumor is

$$N = P(1 - e^{-0.15d})$$

where P is the total population of the community and d is the number of days that have elapsed since the rumor began. In a community of 1,000 students, how many days will elapse before 450 students have heard the rumor?

81. Current in an RL Circuit The equation governing the amount of current I (in amperes) after time t (in seconds) in a simple RL circuit consisting of a resistance R (in ohms), an inductance L (in henrys), and an electromotive force E (in volts) is

$$I = \frac{E}{R}[1 - e^{-(R/L)t}]$$

If $E = 12$ volts, $R = 10$ ohms, and $L = 5$ henrys, how long does it take to obtain a current of 0.5 ampere? Of 1.0 ampere? Graph the equation.

82. Learning Curve Psychologists sometimes use the function

$$L(t) = A(1 - e^{-kt})$$

to measure the amount L learned at time t. The number A represents the amount to be learned, and the number k measures the rate of learning. Suppose that a student has an amount A of 200 vocabulary words to learn. A psychologist determines that the student learned 20 vocabulary words after 5 minutes.

(a) Determine the rate of learning k.
(b) Approximately how many words will the student have learned after 10 minutes?
(c) After 15 minutes?
(d) How long does it take for the student to learn 180 words?

83. Alcohol and Driving The concentration of alcohol in a person's blood is measurable. Suppose that the risk R (given as a percent) of having an accident while driving a car can be modeled by the equation

$$R = 3e^{kx}$$

where x is the variable concentration of alcohol in the blood and k is a constant.

(a) Suppose that a concentration of alcohol in the blood of 0.06 results in a 10% risk ($R = 10$) of an accident. Find the constant k in the equation.
(b) Using this value of k, what is the risk if the concentration is 0.17?
(c) Using the same value of k, what concentration of alcohol corresponds to a risk of 100%?
(d) If the law asserts that anyone with a risk of having an accident of 15% or more should not have driving privileges, at what concentration of alcohol in the blood should a driver be arrested and charged with a DUI?
 (e) Compare this situation with that of Example 9. If you were a lawmaker, which situation would you support? Give your reasons.

84. Is there any function of the form $y = x^{\alpha}, 0 < \alpha < 1$, that increases more slowly than a logarithmic function whose base is greater than 1? Explain.

85. **Constructing a Function** Look back at Figure 12. Assuming that the points (1950, 20) and (1990, 50) are on the graph, find an exponential equation $y = Ae^{bt}$ that fits the data.

[**Hint:** Let $t = 0$ correspond to the year 1950. Then show that $A = 20$. Now find b.]

Is the projection of 102 million in 2015 confirmed by your model? Try to obtain similar data about the United States birthrate and construct a function to fit those data.

86. **Critical Thinking** In buying a new car, one consideration might be how well the price of the car holds up over time. Different makes of cars have different depreciation rates. One way to compute a depreciation rate for a car is given here. Suppose that the current prices of a certain Mercedes automobile are as follows:

New	1 Year Old	2 Years Old	3 Years Old	4 Years Old	5 Years Old
$38,000	$36,600	$32,400	$28,750	$25,400	$21,200

Use the formula New = Old(e^{Rt}) to find R, the annual depreciation rate, for a specific time t. When might be the best time to trade in the car? Consult the NADA ("blue") book and compare two like models that you are interested in. Which has the better depreciation rate?

6.4 | PROPERTIES OF LOGARITHMS; CURVE FITTING

1 Write a Logarithmic Expression as a Sum or Difference of Logarithms
2 Write a Logarithmic Expression as a Single Logarithm
3 Evaluate Logarithms Whose Base Is Neither 10 nor e
4 Draw Scatter Diagrams of Exponential and Logarithmic Data

Logarithms have some very useful properties that can be derived directly from the definition and the laws of exponents.

E X A M P L E 1 Establishing Properties of Logarithms

(a) Show that $\log_a 1 = 0$.
(b) Show that $\log_a a = 1$.

Solution (a) This fact was established when we graphed $y = \log_a x$ (see Figure 20). Algebraically, for $y = \log_a 1$, we have $a^y = 1 = a^0$, so $y = 0$.

(b) For $y = \log_a a$, we have $a^y = a = a^1$, so $y = 1$. ∎

To summarize

$$\log_a 1 = 0 \qquad \log_a a = 1$$

Theorem Properties of Logarithms

In the properties given next, M and a are positive real numbers, with $a \neq 1$, and r is any real number.

The number $\log_a M$ is the exponent to which a must be raised to obtain M. That is,

$$a^{\log_a M} = M \tag{1}$$

The logarithm to the base a of a raised to a power equals that power. That is,

$$\log_a a^r = r \tag{2}$$

Proof of Property (1) Let $x = \log_a M$. Change this logarithmic expression to the equivalent exponential expression:

$a^x = M$

But $x = \log_a M$, so

$a^{\log_a M} = M$

Proof of Property (2) Let $x = a^r$. Change this exponential expression to the equivalent logarithmic expression

$\log_a x = r$

But $x = a^r$, so

$\log_a a^r = r$

E X A M P L E   2

Using Properties (1) and (2)

(a) $2^{\log_2 \pi} = \pi$ (b) $\log_{0.2} 0.2^{-\sqrt{2}} = -\sqrt{2}$ (c) $\ln e^{kt} = kt$

Other useful properties of logarithms are given now.

Theorem   Properties of Logarithms

In the following properties, M, N, and a are positive real numbers, with $a \neq 1$, and r is any real number.

The Log of a Product Equals the Sum of the Logs

$$\log_a (MN) = \log_a M + \log_a N \tag{3}$$

The Log of a Quotient Equals the Difference of the Logs

$$\log_a \left(\frac{M}{N}\right) = \log_a M - \log_a N \tag{4}$$

$$\log_a \left(\frac{1}{N}\right) = -\log_a N \tag{5}$$

$$\log_a M^r = r \log_a M \tag{6}$$

We shall derive properties (3) and (6) and leave the derivations of properties (4) and (5) as exercises (see Problems 76 and 77).

Proof of Property (3)   Let $A = \log_a M$ and let $B = \log_a N$. These expressions are equivalent to the exponential expressions

$$a^A = M \quad \text{and} \quad a^B = N$$

Now

$$
\begin{aligned}
\log_a (MN) = \log_a (a^A a^B) = \log_a a^{A+B} && \text{Law of exponents} \\
= A + B && \text{Property (2) of logarithms} \\
= \log_a M + \log_a N &&
\end{aligned}
$$

Proof of Property (6)   Let $A = \log_a M$. This expression is equivalent to

$$a^A = M$$

Now

$$
\begin{aligned}
\log_a M^r = \log_a (a^A)^r = \log_a a^{rA} && \text{Law of exponents} \\
= rA && \text{Property (2) of logarithms} \\
= r \log_a M &&
\end{aligned}
$$

1   Logarithms can be used to transform products into sums, quotients into differences, and powers into factors. Such transformations prove useful in certain types of calculus problems.

E X A M P L E   3

Writing a Logarithmic Expression as a Sum of Logarithms

Write $\log_a(x\sqrt{x^2 + 1})$ as a sum of logarithms. Express all powers as factors.

Solution

$$
\begin{aligned}
\log_a(x\sqrt{x^2 + 1}) = \log_a x + \log_a \sqrt{x^2 + 1} && \text{Property (3)} \\
= \log_a x + \log_a (x^2 + 1)^{1/2} && \\
= \log_a x + \tfrac{1}{2} \log_a (x^2 + 1) && \text{Property (6)}
\end{aligned}
$$

E X A M P L E   4

Writing a Logarithmic Expression as a Difference of Logarithms

Write

$$\log_a \frac{x^2}{(x - 1)^3}$$

as a difference of logarithms. Express all powers as factors.

Solution

$$\log_a \frac{x^2}{(x - 1)^3} = \log_a x^2 - \log_a (x - 1)^3 = 2 \log_a x - 3 \log_a (x - 1)$$

        ↑   ↑

      Property (4)   Property (6)

 Now work Problem 13.

E X A M P L E   5

Writing a Logarithmic Expression as a Sum and Difference of Logarithms

Write

$$\log_a \frac{x^3 \sqrt{x^2 + 1}}{(x + 1)^4}$$

as a sum and difference of logarithms. Express all powers as factors.

Solution

$$\log_a \frac{x^3 \sqrt{x^2 + 1}}{(x + 1)^4} = \log_a (x^3 \sqrt{x^2 + 1}) - \log_a (x + 1)^4$$

$$= \log_a x^3 + \log_a \sqrt{x^2 + 1} - \log_a (x + 1)^4$$
$$= \log_a x^3 + \log_a (x^2 + 1)^{1/2} - \log_a (x + 1)^4$$
$$= 3 \log_a x + \frac{1}{2} \log_a (x^2 + 1) - 4 \log_a (x + 1)$$

▬

2

Another use of properties (3) through (6) is to write sums and/or differences of logarithms with the same base as a single logarithm.

E X A M P L E   6

Writing Expressions as a Single Logarithm

Write each of the following as a single logarithm:

(a) $\log_a 7 + 4 \log_a 3$
(b) $\frac{2}{3} \log_a 8 - \log_a (3^4 - 8)$
(c) $\log_a x + \log_a 9 + \log_a (x^2 + 1) - \log_a 5$

Solution

(a) $\log_a 7 + 4 \log_a 3 = \log_a 7 + \log_a 3^4$ Property (6)
$$= \log_a 7 + \log_a 81$$
$$= \log_a (7 \cdot 81)$$ Property (3)
$$= \log_a 567$$

(b) $\frac{2}{3} \log_a 8 - \log_a (3^4 - 8) = \log_a 8^{2/3} - \log_a (81 - 8)$ Property (6)
$$= \log_a 4 - \log_a 73$$
$$= \log_a \left(\tfrac{4}{73}\right)$$ Property (4)

(c) $\log_a x + \log_a 9 + \log_a (x^2 + 1) - \log_a 5 = \log_a 9x + \log_a (x^2 + 1) - \log_a 5$
$$= \log_a [9x(x^2 + 1)] - \log_a 5$$
$$= \log_a \left[\frac{9x(x^2 + 1)}{5}\right]$$

▬

Warning: A common error made by some students is to express the logarithm of a sum as the sum of logarithms:

$$\log_a(M + N) \quad \text{is not equal to} \quad \log_a M + \log_a N$$

Correct statement $\log_a (MN) = \log_a M + \log_a N$ Property (3)

Another common error is to express the difference of logarithms as the quotient of logarithms:

$$\log_a M - \log_a N \quad \text{is not equal to} \quad \frac{\log_a M}{\log_a N}$$

Correct statement $\quad \log_a M - \log_a N = \log_a\left(\dfrac{M}{N}\right) \quad$ Property (4)

Now work Problem 23.

There remain two other properties of logarithms that we need to know. They are a consequence of the fact that the logarithmic function $y = \log_a x$ is one-to-one.

Theorem

In the following properties, M, N, and a are positive real numbers, with $a \neq 1$:

$$\text{If } M = N, \quad \text{then} \quad \log_a M = \log_a N. \tag{7}$$
$$\text{If } \log_a M = \log_a N, \quad \text{then} \quad M = N. \tag{8}$$

Properties (7) and (8) are useful for solving *logarithmic equations,* a topic discussed in the next section.

Using a Calculator to Evaluate Logarithms with Bases Other Than e or 10

$\boxed{3}$ Logarithms to the base 10, called **common logarithms,** were used to facilitate arithmetic computations before the widespread use of calculators. (See the Historical Feature at the end of this section.) Natural logarithms, that is, logarithms whose base is the number e, remain very important because they arise frequently in the study of natural phenomena.

Common logarithms are usually abbreviated by writing **log,** with the base understood to be 10, just as natural logarithms are abbreviated by **ln,** with the base understood to be e.

Most calculators have both $\boxed{\log}$ and $\boxed{\ln}$ keys to calculate the common logarithm and natural logarithm of a number. Let's look at an example to see how to calculate logarithms having a base other than 10 or e.

E X A M P L E 7 Evaluating Logarithms Whose Base Is Neither 10 Nor e

Evaluate $\log_2 7$.

Solution   Let $y = \log_2 7$. Then $2^y = 7$, so

$$2^y = 7$$
$$\ln 2^y = \ln 7 \qquad \text{Property (7)}$$
$$y \ln 2 = \ln 7 \qquad \text{Property (6)}$$
$$y = \frac{\ln 7}{\ln 2} \qquad \text{Solve for } y$$
$$\approx 2.8074 \qquad \text{Use calculator (} \boxed{\text{ln}} \text{ key).}$$

Example 7 shows how to change the base from 2 to e. In general, to change from the base b to the base a, we use the **Change-of-Base Formula.**

Theorem   Change-of-Base Formula

If $a \neq 1$, $b \neq 1$, and M are positive real numbers, then

$$\log_a M = \frac{\log_b M}{\log_b a} \tag{9}$$

Proof   We derive this formula as follows: Let $y = \log_a M$. Then $a^y = M$, so

$$\log_b a^y = \log_b M \qquad \text{Property (7)}$$
$$y \log_b a = \log_b M \qquad \text{Property (6)}$$
$$y = \frac{\log_b M}{\log_b a} \qquad \text{Solve for } y$$
$$\log_a M = \frac{\log_b M}{\log_b a} \qquad y = \log_a M$$

Since calculators have only keys for $\boxed{\text{log}}$ and $\boxed{\text{ln}}$, in practice, the Change-of-Base Formula uses either $b = 10$ or $b = e$. Thus,

---

$$\log_a M = \frac{\log M}{\log a} \quad \text{and} \quad \log_a M = \frac{\ln M}{\ln a} \tag{10}$$

Comment   To graph logarithmic functions when the base is different from e or 10 requires the Change-of-Base Formula. For example, to graph $y = \log_2 x$, we would instead graph $y = (\ln x)/(\ln 2)$. Try it.

E X A M P L E   8   Using the Change-of-Base Formula

Calculate:

(a) $\log_5 89$ (b) $\log_{\sqrt{2}} \sqrt{5}$

Solution (a) $\log_5 89 = \dfrac{\log 89}{\log 5} \approx \dfrac{1.94939}{0.69897} = 2.7889$

or

$\log_5 89 = \dfrac{\ln 89}{\ln 5} \approx \dfrac{4.4886}{1.6094} = 2.7889$

(b) $\log_{\sqrt{2}} \sqrt{5} = \dfrac{\log \sqrt{5}}{\log \sqrt{2}} = \dfrac{\frac{1}{2}\log 5}{\frac{1}{2}\log 2} \approx \dfrac{0.69897}{0.30103} = 2.3219$

or

$\log_{\sqrt{2}} \sqrt{5} = \dfrac{\ln \sqrt{5}}{\ln \sqrt{2}} = \dfrac{\frac{1}{2}\ln 5}{\frac{1}{2}\ln 2} \approx \dfrac{1.6094}{0.6931} = 2.3219$

 Now work Problem 37.

Exponential and Logarithmic Curve Fitting

 When placed in a scatter diagram, data sometimes indicate a relation between the variables that is exponential or logarithmic. Figure 25 shows typical scatter diagrams for data that are exponential or logarithmic.

FIGURE 25

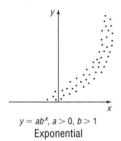

$y = ab^x, a > 0, b > 1$
Exponential

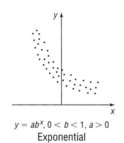

$y = ab^x, 0 < b < 1, a > 0$
Exponential

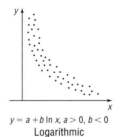

$y = a + b \ln x, a > 0, b < 0$
Logarithmic

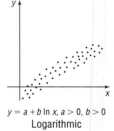

$y = a + b \ln x, a > 0, b > 0$
Logarithmic

E X A M P L E 9

Fitting a Curve to an Exponential Model

Beth is interested in finding a function that explains the relation between the amount of an unknown radioactive material and time. She obtains 100 grams of the radioactive material and every day for 9 days measures the amount of radioactive material in the sample; she obtains the following data:

Day	Weight (in Grams)
0	100
1	91.6
2	84.1
3	74.2
4	68.3
5	63.7
6	58.5
7	53.8
8	50.2
9	47.3

(a) Draw a scatter diagram using day as the independent variable.
(b) The exponential function of best fit to the data is found to be

$$y = 98(0.92)^x$$

Express the function of best fit in the form $A = A_0 e^{kt}$.
(c) Use the solution to (b) to estimate the time it takes until 50 grams of material is left. (Since 50 grams is one-half of the original amount, this time is referred to as the **half-life** of the radioactive material).
(d) Use the solution to (b) to predict how much of the material remains after 25 days.

(e) Use a graphing utility to verify the exponential function of best fit.
(f) Use a graphing utility to draw a scatter diagram of the data and then graph the exponential function of best fit on it.

Solution (a) See Figure 26.

FIGURE 26

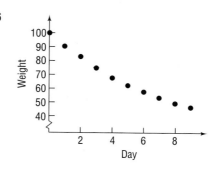

(b) To express $y = ab^x$ in the form $A = A_0e^{kt}$, we proceed as follows:

$$ab^x = A_0e^{kt}$$
$$a = A_0 \qquad b^x = e^{kt} = (e^k)^t$$
$$b = e^k$$

FIGURE 27

Since $y = ab^x = 98(0.92)^x$, we find $a = 98$ and $b = 0.92$. Thus,

$$a = A_0 = 98 \quad \text{and} \quad b = 0.92 = e^k$$
$$k = \ln(0.92) = -0.083$$

As a result, $A = A_0e^{kt} = 98e^{-0.083t}$.

(c) We set $A = 50$ and solve the equation $50 = 98e^{-0.083t}$ for t. The result is that the half-life is $t = 8.1$ days.

FIGURE 28

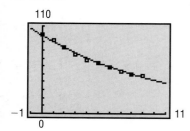

(d) After $t = 25$ days, the amount A of the radioactive material left is

$$A = 98e^{-0.083(25)} = 12.3 \text{ grams}$$

(e) A graphing utility fits the data in Figure 26 to an exponential model of the form $y = ab^x$ by using the EXPonential REGression option.* See Figure 27. Notice that $r = -0.997178162$, so $|r|$ is close to 1, indicating a good fit.

(f) See Figure 28.

 Now work Problem 71.

*If your utility does not have such an option but does have a LINear REGression option, you can transform the exponential model to a linear model by the following technique:

$y = ab^x$	Exponential form
$\ln y = \ln(ab^x)$	Take natural logs of each side.
$\ln y = \ln a + \ln b^x$	Property of logarithms
$\ln y = \ln a + (\ln b)x$	Property of logarithms
$Y = \ln a + (\ln b)x$	

Now apply the LINear REGression techniques discussed in Chapter 3. Be careful though. The dependent variable is now $Y = \ln y$, while the independent variable remains x. Once the line of best fit is obtained, you can find a and b and obtain the exponential curve $y = ab^x$.

Many relations between variables do not follow an exponential model but instead, the independent variable is related to the dependent variable using a logarithmic model.

E X A M P L E   10

Fitting a Curve to a Logarithmic Model

Jodi, a meteorologist, is interested in finding a function that explains the relation between the height of a weather balloon (in kilometers) and the atmospheric pressure (measured in millimeters of mercury) on the balloon. She collects the data in the table to the left:

Atmospheric Pressure, p	Height, h
760	0
740	0.184
725	0.328
700	0.565
650	1.079
630	1.291
600	1.634
580	1.862
550	2.235

(a) Draw a scatter diagram of the data with atmospheric pressure as the independent variable.

(b) The logarithmic function of best fit to the data is found to be

$$h = 45.8 - 6.9 \ln p$$

where h is the height of the weather balloon and p is the atmospheric pressure. Use the logarithmic function of best fit to predict the height of the balloon if the atmospheric pressure is 560 millimeters of mercury.

(c) Use a graphing utility to verify the logarithmic function of best fit.

(d) Use a graphing utility to draw a scatter diagram of the data, and then graph the logarithmic function of best fit on it.

Solution   (a) See Figure 29.

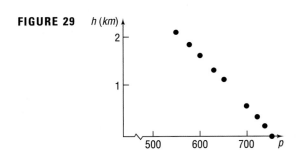

FIGURE 29

(b) Using the given function, Jodi predicts the height of the weather balloon when the atmospheric pressure is 560 to be

$$h = 45.8 - 6.9 \ln 560 \approx 2.14 \text{ kilometers}$$

(c) A graphing utility fits the data in Figure 29 to a logarithmic model of the form $y = a + b \ln x$ by using the Logarithm REGression option. See Figure 30. Notice that $|r|$ is close to 1, indicating a good fit.

(d) Figure 31 shows the graph of $h = 45.8 - 6.9 \ln p$ on the scatter diagram.

FIGURE 30

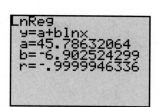

FIGURE 31

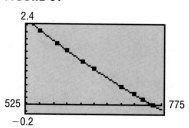

SUMMARY OF PROPERTIES OF LOGARITHMS

In the list that follows, $a > 0$, $a \neq 1$, and $b > 0$, $b \neq 1$; also, $M > 0$ and $N > 0$.

Definition $\qquad\qquad\qquad$ $y = \log_a x$ means $x = a^y$

Properties of logarithms $\qquad$ $\log_a 1 = 0; \quad \log_a a = 1$

$\qquad\qquad\qquad\qquad\qquad$ $a^{\log_a M} = M; \quad \log_a a^r = r$

$\qquad\qquad\qquad\qquad\qquad$ $\log_a (MN) = \log_a M + \log_a N$

$\qquad\qquad\qquad\qquad\qquad$ $\log_a\left(\dfrac{M}{N}\right) = \log_a M - \log_a N$

$\qquad\qquad\qquad\qquad\qquad$ $\log_a\left(\dfrac{1}{N}\right) = -\log_a N$

$\qquad\qquad\qquad\qquad\qquad$ $\log_a M^r = r \log_a M$

Change-of-Base Formula $\qquad$ $\log_a M = \dfrac{\log_b M}{\log_b a}$

HISTORICAL FEATURE Logarithms were invented about 1590 by John Napier (1550–1617) and Jobst Bürgi (1552–1632), working independently. Napier, whose work had the greater influence, was a Scottish lord, a secretive man whose neighbors were inclined to believe him to be in league with the devil. His approach to logarithms was quite different from ours; it was based on the relationship between arithmetic and geometric sequences (see the chapter on induction and sequences), and not on the inverse function relationship of logarithms to exponential functions (described in Section 6.3). Napier's tables, published in 1614, listed what would now be called *natural logarithms* of sines and were rather difficult to use. A London professor, Henry Briggs, became interested in the tables and visited Napier. In their conversations, they developed the idea of common logarithms, which were published in 1617. Their importance for calculation was immediately recognized, and by 1650 they were being printed as far away as China. They remained an important calculation tool until the advent of the inexpensive hand-held calculator about 1972, which has decreased their calculational, but not their theoretical, importance.

$\qquad$A side effect of the invention of logarithms was the popularization of the decimal system of notation for real numbers.

6.4 | EXERCISES

In Problems 1–12, suppose that $\ln 2 = a$ *and* $\ln 3 = b$. *Use properties of logarithms to write each logarithm in terms of a and b.*

1. $\ln 6$

2. $\ln \frac{2}{3}$

3. $\ln 1.5$

4. $\ln 0.5$

5. $\ln 2e$

6. $\ln\left(\dfrac{3}{e}\right)$

7. $\ln 12$

8. $\ln 24$

9. $\ln \sqrt[5]{18}$

10. $\ln \sqrt[4]{48}$

11. $\log_2 3$

12. $\log_3 2$

In Problems 13–22, write each expression as a sum and/or difference of logarithms. Express powers as factors.

13. $\ln(x^2\sqrt{1-x})$

14. $\ln(x^4\sqrt{x^2+1})$

15. $\log_2\left(\dfrac{x^3}{x-3}\right)$

16. $\log_5\left(\dfrac{\sqrt[3]{x^2+1}}{x^2-1}\right)$

17. $\log\left[\dfrac{x(x+2)}{(x+3)^2}\right]$

18. $\log\dfrac{x^3\sqrt{x+1}}{(x-2)^2}$

19. $\ln\left[\dfrac{x^2-x-2}{(x+4)^2}\right]^{1/3}$

20. $\ln\left[\dfrac{(x-4)^2}{x^2-1}\right]^{2/3}$

21. $\ln\dfrac{5x\sqrt{1-3x}}{(x-4)^3}$

22. $\ln\left[\dfrac{5x^2\sqrt[3]{1-x}}{4(x+1)^2}\right]$

In Problems 23–32, write each expression as a single logarithm.

23. $3\log_5 u + 4\log_5 v$

24. $\log_3 u^2 - \log_3 v$

25. $\log_{1/2}\sqrt{x} - \log_{1/2} x^3$

26. $\log_2\left(\dfrac{1}{x}\right) + \log_2\left(\dfrac{1}{x^2}\right)$

27. $\ln\left(\dfrac{x}{x-1}\right) + \ln\left(\dfrac{x+1}{x}\right) - \ln(x^2-1)$

28. $\log\left(\dfrac{x^2+2x-3}{x^2-4}\right) - \log\left(\dfrac{x^2+7x+6}{x+2}\right)$

29. $8\log_2\sqrt{3x-2} - \log_2\left(\dfrac{4}{x}\right) + \log_2 4$

30. $21\log_3\sqrt[3]{x} + \log_3 9x^2 - \log_5 25$

31. $2\log_a 5x^3 - \frac{1}{2}\log_a(2x+3)$

32. $\frac{1}{3}\log(x^3+1) + \frac{1}{2}\log(x^2+1)$

In Problems 33–36, find the exact value of each expression.

33. $3^{\log_3 5 - \log_3 4}$

34. $5^{\log_5 6 + \log_5 7}$

35. $e^{\log_{e^2} 16}$

36. $e^{\log_{e^2} 9}$

In Problems 37–44, use the Change-of-Base Formula and a calculator to evaluate each logarithm. Round your answer to three decimal places.

37. $\log_3 21$

38. $\log_5 18$

39. $\log_{1/3} 71$

40. $\log_{1/2} 15$

41. $\log_{\sqrt{2}} 7$

42. $\log_{\sqrt{5}} 8$

43. $\log_\pi e$

44. $\log_\pi \sqrt{2}$

45. Show that $\log_a(x+\sqrt{x^2-1}) + \log_a(x-\sqrt{x^2-1}) = 0$.

46. Show that $\log_a(\sqrt{x}+\sqrt{x-1}) + \log_a(\sqrt{x}-\sqrt{x-1}) = 0$.

47. Show that $\ln(1+e^{2x}) = 2x + \ln(1+e^{-2x})$.

48. If $f(x) = \log_a x$, show that $\dfrac{f(x+h)-f(x)}{h} = \log_a\left(1+\dfrac{h}{x}\right)^{1/h}$, $h \neq 0$.

49. If $f(x) = \log_a x$, show that $-f(x) = \log_{1/a} x$.

50. If $f(x) = \log_a x$, show that $f(1/x) = -f(x)$.

51. If $f(x) = \log_a x$, show that $f(AB) = f(A) + f(B)$.

52. If $f(x) = \log_a x$, show that $f(x^\alpha) = \alpha f(x)$.

 For Problems 53–56, graph each function using a graphing utility and the Change-of-Base Formula.

53. $y = \log_4 x$

54. $y = \log_5 x$

55. $y = \log_2 (x + 2)$

56. $y = \log_4 (x - 3)$

In Problems 57–66, express y as a function of x. The constant C is a positive number.

57. $\ln y = \ln x + \ln C$

58. $\ln y = \ln(x + C)$

59. $\ln y = \ln x + \ln(x + 1) + \ln C$

60. $\ln y = 2 \ln x - \ln(x + 1) + \ln C$

61. $\ln y = 3x + \ln C$

62. $\ln y = -2x + \ln C$

63. $\ln(y - 3) = -4x + \ln C$

64. $\ln(y + 4) = 5x + \ln C$

65. $3 \ln y = \frac{1}{2} \ln(2x + 1) - \frac{1}{3} \ln(x + 4) + \ln C$

66. $2 \ln y = -\frac{1}{2} \ln x + \frac{1}{3} \ln(x^2 + 1) + \ln C$

67. Find the value of $\log_2 3 \cdot \log_3 4 \cdot \log_4 5 \cdot \log_5 6 \cdot \log_6 7 \cdot \log_7 8$.

68. Find the value of $\log_2 4 \cdot \log_4 6 \cdot \log_6 8$.

69. Find the value of $\log_2 3 \cdot \log_3 4 \cdot \ldots \cdot \log_n (n + 1) \cdot \log_{n+1} 2$.

70. Find the value of $\log_2 2 \cdot \log_2 4 \cdot \ldots \cdot \log_2 2^n$.

71. Chemistry A chemist has a 100-gram sample of a radioactive material. He records the amount of radioactive material every week for 6 weeks and obtains the following data:

Week	Weight (in Grams)
0	100.0
1	88.3
2	75.9
3	69.4
4	59.1
5	51.8
6	45.5

(a) Draw a scatter diagram with week as the independent variable.

(b) The exponential function of best fit to the data is found to be

$$y = 100(0.88)^x$$

Express the function of best fit in the form $A = A_0 e^{kt}$.

(c) Use the solution to (b) to estimate the time it takes until 50 grams of material is left. (Since 50 grams is one-half of the original

amount, this time is referred to as the **half-life** of the radioactive material).

(d) Use the solution to (b) to predict how much radioactive material will be left after 50 weeks.

 (e) Use a graphing utility to verify the exponential function of best fit.

(f) Use a graphing utility to draw a scatter diagram of the data and then graph the exponential function of best fit on it.

72. Chemistry A chemist has a 1,000-gram sample of a radioactive material. She records the amount of radioactive material remaining in the sample every day for a week and obtains the following data:

Day	Weight (in Grams)
0	1000.0
1	897.1
2	802.5
3	719.8
4	651.1
5	583.4
6	521.7
7	468.3

(a) Draw a scatter diagram with day as the independent variable.
(b) The exponential function of best fit to the data is found to be

$$y = 999(0.898)^x$$

Express the function of best fit in the form $A = A_0 e^{kt}$.
(c) Use the solution to (b) to estimate the time it takes until 500 grams of material is left. (Since 500 grams is one-half of the original amount, this time is referred to as the half-life of the radioactive material.)
(d) Use the solution to (b) to predict how much radioactive material is left after 20 days.
 (e) Use a graphing utility to verify the exponential function of best fit.
(f) Use a graphing utility to draw a scatter diagram of the data and then graph the exponential function of best fit on it.

73. Economics and Marketing A store manager collected the following data regarding price and quantity demanded of shoes:

Price ($/Unit)	Quantity Demanded
79	10
67	20
54	30
46	40
38	50
31	60

(a) Draw a scatter diagram with quantity demanded on the x-axis and price on the y-axis.
(b) The exponential function of best fit to the data is found to be

$$y = 96(0.98)^x$$

Express the function of best fit in the form $y = A_0 e^{kx}$.
(c) Use the solution to (b) to predict the quantity demanded if the price is $60.
(d) Use a graphing utility to verify the exponential function of best fit.
(e) Use a graphing utility to draw a scatter diagram of the data and then graph the exponential function of best fit on it.

74. Economics and Marketing A store manager collected the following data regarding price and quantity supplied of dresses:

Price ($/Unit)	Quantity Supplied
25	10
32	20
40	30
46	40
60	50
74	60

(a) Draw a scatter diagram with quantity supplied on the x-axis and price on the y-axis.
(b) The exponential function of best fit to the data is found to be

$$y = 20.5(1.02)^x$$

Express the function of best fit in the form $A = A_0 e^{kx}$.
(c) Use the solution to (b) to predict the quantity supplied if the price is $45.
(d) Use a graphing utility to verify the exponential function of best fit.
(e) Use a graphing utility to draw a scatter diagram of the data and then graph the exponential function of best fit on it.

75. Economics and Marketing The data at the top of page 459 represent the price and quantity demanded in 1997 for IBM personal computers at Best Buy.
(a) Draw a scatter diagram of the data with price as the dependent variable.
(b) The logarithmic function of best fit to the data is found to be

$$y = 32{,}741 - 6071 \ln x$$

Use it to predict the number of IBM personal computers that would be demanded if the price were $1650.
(c) Use a graphing utility to verify the logarithmic function of best fit.
(d) Use a graphing utility to draw a scatter diagram of the data and then graph the logarithmic function of best fit on it.

Price ($/Computer)	Quantity Demanded
2300	152
2000	159
1700	164
1500	171
1300	176
1200	180
1000	189

76. Show that $\log_a (M/N) = \log_a M - \log_a N$, where a, M, and N are positive real numbers, with $a \neq 1$.

77. Show that $\log_a (1/N) = -\log_a N$, where a and N are positive real numbers, with $a \neq 1$.

 78. Find the domain of $f(x) = \log_a x^2$ and the domain of $g(x) = 2 \log_a x$. Since $\log_a x^2 = 2 \log_a x$, how do you reconcile the fact that the domains are not equal? Write a brief explanation.

6.5 | LOGARITHMIC AND EXPONENTIAL EQUATIONS

> 1 Solve Logarithmic Equations
> 2 Solve Exponential Equations
> 3 Solve Logarithmic and Exponential Equations Using a Graphing Utility

Logarithmic Equations

1 Equations that contain terms of the form $\log_a x$, where a is a positive real number, with $a \neq 1$, are often called **logarithmic equations.**

Our practice will be to solve equations, whenever possible, by finding exact solutions using algebraic methods. When algebraic methods cannot be used, approximate solutions will be obtained using a graphing utility. The reader is encouraged to pay particular attention to the form of the equations for which exact solutions are possible.

E X A M P L E   1

Solving a Logarithmic Equation

Solve: $\log_3 (4x - 7) = 2$

Solution   We can obtain an exact solution by changing the logarithm to exponential form.

$$\log_3 (4x - 7) = 2$$
$$4x - 7 = 3^2$$
$$4x - 7 = 9$$
$$4x = 16$$
$$x = 4$$

E X A M P L E   2

Solving a Logarithmic Equation

Solve: $2 \log_5 x = \log_5 9$

Solution
Because each logarithm is to the same base, 5, we can obtain an exact solution, as follows:

$$2 \log_5 x = \log_5 9$$
$$\log_5 x^2 = \log_5 9 \qquad \text{Property (6), Section 6.4}$$
$$x^2 = 9 \qquad \text{Property (8), Section 6.4}$$
$$x = 3 \quad \text{or} \quad \cancel{x = -3} \qquad \text{Recall that logarithms of negative}$$

numbers are not defined, so, in the
expression $2 \log_5 x$, x must be
positive. Therefore, -3 is
extraneous and we discard it.

The equation has only one solution, 3.

Now work Problem 1.

E X A M P L E   3

Solving a Logarithmic Equation

Solve: $\log_4 (x + 3) + \log_4 (2 - x) = 1$

Solution
To obtain an exact solution, we need to express the left side as a single logarithm. Then we will change the expression to exponential form.

$$\log_4 (x + 3) + \log_4 (2 - x) = 1$$
$$\log_4 [(x + 3)(2 - x)] = 1 \qquad \text{Property (3), Section 6.4}$$
$$(x + 3)(2 - x) = 4^1 = 4$$
$$-x^2 - x + 6 = 4$$
$$x^2 + x - 2 = 0$$
$$(x + 2)(x - 1) = 0$$
$$x = -2 \quad \text{or} \quad x = 1$$

You should verify that both of these are solutions.

Now work Problem 11.

Care must be taken when solving logarithmic equations. Be sure to check each apparent solution in the original equation and discard any that are extraneous. In the expression $\log_a M$, remember that a and M are positive and $a \neq 1$.

Exponential Equations

2 Equations that involve terms of the form a^x, $a > 0$, $a \neq 1$, are often referred to as **exponential equations.** Such equations sometimes can be solved by appropriately applying the laws of exponents and equations (1), that is,

$$\text{If } a^u = a^v, \quad \text{then } u = v. \tag{1}$$

To use equation (1), each side of the equation must be written with the same base.

MISSION POSSIBLE

McNewton's Coffee

Your team has been called in to solve a problem encountered by a fast food restaurant. They believe that their coffee should be brewed at 170° Fahrenheit. However, at that temperature it is too hot to drink, and a customer who accidentally spills the coffee on himself might receive third-degree burns.

What they need is a special container that will heat the water to 170°, brew the coffee at that temperature, then cool it quickly to a drinkable temperature, say 140°F, and hold it there, or at least keep it at or above 120°F for a reasonable period of time without further cooking. To cool down the coffee, three companies have submitted proposals with these specifications:

(a) The CentiKeeper Company has a container that will reduce the temperature of a liquid from 200°F to 100°F in 90 minutes by maintaining a constant temperature of 70°F.
(b) The TempControl Company has a container that will reduce the temperature of a liquid from 200°F to 110°F in 60 minutes by maintaining a constant temperature of 60°F.
(c) The Hot'n'Cold, Inc., has a container that will reduce the temperature of a liquid from 210°F to 90°F in 30 minutes by maintaining a constant temperature of 50°F.

Your job is to make a recommendation as to which container to purchase. For this you will need Newton's Law of Cooling, which follows:

$$u(t) = T + (u_0 - T)e^{kt} \quad k < 0$$

In this formula, T represents the temperature of the surrounding medium, u_0 is the initial temperature of the heated object, t is the length of time in minutes, k is a negative constant, and u represents the temperature at time t.

1. Use Newton's Law of Cooling to find the constant k of the formula for each container.
2. Use a graphing utility to graph each relation.
3. How long does it take each container to lower the coffee temperature from 170°F to 140°F?
4. How long will the coffee temperature remain between 120°F and 140°F?
5. On the basis of this information, which company should get the contract with McNewton's? What are your reasons?
6. Define "capital cost" and "operating cost". How might they affect your choice?

E X A M P L E   4   Solving an Exponential Equation

Solve: $3^{x+1} = 81$

Solution   Since $81 = 3^4$, we can write the equation as

$$3^{x+1} = 81 = 3^4$$

Now we have the same base, 3, on each side, so we can apply (1) to obtain

$$x + 1 = 4$$
$$x = 3$$ ▬

 Now work Problem 19.

E X A M P L E   5   Solving an Exponential Equation

Solve: $e^{-x^2} = (e^x)^2 \cdot \dfrac{1}{e^3}$

Solution   We use some Laws of Exponents first to express the right-hand side with the base e.

$$(e^x)^2 \cdot \dfrac{1}{e^3} = e^{2x} \cdot e^{-3} = e^{2x-3}$$

Now we have $e^{-x^2} = e^{2x-3}$, so we can apply (1) to get

$$-x^2 = 2x - 3$$
$$x^2 + 2x - 3 = 0$$
$$(x + 3)(x - 1) = 0$$
$$x = -3 \quad \text{or} \quad x = 1$$ ▬

E X A M P L E   6   Solving an Exponential Equation

Solve: $4^x - 2^x - 12 = 0$

Solution   We note that $4^x = (2^2)^x = 2^{2x} = (2^x)^2$, so the equation is actually quadratic in form, and we can write it as

$$(2^x)^2 - 2^x - 12 = 0 \quad \text{\small If } u = 2^x \text{, then } u^2 - u - 12 = 0.$$

Now we can factor as usual:

$$(2^x - 4)(2^x + 3) = 0 \qquad \text{\small } (u - 4)(u + 3) = 0$$
$$2^x - 4 = 0 \quad \text{or} \quad 2^x + 3 = 0 \qquad \text{\small } u - 4 = 0 \quad \text{or} \quad u + 3 = 0$$
$$2^x = 4 \qquad\qquad 2^x = -3 \quad \text{\small } u = 2^x = 4 \qquad u = 2^x = -3$$

The equation on the left has the solution $x = 2$, since $4 = 2^2$; the equation on the right has no solution, since $2^x > 0$ for all x.   ▬

In each of the preceding three examples, we were able to write each exponential expression using the same base, obtaining exact solutions to the equation. When this is not possible, logarithms can sometimes be used to obtain the solution.

E X A M P L E   7

Solving an Exponential Equation

Solve: $2^x = 5$

Solution

Since we cannot express 5 as a power of 2, we write the exponential equation as the equivalent logarithmic equation:

$$2^x = 5$$

$$x = \log_2 5 = \frac{\ln 5}{\ln 2}$$

↑

Change-of-Base Formula (10)

Alternatively, we can solve the equation $2^x = 5$ by taking the natural logarithm (or common logarithm) of each side. Taking the natural logarithm,

$$2^x = 5$$
$$\ln 2^x = \ln 5$$
$$x \ln 2 = \ln 5$$
$$x = \frac{\ln 5}{\ln 2}$$

Using a calculator, the solution, rounded to three decimal places, is

$$x = \frac{\ln 5}{\ln 2} \approx 2.322$$

Now work Problem 33.

E X A M P L E   8

Solving an Exponential Equation

Solve: $8 \cdot 3^x = 5$

Solution

$8 \cdot 3^x = 5$ Isolate 3^x on the left side.

$3^x = \frac{5}{8}$ Proceed as in Example 7.

$$x = \log_3 \left(\tfrac{5}{8}\right) = \frac{\ln \frac{5}{8}}{\ln 3}$$

Using a calculator, the solution, rounded to three decimal places, is

$$x = \frac{\ln \left(\frac{5}{8}\right)}{\ln 3} \approx -0.428$$

E X A M P L E 9

Solving an Exponential Equation

Solve: $5^{x-2} = 3^{3x+2}$

Solution

Because the bases on each side are different, we take the natural logarithm of each side and apply appropriate properties of logarithms. The result is an equation in x that we can solve.

$$5^{x-2} = 3^{3x+2}$$

$\ln 5^{x-2} = \ln 3^{3x+2}$	Property (7)
$(x - 2)\ln 5 = (3x + 2)\ln 3$	Property (6)
$(\ln 5)x - 2 \ln 5 = (3 \ln 3)x + 2 \ln 3$	Remove parentheses
$(\ln 5 - 3 \ln 3)x = 2 \ln 3 + 2 \ln 5$	Isolate x terms on the left

$$x = \frac{2(\ln 3 + \ln 5)}{\ln 5 - 3 \ln 3} \approx -3.212$$

Now work Problem 41.

Graphing Utility Solutions

The techniques introduced in this section only apply to certain types of logarithmic and exponential equations. Solutions for other types are usually studied in calculus, using numerical methods. However, we can use a graphing utility to approximate the solution of other types.

E X A M P L E 10

Solving Equations Using a Graphing Utility

Solve: $\log_3 x + \log_4 x = 4$
Express the solution(s) correct to two decimal places.

FIGURE 32

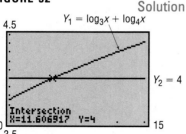

Solution

The solution is found by graphing

$$Y_1 = \log_3 x + \log_4 x = \frac{\log x}{\log 3} + \frac{\log x}{\log 4} \quad \text{and} \quad Y_2 = 4$$

(Remember that you must use the Change-of-Base Formula to graph Y_1.) Y_1 is an increasing function (do you know why?), and so there is only one point of intersection for Y_1 and Y_2. Figure 32 shows the graphs of Y_1 and Y_2. Using the INTERSECT* command, the solution 11.60, correct to two decimal places, is obtained.

Can you discover an algebraic solution to Example 10?

[**Hint:** Factor $\log x$ from Y_1.]

E X A M P L E 11

Solving Equations Using a Graphing Utility

Solve: $x + e^x = 2$
Express the solution(s) correct to two decimal places.

*If your graphing utility does not have an INTERSECT command, you will have to use BOX or ZOOM-IN with TRACE to find the point of intersection.

Solution   The solution is found by graphing $Y_1 = x + e^x$ and $Y_2 = 2$. Y_1 is an increasing function (do you know why?), and so there is only one point of intersection for Y_1 and Y_2. Figure 33 shows the graphs of Y_1 and Y_2. Using the INTERSECT command, the solution 0.44, correct to two decimal places, is obtained.

FIGURE 33

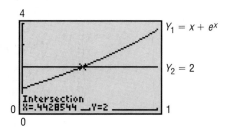

6.5   EXERCISES

In Problems 1–58, solve each equation.

1. $\log_2 (2x + 1) = 3$
2. $\log_3 (3x - 2) = 2$
3. $\log_3 (x^2 + 1) = 2$
4. $\log_5 (x^2 + x + 4) = 2$
5. $\frac{1}{2} \log_3 x = 2 \log_3 2$
6. $-2 \log_4 x = \log_4 9$
7. $2 \log_5 x = 3 \log_5 4$
8. $3 \log_2 x = -\log_2 27$
9. $3 \log_2 (x - 1) + \log_2 4 = 5$
10. $2 \log_3 (x + 4) - \log_3 9 = 2$
11. $\log x + \log (x + 15) = 2$
12. $\log_4 x + \log_4 (x - 3) = 1$
13. $\log_x 4 = 2$
14. $\log_x \left(\frac{1}{8}\right) = 3$
15. $\log_3 (x - 1)^2 = 2$
16. $\log_2 (x + 4)^3 = 6$
17. $\log_{1/2} (3x + 1)^{1/3} = -2$
18. $\log_{1/3} (1 - 2x)^{1/2} = -1$
19. $2^{2x+1} = 4$
20. $5^{1-2x} = \frac{1}{5}$
21. $3^{x^3} = 9^x$
22. $4^{x^2} = 2^x$
23. $8^{x^2-2x} = \frac{1}{2}$
24. $9^{-x} = \frac{1}{3}$
25. $2^x \cdot 8^{-x} = 4^x$
26. $\left(\frac{1}{2}\right)^{1-x} = 4$
27. $2^{2x} + 2^x - 12 = 0$
28. $3^{2x} + 3^x - 2 = 0$
29. $3^{2x} + 3^{x+1} - 4 = 0$
30. $4^x - 2^x = 0$
31. $4^x = 8$
32. $9^{2x} = 27$
33. $2^x = 10$
34. $3^x = 14$
35. $8^{-x} = 1.2$
36. $2^{-x} = 1.5$
37. $3^{1-2x} = 4^x$
38. $2^{x+1} = 5^{1-2x}$
39. $\left(\frac{3}{5}\right)^x = 7^{1-x}$
40. $\left(\frac{4}{3}\right)^{1-x} = 5^x$
41. $1.2^x = (0.5)^{-x}$
42. $(0.3)^{1+x} = 1.7^{2x-1}$
43. $\pi^{1-x} = e^x$
44. $e^{x+3} = \pi^x$
45. $5(2^{3x}) = 8$
46. $0.3(4^{0.2x}) = 0.2$
47. $400e^{0.2x} = 600$
48. $500e^{0.3x} = 600$
49. $\log_a (x - 1) - \log_a (x + 6) = \log_a (x - 2) - \log_a(x + 3)$
50. $\log_a x + \log_a (x - 2) = \log_a (x + 4)$
51. $\log_{1/3} (x^2 + x) - \log_{1/3} (x^2 - x) = -1$
52. $\log_4 (x^2 - 9) - \log_4 (x + 3) = 3$
53. $\log_2 8^x = -3$
54. $\log_3 3^x = -1$
55. $\log_2 (x^2 + 1) - \log_4 x^2 = 1$
56. $\log_2 (3x + 2) - \log_4 x = 3$
      [**Hint:** Change $\log_4 x^2$ to base 2.]
57. $\log_{16} x + \log_4 x + \log_2 x = 7$
58. $\log_9 x + 3 \log_3 x = 14$
59. $(\sqrt[3]{2})^{2-x} = 2^{x^2}$
60. $\log_2 x^{\log_2 x} = 4$

▦ *In Problems 61–76, use a graphing utility to solve each equation. Express your answer correct to two decimal places.*

61. $\log_5 x + \log_3 x = 1$
62. $\log_2 x + \log_6 x = 3$
63. $\log_5 (x + 1) - \log_4 (x - 2) = 1$
64. $\log_2 (x - 1) - \log_6 (x + 2) = 2$
65. $e^x = -x$
66. $e^{2x} = x + 2$
67. $e^x = x^2$
68. $e^x = x^3$
69. $\ln x = -x$
70. $\ln 2x = -x + 2$
71. $\ln x = x^3 - 1$
72. $\ln x = -x^2$
73. $e^x + \ln x = 4$
74. $e^x - \ln x = 4$
75. $e^{-x} = \ln x$
76. $e^{-x} = -\ln x$

6.6 | COMPOUND INTEREST

1 Determine the Future Value of a Lump Sum of Money
2 Calculate Effective Rates of Return
3 Determine the Present Value of a Lump Sum of Money
4 Determine the Time Required to Double or Triple a Lump Sum of Money

1 Interest is money paid for the use of money. The total amount borrowed (whether by an individual from a bank in the form of a loan or by a bank from an individual in the form of a savings account) is called the **principal.** The **rate of interest,** expressed as a percent, is the amount charged for the use of the principal for a given period of time, usually on a yearly (that is, per annum) basis.

If a principal of P dollars is borrowed for a period of t years at a per annum interest rate r, expressed as a decimal, the interest I charged is

> Simple Interest Formula
>
> $$I = Prt \tag{1}$$

Interest charged according to formula (1) is called **simple interest.**

In working with problems involving interest, we use the term **payment period** as follows:

Annually	Once per year	Monthly	12 times per year
Semiannually	Twice per year	Daily	365 times per year*
Quarterly	4 times per year		

When the interest due at the end of a payment period is added to the principal so that the interest computed at the end of the next payment period is based on this new principal amount (old principal + interest), the interest is said to have been **compounded.** Thus, **compound interest** is interest paid on previously earned interest.

E X A M P L E   1

Computing Compound Interest

A credit union pays interest of 8% per annum compounded quarterly on a certain savings plan. If $1000 is deposited in such a plan and the interest is left to accumulate, how much is in the account after 1 year?

Solution We use the simple interest formula, $I = Prt$. The principal P is $1000 and the rate of interest is 8% = 0.08. After the first quarter of a year, the time t is $\frac{1}{4}$ year, so the interest earned is

$$I = Prt = (\$1000)(0.08)(\tfrac{1}{4}) = \$20$$

*Most banks use a 360 day "year." Why do you think they do?

The new principal is $P + I = \$1000 + \$20 = \$1020$. At the end of the second quarter, the interest on this principal is

$$I = (\$1020)(0.08)(\tfrac{1}{4}) = \$20.40$$

At the end of the third quarter, the interest on the new principal of $\$1020 + \$20.40 = \$1040.40$ is

$$I = (\$1040.40)(0.08)(\tfrac{1}{4}) = \$20.81$$

Finally, after the fourth quarter, the interest is

$$I = (\$1061.21)(0.08)(\tfrac{1}{4}) = \$21.22$$

Thus, after 1 year the account contains $\$1082.43$. ∎

The pattern of the calculations performed in Example 1 leads to a general formula for compound interest. To fix our ideas, let P represent the principal to be invested at a per annum interest rate r, which is compounded n times per year. (For computing purposes, r is expressed as a decimal.) The interest earned after each compounding period is the principal times r/n. Thus, the amount A after one compounding period is

$$A = P + P\left(\frac{r}{n}\right) = P\left(1 + \frac{r}{n}\right)$$

After two compounding periods, the amount A, based on the new principal $P(1 + r/n)$, is

$$A = \underbrace{P\left(1 + \frac{r}{n}\right)}_{\substack{\text{New} \\ \text{principal}}} + \underbrace{P\left(1 + \frac{r}{n}\right)\left(\frac{r}{n}\right)}_{\substack{\text{Interest on} \\ \text{new principal}}} = P\left(1 + \frac{r}{n}\right)\left(1 + \frac{r}{n}\right) = P\left(1 + \frac{r}{n}\right)^2$$

After three compounding periods,

$$A = P\left(1 + \frac{r}{n}\right)^2 + P\left(1 + \frac{r}{n}\right)^2\left(\frac{r}{n}\right) = P\left(1 + \frac{r}{n}\right)^2\left(1 + \frac{r}{n}\right) = P\left(1 + \frac{r}{n}\right)^3$$

Continuing this way, after n compounding periods (1 year),

$$A = P\left(1 + \frac{r}{n}\right)^n$$

Because t years will contain $n \cdot t$ compounding periods, after t years we have

$$A = P\left(1 + \frac{r}{n}\right)^{nt}$$

Theorem Compound Interest Formula

The amount A after t years due to a principal P invested at an annual interest rate r compounded n times per year is

$$A = P\left(1 + \frac{r}{n}\right)^{nt} \tag{2}$$

E X A M P L E   2

Comparing Investments Using Different Compounding Periods

Investing $1000 at an annual rate of 10% compounded annually, quarterly, monthly, and daily will yield the following amounts after 1 year:

Annual compounding:

$$A = P(1 + r)$$
$$= (\$1000)(1 + 0.10) = \$1100.00$$

Quarterly compounding:

$$A = P\left(1 + \frac{r}{4}\right)^4$$
$$= (\$1000)(1 + 0.025)^4 = \$1103.81$$

Monthly compounding:

$$A = P\left(1 + \frac{r}{12}\right)^{12}$$
$$= (\$1000)(1 + 0.00833)^{12} = \$1104.71$$

Daily compounding:

$$A = P\left(1 + \frac{r}{365}\right)^{365}$$
$$= (\$1000)(1 + 0.000274)^{365} = \$1105.16$$

Now work Problem 1.

From Example 2, we can see that the effect of compounding more frequently is that the account after 1 year is higher: $1000 compounded 4 times a year at 10% results in $1103.81; $1000 compounded 12 times a year at 10% results in $1104.71; and $1000 compounded 365 times a year at 10% results in $1105.16. This leads to the following question: What would happen to the amount after 1 year if the number of times that the interest is compounded were increased without bound?

Let's find the answer. Suppose that P is the principal, r is the per annum interest rate, and n is the number of times that the interest is compounded each year. The amount after 1 year is

$$A = P\left(1 + \frac{r}{n}\right)^n$$

Now suppose that the number n of times that the interest is compounded per year gets larger and larger; that is, suppose that $n \to \infty$. Then,

$$A = P\left(1 + \frac{r}{n}\right)^n = P\left[1 + \frac{1}{n/r}\right]^n = P\left(\left[1 + \frac{1}{n/r}\right]^{n/r}\right)^r = P\left[\left(1 + \frac{1}{h}\right)^h\right]^r \quad (3)$$

$$\uparrow$$
$$h = \frac{n}{r}$$

In (3), as $n \to \infty$, then $h = n/r \to \infty$, and the expression in brackets equals e. [Refer to equation (2) on p. 430.] Thus, $A \to Pe^r$. Table 6 compares $(1 + r/n)^n$, for large values of n, to e^r for $r = 0.05, r = 0.10, r = 0.15$, and $r = 1$. The larger

n gets, the closer $(1 + r/n)^n$ gets to e^r. Thus, no matter how frequent the compounding, the amount after 1 year has the definite ceiling Pe^r.

TABLE 6

	$\left(1 + \dfrac{r}{n}\right)^n$			
	$n = 100$	$n = 1000$	$n = 10{,}000$	e^r
$r = 0.05$	1.0512579	1.05127	1.051271	1.0512711
$r = 0.10$	1.1051157	1.1051654	1.1051703	1.1051709
$r = 0.15$	1.1617037	1.1618212	1.1618329	1.1618342
$r = 1$	2.7048138	2.7169239	2.7181459	2.7182818

When interest is compounded so that the amount after 1 year is Pe^r, we say the interest is **compounded continuously.**

Theorem Continuous Compounding

The amount A after t years due to a principal P invested at an annual interest rate r compounded continuously is

$$A = Pe^{rt} \qquad (4)$$

E X A M P L E   3   Using Continuous Compounding

The amount A that results from investing a principal P of \$1000 at an annual rate r of 10% compounded continuously for a time t of 1 year is

$$A = \$1000e^{0.10} = (\$1000)(1.10517) = \$1105.17$$

 Now work Problem 9.

 The **effective rate of interest** is the equivalent annual simple rate of interest that would yield the same amount as compounding after 1 year. For example, based on Example 3, a principal of \$1000 will result in \$1105.17 at a rate of 10% compounded continuously. To get this same amount using a simple rate of interest would require that interest of \$1105.17 − \$1000.00 = \$105.17 be earned on the principal. Since \$105.17 is 10.517% of \$1000, a simple rate of interest of 10.517% is needed to equal 10% compounded continuously. Thus, the effective rate of interest of 10% compounded continuously is 10.517%.

Based on the results of Examples 2 and 3, we find the following comparisons:

	Annual Rate	Effective Rate
Annual compounding	10%	10%
Quarterly compounding	10%	10.381%
Monthly compounding	10%	10.471%
Daily compounding	10%	10.516%
Continuous compounding	10%	10.517%

 Now work Problem 21.

E X A M P L E   4   Computing the Value of an IRA

On January 2, 1998, $2000 is placed in an Individual Retirement Account (IRA) that will pay interest of 10% per annum compounded continuously. What will the IRA be worth on January 1, 2018?

Solution   The amount A after 20 years is

$$A = Pe^{rt} = \$2000e^{(0.10)(20)} = \$14{,}778.11$$ ▬

Exploration   How long will it be until $A = \$40{,}000$?

[**Hint:** Graph $Y_1 = 2000e^{0.1x}$ and $Y_2 = 40{,}000$. Use INTERSECT to find x.] ▬

FIGURE 34

Time is money

When people engaged in finance speak of the "time value of money," they are usually referring to the **present value** of money. The present value of A dollars to be received at a future date is the principal you would need to invest now so that it would grow to A dollars in the specified time period. Thus, the present value of money to be received at a future date is always less than the amount to be received, since the amount to be received will equal the present value (money invested now) *plus* the interest accrued over the time period.

We use the compound interest formula (2) to get a formula for present value. If P is the present value of A dollars to be received after t years at a per annum interest rate r compounded n times per year, then, by formula (2),

$$A = P\left(1 + \frac{r}{n}\right)^{nt}$$

To solve for P, we divide both sides by $(1 + r/n)^{nt}$, and the result is

$$\frac{A}{(1 + \frac{r}{n})^{nt}} = P \quad \text{or} \quad P = A\left(1 + \frac{r}{n}\right)^{-nt}$$

Theorem   Present Value Formulas

The present value P of A dollars to be received after t years, assuming a per annum interest rate r compounded n times per year, is

$$P = A\left(1 + \frac{r}{n}\right)^{-nt} \tag{5}$$

If the interest is compounded continuously, then

$$P = Ae^{-rt} \tag{6}$$

To prove (6), solve formula (4) for P.

E X A M P L E   5

Computing the Value of a Zero-Coupon Bond

A zero-coupon (noninterest-bearing) bond can be redeemed in 10 years for $1000. How much should you be willing to pay for it now if you want a return of

(a) 8% compounded monthly?

(b) 7% compounded continuously?

Solution   (a) We are seeking the present value of $1000. Thus, we use formula (5) with $A = \$1000$, $n = 12$, $r = 0.08$, and $t = 10$.

$$P = A\left(1 + \frac{r}{n}\right)^{-nt}$$

$$= \$1000\left(1 + \frac{0.08}{12}\right)^{-12(10)}$$

$$= \$450.52$$

For a return of 8% compounded monthly, you should pay $450.52 for the bond.

(b) Here, we use formula (6) with $A = \$1000$, $r = 0.07$, and $t = 10$:

$$P = Ae^{-rt}$$

$$= \$1000e^{-(0.07)(10)}$$

$$= \$496.59$$

For a return of 7% compounded continuously, you should pay $496.59 for the bond.

 Now work Problem 11.

E X A M P L E   6

 Rate of Interest Required to Double on Investment

What annual rate of interest compounded annually should you seek if you want to double your investment in 5 years?

Solution   If P is the principal and we want P to double, the amount A will be $2P$. We use the compound interest formula with $n = 1$ and $t = 5$ to find r.

$$2P = P(1 + r)^5$$
$$2 = (1 + r)^5$$
$$1 + r = \sqrt[5]{2}$$
$$r = \sqrt[5]{2} - 1 = 1.148698 - 1 = 0.148698$$

The annual rate of interest needed to double the principal in 5 years is 14.87%.

 Now work Problem 23.

E X A M P L E   7   Doubling and Tripling Time for an Investment

(a) How long will it take for an investment to double in value if it earns 5% compounded continuously?

(b) How long will it take to triple at this rate?

Solution   (a) If P is the initial investment and we want P to double, the amount A will be $2P$. We use formula (4) for continuously compounded interest with $r = 0.05$. Then

$$A = Pe^{rt}$$
$$2P = Pe^{0.05t}$$
$$2 = e^{0.05t}$$
$$0.05t = \ln 2$$
$$t = \frac{\ln 2}{0.05} = 13.86$$

It will take about 14 years to double the investment.

(b) To triple the investment, we set $A = 3P$ in formula (4).

$$A = Pe^{rt}$$
$$3P = Pe^{0.05t}$$
$$3 = e^{0.05t}$$
$$0.05t = \ln 3$$
$$t = \frac{\ln 3}{0.05} = 21.97$$

It will take about 22 years to triple the investment.   ■

 Now work Problem 29.

6.6 | EXERCISES

In Problems 1–10, find the amount that results from each investment.

1. $100 invested at 4% compounded quarterly after a period of 2 years

2. $50 invested at 6% compounded monthly after a period of 3 years

3. $500 invested at 8% compounded quarterly after a period of $2\frac{1}{2}$ years

4. $300 invested at 12% compounded monthly after a period of $1\frac{1}{2}$ years

5. $600 invested at 5% compounded daily after a period of 3 years

6. $700 invested at 6% compounded daily after a period of 2 years

7. $10 invested at 11% compounded continuously after a period of 2 years

8. $40 invested at 7% compounded continuously after a period of 3 years

9. $100 invested at 10% compounded continuously after a period of $2\frac{1}{4}$ years

10. $100 invested at 12% compounded continuously after a period of $3\frac{3}{4}$ years

In Problems 11–20, find the principal needed now to get each amount; that is, find the present value.

11. To get $100 after 2 years at 6% compounded monthly

12. To get $75 after 3 years at 8% compounded quarterly

13. To get $1000 after $2\frac{1}{2}$ years at 6% compounded daily

14. To get $800 after $3\frac{1}{2}$ years at 7% compounded monthly

15. To get $600 after 2 years at 4% compounded quarterly

16. To get $300 after 4 years at 3% compounded daily

17. To get $80 after $3\frac{1}{4}$ years at 9% compounded continuously

18. To get $800 after $2\frac{1}{2}$ years at 8% compounded continuously

19. To get $400 after 1 year at 10% compounded continuously

20. To get $1000 after 1 year at 12% compounded continuously

21. Find the effective rate of interest for $5\frac{1}{4}$% compounded quarterly

22. What interest rate compounded quarterly will give an effective interest rate of 7%?

23. What annual rate of interest is required to double an investment in 3 years?

24. What annual rate of interest is required to double an investment in 10 years?

In Problems 25–28, which of the two rates would yield the larger amount in 1 year?
[**Hint:** Start with a principal of $10,000 in each instance.]

25. 6% compounded quarterly or $6\frac{1}{4}$% compounded annually?

26. 9% compounded quarterly or $9\frac{1}{4}$% compounded annually?

27. 9% compounded monthly or 8.8% compounded daily?

28. 8% compounded semiannually or 7.9% compounded daily?

29. How long does it take for an investment to double in value if it is invested at 8% per annum compounded monthly? Compounded continuously?

30. How long does it take for an investment to double in value if it is invested at 10% per annum compounded monthly? Compounded continuously?

31. If Tanisha has $100 to invest at 8% per annum compounded monthly, how long will it be before she has $150? If the compounding is continuous, how long will it be?

32. If Angela has $100 to invest at 10% per annum compounded monthly, how long will it be before she has $175? If the compounding is continuous, how long will it be?

33. How many years will it take for an initial investment of $10,000 to grow to $25,000? Assume a rate of interest of 6% compounded continuously.

34. How many years will it take for an initial investment of $25,000 to grow to $80,000? Assume a rate of interest of 7% compounded continuously.

35. What will a $90,000 house cost 5 years from now if the inflation rate over that period averages 3% compounded annually?

36. Sears charges 1.25% per month on the unpaid balance for customers with charge accounts (in-

terest is compounded monthly). A customer charges $200 and does not pay her bill for 6 months. What is the bill at that time?

37. Jerome will be buying a new car for $15,000 in 3 years. How much money should he ask his parents for now so that, if he invests it at 5% compounded continuously, he will have enough to buy the car?

38. John will require $3000 in 6 months to pay off a loan that has no prepayment privileges. If he has the $3000 now, how much of it should he save in an account paying 3% compounded monthly so that in 6 months he will have exactly $3000?

39. George is contemplating the purchase of 100 shares of a stock selling for $15 per share. The stock pays no dividends. The history of the stock indicates that it should grow at an annual rate of 15% per year. How much will the 100 shares of stock be worth in 5 years?

40. Tracy is contemplating the purchase of 100 shares of stock selling for $15 per share. The stock pays no dividends. Her broker says the stock will be worth $20 per share in 2 years. What is the annual rate of return on this investment?

41. A business purchased for $650,000 in 1995 is sold in 1998 for $850,000. What is the annual rate of return for this investment?

42. Tanya has just inherited a diamond ring appraised at $5000. If diamonds have appreciated in value at an annual rate of 8%, what was the value of the ring 10 years ago when the ring was purchased?

43. Jim places $1000 in a bank account that pays 5.6% compounded continuously. After 1 year, will he have enough money to buy a computer system that costs $1060? If another bank will pay Jim 5.9% compounded monthly, which is the better deal?

44. On January 1, Kim places $1000 in a Certificate of Deposit (CD) that pays 6.8% compounded continuously and matures in 3 months. Then Kim places the $1000 and the interest in a passbook account that pays 5.25% compounded monthly. How much does Kim have in the passbook account on May 1?

45. Will invests $2000 in a bond trust that pays 9% interest compounded semiannually. His friend Henry invests $2000 in a Certificate of Deposit that pays $8\frac{1}{2}$% compounded continuously. Who has more money after 20 years, Will or Henry?

46. Suppose that April has access to an investment that will pay 10% interest compounded continuously. Which is better: To be given $1000 now so that she can take advantage of this investment opportunity or to be given $1325 after 3 years?

47. Colleen and Bill have just purchased a house for $150,000, with the seller holding a second mortgage of $50,000. They promise to pay the seller $50,000 plus all accrued interest 5 years from now. The seller offers them three interest options on the second mortgage:
  (a) Simple interest at 12% per annum
  (b) $11\frac{1}{2}$% interest compounded monthly
  (c) $11\frac{1}{4}$% interest compounded continuously
  Which option is best; that is, which results in the least interest on the loan?

48. The First National Bank advertises that it pays interest on savings accounts at the rate of 4.25% compounded daily. Find the effective rate if the bank uses (a) 360 days or (b) 365 days in determining the daily rate.

Problems 49–52 involve zero-coupon bonds. A zero-coupon bond is a bond that is sold now at a discount and will pay its face value at some time in the future when it matures; no interest payments are made.

49. A zero-coupon bond can be redeemed in 20 years for $10,000. How much should you be willing to pay for it now if you want a return of:
  (a) 10% compounded monthly?
  (b) 10% compounded continuously?

50. A child's grandparents are considering buying a $40,000 face value zero-coupon bond at birth so that she will have enough money for her college education 17 years later. If thay want a rate of return of 8% compounded annually, what should they pay for the bond?

51. How much should a $10,000 face value zero-coupon bond, maturing in 10 years, be sold for now if its rate of return is to be 8% compounded annually?

52. If Pat pays $12,485.52 for a $25,000 face value zero-coupon bond that matures in 8 years, what is his annual rate of return?

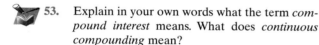 **53.** Explain in your own words what the term *compound interest* means. What does *continuous compounding* mean?

54. Explain in your own words the meaning of present value.

55. Write a program that will calculate the amount after *n* years if a principal *P* is invested at *r*% per annum compounded quarterly. Use it to verify your answers to Problems 1 and 3.

56. Write a program that will calculate the principal needed now to get the amount *A* in *n* years at *r*% per annum compounded daily. Use it to verify your answer to Problem 13.

57. Write a program that will calculate the annual rate of interest required to double an investment in *n* years. Use it to verify your answers to Problems 23 and 24.

58. Write a program that will calculate the number of months required for an initial investment of *x* dollars to grow to *y* dollars at *r*% per annum compounded continuously. Use it to verify your answer to Problems 33 and 34.

59. **Time to Double or Triple an Investment** The formula

$$y = \frac{\ln m}{n \ln\left(1 + \dfrac{r}{n}\right)}$$

can be used to find the number of years *y* required to multiply an investment *m* times when *r* is the per annum interest rate compounded *n* times a year.
  (a) How many years will it take to double the value of an IRA that compounds annually at the rate of 12%?
  (b) How many years will it take to triple the value of a savings account that compounds quarterly at an annual rate of 6%?
  (c) Give a derivation of this formula.

60. Time to Reach an Investment Goal The formula

$$y = \frac{\ln A - \ln P}{r}$$

can be used to find the number of years y required for an investment P to grow to a value A when compounded continuously at an annual rate r.

(a) How long will it take to increase an initial investment of $1000 to $8000 at an annual rate of 10%?

(b) What annual rate is required to increase the value of a $2000 IRA to $30,000 in 35 years?

(c) Give a derivation of this formula.

61. Critical Thinking You have just contracted to buy a house and will seek financing in the amount of $100,000. You go to several banks. Bank 1 will lend you $100,000 at the rate of 8.75% amortized over 30 years with a loan origination fee of 1.75%. Bank 2 will lend you $100,000 at the rate of 8.375% amortized over 15 years with a loan origination fee of 1.5%. Bank 3 will lend you $100,000 at the rate of 9.125% amortized over 30 years with no loan origination fee. Bank 4 will lend you $100,000 at the rate of 8.625% amortized over 15 years with no loan origination fee. Which loan would you take? Why? Be sure to have sound reasons for your choice. If the amount of the monthly payment does not matter to you, which loan would you take? Again, have sound reasons for your choice. Use the information in the table to assist you. Compare your final decision with others in the class. Discuss.

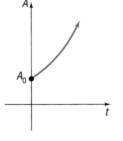

	Monthly Payment	Loan Origination Fee
Bank 1	$786.70	$1,750.00
Bank 2	$977.42	$1,500.00
Bank 3	$813.63	$0.00
Bank 4	$990.68	$0.00

6.7 | GROWTH AND DECAY

1. Find Equations of Populations That Obey the Law of Uninhibited Growth
2. Find Equations of Populations That Obey the Law of Decay
3. Use Newton's Law of Cooling
4. Use Logistic Growth Models

 Many natural phenomena have been found to follow the law that an amount A varies with time t according to

$$A(t) = A_0 e^{kt} \tag{1}$$

where $A_0 = A(0)$ is the original amount ($t = 0$) and $k \neq 0$ is a constant.

If $k > 0$, then equation (1) states that the amount A is increasing over time; if $k < 0$, the amount A is decreasing over time. In either case, when an amount A varies over time according to equation (1), it is said to follow the **exponential law** or the **law of uninhibited growth** ($k > 0$) **or decay** ($k < 0$). See Figure 35.

FIGURE 35

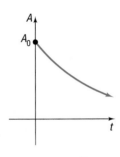

(a) $A(t) = A_0 e^{kt}, k > 0$ **(b)** $A(t) = A_0 e^{kt}, k < 0$

For example, we saw in Section 6.6 that continuously compounded interest follows the law of uninhibited growth. In this section we shall look at three additional phenomena that follow the exponential law.

Biology

Cell division is a universal process in the growth of living organisms such as amoebas, plants, human skin cells, and many others. Based on an ideal situation in which no cells die and no by-products are produced, the number of cells present at a given time follows the law of uninhibited growth. Actually, however, after enough time has passed, growth at an exponential rate will cease due to the influence of factors such as lack of living space, dwindling food supply, and so on. The law of uninhibited growth accurately reflects only the early stages of the cell division process.

The cell division process begins with a culture containing N_0 cells. Each cell in the culture grows for a certain period of time and then divides into two identical cells. We assume that the time needed for each cell to divide in two is constant and does not change as the number of cells increases. These cells then grow, and eventually each divides in two, and so on.

A model that gives the number N of cells in the culture after a time t has passed (in the early stages of growth) is

Uninhibited Growth of Cells
$$N(t) = N_0 e^{kt} \qquad k > 0 \qquad\qquad (2)$$

where k is a positive constant and N_0 is the initial number of cells.

In using equation (2) to model the growth of cells, we are using a function that yields positive real numbers, even though we are counting the number of cells, which must be an integer. This is a common practice in many applications.

E X A M P L E   1

Bacterial Growth

A colony of bacteria increases according to the law of uninhibited growth.

(a) If the number of bacteria doubles in 3 hours, find k in equation (2) and express N as a function of t.

(b) How long will it take for the size of the colony to triple?

(c) How long does it take for the population to double a second time (that is, increase four times).

Solution (a) Using formula (2), the number N of cells at a time t is

$$N(t) = N_0 e^{kt}$$

where N_0 is the initial number of bacteria present and k is a positive number. We seek the number k. The number of cells doubles in 3 hours; thus, we have

$$N(3) = 2N_0$$

But $N(3) = N_0 e^{k(3)}$, so

$$N_0 e^{k(3)} = 2N_0$$
$$e^{3k} = 2$$
$$3k = \ln 2 \qquad \text{Write the exponential equation as a logarithm.}$$
$$k = \tfrac{1}{3} \ln 2 \approx \tfrac{1}{3}(0.6931) = 0.2310$$

Formula (2) for this growth process is therefore

$$N(t) = N_0 e^{0.2310t}$$

(b) The time t needed for the size of the colony to triple requires that $N = 3N_0$. Thus, we substitute $3N_0$ for N to get

$$3N_0 = N_0 e^{0.2310t}$$
$$3 = e^{0.2310t}$$
$$0.2310t = \ln 3$$
$$t = \frac{\ln 3}{0.2310} \approx \frac{1.0986}{0.2310} = 4.756 \text{ hours}$$

It will take about 4.756 hours for the size of the colony to triple.

(c) If the number of bacteria doubles in 3 hours, it will double again in another 3 hours. ▬

Now work Problem 1.

Radioactive Decay

Radioactive materials follow the law of uninhibited decay. Thus, the amount A of a radioactive material present at time t is given by the model

> **Uninhibited Radioactive Decay**
>
> $$A(t) = A_0 e^{kt} \quad k < 0 \tag{3}$$

where $A_0 = A(0)$ is the original amount of radioactive material and k is a negative number.

All radioactive substances have a specific **half-life,** which is the time required for half of the radioactive substance to decay. In **carbon-dating,** we use the fact that all living organisms contain two kinds of carbon, carbon-12 (a stable carbon) and carbon-14 (a radioactive carbon, with a half-life of 5600 years). While an organism is living, the ratio of carbon-12 to carbon-14 is constant. But when an organism dies, the original amount of carbon-12 present remains unchanged, whereas the amount of carbon-14 begins to decrease. This change in the amount of carbon-14 present relative to the amount of carbon-12 present makes it possible to calculate when an organism died.

E X A M P L E   2   Estimating the Age of Ancient Tools

Traces of burned wood found along with ancient stone tools in an archaeo-logical dig in Chile. The wood was found to contain approximately 1.67% of the original amount of carbon-14. If the half-life of carbon-14 is 5600 years, approximately when was the tree cut and burned?

Solution   Using equation (3), the amount A of carbon-14 present at time t is

$$A(t) = A_0 e^{kt}$$

where A_0 is the original amount of carbon-14 present and k is a negative number. We first seek the number k. To find it, we use the fact that after 5600 years half of the original amount of carbon-14 remains. Thus,

$$\frac{1}{2}A_0 = A_0 e^{k(5600)}$$

$$\frac{1}{2} = e^{5600k}$$

$$5600k = \ln\frac{1}{2}$$

$$k = \frac{1}{5600}\ln\frac{1}{2} \approx -0.000124$$

Formula (3) therefore becomes

$$A(t) = A_0 e^{-0.000124t}$$

If the amount A of carbon-14 now present is 1.67% of the original amount, it follows that

$$0.0167A_0 = A_0 e^{-0.000124t}$$
$$0.0167 = e^{-0.000124t}$$
$$-0.000124t = \ln 0.0167$$
$$t = \frac{1}{-0.000124}\ln 0.0167 \approx 33{,}000 \text{ years}$$

The tree was cut and burned about 33,000 years ago. Some archaeologists use this conclusion to argue that humans lived in the Americas 33,000 years ago, much earlier than is generally accepted. ▬

Now work Problem 3.

Newton's Law of Cooling

3   **Newton's Law of Cooling*** states that the temperature of a heated object de-creases exponentially over time toward the temperature of the surrounding medium. That is, the temperature u of a heated object at a given time t can be modeled by the function

*Named after Sir Isaac Newton (1642–1727), one of the cofounders of calculus.

Newton's Law of Cooling

$$u(t) = T + (u_0 - T)e^{kt} \qquad k < 0 \tag{4}$$

where T is the constant temperature of the surrounding medium, u_0 is the initial temperature of the heated object, and k is a negative constant.

E X A M P L E 3   Using Newton's Law of Cooling

An object is heated to 100°C and is then allowed to cool in a room whose air temperature is 30°C.

(a) If the temperature of the object is 80°C after 5 minutes, when will its temperature be 50°C?
(b) Determine the elapsed time before the object is 35°C.
(c) What do you notice about $u(t)$, the temperature, as t, time, increases?

Solution   (a) Using equation (4) with $T = 30$ and $u_0 = 100$, the temperature (in degrees Celsius) of the object at time t (in minutes) is

$$u(t) = 30 + (100 - 30)e^{kt} = 30 + 70e^{kt} \tag{5}$$

where k is a negative constant. To find k, we use the fact that $u = 80$ when $t = 5$. Then

$$80 = 30 + 70e^{k(5)}$$
$$50 = 70e^{5k}$$
$$e^{5k} = \frac{50}{70}$$
$$5k = \ln\frac{5}{7}$$
$$k = \frac{1}{5}\ln\frac{5}{7} \approx -0.0673$$

Formula (5) therefore becomes

$$u(t) = 30 + 70e^{-0.0673t} \tag{6}$$

Now, we want to find t when $u = 50$°C, so

$$50 = 30 + 70e^{-0.0673t}$$
$$20 = 70e^{-0.0673t}$$
$$e^{-0.0673t} = \frac{20}{70}$$
$$-0.0673t = \ln\frac{2}{7}$$
$$t = \frac{1}{-0.0673}\ln\frac{2}{7} \approx 18.6 \text{ minutes}$$

Thus, the temperature of the object will be 50°C after about 18.6 minutes.

(b) If $u = 35°C$, then based on equation (6), we have

$$35 = 30 + 70e^{-0.0673t}$$
$$5 = 70e^{-0.0673t}$$
$$e^{-0.0673t} = \frac{5}{70}$$
$$-0.0673t = \ln\frac{5}{70}$$
$$t = \frac{\ln\frac{5}{70}}{-0.0673} = 39.2 \text{ minutes}$$

The object will reach a temperature of 35°C after about 39.2 minutes.

(c) Refer to equation (6). As t increases, the value of $e^{-0.0673t}$ approaches zero, so the value of $u(t)$ approaches 30°C. ∎

Logistic Models

4 The exponential growth model $A(t) = A_0e^{kt}$, $k > 0$ assumes uninhibited growth, meaning that the value of the function grows without limit. Recall that we stated that cell division could be modeled using this function, assuming that no cells die and no by-products are produced. However, cell division would eventually be limited by factors such as living space, food supply, and so forth. The **logistic growth model** is an exponential function that can model situations where the growth of the dependent variable is limited. The logistic growth model is given by the function

> **Logistic Growth Model**
>
> $$P(t) = \frac{c}{1 + ae^{-bt}}$$

where a, b, and c are constants with $c > 0$ and $b > 0$.

The number c is called the **carrying capacity** because the value $P(t)$ approaches c as t approaches infinity; that is, $\lim_{t \to \infty} P(t) = c$. Thus, c represents the maximum value that the function can attain.

E X A M P L E   4

Fruit Fly Population

Fruit flies are placed in a half-pint milk bottle with a banana (for food) and yeast plants (for food and to provide a stimulus to lay eggs). Suppose that the fruit fly population after t days is given by

$$P(t) = \frac{230}{1 + 56.5e^{-0.37t}}$$

(a) What is the carrying capacity of the half-pint bottle? That is, what is $P(t)$ as $t \to \infty$?

(b) How many fruit flies were initially placed in the half-pint bottle?

(c) When will the population of fruit flies be 180?

(d) Using a graphing utility, graph $P(t)$.

Solution (a) As $t \to \infty$, $e^{-0.37t} \to 0$, and $P(t) \to 230/1$. The carrying capacity of the half-pint bottle is 230 fruit flies.

(b) To find the initial number of fruit flies in the half-pint bottle, we evaluate $P(0)$.

$$P(0) = \frac{230}{1 + 56.5e^{-0.37(0)}}$$

$$= \frac{230}{1 + 56.5}$$

$$= 4$$

Initially, there were four fruit flies in the half-pint bottle.

(c) To determine when the population of fruit flies will be 180, we solve the equation

$$\frac{230}{1 + 56.5e^{-0.37t}} = 180$$

$$230 = 180(1 + 56.5e^{-0.37t})$$

$$1.2778 = 1 + 56.5e^{-0.37t} \qquad \text{Divide each side by 180}$$

$$0.2778 = 56.5e^{-0.37t}$$

$$0.0049 = e^{-0.37t} \qquad \text{Divide each side by 56.5}$$

$$\ln(0.0049) = -0.37t$$

$$t \approx 14.4 \text{ days}$$

It will take approximately 14.4 days for the population to reach 180 fruit flies.

(d) See Figure 36.

FIGURE 36

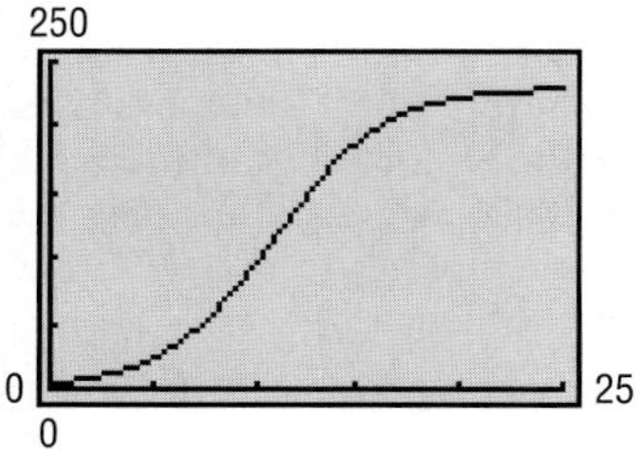

Exploration On the same viewing rectangle, graph $Y_1 = \dfrac{500}{1 + 24e^{-0.03t}}$ and $Y_2 = \dfrac{500}{1 + 24e^{-0.08t}}$. What effect does b have on the logistic growth function?

6.7 | EXERCISES

1. **Growth of an Insect Population** The size P of a certain insect population at time t (in days) obeys the equation $P = 500e^{0.02t}$.
    (a) After how many days will the population reach 1000?
    (b) 2000?

2. **Growth of Bacteria** The number N of bacteria present in a culture at time t (in hours) obeys the equation $N = 1000e^{0.01t}$.
    (a) After how many hours will the population equal 1500?
    (b) 2000?

3. **Radioactive Decay** Strontium-90 is a radioactive material that decays according to the equation $A = A_0e^{-0.0244t}$, where A_0 is the initial

amount present and A is the amount present at time t (in years).
    (a) What is the half-life of strontium-90?
    (b) Determine how long it takes for 100 grams of strontium-90 to decay to 10 grams.

4. **Radioactive Decay** Iodine-131 is a radioactive material that decays according to the equation $A = A_0e^{-0.087t}$, where A_0 is the initial amount present and A is the amount present at time t (in days).
    (a) What is the half-life of iodine-131?
    (b) How long does it take for 100 grams of iodine-131 to decay to 10 grams?

5. **Growth of a Colony of Mosquitoes** The population of a colony of mosquitoes obeys the law

of uninhibited growth. If there are 1000 mosquitoes initially and there are 1800 after 1 day, what is the size of the colony after 3 days? How long is it until there are 10,000 mosquitoes?

6. **Bacterial Growth** A culture of bacteria obeys the law of uninhibited growth. If 500 bacteria are present initially and there are 800 after 1 hour, how many will be present in the culture after 5 hours? How long is it until there are 20,000 bacteria?

7. **Population Growth** The population of a southern city follows an exponential model. If the population doubled in size over an 18-month period and the current population is 10,000, what will the population be 2 years from now?

8. **Population Growth** The population of a midwestern city follows an exponential model. If the population decreased from 900,000 to 800,000 from 1995 to 1997, what will the population be in 1999?

9. **Radioactive Decay** The half-life of radium is 1690 years. If 10 grams are present now, how much will be present in 50 years?

10. **Radioactive Decay** The half-life of radioactive potassium is 1.3 billion years. If 10 grams is present now, how much will be present in 100 years? In 1000 years?

11. **Estimating the Age of a Tree** A piece of charcoal is found to contain 30% of the carbon-14 that it originally had. When did the tree from which the charcoal came die? Use 5600 years as the half-life of carbon-14.

12. **Estimating the Age of a Fossil** A fossilized leaf contains 70% of its normal amount of carbon-14. How old is the fossil?

13. **Cooling Time of a Pizza** A pizza baked at 450°F is removed from the oven at 5:00 PM into a room that is a constant 70°F. After 5 minutes, the pizza is at 300°F.
    (a) At what time can you begin eating the pizza if you want its temperature to be 135°F?
    (b) Determine the time that needs to elapse before the pizza is 160°F?
    (c) What do you notice about the temperature as time passes?

14. **Newton's Law of Cooling** A thermometer reading 72°F is placed in a refrigerator where the temperature is a constant 38°F. If the thermometer reads 60°F after 2 minutes, what will it read after 7 minutes?
    (a) How long will it take before the thermometer reads 39°F?
    (b) Determine the time needed to elapse before the thermometer reads 45°F.
    (c) What do you notice about the temperature as time passes?

15. **Newton's Law of Cooling** A thermometer reading 8°C is brought into a room with a constant temperature of 35°C. If the thermometer reads 15°C after 3 minutes, what will it read after being in the room for 5 minutes? For 10 minutes?
    [**Hint:** You need to construct a formula similar to equation (4).]

16. **Thawing Time of a Steak** A frozen steak has a temperature of 28°F. It is placed in a room with a constant temperature of 70°F. After 10 minutes, the temperature of the steak has risen to 35°F. What will the temperature of the steak be after 30 minutes? How long will it take the steak to thaw to a temperature of 45°F? [See the hint given for Problem 15.]

17. **Decomposition of Salt in Water** Salt (NaCl) decomposes in water into sodium (Na^+) and chloride (Cl^-) ions according to the law of uninhibited decay. If the initial amount of salt is 25 kilograms and, after 10 hours, 15 kilograms of salt is left, how much salt is left after 1 day? How long does it take until $\frac{1}{2}$ kilogram of salt is left?

18. **Voltage of a Conductor** The voltage of a certain conductor decreases over time according to the law of uninhibited decay. If the initial voltage is 40 volts and 2 seconds later it is 10 volts, what is the voltage after 5 seconds?

19. **Radioactivity from Chernobyl** After the release of radioactive material into the atmosphere from a nuclear power plant at Chernobyl (Ukraine) in 1986, the hay in Austria was contaminated by iodine-131 (half-life of 8 years). If it is all right to feed the hay to cows when 10% of the iodine-131 remains, how long do the farmers need to wait to use this hay?

20. **Pig Roasts** The hotel Bora-Bora is having a pig roast. At noon, the chef put the pig in a large earthen oven. The pig's original temperature was 75°F. At 2:00 PM, the chef checked the pig's temperature and was upset because it had reached only 100°F. If the oven's temperature remains a constant 325°F, at what time may the

hotel serve its guests, assuming that pork is done when it reaches 175°F?

21. Proportion of the Population That Owns a VCR The logistic growth model

$$P(t) = \frac{0.9}{1 + 6e^{-0.32t}}$$

relates the proportion of U.S. households that own a VCR to the year. Let $t = 0$ represent 1984, $t = 1$ represent 1985, and so on.
(a) What proportion of U.S. households owned a VCR in 1984?
(b) Determine the maximum proportion of households that will own a VCR.
(c) When will 0.8 (80%) of U.S. households own a VCR?

22. Market Penetration of Intel's Coprocessor The logistic growth model

$$P(t) = \frac{0.90}{1 + 3.5e^{-0.339t}}$$

relates the proportion of new personal computers sold at Best Buy that have Intel's latest coprocessor t months after it has been introduced.
(a) What proportion of new personal computers sold at Best Buy will have Intel's latest coprocessor when it is first introduced (that is, at $t = 0$)?
(b) Determine the maximum proportion of new personal computers sold at Best Buy that will have Intel's latest coprocessor.
(c) When will 0.75 (75%) of new personal computers sold at Best Buy have Intel's latest coprocessor?

23. Population of a Bacteria Culture The logistic growth model

$$P(t) = \frac{1000}{1 + 32.33e^{-0.439t}}$$

represents the population of a bacteria after t hours.
(a) What is the carrying capacity of the environment?
(b) What was the initial amount of bacteria in the population?
(c) When will the amount of bacteria be 800?

24. Population of an Endangered Species Often environmentalists will capture an endangered species and transport the species to a controlled environment where the species can produce offspring and regenerate its population. Suppose that six American Bald Eagles are captured and transported to Montana and set free. Based on experience, the environmentalists expect the population to grow according to the model

$$P(t) = \frac{500}{1 + 83.33e^{-0.162t}}$$

(a) What is the carrying capacity of the environment?
(b) What is the predicted population of the American Bald Eagle in 20 years?
(c) When will the population be 300?

6.8 LOGARITHMIC SCALES

1 Decibels (Loudness of Sound)
2 Richter Scale (Earthquake Magnitude)

Common logarithms often appear in the measurement of quantities, because they provide a way to scale down positive numbers that vary from very small to very large. For example, if a certain quantity can take on values from $0.0000000001 = 10^{-10}$ to $10,000,000,000 = 10^{10}$, the common logarithms of such numbers would be between -10 and 10, respectively.

Loudness of Sound

1 Our first application utilizes a logarithmic scale to measure the loudness of a sound. Physicists define the **intensity of a sound wave** as the amount of energy that the wave transmits through a given area. For example, the least intense sound that a human ear can detect at a frequency of 100 hertz is about 10^{-12} watt per square meter. The **loudness** $L(x)$, measured in **decibels** (named in honor of Alexander Graham Bell), of a sound of intensity x (measured in watts per square meter) is defined as

$$L(x) = 10 \log \frac{x}{I_0} \tag{1}$$

where $I_0 = 10^{-12}$ watt per square meter is the least intense sound that a human ear can detect. If we let $x = I_0$ in equation (1), we get

$$L(I_0) = 10 \log \frac{I_0}{I_0} = 10 \log 1 = 0$$

Thus, at the threshold of human hearing, the loudness is 0 decibels. Figure 37 gives the loudness of some common sounds.

FIGURE 37
Loudness of common sounds (in decibels)

Decibels

Decibels	Sound	Description
140	Shotgun blast, jet 100 feet away at takeoff	Pain
130	Motor test chamber	Human ear pain threshold
120	Firecrackers, severe thunder, pneumatic jackhammer, hockey crowd	Uncomfortably loud
110	Amplified rock music	
100	Textile loom, subway train, elevated train, farm tractor, power lawn mower, newspaper press	Loud
90	Heavy city traffic, noisy factory	
80	Diesel truck going 40 mi/hr 50 feet away, crowded restaurant, garbage disposal, average factory, vacuum cleaner	Moderately loud
70	Passenger car going 50 mi/hr 50 feet away	
60	Quiet typewriter, singing birds, window air conditioner, quiet automobile	Quiet
50	Normal conversation, average office	
40	Household refrigerator, quiet office	Very quiet
30	Average home, dripping faucet, whisper 5 feet away	
20	Light rainfall, rustle of leaves	Average person's threshold of hearing
10	Whisper across room	Just audible
0		Threshold for acute hearing

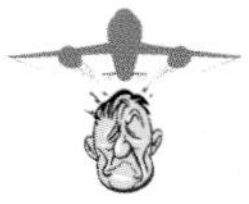

Note that a decibel is not a linear unit like the meter. For example, a noise level of 10 decibels is 10 times as great as a noise level of 0 decibels. [If $L(x) = 10$, then $x = 10I_0$.] A noise level of 20 decibels is 100 times as great as a noise level of 0 decibels. [If $L(x) = 20$, then $x = 100I_0$.] A noise level of 30 decibels is 1000 times as great as a noise level of 0 decibels, and so on.

E X A M P L E   1   Finding the Intensity of a Sound

Use Figure 37 to find the intensity of the sound of a dripping faucet.

Solution   From Figure 37, we see that the loudness of the sound of a dripping faucet is 30 decibels. Thus, by equation (1), its intensity x may be found as follows:

$$30 = 10 \log \left(\frac{x}{I_0}\right)$$

$$3 = \log \left(\frac{x}{I_0}\right) \qquad \text{Divide by 10.}$$

$$\frac{x}{I_0} = 10^3 \qquad \text{Write in exponential form.}$$

$$x = 1000 I_0$$

where $I_0 = 10^{-12}$ watt per square meter. Thus, the intensity of the sound of a dripping faucet is 1000 times as great as a noise level of 0 decibels; that is, such a sound has an intensity of $1000 \cdot 10^{-12} = 10^{-9}$ watt per square meter. ∎

Now work Problem 5.

E X A M P L E   2

Finding the Loudness of a Sound

Use Figure 37 to determine the loudness of a subway train if it is known that this sound is 10 times as intense as the sound due to heavy city traffic.

Solution

The sound due to heavy city traffic has a loudness of 90 decibels. Its intensity, therefore, is the value of x in the equation

$$90 = 10 \log \left(\frac{x}{I_0}\right)$$

A sound 10 times as intense as x has loudness $L(10x)$. Thus, the loudness of the subway train is

$$L(10x) = 10 \log \left(\frac{10x}{I_0}\right) \qquad \text{Substitute } 10x \text{ for } x.$$

$$= 10 \log \left(10 \cdot \frac{x}{I_0}\right)$$

$$= 10 \left[\log 10 + \log \left(\frac{x}{I_0}\right) \right] \qquad \text{Log of product = sum of logs}$$

$$= 10 \log 10 + 10 \log \left(\frac{x}{I_0}\right) \qquad \log 10 = 1; 90 = 10 \log (x/I_0)$$

$$= 10 + 90 = 100 \text{ decibels}$$ ∎

Magnitude of an Earthquake

2   Our second application uses a logarithmic scale to measure the magnitude of an earthquake.

The **Richter scale*** is one way of converting seismographic readings into numbers that provide an easy reference for measuring the magnitude M of an earthquake. All earthquakes are compared to a **zero-level earthquake** whose seismographic reading measures 0.001 millimeter at a distance of 100 kilometers from the epicenter. An earthquake whose seismographic reading measures x millimeters has **magnitude** $M(x)$, given by

*Named after the American scientist, C. F. Richter, who devised it in 1935.

$$M(x) = \log\left(\frac{x}{x_0}\right) \qquad (2)$$

where $x_0 = 10^{-3}$ is the reading of a zero-level earthquake the same distance from its epicenter.

E X A M P L E   3   Finding the Magnitude of an Earthquake

What is the magnitude of an earthquake whose seismographic reading is 0.1 millimeter at a distance of 100 kilometers from its epicenter?

Solution   If $x = 0.1$, the magnitude $M(x)$ of this earthquake is

$$M(0.1) = \log\left(\frac{x}{x_0}\right) = \log\left(\frac{0.1}{0.001}\right) = \log\left(\frac{10^{-1}}{10^{-3}}\right) = \log 10^2 = 2$$

This earthquake thus measures 2.0 on the Richter scale.

Now work Problem 7.

Based on formula (2), we define the **intensity of an earthquake** as the ratio of x to x_0. For example, the intensity of the earthquake described in Example 3 is $\frac{0.1}{0.001} = 10^2 = 100$. That is, it is 100 times as intense as a zero-level earthquake.

E X A M P L E   4   Comparing the Intensity of Two Earthquakes

The devastating San Francisco earthquake of 1906 measured 6.9 on the Richter scale. How did the intensity of that earthquake compare to the Papua, New Guinea, earthquake of 1988, which measured 6.7 on the Richter scale?

FIGURE 38

Solution   Let x_1 and x_2 denote the seismographic readings, respectively, of the 1906 San Francisco earthquake and the 1988 Papua, New Guinea, earthquake. Then, based on formula (2),

$$6.9 = \log\left(\frac{x_1}{x_0}\right) \qquad 6.7 = \log\left(\frac{x_2}{x_0}\right)$$

Consequently,

$$\frac{x_1}{x_0} = 10^{6.9} \qquad \frac{x_2}{x_0} = 10^{6.7}$$

The 1906 San Francisco earthquake was $10^{6.9}$ times as intense as a zero-level earthquake. The Papua, New Guinea, earthquake was $10^{6.7}$ times as intense as a zero-level earthquake. Thus,

$$\frac{x_1}{x_2} = \frac{10^{6.9}x_0}{10^{6.7}x_0} = 10^{0.2} \approx 1.58$$
$$x_1 \approx 1.58 x_2$$

Hence, the San Francisco earthquake was 1.58 times as intense as the Papua, New Guinea, earthquake. ▬

Example 4 demonstrates that the relative intensity of two earthquakes can be found by raising 10 to a power equal to the difference of their readings on the Richter scale.

6.8 | EXERCISES

1. **Loudness of a Dishwasher** Find the loudness of a dishwasher that operates at an intensity of 10^{-5} watt per square meter. Express your answer in decibels.

2. **Loudness of a Diesel Engine** Find the loudness of a diesel engine that operates at an intensity of 10^{-3} watt per square meter. Express your answer in decibels.

3. **Loudness of a Jet Engine** With engines at full throttle, a Boeing 727 jetliner produces noise at an intensity of 0.15 watt per square meter. Find the loudness of the engines in decibels.

4. **Loudness of a Whisper** A whisper produces noise at an intensity of $10^{-9.8}$ watt per square meter. What is the loudness of a whisper in decibels?

5. **Intensity of a Sound at the Threshold of Pain** For humans, the threshold of pain due to sound averages 130 decibels. What is the intensity of such a sound in watts per square meter?

6. **Comparing Sounds** If one sound is 50 times as intense as another, what is the difference in the loudness of the two sounds? Express your answer in decibels.

7. **Magnitude of an Earthquake** Find the magnitude of an earthquake whose seismographic reading is 10.0 millimeters at a distance of 100 kilometers from its epicenter.

8. **Magnitude of an Earthquake** Find the magnitude of an earthquake whose seismographic reading is 1210 millimeters at a distance of 100 kilometers from its epicenter.

9. **Comparing Earthquakes** The Mexico City earthquake of 1985 registered 8.1 on the Richter scale. What would a seismograph 100 kilometers from the epicenter have measured for this earthquake? How does this earthquake compare in intensity to the 1906 San Francisco earthquake, which registered 6.9 on the Richter scale?

10. **Comparing Earthquakes** Two earthquakes differ by 1.0 when measured on the Richter scale. How would the seismographic readings differ at a distance of 100 kilometers from the epicenter? How do their intensities compare?

11. **NBA Finals 1997** In game 5 of the NBA Finals between the Chicago Bulls and the Utah Jazz at the Delta Center, the crowd noise measured 110 decibels. NBA guidelines say sound levels are not to exceed 95 decibels. Compute the ratio of the intensities of these two sounds to determine by how much the crowd noise exceeded guidelines.

CHAPTER REVIEW

THINGS TO KNOW

One-to-one function f

If $x_1 \neq x_2$, then $f(x_1) \neq f(x_2)$ for any choice of x_1 and x_2 in the domain.

Horizontal line test

If horizontal lines intersect the graph of a function f in at most one point, then f is one-to-one.

Inverse function f^{-1} of f

Domain of f = Range of f^{-1};   Range of f = Domain of f^{-1}.

$f^{-1}(f(x)) = x$ and $f(f^{-1}(x)) = x$.

Graphs of f and f^{-1} are symmetric with respect to the line $y = x$.

Properties of the exponential function

$f(x) = a^x, \quad a > 1$ Domain: $(-\infty, \infty)$;   Range: $(0, \infty)$; x-intercepts: none; y-intercept: 1;   horizontal asymptote: x-axis as $x \to -\infty$;   increasing;   one-to-one

See Figure 11 for a typical graph.

$f(x) = a^x, \quad 0 < a < 1$ Domain: $(-\infty, \infty)$;   Range: $(0, \infty)$; x-intercepts: none; y-intercept: 1;   horizontal asymptote: x-axis as $x \to \infty$;   decreasing;   one-to-one

See Figure 14 for a typical graph.

Properties of the logarithmic function

$f(x) = \log_a x, \quad a > 1$
$(y = \log_a x \text{ means } x = a^y)$ Domain: $(0, \infty)$;   Range: $(-\infty, \infty)$; x-intercept: 1; y-intercept: none;   vertical asymptote: y-axis;   increasing;   one-to-one

See Figure 20(b) for a typical graph.

$f(x) = \log_a x, \quad 0 < a < 1$
$(y = \log_a x \text{ means } x = a^y)$ Domain: $(0, \infty)$;   Range: $(-\infty, \infty)$; x-intercept: 1; y-intercept: none;   vertical asymptote: y-axis;   decreasing;   one-to-one

See Figure 20(a) for a typical graph.

Number e

Value approached by the expression $\left(1 + \dfrac{1}{n}\right)^n$ as $n \to \infty$; that is, $\lim\limits_{n \to \infty} \left(1 + \dfrac{1}{n}\right)^n = e$

Natural logarithm

$y = \ln x$ means $x = e^y$

Properties of logarithms

$\log_a 1 = 0 \qquad \log_a a = 1 \qquad a^{\log_a M} = M \qquad \log_a a^r = r$

$\log_a (MN) = \log_a M + \log_a N \qquad \log_a\left(\dfrac{M}{N}\right) = \log_a M - \log_a N \qquad \log_a\left(\dfrac{1}{N}\right) = -\log_a N \qquad \log_a M^r = r \log_a M$

FORMULAS

Change-of-Base Formula $\log_a M = \dfrac{\log_b M}{\log_b a}$

Compound interest $A = P\left(1 + \dfrac{r}{n}\right)^{nt}$

Continuous compounding $A = Pe^{rt}$

Present value $P = A\left(1 + \dfrac{r}{n}\right)^{-nt}$ or $P = Ae^{-rt}$

Growth and decay $A(t) = A_0 e^{kt}$

HOW TO

Find the inverse of certain one-to-one functions (see the procedure given on page 419).

Graph f^{-1} given the graph of f.

Graph exponential and logarithmic functions.

Solve certain exponential equations.

Solve certain logarithmic equations.

Solve problems involving compound interest.

Solve problems involving growth and decay.

Solve problems involving intensity of sound and intensity of earthquakes.

FILL-IN-THE-BLANK ITEMS

1. If every horizontal line intersects the graph of a function f at no more than one point, then f is a(n) _____ function.

2. If f^{-1} denotes the inverse of a function f, then the graphs of f and f^{-1} are symmetric with respect to the line _____.

3. The graph of every exponential function $f(x) = a^x$, $a > 0$, $a \neq 1$, passes through the two points _____.

4. If the graph of an exponential function $f(x) = a^x$, $a > 0$, $a \neq 1$, is decreasing, then its base must be less than _____.

5. If $3^x = 3^4$, then $x =$ _____.

6. The logarithm of a product equals the _____ of the logarithms.

7. For every base, the logarithm of _____ equals 0.

8. If $\log_8 M = \log_5 7/\log_5 8$, then $M =$ _____.

9. The domain of the logarithmic function $f(x) = \log_a x$ consists of _____.

10. The graph of every logarithmic function $f(x) = \log_a x$, $a > 0$, $a \neq 1$, passes through the two points _____.

11. If the graph of a logarithmic function $f(x) = \log_a x, a > 0, a \neq 1$, is increasing, then its base must be larger than _____.

12. If $\log_3 x = \log_3 7$, then $x =$ _____.

TRUE/FALSE ITEMS

T F **1.** If f and g are inverse functions, then the domain of f is the same as the domain of g.

T F **2.** If f and g are inverse functions, then their graphs are symmetric with respect to the line $y = x$.

T F **3.** The graph of every exponential function $f(x) = a^x, a > 0, a \neq 1$, will contain the points $(0, 1)$ and $(1, a)$.

T F **4.** The graphs of $y = 3^{-x}$ and $y = \left(\frac{1}{3}\right)^x$ are identical.

T F **5.** The present value of $1000 to be received after 2 years at 10% per annum compounded continuously is approximately $1205.

T F **6.** If $y = \log_a x$, then $y = a^x$.

T F **7.** The graph of every logarithmic function $f(x) = \log_a x, a > 0, a \neq 1$, will contain the points $(1, 0)$ and $(a, 1)$.

T F **8.** $a^{\log_M a} = M$, where $a > 0, a \neq 1, M > 0$.

T F **9.** $\log_a (M + N) = \log_a M + \log_a N$, where $a > 0, a \neq 1, M > 0, N > 0$.

T F **10.** $\log_a M - \log_a N = \log_a(M/N)$, where $a > 0, a \neq 1, M > 0, N > 0$.

REVIEW EXERCISES

Blue problem numbers indicate the author's suggestions for use in a Practice Test.

In Problems 1–6, the function f is one-to-one. Find the inverse of each function and check your answer. Find the domain and range of f and f^{-1}.

1. $f(x) = \dfrac{2x + 3}{5x - 2}$

2. $f(x) = \dfrac{2 - x}{3 + x}$

3. $f(x) = \dfrac{1}{x - 1}$

4. $f(x) = \sqrt{x - 2}$

5. $f(x) = \dfrac{3}{x^{1/3}}$

6. $f(x) = x^{1/3} + 1$

In Problems 7–12, evaluate each expression. Do not use a calculator.

7. $\log_2\left(\frac{1}{8}\right)$ **8.** $\log_3 81$ **9.** $\ln e^{\sqrt{2}}$ **10.** $e^{\ln 0.1}$ **11.** $2^{\log_2 0.4}$ **12.** $\log_2 2^{\sqrt{3}}$

In Problems 13–18, write each expression as a single logarithm.

13. $3\log_4 x^2 + \frac{1}{2}\log_4 \sqrt{x}$

14. $-2\log_3\left(\frac{1}{x}\right) + \frac{1}{3}\log_3 \sqrt{x}$

15. $\ln\left(\frac{x-1}{x}\right) + \ln\left(\frac{x}{x+1}\right) - \ln(x^2-1)$

16. $\log(x^2-9) - \log(x^2+7x+12)$

17. $2\log 2 + 3\log x - \frac{1}{2}[\log(x+3) + \log(x-2)]$

18. $\frac{1}{2}\ln(x^2+1) - 4\ln\frac{1}{2} - \frac{1}{2}[\ln(x-4) + \ln x]$

In Problems 19–26, find y as a function of x. The constant C is a positive number.

19. $\ln y = 2x^2 + \ln C$

20. $\ln(y-3) = \ln 2x^2 + \ln C$

21. $\frac{1}{2}\ln y = 3x^2 + \ln C$

22. $\ln 2y = \ln(x+1) + \ln(x+2) + \ln C$

23. $\ln(y-3) + \ln(y+3) = x + C$

24. $\ln(y-1) + \ln(y+1) = -x + C$

25. $e^{y+C} = x^2 + 4$

26. $e^{3y-C} = (x+4)^2$

In Problems 27–36, use transformations to graph each function. Determine the domain, range, and any asymptotes.

27. $f(x) = e^{-x}$ **28.** $f(x) = \ln(-x)$ **29.** $f(x) = 1 - e^x$ **30.** $f(x) = 3 + \ln x$

31. $f(x) = 3e^x$ **32.** $f(x) = \frac{1}{2}\ln x$ **33.** $f(x) = e^{|x|}$ **34.** $f(x) = \ln|x|$

35. $f(x) = 3 - e^{-x}$ **36.** $f(x) = 4 - \ln(-x)$

In Problems 37–56, solve each equation.

37. $4^{1-2x} = 2$ **38.** $8^{6+3x} = 4$ **39.** $3^{x^2+x} = \sqrt{3}$

40. $4^{x-x^2} = \frac{1}{2}$ **41.** $\log_x 64 = -3$ **42.** $\log_{\sqrt{2}} x = -6$

43. $5^x = 3^{x+2}$ **44.** $5^{x+2} = 7^{x-2}$ **45.** $9^{2x} = 27^{3x-4}$

46. $25^{2x} = 5^{x^2-12}$ **47.** $\log_3 \sqrt{x-2} = 2$ **48.** $2^{x+1} \cdot 8^{-x} = 4$

49. $8 = 4^{x^2} \cdot 2^{5x}$ **50.** $2^x \cdot 5 = 10^x$ **51.** $\log_6(x+3) + \log_6(x+4) = 1$

52. $\log_{10}(7x-12) = 2\log_{10} x$ **53.** $e^{1-x} = 5$ **54.** $e^{1-2x} = 4$

55. $2^{3x} = 3^{2x+1}$ **56.** $2^{x^3} = 3^{x^2}$

In Problems 57–60, use the following result: If x is the atmospheric pressure (measured in millimeters of mercury), then the formula for the altitude h(x) (measured in meters above sea level) is

$$h(x) = (30T + 8000)\log\left(\frac{P_0}{x}\right)$$

where T is the temperature (in degrees Celsius) and P_0 is the atmospheric pressure at sea level, which is approximately 760 millimeters of mercury.

57. Finding the Altitude of an Airplane At what height is a Piper Cub whose instruments record an outside temperature of 0°C and a barometric pressure of 300 millimeters of mercury?

58. Finding the Height of a Mountain How high is a mountain if instruments placed on its peak record a temperature of 5°C and a barometric pressure of 500 millimeters of mercury?

59. Atmospheric Pressure Outside an Airplane What is the atmospheric pressure outside a Boeing 737 flying at an altitude of 10,000 meters if the outside air temperature is −100°C?

60. Atmospheric Pressure at High Altitudes What is the atmospheric pressure (in millimeters of mercury) on Mt. Everest, which has an altitude of approximately 8900 meters, if the air temperature is 5°C?

61. Amplifying Sound An amplifier's power output P (in watts) is related to its decibel voltage gain d by the formula $P = 25e^{0.1d}$.

(a) Find the power output for a decibel voltage gain of 4 decibels.

(b) For a power output of 50 watts, what is the decibel voltage gain?

62. **Limiting Magnitude of a Telescope** A telescope is limited in its usefulness by the brightness of the star it is aimed at and by the diameter of its lens. One measure of a star's brightness is its *magnitude:* the dimmer the star, the larger its magnitude. A formula for the limiting magnitude L of a telescope, that is, the magnitude of the dimmest star that it can be used to view, is given by

$$L = 9 + 5.1 \log d$$

where d is the diameter (in inches) of the lens.

(a) What is the limiting magnitude of a 3.5-inch telescope?

(b) What diameter is required to view a star of magnitude 14?

63. **Product Demand** The demand for a new product increases rapidly at first and then levels off. The percent P of actual purchases of this product after it has been on the market t months is

$$P = 90 - 80\left(\frac{3}{4}\right)^t$$

(a) What is the percent of purchases of the product after 5 months?

(b) What is the percent of purchases of the product after 10 months?

(c) What is the maximum percent of purchases of the product?

(d) How many months does it take before 40% of purchases occur?

(e) How many months before 70% of purchases occur?

64. **Disseminating Information** A survey of a certain community of 10,000 residents shows that the number of residents N who have heard a piece of information after m months is given by the formula

$$m = 55.3 - 6 \ln(10,000 - N)$$

How many months will it take for half of the citizens to learn about a community program of free blood pressure readings?

65. **Salvage Value** The number of years n for a piece of machinery to depreciate to a known salvage value can be found using the formula

$$n = \frac{\log s - \log i}{\log (1 - d)}$$

where s is the salvage value of the machinery, i is its initial value, and d is the annual rate of depreciation.

(a) How many years will it take for a piece of machinery to decline in value from $90,000 to $10,000 if the annual rate of depreciation is 0.20 (20%)?

(b) How many years will it take for a piece of machinery to lose half of its value if the annual rate of depreciation is 15%?

66. **Funding a College Education** A child's grandparents purchase a $10,000 bond fund that matures in 18 years to be used for her college education. The bond fund pays 4% interest compounded semiannually. How much will the bond fund be worth at maturity?

67. **Funding a College Education** A child's grandparents wish to purchase a bond fund that matures in 18 years to be used for her college education. The bond fund pays 4% interest compounded semiannually. How much should they purchase so that the bond fund will be worth $85,000 at maturity?

68. **Funding an IRA** First Colonial Bankshares Corporation advertised the following IRA investment plans.

Target IRA Plans

For each $5000 Maturity Value Desired	
Deposit:	At a Term of:
$620.17	20 Years
$1045.02	15 Years
$1760.92	10 Years
$2967.26	5 Years

(a) Assuming continuous compounding, what was the annual rate of interest they offered?

(b) First Colonial Bankshares claims that $4000 invested today will have a value of over $32,000 in 20 years. Use the answer found in part (a) to find the actual value of $4000 in 20 years. Assume continuous compounding.

69. **Loudness of a Garbage Disposal** Find the loudness of a garbage disposal unit that operates at an intensity of 10^{-4} watt per square meter. Express your answer in decibels.

70. **Comparing Earthquakes** On September 9, 1985, the western suburbs of Chicago experienced a mild earthquake that registered 3.0 on the Richter scale. How did this earthquake compare in intensity to the great San Francisco earthquake of 1906, which registered 6.9 on the Richter scale?

71. Estimating the Date on Which a Prehistoric Man Died The bones of a prehistoric man found in the desert of New Mexico contain approximately 5% of the original amount of carbon-14. If the half-life of carbon-14 is 5600 years, approximately how long ago did the man die?

72. Temperature of a Skillet A skillet is removed from an oven whose temperature is 450°F and placed in a room whose temperature is 70°F. After 5 minutes, the temperature of the skillet is 400°F. How long will it be until its temperature is 150°F?

73. Biology A certain bacteria initially increases according to the law of uninhibited growth. A biologist collects the following data for this bacteria:

Time (Hours)	Population
0	1000
1	1415
2	2000
3	2828
4	4000
5	5656
6	8000

(a) Draw a scatter diagram.

(b) The exponential function of best fit to the data is found to be

$$y = 1000(\sqrt{2})^x$$

Express the function of best fit in the form $N = N_0 e^{kt}$.

(c) Use the solution to (b) to predict the population at $t = 7$ hours.

 (d) Use a graphing utility to verify the exponential function of best fit.

(e) Use a graphing utility to draw a scatter diagram of the data and then graph the exponential function of best fit on it.

74. Finance The following data represent the amount of money an investor has in an investment account each year for 10 years. She wishes to determine the effective rate of return on her investment.

Year	Value of Account
1985	$10,000
1986	$10,573
1987	$11,260
1988	$11,733
1989	$12,424
1990	$13,269
1991	$13,968
1992	$14,823
1993	$15,297
1994	$16,539

(a) Draw a scatter diagram with time as the independent variable and the value of the account as the dependent variable.

(b) The exponential function of best fit to the data is found to be

$$y = 10,014(1.057)^x$$

Express the function of best fit in the form $A = A_0 e^{kt}$.

(c) Use the solution to (b) to estimate the value of the account in the year 2020.

(d) Use a graphing utility to verify the exponential function of best fit.

(e) Use a graphing utility to draw a scatter diagram of the data and then graph the exponential function of best fit on it.

CHAPTER 7

The Conics

Comets have received quite a bit of attention in the recent years, some of them being visible in the night time sky for many weeks. On the following page, you discover an Unidentified Flying Object (UFO) and need to determine whether or not the next important comet will be named after *you*! You will need to use the Sullivan website at:

www.prenhall.com/sullivan

to help link you to the mathematical software and astronomy resources you will need to determine your claim.

PREPARING FOR THIS CHAPTER

Before getting started on this chapter, review the following concepts:

Rectangular Coordinates *(pp. 163–164)*

Distance Formula *(p. 164)*

Completing the Square *(pp. 110–112)*

Intercepts *(pp. 177–178)*

Symmetry *(pp. 179–180)*

Circles *(pp. 201–204)*

OUTLINE

THE STARDUST MISSION

Aas a budding amateur astronomer you hope that you might discover some new object in space. After many late nights sitting with your telescope you observe an object that you have never seen before, and, according to your list of observable objects, has never been reported! You do not want to let your hopes get too high, since there are many other astronomers that are closely watching the night skies, and yet . . . you cannot help but imagine a comet named after you! You begin to wonder if your space object might even *collide* with the earth.

In order to make an *orbit diagram,* you realize that for the next few weeks you will need to keep careful records of your UFO's position. After only ten evenings, you send your data to the *Minor Planet Center* (MPC) supercomputer. MPC replies with the message that you have discovered an already known comet called Wild2, and in fact this is the comet that has been selected by NASA for *The Stardust Mission,* NASA's newest Discovery Mission, which will fly by a comet and bring back a collection of cosmic history.

Even though this comet has previously been discovered you experienced some of the excitement that comes with exploring space. Since your measurements were found to be very exact, the Jet Propulsion Laboratory (JPL) asks you to be an astrometer for the Stardust Mission.

1. You jump on the Internet and go to the *CBAT* (Central Bureau for Astronomical Telegrams) webpage to see if the ephemeris of your space object matches with any of the known space objects. What information will you need in order to determine if your object will hit the earth?
2. You want to *report* your NEO (Near Earth Object) but you want to do most of the *astrometry* yourself. You plan to use the software package Astrometrica on your computer. The equations of celestial mechanics are still a little bit tedious. What are the units of measurement in the ephemeris?
3. As you collect data on Wild2, you send it in to the Smithsonian Center for Astrophysics. You would like to estimate how fast Wild2 is traveling. Using the fixed stars as a background reference, what other information would you need to estimate the velocity of Wild2?
4. Refer to the table of ephemeris comets. Using the data given there, an astronomer could write an equation that describes the orbit of Wild2:

$$\frac{x^2}{a^2} + \frac{y^2}{b^2} = 1$$

5. Although you have the equation for the path of Wild2, you cannot determine its velocity from this equation alone. What other data would you need in order to calculate Wild2's velocity? Does this velocity change?
6. Review Kepler's laws. In particular, if you knew the angle in degrees at the sun made by the change in position of Wild2 after ten days could you estimate it's velocity? Do you think that the JPL scientists could use your equation to send the Stardust satellite on its intercept?

Historically, Apollonius (200 BC) was among the first to study *conics* and discover some of their interesting properties. Today, conics are still studied because of their many uses. *Paraboloids of revolution* (parabolas rotated about their axes of symmetry) are used as signal collectors (the satellite dishes used with radar and cable TV, for example), as solar energy collectors, and as reflectors (telescopes, light projection, and so on).

The planets circle the Sun in approximately *elliptical* orbits. Elliptical surfaces can be used to reflect signals such as light and sound from one place to another. And *hyperbolas* can be used to determine the positions of ships at sea.

The Greeks used the methods of Euclidean geometry to study conics. We shall use the more powerful methods of analytic geometry, bringing to bear both algebra and geometry, for our study of conics. Thus, we shall give a geometric description of each conic, and then, using rectangular coordinates and the distance formula, we shall find equations that represent conics. We used this same development, you may recall, when we first defined a circle in Section 3.4.

7.1 | PRELIMINARIES

1 Know the names of the conics

1 The word *conic* derives from the word *cone*, which is a geometric figure that can be constructed in the following way: Let *a* and *g* be two distinct lines that intersect at a point *V*. Keep the line *a* fixed. Now rotate the line *g* about *a* while maintaining the same angle between *a* and *g*. The collection of points swept out (generated) by the line *g* is called a **(right circular) cone.** See Figure 1. The fixed line *a* is called the **axis** of the cone; the point *V* is called its **vertex;** the lines that pass through *V* and make the same angle with *a* as *g* are called **generators** of the cone. Thus, each generator is a line that lies entirely on the cone. The cone consists of two parts, called **nappes,** that intersect at the vertex.

Conics, an abbreviation for **conic sections,** are curves that result from the intersection of a (right circular) cone and a plane. The conics we shall study arise when the plane does not contain the vertex, as shown in Figure 2. These conics are **circles** when the plane is perpendicular to the axis of the cone and intersects each generator; **ellipses** when the plane is tilted slightly so that it intersects each generator, but intersects only one nappe of the cone; **parabolas** when the plane is tilted further so that it is parallel to one (and only one) generator and intersects only one nappe of the cone; and **hyperbolas** when the plane intersects both nappes.

If the plane does contain the vertex, the intersection of the plane and the cone is a point, a line, or a pair of intersecting lines. These are usually called **degenerate conics.**

FIGURE 1

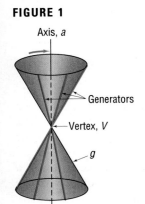

Axis, *a*

Generators

Vertex, *V*

g

FIGURE 2

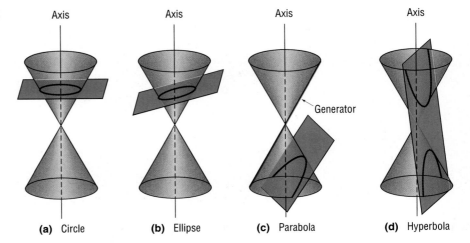

(a) Circle   **(b)** Ellipse   **(c)** Parabola   **(d)** Hyperbola

7.2 | THE PARABOLA

1. Find the Equation of a Parabola
2. Graph Parabolas
3. Discuss the Equation of a Parabola
4. Work with Parabolas with Vertex at (h, k)
5. Solve Applied Problems Involving Properties of Parabolas

We stated earlier (Section 5.1) that the graph of a quadratic function is a parabola. In this section, we begin with a geometric definition of a parabola and use it to obtain an equation.

A **parabola** is defined as the collection of all points P in the plane that are the same distance from a fixed point F as they are from a fixed line D. The point F is called the **focus** of the parabola, and the line D is its **directrix**. As a result, a parabola is the set of points P for which

$$d(F, P) = d(P, D) \qquad (1)$$

FIGURE 3

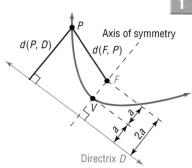

Figure 3 shows a parabola. The line through the focus F and perpendicular to the directrix D is called the **axis of symmetry** of the parabola. The point of intersection of the parabola with its axis of symmetry is called the **vertex** V.

Because the vertex V lies on the parabola, it must satisfy equation (1): $d(F, V) = d(V, D)$. Thus, the vertex is midway between the focus and the directrix. We shall let a equal the distance $d(F, V)$ from F to V. Now we are ready to derive an equation for a parabola. To do this, we use a rectangular system of coordinates positioned so that the vertex V, focus F, and directrix D of the parabola are conveniently located. If we choose to locate the vertex V at the origin $(0, 0)$, then we can conveniently position the focus F on either the x-axis or the y-axis.

First, we consider the case where the focus F is on the positive x-axis, as shown in Figure 4. Because the distance from F to V is a, the coordinates of F will be $(a, 0)$ with $a > 0$. Similarly, because the distance from V to the directrix D is also a and because D must be perpendicular to the x-axis (since the x-axis is the axis of symmetry), the equation of the directrix D must be $x = -a$. Now, if $P = (x, y)$ is any point on the parabola, then P must obey equation (1):

$$d(F, P) = d(P, D)$$

So we have

FIGURE 4
$y^2 = 4ax$

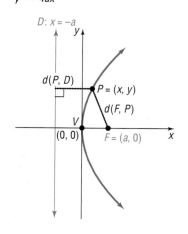

$$\sqrt{(x - a)^2 + y^2} = |x + a| \qquad \text{Use the distance formula.}$$
$$(x - a)^2 + y^2 = (x + a)^2 \qquad \text{Square both sides.}$$
$$x^2 - 2ax + a^2 + y^2 = x^2 + 2ax + a^2 \qquad \text{Simplify.}$$
$$y^2 = 4ax$$

> **Theorem**   Equation of a Parabola; Vertex at (0, 0), Focus at (*a*, 0),
> *a* > 0
>
> The equation of a parabola with vertex at (0, 0), focus at (*a*, 0), and
> directrix *x* = −*a*, *a* > 0, is
>
> $$y^2 = 4ax \qquad\qquad (2)$$

E X A M P L E   1

FIGURE 5
$y^2 = 12x$

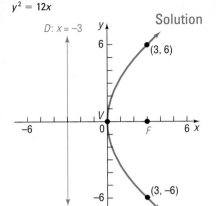

Finding the Equation of a Parabola

Find an equation of the parabola with vertex at (0, 0) and focus at (3, 0).
Graph the equation.

Solution The distance from the vertex (0, 0) to the focus (3, 0) is *a* = 3. Based on equation (2), the equation of this parabola is

$$y^2 = 4ax$$
$$y^2 = 12x \quad a = 3$$

To graph this parabola, it is helpful to plot the two points on the graph above and below the focus. To locate them, we let *x* = 3. Then

$$y^2 = 12x = 36$$
$$y = \pm 6$$

The points on the parabola above and below the focus are (3, −6) and (3, 6). These points help in graphing the parabola because they determine the "opening" of the graph. See Figure 5.

In general, the points on a parabola $y^2 = 4ax$ that lie above and below the focus (*a*, 0) are each at a distance 2*a* from the focus. This follows from the fact that if *x* = *a* then $y^2 = 4ax = 4a^2$, or *y* = ±2*a*. The line segment joining these two points is called the **latus rectum**; its length is 4*a*.

Comment   To graph the parabola $y^2 = 12x$ discussed in Example 1, we need to graph the two functions $Y_1 = \sqrt{12x}$ and $Y_2 = -\sqrt{12x}$. Do this and compare what you see with Figure 5.

Now work Problem 9.

By reversing the steps we used to obtain equation (2), it follows that the graph of an equation of the form of equation (2), $y^2 = 4ax$, is a parabola; its vertex is at (0, 0), its focus is at (*a*, 0), its directrix is the line *x* = −*a*, and its axis of symmetry is the *x*-axis.

For the remainder of this section, the direction "Discuss the equation" will mean to find the vertex, focus, and directrix of the parabola and graph it.

E X A M P L E   2

Discussing the Equation of a Parabola

Discuss the equation: $y^2 = 8x$

Solution The equation $y^2 = 8x$ is of the form $y^2 = 4ax$, where 4*a* = 8. Thus, *a* = 2. Consequently, the graph of the equation is a parabola with vertex at (0, 0) and

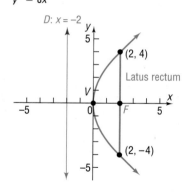

focus on the positive x-axis at $(2, 0)$. The directrix is the vertical line $x = -2$. The two points defining the latus rectum are obtained by letting $x = 2$. Then $y^2 = 16$, or $y = \pm 4$. See Figure 6. ∎

Recall that we arrived at equation (2) after placing the focus on the positive x-axis. If the focus is placed on the negative x-axis, positive y-axis, or negative y-axis, a different form of the equation for the parabola results. The four forms of the equation of a parabola with vertex at $(0, 0)$ and focus on a coordinate axis a distance a from $(0, 0)$ are given in Table 1, and their graphs are given in Figure 7. Notice that each graph is symmetric with respect to its axis of symmetry.

TABLE 1	Equations of a Parabola: Vertex at (0, 0); Focus on Axis; $a > 0$			
Vertex	**Focus**	**Directrix**	**Equation**	**Description**
$(0, 0)$	$(a, 0)$	$x = -a$	$y^2 = 4ax$	Parabola, axis of symmetry is the x-axis, opens to right
$(0, 0)$	$(-a, 0)$	$x = a$	$y^2 = -4ax$	Parabola, axis of symmetry is the x-axis, opens to left
$(0, 0)$	$(0, a)$	$y = -a$	$x^2 = 4ay$	Parabola, axis of symmetry is the y-axis, opens up
$(0, 0)$	$(0, -a)$	$y = a$	$x^2 = -4ay$	Parabola, axis of symmetry is the y-axis, opens down

FIGURE 7

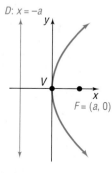

(a) $y^2 = 4ax$

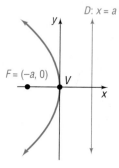

(b) $y^2 = -4ax$

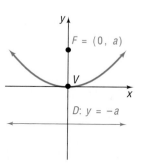

(c) $x^2 = 4ay$

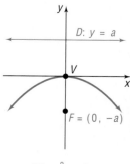

(d) $x^2 = -4ay$

E X A M P L E   3   Discussing the Equation of a Parabola

Discuss the equation: $x^2 = -12y$

Solution   The equation $x^2 = -12y$ is of the form $x^2 = -4ay$, with $a = 3$. Consequently, the graph of the equation is a parabola with vertex at $(0, 0)$, focus at $(0, -3)$, and directrix the line $y = 3$. The parabola opens down, and its axis of symmetry is the y-axis. To obtain the points defining the latus rectum, let $y = -3$. Then $x^2 = 36$, or $x = \pm 6$. See Figure 8.

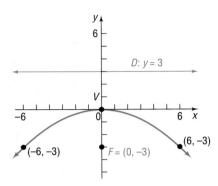

FIGURE 8
$x^2 = -12y$

Now work **Problem 27.**

E X A M P L E   4

Finding the Equation of a Parabola

Find the equation of the parabola with focus at $(0, 4)$ and directrix the line $y = -4$. Graph the equation.

Solution   A parabola whose focus is at $(0, 4)$ and whose directrix is the horizontal line $y = -4$ will have its vertex at $(0, 0)$. (Do you see why? The vertex is mid-way between the focus and the directrix.) Thus, the equation of this parabola is of the form $x^2 = 4ay$, with $a = 4$; that is,

$$x^2 = 16y$$

Figure 9 shows the graph.

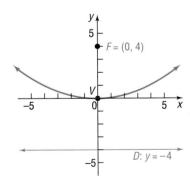

FIGURE 9
$x^2 = 16y$

E X A M P L E   5

Finding the Equation of a Parabola

Find the equation of a parabola with vertex at $(0, 0)$ if its axis of symmetry is the x-axis and its graph passes through the point $(-\frac{1}{2}, 2)$. Find its focus and directrix, and graph the equation.

Solution Because the vertex is at the origin, the axis of symmetry is the x-axis, and the graph passes through a point in the second quadrant, we see from Table 1 that the form of the equation is

$$y^2 = -4ax$$

Because the point $(-\frac{1}{2}, 2)$ is on the parabola, the coordinates $x = -\frac{1}{2}$, $y = 2$ must satisfy the equation. Substituting $x = -\frac{1}{2}$ and $y = 2$ in the equation, we find

$$4 = -4a(-\tfrac{1}{2})$$
$$a = 2$$

Thus, the equation of the parabola is

$$y^2 = -8x$$

The focus is at $(-2, 0)$ and the directrix is the line $x = 2$. Letting $x = -2$, we find $y^2 = 16$ or $y = \pm 4$. The points $(-2, 4)$ and $(-2, -4)$ define the latus rectum. See Figure 10.

FIGURE 10
$y^2 = -8x$

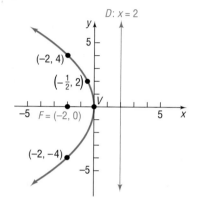

  Now work Problem 19.

Vertex at (h, k)

4 If a parabola with vertex at the origin and axis of symmetry along a coordinate axis is shifted horizontally h units and then vertically k units, the result is a parabola with a vertex at (h, k) and axis of symmetry parallel to a coordinate axis. The equations of such parabolas have the same forms as those in Table 1, but with x replaced by $x - h$ (the horizontal shift) and y replaced by $y - k$ (the vertical shift). Table 2 gives the forms of the equations of such parabolas. Figure 11(a)–(d) illustrates the graphs for $h > 0, k > 0$.

TABLE 2	Parabolas with Vertex at (h, k), Axis of Symmetry Parallel to a Coordinate Axis, $a > 0$			
Vertex	**Focus**	**Directrix**	**Equation**	**Description**
(h, k)	$(h + a, k)$	$x = -a + h$	$(y - k)^2 = 4a(x - h)$	Parabola, axis of symmetry parallel to x-axis, opens to right
(h, k)	$(h - a, k)$	$x = a + h$	$(y - k)^2 = -4a(x - h)$	Parabola, axis of symmetry parallel to x-axis, opens to left
(h, k)	$(h, k + a)$	$y = -a + k$	$(x - h)^2 = 4a(y - k)$	Parabola, axis of symmetry parallel to y-axis, opens up
(h, k)	$(h, k - a)$	$y = a + k$	$(x - h)^2 = -4a(y - k)$	Parabola, axis of symmetry parallel to y-axis, opens down

FIGURE 11

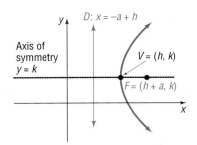
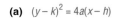

(a) $(y - k)^2 = 4a(x - h)$

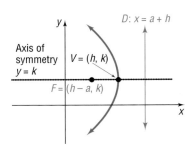
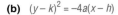

(b) $(y - k)^2 = -4a(x - h)$

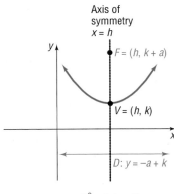

(c) $(x - h)^2 = 4a(y - k)$

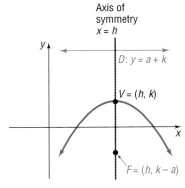

(d) $(x - h)^2 = -4a(y - k)$

E X A M P L E   6

Finding the Equation of a Parabola, Vertex Not at Origin

Find an equation of the parabola with vertex at $(-2, 3)$ and focus at $(0, 3)$. Graph the equation.

Solution   The vertex $(-2, 3)$ and focus $(0, 3)$ both lie on the horizontal line $y = 3$ (the axis of symmetry). The distance a from $(-2, 3)$ to $(0, 3)$ is $a = 2$. Also, because the focus lies to the right of the vertex, we know that the parabola opens to the right. Consequently, the form of the equation is

$$(y - k)^2 = 4a(x - h)$$

where $(h, k) = (-2, 3)$ and $a = 2$. Therefore, the equation is

$$(y - 3)^2 = 4 \cdot 2[x - (-2)]$$
$$(y - 3)^2 = 8(x + 2)$$

If $x = 0$, then $(y - 3)^2 = 16$. Thus, $y - 3 = \pm 4$ and $y = -1$ or $y = 7$. The points $(0, -1)$ and $(0, 7)$ define the latus rectum; the line $x = -4$ is the directrix. See Figure 12 (p. 502).

FIGURE 12
$(y - 3)^2 = 8(x + 2)$

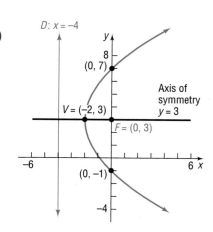

Now work Problem 17.

Polynomial equations define parabolas whenever they involve two variables that are quadratic in one variable and linear in the other. To discuss this type of equation, we first complete the square of the variable that is quadratic.

E X A M P L E   7

Discussing the Equation of a Parabola

Discuss the equation: $x^2 + 4x - 4y = 0$

Solution   To discuss the equation $x^2 + 4x - 4y = 0$, we complete the square involving the variable x. Thus,

FIGURE 13
$x^2 + 4x - 4y = 0$

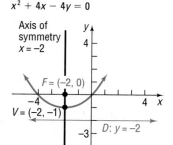

$$
\begin{aligned}
x^2 + 4x - 4y &= 0 \\
x^2 + 4x &= 4y && \text{Isolate the terms involving } x \text{ on the left side.} \\
x^2 + 4x + 4 &= 4y + 4 && \text{Complete the square on the left side.} \\
(x + 2)^2 &= 4(y + 1) && \text{Form: } (x - h)^2 = 4a(y - k).
\end{aligned}
$$

This equation is of the form $(x - h)^2 = 4a(y - k)$, with $h = -2$, $k = -1$, and $a = 1$. The graph is a parabola with vertex at $(h, k) = (-2, -1)$ that opens up. The focus is at $(-2, 0)$, and the directrix is the line $y = -2$. See Figure 13.

Now work Problem 35.

5   Parabolas find their way into many applications. For example, as we discussed in Section 5.1, suspension bridges have cables in the shape of a parabola. Another property of parabolas that is used in applications is their reflecting property.

Reflecting Property

Suppose that a mirror is shaped like a **paraboloid of revolution,** a surface formed by rotating a parabola about its axis of symmetry. If a light (or any other emitting source) is placed at the focus of the parabola, all the rays

FIGURE 14
Searchlight

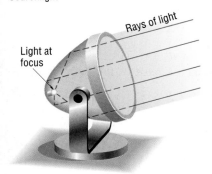

emanating from the light will reflect off the mirror in lines parallel to the axis of symmetry. This principle is used in the design of searchlights, flashlights, certain automobile headlights, and other such devices. See Figure 14.

Conversely, suppose that rays of light (or other signals) emanate from a distant source so that they are essentially parallel. When these rays strike the surface of a parabolic mirror whose axis of symmetry is parallel to these rays, they are reflected to a single point at the focus. This principle is used in the design of some solar energy devices, satellite dishes, and the mirrors used in some types of telescopes. See Figure 15.

FIGURE 15

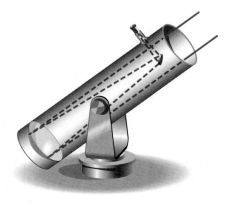

EXAMPLE 8

Satellite Dish

A satellite dish is shaped like a paraboloid of revolution. The signals that emanate from a satellite strike the surface of the dish and are reflected to a single point, where the receiver is located. If the dish is 8 feet across at its opening and is 3 feet deep at its center, at what position should the receiver be placed?

FIGURE 16

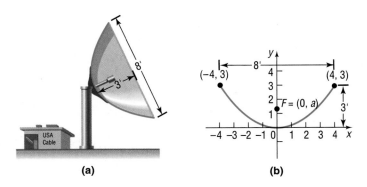

(a)   (b)

Solution   Figure 16(a) shows the satellite dish. We draw the parabola used to form the dish on a rectangular coordinate system so that the vertex of the parabola is at the origin and its focus is on the positive y-axis. See Figure 16(b). The form of the equation of the parabola is

$$x^2 = 4ay$$

and its focus is at $(0, a)$. Since $(4, 3)$ is a point on the graph, we have

$$4^2 = 4a(3)$$

$$a = \frac{4}{3}$$

The receiver should be located $1\frac{1}{3}$ feet from the base of the dish, along its axis of symmetry.

   Now work Problem 51.

7.2 EXERCISES

In Problems 1–8, the graph of a parabola is given. Match each graph to its equation.

A. $y^2 = 4x$ B. $x^2 = 4y$ C. $y^2 = -4x$

D. $x^2 = -4y$ E. $(y - 1)^2 = 4(x - 1)$ F. $(x + 1)^2 = 4(y + 1)$

G. $(y - 1)^2 = -4(x - 1)$ H. $(x + 1)^2 = -4(y + 1)$

1.
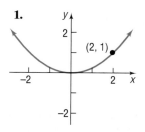
(2, 1)

2.
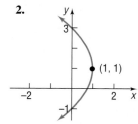
(1, 1)

3.
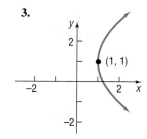
(1, 1)

4.
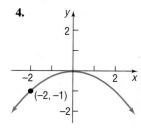
(-2, -1)

5.
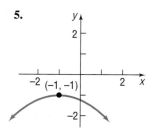
(-1, -1)

6.
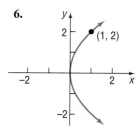
(1, 2)

7.
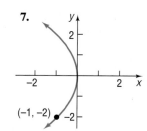
(-1, -2)

8.
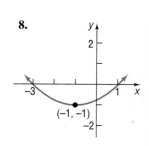
(-1, -1)

In Problems 9–24, find the equation of the parabola described. Find the two points that define the latus rectum, and graph the equation.

9. Focus at $(4, 0)$; vertex at $(0, 0)$

10. Focus at $(0, 2)$; vertex at $(0, 0)$

11. Focus at $(0, -3)$; vertex at $(0, 0)$

12. Focus at $(-4, 0)$; vertex at $(0, 0)$

13. Focus at $(-2, 0)$; directrix the line $x = 2$

14. Focus at $(0, -1)$; directrix the line $y = 1$

15. Directrix the line $y = -\frac{1}{2}$; vertex at $(0, 0)$

16. Directrix the line $x = -\frac{1}{2}$; vertex at $(0, 0)$

17. Vertex at $(2, -3)$; focus at $(2, -5)$

18. Vertex at $(4, -2)$; focus at $(6, -2)$

19. Vertex at $(0, 0)$; axis of symmetry the y-axis; passing through the point $(2, 3)$

20. Vertex at $(0, 0)$; axis of symmetry the x-axis; passing through the point $(2, 3)$

21. Focus at $(-3, 4)$; directrix the line $y = 2$

22. Focus at $(2, 4)$; directrix the line $x = -4$

23. Focus at $(-3, -2)$; directrix the line $x = 1$

24. Focus at $(-4, 4)$; directrix the line $y = -2$

In Problems 25–42, find the vertex, focus, and directrix of each parabola. Graph the equation.

25. $x^2 = 4y$

26. $y^2 = 8x$

27. $y^2 = -16x$

28. $x^2 = -4y$

29. $(y - 2)^2 = 8(x + 1)$

30. $(x + 4)^2 = 16(y + 2)$

31. $(x - 3)^2 = -(y + 1)$

32. $(y + 1)^2 = -4(x - 2)$

33. $(y + 3)^2 = 8(x - 2)$

34. $(x - 2)^2 = 4(y - 3)$

35. $y^2 - 4y + 4x + 4 = 0$

36. $x^2 + 6x - 4y + 1 = 0$

37. $x^2 + 8x = 4y - 8$

38. $y^2 - 2y = 8x - 1$

39. $y^2 + 2y - x = 0$

40. $x^2 - 4x = 2y$

41. $x^2 - 4x = y + 4$

42. $y^2 + 12y = -x + 1$

In Problems 43–50, write an equation for each parabola.

43.
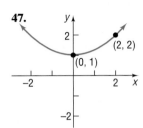

44.
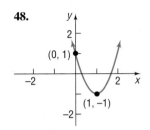

45.
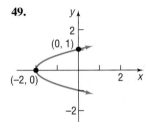

46.
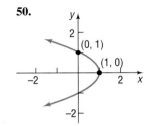

47.
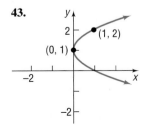

48.
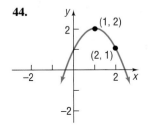

49.
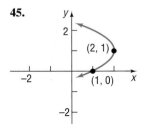

50.
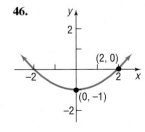

51. Satellite Dish A satellite dish is shaped like a paraboloid of revolution. The signals that emanate from a satellite strike the surface of the dish and are reflected to a single point, where the receiver is located. If the dish is 10 feet across at its opening and is 4 feet deep at its center, at what position should the receiver be placed?

52. Constructing a TV Dish A cable TV receiving dish is in the shape of a paraboloid of revolution. Find the location of the receiver, which is placed at the focus, if the dish is 6 feet across at its opening and 2 feet deep.

53. Constructing a Flashlight The reflector of a flashlight is in the shape of a paraboloid of revolution. Its diameter is 4 inches and its depth is 1 inch. How far from the vertex should the light bulb be placed so that the rays will be reflected parallel to the axis?

54. Constructing a Headlight A sealed-beam headlight is in the shape of a paraboloid of revolution. The bulb, which is placed at the focus, is 1 inch from the vertex. If the depth is to be 2 inches, what is the diameter of the headlight at its opening?

55. Suspension Bridges The cables of a suspension bridge are in the shape of a parabola, as shown in the figure. The towers supporting the cable are 600 feet apart and 80 feet high. If the cables touch the road surface midway between the towers, what is the height of the cable at a point 150 feet from the middle of the bridge?

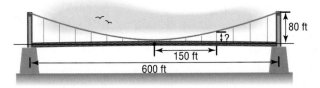
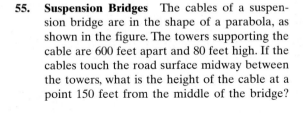

56. Suspension Bridges The cables of a suspension bridge are in the shape of a parabola. The towers supporting the cable are 400 feet apart and 100 feet high. If the cables are at a height of 10 feet midway between the towers, what is the height of the cable at a point 50 feet from a tower?

57. Searchlights A searchlight is shaped like a paraboloid of revolution. If the light source is located 2 feet from the base along the axis of symmetry and the opening is 5 feet across, how deep should the searchlight be?

58. **Searchlights** A searchlight is shaped like a paraboloid of revolution. If the light source is located 2 feet from the base along the axis of symmetry and the depth of the searchlight is 4 feet, what should the width of the opening be?

59. **Solar Heat** A mirror is shaped like a paraboloid of revolution and will be used to concentrate the rays of the sun at its focus, creating a heat source. If the mirror is 20 feet across at its opening and is 6 feet deep, where will the heat source be concentrated?

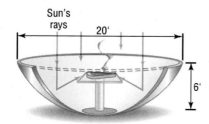

60. **Reflecting Telescopes** A reflecting telescope contains a mirror shaped like a paraboloid of revolution. If the mirror is 4 inches across at its opening and is 3 feet deep, where will the light collected be concentrated?

61. **Parabolic Arch Bridge** A bridge is to be built in the shape of a parabolic arch. The bridge has a span of 120 feet and a maximum height of 25 feet. See the illustration. Choose a suitable rectangular coordinate system and find the height of the arch at distances of 10, 30, and 50 feet from the center.

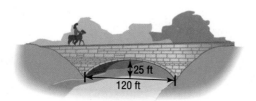

62. **Parabolic Arch Bridge** A bridge is to be built in the shape of a parabolic arch and is to have a span of 100 feet. The height of the arch a distance of 40 feet from the center is to be 10 feet. Find the height of the arch at its center.

63. Show that an equation of the form

$$Ax^2 + Ey = 0 \qquad A \neq 0, E \neq 0$$

is the equation of a parabola with vertex at $(0, 0)$ and axis of symmetry the y-axis. Find its focus and directrix.

64. Show that an equation of the form

$$Cy^2 + Dx = 0 \qquad C \neq 0, D \neq 0$$

is the equation of a parabola with vertex at $(0, 0)$ and axis of symmetry the x-axis. Find its focus and directrix.

65. Show that the graph of an equation of the form

$$Ax^2 + Dx + Ey + F = 0 \qquad A \neq 0$$

(a) Is a parabola if $E \neq 0$.
(b) Is a vertical line if $E = 0$ and $D^2 - 4AF = 0$.
(c) Is two vertical lines if $E = 0$ and $D^2 - 4AF > 0$.
(d) Contains no points if $E = 0$ and $D^2 - 4AF < 0$.

66. Show that the graph of an equation of the form

$$Cy^2 + Dx + Ey + F = 0 \qquad C \neq 0$$

(a) Is a parabola if $D \neq 0$.
(b) Is a horizontal line if $D = 0$ and $E^2 - 4CF = 0$.
(c) Is two horizontal lines if $D = 0$ and $E^2 - 4CF > 0$.

7.3 | THE ELLIPSE

1	Find the Equation of an Ellipse
2	Graph Ellipses
3	Discuss the Equation of an Ellipse
4	Work with Ellipses with Center at (h, k)
5	Solve Applied Problems Involving Properties of Ellipses

An **ellipse** is the collection of all points in the plane the sum of whose distances from two fixed points, called the **foci**, is a constant.

FIGURE 17

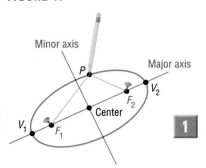

Minor axis

Major axis

V_2

Center

V_1

F_1

F_2

P

FIGURE 18
$d(F_1, P) + d(F_2, P) = 2a$

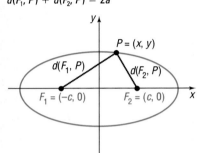

The definition actually contains within it a physical means for drawing an ellipse. Find a piece of string (the length of this string is the constant referred to in the definition). Then take two thumbtacks (the foci) and stick them on a piece of cardboard so that the distance between them is less than the length of the string. Now attach the ends of the string to the thumbtacks and, using the point of a pencil, pull the string taut. Keeping the string taut, rotate the pencil around the two thumbtacks. The pencil traces out an ellipse, as shown in Figure 17.

In Figure 17, the foci are labeled F_1 and F_2. The line containing the foci is called the **major axis.** The midpoint of the line segment joining the foci is called the **center** of the ellipse. The line through the center and perpendicular to the major axis is called the **minor axis.**

The two points of intersection of the ellipse and the major axis are the **vertices,** V_1 and V_2, of the ellipse. The distance from one vertex to the other is called the **length of the major axis.** The ellipse is symmetric with respect to its major axis and with respect to its minor axis.

With these ideas in mind, we are now ready to find the equation of an ellipse in a rectangular coordinate system. First, we place the center of the ellipse at the origin. Second, we position the ellipse so that its major axis coincides with a coordinate axis. Suppose that the major axis coincides with the x-axis, as shown in Figure 18. If c is the distance from the center to a focus, then one focus will be at $F_1 = (-c, 0)$ and the other at $F_2 = (c, 0)$. As we shall see, it is convenient to let $2a$ denote the constant distance referred to in the definition. Then, if $P = (x, y)$ is any point on the ellipse, we have

$d(F_1, P) + d(F_2, P) = 2a$	Sum of the distances from P to the foci equals a constant.
$\sqrt{(x + c)^2 + y^2} + \sqrt{(x - c)^2 + y^2} = 2a$	Use the distance formula.
$\sqrt{(x + c)^2 + y^2} = 2a - \sqrt{(x - c)^2 + y^2}$	Isolate one radical.
$(x + c)^2 + y^2 = 4a^2 - 4a\sqrt{(x - c)^2 + y^2}$ $+ (x - c)^2 + y^2$	Square both sides.
$x^2 + 2cx + c^2 + y^2 = 4a^2 - 4a\sqrt{(x - c)^2 + y^2}$ $+ x^2 - 2cx + c^2 + y^2$	Simplify.
$4cx - 4a^2 = -4a\sqrt{(x - c)^2 + y^2}$	Isolate the radical.
$cx - a^2 = -a\sqrt{(x - c)^2 + y^2}$	Divide each side by 4.
$c^2x^2 - 2a^2cx + a^4 = a^2[(x - c)^2 + y^2]$	Square both sides again.
$c^2x^2 - 2a^2cx + a^4 = a^2(x^2 - 2cx + c^2 + y^2)$	
$(c^2 - a^2)x^2 - a^2y^2 = a^2c^2 - a^4$	
$(a^2 - c^2)x^2 + a^2y^2 = a^2(a^2 - c^2)$	Multiply each side by -1; factor a^2 on the right side. $\quad(1)$

To obtain points on the ellipse off the x-axis, it must be that $a > c$. To see why, look again at Figure 18.

$d(F_1, P) + d(F_2, P) > d(F_1, F_2)$	The sum of the lengths of two sides of a triangle is greater than the length of the third side.
$2a > 2c$	$d(F_1, P) + d(F_2, P) = 2a; \quad d(F_1, F_2) = 2c$
$a > c$	

Since $a > c$, we also have $a^2 > c^2$, so $a^2 - c^2 > 0$. Let $b^2 = a^2 - c^2$, $b > 0$. Then $a > b$ and equation (1) can be written as

$$b^2x^2 + a^2y^2 = a^2b^2$$

$$\frac{x^2}{a^2} + \frac{y^2}{b^2} = 1 \qquad \text{Divide each side by } a^2b^2.$$

Theorem   **Equation of an Ellipse; Center at (0, 0); Foci at ($\pm c$, 0); Major Axis along the x-Axis**

An equation of the ellipse with center at $(0, 0)$ and foci at $(-c, 0)$ and $(c, 0)$ is

$$\frac{x^2}{a^2} + \frac{y^2}{b^2} = 1 \qquad \text{where } a > b > 0 \text{ and } b^2 = a^2 - c^2 \qquad (2)$$

The major axis is the x-axis.

FIGURE 19

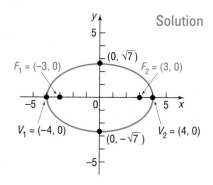

As you can verify, the ellipse defined by equation (2) is symmetric with respect to the x-axis, y-axis, and origin.

To find the vertices of the ellipse defined by equation (2), let $y = 0$. The vertices satisfy the equation $x^2/a^2 = 1$, the solutions of which are $x = \pm a$. Consequently, the vertices of the ellipse given by equation (2) are $V_1 = (-a, 0)$ and $V_2 = (a, 0)$. The y-intercepts of the ellipse, found by letting $x = 0$, have coordinates $(0, -b)$ and $(0, b)$. These four intercepts, $(a, 0)$, $(-a, 0)$, $(0, b)$, and $(0, -b)$, are used to graph the ellipse. See Figure 19.

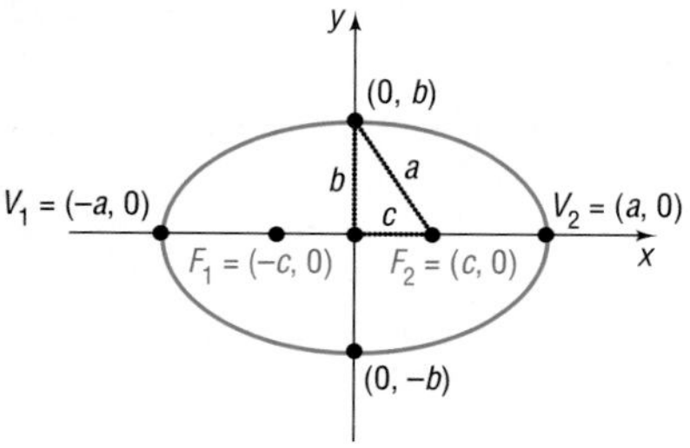

Notice in Figure 19 the right triangle formed with the points $(0, 0)$, $(c, 0)$, and $(0, b)$. Because $b^2 = a^2 - c^2$ (or $b^2 + c^2 = a^2$), the distance from the focus at $(c, 0)$ to the point $(0, b)$ is a.

E X A M P L E   1   **Finding an Equation of an Ellipse**

FIGURE 20
$$\frac{x^2}{16} + \frac{y^2}{7} = 1$$

Find an equation of the ellipse with center at the origin, one focus at $(3, 0)$, and a vertex at $(-4, 0)$. Graph the equation.

Solution   The ellipse has its center at the origin, and the major axis coincides with the x-axis. One focus is at $(c, 0) = (3, 0)$, so $c = 3$. One vertex is at $(-a, 0) = (-4, 0)$, so $a = 4$. From equation (2), it follows that

$$b^2 = a^2 - c^2 = 16 - 9 = 7$$

so an equation of the ellipse is

$$\frac{x^2}{16} + \frac{y^2}{7} = 1$$

Figure 20 shows the graph.

Notice in Figure 20 how we used the intercepts of the equation to graph the ellipse. Following this practice will make it easier for you to obtain an accurate graph of an ellipse.

Comment   The intercepts of the ellipse also provide information about how to set the viewing rectangle. To graph the ellipse

$$\frac{x^2}{16} + \frac{y^2}{7} = 1$$

discussed in Example 1, we would set the viewing rectangle using a square screen that includes the intercepts, perhaps $-6 \le x \le 6$, $-4 \le y \le 4$. Then we would graph the two functions

$$Y_1 = -\sqrt{7}\sqrt{1 - \frac{x^2}{16}}, \qquad Y_2 = \sqrt{7}\sqrt{1 - \frac{x^2}{16}}$$

Do this and compare what you see with Figure 20.   ∎

 Now work Problem 15.

An equation of the form of equation (2), with $a > b$, is the equation of an ellipse with center at the origin, foci on the x-axis at $(-c, 0)$ and $(c, 0)$, where $c^2 = a^2 - b^2$, and major axis along the x-axis.

For the remainder of this section, the direction "Discuss the equation" will mean to find the center, major axis, foci, and vertices of the ellipse and graph it.

E X A M P L E   2

Discussing the Equation of an Ellipse

Discuss the equation: $\dfrac{x^2}{25} + \dfrac{y^2}{9} = 1$

Solution   The given equation is of the form of equation (2), with $a^2 = 25$ and $b^2 = 9$. The equation is that of an ellipse with center $(0, 0)$ and major axis along the x-axis. The vertices are at $(\pm a, 0) = (\pm 5, 0)$. Because $b^2 = a^2 - c^2$, we find

$$c^2 = a^2 - b^2 = 25 - 9 = 16$$

The foci are at $(\pm c, 0) = (\pm 4, 0)$. Figure 21 shows the graph.

FIGURE 21
$\dfrac{x^2}{25} + \dfrac{y^2}{9} = 1$

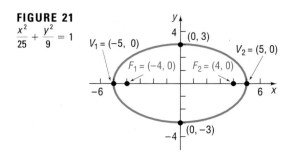

Now work Problem 5.

If the major axis of an ellipse with center at $(0, 0)$ coincides with the y-axis, then the foci are at $(0, -c)$ and $(0, c)$. Using the same steps as before, the definition of an ellipse leads to the following result:

Theorem   Equation of an Ellipse; Center at (0, 0); Foci at (0, ±c); Major Axis along the y-Axis

An equation of the ellipse with center at $(0, 0)$ and foci at $(0, -c)$ and $(0, c)$ is

$$\frac{x^2}{b^2} + \frac{y^2}{a^2} = 1 \qquad \text{where } a > b > 0 \text{ and } b^2 = a^2 - c^2 \qquad (3)$$

The major axis is the y-axis; the vertices are at $(0, -a)$ and $(0, a)$.

FIGURE 22

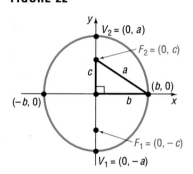

Figure 22 illustrates the graph of such an ellipse. Again, notice the right triangle with the points at $(0, 0)$, $(b, 0)$, and $(0, c)$.

Look closely at equations (2) and (3). Although they may look alike, there is a difference! In equation (2), the larger number, a^2, is in the denominator of the x^2 term, so the major axis of the ellipse is along the x-axis. In equation (3), the larger number, a^2, is in the denominator of the y^2 term, so the major axis is along the y-axis.

E X A M P L E   3

FIGURE 23

$$x^2 + \frac{y^2}{9} = 1$$

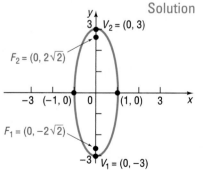

Discussing the Equation of an Ellipse

Discuss the equation: $9x^2 + y^2 = 9$

Solution   To put the equation in proper form, we divide each side by 9:

$$x^2 + \frac{y^2}{9} = 1$$

The larger number, 9, is in the denominator of the y^2 term so, based on equation (3), this is the equation of an ellipse with center at the origin and major axis along the y-axis. Also, we conclude that $a^2 = 9$, $b^2 = 1$, and $c^2 = a^2 - b^2 = 9 - 1 = 8$. The vertices are at $(0, \pm a) = (0, \pm 3)$, and the foci are at $(0, \pm c) = (0, \pm 2\sqrt{2})$. The graph is given in Figure 23.

E X A M P L E   4

Finding an Equation of an Ellipse

Find an equation of the ellipse having one focus at $(0, 2)$ and vertices at $(0, -3)$ and $(0, 3)$. Graph the equation.

FIGURE 24
$$\frac{x^2}{5} + \frac{y^2}{9} = 1$$

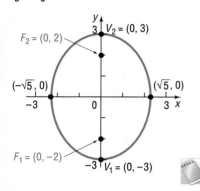

Solution

Because the vertices are at $(0, -3)$ and $(0, 3)$, the center of this ellipse is at the origin. Also, its major axis coincides with the y-axis. The given information also reveals that $c = 2$ and $a = 3$, so $b^2 = a^2 - c^2 = 9 - 4 = 5$. The form of the equation of this ellipse is given by equation (3):

$$\frac{x^2}{b^2} + \frac{y^2}{a^2} = 1$$

$$\frac{x^2}{5} + \frac{y^2}{9} = 1$$

Figure 24 shows the graph. ∎

Now work Problems 9 and 17.

The circle may be considered a special kind of ellipse. To see why, let $a = b$ in equation (2) or in equation (3). Then

$$\frac{x^2}{a^2} + \frac{y^2}{a^2} = 1$$

$$x^2 + y^2 = a^2$$

This is the equation of a circle with center at the origin and radius a. The value of c is

$$c^2 = a^2 - b^2 = 0$$

We conclude that the closer the two foci of an ellipse are the more the ellipse will look like a circle.

Center at (h, k)

4 If an ellipse with center at the origin and major axis coinciding with a coordinate axis is shifted horizontally h units and then vertically k units, the result is an ellipse with center at (h, k) and major axis parallel to a coordinate axis. The equations of such ellipses have the same forms as those given in Equations (2) and (3), except that x is replaced by $x - h$ (the horizontal shift) and y is replaced by $y - k$ (the vertical shift). Table 3 gives the forms of the equations of such ellipses, and Figure 25 shows their graphs.

TABLE 3	Ellipses with Center at (h, k) and Major Axis Parallel to a Coordinate Axis			
Center	**Major Axis**	**Foci**	**Vertices**	**Equation**
(h, k)	Parallel to x-axis	$(h \pm c, k)$	$(h \pm a, k)$	$\dfrac{(x - h)^2}{a^2} + \dfrac{(y - k)^2}{b^2} = 1$, $a > b$ and $b^2 = a^2 - c^2$
(h, k)	Parallel to y-axis	$(h, k \pm c)$	$(h, k \pm a)$	$\dfrac{(x - h)^2}{b^2} + \dfrac{(y - k)^2}{a^2} = 1$, $a > b$ and $b^2 = a^2 - c^2$

FIGURE 25

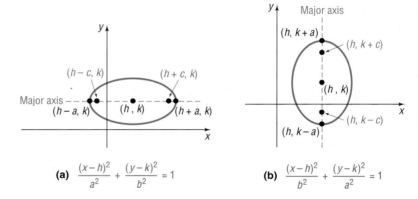

(a) $\dfrac{(x-h)^2}{a^2} + \dfrac{(y-k)^2}{b^2} = 1$ (b) $\dfrac{(x-h)^2}{b^2} + \dfrac{(y-k)^2}{a^2} = 1$

E X A M P L E 5

FIGURE 26

$\dfrac{(x-2)^2}{9} + \dfrac{(y+3)^2}{8} = 1$

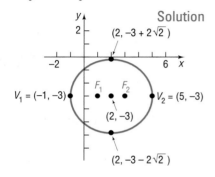

Finding an Equation of an Ellipse, Center Not at the Origin

Find an equation for the ellipse with center at $(2, -3)$, one focus at $(3, -3)$, and one vertex at $(5, -3)$. Graph the equation.

Solution   The center is at $(h, k) = (2, -3)$, so $h = 2$ and $k = -3$. The major axis is parallel to the x-axis. The distance from the center $(2, -3)$ to a focus $(3, -3)$ is $c = 1$; the distance from the center $(2, -3)$ to a vertex $(5, -3)$ is $a = 3$. Thus, $b^2 = a^2 - c^2 = 9 - 1 = 8$. The form of the equation is

$$\frac{(x-h)^2}{a^2} + \frac{(y-k)^2}{b^2} = 1 \qquad \text{where } h = 2, k = -3, a = 3, b = 2\sqrt{2}$$

$$\frac{(x-2)^2}{9} + \frac{(y+3)^2}{8} = 1$$

Figure 26 shows the graph. ∎

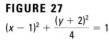 **Now work Problem 41.**

E X A M P L E 6

FIGURE 27

$(x-1)^2 + \dfrac{(y+2)^2}{4} = 1$

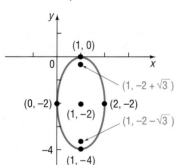

Discussing the Equation of an Ellipse

Discuss the equation: $\quad 4x^2 + y^2 - 8x + 4y + 4 = 0$

Solution   We proceed to complete the squares in x and y:

$$\begin{aligned}
4x^2 + y^2 - 8x + 4y + 4 &= 0 \\
4x^2 - 8x + y^2 + 4y &= -4 \\
4(x^2 - 2x) + (y^2 + 4y) &= -4 \\
4(x^2 - 2x + 1) + (y^2 + 4y + 4) &= -4 + 4 + 4 \\
4(x-1)^2 + (y+2)^2 &= 4 \\
(x-1)^2 + \frac{(y+2)^2}{4} &= 1
\end{aligned}$$

Group like variables; place the constant on the right.

Factor out 4.

Complete each square.

Divide each side by 4.

    This is the equation of an ellipse with center at $(1, -2)$ and major axis parallel to the y-axis. Since $a^2 = 4$ and $b^2 = 1$, we have $c^2 = a^2 - b^2 = 4 - 1 = 3$. The vertices are at $(h, k \pm a) = (1, -2 \pm 2)$ or $(1, 0)$ and $(1, -4)$. The foci are at $(h, k \pm c) = (1, -2 \pm \sqrt{3})$ or $(1, -2 - \sqrt{3})$ and $(1, -2 + \sqrt{3})$. Figure 27 shows the graph. ∎

Now work Problem 29.

Applications

5

Ellipses are found in many applications in science and engineering. For example, the orbits of the planets around the Sun are elliptical, with the Sun's position at a focus. See Figure 28.

FIGURE 28

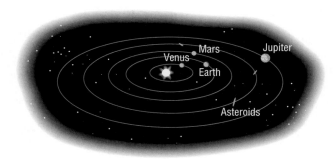

Stone and concrete bridges are often shaped as semielliptical arches. Elliptical gears are used in machinery when a variable rate of motion is required.

Ellipses also have an intersecting reflection property. If a source of light (or sound) is placed at one focus, the waves transmitted by the source will reflect off the ellipse and concentrate at the other focus. This is the principle behind "whispering galleries," which are rooms designed with elliptical ceilings. A person standing at one focus of the ellipse can whisper and be heard by a person standing at the other focus, because all the sound waves that reach the ceiling are reflected to the other person.

E X A M P L E   7   Whispering Galleries

Figure 29 shows the specifications for an elliptical ceiling in a hall designed to be a whispering gallery. In a whispering gallery, a person standing at one focus of the ellipse can whisper and be heard by another person standing at the other focus, because all the sound waves that reach the ceiling from one focus are reflected to the other focus. Where in the hall are the foci located?

FIGURE 29

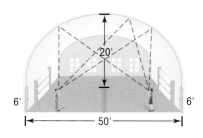

FIGURE 30

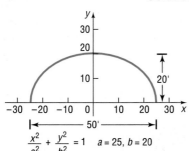

$$\frac{x^2}{a^2} + \frac{y^2}{b^2} = 1 \quad a = 25, b = 20$$

Solution   We set up a rectangular coordinate system so that the center of the ellipse is at the origin and the major axis is along the *x*-axis. See Figure 30. The equation of the ellipse is

$$\frac{x^2}{a^2} + \frac{y^2}{b^2} = 1$$

where $a = 25$ and $b = 20$. Since

$$c^2 = a^2 - b^2 = 25^2 - 20^2 = 625 - 400 = 225$$

we have $c = 15$. Thus, the foci are located 15 feet from the center of the ellipse along the major axis.   ■

    Now work Problem 55.

7.3 | EXERCISES

In Problems 1–4, the graph of an ellipse is given. Match each graph to its equation.

A. $\dfrac{x^2}{4} + y^2 = 1$ B. $x^2 + \dfrac{y^2}{4} = 1$ C. $\dfrac{x^2}{16} + \dfrac{y^2}{4} = 1$ D. $\dfrac{x^2}{4} + \dfrac{y^2}{16} = 1$

1.
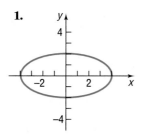

2.
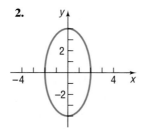

3.
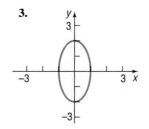

4.
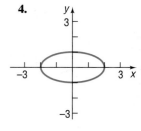

In Problems 5–14, find the vertices and foci of each ellipse. Graph each equation.

5. $\dfrac{x^2}{25} + \dfrac{y^2}{4} = 1$ **6.** $\dfrac{x^2}{9} + \dfrac{y^2}{4} = 1$ **7.** $\dfrac{x^2}{9} + \dfrac{y^2}{25} = 1$ **8.** $x^2 + \dfrac{y^2}{16} = 1$

9. $4x^2 + y^2 = 16$ **10.** $x^2 + 9y^2 = 18$ **11.** $4y^2 + x^2 = 8$ **12.** $4y^2 + 9x^2 = 36$

13. $x^2 + y^2 = 16$ **14.** $x^2 + y^2 = 4$

In Problems 15–24, find an equation for each ellipse. Graph the equation.

15. Center at $(0, 0)$; focus at $(3, 0)$; vertex at $(5, 0)$
16. Center at $(0, 0)$; focus at $(-1, 0)$; vertex at $(3, 0)$
17. Center at $(0, 0)$; focus at $(0, -4)$; vertex at $(0, 5)$
18. Center at $(0, 0)$; focus at $(0, 1)$; vertex at $(0, -2)$
19. Foci at $(\pm 2, 0)$; length of the major axis is 6
20. Focus at $(0, -4)$; vertices at $(0, \pm 8)$
21. Foci at $(0, \pm 3)$; *x*-intercepts are ± 2
22. Foci at $(0, \pm 2)$; length of the major axis is 8
23. Center at $(0, 0)$; vertex at $(0, 4)$; $b = 1$
24. Vertices at $(\pm 5, 0)$; $c = 2$

In Problems 25–28, write an equation for each ellipse.

25.
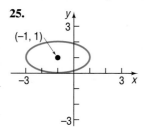

26.
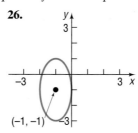

27.
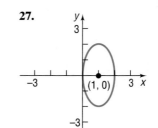

28.
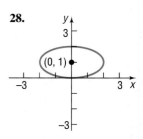

In Problems 29–40, find the center, foci, and vertices of each ellipse. Graph each equation.

29. $\dfrac{(x-3)^2}{4} + \dfrac{(y+1)^2}{9} = 1$

30. $\dfrac{(x+4)^2}{9} + \dfrac{(y+2)^2}{4} = 1$

31. $(x+5)^2 + 4(y-4)^2 = 16$

32. $9(x-3)^2 + (y+2)^2 = 18$

33. $x^2 + 4x + 4y^2 - 8y + 4 = 0$

34. $x^2 + 3y^2 - 12y + 9 = 0$

35. $2x^2 + 3y^2 - 8x + 6y + 5 = 0$

36. $4x^2 + 3y^2 + 8x - 6y = 5$

37. $9x^2 + 4y^2 - 18x + 16y - 11 = 0$

38. $x^2 + 9y^2 + 6x - 18y + 9 = 0$

39. $4x^2 + y^2 + 4y = 0$

40. $9x^2 + y^2 - 18x = 0$

In Problems 41–50, find an equation for each ellipse. Graph the equation.

41. Center at $(2, -2)$; vertex at $(7, -2)$; focus at $(4, -2)$

42. Center at $(-3, 1)$; vertex at $(-3, 3)$; focus at $(-3, 0)$

43. Vertices at $(4, 3)$ and $(4, 9)$; focus at $(4, 8)$

44. Foci at $(1, 2)$ and $(-3, 2)$; vertex at $(-4, 2)$

45. Foci at $(5, 1)$ and $(-1, 1)$; length of the major axis is 8

46. Vertices at $(2, 5)$ and $(2, -1)$; $c = 2$

47. Center at $(1, 2)$; focus at $(4, 2)$; passing through the point $(1, 3)$

48. Center at $(1, 2)$; focus at $(1, 4)$; passing through the point $(2, 2)$

49. Center at $(1, 2)$; vertex at $(4, 2)$; passing through the point $(1, 3)$

50. Center at $(1, 2)$; vertex at $(1, 4)$; passing through the point $(2, 2)$

In Problems 51–54, graph each function.
[**Hint:** Notice that each function is half an ellipse.]

51. $f(x) = \sqrt{16 - 4x^2}$

52. $f(x) = \sqrt{9 - 9x^2}$

53. $f(x) = -\sqrt{64 - 16x^2}$

54. $f(x) = -\sqrt{4 - 4x^2}$

55. Semi-elliptical Arch Bridge An arch in the shape of the upper half of an ellipse is used to support a bridge that is to span a river 20 meters wide. The center of the arch is 6 meters above the center of the river (see the figure). Write an equation for the ellipse in which the x-axis coincides with the water level and the y-axis passes through the center of the arch.

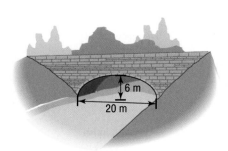

56. Semi-elliptical Arch Bridge The arch of a bridge is a semi-ellipse with a horizontal major axis. The span is 30 feet, and the top of the arch is 10 feet above the major axis. The roadway is horizontal and is 2 feet above the top of the arch. Find the vertical distance from the roadway to the arch at 5 foot intervals along the roadway.

57. Whispering Galleries A hall 100 feet in length is to be designed as a whispering gallery. If the foci are located 25 feet from the center, how high will the ceiling be at the center?

58. Whispering Galleries Jim, standing at one focus of a whispering gallery, is 6 feet from the nearest wall. His friend is standing at the other focus 100 feet away. What is the length of this whispering gallery? How high is its elliptical ceiling at the center?

59. Semi-elliptical Arch Bridge A bridge is built in the shape of a semielliptical arch. The bridge has a span of 120 feet and a maximum height of 25 feet. Choose a suitable rectangular coordinate system and find the height of the arch at distances of 10, 30, and 50 feet from the center.

60. Semi-elliptical Arch Bridge A bridge is built in the shape of a semi-elliptical arch and is to have a span of 100 feet. The height of the arch at a distance of 40 feet from the center is to be 10 feet. Find the height of the arch at its center.

61. Semi-elliptical Arch An arch in the form of half an ellipse is 40 feet wide and 15 feet high at the center. Find the height of the arch at intervals of 10 feet along its width.

62. Semi-elliptical Arch Bridge An arch for a bridge over a highway is in the form of half an ellipse. The top of the arch is 20 feet above the ground level (the major axis). The highway has four lanes, each 12 feet wide; a center safety strip 8 feet wide; and two side strips, each 4 feet wide. What should the span of the bridge be (the length of its major axis) if the height 28 feet from the center is to be 13 feet?

*In Problems 63–66, use the fact that the orbit of a planet about the Sun is an ellipse, with the Sun at one focus. The **aphelion** of a planet is its greatest distance from the Sun and the **perihelion** is its shortest distance. The **mean distance** of a planet from the Sun is the length of the semimajor axis of the elliptical orbit. See the illustration.*

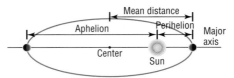

63. Earth The mean distance of Earth from the Sun is 93 million miles. If the aphelion of Earth is 94.5 million miles, what is the perihelion? Write an equation for the orbit of Earth around the Sun.

64. Mars The mean distance of Mars from the Sun is 142 million miles. If the perihelion of Mars is 128.5 million miles, what is the aphelion? Write an equation for the orbit of Mars about the Sun.

65. Jupiter The aphelion of Jupiter is 507 million miles. If the distance from the Sun to the center of its elliptical orbit is 23.2 million miles, what is the perihelion? What is the mean distance? Write an equation for the orbit of Jupiter around the Sun.

66. Pluto The perihelion of Pluto is 4551 million miles and the distance of the Sun from the center of its elliptical orbit is 897.5 million miles. Find the aphelion of Pluto. What is the mean distance of Pluto from the Sun? Write an equation for the orbit of Pluto about the Sun.

67. Racetrack Design Consult the figure. A racetrack is in the shape of an ellipse, 100 feet long and 50 feet wide. What is the width 10 feet from the side?

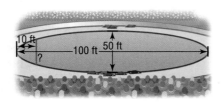

68. Racetrack Design A racetrack is in the shape of an ellipse 80 feet long and 40 feet wide. What is the width 10 feet from the side?

69. Show that an equation of the form

$$Ax^2 + Cy^2 + F = 0 \qquad A \neq 0, C \neq 0, F \neq 0$$

where A and C are of the same sign and F is of opposite sign:
(a) Is the equation of an ellipse with center at $(0, 0)$ if $A \neq C$.
(b) Is the equation of a circle with center $(0, 0)$ if $A = C$.

70. Show that the graph of an equation of the form

$$Ax^2 + Cy^2 + Dx + Ey + F = 0 \qquad A \neq 0, C \neq 0$$

where A and C are of the same sign:
(a) Is an ellipse if $(D^2/4A) + (E^2/4C) - F$ is the same sign as A.
(b) Is a point if $(D^2/4A) + (E^2/4C) - F = 0$.
(c) Contains no points if $(D^2/4A) + (E^2/4C) - F$ is of opposite sign to A.

 71. The **eccentricity** e of an ellipse is defined as the number c/a, where a and c are the numbers given in equation (2). Because $a > c$, it follows that $e < 1$. Write a brief paragraph about the general shape of each of the following ellipses. Be sure to justify your conclusions.
(a) Eccentricity close to 0
(b) Eccentricity $= 0.5$
(c) Eccentricity close to 1

MISSION   POSSIBLE

Building a Bridge over the East River

Your team is working for the transportation authority in New York City. You have been asked to study the construction plans for a new bridge over the East River in New York City. The space between supports needs to be 1050 feet; the height at the center of the arch needs to be 350 feet. One company has suggested that the support be in the shape of a parabola; another company suggests a semi-ellipse. The engineering team will determine the relative strengths of the two plans; your job is to find out if there are any differences in the channel widths.

An empty tanker needs a 280-foot clearance to pass beneath the bridge. You need to find the width of the channel for each of the two different plans.

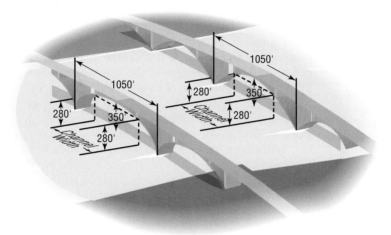

1. To determine the equation of a parabola with these characteristics, first place the parabola on coordinate axes in a convenient location and sketch it.
2. What is the equation of the parabola? (If using a decimal in the equation, you may want to carry six decimal places. If using a fraction, your answer will be more exact.)
3. How wide is the channel that the tanker can pass through if the shape of the support is parabolic?
4. To determine the equation of a semi-ellipse with these characteristics, place the semi-ellipse on coordinate axes in a convenient location and sketch it.
5. What is the equation of the ellipse? How wide is the channel that the tanker can pass through?
6. Now that you know which of the two provides the wider channel, consider some other factors. Your department is also in charge of channel depth, amount of traffic on the river, and various other factors. For example, if the river were to flood and the water level rose by 10 feet, how would the clearances be affected? Make a decision about which plan you think would be the better one as far as your department is concerned, and explain why you think so.

7.4 | THE HYPERBOLA

> 1. Find the Equation of a Hyperbola
> 2. Graph Hyperbolas
> 3. Discuss the Equation of a Hyperbola
> 4. Find the Asymptotes of a Hyperbola
> 5. Work with Hyperbolas with Center at (h, k)
> 6. Solve Applied Problems Involving Properties of Hyperbolas

A **hyperbola** is the collection of all points in the plane the difference of whose distances from two fixed points, called the **foci,** is a constant.

FIGURE 31

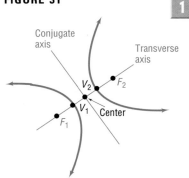

Figure 31 illustrates a hyperbola with foci F_1 and F_2. The line containing the foci is called the **transverse axis.** The midpoint of the line segment joining the foci is called the **center** of the hyperbola. The line through the center and perpendicular to the transverse axis is called the **conjugate axis.** The hyperbola consists of two separate curves, called **branches,** that are symmetric with respect to the transverse axis, conjugate axis, and center. The two points of intersection of the hyperbola and the transverse axis are the **vertices,** V_1 and V_2, of the hyperbola.

With these ideas in mind, we are now ready to find the equation of a hyperbola in a rectangular coordinate system. First, we place the center at the origin. Next, we position the hyperbola so that its transverse axis coincides with a coordinate axis. Suppose that the transverse axis coincides with the x-axis, as shown in Figure 32.

FIGURE 32
$d(F_1, P) - d(F_2, P) = \pm 2a$

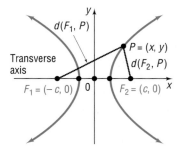

If c is the distance from the center to a focus, then one focus will be at $F_1 = (-c, 0)$ and the other at $F_2 = (c, 0)$. Now we let the constant difference of the distances from any point $P = (x, y)$ on the hyperbola to the foci F_1 and F_2 be denoted by $\pm 2a$. (If P is on the right branch, the $+$ sign is used; if P is on the left branch, the $-$ sign is used.) The coordinates of P must satisfy the equation

$$d(F_1, P) - d(F_2, P) = \pm 2a \qquad \text{Difference of the distances from } P \text{ to the foci equals } \pm 2a.$$

$$\sqrt{(x + c)^2 + y^2} - \sqrt{(x - c)^2 + y^2} = \pm 2a \qquad \text{Use the distance formula.}$$

$$\sqrt{(x + c)^2 + y^2} = \pm 2a + \sqrt{(x - c)^2 + y^2} \qquad \text{Isolate one radical.}$$

$$(x + c)^2 + y^2 = 4a^2 \pm 4a\sqrt{(x - c)^2 + y^2} \qquad \text{Square both sides.}$$
$$+ (x - c)^2 + y^2$$

Next, we remove the parentheses:

$$x^2 + 2cx + c^2 + y^2 = 4a^2 \pm 4a\sqrt{(x-c)^2 + y^2} + x^2 - 2cx + c^2 + y^2$$

$$4cx - 4a^2 = \pm 4a\sqrt{(x-c)^2 + y^2} \qquad \text{Isolate the radical.}$$

$$cx - a^2 = \pm a\sqrt{(x-c)^2 + y^2}. \qquad \text{Divide each side by 4.}$$

$$(cx - a^2)^2 = a^2[(x-c)^2 + y^2] \qquad \text{Square both sides.}$$

$$c^2x^2 - 2ca^2x + a^4 = a^2(x^2 - 2cx + c^2 + y^2)$$

$$c^2x^2 + a^4 = a^2x^2 + a^2c^2 + a^2y^2 \qquad \text{Simplify.}$$

$$(c^2 - a^2)x^2 - a^2y^2 = a^2c^2 - a^4$$

$$(c^2 - a^2)x^2 - a^2y^2 = a^2(c^2 - a^2) \qquad (1)$$

To obtain points on the hyperbola off the x-axis, it must be that $a < c$. To see why, look again at Figure 32.

$$d(F_1, P) < d(F_2, P) + d(F_1, F_2) \qquad \text{Use triangle } F_1PF_2.$$

$$d(F_1, P) - d(F_2, P) < d(F_1, F_2) \qquad \begin{array}{l} P \text{ is on the right branch,} \\ \text{so } d(F_1, P) - d(F_2, P) = 2a. \end{array}$$

$$2a < 2c$$

$$a < c$$

Since $a < c$, we also have $a^2 < c^2$, so $c^2 - a^2 > 0$. Let $b^2 = c^2 - a^2, b > 0$. Then equation (1) can be written as

$$b^2x^2 - a^2y^2 = a^2b^2$$

$$\frac{x^2}{a^2} - \frac{y^2}{b^2} = 1$$

To find the vertices of the hyperbola defined by this equation, let $y = 0$. The vertices satisfy the equation $x^2/a^2 = 1$, the solutions of which are $x = \pm a$. Consequently, the vertices of the hyperbola are $V_1 = (-a, 0)$ and $V_2 = (a, 0)$.

Theorem   Equation of a Hyperbola; Center at (0, 0); Foci at $(\pm c, 0)$; Vertices at $(\pm a, 0)$; Transverse Axis along the x-Axis

An equation of the hyperbola with center at $(0, 0)$, foci at $(-c, 0)$ and $(c, 0)$, and vertices at $(-a, 0)$ and $(a, 0)$ is

$$\frac{x^2}{a^2} - \frac{y^2}{b^2} = 1 \qquad \text{where } b^2 = c^2 - a^2 \qquad (2)$$

The transverse axis is the x-axis.

---

2   As you can verify, the hyperbola defined by equation (2) is symmetric with respect to the x-axis, y-axis, and origin. To find the y-intercepts, if any, let $x = 0$ in equation (2). This results in the equation $y^2/b^2 = -1$, which has no solution. We conclude that the hyperbola defined by equation (2) has no y-intercepts. In fact, since $x^2/a^2 - 1 = y^2/b^2 \geq 0$, it follows that $x^2/a^2 \geq 1$. Thus, there are no points on the graph for $-a < x < a$. See Figure 33.

FIGURE 33

$$\frac{x^2}{a^2} - \frac{y^2}{b^2} = 1, b^2 = c^2 - a^2$$

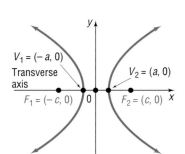

EXAMPLE 1

Finding an Equation of a Hyperbola

Find an equation of the hyperbola with center at the origin, one focus at $(3, 0)$, and one vertex at $(-2, 0)$. Graph the equation.

Solution

The hyperbola has its center at the origin, and the transverse axis coincides with the x-axis. One focus is at $(c, 0) = (3, 0)$, so $c = 3$. One vertex is at $(-a, 0) = (-2, 0)$, so $a = 2$. From equation (2), it follows that $b^2 = c^2 - a^2 = 9 - 4 = 5$, so an equation of the hyperbola is

$$\frac{x^2}{4} - \frac{y^2}{5} = 1$$

See Figure 34.

FIGURE 34

$$\frac{x^2}{4} - \frac{y^2}{5} = 1$$

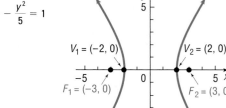

 Comment To graph the hyperbola $(x^2/4) - (y^2/5) = 1$ discussed in Example 1, we need to graph the two functions $Y_1 = \sqrt{5}\sqrt{(x^2/4) - 1}$ and $Y_2 = -\sqrt{5}\sqrt{(x^2/4) - 1}$. Do this and compare what you see with Figure 34.

 Now work Problem 5.

An equation of the form of equation (2) is the equation of a hyperbola with center at the origin, foci on the x-axis at $(-c, 0)$ and $(c, 0)$, where $c^2 = a^2 + b^2$, and transverse axis along the x-axis.

 For the remainder of this section, the direction "Discuss the equation" will mean to find the center, transverse axis, vertices, and foci of the hyperbola and graph it.

EXAMPLE 2

Discussing the Equation of a Hyperbola

Discuss the equation: $\dfrac{x^2}{16} - \dfrac{y^2}{4} = 1$

FIGURE 35
$$\frac{x^2}{16} - \frac{y^2}{4} = 1$$

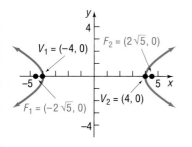

Solution

The given equation is of the form of equation (2), with $a^2 = 16$ and $b^2 = 4$. Thus, the graph of the equation is a hyperbola with center at $(0, 0)$ and transverse axis along the x-axis. Also, we know that $c^2 = a^2 + b^2 = 16 + 4 = 20$. The vertices are at $(\pm a, 0) = (\pm 4, 0)$, and the foci are at $(\pm c, 0) = (\pm 2\sqrt{5}, 0)$. Figure 35 shows the graph.

The next result gives the form of the equation of a hyperbola with center at the origin and transverse axis along the y-axis.

FIGURE 36
$$\frac{y^2}{a^2} - \frac{x^2}{b^2} = 1, b^2 = c^2 - a^2$$

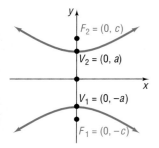

> **Theorem** Equation of a Hyperbola; Center at (0, 0); Foci at (0, ±c); Vertices at (0, ±a); Transverse Axis along the y-Axis
>
> An equation of the hyperbola with center at $(0, 0)$, foci at $(0, -c)$ and $(0, c)$, and vertices at $(0, -a)$ and $(0, a)$ is
>
> $$\frac{y^2}{a^2} - \frac{x^2}{b^2} = 1 \qquad \text{where } b^2 = c^2 - a^2 \qquad (3)$$
>
> The transverse axis is the y-axis.

Figure 36 shows the graph of a typical hyperbola defined by equation (3). Notice the difference in the form of equations (2) and (3). When the y^2 term is subtracted from the x^2 term, the transverse axis is the x-axis. When the x^2 term is subtracted from the y^2 term, the transverse axis is the y-axis.

E X A M P L E   3   Discussing the Equation of a Hyperbola

FIGURE 37
$$\frac{y^2}{4} - x^2 = 1$$

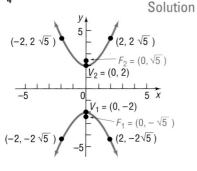

Discuss the equation: $y^2 - 4x^2 = 4$

Solution

To put the equation in proper form, we divide each side by 4:

$$\frac{y^2}{4} - x^2 = 1$$

Since the x^2 term is subtracted from the y^2 term, the equation is that of a hyperbola with center at the origin and transverse axis along the y-axis. Also, comparing the above equation to equation (3), we find $a^2 = 4$, $b^2 = 1$, and $c^2 = a^2 + b^2 = 5$. The vertices are at $(0, \pm a) = (0, \pm 2)$, and the foci are at $(0, \pm c) = (0, \pm \sqrt{5})$. The graph is given in Figure 37.

E X A M P L E   4   Finding an Equation of a Hyperbola

Find an equation of the hyperbola having one vertex at $(0, 2)$ and foci at $(0, -3)$ and $(0, 3)$. Graph the equation.

Solution

Since the foci are at $(0, -3)$ and $(0, 3)$, the center of the hyperbola is at the origin. Also, the transverse axis is along the y-axis. The given information

FIGURE 38
$$\frac{y^2}{4} - \frac{x^2}{5} = 1$$

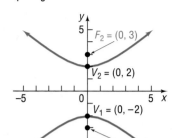

also reveals that $c = 3$, $a = 2$, and $b^2 = c^2 - a^2 = 9 - 4 = 5$. The form of the equation of the hyperbola is given by equation (3):

$$\frac{y^2}{a^2} - \frac{x^2}{b^2} = 1$$

$$\frac{y^2}{4} - \frac{x^2}{5} = 1$$

See Figure 38.   ∎

Now work Problem 7.

Look at the equations of the hyperbolas in Examples 3 and 4. For the hyperbola in Example 3, $a^2 = 4$ and $b^2 = 1$, so $a > b$; for the hyperbola in Example 4, $a^2 = 4$ and $b^2 = 5$, so $a < b$. We conclude that, for hyperbolas, there are no requirements involving the relative sizes of a and b. Contrast this situation to the case of an ellipse, in which the relative sizes of a and b dictate which axis is the major axis. Hyperbolas have another feature to distinguish them from ellipses and parabolas: Hyperbolas have asymptotes.

Asymptotes

Recall from Section 5.3 that a horizontal or oblique asymptote of a graph is a line with the property that the distance from the line to points on the graph approaches 0 as $x \to -\infty$ or as $x \to \infty$. When such asymptotes exist, they give information about the end behavior of the graph.

> **Theorem**   **Asymptotes of a Hyperbola**
>
> The hyperbola $\dfrac{x^2}{a^2} - \dfrac{y^2}{b^2} = 1$ has the two oblique asymptotes
>
> $$y = \frac{b}{a}x \quad \text{and} \quad y = -\frac{b}{a}x$$

Proof   We begin by solving for y in the equation of the hyperbola:

$$\frac{x^2}{a^2} - \frac{y^2}{b^2} = 1$$

$$\frac{y^2}{b^2} = \frac{x^2}{a^2} - 1$$

$$y^2 = b^2\left(\frac{x^2}{a^2} - 1\right)$$

Since $x \neq 0$, we can rearrange the right side in the form

$$y^2 = \frac{b^2 x^2}{a^2}\left(1 - \frac{a^2}{x^2}\right)$$

$$y = \pm\frac{bx}{a}\sqrt{1 - \frac{a^2}{x^2}}$$

Now, as $x \to -\infty$ or as $x \to \infty$, the term a^2/x^2 approaches 0, so the expression under the radical approaches 1. Thus, as $x \to -\infty$ or as $x \to \infty$, the value of y approaches $\pm bx/a$; that is, the graph of the hyperbola approaches the lines

$$y = -\frac{b}{a}x \quad \text{and} \quad y = \frac{b}{a}x$$

These lines are oblique asymptotes of the hyperbola.

FIGURE 39
$$\frac{x^2}{a^2} - \frac{y^2}{b^2} = 1$$

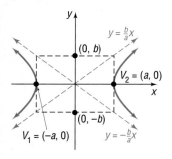

The asymptotes of a hyperbola are not part of the hyperbola, but they do serve as a guide for graphing a hyperbola. For example, suppose that we want to graph the equation

$$\frac{x^2}{a^2} - \frac{y^2}{b^2} = 1$$

We begin by plotting the vertices $(-a, 0)$ and $(a, 0)$. Then we plot the points $(0, -b)$ and $(0, b)$ and use these four points to construct a rectangle, as shown in Figure 39. The diagonals of this rectangle have slopes b/a and $-b/a$, and their extensions are the asymptotes $y = (b/a)x$ and $y = -(b/a)x$ of the hyperbola.

Theorem   Asymptotes of a Hyperbola

The hyperbola $\dfrac{y^2}{a^2} - \dfrac{x^2}{b^2} = 1$ has the two oblique asymptotes

$$y = \frac{a}{b}x \quad \text{and} \quad y = -\frac{a}{b}x$$

You are asked to prove this result in Problem 60.

E X A M P L E   5

FIGURE 40
$$\frac{x^2}{4} - \frac{y^2}{9} = 1$$

Discussing the Equation of a Hyperbola

Discuss the equation: $9x^2 - 4y^2 = 36$

Solution   First, we divide each side by 36 to put the equation in proper form:

$$\frac{x^2}{4} - \frac{y^2}{9} = 1$$

This is the equation of a hyperbola with center at the origin and transverse axis along the x-axis. Using $a^2 = 4$ and $b^2 = 9$, we find $c^2 = a^2 + b^2 = 13$. The vertices are at $(\pm a, 0) = (\pm 2, 0)$, the foci are at $(\pm c, 0) = (\pm \sqrt{13}, 0)$, and the asymptotes have the equations

$$y = \frac{3}{2}x \quad \text{and} \quad y = -\frac{3}{2}x$$

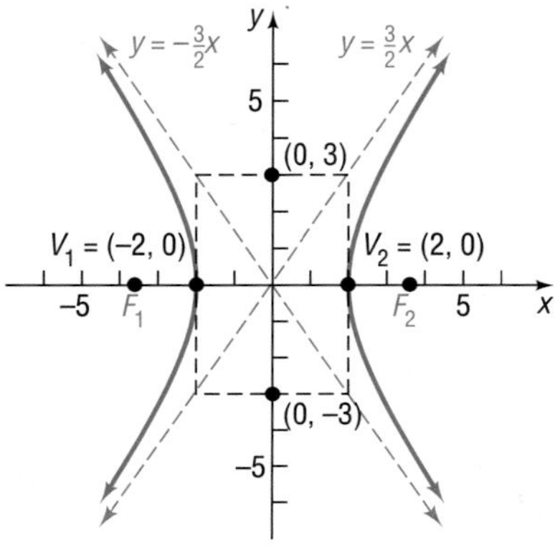

Now form the rectangle containing the points $(\pm a, 0)$ and $(0, \pm b)$, that is, $(-2, 0)$, $(2, 0)$, $(0, -3)$, and $(0, 3)$. The extensions of the diagonals of this rectangle are the asymptotes. See Figure 40 for the graph.

Now work Problem 17.

> **Exploration** Graph the upper portion of the hyperbola $9x^2 - 4y^2 = 36$ discussed in Example 5 and its asymptotes $y = \frac{3}{2}x$ and $y = -\frac{3}{2}x$. Now use ZOOM and TRACE to see what happens as x becomes unbounded in the positive direction. What happens as x becomes unbounded in the negative direction?

Center at (h, k)

5 If a hyperbola with center at the origin and transverse axis coinciding with a coordinate axis is shifted horizontally h units and then vertically k units, the result is a hyperbola with center at (h, k) and transverse axis parallel to a coordinate axis. The equations of such hyperbolas have the same forms as those given in Equations (2) and (3), except that x is replaced by $x - h$ (the horizontal shift) and y is replaced by $y - k$ (the vertical shift). Table 4 gives the forms of the equations of such hyperbolas. See Figure 41 for the graphs.

TABLE 4 Hyperbolas with Center at (h, k) and Transverse Axis Parallel to a Coordinate Axis

Center	Transverse Axis	Foci	Vertices	Equation	Asymptotes
(h, k)	Parallel to x-axis	$(h \pm c, k)$	$(h \pm a, k)$	$\dfrac{(x - h)^2}{a^2} - \dfrac{(y - k)^2}{b^2} = 1,$ $b^2 = c^2 - a^2$	$y - k = \pm\dfrac{b}{a}(x - h)$
(h, k)	Parallel to y-axis	$(h, k \pm c)$	$(h, k \pm a)$	$\dfrac{(y - k)^2}{a^2} - \dfrac{(x - h)^2}{b^2} = 1,$ $b^2 = c^2 - a^2$	$y - k = \pm\dfrac{a}{b}(x - h)$

FIGURE 41

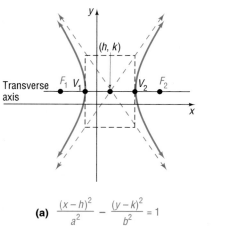

(a) $\dfrac{(x-h)^2}{a^2} - \dfrac{(y-k)^2}{b^2} = 1$

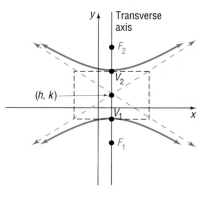

(b) $\dfrac{(y-k)^2}{a^2} - \dfrac{(x-h)^2}{b^2} = 1$

E X A M P L E   6

Finding an Equation of a Hyperbola, Center Not at the Origin

Find an equation for the hyperbola with center at $(1, -2)$, one focus at $(4, -2)$, and one vertex at $(3, -2)$. Graph the equation.

Solution   The center is at $(h, k) = (1, -2)$, so $h = 1$ and $k = -2$. The transverse axis is parallel to the x-axis. The distance from the center $(1, -2)$ to the focus

$(4, -2)$ is $c = 3$; the distance from the center $(1, -2)$ to the vertex $(3, -2)$ is $a = 2$. Thus, $b^2 = c^2 - a^2 = 9 - 4 = 5$. The equation is

$$\frac{(x - h)^2}{a^2} - \frac{(y - k)^2}{b^2} = 1$$

$$\frac{(x - 1)^2}{4} - \frac{(y + 2)^2}{5} = 1$$

See Figure 42.

FIGURE 42

$$\frac{(x - 1)^2}{4} - \frac{(y + 2)^2}{5} = 1$$

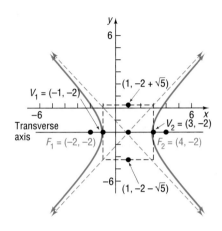

Now work Problem 27.

E X A M P L E   7

Discussing the Equation of a Hyperbola

Discuss the equation: $-x^2 + 4y^2 - 2x - 16y + 11 = 0$

Solution   We complete the squares in x and in y.

$$-x^2 + 4y^2 - 2x - 16y + 11 = 0$$
$$-(x^2 + 2x) + 4(y^2 - 4y) = -11 \qquad \text{Group terms.}$$
$$-(x^2 + 2x + 1) + 4(y^2 - 4y + 4) = -1 + 16 - 11 \qquad \text{Complete each square.}$$
$$-(x + 1)^2 + 4(y - 2)^2 = 4$$
$$(y - 2)^2 - \frac{(x + 1)^2}{4} = 1 \qquad \text{Divide by 4.}$$

FIGURE 43

$$(y - 2)^2 - \frac{(x + 1)^2}{4} = 1$$

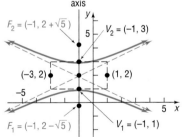

This is the equation of a hyperbola with center at $(-1, 2)$ and transverse axis parallel to the y-axis. Also, $a^2 = 1$ and $b^2 = 4$, so $c^2 = a^2 + b^2 = 5$. The vertices are at $(h, k \pm a) = (-1, 2 \pm 1)$, or $(-1, 1)$ and $(-1, 3)$. The foci are at $(h, k \pm c) = (-1, 2 \pm \sqrt{5})$. The asymptotes are $y - 2 = \frac{1}{2}(x + 1)$ and $y - 2 = -\frac{1}{2}(x + 1)$. Figure 43 shows the graph.

FIGURE 44

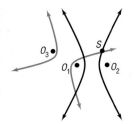

Now work Problem 41.

Applications

See Figure 44. Suppose that a gun is fired from an unknown source S. An observer at O_1 hears the report (sound of gun shot) 1 second after another observer at O_2. Because sound travels at about 1100 feet per second, it follows that the point S must be 1100 feet closer to O_2 than to O_1. Thus, S lies on one branch of a hyperbola with foci at O_1 and O_2. (Do you see why? The difference of the distances from S to O_1 and from S to O_2 is the constant 1100.) If a third observer at O_3 hears the same report 2 seconds after O_1 hears it, then S will lie on a branch of a second hyperbola with foci at O_1 and O_3. The intersection of the two hyperbolas will pinpoint the location of S.

LORAN

In the LOng RAnge Navigation system (LORAN), a master radio sending station and a secondary sending station emit signals that can be received by a ship at sea. (See Figure 45.) Because a ship monitoring the two signals will usually be nearer to one of the two stations, there will be a difference in the distance the two signals travel, which will register as a slight time difference between the signals. As long as the time difference remains constant, the difference of the two distances will also be constant. If the ship follows a path corresponding to the fixed time difference, it will follow the path of a hyperbola whose foci are located at the positions of the two sending stations. So for each time difference a different hyperbolic path results, each bringing the ship to a different shore location. Navigation charts show the various hyperbolic paths corresponding to different time differences.

FIGURE 45

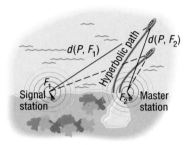

$$d(P, F_1) - d(P, F_2) = \text{constant}$$

E X A M P L E   8   LORAN

Two LORAN stations are positioned 250 miles apart along a straight shore.

(a) A ship records a time difference of 0.00054 second between the LORAN signals. Set up an appropriate rectangular coordinate system to determine where the ship would reach shore if it were to follow the hyperbola corresponding to this time difference.

(b) If the ship wants to enter a harbor located between the two stations 25 miles from the master station, what time difference should it be looking for?

(c) If the ship is 80 miles off shore when the desired time difference is obtained, what is the approximate location of the ship?

[**Note:** The speed of each radio signal is 186,000 miles per second.]

Solution

FIGURE 46

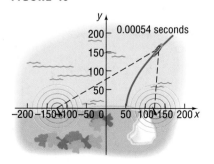

(a) We set up a rectangular coordinate system so that the two stations lie on the x-axis and the origin is midway between them. See Figure 46. The ship lies on a hyperbola whose foci are the locations of the two stations. The reason for this is that the constant time difference of the signals from each station results in a constant difference in the distance of the ship from each station. Since the time difference is 0.00054 second and the speed of the signal is 186,000 miles per second, the difference of the distances from the ship to each station (foci) is

$$\text{Distance} = \text{speed} \times \text{time} = 186{,}000 \times 0.00054 \approx 100 \text{ miles}$$

The difference of the distances from the ship to each station, 100, equals $2a$, so $a = 50$ and the vertex of the corresponding hyperbola is at $(50, 0)$. Since the focus is at $(125, 0)$, following this hyperbola the ship would reach shore 75 miles from the master station.

(b) To reach shore 25 miles from the master station, the ship should follow a hyperbola with vertex at $(100, 0)$. For this hyperbola, $a = 100$, so the constant difference of the distances from the ship to each station is 200. The time difference the ship should look for is

$$\text{Time} = \frac{\text{distance}}{\text{speed}} = \frac{200}{186,000} = 0.001075 \text{ second}$$

(c) To find the exact location of the ship, we need to find the equation of the hyperbola with vertex at $(100, 0)$ and a focus at $(125, 0)$. The form of the equation of this hyperbola is

$$\frac{x^2}{a^2} - \frac{y^2}{b^2} = 1$$

where $a = 100$. Since $c = 125$, we have

$$b^2 = c^2 - a^2 = 125^2 - 100^2 = 5625$$

The equation of the hyperbola is

$$\frac{x^2}{100^2} - \frac{y^2}{5625} = 1$$

Since the ship is 80 miles from shore, we use $y = 80$ in the equation and solve for x.

$$\frac{x^2}{100^2} - \frac{80^2}{5625} = 1$$

$$\frac{x^2}{100^2} = 1 + \frac{80^2}{5625} \approx 2.14$$

$$x^2 = 100^2(2.14)$$

$$x \approx 146$$

The ship is at the position $(146, 80)$. See Figure 47.

FIGURE 47

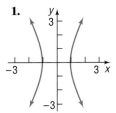

   Now work Problem 53.

| EXERCISES

In Problems 1–4, the graph of a hyperbola is given. Match each graph to its equation.

A. $\dfrac{x^2}{4} - y^2 = 1$ B. $x^2 - \dfrac{y^2}{4} = 1$ C. $\dfrac{y^2}{4} - x^2 = 1$ D. $y^2 - \dfrac{x^2}{4} = 1$

1.

2.
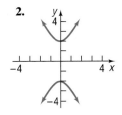

3.
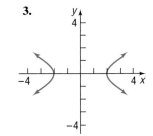

4.
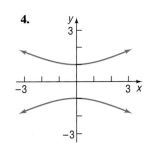

In Problems 5–14, find an equation for the hyperbola described. Graph the equation.

5. Center at $(0, 0)$; focus at $(3, 0)$; vertex at $(1, 0)$

6. Center at $(0, 0)$; focus at $(0, 5)$; vertex at $(0, 3)$

7. Center at $(0, 0)$; focus at $(0, -6)$; vertex at $(0, 4)$

8. Center at $(0, 0)$; focus at $(-3, 0)$; vertex at $(2, 0)$

9. Foci at $(-5, 0)$ and $(5, 0)$; vertex at $(3, 0)$

10. Focus at $(0, 6)$; vertices at $(0, -2)$ and $(0, 2)$

11. Vertices at $(0, -6)$ and $(0, 6)$; asymptote the line $y = 2x$

12. Vertices at $(-4, 0)$ and $(4, 0)$; asymptote the line $y = 2x$

13. Foci at $(-4, 0)$ and $(4, 0)$; asymptote the line $y = -x$

14. Foci at $(0, -2)$ and $(0, 2)$; asymptote the line $y = -x$

In Problems 15–22, find the center, transverse axis, vertices, foci, and asymptotes. Graph each equation.

15. $\dfrac{x^2}{25} - \dfrac{y^2}{9} = 1$

16. $\dfrac{y^2}{16} - \dfrac{x^2}{4} = 1$

17. $4x^2 - y^2 = 16$

18. $4y^2 - x^2 = 16$

19. $y^2 - 9x^2 = 9$

20. $x^2 - y^2 = 4$

21. $y^2 - x^2 = 25$

22. $2x^2 - y^2 = 4$

In Problems 23–26, write an equation for each hyperbola.

23.

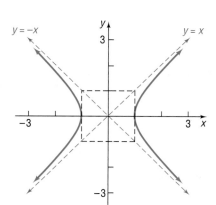

24.

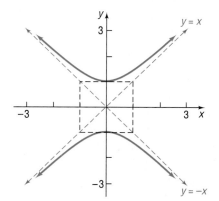

25.

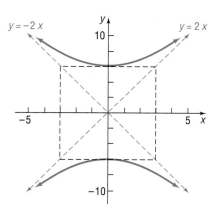

26.

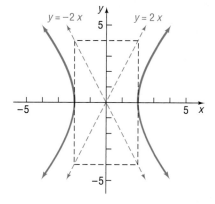

In Problems 27–34, find an equation for the hyperbola described. Graph the equation.

27. Center at $(4, -1)$; focus at $(7, -1)$; vertex at $(6, -1)$

28. Center at $(-3, 1)$; focus at $(-3, 6)$; vertex at $(-3, 4)$

29. Center at $(-3, -4)$; focus at $(-3, -8)$; vertex at $(-3, -2)$

30. Center at $(1, 4)$; focus at $(-2, 4)$; vertex at $(0, 4)$

31. Foci at $(3, 7)$ and $(7, 7)$; vertex at $(6, 7)$

32. Focus at $(-4, 0)$; vertices at $(-4, 4)$ and $(-4, 2)$

33. Vertices at $(-1, -1)$ and $(3, -1)$; asymptote the line $(x - 1)/2 = (y + 1)/3$

34. Vertices at $(1, -3)$ and $(1, 1)$; asymptote the line $(x - 1)/2 = (y + 1)/3$

In Problems 35–48, find the center, transverse axis, vertices, foci, and asymptotes. Graph each equation.

35. $\dfrac{(x-2)^2}{4} - \dfrac{(y+3)^2}{9} = 1$

36. $\dfrac{(y+3)^2}{4} - \dfrac{(x-2)^2}{9} = 1$

37. $(y-2)^2 - 4(x+2)^2 = 4$

38. $(x+4)^2 - 9(y-3)^2 = 9$

39. $(x+1)^2 - (y+2)^2 = 4$

40. $(y-3)^2 - (x+2)^2 = 4$

41. $x^2 - y^2 - 2x - 2y - 1 = 0$

42. $y^2 - x^2 - 4y + 4x - 1 = 0$

43. $y^2 - 4x^2 - 4y - 8x - 4 = 0$

44. $2x^2 - y^2 + 4x + 4y - 4 = 0$

45. $4x^2 - y^2 - 24x - 4y + 16 = 0$

46. $2y^2 - x^2 + 2x + 8y + 3 = 0$

47. $y^2 - 4x^2 - 16x - 2y - 19 = 0$

48. $x^2 - 3y^2 + 8x - 6y + 4 = 0$

In Problems 49–52, graph each function.

[**Hint:** Notice that each function is half a hyperbola.]

49. $f(x) = \sqrt{16 + 4x^2}$

50. $f(x) = -\sqrt{9 + 9x^2}$

51. $f(x) = -\sqrt{-25 + x^2}$

52. $f(x) = \sqrt{-1 + x^2}$

53. LORAN Two LORAN stations are positioned 200 miles apart along a straight shore.
(a) A ship records a time difference of 0.00038 second between the LORAN signals. Set up an appropriate rectangular coordinate system to determine where the ship would reach shore if it were to follow the hyperbola corresponding to this time difference.
(b) If the ship wants to enter a harbor located between the two stations 20 miles from the master station, what time difference should it be looking for?
(c) If the ship is 50 miles off shore when the desired time difference is obtained, what is the approximate location of the ship?
[**Note:** The speed of each radio signal is 186,000 miles per second.]

54. LORAN Two LORAN stations are positioned 100 miles apart along a straight shore.
(a) A ship records a time difference of 0.00032 second between the LORAN signals. Set up an appropriate rectangular coordinate system to determine where the ship would reach shore if it were to follow the hyperbola corresponding to this time difference.
(b) If the ship wants to enter a harbor located between the two stations 10 miles from the master station, what time difference should it be looking for?
(c) If the ship is 20 miles off shore when the desired time difference is obtained, what is the approximate location of the ship?
[**Note:** The speed of each radio signal is 186,000 miles per second.]

55. Calibrating Instruments In a test of their recording devices, a team of seismologists positioned two of the devices 2000 feet apart, with the device at point A to the west of the device at point B. At a point between the devices and 200 feet from point B, a small amount of ex-

plosive was detonated and a note made of the time at which the sound reached each device. A second explosion is to be carried out at a point directly north of point B.
(a) How far north should the site of the second explosion be chosen so that the measured time difference recorded by the devices for the second detonation is the same as that recorded for the first detonation?
 (b) Explain why this experiment can be used to calibrate the instruments.

56. Explain in your own words the LORAN system of navigation.

57. The **eccentricity** e of a hyperbola is defined as the number c/a, where a and c are the numbers given in equation (2). Because $c > a$, it follows that $e > 1$. Describe the general shape of a hyperbola whose eccentricity is close to 1. What is the shape if e is very large?

58. A hyperbola for which $a = b$ is called an **equilateral hyperbola.** Find the eccentricity e of an equilateral hyperbola.
[**Note:** The eccentricity of a hyperbola is defined in Problem 57.]

59. Two hyperbolas that have the same set of asymptotes are called **conjugate.** Show that the hyperbolas

$$\dfrac{x^2}{4} - y^2 = 1 \quad \text{and} \quad y^2 - \dfrac{x^2}{4} = 1$$

are conjugate. Graph each hyperbola on the same set of coordinate axes.

60. Prove that the hyperbola

$$\dfrac{y^2}{a^2} - \dfrac{x^2}{b^2} = 1$$

has the two oblique asymptotes

$$y = \dfrac{a}{b}x \quad \text{and} \quad y = -\dfrac{a}{b}x$$

61. Show that the graph of an equation of the form

$$Ax^2 + Cy^2 + F = 0 \qquad A \neq 0, C \neq 0, F \neq 0$$

where A and C are of opposite sign, is a hyperbola with center at $(0, 0)$.

62. Show that the graph of an equation of the form

$$Ax^2 + Cy^2 + Dx + Ey + F = 0 \quad A \neq 0, C \neq 0$$

where A and C are of opposite sign:

(a) Is a hyperbola if $(D^2/4A) + (E^2/4C) - F \neq 0$.

(b) Is two intersecting lines if

$$\left(\frac{D^2}{4A}\right) + \left(\frac{E^2}{4C}\right) - F = 0$$

CHAPTER REVIEW

THINGS TO KNOW

Equations of Conics

Parabola	See Tables 1 and 2.
Ellipse	See Table 3.
Hyperbola	See Table 4.

HOW TO

Find the vertex, focus, and directrix of a parabola given its equation

Graph a parabola given its equation

Find an equation of a parabola given certain information about the parabola

Find the center, foci, and vertices of an ellipse given its equation

Graph an ellipse given its equation

Find an equation of an ellipse given certain information about the ellipse

Find the center, foci, vertices, and asymptotes of a hyperbola given its equation

Graph a hyperbola given its equation

Find an equation of a hyperbola given certain information about the hyperbola

Solve applied problems involving parabolas, ellipses, and hyperbolas

FILL-IN-THE-BLANK ITEMS

1. A(n) _____ is the collection of all points in the plane such that the distance from each point to a fixed point equals its distance to a fixed line.

2. A(n) _____ is the collection of all points in the plane the sum of whose distances from two fixed points is a constant.

3. A(n) _____ is the collection of all points in the plane the difference of whose distances from two fixed points is a constant.

4. For an ellipse, the foci lie on the _____ axis; for a hyperbola, the foci lie on the _____ axis.

5. For the ellipse $(x^2/9) + (y^2/16) = 1$, the major axis is along the _____.

6. The equations of the asymptotes of the hyperbola $(y^2/9) - (x^2/4) = 1$ are _____ and _____.

TRUE/FALSE ITEMS

T F **1.** On a parabola, the distance from any point to the focus equals the distance from that point to the directrix.

T F **2.** The foci of an ellipse lie on its minor axis.

T F **3.** The foci of a hyperbola lie on its transverse axis.

T F **4.** Hyperbolas always have asymptotes, and ellipses never have asymptotes.

T F **5.** A hyperbola never intersects its conjugate axis.

T F **6.** A hyperbola always intersects its transverse axis.

REVIEW EXERCISES

Blue problem numbers indicate the author's suggestions for use in a Practice Test.

In Problems 1–20, identify each equation. If it is a parabola, give its vertex, focus, and directrix; if it is an ellipse, give its center, vertices, and foci; if it is a hyperbola, give its center, vertices, foci, and asymptotes.

1. $y^2 = -16x$ 2. $16x^2 = y$ 3. $\dfrac{x^2}{25} - y^2 = 1$ 4. $\dfrac{y^2}{25} - x^2 = 1$

5. $\dfrac{y^2}{25} + \dfrac{x^2}{16} = 1$ 6. $\dfrac{x^2}{9} + \dfrac{y^2}{16} = 1$ 7. $x^2 + 4y = 4$ 8. $3y^2 - x^2 = 9$

9. $4x^2 - y^2 = 8$ 10. $9x^2 + 4y^2 = 36$ 11. $x^2 - 4x = 2y$ 12. $2y^2 - 4y = x - 2$

13. $y^2 - 4y - 4x^2 + 8x = 4$ 14. $4x^2 + y^2 + 8x - 4y + 4 = 0$

15. $4x^2 + 9y^2 - 16x - 18y = 11$ 16. $4x^2 + 9y^2 - 16x + 18y = 11$

17. $4x^2 - 16x + 16y + 32 = 0$ 18. $4y^2 + 3x - 16y + 19 = 0$

19. $9x^2 + 4y^2 - 18x + 8y = 23$ 20. $x^2 - y^2 - 2x - 2y = 1$

In Problems 21–36, obtain an equation of the conic described. Graph the equation.

21. Parabola; focus at $(-2, 0)$; directrix the line $x = 2$

22. Ellipse; center at $(0, 0)$; focus at $(0, 3)$; vertex at $(0, 5)$

23. Hyperbola; center at $(0, 0)$; focus at $(0, 4)$; vertex at $(0, -2)$

24. Parabola; vertex at $(0, 0)$; directrix the line $y = -3$

25. Ellipse; foci at $(-3, 0)$ and $(3, 0)$; vertex at $(4, 0)$

26. Hyperbola; vertices at $(-2, 0)$ and $(2, 0)$; focus at $(4, 0)$

27. Parabola; vertex at $(2, -3)$; focus at $(2, -4)$

28. Ellipse; center at $(-1, 2)$; focus at $(0, 2)$; vertex at $(2, 2)$

29. Hyperbola; center at $(-2, -3)$; focus at $(-4, -3)$; vertex at $(-3, -3)$

30. Parabola; focus at $(3, 6)$; directrix the line $y = 8$

31. Ellipse; foci at $(-4, 2)$ and $(-4, 8)$; vertex at $(-4, 10)$

32. Hyperbola; vertices at $(-3, 3)$ and $(5, 3)$; focus at $(7, 3)$

33. Center at $(-1, 2)$; $a = 3$; $c = 4$; transverse axis parallel to the x-axis

34. Center at $(4, -2)$; $a = 1$; $c = 4$; transverse axis parallel to y-axis

35. Vertices at $(0, 1)$ and $(6, 1)$; asymptote the line $3y + 2x - 9 = 0$

36. Vertices at $(4, 0)$ and $(4, 4)$; asymptote the line $y + 2x - 10 = 0$

37. Find an equation of the hyperbola whose foci are the vertices of the ellipse $4x^2 + 9y^2 = 36$ and whose vertices are the foci of this ellipse.

38. Find an equation of the ellipse whose foci are the vertices of the hyperbola $x^2 - 4y^2 = 16$ and whose vertices are the foci of this hyperbola.

39. Describe the collection of points in a plane so that the distance from each point to the point $(3, 0)$ is three-fourths of its distance from the line $x = \frac{16}{3}$.

40. Describe the collection of points in a plane so that the distance from each point to the point $(5, 0)$ is five-fourths of its distance from the line $x = \frac{16}{5}$.

41. **Mirrors**   A mirror is shaped like a paraboloid of revolution. If a light is located 1 foot from the base along the axis of symmetry and the opening is 2 feet across, how deep should the mirror be?

42. **Parabolic Arch Bridge**   A bridge is built in the shape of a parabolic arch. The bridge has a span of 60 feet and a maximum height of 20 feet. Find the height of the arch at distances of 5, 10, and 20 feet from the center.

43. **Semi-elliptical Arch Bridge**   A bridge is built in the shape of a semi-elliptical arch. The bridge has a span of 60 feet and a maximum height of 20 feet. Find the height of the arch at distances of 5, 10, and 20 feet from the center.

44. **Whispering Galleries** The figure shows the specifications for an elliptical ceiling in a hall designed to be a whispering gallery. Where in the hall are the foci located?

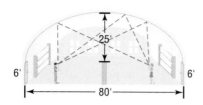

45. **LORAN** Two LORAN stations are positioned 150 miles apart along a straight shore.
    (a) A ship records a time difference of 0.00032 second between the LORAN signals. Set up an appropriate rectangular coordinate system to determine where the ship would reach shore if it were to follow the hyperbola corresponding to this time difference.
    (b) If the ship wants to enter a harbor located between the two stations 15 miles from the master station, what time difference should it be looking for?
    (c) If the ship is 20 miles off shore when the desired time difference is obtained, what is the approximate location of the ship?
    [**Note:** The speed of each radio signal is 186,000 miles per second.]

46. Use the definition of a parabola and the idea of the eccentricity of an ellipse and hyperbola to construct a unifying definition for all three conics. [Refer to Problem 71 in Exercise 7.3 and Problem 57 in Exercise 7.4.]

47. Formulate a strategy for discussing and graphing an equation of the form

$$Ax^2 + Cy^2 + Dx + Ey + F = 0$$

Systems of Equations and Inequalities

Internet usage and the security of the information passing through the Internet is a major issue for corporations and other organizations. On the following page, you and your friends decide to start your own encryption software company. You will need to use the Sullivan website at

www.prenhall.com/sullivan

in order to find the resources available to build your encryption program.

PREPARING FOR THIS CHAPTER

Before getting started on this chapter, review the following concepts:

For Sections 8.1, 8.2, 8.7, 8.8: Lines *(Section 3.3)*

For Section 8.5: Identity *(p. 89)*

Factoring Polynomials *(Section 1.5)*

Proper and Improper Rational Functions *(pp. 347–348)*

Fundamental Theorem of Algebra *(p. 399)*

For Section 8.6: Lines *(Section 3.3)*;

Circles *(pp. 201–204)*

Parabolas *(pp. 496–504)*

Ellipses *(pp. 506–514)*

Hyperbolas *(pp. 518–527)*

OUTLINE

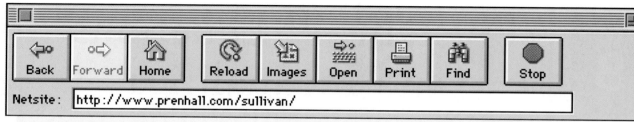

Netsite: http://www.prenhall.com/sullivan/

CRYPTOGRAPHY

Cryptography has a long and fascinating *history*. For example, some believe that Hitler would have won World War II if the German secret codes produced by the *Enigma machine* were not broken!

Today, in the age of the Internet, electronic commerce, and banking, cryptography has once again become indispensable. Companies are investing large sums of money to ensure their security.

You and some of your friends have decided to start your own software company, Phinx-s Security. You want to market your system to small businesses with PCs because this market is wide open at the present moment. Your part of the project is to design the encryption algorithm for the programmer. You have a very clever idea about how to use a dynamic matrix key that will be hardwired into a circuit board.

Now for some of the *basics*.

1. What are some of the current *encryption schemes*?
2. Who are the "cyberpunks"?
3. Are you surprised to see the *$10,000 RSA* and *DES contests*?
4. The kernel of your scheme uses *matrix encryption*. What size matrix might you want to use? 64 by 64? 128 by 128? What are the advantages of using a larger matrix? The disadvantages?
5. How many 64 by 64 matrices are possible mod(26)? Is this equal to the number of possible keys?
6. Would this method work if you were to use mod(71)? Why might you want to use mod(71) over mod(26)? Would you put up a $10,000 challenge?

In this chapter we take up the problem of solving equations and inequalities containing two or more variables. As the section titles suggest, there are various ways to solve such problems.

The *method of substitution* for solving equations in several unknowns goes back to ancient times.

The *method of elimination,* although it had existed for centuries, was put into systematic order by Karl Friedrich Gauss (1777–1855) and by Camille Jordan (1838–1922). This method is now used for solving large systems by computer.

The theory of *matrices* was developed in 1857 by Arthur Cayley (1821–1895), although only later were matrices used as we use them in this chapter. Matrices have become a very flexible instrument, useful in almost all areas of mathematics.

The method of *determinants* was invented by Seki Kōwa (1642–1708) in 1683 in Japan and by Gottfried Wilhelm von Leibniz (1646–1716) in 1693 in Germany. Both used them only in relation to linear equations. *Cramer's Rule* is named after Gabriel Cramer (1704–1752) of Switzerland, who popularized the use of determinants for solving linear systems.

Section 8.5, *partial fraction decomposition,* provides an application of systems of equations. This particular application is one that is used in integral calculus.

Section 8.8 introduces *linear programming,* a modern application of linear inequalities to certain types of problems. This topic is particularly useful for students interested in operations research.

8.1 | SYSTEMS OF LINEAR EQUATIONS: SUBSTITUTION; ELIMINATION

> 1 Solve Systems of Equations by Substitution
> 2 Solve Systems of Equations by Elimination
> 3 Inconsistent Systems and Dependent Equations

We begin with an example.

E X A M P L E   1   **Movie Theater Ticket Sales**

A movie theater sells tickets for $8.00 each, with Seniors receiving a discount of $2.00. One evening the theater took in $3580 in revenue. If x represents the number of tickets sold at $8.00 and y the number of tickets sold at the discounted price of $6.00, write an equation that relates these variables.

Solution   Each nondiscounted ticket brings in $8.00, so x tickets will bring in $8x$ dollars. Similarly, y discounted tickets bring in $6y$ dollars. If the total brought in is $3580, we must have

$$8x + 6y = 3580$$

∎

In Example 1, suppose that we also know that 525 tickets were sold that evening. Then we have another equation relating the variables x and y.

$$x + y = 525$$

The two equations

$$8x + 6y = 3580$$
$$x + y = 525$$

form a *system* of equations.

In general, a **system of equations** is a collection of two or more equations, each containing one or more variables. Example 2 gives some samples of systems of equations.

E X A M P L E   2   **Examples of Systems of Equations**

(a) $\begin{cases} 2x + y = 5 & (1) \\ -4x + 6y = -2 & (2) \end{cases}$ Two equations containing two variables, x and y

(b) $\begin{cases} x + y^2 = 5 & (1) \\ 2x + y = 4 & (2) \end{cases}$ Two equations containing two variables, x and y

(c) $\begin{cases} x + y + z = 6 & (1) \\ 3x - 2y + 4z = 9 & (2) \\ x - y - z = 0 & (3) \end{cases}$ Three equations containing three variables, x, y, and z

(d) $\begin{cases} x + y + z = 5 & (1) \\ x - y = 2 & (2) \end{cases}$ Two equations containing three variables, x, y, and z

(e) $\begin{cases} x + y + z = 6 & (1) \\ 2x + 2z = 4 & (2) \\ y + z = 2 & (3) \\ x = 4 & (4) \end{cases}$ Four equations containing three variables, x, y, and z

∎

We use a brace, as shown, to remind us that we are dealing with a system of equations. We also will find it convenient to number each equation in the system.

A **solution** of a system of equations consists of values for the variables that are solutions of each equation of the system. To **solve** a system of equations means to find all solutions of the system.

For example, $x = 2$, $y = 1$ is a solution of the system in Example 2(a), because

$$2(2) + 1 = 5 \quad \text{and} \quad -4(2) + 6(1) = -2$$

A solution of the system in Example 2(b) is $x = 1$, $y = 2$, because

$$1 + 2^2 = 5 \quad \text{and} \quad 2(1) + 2 = 4$$

Another solution of the system in Example 2(b) is $x = \frac{11}{4}$, $y = -\frac{3}{2}$, which you can check for yourself. A solution of the system in Example 2(c) is $x = 3$, $y = 2$, $z = 1$, because

$$\begin{cases} 3 & + 2 & + 1 & = 6 & \text{(1)} \quad x = 3, y = 2, z = 1 \\ 3(3) & - 2(2) & + 4(1) & = 9 & \text{(2)} \\ 3 & - 2 & - 1 & = 0 & \text{(3)} \end{cases}$$

Note that $x = 3$, $y = 3$, $z = 0$ is not a solution of the system in Example 2(c):

$$\begin{cases} 3 & + 3 & + 0 & = 6 & \text{(1)} \quad x = 3, y = 3, z = 0 \\ 3(3) & - 2(3) & + 4(0) & = 3 \neq 9 & \text{(2)} \\ 3 & - 3 & - 0 & = 0 & \text{(3)} \end{cases}$$

Although these values satisfy equations (1) and (3), they do not satisfy equation (2). Any solution of the system must satisfy *each* equation of the system.

 Now work Problem 3.

When a system of equations has at least one solution, it is said to be **consistent**; otherwise, it is called **inconsistent.**

An equation in n variables is said to be **linear** if it can be written in the form

$$a_1x_1 + a_2x_2 + \cdots + a_nx_n = b$$

where $x_1, x_2, \ldots, x_n$ are n distinct variables, $a_1, a_2, \ldots, a_n$, b are constants, and at least one of the a's is not 0.

Some examples of linear equations are

$$2x + 3y = 2 \qquad 5x - 2y + 3z = 10 \qquad 8x + 8y - 2z + 5w = 0$$

If each equation in a system of equations is linear, then we have a **system of linear equations.** Thus, the systems in Examples 2(a), (c), (d), and (e) are linear, whereas the system in Example 2(b) is nonlinear. We concentrate on solving linear systems in Sections 8.1–8.3, and we will take up nonlinear systems in Section 8.6.

Two Linear Equations Containing Two Variables

We can view the problem of solving a system of two linear equations containing two variables as a geometry problem. The graph of each equation in such a system is a line. Thus, a system of two equations containing two variables represents a pair of lines. The lines either (a) intersect or (b) are parallel or (c) are **coincident** (that is, identical).

(a) If the lines intersect, then the system of equations has one solution, given by the point of intersection. The system is **consistent** and the equations are **independent.**
(b) If the lines are parallel, then the system of equations has no solution, because the lines never intersect. The system is **inconsistent.**
(c) If the lines are coincident, then the system of equations has infinitely many solutions, represented by the totality of points on the line. The system is **consistent** and the equations are **dependent.**

Figure 1 illustrates these conclusions.

FIGURE 1

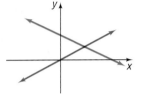
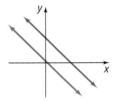
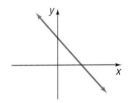

(a) Intersecting lines; system has one solution
(b) Parallel lines; system has no solution
(c) Coincident lines; system has infinitely many solutions

E X A M P L E   3

Graphing a System of Linear Equations

Graph the system: $\begin{cases} 2x + y = 5 & (1) \\ -4x + 6y = 12 & (2) \end{cases}$

Solution   Equation (1) in slope–intercept form is $y = -2x + 5$, which has slope -2 and y-intercept 5. Equation (2) in slope–intercept form is $y = \frac{2}{3}x + 2$, which has slope $\frac{2}{3}$ and y-intercept 2. Figure 2 shows their graph.

FIGURE 2

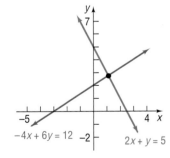

From the graph in Figure 2, we see that the lines intersect, so the system given in Example 3 is consistent. We can also use the graph as a means of approximating the solution. For this system, the solution appears to be close to the point $(1, 3)$. The actual solution, which you should verify, is $\left(\frac{9}{8}, \frac{11}{4}\right)$. To obtain exact solutions we use algebraic methods. The first algebraic method we take up is the *method of substitution.*

Check: Graph the lines $2x + y = 5$ ($Y_1 = -2x + 5$) and $-4x + 6y = 12$ ($Y_2 = \frac{2}{3}x + 2$) and compare what you see with Figure 2. Use INTERSECT to verify that the point of intersection is $(1.125, 2.75)$.

Method of Substitution

We illustrate the **method of substitution** by solving the system given in Example 3.

EXAMPLE 4

Solving a System of Linear Equations by Substitution

Solve: $\begin{cases} 2x + y = 5 & \text{(1)} \\ -4x + 6y = 12 & \text{(2)} \end{cases}$

Solution

We solve the first equation for y, obtaining

$$y = -2x + 5$$

We substitute this result for y in the second equation. The result is an equation containing just the variable x, which we can then solve for.

$$-4x + 6y = 12$$
$$-4x + 6(-2x + 5) = 12$$
$$-4x - 12x + 30 = 12$$
$$-16x = -18$$
$$x = \frac{-18}{-16} = \frac{9}{8}$$

Once we know that $x = \frac{9}{8}$, we can easily find the value of y by **back-substitution,** that is, by substituting $\frac{9}{8}$ for x in one of the previous equations. We use the first one.

$$2x + y = 5$$
$$2\left(\frac{9}{8}\right) + y = 5$$
$$\frac{9}{4} + y = 5$$
$$y = 5 - \frac{9}{4} = \frac{20}{4} - \frac{9}{4} = \frac{11}{4}$$

The solution of the system is $x = \frac{9}{8} = 1.125$, $y = \frac{11}{4} = 2.75$.

The method used to solve the system in Example 4 is called **substitution.** The steps to be used are outlined next.

Steps for Solving by Substitution

STEP 1: Pick one of the equations and solve for one of the variables in terms of the remaining variables.

STEP 2: Substitute the result in the remaining equations.

STEP 3: If one equation in one variable results, solve this equation. Otherwise, repeat Step 1 until a single equation with one variable remains.

STEP 4: Find the values of the remaining variables by back-substitution.

STEP 5: Check the solution found.

E X A M P L E   5

Solving a System of Linear Equations by Substitution

Solve: $\begin{cases} 3x - 2y = 5 & \text{(1)} \\ 5x - y = 6 & \text{(2)} \end{cases}$

Solution   STEP 1: After looking at the two equations, we conclude that it is easiest to solve for the variable y in equation (2).

$$5x - y = 6$$
$$y = 5x - 6$$

STEP 2: We substitute this result into equation (1) and simplify.

$$3x - 2y = 5$$
$$3x - 2(5x - 6) = 5$$
$$-7x + 12 = 5$$
$$-7x = -7$$
$$x = 1$$

STEP 3: Because we now have one solution, $x = 1$, we proceed to Step 4.
STEP 4: Knowing $x = 1$, we can find y from the equation

$$y = 5x - 6 = 5(1) - 6 = -1$$

STEP 5: Check: $\begin{cases} 3(1) - 2(-1) = 3 + 2 = 5 \\ 5(1) - (-1) = 5 + 1 = 6 \end{cases}$

The solution of the system is $x = 1, y = -1$. ■

E X A M P L E   6

Solving a System of Linear Equations by Substitution

Solve: $\begin{cases} 2x - 3y = 7 & \text{(1)} \\ 4x + 5y = 3 & \text{(2)} \end{cases}$

Solution   STEP 1: In looking over the system, we conclude that there is no way to solve for one of the variables without introducing fractions. We solve for the variable x in equation (1).

$$2x - 3y = 7$$
$$2x = 3y + 7$$
$$x = \frac{3}{2}y + \frac{7}{2}$$

STEP 2: We substitute this result for x in equation (2) and simplify.

$$4x + 5y = 3$$
$$4\left(\frac{3}{2}y + \frac{7}{2}\right) + 5y = 3$$
$$6y + 14 + 5y = 3$$
$$11y + 14 = 3$$
$$11y = -11$$

STEP 3: $$y = -1$$

STEP 4: $x = \dfrac{3}{2}y + \dfrac{7}{2} = \dfrac{3}{2}(-1) + \dfrac{7}{2} = \dfrac{4}{2} = 2$

STEP 5: Check: $\begin{cases} 2(2) - 3(-1) = 4 + 3 = 7 \\ 4(2) + 5(-1) = 8 - 5 = 3 \end{cases}$

The solution is $x = 2, y = -1$. ▬

Now use substitution to work Problem 13.

Method of Elimination

A second method for solving a system of linear equations is the *method of elimination.* This method is usually preferred over substitution if substitution leads to fractions or if the system contains more than two variables. Elimination also provides the necessary motivation for solving systems using matrices (the subject of the next section).

The idea behind the method of elimination is to keep replacing the original equations in the system with equivalent equations until a system of equations with an obvious solution is reached. When we proceed in this way, we obtain **equivalent systems of equations.** The rules for obtaining equivalent equations are the same as those studied earlier. However, we may also interchange any two equations of the system and/or replace any equation in the system by the sum (or difference) of that equation and any other equation in the system.

Rules for Obtaining an Equivalent System of Equations

1. Interchange any two equations of the system.
2. Multiply (or divide) each side of an equation by the same nonzero constant.
3. Replace any equation in the system by the sum (or difference) of that equation and any other equation in the system.

An example will give you the idea. As you work through the example, pay particular attention to the pattern being followed.

E X A M P L E 7

Solving a System of Linear Equations by Elimination

Solve: $\begin{cases} 3x + 2y = -1 & (1) \\ -x + y = -3 & (2) \end{cases}$

Solution   The idea behind the method of elimination is to rewrite the system so that adding two of the equations eliminates a variable. For this system, we can accomplish this by multiplying equation (2) by 3. The result is the equivalent system

$$\begin{cases} 3x + 2y = -1 & (1) \\ -3x + 3y = -9 & (2) \end{cases}$$

$5y = -10$ Add equations (1) and (2).

$y = -2$ Solve for y.

We back-substitute by using this value for y in equation (1) and simplify to get

$$3x + 2(-2) = -1$$
$$3x = 3$$
$$x = 1$$

Thus, the solution of the original system is $x = 1$, $y = -2$. We leave it to you to check the solution. ∎

The procedure used in Example 7 is called the **method of elimination.** Notice the pattern of the solution. First, we eliminated the variable x. Then we back-substituted; that is, we substituted the value found for y back into the first equation to find x.

Let's return to the movie theater example (Example 1).

EXAMPLE 8   Movie Theater Ticket Sales

A movie theater sells tickets for $8.00 each, with Seniors receiving a discount of $2.00. One evening the theater sold 525 tickets and took in $3580 in revenue. How many of each type of ticket was sold?

Solution   If x represents the number of tickets sold at $8.00 and y the number of tickets sold at the discounted price of $6.00, then the given information results in the system of equations

$$\begin{cases} 8x + 6y = 3580 \\ x + y = 525 \end{cases}$$

We use elimination and multiply the second equation by -6 and then add the equations.

$$\begin{cases} 8x + 6y = 3580 \\ -6x - 6y = -3150 \end{cases}$$
$$2x = 430$$
$$x = 215$$

Since $x + y = 525$, then $y = 525 - x = 525 - 215 = 310$. Thus, 215 nondiscounted tickets and 310 Senior discount tickets were sold. ∎

Now use elimination to work Problem 13.

The previous examples dealt with consistent systems of equations that had a unique solution. The next two examples deal with two other possibilities that may occur, the first being a system that has no solution.

EXAMPLE 9   An Inconsistent System of Linear Equations

Solve: $\begin{cases} 2x + y = 5 & \text{(1)} \\ 4x + 2y = 8 & \text{(2)} \end{cases}$

Solution   We choose to use the method of substitution and solve equation (1) for y.

$$2x + y = 5$$
$$y = -2x + 5$$

FIGURE 3

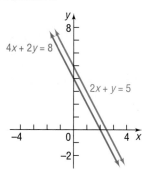

$4x + 2y = 8$

$2x + y = 5$

Substituting in equation (2), we get

$$4x + 2y = 8$$
$$4x + 2(-2x + 5) = 8$$
$$4x - 4x + 10 = 8$$
$$0 \cdot x = -2$$

This equation has no solution. Thus, we conclude that the system itself has no solution and is therefore inconsistent. ▬

Figure 3 illustrates the pair of lines whose equations form the system in Example 9. Notice that the graphs of the two equations are lines, each with slope -2; one has a y-intercept of 5, the other a y-intercept of 4. Thus, the lines are parallel and have no point of intersection. This geometric statement is equivalent to the algebraic statement that the system has no solution.

Check: Graph the lines $2x + y = 5$ ($Y_1 = -2x + 5$) and $4x + 2y = 8$ ($Y_2 = -2x + 4$) and compare what you see with Figure 3. How can you be sure that the lines are parallel? ▬

The next example is an illustration of a system with infinitely many solutions.

EXAMPLE 10

Solving a System of Linear Equations with Infinitely Many Solutions

Solve: $\begin{cases} 2x + y = 4 & (1) \\ -6x - 3y = -12 & (2) \end{cases}$

Solution We choose to use the method of elimination:

$$\begin{cases} 2x + y = 4 & (1) \\ -6x - 3y = -12 & (2) \end{cases}$$

$$\begin{cases} 6x + 3y = 12 & (1) \quad \text{Multiply each side of equation (1) by 3.} \\ -6x - 3y = -12 & (2) \end{cases}$$

$$\begin{cases} 6x + 3y = 12 & (1) \quad \text{Replace equation (2) by the sum} \\ 0 = 0 & (2) \quad \text{of equations (1) and (2).} \end{cases}$$

The original system is thus equivalent to a system containing one equation, so the equations are dependent. This means that any values of x and y for which $6x + 3y = 12$ or, equivalently, $2x + y = 4$ are solutions. For example, $x = 2, y = 0; x = 0, y = 4; x = -1, y = 6; x = 3, y = -2$; and so on, are solutions. There are, in fact, infinitely many values of x and y for which $2x + y = 4$, so the original system has infinitely many solutions. We will write the solutions of the original systems as either

$$y = 4 - 2x$$

where x can be any real number or

$$x = 2 - \frac{1}{2}y$$

where y can be any real number. ▬

FIGURE 4

$\begin{cases} 2x + y = 4 \\ -6x - 3y = -12 \end{cases}$

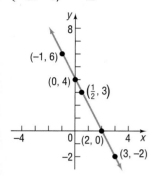

Figure 4 illustrates the situation presented in Example 10. Notice that the graphs of the two equations are lines, each with slope -2 and each with y-intercept 4. Thus, the lines are coincident. Notice also that equation (2) in the original system is just -3 times equation (1), indicating that the two equations are dependent.

For the system in Example 10, we can write down some of the infinite number of solutions by assigning values to x and then finding $y = 4 - 2x$. Thus,

If $x = -1$, then $y = 6$.
If $x = 0$, then $y = 4$.
If $x = 2$, then $y = 0$.

The pairs (x, y) are points on the line in Figure 4.

 Check: Graph the lines $2x + y = 4$ ($Y_1 = -2x + 4$) and $-6x - 3y = -12$ ($Y_2 = -2x + 4$) and compare what you see with Figure 4. How can you be sure that the lines are coincident? ▬

 Now work Problems 19 and 23.

Three Linear Equations Containing Three Variables

Just as with a system of two linear equations containing two variables, a system of three linear equations containing three variables also has either exactly one solution, no solution, or infinitely many solutions.

Let's see how elimination works on a system of three equations containing three variables.

E X A M P L E   11

Solving a System of Three Linear Equations with Three Variables

Use the method of elimination to solve the system of equations:

$$\begin{cases} x + y - z = -1 & (1) \\ 4x - 3y + 2z = 16 & (2) \\ 2x - 2y - 3z = 5 & (3) \end{cases}$$

Solution

For a system of three equations, we attempt to eliminate one variable at a time, using pairs of equations. Our plan of attack on this system will be to first eliminate the variable x from equations (2) and (3). Next, we will eliminate the variable y from equation (3), leaving only the variable z. Back-substitution can then be used to obtain the values of y and then x.

We begin by multiplying each side of equation (1) by -2, in anticipation of eliminating the variable x from equation (3) by adding equations (1) and (3).

$$\begin{cases} -2x - 2y + 2z = 2 & (1) \quad \text{Multiply each side of equation (1)} \\ 4x - 3y + 2z = 16 & (2) \quad \text{by } -2. \\ 2x - 2y - 3z = 5 & (3) \end{cases}$$

$$\begin{cases} -2x - 2y + 2z = 2 & (1) \\ 4x - 3y + 2z = 16 & (2) \\ -4y - z = 7 & (3) \quad \text{Replace equation (3) by the sum of} \\ & \qquad \text{equations (1) and (3).} \end{cases}$$

We now eliminate the variable x from equation (2).

$$\begin{cases} -4x - 4y + 4z = 4 & (1) \\ 4x - 3y + 2z = 16 & (2) \\ -4y - z = 7 & (3) \end{cases}$$
Multiply each side of equation (1) by 2.

$$\begin{cases} -4x - 4y + 4z = 4 & (1) \\ -7y + 6z = 20 & (2) \\ -4y - z = 7 & (3) \end{cases}$$
Replace equation (2) by the sum of equations (1) and (2).

We now eliminate y from equation (3).

$$\begin{cases} -4x - 4y + 4z = 4 & (1) \\ -28y + 24z = 80 & (2) \\ 28y + 7z = -49 & (3) \end{cases}$$
Multiply each side of equation (2) by 4.
Multiply each side of equation (3) by -7.

$$\begin{cases} -4x - 4y + 4z = 4 & (1) \\ -28y + 24z = 80 & (2) \\ 31z = 31 & (3) \end{cases}$$
Replace equation (3) by the sum of equations (2) and (3).

$$\begin{cases} -4x - 4y + 4z = 4 & (1) \\ -28y + 24z = 80 & (2) \\ z = 1 & (3) \end{cases}$$
Multiply each side of equation (3) by $\frac{1}{31}$.

$$\begin{cases} -4x - 4y + 4 = 4 & (1) \\ -28y + 24 = 80 & (2) \\ z = 1 & (3) \end{cases}$$
Back-substitute; replace z by 1 in equations (1) and (2).

$$\begin{cases} -4x - 4y = 0 & (1) \\ y = -2 & (2) \\ z = 1 & (3) \end{cases}$$
Solve equation (2) for y.

$$\begin{cases} -4x + 8 = 0 & (1) \\ y = -2 & (2) \\ z = 1 & (3) \end{cases}$$
Back-substitute; replace y by -2.

$$\begin{cases} x = 2 & (1) \\ y = -2 & (2) \\ z = 1 & (3) \end{cases}$$

The solution of the original system is $x = 2, y = -2, z = 1$. (You should check this.) ∎

Look back over the solution given in Example 11. Note the pattern of making equation (3) contain only the variable z, followed by making equation (2) contain only the variable y and equation (1) contain only the variable x. Although the variables to isolate are your choice, the methodology remains the same for all systems.

8.1 | EXERCISES

In Problems 1–10, verify that the values of the variables listed are solutions of the system of equations.

1. $\begin{cases} 2x - y = 5 \\ 5x + 2y = 8 \end{cases}$
$x = 2, y = -1$

2. $\begin{cases} 3x + 2y = 2 \\ x - 7y = -30 \end{cases}$
$x = -2, y = 4$

3. $\begin{cases} 3x - 4y = 4 \\ \frac{1}{2}x - 3y = -\frac{1}{2} \end{cases}$
$x = 2, y = \frac{1}{2}$

4. $\begin{cases} 2x + \frac{1}{2}y = 0 \\ 3x - 4y = -\frac{19}{2} \end{cases}$
$x = -\frac{1}{2}, y = 2$

5. $\begin{cases} x^2 - y^2 = 3 \\ xy = 2 \end{cases}$
$x = 2, y = 1$

6. $\begin{cases} x^2 - y^2 = 3 \\ xy = 2 \end{cases}$
$x = -2, y = -1$

7. $\begin{cases} \dfrac{x}{1+x} + 3y = 6 \\ x + 9y^2 = 36 \end{cases}$
$x = 0, y = 2$

8. $\begin{cases} \dfrac{x}{x-1} + y = 5 \\ 3x - y = 3 \end{cases}$
$x = 2, y = 3$

9. $\begin{cases} 3x + 3y + 2z = 4 \\ x - y - z = 0 \\ 2y - 3z = -8 \end{cases}$
$x = 1, y = -1, z = 2$

10. $\begin{cases} 4x - z = 7 \\ 8x + 5y - z = 0 \\ -x - y + 5z = 6 \end{cases}$
$x = 2, y = -3, z = 1$

In Problems 11–46, solve each system of equations. If the system has no solution, say it is inconsistent. Use either substitution or elimination.

11. $\begin{cases} x + y = 8 \\ x - y = 4 \end{cases}$

12. $\begin{cases} x + 2y = 5 \\ x + y = 3 \end{cases}$

13. $\begin{cases} 5x - y = 13 \\ 2x + 3y = 12 \end{cases}$

14. $\begin{cases} x + 3y = 5 \\ 2x - 3y = -8 \end{cases}$

15. $\begin{cases} 3x = 24 \\ x + 2y = 0 \end{cases}$

16. $\begin{cases} 4x + 5y = -3 \\ -2y = -4 \end{cases}$

17. $\begin{cases} 3x - 6y = 2 \\ 5x + 4y = 1 \end{cases}$

18. $\begin{cases} 2x + 4y = \frac{2}{3} \\ 3x - 5y = -10 \end{cases}$

19. $\begin{cases} 2x + y = 1 \\ 4x + 2y = 3 \end{cases}$

20. $\begin{cases} x - y = 5 \\ -3x + 3y = 2 \end{cases}$

21. $\begin{cases} 2x - y = 0 \\ 3x + 2y = 7 \end{cases}$

22. $\begin{cases} 3x + 3y = -1 \\ 4x + y = \frac{8}{3} \end{cases}$

23. $\begin{cases} x + 2y = 4 \\ 2x + 4y = 8 \end{cases}$

24. $\begin{cases} 3x - y = 7 \\ 9x - 3y = 21 \end{cases}$

25. $\begin{cases} 2x - 3y = -1 \\ 10x + y = 11 \end{cases}$

26. $\begin{cases} 3x - 2y = 0 \\ 5x + 10y = 4 \end{cases}$

27. $\begin{cases} 2x + 3y = 6 \\ x - y = \frac{1}{2} \end{cases}$

28. $\begin{cases} \frac{1}{2}x + y = -2 \\ x - 2y = 8 \end{cases}$

29. $\begin{cases} \frac{1}{2}x + \frac{1}{3}y = 3 \\ \frac{1}{4}x - \frac{2}{3}y = -1 \end{cases}$

30. $\begin{cases} \frac{1}{3}x - \frac{3}{2}y = -5 \\ \frac{3}{4}x + \frac{1}{3}y = 11 \end{cases}$

31. $\begin{cases} 3x - 5y = 3 \\ 15x + 5y = 21 \end{cases}$

32. $\begin{cases} 2x - y = -1 \\ x + \frac{1}{2}y = \frac{3}{2} \end{cases}$

33. $\begin{cases} x - y = 6 \\ 2x - 3z = 16 \\ 2y + z = 4 \end{cases}$

34. $\begin{cases} 2x + y = -4 \\ -2y + 4z = 0 \\ 3x - 2z = -11 \end{cases}$

35. $\begin{cases} x - 2y + 3z = 7 \\ 2x + y + z = 4 \\ -3x + 2y - 2z = -10 \end{cases}$

36. $\begin{cases} 2x + y - 3z = 0 \\ -2x + 2y + z = -7 \\ 3x - 4y - 3z = 7 \end{cases}$

37. $\begin{cases} x - y - z = 1 \\ 2x + 3y + z = 2 \\ 3x + 2y = 0 \end{cases}$

38. $\begin{cases} 2x - 3y - z = 0 \\ -x + 2y + z = 5 \\ 3x - 4y - z = 1 \end{cases}$

39. $\begin{cases} x - y - z = 1 \\ -x + 2y - 3z = -4 \\ 3x - 2y - 7z = 0 \end{cases}$

40. $\begin{cases} 2x - 3y - z = 0 \\ 3x + 2y + 2z = 2 \\ x + 5y + 3z = 2 \end{cases}$

41. $\begin{cases} 2x - 2y + 3z = 6 \\ 4x - 3y + 2z = 0 \\ -2x + 3y - 7z = 1 \end{cases}$

42. $\begin{cases} 3x - 2y + 2z = 6 \\ 7x - 3y + 2z = -1 \\ 2x - 3y + 4z = 0 \end{cases}$

43. $\begin{cases} x + y - z = 6 \\ 3x - 2y + z = -5 \\ x + 3y - 2z = 14 \end{cases}$

44. $\begin{cases} x - y + z = -4 \\ 2x - 3y + 4z = -15 \\ 5x + y - 2z = 12 \end{cases}$

45. $\begin{cases} x + 2y - z = -3 \\ 2x - 4y + z = -7 \\ -2x + 2y - 3z = 4 \end{cases}$

46. $\begin{cases} x + 4y - 3z = -8 \\ 3x - y + 3z = 12 \\ x + y + 6z = 1 \end{cases}$

47. Solve: $\begin{cases} \dfrac{1}{x} + \dfrac{1}{y} = 8 \\ \dfrac{3}{x} - \dfrac{5}{y} = 0 \end{cases}$

48. Solve: $\begin{cases} \dfrac{4}{x} - \dfrac{3}{y} = 0 \\ \dfrac{6}{x} + \dfrac{3}{2y} = 2 \end{cases}$

[**Hint:** Let $u = 1/x$ and $v = 1/y$, and solve for u and v. Then $x = 1/u$ and $y = 1/v$.]

 In Problems 49–54, use a graphing utility to solve each system of equations. Approximate the solution correct to two decimal places.

49. $\begin{cases} y = \sqrt{2}x - 20\sqrt{7} \\ y = -0.1x + 20 \end{cases}$

50. $\begin{cases} y = -\sqrt{3}x + 100 \\ y = 0.2x + \sqrt{19} \end{cases}$

51. $\begin{cases} \sqrt{2}x + \sqrt{3}y + \sqrt{6} = 0 \\ \sqrt{3}x - \sqrt{2}y + 60 = 0 \end{cases}$

52. $\begin{cases} \sqrt{5}x - \sqrt{6}y + 60 = 0 \\ 0.2x + 0.3y + \sqrt{5} = 0 \end{cases}$

53. $\begin{cases} \sqrt{3}x + \sqrt{2}y = \sqrt{0.3} \\ 100x - 95y = 20 \end{cases}$

54. $\begin{cases} \sqrt{6}x - \sqrt{5}y + \sqrt{1.1} = 0 \\ y = -0.2x + 0.1 \end{cases}$

55. The sum of two numbers is 81. The difference of twice one number and three times the other is 62. Find the two numbers.

56. The difference of two numbers is 40. Six times the smaller number less the larger number is 5. Find the two numbers.

57. The perimeter of a rectangular floor is 90 feet. Find the dimensions of the floor if the length is twice the width.

58. The length of fence required to enclose a rectangular field is 3000 meters. What are the dimensions of the field if it is known that the difference between its length and width is 50 meters?

59. **Cost of Fast Food** Four large cheeseburgers and two chocolate shakes cost a total of $7.90. Two shakes cost 15¢ more than one cheeseburger. What is the cost of a cheeseburger? A shake?

60. **Movie Theater Tickets** A movie theater charges $9.00 for adults and $7.00 for senior citizens. On a day when 325 people paid an admission, the total receipts were $2495. How many who paid were adults? How many were seniors?

61. **Mixing Nuts** A store sells cashews for $5.00 per pound and peanuts for $1.50 per pound. The manager decides to mix 30 pounds of peanuts with some cashews and sell the mixture for $3.00 per pound. How many pounds of cashews should be mixed with the peanuts so that the mixture will produce the same revenue as would selling the nuts separately?

62. **Financial Planning** A recently retired couple need $12,000 per year to supplement their Social Security. They have $150,000 to invest to obtain this income. They have decided on two investment options: AA bonds yielding 10% per annum and a Bank Certificate yielding 5%.
(a) How much should be invested in each to realize exactly $12,000?
(b) If, after two years, the couple requires $14,000 per year in income, how should they reallocate their investment to achieve the new amount?

63. **Computing Wind Speed** With a tailwind, a small Piper aircraft can fly 600 miles in 3 hours. Against this same wind, the Piper can fly the same distance in 4 hours. Find the average wind speed and the average airspeed of the Piper.

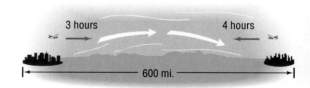

3 hours   4 hours

600 mi.

64. **Computing Wind Speed** The average airspeed of a single-engine aircraft is 150 miles per hour. If the aircraft flew the same distance in 2 hours with the wind as it flew in 3 hours against the wind, what was the wind speed?

65. **Restaurant Management** A restaurant manager wants to purchase 200 sets of dishes. One design costs $25 per set, while another costs $45 per set. If she wants to spend exactly $7400, how many of each design should be ordered?

66. Cost of Fast Food One group of people purchased 10 hot dogs and 5 soft drinks at a cost of $12.50. A second bought 7 hot dogs and 4 soft drinks at a cost of $9.00. What is the cost of a single hot dog? A single soft drink?

We paid $12.50.
How much is one hot dog?
How much is one cola?

We paid $9.00.
How much is one hot dog?
How much is one cola?

67. Computing a Refund The grocery store we use does not mark prices on its goods. My wife went to this store, bought three 1 pound packages of bacon and two cartons of eggs, and paid a total of $7.45. Not knowing she went to the store, I also went to the same store, purchased two 1 pound packages of bacon and three cartons of eggs, and paid a total of $6.45. Now we want to return two 1 pound packages of bacon and two cartons of eggs. How much will be refunded?

68. Finding the Current of a Stream Pamela requires 3 hours to swim 15 miles downstream on the Illinois River. The return trip upstream takes 5 hours. Find Pamela's average speed in still water. How fast is the current? (Assume that Pamela's speed is the same in each direction.)

69. Pharmacy A doctor's prescription calls for a daily intake of liquid containing 40 mg of vitamin C and 30 mg of vitamin D. Your pharmacy stocks two liquids that can be used: one contains 20% vitamin C and 30% vitamin D, the other 40% vitamin C and 20% vitamin D. How many milligrams of each liquid should be mixed to fill the prescription?

70. Pharmacy A doctor's prescription calls for the creation of pills that contain 12 units of vitamin B_{12} and 12 units of vitamin E. Your pharmacy stocks two powders that can be used to make these pills: one contains 20% vitamin B_{12} and 30% vitamin E, the other 40% vitamin B_{12} and 20% vitamin E. How many units of each powder should be mixed in each pill?

71. Electricity: Kirchhoff's Rules An application of *Kirchhoff's Rules* to the circuit shown results in the following system of equations:

$$\begin{cases} I_2 = I_1 + I_3 \\ 5 - 3I_1 - 5I_2 = 0 \\ 10 - 5I_2 - 7I_3 = 0 \end{cases}$$

Find the currents I_1, I_2, and I_3.*

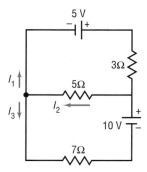

72. Electricity: Kirchhoff's Rules An application of Kirchhoff's Rules to the circuit shown results in the following system of equations:

$$\begin{cases} I_3 = I_1 + I_2 \\ 8 = 4I_3 + 6I_2 \\ 8I_1 = 4 + 6I_2 \end{cases}$$

Find the currents I_1, I_2, and I_3.†

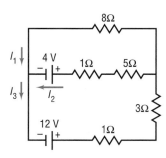

73. Theater Revenues A Broadway theater has 500 seats, divided into orchestra, main, and balcony seating. Orchestra seats sell for $50, main seats for $35, and balcony seats for $25. If all the seats are sold, the gross revenue to the theater is $17,100. If all the main and balcony seats are sold, but only half the orchestra seats are sold, the gross revenue is $14,600. How many are there of each kind of seat?

*Source: Based on Raymond Serway, *Physics,* 3rd ed. (Philadelphia: Saunders, 1990), Prob. 26, p. 790.
†Source: Ibid, Prob. 27, p. 790.

74. Laboratory Work Stations A chemistry laboratory can be used by 38 students at one time. The laboratory has 16 work stations, some set up for 2 students each and the others set up for 3 students each. How many are there of each kind of work station?

 75. Make up a system of two linear equations containing two variables that has
(a) No solution
(b) Exactly one solution
(c) Infinitely many solutions

Give the three systems to a friend to solve and critique.

76. Write a brief paragraph outlining your strategy for solving a system of two linear equations containing two variables.

77. Do you prefer the method of substitution or the method of elimination for solving a system of two linear equations containing two variables? What about for solving a system of three linear equations containing three variables? Give reasons.

78. Curve Fitting Find real numbers b and c such that the parabola $y = x^2 + bx + c$ passes through the points $(1, 3)$ and $(3, 5)$.

79. Curve Fitting Find real numbers b and c such that the parabola $y = x^2 + bx + c$ passes through the points $(1, 2)$ and $(-1, 3)$.

80. Curve Fitting Find real numbers b and c such that the parabola $y = x^2 + bx + c$ passes through the points (x_1, y_1) and (x_2, y_2).

81. Curve Fitting Find real numbers a, b, and c such that the parabola $y = ax^2 + bx + c$ passes through the points $(-1, 4)$, $(2, 3)$, and $(0, 1)$.

82. Curve Fitting Find real numbers a, b, and c such that the parabola $y = ax^2 + bx + c$ passes through the points $(-1, -2)$, $(1, -4)$, and $(2, 4)$.

83. Solve: $\begin{cases} y = m_1 x + b_1 \\ y = m_2 x + b_2 \end{cases}$
where $m_1 \neq m_2$.

84. Solve: $\begin{cases} y = m_1 x + b_1 \\ y = m_2 x + b_2 \end{cases}$
where $m_1 = m_2 = m$ and $b_1 \neq b_2$.

85. Solve: $\begin{cases} y = m_1 x + b_1 \\ y = m_2 x + b_2 \end{cases}$
where $m_1 = m_2 = m$ and $b_1 = b_2 = b$.

8.2 SYSTEMS OF LINEAR EQUATIONS: MATRICES

1 Write the Augmented Matrix of a System of Linear Equations
2 Write the System from the Augmented Matrix
3 Perform Row Operations on a Matrix
4 Solve Systems of Linear Equations Using Matrices

The systematic approach of the method of elimination for solving a system of linear equations provides another method of solution that involves a simplified notation.

Consider the following system of linear equations:

$$\begin{cases} x + 4y = 14 \\ 3x - 2y = 0 \end{cases}$$

If we choose not to write the symbols used for the variables, we can represent this system as

$$\begin{bmatrix} 1 & 4 & | & 14 \\ 3 & -2 & | & 0 \end{bmatrix}$$

where it is understood that the first column represents the coefficients of the variable x, the second column the coefficients of y, and the third column the constants on the right side of the equal signs. The vertical line serves as a reminder of the equal signs. The large square brackets are the traditional symbols used to denote a *matrix* in algebra.

A **matrix** is defined as a rectangular array of numbers,

$$
\begin{array}{cccc}
 & \text{Column 1} & \text{Column 2} & \text{Column } j & \text{Column } n \\
\text{Row 1} \\
\text{Row 2} \\
\vdots \\
\text{Row } i \\
\vdots \\
\text{Row } m
\end{array}
\begin{bmatrix}
a_{11} & a_{12} & \cdots & a_{1j} & \cdots & a_{1n} \\
a_{21} & a_{22} & \cdots & a_{2j} & \cdots & a_{2n} \\
\vdots & \vdots & & \vdots & & \vdots \\
a_{i1} & a_{i2} & \cdots & a_{ij} & \cdots & a_{in} \\
\vdots & \vdots & & \vdots & & \vdots \\
a_{m1} & a_{m2} & \cdots & a_{mj} & \cdots & a_{mn}
\end{bmatrix} \quad (1)
$$

Each number a_{ij} of the matrix has two indices: the **row index** i and the **column index** j. The matrix shown in display (1) has m rows and n columns. The numbers a_{ij} are usually referred to as the **entries** of the matrix.

Now we will use matrix notation to represent a system of linear equations. The matrices used to represent systems of linear equations are called **augmented matrices.** In writing the augmented matrix of a system, the variables of each equation must be on the left side of the equal sign and the constants on the right side. A variable that does not appear in an equation has a coefficient of 0.

E X A M P L E   1   Writing the Augmented Matrix of a System of Linear Equations

Write the augmented matrix of each system of equations.

(a) $\begin{cases} 3x - 4y = -6 & (1) \\ 2x - 3y = -5 & (2) \end{cases}$ (b) $\begin{cases} 2x - y + z = 0 & (1) \\ x + z - 1 = 0 & (2) \\ x + 2y - 8 = 0 & (3) \end{cases}$

Solution   (a) The augmented matrix is

$$
\begin{bmatrix}
3 & -4 & \bigm| & -6 \\
2 & -3 & \bigm| & -5
\end{bmatrix}
$$

(b) Care must be taken that the system is written so that the coefficients of all variables are taken into account (if any variable is missing, its coefficient is 0). Also, all constants must be to the right of the equal sign. Thus, we need to rearrange the given system as follows:

$$
\begin{cases}
2x - y + z = 0 & (1) \\
x + z - 1 = 0 & (2) \\
x + 2y - 8 = 0 & (3)
\end{cases}
$$

$$
\begin{cases}
2x - y + z = 0 & (1) \\
x + 0 \cdot y + z = 1 & (2) \\
x + 2y + 0 \cdot z = 8 & (3)
\end{cases}
$$

The augmented matrix is

$$
\begin{bmatrix}
2 & -1 & 1 & \bigm| & 0 \\
1 & 0 & 1 & \bigm| & 1 \\
1 & 2 & 0 & \bigm| & 8
\end{bmatrix}
$$

If we do not include the constants to the right of the equal sign, that is, to the right of the vertical bar in the augmented matrix of a system of equations, the resulting matrix is called the **coefficient matrix** of the system. For the systems discussed in Example 1, the coefficient matrices are

$$\begin{bmatrix} 3 & -4 \\ 2 & -3 \end{bmatrix} \text{ and } \begin{bmatrix} 2 & -1 & 1 \\ 1 & 0 & 1 \\ 1 & 2 & 0 \end{bmatrix}$$

 Now work Problem 3.

E X A M P L E 2

Writing the System of Linear Equations from the Augmented Matrix

Write the system of linear equations corresponding to each augmented matrix.

(a) $\begin{bmatrix} 5 & 2 & | & 13 \\ -3 & 1 & | & -10 \end{bmatrix}$ (b) $\begin{bmatrix} 3 & -1 & -1 & | & 7 \\ 2 & 0 & 2 & | & 8 \\ 0 & 1 & 1 & | & 0 \end{bmatrix}$

Solution (a) The matrix has two rows and so represents a system of two equations. The two columns to the left of the vertical bar indicate that the system has two variables. If x and y are used to denote these variables, the system of equations is

$$\begin{cases} 5x + 2y = 13 & (1) \\ -3x + y = -10 & (2) \end{cases}$$

(b) This matrix represents a system of three equations containing three variables. If x, y, and z are the three variables, this system is

$$\begin{cases} 3x - y - z = 7 & (1) \\ 2x + 2z = 8 & (2) \\ y + z = 0 & (3) \end{cases}$$

Row Operations on a Matrix

E X A M P L E 3

Solving a System of Equations

Solve the system of equations

$$\begin{cases} 4x - 3y = 11 & (1) \\ 3x + 2y = 4 & (2) \end{cases}$$

Solution We will use a variation of the method of elimination to solve the system. First, multiply each side of equation (2) by -1 and add it to equation (1). Replace equation (1) with the result:

$$\begin{cases} x - 5y = 7 & (1) \\ 3x + 2y = 4 & (2) \end{cases}$$

$$\begin{array}{ll} 4x - 3y = 11 & \text{Equation (1)} \\ -3x - 2y = -4 & \text{Multiply (2) by } -1. \\ \hline x - 5y = 7 & \text{Add; Replace equation (1) by this equation.} \end{array}$$

Multiply each side of equation (1) by -3 and add it to equation (2). Replace equation (2) with the result:

$$\begin{cases} x - 5y = 7 & (1) \\ 17y = -17 & (2) \end{cases}$$

$$\begin{array}{ll} -3x + 15y = -21 & \text{Multiply (1) by } -3. \\ 3x + 2y = 4 & \text{Equation (2)} \\ \hline 17y = -17 & \text{Add; Replace equation (2) by this equation.} \end{array}$$

Multiply each side of equation (2) by $\frac{1}{17}$:

$$\begin{cases} x - 5y = 7 & \text{(1)} \\ y = -1 & \text{(2)} \end{cases}$$

Now we back-substitute $y = -1$ into equation (1) to get

$$x - 5y = 7$$
$$x - 5(-1) = 7$$
$$x = 2$$

The solution of the system is $x = 2$, $y = -1$. ∎

The pattern of solution shown above provides a systematic way to solve any system of equations. The idea is to start with the augmented matrix of the system,

$$\begin{bmatrix} 4 & -3 & | & 11 \\ 3 & 2 & | & 4 \end{bmatrix} \qquad \begin{cases} 4x - 3y = 11 & \text{(1)} \\ 3x + 2y = 4 & \text{(2)} \end{cases}$$

and eventually arrive at the matrix,

$$\begin{bmatrix} 1 & -5 & | & 7 \\ 0 & 1 & | & -1 \end{bmatrix} \qquad \begin{cases} x - 5y = 7 & \text{(1)} \\ y = -1 & \text{(2)} \end{cases}$$

Let's go through the procedure again, this time starting with the augmented matrix and keeping the final augmented matrix given above in mind:

$$\begin{bmatrix} 4 & -3 & | & 11 \\ 3 & 2 & | & 4 \end{bmatrix} \qquad \begin{cases} 4x - 3y = 11 & \text{(1)} \\ 3x + 2y = 4 & \text{(2)} \end{cases}$$

As before, we start by multiplying each side of equation (2) by -1 and adding it to equation (1). This is equivalent to multiplying each entry in the second row of the matrix by -1, adding the result to the corresponding entries in row 1, and replacing row 1 by these entries. The result of this step is that the number 1 appears in row 1, column 1:

$$\begin{bmatrix} (-1)3 + 4 & (-1)2 + (-3) & | & (-1)4 + 11 \\ 3 & 2 & | & 4 \end{bmatrix} = \begin{bmatrix} 1 & -5 & | & 7 \\ 3 & 2 & | & 4 \end{bmatrix} \qquad \begin{cases} x - 5y = 7 & \text{(1)} \\ 3x + 2y = 4 & \text{(2)} \end{cases}$$

Multiply each entry in the first row by -3, add the result to the entries in the second row, and replace the second row by these entries. The result of this step is that the number 0 appears in row 2, column 1:

$$\begin{bmatrix} 1 & -5 & | & 7 \\ (-3)\cdot 1 + 3 & (-3)(-5) + 2 & | & (-3)7 + 4 \end{bmatrix} = \begin{bmatrix} 1 & -5 & | & 7 \\ 0 & 17 & | & -17 \end{bmatrix} \qquad \begin{cases} x - 5y = 7 & \text{(1)} \\ 0 \cdot x + 17y = -17 & \text{(2)} \end{cases}$$

Multiply each entry in the second row by $\frac{1}{17}$. The result of this step is that the number 1 appears in row 2, column 2:

$$\begin{bmatrix} 1 & -5 & | & 7 \\ 0 & 1 & | & -1 \end{bmatrix} \qquad \begin{cases} x - 5y = 7 & \text{(1)} \\ y = -1 & \text{(2)} \end{cases}$$

Now that we know that $y = -1$, we can back-substitute to find that $x = 2$.

The manipulations just performed on the augmented matrix are called **row operations.** There are three basic row operations:

Row Operations

1. Interchange any two rows.
2. Replace a row by a nonzero multiple of that row.
3. Replace a row by the sum of that row and a constant multiple of some other row.

These three row operations correspond to the three rules given earlier for obtaining an equivalent system of equations. Thus, when a row operation is performed on a matrix, the resulting matrix represents a system of equations equivalent to the system represented by the original matrix.

For example, consider the augmented matrix

$$\begin{bmatrix} 1 & 2 & | & 3 \\ 4 & -1 & | & 2 \end{bmatrix}$$

Suppose that we want to apply a row operation to this matrix that results in a matrix whose entry in row 2, column 1 is a 0. The row operation to use is

> Multiply each entry in row 1 by -4 and add the result to the corresponding entries in row 2. $\qquad$ (2)

If we use R_2 to represent the new entries in row 2 and we use r_1 and r_2 to represent the original entries in rows 1 and 2, respectively, then we can represent the row operation in statement (2) by

$$R_2 = -4r_1 + r_2$$

Then

$$\begin{bmatrix} 1 & 2 & | & 3 \\ 4 & -1 & | & 2 \end{bmatrix} \underset{\underset{R_2 = -4r_1 + r_2}{\uparrow}}{\rightarrow} \begin{bmatrix} 1 & 2 & | & 3 \\ -4(1) + 4 & -4(2) + (-1) & | & -4(3) + 2 \end{bmatrix} = \begin{bmatrix} 1 & 2 & | & 3 \\ 0 & -9 & | & -10 \end{bmatrix}$$

As desired, we now have the entry 0 in row 2, column 1.

EXAMPLE 4 $\qquad$ **Applying a Row Operation to an Augmented Matrix**

Apply the row operation $R_2 = -3r_1 + r_2$ to the augmented matrix

$$\begin{bmatrix} 1 & -2 & | & 2 \\ 3 & -5 & | & 9 \end{bmatrix}$$

Solution $\qquad$ The row operation $R_2 = -3r_1 + r_2$ tells us that the entries in row 2 are to be replaced by the entries obtained after multiplying each entry in row 1 by -3 and adding the result to the corresponding entries in row 2. Thus,

$$\begin{bmatrix} 1 & -2 & | & 2 \\ 3 & -5 & | & 9 \end{bmatrix} \underset{\underset{R_2 = -3r_1 + r_2}{\uparrow}}{\rightarrow} \begin{bmatrix} 1 & -2 & | & 2 \\ -3(1) + 3 & (-3)(-2) + (-5) & | & -3(2) + 9 \end{bmatrix} = \begin{bmatrix} 1 & -2 & | & 2 \\ 0 & 1 & | & 3 \end{bmatrix}$$

■

EXAMPLE 5 $\qquad$ **Finding a Particular Row Operation**

Using the matrix

$$\begin{bmatrix} 1 & -2 & | & 2 \\ 0 & 1 & | & 3 \end{bmatrix}$$

find a row operation that will result in a matrix with a 0 in row 1, column 2.

Solution $\qquad$ We want a 0 in row 1, column 2. This result can be accomplished by multiplying a row 2 by 2 and adding the result to row 1. That is, we apply the row operation $R_1 = 2r_2 + r_1$:

$$\begin{bmatrix} 1 & -2 & | & 2 \\ 0 & 1 & | & 3 \end{bmatrix} \underset{\underset{R_1 = 2r_2 + r_1}{\uparrow}}{\rightarrow} \begin{bmatrix} 2(0) + 1 & 2(1) + (-2) & | & 2(3) + 2 \\ 0 & 1 & | & 3 \end{bmatrix} = \begin{bmatrix} 1 & 0 & | & 8 \\ 0 & 1 & | & 3 \end{bmatrix}$$

■

A word about the notation we have introduced. A row operation such as $R_1 = 2r_2 + r_1$ changes the entries in row 1. Note also that for this type of row operation we change the entries in a given row by multiplying the entries in some other row by an appropriate number and adding the results to the original entries of the row to be changed.

Now work Problem 13.

Let's see how we use row operations to solve a system of linear equations.

E X A M P L E   6

Solving a System of Linear Equations Using Matrices

Solve: $\begin{cases} 4x + 3y = 11 & (1) \\ x - 3y = -1 & (2) \end{cases}$

Solution   First, we write the augmented matrix that represents this system:

$$\begin{bmatrix} 4 & 3 & | & 11 \\ 1 & -3 & | & -1 \end{bmatrix}$$

The first step is to get a 1 in row 1, column 1. (This will make the back-substitution that comes later a little easier.) An interchange of rows 1 and 2 is the easiest way to do this.

$$\begin{bmatrix} 1 & -3 & | & -1 \\ 4 & 3 & | & 11 \end{bmatrix}$$

Next we want a 0 under the entry 1 in column 1. (This eliminates the variable x from the second equation.) We use the row operation $R_2 = -4r_1 + r_2$.

$$\begin{bmatrix} 1 & -3 & | & -1 \\ 4 & 3 & | & 11 \end{bmatrix} \underset{\substack{\uparrow \\ R_2 = -4r_1 + r_2}}{\rightarrow} \begin{bmatrix} 1 & -3 & | & -1 \\ 0 & 15 & | & 15 \end{bmatrix}$$

Now we want the entry 1 in row 2, column 2. (This makes it easy to solve for y.) We use $R_2 = \frac{1}{15}r_2$.

$$\begin{bmatrix} 1 & -3 & | & -1 \\ 0 & 15 & | & 15 \end{bmatrix} \underset{\substack{\uparrow \\ R_2 = \frac{1}{15}r_2}}{\rightarrow} \begin{bmatrix} 1 & -3 & | & -1 \\ 0 & 1 & | & 1 \end{bmatrix}$$

The second row of the matrix on the right represents the equation $y = 1$. Thus, using $y = 1$, we back-substitute into the equation $x - 3y = -1$ (from the first row) to get

$$\begin{aligned} x - 3(1) &= -1 \quad y = 1 \\ x &= 2 \end{aligned}$$

The solution of the system is $x = 2$, $y = 1$.   ■

Now work Problem 35.

The steps we used to solve the system of linear equations in Example 6 can be summarized as follows:

Matrix Method for Solving
a System of Linear Equations

STEP 1: Write the augmented matrix that represents the system.
STEP 2: Perform row operations that place the entry 1 in row 1, column 1.
STEP 3: Perform row operations that leave the entry 1 in row 1, column 1 unchanged, while causing 0's to appear below it in column 1.

> STEP 4: Perform row operations that place the entry 1 in row 2, column 2 and leave the entries in columns to the left unchanged. If it is impossible to place a 1 in row 2, column 2, then proceed to place a 1 in row 2, column 3. Once a 1 is in place, perform row operations to place 0's under it.
>
> STEP 5: Now repeat step 4, placing a 1 in the next row, but one column to the right. Continue until the bottom row or the vertical bar is reached.
>
> STEP 6: If any rows are obtained that contain only 0's on the left side of the vertical bar, then place such rows at the bottom of the matrix.

After steps 1 to 6 have been completed, the matrix is said to be in **echelon form.** A little thought should convince you that a matrix is in echelon form when

1. The entry in row 1, column 1 is a 1, and 0's appear below it.
2. The first nonzero entry in each row after the first row is a 1, 0's appear below it, and it appears to the right of the first nonzero entry in any row above.
3. Any rows that contain all 0's to the left of the vertical bar appear at the bottom.

Two advantages of solving a system of equations by writing the augmented matrix in echelon form are the following:

1. The process is algorithmic; that is, it consists of repetitive steps that can be programmed on a computer.
2. The process works on any system of linear equations, no matter how many equations or variables are present.

The next example shows how to write a matrix in echelon form.

E X A M P L E   7   Solving a System of Linear Equations Using Matrices

$$\text{Solve:} \quad \begin{cases} x - y + z = 8 & (1) \\ 2x + 3y - z = -2 & (2) \\ 3x - 2y - 9z = 9 & (3) \end{cases}$$

Solution   STEP 1: The augmented matrix of the system is

$$\begin{bmatrix} 1 & -1 & 1 & 8 \\ 2 & 3 & -1 & -2 \\ 3 & -2 & -9 & 9 \end{bmatrix}$$

STEP 2: Because the entry 1 is already present in row 1, column 1, we can go to step 3.

STEP 3: Perform the row operations $R_2 = -2r_1 + r_2$ and $R_3 = -3r_1 + r_3$.*
Each of these leaves the entry 1 in row 1, column 1 unchanged,
while causing 0's to appear under it.

$$\begin{bmatrix} 1 & -1 & 1 & | & 8 \\ 2 & 3 & -1 & | & -2 \\ 3 & -2 & -9 & | & 9 \end{bmatrix} \rightarrow \begin{bmatrix} 1 & -1 & 1 & | & 8 \\ 0 & 5 & -3 & | & -18 \\ 0 & 1 & -12 & | & -15 \end{bmatrix}$$

$$\uparrow$$
$$R_2 = -2r_1 + r_2$$
$$R_3 = -3r_1 + r_3$$

STEP 4: The easiest way to obtain the entry 1 in row 2, column 2 without
altering column 1 is to interchange rows 2 and 3 (another way
would be to multiply row 2 by $\frac{1}{5}$, but this introduces fractions).

$$\begin{bmatrix} 1 & -1 & 1 & | & 8 \\ 0 & 1 & -12 & | & -15 \\ 0 & 5 & -3 & | & -18 \end{bmatrix}$$

To get 0's under the 1 in row 2, column 2, perform the row operation $R_3 = -5r_2 + r_3$.

$$\begin{bmatrix} 1 & -1 & 1 & | & 8 \\ 0 & 1 & -12 & | & -15 \\ 0 & 5 & -3 & | & -18 \end{bmatrix} \rightarrow \begin{bmatrix} 1 & -1 & 1 & | & 8 \\ 0 & 1 & -12 & | & -15 \\ 0 & 0 & 57 & | & 57 \end{bmatrix}$$

$$\uparrow$$
$$R_3 = -5r_2 + r_3$$

STEP 5: Continuing, we place a 1 in row 3, column 3 by using $R_3 = \frac{1}{57}r_3$.

$$\begin{bmatrix} 1 & -1 & 1 & | & 8 \\ 0 & 1 & -12 & | & -15 \\ 0 & 0 & 57 & | & 57 \end{bmatrix} \rightarrow \begin{bmatrix} 1 & -1 & 1 & | & 8 \\ 0 & 1 & -12 & | & -15 \\ 0 & 0 & 1 & | & 1 \end{bmatrix}$$

$$\uparrow$$
$$R_3 = \frac{1}{57}r_3$$

Because we have reached the bottom row, the matrix is in echelon form and
we can stop.

The system of equations represented by the matrix in echelon form is

$$\begin{cases} x - y + z = 8 \\ y - 12z = -15 \\ z = 1 \end{cases}$$

Using $z = 1$, we back-substitute to get

$$\begin{cases} x - y + 1 = 8 \\ y - 12(1) = -15 \end{cases} \quad \text{or equivalently} \quad \begin{cases} x - y = 7 \\ y = -3 \end{cases}$$

Thus, we get $y = -3$, and back-substituting into $x - y = 7$, we find $x = 4$. The
solution of the system is $x = 4$, $y = -3$, $z = 1$. ∎

Sometimes, it is advantageous to write a matrix in **reduced echelon form.**
In this form, row operations are used to obtain entries that are 0 above (as

*You should convince yourself that doing these two row operations at the same time is the same
as doing one followed by the other.

well as below) the leading 1 in a row. For example, the echelon form obtained in the solution to Example 7 is

$$\begin{bmatrix} 1 & -1 & 1 & | & 8 \\ 0 & 1 & -12 & | & -15 \\ 0 & 0 & 1 & | & 1 \end{bmatrix}$$

To write this matrix in reduced echelon form, we proceed as follows:

$$\begin{bmatrix} 1 & -1 & 1 & | & 8 \\ 0 & 1 & -12 & | & -15 \\ 0 & 0 & 1 & | & 1 \end{bmatrix} \rightarrow \begin{bmatrix} 1 & 0 & -11 & | & -7 \\ 0 & 1 & -12 & | & -15 \\ 0 & 0 & 1 & | & 1 \end{bmatrix} \rightarrow \begin{bmatrix} 1 & 0 & 0 & | & 4 \\ 0 & 1 & 0 & | & -3 \\ 0 & 0 & 1 & | & 1 \end{bmatrix}$$
$$\uparrow \qquad\qquad\qquad\qquad\qquad\qquad \uparrow$$
$$R_1 = r_2 + r_1 \qquad\qquad\qquad\qquad R_1 = 11r_3 + r_1$$
$$R_2 = 12r_3 + r_2$$

The matrix is now written in reduced echelon form. The advantage of writing the matrix in this form is that the solution to the system, $x = 4$, $y = -3$, $z = 1$, is readily found without the need to back-substitute. Another advantage will be seen in Section 8.4, where the inverse of a matrix is discussed.

 Now work Problem 53.

The matrix method for solving a system of linear equations also identifies systems that have infinitely many solutions and systems that are inconsistent. Let's see how.

EXAMPLE 8   Solving a System of Linear Equations Using Matrices

Solve: $\begin{cases} 6x - y - z = 4 & (1) \\ -12x + 2y + 2z = -8 & (2) \\ 5x + y - z = 3 & (3) \end{cases}$

Solution   We start with the augmented matrix of the system:

$$\begin{bmatrix} 6 & -1 & -1 & | & 4 \\ -12 & 2 & 2 & | & -8 \\ 5 & 1 & -1 & | & 3 \end{bmatrix} \rightarrow \begin{bmatrix} 1 & -2 & 0 & | & 1 \\ -12 & 2 & 2 & | & -8 \\ 5 & 1 & -1 & | & 3 \end{bmatrix} \rightarrow \begin{bmatrix} 1 & -2 & 0 & | & 1 \\ 0 & -22 & 2 & | & 4 \\ 0 & 11 & -1 & | & -2 \end{bmatrix}$$
$$\uparrow \qquad\qquad\qquad\qquad\qquad \uparrow$$
$$R_1 = -1r_3 + r_1 \qquad\qquad\qquad R_2 = 12r_1 + r_2$$
$$R_3 = -5r_1 + r_3$$

Obtaining a 1 in row 2, column 2 without altering column 1 can be accomplished by $R_2 = -\frac{1}{22}r_2$ or by $R_3 = \frac{1}{11}r_3$ and interchanging rows or by $R_2 = \frac{23}{11}r_3 + r_2$. We shall use the first of these.

$$\begin{bmatrix} 1 & -2 & 0 & | & 1 \\ 0 & -22 & 2 & | & 4 \\ 0 & 11 & -1 & | & -2 \end{bmatrix} \rightarrow \begin{bmatrix} 1 & -2 & 0 & | & 1 \\ 0 & 1 & -\frac{1}{11} & | & -\frac{2}{11} \\ 0 & 11 & -1 & | & -2 \end{bmatrix} \rightarrow \begin{bmatrix} 1 & -2 & 0 & | & 1 \\ 0 & 1 & -\frac{1}{11} & | & -\frac{2}{11} \\ 0 & 0 & 0 & | & 0 \end{bmatrix}$$
$$\uparrow \qquad\qquad\qquad\qquad\qquad\qquad \uparrow$$
$$R_2 = -\frac{1}{22}r_2 \qquad\qquad\qquad\qquad R_3 = -11r_2 + r_3$$

This matrix is in echelon form. Because the bottom row consists entirely of 0's, the system actually consists of only two equations.

$$\begin{cases} x - 2y = 1 & \text{(1)} \\ y - \frac{1}{11}z = -\frac{2}{11} & \text{(2)} \end{cases}$$

We shall back-substitute the solution for y from the second equation, $y = \frac{1}{11}z - \frac{2}{11}$, into the first equation to get

$$x = 2y + 1 = 2\left(\frac{1}{11}z - \frac{2}{11}\right) + 1 = \frac{2}{11}z + \frac{7}{11}$$

Thus, the original system is equivalent to the system

$$\begin{cases} x = \frac{2}{11}z + \frac{7}{11} & \text{(1)} \\ y = \frac{1}{11}z - \frac{2}{11} & \text{(2)} \end{cases}$$

where z can be any real number.

Let's look at the situation. The original system of three equations is equivalent to a system containing two equations. This means any values of x, y, z that satisfy both

$$x = \frac{2}{11}z + \frac{7}{11} \quad \text{and} \quad y = \frac{1}{11}z - \frac{2}{11}$$

will be solutions. For example, $z = 0, x = \frac{7}{11}, y = -\frac{2}{11}$; $z = 1, x = \frac{9}{11}, y = -\frac{1}{11}$; and $z = -1, x = \frac{5}{11}, y = -\frac{3}{11}$ are some of the solutions of the original system. There are, in fact, infinitely many values of $x, y,$ and z for which the two equations are satisfied. That is, the original system has infinitely many solutions. We will write the solution of the original system as

$$\begin{cases} x = \frac{2}{11}z + \frac{7}{11} \\ y = \frac{1}{11}z - \frac{2}{11} \end{cases}$$

where z can be any real number. ∎

We can also find the solution by writing the augmented matrix in reduced echelon form. Starting with the echelon form, we have

$$\begin{bmatrix} 1 & -2 & 0 & | & 1 \\ 0 & 1 & -\frac{1}{11} & | & -\frac{2}{11} \\ 0 & 0 & 0 & | & 0 \end{bmatrix} \rightarrow \begin{bmatrix} 1 & 0 & -\frac{2}{11} & | & \frac{7}{11} \\ 0 & 1 & -\frac{1}{11} & | & -\frac{2}{11} \\ 0 & 0 & 0 & | & 0 \end{bmatrix}$$
$$\uparrow$$
$$R_1 = 2r_2 + r_1$$

The matrix on the right is in reduced echelon form. The corresponding system of equations is

$$\begin{cases} x - \frac{2}{11}z = \frac{7}{11} & \text{(1)} \\ y - \frac{1}{11}z = -\frac{2}{11} & \text{(2)} \end{cases}$$

or, equivalently

$$\begin{cases} x = \frac{2}{11}z + \frac{7}{11} & (1) \\ y = \frac{1}{11}z - \frac{2}{11} & (2) \end{cases}$$

where z can be any real number.

 Now work Problem 57.

E X A M P L E   9

Solving a System of Linear Equations Using Matrices

$$\text{Solve: } \begin{cases} x + y + z = 6 \\ 2x - y - z = 3 \\ x + 2y + 2z = 0 \end{cases}$$

Solution   We proceed as follows, beginning with the augmented matrix.

$$\begin{bmatrix} 1 & 1 & 1 & | & 6 \\ 2 & -1 & -1 & | & 3 \\ 1 & 2 & 2 & | & 0 \end{bmatrix} \rightarrow \begin{bmatrix} 1 & 1 & 1 & | & 6 \\ 0 & -3 & -3 & | & -9 \\ 0 & 1 & 1 & | & -6 \end{bmatrix} \rightarrow \begin{bmatrix} 1 & 1 & 1 & | & 6 \\ 0 & 1 & 1 & | & -6 \\ 0 & -3 & -3 & | & -9 \end{bmatrix} \rightarrow \begin{bmatrix} 1 & 1 & 1 & | & 6 \\ 0 & 1 & 1 & | & -6 \\ 0 & 0 & 0 & | & -27 \end{bmatrix}$$

$$\begin{array}{c} \uparrow \\ R_2 = -2r_1 + r_2 \\ R_3 = -1r_1 + r_3 \end{array} \qquad \begin{array}{c} \uparrow \\ \text{Interchange rows 2 and 3.} \end{array} \qquad \begin{array}{c} \uparrow \\ R_3 = 3r_2 + r_3 \end{array}$$

This matrix is in echelon form. The bottom row is equivalent to the equation

$$0x + 0y + 0z = -27$$

which has no solution. Hence, the original system is inconsistent.   ■

 Now work Problems 25 and 31.

The matrix method is especially effective for systems of equations for which the number of equations and the number of variables are unequal. Here, too, such a system is either inconsistent or consistent. If it is consistent, it will have either exactly one solution or infinitely many solutions.
Let's look at a system of four equations containing three variables.

E X A M P L E   10

Solving a System of Linear Equations Using Matrices

$$\text{Solve: } \begin{cases} x - 2y + z = 0 & (1) \\ 2x + 2y - 3z = -3 & (2) \\ y - z = -1 & (3) \\ -x + 4y + 2z = 13 & (4) \end{cases}$$

Solution We proceed as follows, beginning with the augmented matrix.

$$
\begin{bmatrix}
1 & -2 & 1 & \bigm| & 0 \\
2 & 2 & -3 & \bigm| & -3 \\
0 & 1 & -1 & \bigm| & -1 \\
-1 & 4 & 2 & \bigm| & 13
\end{bmatrix}
\rightarrow
\begin{bmatrix}
1 & -2 & 1 & \bigm| & 0 \\
0 & 6 & -5 & \bigm| & -3 \\
0 & 1 & -1 & \bigm| & -1 \\
0 & 2 & 3 & \bigm| & 13
\end{bmatrix}
\rightarrow
\begin{bmatrix}
1 & -2 & 1 & \bigm| & 0 \\
0 & 1 & -1 & \bigm| & -1 \\
0 & 6 & -5 & \bigm| & -3 \\
0 & 2 & 3 & \bigm| & 13
\end{bmatrix}
$$

$R_2 = -2r_1 + r_2$
$R_4 = r_1 + r_4$
Interchange rows 2 and 3.

$$
\rightarrow
\begin{bmatrix}
1 & -2 & 1 & \bigm| & 0 \\
0 & 1 & -1 & \bigm| & -1 \\
0 & 0 & 1 & \bigm| & 3 \\
0 & 0 & 5 & \bigm| & 15
\end{bmatrix}
\rightarrow
\begin{bmatrix}
1 & -2 & 1 & \bigm| & 0 \\
0 & 1 & -1 & \bigm| & -1 \\
0 & 0 & 1 & \bigm| & 3 \\
0 & 0 & 0 & \bigm| & 0
\end{bmatrix}
$$

$R_3 = -6r_2 + r_3$
$R_4 = -2r_2 + r_4$
$R_4 = -5r_3 + r_4$

We could stop here, since the matrix is in echelon form, and back-substitute $z = 3$ to find x and y. Or we can continue to obtain the reduced echelon form.

$$
\rightarrow
\begin{bmatrix}
1 & 0 & -1 & \bigm| & -2 \\
0 & 1 & -1 & \bigm| & -1 \\
0 & 0 & 1 & \bigm| & 3 \\
0 & 0 & 0 & \bigm| & 0
\end{bmatrix}
\rightarrow
\begin{bmatrix}
1 & 0 & 0 & \bigm| & 1 \\
0 & 1 & 0 & \bigm| & 2 \\
0 & 0 & 1 & \bigm| & 3 \\
0 & 0 & 0 & \bigm| & 0
\end{bmatrix}
$$

$R_1 = 2r_2 + r_1$
$R_1 = r_3 + r_1$
$R_2 = r_3 + r_2$

The matrix is now in reduced echelon form, and we can see that the solution is $x = 1$, $y = 2$, $z = 3$. ∎

EXAMPLE 11 Mixing Acids

A chemistry laboratory has three containers of nitric acid, HNO_3. One container holds a solution with a concentration of 10% HNO_3, the second holds 20% HNO_3, and the third holds 40% HNO_3. How many liters of each solution should be mixed to obtain 100 liters of a solution whose concentration is 25% HNO_3?

Solution Let x, y, and z represent the number of liters of 10%, 20%, and 40% concentrations of HNO_3, respectively. We want 100 liters in all, and the concentration of HNO_3 from each solution must sum to 25% of 100 liters. Thus, we find that

$$
\begin{cases}
x + y + z = 100 \\
0.10x + 0.20y + 0.40z = 0.25(100)
\end{cases}
$$

We proceed as follows, beginning with the augmented matrix.

$$
\begin{bmatrix}
1 & 1 & 1 & \bigm| & 100 \\
0.10 & 0.20 & 0.40 & \bigm| & 25
\end{bmatrix}
\rightarrow
\begin{bmatrix}
1 & 1 & 1 & \bigm| & 100 \\
0 & 0.10 & 0.30 & \bigm| & 15
\end{bmatrix}
$$

$R_2 = -0.10r_1 + r_2$

$$\underset{\substack{\uparrow \\ R_2 = 10r_2}}{\rightarrow} \begin{bmatrix} 1 & 1 & 1 & | & 100 \\ 0 & 1 & 3 & | & 150 \end{bmatrix} \underset{\substack{\uparrow \\ R_1 = -1r_2 + r_1}}{\rightarrow} \begin{bmatrix} 1 & 0 & -2 & | & -50 \\ 0 & 1 & 3 & | & 150 \end{bmatrix}$$

The matrix is now in reduced echelon form. The final matrix represents the system

$$\begin{cases} x - 2z = -50 & (1) \\ y + 3z = 150 & (2) \end{cases}$$

which has infinitely many solutions given by

$$\begin{cases} x = 2z - 50 & (1) \\ y = -3z + 150 & (2) \end{cases}$$

where z is any real number. However, the practical considerations of this problem require us to restrict the solutions to $x \geq 0$, $y \geq 0$, $z \geq 0$. Therefore, we require $25 \leq z \leq 50$, because otherwise $x < 0$ or $y < 0$. Some of the possible solutions are given in Table 1. The final determination of what solution the laboratory will pick very likely depends on availability, cost differences, and other considerations.

TABLE 1

Liters of 10% Solution	Liters of 20% Solution	Liters of 40% Solution	Liters of 25% Solution
0	75	25	100
10	60	30	100
12	57	31	100
16	51	33	100
26	36	38	100
38	18	44	100
46	6	48	100
50	0	50	100

 Now work Problem 81.

EXERCISES

In Problems 1–12, write the augmented matrix of the given system of equations.

1. $\begin{cases} x - 5y = 5 \\ 4x + 3y = 6 \end{cases}$

2. $\begin{cases} 3x + 4y = 7 \\ 4x - 2y = 5 \end{cases}$

3. $\begin{cases} 2x + 3y - 6 = 0 \\ 4x - 6y + 2 = 0 \end{cases}$

4. $\begin{cases} 9x - y = 0 \\ 3x - y - 4 = 0 \end{cases}$

5. $\begin{cases} 0.01x - 0.03y = 0.06 \\ 0.13x + 0.10y = 0.20 \end{cases}$

6. $\begin{cases} \frac{4}{3}x - \frac{3}{2}y = \frac{3}{4} \\ -\frac{1}{4}x + \frac{1}{3}y = \frac{2}{3} \end{cases}$

7. $\begin{cases} x - y + z = 10 \\ 3x + 2y = 5 \\ x + y + 2z = 2 \end{cases}$

8. $\begin{cases} 5x - y - z = 0 \\ x + y = 5 \\ 2x - 3z = 2 \end{cases}$

9. $\begin{cases} x + y - z = 2 \\ 3x - 2y - 2 = 0 \end{cases}$

10. $\begin{cases} 2x + 3y - 4z = 0 \\ x - 5z + 2 = 0 \end{cases}$

11. $\begin{cases} x - w = 5 + y + z \\ 3x + y + w = 4 + 4z \end{cases}$

12. $\begin{cases} 2x + z = y \\ x + 2w = 1 + y \end{cases}$

In Problems 13–22, perform each row operation in order, (a) followed by (b) followed by (c), on the given augmented matrix.

13. $\begin{bmatrix} 1 & -3 & -5 & | & -2 \\ 2 & -5 & -4 & | & 5 \\ -3 & 5 & 4 & | & 6 \end{bmatrix}$
  (a) $R_2 = -2r_1 + r_2$
  (b) $R_3 = 3r_1 + r_3$
  (c) $R_3 = 4r_2 + r_3$

14. $\begin{bmatrix} 1 & -3 & -3 & | & -3 \\ 2 & -5 & 2 & | & -4 \\ -3 & 2 & 4 & | & 6 \end{bmatrix}$
  (a) $R_2 = -2r_1 + r_2$
  (b) $R_3 = 3r_1 + r_3$
  (c) $R_3 = 7r_2 + r_3$

15. $\begin{bmatrix} 1 & -3 & 4 & | & 3 \\ 2 & -5 & 6 & | & 6 \\ -3 & 3 & 4 & | & 6 \end{bmatrix}$
  (a) $R_2 = -2r_1 + r_2$
  (b) $R_3 = 3r_1 + r_3$
  (c) $R_3 = 6r_2 + r_3$

16. $\begin{bmatrix} 1 & -3 & 3 & | & -5 \\ 2 & -5 & -3 & | & -5 \\ -3 & -2 & 4 & | & 6 \end{bmatrix}$
  (a) $R_2 = -2r_1 + r_2$
  (b) $R_3 = 3r_1 + r_3$
  (c) $R_3 = 11r_2 + r_3$

17. $\begin{bmatrix} 1 & -3 & 2 & | & -6 \\ 2 & -5 & 3 & | & -4 \\ -3 & -6 & 4 & | & 6 \end{bmatrix}$
  (a) $R_2 = -2r_1 + r_2$
  (b) $R_3 = 3r_1 + r_3$
  (c) $R_3 = 15r_2 + r_3$

18. $\begin{bmatrix} 1 & -3 & -4 & | & -6 \\ 2 & -5 & 6 & | & -6 \\ -3 & 1 & 4 & | & 6 \end{bmatrix}$
  (a) $R_2 = -2r_1 + r_2$
  (b) $R_3 = 3r_1 + r_3$
  (c) $R_3 = 8r_2 + r_3$

19. $\begin{bmatrix} 1 & -3 & 1 & | & -2 \\ 2 & -5 & 6 & | & -2 \\ -3 & 1 & 4 & | & 6 \end{bmatrix}$
  (a) $R_2 = -2r_1 + r_2$
  (b) $R_3 = 3r_1 + r_3$
  (c) $R_3 = 8r_2 + r_3$

20. $\begin{bmatrix} 1 & -3 & -1 & | & 2 \\ 2 & -5 & 2 & | & 6 \\ -3 & -6 & 4 & | & 6 \end{bmatrix}$
  (a) $R_2 = -2r_1 + r_2$
  (b) $R_3 = 3r_1 + r_3$
  (c) $R_3 = 15r_2 + r_3$

21. $\begin{bmatrix} 1 & -3 & -2 & | & 3 \\ 2 & -5 & 2 & | & -1 \\ -3 & -2 & 4 & | & 6 \end{bmatrix}$
  (a) $R_2 = -2r_1 + r_2$
  (b) $R_3 = 3r_1 + r_3$
  (c) $R_3 = 11r_2 + r_3$

22. $\begin{bmatrix} 1 & -3 & 5 & | & -3 \\ 2 & -5 & 1 & | & -4 \\ -3 & 3 & 4 & | & 6 \end{bmatrix}$
  (a) $R_2 = -2r_1 + r_2$
  (b) $R_3 = 3r_1 + r_3$
  (c) $R_3 = 6r_2 + r_3$

In Problems 23–34, the reduced echelon form of a system of linear equations is given. Write the system of equations corresponding to the given matrix. Use x, y; or x, y, z; or x_1, x_2, x_3, x_4 as variables. Determine whether the system is consistent or inconsistent. If it is consistent, give the solution.

23. $\begin{bmatrix} 1 & 0 & | & 5 \\ 0 & 1 & | & -1 \end{bmatrix}$

24. $\begin{bmatrix} 1 & 0 & | & -4 \\ 0 & 1 & | & 0 \end{bmatrix}$

25. $\begin{bmatrix} 1 & 0 & 0 & | & 1 \\ 0 & 1 & 0 & | & 2 \\ 0 & 0 & 0 & | & 3 \end{bmatrix}$

26. $\begin{bmatrix} 1 & 0 & 0 & | & 0 \\ 0 & 1 & 0 & | & 0 \\ 0 & 0 & 0 & | & 2 \end{bmatrix}$

27. $\begin{bmatrix} 1 & 0 & 2 & | & -1 \\ 0 & 1 & -4 & | & -2 \\ 0 & 0 & 0 & | & 0 \end{bmatrix}$

28. $\begin{bmatrix} 1 & 0 & 4 & | & 4 \\ 0 & 1 & 3 & | & 2 \\ 0 & 0 & 0 & | & 0 \end{bmatrix}$

29. $\begin{bmatrix} 1 & 0 & 0 & 1 & | & 1 \\ 0 & 1 & 0 & 1 & | & 2 \\ 0 & 0 & 1 & 2 & | & 3 \end{bmatrix}$

30. $\begin{bmatrix} 1 & 0 & 0 & 0 & | & 1 \\ 0 & 1 & 0 & 2 & | & 2 \\ 0 & 0 & 1 & 3 & | & 0 \end{bmatrix}$

31. $\begin{bmatrix} 1 & 0 & 0 & 4 & | & 2 \\ 0 & 1 & 1 & 3 & | & 3 \\ 0 & 0 & 0 & 0 & | & 0 \end{bmatrix}$

32. $\begin{bmatrix} 1 & 0 & 0 & 0 & | & 1 \\ 0 & 1 & 0 & 0 & | & 2 \\ 0 & 0 & 1 & 2 & | & 3 \end{bmatrix}$

33. $\begin{bmatrix} 1 & 0 & 0 & 1 & | & -2 \\ 0 & 1 & 0 & 2 & | & 2 \\ 0 & 0 & 1 & -1 & | & 0 \\ 0 & 0 & 0 & 0 & | & 0 \end{bmatrix}$

34. $\begin{bmatrix} 1 & 0 & 0 & 0 & | & 1 \\ 0 & 1 & 0 & 0 & | & 2 \\ 0 & 0 & 1 & 0 & | & 3 \\ 0 & 0 & 0 & 1 & | & 0 \end{bmatrix}$

In Problems 35–76, solve each system of equations using matrices (row operations). If the system has no solution, say that it is inconsistent.

35. $\begin{cases} x + y = 8 \\ x - y = 4 \end{cases}$

36. $\begin{cases} x + 2y = 5 \\ x + y = 3 \end{cases}$

37. $\begin{cases} x - 5y = -13 \\ 3x + 2y = 12 \end{cases}$

38. $\begin{cases} x + 3y = 5 \\ 2x - 3y = -8 \end{cases}$

39. $\begin{cases} 3x - 6y = 24 \\ 5x + 4y = 12 \end{cases}$

40. $\begin{cases} 2x + 4y = 16 \\ 3x - 5y = -9 \end{cases}$

41. $\begin{cases} 2x + y = 1 \\ 4x + 2y = 6 \end{cases}$

42. $\begin{cases} x - y = 5 \\ -3x + 3y = 2 \end{cases}$

43. $\begin{cases} 2x - 4y = -2 \\ 3x + 2y = 3 \end{cases}$

44. $\begin{cases} 3x + 3y = 3 \\ 4x + 2y = \frac{8}{3} \end{cases}$

45. $\begin{cases} x + 2y = 4 \\ 2x + 4y = 8 \end{cases}$

46. $\begin{cases} 3x - y = 7 \\ 9x - 3y = 21 \end{cases}$

47. $\begin{cases} 2x + 3y = 6 \\ x - y = \frac{1}{2} \end{cases}$

48. $\begin{cases} \frac{1}{2}x + y = -2 \\ x - 2y = 8 \end{cases}$

49. $\begin{cases} 3x - 5y = 3 \\ 15x + 5y = 21 \end{cases}$

50. $\begin{cases} 2x - y = -1 \\ x + \frac{1}{2}y = \frac{3}{2} \end{cases}$

51. $\begin{cases} x - y = 6 \\ 2x - 3z = 16 \\ 2y + z = 4 \end{cases}$

52. $\begin{cases} 2x + y = -4 \\ -2y + 4z = 0 \\ 3x - 2z = -11 \end{cases}$

53. $\begin{cases} x - 2y + 3z = 7 \\ 2x + y + z = 4 \\ -3x + 2y - 2z = -10 \end{cases}$

54. $\begin{cases} 2x + y - 3z = 0 \\ -2x + 2y + z = -7 \\ 3x - 4y - 3z = 7 \end{cases}$

55. $\begin{cases} 2x - 2y - 2z = 2 \\ 2x + 3y + z = 2 \\ 3x + 2y = 0 \end{cases}$

56. $\begin{cases} 2x - 3y - z = 0 \\ -x + 2y + z = 5 \\ 3x - 4y - z = 1 \end{cases}$

57. $\begin{cases} -x + y + z = -1 \\ -x + 2y - 3z = -4 \\ 3x - 2y - 7z = 0 \end{cases}$

58. $\begin{cases} 2x - 3y - z = 0 \\ 3x + 2y + 2z = 2 \\ x + 5y + 3z = 2 \end{cases}$

59. $\begin{cases} 2x - 2y + 3z = 6 \\ 4x - 3y + 2z = 0 \\ -2x + 3y - 7z = 1 \end{cases}$

60. $\begin{cases} 3x - 2y + 2z = 6 \\ 7x - 3y + 2z = -1 \\ 2x - 3y + 4z = 0 \end{cases}$

61. $\begin{cases} x + y - z = 6 \\ 3x - 2y + z = -5 \\ x + 3y - 2z = 14 \end{cases}$

62. $\begin{cases} x - y + z = -4 \\ 2x - 3y + 4z = -15 \\ 5x + y - 2z = 12 \end{cases}$

63. $\begin{cases} x + 2y - z = -3 \\ 2x - 4y + z = -7 \\ -2x + 2y - 3z = 4 \end{cases}$

64. $\begin{cases} x + 4y - 3z = -8 \\ 3x - y + 3z = 12 \\ x + y + 6z = 1 \end{cases}$

65. $\begin{cases} 3x + y - z = \frac{2}{3} \\ 2x - y + z = 1 \\ 4x + 2y = \frac{8}{3} \end{cases}$

66. $\begin{cases} x + y = 1 \\ 2x - y + z = 1 \\ x + 2y + z = \frac{8}{3} \end{cases}$

67. $\begin{cases} x + y + z + w = 4 \\ 2x - y + z = 0 \\ 3x + 2y + z - w = 6 \\ x - 2y - 2z + 2w = -1 \end{cases}$

68. $\begin{cases} x + y + z + w = 4 \\ -x + 2y + z = 0 \\ 2x + 3y + z - w = 6 \\ -2x + y - 2z + 2w = -1 \end{cases}$

69. $\begin{cases} x + 2y + z = 1 \\ 2x - y + 2z = 2 \\ 3x + y + 3z = 3 \end{cases}$

70. $\begin{cases} x + 2y - z = 3 \\ 2x - y + 2z = 6 \\ x - 3y + 3z = 4 \end{cases}$

71. $\begin{cases} x - y + z = 5 \\ 3x + 2y - 2z = 0 \end{cases}$

72. $\begin{cases} 2x + y - z = 4 \\ -x + y + 3z = 1 \end{cases}$

73. $\begin{cases} 2x + 3y - z = 3 \\ x - y - z = 0 \\ -x + y + z = 0 \\ x + y + 3z = 5 \end{cases}$

74. $\begin{cases} x - 3y + z = 1 \\ 2x - y - 4z = 0 \\ x - 3y + 2z = 1 \\ x - 2y = 5 \end{cases}$

75. $\begin{cases} 4x + y + z - w = 4 \\ x - y + 2z + 3w = 3 \end{cases}$

76. $\begin{cases} -4x + y = 5 \\ 2x - y + z - w = 5 \\ z + w = 4 \end{cases}$

77. Curve Fitting Find the parabola $y = ax^2 + bx + c$ that passes through the points $(1, 2)$, $(-2, -7)$, and $(2, -3)$.

78. Curve Fitting Find the parabola $y = ax^2 + bx + c$ that passes through the points $(1, -1)$, $(3, -1)$, and $(-2, 14)$.

79. Curve Fitting Find the function $f(x) = ax^3 + bx^2 + cx + d$ for which $f(-3) = -112, f(-1) = -2, f(1) = 4$, and $f(2) = 13$.

80. Curve Fitting Find the function $f(x) = ax^3 + bx^2 + cx + d$ for which $f(-2) = -10, f(-1) = 3, f(1) = 5$, and $f(3) = 15$.

81. Mixing Acids A chemistry laboratory has three containers of sulfuric acid, H_2SO_4. One container holds a solution with a concentration of 15% H_2SO_4, the second holds 25% H_2SO_4, and the third holds 50% H_2SO_4. How many liters of each solution should be mixed to obtain 100 liters of a solution with a concentration of 40% H_2SO_4? Construct a table similar to Table 1 illustrating some of the possible combinations.

82. Painting a House Three painters, Beth, Bill, and Edie, working together, can paint the exterior of a home in 10 hours. Bill and Edie together have painted a similar house in 15 hours. One day, all three worked on this same kind of house for 4 hours, after which Edie left. Beth and Bill required 8 more hours to finish. Assuming no gain or loss in efficiency, how long should it take each person to complete such a job alone?

83. Prices of Fast Food One group of customers bought 8 deluxe hamburgers, 6 orders of large fries, and 6 large colas for $26.10. A second group ordered 10 deluxe hamburgers, 6 large fries, and 8 large colas and paid $31.60. Is there sufficient information to determine the price of each food item? If not, construct a table showing the various possibilities. Assume that the hamburgers cost between $1.75 and $2.25, the fries between $0.75 and $1.00, and the colas between $0.60 and $0.90.

84. Prices of Fast Food Use the information given in Problem 83, and suppose that a third group purchased 3 deluxe hamburgers, 2 large fries, and 4 large colas for $10.95. Now is there sufficient information to determine the price of each food item?

85. Financial Planning Three retired couples each require an additional annual income of $2000 per year. As their financial consultant, you recommend that they invest some money in Treasury bills that yield 7%, some money in corporate bonds that yield 9%, and some money in junk bonds that yield 11%. Prepare a table for each couple showing the various ways that their goals can be achieved:
(a) If the first couple has $20,000 to invest.
(b) If the second couple has $25,000 to invest.
(c) If the third couple has $30,000 to invest.
(d) What advice would you give each couple regarding the amount to invest and the choices available?

[Higher yields generally carry more risk.]

86. Financial Planning A retired couple has $25,000 to invest. As their financial consultant, you recommend that they invest some money in Treasury bills that yield 7%, some money in corporate bonds that yield 9%, and some money in junk bonds that yield 11%. Prepare a table showing the various ways this couple can achieve the following goals:
(a) The couple wants $1500 per year in income.
(b) The couple wants $2000 per year in income.
(c) The couple wants $2500 per year in income.
(d) What advice would you give this couple regarding the income that they require and the choices available?

[Higher yields generally carry more risk.]

87. Pharmacy A doctor's prescription calls for a daily intake of liquid containing 40 mg of vitamin C and 30 mg of vitamin D. Your pharmacy stocks three liquids that can be used: one contains 20% vitamin C and 30% vitamin D; a second, 40% vitamin C and 20% vitamin D; and a third, 30% vitamin C and 50% vitamin D. Create a table showing the possible combinations that could be used to fill the prescription.

88. Pharmacy A doctor's prescription calls for the creation of pills that contain 12 units of vitamin B_{12} and 12 units of vitamin E. Your pharmacy stocks three powders that can be used to make these pills: one contains 20% vitamin B_{12} and 30% vitamin E; a second, 40% vitamin B_{12} and 20% vitamin E; and a third, 30% vitamin B_{12} and 40% vitamin E. Create a table showing the possible combinations of each powder that could be mixed in each pill.

89. Electricity: Kirchhoff's Rules An application of Kirchhoff's Rules to the circuit shown results in the following system of equations:

$$\begin{cases} I_1 + I_2 = I_3 \\ 16 - 8 - 9I_3 - 3I_1 = 0 \\ 16 - 4 - 9I_3 - 9I_2 = 0 \\ 8 - 4 - 9I_2 + 3I_1 = 0 \end{cases}$$

Find the currents I_1, I_2, and I_3.*

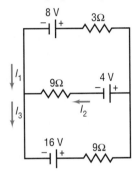

90. Electricity: Kirchhoff's Rules An application of Kirchhoff's Rules to the circuit shown results in the following system of equations:

$$\begin{cases} -4 + 8 - 2I_2 = 0 \\ 8 = 5I_4 + I_1 \\ 4 = 3I_3 + I_1 \\ I_3 + I_4 = I_1 \end{cases}$$

Find the currents I_1, I_2, I_3, and I_4.†

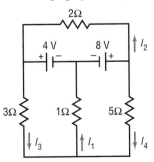

*Source: Based on Raymond Serway, *Physics*, 3rd ed. (Philadelphia: Saunders, 1990), Prob. 31, p. 790.
†*Source: Ibid., Prob. 34, p. 791.

91. Electricity: Kirchhoff's Rules An application of Kirchhoff's Rules to the circuit shown results in the following system of equations:

$$\begin{cases} I_1 = I_3 + I_2 \\ 24 - 6I_1 - 3I_3 = 0 \\ 12 + 24 - 6I_1 - 6I_2 = 0 \end{cases}$$

Find the currents I_1, I_2, and I_3.*

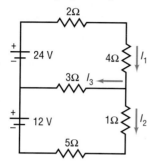

92. Write a brief paragraph or two that outlines your strategy for solving a system of linear equations using matrices.

93. When solving a system of linear equations using matrices, do you prefer to place the augmented matrix in echelon form or in reduced echelon form? Give reasons for your choice.

94. Make up a system of three linear equations containing three variables that has:
(a) No solution
(b) Exactly one solution
(c) Infinitely many solutions

Give the three systems to a friend to solve and critique.

Source: Ibid., Prob., 38, p. 791.

95. Consider the system of equations

$$\begin{cases} a_1 x + b_1 y = c_1 \\ a_2 x + b_2 y = c_2 \end{cases}$$

If $D = a_1 b_2 - a_2 b_1 \neq 0$, use matrices to show that the solution is

$$x = \frac{1}{D}(c_1 b_2 - c_2 b_1), \qquad y = \frac{1}{D}(a_1 c_2 - a_2 c_1)$$

96. For the system in Problem 95, suppose that $D = a_1 b_2 - a_2 b_1 = 0$. Use matrices to show that the system is inconsistent if either $a_1 c_2 \neq a_2 c_1$ or $b_1 c_2 \neq b_2 c_1$ and has infinitely many solutions if both $a_1 c_2 = a_2 c_1$ and $b_1 c_2 = b_2 c_1$.

97. The graph of a linear equation containing three variables is a plane. Give a geometrical argument for what can result when solving a system of two linear equations containing three variables.

[**Hint:** Two planes in a three-dimensional space are either coincident (the same), parallel, or intersect in a line.]

98. Refer to Problem 97. Give a geometrical argument for what can result when solving a system of three linear equations containing three variables.

99. Refer to Problem 97. Give a geometrical argument for what can result when solving a system of four linear equations containing three variables.

100. Explain why no column to the left of the vertical bar of an augmented matrix can consist entirely of zeros.

8.3 SYSTEMS OF LINEAR EQUATIONS: DETERMINANTS

1 Evaluate 2 by 2 Determinants
2 Use Cramer's Rule to Solve a System of Two Equations, Two Variables
3 Evaluate 3 by 3 Determinants
4 Use Cramer's Rule to Solve a System of Three Equations, Three Variables
5 Know Properties of Determinants

1 In the preceding section, we described a method of using matrices to solve any system of linear equations. This section deals with yet another method for solving systems of linear equations; however, it can be used only when the number of equations equals the number of variables. Although the method will work for any system (provided the number of equations equals the number of variables), it is most often used for systems of two equations containing two variables or three equations containing three variables. This method, called *Cramer's Rule*, is based on the concept of a *determinant*.

MISSION POSSIBLE

Markov Chains

The Mission Possible team has just been hired by the Department of Education. The assignment is to determine the percent of the population that will be college educated in the future. The Secretary of Education said he heard someone suggest that Markov chains could be used to solve this problem. Your team is told Markov chains have something to do with matrices. Having just studied matrices, you feel up to the challenge!

A Markov chain (or process) is one in which future outcomes are determined by a current state. Future outcomes are based upon probabilities. The probability of moving to a certain state depends only on the state previously occupied and does not vary with time. An example of a Markov chain would be the maximum education achieved by children based upon the highest education attained of their parents where the states are 1) Earned college degree 2) High-school diploma only 3) Elementary school only. If p_{ij} is the probability of moving from state i to state j, then the **transition matrix** is the $m \times m$ matrix.

$$P = \begin{bmatrix} p_{11} & p_{12} & \cdots & p_{1m} \\ \vdots & \vdots & & \vdots \\ p_{m1} & p_{m2} & \cdots & p_{mm} \end{bmatrix}$$

The following table represents the probabilities of the highest educational level children based on the highest educational level of their parents:

| Highest Educational | Maximum Education Children Achieve | | |
Level of Parents	College	High School	Elementary
College	80%	18%	2%
High School	40%	50%	10%
Elementary	20%	60%	20%

For example, the table shows that the probability is 40% that parents with a high-school education will have children with a college education.

1. Convert the percentages to decimals.
2. What is the transition matrix?
3. Sum across the rows. What do you notice? Why do you think you obtained this result?
4. If P is the transition matrix of a Markov chain, then the (i, j)th entry of P^n (nth power of P) gives the probability of passing from state i to state j in n stages. What is the probability that a grandchild of a college graduate is a college graduate?
5. What is the probability that the grandchild of a high school graduate finishes college?
6. The row vector $v^{(0)} = [0.236 \quad 0.581 \quad 0.183]$ represents the proportion of the U.S. population that have college, high school and elementary school, respectively, as their highest educational level in 1996. (*Source:* U.S. Census Bureau). In a Markov chain the probability distribution $v^{(k)}$ after k stages is $v^{(k)} = v^{(0)}P^k$ where P^k is the kth power of the transition matrix. What will be the distribution of highest educational attainment of the grandchildren of the current population?
7. Calculate P^3, P^4, P^5, . . . Continue until the matrix does not change. This is called the long-run distribution. What is the long-run distribution of highest educational attainment of the population?

2 by 2 Determinants

If a, b, c, and d are four real numbers, the symbol

$$D = \begin{vmatrix} a & b \\ c & d \end{vmatrix}$$

is called a **2 by 2 determinant.** Its value is the number $ad - bc$; that is,

$$D = \begin{vmatrix} a & b \\ c & d \end{vmatrix} = ad - bc \qquad (1)$$

A device that may be helpful for remembering the value of a 2 by 2 determinant is the following:

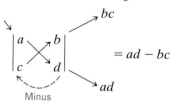

EXAMPLE 1

Evaluating a 2 × 2 Determinant

$$\begin{vmatrix} 3 & -2 \\ 6 & 1 \end{vmatrix} = (3)(1) - (6)(-2) = 3 - (-12) = 15$$

Now work Problem 3.

Let's now see the role that a 2 by 2 determinant plays in the solution of a system of two equations containing two variables. Consider the system

$$\begin{cases} ax + by = s & (1) \\ cx + dy = t & (2) \end{cases} \qquad (2)$$

We shall use the method of elimination to solve this system.

Provided $d \neq 0$ and $b \neq 0$, this system is equivalent to the system

$$\begin{cases} adx + bdy = sd & (1) \quad \text{Multiply by } d. \\ bcx + bdy = tb & (2) \quad \text{Multiply by } b. \end{cases}$$

On subtracting the second equation from the first equation, we get

$$\begin{cases} (ad - bc)x + 0 \cdot y = sd - tb & (1) \\ bcx + bdy = tb & (2) \end{cases}$$

Now, the first equation can be rewritten using determinant notation.

$$\begin{vmatrix} a & b \\ c & d \end{vmatrix} x = \begin{vmatrix} s & b \\ t & d \end{vmatrix}$$

If $D = \begin{vmatrix} a & b \\ c & d \end{vmatrix} = ad - bc \neq 0$, we can solve for x to get

$$x = \frac{\begin{vmatrix} s & b \\ t & d \end{vmatrix}}{\begin{vmatrix} a & b \\ c & d \end{vmatrix}} = \frac{\begin{vmatrix} s & b \\ t & d \end{vmatrix}}{D} \qquad (3)$$

Return now to the original system (2). Provided $a \neq 0$ and $c \neq 0$, the system is equivalent to

$$\begin{cases} acx + bcy = cs & \text{(1)} \quad \text{Multiply by } c. \\ acx + ady = at & \text{(2)} \quad \text{Multiply by } a. \end{cases}$$

On subtracting the first equation from the second equation, we get

$$\begin{cases} acx + \quad bcy \quad = \quad cs & \text{(1)} \\ 0 \cdot x + (ad - bc)y = at - cs & \text{(2)} \end{cases}$$

The second equation can now be rewritten using determinant notation.

$$\begin{vmatrix} a & b \\ c & d \end{vmatrix} y = \begin{vmatrix} a & s \\ c & t \end{vmatrix}$$

If $D = \begin{vmatrix} a & b \\ c & d \end{vmatrix} = ad - bc \neq 0$, we can solve for y to get

$$y = \frac{\begin{vmatrix} a & s \\ c & t \end{vmatrix}}{\begin{vmatrix} a & b \\ c & d \end{vmatrix}} = \frac{\begin{vmatrix} a & s \\ c & t \end{vmatrix}}{D} \tag{4}$$

Equations (3) and (4) lead us to the following result, called **Cramer's Rule.**

Theorem   Cramer's Rule for Two Equations Containing Two Variables

The solution to the system of equations

$$\begin{cases} ax + by = s & \text{(1)} \\ cx + dy = t & \text{(2)} \end{cases} \tag{5}$$

is given by

$$x = \frac{\begin{vmatrix} s & b \\ t & d \end{vmatrix}}{\begin{vmatrix} a & b \\ c & d \end{vmatrix}}, \quad y = \frac{\begin{vmatrix} a & s \\ c & t \end{vmatrix}}{\begin{vmatrix} a & b \\ c & d \end{vmatrix}} \tag{6}$$

provided that

$$D = \begin{vmatrix} a & b \\ c & d \end{vmatrix} = ad - bc \neq 0$$

In the derivation given for Cramer's Rule above, we assumed that none of the numbers a, b, c, and d were 0. In Problem 58 you will be asked to complete the proof under the less stringent conditions that $D = ad - bc \neq 0$.

Now look carefully at the pattern in Cramer's Rule. The denominator in the solution (6) is the determinant of the coefficients of the variables.

$$\begin{cases} ax + by = s \\ cx + dy = t \end{cases} \qquad D = \begin{vmatrix} a & b \\ c & d \end{vmatrix}$$

In the solution for x, the numerator is the determinant, denoted by D_x, formed by replacing the entries in the first column (the coefficients of x) in D by the constants on the right side of the equal sign.

$$D_x = \begin{vmatrix} s & b \\ t & d \end{vmatrix}$$

In the solution for y, the numerator is the determinant, denoted by D_y, formed by replacing the entries in the second column (the coefficients of y) in D by the constants on the right side of the equal sign.

$$D_y = \begin{vmatrix} a & s \\ c & t \end{vmatrix}$$

Cramer's Rule then states that, if $D \neq 0$,

$$x = \frac{D_x}{D}, \qquad y = \frac{D_y}{D} \qquad\qquad (7)$$

E X A M P L E   2   **Solving a System of Linear Equations Using Determinants**

Use Cramer's Rule, if applicable, to solve the system

$$\begin{cases} 3x - 2y = 4 & \text{(1)} \\ 6x + y = 13 & \text{(2)} \end{cases}$$

Solution   The determinant D of the coefficients of the variables is

$$D = \begin{vmatrix} 3 & -2 \\ 6 & 1 \end{vmatrix} = (3)(1) - (6)(-2) = 15$$

Because $D \neq 0$, Cramer's Rule (7) can be used.

$$x = \frac{D_x}{D} = \frac{\begin{vmatrix} 4 & -2 \\ 13 & 1 \end{vmatrix}}{15} = \frac{30}{15} = 2, \qquad y = \frac{D_y}{D} = \frac{\begin{vmatrix} 3 & 4 \\ 6 & 13 \end{vmatrix}}{15} = \frac{15}{15} = 1$$

The solution is $x = 2, y = 1$.   ∎

If, in attempting to use Cramer's Rule, the determinant D of the coefficients of the variables is found to equal 0 (so that Cramer's Rule is not applicable), then the system either is inconsistent or has infinitely many solutions. (Refer to Problem 96 in Exercise 8.2.)

Now work Problem 11.

3 by 3 Determinants

In order to use Cramer's Rule to solve a system of three equations containing three variables, we need to define a 3 by 3 determinant.
  A **3 by 3 determinant** is symbolized by

$$\begin{vmatrix} a_{11} & a_{12} & a_{13} \\ a_{21} & a_{22} & a_{23} \\ a_{31} & a_{32} & a_{33} \end{vmatrix} \qquad\qquad (8)$$

in which $a_{11}, a_{12}, \ldots$ are real numbers.

As with matrices, we use a double subscript to identify an entry by indicating its row and column numbers. For example, the entry a_{23} is in row 2, column 3.

The value of a 3 by 3 determinant may be defined in terms of 2 by 2 determinants by the following formula.

$$\begin{vmatrix} a_{11} & a_{12} & a_{13} \\ a_{21} & a_{22} & a_{23} \\ a_{31} & a_{32} & a_{33} \end{vmatrix} = a_{11}\begin{vmatrix} a_{22} & a_{23} \\ a_{32} & a_{33} \end{vmatrix} \overset{\text{Minus}}{-} a_{12}\begin{vmatrix} a_{21} & a_{23} \\ a_{31} & a_{33} \end{vmatrix} + a_{13}\begin{vmatrix} a_{21} & a_{22} \\ a_{31} & a_{32} \end{vmatrix} \tag{9}$$

2 by 2 determinant left after removing row and column containing a_{11}.

2 by 2 determinant left after removing row and column containing a_{12}.

2 by 2 determinant left after removing row and column containing a_{13}.

Be sure to take note of the minus sign that appears with the second term—it's easy to forget it! Formula (9) is best remembered by noting that each entry in row 1 is multiplied by the 2 by 2 determinant that remains after the row and column containing the entry have been removed, as follows:

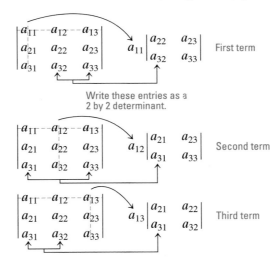

First term

Write these entries as a 2 by 2 determinant.

Second term

Third term

Now insert the minus sign before the middle expression and add.

$$\begin{vmatrix} a_{11} & a_{12} & a_{13} \\ a_{21} & a_{22} & a_{23} \\ a_{31} & a_{32} & a_{33} \end{vmatrix} = a_{11}\begin{vmatrix} a_{22} & a_{23} \\ a_{32} & a_{33} \end{vmatrix} \overset{\text{Minus}}{-} a_{12}\begin{vmatrix} a_{21} & a_{23} \\ a_{31} & a_{33} \end{vmatrix} + a_{13}\begin{vmatrix} a_{21} & a_{22} \\ a_{31} & a_{32} \end{vmatrix}$$

Formula (9) exhibits one way to find the value of a 3 by 3 determinant, *by expanding across row 1*. In fact, the expansion can take place across any row or down any column. The terms to be added or subtracted consist of the row (or column) entry times the value of the 2 by 2 determinant that remains after removing the row and column entry. The value of the determinant is found by adding or subtracting the terms according to the following scheme:

$$\begin{matrix} + & - & + \\ - & + & - \\ + & - & + \end{matrix}$$

For example, if we choose to expand down column 2, we obtain

$$
\begin{vmatrix} a_{11} & a_{12} & a_{13} \\ a_{21} & a_{22} & a_{23} \\ a_{31} & a_{32} & a_{33} \end{vmatrix} = -a_{12}\begin{vmatrix} a_{21} & a_{23} \\ a_{31} & a_{33} \end{vmatrix} + a_{22}\begin{vmatrix} a_{11} & a_{13} \\ a_{31} & a_{33} \end{vmatrix} - a_{32}\begin{vmatrix} a_{11} & a_{13} \\ a_{21} & a_{23} \end{vmatrix}
$$

Expand down column 2 $(-, +, -)$.

If we choose to expand across row 3, we obtain

$$
\begin{vmatrix} a_{11} & a_{12} & a_{13} \\ a_{21} & a_{22} & a_{23} \\ a_{31} & a_{32} & a_{33} \end{vmatrix} = a_{31}\begin{vmatrix} a_{12} & a_{13} \\ a_{22} & a_{23} \end{vmatrix} - a_{32}\begin{vmatrix} a_{11} & a_{13} \\ a_{21} & a_{23} \end{vmatrix} + a_{33}\begin{vmatrix} a_{11} & a_{12} \\ a_{21} & a_{22} \end{vmatrix}
$$

Expand across row 3 $(+, -, +)$.

It can be shown that the value of a determinant does not depend on the choice of the row or column used in the expansion.

EXAMPLE 3

Evaluating a 3 × 3 Determinant

Find the value of the 3 by 3 determinant:
$\begin{vmatrix} 3 & 4 & -1 \\ 4 & 6 & 2 \\ 8 & -2 & 3 \end{vmatrix}$

Solution   We choose to expand across row 1.

Remember the minus sign.
↓

$$
\begin{vmatrix} 3 & 4 & -1 \\ 4 & 6 & 2 \\ 8 & -2 & 3 \end{vmatrix} = 3\begin{vmatrix} 6 & 2 \\ -2 & 3 \end{vmatrix} - 4\begin{vmatrix} 4 & 2 \\ 8 & 3 \end{vmatrix} + (-1)\begin{vmatrix} 4 & 6 \\ 8 & -2 \end{vmatrix}
$$

$$
= 3(18 + 4) - 4(12 - 16) + (-1)(-8 - 48)
$$
$$
= 3(22) - 4(-4) + (-1)(-56)
$$
$$
= 66 + 16 + 56 = 138
$$

We could also find the value of the 3 by 3 determinant in Example 3 by expanding down column 3 (the signs are $+, -, +$).

$$
\begin{vmatrix} 3 & 4 & -1 \\ 4 & 6 & 2 \\ 8 & -2 & 3 \end{vmatrix} = (-1)\begin{vmatrix} 4 & 6 \\ 8 & -2 \end{vmatrix} - 2\begin{vmatrix} 3 & 4 \\ 8 & -2 \end{vmatrix} + 3\begin{vmatrix} 3 & 4 \\ 4 & 6 \end{vmatrix}
$$

$$
= -1(-8 - 48) - 2(-6 - 32) + 3(18 - 16)
$$
$$
= 56 + 76 + 6 = 138
$$

Comment   A graphing utility can be used to evaluate determinants. Check your manual to see how. Then check the answer obtained in Example 3.

 Now work Problem 7.

Systems of Three Equations Containing Three Variables

 Consider the following system of three equations containing three variables:

$$\begin{cases} a_{11}x + a_{12}y + a_{13}z = c_1 \\ a_{21}x + a_{22}y + a_{23}z = c_2 \\ a_{31}x + a_{32}y + a_{33}z = c_3 \end{cases} \tag{10}$$

It can be shown that if the determinant D of the coefficients of the variables is not 0, that is, if

$$D = \begin{vmatrix} a_{11} & a_{12} & a_{13} \\ a_{21} & a_{22} & a_{23} \\ a_{31} & a_{32} & a_{33} \end{vmatrix} \neq 0$$

then the unique solution of system (10) is given by

Cramer's Rule for Three Equations Containing Three Variables

$$x = \frac{D_x}{D}, \qquad y = \frac{D_y}{D}, \qquad z = \frac{D_z}{D}$$

---

where

$$D_x = \begin{vmatrix} c_1 & a_{12} & a_{13} \\ c_2 & a_{22} & a_{23} \\ c_3 & a_{32} & a_{33} \end{vmatrix}, \qquad D_y = \begin{vmatrix} a_{11} & c_1 & a_{13} \\ a_{21} & c_2 & a_{23} \\ a_{31} & c_3 & a_{33} \end{vmatrix}, \qquad D_z = \begin{vmatrix} a_{11} & a_{12} & c_1 \\ a_{21} & a_{22} & c_2 \\ a_{31} & a_{32} & c_3 \end{vmatrix}$$

The similarity of this pattern and the pattern observed earlier for a system of two equations containing two variables should be apparent.

EXAMPLE 4

Using Cramer's Rule

Use Cramer's Rule, if applicable, to solve the following system.

$$\begin{cases} 2x + y - z = 3 & (1) \\ -x + 2y + 4z = -3 & (2) \\ x - 2y - 3z = 4 & (3) \end{cases}$$

Solution   The value of the determinant D of the coefficients of the variables is

$$D = \begin{vmatrix} 2 & 1 & -1 \\ -1 & 2 & 4 \\ 1 & -2 & -3 \end{vmatrix} = 2\begin{vmatrix} 2 & 4 \\ -2 & -3 \end{vmatrix} - 1\begin{vmatrix} -1 & 4 \\ 1 & -3 \end{vmatrix} + (-1)\begin{vmatrix} -1 & 2 \\ 1 & -2 \end{vmatrix}$$

$$= 2(2) - 1(-1) + (-1)(0)$$

$$= 4 + 1 = 5$$

Because $D \neq 0$, we proceed to find the values of D_x, D_y, and D_z.

$$D_x = \begin{vmatrix} 3 & 1 & -1 \\ -3 & 2 & 4 \\ 4 & -2 & -3 \end{vmatrix} = 3\begin{vmatrix} 2 & 4 \\ -2 & -3 \end{vmatrix} - 1\begin{vmatrix} -3 & 4 \\ 4 & -3 \end{vmatrix} + (-1)\begin{vmatrix} -3 & 2 \\ 4 & -2 \end{vmatrix}$$

$$= 3(2) - 1(-7) + (-1)(-2) = 15$$

$$D_y = \begin{vmatrix} 2 & 3 & -1 \\ -1 & -3 & 4 \\ 1 & 4 & -3 \end{vmatrix} = 2\begin{vmatrix} -3 & 4 \\ 4 & -3 \end{vmatrix} - 3\begin{vmatrix} -1 & 4 \\ 1 & -3 \end{vmatrix} + (-1)\begin{vmatrix} -1 & -3 \\ 1 & 4 \end{vmatrix}$$

$$= 2(-7) - 3(-1) + (-1)(-1)$$

$$= -14 + 3 + 1 = -10$$

$$D_z = \begin{vmatrix} 2 & 1 & 3 \\ -1 & 2 & -3 \\ 1 & -2 & 4 \end{vmatrix} = 2\begin{vmatrix} 2 & -3 \\ -2 & 4 \end{vmatrix} - 1\begin{vmatrix} -1 & -3 \\ 1 & 4 \end{vmatrix} + 3\begin{vmatrix} -1 & 2 \\ 1 & -2 \end{vmatrix}$$

$$= 2(2) - 1(-1) + 3(0) = 5$$

As a result,

$$x = \frac{D_x}{D} = \frac{15}{5} = 3, \qquad y = \frac{D_y}{D} = \frac{-10}{5} = -2, \qquad z = \frac{D_z}{D} = \frac{5}{5} = 1$$

The solution is $x = 3$, $y = -2$, $z = 1$. ∎

If the determinant of the coefficients of the variables of a system of three linear equations containing three variables is 0, then Cramer's Rule is not applicable. In such a case, the system either is inconsistent or has infinitely many solutions.

Now work Problem 29.

More about Determinants

5 Determinants have several properties that are sometimes helpful for obtaining their value. We list some of them here.

> ### Theorem
>
> The value of a determinant changes sign if any two rows (or any two columns) are interchanged.   (11)

Proof for 2 by 2 Determinants

$$\begin{vmatrix} a & b \\ c & d \end{vmatrix} = ad - bc \quad \text{and} \quad \begin{vmatrix} c & d \\ a & b \end{vmatrix} = bc - ad = -(ad - bc)$$

EXAMPLE 5   Demonstrating a Theorem (11)

$$\begin{vmatrix} 3 & 4 \\ 1 & 2 \end{vmatrix} = 6 - 4 = 2 \qquad \begin{vmatrix} 1 & 2 \\ 3 & 4 \end{vmatrix} = 4 - 6 = -2$$

Theorem

If all the entries in any row (or any column) equal 0, the value of the determinant is 0. (12)

■

Proof Merely expand across the row (or down the column) containing the 0's.

■

Theorem

If any two rows (or any two columns) of a determinant have corresponding entries that are equal, the value of the determinant is 0. (13)

■

You are asked to prove this result for a 3 by 3 determinant in which the entries in column 1 equal the entries in column 3 in Problem 61.

E X A M P L E   6   Demonstrating a Theorem (13)

$$\begin{vmatrix} 1 & 2 & 3 \\ 1 & 2 & 3 \\ 4 & 5 & 6 \end{vmatrix} = 1\begin{vmatrix} 2 & 3 \\ 5 & 6 \end{vmatrix} - 2\begin{vmatrix} 1 & 3 \\ 4 & 6 \end{vmatrix} + 3\begin{vmatrix} 1 & 2 \\ 4 & 5 \end{vmatrix}$$

$$= 1(-3) - 2(-6) + 3(-3)$$
$$= -3 + 12 - 9 = 0$$

■

Theorem

If any row (or any column) of a determinant is multiplied by a nonzero number k, the value of the determinant is also changed by a factor of k. (14)

■

You are asked to prove this result for a 3 by 3 determinant using row 2 in Problem 60.

E X A M P L E   7   Demonstrating a Theorem (14)

$$\begin{vmatrix} 1 & 2 \\ 4 & 6 \end{vmatrix} = 6 - 8 = -2$$

$$\begin{vmatrix} k & 2k \\ 4 & 6 \end{vmatrix} = 6k - 8k = -2k = k(-2) = k\begin{vmatrix} 1 & 2 \\ 4 & 6 \end{vmatrix}$$

■

Theorem

If the entries of any row (or any column) of a determinant are multiplied by a nonzero number k and the result is added to the corresponding entries of another row (or column), the value of the determinant remains unchanged. (15)

In Problem 62, you are asked to prove this result for a 3 by 3 determinant using rows 1 and 2.

E X A M P L E   8   Demonstrating a Theorem (15)

$$\begin{vmatrix} 3 & 4 \\ 5 & 2 \end{vmatrix} \underset{\uparrow}{=} \begin{vmatrix} -7 & 0 \\ 5 & 2 \end{vmatrix} = -14$$

Multiply row 2 by −2 and add to row 1.

8.3 | EXERCISES

In Problems 1–10, find the value of each determinant.

1. $\begin{vmatrix} 3 & 1 \\ 4 & 2 \end{vmatrix}$ **2.** $\begin{vmatrix} 6 & 1 \\ 5 & 2 \end{vmatrix}$ **3.** $\begin{vmatrix} 6 & 4 \\ -1 & 3 \end{vmatrix}$ **4.** $\begin{vmatrix} 8 & -3 \\ 4 & 2 \end{vmatrix}$ **5.** $\begin{vmatrix} -3 & -1 \\ 4 & 2 \end{vmatrix}$

6. $\begin{vmatrix} -4 & 2 \\ -5 & 3 \end{vmatrix}$ **7.** $\begin{vmatrix} 3 & 4 & 2 \\ 1 & -1 & 5 \\ 1 & 2 & -2 \end{vmatrix}$ **8.** $\begin{vmatrix} 1 & 3 & -2 \\ 6 & 1 & -5 \\ 8 & 2 & 3 \end{vmatrix}$ **9.** $\begin{vmatrix} 4 & -1 & 2 \\ 6 & -1 & 0 \\ 1 & -3 & 4 \end{vmatrix}$ **10.** $\begin{vmatrix} 3 & -9 & 4 \\ 1 & 4 & 0 \\ 8 & -3 & 1 \end{vmatrix}$

In Problems 11–38, solve each system of equations using Cramer's Rule if it is applicable. If Cramer's Rule is not applicable, say so.

11. $\begin{cases} x + y = 8 \\ x - y = 4 \end{cases}$ **12.** $\begin{cases} x + 2y = 5 \\ x - y = 3 \end{cases}$ **13.** $\begin{cases} 5x - y = 13 \\ 2x + 3y = 12 \end{cases}$ **14.** $\begin{cases} x + 3y = 5 \\ 2x - 3y = -8 \end{cases}$

15. $\begin{cases} 3x = 24 \\ x + 2y = 0 \end{cases}$ **16.** $\begin{cases} 4x + 5y = -3 \\ -2y = -4 \end{cases}$ **17.** $\begin{cases} 3x - 6y = 24 \\ 5x + 4y = 12 \end{cases}$ **18.** $\begin{cases} 2x + 4y = 16 \\ 3x - 5y = -9 \end{cases}$

19. $\begin{cases} 3x - 2y = 4 \\ 6x - 4y = 0 \end{cases}$ **20.** $\begin{cases} -x + 2y = 5 \\ 4x - 8y = 6 \end{cases}$ **21.** $\begin{cases} 2x - 4y = -2 \\ 3x + 2y = 3 \end{cases}$ **22.** $\begin{cases} 3x + 3y = 3 \\ 4x + 2y = \frac{8}{3} \end{cases}$

23. $\begin{cases} 2x - 3y = -1 \\ 10x + 10y = 5 \end{cases}$ **24.** $\begin{cases} 3x - 2y = 0 \\ 5x + 10y = 4 \end{cases}$ **25.** $\begin{cases} 2x + 3y = 6 \\ x - y = \frac{1}{2} \end{cases}$ **26.** $\begin{cases} \frac{1}{2}x + y = -2 \\ x - 2y = 8 \end{cases}$

27. $\begin{cases} 3x - 5y = 3 \\ 15x + 5y = 21 \end{cases}$ **28.** $\begin{cases} 2x - y = -1 \\ x - \frac{1}{2}y = \frac{3}{2} \end{cases}$ **29.** $\begin{cases} x + y - z = 6 \\ 3x - 2y + z = -5 \\ x + 3y - 2z = 14 \end{cases}$

30. $\begin{cases} x - y + z = -4 \\ 2x - 3y + 4z = -15 \\ 5x + y - 2z = 12 \end{cases}$ **31.** $\begin{cases} x + 2y - z = -3 \\ 2x - 4y + z = -7 \\ -2x + 2y - 3z = 4 \end{cases}$ **32.** $\begin{cases} x + 4y - 3z = -8 \\ 3x - y + 3z = 12 \\ x + y + 6z = 1 \end{cases}$

33. $\begin{cases} x - 2y + 3z = 1 \\ 3x + y - 2z = 0 \\ 2x - 4y + 6z = 2 \end{cases}$ **34.** $\begin{cases} x - y + 2z = 5 \\ 3x + 2y = 4 \\ -2x + 2y - 4z = -10 \end{cases}$ **35.** $\begin{cases} x + 2y - z = 0 \\ 2x - 4y + z = 0 \\ -2x + 2y - 3z = 0 \end{cases}$

36. $\begin{cases} x + 4y - 3z = 0 \\ 3x - y + 3z = 0 \\ x + y + 6z = 0 \end{cases}$

37. $\begin{cases} x - 2y + 3z = 0 \\ 3x + y - 2z = 0 \\ 2x - 4y + 6z = 0 \end{cases}$

38. $\begin{cases} x - y + 2z = 0 \\ 3x + 2y = 0 \\ -2x + 2y - 4z = 0 \end{cases}$

39. Solve $\begin{cases} \dfrac{1}{x} + \dfrac{1}{y} = 8 \\ \dfrac{3}{x} - \dfrac{5}{y} = 0 \end{cases}$

40. Solve $\begin{cases} \dfrac{4}{x} - \dfrac{3}{y} = 0 \\ \dfrac{6}{x} + \dfrac{3}{2y} = 2 \end{cases}$

[**Hint:** Let $u = 1/x$ and $v = 1/y$ and solve for u and v.]

In Problems 41–46, solve for x.

41. $\begin{vmatrix} x & x \\ 4 & 3 \end{vmatrix} = 5$

42. $\begin{vmatrix} x & 1 \\ 3 & x \end{vmatrix} = -2$

43. $\begin{vmatrix} x & 1 & 1 \\ 4 & 3 & 2 \\ -1 & 2 & 5 \end{vmatrix} = 2$

44. $\begin{vmatrix} 3 & 2 & 4 \\ 1 & x & 5 \\ 0 & 1 & -2 \end{vmatrix} = 0$

45. $\begin{vmatrix} x & 2 & 3 \\ 1 & x & 0 \\ 6 & 1 & -2 \end{vmatrix} = 7$

46. $\begin{vmatrix} x & 1 & 2 \\ 1 & x & 3 \\ 0 & 1 & 2 \end{vmatrix} = -4x$

In Problems 47–54, use properties of determinants to find the value of each determinant if it is known that

$$\begin{vmatrix} x & y & z \\ u & v & w \\ 1 & 2 & 3 \end{vmatrix} = 4$$

47. $\begin{vmatrix} 1 & 2 & 3 \\ u & v & w \\ x & y & z \end{vmatrix}$

48. $\begin{vmatrix} x & y & z \\ u & v & w \\ 2 & 4 & 6 \end{vmatrix}$

49. $\begin{vmatrix} x & y & z \\ -3 & -6 & -9 \\ u & v & w \end{vmatrix}$

50. $\begin{vmatrix} 1 & 2 & 3 \\ x - u & y - v & z - w \\ u & v & w \end{vmatrix}$

51. $\begin{vmatrix} 1 & 2 & 3 \\ x - 3 & y - 6 & z - 9 \\ 2u & 2v & 2w \end{vmatrix}$

52. $\begin{vmatrix} x & y & z - x \\ u & v & w - u \\ 1 & 2 & 2 \end{vmatrix}$

53. $\begin{vmatrix} 1 & 2 & 3 \\ 2x & 2y & 2z \\ u - 1 & v - 2 & w - 3 \end{vmatrix}$

54. $\begin{vmatrix} x + 3 & y + 6 & z + 9 \\ 3u - 1 & 3v - 2 & 3w - 3 \\ 1 & 2 & 3 \end{vmatrix}$

55. Geometry: Equation of a Line An equation of the line containing the two points (x_1, y_1) and (x_2, y_2) may be expressed as the determinant

$$\begin{vmatrix} x & y & 1 \\ x_1 & y_1 & 1 \\ x_2 & y_2 & 1 \end{vmatrix} = 0$$

Prove this result by expanding the determinant and comparing the result to the two-point form of the equation of a line.

56. Geometry: Collinear Points Using the result obtained in Problem 55, show that three distinct points (x_1, y_1), (x_2, y_2), and (x_3, y_3) are collinear (lie on the same line) if and only if

$$\begin{vmatrix} x_1 & y_1 & 1 \\ x_2 & y_2 & 1 \\ x_3 & y_3 & 1 \end{vmatrix} = 0$$

57. Show that $\begin{vmatrix} x^2 & x & 1 \\ y^2 & y & 1 \\ z^2 & z & 1 \end{vmatrix} = (y - z)(x - y)(x - z)$.

58. Complete the proof of Cramer's Rule for two equations containing two variables.
[**Hint:** In system (5), page 567, if $a = 0$, then $b \neq 0$ and $c \neq 0$, since $D = -bc \neq 0$. Now show that equations (6) provide a solution of the system when $a = 0$. There are then three remaining cases: $b = 0$, $c = 0$, and $d = 0$.]

59. Interchange columns 1 and 3 of a 3 by 3 determinant. Show that the value of the new determinant is -1 times the value of the original determinant.

60. Multiply each entry in row 2 of a 3 by 3 determinant by the number k, $k \neq 0$. Show that the value of the new determinant is k times the value of the original determinant.

61. Prove that a 3 by 3 determinant in which the entries in column 1 equal those in column 3 has the value 0.

62. Prove that, if row 2 of a 3 by 3 determinant is multiplied by k, $k \neq 0$, and the result is added to the entries in row 1, then there is no change in the value of the determinant.

8.4 | MATRIX ALGEBRA

1. Work with Equality and Addition of Matrices
2. Know Properties of Matrices
3. Know How to Multiply Matrices
4. Find the Inverse of a Matrix
5. Solve Systems of Equations Using Inverse Matrices

In Section 8.2, we defined a matrix as an array of real numbers and used an augmented matrix to represent a system of linear equations. There is, however, a branch of mathematics, called **linear algebra,** that deals with matrices in such a way that an algebra of matrices is permitted. In this section, we provide a survey of how this **matrix algebra** is developed.

Before getting started, we restate the definition of a matrix.

A **matrix** is defined as a rectangular array of numbers:

$$\begin{array}{c} \\ \text{Row 1} \\ \text{Row 2} \\ \vdots \\ \text{Row } i \\ \vdots \\ \text{Row } m \end{array} \overset{\displaystyle \text{Column 1} \quad \text{Column 2} \qquad \text{Column } j \qquad \text{Column } n}{\begin{bmatrix} a_{11} & a_{12} & \cdots & a_{1j} & \cdots & a_{1n} \\ a_{21} & a_{22} & \cdots & a_{2j} & \cdots & a_{2n} \\ \vdots & \vdots & & \vdots & & \vdots \\ a_{i1} & a_{i2} & \cdots & a_{ij} & \cdots & a_{in} \\ \vdots & \vdots & & \vdots & & \vdots \\ a_{m1} & a_{m2} & \cdots & a_{mj} & \cdots & a_{mn} \end{bmatrix}}$$

Each number a_{ij} of the matrix has two indices: the **row index** i and the **column index** j. The matrix shown above has m rows and n columns. The $m \cdot n$ numbers a_{ij} are usually referred to as the **entries** of the matrix.

Let's begin with an example that illustrates how matrices can be used to conveniently represent an array of information.

E X A M P L E   1   Arranging Data in a Matrix

In a survey of 900 people, the following information was obtained:

200 males	Thought federal defense spending was too high
150 males	Thought federal defense spending was too low
45 males	Had no opinion
315 females	Thought federal defense spending was too high
125 females	Thought federal defense spending was too low
65 females	Had no opinion

We can arrange these data in a rectangular array as follows:

	Too High	Too Low	No Opinion
Male	200	150	45
Female	315	125	65

or as

$$\begin{bmatrix} 200 & 150 & 45 \\ 315 & 125 & 65 \end{bmatrix}$$

This matrix has two rows (representing males and females) and three columns (representing "too high," "too low," and "no opinion"). ▬

The matrix we developed in Example 1 has 2 rows and 3 columns. In general, a matrix with m rows and n columns is called an *m* **by** *n* **matrix.** Thus, the matrix we developed in Example 1 is a 2 by 3 matrix. Notice that an m by n matrix will contain $m \cdot n$ entries.

If an m by n matrix has the same number of rows as columns, that is, if $m = n$, then the matrix is referred to as a **square matrix.**

EXAMPLE 2

Examples of Matrices

(a) $\begin{bmatrix} 5 & 0 \\ -6 & 1 \end{bmatrix}$ A 2 by 2 square matrix   (b) $[1 \quad 0 \quad 3]$ A 1 by 3 matrix

(c) $\begin{bmatrix} 6 & -2 & 4 \\ 4 & 3 & 5 \\ 8 & 0 & 1 \end{bmatrix}$ A 3 by 3 square matrix

▬

Equality and Addition of Matrices

1 We begin our discussion of matrix algebra by first defining what is meant by two matrices being equal and then defining the operations of addition and subtraction. It is important to note that these definitions require each matrix to have the same number of rows *and* the same number of columns as a prerequisite for equality and for addition and subtraction.

We usually represent matrices by capital letters, such as A, B, C, and so on.

Two m by n matrices A and B are said to be **equal**, written as

$$A = B$$

provided each entry a_{ij} in A is equal to the corresponding entry b_{ij} in B.

For example,

$$\begin{bmatrix} 2 & 1 \\ 0.5 & -1 \end{bmatrix} = \begin{bmatrix} \sqrt{4} & 1 \\ \frac{1}{2} & -1 \end{bmatrix} \quad \text{and} \quad \begin{bmatrix} 3 & 2 & 1 \\ 0 & 1 & -2 \end{bmatrix} = \begin{bmatrix} \sqrt{9} & \sqrt{4} & 1 \\ 0 & 1 & \sqrt[3]{-8} \end{bmatrix}$$

$$\begin{bmatrix} 4 & 1 \\ 6 & 1 \end{bmatrix} \neq \begin{bmatrix} 4 & 0 \\ 6 & 1 \end{bmatrix}$$ Because the entries in row 1, column 2 are not equal.

$$\begin{bmatrix} 4 & 1 & 2 \\ 6 & 1 & 2 \end{bmatrix} \neq \begin{bmatrix} 4 & 1 & 2 & 3 \\ 6 & 1 & 2 & 4 \end{bmatrix}$$ Because the matrix on the left is 2 by 3 and the matrix on the right is 2 by 4.

If each of A and B is an m by n matrix (and each therefore contains $m \cdot n$ entries), the statement $A = B$ actually represents a system of $m \cdot n$ ordinary equations. We will make use of this fact a little later.

Suppose that A and B represent two m by n matrices. We define their **sum** $A + B$ to be the m by n matrix formed by adding the corresponding entries a_{ij} of A and b_{ij} of B. The **difference** $A - B$ is defined as the m by n matrix formed by subtracting the entries b_{ij} in B from the corresponding entries a_{ij} in A. Addition and subtraction of matrices are allowed only for matrices having the same number m of rows and the same number n of columns. Thus, for example, a 2 by 3 matrix and a 2 by 4 matrix cannot be added or subtracted.

E X A M P L E   3   Adding and Subtracting Matrices

Suppose that

$$A = \begin{bmatrix} 2 & 4 & 8 & -3 \\ 0 & 1 & 2 & 3 \end{bmatrix} \quad \text{and} \quad B = \begin{bmatrix} -3 & 4 & 0 & 1 \\ 6 & 8 & 2 & 0 \end{bmatrix}$$

Find   (a) $A + B$ (b) $A - B$

Solution   (a) $A + B = \begin{bmatrix} 2 & 4 & 8 & -3 \\ 0 & 1 & 2 & 3 \end{bmatrix} + \begin{bmatrix} -3 & 4 & 0 & 1 \\ 6 & 8 & 2 & 0 \end{bmatrix}$

$= \begin{bmatrix} 2 + (-3) & 4 + 4 & 8 + 0 & -3 + 1 \\ 0 + 6 & 1 + 8 & 2 + 2 & 3 + 0 \end{bmatrix}$ Add corresponding entries.

$= \begin{bmatrix} -1 & 8 & 8 & -2 \\ 6 & 9 & 4 & 3 \end{bmatrix}$

(b) $A - B = \begin{bmatrix} 2 & 4 & 8 & -3 \\ 0 & 1 & 2 & 3 \end{bmatrix} - \begin{bmatrix} -3 & 4 & 0 & 1 \\ 6 & 8 & 2 & 0 \end{bmatrix}$

$= \begin{bmatrix} 2 - (-3) & 4 - 4 & 8 - 0 & -3 - 1 \\ 0 - 6 & 1 - 8 & 2 - 2 & 3 - 0 \end{bmatrix}$ Subtract corresponding entries.

$= \begin{bmatrix} 5 & 0 & 8 & -4 \\ -6 & -7 & 0 & 3 \end{bmatrix}$

FIGURE 5

```
[A]+[B]
  [[-1  8  8  -2]
   [6   9  4   3]]
[A]-[B]
  [[5   0  8  -4]
   [-6 -7  0   3]]
```

Check: Graphing utilities can make the sometimes tedious process of matrix algebra easy. In fact, most graphing calculators can handle matrices as large as 9 by 9, some even larger ones. Enter the matrices into a graphing utility. Name them $[A]$ and $[B]$. Figure 5 shows the results of adding and subtracting $[A]$ and $[B]$.

Now work Problem 1.

Many of the algebraic properties of sums of real numbers are also true for sums of matrices. Suppose that A, B, and C are m by n matrices. Then matrix addition is **commutative.** That is,

Commutative Property

$$A + B = B + A$$

Matrix addition is also **associative.** That is,

Associative Property

$$(A + B) + C = A + (B + C)$$

Although we shall not prove these results, the proofs, as the following example illustrates, are based on the commutative and associative properties for real numbers.

EXAMPLE 4 Demonstrating the Commutative Property

$$\begin{bmatrix} 2 & 3 & -1 \\ 4 & 0 & 7 \end{bmatrix} + \begin{bmatrix} -1 & 2 & 1 \\ 5 & -3 & 4 \end{bmatrix} = \begin{bmatrix} 2 + (-1) & 3 + 2 & -1 + 1 \\ 4 + 5 & 0 + (-3) & 7 + 4 \end{bmatrix}$$

$$= \begin{bmatrix} -1 + 2 & 2 + 3 & 1 + (-1) \\ 5 + 4 & -3 + 0 & 4 + 7 \end{bmatrix}$$

$$= \begin{bmatrix} -1 & 2 & 1 \\ 5 & -3 & 4 \end{bmatrix} + \begin{bmatrix} 2 & 3 & -1 \\ 4 & 0 & 7 \end{bmatrix}$$ ∎

A matrix whose entries are all equal to 0 is called a **zero matrix.** Each of the following matrices is a zero matrix.

$$\begin{bmatrix} 0 & 0 \\ 0 & 0 \end{bmatrix}$$ 2 by 2 square zero matrix $$\begin{bmatrix} 0 & 0 & 0 \\ 0 & 0 & 0 \end{bmatrix}$$ 2 by 3 zero matrix $$\begin{bmatrix} 0 & 0 & 0 \end{bmatrix}$$ 1 by 3 zero matrix

Zero matrices have properties similar to the real number 0. Thus, if A is an m by n matrix and 0 is an m by n zero matrix, then

$$A + 0 = A$$

In other words, the zero matrix is the additive identity in matrix algebra.

We also can multiply a matrix by a real number. If k is a real number and A is an m by n matrix, the matrix kA is the m by n matrix formed by multiplying each entry a_{ij} in A by k. The number k is sometimes referred to as a **scalar,** and the matrix kA is called a **scalar multiple** of A.

EXAMPLE 5 Operations Using Matrices

Suppose that

$$A = \begin{bmatrix} 3 & 1 & 5 \\ -2 & 0 & 6 \end{bmatrix} \quad B = \begin{bmatrix} 4 & 1 & 0 \\ 8 & 1 & -3 \end{bmatrix} \quad C = \begin{bmatrix} 9 & 0 \\ -3 & 6 \end{bmatrix}$$

Find   (a) $4A$ (b) $\frac{1}{3}C$ (c) $3A - 2B$

Solution (a) $4A = 4 \begin{bmatrix} 3 & 1 & 5 \\ -2 & 0 & 6 \end{bmatrix} = \begin{bmatrix} 4 \cdot 3 & 4 \cdot 1 & 4 \cdot 5 \\ 4(-2) & 4 \cdot 0 & 4 \cdot 6 \end{bmatrix} = \begin{bmatrix} 12 & 4 & 20 \\ -8 & 0 & 24 \end{bmatrix}$

(b) $\frac{1}{3}C = \frac{1}{3} \begin{bmatrix} 9 & 0 \\ -3 & 6 \end{bmatrix} = \begin{bmatrix} \frac{1}{3} \cdot 9 & \frac{1}{3} \cdot 0 \\ \frac{1}{3}(-3) & \frac{1}{3} \cdot 6 \end{bmatrix} = \begin{bmatrix} 3 & 0 \\ -1 & 2 \end{bmatrix}$

(c) $3A - 2B = 3\begin{bmatrix} 3 & 1 & 5 \\ -2 & 0 & 6 \end{bmatrix} - 2\begin{bmatrix} 4 & 1 & 0 \\ 8 & 1 & -3 \end{bmatrix}$

$= \begin{bmatrix} 3 \cdot 3 & 3 \cdot 1 & 3 \cdot 5 \\ 3(-2) & 3 \cdot 0 & 3 \cdot 6 \end{bmatrix} - \begin{bmatrix} 2 \cdot 4 & 2 \cdot 1 & 2 \cdot 0 \\ 2 \cdot 8 & 2 \cdot 1 & 2(-3) \end{bmatrix}$

$= \begin{bmatrix} 9 & 3 & 15 \\ -6 & 0 & 18 \end{bmatrix} - \begin{bmatrix} 8 & 2 & 0 \\ 16 & 2 & -6 \end{bmatrix}$

$= \begin{bmatrix} 9 - 8 & 3 - 2 & 15 - 0 \\ -6 - 16 & 0 - 2 & 18 - (-6) \end{bmatrix}$

$= \begin{bmatrix} 1 & 1 & 15 \\ -22 & -2 & 24 \end{bmatrix}$　　　　　　　━

Check:　Enter the matrices $[A], [B]$, and $[C]$ into a graphing utility. Then find $4A, \frac{1}{3}C$, and $3A - 2B$.　　　━

Now work Problem 5.

We list next some of the algebraic properties of scalar multiplication. Let h and k be real numbers, and let A and B be m by n matrices. Then

---

Properties of Scalar Multiplication

$$k(hA) = (kh)A$$
$$(k + h)A = kA + hA$$
$$k(A + B) = kA + kB$$

The proofs of these properties are based on properties of real numbers. For example, if A and B are 2 by 2 matrices, then

$k(A + B) = k\left(\begin{bmatrix} a_{11} & a_{12} \\ a_{21} & a_{22} \end{bmatrix} + \begin{bmatrix} b_{11} & b_{12} \\ b_{21} & b_{22} \end{bmatrix}\right) = k\begin{bmatrix} a_{11} + b_{11} & a_{12} + b_{12} \\ a_{21} + b_{21} & a_{22} + b_{22} \end{bmatrix}$

$= \begin{bmatrix} k(a_{11} + b_{11}) & k(a_{12} + b_{12}) \\ k(a_{21} + b_{21}) & k(a_{22} + b_{22}) \end{bmatrix} = \begin{bmatrix} ka_{11} + kb_{11} & ka_{12} + kb_{12} \\ ka_{21} + kb_{21} & ka_{22} + kb_{22} \end{bmatrix}$

$= \begin{bmatrix} ka_{11} & ka_{12} \\ ka_{21} & ka_{22} \end{bmatrix} + \begin{bmatrix} kb_{11} & kb_{12} \\ kb_{21} & kb_{22} \end{bmatrix} = k\begin{bmatrix} a_{11} & a_{12} \\ a_{21} & a_{22} \end{bmatrix} + k\begin{bmatrix} b_{11} & b_{12} \\ b_{21} & b_{22} \end{bmatrix} = kA + kB$

Multiplication of Matrices

3　Unlike the straightforward definition for adding two matrices, the definition for multiplying two matrices is not what we might expect. In preparation for this definition, we need the following definitions:

　　A **row vector** R is a 1 by n matrix

$$R = [r_1 \quad r_2 \quad \cdots \quad r_n]$$

A **column vector** C is an n by 1 matrix

$$C = \begin{bmatrix} c_1 \\ c_2 \\ \vdots \\ c_n \end{bmatrix}$$

The **product** RC of R times C is defined as the number

$$RC = [r_1 \quad r_2 \quad \cdots \quad r_n] \begin{bmatrix} c_1 \\ c_2 \\ \vdots \\ c_n \end{bmatrix} = r_1 c_1 + r_2 c_2 + \cdots + r_n c_n$$

Notice that a row vector and a column vector can be multiplied only if they contain the same number of entries.

E X A M P L E   6

The Product of a Row Vector by a Column Vector

If $R = [3 \quad -5 \quad 2]$ and $C = \begin{bmatrix} 3 \\ 4 \\ -5 \end{bmatrix}$, then

$$RC = [3 \quad -5 \quad 2] \begin{bmatrix} 3 \\ 4 \\ -5 \end{bmatrix} = 3 \cdot 3 + (-5)4 + 2(-5)$$

$$= 9 - 20 - 10 = -21 \qquad \blacksquare$$

Let's look at an application of the product of a row vector by a column vector.

E X A M P L E   7

Using Matrices to Compute Revenue

A clothing store sells men's shirts for \$25, silk ties for \$8, and wool suits for \$300. Last month, the store had sales consisting of 100 shirts, 200 ties, and 50 suits. What was the total revenue due to these sales?

Solution

We set up a row vector R to represent the prices of each item and a column vector C to represent the corresponding number of items sold.
Then

$$\begin{array}{cc} \text{Prices} & \text{Number} \\ \text{Shirts Ties Suits} & \text{sold} \end{array}$$

$$R = [25 \quad 8 \quad 300] \qquad C = \begin{bmatrix} 100 \\ 200 \\ 50 \end{bmatrix} \begin{array}{l} \text{Shirts} \\ \text{Ties} \\ \text{Suits} \end{array}$$

The total revenue obtained is the product RC. That is,

$$RC = [25 \quad 8 \quad 300] \begin{bmatrix} 100 \\ 200 \\ 50 \end{bmatrix}$$

$$= 25 \cdot 100 + 8 \cdot 200 + 300 \cdot 50 = \$19,100$$

$$\underbrace{\hspace{1.5cm}}_{\text{Shirt revenue}} \underbrace{\hspace{1.5cm}}_{\text{Tie revenue}} \underbrace{\hspace{1.5cm}}_{\text{Suit revenue}} \underbrace{\hspace{1.5cm}}_{\text{Total revenue}} \qquad \blacksquare$$

The definition for multiplying two matrices is based on the definition of a row vector times a column vector.

> Let A denote an m by r matrix and let B denote an r by n matrix. The **product** AB is defined as the m by n matrix whose entry in row i, column j is the product of the ith row of A and the jth column of B.

The definition of the product AB of two matrices A and B, in this order, requires that the number of columns of A equal the number of rows of B; otherwise, no product is defined:

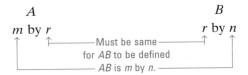

An example will help clarify the definition.

E X A M P L E   8

Multiplying Two Matrices

Find the product AB if

$$A = \begin{bmatrix} 2 & 4 & -1 \\ 5 & 8 & 0 \end{bmatrix} \quad \text{and} \quad B = \begin{bmatrix} 2 & 5 & 1 & 4 \\ 4 & 8 & 0 & 6 \\ -3 & 1 & -2 & -1 \end{bmatrix}$$

Solution

First, we note that A is a 2 by 3 matrix and B is a 3 by 4 matrix, so the product AB is defined and will be a 2 by 4 matrix. Suppose that we want the entry in row 2, column 3 of AB. To find it, we find the product of the row vector from row 2 of A and the column vector from column 3 of B.

$$\underset{\text{Row 2 of } A}{[5 \quad 8 \quad 0]} \overset{\text{Column 3 of } B}{\begin{bmatrix} 1 \\ 0 \\ -2 \end{bmatrix}} = 5 \cdot 1 + 8 \cdot 0 + 0(-2) = 5$$

So far, we have

$$AB = \begin{bmatrix} \underline{\quad} & \underline{\quad} & \overset{\overset{\text{Column 3}}{\downarrow}}{5} & \underline{\quad} \end{bmatrix} \quad \leftarrow \text{Row 2}$$

Now, to find the entry in row 1, column 4 of AB, we find the product of row 1 of A and column 4 of B.

$$\underset{\text{Row 1 of } A}{[2 \quad 4 \quad -1]} \overset{\text{Column 4 of } B}{\begin{bmatrix} 4 \\ 6 \\ -1 \end{bmatrix}} = 2 \cdot 4 + 4 \cdot 6 + (-1)(-1) = 33$$

Continuing in this fashion, we find AB.

$$AB = \begin{bmatrix} 2 & 4 & -1 \\ 5 & 8 & 0 \end{bmatrix} \begin{bmatrix} 2 & 5 & 1 & 4 \\ 4 & 8 & 0 & 6 \\ -3 & 1 & -2 & -1 \end{bmatrix}$$

$$= \begin{bmatrix} \text{Row 1 of } A & \text{Row 1 of } A & \text{Row 1 of } A & \text{Row 1 of } A \\ \text{times} & \text{times} & \text{times} & \text{times} \\ \text{column 1 of } B & \text{column 2 of } B & \text{column 3 of } B & \text{column 4 of } B \\ \text{Row 2 of } A & \text{Row 2 of } A & \text{Row 2 of } A & \text{Row 2 of } A \\ \text{times} & \text{times} & \text{times} & \text{times} \\ \text{column 1 of } B & \text{column 2 of } B & \text{column 3 of } B & \text{column 4 of } B \end{bmatrix}$$

$$= \begin{bmatrix} 2\cdot 2 + 4\cdot 4 + (-1)(-3) & 2\cdot 5 + 4\cdot 8 + (-1)1 & 2\cdot 1 + 4\cdot 0 + (-1)(-2) & 33 \text{ (from earlier)} \\ 5\cdot 2 + 8\cdot 4 + 0(-3) & 5\cdot 5 + 8\cdot 8 + 0\cdot 1 & 5 \text{ (from earlier)} & 5\cdot 4 + 8\cdot 6 + 0(-1) \end{bmatrix}$$

$$= \begin{bmatrix} 23 & 41 & 4 & 33 \\ 42 & 89 & 5 & 68 \end{bmatrix}$$

Check: Enter the matrices A and B. Then find AB. (See what happens if you try to find BA.)

Now work Problem 17.

Notice that for the matrices given in Example 8 the product BA is not defined, because B is 3 by 4 and A is 2 by 3. Try calculating BA on a graphing utility. What do you notice?

Another result that can occur when multiplying two matrices is illustrated in the next example.

E X A M P L E   9   Multiplying Two Matrices

If

$$A = \begin{bmatrix} 2 & 1 & 3 \\ 1 & -1 & 0 \end{bmatrix} \quad \text{and} \quad B = \begin{bmatrix} 1 & 0 \\ 2 & 1 \\ 3 & 2 \end{bmatrix}$$

find   (a) AB (b) BA

Solution   (a) $AB = \underset{\text{2 by 3}}{\begin{bmatrix} 2 & 1 & 3 \\ 1 & -1 & 0 \end{bmatrix}} \underset{\text{3 by 2}}{\begin{bmatrix} 1 & 0 \\ 2 & 1 \\ 3 & 2 \end{bmatrix}} = \underset{\text{2 by 2}}{\begin{bmatrix} 13 & 7 \\ -1 & -1 \end{bmatrix}}$

(b) $BA = \underset{\text{3 by 2}}{\begin{bmatrix} 1 & 0 \\ 2 & 1 \\ 3 & 2 \end{bmatrix}} \underset{\text{2 by 3}}{\begin{bmatrix} 2 & 1 & 3 \\ 1 & -1 & 0 \end{bmatrix}} = \underset{\text{3 by 3}}{\begin{bmatrix} 2 & 1 & 3 \\ 5 & 1 & 6 \\ 8 & 1 & 9 \end{bmatrix}}$

Notice in Example 9 that AB is 2 by 2 and BA is 3 by 3. Thus, it is possible for both AB and BA to be defined, yet be unequal. In fact, even if A

and B are both n by n matrices so that AB and BA are each defined and n by n, AB and BA will usually be unequal.

EXAMPLE 10   Multiplying Two Square Matrices

If

$$A = \begin{bmatrix} 2 & 1 \\ 0 & 4 \end{bmatrix} \quad \text{and} \quad B = \begin{bmatrix} -3 & 1 \\ 1 & 2 \end{bmatrix}$$

find   (a) AB (b) BA

Solution   (a) $AB = \begin{bmatrix} 2 & 1 \\ 0 & 4 \end{bmatrix}\begin{bmatrix} -3 & 1 \\ 1 & 2 \end{bmatrix}$

$$= \begin{bmatrix} 2(-3) + 1\cdot 1 & 2\cdot 1 + 1\cdot 2 \\ 0(-3) + 4\cdot 1 & 0\cdot 1 + 4\cdot 2 \end{bmatrix} = \begin{bmatrix} -5 & 4 \\ 4 & 8 \end{bmatrix}$$

(b) $BA = \begin{bmatrix} -3 & 1 \\ 1 & 2 \end{bmatrix}\begin{bmatrix} 2 & 1 \\ 0 & 4 \end{bmatrix}$

$$= \begin{bmatrix} (-3)2 + 1\cdot 0 & (-3)1 + 1\cdot 4 \\ 1\cdot 2 + 2\cdot 0 & 1\cdot 1 + 2\cdot 4 \end{bmatrix} = \begin{bmatrix} -6 & 1 \\ 2 & 9 \end{bmatrix}$$

The preceding examples demonstrate that an important property of real numbers, the commutative property of multiplication, is not shared by matrices. Thus, in general:

Theorem

Matrix multiplication is not commutative.

 Now work Problems 7 and 9.

Next we give two of the properties of real numbers that are shared by matrices. Assuming that each product and sum is defined, we have:

Associative Property

$$A(BC) = (AB)C$$

Distributive Property

$$A(B + C) = AB + AC$$

The Identity Matrix

For an n by n square matrix, the entries located in row i, column i, $1 \leq i \leq n$, are called the **diagonal entries.** An n by n square matrix whose diagonal entries are 1's, while all other entries are 0's, is called the **identity matrix** I_n. For example,

$$I_2 = \begin{bmatrix} 1 & 0 \\ 0 & 1 \end{bmatrix} \qquad I_3 = \begin{bmatrix} 1 & 0 & 0 \\ 0 & 1 & 0 \\ 0 & 0 & 1 \end{bmatrix}$$

and so on.

E X A M P L E   11

Multiplication with an Identity Matrix

Let

$$A = \begin{bmatrix} -1 & 2 & 0 \\ 0 & 1 & 3 \end{bmatrix} \quad \text{and} \quad B = \begin{bmatrix} 3 & 2 \\ 4 & 6 \\ 5 & 2 \end{bmatrix}$$

Find   (a) AI_3 (b) I_2A (c) BI_2

Solution   (a) $AI_3 = \begin{bmatrix} -1 & 2 & 0 \\ 0 & 1 & 3 \end{bmatrix} \begin{bmatrix} 1 & 0 & 0 \\ 0 & 1 & 0 \\ 0 & 0 & 1 \end{bmatrix} = \begin{bmatrix} -1 & 2 & 0 \\ 0 & 1 & 3 \end{bmatrix} = A$

(b) $I_2A = \begin{bmatrix} 1 & 0 \\ 0 & 1 \end{bmatrix} \begin{bmatrix} -1 & 2 & 0 \\ 0 & 1 & 3 \end{bmatrix} = \begin{bmatrix} -1 & 2 & 0 \\ 0 & 1 & 3 \end{bmatrix} = A$

(c) $BI_2 = \begin{bmatrix} 3 & 2 \\ 4 & 6 \\ 5 & 2 \end{bmatrix} \begin{bmatrix} 1 & 0 \\ 0 & 1 \end{bmatrix} = \begin{bmatrix} 3 & 2 \\ 4 & 6 \\ 5 & 2 \end{bmatrix} = B$

Example 11 demonstrates the following property:

Identity Property

If A is an m by n matrix, then

$$I_mA = A \quad \text{and} \quad AI_n = A$$

If A is an n by n square matrix, then $AI_n = I_nA = A$.

Thus, an identity matrix has properties analogous to those of the real number 1. In other words, the identity matrix is a multiplicative identity in matrix algebra.

4 The Inverse of a Matrix

Let A be a square matrix, n by n. If there exists an n by n matrix A^{-1}, read "A inverse," for which

$$AA^{-1} = A^{-1}A = I_n$$

then A^{-1} is called the **inverse** of the matrix A.

As we shall soon see, not every square matrix has an inverse. When a matrix A does have an inverse A^{-1}, then A is said to be **nonsingular.** If a matrix A has no inverse, it is called **singular.**

E X A M P L E   12   Multiplying a Matrix by Its Inverse

Verify that the inverse of

$$A = \begin{bmatrix} 3 & 1 \\ 2 & 1 \end{bmatrix} \quad \text{is} \quad A^{-1} = \begin{bmatrix} 1 & -1 \\ -2 & 3 \end{bmatrix}$$

Solution   We need to show that $AA^{-1} = A^{-1}A = I_2$.

$$AA^{-1} = \begin{bmatrix} 3 & 1 \\ 2 & 1 \end{bmatrix}\begin{bmatrix} 1 & -1 \\ -2 & 3 \end{bmatrix} = \begin{bmatrix} 3 \cdot 1 + 1(-2) & 3(-1) + 1 \cdot 3 \\ 2 \cdot 1 + 1(-2) & 2(-1) + 1 \cdot 3 \end{bmatrix}$$

$$= \begin{bmatrix} 1 & 0 \\ 0 & 1 \end{bmatrix} = I_2$$

$$A^{-1}A = \begin{bmatrix} 1 & -1 \\ -2 & 3 \end{bmatrix}\begin{bmatrix} 3 & 1 \\ 2 & 1 \end{bmatrix} = \begin{bmatrix} 3 - 2 & 1 - 1 \\ -6 + 6 & -2 + 3 \end{bmatrix} = \begin{bmatrix} 1 & 0 \\ 0 & 1 \end{bmatrix} = I_2 \quad \blacksquare$$

We now show one way to find the inverse of

$$A = \begin{bmatrix} 3 & 1 \\ 2 & 1 \end{bmatrix}$$

Suppose that A^{-1} is given by

$$A^{-1} = \begin{bmatrix} x & y \\ z & w \end{bmatrix} \tag{1}$$

where x, y, z, and w are four variables. Based on the definition of an inverse, if, indeed, A has an inverse, we have

$$AA^{-1} = I_2$$

$$\begin{bmatrix} 3 & 1 \\ 2 & 1 \end{bmatrix}\begin{bmatrix} x & y \\ z & w \end{bmatrix} = \begin{bmatrix} 1 & 0 \\ 0 & 1 \end{bmatrix}$$

$$\begin{bmatrix} 3x + z & 3y + w \\ 2x + z & 2y + w \end{bmatrix} = \begin{bmatrix} 1 & 0 \\ 0 & 1 \end{bmatrix}$$

Because corresponding entries must be equal, it follows that this matrix equation is equivalent to four ordinary equations.

$$\begin{cases} 3x + z = 1 \\ 2x + z = 0 \end{cases} \qquad \begin{cases} 3y + w = 0 \\ 2y + w = 1 \end{cases}$$

The augmented matrix of each system is

$$\begin{bmatrix} 3 & 1 & \Big| & 1 \\ 2 & 1 & \Big| & 0 \end{bmatrix} \qquad \begin{bmatrix} 3 & 1 & \Big| & 0 \\ 2 & 1 & \Big| & 1 \end{bmatrix} \tag{2}$$

The usual procedure would be to transform each augmented matrix into reduced echelon form. Notice, though, that the left sides of the augmented matrices are equal, so the same row operations (see Section 8.2) can be used to reduce each one. Thus, we find it more efficient to combine the two augmented matrices (2) into a single matrix, as shown next, and then transform it into reduced echelon form.

$$\begin{bmatrix} 3 & 1 & \Big| & 1 & 0 \\ 2 & 1 & \Big| & 0 & 1 \end{bmatrix}$$

Now we attempt to transform the left side into an identity matrix.

$$\begin{bmatrix} 3 & 1 & \Big| & 1 & 0 \\ 2 & 1 & \Big| & 0 & 1 \end{bmatrix} \underset{\substack{\uparrow \\ R_1 = -1r_2 + r_1}}{\rightarrow} \begin{bmatrix} 1 & 0 & \Big| & 1 & -1 \\ 2 & 1 & \Big| & 0 & 1 \end{bmatrix}$$

$$\underset{\substack{\uparrow \\ R_2 = -2r_1 + r_2}}{\rightarrow} \begin{bmatrix} 1 & 0 & \Big| & 1 & -1 \\ 0 & 1 & \Big| & -2 & 3 \end{bmatrix} \tag{3}$$

Matrix (3) is in reduced echelon form. Now we reverse the earlier step of combining the two augmented matrices in (2) and write the single matrix (3) as two augmented matrices.

$$\begin{bmatrix} 1 & 0 & \Big| & 1 \\ 0 & 1 & \Big| & -2 \end{bmatrix} \quad \text{and} \quad \begin{bmatrix} 1 & 0 & \Big| & -1 \\ 0 & 1 & \Big| & 3 \end{bmatrix}$$

We conclude from these matrices that $x = 1$, $z = -2$, and $y = -1$, $w = 3$. Substituting these values into matrix (1), we find that

$$A^{-1} = \begin{bmatrix} 1 & -1 \\ -2 & 3 \end{bmatrix}$$

Notice in display (3) that the 2 by 2 matrix to the right of the vertical bar is, in fact, the inverse of A. Also notice that the identity matrix I_2 is the matrix that appears to the left of the vertical bar. These observations and the procedures followed above will work in general.

Procedure for Finding the Inverse of a Nonsingular Matrix

To find the inverse of an n by n nonsingular matrix A, proceed as follows:

STEP 1: Form the matrix $[A|I_n]$.
STEP 2: Transform the matrix $[A|I_n]$ into reduced echelon form.
STEP 3: The reduced echelon form of $[A|I_n]$ will contain the identity matrix I_n on the left of the vertical bar; the n by n matrix on the right of the vertical bar is the inverse of A.

In other words, if A is nonsingular, we begin with the matrix $[A|I_n]$ and, after transforming it into reduced echelon form, we end up with the matrix $[I_n|A^{-1}]$.

Let's look at another example.

E X A M P L E   13   Finding the Inverse of a Matrix

The matrix

$$A = \begin{bmatrix} 1 & 1 & 0 \\ -1 & 3 & 4 \\ 0 & 4 & 3 \end{bmatrix}$$

is nonsingular. Find its inverse.

Solution   First, we form the matrix

$$[A|I_3] = \begin{bmatrix} 1 & 1 & 0 & | & 1 & 0 & 0 \\ -1 & 3 & 4 & | & 0 & 1 & 0 \\ 0 & 4 & 3 & | & 0 & 0 & 1 \end{bmatrix}$$

Next, we use row operations to transform $[A|I_3]$ into reduced echelon form.

$$\begin{bmatrix} 1 & 1 & 0 & | & 1 & 0 & 0 \\ -1 & 3 & 4 & | & 0 & 1 & 0 \\ 0 & 4 & 3 & | & 0 & 0 & 1 \end{bmatrix} \rightarrow \begin{bmatrix} 1 & 1 & 0 & | & 1 & 0 & 0 \\ 0 & 4 & 4 & | & 1 & 1 & 0 \\ 0 & 4 & 3 & | & 0 & 0 & 1 \end{bmatrix} \rightarrow \begin{bmatrix} 1 & 1 & 0 & | & 1 & 0 & 0 \\ 0 & 1 & 1 & | & \frac{1}{4} & \frac{1}{4} & 0 \\ 0 & 4 & 3 & | & 0 & 0 & 1 \end{bmatrix}$$

$$\uparrow \qquad\qquad\qquad\qquad\qquad\qquad \uparrow$$
$$R_2 = r_1 + r_2 \qquad\qquad\qquad\qquad R_2 = \tfrac{1}{4}r_2$$

$$\rightarrow \begin{bmatrix} 1 & 0 & -1 & | & \frac{3}{4} & -\frac{1}{4} & 0 \\ 0 & 1 & 1 & | & \frac{1}{4} & \frac{1}{4} & 0 \\ 0 & 0 & -1 & | & -1 & -1 & 1 \end{bmatrix} \rightarrow \begin{bmatrix} 1 & 0 & -1 & | & \frac{3}{4} & -\frac{1}{4} & 0 \\ 0 & 1 & 1 & | & \frac{1}{4} & \frac{1}{4} & 0 \\ 0 & 0 & 1 & | & 1 & 1 & -1 \end{bmatrix}$$

$$\uparrow \qquad\qquad\qquad\qquad\qquad\qquad \uparrow$$
$$R_1 = -1r_2 + r_1 \qquad\qquad\qquad R_3 = -1r_3$$
$$R_3 = -4r_2 + r_3$$

$$\rightarrow \begin{bmatrix} 1 & 0 & 0 & | & \frac{7}{4} & \frac{3}{4} & -1 \\ 0 & 1 & 0 & | & -\frac{3}{4} & -\frac{3}{4} & 1 \\ 0 & 0 & 1 & | & 1 & 1 & -1 \end{bmatrix}$$

$$\uparrow$$
$$R_1 = r_3 + r_1$$
$$R_2 = -1r_3 + r_2$$

The matrix $[A|I_3]$ is now in reduced echelon form, and the identity matrix I_3 is on the left of the vertical bar. Hence, the inverse of A is

FIGURE 6

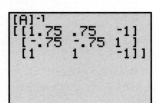

$$A^{-1} = \begin{bmatrix} \frac{7}{4} & \frac{3}{4} & -1 \\ -\frac{3}{4} & -\frac{3}{4} & 1 \\ 1 & 1 & -1 \end{bmatrix}$$

You can (and should) verify that this is the correct inverse by showing that $AA^{-1} = A^{-1}A = I_3$.

Check: Enter the matrix A into a graphing utility. Figure 6 shows A^{-1}.

Now work Problem 23.

If transforming the matrix $[A|I_n]$ into reduced echelon form does not result in the identity matrix I_n to the left of the vertical bar, then A has no inverse. The next example demonstrates such a matrix.

E X A M P L E   14

Showing That a Matrix Has No Inverse

Show that the following matrix has no inverse.

$$A = \begin{bmatrix} 4 & 6 \\ 2 & 3 \end{bmatrix}$$

Solution

Proceeding as in Example 13, we form the matrix

$$[A|I_2] = \begin{bmatrix} 4 & 6 & | & 1 & 0 \\ 2 & 3 & | & 0 & 1 \end{bmatrix}$$

Then we use row operations to transform $[A|I_2]$ into reduced echelon form.

$$[A|I_2] = \begin{bmatrix} 4 & 6 & | & 1 & 0 \\ 2 & 3 & | & 0 & 1 \end{bmatrix} \rightarrow \begin{bmatrix} 1 & \frac{3}{2} & | & \frac{1}{4} & 0 \\ 2 & 3 & | & 0 & 1 \end{bmatrix} \rightarrow \begin{bmatrix} 1 & \frac{3}{2} & | & \frac{1}{4} & 0 \\ 0 & 0 & | & -\frac{1}{2} & 1 \end{bmatrix}$$

$$\uparrow \qquad\qquad\qquad \uparrow$$
$$R_1 = \frac{1}{4}r_1 \qquad\qquad R_2 = -2r_1 + r_2$$

The matrix $[A|I_2]$ is sufficiently reduced for us to see that the identity matrix cannot appear to the left of the vertical bar. We conclude that A has no inverse. ■

Check: Enter the matrix A. Try to find its inverse. What happens? ■

Now work Problem 51.

Solving Systems of Linear Equations

Inverse matrices can be used to solve systems of equations in which the number of equations is the same as the number of variables.

E X A M P L E   15

Using the Inverse Matrix to Solve a System of Linear Equations

Solve the system of equations:
$$\begin{cases} x + y = 3 \\ -x + 3y + 4z = -3 \\ 4y + 3z = 2 \end{cases}$$

Solution

If we let

$$A = \begin{bmatrix} 1 & 1 & 0 \\ -1 & 3 & 4 \\ 0 & 4 & 3 \end{bmatrix}, \qquad X = \begin{bmatrix} x \\ y \\ z \end{bmatrix}, \quad \text{and} \quad B = \begin{bmatrix} 3 \\ -3 \\ 2 \end{bmatrix}$$

then the original system of equations can be written compactly as the matrix equation

$$AX = B \qquad\qquad\qquad\qquad\qquad\qquad\qquad\qquad (4)$$

We know from Example 13 that the matrix A has the inverse A^{-1}, so we multiply each side of equation (4) by A^{-1}.

$$AX = B$$
$$A^{-1}(AX) = A^{-1}B$$
$$(A^{-1}A)X = A^{-1}B \qquad \text{Associative property of multiplication}$$
$$I_3X = A^{-1}B \qquad \text{Definition of inverse matrix}$$
$$X = A^{-1}B \qquad \text{Property of identity matrix} \tag{5}$$

Now we use (5) to find $X = \begin{bmatrix} x \\ y \\ z \end{bmatrix}$.

$$X = \begin{bmatrix} x \\ y \\ z \end{bmatrix} = A^{-1}B = \begin{bmatrix} \frac{7}{4} & \frac{3}{4} & -1 \\ -\frac{3}{4} & -\frac{3}{4} & 1 \\ 1 & 1 & -1 \end{bmatrix} \begin{bmatrix} 3 \\ -3 \\ 2 \end{bmatrix} = \begin{bmatrix} 1 \\ 2 \\ -2 \end{bmatrix}$$

Example 13

Thus, $x = 1$, $y = 2$, $z = -2$. ∎

The method used in Example 15 to solve a system of equations is particularly useful when it is necessary to solve several systems of equations in which the constants appearing to the right of the equal signs change, while the coefficients of the variables on the left side remain the same. See Problems 31–50 for some illustrations.

HISTORICAL FEATURE Matrices were invented in 1857 by Arthur Cayley (1821–1895) as a way of efficiently computing the result of substituting one linear system into another (see Historical Problem 2). The resulting system had incredible richness, in the sense that a very wide variety of mathematical systems could be mimicked by the matrices. Cayley and his friend J. J. Sylvester (1814–1897) spent much of the rest of their lives elaborating the theory. The torch was then passed to G. Frobenius (1849–1917), whose deep investigations established a central place for matrices in modern mathematics. In 1924, rather to the surprise of physicists, it was found that matrices (with complex numbers in them) were exactly the right tool for describing the behavior of atomic systems. Today, matrices are used in a wide variety of applications.

HISTORICAL PROBLEMS 1. *Matrices and Complex Numbers* Frobenius emphasized in his research how matrices could be used to mimic other mathematical systems. Here, we mimic the behavior of complex numbers using matrices. Mathematicians call such a relationship an *isomorphism*.

Complex number ⟷ Matrix

$$a + bi \longleftrightarrow \begin{bmatrix} a & b \\ -b & a \end{bmatrix}$$

Note that the complex number can be read off the top line of the matrix. Thus,

$$2 + 3i \longleftrightarrow \begin{bmatrix} 2 & 3 \\ -3 & 2 \end{bmatrix} \quad \text{and} \quad \begin{bmatrix} 4 & -2 \\ 2 & 4 \end{bmatrix} \longleftrightarrow 4 - 2i.$$

(a) Find the matrices corresponding to $2 - 5i$ and $1 + 3i$.

(b) Multiply the two matrices.

(c) Find the corresponding complex number for the matrix found in part (b).

(d) Multiply $2 - 5i$ by $1 + 3i$. The result should be the same as that found in part (c).

The process also works for addition and subtraction. Try it for yourself.

2. *Cayley's Definition of Matrix Multiplication* Cayley invented matrix multiplication to simplify the following problem:

$$\begin{cases} u = ar + bs \\ v = cr + ds \end{cases} \qquad \begin{cases} x = ku + lv \\ y = mu + nv \end{cases}$$

(a) Find x and y in terms of r and s by substituting u and v from the first system of equations into the second system of equations.

(b) Use the result of part (a) to find the 2 by 2 matrix A in

$$\begin{bmatrix} x \\ y \end{bmatrix} = A \begin{bmatrix} r \\ s \end{bmatrix}$$

(c) Now look at the following way to do it. Write the equations in matrix form.

$$\begin{bmatrix} u \\ v \end{bmatrix} = \begin{bmatrix} a & b \\ c & d \end{bmatrix} \begin{bmatrix} r \\ s \end{bmatrix} \qquad \begin{bmatrix} x \\ y \end{bmatrix} = \begin{bmatrix} k & l \\ m & n \end{bmatrix} \begin{bmatrix} u \\ v \end{bmatrix}$$

So

$$\begin{bmatrix} x \\ y \end{bmatrix} = \begin{bmatrix} k & l \\ m & n \end{bmatrix} \begin{bmatrix} a & b \\ c & d \end{bmatrix} \begin{bmatrix} r \\ s \end{bmatrix}$$

Do you see how Cayley defined matrix multiplication?

8.4 | EXERCISES

In Problems 1–16, use the following matrices to compute the given expression.

$$A = \begin{bmatrix} 0 & 3 & -5 \\ 1 & 2 & 6 \end{bmatrix} \qquad B = \begin{bmatrix} 4 & 1 & 0 \\ -2 & 3 & -2 \end{bmatrix} \qquad C = \begin{bmatrix} 4 & 1 \\ 6 & 2 \\ -2 & 3 \end{bmatrix}$$

1. $A + B$	2. $A - B$	3. $4A$	4. $-3B$
5. $3A - 2B$	6. $2A + 4B$	7. AC	8. BC
9. CA	10. CB	11. $C(A + B)$	12. $(A + B)C$
13. $AC - 3I_2$	14. $CA + 5I_3$	15. $CA - CB$	16. $AC + BC$

In Problems 17–20, compute each product.

17. $\begin{bmatrix} 2 & -2 \\ 1 & 0 \end{bmatrix} \begin{bmatrix} 2 & 1 & 4 & 6 \\ 3 & -1 & 3 & 2 \end{bmatrix}$

18. $\begin{bmatrix} 4 & 1 \\ 2 & 1 \end{bmatrix} \begin{bmatrix} -6 & 6 & 1 & 0 \\ 2 & 5 & 4 & -1 \end{bmatrix}$

19. $\begin{bmatrix} 1 & 0 & 1 \\ 2 & 4 & 1 \\ 3 & 6 & 1 \end{bmatrix} \begin{bmatrix} 1 & 3 \\ 6 & 2 \\ 8 & -1 \end{bmatrix}$

20. $\begin{bmatrix} 4 & -2 & 3 \\ 0 & 1 & 2 \\ -1 & 0 & 1 \end{bmatrix} \begin{bmatrix} 2 & 6 \\ 1 & -1 \\ 0 & 2 \end{bmatrix}$

In Problems 21–30, each matrix is nonsingular. Find the inverse of each matrix. Be sure to check your answer.

21. $\begin{bmatrix} 2 & 1 \\ 1 & 1 \end{bmatrix}$

22. $\begin{bmatrix} 3 & -1 \\ -2 & 1 \end{bmatrix}$

23. $\begin{bmatrix} 6 & 5 \\ 2 & 2 \end{bmatrix}$

24. $\begin{bmatrix} -4 & 1 \\ 6 & -2 \end{bmatrix}$

25. $\begin{bmatrix} 2 & 1 \\ a & a \end{bmatrix}, \quad a \neq 0$

26. $\begin{bmatrix} b & 3 \\ b & 2 \end{bmatrix}, \quad b \neq 0$

27. $\begin{bmatrix} 1 & -1 & 1 \\ 0 & -2 & 1 \\ -2 & -3 & 0 \end{bmatrix}$

28. $\begin{bmatrix} 1 & 0 & 2 \\ -1 & 2 & 3 \\ 1 & -1 & 0 \end{bmatrix}$

29. $\begin{bmatrix} 1 & 1 & 1 \\ 3 & 2 & -1 \\ 3 & 1 & 2 \end{bmatrix}$

30. $\begin{bmatrix} 3 & 3 & 1 \\ 1 & 2 & 1 \\ 2 & -1 & 1 \end{bmatrix}$

In Problems 31–50, use the inverses found in Problems 21–30 to solve each system of equations.

31. $\begin{cases} 2x + y = 8 \\ x + y = 5 \end{cases}$

32. $\begin{cases} 3x - y = 8 \\ -2x + y = 4 \end{cases}$

33. $\begin{cases} 2x + y = 0 \\ x + y = 5 \end{cases}$

34. $\begin{cases} 3x - y = 4 \\ -2x + y = 5 \end{cases}$

35. $\begin{cases} 6x + 5y = 7 \\ 2x + 2y = 2 \end{cases}$

36. $\begin{cases} -4x + y = 0 \\ 6x - 2y = 14 \end{cases}$

37. $\begin{cases} 6x + 5y = 13 \\ 2x + 2y = 5 \end{cases}$

38. $\begin{cases} -4x + y = 5 \\ 6x - 2y = -9 \end{cases}$

39. $\begin{cases} 2x + y = -3 \\ ax + ay = -a \end{cases} \quad a \neq 0$

40. $\begin{cases} bx + 3y = 2b + 3 \\ bx + 2y = 2b + 2 \end{cases} \quad b \neq 0$

41. $\begin{cases} 2x + y = 7/a \\ ax + ay = 5 \end{cases} \quad a \neq 0$

42. $\begin{cases} bx + 3y = 14 \\ bx + 2y = 10 \end{cases} \quad b \neq 0$

43. $\begin{cases} x - y + z = 0 \\ -2y + z = -1 \\ -2x - 3y = -5 \end{cases}$

44. $\begin{cases} x + 2z = 6 \\ -x + 2y + 3z = -5 \\ x - y = 6 \end{cases}$

45. $\begin{cases} x - y + z = 2 \\ -2y + z = 2 \\ -2x - 3y = \frac{1}{2} \end{cases}$

46. $\begin{cases} x + 2z = 2 \\ -x + 2y + 3z = -\frac{3}{2} \\ x - y = 2 \end{cases}$

47. $\begin{cases} x + y + z = 9 \\ 3x + 2y - z = 8 \\ 3x + y + 2z = 1 \end{cases}$

48. $\begin{cases} 3x + 3y + z = 8 \\ x + 2y + z = 5 \\ 2x - y + z = 4 \end{cases}$

49. $\begin{cases} x + y + z = 2 \\ 3x + 2y - z = \frac{7}{3} \\ 3x + y + 2z = \frac{10}{3} \end{cases}$

50. $\begin{cases} 3x + 3y + z = 1 \\ x + 2y + z = 0 \\ 2x - y + z = 4 \end{cases}$

In Problems 51–56, show that each matrix has no inverse.

51. $\begin{bmatrix} 4 & 2 \\ 2 & 1 \end{bmatrix}$

52. $\begin{bmatrix} -3 & \frac{1}{2} \\ 6 & -1 \end{bmatrix}$

53. $\begin{bmatrix} 15 & 3 \\ 10 & 2 \end{bmatrix}$

54. $\begin{bmatrix} -3 & 0 \\ 4 & 0 \end{bmatrix}$

55. $\begin{bmatrix} -3 & 1 & -1 \\ 1 & -4 & -7 \\ 1 & 2 & 5 \end{bmatrix}$

56. $\begin{bmatrix} 1 & 1 & -3 \\ 2 & -4 & 1 \\ -5 & 7 & 1 \end{bmatrix}$

In Problems 57–60, use a graphing utility to find the inverse, if it exists, of each matrix. Round answers to two decimal places.

57. $\begin{bmatrix} 25 & 61 & -12 \\ 18 & -2 & 4 \\ 8 & 35 & 21 \end{bmatrix}$

58. $\begin{bmatrix} 18 & -3 & 4 \\ 6 & -20 & 14 \\ 10 & 25 & -15 \end{bmatrix}$

59. $\begin{bmatrix} 44 & 21 & 18 & 6 \\ -2 & 10 & 15 & 5 \\ 21 & 12 & -12 & 4 \\ -8 & -16 & 4 & 9 \end{bmatrix}$

60. $\begin{bmatrix} 16 & 22 & -3 & 5 \\ 21 & -17 & 4 & 8 \\ 2 & 8 & 27 & 20 \\ 5 & 15 & -3 & -10 \end{bmatrix}$

In Problems 61–64, use the idea behind Example 15 and a graphing utility to solve the following systems of equations. Round answers to two decimal places.

61. $\begin{cases} 25x + 61y - 12z = 10 \\ 18x - 12y + 7z = -9 \\ 3x + 4y - z = -12 \end{cases}$

62. $\begin{cases} 25x + 61y - 12z = 15 \\ 18x - 12y + 7z = -3 \\ 3x + 4y - z = 12 \end{cases}$

63. $\begin{cases} 25x + 61y - 12z = 21 \\ 18x - 12y + 7z = 7 \\ 3x + 4y - z = -2 \end{cases}$

64. $\begin{cases} 25x + 61y - 12z = 25 \\ 18x - 12y + 7z = 10 \\ 3x + 4y - z = -4 \end{cases}$

65. Computing the Cost of Production The Acme Steel Company is a producer of stainless steel and aluminum containers. On a certain day, the following stainless steel containers were manufactured: 500 with 10-gallon capacity, 350 with 5-gallon capacity, and 400 with 1-gallon capacity. On the same day, the following aluminum containers were manufactured: 700 with 10-gallon capacity, 500 with 5-gallon capacity, and 850 with 1-gallon capacity.

(a) Find a 2 by 3 matrix representing the above data. Find a 3 by 2 matrix to represent the same data.

(b) If the amount of material used in the 10-gallon containers is 15 pounds, the amount used in the 5-gallon containers is 8 pounds, and the amount used in the 1-gallon containers is 3 pounds, find a 3 by 1 matrix representing the amount of material.

(c) Multiply the 2 by 3 matrix found in part (a) and the 3 by 1 matrix found in part (b) to get a 2 by 1 matrix showing the day's usage of material.

(d) If stainless steel costs Acme $0.10 per pound and aluminum costs $0.05 per pound, find a 1 by 2 matrix representing cost.

(e) Multiply the matrices found in parts (c) and (d) to determine what the total cost of the day's production was.

66. Computing Profit Rizza Ford has two locations, one in the city and the other in the suburbs. In January, the city location sold 400 subcompacts, 250 intermediate-size cars, and 50 station wagons; in February, it sold 350 subcompacts, 100 intermediates, and 30 station wagons. At the suburban location in January, 450 subcompacts, 200 intermediates, and 140 station wagons were sold. In February, the suburban location sold 350 subcompacts, 300 intermediates, and 100 station wagons.

(a) Find 2 by 3 matrices that summarize the sales data for each location for January and February (one matrix for each month).

(b) Use matrix addition to obtain total sales for the two-month period.

(c) The profit on each kind of car is $100 per subcompact, $150 per intermediate, and $200 per station wagon. Find a 3 by 1 matrix representing this profit.

(d) Multiply the matrices found in parts (b) and (c) to get a 2 by 1 matrix showing the profit at each location.

67. Consider the 2 by 2 square matrix

$$A = \begin{bmatrix} a & b \\ c & d \end{bmatrix}$$

If $D = ad - bc \neq 0$, show that A is nonsingular and that

$$A^{-1} = \frac{1}{D} \begin{bmatrix} d & -b \\ -c & a \end{bmatrix}$$

 68. Make up a situation different from any found in the text that can be represented by a matrix.

8.5 PARTIAL FRACTION DECOMPOSITION

> 1 Decompose P/Q, Where Q Has Only Nonrepeated Linear Factors
> 2 Decompose P/Q, Where Q Has Repeated Linear Factors
> 3 Decompose P/Q, Where Q Has Only Nonrepeated Irreducible Quadratic Factors
> 4 Decompose P/Q, Where Q has Repeated Irreducible Quadratic Factors

Consider the problem of adding two fractions

$$\frac{3}{x+4} \quad \text{and} \quad \frac{2}{x-3}$$

The result is

$$\frac{3}{x+4} + \frac{2}{x-3} = \frac{3(x-3) + 2(x+4)}{(x+4)(x-3)} = \frac{5x-1}{x^2+x-12}$$

The reverse procedure, namely, starting with the rational expression $(5x - 1)/(x^2 + x - 12)$ and writing it as the sum (or difference) of the two simpler fractions $3/(x + 4)$ and $2/(x - 3)$ is referred to as **partial fraction decomposition,** and the two simpler fractions are called **partial fractions.** Decomposing a rational expression into a sum of partial fractions is important in solving certain types of calculus problems. This section presents a systematic way to decompose rational expressions.

We begin by recalling that a rational expression is the ratio of two polynomials, say, P and $Q \neq 0$. We assume P and Q have no common factors. Recall also that a rational expression P/Q is called **proper** if the degree of the polynomial in the numerator is less than the degree of the polynomial in the denominator. Otherwise, the rational expression is termed **improper.**

Because any improper rational expression can be reduced by long division to a mixed form consisting of the sum of a polynomial and a proper rational expression, we shall restrict the discussion that follows to proper rational expressions.

The partial fraction decomposition of the rational expression P/Q depends on the factors of the denominator Q. Recall (from Section 5.5) that any polynomial whose coefficients are real numbers can be factored (over the real numbers) into products of linear and/or irreducible quadratic factors. Thus, the denominator Q of the rational expression P/Q will contain only factors of one or both of the following types:

1. *Linear factors* of the form $x - a$, where a is a real number.
2. *Irreducible quadratic factors* of the form $ax^2 + bx + c$, where a, b, and c are real numbers, $a \neq 0$, and $b^2 - 4ac < 0$ (which guarantees that $ax^2 + bx + c$ cannot be written as the product of two linear factors with real coefficients).

1 As it turns out, there are four cases to be examined. We begin with the case for which Q has only nonrepeated linear factors.

> CASE 1: Q has only nonrepeated linear factors.
>
> Under the assumption that Q has only nonrepeated linear factors, the polynomial Q has the form
>
> $$Q(x) = (x - a_1)(x - a_2) \cdot \ldots \cdot (x - a_n)$$

where none of the numbers $a_1, a_2, \ldots, a_n$ are equal. In this case, the partial fraction decomposition of P/Q is of the form

$$\frac{P(x)}{Q(x)} = \frac{A_1}{x - a_1} + \frac{A_2}{x - a_2} + \cdots + \frac{A_n}{x - a_n} \qquad (1)$$

where the numbers $A_1, A_2, \ldots, A_n$ are to be determined.

We show how to find these numbers in the example that follows.

E X A M P L E 1   **Nonrepeated Linear Factors**

Write the partial fraction decomposition of $\dfrac{x}{x^2 - 5x + 6}$.

Solution   First, we factor the denominator,

$$x^2 - 5x + 6 = (x - 2)(x - 3)$$

and conclude that the denominator contains only nonrepeated linear factors. Then we decompose the rational expression according to equation (1):

$$\frac{x}{x^2 - 5x + 6} = \frac{A}{x - 2} + \frac{B}{x - 3} \qquad (2)$$

where A and B are to be determined. To find A and B, we clear the fractions by multiplying each side by $(x - 2)(x - 3) = x^2 - 5x + 6$. The result is

$$x = A(x - 3) + B(x - 2) \qquad (3)$$

or

$$x = (A + B)x + (-3A - 2B)$$

This equation is an identity in x. Thus, we may equate the coefficients of like powers of x to get

$$\begin{cases} 1 = \quad A + \ B & \text{Equate coefficients of } x: 1x = (A + B)x. \\ 0 = -3A - 2B & \text{Equate coefficients of } x^0, \text{ the constants: } 0x^0 = (-3A - 2B)x^0. \end{cases}$$

This system of two equations containing two variables, A and B, can be solved using whatever method you wish. Solving it, we get

$$A = -2 \qquad B = 3$$

Thus, from equation (2), the partial fraction decomposition is

$$\frac{x}{x^2 - 5x + 6} = \frac{-2}{x - 2} + \frac{3}{x - 3}$$

Check: The decomposition can be checked by adding the fractions.

$$\frac{-2}{x - 2} + \frac{3}{x - 3} = \frac{-2(x - 3) + 3(x - 2)}{(x - 2)(x - 3)} = \frac{x}{(x - 2)(x - 3)}$$

$$= \frac{x}{x^2 - 5x + 6}$$

Now work Problem 9.

The numbers to be found in the partial fraction decomposition can sometimes be found more readily by using suitable choices for x (which may include complex numbers) in the identity obtained after fractions have been cleared. In Example 1, the identity after clearing fractions, equation (3), is

$$x = A(x - 3) + B(x - 2)$$

If we let $x = 2$ in this expression, the term containing B drops out, leaving $2 = A(-1)$, or $A = -2$. Similarly, if we let $x = 3$, the term containing A drops out, leaving $3 = B$. Thus, as before, $A = -2$ and $B = 3$.

We use this method in the next example.

2

CASE 2: Q has repeated linear factors.

If the polynomial Q has a repeated factor, say $(x - a)^n, n \geq 2$ an integer, then, in the partial fraction decomposition of P/Q, we allow for the terms

$$\frac{A_1}{x - a} + \frac{A_2}{(x - a)^2} + \cdots + \frac{A_n}{(x - a)^n}$$

where the numbers $A_1, A_2, \ldots, A_n$ are to be determined.

E X A M P L E   2   **Repeated Linear Factors**

Write the partial fraction decomposition of $\dfrac{x + 2}{x^3 - 2x^2 + x}$.

Solution   First, we factor the denominator,

$$x^3 - 2x^2 + x = x(x^2 - 2x + 1) = x(x - 1)^2$$

and find that the denominator has the nonrepeated linear factor x and the repeated linear factor $(x - 1)^2$. By Case 1, we must allow for the term A/x in the decomposition; and, by Case 2, we must allow for the terms $B/(x - 1) + C/(x - 1)^2$ in the decomposition.

Thus, we write

$$\frac{x + 2}{x^3 - 2x^2 + x} = \frac{A}{x} + \frac{B}{x - 1} + \frac{C}{(x - 1)^2} \tag{4}$$

Again, we clear fractions by multiplying each side by $x^3 - 2x^2 + x = x(x - 1)^2$. The result is the identity

$$x + 2 = A(x - 1)^2 + Bx(x - 1) + Cx \tag{5}$$

If we let $x = 0$ in this expression, the terms containing B and C drop out, leaving $2 = A(-1)^2$, or $A = 2$. Similarly, if we let $x = 1$, the terms containing A and B drop out, leaving $3 = C$. Thus, equation (5) becomes

$$x + 2 = 2(x - 1)^2 + Bx(x - 1) + 3x$$

Now, let $x = 2$ (any choice other than 0 or 1 will work as well). The result is

$$4 = 2(1)^2 + B(2)(1) + 3(2)$$
$$2B = 4 - 2 - 6 = -4$$
$$B = -2$$

Thus, we have $A = 2$, $B = -2$, and $C = 3$.

From equation (4), the partial fraction decomposition is

$$\frac{x + 2}{x^3 - 2x^2 + x} = \frac{2}{x} + \frac{-2}{x - 1} + \frac{3}{(x - 1)^2}$$

■

E X A M P L E 3   Repeated Linear Factors

Write the partial fraction decomposition of $\dfrac{x^3 - 8}{x^2(x - 1)^3}$.

Solution   The denominator contains the repeated linear factor x^2 and the repeated linear factor $(x - 1)^3$. Thus, the partial fraction decomposition takes the form

$$\frac{x^3 - 8}{x^2(x - 1)^3} = \frac{A}{x} + \frac{B}{x^2} + \frac{C}{x - 1} + \frac{D}{(x - 1)^2} + \frac{E}{(x - 1)^3} \qquad (6)$$

As before, we clear fractions and obtain the identity

$$x^3 - 8 = Ax(x - 1)^3 + B(x - 1)^3 + Cx^2(x - 1)^2 + Dx^2(x - 1) + Ex^2 \quad (7)$$

Let $x = 0$. (Do you see why this choice was made?) Then

$$-8 = B(-1)$$
$$B = 8$$

Now let $x = 1$ in equation (7). Then

$$-7 = E$$

Use $B = 8$ and $E = -7$ in equation (7) and collect like terms.

$$x^3 - 8 = Ax(x - 1)^3 + 8(x - 1)^3$$
$$+ Cx^2(x - 1)^2 + Dx^2(x - 1) - 7x^2$$
$$x^3 - 8 - 8(x^3 - 3x^2 + 3x - 1) + 7x^2 = Ax(x - 1)^3 + Cx^2(x - 1)^2 + Dx^2(x - 1)$$
$$-7x^3 + 31x^2 - 24x = x(x - 1)[A(x - 1)^2 + Cx(x - 1) + Dx]$$
$$x(x - 1)(-7x + 24) = x(x - 1)[A(x - 1)^2 + Cx(x - 1) + Dx]$$
$$-7x + 24 = A(x - 1)^2 + Cx(x - 1) + Dx \qquad (8)$$

We now work with equation (8). Let $x = 0$. Then

$$24 = A$$

Now, let $x = 1$ in equation (8). Then

$$17 = D$$

Use $A = 24$ and $D = 17$ in equation (8) and collect like terms.

$$-7x + 24 = 24(x - 1)^2 + Cx(x - 1) + 17x$$
$$-24x^2 + 48x - 24 - 17x - 7x + 24 = Cx(x - 1)$$
$$-24x^2 + 24x = Cx(x - 1)$$
$$-24x(x - 1) = Cx(x - 1)$$
$$-24 = C$$

We now know all the numbers A, B, C, D, and E, so, from equation (6), we have the decomposition

$$\frac{x^3 - 8}{x^2(x - 1)^3} = \frac{24}{x} + \frac{8}{x^2} + \frac{-24}{x - 1} + \frac{17}{(x - 1)^2} + \frac{-7}{(x - 1)^3}$$ ∎

The method employed in Example 3, although somewhat tedious, is still preferable to solving the system of five equations containing five variables that the expansion of equation (7) leads to.

Now work Problem 15.

The final two cases involve irreducible quadratic factors. As mentioned in Section 5.5, a quadratic factor is irreducible if it cannot be factored into linear factors with real coefficients. A quadratic expression $ax^2 + bx + c$ is irreducible whenever $b^2 - 4ac < 0$. For example, $x^2 + x + 1$ and $x^2 + 4$ are irreducible.

CASE 3: Q contains a nonrepeated irreducible quadratic factor.

If Q contains a nonrepeated irreducible quadratic factor of the form $ax^2 + bx + c$, then, in the partial fraction decomposition of P/Q, allow for the term

$$\frac{Ax + B}{ax^2 + bx + c}$$

where the numbers A and B are to be determined.

E X A M P L E   4   Nonrepeated Irreducible Quadratic Factor

Write the partial fraction decomposition of $\dfrac{3x - 5}{x^3 - 1}$.

Solution   We factor the denominator,

$$x^3 - 1 = (x - 1)(x^2 + x + 1)$$

and find that it has a nonrepeated linear factor $x - 1$ and a nonrepeated irreducible quadratic factor $x^2 + x + 1$. Thus, we allow for the term $A/(x - 1)$ by Case 1, and we allow for the term $(Bx + C)/(x^2 + x + 1)$ by Case 3. Hence, we write

$$\frac{3x - 5}{x^3 - 1} = \frac{A}{x - 1} + \frac{Bx + C}{x^2 + x + 1} \qquad (9)$$

We clear fractions by multiplying each side of equation (9) by $x^3 - 1 = (x - 1)(x^2 + x + 1)$ to obtain the identity

$$3x - 5 = A(x^2 + x + 1) + (Bx + C)(x - 1) \qquad (10)$$

Now let $x = 1$. Then equation (10) gives $-2 = A(3)$, or $A = -\frac{2}{3}$. We use this value of A in equation (10) and simplify.

$$3x - 5 = -\frac{2}{3}(x^2 + x + 1) + (Bx + C)(x - 1)$$

$3(3x - 5) = -2(x^2 + x + 1) + 3(Bx + C)(x - 1)$ Multiply each side by 3.

$9x - 15 = -2x^2 - 2x - 2 + 3(Bx + C)(x - 1)$

$2x^2 + 11x - 13 = 3(Bx + C)(x - 1)$ Collect terms.

$(2x + 13)(x - 1) = 3(Bx + C)(x - 1)$ Factor the left side.

$2x + 13 = 3Bx + 3C$

$2 = 3B$ and $13 = 3C$ Equate coefficients.

$B = \dfrac{2}{3}$ $C = \dfrac{13}{3}$

Thus, from equation (9), we see that

$$\frac{3x - 5}{x^3 - 1} = \frac{-\frac{2}{3}}{x - 1} + \frac{\frac{2}{3}x + \frac{13}{3}}{x^2 + x + 1}$$

Now work Problem 17.

<table>
<tr><td>**4**</td><td>**CASE 4:** *Q* contains repeated irreducible quadratic factors.</td></tr>
</table>

If the polynomial Q contains a repeated irreducible quadratic factor $(ax^2 + bx + c)^n$, $n \geq 2$, n an integer, then, in the partial fraction decomposition of P/Q, allow for the terms

$$\frac{A_1 x + B_1}{ax^2 + bx + c} + \frac{A_2 x + B_2}{(ax^2 + bx + c)^2} + \cdots + \frac{A_n x + B_n}{(ax^2 + bx + c)^n}$$

where the numbers $A_1, B_1, A_2, B_2, \ldots, A_n, B_n$ are to be determined.

E X A M P L E   5

Repeated Irreducible Quadratic Factor

Write the partial fraction decomposition of $\dfrac{x^3 + x^2}{(x^2 + 4)^2}$.

Solution   The denominator contains the repeated irreducible quadratic factor $(x^2 + 4)^2$, so we write

$$\frac{x^3 + x^2}{(x^2 + 4)^2} = \frac{Ax + B}{x^2 + 4} + \frac{Cx + D}{(x^2 + 4)^2} \tag{11}$$

We clear fractions to obtain

$$x^3 + x^2 = (Ax + B)(x^2 + 4) + Cx + D$$

Collecting like terms yields

$$x^3 + x^2 = Ax^3 + Bx^2 + (4A + C)x + D + 4B$$

Equating coefficients, we arrive at the system

$$\begin{cases} A = 1 \\ B = 1 \\ 4A + C = 0 \\ D + 4B = 0 \end{cases}$$

The solution is $A = 1$, $B = 1$, $C = -4$, $D = -4$. Hence, from equation (11),

$$\frac{x^3 + x^2}{(x^2 + 4)^2} = \frac{x + 1}{x^2 + 4} + \frac{-4x - 4}{(x^2 + 4)^2}$$

Now work Problem 31.

8.5 EXERCISES

In Problems 1–8, tell whether the given rational expression is proper or improper. If improper, rewrite it as the sum of a polynomial and a proper rational expression.

1. $\dfrac{x}{x^2 - 1}$
2. $\dfrac{5x + 2}{x^3 - 1}$
3. $\dfrac{x^2 + 5}{x^2 - 4}$
4. $\dfrac{3x^2 - 2}{x^2 - 1}$

5. $\dfrac{5x^3 + 2x - 1}{x^2 - 4}$
6. $\dfrac{3x^4 + x^2 - 2}{x^3 + 8}$
7. $\dfrac{x(x - 1)}{(x + 4)(x - 3)}$
8. $\dfrac{2x(x^2 + 4)}{x^2 + 1}$

In Problems 9–42, write the partial fraction decomposition of each rational expression.

9. $\dfrac{4}{x(x - 1)}$
10. $\dfrac{3x}{(x + 2)(x - 1)}$
11. $\dfrac{1}{x(x^2 + 1)}$

12. $\dfrac{1}{(x + 1)(x^2 + 4)}$
13. $\dfrac{x}{(x - 1)(x - 2)}$
14. $\dfrac{3x}{(x + 2)(x - 4)}$

15. $\dfrac{x^2}{(x - 1)^2(x + 1)}$
16. $\dfrac{x + 1}{x^2(x - 2)}$
17. $\dfrac{1}{x^3 - 8}$

18. $\dfrac{2x + 4}{x^3 - 1}$
19. $\dfrac{x^2}{(x - 1)^2(x + 1)^2}$
20. $\dfrac{x + 1}{x^2(x - 2)^2}$

21. $\dfrac{x - 3}{(x + 2)(x + 1)^2}$
22. $\dfrac{x^2 + x}{(x + 2)(x - 1)^2}$
23. $\dfrac{x + 4}{x^2(x^2 + 4)}$

24. $\dfrac{10x^2 + 2x}{(x - 1)^2(x^2 + 2)}$
25. $\dfrac{x^2 + 2x + 3}{(x + 1)(x^2 + 2x + 4)}$
26. $\dfrac{x^2 - 11x - 18}{x(x^2 + 3x + 3)}$

27. $\dfrac{x}{(3x - 2)(2x + 1)}$
28. $\dfrac{1}{(2x + 3)(4x - 1)}$
29. $\dfrac{x}{x^2 + 2x - 3}$

30. $\dfrac{x^2 - x - 8}{(x + 1)(x^2 + 5x + 6)}$
31. $\dfrac{x^2 + 2x + 3}{(x^2 + 4)^2}$
32. $\dfrac{x^3 + 1}{(x^2 + 16)^2}$

33. $\dfrac{7x + 3}{x^3 - 2x^2 - 3x}$
34. $\dfrac{x^5 + 1}{x^6 - x^4}$
35. $\dfrac{x^2}{x^3 - 4x^2 + 5x - 2}$

36. $\dfrac{x^2 + 1}{x^3 + x^2 - 5x + 3}$
37. $\dfrac{x^3}{(x^2 + 16)^3}$
38. $\dfrac{x^2}{(x^2 + 4)^3}$

39. $\dfrac{4}{2x^2 - 5x - 3}$
40. $\dfrac{4x}{2x^2 + 3x - 2}$
41. $\dfrac{2x + 3}{x^4 - 9x^2}$

42. $\dfrac{x^2 + 9}{x^4 - 2x^2 - 8}$

8.6 | SYSTEMS OF NONLINEAR EQUATIONS

> **1** Solve a System of Nonlinear Equations Using Substitution
> **2** Solve a System of Nonlinear Equations Using Elimination

1 There is no general methodology for solving a system of nonlinear equations. There are times when substitution is best; other times, elimination is best; and there are times when neither of these methods works. Experience and a certain degree of imagination are your allies here.

Before we begin, two comments are in order.

1. If the system contains two variables and if the equations in the system are easy to graph, then graph them. By graphing each equation in the system you can get an idea of how many solutions a system has and approximately where they are located.

2. Extraneous solutions can creep in when solving nonlinear systems, so it is imperative that all apparent solutions be checked.

E X A M P L E   1

Solving a System of Nonlinear Equations

Solve the following system of equations:

$$\begin{cases} 3x - y = -2 & \text{(1) A line} \\ 2x^2 - y = 0 & \text{(2) A parabola} \end{cases}$$

Solution Using Substitution

First, we notice that the system contains two variables and that we know how to graph each equation. In Figure 7, we see that the system apparently has two solutions.

We will use substitution to solve the system. Equation (1) is easily solved for y.

FIGURE 7

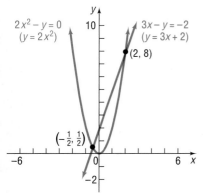

$$3x - y = -2$$
$$y = 3x + 2$$

We substitute this expression for y in equation (2). The result is an equation containing just the variable x, which we can then solve.

$$2x^2 - y = 0$$
$$2x^2 - (3x + 2) = 0$$
$$2x^2 - 3x - 2 = 0$$
$$(2x + 1)(x - 2) = 0$$
$$2x + 1 = 0 \quad \text{or} \quad x - 2 = 0$$
$$x = -\frac{1}{2} \qquad x = 2$$

Using these values for x in $y = 3x + 2$, we find

$$y = 3\left(-\frac{1}{2}\right) + 2 = \frac{1}{2} \quad \text{or} \quad y = 3(2) + 2 = 8$$

The apparent solutions are $x = -\frac{1}{2}, y = \frac{1}{2}$ and $x = 2, y = 8$.

Check: For $x = -\frac{1}{2}, y = \frac{1}{2}$:

$$\begin{cases} 3(-\frac{1}{2}) - \frac{1}{2} = -\frac{3}{2} - \frac{1}{2} = -2 & (1) \\ 2(-\frac{1}{2})^2 - \frac{1}{2} = 2(\frac{1}{4}) - \frac{1}{2} = \ \ \ 0 & (2) \end{cases}$$

For $x = 2, y = 8$:

$$\begin{cases} 3(2) - 8 = \ \ \ \ 6 - 8 = -2 & (1) \\ 2(2)^2 - 8 = 2(4) - 8 = \ \ \ 0 & (2) \end{cases}$$

Each solution checks. Now we know that the graphs in Figure 7 intersect at $(-\frac{1}{2}, \frac{1}{2})$ and at $(2, 8)$.

 Check: Graph $3x - y = -2$ $(Y_1 = 3x + 2)$ and $2x^2 - y = 0$ $(Y_2 = 2x^2)$ and compare what you see with Figure 7. Use INTERSECT (twice) to find the two points of intersection.

Now work Problem 11 using substitution.

 Our next example illustrates how the method of elimination works for nonlinear systems.

E X A M P L E   2   Solving a System of Nonlinear Equations

Solve: $\begin{cases} x^2 + y^2 = 13 & (1) \quad \text{A circle} \\ x^2 - y = 7 & (2) \quad \text{A parabola} \end{cases}$

Solution Using Elimination   First, we graph each equation, as shown in Figure 8. Based on the graph, we expect four solutions. By subtracting equation (2) from equation (1), the variable x is eliminated, leaving

$$y^2 + y = 6$$

FIGURE 8

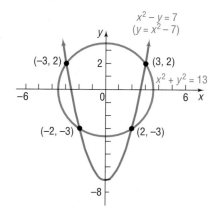

This quadratic equation in y is easily solved by factoring.

$$y^2 + y - 6 = 0$$
$$(y + 3)(y - 2) = 0$$
$$y = -3 \quad \text{or} \quad y = 2$$

We use these values for y in equation (2) to find x. If $y = 2$, then $x^2 = y + 7 = 9$ and $x = 3$ or -3. If $y = -3$, then $x^2 = y + 7 = 4$ and $x = 2$ or -2. Thus, we have four solutions: $x = 3, y = 2; x = -3, y = 2; x = 2, y = -3$; and $x = -2, y = -3$. You should verify that, in fact, these four solutions also satisfy equation (1), so all four are solutions of the system. The four points, $(3, 2)$, $(-3, 2)$, $(2, -3)$, and $(-2, -3)$, are the points of intersection of the graphs. Look again at Figure 8.

 Check: Graph $x^2 + y^2 = 13$ and $x^2 - y = 7$. (Remember that to graph $x^2 + y^2 = 13$ requires two functions: $Y_1 = \sqrt{13 - x^2}$ and $Y_2 = -\sqrt{13 - x^2}$.) Compare what you see with Figure 8. Use INTERSECT to find the four points of intersection.

 Now work Problem 9 using elimination.

E X A M P L E   3   Solving a System of Nonlinear Equations

Solve: $\begin{cases} x^2 - y^2 = 1 & \text{(1)} \\ x^3 - y^2 = x & \text{(2)} \end{cases}$

Solution Using Elimination

Because the second equation is not easy to graph, we omit the graphing step. We use elimination, subtracting equation (2) from equation (1), to obtain

$$x^2 - x^3 = 1 - x$$
$$x^2(1 - x) = 1 - x$$
$$x^2(1 - x) - (1 - x) = 0$$
$$(x^2 - 1)(1 - x) = 0$$
$$x^2 - 1 = 0 \quad \text{or} \quad 1 - x = 0$$
$$x = \pm 1 \qquad\qquad x = 1$$

We now use equation (1) to get y. If $x = 1$, then $1 - y^2 = 1$ and $y = 0$. If $x = -1$, then $1 - y^2 = 1$ and $y = 0$. There are two apparent solutions: $x = 1$, $y = 0$ and $x = -1, y = 0$. Because each of these solutions also satisfies equation (2), the system has two solutions: $x = 1, y = 0$ and $x = -1, y = 0$. ∎

Check: Graph $x^2 - y^2 = 1$ and $x^3 - y^2 = x$. (You will need to graph four functions: $Y_1 = \sqrt{x^2 - 1}$, $Y_2 = -\sqrt{x^2 - 1}$, $Y_3 = \sqrt{x^3 - x}$, and $Y_4 = -\sqrt{x^3 - x}$). See Figure 9. The graphs appear to intersect twice, so the system apparently has two solutions. Using INTERSECT, the solutions to the system of equations are $(-1, 0)$ and $(1, 0)$.

FIGURE 9

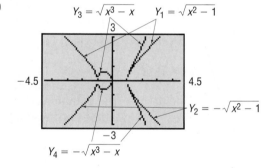

E X A M P L E   4   Solving a System of Nonlinear Equations

Solve: $\begin{cases} x^2 + x + y^2 - 3y + 2 = 0 & \text{(1)} \\ x + 1 + \dfrac{y^2 - y}{x} = 0 & \text{(2)} \end{cases}$

Solution Using Elimination

First, we multiply equation (2) by x to eliminate the fraction. The result is an equivalent system because x cannot be 0 [look at equation (2) to see why].

$$\begin{cases} x^2 + x + y^2 - 3y + 2 = 0 & \text{(1)} \\ x^2 + x + y^2 - y = 0 & \text{(2)} \end{cases}$$

Now subtract equation (2) from equation (1) to eliminate x. The result is

$$-2y + 2 = 0$$
$$y = 1$$

To find x, we back-substitute $y = 1$ in equation (1).

$$x^2 + x + 1 - 3 + 2 = 0$$
$$x^2 + x = 0$$
$$x(x + 1) = 0$$
$$x = 0 \quad \text{or} \quad x = -1$$

Because x cannot be 0, the value $x = 0$ is extraneous, and we discard it. Thus, the solution is $x = -1, y = 1$.

Check: We now check $x = -1, y = 1$.

$$\begin{cases} (-1)^2 + (-1) + 1^2 - 3(1) + 2 = 1 - 1 + 1 - 3 + 2 = 0 & \text{(1)} \\ -1 + 1 + \dfrac{1^2 - 1}{-1} = 0 + \dfrac{0}{-1} = 0 & \text{(2)} \end{cases}$$

Thus, the only solution to the system is $x = -1, y = 1$. ∎

 Now work Problems 25 and 49.

E X A M P L E 5   Solving a System of Nonlinear Equations

Solve: $\begin{cases} x^2 - y^2 = 4 & \text{(1) A hyperbola} \\ y = x^2 & \text{(2) A parabola} \end{cases}$

FIGURE 10

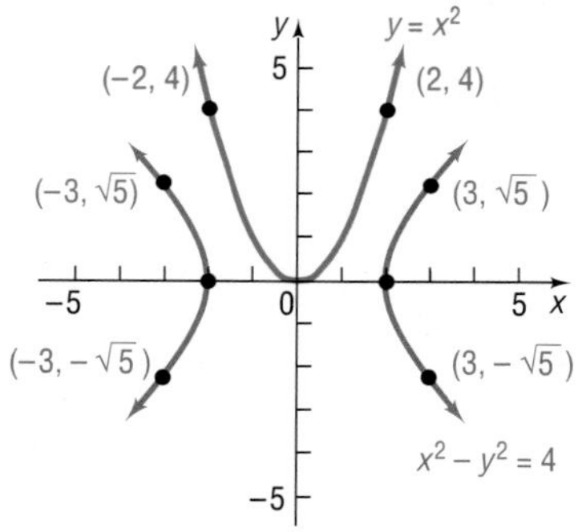

Solution   Either substitution or elimination can be used here. We use substitution and replace x^2 by y in equation (1). The result is

$$y - y^2 = 4$$
$$y^2 - y + 4 = 0$$

This is a quadratic equation whose discriminant is $1 - 4 \cdot 4 = -15 < 0$. Thus, the equation has no real solutions, and hence the system is inconsistent. The graphs of these two equations do not intersect. See Figure 10. ∎

The following examples illustrate two of the more imaginative ways to solve systems of nonlinear equations algebraically.

E X A M P L E 6   Solving a System of Nonlinear Equations

Solve: $\begin{cases} 4x^2 - 9xy - 28y^2 = 0 & \text{(1)} \\ 16x^2 - 4xy = 16 & \text{(2)} \end{cases}$

Solution   We take note of the fact that equation (1) can be factored.

$$4x^2 - 9xy - 28y^2 = 0$$
$$(4x + 7y)(x - 4y) = 0$$

This results in the two equations

$$4x + 7y = 0 \qquad \text{or} \quad x - 4y = 0$$
$$x = -\tfrac{7}{4}y \qquad\qquad x = 4y$$

We substitute each of these values for x in equation (2).

$$16x^2 - 4xy = 16 \qquad\qquad 16x^2 - 4xy = 16$$
$$16(-\tfrac{7}{4}y)^2 - 4(-\tfrac{7}{4}y)y = 16 \qquad 16(4y)^2 - 4(4y)y = 16$$
$$49y^2 + 7y^2 = 16 \qquad\qquad 16(16y^2) - 16y^2 = 16$$
$$56y^2 = 16 \qquad\qquad\qquad 15y^2 = 1$$
$$7y^2 = 2 \qquad\qquad\qquad\quad y^2 = \tfrac{1}{15}$$
$$y^2 = \tfrac{2}{7}$$

Thus, we have

$$y = \pm\sqrt{\frac{2}{7}} = \pm\frac{\sqrt{14}}{7} \qquad\qquad y = \pm\frac{\sqrt{15}}{15}$$

$$x = -\frac{7}{4}y = \mp\frac{\sqrt{14}}{4} \qquad\qquad x = 4y = \pm\frac{4\sqrt{15}}{15}$$

You should verify for yourself that, in fact, the four solutions $x = -\sqrt{14}/4$, $y = \sqrt{14}/7$; $x = \sqrt{14}/4$, $y = -\sqrt{14}/7$; $x = 4\sqrt{15}/15$, $y = \sqrt{15}/15$; and $x = -4\sqrt{15}/15$, $y = -\sqrt{15}/15$ are actually solutions of the system. ∎

E X A M P L E   7

Solving a System of Nonlinear Equations

Solve: $\begin{cases} 3xy - 2y^2 = -2 & (1) \\ 9x^2 + 4y^2 = 10 & (2) \end{cases}$

Solution   We multiply equation (1) by 2 and add the result to equation (2) to eliminate the y^2 terms.

$$\begin{cases} 6xy - 4y^2 = -4 & \text{(1)} \quad \text{Multiply by 2.} \\ 9x^2 + 4y^2 = 10 \end{cases}$$
$$9x^2 + 6xy = 6 \qquad \text{(2)} \quad \text{Add the equations.}$$
$$3x^2 + 2xy = 2 \qquad \text{Divide each side by 3.}$$

Since $x \neq 0$ (do you see why?), we can solve for y in this equation to get

$$y = \frac{2 - 3x^2}{2x}, \qquad x \neq 0 \tag{1}$$

Now substitute for y in equation (2) of the system.

$$9x^2 + 4y^2 = 10$$
$$9x^2 + 4\left(\frac{2 - 3x^2}{2x}\right)^2 = 10$$
$$9x^2 + \frac{4 - 12x^2 + 9x^4}{x^2} = 10$$
$$9x^4 + 4 - 12x^2 + 9x^4 = 10x^2$$
$$18x^4 - 22x^2 + 4 = 0$$
$$9x^4 - 11x^2 + 2 = 0$$

This quadratic equation (in x^2) can be factored.

$$(9x^2 - 2)(x^2 - 1) = 0$$

$$9x^2 - 2 = 0 \qquad \text{or} \quad x^2 - 1 = 0$$

$$x^2 = \frac{2}{9} \qquad\qquad x^2 = 1$$

$$x = \pm \frac{\sqrt{2}}{3} \qquad\qquad x = \pm 1$$

To find y, we use equation (1), $y = \dfrac{2 - 3x^2}{2x}$

If $x = \dfrac{\sqrt{2}}{3}$: $\quad y = \dfrac{2 - 3x^2}{2x} = \dfrac{2 - \frac{2}{3}}{2(\sqrt{2}/3)} = \dfrac{4}{2\sqrt{2}} = \sqrt{2}$

If $x = -\dfrac{\sqrt{2}}{3}$: $\quad y = \dfrac{2 - 3x^2}{2x} = \dfrac{2 - \frac{2}{3}}{-2(\sqrt{2}/3)} = \dfrac{4}{-2\sqrt{2}} = -\sqrt{2}$

If $x = 1$: $\quad y = \dfrac{2 - 3x^2}{2x} = \dfrac{2 - 3}{2} = -\dfrac{1}{2}$

If $x = -1$: $\quad y = \dfrac{2 - 3x^2}{2x} = \dfrac{2 - 3}{-2} = \dfrac{1}{2}$

The system has four solutions. Check them for yourself. ▬

 Now work Problem 45.

E X A M P L E   8

Running a Race

In a 50-mile race, the winner crosses the finish line 1 mile ahead of the second place runner and 4 miles ahead of the third place runner. Assuming that each runner maintains a constant speed throughout the race, by how many miles does the second place runner beat the third place runner?

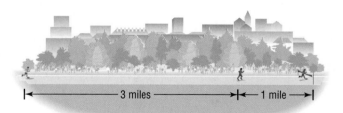

├──────────── 3 miles ────────────┤├── 1 mile ──┤

Solution   Let v_1, v_2, v_3 denote the speeds of the first, second, and third place runners, respectively. Let t_1 and t_2 denote the times (in hours) required for the first place runner and second place runner to finish the race. Then we have the system of equations

$$\begin{cases} 50 = v_1 t_1 & (1) \quad \text{First place runner goes 50 miles in } t_1 \text{ hours.} \\ 49 = v_2 t_1 & (2) \quad \text{Second place runner goes 49 miles in } t_1 \text{ hours.} \\ 46 = v_3 t_1 & (3) \quad \text{Third place runner goes 46 miles in } t_1 \text{ hours.} \\ 50 = v_2 t_2 & (4) \quad \text{Second place runner goes 50 miles in } t_2 \text{ hours.} \end{cases}$$

We seek the distance of the third place runner from the finish at time t_2. That is, we seek

$$50 - v_3 t_2 = 50 - v_3\left(t_1 \cdot \frac{t_2}{t_1}\right)$$

$$= 50 - (v_3 t_1) \cdot \frac{t_2}{t_1}$$

$$= 50 - 46 \cdot \frac{50/v_2}{50/v_1} \qquad \begin{cases} \text{From (3), } v_3 t_1 = 46; \\ \text{from (4), } t_2 = 50/v_2; \\ \text{from (1), } t_1 = 50/v_1. \end{cases}$$

$$= 50 - 46 \cdot \frac{v_1}{v_2}$$

$$= 50 - 46 \cdot \frac{50}{49} \qquad \text{Form the quotient of (1) and (2).}$$

$$\approx 3.06 \text{ miles}$$

HISTORICAL FEATURE Recall that, in the beginning of this section, we said imagination and experience are important in solving simultaneous nonlinear equations. Indeed, these kinds of problems lead into some of the deepest and most difficult parts of modern mathematics. Look again at the graphs in Examples 1 and 2 of this section (Figures 7 and 8). We see that Example 1 has two solutions, and Example 2 has four solutions. We might conjecture that the number of solutions is equal to the product of the degrees of the equations involved. This conjecture was indeed made by Etienne Bezout (1730–1783), but working out the details took about 150 years. It turns out that, to arrive at the correct number of intersections, we must count not only the complex number intersections, but also those intersections that, in a certain sense, lie at infinity. For example, a parabola and a line lying on the axis of the parabola intersect at the vertex and at infinity. This topic is part of the study of algebraic geometry.

HISTORICAL PROBLEM A papyrus dating back to 1950 BC contains the following problem: A given surface area of 100 units of area shall be represented as the sum of two squares whose sides are to each other as $1:\frac{3}{4}$. Solve for the sides by solving the system of equations

$$\begin{cases} x^2 + y^2 = 100 \\ x = \frac{3}{4}y \end{cases}$$

8.6 | EXERCISES

In Problems 1–20, graph each equation of the system. Then solve the system to find the points of intersection.

1. $\begin{cases} y = x^2 + 1 \\ y = x + 1 \end{cases}$

2. $\begin{cases} y = x^2 + 1 \\ y = 4x + 1 \end{cases}$

3. $\begin{cases} y = \sqrt{36 - x^2} \\ y = 8 - x \end{cases}$

4. $\begin{cases} y = \sqrt{4 - x^2} \\ y = 2x + 4 \end{cases}$

5. $\begin{cases} y = \sqrt{x} \\ y = 2 - x \end{cases}$

6. $\begin{cases} y = \sqrt{x} \\ y = 6 - x \end{cases}$

7. $\begin{cases} x = 2y \\ x = y^2 - 2y \end{cases}$

8. $\begin{cases} y = x - 1 \\ y = x^2 - 6x + 9 \end{cases}$

9. $\begin{cases} x^2 + y^2 = 4 \\ x^2 + 2x + y^2 = 0 \end{cases}$

10. $\begin{cases} x^2 + y^2 = 8 \\ x^2 + y^2 + 4y = 0 \end{cases}$

11. $\begin{cases} y = 3x - 5 \\ x^2 + y^2 = 5 \end{cases}$

12. $\begin{cases} x^2 + y^2 = 10 \\ y = x + 2 \end{cases}$

13. $\begin{cases} x^2 + y^2 = 4 \\ y^2 - x = 4 \end{cases}$
14. $\begin{cases} x^2 + y^2 = 16 \\ x^2 - 2y = 8 \end{cases}$
15. $\begin{cases} xy = 4 \\ x^2 + y^2 = 8 \end{cases}$
16. $\begin{cases} x^2 = y \\ xy = 1 \end{cases}$

17. $\begin{cases} x^2 + y^2 = 4 \\ y = x^2 - 9 \end{cases}$
18. $\begin{cases} xy = 1 \\ y = 2x + 1 \end{cases}$
19. $\begin{cases} y = x^2 - 4 \\ y = 6x - 13 \end{cases}$
20. $\begin{cases} x^2 + y^2 = 10 \\ xy = 3 \end{cases}$

In Problems 21–54, solve each system. Use any method you wish.

21. $\begin{cases} 2x^2 + y^2 = 18 \\ xy = 4 \end{cases}$
22. $\begin{cases} x^2 - y^2 = 21 \\ x + y = 7 \end{cases}$
23. $\begin{cases} y = 2x + 1 \\ 2x^2 + y^2 = 1 \end{cases}$

24. $\begin{cases} x^2 - 4y^2 = 16 \\ 2y - x = 2 \end{cases}$
25. $\begin{cases} x + y + 1 = 0 \\ x^2 + y^2 + 6y - x = -5 \end{cases}$
26. $\begin{cases} 2x^2 - xy + y^2 = 8 \\ xy = 4 \end{cases}$

27. $\begin{cases} 4x^2 - 3xy + 9y^2 = 15 \\ 2x + 3y = 5 \end{cases}$
28. $\begin{cases} 2y^2 - 3xy + 6y + 2x + 4 = 0 \\ 2x - 3y + 4 = 0 \end{cases}$
29. $\begin{cases} x^2 - 4y^2 + 7 = 0 \\ 3x^2 + y^2 = 31 \end{cases}$

30. $\begin{cases} 3x^2 - 2y^2 + 5 = 0 \\ 2x^2 - y^2 + 2 = 0 \end{cases}$
31. $\begin{cases} 7x^2 - 3y^2 + 5 = 0 \\ 3x^2 + 5y^2 = 12 \end{cases}$
32. $\begin{cases} x^2 - 3y^2 + 1 = 0 \\ 2x^2 - 7y^2 + 5 = 0 \end{cases}$

33. $\begin{cases} x^2 + 2xy = 10 \\ 3x^2 - xy = 2 \end{cases}$
34. $\begin{cases} 5xy + 13y^2 + 36 = 0 \\ xy + 7y^2 = 6 \end{cases}$
35. $\begin{cases} 2x^2 + y^2 = 2 \\ x^2 - 2y^2 + 8 = 0 \end{cases}$

36. $\begin{cases} y^2 - x^2 + 4 = 0 \\ 2x^2 + 3y^2 = 6 \end{cases}$
37. $\begin{cases} x^2 + 2y^2 = 16 \\ 4x^2 - y^2 = 24 \end{cases}$
38. $\begin{cases} 4x^2 + 3y^2 = 4 \\ 2x^2 - 6y^2 = -3 \end{cases}$

39. $\begin{cases} \dfrac{5}{x^2} - \dfrac{2}{y^2} + 3 = 0 \\ \dfrac{3}{x^2} + \dfrac{1}{y^2} = 7 \end{cases}$
40. $\begin{cases} \dfrac{2}{x^2} - \dfrac{3}{y^2} + 1 = 0 \\ \dfrac{6}{x^2} - \dfrac{7}{y^2} + 2 = 0 \end{cases}$
41. $\begin{cases} \dfrac{1}{x^4} + \dfrac{6}{y^4} = 6 \\ \dfrac{2}{x^4} - \dfrac{2}{y^4} = 19 \end{cases}$

42. $\begin{cases} \dfrac{1}{x^4} - \dfrac{1}{y^4} = 1 \\ \dfrac{1}{x^4} + \dfrac{1}{y^4} = 4 \end{cases}$
43. $\begin{cases} x^2 - 3xy + 2y^2 = 0 \\ x^2 + xy = 6 \end{cases}$
44. $\begin{cases} x^2 - xy - 2y^2 = 0 \\ xy + x + 6 = 0 \end{cases}$

45. $\begin{cases} xy - x^2 + 3 = 0 \\ 3xy - 4y^2 = 2 \end{cases}$
46. $\begin{cases} 5x^2 + 4xy + 3y^2 = 36 \\ x^2 + xy + y^2 = 9 \end{cases}$
47. $\begin{cases} x^3 - y^3 = 26 \\ x - y = 2 \end{cases}$

48. $\begin{cases} x^3 + y^3 = 26 \\ x + y = 2 \end{cases}$
49. $\begin{cases} y^2 + y + x^2 - x - 2 = 0 \\ y + 1 + \dfrac{x - 2}{y} = 0 \end{cases}$
50. $\begin{cases} x^3 - 2x^2 + y^2 + 3y - 4 = 0 \\ x - 2 + \dfrac{y^2 - y}{x^2} = 0 \end{cases}$

51. $\begin{cases} \log_x y = 3 \\ \log_x (4y) = 5 \end{cases}$
52. $\begin{cases} \log_x (2y) = 3 \\ \log_x (4y) = 2 \end{cases}$
53. $\begin{cases} \ln x = 4 \ln y \\ \log_3 x = 2 + 2 \log_3 y \end{cases}$
54. $\begin{cases} \ln x = 5 \ln y \\ \log_2 x = 3 + 2 \log_2 y \end{cases}$

In Problems 55–60, graph each equation and find the point(s) of intersection, if any.

55. The line $x + 2y = 0$ and the circle $(x - 1)^2 + (y - 1)^2 = 5$

56. The line $x + 2y + 6 = 0$ and the circle $(x + 1)^2 + (y + 1)^2 = 5$

57. The circle $(x - 1)^2 + (y + 2)^2 = 4$ and the parabola $y^2 + 4y - x + 1 = 0$

58. The circle $(x + 2)^2 + (y - 1)^2 = 4$ and the parabola $y^2 - 2y - x - 5 = 0$

59. The graph of $y = \dfrac{4}{x - 3}$ and the circle $x^2 - 6x + y^2 + 1 = 0$

60. The graph of $y = \dfrac{4}{x + 2}$ and the circle $x^2 + 4x + y^2 - 4 = 0$

In Problems 61–68, use a graphing utility to solve each system of equations. Express the solution(s) correct to two decimal places.

61. $\begin{cases} y = x^{2/3} \\ y = e^{-x} \end{cases}$

62. $\begin{cases} y = x^{3/2} \\ y = e^{-x} \end{cases}$

63. $\begin{cases} x^2 + y^3 = 2 \\ x^3 y = 4 \end{cases}$

64. $\begin{cases} x^3 + y^2 = 2 \\ x^2 y = 4 \end{cases}$

65. $\begin{cases} x^4 + y^4 = 12 \\ xy^2 = 2 \end{cases}$

66. $\begin{cases} x^4 + y^4 = 6 \\ xy = 1 \end{cases}$

67. $\begin{cases} xy = 2 \\ y = \ln x \end{cases}$

68. $\begin{cases} x^2 + y^2 = 4 \\ y = \ln x \end{cases}$

69. The difference of two numbers is 2 and the sum of their squares is 10. Find the numbers.

70. The sum of two numbers is 7 and the difference of their squares is 21. Find the numbers.

71. The product of two numbers is 4 and the sum of their squares is 8. Find the numbers.

72. The product of two numbers is 10 and the difference of their squares is 21. Find the numbers.

73. The difference of two numbers is the same as their product, and the sum of their reciprocals is 5. Find the numbers.

74. The sum of two numbers is the same as their product, and the difference of their reciprocals is 3. Find the numbers.

75. The ratio of a to b is $\frac{2}{3}$. The sum of a and b is 10. What is the ratio of $a + b$ to $b - a$?

76. The ratio of a to b is $\frac{4}{3}$. The sum of a and b is 14. What is the ratio of $a - b$ to $a + b$?

77. **Geometry**   The perimeter of a rectangle is 16 inches and its area is 15 square inches. What are its dimensions?

78. **Geometry**   An area of 52 square feet is to be enclosed by two squares whose sides are in the ratio of $2 : 3$. Find the sides of the squares.

79. **Geometry**   Two circles have perimeters that add up to 12π centimeters and areas that add up to 20π square centimeters. Find the radius of each circle.

80. **Geometry**   The altitude of an isosceles triangle drawn to its base is 3 centimeters, and its perimeter is 18 centimeters. Find the length of its base.

81. **The Tortoise and the Hare**   In a 21-meter race between a tortoise and a hare, the tortoise leaves 9 minutes before the hare. The hare, by running at an average speed of 0.5 meter per hour faster than the tortoise, crosses the finish line 3 minutes before the tortoise. What are the average speeds of the tortoise and the hare?

82. **Running a Race**   In a 1-mile race, the winner crosses the finish line 10 feet ahead of the second place runner and 20 feet ahead of the third place runner. Assuming that each runner maintains a constant speed throughout the race, by how many feet does the second place runner beat the third place runner?

83. **Constructing a Box**   A rectangular piece of cardboard, whose area is 216 square centimeters, is made into an open box by cutting a 2-centimeter square from each corner and turning up the sides. See the figure. If the box is to have a volume of 224 cubic centimeters, what size cardboard should you start with?

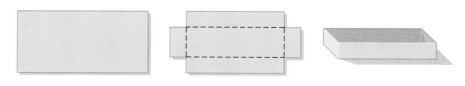

84. Constructing a Cylindrical Tube A rectangular piece of cardboard, whose area is 216 square centimeters, is made into a cylindrical tube by joining together two sides of the rectangle. (See the figure.) If the tube is to have a volume of 224 cubic centimeters, what size cardboard should you start with?

85. Fencing A farmer has 300 feet of fence available to enclose 4500 square feet in the shape of adjoining squares, with sides of length x and y. See the figure. Find x and y.

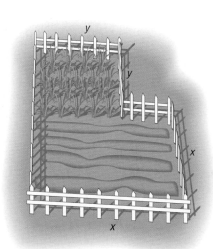

86. Bending Wire A wire 60 feet long is cut into two pieces. Is it possible to bend one piece into the shape of a square and the other into the shape of a circle so that the total area enclosed by the two pieces is 100 square feet? If this is possible, find the length of the side of the square and the radius of the circle.

87. Geometry Find formulas for the length l and width w of a rectangle in terms of its area A and perimeter P.

88. Geometry Find formulas for the base b and one of the equal sides l of an isosceles triangle in terms of its altitude h and perimeter P.

89. Descartes' Method of Equal Roots Descartes's method for finding tangents depends on the idea that, for many graphs, the tangent line at a given point is the *unique* line that intersects the graph at that point only. We will apply his method to find an equation of the tangent line to the parabola $y = x^2$ at the point $(2, 4)$; see the figure. First, we know that the equation of the tangent line must be in the form $y = mx + b$. Using the fact that the point $(2, 4)$ is on the line, we can solve for b in terms of m and get the equation $y = mx + (4 - 2m)$. Now we want $(2, 4)$ to be the *unique* solution to the system

$$\begin{cases} y = x^2 \\ y = mx + 4 - 2m \end{cases}$$

From this system, we get $x^2 = mx + 4 - 2m$ or $x^2 - mx + (2m - 4) = 0$. By using the quadratic formula, we get

$$x = \frac{m \pm \sqrt{m^2 - 4(2m - 4)}}{2}$$

To obtain a unique solution for x, the two roots must be equal; in other words, the discriminant $m^2 - 4(2m - 4)$ must be 0. Complete the work to get m, and write an equation of the tangent line.

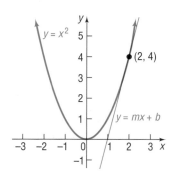

In Problems 90–96, use Descartes's method from Problem 89 to find the equation of the line tangent to each graph at the given point.

90. $x^2 + y^2 = 10$; at $(1, 3)$

91. $y = x^2 + 2$; at $(1, 3)$

92. $x^2 + y = 5$; at $(-2, 1)$

93. $2x^2 + 3y^2 = 14$; at $(1, 2)$

94. $3x^2 + y^2 = 7$; at $(-1, 2)$

95. $x^2 - y^2 = 3$; at $(2, 1)$

96. $2y^2 - x^2 = 14$; at $(2, 3)$

97. If r_1 and r_2 are two solutions of a quadratic equation $ax^2 + bx + c = 0$, then it can be shown that

$$r_1 + r_2 = -\frac{b}{a} \quad \text{and} \quad r_1 r_2 = \frac{c}{a}$$

Solve this system of equations for r_1 and r_2.

 98. A circle and a line intersect at most twice. A circle and a parabola intersect at most four times. Deduce that a circle and the graph of a polynomial of degree 3 intersect at most six times. What do you conjecture about a polynomial of degree 4? What about a polynomial of degree n? Can you explain your conclusions using an algebraic argument?

99. Suppose that you are the manager of a sheet metal shop. A customer asks you to manufacture 10,000 boxes, each box being open on top. The boxes are required to have a square base and a 9-cubic foot capacity. You construct the boxes by cutting a square out from each corner of a square piece of sheet metal and folding along the edges.
(a) What are the dimensions of the square to be cut if the area of the square piece of sheet metal is 100 square feet?
(b) Could you make the box using a smaller piece of sheet metal? Make a list of the dimensions of the box for various pieces of sheet metal.

8.7 | SYSTEMS OF INEQUALITIES

> 1 Graph an Inequality
> 2 Graph a System of Linear Inequalities

In Chapter 2, we discussed inequalities in one variable. In this section, we discuss inequalities in two variables. Samples are given in Example 1.

EXAMPLE 1 Samples of Inequalities in Two Variables

(a) $3x + y - 6 < 0$ (b) $x^2 + y^2 < 4$ (c) $y^2 \leq x$ ▬

1   An inequality in two variables x and y is **satisfied** by an ordered pair (a, b) if, when x is replaced by a and y by b, a true statement results. A **graph of an inequality in two variables** x and y consists of all points (x, y) whose coordinates satisfy the inequality.

Let's look at an example.

EXAMPLE 2 Graphing a Linear Inequality

Graph the linear inequality: $3x + y - 6 \leq 0$

Solution   We begin with the associated problem of the graph of the linear equation

$$3x + y - 6 = 0$$

formed by replacing (for now) the $\leq$ symbol with an $=$ sign. The graph of the linear equation is a line. See Figure 11(a). This line is part of the graph of the inequality that we seek because the inequality is nonstrict. (Do you see why? We are seeking points for which $3x + y - 6$ is less than *or equal to* 0.)

Now, let's test a few randomly selected points to see whether they belong to the graph of the inequality.

	$3x + y - 6$	**Conclusion**
$(4, -1)$	$3(4) + (-1) - 6 = 5 > 0$	Does not belong to graph
$(5, 5)$	$3(5) + 5 - 6 = 14 > 0$	Does not belong to graph
$(-1, 2)$	$3(-1) + 2 - 6 = -7 < 0$	Belongs to graph
$(-2, -2)$	$3(-2) + (-2) - 6 = -14 < 0$	Belongs to graph

Look again at Figure 11(a). Notice that the two points that belong to the graph both lie on the same side of the line, and the two points that do not belong to the graph lie on the opposite side. As it turns out, this is always the case. Thus, the graph we seek consists of all points that lie on the same side of the line as do $(-1, 2)$ and $(-2, -2)$. The graph we seek is the shaded region in Figure 11(b).

FIGURE 11

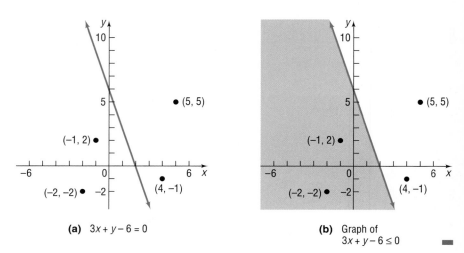

(a) $3x + y - 6 = 0$

(b) Graph of
$3x + y - 6 \leq 0$

The graph of any inequality in two variables may be obtained in a like way. First, the equation corresponding to the inequality is graphed, using dashes if the inequality is strict and solid marks if it is nonstrict. This graph, in almost every case, will separate the xy-plane into two or more regions. In each region either all points satisfy the inequality or no points satisfy the inequality. Thus, the use of a single test point in each region is all that is required to determine whether the points of that region are part of the graph. The steps to follow are given next.

Steps for Graphing an Inequality

STEP 1: Replace the inequality symbol by an equal sign and graph the resulting equation. If the inequality is strict, use dashes; if it is nonstrict, use a solid mark. This graph separates the xy-plane into two or more regions.

STEP 2: In each of the regions, select a test point P.
(a) If the coordinates of P satisfy the inequality, then so do all the points in that region. Indicate this by shading the region.
(b) If the coordinates of P do not satisfy the inequality, then none of the points in that region do.

E X A M P L E   3

FIGURE 12

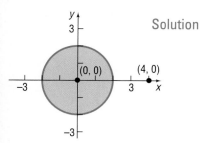

Graphing an Inequality

Graph: $x^2 + y^2 \le 4$

Solution First, we graph the equation $x^2 + y^2 = 4$, a circle of radius 2, center at the origin. A solid circle will be used because the inequality is not strict. We use two test points, one inside the circle, the other outside.

Inside   (0, 0): $x^2 + y^2 = 0^2 + 0^2 = 0 \le 4$ Belongs to the graph
Outside   (4, 0): $x^2 + y^2 = 4^2 + 0^2 = 16 > 4$ Does not belong to graph

All the points inside and on the circle satisfy the inequality. See Figure 12.

Now work Problem 7.

FIGURE 13

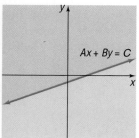

Linear inequalities are inequalities in one of the forms

$$Ax + By < C, \qquad Ax + By > C, \qquad Ax + By \le C, \qquad Ax + By \ge C$$

where A and B are not both zero.

The graph of the corresponding equation of a linear inequality is a line, which separates the xy-plane into two regions, called **half-planes.** See Figure 13.

As shown, if $Ax + By = C$ is the equation of the boundary line, then it divides the plane into two half-planes: one for which $Ax + By < C$ and the other for which $Ax + By > C$. Because of this, for linear inequalities, only one test point is required.

E X A M P L E   4

Graphing Linear Inequalities

Graph: (a) $y < 2$ (b) $y \ge 2x$

Solution (a) The graph of the equation $y = 2$ is a horizontal line and is not part of the graph of the inequality. Since $(0, 0)$ satisfies the inequality, the graph consists of the half-plane below the line $y = 2$. See Figure 14.

(b) The graph of the equation $y = 2x$ is a line and is part of the graph of the inequality. Using $(3, 0)$ as a test point, we find it does not satisfy the inequality $[0 < 2 \cdot 3]$. Thus, points in the half-plane on the opposite side of $y = 2x$ satisfy the inequality. See Figure 15.

FIGURE 14   **FIGURE 15**

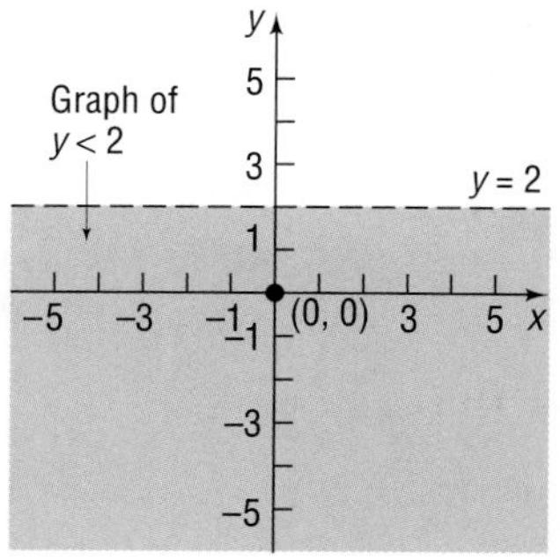

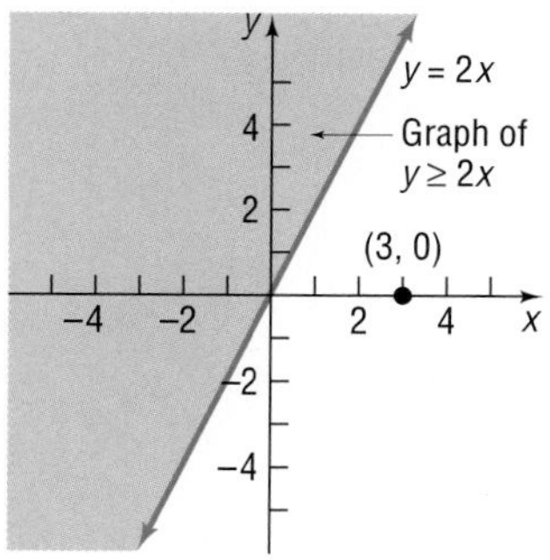

Comment   A graphing utility can be used to graph inequalities. To see how, read Section A.5 in the Appendix.

Now work Problem 3.

Systems of Inequalities in Two Variables

2 The **graph of a system of inequalities** in two variables x and y is the set of all points (x, y) that simultaneously satisfy *each* of the inequalities in the system. Thus, the graph of a system of inequalities can be obtained by graphing each inequality individually and then determining where, if at all, they intersect.

E X A M P L E   5   Graphing a System of Linear Inequalities

Graph the system: $\begin{cases} x + y \geq 2 \\ 2x - y \leq 4 \end{cases}$

Solution   First, we graph the inequality $x + y \geq 2$ as the shaded region in Figure 16(a). Next, we graph the inequality $2x - y \leq 4$ as the shaded region in Figure 16(b). Now, superimpose the two graphs, as shown in Figure 16(c). The points that are in both shaded regions [the overlapping, darker region in Figure 16(c)] are the solutions we seek to the system, because they simultaneously satisfy each linear inequality.

FIGURE 16

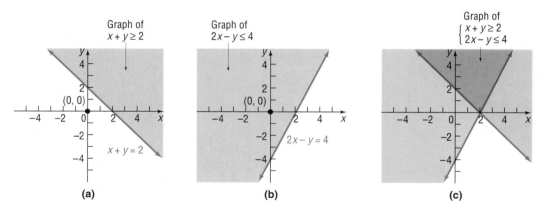

(a)   (b)   (c)

E X A M P L E   6   Graphing a System of Linear Inequalities

Graph the system: $\begin{cases} x + y \leq 2 \\ x + y \geq 0 \end{cases}$

Solution   See Figure 17. The overlapping, darker shaded region between the two boundary lines is the graph of the system.

FIGURE 17 $x + y = 0$ $x + y = 2$

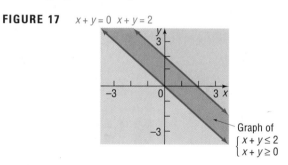

   Now work Problem 25.

E X A M P L E   7

Graphing a System of Linear Inequalities

Graph the system: $\begin{cases} 2x - y \geq 2 \\ 2x - y \geq 0 \end{cases}$

Solution   See Figure 18. The overlappng, darker shaded region is the graph of the system. Note that the graph of the system is identical to the graph of the single inequality $2x - y \geq 2$.

FIGURE 18

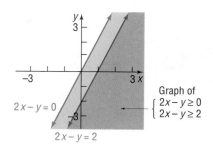

$2x - y = 0$

Graph of
$\begin{cases} 2x - y \geq 0 \\ 2x - y \geq 2 \end{cases}$

$2x - y = 2$

E X A M P L E   8

Graphing a System of Linear Inequalities

Graph the system: $\begin{cases} x + 2y \leq 2 \\ x + 2y \geq 6 \end{cases}$

Solution   See Figure 19. Because no overlapping region results, there are no points in the xy-plane that simultaneously satisfy each inequality. Hence, the system has no solution.

FIGURE 19

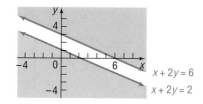

$x + 2y = 6$

$x + 2y = 2$

E X A M P L E   9

Graphing a System of Inequalities

Graph the system: $\begin{cases} y \geq x^2 - 4 \\ x + y \leq 2 \end{cases}$

FIGURE 20

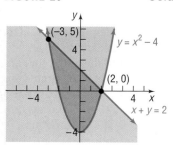

Solution   Figure 20 shows that the graph of the system consists of the region enclosed by the graphs of the parabola $y = x^2 - 4$ and the line $x + y = 2$. The points of intersection of the two equations are found by solving the system of equations

$$\begin{cases} y = x^2 - 4 \\ x + y = 2 \end{cases}$$

Using substitution, we find

$$x + (x^2 - 4) = 2$$
$$x^2 + x - 6 = 0$$
$$(x + 3)(x - 2) = 0$$
$$x = -3, \quad x = 2$$

The two points of intersection are $(-3, 5)$ and $(2, 0)$.

 Now work Problem 19.

E X A M P L E 10   Graphing a System of Four Linear Inequalities

FIGURE 21

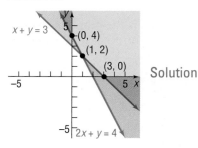

Graph the system: $\begin{cases} x + y \geq 3 \\ 2x + y \geq 4 \\ x \geq 0 \\ y \geq 0 \end{cases}$

Solution   The two inequalities $x \geq 0$ and $y \geq 0$ require the graph of the system to be in quadrant I. Thus, we concentrate on the remaining two inequalities. The intersection of the graphs of these two inequalities is shown in purple in Figure 21. The intersection of this region and quadrant I, shown in gray, is the graph of the system. ∎

E X A M P L E 11   Financial Planning

A retired couple has up to $25,000 to invest. As their financial adviser, you recommend that they place at least $15,000 in Treasury bills yielding 6% and at most $5000 in corporate bonds yielding 9%.

(a) Using x to denote the amount of money invested in Treasury bills and y the amount invested in corporate bonds, write a system of linear inequalities that describes the possible amounts of each investment.

(b) Graph the system.

FIGURE 22

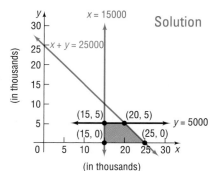

Solution   (a) The system of linear inequalities is

$\begin{cases} x \geq 0 \\ y \geq 0 \\ x + y \leq 25{,}000 \\ x \geq 15{,}000 \\ y \leq 5000 \end{cases}$

x and y are nonnegative variables since they represent money invested.

The total of the two investments, $x + y$, cannot exceed $25,000.

At least $15,000 in Treasury bills.

At most $5000 in corporate bonds.

(b) See the shaded region in Figure 22. Note that the inequalities $x \geq 0$ and $y \geq 0$ again require that the graph of the system be in quadrant I. ∎

The graph of the system of linear inequalities in Figure 22 is said to be **bounded,** because it can be contained within some circle of sufficiently large radius. A graph that cannot be contained in any circle is said to be **unbounded.** For example, the graph of the system of linear inequalities in Figure 21 is unbounded, since it extends indefinitely in a particular direction.

Notice in Figures 21 and 22 that those points belonging to the graph that are also points of intersection of boundary lines have been plotted. Such points are referred to as **vertices** or **corner points** of the graph. Thus, the system graphed in Figure 21 has three corner points: $(0, 4)$, $(1, 2)$, and $(3, 0)$. The system graphed in Figure 22 has four corner points: $(15, 0)$, $(25, 0)$, $(20, 5)$, and $(15, 5)$.

These ideas will be used in the next section in developing a method for solving linear programming problems, an important application of linear inequalities.

Now work Problem 33.

8.7 | EXERCISES

In Problems 1–12, graph each inequality.

1. $x \geq 0$

2. $y \geq 0$

3. $x \geq 4$

4. $y \leq 2$

5. $2x + y \geq 6$

6. $3x + 2y \leq 6$

7. $x^2 + y^2 > 1$

8. $x^2 + y^2 \leq 9$

9. $y \leq x^2 - 1$

10. $y > x^2 + 2$

11. $xy \geq 4$

12. $xy \leq 1$

In Problems 13–30, graph each system of inequalities.

13. $\begin{cases} x + y \leq 2 \\ 2x + y \geq 4 \end{cases}$

14. $\begin{cases} 3x - y \geq 6 \\ x + 2y \leq 2 \end{cases}$

15. $\begin{cases} 2x - y \leq 4 \\ 3x + 2y \geq -6 \end{cases}$

16. $\begin{cases} 4x - 5y \leq 0 \\ 2x - y \geq 2 \end{cases}$

17. $\begin{cases} 2x - 3y \leq 0 \\ 3x + 2y \leq 6 \end{cases}$

18. $\begin{cases} 4x - y \geq 2 \\ x + 2y \geq 2 \end{cases}$

19. $\begin{cases} x^2 + y^2 \leq 9 \\ x + y \geq 3 \end{cases}$

20. $\begin{cases} x^2 + y^2 \geq 9 \\ x + y \leq 3 \end{cases}$

21. $\begin{cases} y \geq x^2 - 4 \\ y \leq x - 2 \end{cases}$

22. $\begin{cases} y^2 \leq x \\ y \geq x \end{cases}$

23. $\begin{cases} xy \geq 4 \\ y \geq x^2 + 1 \end{cases}$

24. $\begin{cases} y + x^2 \leq 1 \\ y \geq x^2 - 1 \end{cases}$

25. $\begin{cases} x - 2y \leq 6 \\ 2x - 4y \geq 0 \end{cases}$

26. $\begin{cases} x + 4y \leq 8 \\ x + 4y \geq 4 \end{cases}$

27. $\begin{cases} 2x + y \geq -2 \\ 2x + y \geq 2 \end{cases}$

28. $\begin{cases} x - 4y \leq 4 \\ x - 4y \geq 0 \end{cases}$

29. $\begin{cases} 2x + 3y \geq 6 \\ 2x + 3y \leq 0 \end{cases}$

30. $\begin{cases} 2x + y \geq 0 \\ 2x + y \geq 2 \end{cases}$

In Problems 31–40, graph each system of linear inequalities. Tell whether the graph is bounded or unbounded, and label the corner points.

31. $\begin{cases} x \geq 0 \\ y \geq 0 \\ 2x + y \leq 6 \\ x + 2y \leq 6 \end{cases}$

32. $\begin{cases} x \geq 0 \\ y \geq 0 \\ x + y \geq 4 \\ 2x + 3y \geq 6 \end{cases}$

33. $\begin{cases} x \geq 0 \\ y \geq 0 \\ x + y \geq 2 \\ 2x + y \geq 4 \end{cases}$

34. $\begin{cases} x \geq 0 \\ y \geq 0 \\ 3x + y \leq 6 \\ 2x + y \leq 2 \end{cases}$

35. $\begin{cases} x \geq 0 \\ y \geq 0 \\ x + y \geq 2 \\ 2x + 3y \leq 12 \\ 3x + y \leq 12 \end{cases}$

36. $\begin{cases} x \geq 0 \\ y \geq 0 \\ x + y \geq 2 \\ x + y \leq 10 \\ 2x + y \leq 3 \end{cases}$

37. $\begin{cases} x \geq 0 \\ y \geq 0 \\ x + y \geq 2 \\ x + y \leq 8 \\ 2x + y \leq 10 \end{cases}$

38. $\begin{cases} x \geq 0 \\ y \geq 0 \\ x + y \geq 2 \\ x + y \leq 8 \\ x + 2y \geq 1 \end{cases}$

39. $\begin{cases} x \geq 0 \\ y \geq 0 \\ x + 2y \geq 1 \\ x + 2y \leq 10 \end{cases}$

40. $\begin{cases} x \geq 0 \\ y \geq 0 \\ x + 2y \geq 1 \\ x + 2y \leq 10 \\ x + y \geq 2 \\ x + y \leq 8 \end{cases}$

In Problems 41–44, write a system of linear inequalities that has the given graph.

41.

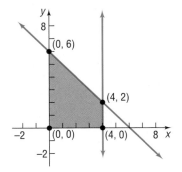

42.

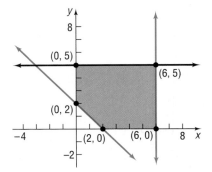

43.

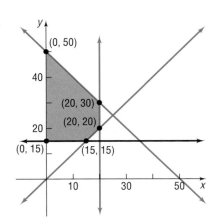

44.

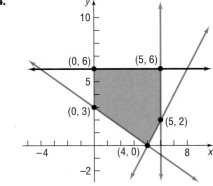

45. Financial Planning A retired couple has up to $50,000 to invest. As their financial adviser, you recommend that they place at least $35,000 in Treasury bills yielding 7% and at most $10,000 in corporate bonds yielding 10%.
(a) Using x to denote the amount of money invested in Treasury bills and y the amount invested in corporate bonds, write a system of linear inequalities that describe the possible amounts of each investment.
(b) Graph the system and label the corner points.

46. Manufacturing Trucks Mike's Toy Truck Company manufacturers two models of toy trucks, a standard model and a deluxe model. Each standard model requires 2 hours for painting and 3 hours for detail work; each deluxe model requires 3 hours for painting and 4 hours for detail work. Two painters and three detail workers are employed by the company, and each works 40 hours per week.
(a) Using x to denote the number of standard model trucks and y to denote the number of deluxe model trucks, write a system of linear inequalities that describes the possi-

ble number of each model of truck that can be manufactured in a week.
(b) Graph the system and label the corner points.

47. Blending Coffee Bill's Coffee House, a store that specializes in coffee, has available 75 pounds of A grade coffee and 120 pounds of B grade coffee. These will be blended into 1 pound packages as follows: An economy blend that contains 4 ounces of A grade coffee and 12 ounces of B grade coffee and a superior blend that contains 8 ounces of A grade coffee and 8 ounces of B grade coffee.

(a) Using x to denote the number of packages of the economy blend and y to denote the number of packages of the superior blend, write a system of linear inequalities that describes the possible number of packages of each kind of blend.

(b) Graph the system and label the corner points.

48. **Mixed Nuts** Nola's Nuts, a store that specializes in selling nuts, has 90 pounds of cashews and 120 pounds of peanuts available. These are to be mixed in 12 ounce packages as follows: a lower priced package containing 8 ounces of peanuts and 4 ounces of cashews and a quality package containing 6 ounces of peanuts and 6 ounces of cashews.

(a) Use x to denote the number of lower priced packages and use y to denote the number

of quality packages. Write a system of linear inequalities that describes the possible number of each kind of package.

(b) Graph the system and label the corner points.

49. **Transporting Goods** A small truck can carry no more than 1600 pounds of cargo or 150 cubic feet of cargo. A printer weighs 20 pounds and occupies 3 cubic feet of space. A microwave weighs 30 pounds and occupies 2 cubic feet of space.

(a) Using x to represent the number of microwave ovens and y to represent the number of printers, write a system of linear inequalities that describes the number of ovens and printers that can be hauled by the truck.

(b) Graph the system and label the corner points.

8.8 LINEAR PROGRAMMING

> 1 Set up a Linear Programming Problem
> 2 Solve a Linear Programming Problem

1 Historically, linear programming evolved as a technique for solving problems involving resource allocation of goods and materials for the U.S. Air Force during World War II. Today, linear programming techniques are used to solve a wide variety of problems, such as optimizing airline scheduling and establishing telephone lines. Although most practical linear programming problems involve systems of several hundred linear inequalities containing several hundred variables, we will limit our discussion to problems containing only two variables, because we can solve such problems using graphing techniques.*

We begin by returning to Example 11 of the previous section.

E X A M P L E   1

Financial Planning

A retired couple has up to $25,000 to invest. As their financial adviser, you recommend that they place at least $15,000 in Treasury bills yielding 6% and at most $5000 in corporate bonds yielding 9%. How much money should be placed in each investment so that income is maximized?

The problem given here is typical of a *linear programming problem*. The problem requires that a certain linear expression, the income, be maximized. If I represents income, x the amount invested in Treasury bills at 6%, and y the amount invested in corporate bonds at 9%, then

$$I = 0.06x + 0.09y$$

---

*The **simplex method** is a way to solve linear programming problems involving many inequalities and variables. This method was developed by George Dantzig in 1946 and is particularly well suited for computerization. In 1984, Narendra Karmarkar of Bell Laboratories discovered a way of solving large linear programming problems that improves on the simplex method.

This linear expression is called the **objective function.** Furthermore, the problem requires that the maximum income be achieved under certain conditions or **constraints,** each of which is a linear inequality involving the variables. (See Example 11 in Section 8.7.) The linear programming problem given in Example 1 may be restated as

Maximize $I = 0.06x + 0.09y$

subject to the conditions that

$$x \geq 0, \qquad y \geq 0$$
$$x + y \leq 25,000$$
$$x \geq 15,000$$
$$y \leq 5000$$

In general, every linear programming problem has two components:

1.   A linear objective function that is to be maximized or minimized.
2.   A collection of linear inequalities that must be satisfied simultaneously.

> A **linear programming problem** in two variables x and y consists of maximizing (or minimizing) a linear objective function
>
> $$z = Ax + By, \quad A \text{ and } B \text{ are real numbers, not both } 0$$
>
> subject to certain conditions, or constraints, expressible as linear inequalities in x and y.

To maximize (or minimize) the quantity $z = Ax + By$, we need to identify points (x, y) that make the expression for z the largest (or smallest) possible. But not all points (x, y) are eligible; only those that also satisfy each linear inequality (constraint) can be used. We refer to each point (x, y) that satisfies the system of linear inequalities (the constraints) as a **feasible point.** Thus, in a linear programming problem, we seek the feasible point(s) that maximizes (or minimizes) the objective function.

Let's look again at the linear programming problem in Example 1.

E X A M P L E   2   Analyzing a Linear Programming Problem

Consider the linear programming problem

Maximize $I = 0.06x + 0.09y$

subject to the conditions that

$$x \geq 0, \qquad y \geq 0$$
$$x + y \leq 25,000$$
$$x \geq 15,000$$
$$y \leq 5000$$

Graph the constraints. Then graph the objective function for $I = 0, 0.9, 1.35, 1.65,$ and 1.8.

Solution   Figure 23 shows the graph of the constraints. We superimpose on this graph the graph of the objective function for the given values of I.

> For $I = 0$, the objective function is the line $0 = 0.06x + 0.09y$.
> For $I = 0.9$, the objective function is the line $0.9 = 0.06x + 0.09y$.
> For $I = 1.35$, the objective function is the line $1.35 = 0.06x + 0.09y$.
> For $I = 1.65$, the objective function is the line $1.65 = 0.06x + 0.09y$.
> For $I = 1.8$, the objective function is the line $1.8 = 0.06x + 0.09y$.

FIGURE 23

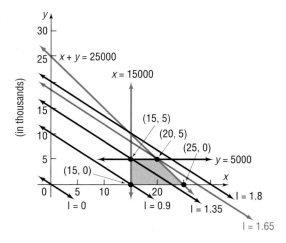

> A **solution** to a linear programming problem consists of a feasible point that maximizes (or minimizes) the objective function, together with the corresponding value of the objective function.

2   One condition for a linear programming problem in two variables to have a solution is that the graph of the feasible points be bounded. (Refer to page 616.)

If none of the feasible points maximizes (or minimizes) the objective function or if there are no feasible points, then the linear programming problem has no solution.

Consider the linear programming problem stated in Example 2, and look again at Figure 23. The feasible points are the points that lie in the shaded region. For example, $(20, 3)$ is a feasible point, as are $(15, 5)$, $(20, 5)$, $(18, 4)$, and so on. To find the solution of the problem requires that we find a feasible point (x, y) that makes $I = 0.06x + 0.09y$ as large as possible. Notice that as I increases in value from $I = 0$ to $I = 0.9$ to $I = 1.35$ to $I = 1.65$ to $I = 1.8$, we obtain a collection of parallel lines. Furthermore, notice that the largest value of I that can be obtained using feasible points is $I = 1.65$, which corresponds to the line $1.65 = 0.06x + 0.09y$. Any larger value of I results in a line that does not pass through any feasible points. Finally, notice that the feasible point that yields $I = 1.65$ is the point $(20, 5)$, a corner point. These observations form the basis of the following result, which we state without proof.

> **Theorem** Location of the Solution of a Linear Programming Problem
>
> If a linear programming problem has a solution, it is located at a corner point of the graph of the feasible points.
>
> If a linear programming problem has multiple solutions, at least one of them is located at a corner point of the graph of the feasible points.
>
> In either case, the corresponding value of the objective function is unique.

We shall not consider here linear programming problems that have no solution. As a result, we can outline the procedure for solving a linear programming problem as follows:

Procedure for Solving a Linear Programming Problem

> STEP 1: Write an expression for the quantity to be maximized (or minimized). This expression is the objective function.
> STEP 2: Write all the constraints as a system of linear inequalities and graph the system.
> STEP 3: List the corner points of the graph of the feasible points.
> STEP 4: List the corresponding values of the objective function at each corner point. The largest (or smallest) of these is the solution.

E X A M P L E   3

Solving a Minimum Linear Programming Problem

Minimize the expression

$$z = 2x + 3y$$

subject to the constraints

$$y \leq 5, \qquad x \leq 6, \qquad x + y \geq 2, \qquad x \geq 0, \qquad y \geq 0$$

Solution   The objective function is $z = 2x + 3y$. We seek the smallest value of z that can occur if x and y are solutions of the system of linear inequalities

$$\begin{cases} y \leq 5 \\ x \leq 6 \\ x + y \geq 2 \\ x \geq 0 \\ y \geq 0 \end{cases}$$

The graph of this system (the feasible points) is shown as the shaded region in Figure 24. We have also plotted the corner points. Table 2 lists the corner points and the corresponding values of the objective function. From the table, we can see that the minimum value of z is 4, and it occurs at the point $(2, 0)$.

FIGURE 24

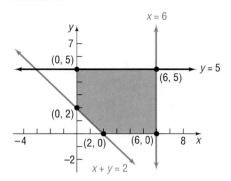

TABLE 2	
Corner Point	**Value of the Objective Function**
(x, y)	$z = 2x + 3y$
$(0, 2)$	$z = 2(0) + 3(2) = 6$
$(0, 5)$	$z = 2(0) + 3(5) = 15$
$(6, 5)$	$z = 2(6) + 3(5) = 27$
$(6, 0)$	$z = 2(6) + 3(0) = 12$
$(2, 0)$	$z = 2(2) + 3(0) = 4$

 Now work Problems 3 and 9.

E X A M P L E   4

Maximizing Profit

At the end of every month, after filling orders for its regular customers, a coffee company has some pure Colombian coffee and some special-blend coffee remaining. The practice of the company has been to package a mixture of the two coffees into 1-pound packages as follows: a low-grade mixture containing 4 ounces of Colombian coffee and 12 ounces of special-blend coffee and a high-grade mixture containing 8 ounces of Colombian and 8 ounces of special-blend coffee. A profit of $0.30 per package is made on the low-grade mixture, whereas a profit of $0.40 per package is made on the high-grade mixture. This month, 120 pounds of special-blend coffee and 100 pounds of pure Colombian coffee remain. How many packages of each mixture should be prepared to achieve a maximum profit? Assume that all packages prepared can be sold.

Solution

We begin by assigning symbols for the two variables.

x = Number of packages of the low-grade mixture
y = Number of packages of the high-grade mixture

If P denotes the profit, then

$$P = \$0.30x + \$0.40y$$

This expression is the objective function. We seek to maximize P subject to certain constraints on x and y. Because x and y represent numbers of packages, the only meaningful values for x and y are nonnegative integers. Thus, we have the two constraints

$$x \geq 0, \qquad y \geq 0 \quad \text{Nonnegative constraints}$$

We also have only so much of each type of coffee available. For example, the total amount of Colombian coffee used in the two mixtures cannot exceed 100 pounds, or 1600 ounces. Because we use 4 ounces in each low-grade package and 8 ounces in each high-grade package, we are led to the constraint

$$4x + 8y \leq 1600 \quad \text{Colombian coffee constraint}$$

Similarly, the supply of 120 pounds, or 1920 ounces, of special-blend coffee leads to the constraint

$$12x + 8y \leq 1920 \qquad \text{Special-blend coffee constraint}$$

The linear programming problem may be stated as

Maximize $\qquad P = 0.3x + 0.4y$

subject to the constraints

$$x \geq 0, \qquad y \geq 0, \qquad 4x + 8y \leq 1600, \qquad 12x + 8y \leq 1920$$

The graph of the constraints (the feasible points) is illustrated in Figure 25. We list the corner points and evaluate the objective function at each. In Table 3, we can see that the maximum profit, $84, is achieved with 40 packages of the low-grade mixture and 180 packages of the high-grade mixture.

FIGURE 25

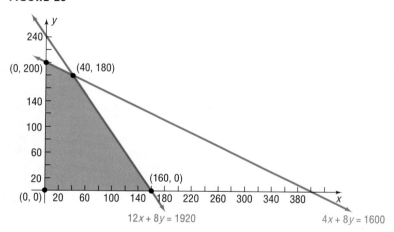

TABLE 3	
Corner Point	**Value of Profit**
(x, y)	$P = 0.3x + 0.4y$
(0, 0)	$P = 0$
(0, 200)	$P = 0.3(0) + 0.4(200) = \$80$
(40, 180)	$P = 0.3(40) + 0.4(180) = \$84$
(160, 0)	$P = 0.3(160) + 0.4(0) = \$48$

 Now work Problem 17.

8.8 EXERCISES

In Problems 1–6, find the maximum and minimum value of the given objective function of a linear programming problem. The figure illustrates the graph of the feasible points.

1. $z = x + y$ **2.** $z = 2x + 3y$ **3.** $z = x + 10y$

4. $z = 10x + y$ **5.** $z = 5x + 7y$ **6.** $z = 7x + 5y$

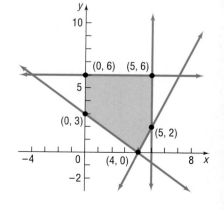

In Problems 7–16, solve each linear programming problem.

7. Maximize $z = 2x + y$ subject to $x \geq 0$, $y \geq 0$, $x + y \leq 6$, $x + y \geq 1$

8. Maximize $z = x + 3y$ subject to $x \geq 0$, $y \geq 0$, $x + y \geq 3$, $x \leq 5$, $y \leq 7$

9. Minimize $z = 2x + 5y$ subject to $x \geq 0$, $y \geq 0$, $x + y \geq 2$, $x \leq 5$, $y \leq 3$

10. Minimize $z = 3x + 4y$ subject to $x \geq 0$, $y \geq 0$, $2x + 3y \geq 6$, $x + y \leq 8$

11. Maximize $z = 3x + 5y$ subject to $x \geq 0$, $y \geq 0$, $x + y \geq 2$, $2x + 3y \leq 12$, $3x + 2y \leq 12$

12. Maximize $z = 5x + 3y$ subject to $x \geq 0$, $y \geq 0$, $x + y \geq 2$, $x + y \leq 8$, $2x + y \leq 10$

13. Minimize $z = 5x + 4y$ subject to $x \geq 0$, $y \geq 0$, $x + y \geq 2$, $2x + 3y \leq 12$, $3x + y \leq 12$

14. Minimize $z = 2x + 3y$ subject to $x \geq 0$, $y \geq 0$, $x + y \geq 3$, $x + y \leq 9$, $x + 3y \geq 6$

15. Maximize $z = 5x + 2y$ subject to $x \geq 0$, $y \geq 0$, $x + y \leq 10$, $2x + y \geq 10$, $x + 2y \geq 10$

16. Maximize $z = 2x + 4y$ subject to $x \geq 0$, $y \geq 0$, $2x + y \geq 4$, $x + y \leq 9$

17. Maximizing Profit A manufacturer of skis produces two types: downhill and cross-country. Use the following table to determine how many of each kind of ski should be produced to achieve a maximum profit. What is the maximum profit? What would the maximum profit be if the maximum time available for manufacturing is increased to 48 hours?

	Downhill	Cross-country	Maximum Time Available
Manufacturing time per ski	2 hours	1 hour	40 hours
Finishing time per ski	1 hour	1 hour	32 hours
Profit per ski	$70	$50	

18. Farm Management A farmer has 70 acres of land available for planting either soybeans or wheat. The cost of preparing the soil, the workdays required, and the expected profit per acre planted for each type of crop are given in the following table:

	Soybeans	Wheat
Preparation cost per acre	$60	$30
Workdays required per acre	3	4
Profit per acre	$180	$100

The farmer cannot spend more than $1800 in preparation costs nor more than a total of 120 workdays. How many acres of each crop should be planted in order to maximize the profit? What is the maximum profit? What is the maximum profit if the farmer is willing to spend no more than $2400 on preparation?

19. Farm Management A small farm in Illinois has 100 acres of land available on which to grow corn and soybeans. The following table shows the cultivation cost per acre, the labor cost per acre, and the expected profit per acre. The column on the right shows the amount of money available for each of these expenses. Find the number of acres of each crop that should be planted in order to maximize profit.

	Soybeans	Corn	Money Available
Cultivation cost per acre	$40	$60	$1800
Labor cost per acre	$60	$60	$2400
Profit per acre	$200	$250	

20. Dietary Requirements A certain diet requires at least 60 units of carbohydrates, 45 units of protein, and 30 units of fat each day. Each ounce of Supplement A provides 5 units of carbohydrates, 3 units of protein, and 4 units of fat. Each ounce of Supplement B provides 2 units of carbohydrates, 2 units of protein, and 1 unit of fat. If Supplement A costs $1.50 per ounce and Supplement B costs $1.00 per ounce, how many ounces of each supplement should be taken daily to minimize the cost of the diet?

21. Production Scheduling In a factory, machine 1 produces 8" plyers at the rate of 60 units per hour and 6" plyers at the rate of 70 units per hour. Machine 2 produces 8" plyers at the rate of 40 units per hour and 6" plyers at the rate of 20 units per hour. It costs $50 per hour to operate machine 1, while machine 2 costs $30 per hour to operate. The production schedule requires that at least 240 units of 8" plyers and at least 140 units of 6" plyers must be produced during each 10 hour day. Which combination of machines will cost the least money to operate?

22. Farm Management An owner of a fruit orchard hires a crew of workers to prune at least 25 of his 50 fruit trees. Each newer tree requires one hour to prune, while each older tree needs one-and-a-half hours. The crew contracts to work for at least 30 hours and charge $15 for each newer tree and $20 for each older tree. To minimize his cost, how many of each kind of tree will the orchard owner have pruned? What will be the cost?

23. Managing a Meat Market A meat market combines ground beef and ground pork in a single package for meat loaf. The ground beef is 75% lean (75% beef, 25% fat) and costs the market $0.75 per pound. The ground pork is 60% lean and costs the market $0.45 per pound. The meat loaf must be at least 70% lean. If the market wants to use at least 50 lb of its available pork, but no more than 200 lb of its available ground beef, how much ground beef should be mixed with ground pork so that the cost is minimized?

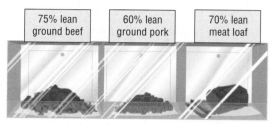

24. Return on Investment An investment broker is instructed by her client to invest up to $20,000, some in a junk bond yielding 9% per annum and some in Treasury bills yielding 7% per annum. The client wants to invest at least $8000 in T-bills and no more than $12,000 in the junk bond.

(a) How much should the broker recommend that the client place in each investment to maximize income if the client insists that the amount invested in T-bills must equal or exceed the amount placed in junk bonds?

(b) How much should the broker recommend that the client place in each investment to maximize income if the client insists that the amount invested in T-bills must not exceed the amount placed in junk bonds?

25. Maximizing Profit on Ice Skates A factory manufacturers two kinds of ice skates: racing skates and figure skates. The racing skates require 6 work-hours in the fabrication department, whereas the figure skates require 4 work-hours there. The racing skates require 1 work-hour in the finishing department, whereas the figure skates require 2 work-hours there. The fabricating department has available at most 120 work-hours per day, and the finishing department has no more than 40 work-hours per

day available. If the profit on each racing skate is $10 and the profit on each figure skate is $12, how many of each should be manufactured each day to maximize profit? (Assume that all skates made are sold.)

26. Financial Planning A retired couple has up to $50,000 to place in fixed-income securities. Their financial adviser suggests two securities to them: one is an AAA bond that yields 8% per annum; the other is a Certificate of Deposit (CD) that yields 4%. After careful consideration of the alternatives, the couple decides to place at most $20,000 in the AAA bond and at least $15,000 in the CD. They also instruct the financial adviser to place at least as much in the CD as in the AAA bond. How should the financial adviser proceed to maximize the return on their investment?

27. Product Design An entrepreneur is having a design group produce at least six samples of a new kind of fastener that he wants to market. It costs $9.00 to produce each metal fastener and $4.00 to produce each plastic fastener. He wants to have at least two of each version of the fastener and needs to have all the samples 24 hours from now. It takes 4 hours to produce each metal sample and 2 hours to produce each plastic sample. To minimize the cost of the samples, how many of each kind should the entrepreneur order? What will be the cost of the samples?

28. Animal Nutrition Kevin's dog Amadeus likes two kinds of canned dog food. "Gourmet Dog" costs 40 cents a can and has 20 units of a vitamin complex; the calorie content is 75 calories. "Chow Hound" costs 32 cents a can and has 35 units of vitamins and 50 calories. Kevin likes Amadeus to have at least 1175 units of vitamins a month and at least 2375 calories during the same time period. Kevin has space to store only 60 cans of dog food at a time. How much of each kind of dog food should Kevin buy each month in order to minimize his cost?

29. Airline Revenue An airline has two classes of service: first class and coach. Management's experience has been that each aircraft should have at least 8 but no more than 16 first-class seats and at least 80 but not more than 120 coach seats.

(a) If management decides that the ratio of first class to coach seats should never exceed 1 : 12, with how many of each type of seat should an aircraft be configured to maximize revenue?

(b) If management decides that the ratio of first class to coach seats should never exceed 1 : 8, with how many of each type of seat should an aircraft be configured to maximize revenue?

(c) If you were management, what would you do?

[**Hint:** Assume that the airline charges $\$C$ for a coach seat and $\$F$ for a first class seat; $C > 0$, $F > C$.]

30. Minimizing Cost A farm that specializes in raising frying chickens supplements the regular chicken feed with four vitamins. The owner wants the supplemental food to contain at least 50 units of vitamin I, 90 units of vitamin II, 60 units of vitamin III, and 100 units of vitamin IV per 100 ounces of feed. Two supplements are available: supplement A, which contains 5 units of vitamin I, 25 units of vitamin II, 10 units of vitamin III, and 35 units of vitamin IV per ounce, and supplement B, which contains 25 units of vitamin I, 10 units of vitamin II, 10 units of vitamin III, and 20 units of vitamin IV per ounce. If supplement A costs $\$0.06$ per ounce and supplement B costs $\$0.08$ per ounce, how much of each supplement should the manager of the farm buy to add to each 100 ounces of feed in order to keep the total cost at a minimum, while still meeting the owner's vitamin specifications?

31. Explain in your own words what a linear programming problem is and how it can be solved.

CHAPTER REVIEW

THINGS TO KNOW

Systems of equations

Systems with no solutions are inconsistent.   Systems with a solution are consistent.

Consistent systems of linear equations have either a unique solution or an infinite number of solutions.

Matrix	Rectangular array of numbers, called entries
m by n matrix	Matrix with m rows and n columns
Identity matrix I	Square matrix whose diagonal entries are 1's, while all other entries are 0's
Inverse of a matrix	A^{-1} is the inverse of A if $AA^{-1} = A^{-1}A = I$
Nonsingular matrix	A matrix that has an inverse

Linear programming problems

Maximize (or minimize) a linear objective function, $z = Ax + By$, subject to certain conditions, or constraints, expressible as linear inequalities in x and y.

Feasible point

A point (x, y) that satisfies the constraints of a linear programming problem

Location of solution

If a linear programming problem has a solution, it is located at a corner point of the graph of the feasible points.

If a linear programming problem has multiple solutions, at least one of them is located at a corner point of the graph of the feasible points.

In either case, the corresponding value of the objective function is unique.

HOW TO

Solve a system of linear equations using the method of substitution

Solve a system of linear equations using the method of elimination

Solve a system of linear equations using matrices

Solve a system of linear equations using determinants

Recognize equal matrices

Add and subtract matrices

Multiply matrices

Find the inverse of a nonsingular matrix

Solve a system of linear equations using the inverse of a matrix

Write the partial fraction decomposition of a rational expression

Solve a system of nonlinear equations

Graph a system of inequalities

Find the corner points of the graph of a system of linear inequalities

Solve linear programming problems

FILL-IN-THE-BLANK ITEMS

1. If a system of equations has no solution, it is said to be _____.
2. An m by n rectangular array of numbers is called a(n) _____.
3. Cramer's Rules uses _____ to solve a system of linear equations.
4. The matrix used to represent a system of linear equations is called a(n) _____ matrix.
5. A matrix B, for which $AB = I_n$, the identity matrix, is called the _____ of A.
6. A matrix that has the same number of rows as columns is called a(n) _____ matrix.
7. In the algebra of matrices, the matrix that has the properties similar to the number 1 is called the _____ matrix.
8. A rational function is called _____ if the degree of its numerator is less than the degree of its denominator.
9. The graph of a linear inequality is called a(n) _____.
10. A linear programming problem requires that a linear expression, called the _____ _____, be maximized or minimized.
11. Each point that satisfies the constraints of a linear programming problem is called a(n) _____ _____.

TRUE/FALSE ITEMS

T F **1.** A system of two linear equations containing two unknowns always has at least one solution.
T F **2.** The augmented matrix of a system of two equations containing three variables has two rows and four columns.
T F **3.** A 3 by 3 determinant can never equal 0.
T F **4.** A consistent system of equations will have exactly one solution.
T F **5.** Every square matrix has an inverse.
T F **6.** Matrix multiplication is commutative.
T F **7.** Any pair of matrices can be multiplied.
T F **8.** The factors of the denominator of a rational expression are used to arrive at the partial fraction decomposition.
T F **9.** The graph of a linear inequality is a half-plane.
T F **10.** The graph of a system of linear inequalities is sometimes unbounded.
T F **11.** If a linear programming problem has a solution, it is located at a corner point of the graph of the feasible points.

REVIEW EXERCISES

Blue problem numbers indicate the author's suggestions for use in a Practice Test.

In Problems 1–20, solve each system of equations using the method of substitution or the method of elimination. If the system has no solution, say that it is inconsistent.

1. $\begin{cases} 2x - y = 5 \\ 5x + 2y = 8 \end{cases}$

2. $\begin{cases} 2x + 3y = 2 \\ 7x - y = 3 \end{cases}$

3. $\begin{cases} 3x - 4y = 4 \\ x - 3y = \frac{1}{2} \end{cases}$

4. $\begin{cases} 2x + y = 0 \\ 5x - 4y = -\frac{13}{2} \end{cases}$

5. $\begin{cases} x - 2y - 4 = 0 \\ 3x + 2y - 4 = 0 \end{cases}$

6. $\begin{cases} x - 3y + 5 = 0 \\ 2x + 3y - 5 = 0 \end{cases}$

7. $\begin{cases} y = 2x - 5 \\ x = 3y + 4 \end{cases}$

8. $\begin{cases} x = 5y + 2 \\ y = 5x + 2 \end{cases}$

9. $\begin{cases} x - y + 4 = 0 \\ \frac{1}{2}x + \frac{1}{6}y + \frac{2}{5} = 0 \end{cases}$

10. $\begin{cases} x + \frac{1}{4}y = 2 \\ y + 4x + 2 = 0 \end{cases}$

11. $\begin{cases} x - 2y - 8 = 0 \\ 2x + 2y - 10 = 0 \end{cases}$

12. $\begin{cases} x - 3y + \frac{7}{2} = 0 \\ \frac{1}{2}x + 3y - 5 = 0 \end{cases}$

13. $\begin{cases} y - 2x = 11 \\ 2y - 3x = 18 \end{cases}$

14. $\begin{cases} 3x - 4y - 12 = 0 \\ 5x + 2y + 6 = 0 \end{cases}$

15. $\begin{cases} 2x + 3y - 13 = 0 \\ 3x - 2y = 0 \end{cases}$

16. $\begin{cases} 4x + 5y = 21 \\ 5x + 6y = 42 \end{cases}$

17. $\begin{cases} 3x - 2y = 8 \\ x - \frac{2}{3}y = 12 \end{cases}$

18. $\begin{cases} 2x + 5y = 10 \\ 4x + 10y = 15 \end{cases}$

19. $\begin{cases} x + 2y - z = 6 \\ 2x - y + 3z = -13 \\ 3x - 2y + 3z = -16 \end{cases}$

20. $\begin{cases} x + 5y - z = 2 \\ 2x + y + z = 7 \\ x - y + 2z = 11 \end{cases}$

In Problems 21–28, use the following matrices to compute each expression.

$$A = \begin{bmatrix} 1 & 0 \\ 2 & 4 \\ -1 & 2 \end{bmatrix} \qquad B = \begin{bmatrix} 4 & -3 & 0 \\ 1 & 1 & -2 \end{bmatrix} \qquad C = \begin{bmatrix} 3 & -4 \\ 1 & 5 \\ 5 & -2 \end{bmatrix}$$

21. $A + C$ **22.** $A - C$ **23.** $6A$ **24.** $-4B$

25. AB **26.** BA **27.** CB **28.** BC

In Problems 29–34, find the inverse of each matrix algebraically, if there is one. If there is not an inverse, say that the matrix is singular.

29. $\begin{bmatrix} 4 & 6 \\ 1 & 3 \end{bmatrix}$ **30.** $\begin{bmatrix} -3 & 2 \\ 1 & -2 \end{bmatrix}$ **31.** $\begin{bmatrix} 1 & 3 & 3 \\ 1 & 2 & 1 \\ 1 & -1 & 2 \end{bmatrix}$

32. $\begin{bmatrix} 3 & 1 & 2 \\ 3 & 2 & -1 \\ 1 & 1 & 1 \end{bmatrix}$ **33.** $\begin{bmatrix} 4 & -8 \\ -1 & 2 \end{bmatrix}$ **34.** $\begin{bmatrix} -3 & 1 \\ -6 & 2 \end{bmatrix}$

In Problems 35–44, solve each system of equations using matrices. If the system has no solution, say that it is inconsistent.

35. $\begin{cases} 3x - 2y = 1 \\ 10x + 10y = 5 \end{cases}$ **36.** $\begin{cases} 3x + 2y = 6 \\ x - y = -\frac{1}{2} \end{cases}$ **37.** $\begin{cases} 5x + 6y - 3z = 6 \\ 4x - 7y - 2z = -3 \\ 3x + y - 7z = 1 \end{cases}$

38. $\begin{cases} 2x + y + z = 5 \\ 4x - y - 3z = 1 \\ 8x + y - z = 5 \end{cases}$ **39.** $\begin{cases} x - 2z = 1 \\ 2x + 3y = -3 \\ 4x - 3y - 4z = 3 \end{cases}$ **40.** $\begin{cases} x + 2y - z = 2 \\ 2x - 2y + z = -1 \\ 6x + 4y + 3z = 5 \end{cases}$

41. $\begin{cases} x - y + z = 0 \\ x - y - 5z - 6 = 0 \\ 2x - 2y + z - 1 = 0 \end{cases}$ **42.** $\begin{cases} 4x - 3y + 5z = 0 \\ 2x + 4y - 3z = 0 \\ 6x + 2y + z = 0 \end{cases}$ **43.** $\begin{cases} x - y - z - t = 1 \\ 2x + y + z + 2t = 3 \\ x - 2y - 2z - 3t = 0 \\ 3x - 4y + z + 5t = -3 \end{cases}$

44. $\begin{cases} x - 3y + 3z - t = 4 \\ x + 2y - z = -3 \\ x + 3z + 2t = 3 \\ x + y + 5z = 6 \end{cases}$

In Problems 45–50, find the value of each determinant.

45. $\begin{vmatrix} 3 & 4 \\ 1 & 3 \end{vmatrix}$ **46.** $\begin{vmatrix} -4 & 0 \\ 1 & 3 \end{vmatrix}$ **47.** $\begin{vmatrix} 1 & 4 & 0 \\ -1 & 2 & 6 \\ 4 & 1 & 3 \end{vmatrix}$ **48.** $\begin{vmatrix} 2 & 3 & 10 \\ 0 & 1 & 5 \\ -1 & 2 & 3 \end{vmatrix}$ **49.** $\begin{vmatrix} 2 & 1 & -3 \\ 5 & 0 & 1 \\ 2 & 6 & 0 \end{vmatrix}$ **50.** $\begin{vmatrix} -2 & 1 & 0 \\ 1 & 2 & 3 \\ -1 & 4 & 2 \end{vmatrix}$

In Problems 51–56, use Cramer's Rule, if applicable, to solve each system.

51. $\begin{cases} x - 2y = 4 \\ 3x + 2y = 4 \end{cases}$ **52.** $\begin{cases} x - 3y = -5 \\ 2x + 3y = 5 \end{cases}$ **53.** $\begin{cases} 2x + 3y - 13 = 0 \\ 3x - 2y = 0 \end{cases}$

54. $\begin{cases} 3x - 4y - 12 = 0 \\ 5x + 2y + 6 = 0 \end{cases}$ **55.** $\begin{cases} x + 2y - z = 6 \\ 2x - y + 3z = -13 \\ 3x - 2y + 3z = -16 \end{cases}$ **56.** $\begin{cases} x - y + z = 8 \\ 2x + 3y - z = -2 \\ 3x - y - 9z = 9 \end{cases}$

In Problems 57–66, write the partial fraction decomposition of each rational expression.

57. $\dfrac{6}{x(x-4)}$

58. $\dfrac{x}{(x+2)(x-3)}$

59. $\dfrac{x-4}{x^2(x-1)}$

60. $\dfrac{2x-6}{(x-2)^2(x-1)}$

61. $\dfrac{x}{(x^2+9)(x+1)}$

62. $\dfrac{3x}{(x-2)(x^2+1)}$

63. $\dfrac{x^3}{(x^2+4)^2}$

64. $\dfrac{x^3+1}{(x^2+16)^2}$

65. $\dfrac{x^2}{(x^2+1)(x^2-1)}$

66. $\dfrac{4}{(x^2+4)(x^2-1)}$

In Problems 67–76, solve each system of equations.

67. $\begin{cases} 2x+y+3=0 \\ x^2+y^2=5 \end{cases}$

68. $\begin{cases} x^2+y^2=16 \\ 2x-y^2=-8 \end{cases}$

69. $\begin{cases} 2xy+y^2=10 \\ 3y^2-xy=2 \end{cases}$

70. $\begin{cases} 3x^2-y^2=1 \\ 7x^2-2y^2-5=0 \end{cases}$

71. $\begin{cases} x^2+y^2=6y \\ x^2=3y \end{cases}$

72. $\begin{cases} 2x^2+y^2=9 \\ x^2+y^2=9 \end{cases}$

73. $\begin{cases} 3x^2+4xy+5y^2=8 \\ x^2+3xy+2y^2=0 \end{cases}$

74. $\begin{cases} 3x^2+2xy-2y^2=6 \\ xy-2y^2+4=0 \end{cases}$

75. $\begin{cases} x^2-3x+y^2+y=-2 \\ \dfrac{x^2-x}{y}+y+1=0 \end{cases}$

76. $\begin{cases} x^2+x+y^2=y+2 \\ x+1=\dfrac{2-y}{x} \end{cases}$

In Problems 77–82, graph each system of inequalities. Tell whether the graph is bounded or unbounded, and label the corner points.

77. $\begin{cases} -2x+y\le 2 \\ x+y\ge 2 \end{cases}$

78. $\begin{cases} x-2y\le 6 \\ 2x+y\ge 2 \end{cases}$

79. $\begin{cases} x\ge 0 \\ y\ge 0 \\ x+y\le 4 \\ 2x+3y\le 6 \end{cases}$

80. $\begin{cases} x\ge 0 \\ y\ge 0 \\ 3x+y\ge 6 \\ 2x+y\ge 2 \end{cases}$

81. $\begin{cases} x\ge 0 \\ y\ge 0 \\ 2x+y\le 8 \\ x+2y\ge 2 \end{cases}$

82. $\begin{cases} x\ge 0 \\ y\ge 0 \\ 3x+y\le 9 \\ 2x+3y\ge 6 \end{cases}$

In Problems 83–86, graph each system of inequalities.

83. $\begin{cases} x^2+y^2\le 16 \\ x+y\ge 2 \end{cases}$

84. $\begin{cases} y^2\le x-1 \\ x-y\le 3 \end{cases}$

85. $\begin{cases} y\le x^2 \\ xy\le 4 \end{cases}$

86. $\begin{cases} x^2+y^2\ge 1 \\ x^2+y^2\le 4 \end{cases}$

In Problems 87–92, solve each linear programming problem.

87. Maximize $z=3x+4y$ subject to $x\ge 0,\ y\ge 0,\ 3x+2y\ge 6,\ x+y\le 8$

88. Maximize $z=2x+4y$ subject to $x\ge 0,\ y\ge 0,\ x+y\le 6,\ x\ge 2$

89. Minimize $z=3x+5y$ subject to $x\ge 0,\ y\ge 0,\ x+y\ge 1,\ 3x+2y\le 12,\ x+3y\le 12$

90. Minimize $z=3x+y$ subject to $x\ge 0,\ y\ge 0,\ x\le 8,\ y\le 6,\ 2x+y\ge 4$

91. Maximize $z=5x+4y$ subject to $x\ge 0,\ y\ge 0,\ x+2y\ge 2,\ 3x+4y\le 12,\ y\ge x$

92. Maximize $z=4x+5y$ subject to $x\ge 0,\ y\ge 0,\ 2x+3y\ge 6,\ x\ge y,\ 2x+y\le 12$

93. Find A so that the system of equations has infinitely many solutions.
$$\begin{cases} 2x + 5y = 5 \\ 4x + 10y = A \end{cases}$$

94. Find A so that the system in Problem 93 is inconsistent.

95. Curve Fitting Find the quadratic function $y = ax^2 + bx + c$ that passes through the three points $(0, 1)$, $(1, 0)$, and $(-2, 1)$.

96. Curve Fitting Find the general equation of the circle that passes through the three points $(0, 1)$, $(1, 0)$, and $(-2, 1)$.

[**Hint:** The general equation of a circle is $x^2 + y^2 + Dx + Ey + F = 0$.]

97. Blending Coffee A coffee distributor is blending a new coffee that will cost $3.90 per pound. It will consist of a blend of $3.00 per pound coffee and $6.00 per pound coffee. What amounts of each type of coffee should be mixed to achieve the desired blend?

[**Hint:** Assume that the weight of the blended coffee is 100 pounds].

$3.00/lb $3.90/lb $6.00/lb

98. Farming A 1000 acre farm in Illinois is used to raise corn and soy beans. The cost per acre for raising corn is $65, and the cost per acre for soybeans is $45. If $54,325 has been budgeted for costs and all the acreage is to be used, how many acres should be allocated for each crop?

99. Cookie Orders A cookie company makes three kinds of cookies, oatmeal raisin, chocolate chip, and shortbread, packaged in small, medium, and large boxes. The small box contains 1 dozen oatmeal raisin and 1 dozen chocolate chip; the medium box has 2 dozen oatmeal raisin, 1 dozen chocolate chip, and 1 dozen shortbread; the large box contains 2 dozen oat-

meal raisin, 2 dozen chocolate chip, and 3 dozen shortbread. If you require exactly 15 dozen oatmeal raisin, 10 dozen chocolate chip, and 11 dozen shortbread, how many of each size box should you buy?

100. Mixed Nuts A store that specializes in selling nuts has 72 pounds of cashews and 120 pounds of peanuts available. These are to be mixed in 12-ounce packages as follows: a lower-priced package containing 8 ounces of peanuts and 4 ounces of cashews and a quality package containing 6 ounces of peanuts and 6 ounces of cashews.

(a) Use x to denote the number of lower-priced packages and use y to denote the number of quality packages. Write a system of linear inequalities that describes the possible number of each kind of package.

(b) Graph the system and label the corner points.

101. A small rectangular lot has a perimeter of 68 feet. If its diagonal is 26 feet, what are the dimensions of the lot?

102. The area of a rectangular window is 4 square feet. If the diagonal measures $2\sqrt{2}$ feet, what are the dimensions of the window?

103. Geometry A certain right triangle has a perimeter of 14 inches. If the hypotenuse is 6 inches long, what are the lengths of the legs?

104. Geometry A certain isosceles triangle has a perimeter of 18 inches. If the altitude is 6 inches, what is the length of the base?

105. Building a Fence How much fence is required to enclose 5000 square feet by two squares whose sides are in the ratio of 1:2?

106. Mixing Acids A chemistry laboratory has three containers of hydrochloric acid, HCl. One container holds a solution with a concentration of 10% HCl, the second holds 25% HCl, and the third holds 40% HCl. How many liters of each should be mixed to obtain 100 liters of a solution with a concentration of 30% HCl? Construct a table showing some of the possible combinations.

107. Calculating Allowances Katy, Mike, Danny, and Colleen agreed to do yard work at home for $45 to be split among them. After they finished, their father determined that Mike deserves twice what Katy gets, Katy and Colleen deserve the same amount, and Danny deserves half of what Katy gets. How much does each one receive?

108. Finding the Speed of the Jet Stream On a flight between Midway Airport in Chicago and Ft. Lauderdale, Florida, a Boeing 737 jet maintains an airspeed of 475 miles per hour. If the trip from Chicago to Ft. Lauderdale takes 2 hours, 30 minutes and the return flight takes 2 hours, 50 minutes, what is the speed of the jet stream? (Assume that the speed of the jet stream remains constant at the various altitudes of the plane.)

109. Constant Rate Jobs If Bruce and Bryce work together for 1 hour and 20 minutes, they will finish a certain job. If Bryce and Marty work together for 1 hour and 36 minutes, the same job can be finished. If Marty and Bruce work together, they can complete this job in 2 hours and 40 minutes. How long will it take each of them working alone to finish the job?

110. Maximizing Profit on Figurines A factory manufactures two kinds of ceramic figurines: a dancing girl and a mermaid, each requiring three processes—molding, painting, and glazing. The daily labor available for molding is no more than 90 work-hours, labor available for painting does not exceed 120 work-hours, and labor available for glazing is no more than 60 work-hours. The dancing girl requires 3 work-hours for molding, 6 work-hours for painting, and 2 work-hours for glazing. The mermaid re-

quires 3 work-hours for molding, 4 work-hours for painting, and 3 work-hours for glazing. If the profit on each figurine is $25 for dancing girls and $30 for mermaids, how many of each should be produced each day to maximize profit? If management decides to produce the number of each figurine that maximizes profit, determine which of these processes has excess work-hours assigned to it.

111. Minimizing Production Cost A factory produces gasoline engines and diesel engines. Each week the factory is obligated to deliver at least 20 gasoline engines and at least 15 diesel engines. Due to physical limitations, however, the factory cannot make more than 60 gasoline engines nor more than 40 diesel engines. Finally, to prevent layoffs, a total of at least 50 engines must be produced. If gasoline engines cost $450 each to produce and diesel engines cost $550 each to produce, how many of each should be produced per week to minimize the cost? What is the excess capacity of the factory; that is, how many of each kind of engine is being produced in excess of the number that the factory is obligated to deliver?

 112. Describe four ways of solving a system of three linear equations containing three variables. Which method do you prefer? Why?

CHAPTER

9

Sequences; Induction; Counting; Probability

Bicycle riding has grown in popularity in the past few years. With the increase of riders also comes an increase of bicycle accidents. On the following page you are placed in the position of being a law student and having to argue a case regarding helmet safety. You will need to use the Sullivan website at

www.prenhall.com/sullivan

to find the relevant data needed to base your case upon.

PREPARING FOR THIS CHAPTER

Before getting started on this chapter, review the following concept:

For Section 9.3: Compound Interest *(Section 6.6)*

```
Netsite: http://www.prenhall.com/sullivan/
```

BICYCLE HELMET SAFETY

Should wearing a bicycle helmet be required by law or would that violate an individual's civil liberty? As a law student you are required to analyze this point of law and defend your position in the people's court. In order to base your case on relevant data you decide to do a statistical analysis of the data on helmet safety.

1. On the radio you hear "If every child wore a helmet, one brain injury could be prevented every five minutes and one death every day." Is this statement true? Should a bicycler be required to wear a helmet? Is this an issue of a civil liberty, as some say? What do you think justifies a mandatory helmet law?
2. Before you analyze the data, do you think that your bias will cause you to be selective in the data that you accept as valid? How can you keep bias from entering into your evaluation?
3. By contrast, do you think that the arguments for bicycle helmets based on the data also apply for motorcycles? Can you find any examples of bias affecting judgment as you look at the apparently contradictory data offered by the *AAOS* and *Opposing Facts*?
4. Review the *statistical data* from various sources. Determine the data sources that you believe are most reliable.
5. Compare the data for bicycles to that for motorcycles. Is there any correlation in the data?
6. What is the probability that you will be involved in a bicycle accident if you ride your bike about two hours per day?
7. Using your most reliable sources of data on bicycle fatalities, could one death every day be prevented?

This chapter introduces topics that are covered in more detail in courses titled *Finite Mathematics* or *Discrete Mathematics*. Applications of these topics can be found in the fields of computer science, engineering, business and economics, the social sciences, and the physical and biological sciences.

The chapter may be divided into four independent parts:

Sections 9.1–9.3, Sequences, which are functions whose domain is the set of natural numbers. Sequences form the basis for the *recursively defined functions* and *recursive procedures* used in computer programming.

Section 9.4, Mathematical Induction, a technique for proving theorems involving the natural numbers.

Section 9.5, the Binomial Theorem, a formula for the expansion of $(x + a)^n$, where n is any natural number.

Sections 9.6–9.8, Counting and Probability. The first two sections deal with techniques and formulas for counting the number of objects in a set, a part of the branch of mathematics called *combinatorics*. These formulas are used in computer science to analyze algorithms and recursive functions and to study stacks and queues. They are also used to determine *probabilities,* the likelihood that a certain outcome of a random experiment will occur.

9.1 | SEQUENCES

> 1 Write the First Several Terms of a Sequence
> 2 Write the Terms of a Sequence Defined by a Recursion Formula
> 3 Find the Sum of a Sequence; Use Summation Notation

> A **sequence** is a function whose domain is the set of positive integers.

Because a sequence is a function, it will have a graph. In Figure 1(a), you will recognize the graph of the function $f(x) = 1/x, x > 0$. If all the points on this graph were removed except those whose x-coordinates are positive integers—that is, if all points were removed except $(1, 1), (2, \frac{1}{2}), (3, \frac{1}{3})$, and so on—the remaining points would be the graph of the sequence $f(n) = 1/n$, as shown in Figure 1(b).

FIGURE 1

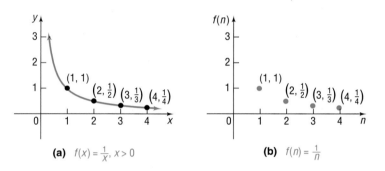

(a) $f(x) = \frac{1}{x}, x > 0$

(b) $f(n) = \frac{1}{n}$

A sequence is usually represented by listing its values in order. For example, the sequence whose graph is given in Figure 1(b) might be represented as

$$f(1), f(2), f(3), f(4), \ldots \quad \text{or} \quad 1, \frac{1}{2}, \frac{1}{3}, \frac{1}{4}, \ldots$$

The list never ends, as the ellipsis dots indicate. The numbers in this ordered list are called the **terms** of the sequence.

In dealing with sequences, we usually use subscripted letters; for example, a_1 to represent the first term, a_2 for the second term, a_3 for the third term, and so on. Thus, for the sequence $f(n) = 1/n$, we write

$$a_1 = f(1) = 1 \quad a_2 = f(2) = \frac{1}{2} \quad a_3 = f(3) = \frac{1}{3} \quad a_4 = f(4) = \frac{1}{4} \ldots a_n = f(n) = \frac{1}{n} \ldots$$

In other words, we usually do not use the traditional function notation $f(n)$ for sequences. For this particular sequence, we have a rule for the nth term, namely, $a_n = 1/n$, so it is easy to find any term of the sequence.

When a formula for the nth term of a sequence is known, rather than write out the terms of the sequence, we usually represent the entire sequence by placing braces around the formula for the nth term. For example, the sequence whose nth term is $b_n = \left(\frac{1}{2}\right)^n$ may be represented as

$$\{b_n\} = \left\{\left(\frac{1}{2}\right)^n\right\}$$

or by

$$b_1 = \frac{1}{2} \quad b_2 = \frac{1}{4} \quad b_3 = \frac{1}{8} \quad \cdots \quad b_n = \left(\frac{1}{2}\right)^n \cdots$$

EXAMPLE 1

FIGURE 2

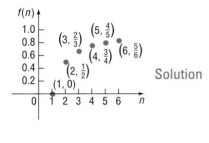

Writing the First Several Terms of a Sequence

Write down the first six terms of the following sequence and graph it.

$$\{a_n\} = \left\{\frac{n-1}{n}\right\}$$

Solution

$$a_1 = 0 \quad a_2 = \frac{1}{2} \quad a_3 = \frac{2}{3} \quad a_4 = \frac{3}{4} \quad a_5 = \frac{4}{5} \quad a_6 = \frac{5}{6}$$

See Figure 2.

Comment Graphing utilities can be used to write the terms of a sequence and graph them. Figure 3 shows the sequence given in Example 1 generated on a TI-83 graphing calculator. We can see the first few terms of the sequence on the viewing window. You need to press the right arrow key to scroll right in order to see the remaining terms of the sequence. Figure 4 shows a graph of the sequence. Notice the first term of the sequence is not visible since it lies on the x-axis.

FIGURE 3

FIGURE 4

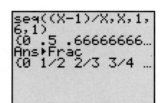

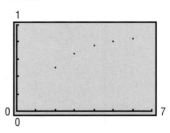

EXAMPLE 2

FIGURE 5

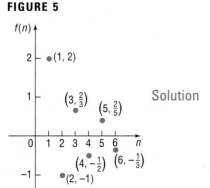

Writing the First Several Terms of a Sequence

Write down the first six terms of the following sequence and graph it.

$$\{b_n\} = \left\{(-1)^{n-1}\left(\frac{2}{n}\right)\right\}$$

Solution

$$b_1 = 2 \quad b_2 = -1 \quad b_3 = \frac{2}{3} \quad b_4 = -\frac{1}{2} \quad b_5 = \frac{2}{5} \quad b_6 = -\frac{1}{3}$$

See Figure 5.

E X A M P L E   3

FIGURE 6

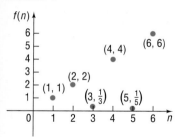

Writing the First Several Terms of a Sequence

Write down the first six terms of the following sequence and graph it.

$$\{c_n\} = \begin{cases} n & \text{if } n \text{ is even} \\ 1/n & \text{if } n \text{ is odd} \end{cases}$$

Solution $c_1 = 1 \quad c_2 = 2 \quad c_3 = \dfrac{1}{3} \quad c_4 = 4 \quad c_5 = \dfrac{1}{5} \quad c_6 = 6$

See Figure 6.

Now work Problems 3 and 5.

Sometimes a sequence is indicated by an observed pattern in the first few terms that makes it possible to infer the makeup of the nth term. In the example that follows, a sufficient number of terms of the sequence is given so that a natural choice for the nth term is suggested.

E X A M P L E   4

Determining a Sequence from a Pattern

(a) $e, \dfrac{e^2}{2}, \dfrac{e^3}{3}, \dfrac{e^4}{4}, \ldots \qquad a_n = \dfrac{e^n}{n}$

(b) $1, \dfrac{1}{3}, \dfrac{1}{9}, \dfrac{1}{27}, \ldots \qquad b_n = \dfrac{1}{3^{n-1}}$

(c) $1, 3, 5, 7, \ldots \qquad c_n = 2n - 1$

(d) $1, 4, 9, 16, 25, \ldots \qquad d_n = n^2$

(e) $1, -\dfrac{1}{2}, \dfrac{1}{3}, -\dfrac{1}{4}, \dfrac{1}{5}, \ldots \qquad e_n = (-1)^{n+1}\left(\dfrac{1}{n}\right)$

Notice in the sequence $\{e_n\}$ in Example 4(e) that the signs of the terms **alternate**. When this occurs, we use factors such as $(-1)^{n+1}$, which equals 1 if n is odd and -1 if n is even, or $(-1)^n$, which equals -1 if n is odd and 1 if n is even.

Now work Problem 13.

The Factorial Symbol

If $n \geq 0$ is an integer, the **factorial symbol** $n!$ is defined as follows:

$$0! = 1 \qquad 1! = 1$$
$$n! = n(n-1) \cdot \ldots \cdot 3 \cdot 2 \cdot 1 \qquad \text{if } n \geq 2$$

For example, $2! = 2 \cdot 1 = 2$, $3! = 3 \cdot 2 \cdot 1 = 6$, $4! = 4 \cdot 3 \cdot 2 \cdot 1 = 24$, and so on. Table 1 lists the values of $n!$ for $0 \le n \le 6$.

TABLE 1							
n	0	1	2	3	4	5	6
$n!$	1	1	2	6	24	120	720

Because

$$n! = n\underbrace{(n-1)(n-2) \cdot \ldots \cdot 3 \cdot 2 \cdot 1}_{(n-1)!}$$

we can use the formula

$$n! = n(n-1)!$$

to find successive factorials. For example, because $6! = 720$, we have

$$7! = 7 \cdot 6! = 7(720) = 5040$$

and

$$8! = 8 \cdot 7! = 8(5040) = 40{,}320$$

> **Comment** Your calculator may have a factorial key. Use it to see how fast factorials increase in value. Find the value of 69!. What happens when you try to find 70!? In fact, 70! is larger than 10^{100} (a *googol*), the largest number most calculators can display. ■

Recursion Formulas

2 A second way of defining a sequence is to assign a value to the first (or the first few) terms and specify the nth term by a formula or equation that involves one or more of the terms preceding it. Sequences defined this way are said to be defined **recursively,** and the rule or formula is called a **recursive formula.**

E X A M P L E 5   Writing the Terms of a Recursively Defined Sequence

Write down the first five terms of the following recursively defined sequence.

$$s_1 = 1, \qquad s_n = 4s_{n-1}$$

Solution   The first term is given as $s_1 = 1$. To get the second term, we use $n = 2$ in the formula to get $s_2 = 4s_1 = 4 \cdot 1 = 4$. To get the third term, we use $n = 3$ in the formula to get $s_3 = 4s_2 = 4 \cdot 4 = 16$. To get a new term requires that we know the value of the preceding term. The first five terms are

$$s_1 = 1$$
$$s_2 = 4 \cdot 1 = 4$$
$$s_3 = 4 \cdot 4 = 16$$
$$s_4 = 4 \cdot 16 = 64$$
$$s_5 = 4 \cdot 64 = 256$$

■

E X A M P L E   6

Writing the Terms of a Recursively Defined Sequence

Write down the first five terms of the following recursively defined sequence.

$$u_1 = 1, \qquad u_2 = 1, \qquad u_{n+2} = u_n + u_{n+1}$$

Solution   We are given the first two terms. To get the third term requires that we know each of the previous two terms. Thus,

$$u_1 = 1$$
$$u_2 = 1$$
$$u_3 = u_1 + u_2 = 2$$
$$u_4 = u_2 + u_3 = 1 + 2 = 3$$
$$u_5 = u_3 + u_4 = 2 + 3 = 5$$

The sequence defined in Example 6 is called a **Fibonacci sequence,** and the terms of this sequence are called **Fibonacci numbers.** These numbers appear in a wide variety of applications (see Problems 67 and 68).

E X A M P L E   7

Writing the Terms of a Recursively Defined Sequence

Write down the first five terms of the following recursively defined sequence.

$$f_1 = 1, \qquad f_{n+1} = (n + 1)f_n$$

Solution   Here

$$f_1 = 1$$
$$f_2 = 2f_1 = 2 \cdot 1 = 2$$
$$f_3 = 3f_2 = 3 \cdot 2 = 6$$
$$f_4 = 4f_3 = 4 \cdot 6 = 24$$
$$f_5 = 5f_4 = 5 \cdot 24 = 120$$

You should recognize the nth term of the sequence in Example 7 as $n!$.

Now work Problems 21 and 29.

Adding the First n Terms of a Sequence; Summation Notation

   It is often important to be able to find the sum of the first n terms of a sequence $\{a_n\}$, namely,

$$a_1 + a_2 + a_3 + \cdots + a_n \qquad (1)$$

Rather than write down all these terms, we introduce a more concise way to express the sum, called **summation notation.** Using summation notation, we would write the sum (1) as

$$a_1 + a_2 + a_3 + \cdots + a_n = \sum_{k=1}^{n} a_k$$

The symbol Σ (a stylized version of the Greek letter sigma, which is an S in our alphabet) is simply an instruction to sum, or add up, the terms. The integer k is called the **index** of the sum; it tells you where to start the sum and where to end it. Therefore, the expression

$$\sum_{k=1}^{n} a_k \qquad (2)$$

is an instruction to add the terms a_k of the sequence $\{a_n\}$ from $k = 1$ through $k = n$. We read expression (2) as "the sum of a_k from $k = 1$ to $k = n$."

Next, we list some properties of sequences using summation notation.

Theorem   Properties of Sequences

If $\{a_n\}$ and $\{b_n\}$ are two sequences and c is a real number, then

1. $\displaystyle\sum_{k=1}^{n} c = c \cdot n$

2. $\displaystyle\sum_{k=1}^{n} ca_k = c \sum_{k=1}^{n} a_k$

3. $\displaystyle\sum_{k=1}^{n} (a_k + b_k) = \sum_{k=1}^{n} a_k + \sum_{k=1}^{n} b_k$

4. $\displaystyle\sum_{k=1}^{n} (a_k - b_k) = \sum_{k=1}^{n} a_k - \sum_{k=1}^{n} b_k$

5. $\displaystyle\sum_{k=1}^{n} a_k = \sum_{k=1}^{j} a_k + \sum_{k=j+1}^{n} a_k, \quad$ when $1 < j < n$

Although we shall not prove these properties, the proofs are based on properties of real numbers.

E X A M P L E   8   Finding the Sum of a Sequence

Find the sum of each sequence.

(a) $\displaystyle\sum_{k=1}^{5} 3k$ (b) $\displaystyle\sum_{k=1}^{4} (k^2 - 7k + 2)$

Solution   (a) $\displaystyle\sum_{k=1}^{5} 3k = 3 \sum_{k=1}^{5} k = 3(1 + 2 + 3 + 4 + 5) = 3(15) = 45$

                ↑

      Property 2

(b) $\displaystyle\sum_{k=1}^{4} (k^2 - 7k + 2) = \sum_{k=1}^{4} k^2 - \sum_{k=1}^{4} 7k + \sum_{k=1}^{4} 2$ Properties 3, 4

$$= \sum_{k=1}^{4} k^2 - 7\sum_{k=1}^{4} k + \sum_{k=1}^{4} 2 \qquad \text{Property 2}$$

$$= (1^2 + 2^2 + 3^2 + 4^2) - 7(1 + 2 + 3 + 4) + 2(4)$$

$$= (1 + 4 + 9 + 16) - 7(10) + 8 \qquad \text{Property 1}$$

$$= 30 - 70 + 8$$

$$= -32$$

Comment A graphing utility can be used to find the sum of a sequence. Figure 7 shows the solution using a TI-83 graphing calculator.

FIGURE 7

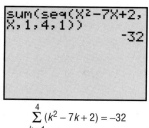

$$\sum_{k=1}^{5} 3k = 45$$

(a)

$$\sum_{k=1}^{4} (k^2 - 7k + 2) = -32$$

(b)

 Now work Problem 39.

E X A M P L E   9   Expanding Summation Notation

Write out each sum.

(a) $\displaystyle\sum_{k=1}^{n} \frac{1}{k}$ (b) $\displaystyle\sum_{k=1}^{n} k!$

Solution (a) $\displaystyle\sum_{k=1}^{n} \frac{1}{k} = \frac{1}{1} + \frac{1}{2} + \frac{1}{3} + \cdots + \frac{1}{n}$ (b) $\displaystyle\sum_{k=1}^{n} k! = 1! + 2! + \cdots + n!$

E X A M P L E   10   Writing a Sum in Summation Notation

Express each sum using summation notation.

(a) $1^2 + 2^2 + 3^2 + \cdots + 9^2$ (b) $1 + \dfrac{1}{2} + \dfrac{1}{4} + \dfrac{1}{8} + \cdots + \dfrac{1}{2^{n-1}}$

Solution (a) The sum $1^2 + 2^2 + 3^2 + \cdots + 9^2$ has 9 terms, each of the form k^2, and starts at $k = 1$ and ends at $k = 9$. Thus,

$$1^2 + 2^2 + 3^2 + \cdots + 9^2 = \sum_{k=1}^{9} k^2$$

(b) The sum

$$1 + \frac{1}{2} + \frac{1}{4} + \frac{1}{8} + \cdots + \frac{1}{2^{n-1}}$$

has n terms, each of the form $1/2^{k-1}$, and starts at $k = 1$ and ends at $k = n$. Thus,

$$1 + \frac{1}{2} + \frac{1}{4} + \frac{1}{8} + \cdots + \frac{1}{2^{n-1}} = \sum_{k=1}^{n} \frac{1}{2^{k-1}}$$

The index of summation need not always begin at 1 or end at n; for example,

$$\sum_{k=0}^{n-1} \frac{1}{2^k} = 1 + \frac{1}{2} + \frac{1}{4} + \cdots + \frac{1}{2^{n-1}}$$

Letters other than k may be used as the index. For example,

$$\sum_{j=1}^{n} j! \quad \text{and} \quad \sum_{i=1}^{n} i!$$

each represent the same sum as the one given in Example 9(b).

 Now work Problems 49 and 59.

9.1 | EXERCISES

In Problems 1–12, write down the first five terms of each sequence.

1. $\{n\}$

2. $\{n^2 + 1\}$

3. $\left\{\dfrac{n}{n + 2}\right\}$

4. $\left\{\dfrac{2n + 1}{2n}\right\}$

5. $\{(-1)^{n+1} n^2\}$

6. $\left\{(-1)^{n-1}\left(\dfrac{n}{2n - 1}\right)\right\}$

7. $\left\{\dfrac{3^n}{2^n + 1}\right\}$

8. $\left\{\left(\dfrac{2}{3}\right)^n\right\}$

9. $\left\{\dfrac{(-1)^n}{(n + 1)(n + 2)}\right\}$

10. $\left\{\dfrac{3^n}{n}\right\}$

11. $\left\{\dfrac{n}{e^n}\right\}$

12. $\left\{\dfrac{n^2}{2^n}\right\}$

In Problems 13–20, the given pattern continues. Write down the nth term of each sequence suggested by the pattern.

13. $\dfrac{1}{2}, \dfrac{2}{3}, \dfrac{3}{4}, \dfrac{4}{5}, \ldots$

14. $\dfrac{1}{1 \cdot 2}, \dfrac{1}{2 \cdot 3}, \dfrac{1}{3 \cdot 4}, \dfrac{1}{4 \cdot 5}, \ldots$

15. $1, \dfrac{1}{2}, \dfrac{1}{4}, \dfrac{1}{8}, \ldots$

16. $\dfrac{2}{3}, \dfrac{4}{9}, \dfrac{8}{27}, \dfrac{16}{81}, \ldots$

17. $1, -1, 1, -1, 1, -1, \ldots$

18. $1, \dfrac{1}{2}, 3, \dfrac{1}{4}, 5, \dfrac{1}{6}, 7, \dfrac{1}{8}, \ldots$

19. $1, -2, 3, -4, 5, -6, \ldots$

20. $2, -4, 6, -8, 10, \ldots$

In Problems 21–34, a sequence is defined recursively. Write the first five terms.

21. $a_1 = 3; a_{n+1} = 2 + a_n$

22. $a_1 = 2; a_{n+1} = 5 - a_n$

23. $a_1 = -2; a_{n+1} = n + a_n$

24. $a_1 = 1; a_{n+1} = n - a_n$

25. $a_1 = 5; a_{n+1} = 2a_n$

26. $a_1 = 2; a_{n+1} = -a_n$

27. $a_1 = 3; a_{n+1} = \dfrac{a_n}{n}$

28. $a_1 = -2; a_{n+1} = n + 3a_n$

29. $a_1 = 1; a_2 = 2; a_{n+2} = a_n a_{n+1}$

30. $a_1 = -1; a_2 = 1; a_{n+2} = a_{n+1} + na_n$

31. $a_1 = A; a_{n+1} = a_n + d$

32. $a_1 = A; a_{n+1} = ra_n; r \neq 0$

33. $a_1 = \sqrt{2}; a_{n+1} = \sqrt{2 + a_n}$

34. $a_1 = \sqrt{2}; a_{n+1} = \sqrt{a_n/2}$

In Problems 35–46, find the sum of each sequence.

35. $\sum_{k=1}^{10} 5$

36. $\sum_{k=1}^{20} 8$

37. $\sum_{k=1}^{6} k$

38. $\sum_{k=1}^{4} (-k)$

39. $\sum_{k=1}^{5} (5k + 3)$

40. $\sum_{k=1}^{6} (3k - 7)$

41. $\sum_{k=1}^{3} (k^2 + 4)$

42. $\sum_{k=0}^{4} (k^2 - 4)$

43. $\sum_{k=1}^{6} (-1)^k 2^k$

44. $\sum_{k=1}^{4} (-1)^k 3^k$

45. $\sum_{k=1}^{4} (k^3 - 1)$

46. $\sum_{k=0}^{3} (k^3 + 2)$

In Problems 47–56, write out each sum.

47. $\sum_{k=1}^{n} (k + 3)$

48. $\sum_{k=1}^{n} (2k + 3)$

49. $\sum_{k=1}^{n} \frac{k^2}{2}$

50. $\sum_{k=1}^{n} (k + 1)^2$

51. $\sum_{k=0}^{n} \frac{1}{3^k}$

52. $\sum_{k=0}^{n} \left(\frac{3}{2}\right)^k$

53. $\sum_{k=0}^{n-1} \frac{1}{3^{k+1}}$

54. $\sum_{k=0}^{n-1} (2k + 1)$

55. $\sum_{k=2}^{n} (-1)^k \ln k$

56. $\sum_{k=3}^{n} (-1)^{k+1} 2^k$

In Problems 57–66, express each sum using summation notation.

57. $1 + 2 + 3 + \cdots + 20$

58. $1^3 + 2^3 + 3^3 + \cdots + 8^3$

59. $\dfrac{1}{2} + \dfrac{2}{3} + \dfrac{3}{4} + \cdots + \dfrac{13}{13 + 1}$

60. $1 + 3 + 5 + 7 + \cdots + [2(12) - 1]$

61. $1 - \dfrac{1}{3} + \dfrac{1}{9} - \dfrac{1}{27} + \cdots + (-1)^6 \left(\dfrac{1}{3^6}\right)$

62. $\dfrac{2}{3} - \dfrac{4}{9} + \dfrac{8}{27} - \cdots + (-1)^{11+1} \left(\dfrac{2}{3}\right)^{11}$

63. $3 + \dfrac{3^2}{2} + \dfrac{3^3}{3} + \cdots + \dfrac{3^n}{n}$

64. $\dfrac{1}{e} + \dfrac{2}{e^2} + \dfrac{3}{e^3} + \cdots + \dfrac{n}{e^n}$

65. $a + (a + d) + (a + 2d) + \cdots + (a + nd)$

66. $a + ar + ar^2 + \cdots + ar^{n-1}$

67. **Growth of a Rabbit Colony** A colony of rabbits begins with one pair of mature rabbits, which will produce a pair of offspring (one male, one female) each month. Assume that all rabbits mature in 1 month and produce a pair of offspring (one male, one female) after 2 months. If no rabbits ever die, how many pairs of mature rabbits are there after 7 months?
[**Hint:** A Fibonacci sequence models this colony. Do you see why?]

68. **Fibonacci Sequence** Let

$$u_n = \frac{(1 + \sqrt{5})^n - (1 - \sqrt{5})^n}{2^n \sqrt{5}}$$

define the nth term of a sequence.
(a) Show that $u_1 = 1$ and $u_2 = 1$.
(b) Show that $u_{n+2} = u_{n+1} + u_n$.
(c) Draw the conclusion that $\{u_n\}$ is a Fibonacci sequence.

1 mature pair

1 mature pair

2 mature pairs

3 mature pairs

In Problems 69 and 70, we use the fact that in some programming languages it is possible to have a function subroutine include a call to itself.

69. **Programming Exercise** Write a program that accepts an integer as input and prints the number and its factorial. Use a recursively defined function, that is, use a function subroutine that calls itself.

70. **Programming Exercise** Write a program that accepts a positive integer N as input and outputs the Nth Fibonacci number. Use a recursively defined subroutine.

71. **Pascal's Triangle** Divide the triangular array shown (called Pascal's triangle) using diagonal lines as indicated. Find the sum of the numbers in each of these diagonal rows. Do you recognize this sequence?

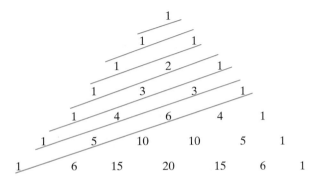

72. **Fibonacci Sequence** Use the result of Problem 68 to do the following problems.
    (a) Write the first 10 terms of the Fibonacci sequence.
    (b) Compute the ratio $\dfrac{u_{n+1}}{u_n}$ for the first 10 terms.
    (c) As n gets large, what number does the ratio approach? This number is referred to as the **golden ratio.** Rectangles whose sides are in this ratio were considered pleasing to the eye by the Greeks. For example, the facade of the Parthenon was constructed using the "golden ratio."
    (d) Compute the ratio $\dfrac{u_n}{u_{n+1}}$ for the first 10 terms.
    (e) As n gets large, what number does the ratio approach? This number is also referred to as the **golden ratio.** This ratio is believed to have been used in the construction of the Great Pyramid in Egypt. The ratio equals the sum of the areas of the four face triangles divided by the total surface area of the Great Pyramid.

73. Investigate various applications that lead to a Fibonacci sequence such as art, architecture, or financial markets. Write an essay on these applications.

9.2 ARITHMETIC SEQUENCES

1 Determine If a Sequence Is Arithmetic
2 Find a Formula for an Arithmetic Sequence
3 Find the Sum of an Arithmetic Sequence

1 When the difference between successive terms of a sequence is always the same number, the sequence is called **arithmetic.** Thus, an **arithmetic sequence*** may be defined recursively as $a_1 = a$, $a_{n+1} - a_n = d$, or as

$$a_1 = a, \qquad a_{n+1} = a_n + d \qquad (1)$$

where $a = a_1$ and d are real numbers. The number a is the first term, and the number d is called the **common difference.**

Thus, the terms of an arithmetic sequence with first term a and common difference d follow the pattern

$$a, \quad a + d, \quad a + 2d, \quad a + 3d, \quad \dots$$

*Sometimes called an **arithmetic progression.**

E X A M P L E   1   Determining If a Sequence Is Arithmetic

The sequence

$$4, 7, 10, 13, \ldots$$

is arithmetic since the difference of successive terms is 3. The first term is 4, and the common difference is 3.   ▬

E X A M P L E   2   Determining If a Sequence Is Arithmetic

Show that the following sequence is arithmetic. Find the first term and the common difference.

$$\{s_n\} = \{3n + 5\}$$

Solution   The first term is $s_1 = 3 \cdot 1 + 5 = 8$. The $(n + 1)$st and nth terms of the sequence $\{s_n\}$ are

$$s_{n+1} = 3(n + 1) + 5 = 3n + 8 \quad \text{and} \quad s_n = 3n + 5$$

Their difference is

$$s_{n+1} - s_n = (3n + 8) - (3n + 5) = 8 - 5 = 3$$

Since, the difference of two successive terms does not depend on n (it always equals 3), the sequence is arithmetic and the common difference is 3.   ▬

E X A M P L E   3   Determining If a Sequence Is Arithmetic

Show that the sequence $\{t_n\} = \{4 - n\}$ is arithmetic. Find the first term and the common difference.

Solution   The first term is $t_1 = 4 - 1 = 3$. The $(n + 1)$st and nth terms are

$$t_{n+1} = 4 - (n + 1) = 3 - n \quad \text{and} \quad t_n = 4 - n$$

Their difference is

$$t_{n+1} - t_n = (3 - n) - (4 - n) = 3 - 4 = -1$$

The difference of two successive terms does not depend on n; it always equals the same number, -1. Hence, $\{t_n\}$ is an arithmetic sequence whose common difference is -1.   ▬

Now work Problem 3.

Suppose that a is the first term of an arithmetic sequence whose common difference is d. We seek a formula for the nth term, a_n. To see the pattern, we write down the first few terms:

$$a_1 = a$$
$$a_2 = a_1 + d = a + 1 \cdot d$$
$$a_3 = a_2 + d = (a + d) + d = a + 2 \cdot d$$
$$a_4 = a_3 + d = (a + 2 \cdot d) + d = a + 3 \cdot d$$
$$a_5 = a_4 + d = (a + 3 \cdot d) + d = a + 4 \cdot d$$
$$\vdots$$
$$a_n = a_{n-1} + d = [a + (n - 2)d] + d = a + (n - 1)d$$

We are led to the following result:

> **Theorem**   *n*th Term of an Arithmetic Sequence
>
> For an arithmetic sequence $\{a_n\}$ whose first term is a and whose common difference is d, the *n*th term is determined by the formula
>
> $$a_n = a + (n-1)d \qquad (2)$$

E X A M P L E   4

Finding a Particular Term of an Arithmetic Sequence

Find the 13th term of the arithmetic sequence: $2, 6, 10, 14, 18, \ldots$.

Solution   The first term of this arithmetic sequence is $a = 2$, and the common difference is 4. By formula (2), the *n*th term is

$$a_n = 2 + (n-1)4$$

Hence, the 13th term is

$$a_{13} = 2 + 12 \cdot 4 = 50$$

 Exploration   Use a graphing utility to find the 13th term of the sequence given in Example 4. Use it to find the 20th term and the 50th term.

E X A M P L E   5

Finding a Recursive Formula for an Arithmetic Sequence

The 8th term of an arithmetic sequence is 75, and the 20th term is 39. Find the first term and the common difference. Give a recursive formula for the sequence.

Solution   By equation (2), we know that $a_n = a + (n-1)d$. As a result,

$$\begin{cases} a_8 = a + 7d = 75 \\ a_{20} = a + 19d = 39 \end{cases}$$

This is a system of two linear equations containing two variables, a and d, which we can solve by elimination. Thus, subtracting the second equation from the first equation, we get

$$-12d = 36$$
$$d = -3$$

With $d = -3$, we find $a = 75 - 7d = 75 - 7(-3) = 96$. The first term is $a = 96$ and the common difference is $d = -3$. A recursive formula for this sequence is found using (1).

$$a_1 = 96, \quad a_{n+1} = a_n - 3$$

Based on formula (2), a formula for the *n*th term of the sequence $\{a_n\}$ in Example 5 is

$$a_n = a + (n-1)d = 96 + (n-1)(-3) = 99 - 3n$$

 Now work Problems 19 and 25.

Adding the First *n* Terms of an Arithmetic Sequence

3   The next result gives a formula for finding the sum of the first *n* terms of an arithmetic sequence.

Theorem Sum of n Terms of an Arithmetic Sequence

Let $\{a_n\}$ be an arithmetic sequence with first term a and common difference d. The sum S_n of the first n terms of $\{a_n\}$ is

$$S_n = \frac{n}{2}[2a + (n-1)d] = \frac{n}{2}(a + a_n) \qquad (3)$$

Proof

$$
\begin{aligned}
S_n &= a_1 + a_2 + a_3 + \cdots + a_n && \text{Sum of first } n \text{ terms} \\
&= a + (a+d) + (a+2d) + \cdots + [a + (n-1)d] && \text{Formula (2)} \\
&= \underbrace{(a + a + \cdots + a)}_{n \text{ terms}} + [d + 2d + \cdots + (n-1)d] \\
&= na + d[1 + 2 + \cdots + (n-1)] \\
&= na + d\left[\frac{n}{2}(n-1)\right] && \text{See Problem 53} \\
&= na + \frac{n}{2}(n-1)d \\
&= \frac{n}{2}[2a + (n-1)d] && \text{Factor out } n/2 \qquad (4) \\
&= \frac{n}{2}[a + a + (n-1)d] \\
&= \frac{n}{2}(a + a_n) && \text{Formula (2)} \qquad (5)
\end{aligned}
$$

■

Formula (3) provides two ways to find the sum of the first n terms of an arithmetic sequence. Notice that one involves the first term and common difference (4), while the other involves the first term and the nth term (5). Use whichever form is easier.

E X A M P L E   6

Finding the Sum of n Terms of an Arithmetic Sequence

Find the sum S_n of the first n terms of the sequence $\{3n + 5\}$; that is, find

$$8 + 11 + 14 + \cdots + (3n + 5)$$

Solution The sequence $\{3n + 5\}$ is an arithmetic sequence with first term $a = 8$ and the nth term $(3n + 5)$. To find the sum S_n we use formula (3) [as given in (5)]:

$$S_n = \frac{n}{2}(a + a_n) = \frac{n}{2}[8 + (3n + 5)] = \frac{n}{2}(3n + 13)$$

■

Now work Problem 33.

E X A M P L E   7   Using a Graphing Utility to Find the Sum of 20 Terms of an Arithmetic Sequence

Use a graphing utility to find the sum S_n of the first 20 terms of the sequence $\{9.5n + 2.6\}$.

Solution   Figure 8 shows the results obtained using a TI-83 graphing calculator.

FIGURE 8

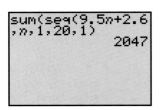

Thus, the sum of the first 20 terms of the sequence $\{9.5n + 2.6\}$ is 2047. ■

   Now work Problem 41.

E X A M P L E   8   Creating a Floor Design

A ceramic tile floor is designed in the shape of a trapezoid 20 feet wide at the base and 10 feet wide at the top. See Figure 9. The tiles, 12 inches by 12 inches, are to be placed so that each successive row contains one less tile than the row below. How many tiles will be required?

FIGURE 9

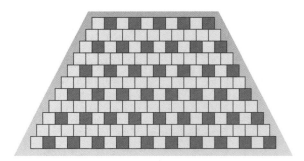

Solution   The bottom row requires 20 tiles and the top row 10 tiles. Since each successive row requires one less tile, the total number of tiles required is

$$S = 20 + 19 + 18 + \cdots + 11 + 10$$

This is the sum of an arithmetic sequence; the common difference is -1. The number of terms to be added is $n = 11$, with the first term $a = 20$ and the last term $a_{11} = 10$. The sum S is

$$S = \frac{n}{2}(a + a_{11}) = \frac{11}{2}(20 + 10) = 165$$

Thus, 165 tiles will be required. ■

9.2 | EXERCISES

In Problems 1–10, an arithmetic sequence is given. Find the common difference and write out the first four terms.

1. $\{n + 4\}$
2. $\{n - 5\}$
3. $\{2n - 5\}$
4. $\{3n + 1\}$
5. $\{6 - 2n\}$

6. $\{4 - 2n\}$
7. $\left\{\dfrac{1}{2} - \dfrac{1}{3}n\right\}$
8. $\left\{\dfrac{2}{3} + \dfrac{n}{4}\right\}$
9. $\{\ln 3^n\}$
10. $\{e^{\ln n}\}$

In Problems 11–18, find the nth term of the arithmetic sequence whose initial term a and common difference d are given. What is the fifth term?

11. $a = 3; d = 2$
12. $a = -2; d = 3$
13. $a = 5; d = -3$

14. $a = 6; d = -2$
15. $a = 0; d = \frac{1}{2}$
16. $a = 1; d = -\frac{1}{3}$

17. $a = \sqrt{2}; d = \sqrt{2}$
18. $a = 0; d = \pi$

In Problems 19–24, find the indicated term in each arithmetic sequence.

19. 11th term of $2, 4, 6, \ldots$
20. 10th term of $-1, 1, 3, \ldots$

21. 10th term of $1, -2, -5, \ldots$
22. 9th term of $5, 0, -5, \ldots$

23. 8th term of $a, a + b, a + 2b, \ldots$
24. 7th term of $2\sqrt{5}, 4\sqrt{5}, 6\sqrt{5}, \ldots$

In Problems 25–32, find the first term and the common difference of the arithmetic sequence described. Give a recursive formula for the sequence.

25. 8th term is 8; 21st term is 47
26. 3rd term is 1; 20th term is 35

27. 9th term is -5; 15th term is 31
28. 8th term is 4; 18th term is -96

29. 15th term is 0; 40th term is -50
30. 5th term is -2; 13th term is 30

31. 14th term is -1; 18th term is -9
32. 12th term is 4; 18th term is 28

In Problems 33–40, find the sum.

33. $1 + 3 + 5 + \cdots + (2n - 1)$
34. $2 + 4 + 6 + \cdots + 2n$

35. $7 + 12 + 17 + \cdots + (2 + 5n)$
36. $-1 + 3 + 7 + \cdots + (4n - 5)$

37. $2 + 4 + 6 + \cdots + 70$
38. $1 + 3 + 5 + \cdots + 59$

39. $5 + 9 + 13 + \cdots + 49$
40. $2 + 5 + 8 + \cdots + 41$

For Problems 41–46, use a graphing utility to find the sum of each sequence.

41. $\{3.45n + 4.12\}\ n = 20$
42. $\{2.67n - 1.23\}\ n = 25$

43. $2.8 + 5.2 + 7.6 + \cdots + 36.4$
44. $5.4 + 7.3 + 9.2 + \cdots + 32$

45. $4.9 + 7.48 + 10.06 + \cdots + 66.82$
46. $3.71 + 6.9 + 10.09 + \cdots + 80.27$

47. Find x so that $x + 3$, $2x + 1$, and $5x + 2$ are terms of an arithmetic sequence.

48. Find x so that $2x$, $3x + 2$, and $5x + 3$ are terms of an arithmetic sequence.

49. **Drury Lane Theater** The Drury Lane Theater has 25 seats in the first row and 30 rows in all. Each successive row contains one additional seat. How many seats are in the theater?

50. **Football Stadium** The corner section of a football stadium has 15 seats in the first row and 40 rows in all. Each successive row contains two additional seats. How many seats are in this section?

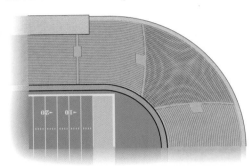

51. **Creating a Mosaic** A mosaic is designed in the shape of an equilateral triangle, 20 feet on each side. Each tile in the mosaic is in the shape of an equilateral triangle, 12 inches to a side. The tiles are to alternate in color as shown in the illustration. How many tiles of each color will be required?

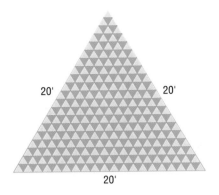

52. **Constructing a Brick Staircase** A brick staircase has a total of 30 steps. The bottom step requires 100 bricks. Each successive step requires two less bricks than the prior step.
   (a) How many bricks are required for the top step?
   (b) How many bricks are required to build the staircase?

53. $1 + 2 + \cdots + (n - 1) = \dfrac{n}{2}(n - 1)$ Let
$S = 1 + 2 + \cdots + (n - 1)$. Then

$$2S = [1 + 2 + \cdots + (n - 1)] + [(n - 1) + (n - 2) + \cdots + 1]$$
$$= \underbrace{[1 + (n - 1)] + [2 + (n - 2)] + \cdots + [(n - 1) + 1]}_{n - 1 \text{ terms in brackets}}$$

Now complete the derivation.

54. Make up an arithmetic sequence. Give it to a friend and ask for its 20th term.

9.3 | GEOMETRIC SEQUENCES; GEOMETRIC SERIES

1	Determine If a Sequence Is Geometric
2	Find a Formula for a Geometric Sequence
3	Find the Sum of a Geometric Sequence
4	Solve Annuity Problems
5	Find the Sum of a Geometric Series

1 When the ratio of successive terms of a sequence is always the same nonzero number, the sequence is called **geometric**. Thus, a **geometric sequence*** may be defined recursively as $a_1 = a$, $a_{n+1}/a_n = r$, or as

$$a_1 = a, \quad a_{n+1} = ra_n \qquad (1)$$

where $a_1 = a$ and $r \neq 0$ are real numbers. The number a is the first term, and the nonzero number r is called the **common ratio.**

*Sometimes called a **geometric progression**.

Thus, the terms of a geometric sequence with first term a and common ratio r follow the pattern

$$a, ar, ar^2, ar^3, \ldots$$

E X A M P L E   1

Determining If a Sequence Is Geometric

The sequence

$$2, 6, 18, 54, 162, \ldots$$

is geometric since the ratio of successive terms is 3. The first term is 2, and the common ratio is 3.   ▬

E X A M P L E   2

Determining If a Sequence Is Geometric

Show that the following sequence is geometric. Find the first term and the common ratio.

$$\{s_n\} = 2^{-n}$$

Solution   The first term is $s_1 = 2^{-1} = \frac{1}{2}$. The $(n + 1)$st and nth terms of the sequence $\{s_n\}$ are

$$s_{n+1} = 2^{-(n+1)} \quad \text{and} \quad s_n = 2^{-n}$$

Their ratio is

$$\frac{s_{n+1}}{s_n} = \frac{2^{-(n+1)}}{2^{-n}} = 2^{-n-1+n} = 2^{-1} = \frac{1}{2}$$

Because the ratio of successive terms is a nonzero number independent of n, the sequence $\{s_n\}$ is geometric with common ratio $\frac{1}{2}$.   ▬

E X A M P L E   3

Determining If a Sequence Is Geometric

Show that the following sequence is geometric. Find the first term and the common ratio.

$$\{t_n\} = \{4^n\}$$

Solution   The first term is $t_1 = 4^1 = 4$. The $(n + 1)$st and nth terms are

$$t_{n+1} = 4^{n+1} \quad \text{and} \quad t_n = 4^n$$

Their ratio is

$$\frac{t_{n+1}}{t_n} = \frac{4^{n+1}}{4^n} = 4$$

Thus, $\{t_n\}$ is a geometric sequence with common ratio 4.   ▬

Now work Problem 3.

Suppose a is the first term of a geometric sequence with common ratio $r \neq 0$. We seek a formula for the nth term a_n. To see the pattern, we write down the first few terms:

$$a_1 = 1 \cdot a = ar^0$$
$$a_2 = ra_1 = ar^1$$
$$a_3 = ra_2 = r(ar) = ar^2$$
$$a_4 = ra_3 = r(ar^2) = ar^3$$
$$a_5 = ra_4 = r(ar^3) = ar^4$$
$$\vdots$$
$$a_n = ra_{n-1} = r(ar^{n-2}) = ar^{n-1}a$$

We are led to the following result:

> **Theorem** nth Term of a Geometric Sequence
>
> For a geometric sequence $\{a_n\}$ whose first term is a and whose common ratio is r, the nth term is determined by the formula
>
> $$a_n = ar^{n-1}, \quad r \neq 0 \tag{2}$$

E X A M P L E   4   **Finding a Particular Term of a Geometric Sequence**

Find the 9th term of the geometric sequence $2, \frac{2}{3}, \frac{2}{9}, \frac{2}{27}, \ldots$.

Solution   The first term of this geometric sequence is $a = 2$, and the common ratio is $\frac{1}{3}$. (Use $\frac{2}{3}/2 = \frac{1}{3}$, or $\frac{2}{9}/\frac{2}{3} = \frac{1}{3}$, or any two successive terms.) By formula (2), the nth term is

$$a_n = 2\left(\frac{1}{3}\right)^{n-1}$$

Hence, the 9th term is

$$a_9 = 2\left(\frac{1}{3}\right)^8 = \frac{2}{3^8} = \frac{2}{6561} \approx 0.0003$$

 Exploration   Use a graphing utility to find the 9th term of the sequence given in Example 4. Use it to find the 20th term and the 50th term.

 Now work Problems 25 and 33.

Adding the First n Terms of a Geometric Sequence

 The next result gives us a formula for finding the sum of the first n terms of a geometric sequence.

> **Theorem** Sum of n Terms of a Geometric Sequence
>
> Let $\{a_n\}$ be a geometric sequence with first term a and common ratio r. The sum S_n of the first n terms of $\{a_n\}$ is
>
> $$S_n = a\frac{1 - r^n}{1 - r}, \quad r \neq 0, 1 \tag{3}$$

Proof

$$S_n = a + ar + \cdots + ar^{n-1} \tag{4}$$

Multiply each side by r to obtain

$$rS_n = ar + ar^2 + \cdots + ar^n \tag{5}$$

Now, subtract (5) from (4). The result is

$$S_n - rS_n = a - ar^n$$
$$(1 - r)S_n = a(1 - r^n)$$

Since $r \neq 1$, we can solve for S_n:

$$S_n = a\frac{1 - r^n}{1 - r}$$

E X A M P L E   5

Finding the Sum of n Terms of a Geometric Sequence

Find the sum S_n of the first n terms of the sequence $\{(\frac{1}{2})^n\}$; that is, find

$$\frac{1}{2} + \frac{1}{4} + \frac{1}{8} + \cdots + \left(\frac{1}{2}\right)^n$$

Solution   The sequence $\{(\frac{1}{2})^n\}$ is a geometric sequence with $a = \frac{1}{2}$ and $r = \frac{1}{2}$. The sum S_n that we seek is the sum of the first n terms of the sequence, so we use formula (3) to get

$$S_n = \sum_{k=1}^{n} \left(\frac{1}{2}\right)^k = \frac{1}{2} + \frac{1}{4} + \frac{1}{8} + \cdots + \left(\frac{1}{2}\right)^n$$

$$= \frac{1}{2}\left[\frac{1 - (\frac{1}{2})^n}{1 - \frac{1}{2}}\right]$$

$$= \frac{1}{2}\left[\frac{1 - (\frac{1}{2})^n}{\frac{1}{2}}\right]$$

$$= 1 - \left(\frac{1}{2}\right)^n$$

Now work Problem 39.

E X A M P L E   6

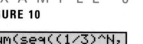

Using a Graphing Utility to Find the Sum of a Geometric Sequence

FIGURE 10

Use a graphing utility to find the sum S_n of the first 15 terms of the sequence $\left\{\left(\frac{1}{3}\right)^n\right\}$; that is, find

$$\frac{1}{3} + \frac{1}{9} + \frac{1}{27} + \cdots + \left(\frac{1}{3}\right)^{15}$$

Solution   Figure 10 shows the result obtained using a TI-83 graphing calculator.

Thus, the sum of the first 15 terms of the sequence $\left\{\left(\frac{1}{3}\right)^n\right\}$ is approximately 0.4999999652.

Now work Problem 45.

Annuities

In Section 6.6 we developed the compound interest formula which gives the future value when a fixed amount of money is deposited in an account that pays interest compounded periodically. Often, though, money is invested in small amounts at periodic intervals. An **annuity** is a sequence of equal periodic deposits. The periodic deposits may be made annually, quarterly, monthly, or daily.

When deposits are made at the same time the interest is credited, the annuity is called **ordinary.** We will only deal with ordinary annuities here. The **amount of an annuity** is the sum of all deposits made plus all interest paid.

Suppose the interest an account earns is i percent per payment period (expressed as a decimal). For example, if an account pays 12% compounded monthly (12 times a year) then $i = 0.12/12 = 0.01$. If an account pays 8% compounded quarterly (4 times a year) then $i = 0.08/4 = 0.02$. To develop a formula for the amount of an annuity, suppose P is deposited each payment period for n payment periods in an account that earns i percent per payment period. When the last deposit is made at the nth payment period, the first deposit of P has earned interest compounded for $n - 1$ payment periods, the second deposit of P has earned interest compounded for $n - 2$ payment periods, and so on. Table 2 shows the value of each deposit after n deposits have been made.

TABLE 2

Deposit	1	2	3	$\cdots$	$n - 1$	n
Amount	$P(1 + i)^{n-1}$	$P(1 + i)^{n-2}$	$P(1 + i)^{n-3}$	$\cdots$	$P(1 + i)$	P

The amount A of the annuity is the sum of the amounts shown in Table 2, namely,

$$A = P(1 + i)^{n-1} + P(1 + i)^{n-2} + \cdots + P(1 + i) + P$$
$$= P[1 + (1 + i) + \cdots + (1 + i)^{n-1}]$$

The expression in brackets is the sum of a geometric sequence with n terms and a common ratio of $(1 + i)$. As a result,

$$A = P[1 + (1 + i) + \cdots + (1 + i)^{n-2} + (1 + i)^{n-1}]$$
$$= P\frac{1 - (1 + i)^n}{1 - (1 + i)} = P\frac{1 - (1 + i)^n}{-i} = P\frac{(1 + i)^n - 1}{i}$$

We have established the following result.

Theorem   Amount of an Annuity

If P represents the deposit in dollars made at each payment period for an annuity at i percent interest per payment period, the amount A of the annuity after n payment periods is

$$A = P\frac{(1 + i)^n - 1}{i} \tag{6}$$

EXAMPLE 7   Determining the Amount of an Annuity

To save for retirement, Brett decides to place $2000 into an Individual Retirement Account (IRA) each year for the next 30 years. What will the value of the IRA be when Brett retires in 30 years if the rate of return of the IRA is assumed to be 10% per annum compounded annually?

Solution   This is an ordinary annuity with 30 annual deposits of $P = \$2000$. The rate of interest per payment period is $i = 0.10/1 = 0.10$. The number of payment periods is $n = 30$. The amount A of the annuity in 30 years is

$$A = 2000\left[\frac{(1 + 0.10)^{30} - 1}{0.10}\right] = \$2000(164.49402) = \$328,988.05$$

EXAMPLE 8   Determining the Amount of an Annuity

To save for his daughter's college education, Mr. McGowen decides to put $50 aside every month in a credit union account paying 10% interest compounded monthly. If he begins this savings program when his daughter is 3 years old, how much will he have saved by the time his daughter is 18 years old?

Solution   When his daughter is 18 years old, Mr. McGowen will have made his 180th payment (15 years × 12 payments per year). This is an annuity with $P = \$50$, $n = 180$, and $i = \frac{0.10}{12}$. The amount A saved is

$$A = 50\left[\frac{\left(1 + \dfrac{0.10}{12}\right)^{180} - 1}{\dfrac{0.10}{12}}\right] = \$50(414.4703) = \$20,723.52$$

Geometric Series

An infinite sum of the form

$$a + ar + ar^2 + \cdots + ar^{n-1} + \cdots$$

with first term a and common ratio r, is called an **infinite geometric series** and is denoted by

$$\sum_{k=1}^{\infty} ar^{k-1}$$

5   Based on formula (3), the sum S_n of the first n terms of a geometric series is

$$S_n = a\frac{1 - r^n}{1 - r} = \frac{a}{1 - r} - \frac{ar^n}{1 - r} \qquad (7)$$

If this finite sum S_n approaches a number L as $n \to \infty$, then we call L the **sum of the infinite geometric series,** and we write

$$L = \sum_{k=1}^{\infty} ar^{k-1}$$

Theorem Sum of an Infinite Geometric Series

If $|r| < 1$, the sum of the infinite geometric series $\sum_{k=1}^{\infty} ar^{k-1}$ is

$$\sum_{k=1}^{\infty} ar^{k-1} = \frac{a}{1-r} \tag{8}$$

Intuitive Proof Since $|r| < 1$, it follows that $|r^n|$ approaches 0 as $n \to \infty$. Then, based on formula (7), the sum S_n approaches $a/(1-r)$ as $n \to \infty$.

E X A M P L E 9

Finding the Sum of a Geometric Series

Find the sum of the geometric series $2 + \frac{4}{3} + \frac{8}{9} + \cdots$.

Solution   The first term is $a = 2$ and the common ratio is

$$r = \frac{\frac{4}{3}}{2} = \frac{4}{6} = \frac{2}{3}$$

Since $|r| < 1$, we use formula (8) to find that

$$2 + \frac{4}{3} + \frac{8}{9} + \cdots = \frac{2}{1 - \frac{2}{3}} = 6$$

 Now work Problem 51.

Exploration   Use a graphing utility to graph $U_n = 2\left(\frac{2}{3}\right)^{n-1} + U_{n-1}$ in sequence mode. TRACE the graph for large values of n. What happens to the value of U_n as n increases without bound? What can you conclude about $\sum_{n=1}^{\infty} 2\left(\frac{2}{3}\right)^{n-1}$?

E X A M P L E 10

Repeating Decimals

Show that the repeating decimal $0.999\ldots$ equals 1.

Solution $0.999\ldots = \frac{9}{10} + \frac{9}{100} + \frac{9}{1000} + \cdots$

Thus, $0.999\ldots$ is a geometric series with first term $\frac{9}{10}$ and common ratio $\frac{1}{10}$. Hence,

$$0.999\ldots = \frac{\frac{9}{10}}{1 - \frac{1}{10}} = \frac{\frac{9}{10}}{\frac{9}{10}} = 1$$

E X A M P L E   11   Pendulum Swings

FIGURE 11

18"

Initially, a pendulum swings through an arc of 18 inches. See Figure 11. On each successive swing, the length of the arc is 0.98 of the previous length.

(a) What is the length of arc after 10 swings?
(b) On which swing is the length of arc first less than 12 inches?
(c) After 15 swings, what total length will the pendulum have swung?
(d) When it stops, what total length will the pendulum have swung?

Solution   (a) The length of the first swing is 18 inches. The length of the second swing is $0.98(18)$ inches; the length of the third swing is $0.98(0.98)(18) = 0.98^2(18)$ inches. The length of arc of the 10th swing is

$$(0.98)^9(18) = 15.007 \text{ inches}$$

(b) The length of arc of the nth swing is $(0.98)^{n-1}(18)$. For this to be exactly 12 inches requires

$$(0.98)^{n-1}(18) = 12$$
$$(0.98)^{n-1} = \frac{12}{18} = \frac{2}{3}$$
$$n - 1 = \log_{0.98}\left(\frac{2}{3}\right)$$
$$n = 1 + \frac{\ln(\frac{2}{3})}{\ln 0.98} = 1 + 20.07 = 21.07$$

The length of arc of the pendulum exceeds 12 inches on the 21st swing and is first less than 12 inches on the 22nd swing.

(c) After 15 swings, the pendulum will have swung the following total length L:

$$L = 18 + 0.98(18) + (0.98)^2(18) + (0.98)^3(18) + \cdots + (0.98)^{14} \quad (18)$$

 1st   2nd   3rd   4th   15th

This is the sum of a geometric sequence. The common ratio is 0.98; the first term is 18. The sum has 15 terms, so

$$L = 18\frac{1 - 0.98^{15}}{1 - 0.98} = 18(13.07) = 235.29 \text{ inches}$$

The pendulum will have swung through 235.29 inches after 15 swings.

(d) When the pendulum stops, it will have swung the following total length T:

$$T = 18 + 0.98(18) + (0.98)^2(18) + (0.98)^3(18) + \cdots$$

This is the sum of a geometric series. The common ratio is $r = 0.98$; the first term is $a = 18$. The sum is

$$T = \frac{a}{1 - r} = \frac{18}{1 - 0.98} = 900$$

The pendulum will have swung a total of 900 inches when it finally stops. ∎

HISTORICAL FEATURE   Sequences are among the oldest objects of mathematical investigation, having been studied for over 3500 years. After the initial steps, however, little progress was made until about 1600.

Arithmetic and geometric sequences appear in the Rhind papyrus, a mathematical text containing 85 problems copied around 1650 BC by the Egyptian scribe Ahmes from an earlier work (see Historical Problems 1). Fibonacci (AD 1220) wrote about problems similar to those found in the Rhind papyrus, leading one to suspect that Fibonacci may have had material available that is now lost. This material would have been in the non-Euclidean Greek tradition of Heron (about AD 75) and Diophantus (about AD 250). One problem, again modified slightly, is still with us in the familiar puzzle rhyme "As I was going to St. Ives . . ." (see Historical Problem 2).

The Rhind papyrus indicates that the Egyptians knew how to add up the terms of an arithmetic or geometric sequence, as did the Babylonians. The rule for summing up a geometric sequence is found in Euclid's *Elements* (book IX, 35, 36), where, like all of Euclid's algebra, it is presented in a geometric form.

Investigations of other kinds of sequences began in the 1500s, when algebra became sufficiently developed to handle the more complicated problems. The development of calculus in the 1600s added a powerful new tool, especially for finding the sum of infinite series, and the subject continues to flourish today.

HISTORICAL PROBLEMS

1. *Arithmetic sequence problem from the Rhind papyrus (statement modified slightly for clarity)* One hundred loaves of bread are to be divided among five people so that the amounts they receive form an arithmetic sequence. The first two together receive one-seventh of what the last three receive. How many does each receive? [*Partial answer:* First person receives $1\frac{2}{3}$ loaves.]

2. The following old English children's rhyme resembles one of the Rhind papyrus problems:

    As I was going to St. Ives

    I met a man with seven wives

    Each wife had seven sacks

    Each sack had seven cats

    Each cat had seven kits [kittens]

    Kits, cats, sacks, wives

    How many were going to St. Ives?

(a) Assuming that the speaker and the cat fanciers met by traveling in opposite directions, what is the answer?

(b) How many kittens are being transported?

(c) Kits, cats, sacks, wives; how many?

[**Hint:** It is easier to include the man, find the sum with the formula, and then subtract 1 for the man.]

9.3 EXERCISES

In Problems 1–10, a geometric sequence is given. Find the common ratio and write out the first four terms.

1. $\{3^n\}$

2. $\{(-5)^n\}$

3. $\left\{-3\left(\frac{1}{2}\right)^n\right\}$

4. $\left\{\left(\frac{5}{2}\right)^n\right\}$

5. $\left\{\frac{2^{n-1}}{4}\right\}$

6. $\left\{\frac{3^n}{9}\right\}$

7. $\{2^{n/3}\}$

8. $\{3^{2n}\}$

9. $\left\{\frac{3^{n-1}}{2^n}\right\}$

10. $\left\{\frac{2^n}{3^{n-1}}\right\}$

In Problems 11–24, determine whether the given sequence is arithmetic, geometric, or neither. If the sequence is arithmetic, find the common difference; if it is geometric, find the common ratio.

11. $\{n + 4\}$

12. $\{3n - 5\}$

13. $\{4n^2\}$

14. $\{5n^2 + 1\}$

15. $\{3 - \frac{2}{3}n\}$

16. $\{8 - \frac{3}{4}n\}$

17. $1, 3, 6, 10, \ldots$

18. $2, 4, 6, 8, \ldots$

19. $\{(\frac{2}{3})^n\}$

20. $\{(\frac{5}{4})^n\}$

21. $-1, -2, -4, -8, \ldots$

22. $1, 1, 2, 3, 5, 8, \ldots$

23. $\{3^{n/2}\}$

24. $\{(-1)^n\}$

In Problems 25–32, find the fifth term and the nth term of the geometric sequence whose initial term a and common ratio r are given.

25. $a = 3; r = 2$

26. $a = -2; r = 3$

27. $a = 5; r = -1$

28. $a = 6; r = -2$

29. $a = 0; r = \frac{1}{2}$

30. $a = 1; r = -\frac{1}{3}$

31. $a = \sqrt{2}; r = \sqrt{2}$

32. $a = 0; r = 1/\pi$

In Problems 33–38, find the indicated term of each geometric sequence.

33. 7th term of $1, \frac{1}{2}, \frac{1}{4}, \ldots$

34. 8th term of $1, 3, 9, \ldots$

35. 9th term of $1, -1, 1, \ldots$

36. 10th term of $-1, 2, -4, \ldots$

37. 8th term of $0.4, 0.04, 0.004, \ldots$

38. 7th term of $0.1, 1.0, 10.0, \ldots$

In Problems 39–44, find the sum.

39. $\dfrac{1}{4} + \dfrac{2}{4} + \dfrac{2^2}{4} + \dfrac{2^3}{4} + \cdots + \dfrac{2^{n-1}}{4}$

40. $\dfrac{3}{9} + \dfrac{3^2}{9} + \dfrac{3^3}{9} + \cdots + \dfrac{3^n}{9}$

41. $\displaystyle\sum_{k=1}^{n} \left(\frac{2}{3}\right)^k$

42. $\displaystyle\sum_{k=1}^{n} 4 \cdot 3^{k-1}$

43. $-1 - 2 - 4 - 8 - \cdots - (2^{n-1})$

44. $2 + \dfrac{6}{5} + \dfrac{18}{25} + \cdots + 2\left(\dfrac{3}{5}\right)^n$

For Problems 45–50, use a graphing utility to find the sum of each geometric sequence.

45. $\dfrac{1}{4} + \dfrac{2}{4} + \dfrac{2^2}{4} + \dfrac{2^3}{4} + \cdots + \dfrac{2^{14}}{4}$

46. $\dfrac{3}{9} + \dfrac{3^2}{9} + \dfrac{3^3}{9} + \cdots + \dfrac{3^{15}}{9}$

47. $\displaystyle\sum_{n=1}^{15} \left(\frac{2}{3}\right)^n$

48. $\displaystyle\sum_{n=1}^{15} 4 \cdot 3^{n-1}$

49. $-1 - 2 - 4 - 8 - \cdots - 2^{14}$

50. $2 + \dfrac{6}{5} + \dfrac{18}{25} + \cdots + 2\left(\dfrac{3}{5}\right)^{15}$

In Problems 51–60, find the sum of each infinite geometric series.

51. $1 + \frac{1}{4} + \frac{1}{16} + \cdots$

52. $2 - \frac{4}{3} + \frac{8}{9} - \frac{16}{27} + \cdots$

53. $8 + 4 + 2 + \cdots$

54. $6 + 2 + \frac{2}{3} + \cdots$

55. $2 - \frac{1}{2} + \frac{1}{8} - \frac{1}{32} + \cdots$

56. $1 - \frac{3}{4} + \frac{9}{16} - \frac{27}{64} + \cdots$

57. $\sum_{k=1}^{\infty} 5(\frac{1}{4})^{k-1}$

58. $\sum_{k=1}^{\infty} 8(\frac{1}{3})^{k-1}$

59. $\sum_{k=1}^{\infty} 6(-\frac{2}{3})^{k-1}$

60. $\sum_{k=1}^{\infty} 4(-\frac{1}{2})^{k-1}$

61. Find x so that x, $x + 2$, and $x + 3$ are terms of a geometric sequence.

62. Find x so that $x - 1$, x, and $x + 2$ are terms of a geometric sequence.

63. **Retirement** Christine contributes $100 each month to her 401(k). What will be the value of Christine's 401(k) in 30 years if the per annum rate of return is assumed to be 12% compounded monthly?

64. **Saving for a Home** Jolene wants to purchase a new home. Suppose she invests $400 per month into a mutual fund. If the per annum rate of return of the mutual fund is assumed to be 10% compounded monthly, how much will Jolene have for a down payment after 3 years?

65. **Tax Sheltered Annuity** Don contributes $500 at the end of each quarter to a Tax Sheltered Annuity (TSA). What will the value of the TSA be in 20 years if the per annum rate of return is assumed to be 8% compounded quarterly?

66. **Retirement** Ray, planning on retiring in 15 years, contributes $1000 to an Individual Retirement Account (IRA) semiannually. What will the value of the IRA be when Ray retires if the per annum rate of return is assumed to be 10% compounded semiannually?

67. **Sinking Fund** Scott and Alice want to purchase a vacation home in 10 years and need $50,000 for a down payment. How much should they place in a savings account each month if the per annum rate of return is assumed to be 6% compounded monthly?

68. **Sinking Fund** For a child born in 1996, a 4-year college education at a public university is projected to be $150,000. Assuming an 8% per annum rate of return compounded monthly, how much must be contributed to a college fund every month in order to have $150,000 in 18 years when the child begins college.

69. **Pendulum Swings** Initially, a pendulum swings through an arc of 2 feet. On each successive swing, the length of arc is 0.9 of the previous length.
(a) What is the length of arc after 10 swings?
(b) On which swing is the length of arc first less than 1 foot?

(c) After 15 swings, what total length will the pendulum have swung?
(d) When it stops, what total length will the pendulum have swung?

70. **Bouncing Balls** A ball is dropped from a height of 30 feet. Each time it strikes the ground, it bounces up to 0.8 of the previous height.

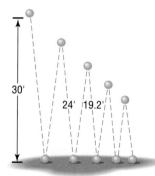

(a) What height will the ball bounce up to after it strikes the ground for the third time?
(b) What is its height after it strikes the ground for the nth time?
(c) How many times does the ball need to strike the ground before its height is less than 6 inches?
(d) What total distance does the ball travel before it stops bouncing?

71. **Salary Increases** Suppose you have just been hired at an annual salary of $18,000 and expect to receive annual increases of 5%. What will your salary be when you begin your fifth year?

72. **Equipment Depreciation** A new piece of equipment cost a company $15,000. Each year, for tax purposes, the company depreciates the value by 15%. What value should the company give the equipment after 5 years?

 73. **Critical Thinking** You have just signed a 7-year professional football league contract with a beginning salary of $2,000,000 per year. Management gives you the following options with regard to your salary over the 7 years.
(1) A bonus of $100,000 each year

(2) An annual increase of 4.5% per year beginning after 1 year

(3) An annual increase of $95,000 per year beginning after 1 year

Which option provides the most money over the 7-year period? Which the least? Which would you choose? Why?

74. **A Rich Man's Promise** A rich man promises to give you $1000 on September 1, 1998. Each day thereafter he will give you $\frac{9}{10}$ of what he gave you the previous day. What is the first date on which the amount you receive is less than 1¢? How much have you received when this happens?

75. **Grains of Wheat on a Chess Board** In an old fable, a commoner who had just saved the king's life was told he could ask the king for any just reward. Being a shrewd man, the commoner said, "A simple wish, sire. Place one grain of wheat on the first square of a chessboard, two grains on the second square, four grains on the third square, continuing until you have filled the board. This is all I seek." Compute the total number of grains needed to do this to see why the request, seemingly simple, could not be granted. (A chessboard consists of $8 \times 8 = 64$ squares.)

76. Can a sequence be both arithmetic and geometric? Give reasons for your answer.

77. Make up a geometric sequence. Give it to a friend and ask for its 20th term.

78. Make up two infinite geometric series, one that has a sum and one that does not. Give them to a friend and ask for the sum of each series.

79. If $x < 1$, then $1 + x + x^2 + x^3 + \cdots + x^n + \cdots = 1/(1-x)$. Make up a table of values using $x = 0.1$, $x = 0.25$, $x = 0.5$, $x = 0.75$, and $x = 0.9$ to compute $1/(1-x)$. Now determine how many terms are needed in the expansion $1 + x + x^2 + x^3 + \cdots + x^n + \cdots$ before it approximates $1/(1-x)$ correct to two decimal places. For example, if $x = 0.1$, then $1/(1-x) = 10/9 = 1.111 \ldots$. The expansion requires three terms.

80. Which of the following choices, A or B, results in more money?

A: To receive $1000 on day 1, $999 on day 2, $998 on day 3, with the process to end after 1000 days

B: To receive $1 on day 1, $2 on day 2, $4 on day 3, for 19 days

81. You are interviewing for a job and receive two offers:

A: $20,000 to start with guaranteed annual increases of 6% for the first 5 years

B: $22,000 to start with guaranteed increases of 3% for the first 5 years

Which offer is best if your goal is to be making as much as possible after 5 years? Which is best if your goal is to make as much money as possible over the contract (5 years)?

82. Look at the figure below. What fraction of the square is eventually shaded if the indicated shading process continues indefinitely?

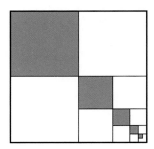

83. **Multiplier** Suppose that, throughout the U.S. economy, individuals spend 90% of every additional dollar that they earn. Economists would say that an individual's **marginal propensity to consume** is 0.90. For example, if Jane earns an additional dollar, she will spend 0.9(1) = $0.90 of it. The individual that earns $0.90 (from Jane) will spend 90% of it or $0.81. This process of spending continues and results in an infinite geometric series as follows:

$1, 0.90, 0.90^2, 0.90^3, 0.90^4, \ldots$

The sum of this infinite geometric series is called the **multiplier.** What is the multiplier if individuals spend 90% of every additional dollar that they earn?

84. **Multiplier** Refer to Problem 83. Suppose that the marginal propensity to consume throughout the U.S. economy is 0.95. What is the multiplier for the U.S. economy?

85. **Stock Price** One method of pricing a stock is to discount the stream of future dividends of the stock. Suppose that a stock pays $P per year

in dividends and, historically, the dividend has been increased $i\%$ per year. If you desire an annual rate of return of $r\%$, this method of pricing a stock states that the price you should pay is the present value of an infinite stream of payments:

$$\text{Price} = P + P\frac{1+i}{1+r} + P\left(\frac{1+i}{1+r}\right)^2 + P\left(\frac{1+i}{1+r}\right)^3 + \cdots$$

Thus, the price of the stock is the sum of an infinite geometric series. Suppose that a stock

pays an annual dividend of $4.00 and, historically, the dividend has been increased 3% per year. You desire an annual rate of return of 9%. What is the most you should pay for the stock?

86. **Stock Price** Refer to Problem 85. Suppose that a stock pays an annual dividend of $2.50 and, historically, the dividend has increased 4% per year. You desire an annual rate of return of 11%. What is the most you should pay for the stock?

9.4 | MATHEMATICAL INDUCTION

1 Prove Statements Using Mathematical Induction

1 *Mathematical induction* is a method for proving that statements involving natural numbers are true for all natural numbers.* For example, the statement "$2n$ is always an even integer" can be proved true for all natural numbers by using mathematical induction. Also, the statement "the sum of the first n positive odd integers equals n^2," that is,

$$1 + 3 + 5 + \cdots + (2n - 1) = n^2 \tag{1}$$

can be proved for all natural numbers n by using mathematical induction.

Before stating the method of mathematical induction, let's try to gain a sense of the power of the method. We shall use the statement in equation (1) for this purpose by restating it for various values of $n = 1, 2, 3, \ldots$:

$n = 1$ The sum of the first positive odd integer is 1^2; $1 = 1^2$.
$n = 2$ The sum of the first 2 positive odd integers is 2^2; $1 + 3 = 4 = 2^2$.
$n = 3$ The sum of the first 3 positive odd integers is 3^2;
         $1 + 3 + 5 = 9 = 3^2$.
$n = 4$ The sum of the first 4 positive odd integers is 4^2;
         $1 + 3 + 5 + 7 = 16 = 4^2$.

Although from this pattern we might conjecture that statement (1) is true for any choice of n, can we really be sure that it does not fail for some choice of n? The method of proof by mathematical induction will, in fact, prove that the statement is true for all n.

> **Theorem** **The Principle of Mathematical Induction**
>
> Suppose the following two conditions are satisfied with regard to a statement about natural numbers:
>
> > **CONDITION I:** The statement is true for the natural number 1.
> > **CONDITION II:** If the statement is true for some natural number k, it is also true for the next natural number $k + 1$.
>
> Then the statement is true for all natural numbers.

*Recall from Chapter 1 that the natural numbers are the numbers $1, 2, 3, 4, \ldots$. In other words, the terms *natural numbers* and *positive integers* are synonymous.

FIGURE 12

We shall not prove this principle. However, we can provide a physical interpretation that will help us to see why the principle works. Think of a collection of natural numbers obeying a statement as a collection of infinitely many dominoes (see Figure 12).

Now, suppose we are told two facts:

1. The first domino is pushed over.
2. If one of the dominoes falls over, say the kth domino, then so will the next one, the $(k + 1)$st domino.

Is it safe to conclude that *all* the dominoes fall over? The answer is yes, because, if the first one falls (Condition I), then the second one does also (by Condition II); and if the second one falls, then so does the third (by Condition II); and so on.

Now let's prove some statements about natural numbers using mathematical induction.

E X A M P L E   1   Using Mathematical Induction

Show that the following statement is true for all natural numbers n:

$$1 + 3 + 5 + \cdots + (2n - 1) = n^2 \tag{2}$$

Solution   We need to show first that statement (2) holds for $n = 1$. Because $1 = 1^2$, statement (2) is true for $n = 1$. Thus, Condition I holds.

Next, we need to show that Condition II holds. Suppose we know for some k that

$$1 + 3 + \cdots + (2k - 1) = k^2 \tag{3}$$

We wish to show that, based on equation (3), statement (2) holds for $k + 1$. Thus, we look at the sum of the first $k + 1$ positive odd integers to determine whether this sum equals $(k + 1)^2$:

$$1 + 3 + \cdots + (2k - 1) + (2k + 1) = \underbrace{[1 + 3 + \cdots + (2k - 1)]}_{= \ k^2 \text{ by equation (3)}} + (2k + 1)$$
$$= k^2 + (2k + 1)$$
$$= k^2 + 2k + 1 = (k + 1)^2$$

Conditions I and II are satisfied; thus, by the Principle of Mathematical Induction, statement (2) is true for all natural numbers. ■

E X A M P L E   2   Using Mathematical Induction

Show that the following statement is true for all natural numbers n:

$$2^n > n$$

Solution   First, we show that the statement $2^n > n$ holds when $n = 1$. Because $2^1 = 2 > 1$, the inequality is true for $n = 1$. Thus, Condition I holds.

Next, we assume, for some natural number k, that $2^k > k$. We wish to show that the formula holds for $k + 1$; that is, we wish to show that $2^{k+1} > k + 1$. Now,

$$2^{k+1} = 2 \cdot 2^k > 2 \cdot k = k + k \geq k + 1$$

$\uparrow$
We know that
$2^k > k.$

$\uparrow$
$k \geq 1.$

Thus, if $2^k > k$, then $2^{k+1} > k + 1$, so Condition II of the Principle of Mathematical Induction is satisfied. Hence, the statement $2^n > n$ is true for all natural numbers n.   ▬

EXAMPLE 3   Using Mathematical Induction

Show that the following formula is true for all natural numbers n:

$$1 + 2 + 3 + \cdots + n = \frac{n(n + 1)}{2} \tag{4}$$

Solution   First, we show that formula (4) is true when $n = 1$. Because

$$\frac{1(1 + 1)}{2} = \frac{1(2)}{2} = 1$$

Condition I of the Principle of Mathematical Induction holds.

Next, we assume that formula (4) holds for some k, and we determine whether the formula then holds for $k + 1$. Thus, we assume that

$$1 + 2 + 3 + \cdots + k = \frac{k(k + 1)}{2} \quad \text{for some } k \tag{5}$$

Now, we need to show that

$$1 + 2 + 3 + \cdots + k + (k + 1) = \frac{(k + 1)(k + 1 + 1)}{2} = \frac{(k + 1)(k + 2)}{2}$$

We do this as follows:

$$1 + 2 + 3 + \cdots + k + (k + 1) = \underbrace{[1 + 2 + 3 + \cdots + k]}_{= \frac{k(k + 1)}{2} \text{ by equation (5)}} + (k + 1)$$

$$= \frac{k(k + 1)}{2} + (k + 1)$$

$$= \frac{k^2 + k + 2k + 2}{2}$$

$$= \frac{k^2 + 3k + 2}{2} = \frac{(k + 1)(k + 2)}{2}$$

Thus, Condition II also holds. As a result, formula (4) is true for all natural numbers.   ▬

 Now work Problem 1.

EXAMPLE 4 Using Mathematical Induction

Show that $3^n - 1$ is divisible by 2 for all natural numbers n.

Solution First, we show that the statement is true when $n = 1$. Because $3^1 - 1 = 3 - 1 = 2$ is divisible by 2, the statement is true when $n = 1$. Thus, Condition I is satisfied.

Next, we assume that the statement holds for some k, and we determine whether the statement then holds for $k + 1$. Thus, we assume that $3^k - 1$ is divisible by 2 for some k. We need to show that $3^{k+1} - 1$ is divisible by 2. Now,

$$3^{k+1} - 1 = 3^{k+1} - 3^k + 3^k - 1$$
$$= 3^k(3 - 1) + (3^k - 1) = 3^k \cdot 2 + (3^k - 1)$$

Because $3^k \cdot 2$ is divisible by 2 and $3^k - 1$ is divisible by 2, it follows that $3^k \cdot 2 + (3^k - 1) = 3^{k+1} - 1$ is divisible by 2. Thus, Condition II is also satisfied. As a result, the statement "$3^n - 1$ is divisible by 2" is true for all natural numbers n. ∎

Warning The conclusion that a statement involving natural numbers is true for all natural numbers is made only after *both* Conditions I and II of the Principle of Mathematical Induction have been satisfied. Problem 27 demonstrates a statement for which only Condition I holds, but the statement is *not* true for all natural numbers. Problem 28 demonstrates a statement for which only Condition II holds, but the statement is *not* true for any natural number.

9.4 EXERCISES

In Problems 1–26, use the Principle of Mathematical Induction to show that the given statement is true for all natural numbers.

1. $2 + 4 + 6 + \cdots + 2n = n(n + 1)$
2. $1 + 5 + 9 + \cdots + (4n - 3) = n(2n - 1)$
3. $3 + 4 + 5 + \cdots + (n + 2) = \frac{1}{2}n(n + 5)$
4. $3 + 5 + 7 + \cdots + (2n + 1) = n(n + 2)$
5. $2 + 5 + 8 + \cdots + (3n - 1) = \frac{1}{2}n(3n + 1)$
6. $1 + 4 + 7 + \cdots + (3n - 2) = \frac{1}{2}n(3n - 1)$
7. $1 + 2 + 2^2 + \cdots + 2^{n-1} = 2^n - 1$
8. $1 + 3 + 3^2 + \cdots + 3^{n-1} = \frac{1}{2}(3^n - 1)$
9. $1 + 4 + 4^2 + \cdots + 4^{n-1} = \frac{1}{3}(4^n - 1)$
10. $1 + 5 + 5^2 + \cdots + 5^{n-1} = \frac{1}{4}(5^n - 1)$
11. $\frac{1}{1 \cdot 2} + \frac{1}{2 \cdot 3} + \frac{1}{3 \cdot 4} + \cdots + \frac{1}{n(n+1)} = \frac{n}{n+1}$
12. $\frac{1}{1 \cdot 2} + \frac{1}{3 \cdot 5} + \frac{1}{5 \cdot 7} + \cdots + \frac{1}{(2n-1)(2n+1)} = \frac{n}{2n+1}$
13. $1^2 + 2^2 + 3^2 + \cdots + n^2 = \frac{1}{6}n(n + 1)(2n + 1)$
14. $1^3 + 2^3 + 3^3 + \cdots + n^3 = \frac{1}{4}n^2(n + 1)^2$
15. $4 + 3 + 2 + \cdots + (5 - n) = \frac{1}{2}n(9 - n)$
16. $-2 - 3 - 4 - \cdots - (n + 1) = -\frac{1}{2}n(n + 3)$
17. $1 \cdot 2 + 2 \cdot 3 + 3 \cdot 4 + \cdots + n(n + 1) = \frac{1}{3}n(n + 1)(n + 2)$
18. $1 \cdot 2 + 3 \cdot 4 + 5 \cdot 6 + \cdots + (2n - 1)(2n) = \frac{1}{3}n(n + 1)(4n - 1)$
19. $n^2 + n$ is divisible by 2.
20. $n^3 + 2n$ is divisible by 3.
21. $n^2 - n + 2$ is divisible by 2.
22. $n(n + 1)(n + 2)$ is divisible by 6.
23. If $x > 1$, then $x^n > 1$.
24. If $0 < x < 1$, then $0 < x^n < 1$.
25. $a - b$ is a factor of $a^n - b^n$.
    [**Hint:** $a^{k+1} - b^{k+1} = a(a^k - b^k) + b^k(a - b)$]
26. $a + b$ is a factor of $a^{2n+1} + b^{2n+1}$.
27. Show that the statement "$n^2 - n + 41$ is a prime number" is true for $n = 1$, but is not true for $n = 41$.

28. Show that the formula
$$2 + 4 + 6 + \cdots + 2n = n^2 + n + 2$$
obeys Condition II of the Principle of Mathematical Induction. That is, show that if the formula is true for some k it is also true for $k + 1$. Then show that the formula is false for $n = 1$ (or for any other choice of n).

29. Use mathematical induction to prove that if $r \neq 1$ then
$$a + ar + ar^2 + \cdots + ar^{n-1} = a\frac{1 - r^n}{1 - r}$$

30. Use mathematical induction to prove that
$$a + (a + d) + (a + 2d) + \cdots$$
$$+ [a + (n - 1)d] = na + d\frac{n(n - 1)}{2}$$

31. Geometry Use mathematical induction to show that the sum of the interior angles of a convex polygon of n sides equals $(n - 2) \cdot 180°$.

32. The Extended Principle of Mathematical Induction The Extended Principle of Mathematical Induction states that if Conditions I and II hold, that is,
(I) a statement is true for a natural number j,
(II) if the statement is true for some natural number $k > j$, then it is also true for the next natural number $k + 1$,
then the statement is true for *all* natural numbers $\geq j$.

Use the Extended Principle of Mathematical Induction to show that the number of diagonals in a convex polygon of n sides is $\frac{1}{2}n(n - 3)$.
[**Hint:** Begin by showing that the result is true when $n = 4$ (Condition I).]

33. How would you explain to a friend the Principle of Mathematical Induction?

9.5   THE BINOMIAL THEOREM

1 Evaluate a Binomial Coefficient
2 Expand a Binomial

In Chapter 1, we listed some special products. Among these were formulas for expanding $(x + a)^n$ for $n = 2$ and $n = 3$. The *Binomial Theorem** is a formula for the expansion of $(x + a)^n$ for n any positive integer. If $n = 1, 2, 3$, and 4, the expansion of $(x + a)^n$ is straightforward:

$$(x + a)^1 = x + a \qquad \text{2 terms, beginning with } x^1 \text{ and ending with } a^1.$$

$$(x + a)^2 = x^2 + 2ax + a^2 \qquad \text{3 terms, beginning with } x^2 \text{ and ending with } a^2.$$

$$(x + a)^3 = x^3 + 3ax^2 + 3a^2x + a^3 \qquad \text{4 terms, beginning with } x^3 \text{ and ending with } a^3.$$

$$(x + a)^4 = x^4 + 4ax^3 + 6a^2x^2 + 4a^3x + a^4 \qquad \text{5 terms, beginning with } x^4 \text{ and ending with } a^4.$$

Notice that each expansion of $(x + a)^n$ begins with x^n and ends with a^n. As you read from left to right, the powers of x are decreasing, while the powers of a are increasing. Also, the number of terms that appear equals $n + 1$. Notice, too, that the degree of each monomial in the expansion equals n. For example, in the expansion of $(x + a)^3$, each monomial ($x^3, 3ax^2, 3a^2x, a^3$) is of degree 3. As a result, we might conjecture that the expansion of $(x + a)^n$ would look like this:

$$(x + a)^n = x^n + \_\_ax^{n-1} + \_\_a^2x^{n-2} + \cdots + \_\_a^{n-1}x + a^n$$

where the blanks are numbers to be found. This is, in fact, the case, as we shall see shortly.

First, we need to introduce a symbol.

---

*The name *binomial* derives from the fact that $x + a$ is a binomial, that is, contains two terms.

Getting the Most Out of Your Contract

You and your band have just signed a recording contract with the nationally acclaimed recording studio NASHBURG TENS. They've promised $200,000 a year for six years, plus a choice of one of the following four options:

(a) a bonus of $10,000 per year
(b) an annual increase of 4.5% per year (starting after the first year)
(c) an annual increase of 6% per year (starting after the second year)
(d) an annual increase of $9500 per year (starting after the first year)

    You have to tell them today which option you want to take. Your agent is out of town and out of cellular phone range, so you'll have to do the math yourselves.

1.   For each option, find out what your payment would be every year for the six years. Round to the nearest dollar.
2.   Identify which options are examples of arithmetic or geometric sequences.
3.   For each option, find out how much the recording studio will have paid in total over the whole six years. Do you have a formula that will shorten your work?
4.   Which option pays your band the most overall? Are there any considerations or circumstances that would make other options better choices even though they pay less money? Are there advantages to being paid more at the beginning of the contract? What are they?

The Symbol $\binom{n}{j}$

1 We define the symbol $\binom{n}{j}$, read "n taken j at a time," as follows:

> If j and n are integers with $0 \le j \le n$, the **symbol** $\binom{n}{j}$ is defined as
>
> $$\binom{n}{j} = \frac{n!}{j!(n-j)!} \qquad (1)$$

E X A M P L E   1

Evaluating $\binom{n}{j}$

Find:

(a) $\binom{3}{1}$ (b) $\binom{4}{2}$ (c) $\binom{8}{7}$ (d) $\binom{65}{15}$

Solution (a) $\binom{3}{1} = \frac{3!}{1!(3-1)!} = \frac{3!}{1!2!} = \frac{3 \cdot 2 \cdot 1}{1(2 \cdot 1)} = \frac{6}{2} = 3$

(b) $\binom{4}{2} = \frac{4!}{2!(4-2)!} = \frac{4!}{2!2!} = \frac{4 \cdot 3 \cdot 2 \cdot 1}{(2 \cdot 1)(2 \cdot 1)} = \frac{24}{4} = 6$

FIGURE 13

```
65 nCr 15
    2.073746998 E14
```

(c) $\binom{8}{7} = \frac{8!}{7!(8-7)!} = \frac{8!}{7!1!} = \frac{8 \cdot 7!}{7! \cdot 1!} = \frac{8}{1} = 8$

$\uparrow$
$8! = 8 \cdot 7!$

(d) We use a calculator.* Figure 13 shows the solution using a TI-83 graphing calculator.

Thus, $\binom{65}{15} = 2.073746998 \times 10^{14}$.

 Now work Problem 1.

Two useful formulas involving the symbol $\binom{n}{j}$ are

> $$\binom{n}{0} = 1 \quad \text{and} \quad \binom{n}{n} = 1$$

*On a calculator, the symbol $\binom{n}{j}$ may be denoted by the key $\boxed{\text{nCr}}$.

Proof

$$\binom{n}{0} = \frac{n!}{0!(n-0)!} = \frac{n!}{0!n!} = \frac{1}{1} = 1$$

You are asked to show that $\binom{n}{n} = 1$ in Problem 41 at the end of this section.

Suppose we arrange the various values of the symbol $\binom{n}{j}$ in a triangular display, as shown next and in Figure 14.

$$\binom{0}{0}$$

$$\binom{1}{0} \quad \binom{1}{1}$$

$$\binom{2}{0} \quad \binom{2}{1} \quad \binom{2}{2}$$

$$\binom{3}{0} \quad \binom{3}{1} \quad \binom{3}{2} \quad \binom{3}{3}$$

$$\binom{4}{0} \quad \binom{4}{1} \quad \binom{4}{2} \quad \binom{4}{3} \quad \binom{4}{4}$$

$$\binom{5}{0} \quad \binom{5}{1} \quad \binom{5}{2} \quad \binom{5}{3} \quad \binom{5}{4} \quad \binom{5}{5}$$

This display is called the **Pascal triangle,** named after Blaise Pascal (1623–1662), a French mathematician.

FIGURE 14
Pascal triangle

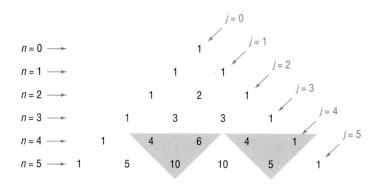

The Pascal triangle has 1's down the sides. To get any other entry, merely add the two nearest entries in the row above it. The shaded triangles in Figure 14 serve to illustrate this feature of the Pascal triangle. Based on this feature, the row corresponding to $n = 6$ is found as follows:

$$\begin{array}{ccccccc} n = 5 \rightarrow & 1 & 5 & 10 & 10 & 5 & 1 \\ n = 6 \rightarrow & 1 & 6 & 15 & 20 & 15 & 6 & 1 \end{array}$$

Later, we shall prove that this addition always works (see the theorem on page 672).

Although the Pascal triangle provides an interesting and organized display of the symbol $\binom{n}{j}$, in practice it is not all that helpful. For example, if you wanted to know the value of $\binom{12}{5}$, you would need to produce twelve rows of the triangle before seeing the answer. It is much faster instead to use the definition (1).

The Binomial Theorem

2 Now we are ready to state the **Binomial Theorem.** A proof is given at the end of this section.

Theorem   Binomial Theorem

Let x and a be real numbers. For any positive integer n, we have

$$(x + a)^n = \binom{n}{0}x^n + \binom{n}{1}ax^{n-1} + \cdots + \binom{n}{j}a^jx^{n-j} + \cdots + \binom{n}{n}a^n$$

$$= \sum_{j=0}^{n}\binom{n}{j}a^jx^{n-j}$$

(2)

Now you know why we needed to introduce the symbol $\binom{n}{j}$; these symbols are the numerical coefficients that appear in the expansion of $(x + a)^n$. Because of this, the symbol $\binom{n}{j}$ is called the **binomial coefficient.**

E X A M P L E   2   Expanding a Binomial

Use the Binomial Theorem to expand $(x + 2)^5$.

Solution   In the Binomial Theorem, let $a = 2$ and $n = 5$. Then

$$(x + 2)^5 = \binom{5}{0}x^5 + \binom{5}{1}2x^4 + \binom{5}{2}2^2x^3 + \binom{5}{3}2^3x^2 + \binom{5}{4}2^4x + \binom{5}{5}2^5$$

    ↑
Use equation (2).

$$= 1 \cdot x^5 + 5 \cdot 2x^4 + 10 \cdot 4x^3 + 10 \cdot 8x^2 + 5 \cdot 16x + 1 \cdot 32$$

    ↑
Use row $n = 5$ of the Pascal triangle or formula (1) for $\binom{n}{j}$.

$$= x^5 + 10x^4 + 40x^3 + 80x^2 + 80x + 32$$

E X A M P L E   3   Expanding a Binomial

Expand $(2y - 3)^4$ using the Binomial Theorem.

Solution   First, we rewrite the expression $(2y - 3)^4$ as $[2y + (-3)]^4$. Now we use the Binomial Theorem with $n = 4$, $x = 2y$, and $a = -3$:

$$[2y + (-3)]^4 = \binom{4}{0}(2y)^4 + \binom{4}{1}(-3)(2y)^3 + \binom{4}{2}(-3)^2(2y)^2$$

$$+ \binom{4}{3}(-3)^3(2y) + \binom{4}{4}(-3)^4$$

$$= 1 \cdot 16y^4 + 4(-3)8y^3 + 6 \cdot 9 \cdot 4y^2 + 4(-27)2y + 1 \cdot 81$$

↑
Use row $n = 4$ of the Pascal triangle or formula (1) for $\binom{n}{j}$.

$$= 16y^4 - 96y^3 + 216y^2 - 216y + 81$$

In this expansion, note that the signs alternate due to the fact that $a = -3 < 0$. ▬

 Now work Problem 17.

E X A M P L E   4   Finding a Particular Coefficient in a Binomial Expansion

Find the coefficient of y^8 in the expansion of $(2y + 3)^{10}$.

Solution   We write out the expansion using the Binomial Theorem:

$$(2y + 3)^{10} = \binom{10}{0}(2y)^{10} + \binom{10}{1}(3)^1(2y)^9 + \binom{10}{2}(3)^2(2y)^8 + \binom{10}{3}(3)^3(2y)^7$$

$$+ \binom{10}{4}(3)^4(2y)^6 + \cdots + \binom{10}{9}(3)^9(2y) + \binom{10}{10}(3)^{10}$$

From the third term in the expansion, the coefficient of y^8 is

$$\binom{10}{2}(3)^2(2)^8 = \frac{10!}{2!8!} \cdot 9 \cdot 2^8 = \frac{10 \cdot 9 \cdot 8!}{2 \cdot 8!} \cdot 9 \cdot 2^8 = 103,680$$ ▬

As this solution demonstrates, we can use the Binomial Theorem to write a particular term in an expansion without writing the entire expansion. Based on the expansion of $(x + a)^n$, the term containing x^j is

$$\binom{n}{n - j}a^{n-j}x^j \qquad (3)$$

For example, we can solve Example 4 by using formula (3) with $n = 10$, $a = 3$, $x = 2y$, and $j = 8$. Then the term containing y^8 is

$$\binom{10}{10 - 8}3^{10-8}(2y)^8 = \binom{10}{2} \cdot 3^2 \cdot 2^8 \cdot y^8 = \frac{10!}{2!8!} \cdot 9 \cdot 2^8 y^8$$

$$= \frac{10 \cdot 9 \cdot 8!}{2!8!} \cdot 9 \cdot 2^8 y^8 = 103{,}680 y^8$$

E X A M P L E   5   Finding a Particular Term in a Binomial Expansion

Find the sixth term in the expansion of $(x + 2)^9$.

Solution A   We expand using the Binomial Theorem until the sixth term is reached:

$$(x + 2)^9 = \binom{9}{0}x^9 + \binom{9}{1}2 \cdot x^8 + \binom{9}{2}2^2 \cdot x^7 + \binom{9}{3}2^3 \cdot x^6 + \binom{9}{4}2^4 \cdot x^5$$

$$+ \binom{9}{5}2^5 \cdot x^4 + \cdots$$

The sixth term is

$$\binom{9}{5}2^5 \cdot x^4 = \frac{9!}{5!4!} \cdot 32 \cdot x^4 = 4032x^4$$

Solution B   The sixth term in the expansion of $(x + 2)^9$, which has 10 terms total, contains x^4. (Do you see why?) Thus, by formula (3), the sixth term is

$$\binom{9}{9 - 4}2^{9-4}x^4 = \binom{9}{5}2^5x^4 = \frac{9!}{5!4!} \cdot 32x^4 = 4032x^4$$

   Now work Problems 25 and 31.

Next we show that the "triangular addition" feature of the Pascal triangle illustrated in Figure 14 always works.

Theorem

If n and j are integers with $1 \le j \le n$, then

$$\binom{n}{j - 1} + \binom{n}{j} = \binom{n + 1}{j} \qquad (4)$$

Proof

$$\binom{n}{j - 1} + \binom{n}{j} = \frac{n!}{(j - 1)![n - (j - 1)]!} + \frac{n!}{j!(n - j)!} \quad \text{Multiply the first term by } j\,/\,j \text{ and the second}$$

$$= \frac{n!}{(j - 1)!(n - j + 1)!} + \frac{n!}{j!(n - j)!} \quad \begin{array}{l}\text{term by } (n - j + 1)/ \\ (n - j + 1).\end{array}$$

$$= \frac{jn!}{j(j - 1)!(n - j + 1)!} + \frac{(n - j + 1)n!}{j!(n - j + 1)(n - j)!}$$

$$= \frac{jn!}{j!(n-j+1)!} + \frac{(n-j+1)n!}{j!(n-j+1)!}$$ Now the denominators are equal.

$$= \frac{jn! + (n-j+1)n!}{j!(n-j+1)!}$$

$$= \frac{n!(j+n-j+1)}{j!(n-j+1)!}$$

$$= \frac{n!(n+1)}{j!(n-j+1)!} = \frac{(n+1)!}{j![(n+1)-j]!} = \binom{n+1}{j}$$ ∎

Proof of the Binomial Theorem   We use mathematical induction to prove the Binomial Theorem. First, we show that formula (2) is true for $n = 1$:

$$(x + a)^1 = x + a = \binom{1}{0}x^1 + \binom{1}{1}a^1$$

Next we suppose that formula (2) is true for some k. That is, we assume that

$$(x+a)^k = \binom{k}{0}x^k + \binom{k}{1}ax^{k-1} + \cdots + \binom{k}{j-1}a^{j-1}x^{k-j+1} + \binom{k}{j}a^j x^{k-j} + \cdots + \binom{k}{k}a^k \qquad (5)$$

Now we calculate $(x+a)^{k+1}$:

$$(x+a)^{k+1} = (x+a)(x+a)^k = x(x+a)^k + a(x+a)^k$$

Use ↑ equation (5).

$$= x\left[\binom{k}{0}x^k + \binom{k}{1}ax^{k-1} + \cdots + \binom{k}{j-1}a^{j-1}x^{k-j+1} + \binom{k}{j}a^j x^{k-j} + \cdots + \binom{k}{k}a^k\right]$$

$$+ a\left[\binom{k}{0}x^k + \binom{k}{1}ax^{k-1} + \cdots + \binom{k}{j-1}a^{j-1}x^{k-j+1} + \binom{k}{j}a^j x^{k-j} + \cdots + \binom{k}{k-1}a^{k-1}x + \binom{k}{k}a^k\right]$$

$$= \binom{k}{0}x^{k+1} + \binom{k}{1}ax^k + \cdots + \binom{k}{j-1}a^{j-1}x^{k-j+2} + \binom{k}{j}a^j x^{k-j+1} + \cdots + \binom{k}{k}a^k x$$

$$+ \binom{k}{0}ax^k + \binom{k}{1}a^2 x^{k-1} + \cdots + \binom{k}{j-1}a^j x^{k-j+1} + \binom{k}{j}a^{j+1}x^{k-j} + \cdots + \binom{k}{k-1}a^k x + \binom{k}{k}a^{k+1}$$

$$= \binom{k}{0}x^{k+1} + \left[\binom{k}{1} + \binom{k}{0}\right]ax^k + \cdots + \left[\binom{k}{j} + \binom{k}{j-1}\right]a^j x^{k-j+1} + \cdots + \left[\binom{k}{k} + \binom{k}{k-1}\right]a^k x + \binom{k}{k}a^{k+1}$$

Because

$$\binom{k}{0} = 1 = \binom{k+1}{0}, \quad \underset{\substack{\uparrow \\ (4)}}{\binom{k}{1} + \binom{k}{0}} = \binom{k+1}{1}, \cdots,$$

$$\underset{\substack{\uparrow \\ (4)}}{\binom{k}{j} + \binom{k}{j-1}} = \binom{k+1}{j}, \cdots, \quad \binom{k}{k} = 1 = \binom{k+1}{k+1}$$

we have

$$(x+a)^{k+1} = \binom{k+1}{0}x^{k+1} + \binom{k+1}{1}ax^k + \cdots + \binom{k+1}{j}a^j x^{k-j+1} + \cdots + \binom{k+1}{k+1}a^{k+1}$$

Thus, Conditions I and II of the Principle of Mathematical Induction are satisfied, and formula (2) is therefore true for all n. ∎

HISTORICAL FEATURE The case $n = 2$ of the Binomial Theorem, $(a + b)^2$, was known to Euclid in 300 BC, but the general law seems to have been discovered by the Persian mathematician and astronomer Omar Khayyám (ca. 1050–1123), who is also well known as the author of the *Rubaiyat*, a collection of four-line poems making observations on the human condition. Omar Khayyám did not state the Binomial Theorem explicitly, but he claimed to have a method for extracting third, fourth, fifth roots, and so on. A little study shows that one must know the Binomial Theorem to create such a method.

The heart of the Binomial Theorem is the formula for the numerical coefficients, and, as we saw, they can be written out in a symmetric triangular form. The Pascal triangle appears first in the books of Yang Hui (about 1270) and Chu Shih-chieh (1303). Pascal's name is attached to the triangle because of the many applications he made of it, especially to counting and probability. In establishing these results, he was one of the earliest users of mathematical induction.

Many people worked on the proof of the Binomial Theorem, which was finally completed for all n (including complex numbers) by Niels Abel (1802–1829).

9.5 | EXERCISES

In Problems 1–12, evaluate each expression.

1. $\binom{5}{2}$ **2.** $\binom{7}{4}$ **3.** $\binom{7}{5}$ **4.** $\binom{9}{7}$ **5.** $\binom{50}{49}$ **6.** $\binom{100}{98}$

7. $\binom{1000}{1000}$ **8.** $\binom{1000}{0}$ **9.** $\binom{55}{23}$ **10.** $\binom{60}{20}$ **11.** $\binom{47}{25}$ **12.** $\binom{37}{19}$

In Problems 13–24, expand each expression using the Binomial Theorem.

13. $(x + 1)^4$ **14.** $(x - 1)^4$ **15.** $(x - 2)^5$ **16.** $(x + 3)^5$

17. $(3x + 1)^4$ **18.** $(2x + 3)^5$ **19.** $(x^2 + y^2)^5$ **20.** $(x^2 - y^2)^6$

21. $(\sqrt{x} + \sqrt{2})^6$ **22.** $(\sqrt{x} - \sqrt{3})^4$ **23.** $(ax + by)^5$ **24.** $(ax - by)^4$

In Problems 25–38, use the Binomial Theorem to find the indicated coefficient or term.

25. The coefficient of x^4 in the expansion of $(x + 3)^{10}$

26. The coefficient of x^5 in the expansion of $(x - 3)^{10}$

27. The coefficient of x^7 in the expansion of $(2x - 1)^{12}$

28. The coefficient of x^3 in the expansion of $(2x + 1)^{12}$

29. The coefficient of x^7 in the expansion of $(2x + 3)^9$

30. The coefficient of x^2 in the expansion of $(2x - 3)^9$

31. The fifth term in the expansion of $(x + 3)^7$

32. The third term in the expansion of $(x - 3)^7$

33. The third term in the expansion of $(3x - 2)^9$

34. The sixth term in the expansion of $(3x + 2)^8$

35. The coefficient of x^0 in the expansion of $\left(x^2 + \dfrac{1}{x}\right)^{12}$

36. The coefficient of x^0 in the expansion of $\left(x - \dfrac{1}{x^2}\right)^9$

37. The coefficient of x^4 in the expansion of $\left(x - \dfrac{2}{\sqrt{x}}\right)^{10}$

38. The coefficient of x^2 in the expansion of $\left(\sqrt{x} + \dfrac{3}{\sqrt{x}}\right)^8$

39. Use the Binomial Theorem to find the numerical value of $(1.001)^5$ correct to five decimal places. [**Hint:** $(1.001)^5 = (1 + 10^{-3})^5$]

40. Use the Binomial Theorem to find the numerical value of $(0.998)^6$ correct to five decimal places.

41. Show that $\binom{n}{n} = 1$.

42. Show that, if n and j are integers with $0 \le j \le n$, then

$$\binom{n}{j} = \binom{n}{n-j}$$

Thus, conclude that the Pascal triangle is symmetric with respect to a vertical line drawn from the topmost entry.

43. If n is a positive integer, show that

$$\binom{n}{0} + \binom{n}{1} + \cdots + \binom{n}{n} = 2^n$$

[**Hint:** $2^n = (1+1)^n$; now use the Binomial Theorem.]

44. If n is a positive integer, show that

$$\binom{n}{0} - \binom{n}{1} + \binom{n}{2} - \cdots + (-1)^n\binom{n}{n} = 0$$

45. $\binom{5}{0}\left(\frac{1}{4}\right)^5 + \binom{5}{1}\left(\frac{1}{4}\right)^4\left(\frac{3}{4}\right) + \binom{5}{2}\left(\frac{1}{4}\right)^3\left(\frac{3}{4}\right)^2$

$+ \binom{5}{3}\left(\frac{1}{4}\right)^2\left(\frac{3}{4}\right)^3 + \binom{5}{4}\left(\frac{1}{4}\right)\left(\frac{3}{4}\right)^4 + \binom{5}{5}\left(\frac{3}{4}\right)^5 = ?$

46. *Stirling's Formula* for approximating $n!$ when n is large is given by

$$n! \approx \sqrt{2n\pi}\left(\frac{n}{e}\right)^n\left(1 + \frac{1}{12n-1}\right)$$

Calculate 12!, 20!, and 25!. Then use Stirling's formula to approximate 12!, 20!, and 25!.

9.6   SETS AND COUNTING

1 Find All the Subsets of a Set
2 Find the Intersection and Union of Sets
3 Find the Complement of a Set
4 Count the Number of Elements in a Set

Sets

A **set** is a well-defined collection of distinct objects. The objects of a set are called its **elements.** By **well-defined,** we mean that there is a rule that enables us to determine whether a given object is an element of the set. If a set has no elements, it is called the **empty set,** or **null set,** and is denoted by the symbol $\emptyset$.

Because the elements of a set are distinct, we never repeat elements. Thus, we would never write {1, 2, 3, 2}; the correct listing is {1, 2, 3}. Furthermore, because a set is a collection, the order in which the elements are listed is immaterial. Thus, {1, 2, 3}, {1, 3, 2}, {2, 1, 3}, and so on, all represent the same set.

EXAMPLE 1   Writing the Elements of a Set

Write the set consisting of the possible outcomes from tossing a coin twice. Use H for "heads" and T for "tails."

Solution   In tossing a coin twice, we can get heads each time, HH; or heads the first time and tails the second, HT; or tails the first time and heads the second, TH; or tails each time, TT. Because no other possibilities exist, the set of outcomes is

[HH, HT, TH, TT]

1   If two sets A and B have precisely the same elements, then we say that A and B are **equal** and write $A = B$.

If each element of a set A is also an element of a set B, then we say that A is a **subset** of B and write $A \subseteq B$.

If $A \subseteq B$ and $A \neq B$, then we say that A is a **proper subset** of B and write $A \subset B$.

Thus, if $A \subseteq B$, every element in set A is also in set B, but B may or may not have additional elements. If $A \subset B$, every element in A is also in B, and B has at least one element not found in A.

Finally, we agree that the empty set is a subset of every set; that is,

$$\varnothing \subseteq A \qquad \text{for any set } A$$

E X A M P L E   2   Finding All the Subsets of a Set

Write down all the subsets of the set $\{a, b, c\}$.

Solution   To organize our work, we write down all the subsets with no elements, then those with one element, then those with two elements, and finally those with three elements. These will give us all the subsets. Do you see why?

0 Elements	1 Element	2 Elements	3 Elements
$\varnothing$	$\{a\}, \{b\}, \{c\}$	$\{a, b\}, \{b, c\}, \{a, c\}$	$\{a, b, c\}$

 Now work Problem 21.

If A and B are sets, the **intersection** of A with B, denoted $A \cap B$, is the set consisting of elements that belong to both A and B. The **union** of A with B, denoted $A \cup B$, is the set consisting of elements that belong to *either A or B,* or both.

E X A M P L E   3   Finding the Intersection and Union of Sets

Let $A = \{1, 3, 5, 8\}$, $B = \{3, 5, 7\}$, and $C = \{2, 4, 6, 8\}$. Find:

(a) $A \cap B$ (b) $A \cup B$ (c) $B \cap (A \cup C)$

Solution   (a) $A \cap B = \{1, 3, 5, 8\} \cap \{3, 5, 7\} = \{3, 5\}$
(b) $A \cup B = \{1, 3, 5, 8\} \cup \{3, 5, 7\} = \{1, 3, 5, 7, 8\}$
(c) $B \cap (A \cup C) = \{3, 5, 7\} \cap [\{1, 3, 5, 8\} \cup \{2, 4, 6, 8\}]$
$\qquad = \{3, 5, 7\} \cap \{1, 2, 3, 4, 5, 6, 8\} = \{3, 5\}$

 Now work Problem 5.

Usually, in working with sets, we designate a **universal set,** the set consisting of all the elements we wish to consider. Once a universal set has been designated, we can consider elements of the universal set not found in a given set.

If A is a set, the **complement** of A, denoted A', is the set consisting of all the elements not in A.

E X A M P L E   4

Finding the Complement of a Set

If the universal set is $U = \{1, 2, 3, 4, 5, 6, 7, 8, 9\}$, and if $A = \{1, 3, 5, 7, 9\}$, then $A' = \{2, 4, 6, 8\}$. ∎

Notice that: $A \cup A' = U$ and $A \cap A' = \emptyset$.

Now work Problem 13.

FIGURE 15

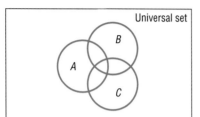

It is often helpful to draw pictures of sets. Such pictures, called **Venn diagrams,** represent sets as circles enclosed in a rectangle, which represents the universal set. Such diagrams often help us to visualize various relationships among sets. See Figure 15.

If we know that $A \subseteq B$, we might use the Venn diagram in Figure 16(a). If we know that A and B have no elements in common, that is, if $A \cap B = \emptyset$, we might use the Venn diagram in Figure 16(b).

FIGURE 16

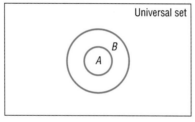 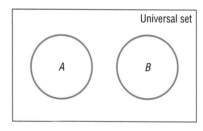

(a) $A \subseteq B$ **(b)** $A \cap B = \emptyset$

Figure 17(a), 17(b), and 17(c) use Venn diagrams to illustrate the definitions of intersection, union, and complement, respectively.

FIGURE 17

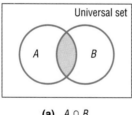 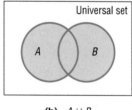 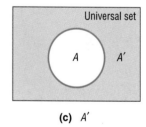

(a) $A \cap B$ **(b)** $A \cup B$ **(c)** A'

Counting

As you count the number of students in a classroom or the number of pennies in your pocket, what you are really doing is matching, on a one-to-one basis, each object to be counted with the counting numbers $1, 2, 3, \ldots, n$, for some number n. If a set A matched up in this fashion with the set $\{1, 2, \ldots, 25\}$, you would conclude that there are 25 elements in the set A. We use the notation $n(A) = 25$ to indicate that there are 25 elements in the set A.

Because the empty set has no elements, we write

$$n(\emptyset) = 0$$

If the number of elements in a set is a nonnegative integer, we say the set is **finite.** Otherwise, it is **infinite.** We shall concern ourselves only with finite sets.

From Example 2, we can see that a set with 3 elements has $2^3 = 8$ subsets. In fact, it can be shown that a set with n elements has exactly 2^n subsets. This fact has an important application to computers, which we take up at the end of this section.

EXAMPLE 5

Analyzing Survey Data

In a survey of 100 college students, 35 were registered in College Algebra, 52 were registered in Introduction to Computer Science, and 18 were in both courses. How many were registered in neither course?

Solution   First, we let A = Set of students in College Algebra
                                    B = Set of students in Introduction to Computer Science

FIGURE 18

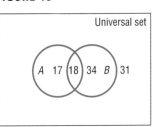

Then the information tells us that

$$n(A) = 35 \qquad n(B) = 52 \qquad n(A \cap B) = 18$$

Refer to Figure 18. Do you see how the numerical entries were determined? Based on the diagram, we conclude that $17 + 18 + 34 = 69$ students were registered in at least one of the two courses. Since 100 students were surveyed, it follows that $100 - 69 = 31$ were registered in neither course. ▬

Now work Problem 35.

The conclusions drawn in Example 5 lead us to formulate a general counting formula. If we count the elements in each of two sets A and B, we necessarily count twice any elements that are in both A and B, that is, those elements in $A \cap B$. Thus, to count correctly the elements that are in A or B, that is, to find $n(A \cup B)$, we need to subtract those in $A \cap B$ from $n(A) + n(B)$.

Theorem   Counting Formula

If A and B are finite sets, then

$$n(A \cup B) = n(A) + n(B) - n(A \cap B) \tag{1}$$

▬

A special case of the counting formula (1) occurs if A and B have no elements in common. In this case, $A \cap B = \varnothing$ so that $n(A \cap B) = 0$.

Theorem   Addition Principle of Counting

If two sets A and B have no elements in common, then

$$n(A \cup B) = n(A) + n(B) \tag{2}$$

▬

E X A M P L E   6   Counting the Number of Possible Codes

A certain code is to consist of either a letter of the alphabet or a digit, but not both. How many codes are possible?

Solution   Let the sets A and B be defined as

A = Set of letters in the alphabet
B = Set of digits $\{0, 1, 2, \ldots, 9\}$

Then

$$n(A) = 26 \qquad n(B) = 10$$

Because letters and digits are different, $A \cap B = \varnothing$. The number of ways either a letter or a digit can be chosen is, therefore,

$$n(A \cup B) = n(A) + n(B) = 26 + 10 = 36$$

Application to Computers

Information stored in a computer may be thought of as a series of switches, which are either on or off and are denoted by either the number 0 (off) or the number 1 (on). These numbers are the binary digits, or **bits.** A **register** holds a certain fixed number of bits. For example, there are 8-bit registers, 16-bit registers, 32-bit registers, and even 64-bit registers. An 8-bit register might hold an entry that looks like this: 01111001 (8 bits). We wish to find out how many different representations are possible in a given register.

We proceed in steps, first looking at a hypothetical 3-bit register. Look again at the solution to Example 2 and arrange all the subsets of $\{a, b, c\}$ as shown in Table 3. As the table illustrates, the number of subsets of a set with 3 elements equals the number of different representations in a 3-bit register. A set with n elements has 2^n subsets; thus, an n-bit register has 2^n representations. So an 8-bit register can hold $2^8 = 256$ different symbols, a 16-bit register can hold $2^{16} = 65,536$ different symbols, and a 32-bit register can hold $2^{32} \approx 4.3 \times 10^9$ different symbols.

TABLE 3

a	b	c	Subset
0	0	0	$\varnothing$
1	0	0	$\{a\}$
0	1	0	$\{b\}$
0	0	1	$\{c\}$
1	1	0	$\{a, b\}$
0	1	1	$\{b, c\}$
1	0	1	$\{a, c\}$
1	1	1	$\{a, b, c\}$

9.6   EXERCISES

In Problems 1–10, use A = $\{1, 3, 5, 7, 9\}$, B = $\{1, 5, 6, 7\}$, *and C* = $\{1, 2, 4, 6, 8, 9\}$ *to find each set.*

1. $A \cup B$ **2.** $A \cup C$ **3.** $A \cap B$ **4.** $A \cap C$

5. $(A \cup B) \cap C$ **6.** $(A \cap C) \cup (B \cap C)$ **7.** $(A \cap B) \cup C$ **8.** $(A \cup B) \cup C$

9. $(A \cup C) \cap (B \cup C)$ **10.** $(A \cap B) \cap C$

In Problems 11–20, use U = Universal set = $\{0, 1, 2, 3, 4, 5, 6, 7, 8, 9\}$, *A* = $\{1, 3, 4, 5, 9\}$, *B* = $\{2, 4, 6, 7, 8\}$, *and C* = $\{1, 3, 4, 6\}$ *to find each set.*

11. A' **12.** C' **13.** $(A \cap B)'$ **14.** $(B \cup C)'$ **15.** $A' \cup B'$

16. $B' \cap C'$ **17.** $(A \cap C')'$ **18.** $(B' \cup C)'$ **19.** $(A \cup B \cup C)'$ **20.** $(A \cap B \cap C)'$

21.   Write down all the subsets of $\{a, b, c, d\}$.

22.   Write down all the subsets of $\{a, b, c, d, e\}$.

23.   If $n(A) = 25$, $n(B) = 30$, and $n(A \cap B) = 15$, find $n(A \cup B)$.

24. If $n(A) = 20$, $n(B) = 30$, and $n(A \cup B) = 45$, find $n(A \cap B)$.

25. If $n(A \cup B) = 50$, $n(A \cap B) = 10$, and $n(B) = 20$, find $n(A)$.

26. If $n(A \cup B) = 60$, $n(A \cap B) = 40$, and $n(A) = n(B)$, find $n(A)$.

In Problems 27–34, use the information given in the figure.

27. How many are in set A?

28. How many are in set B?

29. How many are in A or B?

30. How many are in A and B?

31. How many are in A but not C?

32. How many are not in A?

33. How many are in A and B and C?

34. How many are in A or B or C?

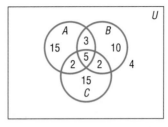

35. Analyzing Survey Data   In a consumer survey of 500 people, 200 indicated that they would be buying a major appliance within the next month; 150 indicated they would buy a car, and 25 said they would purchase both a major appliance and a car. How many will purchase neither? How many will purchase only a car?

36. Analyzing Survey Data   In a student survey, 200 indicated that they would attend Summer Session I and 150 indicated Summer Session II. If 75 students plan to attend both summer sessions and 275 indicated that they would attend neither session, how many students participated in the survey?

37. Analyzing Survey Data   In a survey of 100 investors in the stock market,

50 owned shares in IBM
40 owned shares in AT&T
45 owned shares in GE
20 owned shares in both IBM and GE
15 owned shares in both AT&T and GE
20 owned shares in both IBM and AT&T
5 owned shares in all three

(a) How many of the investors surveyed did not have shares in any of the three companies?
(b) How many owned only IBM shares?
(c) How many owned only GE shares?
(d) How many owned neither IBM nor GE?
(e) How many owned either IBM or AT&T but no GE?

38. Classifying Blood Types   Human blood is classified as either Rh+ or Rh−. Blood is also classified by type: A, if it contains an A antigen; B, if it contains a B antigen; AB, if it contains both A and B antigens; and O, if it contains neither antigen. Draw a Venn diagram illustrating the various blood types. Based on this classification, how many different kinds of blood are there?

 39. Make up a problem different from any found in the text that requires the addition principle of counting to solve. Give it to a friend to solve and critique.

40. Investigate the notion of counting as it relates to infinite sets. Write an essay on your findings.

9.7 | PERMUTATIONS AND COMBINATIONS

> 1 Solve Counting Problems Using the Multiplication Principle
> 2 Solve Counting Problems Using Permutations
> 3 Solve Counting Problems Using Combinations

1 Counting plays a major role in many diverse areas, such as probability, statistics, and computer science. In this section we shall look at special types of counting problems and develop general formulas for solving them.

We begin with an example that will demonstrate a general counting principle.

E X A M P L E   1

Counting the Number of Possible Meals

The fixed-price dinner at a restaurant provides the following choices:

Appetizer: soup or salad
Entree: baked chicken, broiled beef patty, baby beef liver, or roast beef au jus
Dessert: ice cream or cheese cake

How many different meals can be ordered?

Solution   Ordering such a meal requires three separate decisions:

Choose an Appetizer   **Choose an Entree**   **Choose a Dessert**
2 choices   4 choices   2 choices

Look at the **tree diagram** in Figure 19. We see that, for each choice of appetizer, there are 4 choices of entrees. And for each of these $2 \cdot 4 = 8$ choices, there are 2 choices for dessert. Thus, there are a total of

$$2 \cdot 4 \cdot 2 = 16$$

different meals that can be ordered.

FIGURE 19

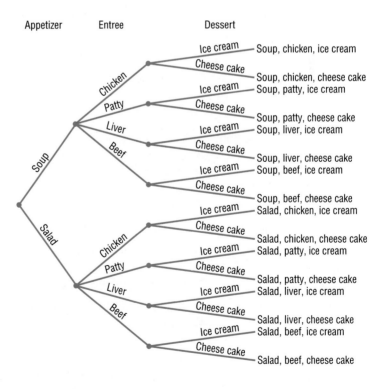

Example 1 illustrates a general counting principle.

> ### Theorem   Multiplication Principle of Counting
>
> If a task consists of a sequence of choices in which there are p selections for the first choice, q selections for the second choice, r selections for the third choice, and so on, then the task of making these selections can be done in
>
> $$p \cdot q \cdot r \cdot \ \ldots$$
>
> different ways.

E X A M P L E   2   Counting Airport Codes

The International Airline Transportation Association (IATA) assigns three-letter codes to represent airport locations. For example, JFK represents Kennedy International in New York. How many different airport codes are possible?

Solution   The task consists of making three selections. Each selection requires choosing a letter of the alphabet (26 choices). Thus, by the Multiplication Principle, there are

$$26 \cdot 26 \cdot 26 = 17,576$$

different airport codes.

In Example 2, we were allowed to repeat a letter. For example, a valid airport code is FLL (Ft. Lauderdale International Airport), in which the letter L appears twice. In the next example, such repetition is not allowed.

E X A M P L E   3   Counting without Repetition

Suppose that we wish to establish a three-letter code using any of the 26 letters of the alphabet, but we require that no letter be used more than once. How many different three-letter codes are there?

Solution   The task consists of making three selections. The first selection requires choosing from 26 letters. Because no letter can be used more than once, the second selection requires choosing from 25 letters. The third selection requires choosing from 24 letters. (Do you see why?) By the Multiplication Principle, there are

$$26 \cdot 25 \cdot 24 = 15,600$$

different three-letter codes with no letter repeated.

 Now work Problems 25 and 29.

Example 3 illustrates a type of counting problem referred to as a *permutation.*

> A **permutation** is an ordered arrangement of n distinct objects without repetitions. The symbol $P(n, r)$ represents the number of permutations of n distinct objects, taken r at a time, where $r \le n$.

For example, the question posed in Example 3 asks for the number of ways the 26 letters of the alphabet can be arranged using three nonrepeated letters. The answer is

$$P(26, 3) = 26 \cdot 25 \cdot 24 = 15,600$$

To arrive at a formula for $P(n, r)$, we note that the task of obtaining an ordered arrangement of n objects in which only $r \le n$ of them are used, without repeating any of them, requires making r selections. For the first selection, there are n choices; for the second selection, there are $n - 1$ choices; for the third selection, there are $n - 2$ choices; ...; for the rth selection, there are $n - (r - 1)$ choices. By the Multiplication Principle, we have

$$
\begin{array}{cccc}
\text{1st} & \text{2nd} & \text{3rd} & r\text{th} \\
\end{array}
$$
$$P(n, r) = n \cdot (n - 1) \cdot (n - 2) \cdot \ldots \cdot [n - (r - 1)]$$
$$= n \cdot (n - 1) \cdot (n - 2) \cdot \ldots \cdot (n - r + 1)$$

This formula for $P(n, r)$ can be compactly written using factorial notation.*

$$P(n, r) = n \cdot (n - 1) \cdot (n - 2) \cdot \ldots \cdot (n - r + 1)$$
$$= n \cdot (n - 1) \cdot (n - 2) \cdot \ldots \cdot (n - r + 1) \cdot \frac{(n - r) \cdot \ldots \cdot 3 \cdot 2 \cdot 1}{(n - r) \cdot \ldots \cdot 3 \cdot 2 \cdot 1} = \frac{n!}{(n - r)!}$$

Theorem **Number of Permutations of n Distinct Objects Taken r at a Time**

The number of different arrangements of n objects using $r \le n$ of them, in which

1. the n objects are distinct,
2. once an object is used it cannot be repeated, and
3. order is important

is given by the formula

$$P(n, r) = \frac{n!}{(n - r)!} \qquad (1)$$

E X A M P L E 4   Computing Permutations

Evaluate:

(a) $P(7, 3)$ (b) $P(6, 1)$ (c) $P(52, 5)$

*Recall that $0! = 1, 1! = 1, 2! = 2 \cdot 1, \ldots, n! = n(n - 1) \cdot \ldots \cdot 3 \cdot 2 \cdot 1$.

Solution   We shall work parts (a) and (b) in two ways.

(a) $P(7, 3) = \underbrace{7 \cdot 6 \cdot 5}_{3 \text{ factors}} = 210$

or

$$P(7, 3) = \frac{7!}{(7 - 3)!} = \frac{7!}{4!} = \frac{7 \cdot 6 \cdot 5 \cdot 4!}{4!} = 210$$

(b) $P(6, 1) = \underbrace{6}_{1 \text{ factor}} = 6$

FIGURE 20

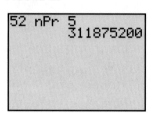

52 nPr 5
          311875200

or

$$P(6, 1) = \frac{6!}{(6 - 1)!} = \frac{6!}{5!} = \frac{6 \cdot 5!}{5!} = 6$$

(c) We use a calculator.* Figure 20 shows the solution using a TI-83 graphing calculator. Thus, $P(52, 5) = 311,875,200$. ∎

 Now work Problem 1.

E X A M P L E   5   Lining Up 5 People

In how many ways can 5 people be lined up?

Solution   The 5 people are obviously distinct. Once a person is in line, that person will
not be repeated elsewhere in the line; and, in lining up people, order is important. Thus, we have a permutation of 5 objects taken 5 at a time. We can
line up 5 people in

$$P(5, 5) = \underbrace{5 \cdot 4 \cdot 3 \cdot 2 \cdot 1}_{5 \text{ factors}} = 5! = 120 \text{ ways}$$

∎

 Now work Problem 31.

Combinations

In a permutation, order is important; for example, the arrangements ABC,
CAB, BAC, . . . are considered different arrangements of the letters A, B,
and C. In many situations, though, order is unimportant. For example, in the
card game of poker, the order in which the cards are received does not matter; it is the *combination* of the cards that matters.

> A **combination** is an arrangement, without regard to order, of n distinct objects
> without repetitions. The symbol $C(n, r)$ represents the number of combinations
> of n distinct objects taken r at a time, where $r \leq n$.

*On most calculators, $P(n, r)$ is represented as nPr.

E X A M P L E   6
Listing Combinations

List all the combinations of the 4 objects a, b, c, d taken 2 at a time. What is $C(4, 2)$?

Solution   One combination of a, b, c, d taken 2 at a time is

ab

The object ba is excluded, because order is not important in a combination. The list of all such combinations (convince yourself of this) is

ab, ac, ad, bc, bd, cd

Thus,

$C(4, 2) = 6$ ▬

We can find a formula for $C(n, r)$ by noting that the only difference between a permutation and a combination is that we disregard order in combinations. Thus, to determine $C(n, r)$, we need only eliminate from the formula for $P(n, r)$ the number of permutations that were simply re-arrangements of a given set of r objects. But that is easily determined from the formula for $P(n, r)$ by calculating $P(r, r) = r!$. So, if we divide $P(n, r)$ by $r!$, we will have the desired formula for $C(n, r)$:

$$C(n, r) = \frac{P(n, r)}{r!} \underset{\substack{\uparrow \\ \text{Use formula (1).}}}{=} \frac{n!/(n - r)!}{r!} = \frac{n!}{(n - r)!r!}$$

We have proved the following result.

Theorem   **Number of Combinations of n Distinct Objects Taken r at a Time**

The number of different arrangements of n objects using $r \leq n$ of them, in which

1. the n objects are distinct,
2. once an object is used, it cannot be repeated, and
3. order is not important

is given by the formula

$$C(n, r) = \frac{n!}{(n - r)!r!} \tag{2}$$

▬

Based on formula (2), we discover that the symbol $C(n, r)$ and the symbol $\binom{n}{r}$ for the binomial coefficients are, in fact, equivalent. Thus, the Pascal triangle (see Section 9.5) can be used to find the value of $C(n, r)$. However, because it is more practical and convenient, we will use formula (2) instead.

E X A M P L E   7
Using Formula (2)

Use formula (2) to find the value of each expression.

(a) $C(3, 1)$ (b) $C(6, 3)$ (c) $C(n, n)$ (d) $C(n, 0)$ (e) $C(52, 5)$

Solution (a) $C(3, 1) = \dfrac{3!}{(3-1)!1!} = \dfrac{3!}{2!1!} = \dfrac{3 \cdot 2 \cdot 1}{2 \cdot 1 \cdot 1} = 3$

FIGURE 21

(b) $C(6, 3) = \dfrac{6!}{(6-3)!3!} = \dfrac{6 \cdot 5 \cdot 4 \cdot 3!}{3! \cdot 3!} = \dfrac{6 \cdot 5 \cdot 4}{6} = 20$

(c) $C(n, n) = \dfrac{n!}{(n-n)!n!} = \dfrac{n!}{0!n!} = \dfrac{1}{1} = 1$

(d) $C(n, 0) = \dfrac{n!}{(n-0)!0!} = \dfrac{n!}{n!0!} = \dfrac{1}{1} = 1$

(e) We use a calculator. Figure 21 shows the solution using a TI-83 graphing calculator. Thus, $C(52, 5) = 2{,}598{,}960$. ∎

 Now work Problem 9.

E X A M P L E   8

Forming Committees

How many different committees of 3 people can be formed from a pool of 7 people?

Solution The 7 people are, of course, distinct. More important, though, is the observation that the order of being selected for a committee is not significant. Thus, the problem asks for the number of combinations of 7 objects taken 3 at a time:

$$C(7, 3) = \frac{7!}{4!3!} = \frac{7 \cdot 6 \cdot 5 \cdot 4!}{4!3!} = \frac{7 \cdot 6 \cdot 5}{6} = 35$$

∎

E X A M P L E   9

Forming Committees

In how many ways can a committee consisting of 2 faculty members and 3 students be formed if there are 6 faculty members and 10 students eligible to serve on the committee?

Solution The problem can be separated into two parts: the number of ways the faculty members can be chosen, $C(6, 2)$, and the number of ways the student members can be chosen, $C(10, 3)$. By the Multiplication Principle, the committee can be formed in

$$C(6, 2) \cdot C(10, 3) = \frac{6!}{4!2!} \cdot \frac{10!}{7!3!} = \frac{6 \cdot 5 \cdot 4!}{4!2!} \cdot \frac{10 \cdot 9 \cdot 8 \cdot 7!}{7!3!}$$

$$= \frac{30}{2} \cdot \frac{720}{6} = 1800 \text{ ways}$$

∎

Now work Problem 47.

Permutations with Repetition

Recall that a permutation involves counting *distinct* objects. A permutation in which some of the objects are repeated is called a **permutation with repetition.** Some books refer to this as a **nondistinguishable permutation.** Let's look at an example.

E X A M P L E   10   Forming Different Words

How many different words can be formed using all the letters in the word REARRANGE?

Solution   Each word formed will have 9 letters: 3 R's, 2 A's, 2 E's, 1 N, and 1 G. To construct each word, we need to fill in 9 positions with the 9 letters:

$$\overline{1}\ \overline{2}\ \overline{3}\ \overline{4}\ \overline{5}\ \overline{6}\ \overline{7}\ \overline{8}\ \overline{9}$$

The process of forming a word consists of five tasks:

Task 1:   Choose the positions for the 3 R's.
Task 2:   Choose the positions for the 2 A's.
Task 3:   Choose the positions for the 2 E's.
Task 4:   Choose the position for the 1 N.
Task 5:   Choose the position for the 1 G.

Task 1 can be done in $C(9, 3)$ ways. There then remain 6 positions to be filled, so Task 2 can be done in $C(6, 2)$ ways. There remain 4 positions to be filled, so Task 3 can be done in $C(4, 2)$ ways. There remain 2 positions to be filled, so Task 4 can be done in $C(2, 1)$ ways. The last position can be filled in $C(1, 1)$ way. Using the Multiplication Principle, the number of possible words that can be formed is

$$C(9, 3) \cdot C(6, 2) \cdot C(4, 2) \cdot C(2, 1) \cdot C(1, 1) = \frac{9!}{3! \cdot \cancel{6}!} \frac{\cancel{6}!}{2! \cdot \cancel{4}!} \frac{\cancel{4}!}{2! \cdot \cancel{2}!} \frac{\cancel{2}!}{1! \cdot \cancel{1}!} \frac{\cancel{1}!}{0! \cdot 1!}$$

$$= \frac{9!}{3! \cdot 2! \cdot 2! \cdot 1! \cdot 1!} = 15,120$$

The form of the answer to Example 10 is suggestive of a general result. Had the letters in REARRANGE each been different, there would have been $P(9, 9) = 9!$ possible words formed. This is the numerator of the answer. The presence of 3 R's, 2 A's, and 2 E's reduces the number of different words, as the entries in the denominator illustrate. We are led to the following result:

Theorem   Permutations with Repetition

The number of permutations of n objects of which n_1 are of one kind, n_2 are of a second kind, ..., and n_k are of a kth kind is given by

$$\frac{n!}{n_1! \cdot n_2! \cdot \ldots \cdot n_k!} \tag{3}$$

where $n = n_1 + n_2 + \cdots + n_k$.

E X A M P L E   11   Arranging Flags

How many different vertical arrangements are there of 8 flags if 4 are white, 3 are blue, and 1 is red?

Solution   We seek the number of permutations of 8 objects, of which 4 are of one kind, 3 of a second kind, and 1 of a third kind. Using formula (3), we find that there are

$$\frac{8!}{4! \cdot 3! \cdot 1!} = \frac{8 \cdot 7 \cdot 6 \cdot 5 \cdot 4!}{4! \cdot 3! \cdot 1!} = 280 \text{ different arrangements:}$$

Now work Problem 53.

9.7 | EXERCISES

In Problems 1–8, find the value of each permutation.

1. $P(6, 2)$ 2. $P(7, 2)$ 3. $P(5, 5)$ 4. $P(4, 4)$

5. $P(8, 0)$ 6. $P(9, 0)$ 7. $P(8, 4)$ 8. $P(8, 3)$

In Problems 9–16, use formula (2) to find the value of each combination.

9. $C(8, 2)$ 10. $C(8, 6)$ 11. $C(6, 4)$ 12. $C(6, 2)$

13. $C(15, 15)$ 14. $C(18, 1)$ 15. $C(26, 13)$ 16. $C(18, 9)$

17. List all the permutations of 5 objects $a, b, c, d,$ and e taken 3 at a time. What is $P(5, 3)$?

18. List all the permutations of 5 objects $a, b, c, d,$ and e taken 2 at a time. What is $P(5, 2)$?

19. List all the permutations of 4 objects $1, 2, 3,$ and 4 taken 3 at a time. What is $P(4, 3)$?

20. List all the permutations of 6 objects $1, 2, 3, 4, 5,$ and 6 taken 3 at a time. What is $P(6, 3)$?

21. List all the combinations of the 5 objects $a, b, c, d,$ and e taken 3 at a time. What is $C(5, 3)$?

22. List all the combinations of the 5 objects $a, b, c, d,$ and e taken 2 at a time. What is $C(5, 2)$?

23. List all the combinations of the 4 objects $1, 2, 3,$ and 4 taken 3 at a time. What is $C(4, 3)$?

24. List all the combinations of the 6 objects $1, 2, 3, 4, 5,$ and 6 taken 3 at a time. What is $C(6, 3)$?

25. A man has 6 shirts and 4 ties. How many different shirt and tie combinations can he wear?

26. A woman has 5 blouses and 4 skirts. How many different outfits can she wear?

27. **Forming Codes** How many two-letter codes can be formed using the letters $A, B, C,$ and D? Repeated letters are allowed.

28. **Forming Codes** How many two-letter codes can be formed using the letters $A, B, C, D,$ and E? Repeated letters are allowed.

29. **Forming Numbers** How many three-digit numbers can be formed using the digits 0 and 1? Repeated digits are allowed.

30. **Forming Numbers** How many three-digit numbers can be formed using the digits $0, 1, 2, 3, 4, 5, 6, 7, 8,$ and 9? Repeated digits are allowed.

31. In how many ways can 4 people be lined up?

32. In how many ways can 5 different boxes be stacked?

33. **Forming Codes** How many different three-letter codes are there if only the letters $A, B, C, D,$ and E can be used and no letter can be used more than once?

34. **Forming Codes** How many different four-letter codes are there if only the letters $A, B, C, D, E,$ and F can be used and no letter can be used more than once?

35. **Arranging Letters** How many arrangements are there of the letters in the word MONEY?

36. **Arranging Digits** How many arrangements are there of the digits in the number 51,342?

37. Establishing Committees In how many ways can a committee of 3 students be formed from a pool of 7 students?

38. Establishing Committees In how many ways can a committee of 4 professors be formed from a department having 8 professors?

39. Possible Answers on a True/False Test How many arrangements of answers are possible for a true/false test with 10 questions?

40. Possible Answers on a Multiple-choice Test How many arrangements of answers are possible in a multiple-choice test with 5 questions, each of which has 4 possible answers?

41. How many four-digit numbers can be formed using the digits 0, 1, 2, 3, 4, 5, 6, 7, 8, and 9 if the first digit cannot be 0? Repeated digits are allowed.

42. How many five-digit numbers can be formed using the digits 0, 1, 2, 3, 4, 5, 6, 7, 8, and 9 if the first digit cannot be 0 or 1? Repeated digits are allowed.

43. Arranging Books Five different mathematics books are to be arranged on a student's desk. How many arrangements are possible?

44. Forming License Plate Numbers How many different license plate numbers can be made using 2 letters followed by 4 digits selected from the digits 0 through 9, if
(a) Letters and digits may be repeated?
(b) Letters may be repeated, but digits are not repeated?
(c) Neither letters nor digits may be repeated?

45. Stock Portfolios As a financial planner, you are asked to select one stock each from the following groups: 8 DOW stocks, 15 NASDAQ stocks, and 4 global stocks. How many different portfolios are possible?

46. Combination Locks A combination lock has 50 numbers on it. To open it, you turn to a number, then rotate clockwise to a second number, and then counterclockwise to the third number. How many different lock combinations are there?

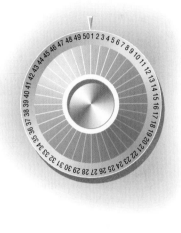

47. A student dance committee is to be formed consisting of 3 boys and 3 girls. If the membership is to be chosen from 6 boys and 8 girls, how many different committees are possible?

48. Baseball Teams A baseball team has 15 members. Five of the players are pitchers, and the remaining 10 members can play any position. How many different teams of 9 players can be formed?

49. The student relations committee of a college consists of 2 administrators, 3 faculty members, and 5 students. There are 4 administrators, 8 faculty members, and 20 students eligible to serve. How many different committees are possible?

50. Football Teams A defensive football squad consists of 25 players. Of these, 10 are linemen, 10 are linebackers, and 5 are safeties. How many different teams of 5 linemen, 3 linebackers, and 3 safeties can be formed?

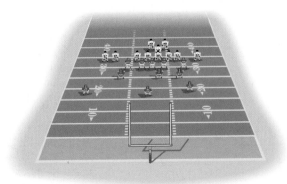

51. Baseball In the American Baseball League, a designated hitter may be used. How many batting orders is it possible for a manager to use? (There are 9 regular players on a team.)

52. Baseball In the National Baseball League, the pitcher usually bats ninth. If this is the case, how many batting orders are possible for a manager to use?

53. Forming Words How many different 9-letter words (real or imaginary) can be formed from the letters in the word ECONOMICS?

54. Forming Words How many different 11-letter words (real or imaginary) can be formed from the letters in the word MATHEMATICS?

55. Senate Committees The U.S. Senate has 100 members. Suppose it is desired to place each senator on exactly 1 of 7 possible committees. The first committee has 22 members, the second has 13, the third has 10, the fourth has 5, the fifth has 16, and the sixth and seventh have 17 apiece. In how many ways can these committees be formed?

56. World Series In the World Series the American League team (*A*) and the National League team (*N*) play until one team wins four games. If the sequence of winners is designated by letters (for example, *NAAAA* means the National League team won the first game and the American League won the next four), how many different sequences are possible?

57. Basketball Teams A basketball team has 6 players who play guard (2 of 5 starting positions). How many different teams are possible, assuming that the remaining 3 positions are filled and it is not possible to distinguish a left guard from a right guard?

58. Basketball Teams On a basketball team of 12 players, 2 only play center, 3 only play guard, and the rest play forward (5 players on a team: 2 forwards, 2 guards, and 1 center). How many different teams are possible, assuming that it is not possible to distinguish left and right guards and left and right forwards?

59. Selecting Objects An urn contains 7 white balls and 3 red balls. Three balls are selected. In how many ways can the 3 balls be drawn from the total of 10 balls:

(a) If 2 balls are white and 1 is red?
(b) If all 3 balls are white?
(c) If all 3 balls are red?

60. Selecting Objects An urn contains 15 red balls and 10 white balls. Five balls are selected. In how many ways can the 5 balls be drawn from the total of 25 balls:

(a) If all 5 balls are red?
(b) If 3 balls are red and 2 are white?
(c) If at least 4 are red balls?

61. Programming Exercise When both *n* and *r* are large, finding *C*(*n*, *r*) on a computer may lead to integers too large to compute. To avoid this, we can approximate the values of *C*(*n*, *r*). One way to do this is the following:

$$C(40, 20) = \frac{40!}{20!20!} = \frac{40 \cdot 39 \cdot 38 \cdot \ldots \cdot 21}{20 \cdot 19 \cdot 18 \cdot \ldots \cdot 1}$$
$$= \frac{40}{20} \cdot \frac{39}{19} \cdot \frac{38}{18} \cdot \ldots \cdot \frac{21}{1}$$
$$\approx 2.000 \cdot 2.053 \cdot 2.1111 \cdot \ldots \cdot 21.000$$
$$= 1.3784652 \times 10^{11}$$

(a) Write a program that inputs two integers *N* and *R* and computes *C*(*N*, *R*) using formula (1).
(b) Use the program to determine where overflow occurs on your computer.
(c) Write a program that inputs two integers *N* and *R* and computes *C*(*N*, *R*) by the approximation technique shown above.
(d) Compare the answers found in parts (a) and (c).

 62. Make up a problem different from any found in the text that requires the Multiplication Principle of counting to solve. Give it to a friend to solve and critique.

63. Make up a problem different from any found in the text that requires a permutation to solve. Give it to a friend to solve and critique.

64. Make up a problem different from any found in the text that requires a combination to solve. Give it to a friend to solve and critique.

65. Explain the difference between a permutation and a combination. Give an example to illustrate your explanation.

9.8 | PROBABILITY

1   Construct Probability Models
2   Utilize the Additive Rule to Find Probabilities
3   Compute Probabilities of Equally Likely Outcomes

Probability is an area of mathematics that deals with experiments that yield random results yet admit a certain regularity. Such experiments do not always produce the same result or outcome, so the result of any one observation is not predictable. However, the results of the experiment over a long period do produce regular patterns that enable us to predict with remarkable accuracy.

EXAMPLE 1

Tossing a Fair Coin

In tossing a fair coin, we know that the outcome is either a head or a tail. On any particular throw, we cannot predict what will happen, but, if we toss the coin many times, we observe that the number of times a head comes up is approximately equal to the number of times we get a tail. It seems reasonable, therefore, to assign a probability of $\frac{1}{2}$ that a head comes up and a probability of $\frac{1}{2}$ that a tail comes up.   ▬

Probability Models

The discussion in Example 1 constitutes the construction of a **probability model** for the experiment of tossing a fair coin once. A probability model has two components: a sample space and an assignment of probabilities. A **sample space** S is a set whose elements represent all the possibilities that can occur as a result of the experiment. Each element of S is called an **outcome.** To each outcome, we assign a number, called the **probability** of that outcome, which has two properties:

1. Each probability is nonnegative.
2. The sum of all the probabilities equals 1.

Thus, if a probability model has the sample space

$$S = \{e_1, e_2, \ldots, e_n\}$$

where $e_1, e_2, \ldots, e_n$ are the possible outcomes, and if $P(e_1), P(e_2), \ldots, P(e_n)$ denote the respective probabilities of these outcomes, then

$$P(e_1) \geq 0, \quad P(e_2) \geq 0, \quad \ldots, \quad P(e_n) \geq 0 \qquad (1)$$

$$P(e_1) + P(e_2) + \cdots + P(e_n) = 1 \qquad (2)$$

Let's look at an example.

EXAMPLE 2

Constructing a Probability Model

An experiment consists of rolling a fair die once.* Construct a probability model for this experiment.

---

*A die is a cube with each face having either 1, 2, 3, 4, 5, or 6 dots on it. See Figure 22.

Solution A sample space S consists of all the possibilities that can occur. Because rolling the die will result in one of six faces showing, the sample space S consists of

FIGURE 22

$$S = \{1, 2, 3, 4, 5, 6\}$$

Because the die is fair, one face is no more likely to occur than another. As a result, our assignment of probabilities is

$$P(1) = \tfrac{1}{6} \qquad P(2) = \tfrac{1}{6}$$
$$P(3) = \tfrac{1}{6} \qquad P(4) = \tfrac{1}{6}$$
$$P(5) = \tfrac{1}{6} \qquad P(6) = \tfrac{1}{6}$$

Suppose that a die is loaded so that the probability assignments are

$$P(1) = 0, \quad P(2) = 0, \quad P(3) = \frac{1}{3}, \quad P(4) = \frac{2}{3}, \quad P(5) = 0, \quad P(6) = 0$$

This assignment would be made if the die was loaded so that only a 3 or 4 could occur and the 4 is twice as likely as the 3 to occur. This assignment is consistent with the definition since each assignment is between 0 and 1, and the sum of all the probability assignments equals 1.

 Now work Problem 13.

E X A M P L E 3

Constructing a Probability Model

An experiment consists of tossing a coin. The coin is weighted so that heads (H) is three times as likely to occur as tails (T). Construct a probability model for this experiment.

Solution The sample space S is $S = \{H, T\}$. If x denotes the probability that a tail occurs, then

$$P(T) = x \quad \text{and} \quad P(H) = 3x$$

Since the sum of the probabilities of the possible outcomes must equal 1, we have

$$P(T) + P(H) = x + 3x = 1$$
$$4x = 1$$
$$x = \frac{1}{4}$$

Thus, we assign the probabilities

$$P(T) = \frac{1}{4} \qquad P(H) = \frac{3}{4}$$

 Now work Problem 17.

E X A M P L E 4

Constructing a Probability Model

An experiment consists of tossing a fair die and then a fair coin. Construct a probability model for this experiment.

Solution
A tree diagram is helpful in listing all the possible outcomes. See Figure 23. The sample space consists of the outcomes

FIGURE 23

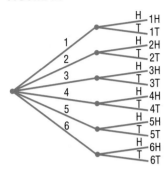

$$S = \{1H, 1T, 2H, 2T, 3H, 3T, 4H, 4T, 5H, 5T, 6H, 6T\}$$

The die and the coin are fair; thus, no one outcome is more likely to occur than another. As a result, we assign the probability $\frac{1}{12}$ to each of the 12 outcomes.   ∎

In working with probability models, the term **event** is used to describe a set of possible outcomes of the experiment. Thus, an event E is some subset of the sample space S. The **probability of an event** E, $E \neq 0$, denoted by $P(E)$, is defined as the sum of the probabilities of the outcomes in E. If $E = \varnothing$, then $P(E) = 0$; if $E = S$, then $P(E) = P(S) = 1$.

E X A M P L E   5

Finding the Probability of an Event

For the experiment described in Example 4, what is the probability that an even number followed by a head occurs?

Solution
The event E, an even number followed by a head, consists of

$$E = \{2H, 4H, 6H\}$$

The probability of E is

$$P(E) = P(2H) + P(4H) + P(6H) = \frac{1}{12} + \frac{1}{12} + \frac{1}{12} = \frac{1}{4} \quad ∎$$

2   The next result, called the **additive rule,** may be used to find the probability of the union of two events.

Theorem   Additive Rule

For any two events E and F,

$$P(E \cup F) = P(E) + P(F) - P(E \cap F) \qquad (3)$$

∎

If E and F are disjoint, so that $E \cap F = \varnothing$, we say that E and F are **mutually exclusive events** and formula (3) takes the form

Mutually Exclusive Events

$$P(E \cup F) = P(E) + P(F) \qquad (4)$$

E X A M P L E   6

Using Formulas (3) and (4)

(a) If $P(E) = 0.2$, $P(F) = 0.3$, and $P(E \cap F) = 0.1$, find $P(E \cup F)$.
(b) If $P(E) = 0.2$, $P(F) = 0.3$, and E, F are mutually exclusive, find $P(E \cup F)$.

Solution   (a) We use the additive rule, formula (3).

$$P(E \cup F) = P(E) + P(F) - P(E \cap F) = 0.2 + 0.3 - 0.1 = 0.4$$

(b) Since E, F are mutually exclusive, we use formula (4).

$$P(E \cup F) = P(E) + P(F) = 0.2 + 0.3 = 0.5$$

A Venn diagram can sometimes be used to obtain probabilities. To construct a Venn diagram representing the information in Example 6(a), we draw two sets E and F. We begin with the fact that $P(E \cap F) = 0.1$. See Figure 24(a). Then, since $P(E) = 0.2$ and $P(F) = 0.3$, we fill in E with $0.2 - 0.1 = 0.1$ and F with $0.3 - 0.1 = 0.2$. See Figure 24(b). Since $P(S) = 1$, we complete the diagram by inserting $1 - [0.1 + 0.1 + 0.2] = 0.6$. See Figure 24(c). Now it is easy to see, for example, that the probability of F, but not E, is 0.2. Also, the probability of neither E nor F is 0.6.

FIGURE 24

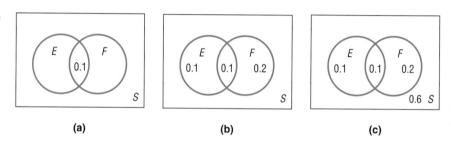

(a)   (b)   (c)

 Now work Problem 23.

Equally Likely Outcomes

3   When the same probability is assigned to each outcome of the sample space, the experiment is said to have **equally likely outcomes.**

Theorem   Probability for Equally Likely Outcomes

If an experiment has n equally likely outcomes, and if the number of ways an event E can occur is m, then the probability of E is

$$P(E) = \frac{\text{Number of ways that } E \text{ can occur}}{\text{Number of all logical possibilities}} = \frac{m}{n} \qquad (5)$$

Thus, if S is the sample space of this experiment, then

$$P(E) = \frac{n(E)}{n(S)} \qquad (6)$$

Based on (6), an alternative method of solution of Example 5 is

$$P(E) = \frac{n(E)}{n(S)} = \frac{3}{12} = \frac{1}{4}$$

E X A M P L E   7

Computing Probabilities for Equally Likely Outcomes

A jar contains 10 marbles; 5 are solid color, 4 are speckled, and 1 is clear.

(a) If one marble is picked at random, what is the probability it is speckled?
(b) If one marble is picked at random, what is the probability it is clear or a solid color?

Solution   The experiment is an example of one in which the outcomes are equally likely; that is, no one marble is more likely to be picked than another. If S is the sample space, then there are 10 possible outcomes in S, so $n(S) = 10$.

(a) Define the event E: Speckled marble is picked. There are 4 ways E can occur. Thus,

$$P(E) = \frac{n(E)}{n(S)} = \frac{4}{10} = 0.4$$

(b) Define the events F: Clear marble is picked and G: Solid color marble is picked. Then there is 1 way for F to occur and 5 ways for G to occur. Thus,

$$P(F) = \frac{n(F)}{n(S)} = \frac{1}{10,} \qquad P(G) = \frac{n(G)}{n(S)} = \frac{5}{10}$$

We seek the probability of the event F or G, that is, $P(F \cup G)$. Since F, G, are mutually exclusive, we use (4).

$$P(F \cup G) = P(F) + P(G) = \frac{1}{10} + \frac{5}{10} = \frac{6}{10} = 0.6$$
■

 Now work Problem 27.

E X A M P L E   8

The Game of Craps

In the game of "craps," two fair dice are rolled. If the total of the faces equals 7 or 11, you win. If the totals are 2, 3, or 12, you have thrown craps and you lose. In all other cases, you throw again.

(a) What is the probability that you will win?
(b) What is the probability that you will lose?
(c) What is the probability that you will need to throw again?

FIGURE 25

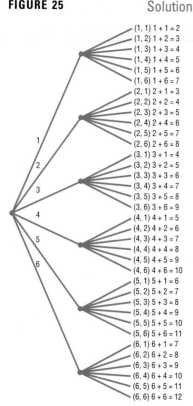

(1, 1) 1 + 1 = 2
(1, 2) 1 + 2 = 3
(1, 3) 1 + 3 = 4
(1, 4) 1 + 4 = 5
(1, 5) 1 + 5 = 6
(1, 6) 1 + 6 = 7
(2, 1) 2 + 1 = 3
(2, 2) 2 + 2 = 4
(2, 3) 2 + 3 = 5
(2, 4) 2 + 4 = 6
(2, 5) 2 + 5 = 7
(2, 6) 2 + 6 = 8
(3, 1) 3 + 1 = 4
(3, 2) 3 + 2 = 5
(3, 3) 3 + 3 = 6
(3, 4) 3 + 4 = 7
(3, 5) 3 + 5 = 8
(3, 6) 3 + 6 = 9
(4, 1) 4 + 1 = 5
(4, 2) 4 + 2 = 6
(4, 3) 4 + 3 = 7
(4, 4) 4 + 4 = 8
(4, 5) 4 + 5 = 9
(4, 6) 4 + 6 = 10
(5, 1) 5 + 1 = 6
(5, 2) 5 + 2 = 7
(5, 3) 5 + 3 = 8
(5, 4) 5 + 4 = 9
(5, 5) 5 + 5 = 10
(5, 6) 5 + 6 = 11
(6, 1) 6 + 1 = 7
(6, 2) 6 + 2 = 8
(6, 3) 6 + 3 = 9
(6, 4) 6 + 4 = 10
(6, 5) 6 + 5 = 11
(6, 6) 6 + 6 = 12

Solution   We begin by constructing a probability model for the experiment. The tree diagram in Figure 25 will help us to see all the possibilities.

Because the dice are fair, no one of the 36 possible outcomes in the sample space S is more likely to occur than any other. Thus, we have equally likely outcomes with $n(S) = 36$.

(a) The event E, "the dice total 7 or 11," consists of the outcomes

$$E = \{(1, 6), (2, 5), (3, 4), (4, 3), (5, 2), (6, 1), (5, 6), (6, 5)\}$$

Because $n(E) = 8$, we have

$$P(E) = \frac{n(E)}{n(S)} = \frac{8}{36} = \frac{2}{9} \approx 0.222$$

(b) The event F, "the dice total 2, 3, or 12," consists of the outcomes

$$F = \{(1, 1), (1, 2), (2, 1), (6, 6)\}$$

Because $n(F) = 4$,

$$P(F) = \frac{n(F)}{n(S)} = \frac{4}{36} = \frac{1}{9} \approx 0.111$$

(c) The number of possibilities that require that you throw again is $36 - n(E) - n(F) = 36 - 8 - 4 = 24$. Thus, the probability that another throw is required is

$$\frac{24}{36} = \frac{2}{3} \approx 0.667$$

Now work Problem 29.

Applications Involving Permutations and Combinations

E X A M P L E   9   Computing Probabilities

Because of a mistake in packaging, 5 defective phones were packaged with 15 good ones. All phones look alike and have equal probability of being chosen. Three phones are selected.

(a) What is the probability that all 3 are defective?
(b) What is the probability that exactly 2 are defective?
(c) What is the probability that at least 2 are defective?

Solution   The sample space S consists of the number of ways 3 objects can be selected from 20 objects, that is, the number of combinations of 20 things taken 3 at a time.

$$n(S) = C(20, 3) = \frac{20!}{17! \cdot 3!} = \frac{20 \cdot 19 \cdot 18}{6} = 1140$$

Each of these outcomes is equally likely to occur.

(a) If E is the event "3 are defective," then the number of elements in E is the number of ways the 3 defective phones can be chosen from the 5 defective phones: $C(5, 3) = 10$. Thus, the probability of E is

$$P(E) = \frac{n(E)}{n(S)} = \frac{10}{1140} \approx 0.0088$$

(b) If F is the event "exactly 2 are defective" and 3 phones are selected, then the number of elements in F is the number of ways to select 2 defective phones from the 5 defective phones and 1 good phone from the 15 good ones. The first of these can be done in $C(5, 2)$ ways and the second in $C(15, 1)$ ways. By the Multiplication Principle, the event F can occur in

$$C(5, 2) \cdot C(15, 1) = \frac{5!}{3! \cdot 2!} \frac{15!}{14! \cdot 1!} = 10 \cdot 15 = 150 \text{ ways}$$

The probability of F is therefore

$$P(F) = \frac{n(F)}{n(S)} = \frac{150}{1140} \approx 0.1316$$

(c) The event G, "at least two are defective," when 3 are chosen, is equivalent to requiring that either exactly 2 defective are chosen or exactly 3 defective are chosen. That is, $G = E \cup F$. Since E and F are mutually exclusive (it is not possible to select 2 defective phones and, at the same time, select 3 defective phones), we find

$$P(G) = P(E) + P(F) \approx 0.0088 + 0.1316 = 0.1404 \qquad \blacksquare$$

 Now work Problem 43.

E X A M P L E   10   Tossing a Coin

A fair coin is tossed 6 times.

(a) What is the probability of obtaining exactly 5 heads and one tail?
(b) What is the probability of obtaining between 4 and 6 heads, inclusive?

Solution   The number of elements in the sample space S is found using the Multiplication Principle. Each toss results in a head (H) or a tail (T). Since the coin is tossed 6 times, we have

$$n(S) = \underbrace{2 \cdot 2 \cdot 2 \cdot 2 \cdot 2 \cdot 2}_{6 \text{ tosses}} = 2^6 = 64$$

The outcomes are equally likely since the coin is fair.

(a) Any sequence that contains 5 heads and 1 tail is determined once the position of the 5 heads (or 1 tail) is known. The number of ways we can position 5 heads in a sequence of 6 slots is $C(6, 5) = 6$. The probability of the event E—exactly 5 heads and one tails—is

$$P(E) = \frac{n(E)}{n(S)} = \frac{C(6, 5)}{2^6} = \frac{6}{64} \approx 0.0938$$

(b) Let F be the event: between 4 and 6 heads, inclusive. To obtain between 4 and 6 heads is equivalent to the event: either 4 heads or 5 heads or 6 heads. Since each of these is mutually exclusive (it is impossible to obtain both 4 heads and 5 heads when tossing a coin 6 times), we have

$$P(F) = P(4 \text{ heads or 5 heads or 6 heads})$$
$$= P(4 \text{ heads}) + P(5 \text{ heads}) + P(6 \text{ heads})$$

The probabilities on the right are obtained as in part (a). Thus

$$P(F) = \frac{C(6, 4)}{2^6} + \frac{C(6, 5)}{2^6} + \frac{C(6, 6)}{2^6} = \frac{15}{64} + \frac{6}{64} + \frac{1}{64} = \frac{22}{64} \approx 0.3438 \qquad \blacksquare$$

HISTORICAL FEATURE Set theory, counting, and probability first took form as a systematic theory in the exchange of letters (1654) between Pierre de Fermat (1601–1665) and Blaise Pascal (1623–1662). They discussed the problem of how to divide the stakes in a game that is interrupted before completion, knowing how many points each player needs to win. Fermat solved the problem by listing all possibilities and counting the favorable ones, whereas Pascal made use of the triangle that now bears his name. As mentioned in the text, the entries in Pascal's triangle are equivalent to $C(n, r)$. This recognition of the role of $C(n, r)$ in counting is the foundation of all further developments.

The first book on probability, the work of Christian Huygens (1629–1695), appeared in 1657. In it, the notion of mathematical expectations is explored. This allows the calculation of the profit or loss a gambler may expect, knowing the probabilities involved in the game (see the Historical Problems that follow).

It is interesting to note that Girolamo Cardano (1501–1576) wrote a treatise on probability, but it was not published until 1663 in Cardano's collected works, and this was too late to have any effect on the development of the theory.

In 1713, the posthumously published *Ars Conjectandi* of Jacob Bernoulli gave the theory the form it would have until 1900. In the current century, both combinatorics (counting) and probability have undergone rapid development due to the use of computers.

A final comment about notation. The notations $C(n, r)$ and $P(n, r)$ are variants of a form of notation developed in England after 1830. The notation $\binom{n}{r}$ for $C(n, r)$ goes back to Leonhard Euler (1707–1783) but is now losing ground because it has no clearly related symbolism of the same type for permutations. The set symbols $\cup$ and $\cap$ were introduced by Giuseppe Peano (1858–1932) in 1888 in a slightly different context. The inclusion symbol $\subset$ was introduced by E. Schroeder (1841–1902) about 1890. The treatment of set theory in the text is due to George Boole (1815–1864), who wrote $A + B$ for $A \cup B$ and AB for $A \cap B$ (statisticians still use AB for $A \cap B$).

HISTORICAL PROBLEMS

1. *The Problem Discussed by Fermat and Pascal* A game between two equally skilled players, A and B, is interrupted when A needs 2 points to win and B needs 3 points. In what proportion should the stakes be divided?

   [**Note:** If each play results in 1 point for either player, at most four more plays will decide the game.]

   (a) *Fermat's solution* List all possible outcomes that will end the game to form the sample space (for example, *ABAA, ABBB,* etc.). The probabilities for A to win and B to win then determine how the stakes should be divided.

   (b) *Pascal's solution* Use combinations to determine the number of ways the 2 points needed for A to win could occur in four plays. Then use combinations to determine the number of ways the 3 points needed for B to win could occur. This is trickier than it looks, since A can win with 2 points in either two plays, three plays, or four plays. Compute the probabilities and compare with the results in part (a).

2. *Huygen's Mathematical Expectation* In a game with n possible outcomes with probabilities $p_1, p_2, \ldots, p_n$, suppose that the *net* winnings are $w_1, w_2, \ldots, w_n$, respectively. Then the mathematical expectation is

   $$E = p_1 w_1 + p_2 w_2 + \cdots + p_n w_n$$

The number E represents the profit or loss per game in the long run. The following problems are a modification of those of Huygens:

(a) A fair die is tossed. A gambler wins $3 if he throws a 6 and $6 if he throws a 5. What is his expectation?

   [**Note:** $w_1 = w_2 = w_3 = w_4 = 0$]

(b) A gambler plays the same game as in part (a), but now the gambler must pay $1 to play. This means $w_5 = \$5$, $w_6 = \$2$, and $w_1 = w_2 = w_3 = w_4 = -\$1$. What is the expectation?

9.8 | EXERCISES

In Problems 1–6, construct a probability model for each experiment.

1. Tossing a fair coin twice

2. Tossing two fair coins once

3. Tossing two fair coins, then a fair die

4. Tossing a fair coin, a fair die, and then a fair coin

5. Tossing three fair coins once

6. Tossing one fair coin three times

In Problems 7–12, use the spinners shown below, and construct a probability model for each experiment.

Spinner I

Spinner II

Spinner III

7. Spin spinner I, then spinner II. What is the probability of getting a 2 or a 4, followed by Red?

8. Spin spinner III, then spinner II. What is the probability of getting Forward, followed by Yellow or Green?

9. Spin spinner I, then II, then III. What is the probability of getting a 1, followed by Red or Green, followed by Backward?

10. Spin spinner II, then I, then III. What is the probability of getting Yellow, followed by a 2 or a 4, followed by Forward?

11. Spin spinner I twice, then spinner II. What is the probability of getting a 2, followed by a 2 or a 4, followed by Red or Green?

12. Spin spinner III, then spinner I twice. What is the probability of getting Forward, followed by a 1 or a 3, followed by a 2 or a 4?

In Problems 13–16, consider the experiment of tossing a coin twice. The table lists six possible assignments of probabilities for this experiment. Using this table, answer the following questions.

13. Which of the assignments of probabilities are consistent with the definition of the probability of an outcome?

14. Which of the assignments of probabilities should be used if the coin is known to be fair?

15. Which of the assignments of probabilities should be used if the coin is known to always come up tails?

16. Which of the assignments of probabilities should be used if tails is twice as likely as heads to occur?

	Sample Space			
Assignments	**HH**	**HT**	**TH**	**TT**
A	$\frac{1}{4}$	$\frac{1}{4}$	$\frac{1}{4}$	$\frac{1}{4}$
B	0	0	0	1
C	$\frac{3}{16}$	$\frac{5}{16}$	$\frac{5}{16}$	$\frac{3}{16}$
D	$\frac{1}{2}$	$\frac{1}{2}$	$-\frac{1}{2}$	$\frac{1}{2}$
E	$\frac{1}{4}$	$\frac{1}{4}$	$\frac{1}{4}$	$\frac{1}{8}$
F	$\frac{1}{9}$	$\frac{2}{9}$	$\frac{2}{9}$	$\frac{4}{9}$

17. Assigning Probabilities A coin is weighted so that heads is four times as likely as tails to occur. What probability should we assign to heads? to tails?

18. Assigning Probabilities A coin is weighted so that tails is twice as likely as heads to occur. What probability should we assign to heads? to tails?

19. Assigning Probabilities A die is weighted so that an odd-numbered face is twice as likely as an even-numbered face. What probability should we assign to each face?

20. Assigning Probabilities A die is weighted so that a six cannot appear. The other faces occur with the same probability. What probability should we assign to each face?

In Problems 21–24, find the probability of the indicated event if $P(A) = 0.25$ and $P(B) = 0.45$.

21. $P(A \cup B)$ if A, B are mutually exclusive

22. $P(A \cap B)$ if A, B are mutually exclusive

23. $P(A \cup B)$ if $P(A \cap B) = 0.15$

24. $P(A \cap B)$ if $P(A \cup B) = 0.6$

In Problems 25–28, a golf ball is selected at random from a container. If the container has 9 white balls, 8 green balls, and 3 orange ones, find the probability of each event.

25. The golf ball is white.

26. The golf ball is green.

27. The golf ball is white or green.

28. The golf ball is not white.

29. What is the probability of throwing a 6 or an 8 in a game of craps? (Consult Example 8.)

30. What is the probability of throwing a 5 or a 9 in a game of craps? (Consult Example 8.)

Problems 31–34 are based on a consumer survey of annual incomes in 100 households. The following table gives the data:

Income	$0–9999	$10,000–19,999	$20,000–29,999	$30,000–39,999	$40,000 or more
Number of households	5	30	35	20	10

31. What is the probability that a household has an annual income of $30,000 or more?

32. What is the probability that a household has an annual income between $10,000 and $29,999, inclusive?

33. What is the probability that a household has an annual income less than $20,000?

34. What is the probability that a household has an annual income of $20,000 or more?

35. Surveys In a survey about the number of TV sets in a house, the following probability table was constructed:
Find the probability of a house having
(a) 1 or 2 TV sets
(b) 1 or more TV sets
(c) 3 or fewer TV sets
(d) 3 or more TV sets
(e) Less than 2 TV sets
(f) Less than 1 TV set
(g) 1, 2, or 3 TV sets
(h) 2 or more TV sets

36. Checkout Lines Through observation it has been determined that the probability for a given number of people waiting in line at the "5 items or less" checkout register of a supermarket is:

Number waiting in line	0	1	2	3	4 or more
Probability	0.10	0.15	0.20	0.24	0.31

Find the probability of
(a) At most 2 people in line
(b) At least 2 people in line
(c) At least 1 person in line

Number of TV sets	0	1	2	3	4 or more
Probability	0.05	0.24	0.33	0.21	0.17

37. Winning a Lottery In a certain lottery, there are ten balls, numbered 1, 2, 3, 4, 5, 6, 7, 8, 9, 10. Of these, five are drawn in order. If you pick five numbers that match those drawn in the correct order, you win $1,000,000. What is the probability of winning such a lottery?

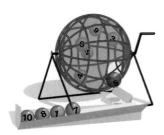

38. A committee of 6 people is to be chosen at random from a group of 14 people consisting of 2 supervisors, 5 skilled laborers, and 7 unskilled laborers. What is the probability that the committee chosen consists of 2 skilled and 4 unskilled laborers?

39. A fair coin is tossed 5 times.
  (a) Find the probability that exactly 3 heads appear.
  (b) Find the probability that no heads appear.

40. A fair coin is tossed 4 times.
  (a) Find the probability that exactly 1 tail appears.
  (b) Find the probability that no more than 1 tail appears.

41. A pair of fair dice is tossed 3 times.
  (a) Find the probability that the sum of 7 appears 3 times.
  (b) Find the probability that a sum of 7 or 11 appears at least twice.

42. A pair of fair dice is tossed 5 times.
  (a) Find the probability that the sum is never 2.
  (b) Find the probability that the sum is never 7.

43. Through a mix-up on the production line, 5 defective TV's were shipped out with 25 good ones. If 5 are selected at random, what is the probability that all 5 are defective? What is the probability that at least 2 of them are defective?

44. In a shipment of 50 transformers, 10 are known to be defective. If 30 transformers are picked at random, what is the probability that all 30 are nondefective? Assume that all transformers look alike and have an equal probability of being chosen.

45. In a promotion, 50 silver dollars are placed in a bag, one of which is valued at more than $10,000. The winner of the promotion is given the opportunity to reach into the bag, while blindfolded, and pull out 5 coins. What is the probability that one of the 5 coins is the one valued at more than $10,000?

 46. Go to the library and look up the "birthday problem" in a book on probability. Write a brief essay about this problem and its solution.

CHAPTER REVIEW

THINGS TO KNOW

Sequence	A function whose domain is the set of positive integers.
Factorials	$0! = 1, 1! = 1, n! = n(n - 1) \cdot \ldots \cdot 3 \cdot 2 \cdot 1$ if $n \geq 2$
Arithmetic sequence	$a_1 = a, a_{n+1} = a_n + d$, where a = first term, d = common difference, $a_n = a + (n - 1)d$
Sum of the first n terms of an arithmetic sequence	$S_n = \dfrac{n}{2}[2a + (n - 1)d] = \dfrac{n}{2}(a + a_n)$
Geometric sequence	$a_1 = a, a_{n+1} = ra_n$; where a = first term, r = common ratio, $a_n = ar^{n-1}, r \neq 0$
Sum of the first n terms of a geometric sequence	$S_n = a\dfrac{1 - r^n}{1 - r}, \quad r \neq 0, 1$
Amount of an annuity	$A = P\dfrac{(1 + i)^n - 1}{i}$
Infinite geometric series	$a + ar + \cdots + ar^{n-1} + \cdots = \displaystyle\sum_{k=1}^{\infty} ar^{k-1}$
Sum of an infinite geometric series	$\displaystyle\sum_{k=1}^{\infty} ar^{k-1} = \dfrac{a}{1 - r}, \quad |r| < 1$

Principle of Mathematical Induction	Condition I: The statement is true for the natural number 1. Condition II: If the statement is true for some natural number k, it is also true for $k + 1$. Then the statement is true for all natural numbers.	

Binomial coefficient	$\binom{n}{j} = \dfrac{n!}{j!(n - j)!}$

Pascal triangle	See Figure 14.

Binomial Theorem	$(x + a)^n = \binom{n}{0}x^n + \binom{n}{1}ax^{n-1} + \cdots + \binom{n}{j}a^j x^{n-j} + \cdots + \binom{n}{n}a^n$

Set		Well-defined collection of distinct objects, called elements
Empty set or Null set	$\varnothing$	Set that has no elements
Equality	$A = B$	A and B have the same elements.
Subset	$A \subseteq B$	Each element of A is also an element of B.
Intersection	$A \cap B$	Set consisting of elements that belong to both A and B
Union	$A \cup B$	Set consisting of elements that belong to either A or B, or both
Universal set	U	Set consisting of all the elements we wish to consider
Complement	A'	Set consisting of elements of the universal set that are not in A
Finite set		The number of elements in the set is a nonnegative integer.
Infinite set		A set that is not finite

Counting formula	$n(A \cup B) = n(A) + n(B) - n(A \cap B)$	
Addition Principle		If $A \cap B = \varnothing$, then $n(A \cup B) = n(A) + n(B)$.
Multiplication Principle		If a task consists of a sequence of choices in which there are p selections for the first choice, q selections for the second choice, and so on, then the task of making these selections can be done in $p \cdot q \cdot \ldots$ different ways.

Permutation	$P(n, r) = n(n - 1) \cdot \ldots \cdot [n - (r - 1)]$ $= \dfrac{n!}{(n - r)!}$	An ordered arrangement of n distinct objects without repetition
Combination	$C(n, r) = \dfrac{P(n, r)}{r!}$ $= \dfrac{n!}{(n - r)!r!}$	An arrangement, without regard to order, of n distinct objects without repetition
Permutations with repetition	$\dfrac{n!}{n_1! n_2! \cdots n_k!}$	The number of permutations of n objects of which n_1 are of one kind, n_2 are of a second kind, . . . , and n_k are of a kth kind, where $n = n_1 + n_2 + \cdots + n_k$
Sample space		Set whose elements represent all the logical possibilities that can occur as a result of an experiment
Probability		A number assigned to each outcome of a sample space; the sum of all the probabilities of the outcomes equals 1
Additive rule	$P(E \cup F) = P(E) + P(F) - P(E \cap F)$	
Equally likely outcomes	$P(E) = \dfrac{n(E)}{n(S)}$	The same probability is assigned to each outcome.

HOW TO:

Write down the terms of a sequence

Use summation notation

Identify an arithmetic sequence

Find the sum of the first n terms of an arithmetic sequence

Identify a geometric sequence

Find the sum of the first n terms of a geometric sequence

Find the sum of an infinite geometric series

Find the sum of arithmetic and geometric sequences using a graphing utility

Solve annuity problems

Prove statements about natural numbers using mathematical induction

Apply the Binomial Theorem

Find unions, intersections, and complements of sets

Use Venn diagrams to illustrate sets

Recognize a permutation problem

Recognize a combination problem

Solve certain probability problems

Count the elements in a sample space

Draw a tree diagram

FILL-IN-THE-BLANK ITEMS

1. A(n) _____ is a function whose domain is the set of positive integers.
2. In a(n) _____ sequence, the difference between successive terms is always the same number.
3. In a(n) _____ sequence, the ratio of successive terms is always the same number.
4. The _____ _____ is a triangular display of the binomial coefficients.
5. $\binom{6}{2} = $ _____
6. The _____ of A with B consists of all elements in either A or B or both; the _____ of A with B consists of all elements in both A and B.
7. $P(5, 2) = $ _____; $C(5, 2) = $ _____.
8. A(n) _____ is an ordered arrangement of n distinct objects.
9. A(n) _____ is an arrangement of n distinct objects without regard to order.
10. When the same probability is assigned to each outcome of a sample space, the experiment is said to have _____ _____ outcomes.

TRUE/FALSE ITEMS

T F **1.** A sequence is a function.

T F **2.** For arithmetic sequences, the difference of successive terms is always the same number.

T F **3.** For geometric sequences, the ratio of successive terms is always the same number.

T F **4.** Mathematical induction can sometimes be used to prove theorems that involve natural numbers.

T F **5.** $\binom{n}{j} = \dfrac{j!}{n!(n-j)!}$

T F **6.** The expansion of $(x + a)^n$ contains n terms.

T F **7.** $\sum\limits_{i=1}^{n+1} i = 1 + 2 + 3 + \cdots + n$

T F **8.** The intersection of two sets is always a subset of their union.

T F **9.** $P(n, r) = \dfrac{n!}{r!}$

T F **10.** In a combination problem, order is not important.

T F **11.** In a permutation problem, once an object is used, it cannot be repeated.

T F **12.** The probability of an event can never equal 0.

REVIEW EXERCISES

Blue problem numbers indicate the author's suggestions for use in a Practice Test.

In Problems 1–8, evaluate each expression.

1. $5!$ **2.** $6!$ **3.** $\binom{5}{3}$ **4.** $\binom{8}{2}$ **5.** $P(8, 3)$ **6.** $P(7, 3)$ **7.** $C(8, 3)$ **8.** $C(7, 3)$

In Problems 9–16, write down the first five terms of each sequence.

9. $\left\{(-1)^n\left(\dfrac{n+3}{n+2}\right)\right\}$ **10.** $\{(-1)^{n+1}(2n+3)\}$ **11.** $\left\{\dfrac{2^n}{n^2}\right\}$

12. $\left\{\dfrac{e^n}{n}\right\}$ **13.** $a_1 = 6; \quad a_{n+1} = \dfrac{2}{3}a_n$ **14.** $a_1 = 8; \quad a_{n+1} = -\dfrac{1}{4}a_n$

15. $a_1 = 2; \quad a_{n+1} = 2 - a_n$ **16.** $a_1 = -3; \quad a_{n+1} = 4 + a_n$

In Problems 17–28, determine whether the given sequence is arithmetic, geometric, or neither. If the sequence is arithmetic, find the common difference and the sum of the first n terms. If the sequence is geometric, find the common ratio and the sum of the first n terms.

17. $\{n + 5\}$ **18.** $\{4n + 3\}$ **19.** $\{2n^3\}$ **20.** $\{2n^2 - 1\}$

21. $\{2^{3n}\}$ **22.** $\{3^{2n}\}$ **23.** $0, 4, 8, 12, \ldots$ **24.** $1, -3, -7, -11, \ldots$

25. $3, \dfrac{3}{2}, \dfrac{3}{4}, \dfrac{3}{8}, \dfrac{3}{16}, \ldots$ **26.** $5, -\dfrac{5}{3}, \dfrac{5}{9}, -\dfrac{5}{27}, \dfrac{5}{81}, \ldots$ **27.** $\dfrac{2}{3}, \dfrac{3}{4}, \dfrac{4}{5}, \dfrac{5}{6}, \ldots$ **28.** $\dfrac{3}{2}, \dfrac{5}{4}, \dfrac{7}{6}, \dfrac{9}{8}, \dfrac{11}{10}, \ldots$

In Problems 29–34, find the sum of each sequence.

29. $\displaystyle\sum_{k=1}^{5} (k^2 + 12)$ **30.** $\displaystyle\sum_{k=1}^{3} (k + 2)^2$ **31.** $\displaystyle\sum_{k=1}^{10} (3k - 9)$ **32.** $\displaystyle\sum_{k=1}^{9} (-2k + 8)$

33. $\displaystyle\sum_{k=1}^{7} \left(\dfrac{1}{3}\right)^k$ **34.** $\displaystyle\sum_{k=1}^{10} (-2)^k$

In Problems 35–40, find the indicated term in each sequence.

35. 9th term of $3, 7, 11, 15, \ldots$ **36.** 8th term of $1, -1, -3, -5, \ldots$

37. 11th term of $1, \dfrac{1}{10}, \dfrac{1}{100}, \ldots$ **38.** 11th term of $1, 2, 4, 8, \ldots$

39. 9th term of $\sqrt{2}, 2\sqrt{2}, 3\sqrt{2}, \ldots$ **40.** 9th term of $\sqrt{2}, 2, 2^{3/2}, \ldots$

In Problems 41–44, find a general formula for each arithmetic sequence.

41. 7th term is 31;   20th term is 96   **42.** 8th term is -20;   17th term is -47

43. 10th term is 0;   18th term is 8   **44.** 12th term is 30;   22nd term is 50

In Problems 45–50, find the sum of each infinite geometric series.

45. $3 + 1 + \dfrac{1}{3} + \dfrac{1}{9} + \cdots$ **46.** $2 + 1 + \dfrac{1}{2} + \dfrac{1}{4} + \cdots$ **47.** $2 - 1 + \dfrac{1}{2} - \dfrac{1}{4} + \cdots$

48. $6 - 4 + \dfrac{8}{3} - \dfrac{16}{9} + \cdots$ **49.** $\displaystyle\sum_{k=1}^{\infty} 4\left(\dfrac{1}{2}\right)^{k-1}$ **50.** $\displaystyle\sum_{k=1}^{\infty} 3\left(-\dfrac{3}{4}\right)^{k-1}$

In Problems 51–56, use the Principle of Mathematical Induction to show that the given statement is true for all natural numbers.

51. $3 + 6 + 9 + \cdots + 3n = \dfrac{3n}{2}(n + 1)$ **52.** $2 + 6 + 10 + \cdots + (4n - 2) = 2n^2$

53. $2 + 6 + 18 + \cdots + 2\cdot 3^{n-1} = 3^n - 1$ **54.** $3 + 6 + 12 + \cdots + 3\cdot 2^{n-1} = 3(2^n - 1)$

55. $1^2 + 4^2 + 7^2 + \cdots + (3n - 2)^2 = \dfrac{1}{2}n(6n^2 - 3n - 1)$

56. $1\cdot 3 + 2\cdot 4 + 3\cdot 5 + \cdots + n(n + 2) = \dfrac{n}{6}(n + 1)(2n + 7)$

In Problems 57–60, expand each expression using the Binomial Theorem.

57. $(x + 3)^5$

58. $(x - 3)^5$

59. $(2x + 3)^4$

60. $(3x - 4)^4$

61. Find the coefficient of x^7 in the expansion of $(x + 2)^9$.

62. Find the coefficient of x^3 in the expansion of $(x - 3)^8$.

63. Find the coefficient of x^3 in the expansion of $(2x + 1)^7$.

64. Find the coefficient of x^3 in the expansion of $(2x + 1)^8$.

In Problems 65–72, use U = Universal set = {1, 2, 3, 4, 5, 6, 7, 8, 9}, A = {1, 3, 5, 7}, B = {3, 5, 6, 7, 8}, and C = {2, 3, 7, 8, 9} to find each set.

65. $A \cup B$

66. $B \cup C$

67. $A \cap C$

68. $A \cap B$

69. $A' \cup B'$

70. $B' \cap C'$

71. $(B \cap C)'$

72. $(A \cup B)'$

73. If $n(A) = 8$, $n(B) = 12$, and $n(A \cap B) = 3$, find $n(A \cup B)$.

74. If $n(A) = 12$, $n(A \cup B) = 30$, and $n(A \cap B) = 6$, find $n(B)$.

In Problems 75–80, use the information supplied in the figure:

75. How many are in A?

76. How many are in A or B?

77. How many are in A and C?

78. How many are not in B?

79. How many are in neither A nor C?

80. How many are in B but not in C?

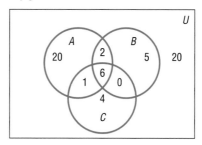

81. A clothing store sells pure wool and polyester/wool suits. Each suit comes in 3 colors and 10 sizes. How many suits are required for a complete assortment?

82. In connecting a certain electrical device, 5 wires are to be connected to 5 different terminals. How many different wirings are possible if 1 wire is connected to each terminal?

83. Baseball On a given day, the American Baseball League schedules 7 games. How many different outcomes are possible, assuming that each game is played to completion?

84. Baseball On a given day, the National Baseball League schedules 6 games. How many different outcomes are possible, assuming that each game is played to completion?

85. If 4 people enter a bus having 9 vacant seats, in how many ways can they be seated?

86. How many different arrangements are there of the letters in the word ROSE?

87. In how many ways can a squad of 4 relay runners be chosen from a track team of 8 runners?

88. A professor has 10 similar problems to put on a test with 3 problems. How many different tests can she design?

89. Baseball In how many different ways can the 14 baseball teams in the American League be paired without regard to which team is at home?

90. Arranging Books on a Shelf There are 5 different French books and 5 different Spanish books. How many ways are there to arrange them on a shelf if
  (a) Books of the same language must be grouped together, French on the left, Spanish on the right?
  (b) French and Spanish books must alternate in the grouping, beginning with a French book?

91. Telephone Numbers Using the digits 0, 1, 2, . . . , 9, how many 7-digit numbers can be formed if the first digit cannot be 0 or 9 and if the last digit is greater than or equal to 2 or less than or equal to 3? Repeated digits are allowed.

92. Home Choices A contractor constructs homes with 5 different choices of exterior finish, 3 different roof arrangements, and 4 different window designs. How many different types of homes can be built?

93. License Plate Possibilities A license plate consists of 1 letter, excluding O and I, followed by a 4-digit number that cannot have a 0 in the lead position. How many different plates are possible?

94. Using the digits 0 and 1, how many different numbers consisting of 8 digits can be formed?

95. Forming Different Words How many different words can be formed using all the letters in the word MISSING?

96. Arranging Flags How many different vertical arrangements are there of 10 flags, if 4 are white, 3 are blue, 2 are green, and 1 is red?

97. Forming Committees A group of 9 people is going to be formed into committees of 4, 3, and 2 people. How many committees can be formed if
(a) A person can serve on any number of committees?
(b) No person can serve on more than one committee?

98. Forming Committees A group consists of 5 men and 8 women. A committee of 4 is to be formed from this group, and policy dictates that at least 1 woman be on this committee.
(a) How many committees can be formed that contain exactly 1 man?
(b) How many committees can be formed that contain exactly 2 women?
(c) How many committees can be formed that contain at least 1 man?

99. From a box containing three 40-watt bulbs, six 60-watt bulbs, and eleven 75-watt bulbs, a bulb is drawn at random. What is the probability that the bulb is 40 watts? What is the probability that it is not a 75 watt bulb?

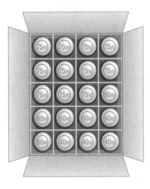

100. You have four $1 bills, three $5 bills, and two $10 bills in your wallet. If you pick a bill at random, what is the probability it will be a $1 bill?

101. Each of the letters in the word ROSE is written on an index card and the cards are then shuffled. What is the probability that, when the cards are dealt, they spell the word ROSE?

102. Each of the numbers $1, 2, \ldots, 100$ is written on an index card and the cards are then shuffled. If a card is selected at random, what is the probability that the number on the card is divisible by 5? What is the probability that the card selected either is a 1 or names a prime number?

103. Computing Probabilities Because of a mistake in packaging, a case of 12 bottles of red wine contained 5 Merlot and 7 Cabernet, each without labels. All the bottles look alike and have equal probability of being chosen. Three bottles are selected.
(a) What is the probability all 3 are Merlot?
(b) What is the probability exactly 2 are Merlot?
(c) What is the probability none is a Merlot?

104. Tossing a Coin A fair coin is tossed 10 times.
(a) What is the probability of obtaining exactly 5 heads?
(b) What is the probability of obtaining all heads?

105. Constructing a Brick Staircase A brick staircase has a total of 25 steps. The bottom step requires 80 bricks. Each successive step requires three less bricks than the prior step.
(a) How many bricks are required for the top step?
(b) How many bricks are required to build the staircase?

106. Creating a Floor Design A mosaic tile floor is designed in the shape of a trapezoid 30 feet wide at the base and 15 feet wide at the top. See Figure 9, page 648. The tiles, 12 inches by 12 inches, are to be placed so that each successive row contains one less tile than the row below. How many tiles will be required?

107. Retirement Planning Chris gets paid once a month and contributes $200 each pay period into his 401(k). If Chris plans on retiring in 20 years, what will the value of his 401(k) be if the per annum rate of return of the 401(k) is 10% compounded monthly?

108. Retirement Planning Jacky contributes $500 every quarter into an IRA. If Jacky plans on retiring in 30 years, what will the value of the IRA be if the per annum rate of return of the IRA is 8% compounded quarterly?

109. Bouncing Balls A ball is dropped from a height of 20 feet. Each time it strikes the ground, it bounces up to $\frac{3}{4}$ of the previous height.
(a) What height will the ball bounce up to after it strikes the ground for the third time?

(b) What is its height after it strikes the ground for the nth time?

(c) How many times does the ball need to strike the ground before its height is less than 6 inches?

(d) What total distance does the ball travel before it stops bouncing?

110. **Salary Increases** Your friend has just been hired at an annual salary of $20,000. If she expects to receive annual increases of 4%, what will her salary be as she begins her fifth year?

111. At the Milex tune-up and brake repair shop, the manager has found that a car will require a tune-up with a probability of 0.6, a brake job with a probability of 0.1, and both with a probability of 0.02.

(a) What is the probability that a car requires either a tune-up or a brake job?

(b) What is the probability that a car requires a tune-up but not a brake job?

(c) What is the probability that a car requires neither type of repair?

3.

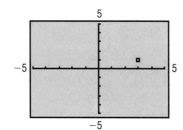

4.

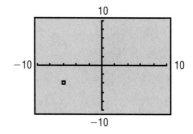

In Problems 5–10, select a RANGE setting so that each of the given points will lie within the viewing rectangle.

5. $(-10, 5), (3, -2), (4, -1)$

6. $(5, 0), (6, 8), (-2, -3)$

7. $(40, 20), (-20, -80), (10, 40)$

8. $(-80, 60), (20, -30), (-20, -40)$

9. $(0, 0), (100, 5), (5, 150)$

10. $(0, -1), (100, 50), (-10, 30)$

In Problems 11–20, determine the RANGE settings used for each viewing rectangle.

11.

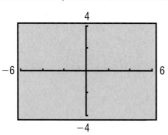

12.

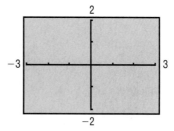

13.

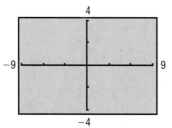

14.

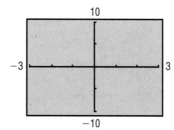

15.

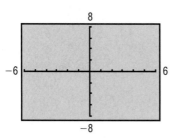

16.

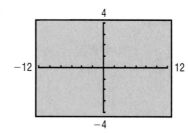

17.

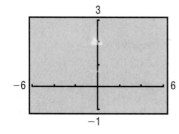

18.

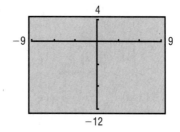

19.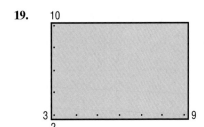

20.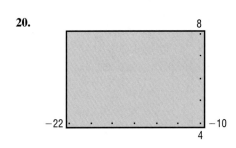

A.2 | USING A GRAPHING UTILITY TO GRAPH EQUATIONS

Examples 1 and 2 of Section 3.2 demonstrate that a graph can be obtained by plotting points in a rectangle coordinate system and connecting them. Graphing utilities perform these same steps when graphing an equation. For example, the TI-83 determines 95 evenly spaced input values,* uses the equation to determine the output values, plots these points on the screen, and, finally (if in the connected mode), draws a line segment between consecutive points.

Most graphing utilities require the following steps in order to obtain the graph of an equation:

Steps for Graphing an Equation Using a Graphing Utility

> STEP 1: Solve the equation for y in terms of x.
>
> STEP 2: Get into the graphing mode of your graphing utility. The screen will usually display $y =$, prompting you to enter the expression involving x that you found in Step 1. (Consult your manual for the correct way to enter the expression; for example, $y = x^2$ might be entered as $x\^2$ or as $x * x$ or as $x\ x^Y\ 2$).
>
> STEP 3: Select the viewing rectangle. Without prior knowledge about the behavior of the graph of the equation, it is common to select the **standard viewing rectangle** initially. The viewing rectangle is then adjusted based on the graph that appears. In this text, the standard viewing rectangle will be
>
> $$X\text{min} = -10 \quad X\text{max} = 10 \quad X\text{scl} = 1$$
> $$Y\text{min} = -10 \quad Y\text{max} = 10 \quad Y\text{scl} = 1$$
>
> STEP 4: Execute (or graph).
>
> STEP 5: Adjust the viewing rectangle until a complete graph is obtained.

EXAMPLE 1

Graphing an Equation Using a Graphing Utility

Graph the equation: $6x^2 + 3y = 36$.

Solution STEP 1: We solve for y in terms of x.

$$6x^2 + 3y = 36$$
$$3y = -6x^2 + 36 \quad \text{Subtract } 6x^2 \text{ from both sides of the equation.}$$
$$y = -2x^2 + 12 \quad \text{Divide both sides of the equation by 3.}$$

*These input values depend on the values of Xmin and Xmax. For example, if Xmin $= -10$ and Xmax $= 10$, then the first input value will be -10 and the next input value will be $-10 + (10 - (-10))/94 \approx -9.7872$, and so on.

STEP 2: From the graphing mode, enter the expression $-2x^2 + 12$ after the prompt $y = \underline{}$.

STEP 3: Set the viewing rectangle to the standard viewing rectangle.

STEP 4: Execute (or graph). The screen should look like Figure 6.

FIGURE 6

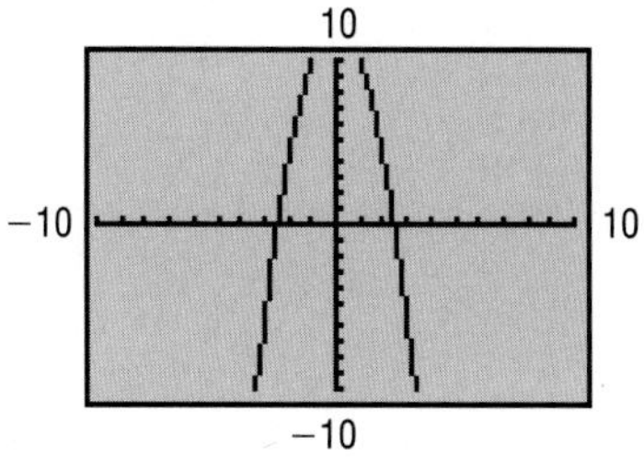

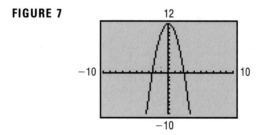

STEP 5: The graph of $y = -2x^2 + 12$ is not complete. The value of Ymax must be increased so that the top portion of the graph is visible. After increasing the value of Ymax to 12, we obtain the graph in Figure 7. The graph is now complete.

FIGURE 7

Look again at Figure 7. Although a complete graph is shown, the graph might be improved by adjusting the values of Xmin and Xmax. Figure 8 shows the graph of $y = -2x^2 + 12$ using Xmin $= -4$ and Xmax $= 4$. Do you think this is a better choice for the viewing rectangle?

FIGURE 8

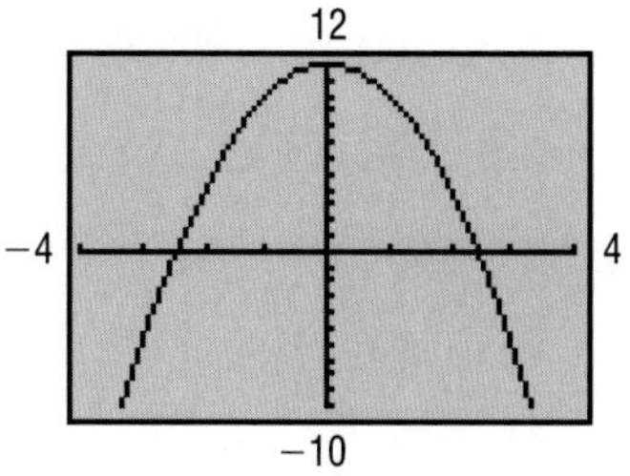

 Now work Problems 11(a)–11(d).

EXAMPLE 2 Creating a Table and Graphing an Equation

Create a table and graph the equation: $y = x^3$

Solution Most graphing utilities have the capability of creating a table of values for an equation. (Check your manual to see if your graphing utility has this capability.) Table 1 illustrates a table of values for $y = x^3$ on a TI-83. See Figure 9 for the graph.

TABLE 1

X	Y1
-3	-27
-2	-8
-1	-1
0	0
1	1
2	8
3	27

Y1=X^3

FIGURE 9

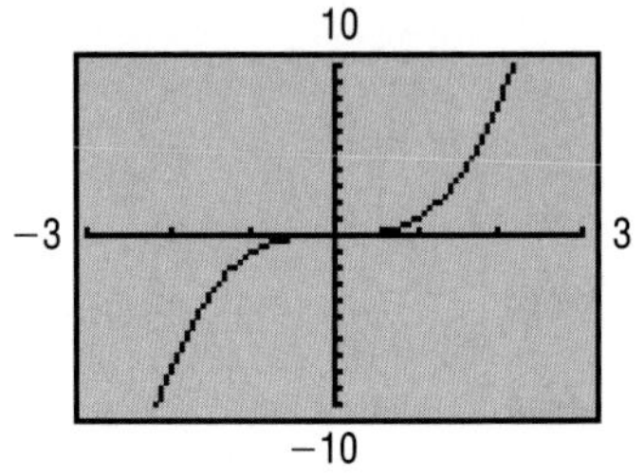

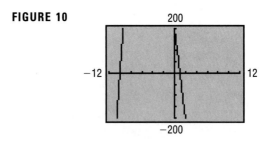

EXAMPLE 3 Using ZOOM-OUT to Obtain a Complete Graph

Graph the equation $y = x^3 - 11x^2 - 190x + 200$ using the following settings for the first viewing rectangle.

Xmin:	-12	Xmax:	12	Xscl:	2
Ymin:	-200	Ymax:	200	Yscl:	50

Solution Figure 10 shows the graph.

FIGURE 10

Notice how ragged the graph looks. The y-scale is clearly not adequate. The graph appears to have two x-intercepts. The graph is not complete. We can use the ZOOM-OUT function to help obtain a complete graph.

After the first ZOOM-OUT, we obtain Figure 11.

FIGURE 11

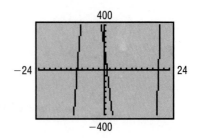

Notice that the min/max settings have been doubled, while the scales remained the same.* Also notice that now there are three x-intercepts, the most a polynomial of degree 3 can have. However, we still cannot see the top or bottom portion of the graph. ZOOM-OUT again to obtain the graph in Figure 12.

FIGURE 12

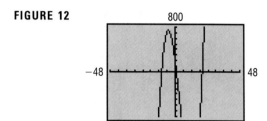

Now we can see the entire graph with the exception of the bottom portion. Therefore, ZOOMing-OUT further may not provide us with a complete graph. To get a complete graph, we instead choose to adjust the RANGE. After some experimentation, we obtain Figure 13, a complete graph.

FIGURE 13
$y = x^3 - 11x^2 - 190x + 200$

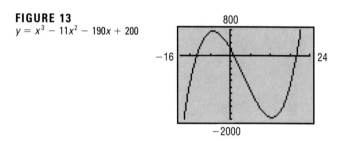

*On some graphing utilities, the default factor for the ZOOM-OUT function is 4, meaning that the original min/max setting will be multiplied by 4. Also, you can set the factor yourself, if you want; furthermore, the factor need not be the same for x and y.

A.2 │ EXERCISES

In Problems 1–20, graph each equation using the following RANGE settings:

(a) Xmin $= -5$ (b) Xmin $= -10$ (c) Xmin $= -10$ (d) Xmin $= -5$
   Xmax $= 5$ Xmax $= 10$ Xmax $= 10$ Xmax $= 5$
   Xscl $= 1$ Xscl $= 1$ Xscl $= 2$ Xscl $= 1$
   Ymin $= -4$ Ymin $= -8$ Ymin $= -8$ Ymin $= -20$
   Ymax $= 4$ Ymax $= 8$ Ymax $= 8$ Ymax $= 20$
   Yscl $= 1$ Yscl $= 1$ Yscl $= 2$ Yscl $= 5$

1. $y = x + 2$ **2.** $y = x - 2$ **3.** $y = -x + 2$ **4.** $y = -x - 2$
5. $y = 2x + 2$ **6.** $y = 2x - 2$ **7.** $y = -2x + 2$ **8.** $y = -2x - 2$
9. $y = x^2 + 2$ **10.** $y = x^2 - 2$ **11.** $y = -x^2 + 2$ **12.** $y = -x^2 - 2$
13. $y = 2x^2 + 2$ **14.** $y = 2x^2 - 2$ **15.** $y = -2x^2 + 2$ **16.** $y = -2x^2 - 2$
17. $3x + 2y = 6$ **18.** $3x - 2y = 6$ **19.** $-3x + 2y = 6$ **20.** $-3x - 2y = 6$

In Problems 21–40, use a graphing utility to graph each equation.

21. $3x + 5y = 75$
22. $3x - 5y = 75$
23. $3x + 5y = -75$
24. $3x - 5y = -75$

25. $y = (x - 10)^2$
26. $y = (x + 10)^2$
27. $y = x^2 - 100$
28. $y = x^2 + 100$

29. $x^2 + y^2 = 100$
30. $x^2 + y^2 = 64$
31. $3x^2 + y^2 = 900$
32. $4x^2 + y^2 = 1600$

33. $x^2 + 3y^2 = 900$
34. $x^2 + 4y^2 = 1600$
35. $y = x^2 - 10x$
36. $y = x^2 + 10x$

37. $y = x^2 - 18x$
38. $y = x^2 + 18x$
39. $y = x^2 - 36x$
40. $y = x^2 + 36x$

A.3 | USING A GRAPHING UTILITY TO LOCATE INTERCEPTS

Four tools that can be used to locate intercepts using a graphing utility are TRACE, BOX, VALUE, and ROOT (or ZERO).

TRACE

Most graphing utilities allow you to move from point to point along the graph, displaying on the screen the coordinates of each point. This feature is called TRACE.

E X A M P L E   1   Using TRACE to Locate Intercepts

Graph the equation: $y = x^3 - 8$. Use TRACE to locate the intercepts.

Solution   Figure 14 shows the graph of $y = x^3 - 8$.

FIGURE 14

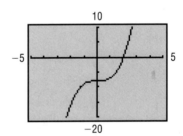

Activate the TRACE feature. As you move the cursor along the graph, you will see the coordinates of each point displayed. Just before you get to the x-axis, the display will look like the one in Figure 15(a). (Due to differences in graphing utilities, your display may be slightly different from the one shown here.)

FIGURE 15

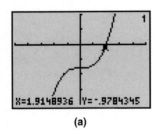
(a)

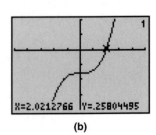
(b)

In Figure 15(a), the negative value of the y-coordinate indicates that we are still below the x-axis. The next position of the cursor is shown in Figure 15(b).

The positive value of the y-coordinate indicates that we are now above the x-axis. This means that between these two points the x-axis was crossed. The x-intercept lies between 1.9148936 and 2.0212766.

Using TRACE, we find that the y-intercept is -8. See Figure 16.

FIGURE 16

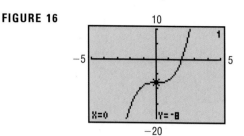

BOX

Most graphing utilities have a BOX feature that allows you to box in a specific part of the graph of an equation.

E X A M P L E   2   Using BOX

Graph the equation $y = x^3 - 8$ and use the BOX feature to improve on the approximation found in Example 1 for the x-intercept.

Solution   Using the viewing rectangle of Figure 14, graph the equation. Activate the BOX feature. (With some graphing utilities, this requires positioning the cursor at one corner of the box and then tracing out the sides of the box to the diagonal corner.) See Figure 17. Once executed, the box becomes the viewing rectangle. The result is shown in Figure 18.

FIGURE 17   **FIGURE 18**

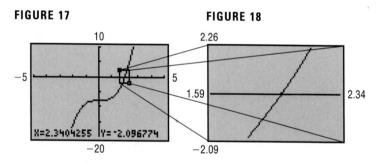

Now you can TRACE to get the approximation shown in Figure 19.

FIGURE 19

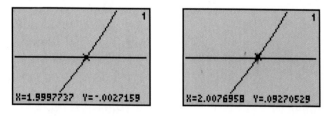

Now we know that the x-intercept lies between 1.9997737 and 2.0076958.

VALUE and ROOT (ZERO)

Most graphing utilities have an eVALUEate feature that, given a value of x, determines the value of y for an equation. We can use this feature to evaluate an equation at $x = 0$ to determine the y-intercept. Using the VALUE feature, we find the y-intercept of $y = x^3 - 8$ is -8.

Most graphing utilities also have a ROOT (or ZERO) feature that can be used to determine the x-intercept of an equation. Using ROOT on a TI-82, we find that the x-intercept of $y = x^3 - 8$ is 2. Consult your owner's manual to determine the appropriate keystrokes for these functions.

If your utility has no ROOT or ZERO feature, you can use BOX to approximate the x-intercepts of an equation.

Approximating *x*-Intercepts Using BOX

To approximate the x-intercepts of an equation, we use an algorithm in which each successive application results in one more decimal place of accuracy.

We begin with an example that illustrates the meaning of writing a number "correct to n decimal places."

E X A M P L E   3   Writing a Number Correct to *n* Decimal Places

Writing the Number	**1/7**	**$\sqrt{2}$**	**π**
Correct to 1 decimal place	0.1	1.4	3.1
Correct to 2 decimal places	0.14	1.41	3.14
Correct to 3 decimal places	0.142	1.414	3.141
Correct to 4 decimal places	0.1428	1.4142	3.1415
Correct to 5 decimal places	0.14285	1.41421	3.14159
Correct to 6 decimal places	0.142857	1.414213	3.141592 ∎

As the example illustrates, **correct to *n* decimal places** means the decimal that results from truncation after the nth decimal.

The following example illustrates the steps to use to approximate the x-intercepts of an equation using a graphing utility.

E X A M P L E   4   Using BOX to Approximate the *x*-Intercepts of an Equation

Find the smaller of the two x-intercepts of the equation $y = x^2 - 6x + 7$. Express the answer correct to two decimal places.

Solution   We begin by graphing the equation using a scale of 1 on each axis. Figure 20 shows the graph.

The smaller of the two x-intercepts lies between 1 and 2. (Remember that Xscl = 1.) Thus, correct to 0 decimal places, the smaller x-intercept is $x = 1$.

Next, we BOX the graph from approximately $x = 1$ to $x = 2$ and from $y = -1$ to $y = 1$. See Figure 21. Adjust the RANGE setting so that Xscl = 0.1 and Yscl = 0.1. Graph the equation again. See Figure 22.

FIGURE 20

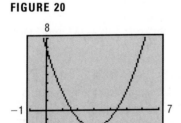

FIGURE 21

FIGURE 22

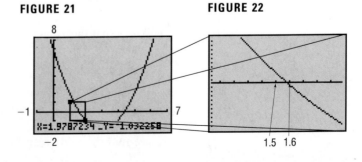

The *x*-intercept lies between 1.5 and 1.6. (Remember that the BOX was from *x* = 1 to *x* = 2 and *X*scl = 0.1.) Thus, correct to one decimal place, the smaller *x*-intercept is *x* = 1.5.

Next, BOX again from *x* = 1.5 to *x* = 1.6 and from *y* = −0.1 to *y* = 0.1. See Figure 23. Adjust the RANGE so that *X*scl = 0.01 and *Y*scl = 0.01 and graph. See Figure 24.

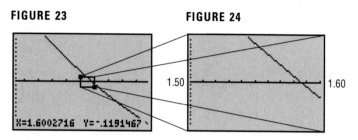

FIGURE 23   **FIGURE 24**

X=1.6002716 Y=-.1191467

1.50   1.60

The *x*-intercept lies between 1.58 and 1.59. Remember that the BOX was from *x* = 1.5 to *x* = 1.6 with *X*scl = 0.01. Thus, correct to two decimal places, the smaller of the two *x*-intercepts is *x* = 1.58.   ▬

The steps to follow for approximating *x*-intercepts of equations correct to any desired number of decimals are given next.

Steps for Approximating the *x*-Intercepts of an Equation

> STEP 1: Write the equation in the form {expression in *x*} = 0.
> STEP 2: Graph the equation *y* = {expression in *x*} using *X*scl = 1 and *Y*scl = 1.
> STEP 3: BOX the graph so that the *x*-intercept is within the BOX. Adjust *X*scl and *Y*scl to one-tenth of their current value and graph *f* again.
> STEP 4: If additional accuracy is desired, repeat step 3.

In these steps, each repetition of step 3 gives one more decimal place of accuracy. Thus, when *X*scl = 0.01, the *x*-intercept will be known correct to two decimal places.

If the *x*-intercept that you are trying to approximate appears to fall on a tick mark, then you must evaluate the equation at that tick mark. If the value is zero, then there is an exact solution. If the value is not zero, then you know whether the graph is above or below the *x*-axis at the tick mark by noting the sign of the value at the tick mark.

A.3 | EXERCISES

In Problems 1–6, write each expression as a decimal correct to three decimal places.

1. $\frac{3}{7}$ **2.** $\frac{2}{11}$ **3.** $\sqrt{5}$ **4.** $\sqrt{6}$ **5.** $\sqrt[3]{2}$ **6.** $\sqrt[3]{3}$

In Problems 7–12, use a graphing utility to approximate the smaller of the two x-intercepts of each equation. Express the answer correct to two decimal places.

7. $y = x^2 + 4x + 2$ **8.** $y = x^2 + 4x - 3$ **9.** $y = 2x^2 + 4x + 1$

10. $y = 3x^2 + 5x + 1$ **11.** $y = 2x^2 - 3x - 1$ **12.** $y = 2x^2 - 4x - 1$

*In Problems 13–20, use a graphing utility to approximate the **positive** x-intercepts of each equation. Express the answer correct to two decimal places.*

13. $y = x^3 + 3.2x^2 - 16.83x - 5.31$

14. $y = x^3 + 3.2x^2 - 7.25x - 6.3$

15. $y = x^4 - 1.4x^3 - 33.71x^2 + 23.94x + 292.41$

16. $y = x^4 + 1.2x^3 - 7.46x^2 - 4.692x + 15.2881$

17. $y = \pi x^3 - (8.88\pi + 1)x^2 - (42.066\pi - 8.88)x + 42.066$

18. $y = \pi x^3 - (5.63\pi + 2)x^2 - (108.392\pi - 11.26)x + 216.784$

19. $y = x^3 + 19.5x^2 - 1021x + 1000.5$

20. $y = x^3 + 14.2x^2 - 4.8x - 12.4$

A.4 SQUARE SCREENS

FIGURE 25

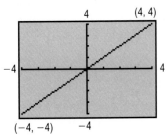

Most graphing utilities have a rectangular screen. Because of this, using the same RANGE for both x and y will result in a distorted view. For example, Figure 25 shows the graph of the line $y = x$ connecting the points $(-4, -4)$ and $(4, 4)$.

We expect the line to bisect the first and third quadrants, but it doesn't. We need to adjust the selections for Xmin, Xmax, Ymin, and Ymax so that a **square screen** results. On most graphing utilities, this is accomplished by setting the ratio of x to y at 3:2.* In other words,

$$2(X\text{max} - X\text{min}) = 3(Y\text{max} - Y\text{min})$$

E X A M P L E   1   Examples of Viewing Rectangles That Result in Square Screens

FIGURE 26

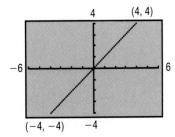

(a) Xmin $= -3$	(b) Xmin $= -6$	(c) Xmin $= -6$
Xmax $=$ 3	Xmax $=$ 6	Xmax $=$ 6
Xscl $=$ 1	Xscl $=$ 1	Xscl $=$ 2
Ymin $= -2$	Ymin $= -4$	Ymin $= -4$
Ymax $=$ 2	Ymax $=$ 4	Ymax $=$ 4
Yscl $=$ 1	Yscl $=$ 1	Yscl $=$ 1

Figure 26 shows the graph of the line $y = x$ on a square screen using the viewing rectangle given in Example 1(b). Notice that the line now bisects the first and third quadrants. Compare this illustration to Figure 25.

*Some graphing utilities have a built-in function that automatically squares the screen. For example, the TI-85 has a ZSQR function that does this. Some graphing utilities require a ratio other than 3:2 to square the screen. For example, the HP 48G requires the ratio of x to y to be 2:1 for a square screen. Consult your manual.

A.4 EXERCISES

In Problems 1–8, determine which of the given RANGE settings result in a square screen.

1. Xmin $= -3$	**2.** Xmin $= -5$	**3.** Xmin $= 0$	**4.** Xmin $= -6$
Xmax $= 3$	Xmax $= 5$	Xmax $= 9$	Xmax $= 6$
Xscl $= 1$	Xscl $= 1$	Xscl $= 3$	Xscl $= 2$
Ymin $= -2$	Ymin $= -4$	Ymin $= -2$	Ymin $= -4$
Ymax $= 2$	Ymax $= 4$	Ymax $= 4$	Ymax $= 4$
Yscl $= 1$	Yscl $= 1$	Yscl $= 2$	Yscl $= 1$

5. Xmin $= -6$
Xmax $= 6$
Xscl $= 1$
Ymin $= -2$
Ymax $= 2$
Yscl $= .5$

6. Xmin $= -6$
Xmax $= 6$
Xscl $= 1$
Ymin $= -4$
Ymax $= 4$
Yscl $= 1$

7. Xmin $= 0$
Xmax $= 9$
Xscl $= 1$
Ymin $= -2$
Ymax $= 4$
Yscl $= 1$

8. Xmin $= -6$
Xmax $= 6$
Xscl $= 2$
Ymin $= -4$
Ymax $= 4$
Yscl $= 2$

9. If Xmin $= -4$, Xmax $= 8$, and Xscl $= 1$, how should Ymin, Ymax, and Yscl be selected so that the viewing rectangle contains the point $(4, 8)$ and the screen is square?

10. If Xmin $= -6$, Xmax $= 12$, and Xscl $= 2$, how should Ymin, Ymax, and Yscl be selected so that the viewing rectangle contains the point $(4, 8)$ and the screen is square?

A.5 | USING A GRAPHING UTILITY TO GRAPH INEQUALITIES

It is easiest to begin with an example.

E X A M P L E   1   Graphing an Inequality Using a Graphing Utility

Use a graphing utility to graph: $3x + y - 6 \leq 0$.

Solution   We begin by graphing the equation $3x + y - 6 = 0$ ($Y_1 = -3x + 6$). See Figure 27.

FIGURE 27

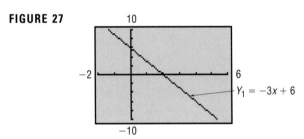

As with graphing by hand, we need to test points selected from each region and determine whether they satisfy the inequality. To test the point $(-1, 2)$, for example, enter $3*-1+2-6\leq0$. See Figure 28(a). The 1 that appears indicates that the statement entered (the inequality) is true. When the point $(5, 5)$ is tested, a 0 appears, indicating that the statement entered is false. Thus, $(-1, 2)$ is a part of the graph of the inequality and $(5, 5)$ is not. Figure 28(b) shows the graph of the inequality on a TI-83.*

FIGURE 28

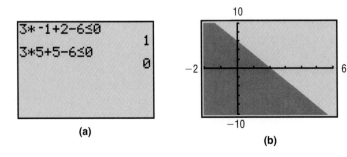

(a)

(b)

*Consult your owner's manual for shading techniques.

The steps to follow to graph an inequality using a graphing utility are given next.

Steps for Graphing an Inequality Using a Graphing Utility

STEP 1: Replace the inequality symbol by an equal sign and graph the resulting equation, $y = \qquad$.

STEP 2: In each of the regions, select a test point P.

(a) Use a graphing utility to determine if the test point P satisfies the inequality. If the test point satisfies the inequality, then so do all the points in the region. Indicate this by using the graphing utility to shade the region.

(b) If the coordinates of P do not satisfy the inequality, then none of the points in that region do.

ANSWERS

CHAPTER 1   Historical Problems

1. (a) 1, 20 **(b)** 2, 50 **2. (a)** $\dfrac{7}{3} = 2.333\ldots$ **(b)** $\dfrac{39}{8} = 4.875$ **(c)** $\dfrac{84,823}{27,000} = 3.141592592\ldots$

1.1 Exercises

1. 0.333... **2.** 0.6 **3.** 0.125 **4.** 3.333... **5.** -1.6 **6.** $-5.333\ldots$ **7.** 0.111... **8.** 0.1 **9.** 0.16 **10.** $-0.8333\ldots$ **11.** $-0.428571428571\ldots$
12. 0.727272... **13.** 7 **14.** 5 **15.** 6 **16.** 0 **17.** 1 **18.** 1 **19.** $\dfrac{13}{3}$ **20.** $\dfrac{3}{2}$ **21.** -11 **22.** -23 **23.** 11 **24.** -11

25. -4 **26.** -12 **27.** $\dfrac{7}{2}$ **28.** $\dfrac{19}{3}$ **29.** $\dfrac{52}{15}$ **30.** $-\dfrac{14}{15}$ **31.** $\dfrac{29}{2}$ **32.** $\dfrac{62}{15}$ **33.** $6x + 24$ **34.** $8x - 4$ **35.** $x^2 - 4x$ **36.** $4x^2 + 12x$
37. $x^2 + 6x + 8$ **38.** $x^2 + 6x + 5$ **39.** $x^2 - x - 2$ **40.** $x^2 - 3x - 4$ **41.** $x^2 - 10x + 16$ **42.** $x^2 - 6x + 8$ **43.** $x^2 - 4$
44. $x^2 - 9$ **49.** No; $\dfrac{1}{3}$; 0.000333... **50.** No; $\dfrac{2}{3}$; 0.000666... **51.** No; $2 - 3 \neq 3 - 2$ **52.** No; $(5 - 4) - 3 \neq 5 - (4 - 3)$

53. No; $\dfrac{2}{3} \neq \dfrac{3}{2}$ **54.** No; $(20 \div 5) \div 2 \neq 20 \div (5 \div 2)$ **55.** Symmetric property

56. $5^2 + 5 = 25 + 5 = 30$ **57.** ──────── **58.** ────────

59. $x > 0$ **60.** $z < 0$ **61.** $x < 3$ **62.** $y > -4$ **63.** $|x| \leq 1$ **64.** $\dfrac{1}{x} \geq 2$ **65.** Positive **66.** Negative **67.** Positive **68.** Negative
69. Positive **70.** Negative **71.** Positive **72.** Positive **73.** 0 **74.** Positive **75.** 1 **76.** 5 **77.** 5 **78.** 1 **79.** 1 **80.** -1 **81.** 22
82. 5 **83.** 2 **84.** 13 **85.** $\{x | x \neq 5\}$ **86.** $\{x | x \neq -4\}$ **87.** $\{x | x \neq -4\}$ **88.** $\{x | x \neq 6\}$ **89.** $\{x | x \neq 0, x \neq 1\}$ **90.** $\{x | x \neq 0, x \neq 4\}$
91. $\{x | x \neq 0, x \neq -2\}$ **92.** $\{x | x \neq -1, x \neq -2\}$ **93** All real numbers **94.** All real numbers
95. $A = lw$; all the variables are positive real numbers **96.** $P = 2(l + w)$; all the variables are positive real numbers
97. $C = \pi d$; d and C are positive real numbers **98.** $A = \dfrac{1}{2}bh$; all the variables are positive real numbers

99. $A = \dfrac{\sqrt{3}}{4} x^2$; x and A are positive real numbers **100.** $P = 3x$; x and P are positive real numbers

101. $V = \dfrac{4}{3}\pi r^3$; r and V are positive real numbers **102.** $S = 4\pi r^2$; r and S are positive real numbers

103. $V = x^3$; x and V are positive real numbers **104.** $S = 6x^2$; x and S are positive real numbers
105. $C = 0°$ **106.** $C = 100°$ **107.** $C = 25°$ **108.** $C = -20°$ **109. (a)** $C = \$6000$ **(b)** $C = \$8000$ **110.** $\$98$ **111.** 4950 **112.** 2500
113. -2450 **114.** 50 **115.** 200 **116.** $2k\pi$ where $k = 1, 2, 3, \ldots$ **117.** No; no **119.** 1 **120.** No **121.** 3.15 or 3.16

1.2 Exercises

1. (a) 18.953 **(b)** 18.952 **2. (a)** 25.861 **(b)** 25.861 **3. (a)** 28.653 **(b)** 28.653 **4. (a)** 99.052 **(b)** 99.052 **5. (a)** 0.063 **(b)** 0.062
6. (a) 0.054 **(b)** 0.053 **7. (a)** 9.999 **(b)** 9.998 **8. (a)** 1.001 **(b)** 1.000 **9. (a)** 0.429 **(b)** 0.428 **10. (a)** 0.556 **(b)** 0.5555
11. (a) 34.733 **(b)** 34.733 **12. (a)** 16.200 **(b)** 16.200 **13.** 72.42 **14.** 92.54 **15.** 30.66 **16.** 40.12 **17.** 133.12 **18.** 112.85 **19.** 11.03
20. 8.30 **21.** -0.49 **22.** 0.03 **23.** 0.59 **24.** 1.22 **25.** 21.58 **26.** 457.56 **27.** 43.90 **28.** -2.24 **29.** 0.30 **30.** 0.29 **31.** 22.68

32. 16.28 **33.** 67.08 **34.** -0.33 **35.** 659.41 **36.** 0.16 **37.** $>$ **38.** $<$ **39.** $>$ **40.** $<$ **41.** $\dfrac{290}{1,000,000}$; 0.00029 **42.** $\dfrac{350}{1,000,000}$; 0.00035
43. 800 **44.** Get an error message

1.3 Exercises

1. 16 **2.** -16 **3.** $\dfrac{1}{16}$ **4.** 16 **5.** $-\dfrac{1}{16}$ **6.** $\dfrac{1}{16}$ **7.** $\dfrac{1}{8}$ **8.** $-\dfrac{1}{8}$ **9.** $\dfrac{1}{4}$ **10.** $\dfrac{2}{9}$ **11.** $\dfrac{1}{9}$ **12.** 4 **13.** $\dfrac{81}{64}$ **14.** 8 **15.** $\dfrac{27}{8}$ **16.** $\dfrac{4}{9}$
17. $\dfrac{81}{2}$ **18.** $\dfrac{25}{81}$ **19.** $\dfrac{4}{81}$ **20.** $\dfrac{125}{216}$ **21.** $\dfrac{1}{12}$ **22.** $\dfrac{1}{18}$ **23.** $-\dfrac{2}{3}$ **24.** $\dfrac{1}{8}$ **25.** y^2 **26.** $\dfrac{y}{x}$ **27.** $\dfrac{x}{y^2}$ **28.** y^4 **29.** $\dfrac{1}{64x^6}$ **30.** $\dfrac{1}{64x^6}$

31. $-\dfrac{4}{x}$ **32.** $-\dfrac{1}{4x}$ **33.** 3 **34.** 1 **35.** $\dfrac{1}{x^3 y}$ **36.** $\dfrac{1}{x^3 y}$ **37.** $\dfrac{1}{xy}$ **38.** $\dfrac{1}{x^3 y^3}$ **39.** $\dfrac{y}{x}$ **40.** $\dfrac{3}{2x}$ **41.** $\dfrac{25x^2}{16y^2}$ **42.** $\dfrac{1}{x^4 y^2}$ **43.** $\dfrac{1}{x^2 y^2}$ **44.** $\dfrac{1}{xy}$

45. $\dfrac{1}{x^3 y^3}$ **46.** $\dfrac{3y^4}{x^6}$ **47.** $-\dfrac{8x^3}{9yz^2}$ **48.** $\dfrac{4z}{25x^6 y^3}$ **49.** $\dfrac{y^3}{x^8}$ **50.** y **51.** $\dfrac{16x^2}{9y^2}$ **52.** $\dfrac{216x^6}{125y^6}$ **53.** $\dfrac{1}{x^3 y}$ **54.** $\dfrac{9x^5}{8y^5}$ **55.** $\dfrac{y^2}{x^2}$ **56.** y^5 **57.** $10; 0$

58. $8; 44$ **59.** 81 **60.** 8 **61.** $304{,}006.671$ **62.** 693.440 **63.** 0.004 **64.** 0.019 **65.** 481.890 **66.** -481.890 **67.** 0.000 **68.** -0.000
69. 4.542×10^2 **70.** 3.214×10^1 **71.** 1.3×10^{-2} **72.** 4.21×10^{-3} **73.** 3.2155×10^4 **74.** 2.121×10^4 **75.** 4.23×10^{-4}
76. 5.14×10^{-2} **77.** $61{,}500$ **78.** 9700 **79.** 0.001214 **80.** 0.000988 **81.** $110{,}000{,}000$ **82.** 411.2 **83.** 0.081 **84.** 0.6453
85. $5.8657 \times 10^{12}\,\text{mi}$ **86.** $5 \times 10^2\,\text{sec}$

1.4 Exercises

1. Yes; 2 **2.** Yes; 1 **3.** Yes; 0 **4.** Yes; 0 **5.** No **6.** No **7.** Yes; 3 **8.** Yes; 2 **9.** No
10. No **11.** $x^2 + 7x + 2$ **12.** $x^3 + 4x^2 - 4x + 6$ **13.** $x^3 - 4x^2 + 9x + 7$ **14.** $-x^3 + 4x^2 - 4x - 9$
15. $6x^5 + 5x^4 + 3x^2 + x$ **16.** $10x^5 + 3x^3 - 10x^2 + 6$ **17.** $9x^2 - 7x - 5$ **18.** $-32x^2 - 8x + 10$ **19.** $-2x^3 + 18x^2 - 18$
20. $8x^3 - 24x^2 - 48x + 4$ **21.** $2x^2 - 4x + 6$ **22.** $-2x^2 + x - 6$ **23.** $15y^2 - 27y + 30$ **24.** $-4y^3 + 4y^2 + 4y + 12$ **25.** $2ax + a^2$
26. $-2ax + a^2$ **27.** $3ax^2 + 3a^2x + a^3$ **28.** $-3ax^2 + 3a^2x - a^3$ **29.** $x^2 + 2x - 15$ **30.** $x^2 + 2x - 8$ **31.** $2x^2 + 17x + 8$
32. $2x^2 + 3x - 2$ **33.** $-3x^2 - 11x + 4$ **34.** $-2x^2 - 3x - 1$ **35.** $12x^2 - 11x + 2$ **36.** $3x^2 - 8x + 4$ **37.** $4x^2 - 25$ **38.** $9x^2 - 1$
39. $4x^2 - 20x + 25$ **40.** $9x^2 + 6x + 1$ **41.** $x^3 - 1$ **42.** $x^3 + 1$ **43.** $2x^3 + 5x^2 - 4x - 10$ **44.** $3x^3 - 5x^2 + 3x - 5$
45. $6x^3 - 13x^2 + 8x - 3$ **46.** $6x^3 + 5x^2 + 2x + 12$ **47.** $9x^2 - 16y^2$ **48.** $4x^2 + 12xy + 9y^2$ **49.** $x^2 - xy - 2y^2$
50. $x^2 - 3xy + 2y^2$ **51.** $x^2 - 2y^2 + 3x + y$ **52.** $-xy + y^2$ **53.** $-4xy$ **54.** $-2xy$ **55.** $2x^2 - 2xy + 13y^2$ **56.** $3x^2 - 2xy$
57. $2x^2 + 2y^2 + 2z^2 + 2xy + 2yz$ **58.** $2x^2 - 2xy + 2y^2 - 2yz + 2z^2$ **59.** $x^3 - y^3$ **60.** $x^3 + y^3$ **61.** $x^3 + 2x^2 - 4x - 8$
62. $x^3 - 3x^2 - 9x + 27$ **63.** $2x^3 - x^2 - 4x + 3$ **64.** $3x^3 + 29x^2 + 65x - 25$ **65.** $x^3 - 6x^2 + 11x - 6$ **66.** $x^3 + 6x^2 + 11x + 6$
67. $6x^2 + 2$ **68.** $-12x^2 - 16$ **69.** $x^4 - 2x^2 + 1$ **70.** $x^4 - 2x^3 + 2x - 1$ **71.** $8x^3 + 36x^2 + 54x + 27$
72. $27x^3 + 54x^2 + 36x + 8$ **73.** $8x^3 - 36ax^2 + 54a^2x - 27a^3$ **74.** $27x^3 - 54ax^2 + 36a^2x - 8a^3$ **75.** $x^2 - y^2 - z^2 - 2yz$
76. $x^2 - y^2 - z^2 + 2yz$ **77.** $4x^2 - 3x + 1$; remainder 1 **78.** $3x^2 - x + 1$; remainder -2 **79.** $4x^2 - 11x + 23$; remainder -45
80. $3x^2 - 7x + 15$; remainder -32 **81.** $4x^2 + 13x + 53$; remainder 213 **82.** $3x^2 + 11x + 45$; remainder 178
83. $4x - 3$; remainder $x + 1$ **84.** $3x - 1$; remainder $x - 2$ **85.** $4x - 3$; remainder $-7x + 7$ **86.** $3x - 1$; remainder $-5x$
87. 2; remainder $-3x^2 + x + 3$ **88.** 1; remainder $-x^2 + x - 1$
89. $2x - \dfrac{5}{2}$; remainder $\dfrac{3}{2}x + \dfrac{7}{2}$ **90.** $x - \dfrac{2}{3}$; remainder $\dfrac{2}{3}x - \dfrac{4}{3}$ **91.** $x - \dfrac{3}{4}$; remainder $\dfrac{7}{4}$ **92.** $x^2 + \dfrac{1}{3}$; remainder $-\dfrac{5}{3}$
93. $x^3 + x^2 + x + 1$; remainder 0 **94.** $x^3 - x^2 + x - 1$; remainder 0
95. $x^2 + 1$; remainder 0 **96.** $x^2 - 1$; remainder 0 **97.** $-4x^2 - 3x - 3$; remainder -7 **98.** $-3x^3 - 3x^2 - 3x - 5$; remainder -6
99. $x^2 - x - 1$; remainder $2x + 2$ **100.** $x^2 + x - 1$; remainder $-2x + 2$ **101.** $-x^2$; remainder 1 **102.** $x^2 - 2$; remainder 3
103. $x^2 + ax + a^2$; remainder 0 **104.** $x^2 - ax + a^2$; remainder 0 **105.** $x^3 + ax^2 + a^2x + a^3$; remainder 0
106. $x^4 + ax^3 + a^2x^2 + a^3x + a^4$; remainder 0
107. The degree of the product equals the degree of the product of the leading terms:
$(a_n x^n + a_{n-1}x^{n-1} + \ldots + a_1 x + a_0)(b_m x^m + b_{m-1}x^{m-1} + \ldots + b_1 x + b_0) = a_n b_m x^{n+m} + (a_n b_{m-1} + a_{n-1} b_m)x^{n+m-1} + \ldots +$
$(a_1 b_0 + a_0 b_1)x + a_0 b_0$, which is of degree $m + n$
108. The degree of the sum equals the degree of the leading term:
If $n > m$, $(a_n x^n + a_{n-1}x^{n-1} + \ldots + a_1 x + a_0) + (b_m x^m + b_{m-1}x^{m-1} + \ldots + b_1 x + b_0) = a_n x^n + \ldots + (a_m + b_m)$
$x^m + \ldots + (a_1 + b_1)x + (a_0 + b_0)$ which is of degree n
109. $(2){:}(x - a)(x + a) = x(x + a) - a(x + a) = x^2 + ax - ax - a^2 = x^2 - a^2$
$(3a){:}(x + a)^2 = (x + a)(x + a) = x(x + a) + a(x + a) = x^2 + 2ax + a^2$
$(3b){:}(x - a)^2 = (x - a)(x - a) = x(x - a) - a(x - a) = x^2 - 2ax + a^2$
$(6){:}(x - a)(x^2 + ax + a^2) = x(x^2 + ax + a^2) - a(x^2 + ax + a^2) = x^3 + ax^2 + a^2x - ax^2 - a^2x - a^3 = x^3 - a^3$
110. $(4a){:}(x + a)(x + b) = x(x + b) + a(x + b) = x^2 + bx + ax + ab = x^2 + (a + b)x + ab$
$(5a){:}(x + a)^3 = (x + a)(x + a)^2 = (x + a)(x^2 + 2ax + a^2) = x^3 + 2ax^2 + a^2x + ax^2 + 2a^2x + a^3 = x^3 + 3ax^2 + 3a^2x + a^3$
$(7){:}(x + a)(x^2 - ax + a^2) = x^3 - ax^2 + a^2x + ax^2 - a^2x + a^3 = x^3 + a^3$
111. $(x + a)^4 = (x + a)^2(x + a)^2 = (x^2 + 2ax + a^2)(x^2 + 2ax + a^2) = x^4 + 4ax^3 + 6a^2x^2 + 4a^3x + a^4$
112. $x^5 + 5ax^4 + 10a^2x^3 + 10a^3x^2 + 5a^4x + a^5$
115. $a = 1, b = -4, c = 11, d = -17; a + b + c + d = -9$

1.5 Exercises

1. $3(x + 2)$ **2.** $7(x - 2)$ **3.** $a(x^2 + 1)$ **4.** $a(x - 1)$ **5.** $x(x^2 + x + 1)$ **6.** $x(x^2 - x + 1)$ **7.** $2x(x - 1)$ **8.** $3x(x - 1)$
9. $3xy(x - 2y + 4)$ **10.** $12xy(5x - 4y + 6x^2)$ **11.** $(x - 1)(x + 1)$ **12.** $(x - 3)(x + 3)$ **13.** $-(2x - 1)(2x + 1)$
14. $-(3x - 1)(3x + 1)$ **15.** $(x + 2)(x + 5)$ **16.** $(x + 4)(x + 1)$ **17.** $(x - 7)(x - 3)$ **18.** $(x - 4)(x - 2)$ **19.** $(x - 4)(x + 2)$
20. $(x - 5)(x + 1)$ **21.** $(x + 1)^2$ **22.** $(x + 2)^2$ **23.** $(x - 2)^2$ **24.** $(x - 1)^2$ **25.** $-(x - 5)(x + 3)$ **26.** Prime
27. $3(x + 2)(x - 6)$ **28.** $x(x + 10)(x - 2)$ **29.** $y^2(y + 5)(y + 6)$ **30.** $3y(y - 8)(y + 2)$ **31.** $(4x + 1)^2$ **32.** $(5x + 1)^2$
33. $(2x + 3)^2$ **34.** $(3x - 2)^2$ **35.** $a(x - 9a)(x + 5a)$ **36.** $b(x + 9b)(x + 5b)$ **37.** $(x - 5)(x^2 + 5x + 25)$
38. $(x + 5)(x^2 - 5x + 25)$ **39.** $-(2x - 3)(4x^2 + 6x + 9)$ **40.** $-8(x - 2)(x^2 + 2x + 4)$ **41.** $2x(x + 2)(x^2 - 2x + 4)$

42. $3x^2(x-1)(x^2+x+1)$ **43.** $(3x+1)(x+1)$ **44.** $(4x-1)(x+1)$ **45.** $(x-3)(x+3)(x^2+9)$
46. $(x-1)(x+1)(x^2+1)$ **47.** $(x-1)^2(x^2+x+1)^2$ **48.** $(x+1)^2(x^2-x+1)^2$ **49.** $x^5(x-1)(x+1)$
50. $x^5(x-1)(x^2+x+1)$ **51.** $(2z+3)(z+1)$ **52.** $(3z+1)(2z-1)$ **53.** $(4x+3)^2$ **54.** $(3x-4)^2$ **55.** $-(4x-5)(4x+1)$
56. $-(x-1)(16x+5)$ **57.** $(2y-5)(2y-3)$ **58.** $(3y+4)(3y-1)$ **59.** $9(2x-3)(x+1)$
60. $2(4x+1)(x-1)$ **61.** $2(4x^2+x+3)$ **62.** $3(3x^2-x+1)$ **63.** Prime   **64.** $(x+3)^2$ **65.** $2x(2x+1)(x-3)$
66. $3x(3x+1)(3x-2)$ **67.** $(x^2+1)^2$ **68.** $(x^2-3)^2$ **69.** $-(3x-1)(3x+1)(x^2+1)$ **70.** $-2(2x-1)(2x+1)(x^2+2)$
71. $-8(2x-1)(4x^2+2x+1)$ **72.** $8(2x+1)(4x^2-2x+1)$ **73.** $(x+3)(x-6)$ **74.** $(3x-7)(x+5)$ **75.** $(x+2)(x-3)$
76. $(x-1)(x-3)$ **77.** $4(x-3)(x+2)$ **78.** $5x(5x-6)$ **79.** $(3x-5)(9x^2-3x+7)$ **80.** $5x(25x^2+15x+3)$
81. $(x+5)(3x+11)$ **82.** $(x-3)(7x-16)$ **83.** $(x-1)(x+1)(x+2)$ **84.** $(x-3)(x-1)(x+1)$
85. $(x-1)(x+1)(x^2-x+1)$ **86.** $(x+1)^2(x^2-x+1)$ **87.** $(x^2+1)(x+2)(x^2-2x+4)$
88. $(x-1)(x+1)(x+2)(x^2-2x+4)$
89. The possibilities are $(x\pm1)(x\pm4)=x^2\pm5x+4$ or $(x\pm2)(x\pm2)=x^2\pm4x+4$, none of which equal x^2+4
90. The only possibilities are $(x\pm1)(x\pm1)=x^2\pm2x+1$, none of which equal x^2+x+1.

1.6 Exercises

1. $\dfrac{3}{x-3}$ **2.** $\dfrac{x}{3}$ **3.** $\dfrac{x}{3}$ **4.** $\dfrac{5x+8}{x}$ **5.** $\dfrac{4x}{2x-1}$ **6.** $\dfrac{(x+2)^2}{(x-4)(x+4)}$ **7.** $\dfrac{y+5}{2(y+1)}$ **8.** $\dfrac{y-1}{y+1}$ **9.** $\dfrac{x+5}{x-1}$ **10.** $-\dfrac{x}{x+2}$

11. $\dfrac{x-2}{x+3}$ **12.** $-\dfrac{x-2}{x-3}$ **13.** $-(x+7)$ **14.** $-(x+3)$ **15.** $\dfrac{2x-5}{x-4}$ **16.** $\dfrac{x-1}{x+5}$ **17.** $\dfrac{x-7}{x+7}$ **18.** 1 **19.** $\dfrac{6(x+1)}{x^2+2x+4}$

20. $\dfrac{3x+11}{2x+3}$ **21.** $\dfrac{3(x+2)(x-3)}{5x^2(x-2)}$ **22.** $-\dfrac{(3x-5)(x+1)}{4}$ **23.** $\dfrac{x(2x-1)}{x+4}$ **24.** $\dfrac{6(x+1)}{x(2x-1)}$ **25.** $\dfrac{8}{3x}$ **26.** $\dfrac{3}{5x}$

27. $\dfrac{x-3}{x+7}$ **28.** $\dfrac{(x-5)(x-2)(x+3)}{(x-3)(x-1)(x+5)}$ **29.** $\dfrac{(x-4)(x+3)}{(x-1)(2x+1)}$ **30.** $\dfrac{(3x+1)(3x-2)(2x+3)}{(3x+2)(3x-1)^2}$ **31.** $\dfrac{4x-1}{4x+1}$

32. $\dfrac{3x-1}{x-2}$ **33.** $\dfrac{x^2(5x-1)^2}{(x-1)(3x+2)^2}$ **34.** $\dfrac{(x-8)(x+1)(x+6)}{(x-3)(x+7)(x+9)}$ **35.** $\dfrac{x+5}{2}$ **36.** $-\dfrac{3}{x}$ **37.** $\dfrac{(x-2)(x+2)}{2x-3}$

38. $\dfrac{3(x^2-3)}{2x-1}$ **39.** $\dfrac{3x-2}{x-3}$ **40.** $\dfrac{3x-1}{3x+2}$ **41.** $\dfrac{x+9}{2x-1}$ **42.** $\dfrac{4x-5}{3x+4}$ **43.** $\dfrac{4-x}{x-2}$ **44.** $\dfrac{x+6}{x-1}$ **45.** $\dfrac{2(x+5)}{(x-1)(x+2)}$

46. $-\dfrac{3x+35}{(x-5)(x+5)}$ **47.** $\dfrac{3x^2-2x-3}{(x+1)(x-1)}$ **48.** $\dfrac{x(5x+1)}{(x-4)(x+3)}$ **49.** $\dfrac{-11x-2}{(x+2)(x-2)}$ **50.** $-\dfrac{2}{(x-1)(x+1)}$

51. $\dfrac{2(x^2-2)}{x(x-2)(x+2)}$ **52.** $\dfrac{x^4+x^3-x^2+x-1}{x^3(x^2+1)}$ **53.** $\dfrac{2x^3-2x^2+2x-1}{x(x-1)^2}$ **54.** $\dfrac{x^2(3x^2-4x-3)}{4(x-1)(x+1)}$

55. $\dfrac{x^3-7x^2+3x+5}{(x+1)(x-1)(x-2)}$ **56.** $\dfrac{2x^3+4x^2-3x-1}{x(x-1)(x+1)}$ **57.** $\dfrac{x^2-2x-1}{x(x-1)(x+1)}$ **58.** $-\dfrac{x^2-2}{x(x-2)(x-1)}$ **59.** $(x-2)(x+2)(x+1)$

60. $(x-4)^2(x+3)$ **61.** $x(x-1)(x+1)$ **62.** $3(x-3)(x+3)(2x+5)$ **63.** $x^3(2x-1)^2$ **64.** $x(x-3)(x+3)$

65. $x(x-1)^2(x+1)(x^2+x+1)$ **66.** $x^2(x+2)^3$ **67.** $\dfrac{5x}{(x-6)(x-1)(x+4)}$ **68.** $\dfrac{x^2+7x-1}{(x-3)(x+8)}$ **69.** $\dfrac{2(2x^2+5x-2)}{(x-2)(x+2)(x+3)}$

70. $\dfrac{3x^2-4x+4}{(x-1)^2}$ **71.** $\dfrac{5x+1}{(x-1)^2(x+1)^2}$ **72.** $\dfrac{-2(2x+7)}{(x-1)^2(x+2)^2}$ **73.** $\dfrac{-x^2+3x+13}{(x-2)(x+1)(x+4)}$ **74.** $\dfrac{x^2-6x+11}{(x+7)(x+1)^2}$

75. $\dfrac{x^3-2x^2+4x+3}{x^2(x+1)(x-1)}$ **76.** $\dfrac{3x^3-5x^2+2x+1}{x^2(x-1)^2}$ **77.** $\dfrac{-1}{x(x+h)}$ **78.** $-\dfrac{2x+h}{x^2(x+h)^2}$ **79.** $\dfrac{x+1}{x-1}$ **80.** $\dfrac{4x^2+1}{3x^2-1}$

81. $\dfrac{(x-1)(x+1)}{x^2+1}$ **82.** $\dfrac{x}{(x+1)^2}$ **83.** $\dfrac{2(5x-1)}{(x-2)(x+1)^2}$ **84.** $\dfrac{-(5x-4)}{(x-2)(x+1)(x+3)}$ **85.** $\dfrac{-2x(x^2-2)}{(x+2)(x^2-x-3)}$

86. $\dfrac{(x+3)(x^2-x-15)}{x(4x^2+5x+3)}$ **87.** $\dfrac{-1}{x-1}$ **88.** $\dfrac{1}{x}$ **89.** $\dfrac{4}{(x+h+2)(x+2)}$ **90.** $\dfrac{-2}{(x-1)(x+h-1)}$

91. a: 1, 2, 3, 5, 8, 13, ...; b: 1, 1, 2, 3, 5, 8, ...; c: 0, 1, 1, 2, 3, 5, ...

Historical Problems

1. (a) $\sqrt[3]{8}$ **(b)** $\sqrt[3]{\sqrt[3]{512}}$ **(c)** $\sqrt{\sqrt[3]{64}}$ **(d)** $\sqrt[3]{\sqrt{64}}$ **(e)** $\sqrt{\sqrt{81}}$ **(f)** $\sqrt{\sqrt{256x^8}}$ **(g)** $\sqrt[3]{-4+\sqrt{8}}$ **2. (a)** $8^{1/3}$ **(b)** $256^{1/4}$ **(c)** $8^{7/3}$
(d) $64^{-5/4}$

1.7 Exercises

1. 5 **2.** 9 **3.** 3 **4.** 5 **5.** -4 **6.** -2 **7.** $\dfrac{1}{3}$ **8.** $\dfrac{3}{2}$ **9.** $5x^2$ **10.** $4x^2$ **11.** $2(1 + x)$ **12.** $2|x + 4|$ **13.** $2\sqrt{2}$ **14.** $2\sqrt[4]{2}$ **15.** $2x\sqrt[3]{2x}$

16. $3x\sqrt{3x}$ **17.** $|x|$ **18.** $|x|\sqrt{x}$ **19.** $\dfrac{5|x|}{3}$ **20.** $\dfrac{1}{2x}$ **21.** $|x^3y^2|$ **22.** x^2y **23.** $\dfrac{x^2}{|y|}$ **24.** $\dfrac{y}{3x}$ **25.** $6\sqrt{x}$ **26.** $3x^2\sqrt{x}$

27. $6x\sqrt{x}$ **28.** $10x^2$ **29.** 1 **30.** $\dfrac{5\sqrt[3]{x^2}}{2y}$ **31.** $\dfrac{4y^2}{3|x|}$ **32.** $\dfrac{3x^2}{4|y^3|}$ **33.** $15\sqrt[3]{3}$ **34.** $300\sqrt[3]{3}$ **35.** $\dfrac{1}{x(2x + 3)}$ **36.** $\dfrac{x - 1}{x + 1}$ **37.** $\sqrt[24]{x}$

38. $\sqrt[13]{x}$ **39.** 1 **40.** $\dfrac{2\sqrt{x}}{(x^2 - 4)}$ **41.** $6\sqrt{2}$ **42.** $3\sqrt{5}$ **43.** $4\sqrt{2}$ **44.** 0 **45.** $\sqrt[3]{2}$ **46.** $15\sqrt[3]{3}$ **47.** $(x + 5)(x - 3)\sqrt{2x}$

48. $-4(2x + 5)\sqrt{y}$ **49.** $(-x - 5y)\sqrt[3]{2xy}$ **50.** $5xy$ **51.** $36\sqrt{2}$ **52.** $-60\sqrt{3}$ **53.** $-9 - \sqrt{3}$

54. $-7 + 4\sqrt{5}$ **55.** $54 + 17\sqrt{7}$ **56.** $33 + 12\sqrt{6}$ **57.** $x - 2\sqrt{x} + 1$ **58.** $2\sqrt{5x} + x + 5$ **59.** $x - 3\sqrt[3]{x^2} + 3\sqrt[3]{x} - 1$

60. $x + 6\sqrt[3]{x^2} + 12\sqrt[3]{x} + 8$ **61.** $4x + 4\sqrt{x} - 15$ **62.** $4x + 9\sqrt{x} - 9$ **63.** $\dfrac{-x^2}{\sqrt{1 - x^2}}$ **64.** $\dfrac{1}{\sqrt{1 - x^2}}$ **65.** $\dfrac{2\sqrt{5}}{5}$ **66.** $\dfrac{\sqrt{15}}{5}$

67. $\dfrac{4\sqrt{6}}{3}$ **68.** $\dfrac{\sqrt{10}}{2}$ **69.** $\dfrac{\sqrt{x}}{x}$ **70.** $\dfrac{x\sqrt{x^2 + 4}}{x^2 + 4}$ **71.** $\dfrac{15 - 3\sqrt{2}}{23}$ **72.** $\dfrac{2\sqrt{7} + 4}{3}$ **73.** $\dfrac{4 - \sqrt{7}}{3}$ **74.** $\dfrac{20 + 5\sqrt{2}}{7}$ **75.** $\dfrac{15 - 2\sqrt{5}}{41}$

76. $2 - \sqrt{3}$ **77.** $5 - 2\sqrt{6}$ **78.** $4 + \sqrt{15}$ **79.** $\dfrac{\sqrt{x} - 2}{x - 4}$ **80.** $\dfrac{\sqrt{x} + 3}{x - 9}$ **81.** $\dfrac{2x + h - 2\sqrt{x(x + h)}}{h}$ **82.** $\dfrac{x + \sqrt{x^2 - h^2}}{h}$

83. $\dfrac{-1}{16 + 7\sqrt{5}}$ **84.** $\dfrac{13}{10\sqrt{3} - 1}$ **85.** $\dfrac{1}{\sqrt{x + h} + \sqrt{x}}$ **86.** $\dfrac{-1}{x\sqrt{x + h} + (x + h)\sqrt{x}}$ **87.** 1.41 **88.** 2.65 **89.** 1.59 **90.** -1.71

91. 4.89 **92.** 0.04 **93.** 2.15 **94.** 1.33

1.8 Exercises

1. 4 **2.** 8 **3.** 9 **4.** 16 **5.** $\dfrac{1}{8}$ **6.** $-\dfrac{1}{32}$ **7.** $\dfrac{1}{27}$ **8.** $\dfrac{1}{3125}$ **9.** $\dfrac{27}{8}$ **10.** $\dfrac{9}{4}$ **11.** $\dfrac{27}{8}$ **12.** $\dfrac{9}{4}$

13. 8 **14.** $\dfrac{1}{64}$ **15.** 8 **16.** $\dfrac{1}{27}$ **17.** 27 **18.** 16 **19.** $\dfrac{1}{5}$ **20.** $\dfrac{1}{9}$ **21.** $x^{1/6}$ **22.** $x^{13/12}$ **23.** x^2y^4 **24.** x^5y^{10}

25. $x^{4/3}y^{5/3}$ **26.** $x^{5/4}y^{5/4}$ **27.** $\dfrac{8x^{3/2}}{y^{1/4}}$ **28.** $\dfrac{8y^{1/2}}{x^{3/2}}$ **29.** $\dfrac{x^{11}}{y^3}$ **30.** $\dfrac{x^4}{y^7}$ **31.** $\dfrac{3x + 2}{(1 + x)^{1/2}}$ **32.** $\dfrac{3x + 1}{2x^{1/2}}$ **33.** $\dfrac{x(3x^2 + 2)}{(x^2 + 1)^{1/2}}$

34. $\dfrac{4x + 3}{3(x + 1)^{2/3}}$ **35.** $\dfrac{22x + 5}{10\sqrt{x} - 5\sqrt{4x + 3}}$ **36.** $\dfrac{65x + 6}{24\sqrt[3]{(x - 2)^2}\sqrt[3]{(8x + 1)^2}}$ **37** $\dfrac{2 + x}{2(1 + x)^{3/2}}$ **38.** $\dfrac{1}{(x^2 + 1)^{3/2}}$ **39.** $\dfrac{4 - x}{(x + 4)^{3/2}}$

40. $\dfrac{9}{(9 - x^2)^{3/2}}$ **41.** $\dfrac{1}{x^2(x^2 - 1)^{1/2}}$ **42.** $\dfrac{4}{(x^2 + 4)^{3/2}}$ **43.** $\dfrac{1 - 3x^2}{2\sqrt{x}(1 + x^2)^2}$ **44.** $\dfrac{6x - 4x^3}{3(1 - x^2)^{4/3}}$

45. $\dfrac{1}{2}(5x + 2)(x + 1)^{1/2}$ **46.** $\dfrac{1}{3}(x^2 + 4)^{1/3}(11x^2 + 12)$ **47.** $2x^{1/2}(3x - 4)(x + 1)$ **48.** $2x^{1/2}(10x + 9)$ **49.** $(x^2 + 4)^{1/3}(11x^2 + 12)$

50. $2x(3x + 4)^{1/3}(5x + 4)$ **51.** $(3x + 5)^{1/3}(2x + 3)^{1/2}(17x + 27)$ **52.** $6(6x + 1)^{1/3}(4x - 3)^{1/2}(10x - 2)$

53. $\sqrt{1 + \left[\dfrac{1}{2}\left(x^3 - \dfrac{1}{x^3}\right)\right]^2} = \sqrt{1 + \dfrac{1}{4}\left(x^6 - 2 + \dfrac{1}{x^6}\right)}$

$= \sqrt{1 + \dfrac{1}{4}\left(\dfrac{x^{12} - 2x^6 + 1}{x^6}\right)}$

$= \sqrt{\dfrac{4x^6 + x^{12} - 2x^6 + 1}{4x^6}}$

$= \dfrac{1}{2}\sqrt{\dfrac{x^{12} + 2x^6 + 1}{x^6}}$

$= \dfrac{1}{2}\sqrt{x^6 + 2 + \dfrac{1}{x^6}}$

$= \dfrac{1}{2}\sqrt{\left(x^3 + \dfrac{1}{x^3}\right)^2}$

$= \dfrac{1}{2}\left(x^3 + \dfrac{1}{x^3}\right)$

54. $1 + (2\sqrt{3} \cdot x\sqrt{3x^2 + 1})^2 = 1 + 4 \cdot 3x^2(3x^2 + 1)$

$= 1 + 12x^2(3x^2 + 1)$

$= 1 + 36x^4 + 12x^2$

$= (1 + 6x^2)(1 + 6x^2)$

$= (6x^2 + 1)^2$

55. $1 + \left[\frac{1}{2}\left(x^2 - \frac{1}{x^2}\right)\right]^2 = 1 + \frac{1}{4}\left(x^4 - 2 + \frac{1}{x^4}\right)$

$\qquad\qquad\qquad\qquad\quad = \frac{4}{4} + \frac{1}{4}\left(x^4 - 2 + \frac{1}{x^4}\right)$

$\qquad\qquad\qquad\qquad\quad = \frac{1}{4}\left(x^4 + 2 + \frac{1}{x^4}\right)$

$\qquad\qquad\qquad\qquad\quad = \frac{1}{4}\left(x^2 + \frac{1}{x^2}\right)^2$

1.9 Exercises

1. 13. **2.** 10 **3.** 26 **4.** 5 **5.** 25 **6.** 50 **7.** Right triangle; 5 **8.** Right triangle; 10 **9.** Not a right triangle **10.** Not a right triangle
11. Right triangle; 25 **12.** Right triangle; 26 **13.** Not a right triangle **14.** Not a right triangle **15.** $A = 15$; $P = 16$
16. $A = 8$; $P = 12$ **17.** $A = \frac{1}{6}$; $P = \frac{5}{3}$ **18.** $A = 1$; $P = \frac{25}{6}$ **19.** 6 **20.** 9 **21.** $\frac{3}{16}$ **22.** $\frac{1}{2}$ **23.** $A \approx 3.14$; $C \approx 6.28$

24. $A \approx 12.56$; $C \approx 12.56$ **25.** $A \approx 7.065$; $C \approx 9.42$ **26.** $A \approx 4.90625$; $C \approx 7.85$ **27.** 48 **28.** 40 **29.** $\frac{10}{3}$ **30.** $\frac{1}{6}$ **31.** About 16.8 ft

32. About 1.6 revolutions **33.** 64 ft^2 **34.** 12 ft^2 **35.** $24 + 2\pi \approx 30.28$ ft^2; $16 + 2\pi \approx 22.28$ ft^2 **36.** $69\pi \approx 216.66$ ft^2; $26\pi \approx 81.64$ ft
37. About 46.6 mi **38.** About 5.5 miles **39.** About 3 mi **40.** About 12.25 miles; about 15 miles
41. $a^2 + b^2 = (m^2 - n^2)^2 + (2mn)^2 = m^4 - 2m^2n^2 + n^4 + 4m^2n^2 = m^4 + 2m^2n^2 + n^4 = (m^2 + n^2)^2 = c^2$
The conclusion follows from the converse of the Pythagorean Theorem.

Fill-in-the blank items

1. symmetric **2.** commutative **3.** distributive **4.** base; exponent **5.** 3; leading **6.** index; radicand **7.** hypotenuse

True/False Items

1. T **2.** T **3.** F **4.** T **5.** F **6.** T

Review Exercises

1. -11 **2.** 10 **3.** $\frac{1}{6}$ **4.** $\frac{3}{8}$ **5.** $\frac{93}{8}$ **6.** $\frac{63}{16}$ **7.** -29 **8.** 44 **9.** $\frac{1}{128}$ **10.** $-\frac{140}{9}$ **11.** 16 **12.** 10 **13.** $\frac{9}{4}$ **14.** $\frac{27}{8}$ **15.** $\frac{1}{2}$

16. $\frac{1}{4\sqrt[4]{2}}$ **17.** 4 **18.** 4 **19.** 4 **20.** -2 **21.** $\frac{y^2}{x^2}$ **22.** $\frac{y^6}{x^2}$ **23.** $\frac{1}{x^5 y}$ **24.** $\frac{x^3}{y^3}$ **25.** xy^4 **26.** $\frac{3y^5}{8x}$ **27.** $\frac{y^2}{x^2 + y^2}$

28. $\frac{y + x}{y - x}$ **29.** $\frac{125}{x^2 y}$ **30.** $\frac{x}{64y^2}$ **31.** $\frac{x^2}{16y^3}$ **32.** $\frac{x}{4y^2}$ **33.** $-8x^2 + 16x - 6$ **34.** $-24x^2 - 38x - 8$ **35.** $9x^3 - 20x^2 + 6x + 13$
36. $16x^4 + 8x^3 - 16x^2 + 4$ **37.** $6x^3 - 15x^2 + 4x - 10$ **38.** $8x^4 - 2x^3 - 4x + 1$ **39.** $x^3 - 7x - 6$ **40.** $x^3 - x^2 - 17x - 15$
41. $3x^2 + 8x + 25$; remainder 79 **42.** $2x^2 + x + 3$; remainder 7 **43.** $-3x^2 + 4$; remainder -2 **44.** $-4x + 1$; remainder $-4x - 1$
45. $8x^2 + 24x + 62$; remainder $167x - 61$ **46.** $3x^2 - 10x + 36$; remainder $-136x + 76$ **47.** $x^4 - x^3 + x^2 - x + 1$; remainder 0
48. $x^4 + x^3 + x^2 + x + 1$; remainder 0 **49.** $3x^4 - 2x^2 + 1$; remainder 0 **50.** $3x^4 - 2x^2 + 1$; remainder 0 **51.** $(x + 7)(x - 2)$
52. $(x - 7)(x - 2)$ **53.** $(3x + 2)(2x - 3)$ **54.** $(3x + 2)(2x - 1)$ **55.** $3(x + 2)(x - 7)$ **56.** $2x(x + 7)(x + 2)$
57. $(2x + 1)(4x^2 - 2x + 1)$ **58.** $(3x - 2)(9x^2 + 6x + 4)$ **59.** $(2x + 3)(x - 1)(x + 1)$ **60.** $(x^2 + 1)(2x + 3)$
61. $(5x - 2)(5x + 2)$ **62.** $(4x + 1)(4x - 1)$ **63.** Prime **64.** Prime **65.** $\frac{2x + 7}{x - 2}$ **66.** $-\frac{x - 7}{x - 2}$ **67.** $\frac{3(3x - 1)}{(x + 3)(3x + 1)}$

68. $-\frac{x + 5}{x^2 - 1}$ **69.** $\frac{4x}{(x + 1)(x - 1)}$ **70.** $-\frac{x^2}{(x + 1)(x + 2)}$ **71.** $\frac{x^2 + 17x + 2}{(x - 2)(x + 2)^2}$

72. $\frac{x^2(2x + 1)}{(x - 2)(x + 3)(2x - 1)} = \frac{2x^3 + x^2}{2x^3 + x^2 - 13x + 6}$ **73.** $\frac{2x + 1}{x + 1}$ **74.** $\frac{1}{1 + x}$ **75.** $\frac{4\sqrt{5}}{5}$ **76.** $-\frac{2\sqrt{3}}{3}$ **77.** $-2(1 + \sqrt{2})$

78. $2 - 2\sqrt{3}$ **79.** $-\frac{3 + \sqrt{5}}{2}$ **80.** 2 **81.** $\frac{2(1 + x^2)}{(2 + x^2)^{1/2}}$ **82.** $\frac{7x^2 + 12}{3(x^2 + 4)^{1/3}}$ **83.** $\frac{x(3x + 16)}{2(x + 4)^{3/2}}$ **84.** $\frac{x^3 + 8x}{(x^2 + 4)^{3/2}}$

85. **(a)** $8000 **(b)** $12,000 **86.** $0.35 per share **87.** About 37.81 ft^2; 25 ft **88.** 216 ft^2; 84 ft **89.** $16\pi \approx 50.24$ ft^2; $10\pi \approx 31.4$ ft
90. Yes; about 229.2 miles

C H A P T E R   2   2.1 Exercises

1. 3 **2.** -4 **3.** -3 **4.** -6 **5.** $\dfrac{3}{2}$ **6.** $-\dfrac{4}{3}$ **7.** $\dfrac{5}{4}$ **8.** $\dfrac{27}{4}$ **9.** -1 **10.** 7 **11.** 3 **12.** -4 **13.** -1 **14.** 1 **15.** $-\dfrac{4}{3}$ **16.** $\dfrac{2}{5}$

17. -18 **18.** $\dfrac{7}{5}$ **19.** -4 **20.** -1 **21.** $-\dfrac{3}{4}$ **22.** 2 **23.** -20 **24.** -10 **25.** 2 **26.** $\dfrac{-5}{2}$ **27.** 0.5 **28.** -10 **29.** $\dfrac{46}{5}$ **30.** 7

31. 2 **32.** $\dfrac{3}{10}$ **33.** 8 **34.** 6 **35.** 2 **36.** $-\dfrac{15}{7}$ **37.** -1 **38.** $\dfrac{1}{3}$ **39.** 3 **40.** 2 **41.** No solution **42.** No solution **43.** $\{0, 9\}$

44. $\{0, 1\}$ **45.** $\{0, 9\}$ **46.** $\{0, 2\}$ **47.** No solution **48.** No solution **49.** 2 **50.** -2 **51.** -31 **52.** 0 **53.** $-\dfrac{20}{39}$ **54.** $-\dfrac{14}{139}$

55. -1 **56.** 3 **57.** $-\dfrac{11}{6}$ **58.** 2 **59.** -6 **60.** -7 **61.** 5.91 **62.** 0.07 **63.** 0.41 **64.** -0.94 **65.** $\{3, 4\}$ **66.** $\{-2, 3\}$ **67.** $\left\{-3, \dfrac{1}{2}\right\}$

68. $\left\{-2, \dfrac{1}{3}\right\}$ **69.** $\{-3, 0, 3\}$ **70.** $\{-1, 0, 1\}$ **71.** $\{-5, 0, 4\}$ **72.** $\{-7, 0, 1\}$ **73.** $\{-1, 1\}$ **74.** $\{-4, -1, 1\}$ **75.** $\{-2, 2, 3\}$ **76.** $\{-1, 1, 3\}$

77. $x = \dfrac{b + c}{a}$ **78.** $x = \dfrac{1 - b}{a}$ **79.** $x = \dfrac{abc}{a + b}$ **80.** $x = \dfrac{a + b}{c}$ **81.** $x = a^2$ **82.** $x = \dfrac{ab}{c}$ **83.** $a = 3$ **84.** $b = 2$ **85.** $R = \dfrac{R_1 R_2}{R_1 + R_2}$

86. $r = \dfrac{A - P}{Pt}$ **87.** $R = \dfrac{mv^2}{F}$ **88.** $T = \dfrac{PV}{nR}$ **89.** $r = \dfrac{S - a}{S}$ **90.** $t = \dfrac{v_0 - v}{g}$

91. In obtaining step (7) we divided by $x - 2$. Since $x = 2$ from step (1), we actually divided by 0.
92. b; Equations are equivalent if they have the same solutions.

2.2 Exercises

1. $A = \pi r^2$; $r =$ Radius, $A =$ Area **2.** $C = 2\pi r$; $r =$ Radius, $C =$ Circumference **3.** $A = s^2$; $A =$ Area, $s =$ Length of a side
4. $P = 4s$; $s =$ length of a side, $p =$ Perimeter **5.** $F = ma$; $F =$ Force, $m =$ Mass, $a =$ Acceleration

6. $P = \dfrac{F}{A}$; $P =$ Pressure, $F =$ Force, $A =$ Area **7.** $W = Fd$; $W =$ Work, $F =$ Force, $d =$ Distance

8. $K = \dfrac{1}{2}mv^2$; $K =$ Kinetic Energy, $m =$ Mass, $v =$ Velocity **9.** $C = 150x$; $C =$ Total cost, $x =$ number of dishwashers

10. $R = 250x$; $R =$ Total Revenue, $x =$ number of dishwashers **11.** \$11,000 will be invested in bonds and \$9000 in CDs.
12. Sean will receive \$6000 and George \$4000. **13.** Scott will receive \$400,000, Alice \$300,000, and Tricia \$200,000.
14. Carter pays \$10.80 and Carole \$7.20. **15.** The regular hourly rate is \$8.50 **16.** Leigh's hourly wage is \$6.00 per hour.
17. The Bears got 5 touchdowns. **18.** The Bulls had 30 field goals. **19.** The length is 19 ft; the width is 11 ft
20. The length is 14 meters; the width is 7 meters. **21.** Invest \$31,250 in bonds and \$18,750 in CDs.
22. Invest \$43,750 in bonds, \$6250 in CDs. **23.** \$11,600 was loaned out at 8%. **24.** He can lend \$333,333.33.
25. Mix 30 cc of 15% HCl with 70 cc of 5% HCl. **26.** 49 pounds of coffee I should be mixed with 51 pounds of Coffee II.

27. Mix 40 lbs of cashews with the peanuts. **28.** Each box should contain 20 carmels and 10 creams. **29.** Add $\dfrac{20}{3}$ ounces of pure water.

30. Add 8 cubic centimeters of pure acid. **31.** There were 3260 adults. **32.** The original price was \$570.
33. The original price was \$147,058.82; purchasing the model saves \$22,058.82.
34. Its list price was \$9411.76; the amount saved was \$1411.76. **35.** The bookstore paid \$44.80.
36. At \$100 over cost, the payment is \$10,300. **37.** Working together, it takes 12 min.
38. April would take 15 hours. **39.** Sarah needs a score of 85. **40.** For a B, Mark needs 78; for an A, he needs 93.
41. The speed of the current is 2.286 mi/hr. **42.** The speed is 9 miles per hour. **43. (a)** The dimensions are 10 ft by 5 ft.
(b) The area is 50 sq ft. **(c)** The dimensions would be 7.5 ft by 7.5 ft. **(d)** The area would be 56.25 sq ft.
44. (a) Its dimensions are 19 ft by 19 ft. **(b)** Its dimensions are 28.5 ft by 9.5 ft. **(c)** The diameter is approximately 25.83 ft.
(d) The circular one has the most. **45.** The defensive back catches up to the tight end at the tight end's 45 yd line.

46. Therese should be allowed 20,000 miles as a business expense. **47.** Add $\dfrac{2}{3}$ gal of water. **48.** 5 liters should be drained.

49. Evaporate 10.67 oz of water. **50.** Evaporate 96 oz of water. **51.** The Metra commuter averages 30 mph; the Amtrak averages
80 mph. **52.** The average speed of the slower car is 60 miles per hour; the average speed of the faster car is 70 miles per hour.
Each traveled 210 miles.
53. 40 grams of 12 karat gold should be mixed with 20 grams of pure gold. **54.** There are 11 atoms of oxygen and 22 atoms of hydrogen.

55. Mike passes Dan $\dfrac{1}{3}$ mile from the start, 2 minutes from the time Mike started to race. **56.** It can fly 742.5 miles.

57. Start the auxiliary pump at 9:45 AM. **58.** 5 pounds must be added. **59.** The tub will fill in 1 hr.
60. It will take an hour and 45 minutes more for the 5 hp pump to empty the pool. **61.** Lewis would beat Burke by 16.75 meters.
62. Set the original price at \$40. At 50% off, there will be no profit.

Historical Problems

1. The area of each shaded square is 9, so the larger square will have area $85 + 4(9) = 121$. The area of the larger square is also given by the expression $(x + 6)^2$, so $(x + 6)^2 = 121$. Taking the positive square root of each side, $x + 6 = 11$ or $x = 5$. **2.** Let $z = -6$, so $z^2 + 12z - 85 = -121$. We get the equation $u^2 - 121 = 0$ or $u^2 = 121$. Thus $u = \pm 11$, so $x = \pm 11 - 6$. $x = -17$ or $x = 5$.

3.
$$\left(x + \frac{b}{2a}\right)^2 = \left(\frac{\sqrt{b^2 - 4ac}}{2a}\right)^2$$
$$\left(x + \frac{b}{2a}\right)^2 - \left(\frac{\sqrt{b^2 - 4ac}}{2a}\right)^2 = 0$$
$$\left(x + \frac{b}{2a} - \frac{\sqrt{b^2 - 4ac}}{2a}\right)\left(x + \frac{b}{2a} + \frac{\sqrt{b^2 - 4ac}}{2a}\right) = 0$$
$$\left(x + \frac{b - \sqrt{b^2 - 4ac}}{2a}\right)\left(x + \frac{b + \sqrt{b^2 - 4ac}}{2a}\right) = 0$$
$$x = \frac{-b + \sqrt{b^2 - 4ac}}{2a} \text{ or } x = \frac{-b - \sqrt{b^2 - 4ac}}{2a}$$

2.3 Exercises

1. $\{0, 9\}$ **2.** $\{-4, 0\}$ **3.** $\{-5, 5\}$ **4.** $\{-3, 3\}$ **5.** $\{-3, 2\}$ **6.** $\{-6, -1\}$ **7.** $\left\{-\frac{1}{2}, 3\right\}$ **8.** $\left\{-1, -\frac{2}{3}\right\}$ **9.** $\{-4, 4\}$ **10.** $\{-5, 5\}$

11. $\{2, 6\}$ **12.** $\{-6, 2\}$ **13.** $\frac{3}{2}$ **14.** $\frac{4}{5}$ **15.** $\left\{-\frac{2}{3}, \frac{3}{2}\right\}$ **16.** $\left\{\frac{1}{2}, \frac{3}{2}\right\}$ **17.** $\left\{-\frac{2}{3}, \frac{3}{2}\right\}$ **18.** $\{3, 4\}$ **19.** $\left\{-\frac{3}{4}, 2\right\}$ **20.** $\left\{-\frac{5}{2}, 1\right\}$

21. $\{2 - \sqrt{2}, 2 + \sqrt{2}\}$ **22.** $\{-2 - \sqrt{2}, -2 + \sqrt{2}\}$ **23.** $\{2 - \sqrt{5}, 2 + \sqrt{5}\}$ **24.** $\{-3 - 2\sqrt{2}, -3 + 2\sqrt{2}\}$ **25.** $\left\{1, \frac{3}{2}\right\}$

26. $\left\{-1, -\frac{3}{2}\right\}$ **27.** No real solution **28.** No real solution **29.** $\left\{\frac{-1 - \sqrt{5}}{4}, \frac{-1 + \sqrt{5}}{4}\right\}$ **30.** $\left\{\frac{-1 - \sqrt{3}}{2}, \frac{-1 + \sqrt{3}}{2}\right\}$

31. $\left\{0, \frac{9}{4}\right\}$ **32.** $\left\{0, \frac{5}{4}\right\}$ **33.** $\frac{1}{3}$ **34.** No real solution **35.** $\left\{-\frac{2}{3}, 1\right\}$ **36.** $\left\{-\frac{3}{2}, 3\right\}$ **37.** $\left\{\frac{1 - \sqrt{33}}{8}, \frac{1 + \sqrt{33}}{8}\right\}$

38. $\left\{\frac{-1 - \sqrt{17}}{8}, \frac{-1 + \sqrt{17}}{8}\right\}$ **39.** No real solution **40.** No real solution **41.** $\{0.63, 3.47\}$ **42.** $\{-3.37, -0.53\}$

43. $\{-2.80, 1.07\}$ **44.** $\{-2.29, 0.87\}$ **45.** $\{-0.85, 1.17\}$ **46.** $\{-1.44, 0.44\}$ **47.** $\{-8.16, -0.22\}$ **48.** $\{1.13, 5.62\}$ **49.** $\{-\sqrt{5}, \sqrt{5}\}$

50. $\{-\sqrt{6}, \sqrt{6}\}$ **51.** $\frac{1}{4}$ **52.** $\frac{1}{3}$ **53.** $\left\{-\frac{3}{5}, \frac{5}{2}\right\}$ **54.** $\left\{-\frac{5}{2}, \frac{4}{3}\right\}$ **55.** $\left\{-\frac{1}{2}, \frac{2}{3}\right\}$ **56.** $\left\{-\frac{2}{3}, \frac{1}{2}\right\}$ **57.** $\left\{\frac{-\sqrt{2} + 2}{2}, \frac{-\sqrt{2} - 2}{2}\right\}$

58. $\{\sqrt{2} - 2, \sqrt{2} + 2\}$ **59.** $\left\{\frac{-1 - \sqrt{17}}{2}, \frac{-1 + \sqrt{17}}{2}\right\}$ **60.** $\left\{\frac{-1 - \sqrt{5}}{2}, \frac{-1 + \sqrt{5}}{2}\right\}$ **61.** No real solution

62. No real solution **63.** Repeated real solution **64.** Repeated real solution **65.** Two unequal real solutions

66. Two unequal real solutions **67.** 16 **68.** 4 **69.** $\frac{1}{16}$ **70.** $\frac{1}{36}$ **71.** $\frac{1}{9}$ **72.** $\frac{1}{25}$ **73.** $\{-7, 3\}$ **74.** $\{3 - \sqrt{22}, 3 + \sqrt{22}\}$

75. $\left\{-\frac{1}{4}, \frac{3}{4}\right\}$ **76.** $\left\{-1, \frac{1}{3}\right\}$ **77.** $\left\{\frac{-1 - \sqrt{7}}{6}, \frac{-1 + \sqrt{7}}{6}\right\}$ **78.** $\left\{\frac{3 - \sqrt{17}}{4}, \frac{3 + \sqrt{17}}{4}\right\}$ **79.** The dimensions are 11 ft by 13 ft

80. The dimensions are 17 cm by 18 cm. **81.** The dimensions are 5 m by 8 m **82.** The shortest radius setting is 25 feet.
83. The dimensions should be 4 ft by 4 ft. **84.** The dimensions should be 6 feet by 3 feet.
85. **(a)** The ball strikes the ground after 6 seconds. **(b)** The ball passes the top of the building on its way down after 5 seconds.
86. The radius is 3 inches. **87.** Mike can paint the house by himself in 10 days.
88. It will take the smaller pump 12 hours and 23 minutes. **89.** 175 boxes were ordered.
90. The patio dimensions will be 36 ft by 18 ft. **91.** The border will be 2.56 ft wide.
92. **(a)** The object will be 15 meters above the ground after 0.99 seconds and after 3.09 seconds.
  (b) It will strike the ground in 4.08 seconds. **(c)** It will never reach 100 meters. **(d)** The maximum height is 20.41 meters.
93. The dimensions should be 11.55 cm by 6.55 cm by 3 cm. **94.** The dimensions should be 11.07 cm by 6.07 cm by 3 cm.
95. The border will be 2.71 ft wide. **96.** The border will be 2.13 ft. wide. **97.** The speed of the current is 5 mi/hr.
98. The lengths of the legs are 5 cm and 12 cm. **99.** The dimensions of the rectangle are 6 in. by 8 in.
100. **(a)** The object will strike ground at 8 seconds. **(b)** The height is 896 feet.

101. $\dfrac{-b + \sqrt{b^2 - 4ac}}{2a} + \dfrac{-b - \sqrt{b^2 - 4ac}}{2a} = \dfrac{-2b}{2a} = -\dfrac{b}{a}$

102. $\left(\dfrac{-b + \sqrt{b^2 - 4ac}}{2a}\right)\left(\dfrac{-b - \sqrt{b^2 - 4ac}}{2a}\right) = \dfrac{b^2 - (b^2 - 4ac)}{4a^2} = \dfrac{4ac}{4a^2} = \dfrac{c}{a}$ **103.** $k = \dfrac{1}{2}$ or $k = -\dfrac{1}{2}$ **104.** $k = 4$ or $k = -4$

105. $ax^2 + bx + c = 0, x = \dfrac{-b \pm \sqrt{b^2 - 4ac}}{2a}$; $ax^2 - bx + c = 0, x = \dfrac{b \pm \sqrt{(-b)^2 - 4ac}}{2a}$

106. For $ax^2 + bx + c = 0$, the solutions are $x_1 = \dfrac{-b + \sqrt{b^2 - 4ac}}{2a}$, $x_2 = \dfrac{-b - \sqrt{b^2 - 4ac}}{2a}$; For $cx^2 + bx + a = 0$, the solutions

are $x_3 = \dfrac{-b + \sqrt{b^2 - 4ac}}{2c}$, $x_4 = \dfrac{-b - \sqrt{b^2 - 4ac}}{2c}$; $x_1 x_4 = \dfrac{b^2 - (b^2 - 4ac)}{4ac} = 1$ and $x_2 x_3 = \dfrac{b^2 - (b^2 - 4ac)}{4ac} = 1$

107. 36 consecutive integers must be added. **108.** 13 diagonals; no **109.** The average speed is 49.5 mi/hr.
110. The tail wind was 137 knots.

2.4 Exercises

1. 1 **2.** 0 **3.** No real solution **4.** No real solution **5.** -13 **6.** 0 **7.** $\{0, 36\}$ **8.** $\{0, 16\}$ **9.** 3 **10.** 3 **11.** 2 **12.** No real solution
13. $-\dfrac{8}{5}$ **14.** No real solution **15.** 8 **16.** 4 **17.** $\{-1, 3\}$ **18.** -2 **19.** $\{1, 5\}$ **20.** 18 **21.** 1 **22.** 25 **23.** 5 **24.** 3 **25.** 2 **26.** -1
27. $\{-4, 4\}$ **28.** $\{-\sqrt{97}, \sqrt{97}\}$ **29.** $\{0, 2\}$ **30.** $\{0, 16\}$ **31.** $\{-2, -1, 1, 2\}$ **32.** $\{-\sqrt{5}, \sqrt{5}\}$ **33.** $\{-1, 1\}$ **34.** $\{-2, 2\}$ **35.** $\{-2, 1\}$
36. $\{-1, 2\}$ **37.** $\{-6, -5\}$ **38.** $\left\{-\dfrac{7}{2}, -1\right\}$ **39.** $-\dfrac{1}{3}$ **40.** $\{-2, 7\}$ **41.** $\left\{-\dfrac{3}{2}, 2\right\}$ **42.** $\left\{\dfrac{5}{3}, 2\right\}$ **43.** $\{0, 16\}$ **44.** 0 **45.** 16 **46.** 4
47. 1 **48.** 16 **49.** $\left\{\left(\dfrac{9 - \sqrt{17}}{8}\right)^4, \left(\dfrac{9 + \sqrt{17}}{8}\right)^4\right\}$ **50.** $\{1, 16\}$ **51.** $\{\sqrt{2}, \sqrt{3}\}$ **52.** $\sqrt{\dfrac{-5 + \sqrt{41}}{2}}$ **53.** $\{-4, 1\}$ **54.** $\{-1, 4\}$
55. $\left\{-2, -\dfrac{1}{2}\right\}$ **56.** $\left\{\dfrac{3}{4}, \dfrac{4}{3}\right\}$ **57.** $\left\{-\dfrac{3}{2}, \dfrac{1}{3}\right\}$ **58.** $\left\{\dfrac{4}{3 - \sqrt{41}}, \dfrac{4}{3 + \sqrt{41}}\right\}$ **59.** $\left\{-\dfrac{1}{8}, 27\right\}$ **60.** $\dfrac{\sqrt{3}}{9}$ **61.** $\left\{-2, -\dfrac{4}{5}\right\}$
62. $\left\{\dfrac{1}{2}, \dfrac{7}{6}\right\}$ **63.** $\{0.34, 11.66\}$ **64.** $\{-39.80, -0.20\}$ **65.** $\{-1.03, 1.03\}$ **66.** $\{-0.93, 0.93\}$ **67.** $\{-1.85, 0.17\}$ **68.** $\{-1.44, 0.44\}$
69. $\left\{\dfrac{3}{2}, 5\right\}$ **70.** $\left\{\dfrac{13}{5}, \dfrac{31}{6}\right\}$ **71.** The depth of the well is 230 ft.

2.5 Exercises

1. $[0, 2]; 0 \le x \le 2$ **2.** $[2, \infty); x \ge 2$ **3.** $(-1, 2); -1 < x < 2$ **4.** $(-\infty, 0]; x \le 0$ **5.** $(-\infty, 0]$ or $(2, \infty); x \le 0$ or $x > 2$
6. $(-\infty, 0]$ or $(1, \infty); x \le 0$ or $x > 1$ **7.** $[0, 3); 0 \le x < 3$ **8.** $(-1, 1]; -1 < x \le 1$ **9.** < **10.** < **11.** > **12.** > **13.** >
14. < **15.** > **16.** <
17. $x \ge -2$　　　　**18.** $x < 4$　　　　**19.** $x \ge 4$ and $x < 6$　　　　**20.** $x > 3$ and $x \le 7$

21. $x \le 0$ or $x < 6$　　　　**22.** $x > 0$ or $x \ge 5$　　　　**23.** There are no numbers for which $x \le -2$ and $x > 1$.
　　　　24. There are no numbers for which $x \ge 4$ and $x < -2$.

25. $x \le -2$ or $x > 1$　　　　**26.** $x \ge 4$ or $x < -2$　　　　**27.** $[0, 4]$　　　　**28.** $(-1, 5)$

29. $[4, 6)$　　　　**30.** $(-2, 0)$　　　　**31.** $[4, \infty)$　　　　**32.** $(-\infty, 5)$

33. $(-\infty, -4)$　　　　**34.** $(1, \infty)$　　　　**35.** $2 \le x \le 5$　　　　**36.** $1 < x < 2$

37. $-3 < x < -2$　　　　**38.** $0 \le x < 1$　　　　**39.** $x \ge 4$　　　　**40.** $x \le 2$

41. $x < -3$　　　　**42.** $x > -8$

43. $a \le b, c > 0;\ a - b \le 0$
　　　　$(a - b)c \le 0(c)$
　　　　$ac - bc \le 0$
　　　　$ac \le bc$

44. $a - b \le 0, c < 0$
　　　$c(a - b) \ge c(0)$
　　　$ca - cb \ge 0$
　　　$ca \ge cb$
　　　$ac \ge bc$

45. $\dfrac{a+b}{2} - a = \dfrac{a+b-2a}{2} = \dfrac{b-a}{2} > 0$; therefore, $a < \dfrac{a+b}{2}$ **46.** $\dfrac{a+b}{2} - a = \dfrac{a+b-2a}{2} = \dfrac{b-a}{2}$

$b - \dfrac{a+b}{2} = \dfrac{2b-a-b}{2} = \dfrac{b-a}{2} > 0$; therefore, $b > \dfrac{a+b}{2}$ $\qquad b - \dfrac{a+b}{2} = \dfrac{2b-a-b}{2} = \dfrac{b-a}{2}$;

thus, $\dfrac{a+b}{2}$ is equidistant from a and b

47. $(\sqrt{ab})^2 - a^2 = ab - a^2 = a(b-a) > 0$; thus, $(\sqrt{ab})^2 > a^2$ and $\sqrt{ab} > a$
$b^2 - (\sqrt{ab})^2 = b^2 - ab = b(b-a) > 0$; thus, $b^2 > (\sqrt{ab})^2$ and $b > \sqrt{ab}$

48. $\dfrac{a+b}{2} - \sqrt{ab} = \dfrac{1}{2}[a - 2\sqrt{ab} + b]$

$= \dfrac{1}{2}(\sqrt{a} - \sqrt{b})^2 > 0$ since $a > b$; thus, $\sqrt{ab} < \dfrac{a+b}{2}$

49. $h - a = \dfrac{2ab}{a+b} - a = \dfrac{ab - a^2}{a+b} = \dfrac{a(b-a)}{a+b} > 0$; thus, $h > a$

$b - h = b - \dfrac{2ab}{a+b} = \dfrac{b^2 - ab}{a+b} = \dfrac{b(b-a)}{a+b} > 0$; thus, $h < b$

50. $\dfrac{1}{h} = \dfrac{1}{2}\left(\dfrac{1}{a} + \dfrac{1}{b}\right)$

$\dfrac{2}{h} = \dfrac{1}{a} + \dfrac{1}{b} = \dfrac{a+b}{ab}$

$\dfrac{h}{2} = \dfrac{ab}{a+b}$

$h = \dfrac{2ab}{a+b}$

$h = \dfrac{(\sqrt{ab})^2}{\dfrac{1}{2}(a+b)}$

51. $21 < \text{Age} < 30$ **52.** $40 \le \text{Age} < 60$
53. (a) Male ≥ 73.4 **(b)** Female ≥ 79.7 **(c)** A female can expect to live at least 6.3 years longer.

2.6 Exercises

1. (a) $0 < 2$ **(b)** $-2 < 0$ **(c)** $9 < 15$ **(d)** $-6 > -10$ **2. (a)** $-1 > -2$ **(b)** $-3 > -4$ **(c)** $6 > 3$ **(d)** $-4 < -2$
3. (a) $2x - 2 < -1$ **(b)** $2x - 4 < -3$ **(c)** $6x + 3 < 6$ **(d)** $-4x - 2 > -4$
4. (a) $-2 - 2x > 2$ **(b)** $-4 - 2x > 0$ **(c)** $3 - 6x > 15$ **(d)** $-2 + 4x < -10$
5. $\{x \mid x < 4\}$ or $(-\infty, 4)$ **6.** $\{x \mid x < 7\}$ or $(-\infty, 7)$ **7.** $\{x \mid x \ge -1\}$ or $[-1, \infty)$ **8.** $\{x \mid x \ge -1\}$ or $[-1, \infty)$

9. $\{x \mid x > 3\}$ or $(3, \infty)$ **10.** $\{x \mid x > -2\}$ or $(-2, \infty)$ **11.** $\{x \mid x \le -2\}$ or $(-\infty, -2]$ **12.** $\{x \mid x \le -5\}$ or $(-\infty, -5]$

13. $\{x \mid x > -7\}$ or $(-7, \infty)$ **14.** $\{x \mid x < 5\}$ or $(-\infty, 5)$ **15.** $\left\{x \mid x \le \dfrac{2}{3}\right\}$ or $\left(-\infty, \dfrac{2}{3}\right]$ **16.** $\{x \mid x \le 0\}$ or $(-\infty, 0]$

17. $\{x \mid x < -20\}$ or $(-\infty, -20)$ **18.** $\left\{x \mid x > -\dfrac{7}{4}\right\}$ or $\left(-\dfrac{7}{4}, \infty\right)$ **19.** $\left\{x \mid x \ge \dfrac{4}{3}\right\}$ or $\left[\dfrac{4}{3}, \infty\right)$ **20.** $\{x \mid x \ge 12\}$ or $[12, \infty)$

21. $\{x \mid 3 \le x \le 5\}$ or $[3, 5]$ **22.** $\{x \mid 1 \le x \le 4\}$ or $[1, 4]$ **23.** $\left\{x \mid \dfrac{2}{3} \le x \le 3\right\}$ or $\left[\dfrac{2}{3}, 3\right]$ **24.** $\{x \mid -3 \le x \le 3\}$ or $[-3, 3]$

25. $\left\{x \mid -4 < x < \dfrac{1}{2}\right\}$ or $\left(-4, \dfrac{1}{2}\right)$ **26.** $\left\{x \mid -\dfrac{2}{3} < x < 4\right\}$ or $\left(-\dfrac{2}{3}, 4\right)$ **27.** $\{x \mid -6 < x < 0\}$ or $(-6, 0)$

28. $\{x \mid 0 < x < 3\}$ or $(0, 3)$ **29.** $\{x \mid x < -5\}$ or $(-\infty, -5)$ **30.** $\{x \mid x < 11\}$ or $(-\infty, 11)$ **31.** $\{x \mid x \geq -1\}$ or $[-1, \infty)$

32. $\{x \mid x \leq 1\}$ or $(-\infty, 1]$ **33.** $\left\{x \mid \dfrac{1}{2} \leq x < \dfrac{5}{4}\right\}$ or $\left[\dfrac{1}{2}, \dfrac{5}{4}\right)$ **34.** $\left\{x \mid -\dfrac{1}{3} < x \leq \dfrac{1}{3}\right\}$ or $\left(-\dfrac{1}{3}, \dfrac{1}{3}\right]$

35. $\left\{x \mid x < -\dfrac{1}{2}\right\}$ or $\left(-\infty, -\dfrac{1}{2}\right)$ **36.** $\left\{x \mid x > \dfrac{1}{2}\right\}$ or $\left(\dfrac{1}{2}, \infty\right)$ **37.** $\left\{x \mid x > \dfrac{10}{3}\right\}$ or $\left(\dfrac{10}{3}, \infty\right)$

38. $\{x \mid x > 6\}$ or $(6, \infty)$ **39.** $\{x \mid x > 3\}$ or $(3, \infty)$ **40.** $\{x \mid x > -1\}$ or $(-1, \infty)$

41. $a = 3, b = 5$ **42.** $a = -9, b = -4$ **43.** $a = -12, b = -8$ **44.** $a = -2, b = 0$ **45.** $a = 3, b = 11$ **46.** $a = -5, b = 7$

47. $a = \dfrac{1}{4}, b = 1$ **48.** $a = -\dfrac{1}{2}, b = -\dfrac{1}{4}$ **49.** $a = 4, b = 16$ **50.** $a = 0, b = 9$ **51.** $\{x \mid -10 < x < 0\}$ **52.** $\{x \mid 3 < x < 6\}$

53. $\{x \mid x \geq -2\}$ **54.** $\{x \mid x \geq -4\}$ **55.** The volume of the gas ranges from 1600 to 2400 cc, inclusive.
56. The interest rates needed are from 9% to 11%, inclusive.
57. The agent's commission ranges from $45,000 to $95,000, inclusive. As a percent of selling price, the commission ranges from 5% to 8.6%, inclusive.   **58.** The commission will vary from $53 to $145, inclusive.
59. The amount withheld varies from $72.14 to $93.14, inclusive.   **60.** The amount withheld varies from $93.14 to $121.14, inclusive.
61. The usage varies from 675.43 kWhr to 2500.86 kWhr, inclusive.   **62.** The water usage varied from 16,000 gallons to 38,000 gallons.
63. The dealer's cost varies from $7457.63 to $7857.14, inclusive.   **64.** The people in the top 2.5% have test scores greater than 123.4.
65. You need at least a 74 on the last test.   **66.** You need at least a 77 on the last test.
67. The amount of gasoline ranged from 12 to 20 gal, inclusive.
68. There were 10 gallons or less of gasoline at the start of the trip.

2.7 Exercises

1. $\{x \mid -2 < x < 5\}$ **2.** $\{x \mid x < -2 \text{ or } x > 5\}$ **3.** $\{x \mid x < 0 \text{ or } x > 4\}$ **4.** $\{x \mid x < -8 \text{ or } x > 0\}$ **5.** $\{x \mid -3 < x < 3\}$

6. $\{x \mid -1 < x < 1\}$ **7.** $\{x \mid x < -2 \text{ or } x > 1\}$ **8.** $\{x \mid -4 < x < -3\}$ **9.** $\left\{x \mid -\dfrac{1}{2} \leq x \leq 3\right\}$ **10.** $\left\{x \mid -\dfrac{2}{3} \leq x \leq \dfrac{3}{2}\right\}$

11. $\{x \mid x < -1 \text{ or } x > 8\}$ **12.** $\{x \mid x < -5 \text{ or } x > 4\}$ **13.** No real solution   **14.** No real solution   **15.** $\left\{x \mid x < -\dfrac{2}{3} \text{ or } x > \dfrac{3}{2}\right\}$

16. All real numbers   **17.** $\{x \mid x > 1\}$ **18.** $\{x \mid x > -2\}$ **19.** $\{x \mid x \leq 1 \text{ or } 2 \leq x \leq 3\}$ **20.** $\{x \mid x \leq -3 \text{ or } -2 \leq x \leq -1\}$
21. $\{x \mid -1 < x < 0 \text{ or } x > 3\}$ **22.** $\{x \mid -3 < x < 0 \text{ or } x > 1\}$ **23.** $\{x \mid x < -1 \text{ or } x > 1\}$ **24.** $\{x \mid -2 < x < 0 \text{ or } 0 < x < 2\}$
25. $\{x \mid x > 4\}$ **26.** $\{x \mid x < 0 \text{ or } 0 < x < 9\}$ **27.** $\{x \mid x < -1 \text{ or } x > 1\}$ **28.** $\{x \mid x > 1\}$ **29.** $\{x \mid x < -1 \text{ or } x > 1\}$
30. $\{x \mid x < -1 \text{ or } x > 3\}$ **31.** $\{x \mid x < -1 \text{ or } 0 < x < 1\}$ **32.** $\{x \mid x < -2 \text{ or } 1 < x < 3\}$ **33.** $\{x \mid x < -1 \text{ or } x > 1\}$
34. $\{x \mid x < -2 \text{ or } x > 2\}$ **35.** $\left\{x \mid x < -\dfrac{2}{3} \text{ or } 0 < x < \dfrac{3}{2}\right\}$ **36.** $\{x \mid x < 0 \text{ or } 3 < x < 4\}$ **37.** $\{x \mid x < 2\}$ **38.** $\{x \mid x > 4\}$
39. $\{x \mid -2 < x \leq 9\}$ **40.** $\{x \mid -8 \leq x < -2\}$ **41.** $\{x \mid x < 2 \text{ or } 3 < x < 5\}$ **42.** $\{x \mid -7 < x < -1 \text{ or } x > 3\}$
43. $\{x \mid x < -3 \text{ or } -1 < x < 1 \text{ or } x > 2\}$ **44.** $\left\{x \mid x < -\dfrac{5}{2} \text{ or } -2 < x < -1\right\}$ **45.** $\{x \mid x < -5 \text{ or } -4 \leq x \leq -3 \text{ or } x = 0 \text{ or } x > 1\}$
46. $\{x \mid x < -1 \text{ or } 0 \leq x < 1 \text{ or } x \geq 2\}$ **47.** $\{x \mid x > 4\}$ **48.** $\{x \mid x > 2\}$ **49.** $\{x \mid x \leq -4 \text{ or } x \geq 4\}$
50. $\{x \mid x = 0 \text{ or } x \geq 3\}$ **51.** $\{x \mid x < -4 \text{ or } x \geq 2\}$ **52.** $\{x \mid x < -4 \text{ or } x \geq 1\}$
53. The ball is more than 96 feet above the ground for time t between 2 and 3 seconds, $2 < t < 3$
54. The ball is less than 64 feet above the ground for time t less than 1 second or t greater than 4 seconds, $0 \leq t < 1 \text{ or } t > 4$
55. For a profit of at least $100, between 10 and 50 watches must be sold, $10 \leq x \leq 50$
56. For a profit of at least $60, between 30 and 40 boxes must be sold, $30 \leq x \leq 40$
57. $b - a = (\sqrt{b} - \sqrt{a})(\sqrt{b} + \sqrt{a})$; since $a \geq 0$ and $b \geq 0$, then $\sqrt{a} \geq 0$ and $\sqrt{b} \geq 0$ so that $\sqrt{b} + \sqrt{a} \geq 0$; thus, $b - a \geq 0$ is equivalent to $\sqrt{b} - \sqrt{a} \geq 0$ and $a \leq b$ is equivalent to $\sqrt{a} \leq \sqrt{b}$ **58.** $-2 < k < 2$ **59.** $k < 1$

2.8 Exercises

1. $\{-3, 3\}$ **2.** $\{-4, 4\}$ **3.** $\{-4, 1\}$ **4.** $\left\{1, -\dfrac{1}{3}\right\}$ **5.** $\left\{-1, \dfrac{3}{2}\right\}$ **6.** $\{-1, 2\}$ **7.** $\{-4, 4\}$ **8.** $\{-1, 1\}$ **9.** 2 **10.** 3 **11.** $\{-12, 12\}$

12. $\{-12, 12\}$ **13.** $\left\{-\dfrac{36}{5}, \dfrac{24}{5}\right\}$ **14.** $\left\{-\dfrac{4}{3}, \dfrac{8}{3}\right\}$ **15.** No real solution **16.** No real solution **17.** $\{-3, 3\}$ **18.** $\{-4, 4\}$ **19.** $\{-1, 3\}$

20. $\{-4, 3\}$ **21.** $\{-2, -1, 0, 1\}$ **22.** $\{-4, -3, 0, 1\}$ **23.** $(-4, 4)$ **24.** $(-5, 5)$ **25.** 0 **26.** $(-1, 0)$ **27.** $\dfrac{\sqrt{5}}{2}$ **28.** $\{5, 41\}$ **29.** $\left\{-\dfrac{1}{8}, 1\right\}$ **30.** $\left\{-\dfrac{1}{8}, \dfrac{1}{8}\right\}$

2.8 Exercises

1. $\{-3,3\}$ **2.** $\{-4,4\}$ **3.** $\{-4,1\}$ **4.** $\left\{1,-\dfrac{1}{3}\right\}$ **5.** $\left\{-1,\dfrac{3}{2}\right\}$ **6.** $\{-1,2\}$ **7.** $\{-4,4\}$ **8.** $\{-1,1\}$ **9.** 2 **10.** 3 **11.** $\{-12,12\}$

12. $\{-12,12\}$ **13.** $\left\{-\dfrac{36}{5},\dfrac{24}{5}\right\}$ **14.** $\left\{-\dfrac{4}{3},\dfrac{8}{3}\right\}$ **15.** No real solution **16.** No real solution **17.** $\{-3,3\}$ **18.** $\{-4,4\}$ **19.** $\{-1,3\}$

20. $\{-4,3\}$ **21.** $\{-2,-1,0,1\}$ **22.** $\{-4,-3,0,1\}$ **23.** $(-4,4)$ **24.** $(-5,5)$ **25.** $(-\infty,-4)$ or $(4,\infty)$ **26.** $(-\infty,-3)$ or $(3,\infty)$ **27.** $(1,3)$

28. $(-6,-2)$ **29.** $\left[-\dfrac{2}{3},2\right]$ **30.** $[-6,1]$ **31.** $(-\infty,1]$ or $[5,\infty)$ **32.** $(-\infty,-6]$ or $[-2,\infty)$ **33.** $\left(-1,\dfrac{3}{2}\right)$ **34.** $(-1,2)$

35. $(-\infty,-1)$ or $(2,\infty)$ **36.** $\left(-\infty,\dfrac{1}{3}\right)$ or $(1,\infty)$ **37.** $(-\infty,-3)$ or $(-3,\infty)$ **38.** $(-\infty,3)$ or $(3,\infty)$ **39.** $(-\infty,\infty)$ **40.** $(-\infty,\infty)$

41. $(0.49,0.51)$ **42.** $(0.66,0.67333...)$ **43.** $a=2,b=8$ **44.** $a=-9,b=1$ **45.** $a=-15,b=-7$ **46.** $a=7,b=13$

47. $a=-1,b=-\dfrac{1}{15}$ **48.** $a=\dfrac{1}{7},b=1$ **49.** $\left|\dfrac{a}{b}\right|=\sqrt{\left(\dfrac{a}{b}\right)^2}=\sqrt{\dfrac{a^2}{b^2}}=\dfrac{\sqrt{a^2}}{\sqrt{b^2}}=\dfrac{|a|}{|b|}$

50. If $a\ge0$, then $a=|a|$ and if $a<0$, then $a<|a|$ so $a\le|a|$.
51. $(a+b)^2=a^2+2ab+b^2\le|a|^2+2|a||b|+|b|^2=(|a|+|b|)^2$; therefore, $\sqrt{(a+b)^2}\le\sqrt{(|a|+|b|)^2}$ or $|a+b|\le|a|+|b|$
52. $|a|=|(a-b)+b|\le|a-b|+|b|$. Thus, $|a|-|b|\le|a-b|+|b|-|b|$ or $|a|-|b|\le|a-b|$. Hence, $|a-b|\ge|a|-|b|$.
53. $|x-3|<\dfrac{1}{2};\dfrac{5}{2}<x<\dfrac{7}{2}$ **54.** $|x+4|<1;-5<x<-3$ **55.** $|x+3|>2;x<-5$ or $x>-1$ **56.** $|x-2|>3;x<-1$ or $x>5$
57. $|x-98.6|\ge1.5;x\le97.1°$F or $x\ge100.1°$F **58.** $|x-115|\le5;110\le x\le120$
59. $x^2-a<0;(x-\sqrt{a})(x+\sqrt{a})<0$; therefore, $-\sqrt{a}<x<\sqrt{a}$
60. $x^2>a$ so $x^2-a>0;(x-\sqrt{a})(x+\sqrt{a})>0$; therefore, $-\infty<x<-\sqrt{a}$ or $\sqrt{a}<x<\infty$ **61.** $-1<x<1$ **62.** $-2<x<2$
63. $x\ge3$ or $x\le-3$ **64.** $x\ge1$ or $x\le-1$ **65.** $-4\le x\le4$ **66.** $-3\le x\le3$ **67.** $x>2$ or $x<-2$ **68.** $x>4$ or $x<-4$ **69.** $\{-1,5\}$
70. $x=0,x=1$

Fill-in-the-Blank Items

1. equivalent **2.** identity **3.** add; $\dfrac{25}{4}$ **4.** discriminant; negative **5.** $-a$ **6.** double; multiplicity 2 **7.** extraneous **8.** negative
9. $-2,2$

True/False Items

1. T **2.** F **3.** F **4.** T **5.** T **6. (a)** T **(b)** F **(c)** T **7. (a)** F **(b)** T **(c)** T

Review Exercises

1. -18 **2.** 24 **3.** 6 **4.** $\dfrac{4}{11}$ **5.** $\dfrac{1}{5}$ **6.** $\dfrac{9}{16}$ **7.** 6 **8.** $\dfrac{11}{18}$ **9.** No real solution **10.** $\{-3,2\}$ **11.** $\dfrac{11}{8}$ **12.** $-\dfrac{27}{13}$ **13.** $\left\{-2,\dfrac{3}{2}\right\}$

14. $\{-4,3\}$ **15.** $\left\{\dfrac{1-\sqrt{13}}{4},\dfrac{1+\sqrt{13}}{4}\right\}$ **16.** $\left\{\dfrac{3-\sqrt{13}}{4},\dfrac{3+\sqrt{13}}{4}\right\}$ **17.** $\{-3,3\}$ **18.** 2 **19.** No real solution **20.** No real solution

21. $\{-2,-1,1,2\}$ **22.** No real solution **23.** 2 **24.** 5 **25.** 0 **26.** $\{-1,0\}$ **27.** $\dfrac{\sqrt{5}}{2}$ **28.** $\{5,41\}$ **29.** $\left\{-\dfrac{1}{8},1\right\}$ **30.** $\left\{-\dfrac{1}{8},\dfrac{1}{8}\right\}$

31. $\left\{-1,\dfrac{1}{2}\right\}$ **32.** $\{4,9\}$ **33.** $\left\{\dfrac{m}{1-n},\dfrac{m}{1+n}\right\}$ **34.** $\left\{-\dfrac{a}{b-1},\dfrac{a}{b+1}\right\}$ **35.** $\left\{-\dfrac{9b}{5a},\dfrac{2b}{a}\right\}$ **36.** $\dfrac{2mn}{n+m}$ **37.** $-\dfrac{9}{5}$

38. No real solution **39.** $\{-5,2\}$ **40.** $\left\{-\dfrac{4}{3},2\right\}$ **41.** $\left\{-\dfrac{5}{3},3\right\}$ **42.** $\{-1,2\}$ **43.** $\{x|x\ge14\}$ **44.** $\left\{x\,\middle|\,x\ge\dfrac{17}{19}\right\}$

45. $\left\{x\,\middle|\,-\dfrac{31}{2}\le x\le\dfrac{33}{2}\right\}$ **46.** $\{x|-5<x<10\}$ **47.** $\{x|-23<x<-7\}$ **48.** $\left\{x\,\middle|\,-\dfrac{7}{3}<x\le\dfrac{11}{3}\right\}$ **49.** $\left\{x|-4<x<\dfrac{3}{2}\right\}$

50. $\left\{x\,\middle|\,x\le-\dfrac{1}{3}\text{ or }x\ge1\right\}$ **51.** $\{x|-3<x\le3\}$ **52.** $\left\{x\,\middle|\,x<\dfrac{1}{3}\text{ or }x>1\right\}$ **53.** $\{x|x<1\text{ or }x>2\}$

54. $\left\{x\,\middle|\,-\dfrac{5}{2}<x\le-\dfrac{7}{6}\right\}$ **55.** $\{x|1<x<2\text{ or }x>3\}$ **56.** $\{x|x\le-1\text{ or }0<x<5\}$ **57.** $\{x|x<-4\text{ or }2<x<4\text{ or }x>6\}$

58. $\{x|x<-5\text{ or }-4<x\le-2\text{ or }0\le x\le1\}$ **59.** $\left\{x\,\middle|\,-\dfrac{3}{2}<x<-\dfrac{7}{6}\right\}$ **60.** $\left\{x\,\middle|\,\dfrac{1}{3}<x<\dfrac{2}{3}\right\}$ **61.** $\{x|x\le-2\text{ or }x\ge7\}$

62. $\left\{x\,\middle|\,x\le-\dfrac{11}{3}\text{ or }x\ge3\right\}$ **63.** 9 **64.** 25 **65.** $\dfrac{4}{9}$ **66.** $\dfrac{4}{25}$ **67.** The storm is 3300 ft away.
68. The range of distances is from 0.5 meters to 0.75 meters, $0.5\le x\le0.75$. **69.** The search plane can go as far as 616 mi.
70. The plane can extend its search 246.4 miles. **71.** The helicopter will reach the life raft in a little less than 1 hr, 35 min.

72. The bees meet for the first time in 18.75 seconds. The bees meet for the second time in 37.5 seconds.
73. It takes Clarissa 10 days by herself. **74.** It will take about 12 hours, 23 minutes.

75. Mix 90 cc of 15% HCl to obtain 150 cc of 25% HCl. **76.** Add $6\frac{2}{3}$ pounds of $8/lb coffee to get $26\frac{2}{3}$ pounds of $5/lb coffee.

77. Add 256 ounces of water. **78.** Evaporate 51.2 ounces of water. **79.** The freight train is 190.67 ft long.

80. (a) 6.5 inches by 6.5 inches; 12.5 inches by 12.5 inches **(b)** $8\frac{2}{3}$ inches by $4\frac{1}{3}$ inches; $14\frac{2}{3}$ inches by $10\frac{1}{3}$ inches

81. It will take the smaller pump two hours. **82.** The length should be approximately 6.47 feet.
83. 36 seniors went on the trip; each one paid $13.40. **84.** It would take the older copier 180 minutes or 3 hours.

85. (a) No **(b)** Todd wins again. **(c)** Todd wins by $\frac{1}{4}$ meter. **(d)** Todd should line up 5.26316 meters behind the start line. **(e)** Yes

86. (a) is an expression, (b) is an equation, and (c) is an inequality. Simplifying the sum is the first step in solving each of the problems. Solving the equality is one step toward solving the inequality.

C H A P T E R 3   3.1 Exercises

1. (a) Quadrant II **(b)** Positive x-axis **(c)** Quadrant III
(d) Quadrant I **(e)** Negative y-axis **(f)** Quadrant IV

2. (a) Quadrant I **(b)** Quadrant III **(c)** Quadrant II
(d) Quadrant I **(e)** positive y-axis **(f)** negative x-axis

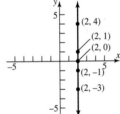

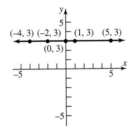

3. The points will be on a vertical line that is 2 units to the right of the y-axis

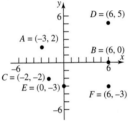

4. The points will be on a horizontal line that is 3 units above the x-axis.

5. $\sqrt{5}$ **6.** $\sqrt{5}$ **7.** $\sqrt{10}$ **8.** $\sqrt{10}$
9. $2\sqrt{17}$ **10.** 5 **11.** $\sqrt{85}$ **12.** $\sqrt{29}$
13. $\sqrt{53}$ **14.** $5\sqrt{5}$ **15.** 2.625 **16.** 1.92
17. $\sqrt{a^2 + b^2}$ **18.** $\sqrt{2}|a|$

19. $d(A, B) = \sqrt{13}$
$d(B, C) = \sqrt{13}$
$d(A, C) = \sqrt{26}$
$(\sqrt{13})^2 + (\sqrt{13})^2 = (\sqrt{26})^2$
Area $= \dfrac{13}{2}$ square units

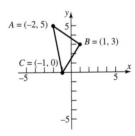

20. $d(A, C) = 20$
$d(A, B) = 10\sqrt{2}$
$d(B, C) = 10\sqrt{2}$
$20^2 = (10\sqrt{2})^2 + (10\sqrt{2})^2$
Area $= 100$ square units

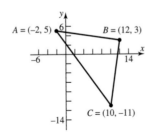

21. $d(A, B) = \sqrt{130}$
$d(B, C) = \sqrt{26}$
$d(A, C) = \sqrt{104}$
$(\sqrt{26})^2 + (\sqrt{104})^2 = (\sqrt{130})^2$
Area $= 26$ square units

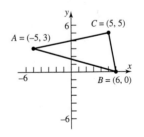

22. $d(A, B) = \sqrt{145}$
$d(A, C) = \sqrt{29}$
$d(B, C) = \sqrt{116}$
$(\sqrt{145})^2 = (\sqrt{29})^2 + (\sqrt{116})^2$
Area = 29 square units

23. $d(A, B) = 4$
$d(A, C) = 5$
$d(B, C) = \sqrt{41}$
$4^2 + 5^2 = 16 + 25 = (\sqrt{41})^2$
Area = 10 square units

24. $d(A, B) = 4$
$d(A, C) = 2\sqrt{5}$
$d(B, C) = 2$
$(2\sqrt{5})^2 = 4^2 + 2^2$
Area = 4 square units

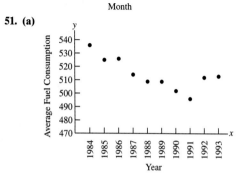

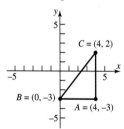

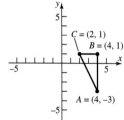

25. $(2, 2); (2, -4)$ **26.** $(13, -3); (-11, -3)$ **27.** $(0, 0); (8, 0)$ **28.** $(0, 1); (0, 7)$ **29.** $(4, -1)$ **30.** $\left(\dfrac{1}{2}, 2\right)$ **31.** $\left(\dfrac{3}{2}, 1\right)$ **32.** $\left(3, -\dfrac{1}{2}\right)$

33. $(5, -1)$ **34.** $\left(-1, -\dfrac{1}{2}\right)$ **35.** $(1.05, 0.7)$ **36.** $(0.45, 1.7)$ **37.** $\left(\dfrac{a}{2}, \dfrac{b}{2}\right)$ **38.** $\left(\dfrac{a}{2}, \dfrac{a}{2}\right)$ **39.** $\sqrt{17}; 2\sqrt{5}; \sqrt{29}$

40. Two triangles are possible. The third vertex is $(2\sqrt{3}, 2)$ or $(-2\sqrt{3}, 2)$.
41. $d(P_1, P_2) = 6; d(P_2, P_3) = 4; d(P_1, P_3) = 2\sqrt{13}$; right triangle
42. $d(P_1, P_2) = \sqrt{53}; d(P_2, P_3) = \sqrt{53}; d(P_1, P_3) = \sqrt{106}$; isosceles right triangle
43. $d(P_1, P_2) = \sqrt{68}; d(P_2, P_3) = \sqrt{34}; d(P_1, P_3) = \sqrt{34}$; isosceles right triangle
44. $d(P_1, P_2) = 5\sqrt{5}; d(P_2, P_3) = 10; d(P_1, P_3) = 5$; right triangle **45.** $4\sqrt{10}$ **46.** $\sqrt{149}$ **47.** $2\sqrt{65}$ **48.** $\sqrt{205}$

49. (a)

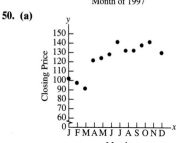

(b)

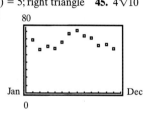

(c) The price of the stock is decreasing, increasing, and then decreasing over time.

50. (a)

(b)

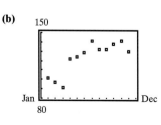

(c) The price of the stock increases with time.

51. (a)

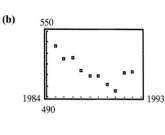

(b)

(c) The average fuel consumption decreases with time.

52. (a)

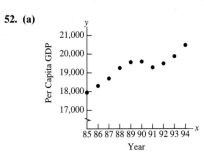

(b)

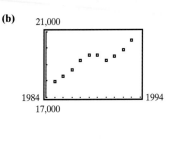

(c) Per capita GDP increases with time

53. (a)

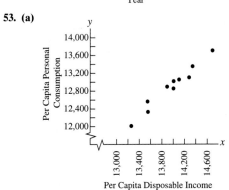

(b)

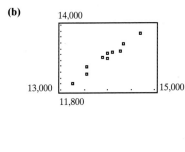

(c) Per capita personal income increases as per capita disposable income increases.

54. (a)

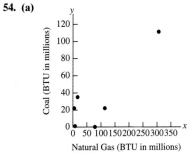

(b)

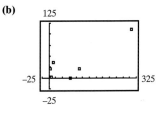

55. (a)

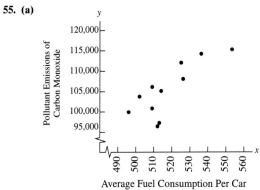

(b)

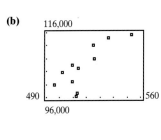

(c) The level of carbon monoxide increases as the average fuel consumption per car increases.

56. (a)

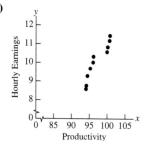

(b)

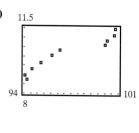

(c) As productivity increases, the average hourly earnings increase.

57. $90\sqrt{2} \approx 127.28$ ft   **58.** $60\sqrt{2} \approx 84.9$ ft   **59. (a)** $(90,0),(90,90),(0,90)$ **(b)** 232.4 ft **(c)** 366.2 ft
60. (a) $(60,0),(60,60),(0,60)$ **(b)** 126.5 ft   **(c)** 272 ft   **61.** $d = 50t$ **62.** $\sqrt{10000 + 484t^2}$

3.2 Exercises

1.

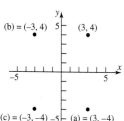

2.

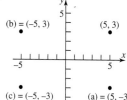

3.

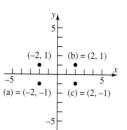

4.

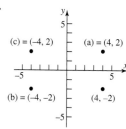

5.

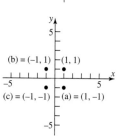

6.

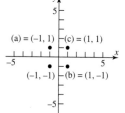

7.

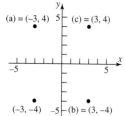

8.

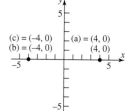

9.

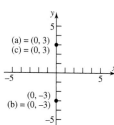

10.

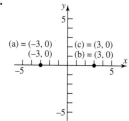

11. (a) $(-1,0),(1,0)$ **(b)** x-axis, y-axis, origin **12. (a)** $(0,1)$ **(b)** none **13. (a)** $\left(-\dfrac{\pi}{2},0\right),(0,1),\left(\dfrac{\pi}{2},0\right)$
(b) y-axis **14. (a)** $(0,0)$ **(b)** origin **15. (a)** $(0,0)$ **(b)** x-axis **16. (a)** $(-2,0),(2,0),(0,-2),(0,2)$ **(b)** x-axis, y-axis, origin
17. (a) $(1,0)$ **(b)** none **18. (a)** $(0,0)$ **(b)** none **19. (a)** $(-3,0),(0,2),(3,0)$ **(b)** y-axis **20. (a)** $(-3,0),(0,2),(2,0)$ **(b)** none
21. (a) $(x,0),0 \le x < 2$ **(b)** none **22. (a)** $(0,1),(2,0),(3.25,0)$ **(b)** none **23. (a)** $(-1.5,0),(0,-2),(1.5,0)$ **(b)** y-axis
24. (a) $(0,0)$ **(b)** origin **25. (a)** none **(b)** origin **26. (a)** none **(b)** x-axis **27.** $(0,0)$ is on the graph.
28. $(0,0)$ and $(1,-1)$ are on the graph. **29.** $(0,3)$ is on the graph. **30.** $(0,1)$ and $(-1,0)$ are on the graph.
31. $(0,2)$ and $(\sqrt{2},\sqrt{2})$ are on the graph.

32. $(0,1)$ and $(2,0)$ are on the graph. **33.** -1 **34.** 12 **35.** $2a + 3b = 6$ **36.** $b = 5; m = -\dfrac{5}{2}$

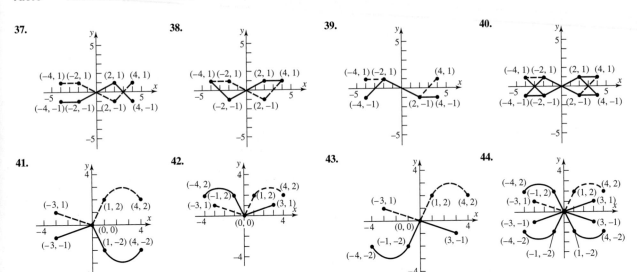

37. **38.** **39.** **40.**

41. **42.** **43.** **44.**

45. $(0,0)$; symmetric with respect to the y-axis   **46.** $(0,0)$; symmetric with respect to the x-axis
47. $(0,0)$; symmetric with respect to the origin   **48.** $(0,0)$; symmetric with respect to the origin
49. $(0,9), (3,0), (-3,0)$; symmetric with respect to the y-axis   **50.** $(-4,0), (0,-2), (0,2)$; symmetric with respect to the x-axis
51. $(-2,0), (2,0), (0,-3), (0,3)$; symmetric with respect to the x-axis, y-axis, and origin
52. $(-1,0), (1,0), (0,-2), (0,2)$; symmetric with respect to the x-axis, y-axis, and origin
53. $(0,-27), (3,0)$; no symmetry   **54.** $(-1,0), (1,0), (0,-1)$; symmetric with respect to the y-axis   **55.** $(0,-4), (4,0), (-1,0)$; no symmetry
56. $(0,4)$; symmetric with respect to the y-axis   **57.** $(0,0)$; symmetric with respect to the origin
58. $(2,0), (-2,0)$; symmetric with respect to the origin
59. (a)

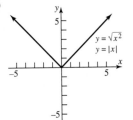 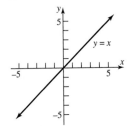 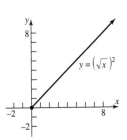

(b) Since $\sqrt{x^2} = |x|$, for all x, the graphs of $y = \sqrt{x^2}$ and $y = |x|$ are the same.
(c) For $y = (\sqrt{x})^2$, the domain of the variable x is $x \geq 0$; for $y = x$, the domain of the variable x is all real numbers. Thus, $(\sqrt{x})^2 = x$ only for $x \geq 0$.
(d) For $y = \sqrt{x^2}$, the range of the variable y is $y \geq 0$; for $y = x$, the range of the variable y is all real numbers. Also, $\sqrt{x^2} = |x|$, which equals x only if $x \geq 0$.

3.3 Exercises

1. (a) $\dfrac{1}{2}$ **(b)** For every 2 unit change in x, y will change 1 unit; if x increases by 2 units, y will increase by 1 unit.

2. (a) -1 **(b)** For every 1 unit change in x, y will decrease by 1 unit.

3. (a) $-\dfrac{1}{3}$ **(b)** For every 3 unit change in x, y will decrease by 1 unit.   **4. (a)** $\dfrac{1}{3}$ **(b)** For every 3 unit change in x, y will change by 1 unit.

5. Slope $= -\dfrac{3}{2}$

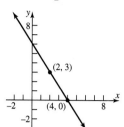

6. Slope $= -2$

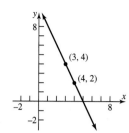

7. Slope $= -\dfrac{1}{2}$

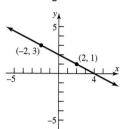

8. Slope $= \dfrac{2}{3}$

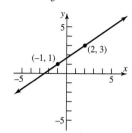

9. Slope $= 0$

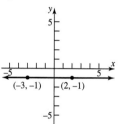

10. Slope $= 0$

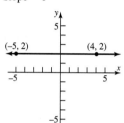

11. Slope undefined

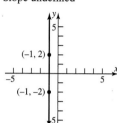

12. Slope undefined

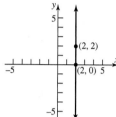

13. Slope $= \dfrac{\sqrt{3}-3}{1-\sqrt{2}} \approx 3.06$

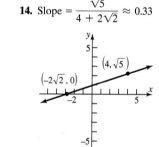

14. Slope $= \dfrac{\sqrt{5}}{4+2\sqrt{2}} \approx 0.33$

15.

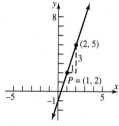

16.

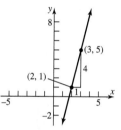

17.

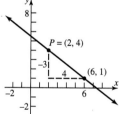

18.

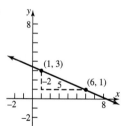

19.

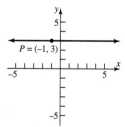

20.

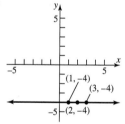

21.

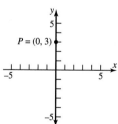

22.

23. $x - 2y = 0$ or $y = \dfrac{1}{2}x$ **24.** $\dfrac{1}{2}x + y = 0$ or $y = -\dfrac{1}{2}x$

25. $x + 3y - 4 = 0$ or $y = -\dfrac{1}{3}x + \dfrac{4}{3}$

26. $x - 3y + 4 = 0$ or $y = \dfrac{1}{3}x + \dfrac{4}{3}$

27. $4x - y + 13 = 0$ or $y = 4x + 13$

28. $3x - y - 9 = 0$ or $y = 3x - 9$

29. $2x + 3y + 1 = 0$ or $y = -\dfrac{2}{3}x - \dfrac{1}{3}$

30. $x - 2y - 1 = 0$ or $y = \dfrac{1}{2}x - \dfrac{1}{2}$ **31.** $x - 2y + 5 = 0$ or $y = \dfrac{1}{2}x + \dfrac{5}{2}$ **32.** $x - 5y + 23 = 0$ or $y = \dfrac{1}{5}x + \dfrac{23}{5}$

33. $4x + y - 3 = 0$ or $y = -4x + 3$ **34.** $2x + y + 3 = 0$ or $y = -2x - 3$ **35.** $x - 2y - 2 = 0$ or $y = \dfrac{1}{2}x - 1$

36. $x - y + 4 = 0$ or $y = x + 4$ **37.** $x - 2 = 0$; no slope–intercept form   **38.** $x - 3 = 0$; no slope–intercept form

39. Slope $= 2$; y-intercept $= 3$

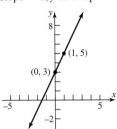

40. Slope $= -3$, y-intercept $= 4$

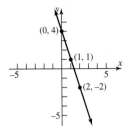

41. $y = 2x - 2$; Slope $= 2$; y-intercept $= -2$

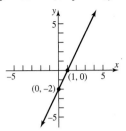

42. $y = -\dfrac{1}{3}x + 2$; Slope $= -\dfrac{1}{3}$, y-intercept $= 2$

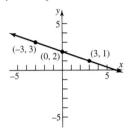

43. Slope $= \dfrac{1}{2}$; y-intercept $= 2$

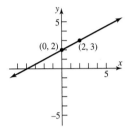

44. Slope $= 2$, y-intercept $= \dfrac{1}{2}$

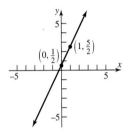

45. $y = -\dfrac{1}{2}x + 2$; Slope $= -\dfrac{1}{2}$; y-intercept $= 2$

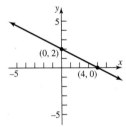

46. $y = \dfrac{1}{3}x + 2$; Slope $= \dfrac{1}{3}$, y-intercept $= 2$

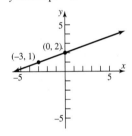

47. $y = \dfrac{2}{3}x - 2$; Slope $= \dfrac{2}{3}$; y-intercept $= -2$

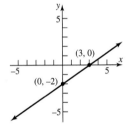

48. $y = -\dfrac{3}{2}x + 3$; Slope $= -\dfrac{3}{2}$, y-intercept $= 3$

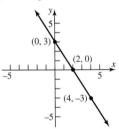

49. $y = -x + 1$; Slope $= -1$; y-intercept $= -1$

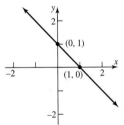

50. $y = x - 2$; Slope $= 1$, y-intercept $= -2$

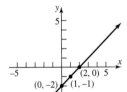

51. Slope undefined; no y-intercept

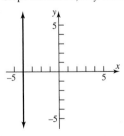

52. Slope $= 0$, y-intercept $= -1$

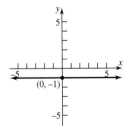

53. Slope $= 0$; y-intercept $= 5$

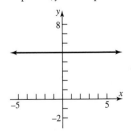

54. Slope undefined; no y-intercept

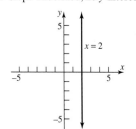

55. $y = x$; Slope $= 1$; y-intercept $= 0$

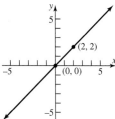

56. $y = -x$, Slope $= -1$, y-intercept $= 0$

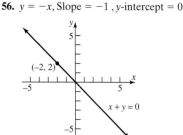

57. $y = \dfrac{3}{2}x$; Slope $= \dfrac{3}{2}$; y-intercept $= 0$

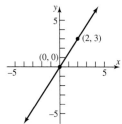

58. $y = -\dfrac{3}{2}x$, Slope $= -\dfrac{3}{2}$, y-intercept $= 0$ **59.** $y = 0$ **60.** $x = 0$

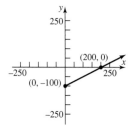

61. $°C = \dfrac{5}{9}(°F - 32)$; approximately 21 °C   **62. (a)** $K = °C + 273$ **(b)** $K = \dfrac{5}{9}°F + \dfrac{2297}{9}$

63. (a) $P = 0.5x - 100$ **(b)** \$400
(c) \$2400

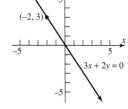

64. (a) $P = 0.55x - 125$ **(b)** \$425
(c) \$2625

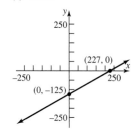

65. $C = 0.06543x + 5.65$; $C = $ \$25.28;
$C = $ \$54.72

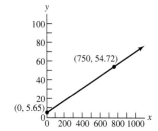

66. $y - 0 = \dfrac{-b}{a}(x - a)$

$y = \dfrac{-bx}{a} + b$

$\dfrac{bx}{a} + y = b$

$\dfrac{x}{a} + \dfrac{y}{b} = 1$

67. (b)   **68. (c)**   **69. (d)**   **70. (a)**   **71.** $x - y + 2 = 0$ or $y = x + 2$

72. $x + y - 1 = 0$ or $y = -x + 1$ **73.** $x + 3y - 3 = 0$ or $y = -\dfrac{1}{3}x + 1$

74. $x + 2y + 2 = 0$ or $y = -\dfrac{1}{2}x - 1$ **75.** $2x + 3y = 0$ or $y = -\dfrac{2}{3}x$ **76.** $2x - 3y = 0$ or $y = \dfrac{2}{3}x$

78. No, if the line is horizontal.   **79.** No, if the intercepts are $(0, 0)$. No, every line crosses at least one axis.   **80.** They are the same line.
81. They are the same line.   **82.** No   **83.** Yes, if the y-intercept $= 0$.

3.4 Exercises

1. (a) 6   **(b)** $-\dfrac{1}{6}$ **2. (a)** -3 **(b)** $\dfrac{1}{3}$ **3. (a)** $-\dfrac{1}{2}$ **(b)** 2   **4. (a)** $\dfrac{2}{3}$ **(b)** $-\dfrac{3}{2}$ **5. (a)** $\dfrac{1}{2}$ **(b)** -2 **6. (a)** -3 **(b)** $\dfrac{1}{3}$

7. (a) $-\dfrac{3}{5}$ **(b)** $\dfrac{5}{3}$ **8. (a)** $\dfrac{4}{3}$ **(b)** $-\dfrac{3}{4}$ **9. (a)** Undefined   **(b)** 0   **10. (a)** 0   **(b)** Undefined   **11.** $2x - y - 3 = 0$ or $y = 2x - 3$

12. $x + y - 3 = 0$ or $y = -x + 3$ **13.** $x + 2y - 5 = 0$ or $y = -\dfrac{1}{2}x + \dfrac{5}{2}$ **14.** $x - y + 2 = 0$ or $y = x + 2$

15. $4x - y + 6 = 0$ or $y = 4x + 6$ **16.** $y = -5x - 3$ or $5x + y + 3 = 0$ **17.** $2x - y = 0$ or $y = 2x$ **18.** $x - 2y = 0$ or $y = \dfrac{1}{2}x$

19. $x - 4 = 0$; no slope–intercept form   **20.** $y - 2 = 0$ or $y = 2$ **21.** $2x + y = 0$ or $y = -2x$ **22.** $x + 2y = 3$ or $y = -\dfrac{1}{2}x - \dfrac{3}{2}$

23. $x - 2y + 3 = 0$ or $y = \dfrac{1}{2}x + \dfrac{3}{2}$ **24.** $2x + y - 4 = 0$ or $y = -2x + 4$ **25.** $y - 4 = 0$ or $y = 4$

26. $x - 3 = 0$; no slope–intercept form **27.** Center $(2, 1)$; Radius 2; $(x - 2)^2 + (x - 1)^2 = 4$

28. Center $(1, 2)$; Radius $= 2$; $(x - 1)^2 + (y - 2)^2 = 4$ **29.** Center $\left(\dfrac{5}{2}, 2\right)$; Radius $\dfrac{3}{2}$; $\left(x - \dfrac{5}{2}\right)^2 + (y - 2)^2 = \dfrac{9}{4}$

30. Center $(1, 2)$; Radius $= \sqrt{2}$; $(x - 1)^2 + (y - 2)^2 = 2$

31. $(x - 1)^2 + (y + 1)^2 = 1$; **32.** $(x + 2)^2 + (y - 1)^2 = 4$; **33.** $x^2 + (y - 2)^2 = 4$; **34.** $(x - 1)^2 + y^2 = 9$;
   $x^2 + y^2 - 2x + 2y + 1 = 0$ $x^2 + y^2 + 4x - 2y + 1 = 0$ $x^2 + y^2 - 4y = 0$ $x^2 + y^2 - 2x - 8 = 0$

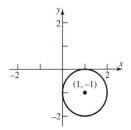 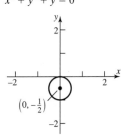 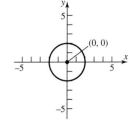 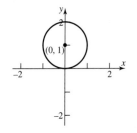

35. $(x - 4)^2 + (y + 3)^2 = 25$; **36.** $(x - 2)^2 + (y + 3)^2 = 16$; **37.** $x^2 + y^2 = 4$; **38.** $x^2 + y^2 = 9$;
   $x^2 + y^2 - 8x + 6y = 0$ $x^2 + y^2 - 4x + 6y - 3 = 0$ $x^2 + y^2 - 4 = 0$ $x^2 + y^2 - 9 = 0$

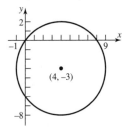 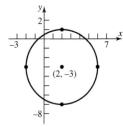 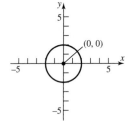 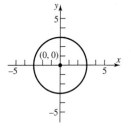

39. $\left(x - \dfrac{1}{2}\right)^2 + y^2 = \dfrac{1}{4}$; **40.** $x^2 + \left(y + \dfrac{1}{2}\right)^2 = \dfrac{1}{4}$; **41.** $r = 2$; $(h, k) = (0, 0)$ **42.** $r = 1$; $(h, k) = (0, 1)$
   $x^2 + y^2 - x = 0$ $x^2 + y^2 + y = 0$

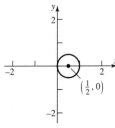 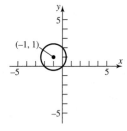 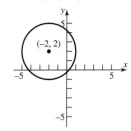 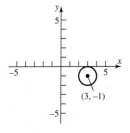

43. $r = 2$; $(h, k) = (3, 0)$ **44.** $r = \sqrt{2}$; $(h, k) = (-1, 1)$ **45.** $r = 3$; $(h, k) = (-2, 2)$ **46.** $r = 1$; $(h, k) = (3, -1)$

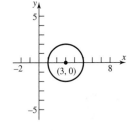 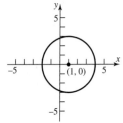

47. $r = \frac{1}{2}; (h, k) = \left(\frac{1}{2}, -1\right)$ **48.** $r = 1; (h, k) = \left(-\frac{1}{2}, -\frac{1}{2}\right)$ **49.** $r = 5; (h, k) = (3, -2)$ **50.** $r = \frac{\sqrt{2}}{2}; (h, k) = (-2, 0)$

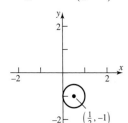

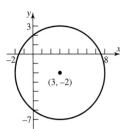

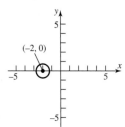

51. $x^2 + y^2 - 13 = 0$ **52.** $x^2 + y^2 - 2x - 17 = 0$ **53.** $x^2 + y^2 - 4x - 6y + 4 = 0$ **54.** $x^2 + y^2 + 6x - 2y + 1 = 0$
55. $x^2 + y^2 + 2x - 6y + 5 = 0$ **56.** $x^2 + y^2 - 4x - 4y + 3 = 0$

57. $P_1 = (-2, 5), P_2 = (1, 3), m_1 = -\frac{2}{3}; P_2 = (1, 3), P_3 = (-1, 0) \ m_2 = \frac{3}{2};$ because $m_1 m_2 = -1$, the lines are perpendicular; the points P_1,
P_2, and P_3 thus form a right triangle.

58. $P_1 = (1, -1), P_3 = (2, 2), m = \frac{3}{1} = 3; P_2 = (4, 1), P_4 = (5, 4), m = \frac{3}{1} = 3; P_3 = (2, 2), P_4 = (5, 4), m = \frac{2}{3}; P_1 = (1, -1), P_2 = (4, 1), m = \frac{2}{3}$
opposite sides are parallel; the points form a parallelogram.

59. $P_1 = (-1, 0), P_2 = (2, 3), m_1 = 1; P_3 = (1, -2), P_4 = (4, 1), m_2 = 1; P_1 = (-1, 0), P_3 = (1, -2), m_3 = -1; P_2 = (2, 3), P_4 = (4, 1), m_4 = -1;$
opposite sides are parallel, and adjacent sides are perpendicular; the points form a rectangle.

60. $P_1 = (0, 0), P_2 = (1, 3), m = 3; P_3 = (4, 2), P_4 = (3, -1), m = 3; P_2 = (1, 3), P_3 = (4, 2), m = -\frac{1}{3}; P_1 = (0, 0), P_4 = (3, -1), m = -\frac{1}{3}$
opposite sides are parallel; adjacent sides are perpendicular;
$d(P_1, P_2) = \sqrt{1^2 + 3^2} = \sqrt{10}; d(P_2, P_3) = \sqrt{(4 - 1)^2 + (2 - 3)^2} = \sqrt{3^2 + (-1)^2} = \sqrt{10};$
$d(P_3, P_4) = \sqrt{(3 - 4)^2 + (-1 - 2)^2} = \sqrt{(-1)^2 + (-3)^2} = \sqrt{10}; d(P_4, P_1) = \sqrt{3^2 + (-1)^2} = \sqrt{10}$
All sides have equal lengths; the quadrilateral is a square.

61. (c) **62.** (d) **63.** (b) **64.** (a) **65.** $(x + 3)^2 + (y - 1)^2 = 16$ **66.** $(x - 4)^2 + (y + 2)^2 = 9$ **67.** $(x - 2)^2 + (y - 2)^2 = 9$
68. $(x - 1)^2 + (y - 3)^2 = 4$
69. Refer to Figure 53; $m_1 m_2 = -1; d(A, B) = \sqrt{(m_2 - m_1)^2}; d(O, A) = \sqrt{1 + m_2^2}; d(O, B) = \sqrt{1 + m_1^2}.$
Now show that $[d(O, B)]^2 + [d(O, A)]^2 = [d(A, B)]^2.$
70. $x^2 + y^2 + 2x + 4y - 4168.16 = 0$
71. **(a)** $x^2 + (mx + b)^2 = r^2$
$(1 + m^2)x^2 + 2mbx + b^2 - r^2 = 0$
One solution if and only if discriminant $= 0$
$(2mb)^2 - 4(1 + m^2)(b^2 - r^2) = 0$
$-4b^2 + 4r^2 + 4m^2 r^2 = 0$
$r^2(1 + m^2) = b^2$
(b) $x = \frac{-2mb}{2(1 + m^2)} = \frac{-2mb}{2b^2/r^2} = -\frac{r^2 m}{b}$
$y = m\left(-\frac{r^2 m}{b}\right) + b = -\frac{r^2 m^2}{b} + b = \frac{-r^2 m^2 + b^2}{b} = \frac{r^2}{b}$
(c) Slope of tangent line $= m$
Slope of line joining center to point of tangency $= \frac{r^2/b}{-r^2 m/b} = -\frac{1}{m}$
72. $\sqrt{2}x + 4y - 9\sqrt{2} = 0$ **73.** $\sqrt{2}x + 4y - 11\sqrt{2} + 12 = 0$ **74.** $(1, 0)$ **75.** $x + 5y + 13 = 0$
76. Slope from (a, b) to (b, a) is $\frac{a - b}{b - a} = -1.$
Slope of the line $y = x$ is 1.
Since $-1 \cdot 1 = -1$, the line containing the points (a, b) and (b, a) is perpendicular to the line $y = x$.
The midpoint of (a, b) and $(b, a) = \left(\frac{a + b}{2}, \frac{b + a}{2}\right).$
Since $\frac{b + a}{2} = \frac{a + b}{2}$, the midpoint lies on the line $y = x$.

77. All have the same slope, 2; the lines are parallel.

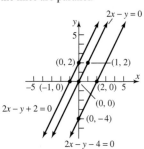

78. The family of lines $Cx + y + 4 = 0$ intersect at the point $(0, -4)$

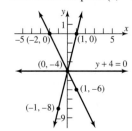

79. $y = 2$

3.5 Exercises

1. Linear **2.** Nonlinear **3.** Linear **4.** Linear **5.** Nonlinear **6.** Nonlinear

7. (a), (c)

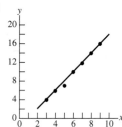

  (b) Using $(3, 4)$ and $(8, 14)$, $y = 2x - 2$
  (d) $y = 2.0357x - 2.3571$ **(e)**

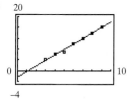

8. (a), (c)

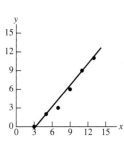

  (b) Using $(5, 2)$ and $(11, 9)$, $y = \dfrac{7}{6}x - \dfrac{23}{6}$.
  (d) $y = 1.129x - 3.862$ **(e)**

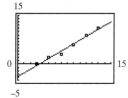

9. (a), (c)

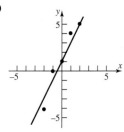

  (b) Using $(0, 1)$ and $(2, 5)$, $y = 2x + 1$.
  (d) $y = 2.2x + 1.2$ **(e)**

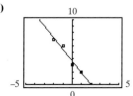

10. (a), (c)

  (b) Using $(-2, 7)$ and $(1, 2)$, $y = -\dfrac{5}{3}x + \dfrac{11}{3}$
  (d) $y = -1.8x + 3.6$ **(e)**

11. (a), (c)

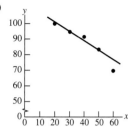

(b) Using $(30, 95)$ and $(50, 83)$, $y = -\dfrac{3}{5}x + 113$.   **(d)** $y = -0.72x + 116.6$

(e)

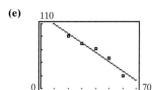

12. (a), (c)

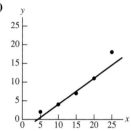

(b) Using $(10, 4)$ and $(20, 11)$, $y = \dfrac{7}{10}x - 3$.   **(d)** $y = 0.78x - 3.3$

(e)

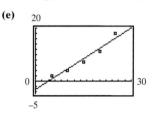

13. (a), (c)

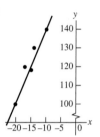

(b) Using $(-20, 100)$ and $(-10, 140)$, $y = 4x + 180$.   **(d)** $y = 3.8613x + 180.292$

(e)

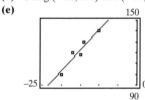

14. (a)

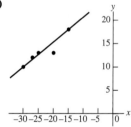

(b) Using $(-30, 10)$ and $(-14, 18)$, $y = \dfrac{1}{2}x + 25$.   **(d)** $y = 0.442x + 23.456$

(e)

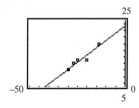

15. (a)

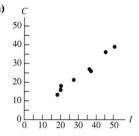

(b) Using $(20, 16)$ and $(50, 39)$, $C - 16 = \dfrac{23}{30}(I - 20)$

(c) As disposable income increases by \$1, consumption increases by about \$0.77.

(d) \$32,201

(e) $C = 0.755I + 0.6266$

16. (a)

S (Savings, thousands of dollars) vs. I (Disposable Income, thousands of dollars) scatter plot.

(b) Using $(20, 4)$ and $(50, 11)$, $S - 4 = \dfrac{7}{30}(I - 20)$

(c) Savings increases by \$233 for every \$1000 extra in disposable income. **(d)** \$9799

(e) $S = 0.245I - 0.627$

17. (a)

Y vs. X scatter plot.

(b) Using $(50, 28)$ and $(65, 15)$, $Y - 28 = -\dfrac{13}{15}(X - 50)$

(c) As speed increases by 1 mph, mpg decreases by about 0.87. **(d)** 18.47 mpg

(e) $Y = -0.8265X + 70.3903$

18. (a)

Weight (pounds) vs. Height (inches) scatter plot.

(b) Using $(60, 105)$ and $(62, 115)$, $y - 105 = 5(x - 60)$

(c) A female's weight increases by 5 lbs for every inch in height. **(d)** 135 pounds

(e) $y = 4.127x - 140.952$

19. (a)

Net Income vs. Sales scatter plot.

(b) Using $(1912.8, 92.6)$ and $(2378.2, 115.6)$, $y - 92.6 = 0.05(x - 1912.8)$

(c) As sales increase by \$1, the net income before taxes increases by \$0.05.

(d) \$119.78 billion

(e) $y = 0.0842x - 88.5776$

20. (a)

Number Employed (millions of people) vs. Civilian Labor Force (millions of people) scatter plot.

(b) Using $(113{,}544, 105{,}005)$ and $(119{,}865, 112{,}440)$, $y - 105{,}005 = 1.176(x - 113{,}544)$

(c) The number of employed increases by 1176 for every 1000 people added to the labor force. **(d)** 115,349,096 people

(e) $y = 1.170x - 28{,}230.774$

21. (a) $y = 1.13x - 2841.69$ **(b)** If per capita disposable income increases \$1, per capita personal consumption increases \$1.13.
(c) \$14,095.88

22. (a) $y = 0.305x + 4.902$
(b) The amount of energy provided by coal increases by 0.305 trillion BTU for every 1 trillion BTU of energy provided by natural gas.
(c) 25.337 trillion BTUs
23. (a) $y = 323.42x - 62,080.03$
(b) If average fuel consumption per car increases by 1 gallon, pollutant emissions of carbon monoxide increase 323.42 thousand tons.
(c) 101,247 thousand tons
24. (a) $y = 0.309x - 19.915$ **(b)** Average hourly earnings increases by $0.309 per hour for each unit increase in productivity.
(c) $11.387 per hour

3.6 Exercises

1. $y = \frac{1}{5}x$ **2.** $v = 8t$ **3.** $A = \pi x^2$ **4.** $V = \frac{4\pi}{3}x^3$ **5.** $F = \frac{250}{d^2}$ **6.** $y = \frac{12}{\sqrt{x}}$ **7.** $z = \frac{1}{5}(x^2 + y^2)$ **8.** $T = d^2\sqrt[3]{x}$

9. $M = \frac{9d^2}{2\sqrt{x}}$ **10.** $z = \frac{1}{17}(x^3 + y^2)$ **11.** $T^2 = \frac{8a^3}{d^2}$ **12.** $z^3 = \frac{8}{97}(x^2 + y^2)$ **13.** $V = \frac{4\pi}{3}r^3$ **14.** $c^2 = a^2 + b^2$ **15.** $A = \frac{1}{2}bh$

16. $p = 2(l + w)$ **17.** $V = \pi r^2 h$ **18.** $V = \frac{\pi}{3}r^2 h$ **19.** $F = 6.67 \times 10^{-11}\left(\frac{mM}{d^2}\right)$ **20.** $T = \frac{2\pi}{\sqrt{32}}\sqrt{l}$ **21.** 144 ft.; 2 sec **22.** 96 ft/sec

23. 2.25 **24.** $\frac{64}{3}$ inches **25.** 54.86 lb **26.** 71.346 lb **27.** $\sqrt[3]{6} \approx 1.82$ in. **28.** 199.9 lb **29.** 900 ft-lb **30.** 282.2 ft

31. 384 psi **32.** 900 lb **33.** $\frac{720}{49} \approx 14.69$ ohms **34.** $V = \frac{kI}{P}; k = 5; 19.4$ atmospheres **35.** $v = \sqrt{gr}$ **36.** 18,775 mph **37.** 18,001 mph

38. 431 mi **39.** 545 mi **40.** 4048 mi **41.** $F = \frac{mv^2}{r}$ **42.** 167 newtons **43.** by 21%

44. 84.8 kilometers per hour **45.** 9 times **46.** Less; half as much

Fill-in-the-Blank Items

1. x-coordinate or abscissa; y-coordinate or ordinate **2.** midpoint **3.** y-axis **4.** circle; radius; center **5.** undefined; 0

6. $m_1 = m_2; m_1 m_2 = -1$ **7.** $\frac{kx^2 y^3}{\sqrt{t}}$

True/False Items

1. F **2.** T **3.** T **4.** F **5.** F **6.** T **7.** T

Review Exercises

1. $2x + y - 5 = 0$ or $y = -2x + 5$ **2.** $y - 4 = 0$ or $y = 4$ **3.** $x + 3 = 0$; no slope-intercept form **4.** $\frac{5}{2}x + y - 5 = 0$ or $y = -\frac{5}{2}x + 5$

5. $x + 5y + 10 = 0$ or $y = -\frac{1}{5}x - 2$ **6.** $5x + y - 11 = 0$ or $y = -5x + 11$ **7.** $2x - 3y + 19 = 0$ or $y = \frac{2}{3}x + \frac{19}{3}$

8. $x + y + 2 = 0$ or $y = -x - 2$ **9.** $-x + y + 7 = 0$ or $y = x - 7$ **10.** $x + 3y - 10 = 0$ or $y = -\frac{1}{3}x + \frac{10}{3}$

11. $4x - 5y + 20 = 0$ **12.** $3x + 4y - 12 = 0$ **13.** $\frac{1}{2}x - \frac{1}{3}y + \frac{1}{6} = 0$

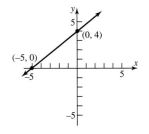

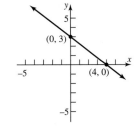

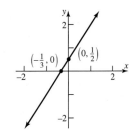

14. $-\dfrac{3}{4}x + \dfrac{1}{2}y = 0$

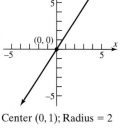

15. $\sqrt{2}x + \sqrt{3}y = \sqrt{6}$

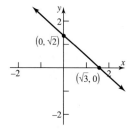

16. $\dfrac{x}{3} + \dfrac{y}{4} = 1$

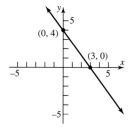

17. Center $(0, 1)$; Radius $= 2$

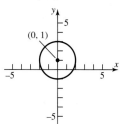

18. Center: $(-2, 0)$; Radius $= 3$

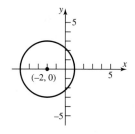

19. Center $(1, -2)$; Radius $= 3$

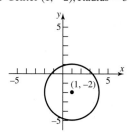

20. Center: $(-2, 2)$; Radius $= 3$

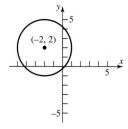

21. Center $(1, -2)$; Radius $= \sqrt{5}$

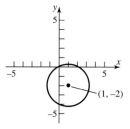

22. Center: $(1, 0)$; Radius $= 1$

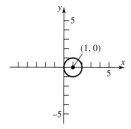

23. Slope $= \dfrac{1}{5}$; distance $= 2\sqrt{26}$; midpoint $= (2, 3)$ **24.** Slope $= -2$; distance $= 4\sqrt{5}$; midpoint $= (4, 1)$

25. Equations in **(c)**   **26.** Equations in **(d)**   **27.** Intercept: $(0, 0)$; symmetric with respect to the x-axis
28. Intercept: $(0, 0)$; symmetric with respect to the origin
29. Intercepts: $(0, -2), (0, 2), (4, 0), (-4, 0)$; symmetric with respect to the x-axis, y-axis, and origin
30. Intercepts: $(1, 0), (-1, 0)$; symmetric with respect to the x-axis, y-axis, and origin.
31. Intercept: $(0, 1)$; symmetric with respect to the y-axis   **32.** Intercepts: $(-1, 0), (0, 0), (1, 0)$; symmetric with respect to the origin
33. Intercepts: $(0, 0), (0, -2), (-1, 0)$; no symmetry
34. Intercepts: $(-4, 0), (0, 0), (0, 2)$; no symmetry   **35.** $\dfrac{8\sqrt[4]{27}}{3}$ **36.** 16 lb   **37.** $a \approx 36$ million miles   **38.** 465 million miles
39. $d(A, B) = \sqrt{13}; d(B, C) = \sqrt{13}$
40. (a) $d(A, B) = \sqrt{[-4 - (-2)]^2 + (4 - 0)^2} = \sqrt{4 + 16} = \sqrt{20} = 2\sqrt{5}$
$\qquad d(B, C) = \sqrt{[8 - (-4)]^2 + (5 - 4)^2} = \sqrt{144 + 1} = \sqrt{145}$
$\qquad d(A, C) = \sqrt{[8 - (-2)]^2 + (5 - 0)^2} = \sqrt{100 + 25} = \sqrt{125} = 5\sqrt{5}$
$\qquad$ Since $d[A, B]^2 + d[A, C]^2 = d[B, C]^2$, by the converse of the Pythagorean
$\qquad$ Theorem, the points $A, B,$ and C are vertices of a right triangle.
$\quad$ **(b)** Slope of $AB = \dfrac{4}{-2} = -2$; Slope of $BC = \dfrac{1}{12}$; Slope of $AC = \dfrac{5}{10} = \dfrac{1}{2}$; Since $(-2)\left(\dfrac{1}{2}\right) = -1$, the lines AB and AC are

$\qquad$ perpendicular and hence form a right angle.
41. Slope using A and B is -1; slope using B and C is -1
42. $(x + 1)^2 + (y - 2)^2 = r^2$
$\qquad$ If $(1, 5)$ lies on the circle, then $r^2 = 13$. Now show that $(2, 4)$ and $(-3, 5)$ lie on the circle $(x + 1)^2 + (y - 2)^2 = 13$
43. Center $(1, -2)$; radius $= 4\sqrt{2}; x^2 + y^2 - 2x + 4y - 27 = 0$ **44.** -4 and 8

45. (a)

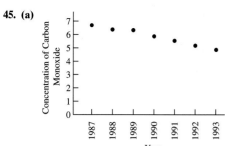

(b) $-\dfrac{0.82}{3} \approx -0.27$

(c) Between the years 1987 and 1990, as x increases by 1 year, the level of carbon monoxide decreases by about 0.27 ppm

(d) $-\dfrac{0.99}{3} = -0.33$

(e) Between the years 1990 and 1993, as x increases by 1 year, the level of carbon monoxide decreases by 0.33 ppm

(f) ≈ -0.308

(g) Between the years 1987 and 1993, as x increases by 1 year, the level of carbon monoxide decreases by about 0.308 ppm

46. (a)

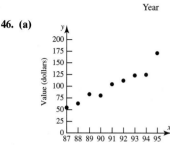

(b) 12.505

(c) Between 1987 and 1991, the value of the portfolio increases by about $12.51 per year

(d) 16.73

(e) Between 1991 and 1995, the value of the portfolio increases by $16.73 per year.

(f) ≈ 12.746

(g) The value of the portfolio increases by about $12.75 per year.

48. (a) y-axis **(b)** x-axis **(c)** line through origin, slope -1 **(d)** the x-axis and y-axis **(e)** the origin

C H A P T E R 4   4.1 Exercises

1. Function **2.** Function **3.** Not a function **4.** Not a function **5.** Function **6.** Function **7.** Function **8.** Function
9. Not a function **10.** Not a function **11.** Function **12.** Function
13. (a) -4 **(b)** -5 **(c)** -9 **(d)** $-3x^2 - 2x - 4$ **(e)** $3x^2 - 2x + 4$ **(f)** $-3x^2 - 4x - 5$ **14. (a)** -1 **(b)** 2 **(c)** 0
(d) $2x^2 - x - 1$ **(e)** $-2x^2 - x + 1$ **(f)** $2x^2 + 5x + 2$ **15. (a)** 0 **(b)** $\dfrac{1}{2}$ **(c)** $-\dfrac{1}{2}$ **(d)** $\dfrac{-x}{x^2+1}$ **(e)** $\dfrac{-x}{x^2+1}$ **(f)** $\dfrac{x+1}{x^2+2x+2}$
16. (a) $-\dfrac{1}{4}$ **(b)** 0 **(c)** 0 **(d)** $-\dfrac{x^2-1}{x-4}$ **(e)** $-\dfrac{x^2-1}{x+4}$ **(f)** $\dfrac{x^2+2x}{x+5}$ **17. (a)** 4 **(b)** 5 **(c)** 5 **(d)** $|x| + 4$ **(e)** $-|x| - 4$
(f) $|x+1| + 4$ **18. (a)** 0 **(b)** $\sqrt{2}$ **(c)** 0 **(d)** $\sqrt{x^2 - x}$ **(e)** $-\sqrt{x^2 + x}$ **(f)** $\sqrt{x^2 + 3x + 2}$
19. (a) $-\dfrac{1}{5}$ **(b)** $-\dfrac{3}{2}$ **(c)** $\dfrac{1}{8}$ **(d)** $\dfrac{-2x+1}{-3x-5}$ **(e)** $\dfrac{-2x-1}{3x-5}$ **(f)** $\dfrac{2x+3}{3x-2}$
20. (a) $\dfrac{3}{4}$ **(b)** $\dfrac{8}{9}$ **(c)** 0 **(d)** $1 - \dfrac{1}{x^2 - 4x + 4}$ **(e)** $-1 + \dfrac{1}{(x+2)^2}$ **(f)** $1 - \dfrac{1}{(x+3)^2}$
21. $f(0) = 3; f(-6) = -3$ **22.** $f(6) = 0; f(11) = 1$ **23.** Positive **24.** Negative **25.** $-3, 6$ and 10 **26.** $-3 < x < 6$ and $10 < x \leq 11$
27. $\{x| -6 \leq x \leq 11\}$ **28.** $\{y| -3 \leq y \leq 4\}$ **29.** $(-3, 0), (6, 0), (10, 0)$ **30.** $(0, 3)$ **31.** 3 times **32.** 2 times
33. (a) No **(b)** $-3; (4, -3)$ **(c)** $14; (14, 2)$ **(d)** $\{x| x \neq 6\}$ **34. (a)** Yes **(b)** $\dfrac{1}{2}; \left(0, \dfrac{1}{2}\right)$ **(c)** 0 or $\dfrac{1}{2}; \left(0, \dfrac{1}{2}\right)$ and $\left(\dfrac{1}{2}, \dfrac{1}{2}\right)$
(d) $\{x|x \neq -4\}$ **35. (a)** Yes **(b)** $\dfrac{8}{17}; \left(2, \dfrac{8}{17}\right)$ **(c)** $-1, 1; (-1, 1), (1, 1)$ **(d)** All real numbers **36. (a)** Yes **(b)** $4; (4, 4)$
(c) $-2; (-2, 1)$ **(d)** $\{x| x \neq 2\}$ **37.** Not a function
38. Function **(a)** Domain: All real numbers; Range: $\{y| 0 < y < \infty\}$ **(b)** Intercept: $(0, 1)$ **(c)** None
39. Function **(a)** Domain: $\{x| -\pi \leq x \leq \pi\}$; Range: $\{y| -1 \leq y \leq 1\}$ **(b)** Intercepts: $\left(-\dfrac{\pi}{2}, 0\right), \left(\dfrac{\pi}{2}, 0\right), (0, 1)$ **(c)** y-axis
40. Function **(a)** Domain: $\left\{x \left| -\dfrac{\pi}{2} < x < \dfrac{\pi}{2}\right.\right\}$; Range: All real numbers **(b)** Intercepts: $(0, 0)$ **(c)** Origin
41. Not a function **42.** Not a function
43. Function **(a)** Domain: $\{x|x > 0\}$; Range: All real numbers **(b)** Intercept: $(1, 0)$ **(c)** None
44. Function **(a)** Domain: $\{x|0 \leq x \leq 4\}$; Range: $\{y|0 \leq y \leq 3\}$ **(b)** Intercept: $(0, 0)$ **(c)** None
45. Function **(a)** Domain: all real numbers; Range: $\{y|y \leq 2\}$ **(b)** Intercept: $(-3, 0), (3, 0), (0, 2)$ **(c)** y-axis
46. Function **(a)** Domain: $\{x|x \geq -3\}$; Range: $\{y|y \geq 0\}$ **(b)** Intercepts: $(-3, 0), (0, 2), (2, 0)$ **(c)** None
47. Function **(a)** Domain: $\{x|x \neq 2\}$; Range: $\{y|y \neq 1\}$ **(b)** Intercept: $(0, 0)$ **(c)** None
48. Function **(a)** Domain: $\{x|x \neq -1\}$; Range: $\{y|y > 0\}$ **(b)** Intercept: $(0, 2)$ **(c)** None
49. All real numbers **50.** All real numbers **51.** All real numbers **52.** All real numbers **53.** $\{x|x \neq -1, x \neq 1\}$ **54.** $\{x|x \neq 1\}$
55. $\{x|x \neq 0\}$ **56.** $\{x|x \neq 0, x \neq -2, x \neq 2\}$ **57.** $\{x|x \geq 4\}$ **58.** $\{x|x \leq 1\}$ **59.** $\{x|x > 9\}$ **60.** $\{x|x > 4\}$ **61.** $\{x|x < 1 \text{ or } x \geq 2\}$
62. $\{x|x \leq 1 \text{ or } x \geq 2\}$ **63. (a)** III **(b)** IV **(c)** I **(d)** V **(e)** II **64. (a)** II **(b)** V **(c)** IV **(d)** III **(e)** I

65.

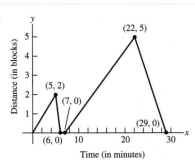

Time (in minutes)

66.

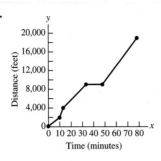

Time (minutes)

67. $A = -\dfrac{7}{2}$ **68.** $B = 5$ **69.** $A = -4$

70. $B = -1$ **71.** $A = 8$; undefined at $x = 3$ **72.** $A = 1, B = 2$

73. (a) No **(b)**

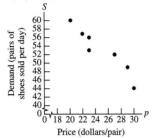

Price (dollars/pair)

(c) For the points $(20, 60)$ and $(30, 44)$, $D = -1.6p + 92$
(d) If the price increases $1, the quantity demanded decreases by 1.6.
(e) $D(p) = -1.6p + 92$
(f) $\{p \mid p > 0\}$
(g) $D(28) = 47.2$; about 47 pairs
(h) $D = -1.34p + 86.20$

74. (a) No **(b)**

Advertising Expenditures
(thousands of dollars)

(c) For the points $(20, 335)$ and $(28.3, 351)$, $S = 1.928A + 296.44$
(d) For every $1000 spent on advertising a product, the sales increase by $1928.
(e) $S(A) = 1.928A + 296.44$
(f) The domain is $\{x \mid x \geq 0\}$
(g) $344,640
(h) $S = 2.07A + 292.887$

75. (a) Yes **(b)**

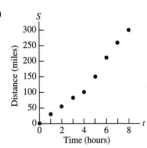

Time (hours)

(c) For the points $(0, 0)$ and $(8, 300)$, $s = \dfrac{75}{2}t$
(d) If t increases by 1 hour, then distance increases by 37.5 miles
(e) $s(t) = \dfrac{75}{2}t$
(f) $\{t \mid t \geq 0\}$
(g) 412.5 miles
(h) $s(t) = 37.78t - 19.13$

76. (a) Yes **(b)**

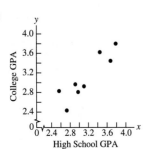

High School GPA

(c) For the points $(2.56, 2.83)$ and $(3.78, 3.81)$, $G = 0.80x + 0.782$.
(d) College GPA increases by 0.80 points for each additional point in high school GPA.
(e) $G(x) = 0.80x + 0.782$
(f) The domain is $\{x \mid 0 \leq x \leq 4\}$
(g) 3.37
(h) $G = 0.964x + 0.0724$

77. (a) 15.1 m, 14.07 m, 12.94 m, 11.72 m **(b)** 1.01 seconds, 1.42 seconds, 1.74 seconds **(c)** 2.02 sec
78. (a) 7m; 4.27m; 1.28 m **(b)** 0.62 seconds, 0.88 seconds, 1.07 seconds **(e)** 1.24 seconds

9. $A(x) = \frac{1}{2}x^2$ **80.** $A(x) = \frac{1}{2}x^2$ **81.** $G(x) = 10x$ **82.** $G(x) = 100 + 10x$

4. The time is least for $x = 1.50$ mi.

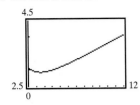

85. **(a)** $C(x) = 10x + 14\sqrt{x^2 - 10x + 29}, 0 \le x \le 5$
(b) $C(1) = \$72.61$ **(c)** $C(3) = \$69.60$
(d) **(e)** Least cost: $x \approx 2.96$ miles

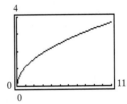

6. **(a)** $T(x) = \frac{\sqrt{9 + x^2}}{12} + \frac{20 - x}{5}$ **(b)** 3.11 hours
(c) 2.63 hrs
(d)

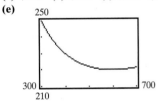

(e) T is least when $x = 20$.

87. **(a)** $A(x) = (8.5 - 2x)(11 - 2x)$
(b) $0 \le x \le 4.25, 0 \le A \le 93.5$
(c) $A(1) = 58.5$ in.^{2}, $A(1.2) = 52.46$ in.^{2}, $A(1.5) = 44$ in.^{2}
(d)

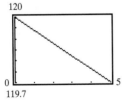

(e) $A(x) = 70$ when $x = 0.64$ in.; $A(x) = 50$ when $x = 1.28$ in.

38. **(a)** $\$222$ **(b)** $\$225$ **(c)** $\$220$ **(d)** $\$230$
(e)

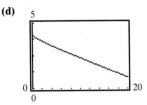

(f) $C(x)$ varies from $\$230$ to $\$220$.

89. **(a)**

(b) The period of T varies from 1.107 seconds to 3.501 seconds.
(c) 81.56 feet

90. **(a)** 119.8 lbs **(b)**

(c) Weight decreases as h approaches 5.
(d) 16.56 miles below sea level.

91. Function **92.** Function **93.** Function **94.** Function
95. Not a function **96.** Not a function **97.** Function
98. Not a function **99.** Only $h(x) = 2x$

4.2 Exercises

1. C **2.** A **3.** E **4.** G **5.** B **6.** D **7.** F **8.** H **9.** **(a)** Domain: $\{x| -3 \le x \le 4\}$; Range: $\{y| 0 \le y \le 3\}$
(b) Increasing on $(-3, 0)$ and on $(2, 4)$; Decreasing on $(0, 2)$ **(c)** Neither **(d)** $(-3, 0), (0, 3), (2, 0)$
10. **(a)** Domain: $\{x|-2 \le x \le 3\}$; Range: $\{y|-1 \le y \le 3\}$ **(b)** Increasing on $(-2, 0)$ and on $(1, 3)$; Decreasing on $(0, 1)$ **(c)** Neither
(d) $\left(-\frac{4}{3}, 0\right), (0, 2), (1, 0)$ **11.** **(a)** Domain: all real numbers; Range: $\{y|y > 0\}$ **(b)** Increasing on $(-\infty, \infty)$ **(c)** Neither **(d)** $(0, 1)$
12. **(a)** Domain: $\{x|x > 0\}$; Range: all real numbers **(b)** Increasing on $(0, \infty)$ **(c)** Neither **(d)** $(1, 0)$
13. **(a)** Domain: $\{x|-\pi \le x \le \pi\}$; Range: $\{y| -1 \le y \le 1\}$ **(b)** Increasing on $\left(-\frac{\pi}{2}, \frac{\pi}{2}\right)$; Decreasing on $\left(-\pi, -\frac{\pi}{2}\right)$ and on $\left(\frac{\pi}{2}, \pi\right)$
(c) Odd **(d)** $(-\pi, 0), (0, 0), (\pi, 0)$
14. **(a)** Domain: $\left\{x\left|-\frac{\pi}{2} < x < \frac{\pi}{2}\right.\right\}$; Range: all real numbers **(b)** Increasing on $\left(-\frac{\pi}{2}, \frac{\pi}{2}\right)$ **(c)** Odd **(d)** $(0, 0)$
15. **(a)** Domain: $\{x| x \ne 2\}$; Range: $\{y| y \ne 1\}$ **(b)** Decreasing on $(-\infty, 2)$ and on $(2, \infty)$ **(c)** Neither **(d)** $(0, 0)$

16. (a) Domain: $\{x|x \neq -1\}$; Range: $\{y|y > 0\}$ **(b)** Increasing on $(-\infty, -1)$; Decreasing on $(-1, \infty)$ **(c)** Neither
(d) $(0, 2)$ **17. (a)** Domain: $\{x|x \neq 0\}$; Range: all real numbers **(b)** Increasing on $(-\infty, 0)$ and on $(0, \infty)$ **(c)** Odd **(d)** $(-1, 0), (1, 0)$
18. (a) Domain: $\{x|x \neq 0\}$; Range: $\{y|y \leq -2 \text{ or } y > 2\}$
(b) Increasing on $(-1, 0)$ and on $(0, 1)$; Decreasing on $(-\infty, -1)$ and on $(1, \infty)$ **(c)** Odd **(d)** None
19. (a) Domain: $\{x|x \neq -2, x \neq 2\}$; Range: $\{y|y \leq 0 \text{ or } y > 1\}$
(b) Increasing on $(-\infty, -2)$ and on $(-2, 0)$; Decreasing on $(0, 2)$ and on $(2, \infty)$ **(c)** Even **(d)** $(0, 0)$
20. (a) Domain: $\{x|x \neq -1, x \neq 1\}$; Range: all real numbers **(b)** Increasing on $(-\infty, -1)$ and on $(-1, 1)$ and on $(1, \infty)$
(c) Odd **(d)** $(0, 0)$
21. (a) Domain: $\{x| -4 \leq x \leq 4\}$; Range: $\{y| 0 \leq y \leq 2\}$ **(b)** Increasing on $(-2, 0)$ and on $(2, 4)$; Decreasing on $(-4, -2)$ and on $(0, 2)$
(c) Even **(d)** $(-2, 0), (0, 2), (2, 0)$
22. (a) Domain: $\{x|-4 \leq x \leq 4\}$; Range: $\{y|-2 \leq y \leq 2\}$ **(b)** Increasing on $(-2, 2)$; Decreasing on $(-4, -2)$ and on $(2, 4)$ **(c)** Odd
(d) $(0, 0)$
23. (a) 4 **(b)** 2 **(c)** 5 **24. (a)** -1 **(b)** 2 **(c)** 5 **25. (a)** 2 **(b)** 3 **(c)** -4 **26. (a)** 0 **(b)** 0 **(c)** -1
27. 3 **28.** -2 **29.** -3 **30.** $x + 1$ **31.** $3x + 1$ **32.** $-2x + 2$ **33.** $x^2 + x$ **34.** $x^2 + x + 2$
35. $-\dfrac{1}{x+1}$ **36.** $-\dfrac{x+1}{x^2}$ **37.** $\dfrac{1}{\sqrt{x}+1} = \dfrac{\sqrt{x}-1}{x-1}$ **38.** $\dfrac{1}{\sqrt{x+3}+2}$ **39.** Odd **40.** Even **41.** Even **42.** Neither
43. Odd **44.** Neither **45.** Neither **46.** Even **47.** Even **48.** Odd **49.** Odd **50.** Odd **51.** At most one **52.** At most one
53. (a) All real numbers **54. (a)** All real numbers **55. (a)** All real numbers **56. (a)** All real numbers
(b) $(0, 0)$ **(b)** $(0, 4)$ **(b)** $(-1, 0), (0, 0)$ **(b)** $(0, 0)$
(c) **(c)** **(c)** **(c)**

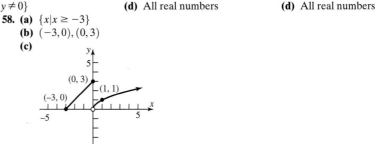

(d) All real numbers **(d)** $\{y|y \neq 0\}$ **(d)** All real numbers **(d)** All real numbers
57. (a) $\{x|x \geq -2\}$ **58. (a)** $\{x|x \geq -3\}$
(b) $(0, 1)$ **(b)** $(-3, 0), (0, 3)$
(c) **(c)**

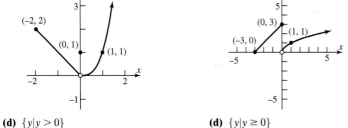

(d) $\{y|y > 0\}$ **(d)** $\{y|y \geq 0\}$
59. (a) $-2x + 5$ **(b)** $-2x - 5$ **(c)** $4x + 5$ **(d)** $2x - 1$ **(e)** $\dfrac{2}{x} + 5$ **(f)** $\dfrac{1}{2x + 5}$
60. (a) $3 + x$ **(b)** $x - 3$ **(c)** $3 - 2x$ **(d)** $6 - x$ **(e)** $3 - \dfrac{1}{x}$ **(f)** $\dfrac{1}{3 - x}$
61. (a) $2x^2 - 4$ **(b)** $-2x^2 + 4$ **(c)** $8x^2 - 4$ **(d)** $2x^2 - 12x + 14$ **(e)** $\dfrac{2}{x^2} - 4$ **(f)** $\dfrac{1}{2x^2 - 4}$
62. (a) $-x^3 + 1$ **(b)** $-x^3 - 1$ **(c)** $8x^3 + 1$ **(d)** $x^3 - 9x^2 + 27x - 26$ **(e)** $\dfrac{1}{x^3} + 1$ **(f)** $\dfrac{1}{x^3 + 1}$
63. (a) $-x^3 + 3x$ **(b)** $-x^3 + 3x$ **(c)** $8x^3 - 6x$ **(d)** $x^3 - 9x^2 + 24x - 18$ **(e)** $\dfrac{1}{x^3} - \dfrac{3}{x}$ **(f)** $\dfrac{1}{x^3 - 3x}$
64. (a) $x^2 - x$ **(b)** $-x^2 - x$ **(c)** $4x^2 + 2x$ **(d)** $x^2 - 5x + 6$ **(e)** $\dfrac{1}{x^2} + \dfrac{1}{x}$ **(f)** $\dfrac{1}{x^2 + x}$
65. (a) $|x|$ **(b)** $-|x|$ **(c)** $2|x|$ **(d)** $|x - 3|$ **(e)** $\dfrac{1}{|x|}$ **(f)** $\dfrac{1}{|x|}$ **66. (a)** $\dfrac{-1}{x}$ **(b)** $\dfrac{-1}{x}$ **(c)** $\dfrac{1}{2x}$ **(d)** $\dfrac{1}{x - 3}$ **(e)** x **(f)** x
67. 2 **68.** -3 **69.** $2x + h + 2$ **70.** $\dfrac{-1}{(x + h)x}$

71. $f(x) = \begin{cases} -x & \text{if } -1 \le x \le 0 \\ \frac{1}{2}x & \text{if } 0 < x \le 2 \end{cases}$ (Other answers are possible.)   **72.** $f(x) = \begin{cases} x & \text{if } -1 \le x \le 0 \\ 1 & \text{if } 0 < x \le 2 \end{cases}$ (Other answers are possible.)

73. $f(x) = \begin{cases} -x & \text{if } x \le 0 \\ -x + 2 & \text{if } 0 < x \le 2 \end{cases}$ (Other answers are possible.)

74. $f(x) = \begin{cases} 2x + 2 & \text{if } -1 \le x \le 0 \\ x & \text{if } x > 0 \end{cases}$ (Other answers are possible.)

75. Not even; $f(2) \ne f(-2)$ **76.** Not even; $f(2) \ne f(-2)$

77. (a), (b), (e)

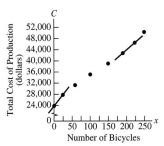

(c) 103 bacteria/day
(d) The population is increasing at an average rate of 103 bacteria per day between day 0 and day 1.
(f) 441 bacteria/day
(g) The population is increasing at an average rate of 441 bacteria per day between day 5 and day 6.
(h) The average rate of change is increasing.

78. (a), (b), (e)

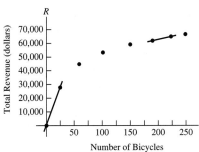

(c) 150 dollars/bicycle
(d) The total cost of producing between 0 and 25 bicycles increases, on average, by $150 for each additional bicycle manufactured.
(f) 113.64 dollars/bicycle
(g) The total cost of producing between 190 and 223 bicycles increases, on average, by $113.64 for each additional bicycle manufactured.

79. (a), (b), (e)

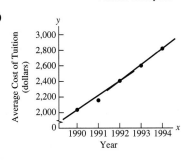

(c) 1120 dollars/bicycle
(d) For each additional bicycle sold between 0 and 25 bicycles, total revenue increases by $1120.
(f) 75 dollars/bicycle
(g) For each additional bicycle sold between 190 and 223 bicycles, total revenue increases by $75.
(h) As the number of bicycles sold increases, marginal revenue decreases.
(i) 150 bicycles

80. (a), (b), (e)

(c) 187.50 dollars/year
(d) The average cost of tuition for the years 1990 to 1992 increased at an average rate of $187.50 per year
(f) 206 dollars/year
(g) The average cost of tuition for the years 1992 to 1994 increased at an average rate of $206 per year.
(h) It is increasing.

81. (a) $43.47 **(b)** $241.49

(c) $C = \begin{cases} 9 + 0.68935x & \text{if } 0 \le x \le 50 \\ 21.465 + 0.44005x & \text{if } x > 50 \end{cases}$

(d)

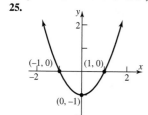

82. (a) $24.71 **(b)** $81.94

(c) $C = \begin{cases} 6 + 0.5125x & \text{if } 0 \le x \le 20 \\ 7.79 + 0.423x & \text{if } 20 < x \le 50 \\ 11.505 + 0.3487x & \text{if } x > 50 \end{cases}$

(d)

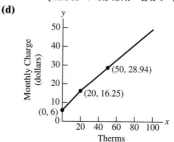

83. Each graph is that of $y = x^2$, but shifted vertically. If $y = x^2 + k, k > 0$, the shift is up k units; if $y = x^2 + k, k < 0$, the shift is down $|k|$ units. **84.** Each graph is that of $y = x^2$, but shifted horizontally. If $y = (x - k)^2$ and $k > 0$, the shift is k units to the right. If $y = (x + k)^2$ and $k > 0$, the shift is k units to the left. **85.** Each graph is that of $y = |x|$, but either compressed or stretched. If $y = k|x|$ and $k > 1$, the graph is stretched vertically; if $y = k|x|, 0 < k < 1$, the graph is compressed vertically. **86.** The graph of $y = -f(x)$ is the reflection about the x-axis of the graph of $y = f(x)$.
87. The graph of $y = f(-x)$ is the reflection about the y-axis of the graph of $y = f(x)$.
88. Shift $y = x^3$ to the right 1 unit and then upward 2 units. **89.** They are all $\cup$-shaped and open upward. All three go through the points $(-1, 1)$, $(0, 0)$ and $(1, 1)$. As the exponent increases, the steepness of the curve increases (except near $x = 0$).
90. They are all increasing functions which pass through the points $(-1, -1)$, $(0, 0)$ and $(1, 1)$. As the exponent increases, the steepness of the curve increases (except near $x = 0$).

4.3 Exercises

1. B **2.** E **3.** H **4.** D **5.** I **6.** A **7.** L **8.** C **9.** F **10.** J **11.** G **12.** K **13.** $y = (x - 4)^3$ **14.** $y = (x + 4)^3$ **15.** $y = x^3 + 4$
16. $y = x^3 - 4$ **17.** $y = -x^3$ **18.** $y = -x^3$ **19.** $y = 4x^3$ **20.** $y = (4x)^3 = 64x^3$ **21.** $y = -\sqrt{-x} - 2$ **22.** $y = -\sqrt{x - 3} - 2$
23. $y = -\sqrt{x + 3} + 2$ **24.** $y = \sqrt{-(x + 3)} + 2$

25.

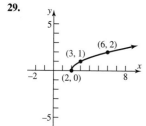

26.

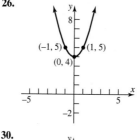

27.

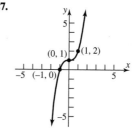

28.

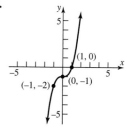

29.

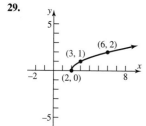

30.

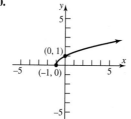

31.

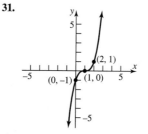

32.

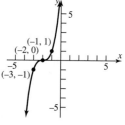

33.

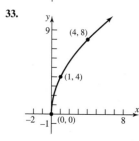

34.

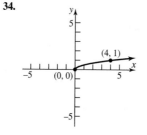

35.

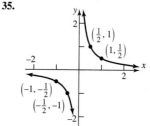

36.

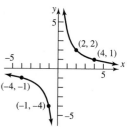

38.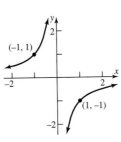

39.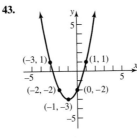

40.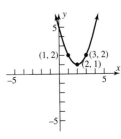

41.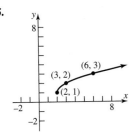

42.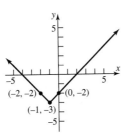

43.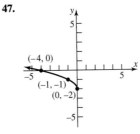

44.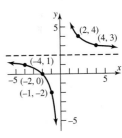

45.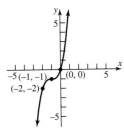

46.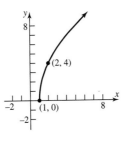

47.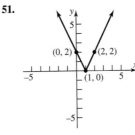

48.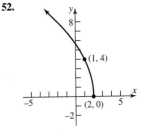

49.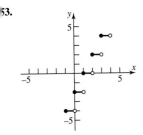

50.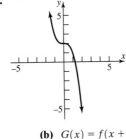

51.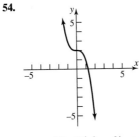

52.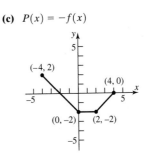

53.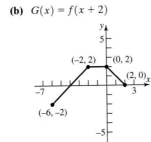

54.

55. (a) $F(x) = f(x) + 3$

(b) $G(x) = f(x + 2)$

(c) $P(x) = -f(x)$

(d) $Q(x) = \frac{1}{2}f(x)$

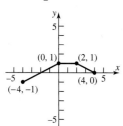

(e) $g(x) = f(-x)$

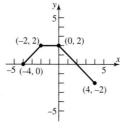

(f) $h(x) = f(2x)$

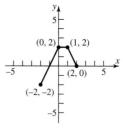

56. (a) $F(x) = f(x) + 3$

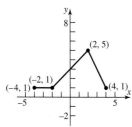

(b) $G(x) = f(x + 2)$

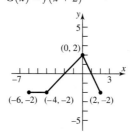

(c) $P(x) = -f(x)$

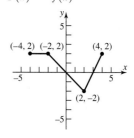

(d) $Q(x) = \frac{1}{2}f(x)$

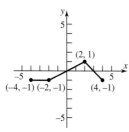

(e) $g(x) = f(-x)$

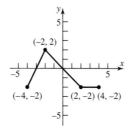

(f) $h(x) = f(2x)$

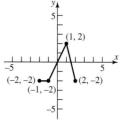

57. (a) $F(x) = f(x) + 3$

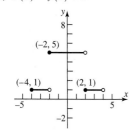

(b) $G(x) = f(x + 2)$

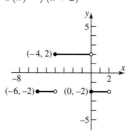

(c) $P(x) = -f(x)$

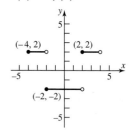

(d) $Q(x) = \frac{1}{2}f(x)$

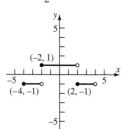

(e) $g(x) = f(-x)$

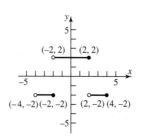

(f) $h(x) = f(2x)$

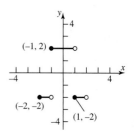

58. (a) $F(x) = f(x) + 3$

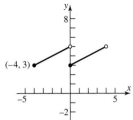

(b) $G(x) = f(x + 2)$

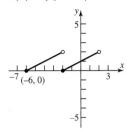

(c) $P(x) = -f(x)$

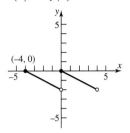

(d) $Q(x) = \dfrac{1}{2}f(x)$

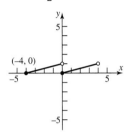

(e) $g(x) = f(-x)$

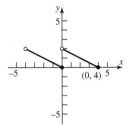

(f) $h(x) = f(2x)$

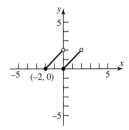

59. (a) $F(x) = f(x) + 3$

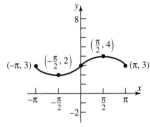

(b) $G(x) = f(x + 2)$

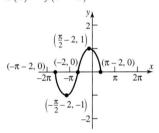

(c) $P(x) = -f(x)$

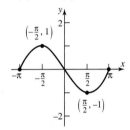

(d) $Q(x) = \dfrac{1}{2}f(x)$

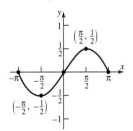

(e) $g(x) = f(-x)$

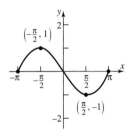

(f) $h(x) = f(2x)$

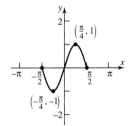

60. (a) $F(x) = f(x) + 3$

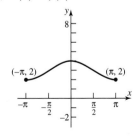

(b) $G(x) = f(x + 2)$

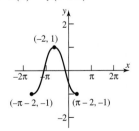

(c) $P(x) = -f(x)$

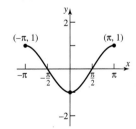

(d) $Q(x) = \dfrac{1}{2}f(x)$

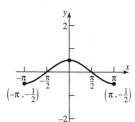

$(-\pi, -\frac{1}{2})$ $(\pi, -\frac{1}{2})$

(e) $g(x) = f(-x)$

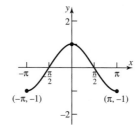

$(-\pi, -1)$ $(\pi, -1)$

(f) $h(x) = f(2x)$

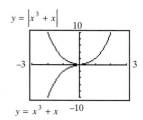

$\left(-\frac{\pi}{2}, -1\right)$ $\left(\frac{\pi}{2}, -1\right)$

61. (a)

$y = |x + 1|$

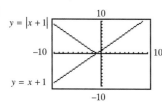

$y = x + 1$

(b) $y = |4 - x^2|$

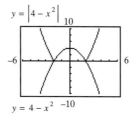

$y = 4 - x^2$

(c) $y = |x^3 + x|$

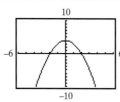

$y = x^3 + x$

(d) Any part of the graph of $y = f(x)$ that lies below the x-axis is reflected about the x-axis to obtain the graph of $y = |f(x)|$.

62. (a)

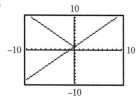

(b)

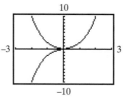

(c)

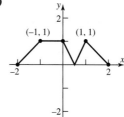

(d) For the graph of $y = f(|x|)$, the graph of $y = f(x)$ to the left of the y-axis is replaced by the reflection about the y-axis of the graph of $y = f(x)$ to the right of the y-axis. To the right of the y-axis, both graphs are the same.

63. (a)

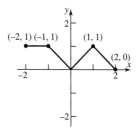

(b)

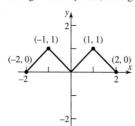

64. (a)

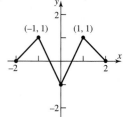

(b)

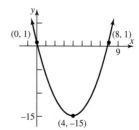

65. $f(x) = (x + 1)^2 - 1$

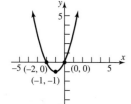

66. $f(x) = (x - 3)^2 - 9$

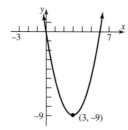

67. $f(x) = (x - 4)^2 - 15$

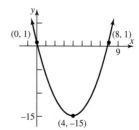

68. $f(x) = (x + 2)^2 - 2$

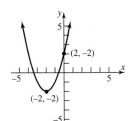

(2, −2)

(−2, −2)

69. $f(x) = \left(x + \dfrac{1}{2}\right)^2 + \dfrac{3}{4}$

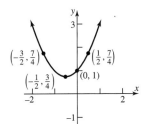

$\left(-\dfrac{3}{2}, \dfrac{7}{4}\right)$ $\left(\dfrac{1}{2}, \dfrac{7}{4}\right)$

$\left(-\dfrac{1}{2}, \dfrac{3}{4}\right)$ (0, 1)

70. $f(x) = \left(x - \dfrac{1}{2}\right)^2 + \dfrac{3}{4}$

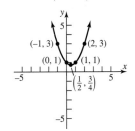

(−1, 3)   (2, 3)

(0, 1)   (1, 1)

$\left(\dfrac{1}{2}, \dfrac{3}{4}\right)$

71.

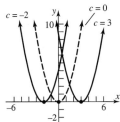

$c = -2$ $c = 0$ $c = 3$

72.

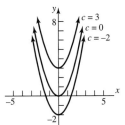

$c = 3$ $c = 0$ $c = -2$

73.

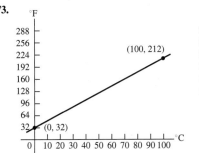

(100, 212)

(0, 32)

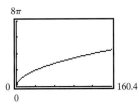

(373, 212)

(273, 32)

74. (a) 8π

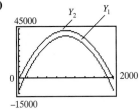

0   160.4

(b) 2π

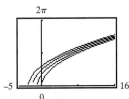

−5   16.1

(c) An increase in l increases the period T.
(d)

8π

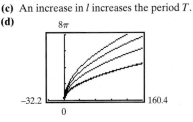

−32.2   160.4

(e) Multiplying the length l by a factor k, increases the period T by a factor of $\sqrt{k}$.

75. (a)

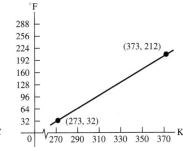

45000 Y_2 Y_1

0   2000

−15000

(b) 10% tax
(c) Y_1 is the graph of $p(x)$ shifted down vertically 10,000 units. Y_2 is the graph of $p(x)$ vertically compressed by a factor of 0.9.
(d) 10% tax

4.4 Exercises

1. (a) $(f + g)(x) = 5x + 1$; All real numbers   **(b)** $(f - g)(x) = x + 7$; All real numbers
(c) $(f \cdot g)(x) = 6x^2 - x - 12$; All real numbers   **(d)** $\left(\dfrac{f}{g}\right)(x) = \dfrac{3x + 4}{2x - 3}; \left\{x \,\middle|\, x \neq \dfrac{3}{2}\right\}$
2. (a) $(f + g)(x) = 5x - 1$; All real numbers   **(b)** $(f - g)(x) = -x + 3$; All real numbers
(c) $(f \cdot g)(x) = 6x^2 - x - 2$; All real numbers   **(d)** $\left(\dfrac{f}{g}\right)(x) = \dfrac{2x + 1}{3x - 2}; \left\{x \,\middle|\, x \neq \dfrac{2}{3}\right\}$

3. (a) $(f + g)(x) = 2x^2 + x - 1$; All real numbers **(b)** $(f - g)(x) = -2x^2 + x - 1$; All real numbers

(c) $(f \cdot g)(x) = 2x^3 - 2x^2$; All real numbers **(d)** $\left(\dfrac{f}{g}\right)(x) = \dfrac{x - 1}{2x^2}$; $\{x | x \neq 0\}$ **4. (a)** $(f + g)(x) = 4x^3 + 2x^2 + 4$; All real numbers

(b) $(f - g)(x) = -4x^3 + 2x^2 + 2$; All real numbers **(c)** $(f \cdot g)(x) = 8x^5 + 12x^3 + 2x^2 + 3$; All real numbers

(d) $\left(\dfrac{f}{g}\right)(x) = \dfrac{2x^2 + 3}{4x^3 + 1}$; $\left\{x \middle| x \neq -\dfrac{1}{\sqrt[3]{4}}\right\}$ **5. (a)** $(f + g)(x) = \sqrt{x} + 3x - 5$; $\{x | x \geq 0\}$

(b) $(f - g)(x) = \sqrt{x} - 3x + 5$; $\{x | x \geq 0\}$ **(c)** $(f \cdot g)(x) = 3x\sqrt{x} - 5\sqrt{x}$; $\{x | x \geq 0\}$

(d) $\left(\dfrac{f}{g}\right)(x) = \dfrac{\sqrt{x}}{3x - 5}$; $\left\{x \middle| x \geq 0, x \neq \dfrac{5}{3}\right\}$ **6. (a)** $(f + g)(x) = |x| + x$; All real numbers **(b)** $(f - g)(x) = |x| - x$; All real numbers

(c) $(f \cdot g)(x) = |x|x$; All real numbers **(d)** $\left(\dfrac{f}{g}\right)(x) = \dfrac{|x|}{x}$; $\{x | x \neq 0\}$ **7. (a)** $(f + g)(x) = 1 + \dfrac{2}{x}$; $\{x | x \neq 0\}$

(b) $(f - g)(x) = 1$; $\{x | x \neq 0\}$ **(c)** $(f \cdot g)(x) = \dfrac{1}{x} + \dfrac{1}{x^2}$; $\{x | x \neq 0\}$ **(d)** $\left(\dfrac{f}{g}\right)(x) = x + 1$; $\{x | x \neq 0\}$

8. (a) $(f + g)(x) = 4x^2$; All real numbers **(b)** $(f - g)(x) = -2x$; All real numbers **(c)** $(f \cdot g)(x) = 4x^4 - x^2$; All real numbers

(d) $\left(\dfrac{f}{g}\right)(x) = \dfrac{2x^2 - x}{2x^2 + x}$; $\left\{x \middle| x \neq 0, x \neq -\dfrac{1}{2}\right\}$ **9. (a)** $(f + g)(x) = \dfrac{6x + 3}{3x - 2}$; $\left\{x \middle| x \neq \dfrac{2}{3}\right\}$

(b) $(f - g)(x) = \dfrac{-2x + 3}{3x - 2}$; $\left\{x \middle| x \neq \dfrac{2}{3}\right\}$ **(c)** $(f \cdot g)(x) = \dfrac{8x^2 + 12x}{(3x - 2)^2}$; $\left\{x \middle| x \neq \dfrac{2}{3}\right\}$ **(d)** $\left(\dfrac{f}{g}\right)(x) = \dfrac{2x + 3}{4x}$; $\left\{x \middle| x \neq 0, x \neq \dfrac{2}{3}\right\}$

10. (a) $(f + g)(x) = \sqrt{x + 1} + \dfrac{2}{x}$; $\{x | x \geq -1, x \neq 0\}$ **(b)** $(f - g)(x) = \sqrt{x + 1} - \dfrac{2}{x}$; $\{x | x \geq -1, x \neq 0\}$

(c) $(f \cdot g)(x) = \dfrac{2\sqrt{x + 1}}{x}$; $\{x | x \geq -1, x \neq 0\}$ **(d)** $\left(\dfrac{f}{g}\right)(x) = \dfrac{x\sqrt{x + 1}}{2}$; $\{x | x \geq -1, x \neq 0\}$ **11.** $g(x) = 5 - \dfrac{7}{2}x$ **12.** $g(x) = \dfrac{x - 1}{x + 1}$

13. (a) 98 **(b)** 49 **(c)** 4 **(d)** 4 **14. (a)** 95 **(b)** 127 **(c)** 17 **(d)** 1

15. (a) 97 **(b)** $-\dfrac{163}{2}$ **(c)** 1 **(d)** $-\dfrac{3}{2}$ **16. (a)** 4418 **(b)** -191 **(c)** 8 **(d)** -2

17. (a) $2\sqrt{2}$ **(b)** $2\sqrt{2}$ **(c)** 1 **(d)** 0 **18. (a)** $\sqrt{13}$ **(b)** $3\sqrt{3}$ **(c)** $\sqrt{\sqrt{2} + 1}$ **(d)** 0

19. (a) $\dfrac{1}{17}$ **(b)** $\dfrac{1}{5}$ **(c)** 1 **(d)** $\dfrac{1}{2}$ **20. (a)** $\dfrac{11}{6}$ **(b)** $\dfrac{3}{2}$ **(c)** 1 **(d)** $\dfrac{12}{17}$

21. (a) $\dfrac{3}{5}$ **(b)** $\dfrac{\sqrt{15}}{5}$ **(c)** $\dfrac{12}{13}$ **(d)** 0 **22. (a)** $\dfrac{8}{4913}$ **(b)** $\dfrac{2}{65}$ **(c)** 1 **(d)** $\dfrac{2}{5}$

23. $\{x | x \neq 0, x \neq 2\}$ **24.** $\left\{x \middle| x \neq 0, x \neq \dfrac{2}{3}\right\}$ **25.** $\{x | x \neq -4, x \neq 0\}$ **26.** $\left\{x \middle| x \neq 0, x \neq -\dfrac{2}{3}\right\}$

27. $\left\{x \middle| x \geq -\dfrac{3}{2}\right\}$ **28.** $\left\{x \middle| x \leq -\dfrac{1}{2}\right\}$ **29.** $\{x | x \leq -1 \text{ or } x > 1\}$ **30.** $\left\{x \middle| x < 2 \text{ or } x \geq \dfrac{8}{3}\right\}$

31. (a) $(f \circ g)(x) = 6x + 3$; All real numbers **(b)** $(g \circ f)(x) = 6x + 9$; All real numbers **(c)** $(f \circ f)(x) = 4x + 9$; All real numbers

(d) $(g \circ g)(x) = 9x$; All real numbers **32. (a)** $(f \circ g)(x) = 4 - 2x$; All real numbers **(b)** $(g \circ f)(x) = -2x - 4$; All real numbers

(c) $(f \circ f)(x) = x$; All real numbers **(d)** $(g \circ g)(x) = 4x - 12$; All real numbers **33. (a)** $(f \circ g)(x) = 3x^2 + 1$; All real numbers

(b) $(g \circ f)(x) = 9x^2 + 6x + 1$; All real numbers **(c)** $(f \circ f)(x) = 9x + 4$; All real numbers **(d)** $(g \circ g)(x) = x^4$; All real numbers

34. (a) $(f \circ g)(x) = x^2 + 5$; All real numbers **(b)** $(g \circ f)(x) = x^2 + 2x + 5$; All real numbers **(c)** $(f \circ f)(x) = x + 2$; All real numbers

(d) $(g \circ g)(x) = x^4 + 8x^2 + 20$; All real numbers **35. (a)** $(f \circ g)(x) = x^4 + 8x^2 + 16$; All real numbers

(b) $(g \circ f)(x) = x^4 + 4$; All real numbers **(c)** $(f \circ f)(x) = x^4$; All real numbers **(d)** $(g \circ g)(x) = x^4 + 8x^2 + 20$; All real numbers

36. (a) $(f \circ g)(x) = 4x^4 + 12x^2 + 10$; All real numbers **(b)** $(g \circ f)(x) = 2x^4 + 4x^2 + 5$; All real numbers

(c) $(f \circ f)(x) = x^4 + 2x^2 + 2$; All real numbers **(d)** $(g \circ g)(x) = 8x^4 + 24x^2 + 21$; All real numbers

37. (a) $(f \circ g)(x) = \dfrac{3x}{2 - x}$; $\{x | x \neq 0, x \neq 2\}$ **(b)** $(g \circ f)(x) = \dfrac{2(x - 1)}{3}$; $\{x | x \neq 1\}$ **(c)** $(f \circ f)(x) = \dfrac{3(x - 1)}{4 - x}$; $\{x | x \neq 1, x \neq 4\}$

(d) $(g \circ g)(x) = x$; $\{x | x \neq 0\}$ **38. (a)** $(f \circ g)(x) = \dfrac{x}{-2 + 3x}$; $\left\{x \middle| x \neq 0, x \neq \dfrac{2}{3}\right\}$ **(b)** $(g \circ f)(x) = -2x - 6$; $\{x | x \neq -3\}$

(c) $(f \circ f)(x) = \dfrac{x + 3}{3x + 10}$; $\left\{x \middle| x \neq -3, x \neq -\dfrac{10}{3}\right\}$ **(d)** $(g \circ g)(x) = x$; $\{x | x \neq 0\}$

39. (a) $(f \circ g)(x) = \dfrac{4}{4 + x}$; $\{x | x \neq -4, x \neq 0\}$ **(b)** $(g \circ f)(x) = \dfrac{-4(x - 1)}{x}$; $\{x | x \neq 0, x \neq 1\}$ **(c)** $(f \circ f)(x) = x$; $\{x | x \neq 1\}$

(d) $(g \circ g)(x) = x$; $\{x | x \neq 0\}$ **40. (a)** $(f \circ g)(x) = \dfrac{2}{2 + 3x}$; $\left\{x \middle| x \neq 0, x \neq -\dfrac{2}{3}\right\}$ **(b)** $(g \circ f)(x) = \dfrac{2x + 6}{x}$; $\{x | x \neq -3, x \neq 0\}$

(c) $(f \circ f)(x) = \dfrac{x}{4x + 9}$; $\left\{x \middle| x \neq -3, x \neq -\dfrac{9}{4}\right\}$ **(d)** $(g \circ g)(x) = x$; $\{x | x \neq 0\}$

41. (a) $(f \circ g)(x) = \sqrt{2x + 3}$; $\left\{x \middle| x \geq -\dfrac{3}{2}\right\}$ **(b)** $(g \circ f)(x) = 2\sqrt{x} + 3$; $\{x | x \geq 0\}$ **(c)** $(f \circ f)(x) = \sqrt[4]{x}$; $\{x | x \geq 0\}$

(d) $(g \circ g)(x) = 4x + 9$; All real numbers   **42. (a)** $(f \circ g)(x) = \sqrt{-1 - 2x}$; $\left\{x \middle| x \leq -\dfrac{1}{2}\right\}$ **(b)** $(g \circ f)(x) = 1 - 2\sqrt{x - 2}$; $\{x | x \geq 2\}$

(c) $(f \circ f)(x) = \sqrt{\sqrt{x - 2} - 2}$; $\{x | x \geq 6\}$ **(d)** $(g \circ g)(x) = -1 + 4x$; All real numbers

43. (a) $(f \circ g)(x) = \sqrt{\dfrac{1 + x}{x - 1}}$; $\{x | x \leq -1 \text{ or } x > 1\}$ **(b)** $(g \circ f)(x) = \dfrac{2}{\sqrt{x + 1} - 1}$; $\{x | x \geq -1, x \neq 0\}$

(c) $(f \circ f)(x) = \sqrt{\sqrt{x + 1} + 1}$; $\{x | x \geq -1\}$ **(d)** $(g \circ g)(x) = \dfrac{2(x - 1)}{3 - x}$; $\{x | x \neq 1, x \neq 3\}$ **44. (a)** $(f \circ g)(x) = \sqrt{\dfrac{3x - 8}{x - 2}}$;

$\left\{x \middle| x < 2 \text{ or } x \geq \dfrac{8}{3}\right\}$ **(b)** $(g \circ f)(x) = \dfrac{2}{\sqrt{3 - x} - 2}$; $\{x | x \leq 3, x \neq -1\}$ **(c)** $(f \circ f)(x) = \sqrt{3 - \sqrt{3 - x}}$; $\{x | -6 \leq x \leq 3\}$

(d) $(g \circ g)(x) = -\dfrac{x - 2}{x - 3}$; $\{x | x \neq 2, x \neq 3\}$ **45. (a)** $(f \circ g)(x) = acx + ad + b$; All real numbers   **(b)** $(g \circ f)(x) = acx + bc + d$;

All real numbers   **(c)** $(f \circ f)(x) = a^2 x + ab + b$; All real numbers   **(d)** $(g \circ g)(x) = c^2 x + cd + d$; All real numbers

46. (a) $(f \circ g)(x) = \dfrac{amx + b}{cmx + d}$; $\left\{x \middle| x \neq -\dfrac{d}{cm}\right\}$ **(b)** $(g \circ f)(x) = \dfrac{max + mb}{cx + d}$; $\left\{x \middle| x \neq -\dfrac{d}{c}\right\}$

(c) $(f \circ f)(x) = \dfrac{a^2 x + ab + bcx + bd}{cax + cb + dcx + d^2}$; $\left\{x \middle| x \neq -\dfrac{d}{c}, x \neq -\dfrac{cb + d^2}{ca + dc}\right\}$ **(d)** $(g \circ g)(x) = m^2 x$; All real numbers

47. $(f \circ g)(x) = f(g(x)) = f\left(\dfrac{1}{2}x\right) = 2\left(\dfrac{1}{2}x\right) = x$; $(g \circ f)(x) = g(f(x)) = g(2x) = \dfrac{1}{2}(2x) = x$

48. $(f \circ g)(x) = f(g(x)) = f\left(\dfrac{1}{4}x\right) = 4\left(\dfrac{1}{4}x\right) = x$; $(g \circ f)(x) = g(f(x)) = g(4x) = \dfrac{1}{4}(4x) = x$

49. $(f \circ g)(x) = f(g(x)) = f(\sqrt[3]{x}) = (\sqrt[3]{x})^3 = x$; $(g \circ f)(x) = g(f(x)) = g(x^3) = \sqrt[3]{x^3} = x$

50. $(f \circ g)(x) = f(g(x)) = f(x - 5) = (x - 5) + 5 = x$; $(g \circ f)(x) = g(f(x)) = g(x + 5) = (x + 5) - 5 = x$

51. $(f \circ g)(x) = f(g(x)) = f\left(\dfrac{1}{2}(x + 6)\right) = 2\left[\dfrac{1}{2}(x + 6)\right] - 6 = x + 6 - 6 = x$;

$(g \circ f)(x) = g(f(x)) = g(2x - 6) = \dfrac{1}{2}(2x - 6 + 6) = x$

52. $(f \circ g)(x) = f(g(x)) = f\left(\dfrac{1}{3}(4 - x)\right) = 4 - 3\left(\dfrac{1}{3}(4 - x)\right) = x$; $(g \circ f)(x) = g(f(x)) = g(4 - 3x) = \dfrac{1}{3}(4 - (4 - 3x)) = x$

53. $(f \circ g)(x) = f(g(x)) = f\left(\dfrac{1}{a}(x - b)\right) = a\left[\dfrac{1}{a}(x - b)\right] + b = x$; $(g \circ f)(x) = g(f(x)) = g(ax + b) = \dfrac{1}{a}(ax + b - b) = x$

54. $(f \circ g)(x) = f(g(x)) = f\left(\dfrac{1}{x}\right) = \dfrac{1}{\frac{1}{x}} = x$; $(g \circ f)(x) = g(f(x)) = g\left(\dfrac{1}{x}\right) = \dfrac{1}{\frac{1}{x}} = x$

55. $F = f \circ g$ **56.** $G = g \circ f$ **57.** $H = h \circ f$ **58.** $p = g \circ h$ **59.** $q = f \circ h$ **60.** $R = g \circ g$ **61.** $P = f \circ f$ **62.** $Q = h \circ h$

63. $f(x) = x^4$; $g(x) = 2x + 3$ (Other answers are possible.)   **64.** $f(x) = x^{3/2}$; $g(x) = 1 + x^2$ (Other answers are possible.)

65. $f(x) = \sqrt{x}$; $g(x) = x^2 + x + 1$ (Other answers are possible.)   **66.** $f(x) = \dfrac{1}{x}$; $g(x) = 1 + x^2$ (Other answers are possible.)

67. $f(x) = x^2$; $g(x) = 1 - \dfrac{1}{x^2}$ (Other answers are possible.)   **68.** $f(x) = |x|$; $g(x) = 2x^2 + 3$ (Other answers are possible.)

69. $f(x) = \text{int}(x)$; $g(x) = x^2 + 1$ (Other answers are possible.)   **70.** $f(x) = x^{-4}$; $g(x) = 4 - x^2$ (Other answers are possible.)

71. $(f \circ g)(x) = 11$; $(g \circ f)(x) = 2$ **72.** $(f \circ f)(x) = x$ **73.** $-3, 3$ **74.** $-5, 5$ **75.** $S(r(t)) = \dfrac{16}{9}\pi t^6$ **76.** $V(r(t)) = \dfrac{32}{81}\pi t^9$

77. $C(N(t)) = 15{,}000 + 800{,}000t - 40{,}000t^2$ **78.** $A(r(t)) = 40{,}000\pi t$ **79.** $C = \dfrac{2\sqrt{100 - p}}{25} + 600$ **80.** $C = \dfrac{\sqrt{5(200 - p)}}{10} + 400$

81. Since f and g are odd, $f(-x) = -f(x)$ and $g(-x) = -g(x)$; $(f \circ g)(-x) = f(g(-x)) = f(-g(x)) = -f(g(x)) = -(f \circ g)(x)$;
thus, $f \circ g$ is odd   **82.** f is odd so $f(-x) = -f(x)$; g is even so $g(-x) = g(x)$
$(f \circ g)(-x) = f(g(-x)) = f(g(x)) = (f \circ g)(x)$ so $(f \circ g)$ is even
$(g \circ f)(-x) = g(f(-x)) = g(-f(x)) = g(f(x))$ so $g \circ f$ is even

4.5 Exercises

1. $V(r) = 2\pi r^3$ **2.** $V(r) = \dfrac{2}{3}\pi r^3$

3. (a) $R(x) = -\dfrac{1}{6}x^2 + 100x$

   (b) \$13,333.33

   (c)

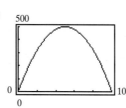

   (d) $x = 300; \$15,000$ **(e)** \$50

5. (a) $R(x) = -\dfrac{1}{5}x^2 + 20x$ **(b)** \$255 **(c)**

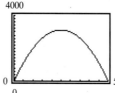

4. (a) $R(x) = -\dfrac{1}{3}x^2 + 100x$

   (b) \$6666.67

   (c)

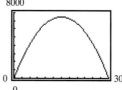

   (d) $x = 150; \$7500$ **(e)** \$50 per unit

   (d) $x = 50; \$500$ **(e)** \$10

6. (a) $R(x) = -\dfrac{1}{20}x^2 + 25x$ **(b)** \$480 **(c)** 4000

(d) $R(x)$ is maximum when $x = 250; \$3125$

(e) \$12.5 per unit

7. (a) $A(x) = -x^2 + 200x$ **(b)** Domain: $\{x | 0 < x < 200\}$ **(c)** 10,000

A is largest when $x = 100$ yards

8. (a) $A(x) = -\dfrac{x^2}{2} + 1500x$ **(b)** A is largest when $x = 1500$ ft

9. (a) $C(x) = x$ **(b)** $A(x) = \dfrac{x^2}{4\pi}$ **10. (a)** $p(x) = x$ **(b)** $A(x) = \dfrac{x^2}{16}$ **11.** $A(x) = \dfrac{1}{2}x^4$ **12.** $A(x) = -\dfrac{1}{2}x^3 + \dfrac{9}{2}x$

13. (a) $d(x) = \sqrt{x^4 - 15x^2 + 64}$ **(b)** $d(0) = 8$

   (c) $d(1) = \sqrt{50} \approx 7.07$

   (d)

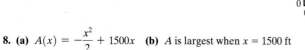

   (e) d is smallest when $x \approx -2.74$ or $x \approx 2.74$

14. (a) $d(x) = \sqrt{x^4 - 13x^2 + 49}$ **(b)** $d(0) = 7$

   (c) $d(-1) = \sqrt{37} \approx 6.08$

   (d)

   (e) d is smallest when $x \approx -2.55$ or $x \approx 2.55$

15. (a) $d(x) = \sqrt{x^2 - x + 1}$

(b)

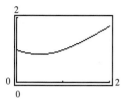

(c) d is smallest when $x = 0.50$

17. $d(t) = 50t$

18. (a) $d(t) = \sqrt{2500t^2 - 360t + 13}$
(b) d is smallest when $t = 0.072$ hrs

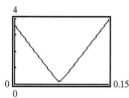

20. (a) $A(x) = x^2 + \dfrac{40}{x}$ **(b)** 41 ft² **(c)** 24 ft²

(d) A is smallest if $x \approx 2.71$ ft

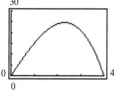

22. $V(S) = \dfrac{S}{6}\sqrt{\dfrac{S}{\pi}}$. If S doubles, V increases by a factor of $2\sqrt{2}$.

23. (a) $A(x) = x(16 - x^2)$
(b) Domain: $\{x \mid 0 < x < 4\}$
(c) The area is largest when $x \approx 2.31$

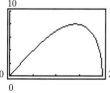

25. (a) $A(x) = 4x\sqrt{4 - x^2}$ **(b)** $p(x) = 4x + 4\sqrt{4 - x^2}$
(c) The area is largest when $x \approx 1.41$ **(d)** The perimeter is largest when $x \approx 1.41$

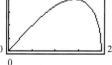

26. (a) $A(r) = 4r^2$ **(b)** $p(r) = 8r$

16. (a) $d(x) = \sqrt{\dfrac{x^4 + 1}{x^2}}$

(b)

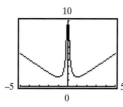

(c) d is smallest when $x = -1$ or $x = 1$

19. (a) $V(x) = x(24 - 2x)^2$
(b) 972 in³ **(c)** 160 in³
(d) The volume is largest when $x = 4$

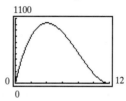

21. (a) $A(x) = 2x^2 + \dfrac{40}{x}$ **(b)** 42 ft² **(c)** 28 ft²
(d) The area is smallest when $x \approx 2.15$

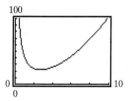

24. (a) $A(x) = 2x\sqrt{4 - x^2}$ **(b)** $p(x) = 4x + 2\sqrt{4 - x^2}$
(c) A is largest when $x \approx 1.41$ **(d)** p is largest when $x \approx 1.79$

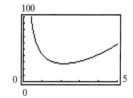 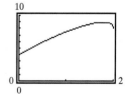

27. (a) $A(x) = x^2 + \dfrac{25 - 20x + 4x^2}{\pi}$

(b) Domain: $\{x \mid 0 < x < 2.5\}$

(c) The area is smallest when $x \approx 1.40$ meters

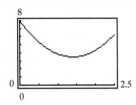

28. (a) $A(x) = \dfrac{(10 - 3x)^2}{4\pi} + \dfrac{\sqrt{3}}{4}x^2$

(b) $\left\{ x \mid 0 < x < \dfrac{10}{3} \right\}$

(c) A is smallest when $x \approx 2.08$

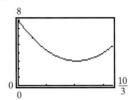

29. (a) $A(r) = 2r^2$ **(b)** $p(r) = 6r$ **30. (a)** $C(x) = \dfrac{2\pi x}{\sqrt{3}}$ **31.** $A(x) = \left(\dfrac{\pi}{3} - \dfrac{\sqrt{3}}{4} \right)x^2$

32. (a)

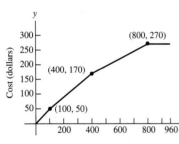

(b) For hauls between 100 and 400 miles, the cost as a function of mileage is
$C(x) = 10 + 0.40x$.

(c) For hauls between 400 and 800 miles, the cost as a function of mileage is
$C(x) = 70 + 0.25x$.

33. $C = \begin{cases} 95 & \text{if } x = 7 \\ 119 & \text{if } 7 < x \le 8 \\ 143 & \text{if } 8 < x \le 9 \\ 167 & \text{if } 9 < x \le 10 \\ 190 & \text{if } 10 < x \le 14 \end{cases}$

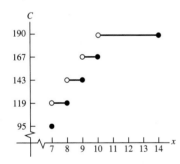

34. $C = \begin{cases} 219 & \text{if } x = 7 \\ 264 & \text{if } 7 < x \le 8 \\ 309 & \text{if } 8 < x \le 9 \\ 354 & \text{if } 9 < x \le 10 \\ 399 & \text{if } 10 < x \le 11 \\ 438 & \text{if } 11 < x \le 14 \end{cases}$

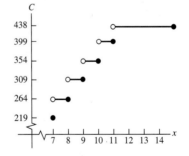

35. $V(h) = \dfrac{\pi}{48}h^3$

36. Schedule X: $f(x) = \begin{cases} 0.15x & \text{if } 0 < x \le 24{,}650 \\ 3{,}698 + 0.28(x - 24{,}650) & \text{if } 24{,}650 < x \le 59{,}750 \\ 13{,}526 + 0.31\,(x - 59{,}750) & \text{if } 59{,}750 < x \le 124{,}650 \\ 33{,}645 + 0.36\,(x - 124{,}650) & \text{if } 124{,}650 < x \le 271{,}050 \\ 86{,}349 + 0.396\,(x - 271{,}050) & \text{if } x > 271{,}050 \end{cases}$

Schedule Y − 1: $f(x) = \begin{cases} 0.15x & \text{if } 0 < x \le 41{,}200 \\ 6{,}180 + 0.28(x - 41{,}200) & \text{if } 41{,}200 < x \le 99{,}600 \\ 22{,}532 + 0.31(x - 99{,}600) & \text{if } 99{,}600 < x \le 151{,}750 \\ 38{,}699 + 0.36(x - 151{,}750) & \text{if } 151{,}750 < x \le 271{,}050 \\ 81{,}647 + 39.6(x - 271{,}050) & \text{if } x > 271{,}050 \end{cases}$

37. $V(h) = \pi h \left(R^2 - \dfrac{h^2}{4} \right)$ **38.** $V(r) = \pi H r^2 \left(1 - \dfrac{r}{R} \right)$

Fill-in-the-Blank Items

1. independent; dependent   **2.** vertical   **3.** even; odd   **4.** horizontal; right   **5.** $g(f(x)) = (g \circ f)(x)$ **6.** $g(f(x))$

True/False Items

1. F   **2.** T   **3.** T   **4.** F   **5.** F   **6.** T   **7.** F   **8.** F

Review Exercises

1. $f(x) = -2x + 3$ **2.** $g(x) = -4x - 6$ **3.** $A = 11$ **4.** $A = 8$ **5.** b, c, d

6. (a) Domain: $\{x | -5 \le x \le 4\}$; Range: $\{y | -3 \le y \le 1\}$ **(b)** Increasing on $(3, 4)$ **(c)** Constant on $(-5, -1)$ **(d)** $(0, 0), (4, 0)$

7. (a) $f(-x) = \dfrac{-3x}{x^2 - 4}$ **(b)** $-f(x) = \dfrac{-3x}{x^2 - 4}$ **(c)** $f(x + 2) = \dfrac{3x + 6}{x^2 + 4x}$ **(d)** $f(x - 2) = \dfrac{3x - 6}{x^2 - 4x}$

8. (a) $f(-x) = \dfrac{x^2}{-x + 2}$ **(b)** $-f(x) = \dfrac{-x^2}{x + 2}$ **(c)** $f(x + 2) = \dfrac{x^2 + 4x + 4}{x + 4}$ **(d)** $f(x - 2) = \dfrac{x^2 - 4x + 4}{x}$

9. (a) $f(-x) = \sqrt{x^2 - 4}$ **(b)** $-f(x) = -\sqrt{x^2 - 4}$ **(c)** $f(x + 2) = \sqrt{x^2 + 4x}$ **(d)** $f(x - 2) = \sqrt{x^2 - 4x}$

10. (a) $f(-x) = |x^2 - 4|$ **(b)** $-f(x) = -|x^2 - 4|$ **(c)** $f(x + 2) = |x^2 + 4x|$ **(d)** $f(x - 2) = |x^2 - 4x|$

11. (a) $f(-x) = \dfrac{x^2 - 4}{x^2}$ **(b)** $-f(x) = -\dfrac{x^2 - 4}{x^2}$ **(c)** $f(x + 2) = \dfrac{x^2 + 4x}{x^2 + 4x + 4}$ **(d)** $f(x - 2) = \dfrac{x^2 - 4x}{x^2 - 4x + 4}$

12. (a) $f(-x) = \dfrac{-x^3}{x^2 - 4}$ **(b)** $-f(x) = \dfrac{-x^3}{x^2 - 4}$ **(c)** $f(x + 2) = \dfrac{x^3 + 6x^2 + 12x + 8}{x^2 + 4x}$ **(d)** $f(x - 2) = \dfrac{x^3 - 6x^2 + 12x - 8}{x^2 - 4x}$

13. Odd   **14.** Even   **15.** Even   **16.** Neither   **17.** Neither   **18.** Neither   **19.** $\{x | x \ne -3, x \ne 3\}$ **20.** $\{x | x \ne 2\}$

21. $\{x | x \le 2\}$ **22.** $\{x | x \ge -2\}$ **23.** $\{x | x > 0\}$ **24.** $\{x | x \ne 0\}$ **25.** $\{x | x \ne -3, x \ne 1\}$ **26.** $\{x | x \ne -1, x \ne 4\}$

27. $\{x | x \ge -1\}$ **28.** $\{x | 0 < x \le 8\}$ **29.** $\{x | x \ge 0\}$ **30.** $\{x | x \le 3\}$ **31.** -5 **32.** $2x$ **33.** $3 - 4x$ **34.** $x - 3$

35.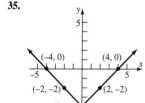

Intercepts: $(-4, 0), (4, 0)$
Domain: all real numbers
Range: $\{y | y \ge -4\}$

36.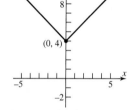

Intercept: $(0, 4)$
Domain: all real numbers
Range: $\{y | y \ge 4\}$.

37.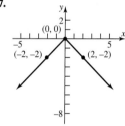

Intercept: $(0, 0)$
Domain: all real numbers
Range: $\{y | y \le 0\}$

38.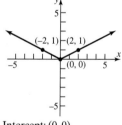

Intercept: $(0, 0)$
Domain: all real numbers
Range: $\{y | y \ge 0\}$

39.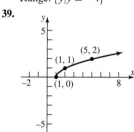

Intercept: $(1, 0)$
Domain: $\{x | x \ge 1\}$
Range: $\{y | y \ge 0\}$

40.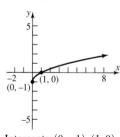

Intercepts: $(0, -1), (1, 0)$
Domain: $\{x | x \ge 0\}$
Range: $\{y | y \ge -1\}$

41.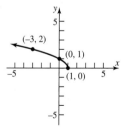

Intercepts: $(0, 1), (1, 0)$
Domain: $\{x | x \le 1\}$
Range: $\{y | y \ge 0\}$

42.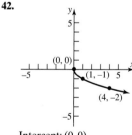

Intercept: $(0, 0)$
Domain: $\{x | x \ge 0\}$
Range: $\{y | y \le 0\}$

43.

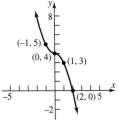

Intercepts: $(0, 4)$, $(2, 0)$
Domain: all real numbers
Range: all real numbers

44.

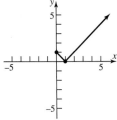

Intercepts: $(0, 1)$, $(1, 0)$
Domain: $\{x | x \geq 0\}$
Range: $\{y | y \geq 0\}$

45.

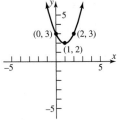

Intercept: $(0, 3)$
Domain: all real numbers
Range: $\{y | y \geq 2\}$

46.

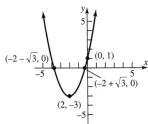

Intercepts: $(-2 - \sqrt{3}, 0)$, $(-2 + \sqrt{3}, 0)$,
$(0, 1)$
Domain: all real numbers
Range: $\{y | y \geq -3\}$

47.

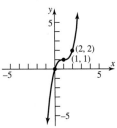

Intercept: $(0, 0)$
Domain: all real numbers
Range: all real numbers

48.

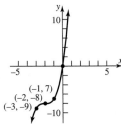

Intercept: $(0, 0)$
Domain: all real numbers.
Range: all real numbers

49.

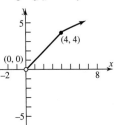

Intercept: none
Domain: $\{x | x > 0\}$
Range: $\{y | y > 0\}$

50.

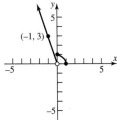

Intercepts: $(0, 1)$, $(1, 0)$
Domain: $\{x | x \leq 1\}$
Range: $\{y | y \geq 0\}$

51.

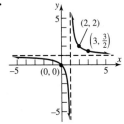

Intercept: $(0, 0)$
Domain: $\{x | x \neq 1\}$
Range: $\{y | y \neq 1\}$

52.

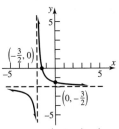

Intercept: $\left(-\frac{3}{2}, 0\right)$, $\left(0, -\frac{3}{2}\right)$

Domain: $\{x | x \neq -2\}$
Range: $\{y | y \neq -2\}$

53.

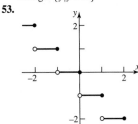

Intercepts: $(x, 0)$, where $-1 < x \leq 0$

Domain: all real numbers
Range: the set of integers

54.

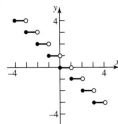

Intercepts: $(x, 0)$, where $0 \leq x < 1$

Domain: all real numbers
Range: the set of integers

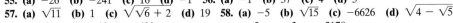

55. (a) -26 **(b)** -241 **(c)** $\dfrac{16}{9}$ **(d)** -1 **56. (a)** -1 **(b)** 37 **(c)** 4 **(d)** 5

57. (a) $\sqrt{11}$ **(b)** 1 **(c)** $\sqrt{\sqrt{6} + 2}$ **(d)** 19 **58. (a)** -5 **(b)** $\sqrt{15}$ **(c)** -6626 **(d)** $\sqrt{4 - \sqrt{5}}$

59. (a) $\dfrac{1}{20}$ **(b)** $-\dfrac{13}{8}$ **(c)** $\dfrac{400}{1601}$ **(d)** -17 **60. (a)** $\dfrac{2}{73}$ **(b)** $\dfrac{2}{3}$ **(c)** $\dfrac{2178}{1097}$ **(d)** -9

61. $(f \circ g)(x) = 1 - 3x$; All real numbers; $(g \circ f)(x) = 7 - 3x$; All real numbers; $(f \circ f)(x) = x$; All real numbers; $(g \circ g)(x) = 9x + 4$;
All real numbers

62. $(f \circ g)(x) = 4x + 1$; All real numbers; $(g \circ f)(x) = 4x - 1$; All real numbers; $(f \circ f)(x) = 4x - 3$; All real numbers;
$(g \circ g)(x) = 4x + 3$; All real numbers

63. $(f \circ g)(x) = 27x^2 + 3|x| + 1$; All real numbers; $(g \circ f)(x) = 3|3x^2 + x + 1|$; All real numbers;
$(f \circ f)(x) = 3(3x^2 + x + 1)^2 + 3x^2 + x + 2$; All real numbers; $(g \circ g)(x) = 9|x|$; All real numbers
64. $(f \circ g)(x) = \sqrt{3 + 3x + 3x^2}$; All real numbers; $(g \circ f)(x) = 1 + \sqrt{3x} + 3x$; $\{x|x \geq 0\}$;
$(f \circ f)(x) = \sqrt{3\sqrt{3x}}$; $\{x|x \geq 0\}$; $(g \circ g)(x) = 3 + 3x + 4x^2 + 2x^3 + x^4$; All real numbers
65. $(f \circ g)(x) = \dfrac{1 + x}{1 - x}$; $\{x \mid x \neq 0, x \neq 1\}$; $(g \circ f)(x) = \dfrac{x - 1}{x + 1}$; $\{x \mid x \neq -1, x \neq 1\}$; $(f \circ f)(x) = x$; $\{x|x \neq 1\}$; $(g \circ g)(x) = x$; $\{x|x \neq 0\}$
66. $(f \circ g)(x) = \sqrt{\dfrac{3}{x} - 3}$; $\{x|0 < x \leq 1\}$; $(g \circ f)(x) = \dfrac{3}{\sqrt{x - 3}}$; $\{x|x > 3\}$; $(f \circ f)(x) = \sqrt{\sqrt{x - 3} - 3}$; $\{x|x \geq 12\}$;
$(g \circ g)(x) = x$; $\{x|x > 0\}$

67. (a)

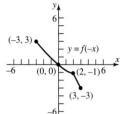

(b)

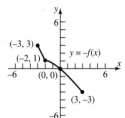

(c)

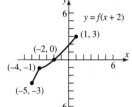

(d)

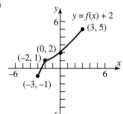

(e)

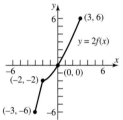

(f)

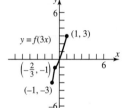

68. (a)

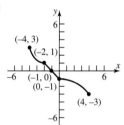

(b)

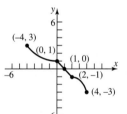

(c)

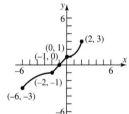

(d)

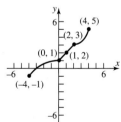

(e)

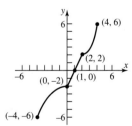

(f)

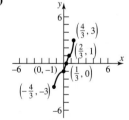

69. $T(h) = -0.0025h + 30, 0 \leq x \leq 10,000$ **70.** 130 ft/sec; $v(t) = 5t - 20$ **71.** $S(x) = kx(36 - x^2)^{3/2}$; Domain: $\{x \mid 0 < x < 6\}$
72. (a), (b), (e)

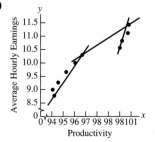

(c) 0.624 dollars/output
(d) For each 1-unit increase in output, earnings increase by $0.62.
(f) 0.273 dollars/output unit
(g) For each 1-unit increase in output, earnings increased by an average of $0.27.
(h) It is decreasing.

73. (a), (b), (e)

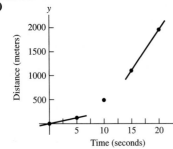

(c) 22.5 ft/sec
(d) Between 0 and 5 seconds, the average speed of the parachutist is 22.5 ft/sec.
(f) 171.5 ft/sec
(g) Between 15 and 20 seconds, the average speed of the parachutist is 171.5 ft/sec
(h) It is increasing.

74. (a) $A = 2\pi r^2 + \dfrac{200}{r}$ (b) 123.22 square feet
(c) 150.53 square feet (d) 197.08 square feet
(e) A is smallest when $r = 2.52$ ft.

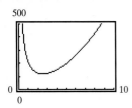

75. (a) $C(r) = 0.12\pi r^2 + \dfrac{40}{r}$ (b) $C(4) = \$16.03$
(c) $C(8) = \$29.13$
(d) The cost in least for r about 3.76 cm.

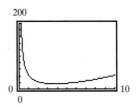

C H A P T E R   5   5.1 Exercises

1. D **2.** F **3.** A **4.** H **5.** B **6.** C **7.** E **8.** G

9.

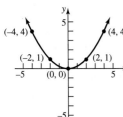

10.

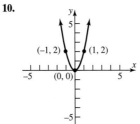

11.

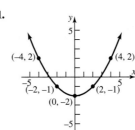

12.

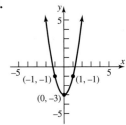

13.

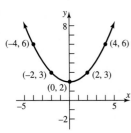

14.

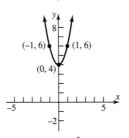

15.

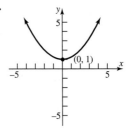

16.

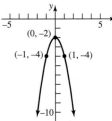
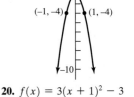

17. $f(x) = (x + 2)^2 - 2$

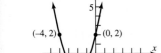

18. $f(x) = (x - 3)^2 - 10$

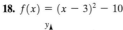

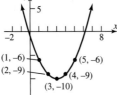

19. $f(x) = 2(x - 1)^2 - 1$

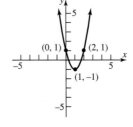

20. $f(x) = 3(x + 1)^2 - 3$

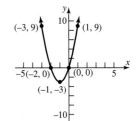

21. $f(x) = -(x + 1)^2 + 1$

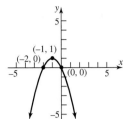

22. $f(x) = -2\left(x - \dfrac{3}{2}\right)^2 + \dfrac{13}{2}$

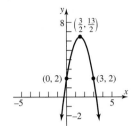

23. $f(x) = \dfrac{1}{2}(x + 1)^2 - \dfrac{3}{2}$

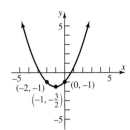

24. $f(x) = \dfrac{2}{3}(x + 1)^2 - \dfrac{5}{3}$

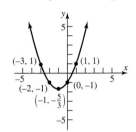

25.

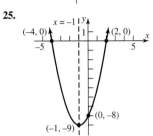

26.

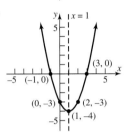

27.

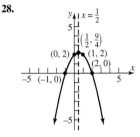

28.

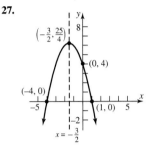

29.

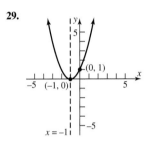

30.

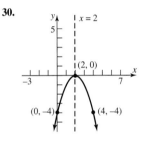

31.

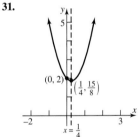

32.

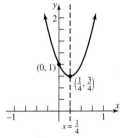

33.

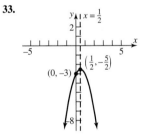

34.

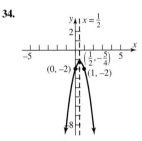

35.

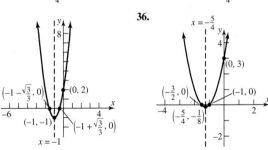

36.

37.

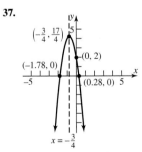

38.

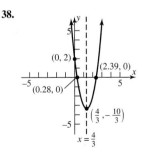

39. Minimum value; -21 **40.** Minimum value; -1

41. Maximum value; 21 **42.** Maximum value; 11

43. Maximum value; 13 **44.** Minimum value; -1

45. $a = 6, b = 0, c = 2, f(x) = 6x^2 + 2$

46. $a = -3, b = 6, c = 1; f(x) = -3x^2 + 6x + 1$

47. Price: \$500; maximum revenue: \$1,000,000

48. Price: \$1900; Maximum revenue: \$1,805,000

49. 10,000 ft²; 100 ft by 100 ft   **50.** $\dfrac{P}{4}$ by $\dfrac{P}{4}$ **51.** 2,000,000 m²

52. 500,000 m²   **53.** 4,166,666.7 m²   **54.** 3,125,000 m²

55. (a) $\dfrac{625}{16} \approx 39$ ft

(b) 219.5 ft

(c) 170 ft

(d)

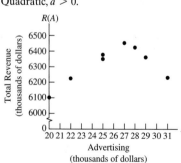

220
0
0   200

(f) When the height is 100 ft, the projectile is 135.7 ft from the cliff.

56. (a) 156.25 ft

(b) 78.125 ft

(c) 312.5 ft

(d)

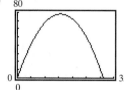

80
0
0   350

(f) When the height is 50 ft, the projectile has traveled horizontally 62.5 and 250 ft

57. 8 PM **58.** 5.3 miles **59.** 18.75 m **60.** $y = -\dfrac{1}{144}x^2 + 25$; $x = 10$, $y \approx 24.3$ ft; $x = 20$, $y \approx 22.2$ ft; $x = 40$, $y \approx 13.9$ ft **61.** 3 in.

62. At approximately 9:17 PM, the ships are approximately 11.04 nautical miles apart. **63.** Width $= \dfrac{40}{\pi + 4} \approx 5.6$ ft; length ≈ 2.8 ft

64. 375 m by 238.7 m **65.** $x = \dfrac{16}{6 - \sqrt{3}} \approx 3.75$ ft; other side ≈ 2.38 ft **66. (a)** $\dfrac{v_0^2}{128}$ ft **(b)** Maximum height increases by a multiple of 4

(c) 128 ft **67.** 70 members **68.** price $= \$28$ or $\$30$

69. $\left.\begin{array}{r} ah^2 - bh + c = y_0 \\ c = y_1 \\ ah^2 + bh + c = y_2 \end{array}\right\}$ $\left.\begin{array}{r} y_0 + y_2 = 2ah^2 + 2c \\ 4y_1 = 4c \end{array}\right\}$ Area $= \dfrac{h}{3}(2ah^2 + 6c) = \dfrac{h}{3}(y_0 + 4y_1 + y_2)$

70. If x is an even integer, so is x^2. The product ax^2 is even as is the product bx. Since c is odd, the sum $ax^2 + bx + c$ will be odd. If x is an odd integer, so is x^2. The product ax^2 is odd as is the product bx. Since c is odd, the sum $ax^2 + bx + c$ will be odd.

71. (a) Quadratic, $a < 0$.

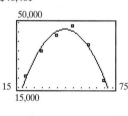

I

Median Income (dollars)

50,000
40,000
30,000
20,000
10,000
0

15-24 25-34 35-44 45-54 55-64 65+

Age

(b) 44.7 years old

(c) \$46,461

(e)

50,000

15   75
15,000

72. (a) Quadratic, $a > 0$.

R(A)

Total Revenue
(thousands of dollars)

6500
6400
6300
6200
6100
6000
0

20 21 22 23 24 25 26 27 28 29 30 31   *A*

Advertising
(thousands of dollars)

(b) Approximately \$26,500

(c) Approximately \$6,408,000

(e)

6460

20   32
6100

73. (a) Quadratic, $a > 0$.

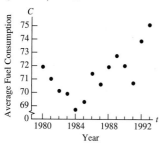

(b) 1984
(c) 67.3 billion gallons
(e)

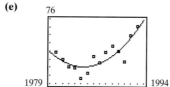

74. (a) Quadratic, $a < 0$.

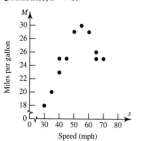

(b) Approximately 53.61 mph
(c) Approximately 24.81 miles per gallon
(e)

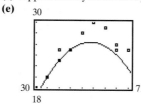

75. (a) Quadratic, $a < 0$.

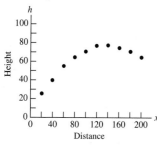

(b) 139.2 ft
(c) 77.4 ft
(e)

76. (a) Quadratic, $a < 0$.

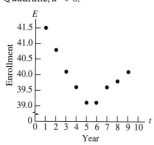

(b) During the academic year 1984–1985.
(d)

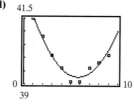

77. $x = \dfrac{a}{2}$

5.2 Exercises

1. Yes; degree 3   **2.** Yes; degree 4   **3.** Yes; degree 2   **4.** Yes; degree 1   **5.** No; x is raised to the -1 power.   **6.** Yes; degree 2

7. No; x is raised to the $\dfrac{3}{2}$ power.   **8.** No; x is raised to the $\dfrac{1}{2}$ power.   **9.** Yes; degree 4

10. No; it is the ratio of two polynomials and the polynomial in the denominator is of positive degree.

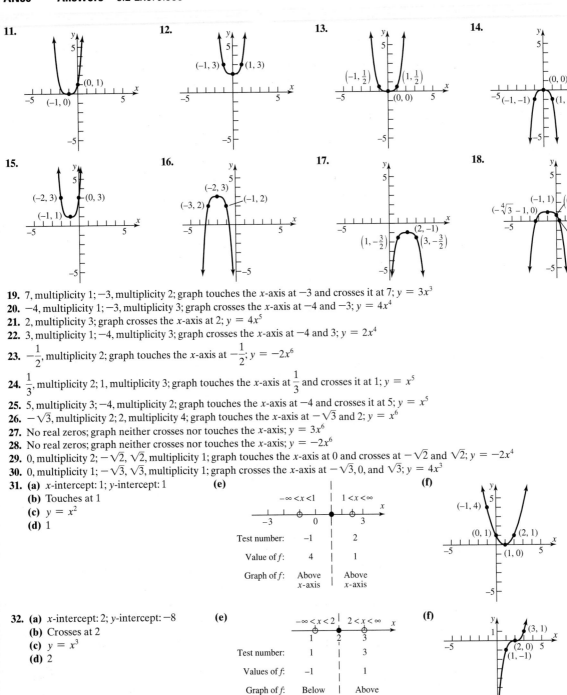

19. 7, multiplicity 1; -3, multiplicity 2; graph touches the x-axis at -3 and crosses it at 7; $y = 3x^3$

20. -4, multiplicity 1; -3, multiplicity 3; graph crosses the x-axis at -4 and -3; $y = 4x^4$

21. 2, multiplicity 3; graph crosses the x-axis at 2; $y = 4x^5$

22. 3, multiplicity 1; -4, multiplicity 3; graph crosses the x-axis at -4 and 3; $y = 2x^4$

23. $-\dfrac{1}{2}$, multiplicity 2; graph touches the x-axis at $-\dfrac{1}{2}$; $y = -2x^6$

24. $\dfrac{1}{3}$, multiplicity 2; 1, multiplicity 3; graph touches the x-axis at $\dfrac{1}{3}$ and crosses it at 1; $y = x^5$

25. 5, multiplicity 3; -4, multiplicity 2; graph touches the x-axis at -4 and crosses it at 5; $y = x^5$

26. $-\sqrt{3}$, multiplicity 2; 2, multiplicity 4; graph touches the x-axis at $-\sqrt{3}$ and 2; $y = x^6$

27. No real zeros; graph neither crosses nor touches the x-axis; $y = 3x^6$

28. No real zeros; graph neither crosses nor touches the x-axis; $y = -2x^6$

29. 0, multiplicity 2; $-\sqrt{2}$, $\sqrt{2}$, multiplicity 1; graph touches the x-axis at 0 and crosses at $-\sqrt{2}$ and $\sqrt{2}$; $y = -2x^4$

30. 0, multiplicity 1; $-\sqrt{3}$, $\sqrt{3}$, multiplicity 1; graph crosses the x-axis at $-\sqrt{3}$, 0, and $\sqrt{3}$; $y = 4x^3$

31. (a) x-intercept: 1; y-intercept: 1
 (b) Touches at 1
 (c) $y = x^2$
 (d) 1

32. (a) x-intercept: 2; y-intercept: -8
 (b) Crosses at 2
 (c) $y = x^3$
 (d) 2

33. (a) x-intercepts: $0, 3$; y-intercept: 0
(b) Touches at 0; crosses at 3
(c) $y = x^3$
(d) 2

(e)

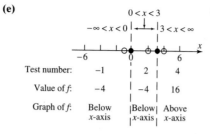

Test number:	-1	2	4
Value of f:	-4	-4	16
Graph of f:	Below x-axis	Below x-axis	Above x-axis

(f)

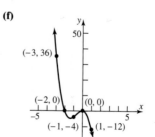

34. (a) x-intercepts: $-2, 0$; y-intercept: 0
(b) Touches at -2; crosses at 0
(c) $y = x^3$
(d) 2
(e)

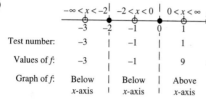

Test number:	-3	-1	1
Values of f:	-3	-1	9
Graph of f:	Below x-axis	Below x-axis	Above x-axis

(f)

35. (a) x-intercepts: $-4, 0$; y-intercept: 0
(b) Crosses at $-4, 0$
(c) $y = 6x^4$
(d) 3
(e)

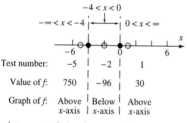

Test number:	-5	-2	1
Value of f:	750	-96	30
Graph of f:	Above x-axis	Below x-axis	Above x-axis

(f)

36. (a) x-intercepts: $0, 1$; y-intercept: 0
(b) Crosses at 0 and 1
(c) $y = 5x^4$
(d) 3
(e)

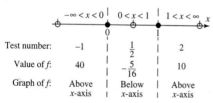

Test number:	-1	$\frac{1}{2}$	2
Value of f:	40	$-\frac{5}{16}$	10
Graph of f:	Above x-axis	Below x-axis	Above x-axis

(f)

37. (a) x-intercepts: $-2, 0$; y-intercept: 0
(b) Crosses at -2; touches at 0
(c) $y = -4x^3$
(d) 2
(e)

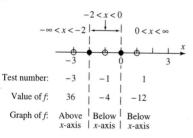

Test number:	-3	-1	1
Value of f:	36	-4	-12
Graph of f:	Above x-axis	Below x-axis	Below x-axis

(f)

38. (a) x-intercepts: $-4, 0$; y-intercept: 0
(b) Crosses at -4 and 0
(c) $y = -\dfrac{1}{2}x^4$
(d) 3
(e)

$-\infty < x < -4$	$-4 < x < 0$	$0 < x < \infty$

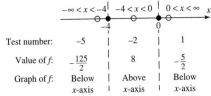

	$-\infty < x < -4$	$-4 < x < 0$	$0 < x < \infty$
Test number:	-5	-2	1
Value of f:	$-\dfrac{125}{2}$	8	$-\dfrac{5}{2}$
Graph of f:	Below x-axis	Above x-axis	Below x-axis

(f)

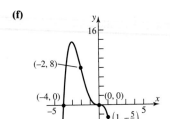

39. (a) x-intercepts: $-4, 0, 2$; y-intercept: 0
(b) Crosses at $-4, 0, 2$
(c) $y = x^3$
(d) 2
(e)

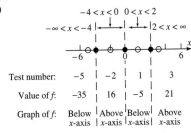

	$-\infty < x < -4$	$-4 < x < 0$	$0 < x < 2$	$2 < x < \infty$
Test number:	-5	-2	1	3
Value of f:	-35	16	-5	21
Graph of f:	Below x-axis	Above x-axis	Below x-axis	Above x-axis

(f)

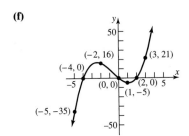

40. (a) x-intercepts: $-4, 0, 3$; y-intercept: 0
(b) Crosses at $-4, 0$, and 3
(c) $y = x^3$
(d) 2
(e)

	$-\infty < x < -4$	$-4 < x < 0$	$0 < x < 3$	$3 < x < \infty$
Test number:	-5	-2	1	4
Values of f:	-40	20	-10	32
Graph of f:	Below x-axis	Above x-axis	Below x-axis	Above x-axis

(f)

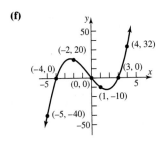

41. $f(x) = 4x - x^3 = -x(x^2 - 4)$
$\quad = -x(x + 2)(x - 2)$
(a) x-intercepts: $-2, 0, 2$; y-intercept: 0
(b) Crosses at $-2, 0, 2$
(c) $y = -x^3$
(d) 2
(e)

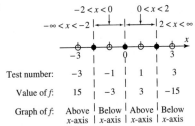

	$-\infty < x < -2$	$-2 < x < 0$	$0 < x < 2$	$2 < x < \infty$
Test number:	-3	-1	1	3
Value of f:	15	-3	3	-15
Graph of f:	Above x-axis	Below x-axis	Above x-axis	Below x-axis

(f)

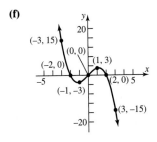

42. $f(x) = x - x^3 = -x(x^2 - 1)$
$\qquad = -x(x-1)(x+1)$
(a) x-intercepts: $-1, 0, 1$; y-intercept: 0
(b) Crosses at $-1, 0,$ and 1
(c) $y = -x^3$
(d) 2
(e)

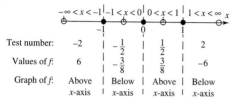

Test number:	-2	$-\frac{1}{2}$	$\frac{1}{2}$	2
Values of f:	6	$-\frac{3}{8}$	$\frac{3}{8}$	-6
Graph of f:	Above x-axis	Below x-axis	Above x-axis	Below x-axis

(f)

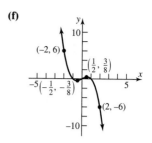

43. (a) x-intercepts: $-2, 0, 2$; y-intercept: 0
(b) Crosses at $-2, 2$; touches at 0
(c) $y = x^4$
(d) 3
(e)

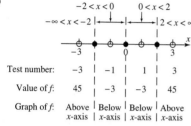

Test number:	-3	-1	1	3
Value of f:	45	-3	-3	45
Graph of f:	Above x-axis	Below x-axis	Below x-axis	Above x-axis

(f)

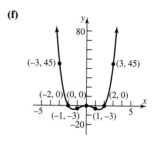

44. (a) x-intercepts: $-4, 0, 3$; y-intercept: 0
(b) Crosses at -4 and 3; touches at 0
(c) $y = x^4$
(d) 3
(e)

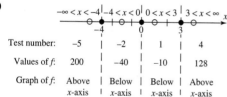

Test number:	-5	-2	1	4
Values of f:	200	-40	-10	128
Graph of f:	Above x-axis	Below x-axis	Below x-axis	Above x-axis

(f)

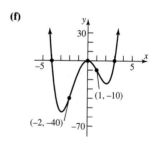

45. (a) x-intercepts: $0, 2$; y-intercept: 0
(b) Touches at $0, 2$
(c) $y = x^4$
(d) 3
(e)

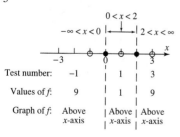

Test number:	-1	1	3
Values of f:	9	1	9
Graph of f:	Above x-axis	Above x-axis	Above x-axis

(f)

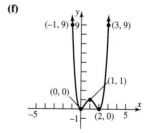

46. (a) x-intercepts: $0, 3$; y-intercept: 0
(b) Crosses at 0 and 3
(c) $y = x^4$
(d) 3
(e)

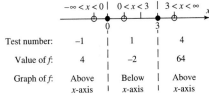

	$-\infty < x < 0$	$0 < x < 3$	$3 < x < \infty$
Test number:	-1	1	4
Value of f:	4	-2	64
Graph of f:	Above x-axis	Below x-axis	Above x-axis

(f)

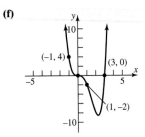

47. (a) x-intercepts: $-1, 0, 3$; y-intercept: 0
(b) Crosses at $-1, 3$; touches at 0
(c) $y = x^4$
(d) 3
(e)

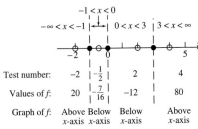

	$-\infty < x < -1$	$-1 < x < 0$	$0 < x < 3$	$3 < x < \infty$
Test number:	-2	$-\frac{1}{2}$	2	4
Values of f:	20	$\left|-\frac{7}{16}\right|$	-12	80
Graph of f:	Above x-axis	Below x-axis	Below x-axis	Above x-axis

(f)

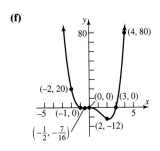

48. (a) x-intercepts: $0, 1$, and 3;
y-intercept: 0
(b) Touches at 0; crosses at 1 and 3
(c) $y = x^4$
(d) 3
(e)

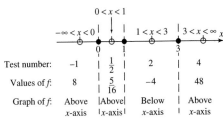

	$-\infty < x < 0$	$0 < x < 1$	$1 < x < 3$	$3 < x < \infty$
Test number:	-1	$\frac{1}{2}$	2	4
Values of f:	8	$\frac{5}{16}$	-4	48
Graph of f:	Above x-axis	Above x-axis	Below x-axis	Above x-axis

(f)

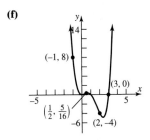

49. (a) x-intercepts: $-2, 0, 4, 6$;
y-intercept: 0
(b) Crosses at $-2, 0, 4, 6$
(c) $y = x^4$
(d) 3
(e)

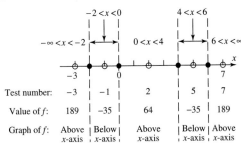

	$-\infty < x < -2$	$-2 < x < 0$	$0 < x < 4$	$4 < x < 6$	$6 < x < \infty$
Test number:	-3	-1	2	5	7
Value of f:	189	-35	64	-35	189
Graph of f:	Above x-axis	Below x-axis	Above x-axis	Below x-axis	Above x-axis

(f)

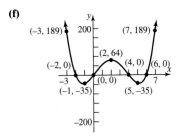

50. (a) x-intercepts: $-4, -2, 0, 2$; y-intercept: 0
(b) Crosses at $-4, -2, 0,$ and 2
(c) $y = x^4$
(d) 3
(e)

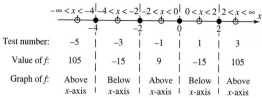

	$-\infty < x < -4$	$-4 < x < -2$	$-2 < x < 0$	$0 < x < 2$	$2 < x < \infty$
Test number:	-5	-3	-1	1	3
Value of f:	105	-15	9	-15	105
Graph of f:	Above x-axis	Below x-axis	Above x-axis	Below x-axis	Above x-axis

(f)

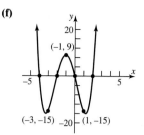

51. (a) x-intercepts: $0, 2$; y-intercept: 0
(b) Touches at 0; crosses at 2
(c) $y = x^5$
(d) 4
(e)

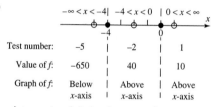

	$-\infty < x < 0$	$0 < x < 2$	$2 < x < \infty$
Test number:	-1	1	3
Value of f:	-12	-4	108
Graph of f:	Below x-axis	Below x-axis	Above x-axis

(f)

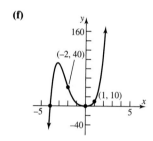

52. (a) x-intercepts: $-4, 0$; y-intercept: 0
(b) Crosses at -4; touches at 0
(c) $y = x^5$
(d) 4
(e)

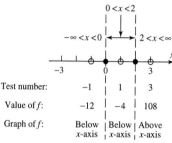

	$-\infty < x < -4$	$-4 < x < 0$	$0 < x < \infty$
Test number:	-5	-2	1
Value of f:	-650	40	10
Graph of f:	Below x-axis	Above x-axis	Above x-axis

(f)

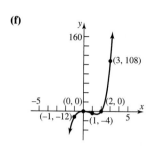

53. (a) x-intercepts: $-1, 0, 1$; y-intercept: 0
(b) Crosses at 1; touches at -1 and 0
(c) $y = -x^5$
(d) 4
(e)

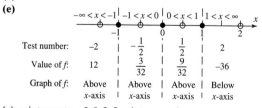

	$-\infty < x < -1$	$-1 < x < 0$	$0 < x < 1$	$1 < x < \infty$
Test number:	-2	$-\dfrac{1}{2}$	$\dfrac{1}{2}$	2
Value of f:	12	$\dfrac{3}{32}$	$\dfrac{9}{32}$	-36
Graph of f:	Above x-axis	Above x-axis	Above x-axis	Below x-axis

(f)

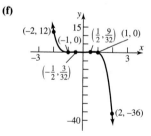

54. (a) x-intercepts: $-2, 0, 2, 5$; y-intercept: 0
(b) Touches at 0; crosses at $-2, 2,$ and 5
(c) $y = x^5$
(d) 4
(e)

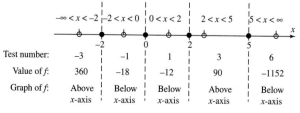

	$-\infty < x < -2$	$-2 < x < 0$	$0 < x < 2$	$2 < x < 5$	$5 < x < \infty$
Test number:	-3	-1	1	3	6
Value of f:	360	-18	-12	90	-1152
Graph of f:	Above x-axis	Below x-axis	Below x-axis	Above x-axis	Below x-axis

(f)

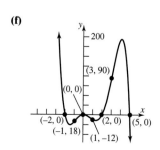

55. c, e, f **56.** c, e, f **57.** c, e **58.** d, e, f

59. x-intercepts: $-1.26, -0.20, 1.26$
Turning points: $(0.66, -0.99)$; $(-0.79, 0.56)$

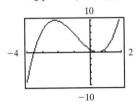

60. x-intercepts: $-2.16, 0.80, 2.16$
Turning points: $(-1.01, 6.60)$; $(1.54, -1.70)$

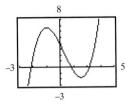

61. x-intercepts: $-3.56, 0.50$
Turning points: $(0.50, 0)$; $(-2.20, 9.91)$

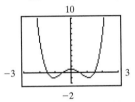

62. x-intercepts: $-0.90, 4.71$
Turning points: $(-0.90, 0)$; $(2.84, -26.16)$

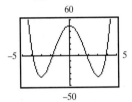

63. x-intercepts: $-1.50, -0.50, 0.50, 1.50$
Turning points: $(-1.11, -1)$, $(1.11, -1)$, $(0, 0.5625)$

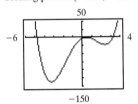

64. x-intercepts: $-3.90, -1.82, 1.82, 3.90$
Turning points: $(-3.04, -35.30)$; $(0, 50.26)$; $(3.04, -35.30)$

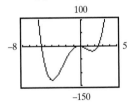

65. x-intercepts: $-4.78, 0.45, 3.23$
Turning points: $(-3.31, -135.91)$, $(2.37, -22.66)$; $(0.45, 0)$

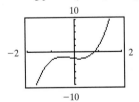

66. x-intercepts: $-5.41, -0.23, 2.42$
Turning points: $(-3.97, -128.71)$; $(-0.23, 0)$; $(1.61, -19.25)$

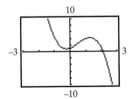

67. x-intercept: 0.83
Turning points: $(-0.50, -1.54)$; $(0.21, -2.12)$

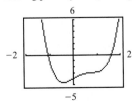

68. x-intercept: 2.10
Turning points: $(-0.23, 0.79)$; $(1.27, 4.17)$

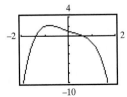

69. x-intercepts: $-1.06, 1.61$
Turning point: $(-0.41, -4.64)$

70. x-intercepts: $-1.47, 0.91$
Turning point: $(-0.81, 3.21)$

71. *x*-intercept: −0.97
No turning points

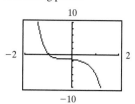

72. *x*-intercept: −0.71
No turning points

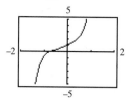

73. **(a)** Cubic, $a < 0$

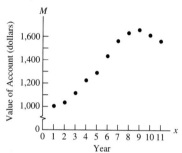

(b) ≈1370 motor vehicle thefts
(d)

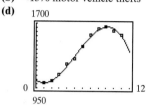

74. **(a)** Cubic, $a < 0$

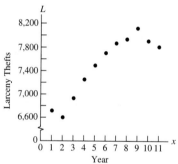

(b) 7396 thousand thefts
(d)

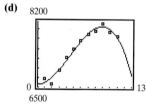

75. **(a)** Cubic, $a > 0$

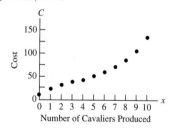

(b) 7 thousand dollars
(c) 20 thousand dollars
(d) ≈$155,000
(f)

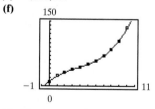

(g) Fixed costs of $10,200.

76. **(a)** Cubic, $a > 0$

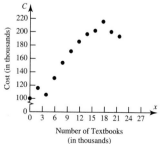

(b) Approximately 3.17 dollars/textbook
(c) 1.85 dollars/textbook
(d) $171,040
(f)

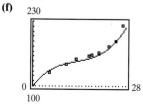

(g) Fixed costs of $100,000.

82. a, b, c, d

5.3 Exercises

1. All real numbers except 3.   **2.** All real numbers except -3.   **3.** All real numbers except 2 and -4.

4. All real numbers except -3 and 4.   **5.** All real numbers except $-\dfrac{1}{2}$ and 3.   **6.** All real numbers except $\dfrac{1}{3}$ and -2.

7. All real numbers except 2.   **8.** All real numbers except -1 and 1.   **9.** All real numbers   **10.** All real numbers

11. All real numbers except -3 and 3.   **12.** All real numbers except -2.

13. (a) Domain: $\{x \mid x \neq 2\}$; Range: $\{y \mid y \neq 1\}$ **(b)** $(0,0)$ **(c)** $y = 1$ **(d)** $x = 2$ **(e)** None

14. (a) Domain: $\{x \mid x \neq -1\}$; Range: $\{y \mid y > 0\}$ **(b)** $(0,2)$ **(c)** $y = 0$ **(d)** $x = -1$ **(e)** None

15. (a) Domain: $\{x \mid x \neq 0\}$; Range: all real numbers   **(b)** $(-1,0), (1,0)$ **(c)** None   **(d)** None   **(e)** $y = 2x$

16. (a) Domain: $\{x \mid x \neq 0\}$; Range: $\{y \mid y \leq -2, y \geq 2\}$ **(b)** None   **(c)** None   **(d)** $x = 0$ **(e)** $y = -x$

17. (a) Domain: $\{x \mid x \neq -2, x \neq 2\}$; Range: $\{y \mid y \leq 0, y > 1\}$ **(b)** $(0,0)$ **(c)** $y = 1$ **(d)** $x = -2, x = 2$ **(e)** None

18. (a) Domain: $\{x \mid x \neq -1, x \neq 1\}$; Range: All real numbers   **(b)** $(0,0)$ **(c)** $y = 0$ **(d)** $x = -1, x = 1$ **(e)** None

19.
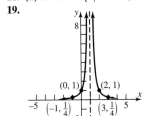

20.
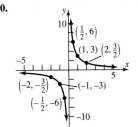

21.
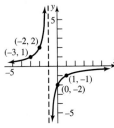

22.
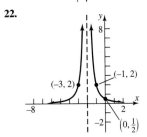

23.
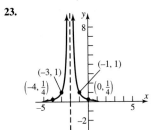

24.
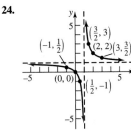

25.
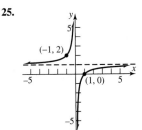

26.
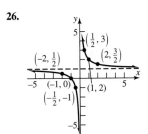

27.
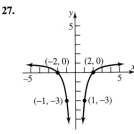

28.
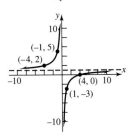

29.
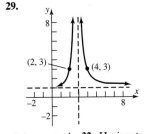

30.
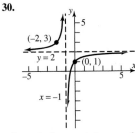

31. Horizontal asymptote: $y = 3$; vertical asymptote: $x = -4$ **32.** Horizontal asymptote: $y = 3$; vertical asymptote: $x = 6$

33. No asymptotes   **34.** Oblique asymptote: $y = -x + 5$; vertical asymptote: $x = -5$

35. Horizontal asymptote: $y = 0$; vertical asymptotes: $x = 1, x = -1$ **36.** Vertical asymptote: $x = 1$

37. Horizontal asymptote: $y = 0$; vertical asymptote: $x = 0$

38. Horizontal asymptote: $y = 0$; vertical asymptote: $x = 0, x = -2$ **39.** Oblique asymptote: $y = 3x$; vertical asymptote: $x = 0$

40. Vertical asymptotes: $x = -\dfrac{1}{3}, x = 2$; horizontal asymptote: $y = 2$

41. Oblique asymptote: $y = -x - 1$; vertical asymptote: $x = 0, x = 1$

42. Vertical asymptotes: $x = -1, x = 0$, and $x = 1$; horizontal asymptote: $y = 0$

43. 1. Domain: $\{x|x \neq 0, x \neq -4\}$
    2. x-intercept: -1; no y-intercept
    3. No symmetry
    4. Vertical asymptotes: $x = 0, x = -4$
    5. Horizontal asymptote: $y = 0$, intersected at $(-1, 0)$
    6. $x < -4$: below x-axis
      $-4 < x < -1$: above x-axis
      $-1 < x < 0$: below x-axis
      $x > 0$: above x-axis

7.
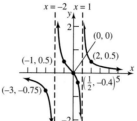

44. 1. Domain: $\{x|x \neq 1, x \neq -2\}$
    2. x-intercept: 0; y-intercept: 0
    3. No symmetry
    4. Vertical asymptotes: $x = 1, x = -2$
    5. Horizontal asymptote: $y = 0$, intersected at $(0, 0)$
    6. $x < -2$: below x-axis
      $-2 < x < 0$: above x-axis
      $0 < x < 1$: below x-axis
      $x > 1$: above x-axis

7.
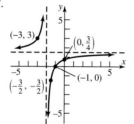

45. 1. Domain: $\{x|x \neq -2\}$
    2. x-intercept: -1; y-intercept: $\dfrac{3}{4}$
    3. No symmetry
    4. Vertical asymptote: $x = -2$
    5. Horizontal asymptote: $y = \dfrac{3}{2}$, not intersected
    6. $x < -2$: above x-axis
      $-2 < x < -1$: below x-axis
      $x > -1$: above x-axis

7.
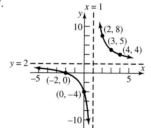

46. 1. Domain: $\{x|x \neq 1\}$
    2. x-intercept: -2; y-intercept: -4
    3. No symmetry
    4. Vertical asymptotes: $x = 1$
    5. Horizontal asymptote: $y = 2$, not intersected
    6. $x < -2$: above x-axis
      $-2 < x < 1$: below x-axis
      $x > 1$: above x-axis

7.
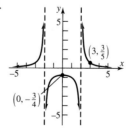

47. 1. Domain: $\{x|x \neq -2, x \neq 2\}$
    2. No x-intercept; y-intercept: $-\dfrac{3}{4}$
    3. Symmetric with respect to y-axis
    4. Vertical asymptotes: $x = 2, x = -2$
    5. Horizontal asymptote: $y = 0$, not intersected
    6. $x < -2$: above x-axis
      $-2 < x < 2$: below x-axis
      $x > 2$: above x-axis

48. 1. Domain: $\{x|x \neq -2, x \neq 3\}$
    2. No x-intercept; y-intercept: -1
    3. No symmetry
    4. Vertical asymptotes: $x = -2, x = 3$
    5. Horizontal asymptotes: $y = 0$, not intersected
    6. $x < -2$: above x-axis
      $-2 < x < 3$: below x-axis
      $x > 3$: above x-axis

7.
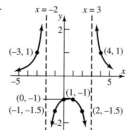

49. 1. Domain: $\{x \mid x \neq -1, x \neq 1\}$
2. No x-intercept; y-intercept: -1
3. Symmetric with respect to y-axis
4. Vertical asymptotes: $x = -1$, $x = 1$
5. No horizontal or oblique asymptotes
6. $x < -1$: above x-axis
$-1 < x < 1$: below x-axis
$x > 1$: above x-axis

7.
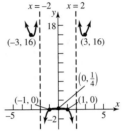

50. 1. Domain: $\{x \mid x \neq -2, x \neq 2\}$
2. x-intercepts: $1, -1$; y-intercept: $\dfrac{1}{4}$
3. Symmetric with respect to y-axis
4. Vertical asymptotes: $x = -2$, $x = 2$
5. No horizontal or oblique asymptotes
6. $x < -2$: above x-axis
$-2 < x < -1$: below x-axis
$-1 < x < 1$: above x-axis
$1 < x < 2$: below x-axis
$x > 2$: above x-axis

7.
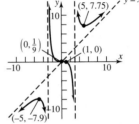

51. 1. Domain: $\{x \mid x \neq -3, x \neq 3\}$
2. x-intercept: 1; y-intercept: $\dfrac{1}{9}$
3. No symmetry
4. Vertical asymptotes: $x = 3$, $x = -3$
5. Oblique asymptote: $y = x$, intersected at $\left(\dfrac{1}{9}, \dfrac{1}{9}\right)$
6. $x < -3$: below x-axis
$-3 < x < 1$: above x-axis
$1 < x < 3$: below x-axis
$x > 3$: above x-axis

7.
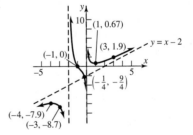

52. 1. Domain: $\{x \mid x \neq 0, x \neq -2\}$
2. x-intercept: -1; no y-intercept
3. No symmetry
4. Vertical asymptotes: $x = 0$, $x = -2$
5. Oblique asymptote: $y = x - 2$, intersected at $\left(-\dfrac{1}{4}, -\dfrac{9}{4}\right)$
6. $x < -2$: below x-axis
$-2 < x < -1$: above x-axis
$-1 < x < 0$: below x-axis
$x > 0$: above x-axis

7.
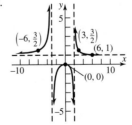

53. 1. Domain: $\{x \neq -3, x \neq 2\}$
2. Intercept: $(0, 0)$
3. No symmetry
4. Vertical asymptotes: $x = 2$, $x = -3$
5. Horizontal asymptote: $y = 1$, intersected at $(6, 1)$
6. $x < -3$: above x-axis
$-3 < x < 0$: below x-axis
$0 < x < 2$: below x-axis
$x > 2$: above x-axis

7.

54. 1. Domain: $\{x|x \neq -2, x \neq 2\}$
2. x-intercepts: $-4, 3$; y-intercepts: 3
3. No symmetry
4. Vertical asymptotes: $x = -2, x = 2$
5. Horizontal asymptote: $y = 1$, not intersected
6. $x < -4$: above x-axis
   $-4 < x < -2$: below x-axis
   $-2 < x < 2$: above x-axis
   $2 < x < 3$: below x-axis
   $x > 3$: above x-axis

7.
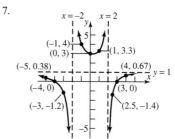

55. 1. Domain: $\{x|x \neq -2, x \neq 2\}$
2. Intercept: $(0, 0)$
3. Symmetry with respect to origin
4. Vertical asymptotes: $x = -2, x = 2$
5. Horizontal asymptote: $y = 0$, intersected at $(0, 0)$
6. $x < -2$: below x-axis
   $-2 < x < 0$: above x-axis
   $0 < x < 2$: below x-axis
   $x > 2$: above x-axis

7.
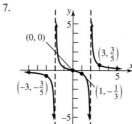

56. 1. Domain: $\{x|x \neq -1, x \neq 1\}$
2. x-intercept: 0; y-intercept: 0
3. Symmetric with respect to the origin
4. Vertical asymptotes: $x = 1, x = -1$
5. Horizontal asymptote: $y = 0$, intersected at $(0, 0)$
6. $x < -1$: below x-axis
   $-1 < x < 0$: above x-axis
   $0 < x < 1$: below x-axis
   $x > 1$: above x-axis

7.
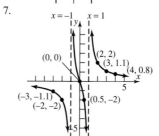

57. 1. Domain: $\{x|x \neq 1, x \neq -2, x \neq 2\}$
2. No x-intercept; y-intercept: $\dfrac{3}{4}$
3. No symmetry
4. Vertical asymptotes: $x = -2, x = 1, x = 2$
5. Horizontal asymptote: $y = 0$, not intersected
6. $x < -2$: below x-axis
   $-2 < x < 1$: above x-axis
   $1 < x < 2$: below x-axis
   $x > 2$: above x-axis

7.
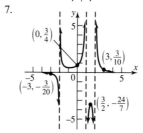

58. 1. Domain: $\{x|x \neq -3, x \neq -1, x \neq 3\}$
2. No x-intercept; y-intercept: $\dfrac{4}{9}$
3. No symmetry
4. Vertical asymptotes: $x = -1, x = 3, x = -3$
5. Horizontal asymptote: $y = 0$, not intersected
6. $x < -3$: above x-axis
   $-3 < x < -1$: below x-axis
   $-1 < x < 3$: above x-axis
   $x > 3$: below x-axis

7.
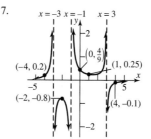

59. 1. Domain: $\{x|x \neq -2, x \neq 2\}$
2. x-intercepts: $-1, 1$; y-intercept: $\dfrac{1}{4}$
3. Symmetric with respect to y-axis
4. Vertical asymptotes: $x = -2, x = 2$
5. Horizontal asymptote: $y = 0$, intersected at $(-1, 0)$ and $(1, 0)$
6. $x < -2$: above x-axis
   $-2 < x < -1$: below x-axis
   $-1 < x < 1$: above x-axis
   $1 < x < 2$: below x-axis
   $x > 2$: above x-axis

7.
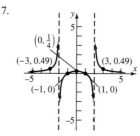

60. 1. Domain: $\{x|x \neq -1, x \neq 1\}$
2. No x-intercept; y-intercept: -4
3. Symmetric about the y-axis
4. Vertical asymptotes: $x = 1$, $x = -1$
5. Horizontal asymptote: $y = 0$, not intersected
6. $x < -1$: above x-axis
$-1 < x < 1$: below x-axis
$x > 1$: above x-axis

7.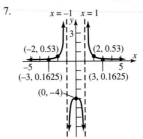

61. 1. Domain: $\{x|x \neq -2\}$
2. x-intercepts: -1, 4; y-intercept: -2
3. No symmetry
4. Vertical asymptote: $x = -2$
5. Oblique asymptote: $y = x - 5$, not intersected
6. $x < -2$: below x-axis
$-2 < x < -1$: above x-axis
$-1 < x < 4$: below x-axis
$x > 4$: above x-axis

7.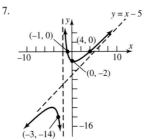

62. 1. Domain: $\{x|x \neq 1\}$
2. x-intercepts: -2, -1; y-intercepts: -2
3. No symmetry
4. Vertical asymptote: $x = 1$
5. Oblique asymptote: $y = x + 4$, not intersected
6. $x < -2$: below x-axis
$-2 < x < -1$: above x-axis
$-1 < x < 1$: below x-axis
$x > 1$: above x-axis

7.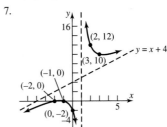

63. 1. Domain: $\{x|x \neq 4\}$
2. x-intercepts: -4, 3; y-intercept: 3
3. No symmetry
4. Vertical asymptote: $x = 4$
5. Oblique asymptote: $y = x + 5$, not intersected
6. $x < -4$: below x-axis
$-4 < x < 3$: above x-axis
$3 < x < 4$: below x-axis
$x > 4$: above x-axis

7.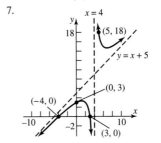

64. 1. Domain: $\{x|x \neq -5\}$
2. x-intercepts: 4, -3; y-intercept: $-\dfrac{12}{5}$
3. No symmetry
4. Vertical asymptote: $x = -5$
5. Oblique asymptote: $y = x - 6$, not intersected
6. $x < -5$: below x-axis
$-5 < x < -3$: above x-axis
$-3 < x < 4$: below x-axis
$x > 4$: above x-axis

7.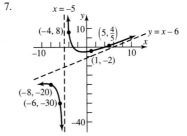

65. 1. Domain: $\{x|x \neq -2\}$
2. x-intercepts: -4, 3; y-intercept: -6
3. No symmetry
4. Vertical asymptote: $x = -2$
5. Oblique asymptote: $y = x - 1$, not intersected
6. $x < -4$: below x-axis
$-4 < x < -2$: above x-axis
$-2 < x < 3$: below x-axis
$x > 3$: above x-axis

7.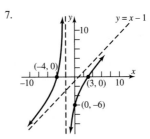

66. 1. Domain: $\{x|x \neq -1\}$
2. x-intercepts: $4, -3$; y-intercept: -12
3. No symmetry
4. Vertical asymptote: $x = -1$
5. Oblique asymptote: $y = x - 2$, not intersected
6. $x < -3$: below x-axis
   $-3 < x < -1$: above x-axis
   $-1 < x < 4$: below x-axis
   $x > 4$: above x-axis

7.

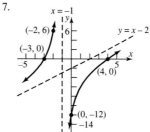

67. 1. Domain: $\{x|x \neq -3\}$
2. x-intercepts: $0, 1$; y-intercept: 0
3. No symmetry
4. Vertical asymptote: $x = -3$
5. Horizontal asymptote: $y = 1$, not intersected
6. $x < -3$: above x-axis
   $-3 < x < 0$: below x-axis
   $0 < x < 1$: above x-axis
   $x > 1$: above x-axis

7.

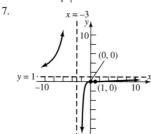

68. 1. Domain: $\{x|x \neq 0, x \neq 4\}$
2. x-intercepts: $1, -2, 3$; no y-intercept
3. No symmetry
4. Vertical asymptotes: $x = 0$; $x = 4$
5. Horizontal asymptote: $y = 1$,
   intersected at $\left(\dfrac{7 + \sqrt{33}}{4}, 1\right)$ and $\left(\dfrac{7 - \sqrt{33}}{4}, 1\right)$
6. $x < -2$: above x-axis
   $-2 < x < 0$: below x-axis
   $0 < x < 1$: above x-axis
   $1 < x < 3$: below x-axis
   $3 < x < 4$: above x-axis
   $x > 4$: above x-axis

7.

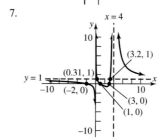

69. 1. Domain: $\{x|x \neq -2, x \neq 3\}$
2. x-intercept: -4; y-intercept: 2
3. No symmetry
4. Vertical asymptote: $x = -2$; hole at $\left(3, \dfrac{7}{5}\right)$
5. Horizontal asymptote: $y = 1$, not intersected
6. $x < -4$: above x-axis
   $-4 < x < -2$: below x-axis
   $-2 < x < 3$: above x-axis
   $x > 3$: above x-axis

7.

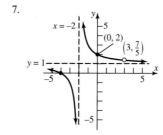

70. 1. Domain: $\{x|x \neq -3, x \neq -5\}$
2. x-intercept: 2; y-intercept: $-\dfrac{2}{3}$
3. No symmetry
4. Vertical asymptote: $x = -3$
5. Horizontal asymptote: $y = 1$, not intersected
6. $x < -3$: above x-axis
   $-3 < x < 2$: below x-axis
   $x > 2$: above x-axis

7.

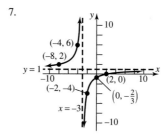

71. 1. Domain: $\left\{ x \,\middle|\, x \neq \dfrac{3}{2}, x \neq 2 \right\}$

2. x-intercept: $-\dfrac{1}{3}$; y-intercept: $-\dfrac{1}{2}$

3. No symmetry

4. Vertical asymptote: $x = 2$; hole at $\left(\dfrac{3}{2}, -11 \right)$

5. Horizontal asymptote: $y = 3$, not intersected

6. $x < -\dfrac{1}{3}$: above x-axis

$-\dfrac{1}{3} < x < \dfrac{3}{2}$: below x-axis

$\dfrac{3}{2} < x < 2$: below x-axis

$x > 2$: above x-axis

7.

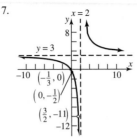

72. 1. Domain: $\left\{ x \,\middle|\, x \neq -\dfrac{5}{2}, x \neq 3 \right\}$

2. x-intercept: $-\dfrac{3}{4}$; y-intercept: -1

3. No symmetry

4. Vertical asymptote: $x = 3$; hole at $\left(-\dfrac{5}{2}, \dfrac{14}{11} \right)$

5. Horizontal asymptote: $y = 4$, not intersected

6. $x < -\dfrac{5}{2}$: above x-axis

$-\dfrac{5}{2} < x < -\dfrac{3}{4}$: above x-axis

$-\dfrac{3}{4} < x < 3$: below x-axis

$x > 3$: above x-axis

7.

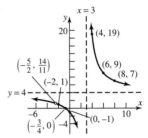

73. 4 must be a zero of the denominator, hence, $x - 4$ must be a factor.   **74.** If $y = 2$ is a horizontal asymptote, then the result of long division is a constant. This is true when the degree of the numerator equals the degree of the denominator.   **75.** c, d   **76.** b, e

77. (a) 9.82 m/sec^3 **(b)** 9.8195 m/sec^2 **(c)** 9.7936 m/sec^2 **(d)** h-axis   **(e)**

78. (a) 25   **(b)** Approximately 596   **(c)** 2500   **(d)**

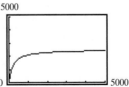

79. (a) t-axis; $C(t) \to 0$ **(b)**   **(c)** 0.70 hours

80. (a) Horizontal asymptote: $y = 0$; the concentration C approaches 0 as t increases. **(b)**
(c) The concentration is highest 5 minutes after injection.

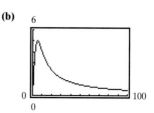

81. (a) $\overline{C}(x) = \dfrac{0.2156x^3 - 2.3473x^2 + 14.3275x + 10.2238}{x}$ **(b)** $\overline{C}(6) = \$9709$ **(c)** $\overline{C}(9) = \$11{,}801$ **(d)**

(e) 6 **(f)** \$9709

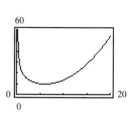

82. (a) $\dfrac{0.0155x^3 - 0.5951x^2 + 9.1502x + 98.4327}{x}$ **(d)**

(b) Approximately \$11.61 **(c)** Approximately \$7.90
(e) Approximately 25,000 books **(f)** Approximately \$7.90

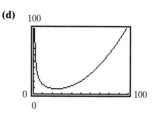

83. (a) $S(x) = 2x^2 + \dfrac{40{,}000}{x}$ **(b)**

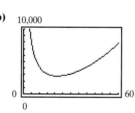

(c) 2784.95 sq in. **(d)** 21.54 in. × 21.54 in. × 21.54 in.

(e) To minimize the cost of material needed for construction.

84. (a) $A(x) = 2x^2 + \dfrac{20{,}000}{x}$ **(b)**

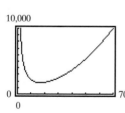

(c) 1754.41 sq in. **(d)** 17.10 in. × 17.10 in. × 17.10 in.
(e) To minimize the cost of material needed for construction.

85. (a) $C(r) = 12\pi r^2 + \dfrac{4000}{r}$ **(b)** The cost is least for r about 3.75 cm

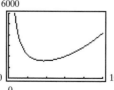

86. (a) $A(r) = 2\pi r^2 + \dfrac{200}{r}$ **(b)** Approximately 123.22 sq ft **(e)**
(c) Approximately 150.53 sq ft **(d)** Approximately 197.08 sq ft

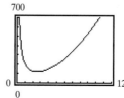

A is smallest when $r = 2.52$ ft.

87. No. Each of the functions is a quotient of polynomials, but not written in lowest terms. Each function is defined for $x = 1$.

88. All four graphs have a vertical asymptote at $x = 1$, $y = \dfrac{x^2}{x - 1}$ has an oblique asymptote as $y = x$.

89. Minimum value: 2.00 at $x = 1.00$

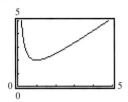

90. Minimum value: 8.48 at $x = 2.12$

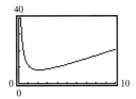

91. Minimum value: 1.88 at $x = 0.79$

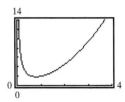

92. Minimum value: 10.30 at $x = 1.31$

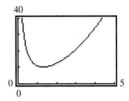

93. Minimum value: 1.75 at $x = 1.31$

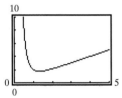

94. Minimum value: 5.11 at $x = 1.91$

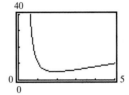

96. No

5.4 Exercises

1. $q(x) = x^2 + x + 4$; $R = 12$ **2.** $q(x) = x^2 + x - 4$; $R = 5$ **3.** $q(x) = 3x^2 + 11x + 32$; $R = 99$
4. $q(x) = -4x^2 + 10x - 21$; $R = 43$ **5.** $q(x) = x^4 - 3x^3 + 5x^2 - 15x + 46$; $R = -138$ **6.** $q(x) = x^3 + 2x^2 + 5x + 10$; $R = 22$
7. $q(x) = 4x^5 + 4x^4 + x^3 + x^2 + 2x + 2$; $R = 7$ **8.** $q(x) = x^4 - x^3 + 6x^2 - 6x + 6$; $R = -16$
9. $q(x) = 0.1x^2 - 0.11x + 0.321$; $R = -0.3531$ **10.** $q(x) = 0.1x - 0.21$; $R = 0.241$ **11.** $q(x) = x^4 + x^3 + x^2 + x + 1$; $R = 0$
12. $q(x) = x^4 - x^3 + x^2 - x + 1$; $R = 0$ **13.** No   **14.** No   **15.** Yes   **16.** Yes   **17.** Yes   **18.** Yes   **19.** No   **20.** Yes   **21.** Yes
22. No

Historical Problems

1.
$$\left(x - \frac{b}{3}\right)^3 + b\left(x - \frac{b}{3}\right)^2 + c\left(x - \frac{b}{3}\right) + d = 0$$
$$x^3 - bx^2 + \frac{b^2 x}{3} - \frac{b^3}{27} + bx^2 - \frac{2b^2 x}{3} + \frac{b^3}{9} + cx - \frac{bc}{3} + d = 0$$
$$x^3 + \left(c - \frac{b^2}{3}\right)x + \left(\frac{2b^3}{27} - \frac{bc}{3} + d\right) = 0$$
Let $p = c - \dfrac{b^2}{3}$, $q = \dfrac{2b^3}{27} - \dfrac{bc}{3} + d$, then $x^3 + px + q = 0$

2.
$$(H + K)^3 + p(H + K) + q = 0$$
$$H^3 + 3H^2 K + 3HK^2 + K^3 + pH + pK + q = 0$$
Let $3HK = -p$:
$$H^3 - pH - pK + K^3 + pH + pK + q = 0$$
$$H^3 + K^3 = -q$$

3.
$$3HK = -p$$
$$K = -\frac{p}{3H}$$
$$H^3 + \left(-\frac{p}{3H}\right)^3 = -q$$
$$H^3 - \frac{p^3}{27H^3} = -q$$
$$27H^6 - p^3 = -27qH^3$$
$$27H^6 + 27qH^3 - p^3 = 0$$
$$H^3 = \frac{-27q \pm \sqrt{(27q)^2 - 4(27)(-p^3)}}{2 \cdot 27}$$
$$H^3 = \frac{-q}{2} \pm \sqrt{\frac{27^2 q^2}{2^2(27^2)} + \frac{4(27)p^3}{2^2(27^2)}}$$
$$H^3 = \frac{-q}{2} \pm \sqrt{\frac{q^2}{4} + \frac{p^3}{27}} \qquad \text{Choose the positive root for now.}$$
$$H = \sqrt[3]{\frac{-q}{2} + \sqrt{\frac{q^2}{4} + \frac{p^3}{27}}}$$

4. $H^3 + K^3 = -q$
$$K^3 = -q - H^3$$
$$K^3 = -q - \left[\frac{-q}{2} + \sqrt{\frac{q^2}{4} + \frac{p^3}{27}}\right]$$
$$K^3 = -\frac{q}{2} - \sqrt{\frac{q^2}{4} + \frac{p^3}{27}}$$
$$K = \sqrt[3]{-\frac{q}{2} - \sqrt{\frac{q^2}{4} + \frac{p^3}{27}}}$$

5. $x = H + K$
$$x = \sqrt[3]{\frac{-q}{2} + \sqrt{\frac{q^2}{4} + \frac{p^3}{27}}} + \sqrt[3]{\frac{-q}{2} - \sqrt{\frac{q^2}{4} + \frac{p^3}{27}}} \text{ (Note that if we had used the negative root in 3, the result would be the same.)}$$

6. $x = 3$ **7.** $x = 2$ **8.** $x = 2$

5.5 Exercises

1. No; $f(2) = 8$ **2.** No; $f(-3) = 161$ **3.** Yes; $f(2) = 0$ **4.** Yes; $f(2) = 0$ **5.** Yes; $f(-3) = 0$ **6.** Yes; $f(-3) = 0$

7. No; $f(-4) = 1$ **8.** Yes; $f(-4) = 0$ **9.** Yes; $f\left(\frac{1}{2}\right) = 0$ **10.** No; $f\left(-\frac{1}{3}\right) = 2$ **11.** 7; 3 or 1 positive; 2 or 0 negative

12. 4; 1 positive; 1 negative **13.** 6; 2 or 0 positive; 2 or 0 negative **14.** 5; 1 positive; 0 negative **15.** 3; 2 or 0 positive; 1 negative
16. 3, 1 positive; 2 or 0 negative **17.** 4; 2 or 0 positive; 2 or 0 negative **18.** 4; 1 positive; 1 negative **19.** 5; 0 positive; 3 or 1 negative

20. 5; 5, 3 or 1 positive; 0 negative **21.** 6; 1 positive, 1 negative **22.** 6; no positive; no negative **23.** $\pm 1, \pm\frac{1}{3}$ **24.** $\pm 1, \pm 3$ **25.** $\pm 1, \pm 3$

26. $\pm 1, \pm\frac{1}{2}$ **27.** $\pm 1, \pm 2, \pm\frac{1}{4}, \pm\frac{1}{2}$ **28.** $\pm 1, \pm 2, \pm\frac{1}{2}, \pm\frac{1}{3}, \pm\frac{1}{6}, \pm\frac{2}{3}$ **29.** $\pm 1, \pm 3, \pm 9, \pm\frac{1}{2}, \pm\frac{1}{3}, \pm\frac{1}{6}, \pm\frac{3}{2}, \pm\frac{9}{2}$

30. $\pm 1, \pm 2, \pm 3, \pm 6, \pm\frac{1}{4}, \pm\frac{1}{2}, \pm\frac{3}{4}, \pm\frac{3}{2}$ **31.** $\pm 1, \pm 2, \pm 3, \pm 4, \pm 6, \pm 12, \pm\frac{1}{2}, \pm\frac{3}{2}$ **32.** $\pm 1, \pm 2, \pm 3, \pm 6, \pm 9, \pm 18, \pm\frac{1}{3}, \pm\frac{2}{3}$

33. $\pm 1, \pm 2, \pm 4, \pm 5, \pm 10, \pm 20, \pm\frac{1}{2}, \pm\frac{5}{2}, \pm\frac{1}{3}, \pm\frac{2}{3}, \pm\frac{4}{3}, \pm\frac{5}{3}, \pm\frac{10}{3}, \pm\frac{20}{3}, \pm\frac{1}{6}, \pm\frac{5}{6}$ **34.** $\pm 1, \pm 2, \pm 5, \pm 10, \pm\frac{1}{2}, \pm\frac{1}{3}, \pm\frac{1}{6}, \pm\frac{2}{3}, \pm\frac{5}{3}, \pm\frac{5}{2}, \pm\frac{5}{6}, \pm\frac{10}{3}$

35. $-3, -1, 2; f(x) = (x + 3)(x + 1)(x - 2)$ **36.** $-4, -5, 1; f(x) = (x - 1)(x + 5)(x + 4)$ **37.** $\frac{1}{2}; f(x) = 2\left(x - \frac{1}{2}\right)(x^2 + 1)$

38. $-\frac{1}{2}; f(x) = 2\left(x + \frac{1}{2}\right)(x^2 + 1)$ **39.** $-1, 1; f(x) = (x + 1)(x - 1)(x^2 + 2)$ **40.** $2, -2; f(x) = (x - 2)(x + 2)(x^2 + 1)$

41. $-\frac{1}{2}, \frac{1}{2}; f(x) = 4\left(x + \frac{1}{2}\right)\left(x - \frac{1}{2}\right)(x^2 + 2)$ **42.** $\frac{1}{2}, -\frac{1}{2}; f(x) = (2x - 1)(2x + 1)(x^2 + 4)$

43. $-2, -1, 1, 1; f(x) = (x + 2)(x + 1)(x - 1)^2$ **44.** $-1, -2, 2, 2; f(x) = (x + 1)(x + 2)(x - 2)^2$

45. $-\sqrt{2}/2, \sqrt{2}/2, 2; f(x) = 4(x + \sqrt{2}/2)(x - \sqrt{2}/2)(x - 2)\left(x^2 + \frac{1}{2}\right)$

46. $-3, -\frac{\sqrt{2}}{2}, \frac{\sqrt{2}}{2}; f(x) = 4(x + 3)\left(x + \frac{\sqrt{2}}{2}\right)\left(x - \frac{\sqrt{2}}{2}\right)\left(x^2 + \frac{1}{2}\right)$ **47.** $\{-1, 2\}$ **48.** $-\frac{3}{2}$ **49.** $\left\{\frac{2}{3}, -1 + \sqrt{2}, -1 - \sqrt{2}\right\}$

50. $\dfrac{5}{2}$ **51.** $\left\{\dfrac{1}{3}, \sqrt{5}, -\sqrt{5}\right\}$ **52.** $\left\{-\dfrac{1}{2}, 2, 4\right\}$ **53.** $\{-3, -2\}$ **54.** 1 **55.** $-\dfrac{1}{3}$ **56.** $\dfrac{1}{2}$ **57.** $\left\{\dfrac{1}{2}, 2, 5\right\}$ **58.** $\left\{2, -\dfrac{1}{2}, -4\right\}$

59. $(-3, 0)$; $(-1, 0)$; $(2, 0)$; $(0, -6)$
$-\infty < x < -3$, $f(-4) = -18$, below x-axis
$-3 < x < -1$, $f(-2) = 4$, above x-axis
$-1 < x < 2$, $f(0) = -6$, below x-axis
$2 < x < \infty$, $f(3) = 24$, above x-axis

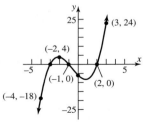

60. $(0, -20)$; $(1, 0)$; $(-5, 0)$; $(-4, 0)$
$-\infty < x < -5$, $f(-6) = -14$, below x-axis
$-5 < x < -4$, $f(-4.5) = 1.375$, above x-axis
$-4 < x < 1$, $f(0) = -20$, below x-axis
$1 < x < \infty$, $f(2) = 42$, above x-axis

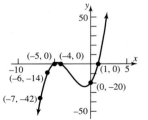

61. $\left(\dfrac{1}{2}, 0\right)$; $(0, -1)$
$-\infty < x < \dfrac{1}{2}$, $f(0) = -1$, below x-axis
$\dfrac{1}{2} < x < \infty$, $f(1) = 2$, above x-axis

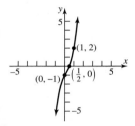

62. $\left(-\dfrac{1}{2}, 0\right)$; $(0, 1)$
$-\infty < x < -\dfrac{1}{2}$, $f(-1) = -2$, below x-axis
$-\dfrac{1}{2} < x < \infty$, $f(0) = 1$, above x-axis

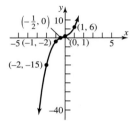

63. $(-1, 0)$; $(1, 0)$; $(0, -2)$
$-\infty < x < -1$, $f(-2) = 18$, above x-axis
$-1 < x < 1$, $f(0) = -2$, below x-axis
$1 < x < \infty$, $f(2) = 18$, above x-axis
(symmetric with respect to the y-axis)

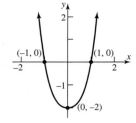

64. $(-2, 0)$, $(2, 0)$; $(0, -4)$
$-\infty < x < -2$, $f(-3) = 50$, above x-axis
$-2 < x < 2$, $f(0) = -4$, below x-axis
$2 < x < \infty$, $f(3) = 50$, above x-axis
(symmetric with respect to the y-axis)

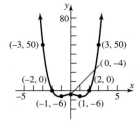

65. $\left(-\dfrac{1}{2}, 0\right); \left(\dfrac{1}{2}, 0\right); (0, -2)$

$-\infty < x < -\dfrac{1}{2}, f(-1) = 9$, above x-axis

$-\dfrac{1}{2} < x < \dfrac{1}{2}, f(0) = -2$, below x-axis

$\dfrac{1}{2} < x < \infty, f(1) = 9$, above x-axis

(symmetric with respect to the y-axis)

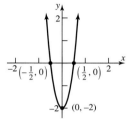

66. $\left(-\dfrac{1}{2}, 0\right); \left(\dfrac{1}{2}, 0\right); (0. -4)$

$-\infty < x < -\dfrac{1}{2}, f(-1) = 15$, above x-axis

$-\dfrac{1}{2} < x < \dfrac{1}{2}, f(0) = -4$, below x-axis

$\dfrac{1}{2} < x < \infty, f(1) - 15$, above x-axis

(symmetric with respect to the y-axis)

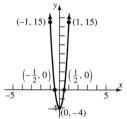

67. $(-1, 0); (-2, 0); (1, 0); (0, 2)$

$-\infty < x < -2, f(-3) = 32$, above x-axis

$-2 < x < -1, f\left(-\dfrac{3}{2}\right) = -\dfrac{25}{16}$, below x-axis

$-1 < x < 1, f(0) = 2$, above x-axis

$1 < x < \infty, f(2) = 12$, above x-axis

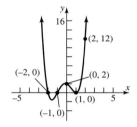

68. $(-1, 0); (-2, 0); (2, 0); (0, 8)$

$-\infty < x < -2, f(-3) = 50$, above x-axis

$-2 < x < -1, f(-1.5) \approx -3$, below x-axis

$-1 < x < 2, f(0) = 8$, above x-axis

$2 < x < \infty, f(3) = 20$, above x-axis

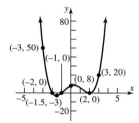

69. $(2, 0); (-\sqrt{2}/2, 0); (\sqrt{2}/2, 0); (0, 2)$

$-\infty < x < -\sqrt{2}/2, f(-1) = -9$, below x-axis

$-\sqrt{2}/2 < x < \sqrt{2}/2, f(0) = 2$, above x-axis

$\sqrt{2}/2 < x < 2, f(1) = -3$, below x-axis

$2 < x < \infty, f(3) = 323$, above x-axis

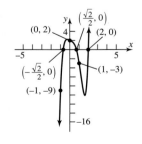

70. $(-3, 0); \left(-\dfrac{\sqrt{2}}{2}, 0\right); \left(\dfrac{\sqrt{2}}{2}, 0\right); (0, -3)$

$-\infty < x < -3, f(-3.5) \approx -300$, below x-axis

$-3 < x < -\dfrac{\sqrt{2}}{2}, f(-2) = 63$, above x-axis

$-\dfrac{\sqrt{2}}{2} < x < \dfrac{\sqrt{2}}{2}, f(0) = -3$, below x-axis

$\dfrac{\sqrt{2}}{2} < x < \infty, f(1) = 12$, above x-axis

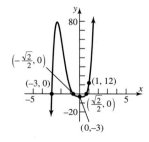

71. -5 to 5 **72.** -37 to 37 **73.** -2 to 2 **74.** -2 to 2 **75.** -5 to 5 **76.** -13 to 13 **77.** $-\dfrac{7}{4}$ to $\dfrac{7}{4}$ **78.** $-\dfrac{3}{2}$ to $\dfrac{3}{2}$

79. $f(0) = -1; f(1) = 10$ **80.** $f(-1) = -6; f(0) = 2$ **81.** $f(-5) = -58; f(-4) = 2$ **82.** $f(-3) = -42; f(-2) = 5$
83. $f(1.4) = -0.17536; f(1.5) = 1.40625$ **84.** $f(1.7) = 0.35627; f(1.8) = -1.02112$ **85.** 0.21 **86.** -0.60 **87.** -4.04 **88.** -2.17

89. 1.15 **90.** 0.70 **91.** 2.53 **92.** 2.13 **93.** $k = 5$ **94.** $k = -\dfrac{17}{12}$ **95.** -7 **96.** 1

97. If $f(x) = x^n - c^n$, then $f(c) = c^n - c^n = 0$ so $x - c$ is a factor of f. **98.** $f(-c) = 0; (-c)^n + c^n = 0$ if $n \geq 1$ is an odd integer

99. 5 **100.** -3 **101.** No (use the Rational Zeros Theorem) **102.** No (use the Rational Zeros Theorem)

103. No (use the Rational Zeros Theorem) **104.** No (use the Rational Zeros Theorem) **105.** 7 in. **106.** 6 cm

107. All the potential rational zeroes are integers. Hence, r either is an integer or is not a rational root (and is therefore irrational).

108. Let $\dfrac{p}{q}$, where p and q have no common factors except 1 and -1, be a solution of the polynomial

$$f(x) = a_n x^n + a_{n-1} x^{n-1} + \cdots + a_1 x + a_0$$

whose coefficients are all integers. Then

$$f\left(\frac{p}{q}\right) = a_n\left(\frac{p}{q}\right)^n + a_{n-1}\left(\frac{p}{q}\right)^{n-1} + \cdots + a_1\left(\frac{p}{q}\right) + a_0 = 0$$

$$a_n p^n + a_{n-1} p^{n-1} q + \cdots + a_1 p q^{n-1} + a_0 q^n = 0$$

Because p is a factor of the first n terms of this equation, p must also be a factor of $a_0 q^n$. Since p is not a factor of q (p and q have no common factors except 1 and -1), p must be a factor of a_0. Similarly, q must be a factor of a_n.

5.6 Exercises

1. $8 + 5i$ **2.** $-4 + 7i$ **3.** $-7 + 6i$ **4.** 6 **5.** $-6 - 11i$ **6.** $-10 + 6i$ **7.** $6 - 18i$ **8.** $-8 - 32i$ **9.** $6 + 4i$ **10.** $-12 - 9i$

11. $10 - 5i$ **12.** $13 + i$ **13.** 37 **14.** -10 **15.** $\dfrac{6}{5} + \dfrac{8}{5}i$ **16.** $\dfrac{5}{13} + \dfrac{12}{13}i$ **17.** $1 - 2i$ **18.** $\dfrac{1}{2} + i$ **19.** $\dfrac{5}{2} - \dfrac{7}{2}i$ **20.** $-\dfrac{1}{2} + \dfrac{5}{2}i$

21. $-\dfrac{1}{2} + \dfrac{\sqrt{3}}{2}i$ **22.** $\dfrac{1}{2} - \dfrac{\sqrt{3}}{2}i$ **23.** $2i$ **24.** $-2i$ **25.** $-i$ **26.** -1 **27.** i **28.** i **29.** -6 **30.** $4 - i$ **31.** $-10i$ **32.** $3 - 4i$

33. $-2 + 2i$ **34.** 82 **35.** 0 **36.** 0 **37.** 0 **38.** 0 **39.** $2i$ **40.** $3i$ **41.** $5i$ **42.** $8i$ **43.** $5i$ **44.** $5i$ **45.** $\{-2i, 2i\}$ **46.** $\{-2, 2\}$

47. $\{-4, 4\}$ **48.** $\{-5i, 5i\}$ **49.** $\{3 - 2i, 3 + 2i\}$ **50.** $\{-2 + 2i, -2 - 2i\}$ **51.** $\{3 - i, 3 + i\}$ **52.** $\{1 - 2i, 1 + 2i\}$

53. $\left\{\dfrac{1}{4} - \dfrac{1}{4}i, \dfrac{1}{4} + \dfrac{1}{4}i\right\}$ **54.** $\left\{-\dfrac{3}{10} - \dfrac{i}{10}, -\dfrac{3}{10} + \dfrac{i}{10}\right\}$ **55.** $\left\{\dfrac{1}{5} - \dfrac{2}{5}i, \dfrac{1}{5} + \dfrac{2}{5}i\right\}$ **56.** $\left\{\dfrac{3}{13} - \dfrac{2}{13}i, \dfrac{3}{13} + \dfrac{2}{13}i\right\}$

57. $\left\{-\dfrac{1}{2} - \dfrac{\sqrt{3}}{2}i, -\dfrac{1}{2} + \dfrac{\sqrt{3}}{2}i\right\}$ **58.** $\left(\dfrac{1}{2} - \dfrac{\sqrt{3}i}{2}, \dfrac{1}{2} + \dfrac{\sqrt{3}i}{2}\right)$ **59.** $\{2, -1 - \sqrt{3}i, -1 + \sqrt{3}i\}$ **60.** $\left\{-3, 3 - \dfrac{3\sqrt{3}}{2}i, 3 + \dfrac{3\sqrt{3}}{2}i\right\}$

61. $\{-2, 2, -2i, 2i\}$ **62.** $\{-1, 1 - i, i\}$ **63.** $\{-3i, -2i, 2i, 3i\}$ **64.** $\{-1, 1, -2i, 2i\}$ **65.** Two complex solutions.

66. Two unequal real solutions. **67.** Two unequal real solutions. **68.** Two complex solutions that are conjugates of each other.

69. A repeated real solution. **70.** A repeated real solution. **71.** $2 - 3i$ **72.** $4 + i$ **73.** 6 **74.** $6i$ **75.** 25 **76.** $-5 + 7i$

77. $z + \bar{z} = (a + bi) + (a - bi) = 2a; z - \bar{z} = (a + bi) - (a - bi) = 2bi$ **78.** $\bar{\bar{z}} = \overline{a + bi} = \overline{a - bi} = a + bi = z$

79. $\overline{z + w} = \overline{(a + bi) + (c + di)} = \overline{(a + c) + (b + d)i} = (a + c) - (b + d)i = (a - bi) + (c - di) = \bar{z} + \bar{w}$

80. $\overline{z \cdot w} = \overline{(a + bi) \cdot (c + di)}$
$= \overline{(ac - bd) + (ad + bc)i}$
$= (ac - bd) - (ad + bc)i$
$= (ac - bd) + (-ad - bc)i$
$= (a - bi) \cdot (c - di)$
$= \bar{z} \cdot \bar{w}$

5.7 Exercises

1. $4 + i$ **2.** $3 - i$ **3.** $-i, 1 - i$ **4.** $2 - i$ **5.** $-i, -2i$ **6.** $-i$ **7.** $-i$ **8.** $2 + i, i$ **9.** $2 - i, -3 + i$ **10.** $-i, 3 + 2i, -2 - i$

11. $f(x) = x^4 - 14x^3 + 77x^2 - 200x + 208; a = 1$ **12.** $f(x) = x^4 - 2x^3 + 6x^2 - 2x + 5; a = 1$

13. $f(x) = x^5 - 4x^4 + 7x^3 - 8x^2 + 6x - 4; a = 1$ **14.** $f(x) = x^6 - 12x^5 + 55x^4 - 120x^3 + 139x^2 - 108x + 85; a = 1$

15. $f(x) = x^4 - 6x^3 + 10x^2 - 6x + 9; a = 1$ **16.** $f(x) = x^5 - 5x^4 + 11x^3 - 13x^2 + 8x - 2; a = 1$

17. $-2i, 4$ **18.** $5i, -3$ **19.** $2i, -3, \dfrac{1}{2}$ **20.** $3i, -2, \dfrac{1}{3}$ **21.** $3 + 2i, -2, 5$ **22.** $1 - 3i, -1, 6$ **23.** $4i, -\sqrt{11}, \sqrt{11}, -\dfrac{2}{3}$

24. $-3i, -3, 4, \dfrac{1}{2}$ **25.** $1, -\dfrac{1}{2} - \dfrac{\sqrt{3}}{2}i, -\dfrac{1}{2} + \dfrac{\sqrt{3}}{2}i$ **26.** $-i, i, -1, 1$ **27.** $2, 3 - 2i, 3 + 2i$ **28.** $-5, -4 + i, -4 - i$ **29.** $-i, i, -2i, 2i$

30. $-2i, 2i, -3i, 3i$ **31.** $-5i, 5i, -3, 1$ **32.** $-3i, 3i, -7, 4$ **33.** $-4, \dfrac{1}{3}, 2 - 3i, 2 + 3i$ **34.** $-3 - 2i, -3 + 2i, \dfrac{1}{2}, 5$

35. Zeros that are complex numbers must occur in conjugate pairs; or a polynomial with real coefficients of odd degree must have at least one real zero. **36.** Zeros that are complex numbers must occur in conjugate pairs.

37. If the remaining zero were a complex number, then its conjugate would also be zero, creating a polynomial of degree 5.

38. A missing zero is $4 + i$. If the remaining zero were a complex number, then its conjugate would also be a zero, creating a polynomial of degree 5.

Fill-in-the-Blank Items

1. parabola; vertex   **2.** remainder; dividend   **3.** $f(c)$ **4.** $f(c) = 0$ **5.** zero   **6.** three; one; two; no   **7.** $\pm 1, \pm\dfrac{1}{2}$ **8.** $y = 1$
9. $x = -1$ **10.** real; imaginary; imaginary unit   **11.** $3 - 4i$ **12.** $-2i; 2i; -2; 2$

True/False Items

1. F   **2.** F   **3.** T   **4.** T   **5.** T   **6.** F   **7.** T

Review Exercises

1.

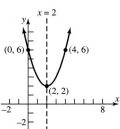

2.

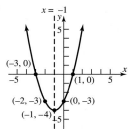

3.

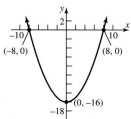

4.

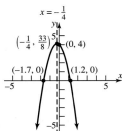

5.

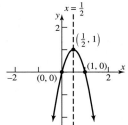

6.

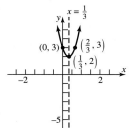

7.

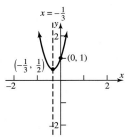

8.

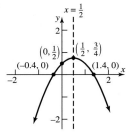

9.

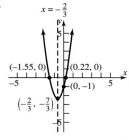

10.

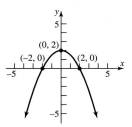

11.

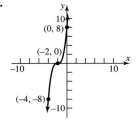

12.

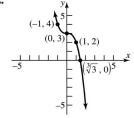

13.

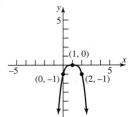

14.

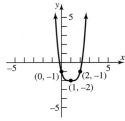

15.

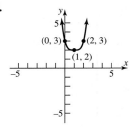

16.

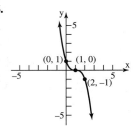

17. Minimum value; 1
18. Minimum value; -3
19. Maximum value; 12
20. Maximum value; 22
21. Maximum value; 16
22. Maximum value; 4

23. **(a)** x-intercepts: $-4, -2, 0$; y-intercept: 0   **(e)**
(b) Crosses at $-4, -2, 0$
(c) $y = x^3$
(d) 2

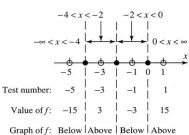

(f)

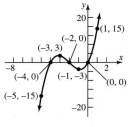

24. **(a)** x-intercepts: 0, 2, 4; y-intercept: 0
(b) Crosses at 0, 2, and 4
(c) $y = x^3$
(d) 2
(e)

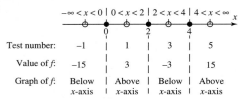

(f)

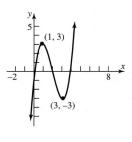

25. **(a)** x-intercepts: $-4, 2$; y-intercept: 16
(b) Crosses at -4; touches at 2
(c) $y = x^3$
(d) 2
(e)

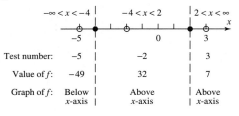

(f)

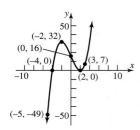

26. (a) x-intercepts: $-4, 2$; y-intercept: -32
   (b) Touches at -4; crosses at 2
   (c) $y = x^3$
   (d) 2
   (e)

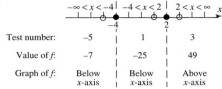

Test number:	-5		1		3
Value of f:	-7		-25		49
Graph of f:	Below x-axis		Below x-axis		Above x-axis

(f)

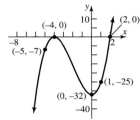

27. (a) x-intercepts: $0, 4$; y-intercept: 0
   (b) Touches at 0; crosses at 4
   (c) $y = x^3$
   (d) 2
   (e)

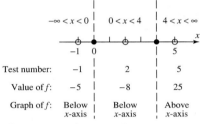

Test number:	-1		2		5
Value of f:	-5		-8		25
Graph of f:	Below x-axis		Below x-axis		Above x-axis

(f)

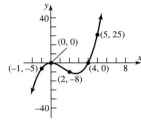

28. (a) x-intercept: 0; y-intercept: 0
   (b) Crosses at 0
   (c) $y = x^3$
   (d) 2
   (e)

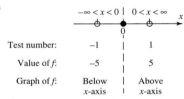

Test number:	-1		1
Value of f:	-5		5
Graph of f:	Below x-axis		Above x-axis

(f)

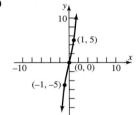

29. (a) x-intercepts: $-3, -1, 1$; y-intercept: 3
   (b) Crosses at $-3, -1$; touches at 1
   (c) $y = x^4$
   (d) 3
   (e)

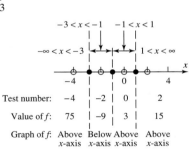

Test number:	-4	-2	0	2
Value of f:	75	-9	3	15
Graph of f:	Above x-axis	Below x-axis	Above x-axis	Above x-axis

(f)

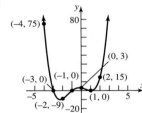

30. (a) x-intercepts: $-2, 2, 4$; y-intercept: 32

(b) Crosses at 2 and 4; touches at -2

(c) $y = x^4$

(d) 3

(e)

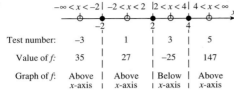

Test number:	-3	1	3	5
Value of f:	35	27	-25	147
Graph of f:	Above x-axis	Above x-axis	Below x-axis	Above x-axis

(f)

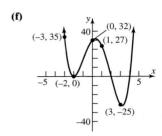

31. 1. Domain: $\{x \mid x \neq 0\}$

2. x-intercept: 3; no y-intercept

3. No symmetry

4. Vertical asymptote: $x = 0$

5. Horizontal asymptote: $y = 2$; not intersected

6. $x < 0$: above x-axis

     $0 < x < 3$: below x-axis

     $x > 3$: above x-axis

7.

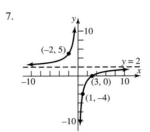

32. 1. Domain: $\{x \mid x \neq 0\}$

2. x-intercept: 4; no y-intercept

3. No symmetry

4. Vertical asymptote: $x = 0$

5. Horizontal asymptote: $y = -1$; not intersected

6. $x < 0$: below x-axis

     $0 < x < 4$: above x-axis

     $x > 4$: below x-axis

7.

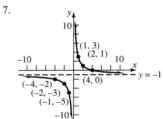

33. 1. Domain: $\{x \mid x \neq 0, x \neq 2\}$

2. x-intercept: -2; no y-intercept

3. No symmetry

4. Vertical asymptotes: $x = 0, x = 2$

5. Horizontal asymptote: $y = 0$; intersected at $(-2, 0)$

6. $x < -2$: below x-axis

     $-2 < x < 0$: above x-axis

     $0 < x < 2$: below x-axis

     $x > 2$: above x-axis

7.

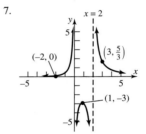

34. 1. Domain: $\{x \mid x \neq -1, x \neq 1\}$

2. x-intercept: 0; y-intercept: 0

3. Symmetric with respect to the origin

4. Vertical asymptotes: $x = -1$ and $x = 1$

5. Horizontal asymptote: $y = 0$; intersected at $(0, 0)$

6. $x < -1$: below x-axis

     $-1 < x < 0$: above x-axis

     $0 < x < 1$: below x-axis

     $x > 1$: above x-axis

7.

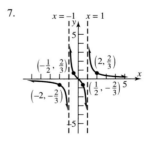

35. 1. Domain: $\{x \mid x \neq -2, x \neq 3\}$
2. x-intercepts: $-3, 2$; y-intercept: 1
3. No symmetry
4. Vertical asymptote: $x = -2, x = 3$
5. Horizontal asymptote: $y = 1$; intersected at $(0, 1)$
6. $x < -3$: above x-axis
   $-3 < x < -2$: below x-axis
   $-2 < x < 2$: above x-axis
   $2 < x < 3$: below x-axis
   $x > 3$: above x-axis

7.
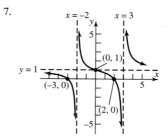

36. 1. Domain: $\{x \mid x \neq 0\}$
2. x-intercept: 3; no y-intercept
3. No symmetry
4. Vertical asymptote: $x = 0$
5. Horizontal asymptote: $y = 1$; intersected at $\left(\dfrac{3}{2}, 1\right)$
6. $x < 0$: above x-axis
   $0 < x < 3$: above x-axis
   $x > 3$: above x-axis

7.
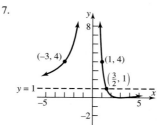

37. 1. Domain: $\{x \mid x \neq -2, x \neq 2\}$
2. x-intercept: 0; y-intercept: 0
3. Symmetric with respect to the origin
4. Vertical asymptotes: $x = -2, x = 2$
5. Oblique asymptote: $y = x$; intersected at $(0, 0)$
6. $x < -2$: below x-axis
   $-2 < x < 0$: above x-axis
   $0 < x < 2$: below x-axis
   $x > 2$: above x-axis

7.
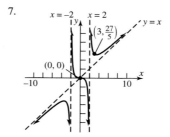

38. 1. Domain: $\{x \mid x \neq 1\}$
2. x-intercept: 0; y-intercept: 0
3. No symmetry
4. Vertical asymptote: $x = 1$
5. Oblique asymptote: $y = 3x + 6$; intersected at $\left(\dfrac{2}{3}, 8\right)$
6. $x < 0$: below x-axis
   $0 < x < 1$: above x-axis
   $x > 1$: above x-axis

7.
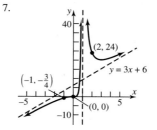

39. 1. Domain: $\{x \mid x \neq 1\}$
2. x-intercept: 0; y-intercept: 0
3. No symmetry
4. Vertical asymptote: $x = 1$
5. No oblique or horizontal asymptote
6. $x < 0$: above x-axis
   $0 < x < 1$: above x-axis
   $x > 1$: above x-axis

7.
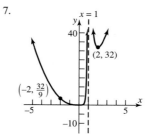

40. 1. Domain: $\{x | x \neq -3, x \neq 3\}$
2. x-intercept: 0; y-intercept: 0
3. Symetric with respect to the y-axis
4. Vertical asymptotes: $x = 3$ and $x = -3$
5. No horizontal asymptotes and no oblique asymptotes
6. $x < -3$: above x-axis
$-3 < x < 0$: below x-axis
$0 < x < 3$: below x-axis
$x > 3$: above x-axis

7.
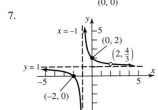

41. 1. Domain: $\{x | x \neq -1, x \neq 2\}$
2. x-intercept: -2; y-intercept: 2
3. No symmetry
4. Vertical asymptote: $x = -1$; hole at $\left(2, \dfrac{4}{3}\right)$
5. Horizontal asymptote: $y = 1$
6. $x < -2$: above x-axis
$-2 < x < -1$: below x-axis
$-1 < x < 2$: above x-axis
$x > 2$: above x-axis

7.
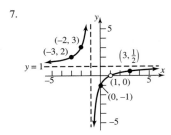

42. 1. Domain: $\{x | x \neq -1, x \neq 1\}$
2. No x-intercepts; y-intercept: -1
3. No symmetry
4. Vertical asymptote: $x = -1$; hole at $(1, 0)$
5. Horizontal asymptote: $y = 1$, not intersected
6. $-\infty < x < -1$: above x-axis
$-1 < x < 1$: below x-axis
$1 < x < \infty$: above x-axis

7.

43. $q(x) = 8x^2 + 5x + 6$; $R = 10$ **44.** $q(x) = 2x^2 + 12x + 19$; $R = 43$ **45.** $q(x) = x^3 - 4x^2 + 8x - 15$; $R = 29$
46. $q(x) = x^3 - x^2 + 3$; $R = -3$ **47.** $f(4) = 47{,}105$ **48.** $f(-2) = 204$ **49.** 4, 2, or 0 positive; 2 or 0 negative

50. 1 positive; 2 or 0 negative **51.** $\pm\dfrac{1}{12}, \pm\dfrac{1}{6}, \pm\dfrac{1}{4}, \pm\dfrac{1}{3}, \pm\dfrac{1}{2}, \pm\dfrac{3}{4}, \pm1, \pm\dfrac{3}{2}; \pm3$ **52.** $\pm1, \pm\dfrac{1}{2}, \pm\dfrac{1}{3}, \pm\dfrac{1}{6}$

53. $-2, 1, 4$; $f(x) = (x + 2)(x - 1)(x - 4)$ **54.** $-1, 4, -2$; $f(x) = (x + 1)(x - 4)(x + 2)$

55. $\dfrac{1}{2}$, multiplicity 2; -2; $f(x) = 4\left(x - \dfrac{1}{2}\right)^2(x + 2)$ **56.** $2, -\dfrac{1}{2}$; $f(x) = (x - 2)(2x + 1)^2$

57. 2, multiplicity 2; $f(x) = (x - 2)^2(x^2 + 5)$ **58.** -3, multiplicity 2; $f(x) = (x + 3)^2(x^2 + 2)$ **59.** $\{-3, 2\}$ **60.** $\{2, -3\}$

61. $\left\{-3, -1, -\dfrac{1}{2}, 1\right\}$ **62.** $\left\{2, -2, -\dfrac{1}{2}, -3\right\}$

63. x-intercepts: $-2, 1, 4$
y-intercept: 8
Above x-axis: $-2 < x < 1, 4 < x < \infty$
Below x-axis: $-\infty < x < -2, 1 < x < 4$

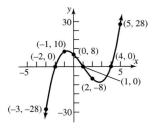

64. x-intercepts: $-2, -1, 4$
y-intercept: -8
Above x-axis: $-2 < x < -1, 4 < x < \infty$
Below x-axis: $-\infty < x < -2, 1 < x < 4$

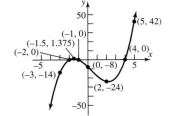

65. x-intercepts: $-2, \dfrac{1}{2}$

y-intercept: 2
Above x-axis: $-2 < x < \infty$
Below x-axis: $-\infty < x < -2$

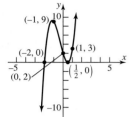

66. x-intercepts: $-\dfrac{1}{2}, 2$

y-intercept: -2
Above x-axis: $2 < x < \infty$
Below x-axis: $-\infty < x < 2$

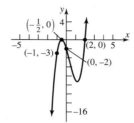

67. x-intercept: 2
y-intercept: 20
Above x-axis: All x

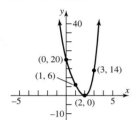

68. x-intercept: -3
y-intercept: 18
Above x-axis: All x

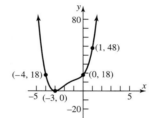

69. x-intercepts: $-3, 2$
y-intercept: -6
Above x-axis: $-\infty < x < -3, 2 < x < \infty$
Below x-axis: $-3 < x < 2$

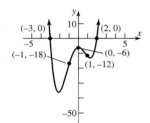

70. x-intercepts: $-3, 2$
y-intercept: -6
Above x-axis: $-\infty < x < -3, 2 < x < \infty$
Below x-axis: $-3 < x < 2$

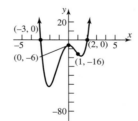

71. x-intercepts: $-3, -1, -\dfrac{1}{2}, 1$

y-intercept: -3

Above x-axis: $-\infty < x < -3, -1 < x < -\dfrac{1}{2}, 1 < x < \infty$

Below x-axis: $-3 < x < -1, -\dfrac{1}{2} < x < 1$

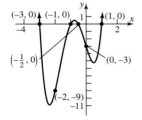

72. x-intercepts: $-3, -2, -\dfrac{1}{2}, 2$

y-intercept: -12

Above x-axis: $-\infty < x < -3, -2 < x < -\dfrac{1}{2}, 2 < x < \infty$

Below x-axis: $-3 < x < -2, -\dfrac{1}{2} < x < 2$

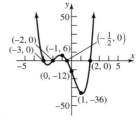

73. between -5 and 5   **74.** between -11 and 11   **75.** between $-\dfrac{37}{2}$ and $\dfrac{37}{2}$ **76.** between $-\dfrac{17}{3}$ and $\dfrac{17}{3}$ **77.** $f(0) = -1; f(1) = 1$

78. $f(1) = -2, f(2) = 9$ **79.** $f(0) = -1; f(1) = 1$ **80.** $f(1) = -3, f(2) = 62$ **81.** 1.52   **82.** 1.33   **83.** 0.93   **84.** 1.14   **85.** $4 + 7i$

86. $2 - i$ **87.** $-3 + 2i$ **88.** $-4 + 11i$ **89.** $\dfrac{9}{10} - \dfrac{3}{10}i$ **90.** $\dfrac{8}{5} + \dfrac{4}{5}i$ **91.** -1 **92.** i **93.** $-46 + 9i$ **94.** $-9 - 46i$ **95.** $4 - i$

96. $3 - 4i$ **97.** $-i, 1 - i$ **98.** $1 - i$ **99.** $\left\{ -\dfrac{1}{2} - \dfrac{\sqrt{3}}{2}i, -\dfrac{1}{2} + \dfrac{\sqrt{3}}{2}i \right\}$ **100.** $\left\{ \dfrac{1 - \sqrt{3}i}{2}, \dfrac{1 + \sqrt{3}i}{2} \right\}$ **101.** $\left\{ \dfrac{-1 - \sqrt{17}}{4}, \dfrac{-1 + \sqrt{17}}{4} \right\}$

102. $\left\{ -\dfrac{1}{3}, 1 \right\}$ **103.** $\left\{ \dfrac{1}{2} - \dfrac{\sqrt{11}i}{2}, \dfrac{1}{2} + \dfrac{\sqrt{11}i}{2} \right\}$ **104.** $\left\{ \dfrac{1}{2} - \dfrac{1}{2}i, \dfrac{1}{2} + \dfrac{1}{2}i \right\}$ **105.** $\left\{ \dfrac{1}{2} - \dfrac{\sqrt{23}}{2}i, \dfrac{1}{2} + \dfrac{\sqrt{23}}{2}i \right\}$ **106.** $\{-2, 1\}$

107. $\{-\sqrt{2}, \sqrt{2}, -2i, 2i\}$ **108.** $\{-3i, 3i, -1, 1\}$ **109.** $\{-3, 2\}$ **110.** $\{-2, 2, 3\}$ **111.** $\left\{ \dfrac{1}{3}, 1, -i, i \right\}$ **112.** $\{-2, \sqrt{2}, -\sqrt{2}\}$ **113.** $(2, 2)$

114. (2.3) **115.** 50 ft by 50 ft **116.** 2 ft by 8 ft **117.** 25 sq units **118. (a)** 62 clubs **(b)** approximately \$12.58 per club **(c)** \$20,000

119. (a) 5000 **(b)** $\dfrac{C(5000)}{5000} = \$7.36$ **(c)** about \$111,770 **120.** 5:38 PM

121. The side with the semi-circles should be $\dfrac{50}{\pi}$ ft; the other side should be 25 ft. **122.** 3.6 ft

123. (a)

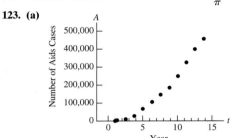

(b) 544,721 **(d)**

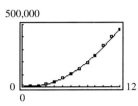

124. (a)

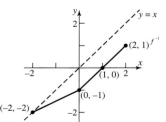

(b) -10 ft per second
(c) When $t = 17.5$ seconds
(f) -5.4 ft per second^2

(e)

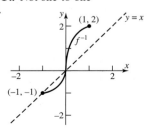

C H A P T E R 6 6.1 Exercises

1. (a)

Domain	Range
\$200 →	20 hours
\$300 →	25 hours
\$350 →	30 hours
\$425 →	40 hours

2. (a)

Domain	Range
Beth →	Dave
Diane →	Bob
Linda →	John
Marcia →	Chuck

3. (a)

Domain	Range
\$200	20 hours
	25 hours
\$350	30 hours
\$425	40 hours

4. (a)

Domain	Range
Beth	Bob
Diane	Dave
Marcia	John
	Chuck

(b) Inverse is a function   **(b)** Inverse is a function   **(b)** Inverse is not a function   **(b)** Inverse is not a function

5. (a) $\{(6, 2), (6, -3), (9, 4), (10, 1)\}$ **(b)** Inverse is not a function   **6. (a)** $\{(5, -2), (3, -1), (7, 3), (12, 4)\}$ **(b)** Inverse is a function

7. (a) $\{(0, 0), (1, 1), (16, 2), (81, 3)\}$ **(c)** Inverse is a function   **8. (a)** $\{(2, 1), (8, 2), (18, 3), (32, 4)\}$ **(b)** Inverse is a function

9. One-to-one **10.** One-to-one **11.** Not one-to-one **12.** Not one-to-one **13.** One-to-one **14.** Not one-to-one

15.

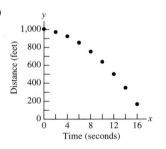

16.

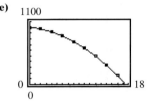

17.

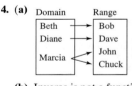

18.

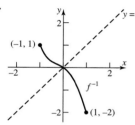

19.

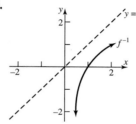

20.

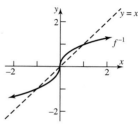

21. $f(g(x)) = f\left(\frac{1}{3}(x-4)\right) = 3\left[\frac{1}{3}(x-4)\right] + 4 = x; g(f(x)) = g(3x+4) = \frac{1}{3}[(3x+4)-4] = x$

22. $f(g(x)) = f\left(-\frac{1}{2}(x-3)\right) = 3 - 2\left[-\frac{1}{2}(x-3)\right] = 3 + x - 3 = x; g(f(x)) = g(3-2x) = -\frac{1}{2}[(3-2x)-3] = x$

23. $f(g(x)) = 4\left[\frac{x}{4}+2\right] - 8 = x; g(f(x)) = \frac{4x-8}{4} + 2 = x$

24. $f(g(x)) = f\left(\frac{1}{2}(x-3)\right) = 2\left(\frac{1}{2}x-3\right) + 6 = x - 6 + 6 = x; g(f(x)) = g(2x+6) = \frac{1}{2}(2x+6) - 3 = x$

25. $f(g(x)) = (\sqrt[3]{x+8})^3 - 8 = x; g(f(x)) = \sqrt[3]{(x^3-8)+8} = x$

26. $f(g(x)) = f(\sqrt{x}+2) = [(\sqrt{x}+2)-2]^2 = (\sqrt{x})^2 = x$

$g(f(x)) = g((x-2)^2) = \sqrt{(x-2)^2} + 2 = |x-2| + 2 = x, x \geq 2$

27. $f(g(x)) = \dfrac{1}{\left(\dfrac{1}{x}\right)} = x; g(f(x)) = \dfrac{1}{\left(\dfrac{1}{x}\right)} = x$ **28.** $f(g(x)) = f(x) = x; g(f(x)) = g(x) = x$

29. $f(g(x)) = \dfrac{2\left(\dfrac{4x-3}{2-x}\right) + 3}{\dfrac{4x-3}{2-x} + 4} = x; g(f(x)) = \dfrac{4\left(\dfrac{2x+3}{x+4}\right) - 3}{2 - \dfrac{2x+3}{x+4}} = x$

30. $f(g(x)) = f\left(\dfrac{3x+5}{1-2x}\right) = \dfrac{\dfrac{3x+5}{1-2x} - 5}{2\left(\dfrac{3x+5}{1-2x}\right) + 3} = x; g(f(x)) = g\left(\dfrac{x-5}{2x-3}\right) = \dfrac{3\left(\dfrac{x-5}{2x-3}\right) + 5}{1 - 2\left(\dfrac{x-5}{2x+3}\right)} = x$

31. $f^{-1}(x) = \frac{1}{3}x$

$f(f^{-1}(x)) = 3\left(\frac{1}{3}x\right) = x$

$f^{-1}(f(x)) = \frac{1}{3}(3x) = x$

Domain f = Range $f^{-1} = (-\infty, \infty)$
Range f = Domain $f^{-1} = (-\infty, \infty)$

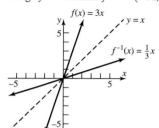

32. $f^{-1}(x) = \dfrac{-x}{4}$

$f(f^{-1}(x)) = -4\left(\dfrac{-x}{4}\right) = x$

$f^{-1}(f(x)) = \dfrac{-(-4x)}{4} = x$

Domain f = Range $f^{-1} = (-\infty, \infty)$
Range f = Domain $f^{-1} = (-\infty, \infty)$

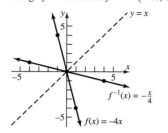

33. $f^{-1}(x) = \dfrac{x}{4} - \dfrac{1}{2}$

$f(f^{-1}(x)) = 4\left(\dfrac{x}{4} - \dfrac{1}{2}\right) + 2 = x$

$f^{-1}(f(x)) = \dfrac{4x+2}{4} - \dfrac{1}{2} = x$

Domain f = Range $f^{-1} = (-\infty, \infty)$
Range f = Domain $f^{-1} = (-\infty, \infty)$

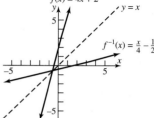

34. $f^{-1}(x) = \dfrac{-x + 1}{3}$

$f(f^{-1}(x)) = 1 - 3\left(\dfrac{-x + 1}{3}\right) = x$

$f^{-1}(f(x)) = \dfrac{-(1 - 3x) + 1}{3} = x$

Domain f = Range f^{-1} = $(-\infty, \infty)$

Range f = Domain f^{-1} = $(-\infty, \infty)$

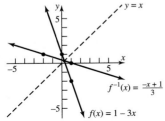

35. $f^{-1}(x) = \sqrt[3]{x + 1}$

$f(f^{-1}(x)) = (\sqrt[3]{x + 1})^3 - 1 = x$

$f^{-1}(f(x)) = \sqrt[3]{(x^3 - 1) + 1} = x$

Domain f = Range f^{-1} = $(-\infty, \infty)$

Range f = Domain f^{-1} = $(-\infty, \infty)$

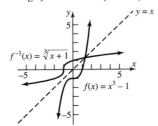

36. $f^{-1}(x) = \sqrt[3]{x - 1}$

$f(f^{-1}(x)) = (\sqrt[3]{x - 1})^3 + 1 = x$

$f^{-1}(f(x)) = \sqrt[3]{(x^3 + 1) - 1} = x$

Domain f = Range f^{-1} = $(-\infty, \infty)$

Range f = Domain f^{-1} = $(-\infty, \infty)$

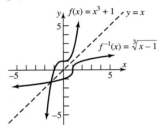

37. $f^{-1}(x) = \sqrt{x - 4}$

$f(f^{-1}(x)) = (\sqrt{x - 4})^2 + 4 = x$

$f^{-1}(f(x)) = \sqrt{(x^2 + 4) - 4} = \sqrt{x^2} = |x| = x$

Domain f = Range f^{-1} = $[0, \infty)$

Range f = Domain f^{-1} = $[4, \infty)$

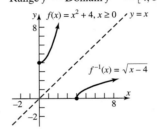

38. $f^{-1}(x) = \sqrt{x - 9}$

$f(f^{-1}(x)) = (\sqrt{x - 9})^2 + 9 = x$

$f^{-1}(f(x)) = \sqrt{(x^2 + 9) - 9} = \sqrt{x^2} = |x| = x, \ x \geq 0$

Domain f = Range f^{-1} = $[0, \infty)$

Range f = Domain f^{-1} = $[9, \infty)$

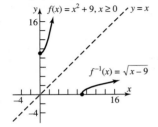

39. $f^{-1}(x) = \dfrac{4}{x}$

$f(f^{-1}(x)) = \dfrac{4}{\dfrac{4}{x}} = x$

$f^{-1}(f(x)) = \dfrac{4}{\dfrac{4}{x}} = x$

Domain f = Range f^{-1} = All real numbers except 0

Range f = Domain f^{-1} = All real numbers except 0

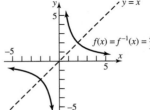

40. $f^{-1}(x) = -\dfrac{3}{x}$

$f(f^{-1}(x)) = -\dfrac{3}{-\dfrac{3}{x}} = x$

$f^{-1}(f(x)) = -\dfrac{3}{-\dfrac{3}{x}} = x$

Domain f = Range f^{-1} = Domain f^{-1} = Range f
= All real numbers except 0

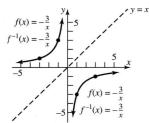

$f(x) = -\dfrac{3}{x}$
$f^{-1}(x) = -\dfrac{3}{x}$

$y = x$

$f(x) = -\dfrac{3}{x}$
$f^{-1}(x) = -\dfrac{3}{x}$

41. $f^{-1}(x) = \dfrac{2x + 1}{x}$

$f(f^{-1}(x)) = \dfrac{1}{\dfrac{2x + 1}{x} - 2} = x$

$f^{-1}(f(x)) = \dfrac{2\left(\dfrac{1}{x - 2}\right) + 1}{\dfrac{1}{x - 2}} = x$

Domain f = Range f^{-1} = All real numbers except 2
Range f = Domain f^{-1} = All real numbers except 0

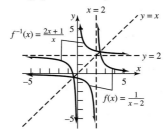

$f^{-1}(x) = \dfrac{2x + 1}{x}$

$x = 2$
$y = x$
$y = 2$
$f(x) = \dfrac{1}{x - 2}$

42. $f^{-1}(x) = \dfrac{4 - 2x}{x}$

$f(f^{-1}(x)) = \dfrac{4}{\dfrac{4 - 2x}{x} + 2} = x$

$f^{-1}(f(x)) = \dfrac{4 - 2\left(\dfrac{4}{x + 2}\right)}{\dfrac{4}{x + 2}} = x$

Domain f = Range f^{-1} = All real numbers except -2
Range f = Domain f^{-1} = All real numbers except 0

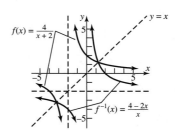

$f(x) = \dfrac{4}{x + 2}$

$y = x$

$f^{-1}(x) = \dfrac{4 - 2x}{x}$

43. $f^{-1}(x) = \dfrac{2 - 3x}{x}$

$f(f^{-1}(x)) = \dfrac{2}{3 + \dfrac{2 - 3x}{x}} = x$

$f^{-1}(f(x)) = \dfrac{2 - 3\left(\dfrac{2}{3 + x}\right)}{\dfrac{2}{3 + x}} = x$

Domain f = Range f^{-1} = All real numbers except -3
Range f = Domain f^{-1} = All real numbers except 0

44. $f^{-1}(x) = \dfrac{2x - 4}{x}$

$f(f^{-1}(x)) = \dfrac{4}{2 - \left(\dfrac{2x - 4}{x}\right)} = x$

$f^{-1}(f(x)) = \dfrac{2\left(\dfrac{4}{2 - x}\right) - 4}{\dfrac{4}{2 - x}} = x$

Domain f = Range f^{-1} = All real numbers except 2
Range f = Domain f^{-1} = All real numbers except 0

45. $f^{-1}(x) = \sqrt{x} - 2$
$f(f^{-1}(x)) = (\sqrt{x} - 2 + 2)^2 = x$
$f^{-1}(f(x)) = \sqrt{(x + 2)^2} - 2 = |x + 2| - 2 = x, x \geq -2$
Domain f = Range f^{-1} = $[-2, \infty)$
Range f = Domain f^{-1} = $[0, \infty)$

46. $f^{-1}(x) = \sqrt{x} + 1$
$f(f^{-1}(x)) = (\sqrt{x} + 1 - 1)^2 = x$
$f^{-1}(f(x)) = \sqrt{(x - 1)^2} + 1 = |x - 1| + 1 = x, x \geq 1$
Domain f = Range f^{-1} = $[1, \infty)$
Range f = Domain f^{-1} = $[0, \infty)$

47. $f^{-1}(x) = \dfrac{x}{x-2}$

$$f(f^{-1}(x)) = \dfrac{2\left(\dfrac{x}{x-2}\right)}{\dfrac{x}{x-2}-1} = x$$

$$f^{-1}(f(x)) = \dfrac{\dfrac{2x}{x-1}}{\dfrac{2x}{x-1}-2} = x$$

Domain f = Range f^{-1} = All real numbers except 1
Range f = Domain f^{-1} = All real numbers except 2

48. $f^{-1}(x) = \dfrac{1}{x-3}$

$$f(f^{-1}(x)) = \dfrac{3\left(\dfrac{1}{x-3}\right)+1}{\dfrac{1}{x-3}} = x$$

$$f^{-1}(f(x)) = \dfrac{1}{\dfrac{3x+1}{x}-3} = x$$

Domain f = Range f^{-1} = All real numbers except 0
Domain f^{-1} = Range f = All real numbers except 3

49. $f^{-1}(x) = \dfrac{3x+4}{2x-3}$

$$f(f^{-1}(x)) = \dfrac{3\left(\dfrac{3x+4}{2x-3}\right)+4}{2\left(\dfrac{3x+4}{2x-3}\right)-3} = x$$

$$f^{-1}(f(x)) = \dfrac{3\left(\dfrac{3x+4}{2x-3}\right)+4}{2\left(\dfrac{3x+4}{2x-3}\right)-3} = x$$

Domain f = Range f^{-1} = All real numbers except $\dfrac{3}{2}$

Range f = Domain f^{-1} = All real numbers except $\dfrac{3}{2}$

50. $f^{-1}(x) = \dfrac{-4x-3}{x-2}$

$$f(f^{-1}(x)) = \dfrac{2\left(\dfrac{-4x-3}{x-2}\right)-3}{\left(\dfrac{-4x-3}{x-2}\right)+4} = x$$

$$f^{-1}(f(x)) = \dfrac{-4\left(\dfrac{2x-3}{x+4}\right)-3}{\left(\dfrac{2x-3}{x+4}\right)-2} = x$$

Domain f = Range f^{-1} = All real numbers except -4

Range f = Domain f^{-1} = All real numbers except 2

51. $f^{-1}(x) = \dfrac{-2x+3}{x-2}$

$$f(f^{-1}(x)) = \dfrac{2\left(\dfrac{-2x+3}{x-2}\right)+3}{\dfrac{-2x+3}{x-2}+2} = x$$

$$f^{-1}(f(x)) = \dfrac{-2\left(\dfrac{2x+3}{x+2}\right)+3}{\dfrac{2x+3}{x+2}-2} = x$$

Domain f = Range f^{-1} = All real numbers except -2
Range f = Domain f^{-1} = All real numbers except 2

52. $f^{-1}(x) = \dfrac{2x-4}{x+3}$

$$f(f^{-1}(x)) = \dfrac{-3\left(\dfrac{2x-4}{x+3}\right)-4}{\left(\dfrac{2x-4}{x+3}\right)-2} = x$$

$$f^{-1}(f(x)) = \dfrac{2\left(\dfrac{-3x-4}{x-2}\right)-4}{\left(\dfrac{-3x-4}{x-2}\right)+3} = x$$

Domain f = Range f^{-1} = All real numbers except 2
Range f = Domain f^{-1} = All real numbers except -3

53. $f^{-1}(x) = \dfrac{x^3}{8}$

$$f(f^{-1}(x)) = 2\sqrt[3]{\dfrac{x^3}{8}} = x$$

$$f^{-1}(f(x)) = \dfrac{(2\sqrt[3]{x})^3}{8} = x$$

Domain f = Range f^{-1} = $(-\infty, \infty)$
Range f = Domain f^{-1} = $(-\infty, \infty)$

54. $f^{-1}(x) = \dfrac{16}{x^2}, x > 0$

$$f(f^{-1}(x)) = \dfrac{4}{\sqrt{\dfrac{16}{x^2}}} = \dfrac{4}{\left|\dfrac{4}{x}\right|} = x, x > 0$$

$$f^{-1}(f(x)) = \dfrac{16}{\left(\dfrac{4}{\sqrt{x}}\right)^2} = \dfrac{16}{\dfrac{16}{x}} = x$$

Domain f = Range f^{-1} = $(0, \infty)$
Range f = Domain f^{-1} = $(0, \infty)$

55. $f^{-1}(x) = \dfrac{1}{m}(x-b), m \neq 0$ **56.** $f^{-1}(x) = \sqrt{r^2 - x^2}, 0 \le x \le r$

59. Quadrant I **60.** Quadrant IV **61.** $f(x) = |x|, x \ge 0$, is one-to-one; $f^{-1}(x) = x, x \ge 0$

62. $f(x) = x^4, x \ge 0$, is one-to-one; $f^{-1}(x) = \sqrt[4]{x}$ **63.** $f(g(x)) = \dfrac{9}{5}\left[\dfrac{5}{9}(x-32)\right] + 32 = x; g(f(x)) = \dfrac{5}{9}\left[\left(\dfrac{9}{5}x+32\right)-32\right] = x$

64. $x(p) = \dfrac{1}{50}(300-p)$ **65.** $l(T) = \dfrac{gT^2}{4\pi^2}, T > 0$ **66.** $f(x) = \begin{cases} x & \text{if } x \text{ is rational} \\ -x & \text{if } x \text{ is irrational} \end{cases}$ (Other answers are possible.)

67. $f^{-1}(x) = \dfrac{-dx + b}{cx - a}$; $f = f^{-1}$ if $a = -d$ **68. (a)** $f^{-1}(x) = \dfrac{3x + 5}{x - 2}$; The range of f is $\{y|y \neq 2\}$

(b) $g^{-1}(x) = \dfrac{-2}{x - 4}$; The range of g is $\{y|y \neq 4\}$ **(c)** $F^{-1}(x) = \dfrac{4x - 3}{x}$; The range of F is $\{y|y \neq 0\}$

6.2 Exercises

1. (a) 11.212 **(b)** 11.587 **(c)** 11.664 **(d)** 11.665 **2. (a)** 15.426 **(b)** 16.189 **(c)** 16.241 **(d)** 16.242
3. (a) 8.815 **(b)** 8.821 **(c)** 8.824 **(d)** 8.825 **4. (a)** 6.498 **(b)** 6.543 **(c)** 6.580 **(d)** 6.581
5. (a) 21.217 **(b)** 22.217 **(c)** 22.440 **(d)** 22.459 **6. (a)** 21.738 **(b)** 22.884 **(c)** 23.119 **(d)** 23.141
7. 3.320 **8.** 0.273 **9.** 0.427 **10.** 8.166 **11.** B **12.** F **13.** D **14.** H **15.** A **16.** C **17.** E **18.** G

19.

Domain: $(-\infty, \infty)$
Range: $(0, \infty)$
Horizontal asymptote: $y = 0$

20.

Domain: $(-\infty, \infty)$
Range: $(-\infty, 0)$
Horizontal asymptote: $y = 0$

21.

Domain: $(-\infty, \infty)$
Range: $(0, \infty)$
Horizontal asymptote: $y = 0$

22.

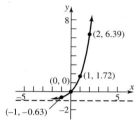

Domain: $(-\infty, \infty)$
Range: $(-1, \infty)$
Horizontal asymptote:
$y = -1$

23.

Domain: $(-\infty, \infty)$
Range: $(-\infty, 5)$
Horizontal asymptote: $y = 5$

24.

Domain: $(-\infty, \infty)$
Range: $(-\infty, 9)$
Horizontal asymptote: $y = 9$

25.

Domain: $(-\infty, \infty)$
Range: $(-\infty, 2)$
Horizontal asymptote: $y = 2$

26.

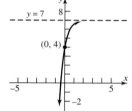

Domain: $(-\infty, \infty)$
Range: $(-\infty, 7)$
Horizontal asymptote: $y = 7$

27.

Domain: $(-\infty, \infty)$
Range: $[1, \infty)$
Intercept: $(0, 1)$

28.

Domain: $(-\infty, \infty)$
Range: $(0, 1]$
Intercept: $(0, 1)$

29.

Domain: $(-\infty, \infty)$
Range: $[-1, 0)$
Intercept: $(0, -1)$

30.

Domain: $(-\infty, \infty)$
Range: $(-\infty, -1]$
Intercept: $(0, -1)$

31. $\dfrac{1}{49}$ **32.** $\dfrac{1}{9}$ **33.** $\dfrac{1}{4}$ **34.** $\dfrac{1}{27}$ **35. (a)** 74% **(b)** 47% **36. (a)** 568.68 mm Hg **(b)** 178.27 mm Hg **37. (a)** 44 watts **(b)** 11.6 watts
38. (a) 34.99 cm^2 **(b)** 3.02 cm^2 **39.** 3.35 milligrams; 0.45 milligrams **40.** About 362 students

41. (a) 0.63   **(b)** 0.98
(c)

(d) About 7 minutes   **(e)** 1

42. (a) 89.5%   **(b)** 98.9%
(c)

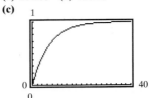

(d) About 6 minutes   **(e)** 1

43. (a) 60.5%   **(b)** 68.7%   **(c)** 70%   **(d)** About $4\frac{1}{4}$ days

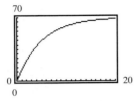

44. (a) \$98,125   **(b)** \$99,941.41   **(c)** \$100,000   **(d)** About $\frac{3}{4}$ year

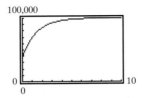

45. (a) 5.414 amperes, 7.585 amperes, 10.376 amperes   **(b)** 12 amperes
(d) 3.343 amperes, 5.309 amperes, 9.443 amperes   **(e)** 24 amperes
(c), (f)

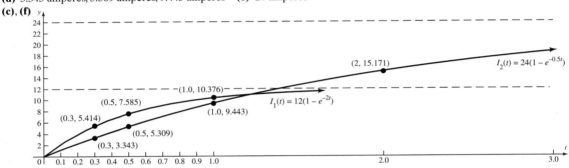

46. (a) 0.06 milliamperes, 0.0364 milliamperes, 0.0134 milliamperes   **(b)** 0.06 milliamperes
(d) 0.12 milliamperes, 0.0728 milliamperes, 0.0268 milliamperes   **(e)** 0.12 milliamperes
(c),(f)

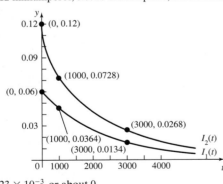

(f)

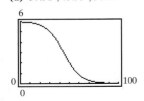

47. (a) 9.23×10^{-3}, or about 0
(b) 0.81, or about 1   **(c)** 5.01, or about 5
(d) 57.91°, 43.99°, 30.07°

48. (a) About 3317 stamps
(b) About 3487 stamps
(c)

Year	Actual	Predicted
1848	2	78
1868	88	129
1888	218	212
1908	341	350

The predictions appear to get better with time.

49. $n = 4: 2.7083; n = 6: 2.7181; n = 8: 2.7182788; n = 10: 2.7182818$
50. $3.0, 2.666666, 2.727272, 2.71698, 2.71845, 2.71826;$ The expression gets closer to e, 2.71828182846.
51. $\dfrac{f(x + h) - f(x)}{h} = \dfrac{a^{x+h} - a^x}{h} = \dfrac{a^x a^h - a^x}{h} = \dfrac{a^x(a^h - 1)}{h}$ **52.** $f(A + B) = a^{A+B} = a^A \cdot a^B = f(A) \cdot f(B)$
53. $f(-x) = a^{-x} = \dfrac{1}{a^x} = \dfrac{1}{f(x)}$ **54.** $f(\alpha x) = a^{\alpha x} = (a^x)^{\alpha} = [f(x)]^{\alpha}$
55. $f(1) = 5, f(2) = 17, f(3) = 257, f(4) = 65,537, f(5) = 4,294,967,297 = 641 \times 6,700,417$ **56.** 59 min **58.** No

6.3 Exercises

1. $2 = \log_3 9$ **2.** $2 = \log_4 16$ **3.** $2 = \log_a 1.6$ **4.** $3 = \log_a 2.1$ **5.** $2 = \log_{1.1} M$ **6.** $3 = \log_{2.2} N$ **7.** $x = \log_2 7.2$ **8.** $x = \log_3 4.6$
9. $\sqrt{2} = \log_x \pi$ **10.** $\pi = \log_x e$ **11.** $x = \ln 8$ **12.** $2.2 = \ln M$ **13.** $2^3 = 8$ **14.** $3^{-2} = \dfrac{1}{9}$ **15.** $a^6 = 3$ **16.** $b^2 = 4$ **17.** $3^x = 2$
18. $2^x = 6$ **19.** $2^{1.3} = M$ **20.** $3^{2.1} = N$ **21.** $(\sqrt{2})^x = \pi$ **22.** $\pi^{1/2} = x$ **23.** $e^x = 4$ **24.** $e^4 = x$ **25.** 0 **26.** 1 **27.** 2 **28.** -2
29. -4 **30.** -2 **31.** $\dfrac{1}{2}$ **32.** $\dfrac{2}{3}$ **33.** 4 **34.** 4 **35.** $\dfrac{1}{2}$ **36.** 3 **37.** $\{x | x < 3\}; (0, \ln 3), (2, 0)$ **38.** $\{x | x > 1\}; (2, 0)$
39. All real numbers except 0; $(-1, 0), (1, 0)$ **40.** $\{x | x > 0\}; (1, 0)$ **41.** All real numbers except 1; $(0, 0), (2, 0)$
42. $\{x | x < -1 \text{ or } x > 1\}; (-\sqrt{2}, 0), (\sqrt{2}, 0)$ **43.** $\{x | x < -1 \text{ or } x > 0\};$ No intercepts **44.** $\{x | x < 0 \text{ or } x > 1\};$ No intercepts
45. 0.511 **46.** 0.536 **47.** 30.099 **48.** 4.055 **49.** $\sqrt{2}$ **50.** $\sqrt[4]{2}$ **51.** B **52.** F **53.** D **54.** H **55.** A **56.** C **57.** E **58.** G

59.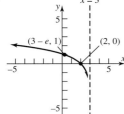
Domain: $(-4, \infty)$
Range: $(-\infty, \infty)$
Vertical asymptote: $x = -4$

60.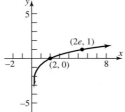
Domain: $(3, \infty)$
Range: $(-\infty, \infty)$
Vertical asymptote: $x = 3$

61.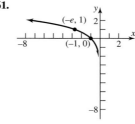
Domain: $(-\infty, 0)$
Range: $(-\infty, \infty)$
Vertical asymptote: $x = 0$

62.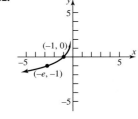
Domain: $(-\infty, 0)$
Range: $(-\infty, \infty)$
Vertical asymptote: $x = 0$

63.
Domain: $(0, \infty)$
Range: $(-\infty, \infty)$
Vertical asymptote: $x = 0$

64.
Domain: $(0, \infty)$
Range: $(-\infty, \infty)$
Vertical asymptote: $x = 0$

65.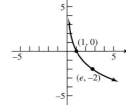
Domain: $(0, \infty)$
Range: $(-\infty, \infty)$
Vertical asymptote: $x = 0$

66.
Domain: $(0, \infty)$
Range: $(-\infty, \infty)$
Vertical asymptote: $x = 0$

67.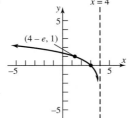
Domain: $(-\infty, 3)$
Range: $(-\infty, \infty)$
Vertical asymptote: $x = 3$

68.
Domain: $(-\infty, 4)$
Range: $(-\infty, \infty)$
Vertical asymptote: $x = 4$

69.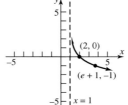
Domain: $(1, \infty)$
Range: $(-\infty, \infty)$
Vertical asymptote: $x = 1$

70.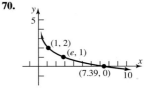
Domain: $(0, \infty)$
Range: $(-\infty, \infty)$
Vertical asymptote: $x = 0$

71.

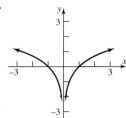

Domain: $\{x|x \neq 0\}$
Range: $(-\infty, \infty)$
Intercepts: $(-1, 0)$, $(1, 0)$

72.

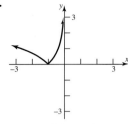

Domain: $\{x|x < 0\}$
Range: $\{y|y \geq 0\}$
Intercept: $(-1, 0)$

73.

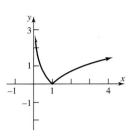

Domain: $\{x|x > 0\}$
Range: $\{y|y \geq 0\}$
Intercept: $(1, 0)$

74.

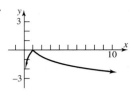

Domain: $\{x|x > 0\}$
Range: $\{y|y \leq 0\}$
Intercept: $(1, 0)$

75. (a) $n \approx 6.93$ so 7 panes are necessary **(b)** $n \approx 13.86$ so 14 panes are necessary **76. (a)** 7 **(b)** 0.0000631 moles per liter

77. (a) $d \approx 127.7$ so it takes about 128 days **(b)** $d \approx 575.6$ so it takes about 576 days

78. (a) $n \approx 1.98$ so it takes about 2 days **(b)** $n \approx 6.58$ so it takes about $6\frac{1}{2}$ days

79. $h \approx 2.29$ so the time between injections is about 2 hours, 17 minutes **80.** $d \approx 3.99$ so it takes about 4 days

81. 0.2695 sec
0.8959 sec

82. (a) 0.0211
(b) 38 words
(c) 54 words
(d) About 1 hr 49 min

83. (a) $k = 20.07$
(b) 91%
(c) 0.175
(d) 0.08

84. No **85.** $y = 20e^{0.023t}$; $y = 89.2$ is predicted **86.** after $t = 1$ year, $R \approx 0.0375$; after $t = 2$ years, $R \approx 0.0797$; after $t = 3$ years, $R \approx 0.0930$; after $t = 4$ years, $R \approx 0.1007$; after $t = 5$ years, $R \approx 0.1167$

6.4 Exercises

1. $a + b$ **2.** $a - b$ **3.** $b - a$ **4.** $-a$ **5.** $a + 1$ **6.** $b - 1$ **7.** $2a + b$ **8.** $b + 3a$ **9.** $\frac{1}{5}(a + 2b)$ **10.** $\frac{1}{4}b + a$ **11.** $\frac{b}{a}$ **12.** $\frac{a}{b}$

13. $2 \ln x + \frac{1}{2}\ln(1 - x)$ **14.** $4 \ln x + \frac{1}{2}\ln(x^2 + 1)$ **15.** $3 \log_2 x - \log_2(x - 3)$ **16.** $\frac{1}{3}\log_5(x^2 + 1) - \log_5(x^2 - 1)$

17. $\log x + \log(x + 2) - 2\log(x + 3)$ **18.** $3 \log x + \frac{1}{2}\log(x + 1) - 2\log(x - 2)$ **19.** $\frac{1}{3}\ln(x - 2) + \frac{1}{3}\ln(x + 1) - \frac{2}{3}\ln(x + 4)$

20. $\frac{4}{3}\ln(x - 4) - \frac{2}{3}\ln(x^2 - 1)$ **21.** $\ln 5 + \ln x + \frac{1}{2}\ln(1 - 3x) - 3\ln(x - 4)$

22. $\ln 5 + 2\ln x + \frac{1}{3}\ln(1 - x) - \ln 4 - 2\ln(x + 1)$ **23.** $\log_5 u^3 v^4$ **24.** $\log_3 \frac{u^2}{v}$ **25.** $-\frac{5}{2}\log_{1/2} x$ **26.** $-3 \log_2 x$

27. $-2 \ln (x - 1)$ **28.** $\log \dfrac{(x + 3)(x - 1)}{(x - 2)(x + 6)(x + 1)}$ **29.** $\log_2[x(3x - 2)^4]$ **30.** $9 \log_3 x$ **31.** $\log_a\left(\dfrac{25x^6}{\sqrt{2x + 3}}\right)$

32. $\log (\sqrt[3]{x^3 + 1}\sqrt{x^2 + 1})$ **33.** $\frac{5}{4}$ **34.** 42 **35.** 4 **36.** 3 **37.** 2.771 **38.** 1.796 **39.** -3.880 **40.** -3.907 **41.** 5.615 **42.** 2.584

43. 0.874 **44.** 0.303

45. $\log_a(x + \sqrt{x^2 - 1}) + \log_a(x - \sqrt{x^2 - 1}) = \log_a[(x + \sqrt{x^2 - 1})(x - \sqrt{x^2 - 1})] = \log_a[x^2 - (x^2 - 1)] = \log_a 1 = 0$

46. $\log_a(\sqrt{x} + \sqrt{x - 1}) + \log_a(\sqrt{x} - \sqrt{x - 1}) = \log_a[(\sqrt{x} + \sqrt{x - 1})(\sqrt{x} - \sqrt{x - 1})] = \log_a[x - (x - 1)] = \log_a 1 = 0$

47. $\ln(1 + e^{2x}) = \ln(e^{2x} \cdot e^{-2x} + e^{2x}) = \ln [e^{2x}(e^{-2x} + 1)] = \ln e^{2x} + \ln(e^{-2x} + 1) = 2x + \ln(1 + e^{-2x})$

48. $\dfrac{f(x+h)-f(x)}{h} = \dfrac{\log_a(x+h)-\log_a x}{h} = \dfrac{\log_a\left(\dfrac{x+h}{x}\right)}{h} = \dfrac{1}{h}\log_a\left(1+\dfrac{h}{x}\right) = \log_a\left(1+\dfrac{h}{x}\right)^{1/h}, h \ne 0$

49. If $y = \log_a x$, then $x = a^y = \dfrac{1}{a^{-y}} = \left(\dfrac{1}{a}\right)^{-y}$ so $-y = \log_{1/a} x$ **50.** $f\left(\dfrac{1}{x}\right) = \log_a\left(\dfrac{1}{x}\right) = \log_a x^{-1} = -\log_a x = -f(x)$

51. $f(AB) = \log_a(AB) = \log_a A + \log_a B = f(A) + f(B)$ **52.** $f(x^\alpha) = \log_a x^\alpha = \alpha \log_a x = \alpha f(x)$

53.   **54.** $y = \left(\dfrac{\log x}{\log 5}\right)$ **55.**   **56.** $y = \dfrac{\log(x-3)}{\log 4}$

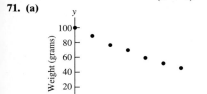

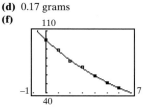

57. $y = Cx$ **58.** $y = x + C$ **59.** $y = Cx(x+1)$ **60.** $\dfrac{Cx^2}{x+1}$ **61.** $y = Ce^{3x}$ **62.** $y = Ce^{-2x}$ **63.** $y = Ce^{-4x} + 3$

64. $y = Ce^{5x} - 4$ **65.** $y = \dfrac{\sqrt[3]{C}(2x+1)^{1/6}}{(x+4)^{1/9}}$ **66.** $y = \sqrt{C}x^{-1/4}(x^2+1)^{1/6}$ **67.** 3   **68.** 3   **69.** 1   **70.** $n!$

71. (a)
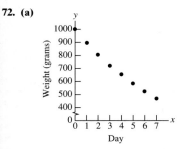

(b) $A = 100e^{-0.128t}$
(c) 5.4 weeks
(d) 0.17 grams
(f)
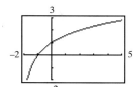

72. (a)
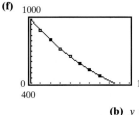

(b) $A = 999e^{-0.1076t}$
(c) $t = 6.4$ days
(d) 116 grams
(f)

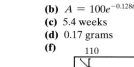

73. (a)
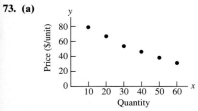

(b) $y = 96e^{-0.02x}$
(c) 23.3 units
(e)

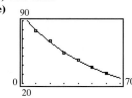

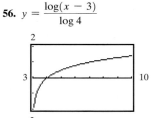

74. (a)

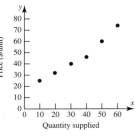

(b) $A = 20.5e^{0.0198x}$

(c) 39.7 units

(e)

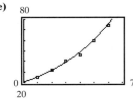

75. (a)

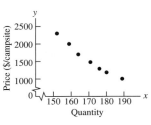

(b) 168 computers

(d)

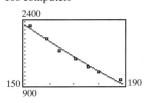

76. If $A = \log_a M$ and $B = \log_a N$, then $a^A = M$ and $a^B = N$.

Then $\log_a\left(\dfrac{M}{N}\right) = \log_a\left(\dfrac{a^A}{a^B}\right) = \log_a a^{A-B} = A - B = \log_a M - \log_a N$.   **77.** $\log_a \dfrac{1}{N} = \log_a 1 - \log_a N = -\log_a N$

6.5 Exercises

1. $\dfrac{7}{2}$ **2.** $\dfrac{11}{3}$ **3.** $\{-2\sqrt{2}, 2\sqrt{2}\}$ **4.** $\left\{\dfrac{-1 - \sqrt{85}}{2}, \dfrac{-1 + \sqrt{85}}{2}\right\}$ **5.** 16   **6.** $\dfrac{1}{3}$ **7.** 8   **8.** $\dfrac{1}{3}$ **9.** 3   **10.** 5   **11.** 5   **12.** 4   **13.** 2   **14.** $\dfrac{1}{2}$

15. $\{-2, 4\}$ **16.** 0   **17.** 21   **18.** -4 **19.** $\dfrac{1}{2}$ **20.** 1   **21.** $\{-\sqrt{2}, 0, \sqrt{2}\}$ **22.** $\left\{0, \dfrac{1}{2}\right\}$ **23.** $\left\{1 - \dfrac{\sqrt{6}}{3}, 1 + \dfrac{\sqrt{6}}{3}\right\}$ **24.** $\dfrac{1}{2}$ **25.** 0   **26.** 3

27. $\dfrac{\ln 3}{\ln 2} \approx 1.585$ **28.** 0   **29.** 0   **30.** 0   **31.** $\dfrac{3}{2}$ **32.** $\dfrac{3}{4}$ **33.** $\dfrac{\ln 10}{\ln 2} \approx 3.322$ **34.** $\dfrac{\ln 14}{\ln 3} \approx 2.402$ **35.** $-\dfrac{\ln 1.2}{\ln 8} \approx -0.088$

36. $-\dfrac{\ln 1.5}{\ln 2} \approx -0.585$ **37.** $\dfrac{\ln 3}{2\ln 3 + \ln 4} \approx 0.307$ **38.** $\dfrac{\ln 5 - \ln 2}{\ln 2 + 2\ln 5} \approx 0.234$ **39.** $\dfrac{\ln 7}{\ln 0.6 + \ln 7} \approx 1.356$ **40.** $\dfrac{\ln (4/3)}{\ln (4/3) + \ln 5} \approx 0.152$

41. 0   **42.** $\dfrac{\ln 0.3 + \ln 1.7}{2\ln 1.7 - \ln 0.3} \approx -0.297$ **43.** $\dfrac{\ln \pi}{1 + \ln \pi} \approx 0.534$ **44.** $\dfrac{-3}{1 - \ln \pi} \approx 20.728$ **45.** $\dfrac{\ln 1.6}{3\ln 2} \approx 0.226$ **46.** $\dfrac{5\ln (2/3)}{\ln 4} \approx -1.462$

47. $5\ln 1.5 \approx 2.027$ **48.** $\dfrac{\ln (6/5)}{0.3} \approx 0.608$ **49.** $\dfrac{9}{2}$ **50.** 4   **51.** 2   **52.** 67   **53.** -1 **54.** -1 **55.** 1   **56.** $\left\{\dfrac{26 - 8\sqrt{10}}{9}, \dfrac{26 + 8\sqrt{10}}{9}\right\}$

57. 16   **58.** 81   **59.** $\left\{-1, \dfrac{2}{3}\right\}$ **60.** 4   **61.** 1.92   **62.** 4.47   **63.** 2.78   **64.** 12.14   **65.** -0.56 **66.** $\{0.44, -1.98\}$ **67.** -0.70

68. $\{1.85, 4.53\}$ **69.** 0.56   **70.** 1.15   **71.** $\{0.39, 1.00\}$ **72.** 0.65   **73.** 1.31   **74.** $\{0.05, 1.47\}$ **75.** 1.30   **76.** 0.56

6.6 Exercises

1. $108.29   **2.** $59.83   **3.** $609.50   **4.** $358.84   **5.** $697.09   **6.** $789.24   **7.** $12.46   **8.** $49.35   **9.** $125.23   **10.** $156.83   **11.** $88.72
12. $59.14   **13.** $860.72   **14.** $626.61   **15.** $554.09   **16.** $266.08   **17.** $59.71   **18.** $654.98   **19.** $361.93   **20.** $886.92   **21.** 5.35%

22. 6.82%   **23.** 26%   **24.** 7.18%   **25.** $6\frac{1}{4}$% compounded annually   **26.** 9% compounded quarterly

27. 9% compounded monthly   **28.** 7.9% compounded daily   **29.** 104.32 mo; 103.97 mo   **30.** 83.52 mo; 83.18 mo   **31.** 61.02 mo; 60.82 mo

32. 67.43 mo; 67.15 mo   **33.** 15.27 yrs or 15 yrs, 4 mo   **34.** 16.62 yrs or 16 yrs, 8 mo   **35.** $104,335   **36.** $212.82   **37.** $12,910.62

38. $2955.39   **39.** About $30.17 per share or $3017   **40.** 15.47%   **41.** 9.35%   **42.** $2315.97
43. Not quite. Jim will have $1057.60. The second bank gives a better deal, since Jim will have $1060.62 after 1 year.   **44.** $1021.60
45. Will has $11,632.73; Henry has $10,947.89.   **46.** Take $1000 now; it will be worth $1349.86 in 3 years.
47. (a) Interest is $30,000   **(b)** Interest is $38,613.59   **(c)** Interest is $37,752.73. Simple interest at 12% is best.
48. (a) 4.341344%   **(b)** 4.341347%   **49. (a)** $1364.62   **(b)** $1353.35   **50.** $10,810.76   **51.** $4631.93   **52.** 9.07%

59. (a) 6.1 yrs   **(b)** 18.45 yrs

(c) $mP = P\left(1 + \dfrac{r}{n}\right)^{nt}$

$m = \left(1 + \dfrac{r}{n}\right)^{nt}$

$\ln m = \ln\left(1 + \dfrac{r}{n}\right)^{nt} = nt\ln\left(1 + \dfrac{r}{n}\right)$

$t = \dfrac{\ln m}{n\ln\left(1 + \dfrac{r}{n}\right)}$

60. (a) 20.79 yrs   **(b)** 7.74%

(c) $A = Pe^{rt}$

$\dfrac{A}{P} = e^{rt}$

$\ln\left(\dfrac{A}{P}\right) = rt$

$\ln A - \ln P = rt$

$t = \dfrac{\ln A - \ln P}{r}$

6.7 Exercises

1. (a) 34.7 days   **(b)** 69.3 days   **2.** 40.55 hrs; 69.31 hrs   **3. (a)** 28.4 yrs   **(b)** 94.4 yrs   **4. (a)** 7.97 days   **(b)** 26.47 days
5. 5832; 3.9 days   **6.** 5243 bacteria cells; 7.85 hrs   **7.** 25,198   **8.** 711,111   **9.** 9.797 g   **10.** 9.99999947 grams; 9.9999947 grams
11. 9727 yrs ago   **12.** 2882 yrs old
13. (a) 5:18 PM   **(b)** 14.3 min   **(d)** As time passes, the temperature of the pizza gets closer to 70°F.
14. (a) 16.2 min   **(b)** 7.3   **(c)** As time passes, the temperature of the thermometer gets closer to 38°F.
15. 18.63°C; 25.1°C   **16.** 45.69°F; 28.3 min   **17.** 7.34 kg; 76.6 hr   **18.** 1.25 volts   **19.** 26.5 days   **20.** 9:42 PM
21. (a) 0.1286   **(b)** 0.9   **(c)** 1996   **22. (a)** 0.2   **(b)** 0.9   **(c)** 8.4 months after it has been introduced
23. (a) 1000   **(b)** 30   **(c)** After 11.076 hrs   **24. (a)** 500   **(b)** 117   **(c)** After 29.8 yrs

6.8 Exercises

1. 70 decibels   **2.** 90 decibels   **3.** 111.76 decibels   **4.** 22 decibels   **5.** 10 watts/m²   **6.** 16.99 decibels   **7.** 4.0 on the Richter scale
8. 6.0827 on the Richter scale   **9.** 125,892.54 mm; the Mexico City earthquake was 15.85 times as intense as the one in San Francisco.
10. $x_1 = 10x_2$ so x_1 is 10 times as intense as x_2.   **11.** Crowd noise was 31.6 times as intense as guidelines allow.

Fill-in-the-Blank Items

1. one-to-one   **2.** $y = x$ **3.** $(0, 1)$ and $(1, a)$ **4.** 1   **5.** 4   **6.** sum   **7.** 1   **8.** 7   **9.** $\{x | x > 0\}$ **10.** $(1, 0)$ and $(a, 1)$ **11.** 1   **12.** 7

True/False Items

1. F   **2.** T   **3.** T   **4.** T   **5.** F   **6.** F   **7.** T   **8.** F   **9.** F   **10.** T

Review Exercises

1. $f^{-1}(x) = \dfrac{2x + 3}{5x - 2}$; $f(f^{-1}(x)) = \dfrac{2\left(\dfrac{2x + 3}{5x - 2}\right) + 3}{5\left(\dfrac{2x + 3}{5x - 2}\right) - 2} = x$; $f^{-1}(f(x)) = \dfrac{2\left(\dfrac{2x + 3}{5x - 2}\right) + 3}{5\left(\dfrac{2x + 3}{5x - 2}\right) - 2} = x$;

Domain f = Range f^{-1} = All real numbers except $\dfrac{2}{5}$; Range f = Domain f^{-1} = All real numbers except $\dfrac{2}{5}$

2. $f^{-1}(x) = \dfrac{2 - 3x}{x + 1}$; $f(f^{-1}(x)) = \dfrac{2 - \dfrac{2 - 3x}{x + 1}}{3 + \dfrac{2 - 3x}{x + 1}} = x$; $f^{-1}(f(x)) = \dfrac{2 - 3\left(\dfrac{2 - x}{3 + x}\right)}{\dfrac{2 - x}{3 + x} + 1} = x$;

Domain f = Range f^{-1} = All real numbers except -3; Range f = Domain f^{-1} = All real numbers except -1

3. $f^{-1}(x) = \dfrac{x + 1}{x}$; $f(f^{-1}(x)) = \dfrac{1}{\dfrac{x + 1}{x} - 1} = x$; $f^{-1}(f(x)) = \dfrac{\dfrac{1}{x - 1} + 1}{\dfrac{1}{x - 1}} = x$;

Domain f = Range f^{-1} = All real numbers except 1; Range f = Domain f^{-1} = All real numbers except 0
4. $f^{-1}(x) = x^2 + 2$, $x \geq 0$; $f(f^{-1}(x)) = (\sqrt{x - 2})^2 + 2 = x$; $f^{-1}(f(x)) = \sqrt{(x^2 + 2) - 2} = |x| = x$, $x \geq 0$;
Domain f = Range f^{-1} = $\{x | x \geq 2\}$; Range f = Domain f^{-1} = $\{x | x \geq 0\}$

5. $f^{-1}(x) = \dfrac{27}{x^3}$; $f(f^{-1}(x)) = \dfrac{3}{\left(\dfrac{27}{x^3}\right)^{1/3}} = x$; $f^{-1}(f(x)) = \dfrac{27}{\left(\dfrac{3}{x^{1/3}}\right)^3} = x$;

Domain f = Range f^{-1} = All real numbers except 0; Range f = Domain f^{-1} = All real numbers except 0

6. $f^{-1}(x) = (x-1)^3$; $f(f^{-1}(x)) = [(x-1)^3]^{1/3} + 1 = x$; $f^{-1}(f(x)) = [(x^{1/3} + 1) - 1]^3 = x$;
Domain f = Range f^{-1} = All real numbers; Range f = Domain f^{-1} = All real numbers

7. -3 **8.** 4 **9.** $\sqrt{2}$ **10.** 0.1 **11.** 0.4 **12.** $\sqrt{3}$ **13.** $\dfrac{25}{4}\log_4 x$ **14.** $\log_3 x^{13/6}$ **15.** $-2\ln(x+1)$ **16.** $\log\dfrac{x-3}{x+4}$

17. $\log\left(\dfrac{4x^3}{[(x+3)(x-2)]^{1/2}}\right)$ **18.** $\ln\left(16\sqrt{\dfrac{x^2+1}{x(x-4)}}\right)$ **19.** $y = Ce^{2x^2}$ **20.** $y = 3 + 2Cx^2$ **21.** $y = (Ce^{3x^2})^2$

22. $y = \dfrac{C(x^2 + 3x + 2)}{2}$ **23.** $y = \sqrt{e^{x+C} + 9}$ **24.** $y = \sqrt{e^{-x+c} + 1}$ **25.** $y = \ln(x^2 + 4) - C$ **26.** $y = \dfrac{C + \ln(x+4)^2}{3}$

27.

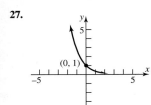

Domain: $(-\infty, \infty)$
Range: $(0, \infty)$
Asymptote: $y = 0$

28.

Domain: $(-\infty, 0)$
Range: $(-\infty, \infty)$
Asymptote: $x = 0$

29.

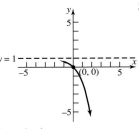

Domain: $(-\infty, \infty)$
Range: $(-\infty, 1)$
Asymptote: $y = 1$

30.

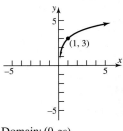

Domain: $(0, \infty)$
Range: $(-\infty, \infty)$
Asymptote: $x = 0$

31.

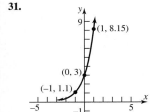

Domain: $(-\infty, \infty)$
Range: $(0, \infty)$
Asymptote: $y = 0$

32.

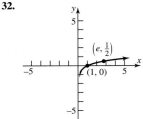

Domain: $(0, \infty)$
Range: $(-\infty, \infty)$
Asymptote: $x = 0$

33.

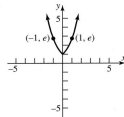

Domain: $(-\infty, \infty)$
Range: $[1, \infty)$
Asymptote: None

34.

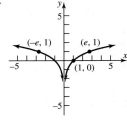

Domain: $\{x \mid x \neq 0\}$
Range: $(-\infty, \infty)$
Asymptote: $x = 0$

35.

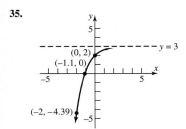

Domain: $(-\infty, \infty)$
Range: $(-\infty, 3)$
Asymptote: $y = 3$

36.

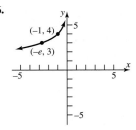

Domain: $(-\infty, 0)$
Range: $(-\infty, \infty)$
Asymptote: $x = 0$

37. $\dfrac{1}{4}$ **38.** $-\dfrac{16}{9}$

39. $\left\{\dfrac{-1 - \sqrt{3}}{2}, \dfrac{-1 + \sqrt{3}}{2}\right\}$

40. $\left\{\dfrac{1 + \sqrt{3}}{2}, \dfrac{1 - \sqrt{3}}{2}\right\}$ **41.** $\dfrac{1}{4}$ **42.** $\dfrac{1}{8}$

43. $\dfrac{2\ln 3}{\ln 5 - \ln 3} \approx 4.301$

44. $\dfrac{2(\ln 7 + \ln 5)}{\ln 7 - \ln 5} \approx 21.1331$

45. $\dfrac{12}{5}$ **46.** $\{-2, 6\}$ **47.** 83 **48.** $-\dfrac{1}{2}$ **49.** $\left\{-3, \dfrac{1}{2}\right\}$ **50.** 1 **51.** -1 **52.** $\{3, 4\}$ **53.** $1 - \ln 5 \approx -0.609$

54. $\dfrac{1}{2} - \ln 2 \approx -0.193$ **55.** $\dfrac{\ln 3}{3\ln 2 - 2\ln 3} \approx -9.327$ **56.** $\left\{0, \dfrac{\ln 3}{\ln 2}\right\}$ or $\{0, 1.585\}$ **57.** 3229.5 m **58.** 1482 m **59.** 7.6 mm of mercury

60. 61.5 mm Hg **61. (a)** 37.3 watts **(b)** 6.9 decibels **62. (a)** 11.8 **(b)** 9.56 in **63. (a)** 71% **(b)** 85.5% **(c)** 90%
(d) About 1.6 mo **(e)** About 4.8 mo **64.** About 4.2 mo **65. (a)** 9.85 yrs **(b)** 4.27 yrs **66.** $20,399 **67.** $41,669
68. (a) $r = 10.436\%$ **(b)** $32,249.24 **69.** 80 decibels

70. The San Francisco earthquake was 7943 times as intense as the Chicago earthquake.   **71.** 24,203 yrs ago **72.** 55.2 min
73. (a)

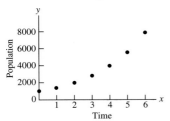

(b) $N = 1000e^{0.346574t}$
(c) 11,314
(e)

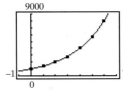
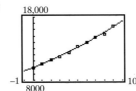

74. (a)

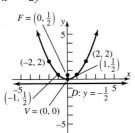

(b) $A = 10,014e^{0.055435t}$
(c) \$69,700
(e)

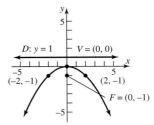

C H A P T E R   7   7.2 Exercises

1. B **2.** G **3.** E **4.** D **5.** H **6.** A **7.** C **8.** F
9. $y^2 = 16x$ **10.** $x^2 = 8y$ **11.** $x^2 = -12y$

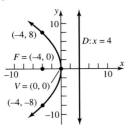   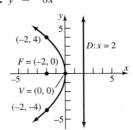   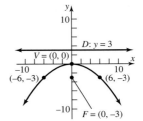

12. $y^2 = -16x$ **13.** $y^2 = -8x$ **14.** $x^2 = -4y$

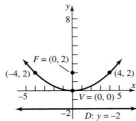

15. $x^2 = 2y$ **16.** $y^2 = 2x$ **17.** $(x - 2)^2 = -8(y + 3)$

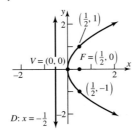   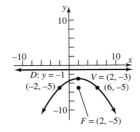

18. $(y + 2)^2 = 8(x - 4)$

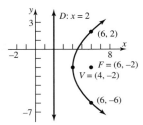

19. $x^2 = \frac{4}{3}y$

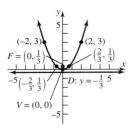

20. $y^2 = \frac{9}{2}x$

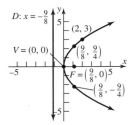

21. $(x + 3)^2 = 4(y - 3)$

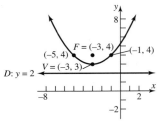

22. $(y - 4)^2 = 12(x + 1)$

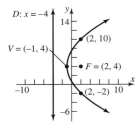

23. $(y + 2)^2 = -8(x + 1)$

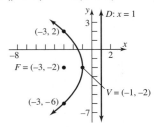

24. $(x + 4)^2 = 12(y - 1)$

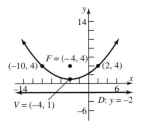

25. Vertex: $(0, 0)$; Focus: $(0, 1)$; Directrix: $y = -1$

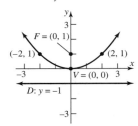

26. Vertex: $(0, 0)$; Focus: $(2, 0)$; Directrix: $x = -2$

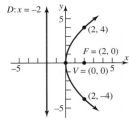

27. Vertex: $(0, 0)$; Focus: $(-4, 0)$; Directrix: $x = 4$

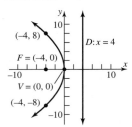

28. Vertex: $(0, 0)$; Focus: $(0, -1)$; Directrix: $y = 1$

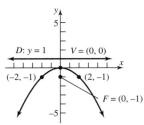

29. Vertex: $(-1, 2)$; Focus: $(1, 2)$; Directrix: $x = -3$

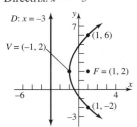

30. Vertex: $(-4, -2)$; Focus: $(-4, 2)$; Directrix: $y = -6$

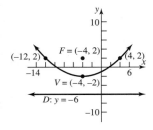

31. Vertex: $(3, -1)$; Focus: $\left(3, -\frac{5}{4}\right)$; Directrix: $y = -\frac{3}{4}$

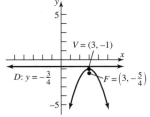

32. Vertex: $(2, -1)$; Focus: $(1, -1)$; Directrix: $x = 3$

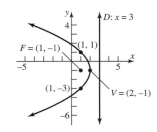

33. Vertex: $(2, -3)$; Focus: $(4, -3)$;
Directrix: $x = 0$

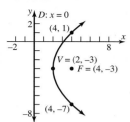

34. Vertex: $(2, 3)$; Focus: $(2, 4)$;
Directrix: $y = 2$

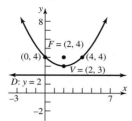

35. Vertex: $(0, 2)$; Focus: $(-1, 2)$;
Directrix: $x = 1$

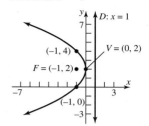

36. Vertex: $(-3, -2)$; Focus: $(-3, -1)$;
Directrix: $y = -3$

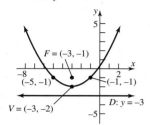

37. Vertex: $(-4, -2)$; Focus: $(-4, -1)$;
Directrix: $y = -3$

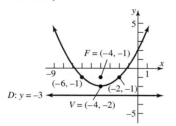

38. Vertex: $(0, 1)$; focus: $(2, 1)$;
Directrix: $x = -2$

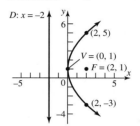

39. Vertex: $(-1, -1)$; Focus: $\left(-\dfrac{3}{4}, -1\right)$;

Directrix: $x = -\dfrac{5}{4}$

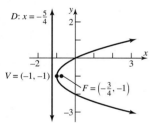

40. Vertex: $(2, -2)$; Focus: $\left(2, -\dfrac{3}{2}\right)$;

Directrix: $y = -\dfrac{5}{2}$

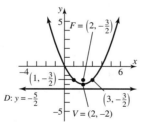

41. Vertex: $(2, -8)$; Focus: $\left(2, -\dfrac{31}{4}\right)$;

Directrix: $y = -\dfrac{33}{4}$

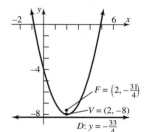

42. Vertex: $(37, -6)$; Focus: $\left(\dfrac{147}{4}, -6\right)$;

Directrix: $x = \dfrac{149}{4}$

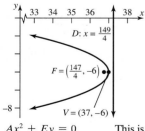

43. $(y - 1)^2 = x$ **44.** $(x - 1)^2 = -(y - 2)$ **45.** $(y - 1)^2 = -(x - 2)$

46. $x^2 = 4(y + 1)$ **47.** $x^2 = 4(y - 1)$ **48.** $(x - 1)^2 = \dfrac{1}{2}(y + 1)$ **49.** $y^2 = \dfrac{1}{2}(x + 2)$

50. $y^2 = -(x - 1)$ **51.** 1.5625 ft from the base of the dish, along the axis of symmetry
52. 1.125 ft or 13.5 in. from the base of the dish **53.** 1 in. from the vertex **54.** $4\sqrt{2}$ in.
55. 20 ft **56.** 60.625 ft **57.** 0.78125 ft **58.** $8\sqrt{2}$ ft
59. 4.17 ft from the base along the axis of symmetry
60. 0.0278 in. from the base of the mirror **61.** 24.31 ft, 18.75 ft, 7.64 ft **62.** 27.78 ft

63. $Ax^2 + Ey = 0$ This is the equation of a parabola with vertex at $(0, 0)$ and axis of symmetry the y-axis. The focus is

$x^2 = -\dfrac{E}{A}y$ $\left(0, -\dfrac{E}{4A}\right)$; the directrix is the line $y = \dfrac{E}{4A}$. The parabola opens up if $-\dfrac{E}{A} > 0$ and down if $-\dfrac{E}{A} < 0$.

64. $Cy^2 + Dx = 0$ $C \neq 0, D \neq 0$

$Cy^2 = -Dx$ This is the equation of a parabola with vertex at $(0, 0)$ and axis of symmetry the x-axis. The focus is $\left(-\dfrac{D}{4C}, 0\right)$;

$y^2 = -\dfrac{D}{C}x$ the directrix is the line $x = \dfrac{D}{4C}$. The parabola opens to the right if $-\dfrac{D}{C} > 0$ and to the left if $-\dfrac{D}{C} < 0$.

65. $Ax^2 + Dx + Ey + F = 0, A \neq 0$

$$Ax^2 + Dx = -Ey - F$$

$$x^2 + \frac{D}{A}x = -\frac{E}{A}y - \frac{F}{A}$$

$$\left(x + \frac{D}{2A}\right)^2 = -\frac{E}{A}y - \frac{F}{A} + \frac{D^2}{4A^2}$$

$$\left(x + \frac{D}{2A}\right)^2 = -\frac{E}{A}y + \frac{D^2 - 4AF}{4A^2}$$

(a) If $E \neq 0$, then the equation may be written as

$$\left(x + \frac{D}{2A}\right)^2 = -\frac{E}{A}\left(y - \frac{D^2 - 4AF}{4AE}\right)$$

This is the equation of a parabola with vertex at

$$\left(-\frac{D}{2A}, \frac{D^2 - 4AF}{4AE}\right)$$ and axis of symmetry parallel to the y-axis.

(b)–(d) If $E = 0$, the graph of the equation contains no points if $D^2 - 4AF < 0$, is a single vertical line if $D^2 - 4AF = 0$, and is two vertical lines if $D^2 - 4AF > 0$.

66. $Cy^2 + Dx + Ey + F = 0, C \neq 0$

$$Cy^2 + Ey = -Dx - F$$

$$y^2 + \frac{E}{C}y = -\frac{D}{C}x - \frac{F}{C}$$

$$\left(y + \frac{E}{2C}\right)^2 = -\frac{D}{C}x - \frac{F}{C} + \frac{E^2}{4C^2}$$

$$\left(y + \frac{E}{2C}\right)^2 = -\frac{D}{C}x + \frac{E^2 - 4CF}{4C^2}$$

(a) If $D \neq 0$, then the equation may be written as

$$\left(y + \frac{E}{2C}\right)^2 = -\frac{D}{C}\left(x - \frac{E^2 - 4CF}{4CD}\right).$$

This is the equation of a parabola with vertex at

$$\left(\frac{E^2 - 4CF}{4CD}, -\frac{E}{2C}\right)$$ and axis of symmetry parallel to the x-axis.

(b)–(d) If $D \neq 0$, the graph of the equation contains no points if $E^2 - 4CF < 0$, is a single vertical line if $E^2 - 4CF = 0$, and is two horizontal lines if $E^2 - 4CF > 0$.

7.3 Exercises

1. C **2.** D **3.** B **4.** A

5. Vertices: $(-5, 0)$, $(5, 0)$
Foci: $(-\sqrt{21}, 0)$, $(\sqrt{21}, 0)$

6. Vertices: $(-3, 0)$, $(3, 0)$
Foci: $(-\sqrt{5}, 0)$, $(\sqrt{5}, 0)$

7. Vertices: $(0, -5)$, $(0, 5)$
Foci: $(0, -4)$, $(0, 4)$

8. Vertices: $(0, -4)$, $(0, 4)$
Foci: $(0, -\sqrt{15})$, $(0, \sqrt{15})$

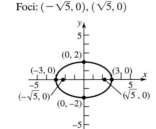

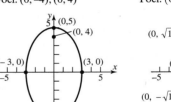

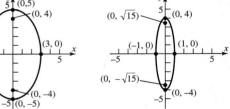

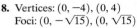

9. $\dfrac{x^2}{4} + \dfrac{y^2}{16} = 1$
Vertices: $(0, -4)$, $(0, 4)$
Foci: $(0, -2\sqrt{3})$, $(0, 2\sqrt{3})$

10. $\dfrac{x^2}{18} + \dfrac{y^2}{2} = 1$
Vertices: $(-3\sqrt{2}, 0)$, $(3\sqrt{2}, 0)$
Foci: $(-4, 0)$, $(4, 0)$

11. $\dfrac{x^2}{8} + \dfrac{y^2}{2} = 1$
Vertices: $(-2\sqrt{2}, 0)$, $(2\sqrt{2}, 0)$
Foci: $(-\sqrt{6}, 0)$, $(\sqrt{6}, 0)$

12. $\dfrac{x^2}{4} + \dfrac{y^2}{9} = 1$
Vertices: $(0, -3)$, $(0, 3)$
Foci: $(0, -\sqrt{5})$, $(0, \sqrt{5})$

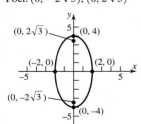

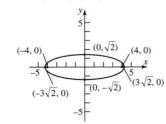

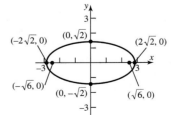

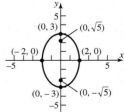

13. Vertices: $(-4, 0), (4, 0), (0, -4), (0, 4)$

Focus: $(0, 0)$

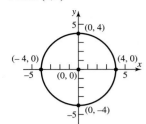

14. Vertices: $(-2, 0), (2, 0), (0, -2), (0, 2)$

Focus: $(0, 0)$

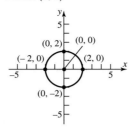

15. $\dfrac{x^2}{25} + \dfrac{y^2}{16} = 1$

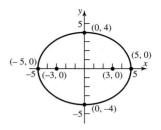

16. $\dfrac{x^2}{9} + \dfrac{y^2}{8} = 1$

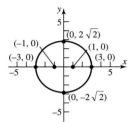

17. $\dfrac{x^2}{9} + \dfrac{y^2}{25} = 1$

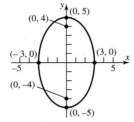

18. $\dfrac{x^2}{3} + \dfrac{y^2}{4} = 1$

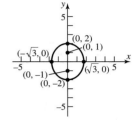

19. $\dfrac{x^2}{9} + \dfrac{y^2}{5} = 1$

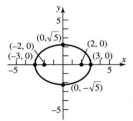

20. $\dfrac{x^2}{48} + \dfrac{y^2}{64} = 1$

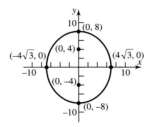

21. $\dfrac{x^2}{4} + \dfrac{y^2}{13} = 1$

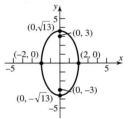

22. $\dfrac{x^2}{12} + \dfrac{y^2}{16} = 1$

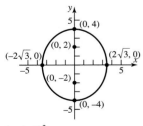

23. $x^2 + \dfrac{y^2}{16} = 1$

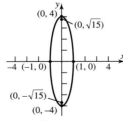

24. $\dfrac{x^2}{25} + \dfrac{y^2}{21} = 1$

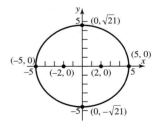

25. $\dfrac{(x + 1)^2}{4} + (y - 1)^2 = 1$ **26.** $(x + 1)^2 + \dfrac{(y + 1)^2}{4} = 1$ **27.** $(x - 1)^2 + \dfrac{y^2}{4} = 1$ **28.** $\dfrac{x^2}{4} + (y - 1)^2 = 1$

29. Center: $(3, -1)$; Vertices: $(3, -4)$, $(3, 2)$
Foci: $(3, -1 - \sqrt{5})$, $(3, -1 + \sqrt{5})$

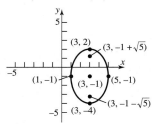

30. Center: $(-4, -2)$; Vertices: $(-7, -2)$, $(-1, -2)$;
Foci: $(-4 - \sqrt{5}, -2)$, $(-4 + \sqrt{5}, -2)$

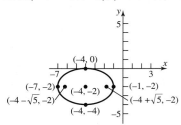

31. $\dfrac{(x + 5)^2}{16} + \dfrac{(y - 4)^2}{4} = 1$
Center: $(-5, 4)$; Vertices: $(-9, 4)$, $(-1, 4)$
Foci: $(-5 - 2\sqrt{3}, 4)$, $(-5 + 2\sqrt{3}, 4)$

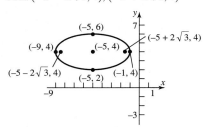

32. $\dfrac{(x - 3)^2}{2} + \dfrac{(y + 2)^2}{18} = 1$
Center: $(3, -2)$; Vertices: $(3, -2 - 3\sqrt{2})$, $(3, -2 + 3\sqrt{2})$
Foci: $(3, -6)$, $(3, 2)$

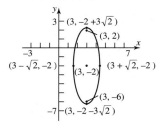

33. $\dfrac{(x + 2)^2}{4} + (y - 1)^2 = 1$
Center: $(-2, 1)$; Vertices: $(-4, 1)$, $(0, 1)$
Foci: $(-2 - \sqrt{3}, 1)$, $(-2 + \sqrt{3}, 1)$

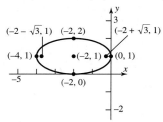

34. $\dfrac{x^2}{3} + (y - 2)^2 = 1$
Center: $(0, 2)$; Vertices: $(-\sqrt{3}, 0)$, $(\sqrt{3}, 0)$
Foci: $(-\sqrt{2}, 2)$, $(\sqrt{2}, 2)$

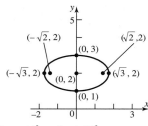

35. $\dfrac{(x - 2)^2}{3} + \dfrac{(y + 1)^2}{2} = 1$,
Center: $(2, -1)$; Vertices: $(2 - \sqrt{3}, -1)$, $(2 + \sqrt{3}, -1)$; Foci: $(1, -1)$, $(3, -1)$

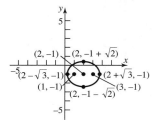

36. $\dfrac{(x + 1)^2}{3} + \dfrac{(y - 1)^2}{4} = 1$
Center: $(-1, 1)$; Vertices: $(-1, -1)$, $(-1, 3)$
Foci: $(-1, 0)$, $(-1, 2)$

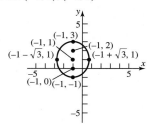

37. $\dfrac{(x - 1)^2}{4} + \dfrac{(y + 2)^2}{9} = 1$
Center: $(1, -2)$; Vertices: $(1, -5)$, $(1, 1)$
Foci: $(1, -2 - \sqrt{5})$, $(1, -2 + \sqrt{5})$

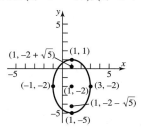

38. $\dfrac{(x + 3)^2}{9} + (y - 1)^2 = 1$
Center: $(-3, 1)$; Vertices: $(-6, 1)$, $(0, 1)$
Foci: $(-3 - \sqrt{8}, 1)$, $(-3 + \sqrt{8}, 1)$

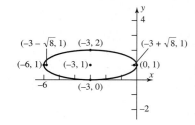

39. $x^2 + \dfrac{(y + 2)^2}{4} = 1$

Center: $(0, -2)$; Vertices: $(0, -4)$, $(0, 0)$
Foci: $(0, -2 - \sqrt{3})$, $(0, -2 + \sqrt{3})$

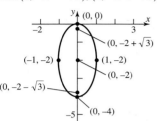

40. $(x - 1)^2 + \dfrac{y^2}{9} = 1$

Center: $(1, 0)$; Vertices: $(1, -3)$, $(1, 3)$
Foci: $(1, -\sqrt{8})$, $(1, \sqrt{8})$

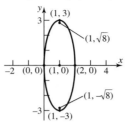

41. $\dfrac{(x - 2)^2}{25} + \dfrac{(y + 2)^2}{21} = 1$

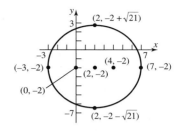

42. $\dfrac{(x + 3)^2}{3} + \dfrac{(y - 1)^2}{4} = 1$

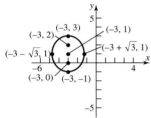

43. $\dfrac{(x - 4)^2}{5} + \dfrac{(y - 6)^2}{9} = 1$

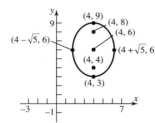

44. $\dfrac{(x + 1)^2}{9} + \dfrac{(y - 2)^2}{5} = 1$

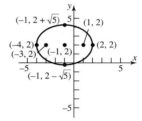

45. $\dfrac{(x - 2)^2}{16} + \dfrac{(y - 1)^2}{7} = 1$

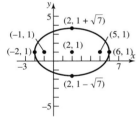

46. $\dfrac{(x - 2)^2}{5} + \dfrac{(y - 2)^2}{9} = 1$

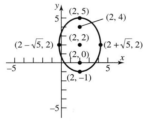

47. $\dfrac{(x - 1)^2}{10} + (y - 2)^2 = 1$

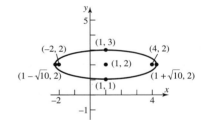

48. $\dfrac{(x - 1)^2}{1} + \dfrac{(y - 2)^2}{5} = 1$

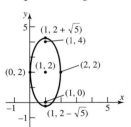

49. $\dfrac{(x - 1)^2}{9} + (y - 2)^2 = 1$

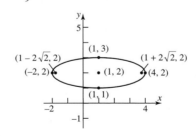

50. $\dfrac{(x - 1)^2}{1} + \dfrac{(y - 2)^2}{4} = 1$

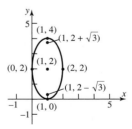

51.

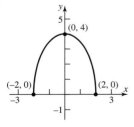

52.

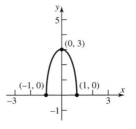

53.

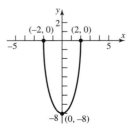

54.

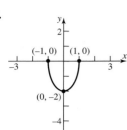

55. $\dfrac{x^2}{100} + \dfrac{y^2}{36} = 1$ **56.** at $x = 5$, distance is 2.57 ft; at $x = 10$, distance is 4.55 ft; at $x = 15$, distance is 12 ft **57.** 43.3 ft **58.** 112 ft; 25.2 ft **59.** 24.65 ft, 21.65 ft, 13.82 ft

60. 16.67 ft **61.** 0 ft, 12.99 ft, 15 ft, 12.99 ft, 0 ft **62.** 73.68 ft

63. 91.5 million miles; $\dfrac{x^2}{(93)^2} + \dfrac{y^2}{8646.75} = 1$ **64.** 155.5 million miles; $\dfrac{x^2}{(142)^2} + \dfrac{y^2}{19,981.75} = 1$

65. perihelion: 460.6 million miles; mean distance: 483.8 million miles; $\dfrac{x^2}{(483.8)^2} + \dfrac{y^2}{233,524.2} = 1$

66. aphelion: 6346 million miles; mean distance: 5448.5 million miles; $\dfrac{x^2}{5448.5^2} + \dfrac{y^2}{28,880,646} = 1$ **67.** 30 ft **68.** $10\sqrt{7}$ ft

69. (a) $Ax^2 + Cy^2 + F = 0$ If A and C are of the same and F is of opposite sign, then the equation takes the form
$Ax^2 + Cy^2 = -F$ $\dfrac{x^2}{\left(-\dfrac{F}{A}\right)} + \dfrac{y^2}{\left(-\dfrac{F}{C}\right)} = 1$, where $-\dfrac{F}{A}$ and $-\dfrac{F}{C}$ are positive. This is the equation of an ellipse with center at $(0,0)$.

(b) If $A = C$, the equation may be written as $x^2 + y^2 = -\dfrac{F}{A}$. This is the equation of a circle with center at $(0,0)$ and radius equal to $\sqrt{-\dfrac{F}{A}}$.

70. $Ax^2 + Cy^2 + Dx + Ey + F = 0$ $A \neq 0, C \neq 0$
$Ax^2 + Dx + Cy^2 + Ey = -F$
$A\left(x^2 + \dfrac{D}{A}x\right) + C\left(y^2 + \dfrac{E}{C}y\right) = -F$
$A\left(x + \dfrac{D}{2A}\right)^2 + C\left(y + \dfrac{E}{2C}\right)^2 = -F + \dfrac{D^2}{4A} + \dfrac{E^2}{4C}$
(a) If $\dfrac{D^2}{4A} + \dfrac{E^2}{4C} - F$ is of the same sign as A (and C), this is the equation of an ellipse with center at $\left(\dfrac{-D}{2A}, \dfrac{-E}{2C}\right)$.
(b) If $\dfrac{D^2}{4A} + \dfrac{E^2}{4C} - F = 0$, the graph is the single point $\left(\dfrac{-D}{2A}, \dfrac{-E}{2C}\right)$.
(c) If $\dfrac{D^2}{4A} + \dfrac{E^2}{4C} - F$ is of the opposite sign to A (and C), the graph contains no points since, in this case, the left side has a sign opposite that of the right side.

7.4 Exercises

1. B **2.** C **3.** A **4.** D

5. $x^2 - \dfrac{y^2}{8} = 1$ **6.** $\dfrac{y^2}{9} - \dfrac{x^2}{16} = 1$ **7.** $\dfrac{y^2}{16} - \dfrac{x^2}{20} = 1$

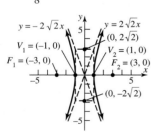 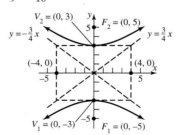 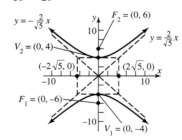

8. $\dfrac{x^2}{4} - \dfrac{y^2}{5} = 1$

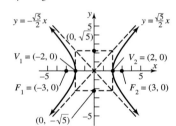

9. $\dfrac{x^2}{9} - \dfrac{y^2}{16} = 1$

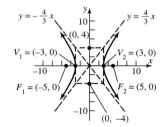

10. $\dfrac{y^2}{4} - \dfrac{x^2}{32} = 1$

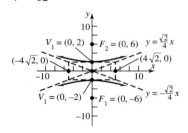

11. $\dfrac{y^2}{36} - \dfrac{x^2}{9} = 1$

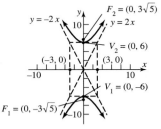

12. $\dfrac{x^2}{16} - \dfrac{y^2}{64} = 1$

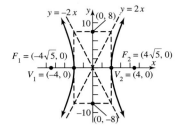

13. $\dfrac{x^2}{8} - \dfrac{y^2}{8} = 1$

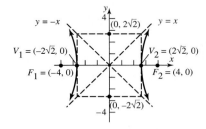

14. $\dfrac{y^2}{2} - \dfrac{x^2}{2} = 1$

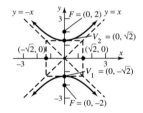

15. $\dfrac{x^2}{25} - \dfrac{y^2}{9} = 1$

Center: $(0, 0)$
Transverse axis: x-axis
Vertices: $(-5, 0)$, $(5, 0)$
Foci: $(-\sqrt{34}, 0)$, $(\sqrt{34}, 0)$

Asymptotes: $y = \pm \dfrac{3}{5}x$

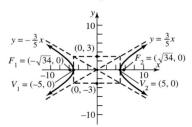

16. $\dfrac{y^2}{16} - \dfrac{x^2}{4} = 1$

Center: $(0, 0)$
Transverse axis: y-axis
Vertices: $(0, -4)$, $(0, 4)$
Foci: $(0, -2\sqrt{5})$, $(0, 2\sqrt{5})$

Asymptotes: $y = \pm 2x$

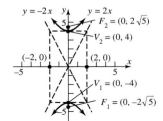

17. $\dfrac{x^2}{4} - \dfrac{y^2}{16} = 1$

Center: $(0,0)$
Transverse axis: x-axis
Vertices: $(-2, 0)$, $(2, 0)$
Foci: $(-2\sqrt{5}, 0)$, $(2\sqrt{5}, 0)$

Asymptotes: $y = \pm 2x$

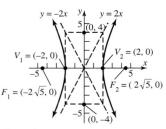

18. $\dfrac{y^2}{4} - \dfrac{x^2}{16} = 1$

Center: $(0,0)$
Transverse axis: y-axis
Vertices: $(0, -2)$, $(0, 2)$
Foci: $(0, -2\sqrt{5})$, $(0, 2\sqrt{5})$

Asymptotes: $y = \pm \dfrac{1}{2}x$

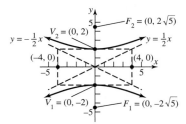

19. $\dfrac{y^2}{9} - x^2 = 1$

Center: $(0,0)$
Transverse axis: y-axis
Vertices: $(0, -3)$, $(0, 3)$
Foci: $(0, -\sqrt{10})$, $(0, \sqrt{10})$

Asymptotes: $y = \pm 3x$

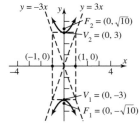

20. $\dfrac{x^2}{4} - \dfrac{y^2}{4} = 1$

Center: $(0,0)$
Transverse axis: x-axis
Vertices: $(-2, 0)$, $(2, 0)$
Foci: $(-2\sqrt{2}, 0)$, $(2\sqrt{2}, 0)$
Asymptotes: $y = \pm x$

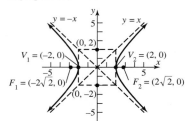

21. $\dfrac{y^2}{25} - \dfrac{x^2}{25} = 1$

Center: $(0,0)$
Transverse axis: y-axis
Vertices: $(0, -5)$, $(0, 5)$
Foci: $(0, -5\sqrt{2})$, $(0, 5\sqrt{2})$
Asymptotes: $y = \pm x$

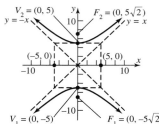

22. $\dfrac{x^2}{2} - \dfrac{y^2}{4} = 1$

Center: $(0,0)$
Transverse axis: x-axis
Vertices: $(-\sqrt{2}, 0)$, $(\sqrt{2}, 0)$
Foci: $(-\sqrt{6}, 0)$, $(\sqrt{6}, 0)$
Asymptotes: $y = \pm\sqrt{2}x$

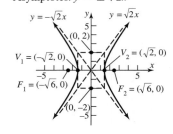

23. $x^2 - y^2 = 1$ **24.** $y^2 - x^2 = 1$ **25.** $\dfrac{y^2}{36} - \dfrac{x^2}{9} = 1$ **26.** $\dfrac{x^2}{4} - \dfrac{y^2}{16} = 1$

27. $\dfrac{(x-4)^2}{4} - \dfrac{(y+1)^2}{5} = 1$

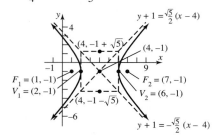

28. $\dfrac{(y-1)^2}{9} - \dfrac{(x+3)^2}{16} = 1$

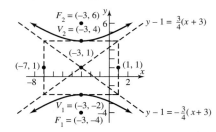

29. $\dfrac{(y+4)^2}{4} - \dfrac{(x+3)^2}{12} = 1$

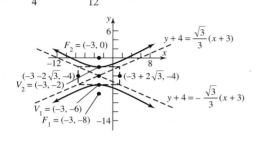

30. $\dfrac{(x-1)^2}{1} - \dfrac{(y-4)^2}{8} = 1$

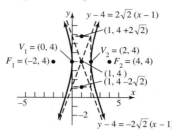

31. $(x-5)^2 - \dfrac{(y-7)^2}{3} = 1$

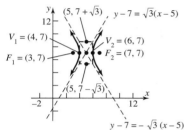

32. $\dfrac{(y-3)^2}{1} - \dfrac{(x+4)^2}{8} = 1$

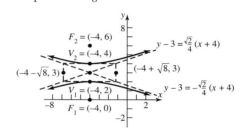

33. $\dfrac{(x-1)^2}{4} - \dfrac{(y+1)^2}{9} = 1$

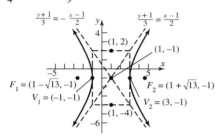

34. $\dfrac{(y+1)^2}{4} - \dfrac{9(x-1)^2}{16} = 1$

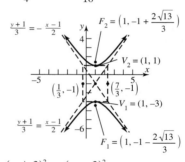

35. $\dfrac{(x-2)^2}{4} - \dfrac{(y+3)^2}{9} = 1$

Center: $(2, -3)$

Transverse axis: Parallel to x-axis

Vertices: $(0, -3), (4, -3)$

Foci: $(2 - \sqrt{13}, -3), (2 + \sqrt{13}, -3)$

Asymptotes: $y + 3 = \pm\dfrac{3}{2}(x - 2)$

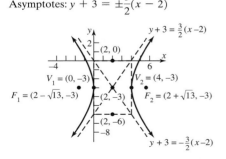

36. $\dfrac{(y+3)^2}{4} - \dfrac{(x-2)^2}{9} = 1$

Center: $(2, -3)$

Transverse axis: Parallel to x-axis

Vertices: $(2, -5), (2, -1)$

Foci: $(2, -3 + \sqrt{13}), (2, -3 - \sqrt{13})$

Asymptotes: $y + 3 = \pm\dfrac{2}{3}(x - 2)$

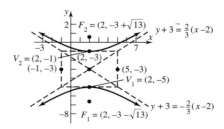

37. $\dfrac{(y-2)^2}{4} - (x+2)^2 = 1$

Center: $(-2, 2)$
Transverse axis: Parallel to y-axis
Vertices: $(-2, 0)$, $(-2, 4)$
Foci: $(-2, 2 - \sqrt{5})$, $(-2, 2 + \sqrt{5})$

Asymptotes: $y - 2 = \pm 2(x + 2)$

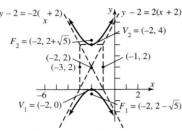

38. $\dfrac{(x+4)^2}{9} - (y-3)^2 = 1$

Center: $(-4, 3)$
Transverse axis: Parallel to x-axis
Vertices: $(-7, 3)$, $(-1, 3)$
Foci: $(-4 - \sqrt{10}, 3)$, $(-4 + \sqrt{10}, 3)$

Asymptotes: $y - 3 = \pm\dfrac{1}{3}(x + 4)$

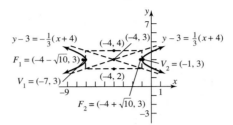

39. $\dfrac{(x+1)^2}{4} - \dfrac{(y+2)^2}{4} = 1$

Center: $(-1, -2)$
Transverse axis: Parallel to x-axis
Vertices: $(-3, -2)$, $(1, -2)$
Foci: $(-1 - 2\sqrt{2}, -2)$, $(-1 + 2\sqrt{2}, -2)$
Asymptotes: $y + 2 = \pm(x + 1)$

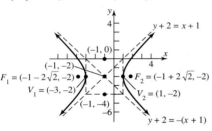

40. $\dfrac{(y-3)^2}{4} - \dfrac{(x+2)^2}{4} = 1$

Center: $(-2, 3)$
Transverse axis: Parallel to y-axis
Vertices: $(-2, 5)$, $(-2, 1)$
Foci: $(-2, 3 + 2\sqrt{2})$, $(-2, 3 - 2\sqrt{2})$
Asymptotes: $y - 3 = \pm(x + 2)$

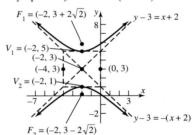

41. $(x-1)^2 - (y+1)^2 = 1$
Center: $(1, -1)$
Transverse axis: Parallel to x-axis
Vertices: $(0, -1)$, $(2, -1)$
Foci: $(1 - \sqrt{2}, -1)$, $(1 + \sqrt{2}, -1)$
Asymptotes: $y + 1 = \pm(x - 1)$

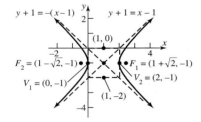

42. $(y-2)^2 - (x-2)^2 = 1$
Center: $(2, 2)$
Transverse axis: Parallel to y-axis
Vertices: $(2, 1)$, $(2, 3)$
Foci: $(2, 2 - \sqrt{2})$, $(2, 2 + \sqrt{2})$
Asymptotes: $y - 2 = \pm(x - 2)$

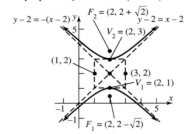

43. $\dfrac{(y-2)^2}{4} - (x+1)^2 = 1$

Center: $(-1, 2)$
Transverse axis: Parallel to y-axis
Vertices: $(-1, 0), (-1, 4)$
Foci: $(-1, 2 - \sqrt{5}), (-1, 2 + \sqrt{5})$
Asymptotes: $y - 2 = \pm 2(x + 1)$

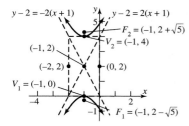

44. $(x+1)^2 - \dfrac{(y-2)^2}{2} = 1$

Center: $(-1, 2)$
Transverse axis: Parallel to x-axis
Vertices: $(-2, 2), (0, 2)$
Foci: $(-1 - \sqrt{3}, 2), (-1 + \sqrt{3}, 2)$
Asymptotes: $y - 2 = \pm\sqrt{2}(x + 1)$

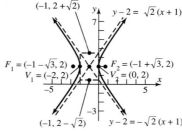

45. $\dfrac{(x-3)^2}{4} - \dfrac{(y+2)^2}{16} = 1$

Center: $(3, -2)$
Transverse axis: Parallel to x-axis
Vertices: $(1, -2), (5, -2)$
Foci: $(3 - 2\sqrt{5}, -2), (3 + 2\sqrt{5}, -2)$

Asymptotes: $y + 2 = \pm 2(x - 3)$

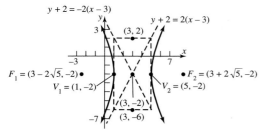

46. $\dfrac{(y+2)^2}{2} - \dfrac{(x-1)^2}{4} = 1$

Center: $(-1, 2)$
Transverse axis: Parallel to y-axis
Vertices: $(1, -2 - \sqrt{2}), (1, -2 + \sqrt{2})$
Foci: $(1, -2 - \sqrt{6}), (1, -2 + \sqrt{6})$

Asymptotes: $y + 2 = \pm\dfrac{\sqrt{2}}{2}(x - 1)$

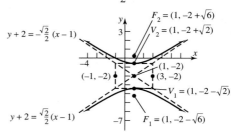

47. $\dfrac{(y-1)^2}{4} - (x+2)^2 = 1$

Center: $(-2, 1)$
Transverse axis: Parallel to y-axis
Vertices: $(-2, -1), (-2, 3)$
Foci: $(-2, 1 - \sqrt{5}), (-2, 1 + \sqrt{5})$

Asymptotes: $y - 1 = \pm 2(x + 2)$

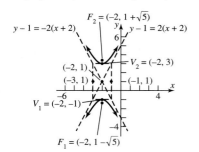

48. $\dfrac{(x+4)^2}{9} - \dfrac{(y+1)^2}{3} = 1$

Center: $(-4, -1)$
Transverse axis: Parallel to x-axis
Vertices: $(-7, 1), (-1, 1)$
Foci: $(-4 - 2\sqrt{3}, -1), (-4 + 2\sqrt{3}, -1)$

Asymptotes: $y + 1 = \pm\dfrac{\sqrt{3}}{3}(x + 4)$

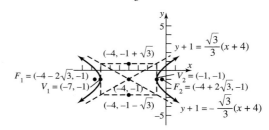

49.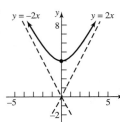

50.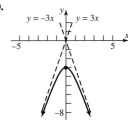

51.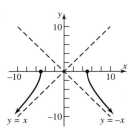

52.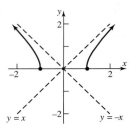

53. (a) The ship will reach shore at a point 64.66 miles from the master station.   **(b)** 0.00086 sec   **(c)** $(104, 50)$
54. (a) The ship would reach shore 20.24 miles from the master station.   **(b)** 0.0004301 sec   **(c)** $(48, 20)$
55. (a) 450 ft   **57.** If e is close to 1, narrow hyperbola; if e is very large, wide hyperbola   **58.** $\sqrt{2}$
59. $\dfrac{x^2}{4} - y^2 = 1$; asymptotes $y = \pm\dfrac{1}{2}x$, $y^2 - \dfrac{x^2}{4} = 1$; asymptotes $y = \pm\dfrac{1}{2}x$

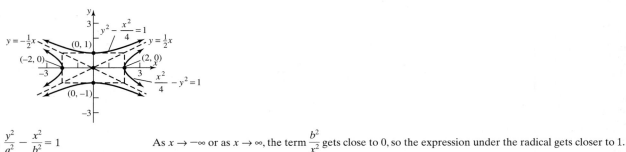

60. $\dfrac{y^2}{a^2} - \dfrac{x^2}{b^2} = 1$

As $x \to -\infty$ or as $x \to \infty$, the term $\dfrac{b^2}{x^2}$ gets close to 0, so the expression under the radical gets closer to 1.

$\dfrac{y^2}{a^2} = 1 + \dfrac{x^2}{b^2}$

Thus, the graph of the hyperbola gets closer to the lines

$y^2 = a^2\left(1 + \dfrac{x^2}{b^2}\right)$

$y = -\dfrac{a}{b}x$ and $y = \dfrac{a}{b}x$

$y^2 = \dfrac{a^2 x^2}{b^2}\left(\dfrac{b^2}{x^2} + 1\right)$

The lines are asymptotes of the hyperbola.

$y = \pm\dfrac{ax}{b}\sqrt{\dfrac{b^2}{x^2} + 1}$

61. $Ax^2 + Cy^2 + F = 0$

If A and C are of opposite sign and $F \neq 0$, this equation may be written as $\dfrac{x^2}{\left(-\dfrac{F}{A}\right)} + \dfrac{y^2}{\left(-\dfrac{F}{C}\right)} = 1$,

$Ax^2 + Cy^2 = -F$

where $-\dfrac{F}{A}$ and $-\dfrac{F}{C}$ are opposite in sign. This is the equation of a hyperbola with center $(0, 0)$.

The transverse axis is the x-axis if $-\dfrac{F}{A} > 0$; the transverse axis is the y-axis if $-\dfrac{F}{A} < 0$.

62. $Ax^2 + Cy^2 + Dx + Ey + F = 0$ where A and C

are of opposite sign

$Ax^2 + Dx + Cy^2 + Ey = -F$

$A\left(x^2 + \dfrac{D}{A}x\right) + C\left(y^2 + \dfrac{E}{C}y\right) = -F$

$A\left(x + \dfrac{D}{2A}\right)^2 + C\left(y + \dfrac{E}{2C}\right)^2 = -F + \dfrac{D^2}{4A} + \dfrac{E^2}{4C}$

(a) If $\dfrac{D^2}{4A} + \dfrac{E^2}{4C} - F \neq 0$, this is the equation of hyperbola with

center at $\left(-\dfrac{D}{2A}, -\dfrac{E}{2C}\right)$.

(b) If $\dfrac{D^2}{4A} + \dfrac{E^2}{4C} - F = 0$, the graph is intersecting lines through the

point $\left(-\dfrac{D}{2A}, -\dfrac{E}{2C}\right)$ with slopes $\pm\sqrt{\left|\dfrac{A}{C}\right|}$.

Fill-in-the-Blank Items

1. parabola   **2.** ellipse   **3.** hyperbola   **4.** major; transverse   **5.** y-axis   **6.** $\dfrac{y}{3} = \dfrac{x}{2}; \dfrac{y}{3} = \dfrac{-x}{2}$

True/False Items

1. T **2.** F **3.** T **4.** T **5.** T **6.** T

Review Exercises

1. Parabola; vertex $(0, 0)$, focus $(-4, 0)$, directrix $x = 4$ **2.** Parabola; vertex $(0, 0)$, focus $\left(0, \dfrac{1}{64}\right)$, directrix: $y = -\dfrac{1}{64}$

3. Hyperbola; center $(0, 0)$, vertices $(5, 0)$ and $(-5, 0)$, foci $(\sqrt{26}, 0)$ and $(-\sqrt{26}, 0)$, asymptotes $y = \dfrac{1}{5}x$ and $y = -\dfrac{1}{5}x$

4. Hyperbola; center $(0, 0)$, vertices $(0, 5)$ and $(0, -5)$, foci $(0, \sqrt{26})$ and $(0, -\sqrt{26})$, asymptotes $y = 5x$ and $y = -5x$

5. Ellipse; center $(0, 0)$, vertices $(0, 5)$ and $(0, -5)$, foci $(0, 3)$ and $(0, -3)$
6. Ellipse; center $(0, 0)$, vertices $(0, 4)$ and $(0, -4)$, foci $(0, \sqrt{7})$ and $(0, -\sqrt{7})$
7. $x^2 = -4(y - 1)$: Parabola; vertex $(0, 1)$, focus $(0, 0)$, directrix $y = 2$

8. $\dfrac{y^2}{3} - \dfrac{x^2}{9} = 1$: Hyperbola; center $(0, 0)$, vertices $(0, \sqrt{3})$ and $(0, -\sqrt{3})$, foci $(0, 2\sqrt{3})$ and $(0, -2\sqrt{3})$; asymptotes $y = \dfrac{\sqrt{3}}{3}x$, $y = -\dfrac{\sqrt{3}}{3}x$

9. $\dfrac{x^2}{2} - \dfrac{y^2}{8} = 1$: Hyperbola; center $(0, 0)$, vertices $(\sqrt{2}, 0)$ and $(-\sqrt{2}, 0)$, foci $(\sqrt{10}, 0)$ and $(-\sqrt{10}, 0)$, asymptotes $y = 2x$ and $y = -2x$

10. $\dfrac{x^2}{4} + \dfrac{y^2}{9} = 1$: Ellipse; center $(0, 0)$, vertices $(0, 3)$ and $(0, -3)$, foci $(0, \sqrt{5})$ and $(0, -\sqrt{5})$

11. $(x - 2)^2 = 2(y + 2)$: Parabola; vertex $(2, -2)$, focus $\left(2, -\dfrac{3}{2}\right)$, directrix $y = -\dfrac{5}{2}$

12. $(y - 1)^2 = \dfrac{1}{2}x$: Parabola; vertex $(0, 1)$, focus $\left(\dfrac{1}{8}, 1\right)$, directrix: $y = -\dfrac{1}{8}$

13. $\dfrac{(y - 2)^2}{4} - (x - 1)^2 = 1$: Hyperbola; center $(1, 2)$, vertices $(1, 4)$ and $(1, 0)$, foci $(1, 2 + \sqrt{5})$ and $(1, 2 - \sqrt{5})$, asymptotes $y - 2 = \pm 2(x - 1)$

14. $(x + 1)^2 + \dfrac{(y - 2)^2}{4} = 1$: Ellipse; center $(-1, 2)$, vertices $(-1, 0)$ and $(-1, 4)$, foci $(-1, 2 \pm \sqrt{3})$

15. $\dfrac{(x - 2)^2}{9} + \dfrac{(y - 1)^2}{4} = 1$: Ellipse; center $(2, 1)$, vertices $(5, 1)$ and $(-1, 1)$, foci $(2 + \sqrt{5}, 1)$ and $(2 - \sqrt{5}, 1)$

16. $\dfrac{(x - 2)^2}{9} + \dfrac{(y + 1)^2}{4} = 1$: Ellipse; center $(2, -1)$; vertices $(5, -1)$ and $(-1, -1)$; foci $(2 \pm \sqrt{5}, -1)$

17. $(x - 2)^2 = -4(y + 1)$: Parabola; vertex $(2, -1)$, focus $(2, -2)$, directrix $y = 0$

18. $(y - 2)^2 = -\dfrac{3}{4}(x + 1)$: Parabola; vertex $(-1, 2)$; focus $\left(-\dfrac{19}{16}, 2\right)$; directrix: $x = -\dfrac{13}{16}$

19. $\dfrac{(x - 1)^2}{4} + \dfrac{(y + 1)^2}{9} = 1$: Ellipse; center $(1, -1)$, vertices $(1, 2)$ and $(1, -4)$, foci $(1, -1 + \sqrt{5})$ and $(1, -1 - \sqrt{5})$

20. Hyperbola; center $(1, -1)$; vertices $(0, -1)$ and $(2, -1)$; foci $(1 \pm \sqrt{2}, -1)$; asymptotes: $y + 1 = \pm(x - 1)$

21. $y^2 = -8x$ **22.** $\dfrac{x^2}{16} + \dfrac{y^2}{25} = 1$ **23.** $\dfrac{y^2}{4} - \dfrac{x^2}{12} = 1$

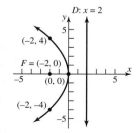

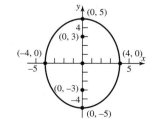

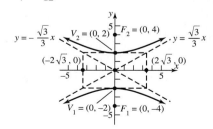

24. $x^2 = 12y$

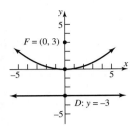

25. $\dfrac{x^2}{16} + \dfrac{y^2}{7} = 1$

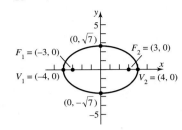

26. $\dfrac{x^2}{4} - \dfrac{y^2}{12} = 1$

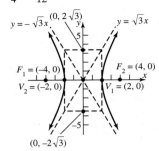

27. $(x - 2)^2 = -4(y + 3)$

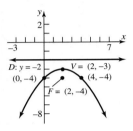

28. $\dfrac{(x + 1)^2}{9} + \dfrac{(y - 2)^2}{8} = 1$

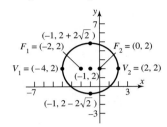

29. $(x + 2)^2 - \dfrac{(y + 3)^2}{3} = 1$

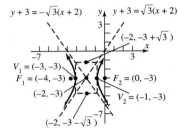

30. $(x - 3)^2 = -4(y - 7)$

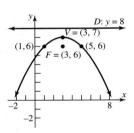

31. $\dfrac{(x + 4)^2}{16} + \dfrac{(y - 5)^2}{25} = 1$

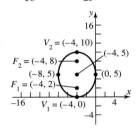

32. $\dfrac{(x - 1)^2}{16} - \dfrac{(y - 3)^2}{20} = 1$

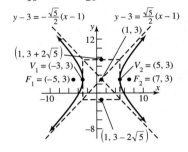

33. $\dfrac{(x + 1)^2}{9} - \dfrac{(y - 2)^2}{7} = 1$

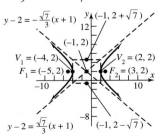

34. $(y + 2)^2 - \dfrac{(x - 4)^2}{15} = 1$

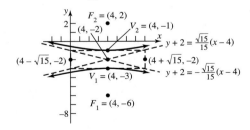

35. $\dfrac{(x-3)^2}{9} - \dfrac{(y-1)^2}{4} = 1$

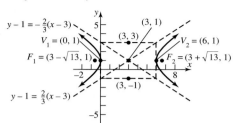

36. $\dfrac{(y-2)^2}{4} - \dfrac{(x-4)^2}{1} = 1$

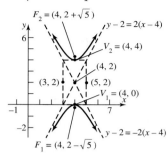

37. $\dfrac{x^2}{5} - \dfrac{y^2}{4} = 1$ **38.** $\dfrac{x^2}{4} + \dfrac{y^2}{20} = 1$ **39.** The ellipse $\dfrac{x^2}{16} + \dfrac{y^2}{7} = 1$ **40.** The hyperbola $\dfrac{x^2}{16} - \dfrac{y^2}{9} = 1$ **41.** $\dfrac{1}{4}$ ft or 3 in.

42. 19.44 ft, 17.78 ft, 11.11 ft   **43.** 19.72 ft, 18.86 ft, 14.91 ft   **44.** The foci are located 8.8 ft in from each wall.

45. **(a)** 45.24 miles from the master station   **(b)** 0.000645 sec   **(c)** $(66, 20)$

C H A P T E R 8   8.1 Exercises

1. $2(2) - (-1) = 5$ and $5(2) + 2(-1) = 8$ **2.** $3(-2) + 2(4) = 2$ and $(-2) - 7(4) = -30$

3. $3(2) - 4\left(\dfrac{1}{2}\right) = 4$ and $\dfrac{1}{2}(2) - 3\left(\dfrac{1}{2}\right) = -\dfrac{1}{2}$ **4.** $2\left(-\dfrac{1}{2}\right) + \dfrac{1}{2}(2) = 0$ and $3\left(-\dfrac{1}{2}\right) - 4(2) = -\dfrac{19}{2}$ **5.** $2^2 - 1^2 = 3$ and $(2)(1) = 2$

6. $(-2)^2 - (-1)^2 = 3$ and $(-2)(-1) = 2$ **7.** $\dfrac{0}{1+0} + 3(2) = 6$ and $0 + 9(2)^2 = 36$ **8.** $\dfrac{2}{2-1} + 3 = 5$ and $3(2) - 3 = 3$

9. $3(1) + 3(-1) + 2(2) = 4$, $1 - (-1) - 2 = 0$, and $2(-1) - 3(2) = -8$

10. $4(2) - 1 = 7$, $8(2) + 5(-3) - 1 = 0$, and $-2 - (-3) + 5(1) = 6$ **11.** $x = 6, y = 2$ **12.** $x = 1, y = 2$ **13.** $x = 3, y = 2$

14. $x = -1, y = 2$ **15.** $x = 8, y = -4$ **16.** $x = -\dfrac{13}{4}, y = 2$ **17.** $x = \dfrac{1}{3}, y = -\dfrac{1}{6}$ **18.** $x = -\dfrac{5}{3}, y = 1$ **19.** Inconsistent

20. Inconsistent   **21.** $x = 1, y = 2$ **22.** $x = 1, y = -\dfrac{4}{3}$ **23.** $x = 4 - 2y$, y is any real number   **24.** $y = 3x - 7$, x is any real number

25. $x = 1, y = 1$ **26.** $x = \dfrac{1}{5}, y = \dfrac{3}{10}$ **27.** $x = \dfrac{3}{2}, y = 1$ **28.** $x = 2, y = -3$ **29.** $x = 4, y = 3$ **30.** $x = 12, y = 6$

31. $x = \dfrac{4}{3}, y = \dfrac{1}{5}$ **32.** $x = \dfrac{1}{2}, y = 2$ **33.** $x = 8, y = 2, z = 0$ **34.** $x = -3, y = 2, z = 1$ **35.** $x = 2, y = -1, z = 1$

36. $x = \dfrac{56}{13}, y = -\dfrac{7}{13}, z = \dfrac{35}{13}$ **37.** Inconsistent   **38.** Inconsistent   **39.** $x = 5z - 2, y = 4z - 3$ where z is any real number,

or $x = \dfrac{5}{4}y + \dfrac{7}{4}, z = \dfrac{1}{4}y + \dfrac{3}{4}$ where y is any real number, or $y = \dfrac{4}{5}x - \dfrac{7}{5}, z = \dfrac{1}{5}x + \dfrac{2}{5}$ where x is any real number

40. $x = -\dfrac{4}{13}z + \dfrac{6}{13}, y = -\dfrac{7}{13}z + \dfrac{4}{13}$, where z is any real number, or $x = \dfrac{4}{7}y + \dfrac{2}{7}, z = -\dfrac{13}{7}y + \dfrac{4}{7}$, where y is any real number, or

$y = \dfrac{7}{4}x - \dfrac{1}{2}, z = -\dfrac{13}{4}x + \dfrac{3}{2}$, where x is any real number   **41.** Inconsistent   **42.** Inconsistent   **43.** $x = 1, y = 3, z = -2$

44. $x = 1, y = 3, z = -2$ **45.** $x = -3, y = \dfrac{1}{2}, z = 1$ **46.** $x = 3, y = -\dfrac{8}{3}, z = \dfrac{1}{9}$ **47.** $x = \dfrac{1}{5}, y = \dfrac{1}{3}$ **48.** $x = 4, y = 3$

49. $x = 48.15, y = 15.18$ **50.** $x = 49.50, y = 14.25$ **51.** $x = -21.47, y = 16.12$ **52.** $x = -20.22, y = 6.03$ **53.** $x = 0.26, y = 0.06$

54. $x = -0.28, y = 0.15$ **55.** 20 and 61   **56.** 49 and 9   **57.** Length 30 ft; width 15 ft   **58.** 725 m by 775 m

59. Cheeseburger $1.55; shake $0.85   **60.** 110 adults, 215 senior citizens   **61.** 22.5 lbs

62. **(a)** $90,000 in AA bonds and $60,000 in a Bank Certificate   **(b)** $130,000 in AA bonds and $20,000 in a Bank Certificate

63. Average wind speed 25 mph; average airspeed 175 mph   **64.** wind speed was 30 mph   **65.** 80 $25 sets and 120 $45 sets

66. Hot dog $1.00, soft drink $0.50   **67.** $5.56   **68.** Average speed of the swimmer is 4 mph; speed of current is 1 mph

69. Mix 50 mg of the first liquid with 75 mg of the second.   **70.** 30 units of the first powder and 15 units of the second powder.

71. $I_1 = \dfrac{10}{71}, I_2 = \dfrac{65}{71}, I_3 = \dfrac{55}{71}$ **72.** $I_1 = \dfrac{11}{13}, I_2 = \dfrac{6}{13}, I_3 = \dfrac{17}{13}$ **73.** 100 orchestra, 210 main, and 190 balcony seats

74. 10 work stations for 2 students; 6 work stations for 3 students   **78.** $b = -3, c = 5$ **79.** $b = -\dfrac{1}{2}; c = \dfrac{3}{2}$

80. $b = \dfrac{y_2 - y_1 + x_1^2 - x_2^2}{x_2 - x_1}, c = y_1 - x_1^2 - \dfrac{(y_2 - y_1 + x_1^2 - x_2^2)}{(x_2 - x_1)}x_1$ 81. $a = \dfrac{4}{3}, b = -\dfrac{5}{3}, c = 1$ 82. $a = 3, b = -1, c = -6$

83. $x = \dfrac{b_1 - b_2}{m_2 - m_1}, y = \dfrac{m_2 b_1 - m_1 b_2}{m_2 - m_1}$ 84. No solution   85. $y = mx + b, x$ is any real number

8.2 Exercises

1. $\left[\begin{array}{rr|r} 1 & -5 & 5 \\ 4 & 3 & 6 \end{array}\right]$ 2. $\left[\begin{array}{rr|r} 3 & 4 & 7 \\ 4 & -2 & 5 \end{array}\right]$ 3. $\left[\begin{array}{rr|r} 2 & 3 & 6 \\ 4 & -6 & -2 \end{array}\right]$ 4. $\left[\begin{array}{rr|r} 9 & -1 & 0 \\ 3 & -1 & 4 \end{array}\right]$ 5. $\left[\begin{array}{rr|r} 0.01 & -0.03 & 0.06 \\ 0.13 & 0.10 & 0.20 \end{array}\right]$ 6. $\left[\begin{array}{rr|r} \frac{4}{3} & -\frac{3}{2} & \frac{3}{4} \\ -\frac{1}{4} & \frac{1}{3} & \frac{2}{3} \end{array}\right]$

7. $\left[\begin{array}{rrr|r} 1 & -1 & 1 & 10 \\ 3 & 2 & 0 & 5 \\ 1 & 1 & 2 & 2 \end{array}\right]$ 8. $\left[\begin{array}{rrr|r} 5 & -1 & -1 & 0 \\ 1 & 1 & 0 & 5 \\ 2 & 0 & -3 & 2 \end{array}\right]$ 9. $\left[\begin{array}{rrr|r} 1 & 1 & -1 & 2 \\ 3 & -2 & 0 & 2 \end{array}\right]$ 10. $\left[\begin{array}{rrr|r} 2 & 3 & -4 & 0 \\ 1 & 0 & -5 & -2 \end{array}\right]$ 11. $\left[\begin{array}{rrrr|r} 1 & -1 & -1 & -1 & 5 \\ 3 & 1 & -4 & 1 & 4 \end{array}\right]$

12. $\left[\begin{array}{rrrr|r} 2 & -1 & 1 & 0 & 0 \\ 1 & -1 & 0 & 2 & 1 \end{array}\right]$ 13. $\left[\begin{array}{rrr|r} 1 & -3 & -5 & -2 \\ 0 & 1 & 6 & 9 \\ 0 & 0 & 13 & 36 \end{array}\right]$ 14. $\left[\begin{array}{rrr|r} 1 & -3 & -3 & -3 \\ 0 & 1 & 8 & 2 \\ 0 & 0 & 51 & 11 \end{array}\right]$ 15. $\left[\begin{array}{rrr|r} 1 & -3 & 4 & 3 \\ 0 & 1 & -2 & 0 \\ 0 & 0 & 4 & 15 \end{array}\right]$ 16. $\left[\begin{array}{rrr|r} 1 & -3 & -3 & -5 \\ 0 & 1 & 3 & 5 \\ 0 & 0 & 28 & 46 \end{array}\right]$

17. $\left[\begin{array}{rrr|r} 1 & -3 & 2 & -6 \\ 0 & 1 & -1 & 8 \\ 0 & 0 & -5 & 108 \end{array}\right]$ 18. $\left[\begin{array}{rrr|r} 1 & -3 & -4 & -6 \\ 0 & 1 & 14 & 6 \\ 0 & 0 & 104 & 36 \end{array}\right]$ 19. $\left[\begin{array}{rrr|r} 1 & -3 & 1 & -2 \\ 0 & 1 & 4 & 2 \\ 0 & 0 & 39 & 16 \end{array}\right]$ 20. $\left[\begin{array}{rrr|r} 1 & -3 & -1 & 2 \\ 0 & 1 & 4 & 2 \\ 0 & 0 & 61 & 42 \end{array}\right]$

21. $\left[\begin{array}{rrr|r} 1 & -3 & -2 & 3 \\ 0 & 1 & 6 & -7 \\ 0 & 0 & 64 & -62 \end{array}\right]$ 22. $\left[\begin{array}{rrr|r} 1 & -3 & 5 & -3 \\ 0 & 1 & -9 & 2 \\ 0 & 0 & -35 & 9 \end{array}\right]$

23. $\begin{cases} x = 5 \\ y = -1 \end{cases}$
consistent; $x = 5, y = -1$

24. $\begin{cases} x = -4 \\ y = 0 \end{cases}$
consistent; $x = -4, y = 0$

25. $\begin{cases} x = 1 \\ y = 2 \\ 0 = 3 \end{cases}$
inconsistent

26. $\begin{cases} x = 0 \\ y = 0 \\ 0 = 2 \end{cases}$
inconsistent

27. $\begin{cases} x + 2z = -1 \\ y - 4z = -2 \\ \quad\ 0 = 0 \end{cases}$
consistent;
$x = -1 - 2z,$
$y = -2 + 4z,$
z is any real number

28. $\begin{cases} x + 4z = 4 \\ y + 3z = 2 \\ \quad\ 0 = 0 \end{cases}$
consistent;
$x = 4 - 4z,$
$y = 2 - 3z,$
z is any real number

29. $\begin{cases} x_1 = 1 \\ x_2 + x_4 = 2 \\ x_3 + 2x_4 = 3 \end{cases}$
consistent;
$x_1 = 1, x_2 = 2 - x_4,$
$x_3 = 3 - 2x_4,$
x_4 is any real number

30. $\begin{cases} x_1 = 1 \\ x_2 + 2x_4 = 2 \\ x_3 + 3x_4 = 0 \end{cases}$
consistent;
$x_1 = 1,$
$x_2 = 2 - 2x_4,$
$x_3 = -3x_4,$
x_4 is any real number

31. $\begin{cases} x_1 + 4x_4 = 2 \\ x_2 + x_3 + 3x_4 = 3 \\ \qquad\quad 0 = 0 \end{cases}$
consistent;
$x_1 = 2 - 4x_4,$
$x_2 = 3 - x_3 - 3x_4,$
x_3, x_4 are any real numbers

32. $\begin{cases} x_1 = 1 \\ x_2 = 2 \\ x_3 + 2x_4 = 3 \end{cases}$
consistent;
$x_1 = 1,$
$x_2 = 2,$
$x_3 = 3 - 2x_4,$
x_4 is any real number

33. $\begin{cases} x_1 + x_4 = -2 \\ x_2 + 2x_4 = 2 \\ x_3 - x_4 = 0 \end{cases}$
consistent;
$x_1 = -2 - x_4,$
$x_2 = 2 - 2x_4,$
$x_3 = x_4,$
x_4 is any real number

34. $\begin{cases} x_1 = 1 \\ x_2 = 2 \\ x_3 = 3 \\ x_4 = 0 \end{cases}$
consistent;
$x_1 = 1,$
$x_2 = 2,$
$x_3 = 3,$
$x_4 = 0$

35. $x = 6, y = 2$ 36. $x = 1, y = 2$ 37. $x = 2, y = 3$ 38. $x = -1, y = 2$ 39. $x = 4, y = -2$ 40. $x = 2, y = 3$ 41. Inconsistent

42. Inconsistent   43. $x = \dfrac{1}{2}, y = \dfrac{3}{4}$ 44. $x = \dfrac{1}{3}, y = \dfrac{2}{3}$ 45. $x = 4 - 2y, y$ is any real number   46. $y = 3x - 7, x$ is any real number

47. $x = \dfrac{3}{2}, y = 1$ 48. $x = 2, y = -3$ 49. $x = \dfrac{4}{3}, y = \dfrac{1}{5}$ 50. $x = \dfrac{1}{2}, y = 2$ 51. $x = 8, y = 2, z = 0$ 52. $x = -3, y = 2, z = 1$

53. $x = 2, y = -1, z = 1$ 54. $x = \dfrac{56}{13}, y = -\dfrac{7}{13}, z = \dfrac{35}{13}$ 55. Inconsistent   56. Inconsistent   57. $x = 5z - 2, y = 4z - 3$, where z is

any real number, or $x = \dfrac{5}{4}y + \dfrac{7}{4}, z = \dfrac{1}{4}y + \dfrac{3}{4},$ where y is any real number, or $y = \dfrac{4}{5}x - \dfrac{7}{5}, z = \dfrac{1}{5}x + \dfrac{2}{5},$ where x is any real number

58. $x = -\dfrac{4}{13}z + \dfrac{6}{13}, y = -\dfrac{7}{13}z + \dfrac{4}{13},$ where z is any real number   59. Inconsistent   60. Inconsistent   61. $x = 1, y = 3, z = -2$

62. $x = 1, y = 3, z = -2$ 63. $x = -3, y = \dfrac{1}{2}, z = 1$ 64. $x = 3, y = -\dfrac{8}{3}, z = \dfrac{1}{9}$ 65. $x = \dfrac{1}{3}, y = \dfrac{2}{3}, z = 1$ 66. $x = \dfrac{1}{3}, y = \dfrac{2}{3}, z = 1$

67. $x = 1, y = 2, z = 0, w = 1$ 68. $x = 2, y = 1, z = 0, w = 1$ 69. $y = 0, z = 1 - x, x$ is any real number   70. Inconsistent

71. $x = 2, y = z - 3, z$ is any real number   **72.** $x = 1 + \frac{4}{3}z, y = 2 - \frac{5}{3}z$, where z is any real number   **73.** $x = \frac{13}{9}, y = \frac{7}{18}, z = \frac{19}{18}$

74. Inconsistent   **75.** $x = \frac{7}{5} - \frac{3}{5}z - \frac{2}{5}w, y = -\frac{8}{5} + \frac{7}{5}z + \frac{13}{5}w$, where z and w are any real numbers

76. $x = -3 - w, y = -7 - 4w, z = 4 - w$, where w is any real number   **77.** $y = -2x^2 + x + 3$ **78.** $y = x^2 - 4x + 2$
79. $f(x) = 3x^3 - 4x^2 + 5$ **80.** $f(x) = x^3 - 2x^2 + 6$

81. $x = $ liters of 15% $H_2SO_4, y = $ liters of 25% $H_2SO_4, z = $ liters of 50% H_2SO_4: $\begin{cases} x = \frac{5}{2}z - 150 \\ y = 250 - \frac{7}{2}z \end{cases}$

15%	25%	50%	40%
0	40	60	100
10	26	64	100
20	12	68	100

82. It will take Beth 30 hrs, Bill 24 hrs, and Edie 40 hrs.

83. If $x = $ Price of hamburgers, $y = $ Price of fries, $z = $ Price of colas, then $x = 2.75 - z, y = 0.68 + \frac{1}{3}z, z$ is any real number.

There is not sufficient information:

x	$2.15	$2.00	$1.85
y	$0.88	$0.93	$0.98
z	$0.60	$0.75	$0.90

84. Yes; hamburgers are $1.95, fries are $.95, cola is $.80.
85. (a)

Amount Invested At

7%	9%	11%
0	10,000	10,000
1000	8000	11,000
2000	6000	12,000
3000	4000	13,000
4000	2000	14,000
5000	0	15,000

(b)

Amount Invested At

7%	9%	11%
12,500	12,500	0
14,500	8500	2000
16,500	4500	4000
18,750	0	6250

(c) All the money invested at 7% provides $2100, more than what is required.

86. (a) Investing all the money at 7% yields more than $1500
(b)

7%	9%	11%
12,500	12,500	0
15,500	6500	3000
18,750	0	6250

(c)

7%	9%	11%
0	12,500	12,500
1000	10,500	13,500
6250	0	18,750

87.

First Liquid	Second Liquid	Third Liquid
50 mg	75 mg	0 mg
36 mg	76 mg	8 mg
22 mg	77 mg	16 mg
8 mg	78 mg	24 mg

88.

First Powder	Second Powder	Third Powder
30 units	15 units	0 units
20 units	14 units	8 units
10 units	13 units	16 units
0 units	12 units	24 units

89. $I_1 = \frac{4}{15}, I_2 = \frac{8}{15}, I_3 = \frac{4}{5}$ **90.** $I_1 = \frac{44}{23}, I_2 = 2, I_3 = \frac{16}{23}, I_4 = \frac{28}{23}$ **91.** $I_1 = 3.5, I_2 = 2.5, I_3 = 1$

95. If $a_1 \neq 0$,

$$\begin{bmatrix} a_1 & b_1 & | & c_1 \\ a_2 & b_2 & | & c_2 \end{bmatrix} \rightarrow \begin{bmatrix} 1 & \frac{b_1}{a_1} & | & \frac{c_1}{a_1} \\ a_2 & b_2 & | & c_2 \end{bmatrix} \rightarrow \begin{bmatrix} 1 & \frac{b_1}{a_1} & | & \frac{c_1}{a_1} \\ 0 & \frac{-a_2 b_1}{a_1} + b_2 & | & \frac{-a_2 c_1}{a_1} + c_2 \end{bmatrix} \rightarrow \begin{bmatrix} 1 & \frac{b_1}{a_1} & | & \frac{c_1}{a_1} \\ 0 & \frac{-a_2 b_1 + b_2 a_1}{a_1} & | & \frac{-a_2 c_1 + c_2 a_1}{a_1} \end{bmatrix}$$

$$\rightarrow \begin{bmatrix} 1 & \frac{b_1}{a_1} & | & \frac{c_1}{a_1} \\ 0 & 1 & | & \frac{-a_2 c_1 + c_2 a_1}{a_1} \cdot \frac{a_1}{-a_2 b_1 + b_2 a_1} \end{bmatrix} \rightarrow \begin{bmatrix} 1 & \frac{b_1}{a_1} & | & \frac{c_1}{a_1} \\ 0 & 1 & | & \frac{-a_2 c_1 + c_2 a_1}{-a_2 b_1 + b_2 a_1} \end{bmatrix} \rightarrow \begin{bmatrix} 1 & 0 & | & \frac{-b_1 c_2 + b_2 c_1}{-a_2 b_1 + b_2 a_1} \\ 0 & 1 & | & \frac{-a_2 c_1 + c_2 a_1}{-a_2 b_1 + b_2 a_1} \end{bmatrix}$$

$$x = \frac{1}{a_1 b_2 - a_2 b_1}(c_1 b_2 - c_2 b_1) = \frac{1}{D}(c_1 b_2 - c_2 b_1), \quad y = \frac{1}{a_1 b_2 - a_2 b_1}(a_1 c_2 - a_2 c_1) = \frac{1}{D}(a_1 c_2 - a_2 c_1)$$

If $a_1 = 0$, then $a_2 \neq 0, b_1 \neq 0$, and

$$\begin{bmatrix} 0 & b_1 & | & c_1 \\ a_2 & b_2 & | & c_2 \end{bmatrix} \rightarrow \begin{bmatrix} a_2 & b_2 & | & c_2 \\ 0 & b_1 & | & c_1 \end{bmatrix} \rightarrow \begin{bmatrix} 1 & \frac{b_2}{a_2} & | & \frac{c_2}{a_2} \\ 0 & b_1 & | & c_1 \end{bmatrix} \rightarrow \begin{bmatrix} 1 & \frac{b_2}{a_2} & | & \frac{c_2}{a_2} \\ 0 & 1 & | & \frac{c_1}{b_1} \end{bmatrix} \rightarrow \begin{bmatrix} 1 & 0 & | & \frac{c_2}{a_2} - \frac{b_2 c_1}{a_2 b_1} = \frac{c_1 b_2 - c_2 b_1}{-a_2 b_1} \\ 0 & 1 & | & \frac{c_1}{b_1} = \frac{-a_2 c_1}{-a_2 b_1} \end{bmatrix}$$

96. Sketch of proof: Following the solution for Problem 95, if $a_1 c_2 \neq a_2 c_1$ and $b_1 c_2 \neq b_2 c_1$ then using row operations we can obtain a row of zeros on the left side of the matrix with a nonzero entry on the right. If both $a_1 c_2 = a_2 c_1$ and $b_1 c_2 = b_2 c_1$ then if $a_1 \neq 0$,

$$\begin{bmatrix} a_1 & b_1 & | & c_1 \\ a_2 & b_2 & | & c_2 \end{bmatrix} \rightarrow \begin{bmatrix} 1 & \frac{b_1}{a_1} & | & \frac{c_1}{a_1} \\ a_2 & b_2 & | & c_2 \end{bmatrix} \rightarrow \begin{bmatrix} 1 & \frac{b_1}{a_1} & | & \frac{c_1}{a_1} \\ 0 & \frac{-a_2 b_1}{a_1} + b_2 & | & \frac{-a_2 c_1}{a_1} + c_2 \end{bmatrix} \rightarrow \begin{bmatrix} 1 & \frac{b_1}{a_1} & | & \frac{c_1}{a_1} \\ 0 & \frac{-a_2 b_1 + b_2 a_1}{a_1} & | & \frac{-a_2 c_1 + c_2 a_1}{a_1} \end{bmatrix}$$

$$\rightarrow \begin{bmatrix} 1 & \frac{b_1}{a_1} & | & \frac{c_1}{a_1} \\ 0 & 0 & | & 0 \end{bmatrix} \rightarrow \begin{bmatrix} a_1 & b_1 & | & c_1 \\ 0 & 0 & | & 0 \end{bmatrix}$$

If $a_1 = 0$, then $a_2 = 0$ or $b_1 = 0$, and $a_2 c_1 = 0$ and $a_2 b_1 = 0$. If $a_2 \neq 0$, then $c_1 = 0$ and $b_1 = 0$ so we have a row of zeros. If $a_2 = 0$ and $b_1 \neq 0$ then $y = \frac{c_1}{b_1} = \frac{c_2}{b_2}$ and x can be any real number.

8.3 Exercises

1. 2 **2.** 7 **3.** 22 **4.** 28 **5.** -2 **6.** -2 **7.** 10 **8.** -169 **9.** -26 **10.** -119 **11.** $x = 6, y = 2$ **12.** $x = \frac{11}{3}, y = \frac{2}{3}$

13. $x = 3, y = 2$ **14.** $x = -1, y = 2$ **15.** $x = 8, y = -4$ **16.** $x = -\frac{13}{4}, y = 2$ **17.** $x = 4, y = -2$ **18.** $x = 2, y = 3$

19. Not applicable **20.** Not applicable **21.** $x = \frac{1}{2}, y = \frac{3}{4}$ **22.** $x = \frac{1}{3}, y = \frac{2}{3}$ **23.** $x = \frac{1}{10}, y = \frac{2}{5}$ **24.** $x = \frac{1}{5}, y = \frac{3}{10}$

25. $x = \frac{3}{2}, y = 1$ **26.** $x = 2, y = -3$ **27.** $x = \frac{4}{3}, y = \frac{1}{5}$ **28.** Not applicable **29.** $x = 1, y = 3, z = -2$ **30.** $x = 1, y = 3, z = -2$

31. $x = -3, y = \frac{1}{2}, z = 1$ **32.** $x = 3, y = -\frac{8}{3}, z = \frac{1}{9}$ **33.** Not applicable **34.** Not applicable **35.** $x = 0, y = 0, z = 0$

36. $x = 0, y = 0, z = 0$ **37.** Not applicable **38.** Not applicable **39.** $x = \frac{1}{5}, y = \frac{1}{3}$ **40.** $x = 4, y = 3$ **41.** -5 **42.** 1 or -1 **43.** $\frac{13}{11}$

44. $-\frac{7}{6}$ **45.** 0 or -9 **46.** 0 or $-\frac{1}{2}$ **47.** -4 **48.** 8 **49.** 12 **50.** 4 **51.** 8 **52.** 4 **53.** 8 **54.** 12

55. $(y_1 - y_2)x - (x_1 - x_2)y + (x_1 y_2 - x_2 y_1) = 0$
$(y_1 - y_2)x + (x_2 - x_1)y = x_2 y_1 - x_1 y_2$
$(x_2 - x_1)y - (x_2 - x_1)y_1 = (y_2 - y_1)x + x_2 y_1 - x_1 y_2 - (x_2 - x_1)y_1$
$(x_2 - x_1)(y - y_1) = (y_2 - y_1)x - (y_2 - y_1)x_1$
$y - y_1 = \frac{y_2 - y_1}{x_2 - x_1}(x - x_1)$

56. This result follows from Problem 55 by exchanging rows.

57. $\begin{vmatrix} x^2 & x & 1 \\ y^2 & y & 1 \\ z^2 & z & 1 \end{vmatrix} = x^2 \begin{vmatrix} y & 1 \\ z & 1 \end{vmatrix} - x \begin{vmatrix} y^2 & 1 \\ z^2 & 1 \end{vmatrix} + \begin{vmatrix} y^2 & y \\ z^2 & z \end{vmatrix} = x^2(y - z) - x(y^2 - z^2) + yz(y - z)$

$$= (y - z)[x^2 - x(y + z) + yz] = (y - z)[(x^2 - xy) - (xz - yz)] = (y - z)[x(x - y) - z(x - y)]$$
$$= (y - z)(x - y)(x - z)$$

58. Follow hint given in problem statement.

59. $\begin{vmatrix} a_{13} & a_{12} & a_{11} \\ a_{23} & a_{22} & a_{21} \\ a_{33} & a_{32} & a_{31} \end{vmatrix} = a_{13}(a_{22}a_{31} - a_{32}a_{21}) - a_{12}(a_{23}a_{31} - a_{33}a_{21}) + a_{11}(a_{23}a_{32} - a_{33}a_{22})$

$$= -[a_{11}(a_{22}a_{33} - a_{32}a_{23}) - a_{12}(a_{21}a_{33} - a_{31}a_{23}) + a_{13}(a_{21}a_{32} - a_{31}a_{22})] = -\begin{vmatrix} a_{11} & a_{12} & a_{13} \\ a_{21} & a_{22} & a_{23} \\ a_{31} & a_{32} & a_{33} \end{vmatrix}$$

60. $\begin{vmatrix} a_{11} & a_{12} & a_{13} \\ ka_{21} & ka_{22} & ka_{23} \\ a_{31} & a_{32} & a_{33} \end{vmatrix} = a_{11} \begin{vmatrix} ka_{22} & ka_{23} \\ a_{32} & a_{33} \end{vmatrix} - a_{12} \begin{vmatrix} ka_{21} & ka_{23} \\ a_{31} & a_{33} \end{vmatrix} + a_{13} \begin{vmatrix} ka_{21} & ka_{22} \\ a_{31} & a_{32} \end{vmatrix}$

$$= a_{11}(ka_{22}a_{33} - ka_{23}a_{32}) - a_{12}(ka_{21}a_{33} - ka_{23}a_{31}) + a_{13}(ka_{21}a_{32} - ka_{22}a_{31})$$
$$= ka_{11}(a_{22}a_{33} - a_{23}a_{32}) - ka_{12}(a_{21}a_{33} - a_{23}a_{31}) + ka_{13}(a_{21}a_{32} - a_{22}a_{31})$$

$$= ka_{11} \begin{vmatrix} a_{22} & a_{23} \\ a_{32} & a_{33} \end{vmatrix} - ka_{12} \begin{vmatrix} a_{21} & a_{23} \\ a_{31} & a_{33} \end{vmatrix} + ka_{13} \begin{vmatrix} a_{21} & a_{22} \\ a_{31} & a_{32} \end{vmatrix} = k \cdot \begin{vmatrix} a_{11} & a_{12} & a_{13} \\ a_{21} & a_{22} & a_{23} \\ a_{31} & a_{32} & a_{33} \end{vmatrix}$$

61. $\begin{vmatrix} a_{11} & a_{12} & a_{11} \\ a_{21} & a_{22} & a_{21} \\ a_{31} & a_{32} & a_{31} \end{vmatrix} = a_{11}(a_{22}a_{31} - a_{32}a_{21}) - a_{12}(a_{21}a_{31} - a_{31}a_{21}) + a_{11}(a_{21}a_{32} - a_{31}a_{22})$

$$= a_{11}a_{22}a_{31} - a_{11}a_{32}a_{21} - a_{12}(0) + a_{11}a_{21}a_{32} - a_{11}a_{31}a_{22} = 0$$

62. $\begin{vmatrix} ka_{21} + a_{11} & ka_{22} + a_{12} & ka_{23} + a_{13} \\ a_{21} & a_{22} & a_{23} \\ a_{31} & a_{32} & a_{33} \end{vmatrix} = (ka_{21} + a_{11}) \begin{vmatrix} a_{22} & a_{23} \\ a_{32} & a_{33} \end{vmatrix} - (ka_{22} + a_{12}) \begin{vmatrix} a_{21} & a_{23} \\ a_{31} & a_{33} \end{vmatrix} + (ka_{23} + a_{13}) \begin{vmatrix} a_{21} & a_{22} \\ a_{31} & a_{32} \end{vmatrix}$

$$= (ka_{21} + a_{11})(a_{22}a_{33} - a_{23}a_{32}) - (ka_{22} + a_{12})(a_{21}a_{33} - a_{23}a_{31})$$
$$+ (ka_{23} + a_{13})(a_{21}a_{32} - a_{22}a_{31})$$
$$= ka_{21}(a_{22}a_{33} - a_{23}a_{32}) - ka_{22}(a_{21}a_{33} - a_{23}a_{31}) + ka_{23}(a_{21}a_{32} - a_{22}a_{31})$$
$$+ a_{11}(a_{22}a_{33} - a_{23}a_{32}) - a_{12}(a_{21}a_{33} - a_{23}a_{31}) + a_{13}(a_{21}a_{32} - a_{22}a_{31})$$
$$= ka_{21}a_{22}a_{33} - ka_{21}a_{23}a_{32} - ka_{22}a_{21}a_{33} + ka_{22}a_{23}a_{31} + ka_{23}a_{21}a_{32} - ka_{23}a_{22}a_{31}$$

$$+ a_{11} \begin{vmatrix} a_{22} & a_{23} \\ a_{32} & a_{33} \end{vmatrix} - a_{12} \begin{vmatrix} a_{21} & a_{23} \\ a_{31} & a_{33} \end{vmatrix} + a_{13} \begin{vmatrix} a_{21} & a_{22} \\ a_{31} & a_{32} \end{vmatrix}$$

$$= 0 + \begin{vmatrix} a_{11} & a_{12} & a_{13} \\ a_{21} & a_{22} & a_{23} \\ a_{31} & a_{32} & a_{33} \end{vmatrix}$$

Historical Problems

1. (a) $2 - 5i \longleftrightarrow \begin{bmatrix} 2 & -5 \\ 5 & 2 \end{bmatrix}$, $1 + 3i \longleftrightarrow \begin{bmatrix} 1 & 3 \\ -3 & 1 \end{bmatrix}$ **(b)** $\begin{bmatrix} 2 & -5 \\ 5 & 2 \end{bmatrix} \begin{bmatrix} 1 & 3 \\ -3 & 1 \end{bmatrix} = \begin{bmatrix} 17 & 1 \\ -1 & 17 \end{bmatrix}$ **(c)** $17 + i$ **(d)** $17 + i$

2. (a) $x = (ka + lc)r + (kb + ld)s$; $y = (ma + nc)r + (mb + nd)s$ **(b)** $A = \begin{bmatrix} ka + lc & kb + ld \\ ma + nc & mb + nd \end{bmatrix}$

8.4 Exercises

1. $\begin{bmatrix} 4 & 4 & -5 \\ -1 & 5 & 4 \end{bmatrix}$ **2.** $\begin{bmatrix} -4 & 2 & -5 \\ 3 & -1 & 8 \end{bmatrix}$ **3.** $\begin{bmatrix} 0 & 12 & -20 \\ 4 & 8 & 24 \end{bmatrix}$ **4.** $\begin{bmatrix} -12 & -3 & 0 \\ 6 & -9 & 6 \end{bmatrix}$ **5.** $\begin{bmatrix} -8 & 7 & -15 \\ 7 & 0 & 22 \end{bmatrix}$ **6.** $\begin{bmatrix} 16 & 10 & -10 \\ -6 & 16 & 4 \end{bmatrix}$

7. $\begin{bmatrix} 28 & -9 \\ 4 & 23 \end{bmatrix}$ **8.** $\begin{bmatrix} 22 & 6 \\ 14 & -2 \end{bmatrix}$ **9.** $\begin{bmatrix} 1 & 14 & -14 \\ 2 & 22 & -18 \\ 3 & 0 & 28 \end{bmatrix}$ **10.** $\begin{bmatrix} 14 & 7 & -2 \\ 20 & 12 & -4 \\ -14 & 7 & -6 \end{bmatrix}$ **11.** $\begin{bmatrix} 15 & 21 & -16 \\ 22 & 34 & -22 \\ -11 & 7 & 22 \end{bmatrix}$ **12.** $\begin{bmatrix} 50 & -3 \\ 18 & 21 \end{bmatrix}$ **13.** $\begin{bmatrix} 25 & -9 \\ 4 & 20 \end{bmatrix}$

14. $\begin{bmatrix} 6 & 14 & -14 \\ 2 & 27 & -18 \\ 3 & 0 & 33 \end{bmatrix}$ **15.** $\begin{bmatrix} -13 & 7 & -12 \\ -18 & 10 & -14 \\ 17 & -7 & 34 \end{bmatrix}$ **16.** $\begin{bmatrix} 50 & -3 \\ 18 & 21 \end{bmatrix}$ **17.** $\begin{bmatrix} -2 & 4 & 2 & 8 \\ 2 & 1 & 4 & 6 \end{bmatrix}$ **18.** $\begin{bmatrix} -22 & 29 & 8 & -1 \\ -10 & 17 & 6 & -1 \end{bmatrix}$ **19.** $\begin{bmatrix} 9 & 2 \\ 34 & 13 \\ 47 & 20 \end{bmatrix}$

20. $\begin{bmatrix} 6 & 32 \\ 1 & 3 \\ -2 & -4 \end{bmatrix}$ **21.** $\begin{bmatrix} 1 & -1 \\ -1 & 2 \end{bmatrix}$ **22.** $\begin{bmatrix} 1 & 1 \\ 2 & 3 \end{bmatrix}$ **23.** $\begin{bmatrix} 1 & -\frac{5}{2} \\ -1 & 3 \end{bmatrix}$ **24.** $\begin{bmatrix} -1 & -\frac{1}{2} \\ -3 & -2 \end{bmatrix}$ **25.** $\begin{bmatrix} 1 & \frac{-1}{a} \\ -1 & \frac{2}{a} \end{bmatrix}$ **26.** $\begin{bmatrix} -\frac{2}{b} & \frac{3}{b} \\ 1 & -1 \end{bmatrix}$

27. $\begin{bmatrix} 3 & -3 & 1 \\ -2 & 2 & -1 \\ -4 & 5 & -2 \end{bmatrix}$ **28.** $\begin{bmatrix} 3 & -2 & -4 \\ 3 & -2 & -5 \\ -1 & 1 & 2 \end{bmatrix}$ **29.** $\begin{bmatrix} -\frac{5}{7} & \frac{1}{7} & \frac{3}{7} \\ \frac{9}{7} & \frac{1}{7} & -\frac{4}{7} \\ \frac{3}{7} & -\frac{2}{7} & \frac{1}{7} \end{bmatrix}$ **30.** $\begin{bmatrix} \frac{3}{7} & -\frac{4}{7} & \frac{1}{7} \\ \frac{1}{7} & \cdot \frac{1}{7} & -\frac{2}{7} \\ -\frac{5}{7} & \frac{9}{7} & \frac{3}{7} \end{bmatrix}$ **31.** $x = 3, y = 2$ **32.** $x = 12, y = 28$

33. $x = -5, y = 10$ **34.** $x = 9, y = 23$ **35.** $x = 2, y = -1$ **36.** $x = -7, y = -28$ **37.** $x = \frac{1}{2}, y = 2$ **38.** $x = -\frac{1}{2}, y = 3$

39. $x = -2, y = 1$ **40.** $x = 2, y = 1$ **41.** $x = \frac{2}{a}, y = \frac{3}{a}$ **42.** $x = \frac{2}{b}, y = 4$ **43.** $x = -2, y = 3, z = 5$ **44.** $x = 4, y = -2, z = 1$

45. $x = \frac{1}{2}, y = -\frac{1}{2}, z = 1$ **46.** $x = 1, y = -1, z = \frac{1}{2}$ **47.** $x = -\frac{34}{7}, y = \frac{85}{7}, z = \frac{12}{7}$ **48.** $x = \frac{8}{7}, y = \frac{5}{7}, z = \frac{17}{7}$

49. $x = \frac{1}{3}, y = 1, z = \frac{2}{3}$ **50.** $x = 1, y = -1, z = 1$ **51.** $\left[\begin{array}{cc|cc} 4 & 2 & 1 & 0 \\ 2 & 1 & 0 & 1 \end{array}\right] \rightarrow \left[\begin{array}{cc|cc} 1 & \frac{1}{2} & \frac{1}{4} & 0 \\ 2 & 1 & 0 & 1 \end{array}\right] \rightarrow \left[\begin{array}{cc|cc} 1 & \frac{1}{2} & \frac{1}{4} & 0 \\ 0 & 0 & -\frac{1}{2} & 1 \end{array}\right]$

52. $\left[\begin{array}{cc|cc} -3 & \frac{1}{2} & 1 & 0 \\ 6 & -1 & 0 & 1 \end{array}\right] \rightarrow \left[\begin{array}{cc|cc} 1 & -\frac{1}{6} & -\frac{1}{3} & 0 \\ 6 & -1 & 0 & 1 \end{array}\right] \rightarrow \left[\begin{array}{cc|cc} 1 & -\frac{1}{6} & -\frac{1}{3} & 0 \\ 0 & 0 & 2 & 1 \end{array}\right]$ **53.** $\left[\begin{array}{cc|cc} 15 & 3 & 1 & 0 \\ 10 & 2 & 0 & 1 \end{array}\right] \rightarrow \left[\begin{array}{cc|cc} 1 & \frac{1}{5} & \frac{1}{15} & 0 \\ 10 & 2 & 0 & 1 \end{array}\right] \rightarrow \left[\begin{array}{cc|cc} 1 & \frac{1}{5} & \frac{1}{15} & 0 \\ 0 & 0 & -\frac{2}{3} & 1 \end{array}\right]$

54. $\left[\begin{array}{cc|cc} -3 & 0 & 1 & 0 \\ 4 & 0 & 0 & 1 \end{array}\right] \rightarrow \left[\begin{array}{cc|cc} 1 & 0 & -\frac{1}{3} & 0 \\ 4 & 0 & 0 & 1 \end{array}\right] \rightarrow \left[\begin{array}{cc|cc} 1 & 0 & -\frac{1}{3} & 0 \\ 0 & 0 & \frac{4}{3} & 1 \end{array}\right]$

55. $\left[\begin{array}{ccc|ccc} -3 & 1 & -1 & 1 & 0 & 0 \\ 1 & -4 & -7 & 0 & 1 & 0 \\ 1 & 2 & 5 & 0 & 0 & 1 \end{array}\right] \rightarrow \left[\begin{array}{ccc|ccc} 1 & 2 & 5 & 0 & 0 & 1 \\ 1 & -4 & -7 & 0 & 1 & 0 \\ -3 & 1 & -1 & 1 & 0 & 0 \end{array}\right] \rightarrow \left[\begin{array}{ccc|ccc} 1 & 2 & 5 & 0 & 0 & 1 \\ 0 & -6 & -12 & 0 & 1 & -1 \\ 0 & 7 & 14 & 1 & 0 & 3 \end{array}\right] \rightarrow \left[\begin{array}{ccc|ccc} 1 & 2 & 5 & 0 & 0 & 1 \\ 0 & 1 & 2 & 0 & -\frac{1}{6} & \frac{1}{6} \\ 0 & 1 & 2 & \frac{1}{7} & 0 & \frac{3}{7} \end{array}\right]$

$\rightarrow \left[\begin{array}{ccc|ccc} 1 & 2 & 5 & 0 & 0 & 1 \\ 0 & 1 & 2 & 0 & -\frac{1}{6} & \frac{1}{6} \\ 0 & 0 & 0 & \frac{1}{7} & \frac{1}{6} & \frac{11}{42} \end{array}\right]$

56. $\left[\begin{array}{ccc|ccc} 1 & 1 & -3 & 1 & 0 & 0 \\ 2 & -4 & 1 & 0 & 1 & 0 \\ -5 & 7 & 1 & 0 & 0 & 1 \end{array}\right] \rightarrow \left[\begin{array}{ccc|ccc} 1 & 1 & -3 & 1 & 0 & 0 \\ 0 & -6 & 7 & -2 & 1 & 0 \\ 0 & 12 & -14 & 5 & 0 & 1 \end{array}\right] \rightarrow \left[\begin{array}{ccc|ccc} 1 & 1 & -3 & 1 & 0 & 0 \\ 0 & 1 & -\frac{7}{6} & \frac{1}{3} & -\frac{1}{6} & 0 \\ 0 & 12 & -14 & 5 & 0 & 1 \end{array}\right] \rightarrow \left[\begin{array}{ccc|ccc} 1 & 0 & -\frac{11}{6} & \frac{2}{3} & \frac{1}{6} & 0 \\ 0 & 1 & -\frac{7}{6} & \frac{1}{3} & -\frac{1}{6} & 0 \\ 0 & 0 & 0 & 1 & 2 & 1 \end{array}\right]$

57. $\begin{bmatrix} 0.01 & 0.05 & -0.01 \\ 0.01 & -0.02 & 0.01 \\ -0.02 & 0.01 & 0.03 \end{bmatrix}$ **58.** $\begin{bmatrix} 0.26 & -0.29 & -0.20 \\ -1.21 & 1.63 & 1.20 \\ -1.84 & 2.53 & 1.80 \end{bmatrix}$ **59.** $\begin{bmatrix} 0.02 & -0.04 & -0.01 & 0.01 \\ -0.02 & 0.05 & 0.03 & -0.03 \\ 0.02 & 0.01 & -0.04 & 4.88 \\ -0.02 & 0.06 & 0.07 & 0.06 \end{bmatrix}$

60. $\begin{bmatrix} 0.01 & 0.04 & 0.00 & 0.03 \\ 0.02 & -0.02 & 0.01 & 0.01 \\ -0.04 & 0.02 & 0.04 & 0.06 \\ 0.05 & -0.02 & 0.00 & -0.09 \end{bmatrix}$ **61.** $x = -5.38, y = 7.30, z = 25.05$ **62.** $x = 4.56, y = -6.06, z = -22.55$

63. $x = -1.19, y = 2.46, z = 8.27$ **64.** $x = -2.05, y = 3.88, z = 13.36$

65. (a) $\begin{bmatrix} 500 & 350 & 400 \\ 700 & 500 & 850 \end{bmatrix}$; $\begin{bmatrix} 500 & 700 \\ 350 & 500 \\ 400 & 850 \end{bmatrix}$ **(b)** $\begin{bmatrix} 15 \\ 8 \\ 3 \end{bmatrix}$ **(c)** $\begin{bmatrix} 11,500 \\ 17,050 \end{bmatrix}$ **(d)** $\begin{bmatrix} 0.10 & 0.05 \end{bmatrix}$ **(e)** $2002.50

66. (a) January: $\begin{bmatrix} 400 & 250 & 50 \\ 450 & 200 & 140 \end{bmatrix}$ February: $\begin{bmatrix} 350 & 100 & 30 \\ 350 & 300 & 100 \end{bmatrix}$ **(b)** $\begin{bmatrix} 750 & 350 & 80 \\ 800 & 500 & 240 \end{bmatrix}$ **(c)** $\begin{bmatrix} 100 \\ 150 \\ 200 \end{bmatrix}$ **(d)** $\begin{bmatrix} 143,500 \\ 203,000 \end{bmatrix}$

67. If $a \neq 0$, $\left[\begin{array}{cc|cc} a & b & 1 & 0 \\ c & d & 0 & 1 \end{array}\right] \rightarrow \left[\begin{array}{cc|cc} 1 & \frac{b}{a} & \frac{1}{a} & 0 \\ c & d & 0 & 1 \end{array}\right] \rightarrow \left[\begin{array}{cc|cc} 1 & \frac{b}{a} & \frac{1}{a} & 0 \\ 0 & \frac{-cb + da}{a} & -\frac{c}{a} & 1 \end{array}\right] \rightarrow \left[\begin{array}{cc|cc} 1 & \frac{b}{a} & \frac{1}{a} & 0 \\ 0 & 1 & \frac{-c}{-cb + da} & \frac{a}{-cb + da} \end{array}\right] \rightarrow \left[\begin{array}{cc|cc} 1 & 0 & \frac{d}{ad - bc} & \frac{-b}{ad - bc} \\ 0 & 1 & \frac{-c}{ad - bc} & \frac{a}{ad - bc} \end{array}\right]$.

Therefore, $A^{-1} = \frac{1}{D}\begin{bmatrix} d & -b \\ -c & a \end{bmatrix}$. If $a = 0$, $\left[\begin{array}{cc|cc} 0 & b & 1 & 0 \\ c & d & 0 & 1 \end{array}\right] \rightarrow \left[\begin{array}{cc|cc} c & d & 0 & 1 \\ 0 & b & 1 & 0 \end{array}\right] \rightarrow \left[\begin{array}{cc|cc} 1 & \frac{d}{c} & 0 & \frac{1}{c} \\ 0 & b & 1 & 0 \end{array}\right] \rightarrow \left[\begin{array}{cc|cc} 1 & 0 & \frac{-d}{cb} & \frac{1}{c} \\ 0 & 1 & \frac{1}{b} & 0 \end{array}\right]$

$\rightarrow \left[\begin{array}{cc|cc} 1 & 0 & \frac{d}{-bc} & \frac{-b}{-bc} \\ 0 & 1 & \frac{-c}{-bc} & 0 \end{array}\right]$. Since $a = 0$, $D = ad - bc = -bc$, so $A^{-1} = \frac{1}{D}\begin{bmatrix} d & -b \\ -c & a \end{bmatrix}$.

8.5 Exercises

1. Proper **2.** Proper **3.** Improper: $1 + \dfrac{9}{x^2 - 4}$ **4.** Improper; $3 + \dfrac{1}{x^2 - 1}$ **5.** Improper; $5x + \dfrac{22x - 1}{x^2 - 4}$

6. Improper; $3x + \dfrac{x^2 - 24x - 2}{x^3 + 8}$ **7.** Improper; $1 + \dfrac{-2(x - 6)}{(x + 4)(x - 3)}$ **8.** Improper; $2x + \dfrac{6x}{x^2 + 1}$ **9.** $\dfrac{-4}{x} + \dfrac{4}{x - 1}$

10. $\dfrac{2}{x+2} + \dfrac{1}{x-1}$ **11.** $\dfrac{1}{x} + \dfrac{-x}{x^2+1}$ **12.** $\dfrac{\frac{1}{5}}{x+1} + \dfrac{-\frac{1}{5}x+\frac{1}{5}}{x^2+4}$ **13.** $\dfrac{-1}{x-1} + \dfrac{2}{x-2}$ **14.** $\dfrac{1}{x+2} + \dfrac{2}{x-4}$

15. $\dfrac{\frac{1}{4}}{x+1} + \dfrac{\frac{3}{4}}{x-1} + \dfrac{\frac{1}{2}}{(x-1)^2}$ **16.** $\dfrac{-\frac{3}{4}}{x} + \dfrac{-\frac{1}{2}}{x^2} + \dfrac{\frac{3}{4}}{x-2}$ **17.** $\dfrac{\frac{1}{12}}{x-2} + \dfrac{-\frac{1}{12}(x+4)}{x^2+2x+4}$ **18.** $\dfrac{2}{x-1} + \dfrac{-2x-2}{x^2+x+1}$

19. $\dfrac{\frac{1}{4}}{x-1} + \dfrac{\frac{1}{4}}{(x-1)^2} + \dfrac{-\frac{1}{4}}{x+1} + \dfrac{\frac{1}{4}}{(x+1)^2}$ **20.** $\dfrac{\frac{1}{2}}{x} + \dfrac{\frac{1}{4}}{x^2} + \dfrac{-\frac{1}{2}}{x-2} + \dfrac{\frac{3}{4}}{(x-2)^2}$ **21.** $\dfrac{-5}{x+2} + \dfrac{5}{x+1} + \dfrac{-4}{(x+1)^2}$

22. $\dfrac{\frac{2}{9}}{x+2} + \dfrac{\frac{7}{9}}{x-1} + \dfrac{\frac{2}{3}}{(x-1)^2}$ **23.** $\dfrac{\frac{1}{4}}{x} + \dfrac{1}{x^2} + \dfrac{-\frac{1}{4}(x+4)}{x^2+4}$ **24.** $\dfrac{\frac{14}{3}}{x-1} + \dfrac{4}{(x-1)^2} + \dfrac{-\frac{14}{3}x+\frac{4}{3}}{x^2+2}$ **25.** $\dfrac{\frac{2}{3}}{x+1} + \dfrac{\frac{1}{3}(x+1)}{x^2+2x+4}$

26. $\dfrac{-6}{x} + \dfrac{7x+7}{x^2+3x+3}$ **27.** $\dfrac{\frac{2}{7}}{3x-2} + \dfrac{\frac{1}{7}}{2x+1}$ **28.** $\dfrac{-\frac{1}{7}}{2x+3} + \dfrac{\frac{2}{7}}{4x-1}$ **29.** $\dfrac{\frac{3}{4}}{x+3} + \dfrac{\frac{1}{4}}{x-1}$ **30.** $\dfrac{-3}{x+1} + \dfrac{2}{x+2} + \dfrac{2}{x+3}$

31. $\dfrac{1}{x^2+4} + \dfrac{2x-1}{(x^2+4)^2}$ **32.** $\dfrac{x}{x^2+16} + \dfrac{-16x+1}{(x^2+16)^2}$ **33.** $\dfrac{-1}{x} + \dfrac{2}{x-3} + \dfrac{-1}{x+1}$ **34.** $\dfrac{-1}{x^2} - \dfrac{1}{x^4} + \dfrac{1}{x-1}$

35. $\dfrac{4}{x-2} + \dfrac{-3}{x-1} + \dfrac{-1}{(x-1)^2}$ **36.** $\dfrac{\frac{3}{8}}{x-1} + \dfrac{\frac{1}{2}}{(x-1)^2} + \dfrac{\frac{5}{8}}{x+3}$ **37.** $\dfrac{x}{(x^2+16)^2} + \dfrac{-16x}{(x^2+16)^3}$ **38.** $\dfrac{1}{(x^2+4)^2} - \dfrac{4}{(x^2+4)^3}$

39. $\dfrac{-\frac{8}{7}}{2x+1} + \dfrac{\frac{4}{7}}{x-3}$ **40.** $\dfrac{\frac{4}{5}}{2x-1} + \dfrac{\frac{8}{5}}{x+2}$ **41.** $\dfrac{-\frac{2}{9}}{x} + \dfrac{-\frac{1}{3}}{x^2} + \dfrac{\frac{1}{6}}{x-3} + \dfrac{\frac{1}{18}}{x+3}$ **42.** $\dfrac{\frac{13}{24}}{x-2} + \dfrac{-\frac{13}{24}}{x+2} + \dfrac{-\frac{7}{6}}{x^2+2}$

Historical Problem

1. $x = 6$ units, $y = 8$ units

8.6 Exercises

1.
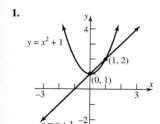

2.
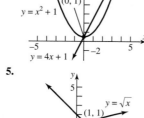

3.
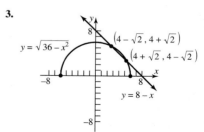

4.
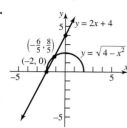

5.
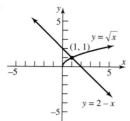

6.
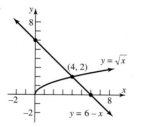

7.
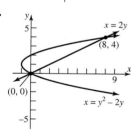

8.
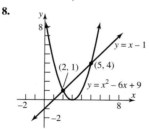

9.
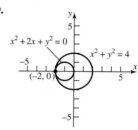

10.

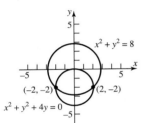

11.

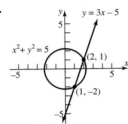

12.

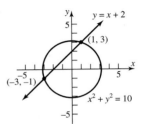

13.

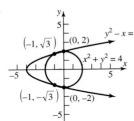

14.

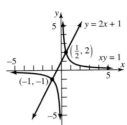

15.

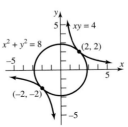

16.

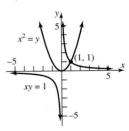

17. No points of intersection.

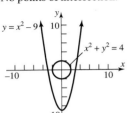

18.

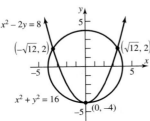

19.

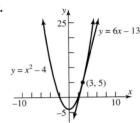

20.

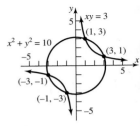

21. $x = 1, y = 4; x = -1, y = -4; x = 2\sqrt{2}, y = \sqrt{2}; x = -2\sqrt{2}, y = -\sqrt{2}$ **22.** $x = 5, y = 2$ **23.** $x = 0, y = 1; x = -\dfrac{2}{3}, y = -\dfrac{1}{3}$

24. $x = -5, y = -\dfrac{3}{2}$ **25.** $x = 0, y = -1; x = \dfrac{5}{2}, y = -\dfrac{7}{2}$

26. $x = -2, y = -2; x = -\sqrt{2}, y = -2\sqrt{2}; x = \sqrt{2}, y = 2\sqrt{2}; x = 2, y = 2$ **27.** $x = 2, y = \dfrac{1}{3}; x = \dfrac{1}{2}, y = \dfrac{4}{3}$

28. $x = -2, y = 0; x = 7, y = 6$ **29.** $x = 3, y = 2; x = 3, y = -2; x = -3, y = 2; x = -3, y = -2$

30. $x = -1, y = -2; x = -1, y = 2; x = 1, y = -2; x = 1, y = 2$

31. $x = \dfrac{1}{2}, y = \dfrac{3}{2}; x = \dfrac{1}{2}, y = -\dfrac{3}{2}; x = -\dfrac{1}{2}, y = \dfrac{3}{2}; x = -\dfrac{1}{2}, y = -\dfrac{3}{2}$

32. $x = -2\sqrt{2}, y = \sqrt{3}; x = 2\sqrt{2}, y = \sqrt{3}; x = 2\sqrt{2}, y = -\sqrt{3}; x = -2\sqrt{2}, y = -\sqrt{3}$

33. $x = \sqrt{2}, y = 2\sqrt{2}; x = -\sqrt{2}, y = -2\sqrt{2}$ **34.** $x = -5\sqrt{3}, y = \sqrt{3}; x = 5\sqrt{3}, y = -\sqrt{3}$

35. No real solution exists   **36.** No real solution exists

37. $x = \dfrac{8}{3}, y = \dfrac{2\sqrt{10}}{3}; x = -\dfrac{8}{3}, y = \dfrac{2\sqrt{10}}{3}; x = \dfrac{8}{3}, y = -\dfrac{2\sqrt{10}}{3}; x = -\dfrac{8}{3}, y = -\dfrac{2\sqrt{10}}{3}$

38. $x = \dfrac{-\sqrt{2}}{2}, y = \dfrac{-\sqrt{6}}{3}; x = \dfrac{-\sqrt{2}}{2}, y = \dfrac{\sqrt{6}}{3}; x = \dfrac{\sqrt{2}}{2}, y = \dfrac{-\sqrt{6}}{3}; x = \dfrac{\sqrt{2}}{2}, y = \dfrac{\sqrt{6}}{3}$

39. $x = 1, y = \dfrac{1}{2}; x = -1, y = \dfrac{1}{2}; x = 1, y = -\dfrac{1}{2}; x = -1, y = -\dfrac{1}{2}$

40. $x = -2, y = -\sqrt{2}; x = -2, y = \sqrt{2}; x = 2, y = -\sqrt{2}; x = 2, y = \sqrt{2}$ **41.** No real solution exists

42. $x = -\sqrt[4]{\dfrac{2}{5}}, y = -\sqrt[4]{\dfrac{2}{3}}; x = -\sqrt[4]{\dfrac{2}{5}}, y = \sqrt[4]{\dfrac{2}{3}}; x = \sqrt[4]{\dfrac{2}{5}}, y = -\sqrt[4]{\dfrac{2}{3}}; x = \sqrt[4]{\dfrac{2}{5}}, y = \sqrt[4]{\dfrac{2}{3}}$

43. $x = \sqrt{3}, y = \sqrt{3}; x = -\sqrt{3}, y = -\sqrt{3}; x = 2, y = 1; x = -2, y = -1$ **44.** $x = -2, y = 2; x = 3, y = -3$

45. $x = 3, y = 2; x = -3, y = -2; x = 2, y = \dfrac{1}{2}; x = -2, y = -\dfrac{1}{2}$

46. $x = -3, y = 3; x = -\sqrt{3}, y = -\sqrt{3}; x = \sqrt{3}, y = \sqrt{3}; x = 3, y = -3$ **47.** $x = 3, y = 1; x = -1, y = -3$

48. $x = -1, y = 3; x = 3, y = -1$ **49.** $x = 0, y = -2; x = 0, y = 1; x = 2, y = -1$ **50.** $x = 2, y = 1$ **51.** $x = 2, y = 8$

52. $x = \dfrac{1}{2}, y = \dfrac{1}{16}$ **53.** $x = 81, y = 3$ **54.** $x = 32, y = 2$

55.

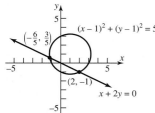

56.

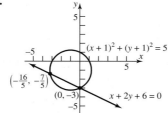

57.

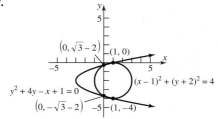

58.

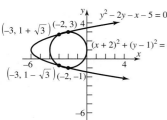

59.

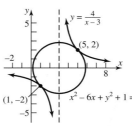

60.

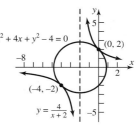

61. $x = 0.48$, $y = 0.61$ **62.** $x = 0.64$, $y = 0.52$ **63.** $x = -1.64$, $y = -0.89$ **64.** $x = -1.36$, $y = 2.13$
65. $x = 0.58$, $y = 1.85$; $x = 1.81$, $y = 1.05$; $x = 0.58$, $y = -1.85$; $x = 1.81$, $y = -1.05$ **66.** $x = 0.64$, $y = 1.55$; $x = -0.64$, $y = -1.55$
67. $x = 2.34$, $y = 0.85$ **68.** $x = 1.89$, $y = 0.63$; $x = 0.13$, $y = -1.99$ **69.** 3 and 1; -3 and -1 **70.** 2 and 5 **71.** 2 and 2; -2 and -2

72. 2 and 5; -2 and -5 **73.** $\frac{1}{2}$ and $\frac{1}{3}$ **74.** $\frac{1}{2}$ and -1 **75.** 5 **76.** $\frac{1}{7}$ **77.** 5 in. by 3 in. **78.** 4 ft , 6 ft **79.** 2 cm and 4 cm **80.** 8 cm

81. tortoise: 7 m/hr, hare: $7\frac{1}{2}$ m/hr **82.** 10.02 ft **83.** 12 cm by 18 cm **84.** 16.57 cm by 13.03 cm **85.** $x = 60$ ft; $y = 30$ ft

86. It is not possible. **87.** $l = \dfrac{P + \sqrt{P^2 - 16A}}{4}$; $w = \dfrac{P - \sqrt{P^2 - 16A}}{4}$ **88.** $b = P - \dfrac{4h^2 + P^2}{2P}$, $l = \dfrac{4h^2 + P^2}{4P}$

89. $y = 4x - 4$ **90.** $y = -\dfrac{1}{3}x + \dfrac{10}{3}$ **91.** $y = 2x + 1$ **92.** $y = 4x + 9$ **93.** $y = -\dfrac{1}{3}x + \dfrac{7}{3}$ **94.** $y = \dfrac{3}{2}x + \dfrac{7}{2}$ **95.** $y = 2x - 3$

96. $y = \dfrac{1}{3}x + \dfrac{7}{3}$ **97.** $r_1 = \dfrac{-b + \sqrt{b^2 - 4ac}}{2a}$; $r_2 = \dfrac{-b - \sqrt{b^2 - 4ac}}{2a}$ **99. (a)** 3.5 ft by 3.5 ft

8.7 Exercises

1.

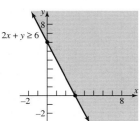

2.

3.

4.

5.

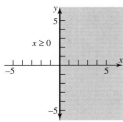

6.

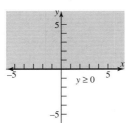

7.

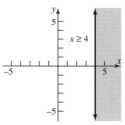

8.

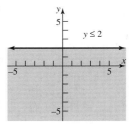

9.

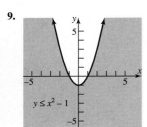

$y \le x^2 - 1$

10.

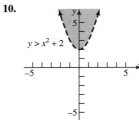

$y > x^2 + 2$

11.

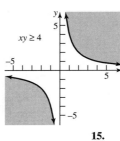

$xy \ge 4$

12.

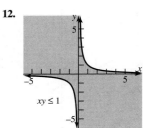

$xy \le 1$

13.

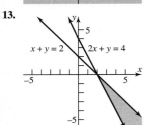

$x + y = 2$ $2x + y = 4$

14.

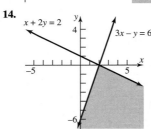

$x + 2y = 2$ $3x - y = 6$

15.

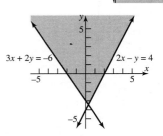

$3x + 2y = -6$ $2x - y = 4$

16.

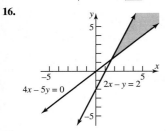

$4x - 5y = 0$ $2x - y = 2$

17.

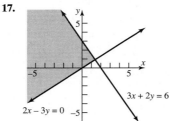

$2x - 3y = 0$ $3x + 2y = 6$

18.

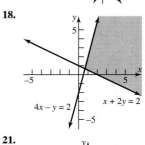

$4x - y = 2$ $x + 2y = 2$

19.

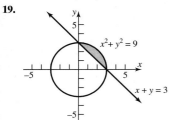

$x^2 + y^2 = 9$ $x + y = 3$

20.

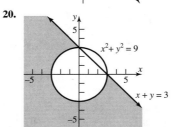

$x^2 + y^2 = 9$ $x + y = 3$

21.

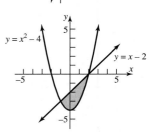

$y = x^2 - 4$ $y = x - 2$

22.

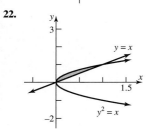

$y = x$ $y^2 = x$

23.

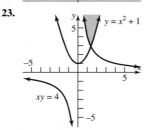

$y = x^2 + 1$ $xy = 4$

24.

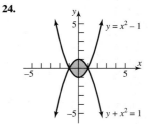

$y = x^2 - 1$ $y + x^2 = 1$

25.

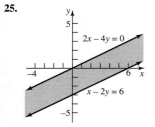

$2x - 4y = 0$ $x - 2y = 6$

26.

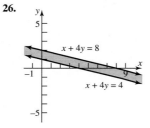

$x + 4y = 8$ $x + 4y = 4$

27.

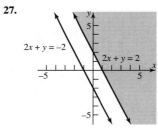

$2x + y = -2$ $2x + y = 2$

28.

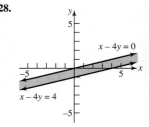

$x - 4y = 0$ $x - 4y = 4$

29. No solution

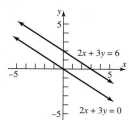

30.

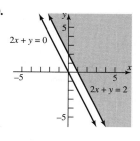

31. Bounded; corner points $(0, 0)$, $(3, 0)$, $(2, 2)$, $(0, 3)$

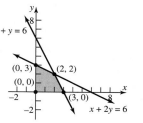

32. Unbounded; corner points $(0, 4)$, $(4, 0)$

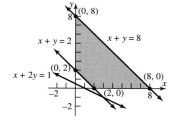

33. Unbounded; corner points $(2, 0)$, $(0, 4)$

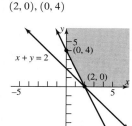

34. Bounded; corner points $(0, 0)$, $(0, 2)$, $(1, 0)$

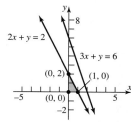

35. Bounded; corner points $(2, 0)$, $(4, 0)$, $\left(\dfrac{24}{7}, \dfrac{12}{7}\right)$, $(0, 4)$, $(0, 2)$

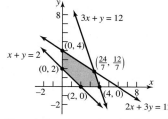

36. Bounded; corner points $(0, 2)$, $(0, 3)$, $(1, 1)$

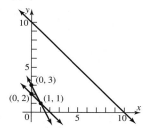

37. Bounded; corner points $(2, 0)$, $(5, 0)$, $(2, 6)$, $(0, 8)$, $(0, 2)$

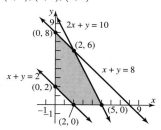

38. Bounded; corner points $(2, 0)$, $(0, 2)$, $(0, 8)$, $(8, 0)$

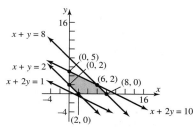

39. Bounded; corner points $(1, 0)$, $(10, 0)$, $(0, 5)$, $\left(0, \dfrac{1}{2}\right)$

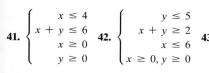

40. Bounded, corner points $(0, 2)$, $(0, 5)$, $(6, 2)$, $(8, 0)$, $(2, 0)$

41. $\begin{cases} x \le 4 \\ x + y \le 6 \\ x \ge 0 \\ y \ge 0 \end{cases}$

42. $\begin{cases} y \le 5 \\ x + y \ge 2 \\ x \le 6 \\ x \ge 0, y \ge 0 \end{cases}$

43. $\begin{cases} x \le 20 \\ y \ge 15 \\ x + y \le 50 \\ x - y \le 0 \\ x \ge 0 \end{cases}$

44. $\begin{cases} y \le 6 \\ x \le 5 \\ 3x + 4y \ge 12 \\ y \ge 2x - 8 \\ x \ge 0, y \ge 0 \end{cases}$

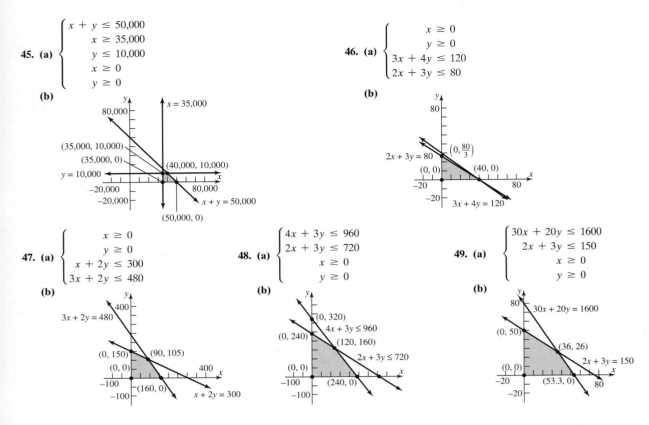

45. (a) $\begin{cases} x + y \leq 50{,}000 \\ x \geq 35{,}000 \\ y \leq 10{,}000 \\ x \geq 0 \\ y \geq 0 \end{cases}$

(b)

46. (a) $\begin{cases} x \geq 0 \\ y \geq 0 \\ 3x + 4y \leq 120 \\ 2x + 3y \leq 80 \end{cases}$

(b)

47. (a) $\begin{cases} x \geq 0 \\ y \geq 0 \\ x + 2y \leq 300 \\ 3x + 2y \leq 480 \end{cases}$

(b)

48. (a) $\begin{cases} 4x + 3y \leq 960 \\ 2x + 3y \leq 720 \\ x \geq 0 \\ y \geq 0 \end{cases}$

(b)

49. (a) $\begin{cases} 30x + 20y \leq 1600 \\ 2x + 3y \leq 150 \\ x \geq 0 \\ y \geq 0 \end{cases}$

(b)

8.8 Exercises

1. Maximum value is 11; minimum value is 3   **2.** Maximum value is 28; minimum value is 8   **3.** Maximum value is 65; minimum value is 4
4. Maximum value is 56; minimum value is 3   **5.** Maximum value is 67; minimum value is 20
6. Maximum value is 65; minimum value is 15   **7.** The maximum value of z is 12, and it occurs at the point $(6, 0)$.
8. Maximum value is 26, and it occurs at the point $(5, 7)$.   **9.** The minimum value of z is 4, and it occurs at the point $(2, 0)$.
10. The minimum value of z is 8, and it occurs at the point $(0, 2)$.   **11.** The maximum value of z is 20, and it occurs at the point $(0, 4)$.
12. The maximum value of z is 28, and it occurs at the point $(2, 6)$.   **13.** The minimum value of z is 8, and it occurs at the point $(0, 2)$.
14. The minimum value of z is $\dfrac{15}{2}$, and it occurs at the point $\left(\dfrac{3}{2}, \dfrac{3}{2}\right)$.
15. The maximum value of z is 50, and it occurs at the point $(10, 0)$.   **16.** The maximum value of z is 36, and it occurs at the point $(0, 9)$.
17. 8 downhill, 24 cross-country; $1760; $1920   **18.** 24 acres of soybeans and 12 acres of wheat for a maximum profit of $5520; $7200.
19. 30 acres of soybeans and 10 acres of corn.
20. 15 oz of supplement A and 0 oz of supplement B or 7.5 oz of supplement A and 11.25 oz of supplement B, minimum cost $22.50
21. $\dfrac{1}{2}$ hr on machine 1; $5\dfrac{1}{4}$ hrs on machine 2   **22.** 15 newer trees and 10 older trees; minimum cost $425
23. 100 lbs of ground beef and 50 lbs of pork   **24. (a)** $10,000 in a junk bond and $10,000 in treasury bills
(b) $12,000 in junk bond and $8,000 in treasury bills   **25.** 10 racing skates, 15 figure skates
26. $20,000 in a AAA bond and $30,000 in a CD   **27.** 2 metal samples, 4 plastic samples; $34
28. 15 cans of "Gourmet Dog" and 25 cans of "Chow Hound"
29. (a) 10 first class, 120 coach   **(b)** 15 first class, 120 coach   **30.** 5 oz of supplement A and 1 oz of supplement B; minimum cost $0.38

Fill-in-the-Blank Items

1. inconsistent   **2.** m by n matrix   **3.** determinants   **4.** augmented   **5.** inverse   **6.** square   **7.** identity   **8.** proper
9. half-plane   **10.** objective function   **11.** feasible point

True/False Items

1. F **2.** T **3.** F **4.** F **5.** F **6.** F **7.** F **8.** T **9.** T **10.** T **11.** T

Review Exercises

1. $x = 2, y = -1$ **2.** $x = \dfrac{11}{23}, y = \dfrac{8}{23}$ **3.** $x = 2, y = \dfrac{1}{2}$ **4.** $x = -\dfrac{1}{2}, y = 1$ **5.** $x = 2, y = -1$ **6.** $x = 0, y = \dfrac{5}{3}$

7. $x = \dfrac{11}{5}, y = -\dfrac{3}{5}$ **8.** $x = -\dfrac{1}{2}, y = -\dfrac{1}{2}$ **9.** $x = -\dfrac{8}{5}, y = \dfrac{12}{5}$ **10.** Inconsistent **11.** $x = 6, y = -1$ **12.** $x = 1, y = \dfrac{3}{2}$

13. $x = -4, y = 3$ **14.** $x = 0, y = -3$ **15.** $x = 2, y = 3$ **16.** $x = 84, y = -63$ **17.** Inconsistent **18.** Inconsistent

19. $x = -1, y = 2, z = -3$ **20.** $x = -1, y = 2, z = 7$ **21.** $\begin{bmatrix} 4 & -4 \\ 3 & 9 \\ 4 & 0 \end{bmatrix}$ **22.** $\begin{bmatrix} -2 & 4 \\ 1 & -1 \\ -6 & 4 \end{bmatrix}$ **23.** $\begin{bmatrix} 6 & 0 \\ 12 & 24 \\ -6 & 12 \end{bmatrix}$ **24.** $\begin{bmatrix} -16 & 12 & 0 \\ -4 & -4 & 8 \end{bmatrix}$

25. $\begin{bmatrix} 4 & -3 & 0 \\ 12 & -2 & -8 \\ -2 & 5 & -4 \end{bmatrix}$ **26.** $\begin{bmatrix} -2 & -12 \\ 5 & 0 \end{bmatrix}$ **27.** $\begin{bmatrix} 8 & -13 & 8 \\ 9 & 2 & -10 \\ 18 & -17 & 4 \end{bmatrix}$ **28.** $\begin{bmatrix} 9 & -31 \\ -6 & 5 \end{bmatrix}$ **29.** $\begin{bmatrix} \frac{1}{2} & -1 \\ -\frac{1}{6} & \frac{2}{3} \end{bmatrix}$ **30.** $\begin{bmatrix} -\frac{1}{2} & -\frac{1}{2} \\ -\frac{1}{4} & -\frac{3}{4} \end{bmatrix}$

31. $\begin{bmatrix} -\frac{5}{7} & \frac{9}{7} & \frac{3}{7} \\ \frac{1}{7} & \frac{1}{7} & -\frac{2}{7} \\ \frac{3}{7} & -\frac{4}{7} & \frac{1}{7} \end{bmatrix}$ **32.** $\begin{bmatrix} \frac{3}{7} & \frac{1}{7} & -\frac{5}{7} \\ -\frac{4}{7} & \frac{1}{7} & \frac{9}{7} \\ -\frac{1}{7} & -\frac{2}{7} & \frac{3}{7} \end{bmatrix}$ **33.** Singular **34.** Singular **35.** $x = \dfrac{2}{5}, y = \dfrac{1}{10}$ **36.** $x = 1, y = \dfrac{3}{2}$

37. $x = \dfrac{1}{2}, y = \dfrac{2}{3}, z = \dfrac{1}{6}$ **38.** Inconsistent **39.** $x = -\dfrac{1}{2}, y = -\dfrac{2}{3}, z = -\dfrac{3}{4}$ **40.** $x = \dfrac{1}{3}, y = \dfrac{4}{5}, z = -\dfrac{1}{15}$

41. $z = -1, x = y + 1, y$ is any real number **42.** x is any real number, $y = -2x, z = -2x$ **43.** $x = 1, y = 2, z = -3, t = 1$

44. $x = -\dfrac{17}{15}, y = -\dfrac{1}{15}, z = \dfrac{22}{15}, t = -\dfrac{2}{15}$ **45.** 5 **46.** -12 **47.** 108 **48.** -19 **49.** -100 **50.** 11 **51.** $x = 2, y = -1$

52. $x = 0, y = \dfrac{5}{3}$ **53.** $x = 2, y = 3$ **54.** $x = 0, y = -3$ **55.** $x = -1, y = 2, z = -3$ **56.** $x = \dfrac{37}{9}, y = -\dfrac{19}{6}, z = \dfrac{13}{18}$

57. $\dfrac{-\frac{3}{2}}{x} + \dfrac{\frac{3}{2}}{x - 4}$ **58.** $\dfrac{\frac{2}{5}}{x + 2} + \dfrac{\frac{3}{5}}{x - 3}$ **59.** $\dfrac{-3}{x - 1} + \dfrac{3}{x} + \dfrac{4}{x^2}$ **60.** $\dfrac{4}{x - 2} + \dfrac{-2}{(x - 2)^2} + \dfrac{-4}{x - 1}$ **61.** $\dfrac{-\frac{1}{10}}{x + 1} + \dfrac{\frac{1}{10}x + \frac{9}{10}}{x^2 + 9}$

62. $\dfrac{\frac{6}{5}}{x - 2} + \dfrac{-\frac{6}{5}x + \frac{3}{5}}{x^2 + 1}$ **63.** $\dfrac{x}{x^2 + 4} + \dfrac{-4x}{(x^2 + 4)^2}$ **64.** $\dfrac{x}{x^2 + 16} + \dfrac{-16x + 1}{(x^2 + 16)^2}$ **65.** $\dfrac{\frac{1}{2}}{x^2 + 1} + \dfrac{\frac{1}{4}}{x - 1} + \dfrac{-\frac{1}{4}}{x + 1}$

66. $\dfrac{-\frac{4}{5}}{x^2 + 4} + \dfrac{\frac{2}{5}}{x - 1} + \dfrac{-\frac{2}{5}}{x + 1}$ **67.** $x = -\dfrac{2}{5}, y = -\dfrac{11}{5}; x = -2, y = 1$ **68.** $x = 2, y = -2\sqrt{3}; x = 2, y = 2\sqrt{3}; x = -4, y = 0$

69. $x = 2\sqrt{2}, y = \sqrt{2}; x = -2\sqrt{2}, y = -\sqrt{2}$

70. $x = -\sqrt{3}, y = -2\sqrt{2}; x = -\sqrt{3}, y = 2\sqrt{2}; x = \sqrt{3}, y = -2\sqrt{2}; x = \sqrt{3}, y = 2\sqrt{2}$

71. $x = 0, y = 0; x = -3, y = 3; x = 3, y = 3$ **72.** $x = 0, y = -3; x = 0, y = 3$

73. $x = \sqrt{2}, y = -\sqrt{2}; x = -\sqrt{2}, y = \sqrt{2}; x = \dfrac{4}{3}\sqrt{2}, y = -\dfrac{2}{3}\sqrt{2}; x = -\dfrac{4}{3}\sqrt{2}, y = \dfrac{2}{3}\sqrt{2}$

74. $x = -\dfrac{5}{21}\sqrt{42}, y = -\dfrac{2}{7}\sqrt{42}; x = \dfrac{5}{21}\sqrt{42}, y = \dfrac{2}{7}\sqrt{42}; x = -2, y = 1; x = 2, y = -1$ **75.** $x = 1, y = -1$

76. $x = -2, y = 0; x = 1, y = 0; x = -1, y = 2$

77. Unbounded; corner point $(0, 2)$

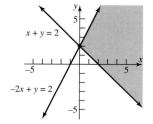

78. Unbounded; corner point $(2, -2)$

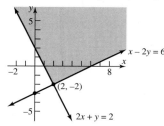

79. Bounded; corner points $(0, 0), (0, 2), (3, 0)$

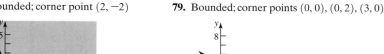
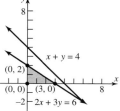

80. Unbounded; corner points (2, 0), (0, 6)

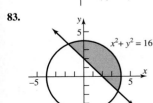

81. Bounded; corner points (0, 1), (0, 8), (4, 0), (2, 0)

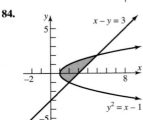

82. Bounded; corner points (0, 2), (0, 9), (3, 0)

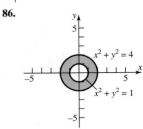

83.

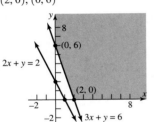

84.

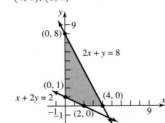

85.

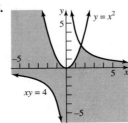

86.

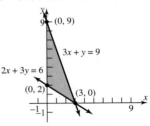

87. The maximum value is 32 when $x = 0$ and $y = 8$. **88.** The maximum value is 20, which occurs at (2, 4).
89. The minimum value is 3 when $x = 1$ and $y = 0$. **90.** The minimum value is 4, which occurs at (0, 4).
91. The maximum value is $\dfrac{108}{7}$ when $x = \dfrac{12}{7}$ and $y = \dfrac{12}{7}$. **92.** The maximum value is 36, which occurs at (4, 4). **93.** 10

94. Inconsistent if $A \neq 10$ **95.** $y = -\dfrac{1}{3}x^2 - \dfrac{2}{3}x + 1$ **96.** $x^2 + y^2 + 2x + 2y - 3 = 0$ **97.** 70 lbs of \$3 coffee and 30 lbs of \$6 coffee

98. Corn—466.25 acres, soybeans—533.75 acres **99.** 1 small, 5 medium, 2 large

100. (a) $\begin{cases} \dfrac{1}{2}x + \dfrac{3}{8}y \leq 120 \\ \dfrac{1}{4}x + \dfrac{3}{8}y \leq 72 \\ x \geq 0 \\ y \geq 0 \end{cases}$

(b)

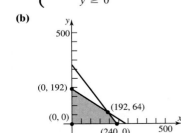

101. 24 ft by 10 ft **102.** 2 ft by 2 ft **103.** $4 + \sqrt{2}$ in. and $4 - \sqrt{2}$ in. **104.** 5 in. **105.** $100\sqrt{10}$ ft
106. $x =$ amount of 10% HCl
$y =$ amount of 25% HCl
$z =$ amount of 40% HCl

X	Y	Z
0	66.7	33.3
16.7	33.3	50.0
33.3	0	66.7

107. Katy gets \$10, Mike gets \$20, Danny gets \$5, Colleen gets \$10 **108.** Jet-stream speed is 29.69 mph
109. Bruce: 4 hr; Bryce: 2 hr; Marty: 8 hr
110. 12 dancing girls and 12 mermaids; maximum profit \$660. The process of molding has excess work-hours assigned.
111. 35 gasoline engines, 15 diesel engines; 15 gasoline engines, 0 diesel engines

CHAPTER 9   9.1 Exercises

1. $1, 2, 3, 4, 5$ **2.** $2, 5, 10, 17, 26$ **3.** $\dfrac{1}{3}, \dfrac{2}{4} = \dfrac{1}{2}, \dfrac{3}{5}, \dfrac{4}{6} = \dfrac{2}{3}, \dfrac{5}{7}$ **4.** $\dfrac{3}{2}, \dfrac{5}{4}, \dfrac{7}{6}, \dfrac{9}{8}, \dfrac{11}{10}$ **5.** $1, -4, 9, -16, 25$ **6.** $1, -\dfrac{2}{3}, \dfrac{3}{5}, -\dfrac{4}{7}, \dfrac{5}{9}$ **7.** $1, \dfrac{9}{5}, 3, \dfrac{81}{17}, \dfrac{81}{11}$

8. $\dfrac{2}{3}, \dfrac{4}{9}, \dfrac{8}{27}, \dfrac{16}{81}, \dfrac{32}{243}$ **9.** $-\dfrac{1}{6}, \dfrac{1}{12}, -\dfrac{1}{20}, \dfrac{1}{30}, -\dfrac{1}{42}$ **10.** $3, \dfrac{9}{2}, 9, \dfrac{81}{4}, \dfrac{243}{5}$ **11.** $\dfrac{1}{e}, \dfrac{2}{e^2}, \dfrac{3}{e^3}, \dfrac{4}{e^4}, \dfrac{5}{e^5}$ **12.** $\dfrac{1}{2}, 1, \dfrac{9}{8}, 1, \dfrac{25}{32}$ **13.** $\dfrac{n}{n+1}$

14. $\dfrac{1}{n(n+1)}$ **15.** $\dfrac{1}{2^{n-1}}$ **16.** $\dfrac{2^n}{3^n} = \left(\dfrac{2}{3}\right)^n$ **17.** $(-1)^{n+1}$ **18.** $\left(\dfrac{1}{n}\right)^{(-1)^n}$ or $n^{(-1)^{n+1}}$ **19.** $(-1)^{n+1}n$ **20.** $(-1)^{n+1} \cdot 2n$

21. $a_1 = 3, a_2 = 5, a_3 = 7, a_4 = 9, a_5 = 11$ **22.** $a_1 = 2, a_2 = 3, a_3 = 2, a_4 = 3, a_5 = 2$ **23.** $a_1 = -2, a_2 = -1, a_3 = 1, a_4 = 4, a_5 = 8$
24. $a_1 = 1, a_2 = 0, a_3 = 2, a_4 = 1, a_5 = 3$ **25.** $a_1 = 5, a_2 = 10, a_3 = 20, a_4 = 40, a_5 = 80$

26. $a_1 = 2, a_2 = -2, a_3 = 2, a_4 = -2, a_5 = 2$ **27.** $a_1 = 3, a_2 = 3, a_3 = \dfrac{3}{2}, a_4 = \dfrac{1}{2}, a_5 = \dfrac{1}{8}$

28. $a_1 = -2, a_2 = -5, a_3 = -13, a_4 = -36, a_5 = -104$ **29.** $a_1 = 1, a_2 = 2, a_3 = 2, a_4 = 4, a_5 = 8$
30. $a_1 = -1, a_2 = 1, a_3 = 0, a_4 = 2, a_5 = 2$ **31.** $a_1 = A, a_2 = A + d, a_3 = A + 2d, a_4 = A + 3d, a_5 = A + 4d$
32. $a_1 = A, a_2 = rA, a_3 = r^2A, a_4 = r^3A, a_5 = r^4A$

33. $a_1 = \sqrt{2}, a_2 = \sqrt{2 + \sqrt{2}}, a_3 = \sqrt{2 + \sqrt{2 + \sqrt{2}}}, a_4 = \sqrt{2 + \sqrt{2 + \sqrt{2 + \sqrt{2}}}}, a_5 = \sqrt{2 + \sqrt{2 + \sqrt{2 + \sqrt{2 + \sqrt{2}}}}}$

34. $a_1 = \sqrt{2}, a_2 = \dfrac{1}{2^{1/4}}, a_3 = \dfrac{1}{2^{5/8}}, a_4 = \dfrac{1}{2^{13/16}}, a_5 = \dfrac{1}{2^{29/32}}$ **35.** 50   **36.** 160   **37.** 21   **38.** −10   **39.** 90   **40.** 21   **41.** 26   **42.** 10   **43.** 42

44. 60   **45.** 96   **46.** 44   **47.** $4 + 5 + \cdots + (n + 3)$ **48.** $5 + 7 + 9 + \cdots + (2n + 3)$ **49.** $\dfrac{1}{2} + 2 + \dfrac{9}{2} + \cdots + \dfrac{n^2}{2}$

50. $4 + 9 + 16 + \cdots + (n + 1)^2$ **51.** $1 + \dfrac{1}{3} + \dfrac{1}{9} + \cdots + \dfrac{1}{3^n}$ **52.** $1 + \dfrac{3}{2} + \dfrac{9}{4} + \cdots + \left(\dfrac{3}{2}\right)^n$ **53.** $\dfrac{1}{3} + \dfrac{1}{9} + \cdots + \dfrac{1}{3^n}$

54. $1 + 3 + 5 + \cdots + (2n - 1)$ **55.** $\ln 2 - \ln 3 + \ln 4 - \cdots + (-1)^n \ln n$ **56.** $2^3 - 2^4 + 2^5 - 2^6 + \cdots + (-1)^{n+1} \cdot 2^n$

57. $\displaystyle\sum_{k=1}^{20} k$ **58.** $\displaystyle\sum_{k=1}^{8} k^3$ **59.** $\displaystyle\sum_{k=1}^{13} \dfrac{k}{k+1}$ **60.** $\displaystyle\sum_{k=1}^{12} 2k - 1$ **61.** $\displaystyle\sum_{k=0}^{6} (-1)^k \left(\dfrac{1}{3^k}\right)$ **62.** $\displaystyle\sum_{k=1}^{11} (-1)^{k+1} \left(\dfrac{2}{3}\right)^k$ **63.** $\displaystyle\sum_{k=1}^{n} \dfrac{3^k}{k}$ **64.** $\displaystyle\sum_{k=1}^{n} \dfrac{k}{e^k}$

65. $\displaystyle\sum_{k=0}^{n} (a + kd)$ **66.** $\displaystyle\sum_{k=1}^{n} ar^{k-1}$ **67.** 21

68. (a) $u_1 = \dfrac{(1 + \sqrt{5})^1 - (1 - \sqrt{5})^1}{2^1\sqrt{5}} = \dfrac{1 + \sqrt{5} - 1 + \sqrt{5}}{2\sqrt{5}} = \dfrac{2\sqrt{5}}{2\sqrt{5}} = 1$

$u_2 = \dfrac{(1 + \sqrt{5})^2 - (1 - \sqrt{5})^2}{2^2\sqrt{5}} = \dfrac{1 + 2\sqrt{5} + 5 - (1 - 2\sqrt{5} + 5)}{4\sqrt{5}}$

$= \dfrac{1 + 2\sqrt{5} + 5 - 1 + 2\sqrt{5} - 5}{4\sqrt{5}} = \dfrac{4\sqrt{5}}{4\sqrt{5}} = 1$

(b) $u_{n+1} + u_n = \dfrac{(1 + \sqrt{5})^{n+1} - (1 - \sqrt{5})^{n+1}}{2^{n+1}\sqrt{5}} + \dfrac{(1 + \sqrt{5})^n - (1 - \sqrt{5})^n}{2^n\sqrt{5}}$

$= \dfrac{(1 + \sqrt{5})^{n+1} - (1 - \sqrt{5})^{n+1} + 2(1 + \sqrt{5})^n - 2(1 - \sqrt{5})^n}{2^{n+1}\sqrt{5}}$

$= \dfrac{(1 + \sqrt{5})^n[1 + \sqrt{5} + 2] - (1 - \sqrt{5})^n[1 - \sqrt{5} + 2]}{2^{n+1}\sqrt{5}}$

$= \dfrac{(1 + \sqrt{5})^n(3 + \sqrt{5}) - (1 - \sqrt{5})^n - [3 - \sqrt{5}]}{2^{n+1}\sqrt{5}}$

$= \dfrac{(1 + \sqrt{5})^{n+2}\dfrac{(3 + \sqrt{5})}{(1 + \sqrt{5})^2} - (1 - \sqrt{5})^{n+2}\dfrac{(3 - \sqrt{5})}{(1 - \sqrt{5})^2}}{2^{n+1}\sqrt{5}}$

$= \dfrac{(1 + \sqrt{5})^{n+2}\dfrac{(3 + \sqrt{5})}{(6 + 2\sqrt{5})} - (1 - \sqrt{5})^{n+2}\dfrac{(3 - \sqrt{5})}{(6 - 2\sqrt{5})}}{2^{n+1}\sqrt{5}}$

$= \dfrac{(1 + \sqrt{5})^{n+2}\dfrac{1}{2} - (1 - \sqrt{5})^{n+2}\left(\dfrac{1}{2}\right)}{2^{n+1}\sqrt{5}}$

$= \dfrac{(1 + \sqrt{5})^{n+2} - (1 - \sqrt{5})^{n+2}}{2^{n+2}\sqrt{5}}$

$= u_{n+2}$

(c) Since $u_1 = 1, u_2 = 1, u_{n+2} = u_n + u_{n+1}$, $\{u_n\}$ is the Fibonacci sequence.
71. The Fibonacci sequence

72. (a) $1, 1, 2, 3, 5, 8, 13, 21, 34, 55$ **(b)** $1, 2, \dfrac{3}{2}, \dfrac{5}{3}, \dfrac{8}{5}, \dfrac{13}{8}, \dfrac{21}{13}, \dfrac{34}{21}, \dfrac{55}{34}$ **(c)** 1.6   **(d)** $1, \dfrac{1}{2}, \dfrac{2}{3}, \dfrac{3}{5}, \dfrac{5}{8}, \dfrac{8}{13}, \dfrac{13}{21}, \dfrac{21}{34}, \dfrac{34}{55}$ **(e)** 0.6

9.2 Exercises

1. $d = 1; 5, 6, 7, 8$ **2.** $d = 1; -4, -3, -2, -1$ **3.** $d = 2; -3, -1, 1, 3$ **4.** $d = 3; 4, 7, 10, 13$ **5.** $d = -2; 4, 2, 0, -2$

6. $d = -2; 2, 0, -2, -4$ **7.** $d = -\dfrac{1}{3}; \dfrac{1}{6}, -\dfrac{1}{6}, -\dfrac{1}{2}, -\dfrac{5}{6}$ **8.** $d = \dfrac{1}{4}; \dfrac{11}{12}, \dfrac{7}{6}, \dfrac{17}{12}, \dfrac{5}{3}$ **9.** $d = \ln 3; \ln 3, 2 \ln 3, 3, \ln 3, 4 \ln 3$ **10.** $d = 1; 1, 2, 3, 4$

11. $a_n = 2n + 1; a_5 = 11$ **12.** $a_n = 3n - 5; a_5 = 10$ **13.** $a_n = 8 - 3n; a_5 = -7$ **14.** $a_n = 8 - 2n; a_5 = -2$

15. $a_n = \dfrac{1}{2}(n - 1); a_5 = 2$ **16.** $a_n = \dfrac{4 - n}{3}; a_5 = -\dfrac{1}{3}$ **17.** $a_n = \sqrt{2}n; a_5 = 5\sqrt{2}$ **18.** $a_n = \pi(n - 1); a_5 = 4\pi$ **19.** $a_{11} = 22$

20. $a_{10} = 17$ **21.** $a_{10} = -26$ **22.** $a_9 = -35$ **23.** $a_8 = a + 7b$ **24.** $a_7 = 14\sqrt{5}$ **25.** $a_1 = -13; d = 3; a_1 = -13, a_{n+1} = a_n + 3$

26. $a_1 = -3; d = 2; a_1 = -3, a_{n+1} = a_n + 2$ **27.** $a_1 = -53; d = 6; a_1 = -53, a_{n+1} = a_n + 6$

28. $a_1 = 74; d = -10; a_1 = 74, a_{n+1} = a_n - 10$ **29.** $a_1 = 28; d = -2; a_1 = 28, a_{n+1} = a_n - 2$

30. $a_1 = -18; d = 4; a_1 = -18, a_{n+1} = a_n + 4$ **31.** $a_1 = 25; d = -2; a_1 = 25, a_{n+1} = a_n - 2$

32. $a_1 = -40; d = 4; a_1 = -40, a_{n+1} = a_n + 4$ **33.** n^2 **34.** $n + n^2$ **35.** $\dfrac{n}{2}(9 + 5n)$ **36.** $2n^2 - 3n$ **37.** 1260   **38.** 900   **39.** 324

40. 301

41.
```
sum(seq(3.45n+4.
12,n,1,20,1))
            806.9
```

42.
```
sum(seq(2.67n-1.
23,n,1,25,1)
             837
```

43.
```
sum(seq(2.4n+.4,
n,1,15,1))
             294
```

44.
```
sum(seq(1.9n+3.5
,n,1,15,1))
           280.5
```

45.
```
sum(seq(2.58n+2.
32,n,1,25,1))
           896.5
```

46.
```
sum(seq(3.19n+.5
2,n,1,25,1))
          1049.75
```

47. $-\dfrac{3}{2}$ **48.** 1   **49.** 1185 seats   **50.** 2160 seats

51. 210 of (1st colors name) and 190 (2nd colors name)

52. **(a)** 42 bricks   **(b)** 2130 bricks

53. $2S = n + n + \cdots + n \ (n - 1 \text{ terms})$

$2S = n(n - 1)$

$S = \dfrac{n}{2}(n - 1)$

Historical Problems

1. $1\dfrac{2}{3}$ loaves, $10\dfrac{5}{6}$ loaves, 20 loaves, $29\dfrac{1}{6}$ loaves, $38\dfrac{1}{3}$ loaves   **2.** **(a)** 1   **(b)** 2401   **(c)** 2800

9.3 Exercises

1. $r = 3; 3, 9, 27, 81$ **2.** $r = -5; -5, 25, -125, 625$ **3.** $r = \dfrac{1}{2}; -\dfrac{3}{2}, -\dfrac{3}{4}, -\dfrac{3}{8}, -\dfrac{3}{16}$ **4.** $r = \dfrac{5}{2}; \dfrac{5}{2}, \dfrac{25}{4}, \dfrac{125}{8}, \dfrac{625}{16}$ **5.** $r = 2; \dfrac{1}{4}, \dfrac{1}{2}, 1, 2$

6. $r = 3; \dfrac{1}{3}, 1, 3, 9$ **7.** $r = 2^{1/3}; 2^{1/3}, 2^{2/3}, 2, 2^{4/3}$ **8.** $r = 9; 9, 81, 729, 6561$ **9.** $r = \dfrac{3}{2}; \dfrac{1}{2}, \dfrac{3}{4}, \dfrac{9}{8}, \dfrac{27}{16}$ **10.** $r = \dfrac{2}{3}; 2, \dfrac{4}{3}, \dfrac{8}{9}, \dfrac{16}{27}$

11. Arithmetic; $d = 1$ **12.** Arithmetic; $d = 3$ **13.** Neither   **14.** Neither   **15.** Arithmetic; $d = -\dfrac{2}{3}$ **16.** Arithmetic; $d = -\dfrac{3}{4}$

17. Neither   **18.** Arithmetic; $d = 2$ **19.** Geometric; $r = \dfrac{2}{3}$ **20.** Geometric; $r = \dfrac{5}{4}$ **21.** Geometric; $r = 2$ **22.** Neither

23. Geometric; $r = 3^{1/2}$ **24.** Geometric; $r = -1$ **25.** $a_5 = 48; a_n = 3 \cdot 2^{n-1}$ **26.** $a_5 = -162; a_n = (-2)3^{n-1}$

27. $a_5 = 5; a_n = 5 \cdot (-1)^{n-1}$ **28.** $a_5 = 96; a_n = (6)(-2)^{n-1}$ **29.** $a_5 = 0; a_n = 0$ **30.** $a_5 = \dfrac{1}{81}; a_n = \left(-\dfrac{1}{3}\right)^{n-1}$

31. $a_5 = 4\sqrt{2}; a_n = (\sqrt{2})^n$ **32.** $a_5 = 0; a_n = 0$ **33.** $a_7 = \dfrac{1}{64}$ **34.** $a_8 = 2187$ **35.** $a_9 = 1$ **36.** $a_{10} = 512$ **37.** $a_8 = 0.00000004$

38. $a_7 = 100,000$ **39.** $-\dfrac{1}{4}(1 - 2^n)$ **40.** $-\dfrac{1}{6}(1 - 3^n)$ **41.** $2\left[1 - \left(\dfrac{2}{3}\right)^n\right]$ **42.** $-2(1 - 3^n)$ **43.** $1 - 2^n$ **44.** $5\left[1 - \left(\dfrac{3}{5}\right)^n\right]$

45.
```
(1/4)sum(seq(2^n
,n,0,14,1))
          8191.75
```

46.
```
(1/9)sum(seq(3^n
,n,1,15,1)
       2391484.333
```

47.
```
sum(seq((2/3)^n,
n,1,15,1))
       1.995432683
```

48.
```
4sum(seq(3^(n-1)
,n,1,15,1))
         28697812
```

49.
```
-1sum(seq(2^n,n,
0,14,1))
            -32767
```

50.
```
2sum(seq((3/5)^n
,n,0,15,1))
         4.998589445
```

51. $\dfrac{4}{3}$ **52.** $\dfrac{6}{5}$ **53.** 16 **54.** 9 **55.** $\dfrac{8}{5}$ **56.** $\dfrac{4}{7}$ **57.** $\dfrac{20}{3}$ **58.** 12 **59.** $\dfrac{18}{5}$ **60.** $\dfrac{8}{3}$ **61.** -4 **62.** 2 **63.** \$349,496.41 **64.** \$16,712.73

65. \$96,885.98 **66.** \$66,438.85 **67.** \$305.10 **68.** \$312.44 **69. (a)** 0.775 ft **(b)** 8th **(c)** 15.88 ft **(d)** 20 ft

70. (a) 15.36 ft **(b)** $30(0.8)^n$ **(c)** After the 19th strike **(d)** 270 ft **71.** \$21,879.11 **72.** \$6655.58

73. Option 2 results in the most: \$16,038, 304; Option 1 results in the least: \$14,700,000 **74.** December 20, 1998 (111 days); \$9999.91

75. 1.845×10^{19} **76.** Yes. A constant function is both arithmetic and geometric. For example, $3, 3, 3, \ldots$ is an arithmetic sequence with $a_1 = 3$ and $d = 0$ and is a geometric sequence with $a_1 = 3$ and $r = 1$. **79.** $3, 5$, never, never, never

80. Option B results in more money (\$500,500 versus \$524,287)

81. A: \$25,250 per year in 5th year, \$112,742 total; B: \$24,761 per year in 5th year, \$116,801 total **82.** $\dfrac{1}{3}$ **83.** 10 **84.** 20

85. \$72.67 per share **86.** \$39.64

9.4 Exercises

1. (I) $n = 1: 2 \cdot 1 = 2$ and $1(1 + 1) = 2$

   (II) If $2 + 4 + 6 + \cdots + 2k = k(k + 1)$, then $2 + 4 + 6 + \cdots + 2k + 2(k + 1) = (2 + 4 + 6 + \cdots + 2k) + 2(k + 1)$
$$= k(k + 1) + 2(k + 1) = k^2 + 3k + 2 = (k + 1)(k + 2).$$
   Thus, by I and II, it is true for all natural numbers.

2. (I) $n = 1: 4 \cdot 1 - 3 = 1$ and $1(2 \cdot 1 - 1) = 1$

   (II) If $1 + 5 + 9 + \cdots + (4k - 3) = k(2k - 1)$, then $1 + 5 + 9 + \cdots + (4k - 3) + (4k + 1)$
$$= [1 + 5 + 9 + \cdots + (4k - 3)] + (4k + 1)$$
$$= k(2k - 1) + (4k + 1) = 2k^2 - k + 4k + 1 = 2k^2 + 3k + 1$$
$$= (k + 1)(2k + 1).$$
   Thus, by I and II, it is true for all natural numbers.

3. (I) $n = 1: 1 + 2 = 3$ and $\dfrac{1}{2}(1)(1 + 5) = \dfrac{1}{2}(6) = 3$

   (II) If $3 + 4 + 5 + \cdots + (k + 2) = \dfrac{1}{2}k(k + 5)$, then $3 + 4 + 5 + \cdots + (k + 2) + [(k + 1) + 2]$
$$= [3 + 4 + 5 + \cdots + (k + 2)] + (k + 3) = \dfrac{1}{2}k(k + 5) + k + 3 = \dfrac{1}{2}(k^2 + 7k + 6) = \dfrac{1}{2}(k + 1)(k + 6)$$
   Thus, by I and II, it is true for all natural numbers.

4. (I) $n = 1: 2 \cdot 1 + 1 = 3$ and $1(1 + 2) = 3$

   (II) If $3 + 5 + 7 \cdots + (2k + 1) = k(k + 2)$, then $3 + 5 + 7 + \cdots + (2k + 1) + (2k + 3)$
$$= [3 + 5 + 7 + \cdots + (2k + 1)] + (2k + 3)$$
$$= k(k + 2) + (2k + 3) = k^2 + 2k + 2k + 3 = k^2 + 4k + 3$$
$$= (k + 1)(k + 3).$$
   Thus, by I and II, it is true for all natural numbers.

5. (I) $n = 1: 3 \cdot 1 - 1 = 2$ and $\dfrac{1}{2}(1)[3(1) + 1] = \dfrac{1}{2}(4) = 2$

   (II) If $2 + 5 + 8 + \cdots + (3k - 1) = \dfrac{1}{2}k(3k + 1)$, then $2 + 5 + 8 + \cdots + (3k - 1) + [3(k + 1) - 1]$
$$= [2 + 5 + 8 + \cdots + (3k - 1)] + 3k + 2 = \dfrac{1}{2}k(3k + 1) + (3k + 2) = \dfrac{1}{2}(3k^2 + 7k + 4) = \dfrac{1}{2}(k + 1)(3k + 4).$$
   Thus, by I and II, it is true for all natural numbers.

6. (I) $n = 1: 3 \cdot 1 - 2 = 1$ and $\dfrac{1}{2} \cdot 1(3 \cdot 1 - 1) = 1$

   (II) If $1 + 4 + 7 \cdots + (3k - 2) = \dfrac{1}{2}k(3k - 1)$, then $1 + 4 + 7 + \cdots + (3k - 2) + (3k + 1)$
$$= [1 + 4 + 7 + \cdots + (3k - 2)] + (3k + 1)$$
$$= \dfrac{1}{2}k(3k - 1) + (3k + 1) = \dfrac{3}{2}k^2 - \dfrac{1}{2}k + 3k + 1$$
$$= \dfrac{3}{2}k^2 + \dfrac{5}{2}k + 1 = \dfrac{1}{2}(3k^2 + 5k + 2) = \dfrac{1}{2}(k + 1)(3k + 2).$$
   Thus, by I and II, it is true for all natural numbers.

7. (I) $n = 1: 2^{1-1} = 1$ and $2^1 - 1 = 1$

(II) If $1 + 2 + 2^2 + \cdots + 2^{k-1} = 2^k - 1$, then $1 + 2 + 2^2 + \cdots + 2^{k-1} + 2^{(k+1)-1} = (1 + 2 + 2^2 + \cdots + 2^{k-1}) + 2^k$
$= 2^k - 1 + 2^k = 2(2^k) - 1 = 2^{k+1} - 1.$

Thus, by I and II, it is true for all natural numbers.

8. (I) $n = 1: 3^{1-1} = 1$ and $\frac{1}{2}(3^1 - 1) = 1$

(II) If $1 + 3 + 3^2 \cdots + 3^{k-1} = \frac{1}{2}(3^k - 1)$, then $1 + 3 + 3^2 + \cdots + 3^{k-1} + 3^k$
$= [1 + 3 + 3^2 + \cdots + 3^{k-1}] + 3^k$
$= \frac{1}{2}(3^k - 1) + 3^k = \frac{3^k}{2} - \frac{1}{2} + 3^k = \frac{3^k}{2} + 3^k - \frac{1}{2}$
$= 3^k\left(\frac{1}{2} + 1\right) - \frac{1}{2} = 3^k\left(\frac{3}{2}\right) - \frac{1}{2} = \frac{1}{2}[3^k(3) - 1] = \frac{1}{2}(3^{k+1} - 1).$

Thus, by I and II, it is true for all natural numbers.

9. (I) $n = 1: 4^{1-1} = 1$ and $\frac{1}{3}(4^1 - 1) = \frac{1}{3}(3) = 1$

(II) If $1 + 4 + 4^2 + \cdots + 4^{k-1} = \frac{1}{3}(4^k - 1)$, then $1 + 4 + 4^2 + \cdots + 4^{k-1} + 4^{(k+1)-1} = (1 + 4 + 4^2 + \cdots + 4^{k-1}) + 4^k$
$= \frac{1}{3}(4^k - 1) + 4^k = \frac{1}{3}[4^k - 1 + 3(4^k)] = \frac{1}{3}[4(4^k) - 1] = \frac{1}{3}(4^{k+1} - 1).$

Thus, by I and II, it is true for all natural numbers.

10. (I) $n = 1: 5^{1-1} = 1$ and $\frac{1}{4}(5^1 - 1) = 1$

(II) If $1 + 5 + 5^2 \cdots + 5^{k-1} = \frac{1}{4}(5^k - 1)$, then $1 + 5 + 5^2 + \cdots + 5^{k-1} + 5^k$
$= [1 + 5 + 5^2 + \cdots + 5^{k-1}] + 5^k$
$= \frac{1}{4}[5^k - 1] + 5^k = \frac{5^k}{4} - \frac{1}{4} + 5^k = 5^k\left(\frac{1}{4} + 1\right) - \frac{1}{4}.$
$5^k\left(\frac{5}{4}\right) - \frac{1}{4} = \frac{1}{4}(5^k \cdot 5 - 1) = \frac{1}{4}(5^{k+1} - 1)$

Thus, by I and II, it is true for all natural numbers.

11. (I) $n = 1: \frac{1}{1 \cdot 2} = \frac{1}{2}$ and $\frac{1}{1 + 1} = \frac{1}{2}$

(II) If $\frac{1}{1 \cdot 2} + \frac{1}{2 \cdot 3} + \frac{1}{3 \cdot 4} + \cdots + \frac{1}{k(k + 1)} = \frac{k}{k + 1}$, then $\frac{1}{1 \cdot 2} + \frac{1}{2 \cdot 3} + \frac{1}{3 \cdot 4} + \cdots + \frac{1}{k(k + 1)} + \frac{1}{(k + 1)[(k + 1) + 1]}$
$= \left[\frac{1}{1 \cdot 2} + \frac{1}{2 \cdot 3} + \frac{1}{3 \cdot 4} + \cdots + \frac{1}{k(k + 1)}\right] + \frac{1}{(k + 1)(k + 2)} = \frac{k}{k + 1} + \frac{1}{(k + 1)(k + 2)}$
$= \frac{k(k + 2) + 1}{(k + 1)(k + 2)} = \frac{k^2 + 2k + 1}{(k + 1)(k + 2)} = \frac{(k + 1)(k + 1)}{(k + 1)(k + 2)} = \frac{k + 1}{k + 2}.$

Thus, by I and II, it is true for all natural numbers.

12. (I) $n = 1: \frac{1}{(2 \cdot 1 - 1)(2 \cdot 1 + 1)} = \frac{1}{3}$ and $\frac{1}{2 \cdot 1 + 1} = \frac{1}{3}$

(II) If $\frac{1}{1 \cdot 3} + \frac{1}{3 \cdot 5} + \frac{1}{5 \cdot 7} + \cdots + \frac{1}{(2k - 1)(2k + 1)} = \frac{k}{2k + 1}$, then $\left[\frac{1}{1 \cdot 3} + \frac{1}{3 \cdot 5} + \frac{1}{5 \cdot 7} + \cdots + \frac{1}{(2k - 1)(2k + 1)}\right]$
$+ \frac{1}{(2k + 1)(2k + 3)} = \frac{k}{2k + 1} + \frac{1}{(2k + 1)(2k + 3)} = \frac{k(2k + 3) + 1}{(2k + 1)(2k + 3)}$
$= \frac{2k^2 + 3k + 1}{(2k + 1)(2k + 3)} = \frac{(2k + 1)(k + 1)}{(2k + 1)(2k + 3)} = \frac{k + 1}{2k + 3}.$

Thus, by I and II, it is true for all natural numbers.

13. (I) $n = 1: 1^2 = 1$ and $\frac{1}{6} \cdot 1 \cdot 2 \cdot 3 = 1$

(II) If $1^2 + 2^2 + 3^2 + \cdots + k^2 = \frac{1}{6}k(k + 1)(2k + 1)$, then $1^2 + 2^2 + 3^2 + \cdots + k^2 + (k + 1)^2$
$= (1^2 + 2^2 + 3^2 + \cdots + k^2) + (k + 1)^2 = \frac{1}{6}k(k + 1)(2k + 1) + (k + 1)^2 = \frac{1}{6}(2k^3 + 9k^2 + 13k + 6)$
$= \frac{1}{6}(k + 1)(k + 2)(2k + 3).$

Thus, by I and II, it is true for all natural numbers.

14. (I) $n = 1: 1^3 = 1$ and $\frac{1}{4} \cdot 1^2(1 + 1)^2 = 1$

(II) If $1^3 + 2^3 + 3^3 \cdots + k^3 = \frac{1}{4}k^2(k + 1)^2$, then $1^3 + 2^3 + 3^3 + \cdots + k^3 + (k + 1)^3$

$$= [1^3 + 2^3 + 3^3 + \cdots k^3] + (k + 1)^3$$

$$= \frac{1}{4}k^2(k + 1)^2 + (k + 1)^3 = (k + 1)^2 \left[\frac{1}{4}k^2 + k + 1 \right]$$

$$= (k + 1)^2 \cdot \frac{1}{4}(k^2 + 4k + 4) = (k + 1)^2 \cdot \frac{1}{4}(k + 2)^2$$

$$= \frac{1}{4}(k + 1)^2(k + 2)^2$$

Thus, by I and II, it is true for all natural numbers.

15. (I) $n = 1: 5 - 1 = 4$ and $\frac{1}{2}(1)(9 - 1) = \frac{1}{2} \cdot 8 = 4$

(II) If $4 + 3 + 2 + \cdots + (5 - k) = \frac{1}{2}k(9 - k)$, then $4 + 3 + 2 + \cdots + (5 - k) + 5 - (k + 1)$

$$= [4 + 3 + 2 + \cdots + (5 - k)] + 5 - (k + 1) = \frac{1}{2}k(9 - k) + 4 - k = \frac{1}{2}(-k^2 + 7k + 8) = \frac{1}{2}(8 - k)(k + 1)$$

$$= \frac{1}{2}(k + 1)[9 - (k + 1)].$$

Thus, by I and II, it is true for all natural numbers.

16. (I) $n = 1: -(1 + 1) = -2$ and $-\frac{1}{2} \cdot 1(1 + 3) = -2$

(II) If $-2 - 3 - 4 - \cdots - (k + 1) = -\frac{1}{2}k(k + 3)$, then $-2 - 3 - 4 - \cdots - (k + 1) - (k + 2)$

$$= [-2 - 3 - 4 - \cdots - (k + 1)] - (k + 2)$$

$$= -\frac{1}{2}k(k + 3) - (k + 2) = -\frac{1}{2}k^2 - \frac{3}{2}k - k - 2$$

$$= -\frac{1}{2}k^2 - \frac{5}{2}k - 2$$

$$= -\frac{1}{2}(k^2 + 5k + 4) = -\frac{1}{2}(k + 1)(k + 4).$$

Thus, by I and II, it is true for all natural numbers.

17. (I) $n = 1: 1 \cdot (1 + 1) = 2$ and $\frac{1}{3} \cdot 1 \cdot 2 \cdot 3 = 2$

(II) If $1 \cdot 2 + 2 \cdot 3 + 3 \cdot 4 + \cdots + k(k + 1) = \frac{1}{3}k(k + 1)(k + 2)$, then

$$1 \cdot 2 + 2 \cdot 3 + 3 \cdot 4 + \cdots + k(k + 1) + (k + 1)(k + 2) = [1 \cdot 2 + 2 \cdot 3 + 3 \cdot 4 + \cdots + k(k + 1)] + (k + 1)(k + 2)$$

$$= \frac{1}{3}k(k + 1)(k + 2) + (k + 1)(k + 2) = \frac{1}{3}(k + 1)(k + 2)(k + 3)$$

Thus, by I and II, it is true for all natural numbers.

18. (I) $n = 1: (2 \cdot 1 - 1)(2 \cdot 1) = 2$ and $\frac{1}{3} \cdot 1(1 + 1)(4 \cdot 1 - 1) = 2$

(II) If $1 \cdot 2 + 3 \cdot 4 + 5 \cdot 6 + \cdots + (2k - 1)(2k) = \frac{1}{3}k(k + 1)(4k - 1)$, then

$$1 \cdot 2 + 3 \cdot 4 + 5 \cdot 6 + \cdots + (2k - 1)(2k) + (2k + 1)(2k + 2)$$
$$= [1 \cdot 2 + 3 \cdot 4 + 5 \cdot 6 + \cdots + (2k - 1)(2k)] + (2k + 1)(2k + 2)$$

$$= \frac{1}{3}k(k + 1)(4k - 1) + (2k + 1)(2k + 2)$$

$$= \frac{1}{3}k(4k^2 + 3k - 1) + (2k + 1)(2k + 2)$$

$$= \frac{4}{3}k^3 + k^2 - \frac{1}{3}k + 4k^2 + 6k + 2$$

$$= \frac{4}{3}k^3 + 5k^2 + \frac{17}{3}k + 2 = \frac{1}{3}(4k^3 + 15k^2 + 17k + 6)$$

$$= \frac{1}{3}(k + 1)(4k^2 + 11k + 6) = \frac{1}{3}(k + 1)(k + 2)(4k + 3)$$

Thus, by I and II, it is true for all natural numbers.

19. (I) $n = 1: 1^2 + 1 = 2$ is divisible by 2.

(II) If $k^2 + k$ is divisible by 2, then $(k + 1)^2 + (k + 1) = k^2 + 2k + 1 + k + 1 = (k^2 + k) + 2k + 2$. Since $k^2 + k$ is divisible by 2 and $2k + 2$ is divisible by 2, therefore, $(k + 1)^2 + k + 1$ is divisible by 2.

Thus, by I and II, it is true for all natural numbers.

20. (I) $n = 1: 1^3 + 2 \cdot 1 = 3$ is divisible by 3.

(II) If $k^3 + 2k$ is divisible by 3, then $(k + 1)^3 + 2(k + 1) = k^3 + 3k^2 + 3k + 1 + 2k + 2 = k^3 + 2k + 3k^2 + 3k + 3$. Since $k^3 + 2k$ is divisible by 3 and $3k^2 + 3k + 3$ is divisible by 3, therefore, $(k + 1)^3 + 2(k + 1)$ is divisible by 3.

Thus, by I and II, it is true for all natural numbers.

21. (I) $n = 1: 1^2 - 1 + 2 = 2$ is divisible by 2.

(II) If $k^2 - k + 2$ is divisible by 2, then $(k + 1)^2 - (k + 1) + 2 = k^2 + 2k + 1 - k - 1 + 2 = (k^2 - k + 2) + 2k$. Since $k^2 - k + 2$ is divisible by 2 and $2k$ is divisible by 2, therefore, $(k + 1)^2 - (k + 1) + 2$ is divisible by 2.

Thus, by I and II, it is true for all natural numbers.

22. (I) $n = 1: 1(1 + 2)(1 + 2) = 6$ is divisible by 6.

(II) If $k(k + 1)(k + 2)$ is divisible by 6, then

$(k + 1)(k + 2)(k + 3) = k^3 + 6k^2 + 11k + 6 = k^3 + 3k^2 + 2k + 3k^2 + 9k + 6 = k(k^2 + 3k + 2) + 3(k^2 + 3k + 2)$
$= k(k + 1)(k + 2) + 3(k + 1)(k + 2)$. We know that $k(k + 1)(k + 2)$ is divisible by 6 and $3 (k + 1)(k + 2)$ is also divisible by 6 since either $k + 1$ or $k + 2$ must be even (divisible by 2).

Thus, by I and II, it is true for all natural numbers.

23. (I) $n = 1$: If $x > 1$, then $x^1 = x > 1$.

(II) Assume, for any natural number k, that if $x > 1$, then $x^k > 1$. Show that if $x > 1$, then $x^{k+1} > 1$:

$$x^{k+1} = \underset{\underset{\displaystyle x^k > 1}{\uparrow}}{x^k \cdot x^1} > 1 \cdot x = x > 1$$

Thus, by I and II, it is true for all natural numbers.

24. (I) If $0 < x < 1$, then $0 < x^1 = x < 1$.

(II) Assume that if $0 < x < 1$, then $0 < x^k < 1$ for any natural number k.

Show that, if $0 < x < 1$, then $0 < x^{k+1} < 1$: $0 < x^{k+1} = \underset{\underset{\displaystyle x^k < 1}{\uparrow}}{x^k \cdot x} < 1 \cdot x = x < 1$

Thus, by I and II, it is true for all natural numbers.

25. (I) $n = 1: a - b$ is a factor of $a^1 - b^1 = a - b$.

(II) If $a - b$ is a factor of $a^k - b^k$, show that $a - b$ is a factor of $a^{k+1} - b^{k+1}$: $a^{k+1} - b^{k+1} = a(a^k - b^k) + b^k(a - b)$. Since $a - b$ is a factor of $a^k - b^k$ and $a - b$ is a factor of $a - b$, therefore, $a - b$ is a factor of $a^{k+1} - b^{k+1}$.

Thus, by I and II, it is true for all natural numbers.

26. (I) $n = 1: a + b$ is a factor of $a^{2 \cdot 1 + 1} + b^{2 \cdot 1 + 1} = a^3 + b^3$

(II) Suppose for some natural number k that $a + b$ is a factor of $a^{2k+1} + b^{2k+1}$. Show that $a + b$ is a factor of $a^{2k+3} + b^{2k+3}$:

$a^{2k+3} + b^{2k+3} = a^2 \cdot a^{2k+1} + a^2 b^{2k+1} - a^2 b^{2k+1} + b^{2k+3}$
$= a^2(a^{2k+1} + b^{2k+1}) - b^{2k+1}(a^2 - b^2)$

Since $a + b$ divides $a^{2k+1} + b^{2k+1}$ and $a + b$ divides $a^2 - b^2$, then $a + b$ divides $a^{2k+3} + b^{2k+3}$.

Thus, by I and II, it is true for all natural numbers.

27. $n = 1: 1^2 - 1 + 41 = 41$ is a prime number.

$n = 41: 41^2 - 41 + 41 = 1681 = 41^2$ is not prime.

28. (II) If $2 + 4 + 6 + \cdots + 2k = k^2 + k + 2$, then $[2 + 4 + 6 + \cdots + 2k] + 2k + 2 = k^2 + k + 2 + 2k + 2$
$= k^2 + 3k + 4 = k^2 + 2k + k + 1 + 1 + 2 = (k^2 + 2k + 1) + k + 1 + 2 = (k + 1)^2 + (k + 1) + 2$

(I) $n = 1: 2 \cdot 1 = 2$ but $1^2 + 1 + 2 = 4 \neq 2$

29. (I) $n = 1: ar^{1-1} = a \cdot 1 = a$ and $a \cdot \dfrac{1 - r^1}{1 - r} = a$, because $r \neq 1$.

(II) If $a + ar + ar^2 + \cdots + ar^{k-1} = a\left(\dfrac{1 - r^k}{1 - r}\right)$, then

$$a + ar + ar^2 + \cdots + ar^{k-1} + ar^{(k+1)-1} = (a + ar + ar^2 + \cdots + ar^{k-1}) + ar^k$$

$$= a\left(\frac{1 - r^k}{1 - r}\right) + ar^k = \frac{a(1 - r^k) + ar^k(1 - r)}{1 - r} = \frac{a - ar^k + ar^k - ar^{k+1}}{1 - r} = a\left(\frac{1 - r^{k+1}}{1 - r}\right)$$

Thus, by I and II, it is true for all natural numbers.

30. (I) $n = 1$: $[a + (1 - 1)d] = a$ and $1 \cdot a + d\dfrac{1(1 - 1)}{2} = a$

(II) If $a + (a + d) + (a + 2d) + \cdots + [a + (k - 1)d] = ka + d\dfrac{k(k - 1)}{2}$, then

$a + (a + d) + (a + 2d) + \cdots + [a + (k - 1)d] + [a + kd]$
$= \{a + (a + d) + (a + 2d) + \cdots + [a + (k - 1)d]\} + [a + kd]$
$= ka + d\dfrac{k(k - 1)}{2} + a + kd = ka + \dfrac{kd}{2}(k - 1) + a + kd$
$= ka + \dfrac{k^2 d}{2} - \dfrac{kd}{2} + a + kd$
$= (k + 1)a + \dfrac{d}{2}k^2 - \dfrac{d}{2}k + dk$
$= (k + 1)a + \dfrac{d}{2}k^2 + \dfrac{d}{2}k = (k + 1)a + \dfrac{dk(k + 1)}{2}$

Thus, by I and II, it is true for all natural numbers.

31. (I) $n = 3$: The sum of the angles of a triangle is $(3 - 2) \cdot 180° = 180°$.

(II) Assume for any k that the sum of the angles of a convex polygon of k sides is $(k - 2) \cdot 180°$. A convex polygon of $k + 1$ sides consists of a convex polygon of k sides plus a triangle (see the illustration). The sum of the angle is $(k - 2) \cdot 180° + 180°$ $= (k - 1) \cdot 180°$. Since Conditions I and II have been met, the result follows.

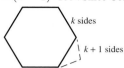

k sides

$k + 1$ sides

32. (I) $n = 4$: The number of diagonals of a quadrilateral is

$$\frac{1}{2} \cdot 4(4 - 3) = 2$$

(II) Assume that for any k the number of diagonals of a convex polygon of k sides (k vertices) is $\dfrac{1}{2}k(k - 3)$. A convex polygon of

$k + 1$ sides ($k + 1$ vertices) consists of a convex polygon of k sides (k vertices) plus a triangle for a total of $k + 1$ vertices; see the illustration. The number of diagonals of this convex polygon consists of all the original ones plus $k - 1$ additional ones, namely,

$$\frac{1}{2}k(k - 3) + (k - 1) = \frac{1}{2}k^2 - \frac{3}{2}k + k - 1$$
$$= \frac{1}{2}k^2 - \frac{1}{2}k - 1$$
$$= \frac{1}{2}(k + 1)(k - 2)$$

Since Conditions I and II have been met, the result follows.

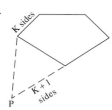

k sides

$k + 1$ sides

P

9.5 Exercises

1. 10 **2.** 35 **3.** 21 **4.** 36 **5.** 50 **6.** 4950 **7.** 1 **8.** 1 **9.** 1.866×10^{15} **10.** 4.192×10^{15} **11.** 1.483×10^{13} **12.** 1.767×10^{10}
13. $x^4 + 4x^3 + 6x^2 + 4x + 1$ **14.** $x^4 - 4x^3 + 6x^2 - 4x + 1$ **15.** $x^5 - 10x^4 + 40x^3 - 80x^2 + 80x - 32$
16. $x^5 + 15x^4 + 90x^3 + 270x^2 + 405x + 243$ **17.** $81x^4 + 108x^3 + 54x^2 + 12x + 1$
18. $32x^5 + 240x^4 + 720x^3 + 1080x^2 + 810x + 243$ **19.** $x^{10} + 5y^2 x^8 + 10y^4 x^6 + 10y^6 x^4 + 5y^8 x^2 + y^{10}$
20. $x^{12} - 6y^2 x^{10} + 15y^4 x^8 - 20y^6 x^6 + 15y^8 x^4 - 6y^{10} x^2 + y^{12}$ **21.** $x^3 + 6\sqrt{2}x^{5/2} + 30x^2 + 40\sqrt{2}x^{3/2} + 60x + 24\sqrt{2}x^{1/2} + 8$
22. $x^2 - 4\sqrt{3}x^{3/2} + 18x - 12\sqrt{3}x^{1/2} + 9$ **23.** $(ax)^5 + 5by(ax)^4 + 10(by)^2(ax)^3 + 10(by)^3(ax)^2 + 5(by)^4(ax) + (by)^5$
24. $(ax)^4 - 4by(ax)^3 + 6(by)^2(ax)^2 - 4(by)^3(ax) + (by)^4$ **25.** 153,090 **26.** −61,236 **27.** −101,376 **28.** 1760 **29.** 41,472
30. −314,928 **31.** $2835x^3$ **32.** $189x^5$ **33.** $314,928x^7$ **34.** $48,384x^3$ **35.** 495 **36.** −84 **37.** 3360 **38.** 252 **39.** 1.00501
40. 0.98805 **41.** $\dbinom{n}{n} = \dfrac{n!}{n!(n - n)!} = \dfrac{n!}{n!0!} = \dfrac{n!}{n!} = 1$

42. If n and j are integers with $0 \le j \le n$, then
$$\binom{n}{j} = \left(\frac{n!}{j!(n-j)!}\right) = \left(\frac{n!}{(n-j)!\,(n-(n-j))!}\right) = \binom{n}{n-j}$$

43. $2^n = (1+1)^n = \binom{n}{0}1^n + \binom{n}{1}(1)(1)^{n-1} + \cdots + \binom{n}{n}1^n = \binom{n}{0} + \binom{n}{1} + \cdots + \binom{n}{n}$

44. Let $0 = (1-1)^n = \binom{n}{0}1^n + \binom{n}{1}(-1)1^{n-1} + \binom{n}{2}(-1)^2 1^{n-2} + \cdots + \binom{n}{n}(-1)^n$

Therefore, $\binom{n}{0} - \binom{n}{1} + \binom{n}{2} - \cdots + (-1)^n \binom{n}{n} = 0$

45. 1

46. $12! = 4.790016 \times 10^8$
$20! = 2.432902 \times 10^{18}$
$25! = 1.551121 \times 10^{25}$
Using Sterlings Formula
$12! \approx 4.79013972 \times 10^8$
$20! \approx 2.43292403 \times 10^{18}$
$25! \approx 1.55112992 \times 10^{25}$

9.6 Exercises

1. $\{1, 3, 5, 6, 7, 9\}$ **2.** $\{1, 2, 3, 4, 5, 6, 7, 8, 9\}$ **3.** $\{1, 5, 7\}$ **4.** $\{1, 9\}$ **5.** $\{1, 6, 9\}$ **6.** $\{1, 6, 9\}$ **7.** $\{1, 2, 4, 5, 6, 7, 8, 9\}$
8. $\{1, 2, 3, 4, 5, 6, 7, 8, 9\}$ **9.** $\{1, 2, 4, 5, 6, 7, 8, 9\}$ **10.** $\{1\}$ **11.** $\{0, 2, 6, 7, 8\}$ **12.** $\{0, 2, 5, 7, 8, 9\}$ **13.** $\{0, 1, 2, 3, 5, 6, 7, 8, 9\}$
14. $\{0, 5, 9\}$ **15.** $\{0, 1, 2, 3, 5, 6, 7, 8, 9\}$ **16.** $\{0, 5, 9\}$ **17.** $\{0, 1, 2, 3, 4, 6, 7, 8\}$ **18.** $\{2, 7, 8\}$ **19.** $\{0\}$ **20.** $\{0, 1, 2, 3, 5, 6, 7, 8, 9\}$
21. $\varnothing, \{a\}, \{b\}, \{c\}, \{d\}, \{a, b\}, \{a, c\}, \{a, d\}, \{b, c\}, \{b, d\}, \{c, d\}, \{a, b, c\}, \{b, c, d\}, \{a, c, d\}, \{a, b, d\}, \{a, b, c, d\}$
22. $\varnothing, \{a\}, \{b\}, \{c\}, \{d\}, \{e\}, \{a, b\}, \{a, c\}, \{a, d\}, \{a, e\}, \{b, c\}, \{b, d\}, \{b, e\}, \{c, d\}, \{c, e\}, \{d, e\}, \{a, b, c\}, \{a, b, d\}, \{a, b, e\}, \{a, c, d\},$
$\{a, c, e\}, \{a, d, e\}, \{b, c, d\}, \{b, c, e\}, \{b, d, e\}, \{c, d, e\}, \{a, b, c, d\}, \{a, b, c, e\}, \{a, b, d, e\}, \{a, c, d, e\}, \{b, c, d, e\}, \{a, b, c, d, e\}$
23. 40 **24.** 5 **25.** 40 **26.** 50 **27.** 25 **28.** 20 **29.** 37 **30.** 8 **31.** 18 **32.** 31 **33.** 5 **34.** 52 **35.** 175; 125 **36.** 550
37. (a) 15 **(b)** 15 **(c)** 15 **(d)** 25 **(e)** 40 **38.** 8

9.7 Exercises

1. 30 **2.** 42 **3.** 120 **4.** 24 **5.** 1 **6.** 1 **7.** 1680 **8.** 336 **9.** 28 **10.** 28 **11.** 15 **12.** 15 **13.** 1 **14.** 18 **15.** 10,400,600 **16.** 48,620
17. $\{abc, abd, abe, acb, acd, ace, adb, adc, ade, aeb, aec, aed$
    $bac, bad, bae, bca, bcd, bce, bda, bdc, bde, bea, bec, bed$
    $cab, cad, cae, cba, cbd, cbe, cda, cdb, cde, cea, ceb, ced$
    $dab, dac, dae, dba, dbc, dbe, dca, dcb, dce, dea, deb, dec$
    $eab, eac, ead, eba, ebc, ebd, eca, ecb, ecd, eda, edb, edc\}$; 60
18. $\{ab, ac, ad, ae, ba, bc, bd, be, ca, cb, cd, ce, da, db, dc, de, ea, eb, ec, ed\}$; 20
19. $\{123, 124, 132, 134, 142, 143, 213, 214, 231, 234, 241, 243, 312, 314, 321, 324, 341, 342, 412, 413, 421, 423, 431, 432\}$; 24
20. $\{123, 124, 125, 126, 132, 134, 135, 136, 142, 143, 145, 146, 152, 153, 154, 156, 162, 163, 164, 165, 213, 214, 215, 216, 231, 234, 235, 236,$
    $241, 243, 245, 246, 251, 253, 254, 256, 261, 263, 264, 265, 312, 314, 315, 316, 321, 324, 325, 326, 341, 342, 345, 346, 351, 352, 354, 356,$
    $361, 362, 364, 365, 412, 413, 415, 416, 421, 423, 425, 426, 431, 432, 435, 436, 451, 452, 453, 456, 461, 462, 463, 465, 512, 513, 514, 516,$
    $521, 523, 524, 526, 531, 532, 534, 536, 541, 542, 543, 546, 561, 562, 563, 564, 612, 613, 614, 615, 621, 623, 624, 625, 631, 632, 634, 635,$
    $641, 642, 643, 645, 651, 652, 653, 654\}$; 120 **21.** $\{abc, abd, abe, acd, ace, ade, bcd, bce, bde, cde\}$; 10
22. $\{ab, ac, ad, ae, bc, bd, be, cd, ce, de\}$; 10 **23.** $\{123, 234, 124, 134\}$; 4
24. $\{123, 124, 125, 126, 234, 235, 236, 345, 346, 456, 134, 135, 136, 245, 246, 356, 145, 146, 256, 156\}$; 20 **25.** 24 **26.** 20 **27.** 16 **28.** 25
29. 8 **30.** 1000 **31.** 24 **32.** 120 **33.** 60 **34.** 360 **35.** 120 **36.** 120 **37.** 35 **38.** 70 **39.** 1024 **40.** 1024 **41.** 9000 **42.** 80,000
43. 120 **44. (a)** 6,760,000 **(b)** 3,407,040 **(c)** 3,276,000 **45.** 480 **46.** 125,000 **47.** 1120 **48.** 225 **49.** 5,209,344
50. 302,400 **51.** 362,880 **52.** 40,320 **53.** 90,720 **54.** 4,989,600 **55.** 1.157×10^{76} **56.** 70 **57.** 15 **58.** 126
59. (a) 63 **(b)** 35 **(c)** 1 **60. (a)** 3003 **(b)** 20,475 **(c)** 16,653

Historical Problems

1. (a) $P(A \text{ wins}) = \dfrac{11}{16}$; $P(B \text{ wins}) = \dfrac{5}{16}$; Player A gets $\dfrac{11}{16}$ of the stakes; player B gets $\dfrac{5}{16}$ of the stakes.

(b) Player A has $C(4, 2) + C(4, 3) + C(4, 4) = 6 + 4 + 1 = 11$ chances to win. Player B has $C(4, 3) + C(4, 4) = 4 + 1 = 5$ Chances
to win. $P(A \text{ wins}) = \dfrac{11}{16}$; $P(B \text{ wins}) = \dfrac{5}{16}$; The results are the same. **2. (a)** \$1.50 **(b)** \$0.50

9.8 Exercises

1. $S = \{HH, HT, TH, TT\}$; $P(HH) = \frac{1}{4}$, $P(HT) = \frac{1}{4}$, $P(TH) = \frac{1}{4}$, $P(TT) = \frac{1}{4}$

2. $S = \{HH, HT, TH, TT\}$

$P(HH) = \frac{1}{4}$, $P(HT) = \frac{1}{4}$, $P(TH) = \frac{1}{4}$, $P(TT) = \frac{1}{4}$

3. $S = \{HH1, HH2, HH3, HH4, HH5, HH6, HT1, HT2, HT3, HT4, HT5, HT6, TH1, TH2, TH3, TH4, TH5, TH6, TT1, TT2, TT3,$
$TT4, TT5, TT6\}$; each outcome has the probability of $\frac{1}{24}$.

4. $S = \{H1H, H2H, H3H, H4H, H5H, H6H, H1T, H2T, H3T, H4T, H5T, H6T, T1H, T2H, T3H, T4H, T5H, T6H, T1T, T2T, T3T,$
$T4T, T5T, T6T\}$; each outcome has the probability of $\frac{1}{24}$.

5. $S = \{HHH, HHT, HTH, HTT, THH, THT, TTH, TTT\}$; each outcome has the probability of $\frac{1}{8}$.

6. $S = \{HHH, HHT, HTH, HTT, THH, THT, TTH, TTT\}$; each outcome has the probability of $\frac{1}{8}$.

7. $S = \{1$ Yellow, 1 Red, 1 Green, 2 Yellow, 2 Red, 2 Green, 3 Yellow, 3 Red, 3 Green, 4 Yellow, 4 Red, 4 Green$\}$; each outcome has the
probability of $\frac{1}{12}$; thus, $P(2 \text{ Red}) + P(4 \text{ Red}) = \frac{1}{12} + \frac{1}{12} = \frac{1}{6}$.

8. $S = \{$Forward Yellow, Forward Green, Forward Red, Backward Yellow, Backward Green, Backward Red$\}$; each outcome has the
probability of $\frac{1}{6}$; thus, $P(\text{Forward Yellow}) + P(\text{Forward Green}) = \frac{1}{6} + \frac{1}{6} = \frac{1}{3}$.

9. $S = \{1$ Yellow Forward, 1 Yellow Backward, 1 Red Forward, 1 Red Backward, 1 Green Forward, 1 Green Backward, 2 Yellow Forward,
2 Yellow Backward, 2 Red Forward, 2 Red Backward, 2 Green Forward, 2 Green Backward, 3 Yellow Forward, 3 Yellow Backward,
3 Red Forward, 3 Red Backward, 3 Green Forward, 3 Green Backward, 4 Yellow Forward, 4 Yellow Backward, 4 Red Forward, 4 Red
Backward, 4 Green Forward, 4 Green Backward$\}$; each outcome has the probability of $\frac{1}{24}$; thus,

$P(1 \text{ Red Backward}) + P(1 \text{ Green Backward}) = \frac{1}{24} + \frac{1}{24} = \frac{1}{12}$.

10. $S = \{$Yellow 1 Forward, Yellow 1 Backward, Yellow 2 Forward, Yellow 2 Backward, Yellow 3 Forward, Yellow 3 Backward, Yellow 4
Forward, Yellow 4 Backward, Green 1 Forward, Green 1 Backward, Green 2 Forward, Green 2 Backward, Green 3 Forward, Green 3
Backward, Green 4 Forward, Green 4 Backward, Red 1 Forward, Red 1 Backward, Red 2 Forward, Red 2 Backward, Red 3 Forward,
Red 3 Backward, Red 4 Forward, Red 4 Backward$\}$; each outcome has the probability of $\frac{1}{24}$; thus,

$P(\text{Yellow 2 Forward}) + P(\text{Yellow 4 Forward}) = \frac{1}{24} + \frac{1}{24} = \frac{1}{12}$.

11. $S = \{11$ Red, 11 Yellow, 11 Green, 12 Red, 12 Yellow, 12 Green, 13 Red, 13 Yellow, 13 Green, 14 Red, 14 Yellow, 14 Green, 21 Red,
21 Yellow, 21 Green, 22 Red, 22 Yellow, 22 Green, 23 Red, 23 Yellow, 23 Green, 24 Red, 24 Yellow, 24 Green, 31 Red, 31 Yellow,
31 Green, 32 Red, 32 Yellow, 32 Green, 33 Red, 33 Yellow, 33 Green, 34 Red, 34 Yellow, 34 Green, 41 Red, 41 Yellow, 41 Green, 42 Red,
42 Yellow, 42 Green, 43 Red, 43 Yellow, 43 Green, 44 Red, 44 Yellow, 44 Green$\}$; each outcome has the probability of $\frac{1}{48}$;

thus, $E = \{22$ Red, 22 Green, 24 Red, 24 Green$\}$; $P(E) = \frac{n(E)}{n(S)} = \frac{4}{48} = \frac{1}{12}$.

12. $S = \{$Forward 11, Forward 12, Forward 13, Forward 14, Forward 21, Forward 22, Forward 23, Forward 24, Forward 31, Forward 32,
Forward 33, Forward 34, Forward 41, Forward 42, Forward 43, Forward 44, Backward 11, Backward 12, Backward 13, Backward 14,
Backward 21, Backward 22, Backward 23, Backward 24, Backward 31, Backward 32, Backward 33, Backward 34, Backward 41,
Backward 42, Backward 43, Backward 44$\}$; each outcome has the probability of $\frac{1}{32}$; thus,

$P(\text{Forward 12}) + P(\text{Forward 14}) + P(\text{Forward 32}) + P(\text{Forward 34}) = \frac{1}{32} + \frac{1}{32} + \frac{1}{32} + \frac{1}{32} = \frac{1}{8}$

13. A, B, C, F **14.** A **15.** B **16.** F **17.** $\frac{4}{5}; \frac{1}{5}$ **18.** $P(H) = \frac{1}{3}; P(T) = \frac{2}{3}$ **19.** $P(1) = P(3) = P(5) = \frac{2}{9}; P(2) = P(4) = P(6) = \frac{1}{9}$

20. $P(1) = P(2) = P(3) = P(4) = P(5) = \frac{1}{5}; P(6) = 0$ **21.** 0.7 **22.** 0 **23.** 0.55 **24.** 0.1 **25.** $\frac{9}{20}$ **26.** $\frac{2}{5}$ **27.** $\frac{17}{20}$ **28.** $\frac{11}{20}$ **29.** $\frac{5}{18}$

30. $\frac{2}{9}$ **31.** 0.3 **32.** 0.65 **33.** 0.35 **34.** 0.65 **35. (a)** 0.57 **(b)** 0.95 **(c)** 0.83 **(d)** 0.38 **(e)** 0.29 **(f)** 0.05 **(g)** 0.78 **(h)** 0.71

36. (a) 0.45 **(b)** 0.75 **(c)** 0.9 **37.** $\dfrac{1}{30,240} \approx 0.000033069$ **38.** $\dfrac{350}{3003} \approx 0.1166$ **39. (a)** $\dfrac{10}{32}$ **(b)** $\dfrac{1}{32}$ **40. (a)** $\dfrac{1}{4}$ **(b)** $\dfrac{5}{16}$

41. (a) 0.00463 **(b)** 0.049 **42. (a)** $\left(\dfrac{35}{36}\right)^5 \approx 0.869$ **(b)** $\left(\dfrac{5}{6}\right)^5 \approx 0.402$ **43.** $\dfrac{1}{C(30,\,5)} \approx 7.02 \times 10^{-6}; 0.183$ **44.** 1.80×10^{-5} **45.** 0.1

Fill-in-the-Blank Items

1. sequence **2.** arithmetic **3.** geometric **4.** Pascal triangle **5.** 15 **6.** union; intersection **7.** 20; 10 **8.** permutation
9. combination **10.** equally likely

True/False Items

1. T **2.** T **3.** T **4.** T **5.** F **6.** F **7.** F **8.** T **9.** F **10.** T **11.** T **12.** F

Review Exercises

1. 120 **2.** 720 **3.** 10 **4.** 28 **5.** 336 **6.** 210 **7.** 56 **8.** 35 **9.** $-\dfrac{4}{3}, \dfrac{5}{4}, -\dfrac{6}{5}, \dfrac{7}{6}, -\dfrac{8}{7}$ **10.** $5, -7, 9, -11, 13$ **11.** $2, 1, \dfrac{8}{9}, 1, \dfrac{32}{25}$

12. $e, \dfrac{e^2}{2}, \dfrac{e^3}{3}, \dfrac{e^4}{4}, \dfrac{e^5}{5}$ **13.** $6, 4, \dfrac{8}{3}, \dfrac{16}{9}, \dfrac{32}{27}$ **14.** $8, -2, \dfrac{1}{2}, -\dfrac{1}{8}, \dfrac{1}{32}$ **15.** $2, 0, 2, 0, 2$ **16.** $-3, 1, 5, 9, 13$ **17.** Arithmetic; $d = 1; \dfrac{n}{2}(n + 11)$

18. Arithmetic; $d = 4; 2n^2 + 5n$ **19.** Neither **20.** Neither **21.** Geometric; $r = 8; \dfrac{8}{7}(8^n - 1)$ **22.** Geometric; $r = 9; \dfrac{9}{8}(9^n - 1)$

23. Arithmetic; $d = 4; 2n(n - 1)$ **24.** Arithmetic; $d = -4; 3n - 2n^2$ **25.** Geometric; $r = \dfrac{1}{2}; 6\left[1 - \left(\dfrac{1}{2}\right)^n\right]$

26. Geometric; $r = -\dfrac{1}{3}; \dfrac{15}{4}\left(1 - \left(-\dfrac{1}{3}\right)^n\right)$ **27.** Neither **28.** Neither **29.** 115 **30.** 50 **31.** 75 **32.** -18 **33.** 0.49977 **34.** 682

35. 35 **36.** -13 **37.** $\dfrac{1}{10^{10}}$ **38.** 1024 **39.** $9\sqrt{2}$ **40.** $2^{9/2} = 16\sqrt{2}$ **41.** $5n - 4$ **42.** $4 - 3n$ **43.** $n - 10$ **44.** $6 + 2n$ **45.** $\dfrac{9}{2}$ **46.** 4

47. $\dfrac{4}{3}$ **48.** $\dfrac{18}{5}$ **49.** 8 **50.** $\dfrac{12}{7}$

51. (I) $n = 1: 3 \cdot 1 = 3$ and $\dfrac{3 \cdot 1}{2}(2) = 3$

(II) If $3 + 6 + 9 + \cdots + 3k = \dfrac{3k}{2}(k + 1)$, then $3 + 6 + 9 + \cdots + 3k + 3(k + 1) = (3 + 6 + 9 + \cdots + 3k) + (3k + 3)$

$= \dfrac{3k}{2}(k + 1) + (3k + 3) = \dfrac{3k^2}{2} + \dfrac{9k}{2} + \dfrac{6}{2} = \dfrac{3}{2}(k^2 + 3k + 2) = \dfrac{3}{2}(k + 1)(k + 2)$.

52. (I) $n = 1: 4 \cdot 1 - 2 = 2$ and $2 \cdot 1^2 = 2$
(II) If $2 + 6 + 10 + \cdots + (4k - 2) = 2k^2$, then $2 + 6 + 10 + \cdots + (4k - 2) + (4k + 2)$
$= [2 + 6 + 10 + \cdots + (4k - 2)] + (4k + 2)$
$= 2k^2 + (4k + 2)$
$= 2(k^2 + 2k + 1) = 2(k + 1)^2$

53. (I) $n = 1: 2 \cdot 3^{1-1} = 2$ and $3^1 - 1 = 2$
(II) If $2 + 6 + 18 + \cdots + 2 \cdot 3^{k-1} = 3^k - 1$, then $2 + 6 + 18 + \cdots + 2 \cdot 3^{k-1} + 2 \cdot 3^{(k+1)-1}$
$= (2 + 6 + 18 + \cdots + 2 \cdot 3^{k-1}) + 2 \cdot 3^k = 3^k - 1 + 2 \cdot 3^k = 3 \cdot 3^k - 1 = 3^{k+1} - 1$.

54. (I) $n = 1: 3 \cdot 2^{1-1} = 3$ and $3(2^1 - 1) = 3$
(II) If $3 + 6 + 12 + \cdots + 3 \cdot 2^{k-1} = 3(2^k - 1)$, then $3 + 6 + 12 + \cdots + 3 \cdot 2^{k-1} + 3 \cdot 2^k$
$= [3 + 6 + 12 + \cdots + 3 \cdot 2^{k-1}] + 3 \cdot 2^k$
$= 3(2^k - 1) + 3 \cdot 2^k$
$= 3 \cdot 2^k - 3 + 3 \cdot 2^k = 3(2^{k+1} - 1)$.

55. (I) $n = 1: 1^2 = 1$ and $\dfrac{1}{2}(6 - 3 - 1) = \dfrac{1}{2}(2) = 1$

(II) If $1^2 + 4^2 + 7^2 + \cdots + (3k - 2)^2 = \dfrac{1}{2}k(6k^2 - 3k - 1)$, then

$1^2 + 4^2 + 7^2 + \cdots + (3k - 2)^2 + [3(k + 1) - 2]^2 = [1^2 + 4^2 + 7^2 + \cdots + (3k - 2)^2] + (3k + 1)^2$

$= \dfrac{1}{2}k(6k^2 - 3k - 1) + (3k + 1)^2$

$= \dfrac{1}{2}(6k^3 + 15k^2 + 11k + 2) = \dfrac{1}{2}(k + 1)(6k^2 + 9k + 2) = \dfrac{1}{2}(k + 1)[6(k + 1)^2 - 3(k + 1) - 1]$.

56. (I) $n = 1: 1(1 + 2) = 3$ and $\frac{1}{6}(1 + 1)(2 \cdot 1 + 7) = 3$

(II) If $1 \cdot 3 + 2 \cdot 4 + 3 \cdot 5 + \cdots + k(k + 2) = \frac{k}{6}(k + 1)(2k + 7)$, then $1 \cdot 3 + 2 \cdot 4 + 3 \cdot 5 + \cdots + k(k + 2) + (k + 1)(k + 3)$

$= (1 \cdot 3 + 2 \cdot 4 + 3 \cdot 5 + \cdots + k(k + 2)] + (k + 1)(k + 3)$

$= \frac{k}{6}(k + 1)(2k + 7) + (k + 1)(k + 3)$

$= (k + 1)\left[\frac{k}{6}(2k + 7) + (k + 3)\right]$

$= (k + 1)\left[\frac{2k^2}{6} + \frac{7k}{6} + k + 3\right] = (k + 1)\left[\frac{2k^2 + 7k + 6k + 18}{6}\right]$

$= \frac{k + 1}{6}(2k^2 + 13k + 18) = \frac{k + 1}{6}(k + 2)(2k + 9).$

57. $x^5 + 15x^4 + 90x^3 + 270x^2 + 405x + 243$ **58.** $x^5 - 15x^4 + 90x^3 - 270x^2 + 405x - 243$ **59.** $16x^4 + 96x^3 + 216x^2 + 216x + 81$
60. $81x^4 - 432x^3 + 864x^2 - 768x + 256$ **61.** 144   **62.** $-13{,}608$ **63.** 280   **64.** 448   **65.** $\{1, 3, 5, 6, 7, 8\}$ **66.** $\{2, 3, 5, 6, 7, 8, 9\}$
67. $\{3, 7\}$ **68.** $\{3, 5, 7\}$ **69.** $\{1, 2, 4, 6, 8, 9\}$ **70.** $\{1, 4\}$ **71.** $\{1, 2, 4, 5, 6, 9\}$ **72.** $\{2, 4, 9\}$ **73.** 17   **74.** 24   **75.** 29   **76.** 34   **77.** 7
78. 45   **79.** 25   **80.** 7   **81.** 60   **82.** 120   **83.** 128   **84.** 64   **85.** 3024   **86.** 24   **87.** 70   **88.** 120   **89.** 91   **90. (a)** 14,400
(b) 14,400   **91.** 1,600,000   **92.** 60   **93.** 216,000   **94.** 256   **95.** 1260   **96.** 12,600   **97. (a)** 381,024   **(b)** 1260   **98. (a)** 280   **(b)** 280
(c) 640   **99.** $\frac{3}{20}; \frac{9}{20}$ **100.** $\frac{4}{9}$ **101.** $\frac{1}{24}$ **102.** $0.2; 0.26$ **103. (a)** $\frac{10}{220} \approx 0.045$ **(b)** $\frac{70}{220} \approx 0.318$ **(c)** $\frac{35}{220} \approx 0.159$
104. (a) $\frac{252}{1024} \approx 0.246$ **(b)** $\frac{1}{1024}$ **105. (a)** 8   **(b)** 1100   **106.** 360   **107.** \$151,873.77   **108.** \$244,129.08
109. (a) $\left(\frac{3}{4}\right)^3 \cdot 20 = \frac{135}{16}$ ft   **(b)** $20\left(\frac{3}{4}\right)^n$ ft   **(c)** after the 13th time   **(d)** 140 ft   **110.** \$23,397.17   **111. (a)** 0.68   **(b)** 0.58   **(c)** 0.32

A P P E N D I X   Exercise 1

1. $(-1, 4)$ **2.** $(3, 4)$ **3.** $(3, 1)$ **4.** $(-6, -4)$ **5.** $X\min = -11, X\max = 5, X\text{scl} = 1, Y\min = -3, Y\max = 6, Y\text{scl} = 1$
6. $X\min = -3, X\max = 7, X\text{scl} = 1, Y\min = -4, Y\max = 9, Y\text{scl} = 1$ **7.** $X\min = -30, X\max = 50, X\text{scl} = 10, Y\min = -90,$
$Y\max = 50, Y\text{scl} = 10$ **8.** $X\min = -90, X\max = 30, X\text{scl} = 10, Y\min = -50, Y\max = 70, Y\text{scl} = 10$ **9.** $X\min = -10,$
$X\max = 110, X\text{scl} = 10, Y\min = -10, Y\max = 160, Y\text{scl} = 10$ **10.** $X\min = -20, X\max = 110, X\text{scl} = 10, Y\min = -10,$
$Y\max = 60, Y\text{scl} = 10$ **11.** $X\min = -6, X\max = 6, X\text{scl} = 2, Y\min = -4, Y\max = 4, Y\text{scl} = 2$ **12.** $X\min = -3, X\max = 3,$
$X\text{scl} = 1, Y\min = -2, Y\max = 2, Y\text{scl} = 1$ **13.** $X\min = -9, X\max = 9, X\text{scl} = 3, Y\min = -4, Y\max = 4, Y\text{scl} = 2$
14. $X\min = -3, X\max = 3, X\text{scl} = 1, Y\min = -10, Y\max = 10, Y\text{scl} = 5$ **15.** $X\min = -6, X\max = 6, X\text{scl} = 1, Y\min = -8,$
$Y\max = 8, Y\text{scl} = 2$ **16.** $X\min = -12, X\max = 12, X\text{scl} = 2, Y\min = -4, Y\max = 4, Y\text{scl} = 1$ **17.** $X\min = -6, X\max = 6,$
$X\text{scl} = 2, Y\min = -1, Y\max = 3, Y\text{scl} = 1$ **18.** $X\min = -9, X\max = 9, X\text{scl} = 3, Y\min = -12, Y\max = 4, Y\text{scl} = 4$
19. $X\min = 3, X\max = 9, X\text{scl} = 1, Y\min = 2, Y\max = 10, Y\text{scl} = 2$ **20.** $X\min = -22, X\max = -10, X\text{scl} = 2, Y\min = 4,$
$Y\max = 8, Y\text{scl} = 1$

Exercise 2

1. (a)   **(b)**   **(c)**   **(d)**

2. (a)   **(b)**   **(c)**   **(d)**

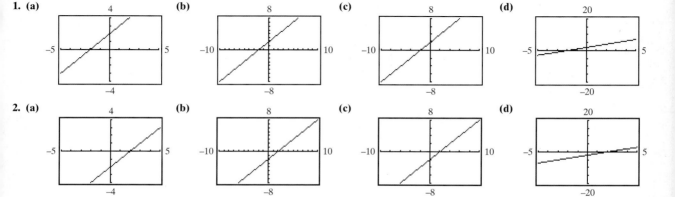

3. (a) **(b)** **(c)** **(d)**

4. (a) **(b)** **(c)** **(d)**

5. (a) **(b)** **(c)** **(d)**

6. (a) **(b)** **(c)** **(d)**

7. (a) **(b)** **(c)** **(d)**

8. (a) **(b)** **(c)** **(d)**

9. (a) **(b)** **(c)** **(d)**

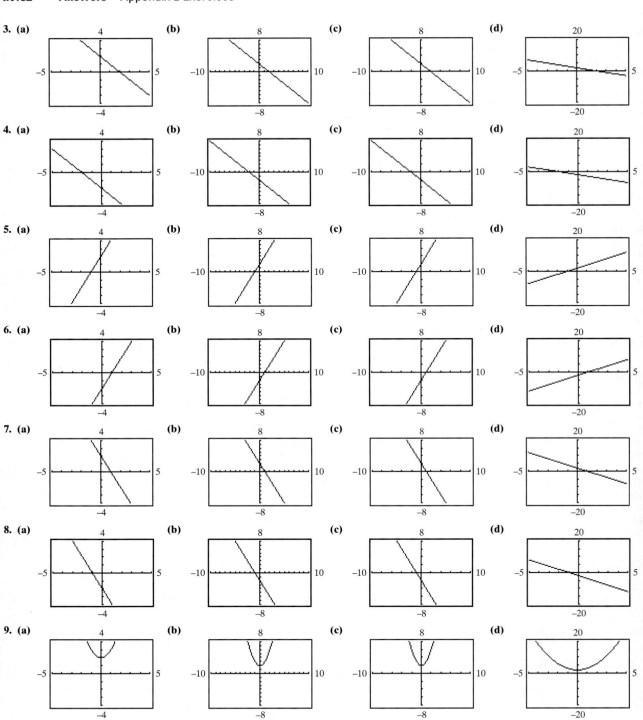

10. (a) 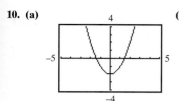 **(b)** 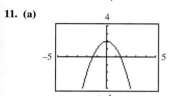 **(c)** 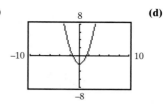 **(d)**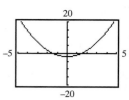

11. (a) 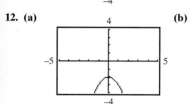 **(b)** 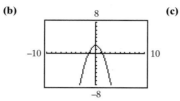 **(c)** 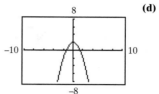 **(d)**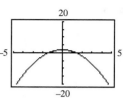

12. (a) 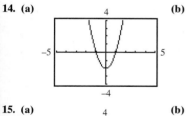 **(b)** 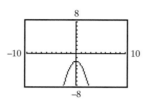 **(c)** 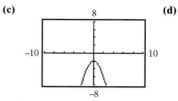 **(d)**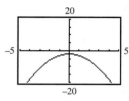

13. (a) 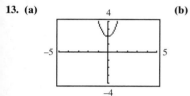 **(b)** 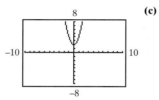 **(c)** 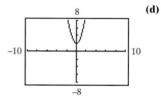 **(d)**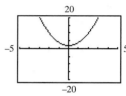

14. (a) 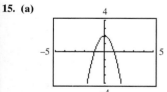 **(b)** 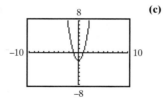 **(c)** 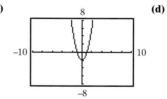 **(d)**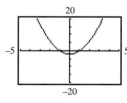

15. (a) 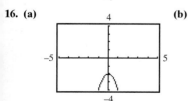 **(b)** 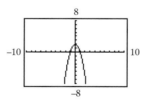 **(c)** 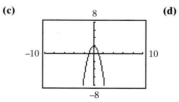 **(d)**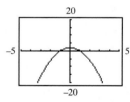

16. (a) 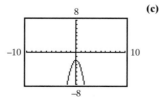 **(b)** 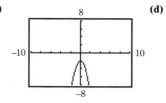 **(c)** 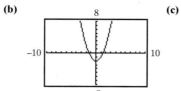 **(d)**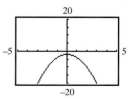

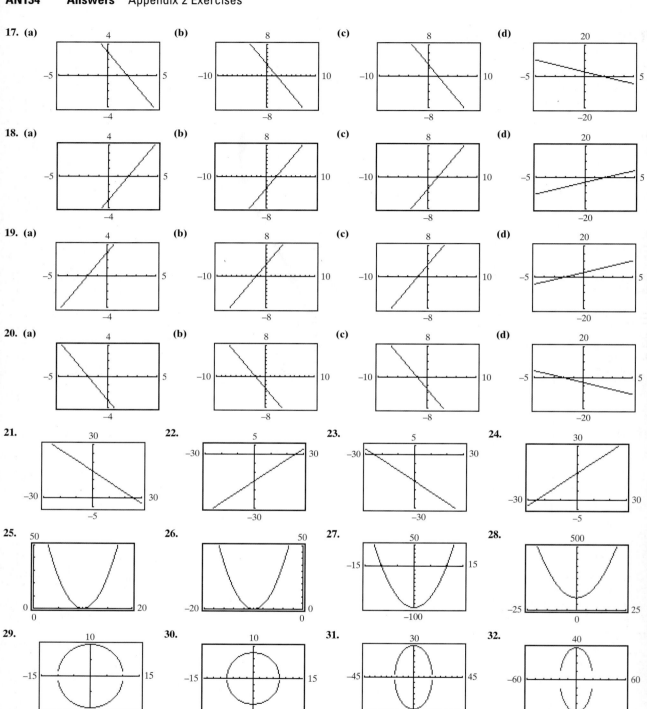

33.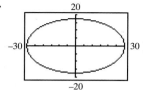

34.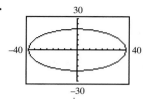

35.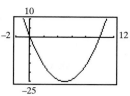

36.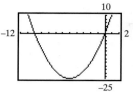

37.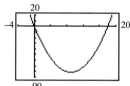

38.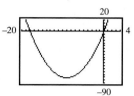

39.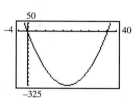

40.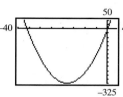

Exercise 3

1. 0.428   **2.** .181   **3.** 2.236   **4.** 2.449   **5.** 1.259   **6.** 1.442   **7.** −3.41   **8.** −4.64   **9.** −1.70   **10.** −1.43   **11.** −0.28   **12.** −0.22   **13.** 3.00
14. 2.00   **15.** 4.50   **16.** 1.69   **17.** 0.31, 12.30   **18.** 0.63, 13.60   **19.** 1.00, 23.00   **20.** 1.07

Exercise 4

1. Yes   **2.** No   **3.** Yes   **4.** Yes   **5.** No   **6.** Yes   **7.** Yes   **8.** Yes

9. $y\min = 4$ Other answers are possible
   $y\max = 12$
   $y\text{scl} = 1$

10. $y\min = -2$ Other answers are possible
    $y\max = 10$
    $y\text{scl} = 2$

INDEX

Conics

Parabola

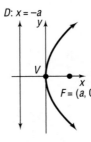 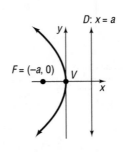 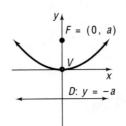 

$$y^2 = 4ax \qquad y^2 = -4ax \qquad x^2 = 4ay \qquad x^2 = -4ay$$

Ellipse

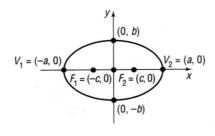 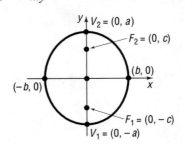

$$\frac{x^2}{a^2} + \frac{y^2}{b^2} = 1, \quad c^2 = a^2 - b^2 \qquad\qquad \frac{x^2}{b^2} + \frac{y^2}{a^2} = 1, \quad c^2 = a^2 - b^2$$

Hyperbola

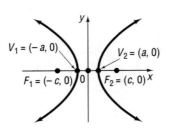 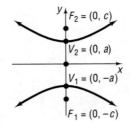

$$\frac{x^2}{a^2} - \frac{y^2}{b^2} = 1, \quad c^2 = a^2 + b^2 \qquad\qquad \frac{y^2}{a^2} - \frac{x^2}{b^2} = 1, \quad c^2 = a^2 + b^2$$

$$\text{Asymptotes:} \quad y = \frac{b}{a}x, \quad y = -\frac{b}{a}x \qquad\qquad \text{Asymptotes:} \quad y = \frac{a}{b}x, \quad y = -\frac{a}{b}x$$